CATECHOLAMINES

BRIDGING BASIC SCIENCE WITH CLINICAL MEDICINE

Edited by

David S. Goldstein
Graeme Eisenhofer
Clinical Neuroscience Branch
National Institute of Neurological Disorders and Stroke
National Institutes of Health
Bethesda, Maryland

Richard McCarty
Department of Psychology
University of Virginia
Charlottesville, Virginia

ACADEMIC PRESS
San Diego London Boston New York Sydney Tokyo Toronto

This book is printed on acid-free paper. ∞

Academic Press
a division of Harcourt Brace & Company
525 B Street, Suite 1900, San Diego, California 92101-4495, USA
http://www.apnet.com

Academic Press Limited
24-28 Oval Road, London NW1 7DX, UK
http://www.hbuk.co.uk/ap/

International Standard Book Number: 0-12-032943-3

PRINTED IN THE UNITED STATES OF AMERICA
97 98 99 00 01 02 MM 9 8 7 6 5 4 3 2 1

Contents

Other Catecholamine-Synthesizing Enzymes

PART B CATECHOLAMINE REUPTAKE AND STORAGE

The Plasma Membrane Transporters

Intracellular Mechanisms

Pharmacology

Catecholamine Receptors in Physiology and Behavior

Pathophysiological States

PART E CATECHOLAMINES IN THE PERIPHERY

PART F CATECHOLAMINES IN THE CENTRAL NERVOUS SYSTEM

PART I DRUG ABUSE AND ALCOHOLISM

* These chapters present current findings in catecholamine research by the many scientists whose lives have been touched by Dr. Irwin J. Kopin.

Contributors

Abercrombie, Elizabeth D.
Acquas, Elio
Aebischer, P.
Aicher, Sue A.
Akhter, Shahab A.
Albillos, Almundena
Aloe, Luigi
Altar, C. Anthony
Amara, S. G.
Andreasen, Jan
Andringa, G.
Aoki, Chiye
Anouar, Youssef
Aperia, Anita
Apparsundaram, Subramaniam
Arai, R.
Arbuthnott, G. W.
Arnsten, A. F. T.
Aroda, Vanita R.
Arrondo, José Luis
Aston-Jones, G.
Axelrod, F. B.
Badiani, Aldo
Baffi, Judit S.
Baik, Ja-Hyun
Balboa, Maria A.
Bankiewicz, K. S.
Barak, Larry S.
Barker, E. L.
Barnes, R. D.
Bartholomé, Klaus
Bassareo, Valentina
Bauerfeind, Rudolf
Belknap, John K.
Bellinger, D. L.
Bellini, F.
Bennett, M. R.
Benovic, Jeffrey L.
Berlan, M.
Bernstein, D.
Berridge, Craig W.
Betuing, S.
Biddlecome, Gloria H.
Black, Ira B.

Blakely, Randy D.
Bloch, B.
Bolli, Geremia B.
Bonci, A.
Bönisch, Heinz
Bordet, R.
Borrelli, Emiliana
Bossé, Roger
Botton, D.
Bouloumié, A.
Boulton, Alan A.
Bouvier, Michel
Bozzi, Yuri
Breidert, Tilo
Breier, Alan
Brierley, C. H.
Bringas, J. R.
Bronner-Fraser, Marianne
Brühl, Barbara
Brunner, Han G.
Bruns, Dieter
Brüss, M.
Bruyette, D. S.
Bryan-Lluka, Lesley J.
Buck, Kari J.
Buckley, K. M.
Bunin, Melissa
Burchell, B.
Burris, Kevin D.
Cadoni, Cristina
Cahill, Larry
Cannon, Richard O., III
Cano-Abad, Mariá F.
Caohuy, Hung
Caplan, M. J.
Carboni, Ezio
Cardinaud, Bruno
Carey, Robert M.
Caron, Marc G.
Carpéné, C.
Cavalieri, E. L.
Celada, P.
Cepeda, Carlos
Cerruti, C.
Chan, June
Chen, B-H.
Chen, Kevin
Chen, Xiaoli
Cheng, Peter Y.
Cheng, Z.
Chesselet, M-F.
Chio, Chris L.
Chiueh, Chuang C.
Cho, S.
Christensen, Niels Juel
Chritton, S. L.
Chrousos, George P.
Chruscinski, A.
Cimini, V.
Clarke, D. J.
Clarke, David E.
Clawson, G. A.
Clifford, J.
Coffman, B. L.
Cohen, J.
Condron, Barry
Conti, B.
Cools, A. R.
Corbett, R.
Coughtrie, Michael W. H.

Cox, H. S.
Crabbe, John C.
Creese, Ian
Creveling, C. R.
Croke, D. T.
Cryer, Philip E.
Cummings, B. J.
Cunnane, T. C.
Curtis, Andre L.
Daaka, Yehia
Danek, K. S.
Daniels, G. M.
Das, S. K.
Davenport, R.
David, Carol
Davis, Bruce A.
DeBoer, Peter
De Camilli, Pietro
DeFelice, Louis J.
Déglon, N.
Deimling, Frauke
Delfs, J. M.
Della Rocca, Gregory J.
Desai, K. H.
Deveney, A. M.
Devilbiss, David
Dey, S. K.
Diana, Marco
Diaz, J.
Di Chiara, Gaetano
Ding, Y-S.
Dirkx, R.
Drago, J.
Driessen, Bernd
Drijfhout, W. J.
Drukarch, B.
Duerr, Janet S.
Dumas, Sylvie
Durden, David A.
Dyck, Lillian E.
Ebert, Steven N.
Edwards, Robert H.
Egan, Terry
Eiden, Lee E.
Eisenhofer, Graeme
Eisner, Gilbert M.
Ekblom, Junas
Eklöf, Ann-Christine
Eldrup, Ebbe
Elman, Igor
Elsworth, John D.
Emborg, M. E.
Ennis, E.
Erickson, J. D.
Ernsberger, Uwe
Esler, M. D.
Evinger, M. J.
Fagerquist, M.
Fausing, Susanne M.
Felder, Robin A.
Felten, D. L.
Felten, S. Y.
Ferguson, Stephen S. G.
Ferrier, C.
Feuerstein, Gloria
Finn, J. P.
Finotto, Susetta
Fiorillo, C. D.
Flatmark, Torgier
Flemström, Gunnar

Fleischer, Daniel
Florin-Lechner, Sandra M.
Ford, Anthony P. D. W.
Fowler, J. S.
Fraenkel, Yigal
Freed, C.
Frey, Kirk A.
Friedgen, Bernd
Frisby, Dennis L.
Fritsche, Michelle
Fujii, Y.
Fumagalli, Fabio
Gagnon, Alison W.
Galitzky, J.
Galli, Aurelio
Gandia, Luis
García, Antonio G.
Garcia-Arraras, Jose E.
García-Palomero, Esther
Garpenstrand, Håkan
Garris, Paul A.
Gash, Don M.
Gasnier, B.
Gassen, M.
Gatley, S. J.
Gerfen, Charles R.
Gerhardt, Greg A.
Gether, Ulrik
Gibb, James W.
Gibert, J-M.
Gioanni, Y.
Giros, Bruno
Gjedde, A.
Gliese, Martin
Glowinski, J.
Goldman-Rakic, Patricia S.
Goldstein, David S.
Golomb, Eliahu
Gonçalves, Jorge
Gonon, F.
Goodman, Linda V.
Goodman, Murray
Goodman, Oscar B., Jr.
Gorelick, David A.
Goridis, Christo
Goshima, Y.
Grabs, Detlev
Grace, Anthony A.
Graefe, Karl-Heinz
Graybiel, Ann M.
Green, M. D.
Greengard, Paul
Grisel, Judith E.
Gründemann, Dirk
Gründer, T.
Gu, H.
Guice, Karen S.
Guido, Margaret A.
Guidry, Guy L.
Guix, T.
Guo, T. Z.
Gurevich, Vsevolod V.
Guyenet, Patrice G.
Haass, Markus
Haavik, Jan
Hagström-Toft, Eva
Hahm, Sung Ho
Hale, Nannette
Hallman, Jarmila
Hammermann, R.

Hammond, H. Kirk
Han, Song-Ping
Hanson, Glen R.
Hansson, Stefan R.
Harley, Carolyn W.
Harris, J.
Hart, B.
Hastings, Teresa G.
Hawes, Brian E.
Hawrylak, Valerie A.
Head, E.
Heeringa, Marten J.
Hein, L.
Henry, J. P.
Hermel, J-M.
Hernández Guijo, M.
Herrero, Carlos J.
Hetland, Merete L.
Hilsted, Jannik
Hiremagalur, B.
Hiroi, N.
Hirsch, Jay
Hoffer, Barry J.
Hoffman, A. F.
Hoffman, Beth J.
Holmes, Courtney S.
Holtbäck, Ulla
Hoshino, Masato
Houts, Frederick W.
Hsu, Chang-Mei
Hu, Xiu-Ti
Huangfu, Donghai
Huff, Rita M.
Huff, Robin A.
Hundal, R.
Hunter, L. W.
Hurd, Y. L.
Hwa, John
Ianni, L.
Ichinose, H.
Illi, Ari
Ingham, C. A.
Innis, Robert B.
Insel, Paul A.
Isaacs, K. R.
Isambert, M. F.
Ivanova, S.
Izumi, Futoshi
Jaber, Mohamed
Jacobowitz, D. M.
Jacobs, M. C.
Jahn, Reinhard
James, Kristy M.
Javitch, Jonathan A.
Jennings, G. L.
Jentsch, D. J.
Jentsch, J. David
Jiang, Dong
Joh, T. H.
Johnson, Steven W.
Jones, Sara
Jose, Pedro A.
Joseph, J. M.
Juorio, Augusto V.
Justice, J. B.
Kable, J. W.
Kaler, Stephen G.
Karasawa, N.
Karoor, Vijaya
Kaufman, Seymour

Kawahara, H.
Kawahara, Y.
Kaye, D. M.
Keefe, Kristen A.
Keen, James H.
Kennedy, Brian P.
Khan, Naseem
Kikuchi, Y.
Kilbourn, Michael R.
Kim, G. D.
Kim, S. J.
King, C.
Kirk, K. L.
Kitai, S. T.
Knudsen, Jens H.
Knutson, Lars
Kobilka, Brian K.
Koch, Walter J.
Konishi, Shiro
Konradi, Christine
Koob, George F.
Kopin, Irwin J.
Koulu, Markku
Krieglstein, Kerstin
Krueger, Kathleen
Krupnick, Jason G.
Kruszewska, B.
Kubiak, P.
Kubo, T.
Kuhar, Michael J.
Kunikata-Sumitomo, Mayuko
Kurose, Hitoshi
Kvetňanský, Richard
Kwatra, Madan M.
Lafontan, M.
LaHoste, Gerald J.
Lajiness, Mary E.
Lakhlani, P.
Lambert, G. W.
Landis, Story C.
Lanier, Stephen M.
Lanzillotti, R.
Laslop, A.
Lefkowitz, Robert J.
Leitner, B.
Lenders, Jacques W. M.
Leon, Michael
Levey, Allan I.
Levine, Eric S.
Levine, Michael S.
Lewis, David A.
Li, B. M.
Li, Xi-Min
Li, Yu Wen
Liehr, Joachim G.
Liggett, Stephen B.
Limbird, L. E.
Lindner, Karin
Lindsay, Ronald M.
Lipska, Barbara K.
Liu, Fang
Liu, Fu-Chin
Liu, Yongjian
Livsey, Carolyn V.
Logan, J.
Lomax, Richard B.
Lovinger, D.
Lüdecke, Barbara
Lupica, C. R.
Luttrell, Deirdre K.

Luttrell, Louis M.
MacArthur, Heather
MacArthur, Linda
MacEwan, D.
MacGregor, A.
MacGregor, R. R.
Mackenzie, L.
MacMillan, L. B.
Madden, K. S.
Maggs, D. G.
Mahata, Manjula
Mahata, Sushil K.
Maisel, Alan S.
Mak, Chun
Malbon, Craig C.
Mallet, Jacques
Mannelli, M.
Männistö, Pekka T.
Marcus, M.
Marshall, John F.
Martel, Fátima
Martínez, Aurora
Mathé, J. M.
Maze, M.
McCullough, Laura A.
McGinty, Jacqueline F.
McLaughlin, W.
McPherson, Peter S.
Meloni, Rolando
Meltzer, H. Y.
Merchant, Kalpana
Meredith, I. T.
Merickel, A.
Mertes, P. M.
Metten, Pamela
Mezey, Eva
Michel, Martin C.
Mickelson, George E.
Milano, Carmelo A.
Millan, Mark J.
Milligan, G.
Mills, Paul J.
Miner, Lucinda
Minton, Allen P.
Missale, Cristina
Misu, Y.
Miyamae, T.
Miyamoto, Eishichi
Mochizuki, Naoki
Mohney, R.
Molinoff, Perry B.
Møller, Svend E.
Monaghan, G.
Moore, Holly
Morita, Kyoji
Moynihan, J. A.
Muga, Arturo
Mullaney, I.
Murphy, Dennis L.
Naes, Linda
Nagatsu, I.
Nagatsu, Toshi
Nagy, D.
Naidu, S.
Nakamura, S.
Nankova, Bistra B.
Natali, A.
Neilson, J.
Nemoto, Yasuo
Nestler, E. J.

Neve, R. L.
Nicholls, David G.
Nie, J-Y.
Nirenburg, Melissa J.
Nishiike, S.
Nishikimi, Toshio
Niznik, Hyman B.
Nomikos, G. G.
Nowick, Susana
Nutt, John G.
O'Connor, Daniel T.
O'Donnell, Patricio
Ojala-Karlsson, Pirjo
Oka, Motoo
Oldham, Keith T.
Oreland, Lars
Osterhout, Cheryl A.
Pacák, Karel
Page, Michelle E.
Paladini, C. A.
Palkovits, Miklós
Palmiter, Richard D.
Pappas, N.
Paria, B. C.
Parini, Angelo
Parmer, Robert J.
Paterson, I. Alick
Pavkovich, Luis A.
Penn, Raymond B.
Perez, Dianne M.
Pert, Agu
Peter, D.
Piascik, M.
Picetti, Roberto
Pickel, Virginia M.
Piech, Kristen M.
Pike, V. W.
Pinchasi, B.
Ping, Peipei
Pirot, S.
Pitcher, Julie
Pivirotto, P.
Plummer, Mark R.
Pollard, Harvey B.
Pontén, M.
Porter, James
Pörzgen, Peter
Post, Steven R.
Pothos, Emmanuel N.
Povlock, S. L.
Provoda, C.
Przywara, Dennis
Pupilli, C.
Rabin, D. U.
Raddatz, Rita
Raja, Srinivasa N.
Rajkowski, J.
Rand, James B.
Rauhala, Pekka
Rea, Robert F.
Rebec, George V.
Redmond, Eugene, Jr.
Regunathan, S.
Rehman, Jalees
Reis, D. J.
Revay, R.
Ricaurte, G.
Richter, Erik A.
Ridray, S.
Rios, G.

Robertson, David
Robinson, Terry E.
Rockman, Howard A.
Rodriguez, Lawrence A.
Rogan, E. G.
Rohrer, D. K.
Rohrer, Hermann
Roig-Lopez, Jose L.
Rorie, D. K.
Rosenberg, David R.
Rosenthal, Arnon
Ross, Elliott M.
Roth, Bryan L.
Roth, Robert H.
Rothman, R.
Rousseau, Guy
Rudnick, G.
Ruehl, W.
Ruffolo, Robert R., Jr.
Ruskin, David N.
Rusnak, M.
Russ, Hermann
Sabban, Esther L.
Saez, E.
Säfsten, Bengt
Sagen, Jacqueline
Sagné, C.
Saiardi, Adolfo
Sakai, M.
Samad, Tarek Abdel
Santini, Francesca
Saulnier-Blache, J-S.
Sawaguchi, T.
Schäfer, M. K-H.
Schäfers, Rafael S.
Schauble, E.
Scheinin, Mika
Schilström, B.
Schömig, Edgar
Schreiber, R.
Schreihofer, Ann M.
Schuldiner, Shimon
Schultz, Wolfram
Schütz, B.
Schwartz, J-C.
Schwartz, Joan P.
Schwinn, Debra A.
Seeman, P.
Serova, L. I.
Sesack, Susan R.
Sessler, Francis M.
Shadiack, A.
Shea, C.
Sherwin, R. S.
Shih, Jean Chen
Shipley, M. T.
Sibley, David R.
Simpson, Kimberly L.
Singer, H.
Singh, M.
Siragy, Helmy M.
Slepnev, Vladimir I.
Smith, A.
Soares-da-Silva, P.
Sokoloff, P.
Solimena, M.
Son, J. H.
Sonders, M. S.
Song, Si-Young
Song, Wen-Jie

Spielgeman, B.
Spitzenberger, Folker
Srivastava, Meera
Stack, D. E.
Starke, Klaus
Staudt, Kerstin
Steere, J. C.
Steiner, Heinz
Steiner-Mordoch, Sonia
Sterling, Carol
Stjärne, Lennart
Stoof, J. C.
Strosberg, A. Donny
Sugamori, K. S.
Sulzer, David
Sun, Baoyong
Sun, Y.
Sundlöf, Martin
Surmeier, D. James
Svensson, T. H.
Szymanski, S.
Taguchi, Katsunari
Takahashi, Nobuyuki
Takeda, N.
Takei, Kohji
Takeuchi, T.
Tallman, John F.
Tanda, Gianluigi
Tank, A. William
Tao-Cheng, J. H.
Tassin, Jean-Pol
Tatton, W. G.
Taulane, Joseph P.
Taupenot, Laurent
Taylor, Jane R.
Tephly, T. R.
Tepper, J. M.
Thien, Th.
Thierry, A. M.
Thomas, Steven A.
Thompson, J. M.
Tighe, O.
Tinti, C.
Tóth, Z.
Tseng, J. L.
Tu, Yaping
Turner, A. G.
Tyce, G. M.
Uhl, George R.
Ungerstedt, U.
Unsicker, Klaus
Urasawa, Kazushi
Usher, M.
Vaccariello, S.
Valentino, Rita J.
Valet, P.
van Biesen, Tim
Vandenbergh, David J.
vanGalen, M.
Van Tol, Hubert H. M.
Vaughan, Roxanne A.
Vaz, M.
Venkatesan, Charu
Ventura, Ana L. M.
Vermeulen, R. J.
Vernier, Phillipe
Vickery, Lillian
Vidgren, Jukka
Vieira-Coelho, M. A.
Villemagne, V.

Vincent, J-D.
Volkow, N. D.
von Kügelgen, Ivar
Waddington, J. L.
Wagstaff, John D.
Wakade, Arun R.
Wakade, Taruna D.
Wang, G-J.
Wang, John Q.
Wang, Jun
Wang, Zhi-Qin
Ward, L. E.
Waterhouse, Barry D.
Weihe, E.
Wieland, Donald M.
Weinberger, Daniel R.
Weiss, C.
Weisz, J.
West, A. E.
Westerink, B. H. C.
Westfall, Thomas C.
Weston, James A.
White, Francis J.
Wickens, J. R.
Wightman, R. Mark
Wilkinson, D.
Williams, J. T.
Williams, Timothy J.
Winkler, H.
Wölfel, Reinhard
Wong, D. F.
Wong, Dona L.
Wu, Hongjiang
Xing, Mingxhao
Yadid, Gal
Yamada, K.
Yamakuni, Tohru
Yamamoto, Hideko
Yamamoto, Toshibumi
Yan, Zhen
Yanagihara, Nobuyuki
Yang, B.
Yang, Chun Lian
Yelin, Rodrigo
Yokoi, F.
Yoo, Seung Hyun
Youdim, M. B. H.
Yu, Peter H.
Yue, J-L.
Zahniser, N. R.
Zhang, Jie
Zhang, Xu-Feng
Zhao, Ming-Ming
Zhou, Y.
Ziegler, Michael G.
Zigmond, Michael J.
Zigmond, Richard E.
Zinn, Kai
Zukowska-Grojec, Zofia
Zurn, A. D.

Preface

The catecholamines dopamine, norepinephrine, and epinephrine participate in a wide range of behaviors, physiological mechanisms, drug actions, and neurological, psychiatric, endocrine, and cardiovascular diseases. One hundred years of increasingly intensive investigation of this family of compounds have yielded an enormous body of knowledge.

Indeed, much of the recent history of scientific medicine can be written in terms of milestone discoveries based on catecholamine research. Many Nobel Prizes in Physiology or Medicine have depended on catecholaminergic systems—the 1936 prize for the theory of chemical neurotransmission; the 1938 prize for studies of cardiorespiratory reflexes; the 1948 prize for the description of the role of the hypothalamus in vegetative and emotional behaviors; the 1970 prize for the identification of norepinephrine as the sympathetic neurotransmittor and neuronal reuptake as the main means of inactivation of norepinephrine; the 1971 prize for the discovery of cAMP; the 1986 prize for the discovery of nerve growth factor; the 1988 prize for the development of β-adrenoceptor blockers; the 1992 prize for the elucidation of intracellular phosphorylation; and the 1994 prize for the discovery of G proteins.

There are several reasons for this remarkable history. First, catecholamines constitute the only neurochemical messengers for which virtually all steps in an entire functional cycle are amenable to detailed scientific study—from central neural changes to nerve impulses to transmitter release to transmitter deactivation to receptor function to cellular activation to afferent information back to the central nervous system. Second, numerous genes encoding catecholamine-synthesizing and catecholamine-metabolizing enzymes, catecholamine transporters, and catecholamine receptors have been identified, and levels of the precursor, catecholamines, and many metabolites can be measured in body fluids, enabling studies of how genotypic changes produce specific neurochemical phenotypes and clinical diseases. Third, the adrenomedullary hormonal and sympathetic neural systems, two of the most powerful and rapidly acting of the body's "stress" systems, use the catecholamines epinephrine and norepinephrine as the main effector biochemicals. Fourth, norepinephrine and dopamine are classical central neurotransmitters, thought to participate in movement, attention, memory, ideation, and neuroendocrine manifestations of distress. Fifth, catecholaminergic systems exemplify the major known means by which chemical messengers act on cells, since norepinephrine is the neurotransmitter released from nerve terminals of the sympathetic nervous system, epinephrine is the main hormone released from the adrenal medulla into the systemic circulation, and dopamine, the natriuretic catecholamine, is synthesized, is released, and acts locally in the kidneys and probably elsewhere as an autocrine/paracrine substance. And sixth, measurements of levels of endogenous DOPA, catecholamines, and catecholamine metabolites can provide information important in the diagnosis, assessment of treatment, and mechanisms of drug action in a variety of both common and rare diseases.

The goals of this book are to disseminate the latest important basic scientific and clinical medical advances about catecholamines; to bridge molecular biologic, genetic, neurochemical, psychological, and clinical interests, via research about catecholaminergic systems; to facilitate application of findings from basic research about catecholamines to issues in clinical pathogenesis and treatment; to emphasize integrative approaches for understanding the roles of catecholamines in cellular, organ, systemic, and organismic homeostasis; and to foster synthesis of molecular genetics with integrative physiology, based on an evolving understanding of catecholaminergic systems.

The contents of this book include approximately 250 mini-chapters, based on presentations at the Eighth International Catecholamine Symposium, which took place at the Asilomar Conference Center in Pacific Grove, California, October 13–18, 1996. This book provides the first comprehensive review of the general field of catecholamine research in the past eight years. During this interval, a tremendous amount of new information, obtained using molecular, genetic, neurochemical, and nuclear scanning techniques, has substantiated the important roles of catecholaminergic systems in development, cardiovascular function, metabolism, psychiatric disorders, neurodegenerative diseases, and neurocardiological disorders. The Editors attempted to foster an emerging sense of increasing integration of molecular science with clinical physiology and pathophysiology.

The nine parts were chosen to reflect continuity with previous overviews, yet incorporate new developments relevant to catecholamine research. Thus, eight of the nine parts deal with catecholamine synthesis, release, recycling, and metabolism; adrenoceptors; peripheral, central, and novel catecholaminergic systems; and development and plasticity. Because of the importance of drug abuse and alcoholism in modern society and the putative roles of central catecholamines in addictive behaviors, this part has been added.

In contrast with previous publications, within a given part the presentations range from basic to clinical. This should enable readers interested in a basic topic, such as the molecular genetics of catecholamine-synthesizing enzymes, to learn about neurogenetic diseases involving catecholamine biosynthesis; those interested in new clinical neurochemical and nuclear imaging techniques to learn about basic aspects of catecholamine release and recycling; those interested in the molecular biology of adrenoceptors to learn about adrenoceptor-based pharmacotherapy; those interested in movement disorders to learn about central neural dopaminergic function; and those interested in basic research about development to learn about ontogenetic diseases such as familial dysautonomia.

The Symposium also featured a Festschrift to honor Dr. Irwin J. Kopin. In 1957, Irv obtained a research position in the Laboratory of Clinical Science at the National Institute of Mental Health (NIMH). Over the next 10 years he published over 85 peer-reviewed papers, dealing largely with the disposition and metabolism of catecholamines. After a brief period in New York, Irv returned to the NIMH, where he headed the Laboratory of Clinical Science with distinction until 1983. He was then appointed Scientific Director of the National Institute of Neurological Disorders and Stroke, a position he held for more than a decade.

Throughout his career, Irv has demonstrated an unstinting commitment to the mission of the NIH and more generally to advancing medical scientific knowledge. He has served on numerous committees, won several medals and awards, participated on numerous scientific advisory boards, and been a co-editor or an editorial board member of more than 20 scientific journals. To date, Irv has authored or co-authored more than 650 articles, reviews, and book chapters that together constitute a major part of current scientific knowledge about catecholamines. His collaborations with many of today's leaders in the field of catecholamine research have provided the growing points that have shaped the direction of much of present-day research into catecholaminergic systems. Perhaps most importantly, Irv has been a mentor and role model for scores of postdoctoral researchers, many of whom now occupy key positions in academic medicine and the pharmaceutical industry. As a way of honoring these achievements, an asterisk (*) in the Table of Contents indicates chapters that present current findings in catecholamine research by the many scientists whose lives Irv has touched.

David S. Goldstein

PART A

CATECHOLAMINE SYNTHESIS AND RELEASE

Toshi Nagatsu* and Lennart Stjärne†

* Institute for Comprehensive Medical Science
School of Medicine
Fujita Health University
Toyoake, Aichi, Japan

† Department of Physiology and Pharmacology
Karolinska Institute
Stockholm, Sweden

Overview

Consideration of the roles of catecholaminergic systems in physiology and pathophysiology begins with mechanisms and regulation of catecholamine biosynthesis and release. Both functions are linked in that changes in release must be balanced by changes in synthesis, so it is appropriate that they are considered together. The regulation of synthesis relative to release is, however, influenced importantly by functions of catecholaminergic systems that affect or contribute to transmitter turnover independently of transmitter release. Removal of catecholamines from sites of action by membrane transporters not only operates to inactivate released transmitter, but also serves—in sequence with the vesicular transporter—to return catecholamines back into storage vesicles. This minimizes the impact of transmitter release on turnover and the requirement for ongoing synthesis. Vesicular transporters also function to counteract the considerable leakage of transmitter from stores into the axoplasm, this representing

1054-3589/98 $25.00

the main determinant of transmitter turnover at rest (1, 2). By maintaining a high concentration gradient between storage vesicles and the axoplasm, the vesicular transporter competes with the metabolizing enzyme, monoamine oxidase, within the same compartment. This enzyme is responsible for most catecholamine turnover. Thus, the processes of catecholamine biosynthesis and release that are covered in this section should also be considered in relation to the catecholamine transporter and metabolizing systems that are covered in the next two sections of this volume.

I. Regulation and Expression of Tyrosine Hydroxylase

As the first and rate-limiting enzyme in catecholamine biosynthesis, tyrosine hydroxylase (TH) is subject to multiple controls. Short-term regulation includes activation of TH in response to increased nerve traffic and negative feedback control through end-product inhibition. Long-term regulation involving production of new TH enzyme occurs at the levels of transcription and translation. Considerable research continues to focus on these multiple levels of control.

Short-term activation of TH involves phosphorylation at four sites, corresponding to serine residues 8, 19, 31, and 40. Muga and colleagues (p. 15) report that after phosphorylation of Ser-40 of human TH type 1, the dissociation rate of the endogenous inhibitor, dopamine, increased due to a conformational change in the enzyme. Thus, phosphorylation of Ser-40 appears to activate TH by dissociation of dopamine at the active site iron.

Yanagihara and coworkers (p. 18) propose that a single signal transduction mechanism may regulate acetylcholine-induced synthesis and secretion of catecholamines in adrenal medullary cells. Ca^{2+}/calmodulin-dependent protein kinase II stimulates both catecholamine synthesis and secretion by phosphorylating several proteins, including TH, chromogranin A, and proteins in chromaffin granule membranes.

In the rat superior cervical ganglion, increased preganglionic nerve activity leads to both acute and long-term increases in TH activity in postganglionic neurons. Both cholinergic and noncholinergic neurotransmitters participate in this activation. As noted by Zigmond (p. 21), among noncholinergic preganglionic neurotransmitters, vasoactive intestinal peptide, peptide histidine isoleucineamide, and pituitary adenylyl cyclase–activating peptide have received attention. These peptides may acutely activate TH via phosphorylation or may be involved in triggering TH induction. Substance P selectively blocks nicotinic activation of TH; the physiological significance remains unknown.

Alterations in the rate of TH gene transcription, TH mRNA stabilization, and other post transcriptional mechanisms all represent potential long-term mechanisms for regulating TH expression in response to various stimuli. As shown by Tank *et al.* (p. 25), among these mechanisms, transsynaptic regulation and regulation by increased contact between neighboring catecholamine cells lead to long-term induction of TH. Transsynaptic mechanisms appear responsible for persistent stimulation of TH gene transcription in the adrenal medulla. Transsynaptic regulation occurs via both cholinergic and noncholinergic receptors, with multiple signaling pathways that include three major second messengers for regulating the TH gene: cyclic adenosine monophosphate (cAMP), diacylglycerol, and calcium. Increased cell–cell contact increases the rate of TH gene transcription in cultured rat pheochromocytoma cells. Cyclic AMP does not appear to mediate this response.

Tank and coworkers (p. 25) provide evidence that the cAMP response element (CRE) may participate in protein kinase A- and protein kinase C-mediated induction of TH gene expression. In rat pheochromocytoma cells, both cAMP and phorbol ester increase the transcription rate of TH gene via CRE, but through distinct signaling pathways.

Yamakuni and colleagues (p. 30) report a new regulatory protein of catecholamine synthesizing-enzyme expression: V-1 protein. This occurs in various central and peripheral catecholaminergic neurons. The V-1 protein consists of 117 amino acids, with 2.5 contiguous repeats of the cdc 10/SW16 motif. V-1-overexpressing PC12 cells have increased dopamine contents, TH mRNA and activity, mRNA for aromatic-amino-acid decarboxylase (AAAD), dopamine-β-hydroxylase (DBH), but not phenylethanolamine-*N*-methyltransferase (PNMT). These results suggest that the V-1 protein plays a role in the control of catecholamine synthesis, altering coordinated transcriptional regulation of genes for TH, AADC, and DBH.

Although several cis-acting DNA motifs in the upstream sequence of the TH gene have been characterized to control TH gene expression, some controversial results remain. Joh and coworkers (p. 33) note that each cell type expresses and regulates the TH gene specifically, depending on differences in cell–type–specific transcriptional factors. Previous data from *in vitro* studies suggest that relatively small DNA sequences upstream of the TH gene, such as the cAMP-responsive element at approximately −40 bp and the activator protein-1 (AP1) element at approximately -200 bp, are sufficient to regulate cell-type–specific expression of TH. In contrast, Joh *et al.* show that at least 9.1 kb upstream of the rat TH gene is necessary for tissue-specific and developmentally correct expression of TH in transgenic mice and transcriptional induction of TH in the brain by reserpine administration or cold exposure. Also, for the strictly region-specific and developmental stage-specific expression of human TH gene, not only the 5′ upstream but also the structural portion of the TH gene or 3′-downstream regions may be required (3, 4).

II. Tyrosine Hydroxylase Transgenics and Deficiency Syndromes

Transgenic approaches have provided useful information about effects of TH deficiency (5, 6, 7). Targeted disruption of TH causes severe catecholamine depletion and perinatal death in homozygote TH −/− mice (5, 6), probably via changes in cardiac functions that depend on catecholamines (6). The lethality can be prevented either pharmacologically with L-dopa (5) or by transfer of a human TH transgene into the homozygous mice (6). The preceding findings probably indicate why there have been no reports of complete TH deficiencies in the clinical literature. Rather, clinical syndromes of TH deficiency invariably present as a partial loss of activity, often due to deficiencies in the availability of crucial cofactors such as tetrahydrobiopterin (BH_4).

The three aromatic amino acid hydroxylases, phenylalanine hydroxylase, tryptophan hydroxylase, and TH all depend on BH_4 as a cofactor. Homeostasis of BH_4 requires the three BH_4-synthesizing enzymes and two BH_4-regenerating enzymes. As pointed out by Kaufman, (p. 41) mutations of these BH_4-related enzymes causes variant forms of phenylketonuria (PKU), with deficient synthesis of catecholamines and serotonin and with hyperphenylalaninemia. Part of the treatment for these PKU variants consists of administration of the catecholamine precursor, L-dopa, and the serotonin precursor, L-5-hydroxytryptophan, bypassing the metabolic blocks. In addition, for defects in BH_4 synthesis, administration of BH_4 constitutes an essential part of the treatment.

Guanosine triphosphate (GTP) cyclohydrolase I (GCH) contributes to the regulation of TH activity as the first and rate-limiting enzyme in BH_4 synthesis. Human and mouse GCH genes have been cloned, and the human gene has been mapped to chromosome 14q22.1-q22.2. As noted by Nagatsu and Ichinose (p. 44), dominant dystonia, an hereditary progressive dystonia with marked diurnal fluctuation (also termed Segawa's disease and dopa-responsive dystonia) results from mutations of the GCH gene. The patients are heterozygous for these mutations, which would cause decreased TH activity and dopamine deficiency in the substantia nigra. In contrast with the dominant dystonia, recessive GCH deficiency, in homozygous patients, results from missense mutations of the GCH gene, producing complete loss of GCH activity and BH_4 and severe neurological symptoms, probably from a combination of hyperphenylalaninemia (due to low phenylalanine hydroxylase activity), catecholamine deficiency (due to low TH activity), and serotonin deficiency (due to low tryptophan hydroxylase activity).

Bartholomé and Lüdecke (p. 48) report that recessive dopa-responsive dystonia can also result from mutations of the TH gene itself. The mutation, Gln 381 Lys, of human TH type 1 causes a partial decrease in TH activity to about 15% of the wild-type form, causing a mild form of dopa-responsive dystonia. Another mutation, Leu 205 Pro, causes a severe decrease in TH activity and a more severe from of dopa-responsive dystonia.

Genetic factors may play an important role in the pathogenesis of neuropsychiatric conditions such as manic-depressive illness and schizophrenia. Human TH generates four different mRNA species by alternative splicing. Three new species of mRNA derived from alternative splicing of exon 3 have been found. Populations of novel TH mRNA are increased in adrenal medullary tissue of patients with progressive supranuclear palsy, which may have a genetic etiology. For the genetic analysis of psychiatric diseases, the detection of allelic association (linkage disequilibrium) between a marker allele and the disease phenotype is a useful strategy. A significant association has been found between a microsatellite localized in the first intron of the TH gene and manic-depressive illness as well as schizophrenia. A rare variant (T10p) of a common allele (T10i) of this repeated sequence is found among schizophrenic patients but is absent among controls free of personal and familial history of psychiatric diseases, in three different populations. The T10i and T10p alleles are able to bind nuclear proteins and may act as transcriptional regulators. As pointed out by Meloni and coworkers (p. 50), these findings could have important implications for our understanding of the genetic predisposition to schizophrenia.

III. Other Catecholamine-Synthesizing Enzymes

Regulation of other catecholamine synthesizing-enzymes, such as DBH and PNMT, is controlled importantly at the transcriptional level. Sabban and Nankoba (p. 53) describe increased binding of AP1-like factors to a composite regulatory element in the promoter (DBH-1, position -175-145) containing two adjacent cAMP regulatory elements (CRE1, CRE2) in nuclear extracts from repeatedly immobilized animals. The rat DBH CRE1 contains a core CRE/AP1 sequence indentical to CRE2 in the enkephalin gene. Different sets of transcription factors, with different kinetics of induction, interact with the DBH-1 composite enhancer and are likely candidates for modulating DBH transcription in response to various physiologically relevant stimuli.

Using targeted disruption of the DBH gene, Thomas and Palmiter (p. 57) show the importance of DBH in development, physiology, and behavior. Most DBH −/− mice die *in utero* (8). However, survival can be enhanced by perinatal administration of dihydroxyphenylserine, a direct precursor of norepinephrine. DBH −/− mice have impaired adaptation to cold and elevated basal metabolic rates without abnormalities in thyroid hormone levels. This suggests that norepinephrine (NE) plays a fundamental role in determining basal metabolic rate independent of thyroid activity. A deficit in maternal behavior in DBH −/− females may also exist. The results of cross-fostering between DBH −/− females and DBH +/− females suggest that an important interaction mediated by NE occurs between dam and neonate during the first 24 hr for establishing maternal behavior. The results of behavioral tests indicate that NE plays an important role in motor learning and performance and in the retention of several behaviors. Restoration of NE with dihydroxyphenylserine eliminated the motor deficits and improved fertility in males, indicating that these differences are due to physiological absence of NE rather than lasting developmental defects secondary to NE deficiency.

As noted by Robertson and Hale (p. 61), postural hypotension is the most pronounced clinical finding in patients with complete DBH deficiency. This can be successfully treated with dihydroxyphenylserine, providing encouragement for development of treatments for other autonomic disorders.

Kaler *et al.* (p. 66) describe a partial deficiency of DBH in patients with Menkes disease or occipital horn syndrome. These X-linked recessive disorders of copper metabolism produce a spectrum of neurological symptoms, ranging from fatal infantile neurodegeneration to mild dysautonomia. The basis for DBH deficiency in these syndromes is due to to impaired copper incorporation into the DBH apoenzyme.

Expression of the PNMT gene involves coordination of three major regulatory mechanisms: hormonal regulation by glucocorticoids, neural regulation by cholinergic stimuli, and cell-specific determinants of expression. A glucocorticoid response element (GRE) encoded within 5′ regulatory region mediates hormonal regulation of the PNMT gene. Mannelli and colleagues (p. 69) report the clinical relevance on the hormonal control of PNMT from studies of patients with hyper- and hypo-cortisolism. Their results show that normal cortisol secretion from the adrenal cortex is necessary for PNMT activation and normal epinephrine secretion; however, in patients with Cushing's syndrome, adrenal PNMT activity and overall sympathetic-adrenal activity are reduced, for as yet unexplained reasons.

As noted by Evinger (p. 73), neural regulation of the PNMT gene is modulated through separate muscarinic and nicotinic mechanisms, involving not only different transduction pathways, but also distinctive response regions of the gene. Wong and colleagues (p. 77) describe how Egr-1, a member of the immediate early gene family, is a transcription factor mediating neural control of the PNMT gene. Muscarinic receptor activation of PNMT gene transcription in RS1 cells appears to result from induction of the Egr-1 transcription factor, secondary to mobilization of intracellular calcium.

Cell-specific determinants play critical roles in specifying the highly restricted expression of the PNMT gene. Some of these regulatory elements of the PNMT gene appear to share sequence similarities with previously characterized neural-specific silencers.

IV. Exocytosis

A. Caveats

In considering research about exocytosis, a few caveats are in order. Hormones and neurotransmitters (with the exception of nitric oxide) are stored in vesicles at extremely high concentrations (100 mM in nerves and 600 mM in chromaffin cells) but generally also occur in the cytosol at low micromolar concentrations. It is important, therefore, to use methods that distinguish release by exocytosis of vesicular contents from release by carrier-mediated molecular leakage from the cytosol (9).

No current method measures exocytosis per se. What is measured are signals triggered by interaction of released transmitter with a (biological or artificial) sensor. At best, the amplitude and time course of the signal reflects the rise and fall in the transmitter concentration at the sensor (i.e., exocytosis minus clearance) (10). Even this may not apply if the sensor is saturated or desensitized by released transmitter (11). The signal is thus not always a linear function of the rate of exocytosis.

Much progress in the field has been made by using simple preparations to study in detail individual steps in transmitter exocytosis. Findings in these systems, however, may not necessarily apply to intact adult endocrine cells or neurons of adult tissue *in vivo*.

Methods now exist that can resolve release of single transmitter quanta from individual visualized boutons or sympathetic nerve varicosities. Such studies focus on high probability release sites, which may be exceptions. Methods with less resolution probably would give a better overview, because most of the approximately 20,000 varicosities in the terminals of a single sympathetic neuron are silent most of the time (9). Moreover, slow, long-term changes in catecholaminergic outflows (e.g., diurnal rhythms) require methods with low temporal resolution (e.g., microdialysis in freely moving rats), reported by Westerink and coworkers (p. 136).

B. Cellular and Molecular Mechanisms of Exocytosis

Constitutive exocytosis, which goes on continuously in all eukaryotic cells, utilizes in principle the same molecular mechanisms as Ca^{2+}-triggered exocytosis of hormones or neurotransmitters (12). The exact sequence of events remains to be worked out. According to the simplest version of the SNARE hypothesis, all vesicles generated in the endoplasmic reticulum, Golgi complex, lysosomal system, or plasmalemma carry a v-SNARE (*v* for vesicle) that interacts with a cognate t-SNARE (*t* for target) on the respective acceptor membrane.

In this hypothesis, eight proteins or protein complexes regulate exocytosis: the v-SNARE synaptobrevin, the plasmalemmal t-SNAREs synaptosome-associated protein 25 (SNAP-25) and syntaxin 1, two soluble proteins, NSF and α,β,γ-SNAP (soluble NSF-accepting protein), a voltage-gated Ca^{2+} channel that tends to complex with syntaxin on the plasma membrane, and synaptotagmin, an integral vesicle membrane protein whose N-terminal spans the membrane while the C2a domain on the outside binds Ca^{2+} with fourfold cooperativity.

The SNARE hypothesis helps explain how vesicles dock at specific release sites and the need for priming to render them "exocytosis-competent." A still unsolved problem is how these steps lead to fusion of the lipid membranes and formation of a fusion pore (13, 14). This question is addressed in the chapter by Pollard and coworkers (p. 81), who propose that the fusion protein may be synexin, a member of a family of Ca^{2+}- and phospholipid-binding proteins that stimulate exocytosis in permeabilized chromaffin cells. The hypothesis seems plausible for exocytosis of large vesicles (in chromaffin cell and nerve) but may not apply for small synaptic vesicles. Direct evidence is lacking that removal of synexin prevents fusion.

C. Quantal Release

The first direct evidence of quantal release was obtained by amperometric recording of the evoked catecholamine oxidation current spikes at a small carbon fiber electrode at the surface of a chromaffin cell, in combination with membrane capacitance measurements (15, 16). The results indicated that chromaffin cells sometimes release all, sometimes only part, of the catecholamine content of single vesicles. A corresponding analysis from sympathetic nerve has not been reported. As reported by Bruns and Jahn (p. 87), from a study of serotonergic leech Retzius cells grown in culture, nerve impulses in the axon give rise to either "small" or "large" serotonin oxidation current spikes. Small spikes were never preceded by a "foot," but large spikes often were. The small and large spikes were thought to represent all-or-none exocytosis of the serotonin contents of small or large synaptic vesicles, the "abortive foot" partial exocytosis of large vesicles. Exocytosis from small vesicles may thus always be all-or-none, whereas exocytosis from large vesicles may be all-or-none or graded.

D. Calcium Dependence of Exocytosis

Endocrine cells and nerve terminals possess a wide range of voltage-, receptor-, G-protein-, or second messenger-operated Ca^{2+} channels. Intracellular Ca^{2+} is distributed in an extremely nonuniform manner and controls many different functions related to hormone or transmitter exocytosis.

At the high concentrations ($\geq$100 μM) that occur for tens of microseconds within tens of nanometers from the mouths of open N-, P- or Q-type channels at active zones, binding of Ca^{2+} to low-affinity receptors triggers exocytosis. At much lower concentrations ($\leq$1 μM), Ca^{2+} binds to high-affinity receptors throughout the cell, probably controlling a wide range of "supporting" functions (e.g., phosphorylation of synapsins and thereby the size of the releasable vesicle pool) (17). The chapter by Garcia and coworkers (p. 91) reports results of patch-clamp analyses of the effects of selective channel blocking agents on the K^+-pulse-evoked secretory activity of single chromaffin cells. Ca^{2+} was found to enter the cells through L-, N-, P-, Q-, and R-type channels. However, not all of these channels played a role in evoked catecholamine secretion. In noradrenergic cells, secretion was mediated mainly by L-type channels, whereas in adrenergic cells, secretion was mediated mainly by Q-type channels. These findings do not apply to sympathetic nerves, where L-type channel blockers inhibit the secretory response to high K^+ but not to electrical nerve stimulation, and may not apply to the role of L-channels in mediating nerve stimulation-induced hormone release from noradrenergic chromaffin cells *in situ*.

Moreover, the channels through which Ca^{2+} enters the cytosol to trigger exocytosis may vary with the stimulation parameters. Thus, as reported by Cunnane and Smith (p. 94), blockade of N-type channels abolishes the secretory response of sympathetic nerves at low but not at high frequency. The residual release was still triggered by Ca^{2+} entry, but now through non-L, -N, -P, or -Q voltage-activated channels. With time and repeated activation, the response became increasingly dependent on Ca^{2+} entry from intracellular stores via ryanodine-sensitive channels (Ca^{2+}-induced Ca^{2+} release).

E. Probability of Monoquantal Release

As a rule, individual release sites in nerves either ignore the action potential or release a single transmitter quantum; individual release sites vary greatly in the probability of monoquantal release. Electrophysiological analyses in guinea pig and mouse vas deferens or rat tail artery indicate a probability in the average varicosity of less than or equal to .03 (18, 19). However, as reported by Bennett (p. 98) both electrophysiological and optical analyses of secretory activity of sympathetic nerve terminals indicate highly nonuniform p values ranging from less than or equal to .03 to greater than or equal to .5 in individual varicosities.

Monoquantal release may be the rule, but as noted by Mickelson and colleagues (p. 144), field stimulation of brain slices with single pulses at high voltage apparently releases no more than 1 quantum from single sites. Studies of NE release from sympathetic nerves have reported analogous findings (20).

F. The Secretosome Hypothesis

Differences in size of active zones complicate comparison of values for p in individual sites in different neurons. In some nerve terminals (e.g., amphibian neuromuscular junction), active zones are large with 50 or more docked synaptic vesicles (by electron microscope criteria); in others (e.g., sympathetic varicosites), EM criteria for active zones are not always visible. In the search for a common denominator, Bennett (p. 98) found that the decay of Ca^{2+} in single boutons or varicosities of adult pre- or postganglionic sympathetic nerve terminals, just as in amphibian neuromuscular junction, paralleled the changes in p induced by stimulation with single pulses or short high-frequency trains, and that the fourfold Ca^{2+} cooperativity characteristic of transmitter release did not exist in immature neurons but appeared in parallel with the C2a domain of the integral vesicle membrane protein synaptotagmin, which binds Ca^{2+} with fourfold cooperativity. Based on these findings, Bennett proposed a hypothesis in which the basic secretory unit in nerve terminals is the secretosome. This unit, consisting of a "primed" synaptic vesicle docked at the plasma membrane by binding to the SNARE receptor, syntaxin, and thereby complexed with a voltage-gated Ca^{2+} channel, uses the C2a domain of synaptotagmin at the vesicle surface as its low-affinity Ca^{2+} receptor. In its fully loaded form, the secretosome would always respond to Ca^{2+} entry by exocytosis of the vesicle contents. The Ca^{2+} concentration elsewhere in the terminal would regulate the availability of primed vesicles, which would in turn determine the value for p.

In apparent contrast to Bennett's findings in intact organs, Wakade and coworkers (p. 102) report that "immature" axons of cultured sympathetic neurons grown in isolation had an abnormally high N-type Ca^{2+} channel density and responded to electrical field stimulation with an abnormally high per pulse [^{3}H]-NE secretion, with negative frequency-dependence. Coculture with appropriate target cells normalized these features. In the cultured neurons, "maturation" of the secretory machinery may thus involve downregulation of Ca^{2+} channels.

G. Autoregulation of the Release Probability

Most chromaffin cells and sympathetic varicosities possess autoreceptors for endogenous ligands. As described by Westfall, and coworkers (p. 106), exogenous neuropeptide Y (NPY) depresses catecholamine synthesis via Y_3 and catecholamine release via Y_2 presynaptic receptors. It is unclear, however, to what extent hormones and transmitters influence either synthesis, or values for p of exocytosis, of the contents of different vesicles. The physiological role of this presynaptic modulation remains poorly understood.

In some but not all sympathetic nerves, released NE autoregulates p both by positive feedback, via β_2-adrenoceptors and negative feedback via α_2-

adrenoceptors (9). Nicholls (p. 110) reports that glutamate released from rat brain synaptosomes exerts both positive and negative feedback control of the release machinery, via different G-protein–coupled glutamate receptors. Positive feedback, reinforced by cooperative interaction with arachidonic acid released from the postjunctional cell, operated via protein kinase C–induced inhibition of K^+ channels. Negative feedback operated via G-protein–mediated inhibition of Ca^{2+} channels; this effect was counteracted by activation of protein kinase C.

The finding that chromaffin vesicles both store and release catecholamines and ATP at a fixed molar ratio raises the question of whether released ATP itself exerts effects on extracellular targets. The chapter by Garcia and coworkers (p. 91) reports that both ATP and enkephalins released from chromaffin cells autoinhibited the secretory machinery by inhibiting Ca^{2+} entry through non-L channels. In line with this is confirmation that exogenous ATP and opioids inhibit non-L Ca^{2+} channels in these cells, but only released ATP may produce autoinhibition (21).

H. Cotransmission: Are NE, ATP, and NPY Stored in and Released from the Same Vesicles?

NE and ATP are stored both in small and large dense-core vesicles (SDVs, LDVs), but neuropeptides only in LDVs. Release of NPY, therefore, does not parallel that of NE and ATP, as exocytosis from small and large vesicles is regulated differentially (22). It is unclear if all SDVs and LDVs store and release NE and ATP in the same fixed molar ratios. For example, as reported by Stjärne and von Kügelgen *et al.* (p. 120), values for the parameters used to estimate the release of NE and ATP per pulse often vary in parallel. Results of experiments using overflow techniques to measure exocytosis of NE and ATP in guinea pig vas deferens or myenteric plexus synaptosomes have, however, suggested differential release of NE and ATP; thus, the issue remains open.

I. Cotransmission: Postsynaptic Actions and Interactions

Released sympathetic cotransmitters can inhibit or potentiate each other's actions. As reported by Zukowska-Grojec (p. 125), NPY exerts vasoconstrictor effects only at larger nanomolar concentrations; at resting subnanomolar levels, NPY acts as a trophic factor (e.g., for endothelial cells). Enzymatic cleavage of NPY inactivates Y_1 but not Y_2 functions; just as intact NPY, the metabolite stimulates vascular smooth-muscle proliferation, growth of endothelial cells, and angiogenesis.

Jacobowitz and coworkers (p. 37) reported triple colocalization of TH, calretinin (CR), and calbindin (CB) in dopaminergic fibers of the intermediate lobe of the pituitary. The calcium-binding proteins CR and CB may function as modulatory factors in the dopaminergic inhibitory action on release of α-melanocyte stimulating hormone (α-MSH) in the intermediate lobe.

J. Geometry of Noradrenergic Neuromuscular Transmission

Release of single quanta causes the norepinephrine concentration at the release site to reach a theoretical maximum near 100 mM (10). This is four to five orders of magnitude higher than the concentration of exogenous NE required to trigger a maximal contraction in preparations such as rat tail artery (23). The size of the α_2-adrenoceptor–mediated neurogenic contraction triggered by release of single quanta from ≤1% or less of the varicosities is mediated mainly by α_2-adrenoceptors and amounts to 0.1% or less of the neurogenic maximum. However, a 15-fold increase in the number of pulses (at 20 Hz) causes the response to be driven mainly by α_1-adrenoceptors and increases the response amplitude by 500-fold. The steep increase in contractile amplitude during high-frequency trains thus reflects an increasing recruitment of previously unstimulated receptors. As suggested by the work of Bennett (p. 98), the presumed explanation for this phenomenon is that the value of p for quantal release is non-uniform. Pulses in high-frequency trains are thus likely to release quanta repeatedly from the same high-probability sites. This would saturate local NE reuptake and cause released NE to diffuse further away (i.e., increase the radius of the hotspot) and increasingly activate extrajunctional, high-affinity α_1-adrenoceptors (23).

K. Advantages and Disadvantages of Microdialysis and Voltammetry

The main strength of *in vivo* microdialysis is its high sensitivity and specificity, and the main weakness, its limited spatial and temporal resolution. The method is particularly useful for study of slow, widespread changes. As noted by Abercrombie *et al.* (p. 133), this method may be used to characterize a number of similarities as well as differences between dopamine release (probably by exocytosis) in dendrites (of dopaminergic neurons in substantia nigra) and terminals (in striatum). The paper by Westerink and coworkers (p. 136) illustrates the usefulness of this method to sample neurotransmitters, administer drugs, or examine diffusion in the brain. Their study of the 24-hr time course of NE release from nerve terminals in the pineal gland that drive melatonin release provides a striking example of the usefulness of this "slow" method for attacking selected problems.

In contrast to microdialysis, the electrochemical oxidation current at a small carbon fiber electrode resolves dopamine (DA) or NE release on a millisecond time scale, with micrometer spatial resolution. As reported by Gonon and Bloch (p. 140), the method makes it possible to study *in vivo* in real-time the relationship between the concentration of DA in the nucleus accumbens (released by nerve impulses or by chemical stimulation of the DA cell bodies) and the extracellularly recorded D_1-receptor–mediated increase in firing rate of target cells. The results suggest that DA-ergic transmission at this synapse is excitatory and probably mediated via perisynaptic D_1-receptors. Mickelson and colleagues (p. 144) describe how the same method can be used to study per-pulse quantal release and clearance of DA in rat brain. In the striatum *in vivo*, nerve impulses apparently normally released single DA quanta (approximately 1000 molecules). The released DA diffused rapidly out of the narrow synaptic cleft to drive effector responses via extrasynaptic receptors located within a few micrometers of the margin of the cleft. DA release may not be monoquantal under all conditions, however. In field-stimulated brain slices (biphasic pulses, 2 msec in each phase, variable voltage), per-pulse DA release increased biphasically (at high but not at low Ca^{2+}) as a function of stimulus voltage. The proposed explanation is that, in contrast to the "normal" nerve impulse, each high-voltage pulse of Ca^{2+} often caused individual sites to release two quanta.

V. Future Directions

The considerable progress at molecular levels of control of catecholamine synthesis has already shown direct relevance to the identification and treatment of a number of clinical abnormalities of catecholamine biosynthesis. Typically these have involved gross and rare abnormalities of catecholamine biosynthesis, but future progress is likely to lead to identification of more subtle abnormalities that may be involved in more common disease processes. For example, the elucidation of cell-specific transcription factors makes it possible that some psychiatric or neurological disorders may involve abnormalites of catecholamine synthesis within specific cell types. Progress in this area can benefit from the development of novel "slow or chronic knockout" methods that target specific cell types to create animal models of neurodegenerative diseases (e.g., Parkinson's disease). Also, the recently developed "conditional knockout" experiments of a specific protein of a specific cell type at a specific postnatal period may be useful to elucidate the functional roles of the protein in catecholamine synthesis or release and of the specific cell type (7, 24).

Sensitivities of modern molecular approaches are suggesting that catecholamine biosynthesis may occur in novel cell types not previously considered in the investigation of catecholamine systems (see section on Novel Catecholamine Systems). Elucidation of cell-specific transcription factors for these systems may warrant attention should these novel catecholamine systems prove important in physiology.

Over the last few years, much has been learned about the molecular and cellular mechanisms that control exocytosis in chromaffin cells, but little in monoaminergic neurons. Still poorly understood are the molecular mechanisms that cause transmitter release from sympathetic varicosities to be mono- (or at the most, oligo-) quantal and 90% or more of the terminal branches to be virtually silent while the remaining 10% or less are "active" with p in individual varicosities ranging from .1 to .5 or higher. This field will almost certainly develop rapidly during the next few years through, for example, development of knockouts of docking or fusion proteins and by recording of quantal release from individual visualized varicosities. Progress in this area is important both in its own right and because it may help to elucidate the control of transmitter release in other peripheral or central neurons with terminals of similar morphology, regardless of the nature of the transmitter.

It can be anticipated that future research in the area of exocytosis will provide improved understanding of the molecular and cellular basis of the malfunctioning catecholamine release that almost certainly underlies a wide variety of clinical conditions, ranging from arterial hypertension to schizophrenia. Although much research in this field already exists, little progress has been made. Improved understanding of the molecular and cellular bases of catecholamine release will ensure that the relevant questions are asked and the appropriate parameters measured.

References

1. Eisenhofer, G., Esler, M. D., Meredith, I. T., Dart, A., Cannon, R. O., Quyyumi, A. A., Lambert, G., Chin, J., Jennings, G. L., and Goldstein, D. S. (1992). Sympathetic nervous function in the human heart as assessed by cardiac spillovers of dihydroxyphenylglycol and norepinephrine. *Circulation* **85**, 1775–1785.
2. Eisenhofer, G., Friberg, P., Rundqvist, B., Quyyumi, A. A., Lambert, G., Kaye, D. M., Kopin, I. J., Goldstein, D. S., and Esler, M. D. (1996). Cardiac sympathetic nerve function in congestive heart failure. *Circulation* **93**, 1667–1676.
3. Kaneda, N., Sasaoka, T., Kobayashi, K., Kiuchi, K., Nagatsu, I., Kurosawa, Y., Fujita, K., Yokoyama, M., Nomura, T., Katsuki, M., and Nagatsu, T. (1991). Tissue-specific and high-level expression of the human tyrosine hydroxylase gene in transgenic mice. *Neuron* **6**, 583–594.
4. Sasaoka, T., Kobayashi, K., Nagatsu, I., Takahashi, R., Kimura, M., Yokoyama, M., Nomura, T., Katsuki, M., and Nagatsu, T. (1992). Analysis of the human tyrosine hydroxylase promoter-chloramphenicol acetyltransferase chimeric gene expression in transgenic mice. *Mol. Brain Res.* **16**, 274–286.
5. Zhou, Q. Y., Quaife, C. J., and Palmiter, R. D. (1995). Targeted disruption of the tyrosine hydroxylase gene reveals that catecholamines are required for mouse fetal development. *Nature* **374**, 640–643.
6. Kobayashi, K., Morita, S., Sawada, H., Mizuguchi, T., Yamada, K., Nagatsu, I., Hata, T., Watanabe, Y., Fujita, K., and Nagatsu, T. (1995). Targeted disruption of the tyrosine hydroxylase locus results in severe catecholamine depletion and perinatal lethality in mice. *J. Biol. Chem.* **270**, 27235–27243.
7. Kobayashi, K., Morita, S., Sawada, H., Mizuguchi, T., Yamada, K., Nagatsu, I., Fujita, K., Kreitman, R. J., Pastan, I., and Nagatsu, T. (1995). Immunotoxin-mediated conditional disruption of specific neurons in transgenic mice. *Proc. Natl. Acad. Sci. U.S.A.* **92**, 1132–1136.
8. Thomas, S. A., Matsumoto, A. M., and Palmiter, R. D. (1995). Noradrenaline is essential for mouse fetal development. *Nature* **374**, 643–646.
9. Stjärne, L. (1989). Basic mechanisms and local modulatin of nerve impulse-induced secretion of neurotransmitters from individual sympathetic nerve varicosities. *Rev. Physiol. Biochem. Pharmacol.* **112**, 1–137.
10. Clements, J. D. (1996). Transmitter timecourse in the synaptic cleft: Its role in central synaptic function. *Trends Neurosci.* **19**, 163–171.
11. Jones, M. V., and Westbrook, G. L. (1996). The impact of receptor desensitization on fast synaptic transmission. *Trends Neurosci.* **19**, 96–101.
12. Südhoff, T. C. (1995). The synaptic vesicle cycle: A cascade of protein-protein interactions. *Nature* **375**, 645–653.
13. Burgoyne, R. D., and Morgan, A. (1995). Ca^{2+} and secretory-vesicle dynamics. *Trends Neurosci.* **18**, 191–196.

14. Gillis, K. D., Mössner, R., and Neher, E. (1996). Protein kinase C enhances exocytosis from chromaffin cells by increasing the size of the readily releasable pool of secretory granules. *Neuron* **16,** 1209–1220.
15. Wightman, R. M., Jankowski, J. A., Kennedy, R. T., Kawagoe, K. T., Schroeder, T. J., Leszczyszyn, D. J., Near, J. A., Diliberto, E. J., and Viveros, O. H. (1991). Temporally resolved catecholamine spikes correspond to single vesicle release from individual chromaffin cells. *Proc. Natl. Acad. Sci. U.S.A.* **88,** 10754–10758.
16. Chow, R. H., von Rüden, L., and Neher, E. (1992). Delay in vesicle fusion revealed by electrochemical monitoring of single secretory events in adrenal chromaffin cells. *Nature* **356,** 60–63.
17. Augustine, G. J., Betz, H., Bommert, K., Charlton, M. P., DeBello, W. M., Hans, M., and Swandulla, D. (1994). Molecular pathways for presynaptic calcium signaling. *In* Molecular and Cellular Mechanisms of Neurotransmitter Release. (L. Stjärne, P. Greengard, S. Grillner, T. Hökfelt, and D. Ottoson, eds.) pp. 139–154. Raven Press, New York.
18. Cunnane, T. C., and Stjärne, L. (1984). Transmitter secretion from individual varicosities of guinea-pig and mouse vas deferens. *Neuroscience* **13,** 1–20.
19. Åstrand, P., and Stjärne, L. (1989). On the secretory activity of single varicosities in the sympathetic nerves innervating the rat tail artery. *J. Physiol.* (*Lond.*) **409,** 207–220.
20. Alberts, P., Bartfai, T., and Stjärne, L. (1981). Sites and ionic basis of α-autoinhibition and facilitation of [^{3}H]noradrenaline secretion in guinea-pig vas deferens. *J. Physiol.* (*Lond.*) **312,** 297–334.
21. Currie, K. P. M., and Fox, A. P. (1996). ATP serves as a negative feedback inhibitor of voltage-gated Ca^{2+} channel currents in cultured bovine adrenal chromaffin cells. *Neuron* **16,** 1027–1036.
22. Lundberg, J. M., and Hökfelt, T. (1986). Multiple co-existence of peptides and classical transmitter in peripheral autonomic and sensory neurons: Functional and pharmacological implications. *Prog. Brain Res.* **68,** 241–262.
23. Stjärne, L., and Stjärne, E. (1995). Geometry, kinetics and plasticity of release and clearance of ATP and noradrenaline as sympathetic cotransmitters: Roles for the neurogenic contraction. *Prog. Neurobiol.* **47,** 45–94.
24. Kuhn, R., Schwenk, F., Aguet, M., and Rajewsky, K. (1995). Inducible gene targeting in mice. *Science* **269,** 1427–1429.

Arturo Muga,* José Luis R. Arrondo,* Aurora Martínez,†
Torgeir Flatmark,† and Jan Haavik†

* Department of Biochemistry and Molecular Biology
Faculty of Science
University of the Basque Country
E-48080 Bilbao, Spain

† Department of Biochemistry and Molecular Biology
University of Bergen
N-5009 Bergen, Norway

The Effect of Phosphorylation at Ser-40 on the Structure and Thermal Stability of Tyrosine Hydroxylase

The short-term activation of tyrosine hydroxylase (TH) is believed to be mediated, at least in part, by phosphorylation. Four phosphorylation sites have been identified, corresponding to serine residues (Ser) 8, 19, 31, and 40 in the rat TH sequence. Several of the common multiprotein kinases phosphorylate one or more of these sites, and phosphorylation of Ser 31 and 40 is accompanied by a significant enzyme activation *in vitro*. It has been proposed that activation of TH by phosphorylation of Ser-40 is due to a conformational change that triggers the release of inhibitory catecholamines bound to the active site iron (1). The purpose of this study was to further investigate the conformational and stability changes induced by phosphorylation.

Infrared spectroscopy can provide qualitative as well as quantitative information on the secondary structure of proteins in aqueous environments (2). The infrared spectra are dominated by the amide I band, which appears between 1700 and 1600 cm^{-1}. Specific secondary structures within proteins are associated with particular hydrogen-bonding patterns, which give rise to characteristic amide I bands. Isoform 1 of human TH (hTH1) was expressed in *Escherichia coli* and purified to homogeneity in the nonphosphorylated state (3). Bovine adrenal TH (bTH) was isolated as previously described (1). The spectroscopic properties of the enzymes were examined before and after phosphorylation of Ser-40 by the catalytic subunit of bovine heart cyclic AMP (cAMP)–dependent protein kinase (PKA) (1).

Figure 1A shows the amide I band region of the deconvoluted infrared spectra of recombinant hTH1 and bTH, recorded in D_2O medium. The spectra

Advances in Pharmacology, Volume 42

1054-3589/98 $25.00

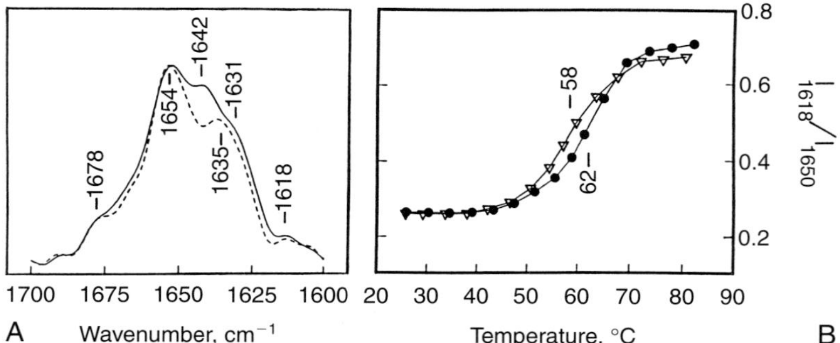

FIGURE 1 Infrared spectra and effect of phosphorylation on the thermal stability of bovine tyrosine hydroxylase. (*A*) Deconvoluted infrared spectra of bovine (*continuous line*) and recombinant human (*broken line*) TH in D_2O buffer. Fourier self-deconvolution was performed using a lorentzian line shape of 18 cm^{-1} half-width and a resolution enhancement factor of 2. (*B*) Ratio of amide I band intensity at 1618 cm^{-1} to that at 1650 nm^{-1} as a function of temperature for nonphosphorylated (•) and phosphorylated (∇) bTH.

of both proteins in this region show several components, which have been assigned to specific types of secondary structure as reported (2), and can be summarized as follows. The intense band at 1654 cm^{-1} is characteristic of helical structures. Those appearing in the spectrum of bTH at 1642 and 1631 cm^{-1} represent nonstructured and β-sheet conformations, respectively. These components are not resolved in the spectrum of hTH1, and instead a broad band is observed at 1635 cm^{-1} that contains contributions from both types of secondary structures (3). This difference is not observable in H_2O medium (data not shown), where the amide I band of both samples exhibits the same number of components at identical positions. Minor components are seen in both spectra at around 1668 and 1678 cm^{-1}, which mainly arise from β-turns. Quantification of the secondary structure, following a procedure described by Arrondo *et al.*, (4), shows that the estimated content of α-helix (~43%), extended structures (including β-sheet; 35%), β-turns (14%), and nonstructured (8%) conformations, is very similar for the two proteins. Therefore, the nature of the difference observed in D_2O, which affects the β and nonstructured conformations, is mainly qualitative and may reflect a different degree of accessibility of these substructures to solvent exchange.

The spectra of the phosphorylated proteins look similar to those described for the nonphosphorylated samples, including the number and position of the amide I component (data not shown). A quantitative estimate of the secondary structure of the phosphorylated hTH1 revealed an increase of about 10% in the α-helix content, concomitant with a loss of nonstructured conformation, as compared with the nonphosphorylated form of the protein (3). However, the same phosphorylation-induced structural modification was not observed for bTH. This apparent discrepancy may be caused by the microheterogeneity of bTH, which is isolated in a partially phosphorylated state and incorporates only 0.6 mol phosphate per subunit after treatment with PKA. By contrast, the purified recombinant hTH1 probably does not contain any covalently bound phosphate and is phosphorylated by PKA to a stoichiometry of 1.0 phosphate per subunit at Ser-40 (3). Therefore, the lack of significant differences between phosphorylated and nonphosphorylated bTH may be due to the difficulty in obtaining homogenous protein populations and stresses the advantage of using recombinant enzymes to study the conformational consequences of regulatory chemical modifications, such as phosphorylation, by biophysical techniques. Nevertheless, it should be noted that the spectral differences between the two proteins in D_2O buffer are preserved under all of the experimental conditions studied (e.g., nonphosphorylated, phosphorylated, and in the presence or absence of catecholamines), suggesting that in spite of their high degree of homology (84%) (5) the arrangement of their nonstructured and β-sheet conformations is distinct.

Infrared spectroscopy was also employed to study the thermal stability of phosphorylated and nonphosphorylated bovine and human TH. The temperature-induced changes observed in the amide I band of these samples are similar to those reported previously (3, 4), that is, a loss of regular secondary substructures and the emergence of band components at 1618 and 1685 cm^{-1}, which have been assigned to hydrogen-bonded extended structures formed upon aggregation of thermally denatured proteins (3) (data not shown). Therefore, protein unfolding can be followed by the temperature dependence of the intensity ratio of the amide I band at 1618 cm^{-1} and 1650 cm^{-1} (I1618/I1650). As indicated by this parameter, the midpoint transition temperature (Tm) for bTH (62°C) is significantly higher than that found for hTH1 (54°C) (Fig. 1B,3), indicating that the structure of the former is more resistant against thermal denaturation. However, the same Tm value (62°C; Fig. 1B) can be obtained for hTH1 after incubation with equimolar amounts of dopamine, an enzyme inhibitor that coordinates to iron (III) at the active site (1, 3). Thus, the higher thermal stability of bTH, as compared with that of hTH1, may be due to the presence of substoichiometrical amounts of catecholamines bound to the active site of bTH (1).

Phosphorylation has an opposite effect on these proteins; it decreases T_m for bTH by 4°C (Fig. 1B), whereas it increases its value by 3°C for hTH1 (3). Because phosphorylation lowers the affinity of catecholamine binding to both enzymes, the observed decreased thermostability of bTH may be due to a phosphorylation-induced catecholamine release from its active site, masking in this way its real effect on bTH stability. By contrast, hTH1 is isolated without bound catecholamines, and therefore the stabilizing effect of phosphorylation can be directly related to a protein conformational change, including modifications of secondary and tertiary structural features.

References

1. Haavik, J., Martínez, A., and Flatmark, T. (1990). pH-dependent release of catecholamines from tyrosine hydroxylase and the effect of phosphorylation of Ser-40. *FEBS Lett.* **262**, 363–365.

2. Arrondo, J. L. R., Muga, A., Castresana, J., and Goñi, F. M. (1993). Quantitative studies of the structure of proteins in solution by Fourier-transformed infrared spectroscopy. *Prog. Biophys. Mol. Biol.* **59**, 23–56.
3. Martínez, A., Haavik, J., Flatmark, T., Arrondo, J. L. R., and Muga, A. (1996). Conformational properties and stability of tyrosine hydroxylase studied by infrared spectroscopy. *J. Biol. Chem.* **271**, 19737–19742.
4. Arrondo, J. L. R., Castresana, J., Valpuesta, J. M., and Goñi, F. M. (1994). Structure and thermal denaturation of crystalline and noncrystalline cytochrome oxidase as studied by infrared spectroscopy. *Biochemistry* **33**, 11650–11655.
5. Saadat, S., Stehle, A. D., Lamouroux, A., Mallet, J., and Thoenen, H. (1988). Predicted amino acid sequence of bovine tyrosine hydroxylase and its similarity to tyrosine hydroxylases from other species. *J. Neurochem.* **51**, 572–578.

Nobuyuki Yanagihara,* Futoshi Izumi,*
Eishichi Miyamoto,† and Motoo Oka‡

* Department of Pharmacology
University of Occupational and Environmental Health
School of Medicine, Kitakyusyu 807, Japan

† Department of Pharmacology
University of Kumamoto School of Medicine
Kumamoto 860, Japan

‡ Osaka University of Pharmaceutical Sciences
Takatsuki 569, Japan

Factors Affecting Adrenal Medullary Catecholamine Biosynthesis and Release

In adrenal medullary cells, stimulation of acetylcholine (ACh) receptors leads to the generation of intracellular second messengers such as cyclic AMP (cAMP), Ca^{2+}, and diacyl glycerol, which causes the activation of a variety of protein kinases. These protein kinases, including cAMP-dependent protein kinase (PKA), protein kinase C (PKC), and Ca^{2+}/calmodulin–dependent protein kinase II (CaM kinase II), have broad substrate specificities and therefore are considered as multifunctional protein kinases that are involved in stimulus–secretion coupling and the regulation of catecholamine biosynthesis. Among these protein kinases, CaM kinase II is noted as a possible candidate that mediates the Ca^{2+} dependent secretion and synthesis of catecholamines. Indeed,

Advances in Pharmacology, Volume 42

1054-3589/98 $25.00

our previous study has demonstrated that the tryptic peptide maps of tyrosine hydroxylase (TH) phosphorylated in rat pheochromocytoma PC12 cells after treatment with 56 mM K^+ or ionomycin are similar to those obtained after the enzyme is phosphorylated by CaM kinase II *in vitro* (1). Furthermore, injection of activated CaM kinase II into brain synaptosomes increased the release of noradrenaline (2). However, the existence and characterization of CaM kinase II have not been well documented in adrenal medulla. In the present study, we have purified an isozyme of CaM kinase II and its endogenous substrate from cultured bovine adrenal medullary cells. Furthermore, the activation of CaM kinase II has been investigated in association with the functional responses in the intact cells.

Bovine adrenal medullary cells were isolated, purified, and maintained as previously reported (3). The Ca^{2+}-independent activity of CaM kinase II in the cell supernatant was measured by phosphorylation of syntide-2 (4). TH activity and catecholamines secreted were assayed (4). CaM kinase II and its endogenous substrate were purified from adrenal medullary cells by a series of chromatographic steps using DEAE-cellulose, calmodulin affinity, and Sephacryl S-300 columns (3). Chromogranin A (CgA) and its antisera were prepared, as described previously (5).

Previous *in vitro* studies have shown a unique activation property of CaM kinase II; that is, autophosphorylation of the kinase converts it from the Ca^{2+}-dependent form to the Ca^{2+}-independent (autonomous) form (6). Therefore, the autophosphorylation, as well as the Ca^{2+}-independent activity, has been used as a useful index of CaM kinase II activation in various intact cells, such as cerebellar granule cells, GH_3 pituitary cells, and PC12 cells. Therefore, first, we examined the effects of ACh on the autophosphorylation and the Ca^{2+}-independent activity of CaM kinase II in cultured adrenal medullary cells. After labeling the cells with $^{32}P\text{-}PO_4$, they were stimulated with 0.3 mM ACh for 3 min. ACh enhanced the phosphorylation of CaM kinase II by two- to-three-fold and the Ca^{2+}-independent activity of CaM kinase II by 1.5 to two-fold. The Ca^{2+}-independent activity increased rapidly at 20 sec and reached a plateau at 3–5 min. ACh also stimulated catecholamine secretion and TH activity in a time-dependent manner similar to that of the Ca^{2+}-independent activity of CaM kinase II. The concentration-dependent increase in Ca^{2+}-independent activity produced by ACh was also correlated with those of catecholamine secretion and TH activity. The removal of extracellular Ca^{2+} completely diminished the ACh-induced increase in Ca^{2+}-independent activity as well as catecholamine secretion and TH activity. These results suggest that ACh stimulates the autophosphorylation and the Ca^{2+}-independent activity of CaM kinase II, which is associated with catecholamine secretion and TH activation.

We isolated and characterized calmodulin-binding proteins, including an apparent isoform of CaM kinase II, from cultured bovine adrenal medullary cells. The supernatant fraction from the cells was subjected to several chromatographic steps. Two peaks of CaM kinase activity (peaks I and III) and one major peak of a protein (peak II) were obtained on the Sephacryl S-300 column. Apparent molecular masses of peaks I, II, and III were estimated to be approximately 650, 300, and 200 kDa, respectively. Using SDS-PAGE, apparent molecular masses of peaks I and II were approximately 50 and 70 kDa, respectively.

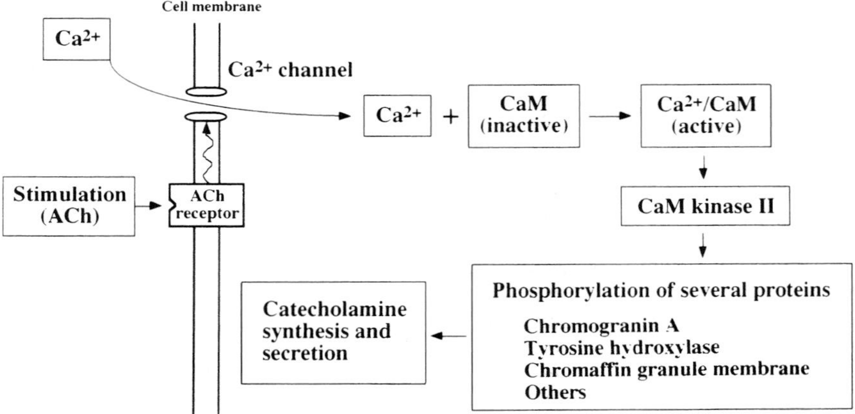

FIGURE 1 Signal transduction via Ca^{2+}/calmodulin-dependent pathway in acetylcholine-stimulated adrenal medullary cells. ACh, acetylcholine; CaM, calmodulin.

Peak I but not peak III was immunoprecipitated by an antibrain CaM kinase II antibody, suggesting that peak I kinase is an isozyme of CaM kinase II in adrenal medullary cells. The peak I kinase (CaM kinase II) phosphorylated TH purified from PC12 cells *in vitro*. Furthermore, this kinase phosphorylated several proteins in chromaffin granule membranes. These proteins were estimated to be 121, 70, 58, 49, and 26 kDa. To investigate whether the protein in peak II is phosphorylated by the CaM kinase activities in peaks I and III, we incubated peak II with peak I or peak III. Interestingly, peak I kinase (CaM kinase II) but not peak III kinase enhanced the phosphorylation of the peak II protein. This result indicates that the peak II protein is an endogenous substrate for CaM kinase II in adrenal medulla.

We analyzed the partial amino acid sequence of the peak II protein. The sequences of N-terminal and another four peptides derived from the peak II protein exhibited a complete homology with the published sequences 1–8, 173–179, 195–206, 244–259, and 315–318 of the bovine adrenal chromogranin A (CgA). Furthermore, CgA isolated from bovine adrenal chromaffin granules was also phosphorylated by CaM kinase II *in vitro*.

In the signal transduction of adrenal medullary cells (Fig. 1), ACh causes the Ca^{2+} channels to open and produces the influx of Ca^{2+}, which, in turn, stimulates CaM kinase II. The activation of CaM kinase II results in the phosphorylation of several proteins, such as TH, chromaffin granule membranes, and probably CgA, which may produce catecholamine synthesis and secretion in the cells.

References

1. Tachikawa, E., Tank, A. W., Yanagihara, N., Mosimann, W., and Weiner, N. (1986). Phosphorylation of tyrosine hydroxylase on at least three sites in rat pheochromocytoma

PC12 cells treated with 56mM K^+: Determination of the sites on tyrosine hydroxylase phosphorylated by cyclic AMP-dependent and calcium/calmodulin-dependent protein kinases. *Mol. Pharmacol.* **30**, 476–485.

2. Nicols, R. A., Sihra, T. S., Czernik, A. J., Nairn, A. C., and Greengard, P. (1990). Calcium/calmodulin-dependent protein kinase II increases glutamate and noradrenaline release from synaptosomes. *Nature* **343**, 647–651.
3. Yanagihara, N., Toyohira, Y., Yamamoto, H., Ohta, Y., Tsutsui, M., Miyamoto, E., and Izumi, F. (1994). Occurrence and activation of Ca^{2+}/calmodulin-dependent protein kinase II and its endogenous substrates in bovine adrenal medullary cells. *Mol. Pharmacol.* **46**, 423–430.
4. Tsutsui, M., Yanagihara, N., Miyamoto, E., Kuroiwa, A., and Izumi, F. (1994). Correlation of activation of Ca^{2+}/calmodulin-dependent protein kinase II with catecholamine secretion and tyrosine hydroxylase activation in cultured bovine adrenal medullary cells. *Mol. Pharmacol.* **46**, 1041–1047.
5. Yanagihara, N., Oishi, Y., Yamamoto, H., Tsutsui, M., Kondoh, J., Sugiura, T., Miyamoto, E., and Izumi, F. (1996). Phosphorylation of chromogranin A and catecholamine secretion stimulated by elevation of intracellular Ca^{2+} in cultured bovine adrenal medullar cells. *J. Biol. Chem.* **271**, 17463–17468.
6. Hanson, P. I., and Schulman, H. (1992). Neuronal Ca^{2+}/calmodulin-dependent protein kinases. *Annu. Rev. Biochem.* **61**, 559–601.

Richard E. Zigmond

Case Western Reserve University
Cleveland, Ohio 44106

Regulation of Tyrosine Hydroxylase by Neuropeptides

Neurotransmitter biosynthesis is a regulated process in many (if not all) neurons. The mechanisms underlying such regulation are best understood for catecholaminergic neurons. The rate-limiting step in catecholamine biosynthesis, the hydroxylation of tyrosine (TH), is regulated primarily by two processes, one involving activation of preexisting TH molecules and the other involving the synthesis of new enzyme molecules (1, 2). Activation of TH occurs rapidly after a stimulus (i.e., within minutes), is rapidly reversible, and involves post-translational modification of preexisting TH molecules (1). Induction of TH occurs slowly (i.e., over a period of hours), is long-lasting, and involves increases in TH mRNA and TH protein (2).

One of the stimuli that lead to both activation and induction of TH is increased synaptic stimulation (1, 2). Thus, for example, stimulation of the preganglionic cervical sympathetic trunk at 10 Hz either *in vivo* or *in vitro* leads to an activation of TH in the rat superior cervical ganglion (SCG). Given

Advances in Pharmacology, Volume 42

1054-3589/98 $25.00

that cholinergic agonists also lead to an activation of TH in this preparation, it was reasonable to assume that the effects of preganglionic nerve stimulation on TH would be blocked entirely by cholinergic antagonists. However, when ganglia are exposed *in vitro* to high concentrations of hexamethonium and atropine and then stimulated at 10 Hz, the activation of TH is only inhibited by 50% (1). Increasing both cholinergic antagonists by an order of magnitude does not further diminish the increase in TH activity.

This unexpected finding eventually led to the discovery that certain neuropeptides cause the activation of TH. Of the large number of neuropeptides that were tested as possible mediators of the noncholinergic transsynaptic regulation of TH, only a subgroup of the peptides of the secretin-glucagon family were found to be candidates. Five peptides from this family, namely secretin, vasoactive intestinal peptide (VIP), peptide histidine isoleucineamide (PHI), rat growth hormone–releasing hormone, and helodermin, increase TH activity acutely, whereas five other members of this same peptide family (glucagon, glucagon-like peptides 1 and 2, human growth hormone–releasing hormone, and gastric inhibitory peptide) have no effect (1).

With regard to the mechanisms underlying this peptidergic activation of TH, an important observation was the finding that those peptides that caused TH activation also cause increased levels of cyclic adenosine monophosphate (cAMP) in the SCG, whereas those peptides that have no effect on TH activity have no effect on cAMP levels (1). Many earlier studies on a variety of peripheral and central adrenergic cell types had established that cAMP leads to an activation of TH via phosphorylation of the enzyme by protein kinase A. However, prior to the discovery that secretin and related peptides activate TH, first messengers that use this pathway were not known. Among the several serine residues that can be phosphorylated on TH, it was found that cAMP, via activation of protein kinase A, caused the phosphorylation of Ser-40. This same residue is phosphorylated when bovine chromaffin cells are incubated with ^{32}P-labeled inorganic phosphate (3).

Immunohistochemical studies established that a small number of neural processes in the SCG exhibit VIP- or PHI-like immunoreactivity, as do a number of preganglionic neuronal cell bodies that project to the SCG from the spinal cord (1). Secretin-like immunoreactivity was not detectable in either place, although this negative result may be the result of methodological problems. Unfortunately, the most crucial experiment for determining if VIP mediates the noncholinergic effect of preganglionic nerve stimulation on TH activation can not be done because no effective VIP antagonist has yet been found for this system. VIP was subsequently shown, however, to be a likely candidate mediating a noncholinergic component to transsynaptic stimulation of catecholamine release and TH activation in the rat adrenal gland. In the case of the adrenal gland, it was shown by Wakade and colleagues that a VIP antagonist ([Ac-Tyr^1,D-Phe^2] growth hormone–releasing hormone [1-29] amide) was effective in reducing catecholamine release caused by both exogenous VIP and preganglionic nerve stimulation. Recently, pituitary adenylate cyclase–activating peptide (PACAP), a newly identified member of the secretin-glucagon family, has become another candidate for the noncholinergic transmitter–regulating catecholamine biosynthesis in the sympathoadrenal system. The resolution of the identity of the endogenous transmitter or transmitters must await the discovery of selective antagonists for VIP and PACAP.

That TH activation following preganglionic nerve stimulation is mediated by both cholinergic and noncholinergic neurotransmitters raised a question that is relevant to any neural system exhibiting cotransmission, namely, what is the functional significance of the presence of multiple transmitters? Does the noncholinergic component of TH activation simply represent a redundant mechanism or does it serve some unique function? Experiments in which the effects of different patterns of nerve stimulation were examined demonstrated that the proportion of the cholinergic and noncholinergic components can vary depending on the pattern of preganglionic stimulation (1). When the cervical sympathetic trunk was stimulated continuously at 10 Hz for 30 min, approximately half of the activation of TH was sensitive to cholinergic antagonists and half was not. The same relative proportions were found if the nerve was stimulated at 10 Hz for 5 min or at 1.67 Hz for 30 min. However, if the nerve was stimulated at 10 Hz for 1 out of every 6 seconds for 30 min, TH activation was mediated entirely by a mechanism that is insensitive to cholinergic antagonists. It should be emphasized that synaptic transmission in the SCG, monitored by measuring the compound action potential from the surface of the ganglion, is entirely dependent on cholinergic transmission under all stimulation conditions examined. Thus, with a discontinuous pattern of stimulation, acetylcholine and the noncholinergic transmitter have two distinct effects on the postganglionic neurons: The former triggers action potentials and the latter triggers the activation of TH.

To determine whether neuropeptides of the secretin–glucagon family might influence catecholamine biosynthesis not only in adrenergic cells but also in adrenergic nerve terminals, tissues innervated by the SCG were examined, namely, the iris, pineal gland, and submaxillary gland. Incubation of these target tissues with secretin and VIP leads to TH activation (5). Whether these peptidergic effects on sympathetic nerve terminals mimic an endogenous mechanism is not certain at this point; however, a potential source of peptides that could act on sympathetic nerve terminals is parasympathetic nerve endings, many of which are known to contain VIP.

Given that neuropeptides can activate TH, can they also cause an induction of the enzyme? An effect of VIP on TH induction was first demonstrated in 1991 by Strong and his collaborators in PC12 cells. Subsequently, VIP and secretin were both found to produce small but significant long-term increases in TH activity in the SCG, though glucagon had no effect (5). Although it had previously been shown that administration of nicotinic ganglionic antagonists did not completely block the induction of TH produced by preganglionic nerve stimulation *in vivo,* this effect could have been caused either by incomplete cholinergic blockade or by the action of an endogenous noncholinergic neurotransmitter.

In addition to having VIPergic fibers, the rat SCG is also known to contain fibers exhibiting substance P–like immunoreactivity. Because earlier work by Stallcup and Patrick in 1980 had established that substance P can inhibit the ion fluxes caused by stimulation of nicotinic receptors, the effect of substance P on the nicotinic activation of TH was examined. Substance P selectively blocks

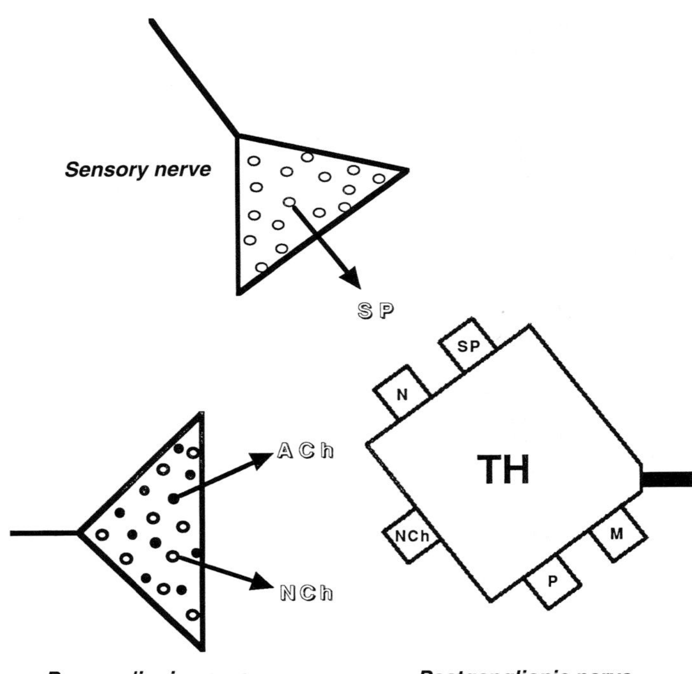

FIGURE 1 Model for the regulation of in SCG by multiple neruotransmitters. ACh, acetylcholine; NCh, noncholinergic transmitter; SP, substance P; N, nicotinic; M, muscarinic.

the activation of TH by the nicotinic agonist dimethylphenylpiperazinium while having no effect on the activation produced by VIP. This effect of substance P can not be mimicked by the deamidated analogue of the peptide. The physiological significance of this inhibitory effect of substance P remains to be determined.

A model for the regulation of TH in the SCG by multiple neurotransmitters is shown in Figure 1. Four important questions that remain to be resolved concerning this model are: (1) whether VIP is the noncholinergic neurotransmitter released by the preganglionic nerve; (2) under what condition sensory neurons in the SCG release substance P; (3) whether, in addition to blocking the effects of exogenous nicotinic agonists, substance P also blocks the nicotinic effects of neurally released acetylcholine; and (4) whether muscarinic receptors are stimulated by neurally released acetylcholine or only by bath-applied muscarinic agonists.

References

1. Zigmond, R. E., Schwarzschild, M. A., and Rittenhouse, A. R. (1989). Acute regulation of tyrosine hydroxylase by nerve activity and by neurotransmitters via phosphorylation. *Annu. Rev. Neurosci.* 2, 415–461.

2. Zigmond, R. E. (1989). A comparison of the long-term and short-term regulations of tyrosine hydroxylase activity. *J. Physiol. Paris* **83**, 267–271.
3. Waymire, J. C., Craviso, G. L., Lichteig, K., Johnston, J. P., Baldwin, C., and Zigmond, R. E. (1991). Vasoactive intestinal peptide stimulates catecholamine biosynthesis in isolated adrenal chromaffin cells: Evidence for a cyclic AMP–dependent phosphorylation and activation of tyrosine hydroxylase. *J. Neurochem.* **57**, 1313–1324.
4. Wakade, T. D., Blank, M. A., Malhotra, R. K., Pourcho, R., and Wakade, A. R. (1991). The peptide VIP is a neurotransmitter in rat adrenal medulla: physiological role in controlling catecholamine secretion. *J. Physiol.* **444**, 349–362.
5. Schwarzschild, M. A., and Zigmond, R. E. (1989). Secretin and vasoactive intestinal peptide activate tyrosine hydroxylase in sympathetic nerve endings. *J. Neurosci.* **9**, 160–166.
6. Wessels-Reiker, M., Haycock, J. W., Howlett, A. C., and Strong, R. (1991). Vasoactive intestinal polypeptide induces tyrosine hydroxylase in PC12 cells. *J. Biol. Chem.* **266**, 9347–9350.
7. McKeon, T. W., and Zigmond, R. E. (1993). Vasoactive intestinal peptide and secretin produce long-term increases in tyrosine hydroxylase activity in the superior cervical ganglion. *Brain Res.* **607**, 345–348.
8. Stallcup, W. B., and Patrick, J. (1980). Substance P enhances cholinergic receptor desensitization in a clonal nerve cell line. *Proc. Natl. Acad. Sci. U.S.A.* **77**(1), 634–638.

A. William Tank, Kristen M. Piech, Cheryl A. Osterhout, Baoyong Sun, and Carol Sterling

Department of Pharmacology and Physiology
University of Rochester Medical Center
Rochester, New York 14642

Regulation of Tyrosine Hydroxylase Gene Expression by Transsynaptic Mechanisms and Cell–Cell Contact

Numerous stimuli induce the expression of tyrosine hydroxylase (TH) in catecholaminergic neurons and adrenal medulla. These stimuli include neurotransmitters released from presynaptic neurons (transsynaptic regulation), glucocorticoids, nerve growth factor (and other neurotrophic factors), cell–cell contact, and aging. Some of these stimuli are important during development and differentiation to promote basal expression of the enzyme and to maintain the catecholaminergic phenotype; these include neurotrophic factors and cell–cell contact. In contrast, glucocorticoids and transsynaptic regulation mediate the induction of TH above its basal expression in response to environmental changes such as stress, hypoxia, or exposure to certain drugs, such as reserpine, nicotine, or cocaine. All of these stimuli lead to increases in TH protein in the appropriate catecholaminergic cells. The mechanisms responsible for these

Advances in Pharmacology, Volume 42

1054-3589/98 $25.00

increases in TH protein expression have been intensely studied over the past 25 years (for review, see Kumer and Vrana [1]). With the introduction of molecular biological techniques, it has been clearly demonstrated that most, if not all, of these stimuli induce TH mRNA prior to the observed changes in TH protein. In more recent years, a major question that has been investigated is whether these increases in TH mRNA are due to stimulation of TH gene transcription rate and/or stabilization of TH mRNA. Most of the stimuli already noted stimulate TH gene transcription rate.

In this chapter we focus on the regulation of TH expression by stimuli that activate transsynaptic mechanisms and by cell–cell contact. We discuss the evidence supporting the hypothesis that these stimuli activate TH gene transcription rate and summarize some of the available information concerning the mechanisms responsible for this transcriptional regulation. However, it is important to note that TH mRNA stabilization also plays an important role in regulating TH expression during hypoxia (work from Drs. Czyzyk-Krzeska and Millhorn) and aging (work from Dr. Strong and coworkers), and that other posttranscriptional mechanisms may also influence TH expression in response to certain stimuli (1).

I. Transsynaptic Regulation of TH Gene Expression Is Mediated via Transcriptional Mechanisms

Many years ago, Mueller, Thoenen, and Axelrod published a series of papers demonstrating that stress or catecholamine-depleting drugs, such as reserpine, induce TH in adrenal medulla and sympathetic ganglia, and that this induction was dependent on innervation of these catecholaminergic structures by cholinergic presynaptic nerves (1). Subsequent studies from numerous laboratories showed that this transsynaptic induction of TH required 2–4 hr of prolonged stress, occurred slowly over 12–24 hr, was due to an increase in TH protein and TH mRNA, and was mediated by nicotinic cholinergic receptors. More recently, our laboratory used nuclear run-on assays to test whether these transsynaptic increases in TH expression were due to initial increases in TH gene transcription rate. In our first experiments, we showed that adrenal TH gene transcription rates were elevated two- to threefold, when measured 3 hr after reserpine administration (2). This result supports the hypothesis that transsynaptic induction of adrenal TH is mediated via transcriptional mechanisms.

In subsequent studies we used systemic nicotine administration as a model to study in more detail this transsynaptic induction in the adrenal medulla (2, 3). Subcutaneous injection of nicotine rapidly stimulates TH gene transcription rate in the adrenal medulla; transcription rate remains elevated for at least 3 hr after multiple injections of the drug. Interestingly, we found that TH mRNA and TH protein were induced only when the transcription rate was elevated for at least 3 hr. Moreover, surprisingly, we also found that the nicotinic receptor antagonist, hexamethonium, did not inhibit the effect of nicotine on TH gene transcription rate or TH enzymatic activity unless the adrenal gland was denervated. Finally, we performed studies demonstrating that muscarinic agonists stimulated TH gene transcription rate in both innervated and denervated adrenal glands. These results, coupled with other findings, led us to the following conclusions: (1) Direct interaction of nicotine with nicotinic cholinergic receptors on rat adrenal chromaffin cells activates signaling mechanisms, leading to stimulation of TH gene transcription rate. (2) Systemic administration of nicotine also stimulates adrenal TH gene transcription by activation of CNS pathways, leading to increased sympathetic outflow and transsynaptic regulation of the gene via multiple chromaffin cell receptors. (3) Muscarinic cholinergic receptors in the adrenal medulla are linked to signal transduction pathways that stimulate the TH gene. (4) Noncholinergic receptors participate in the transsynaptic regulation of the gene.

To test whether the transsynaptic induction of TH elicited by stress was also mediated via transcriptional mechanisms, in collaboration with Dr. Dona Chikaraishi we used a transgenic mouse model (4). These mice express a transgene containing 4.8 kb of TH gene 5′ flanking region directing the expression of the reporter gene, chloramphenicol acetyltransferase (CAT). Cold stress for 2–3 days increased TH protein and TH activity in the mouse adrenal gland. In addition, cold stress induced the expression of the transgene. Similarly, immobilization stress induced both endogenous adrenal TH enzyme and the TH-CAT transgene. These results provide evidence that the induction of adrenal TH elicited by stress is also mediated by transcriptional mechanisms.

II. Cell–Cell Contact Induces TH Expression via Transcriptional Mechanisms

During development of the autonomic nervous system, neuroblasts migrate from the neural crest to the sites of primitive sympathetic ganglia, where they aggregate and TH expression is first observed. Many environmental cues are thought to participate in this initial expression of TH and in maintaining its basal expression; one of these cues is increased cell–cell contact. Using cultured catecholaminergic cells, various researchers have shown that TH expression increases dramatically at high cell density. This phenomenon has been observed in tumor cell lines derived from neuroblastoma or pheochromocytoma, as well as in bovine adrenal chromaffin cells. Using these cell culture models, it has been shown that the increase in TH activity is due to an increase in enzyme protein and is preceded by an increase in TH mRNA. Similarly, TH mRNA levels increase in primitive sympathetic ganglia soon after the neuroblasts aggregate. We have used nuclear run-on assays to demonstrate that the induction of TH mRNA in rat pheochromocytoma PC18 cells is mediated primarily by an increase in TH gene transcription rate (5). These results support the hypothesis that increased cell–cell contact stimulates the TH gene.

III. Multiple Signaling Pathways and TH Gene Promoter Sites Are Responsible for Regulating the Gene

Transsynaptic regulation of the TH gene is mediated by both cholinergic and noncholinergic receptors; hence, multiple signaling pathways also regulate the gene. Three major second messengers have been implicated in this response: cyclic adenosine monophosphate (cAMP), diacylglycerol, and calcium. Each of these second messengers stimulates TH gene transcription rate and induces TH mRNA and TH protein in cultured cells. Several laboratories have studied the promoter elements within the TH gene 5′ flanking region that mediate the responses to these second messengers. Two major response elements within the proximal promoter region have been identified: the cAMP-responsive element (CRE; at approximately −40 bp) and the activating protein-1 element (AP1; at approximately -200 bp). A third modulatory element (THCRE2; at approximately −100 bp) has recently been identified in the rat TH gene promoter. All of these promoter elements are likely to participate in the response of the gene to transsynaptic stimuli.

There is a dearth of information concerning the signaling pathways responsible for the cell contact–mediated regulation of the gene. In their cited study, Fader and Lewis (6) have shown that high cell density does not stimulate the TH gene proximal promoter. Because this proximal promoter responds robustly to cAMP, diacylglycerol, and Ca^{2+}, it is unlikely that these second messengers mediate this response. There is evidence that this effect is mediated by membrane proteins that interact on adjacent cells; however, the signal transduction pathways activated by this interaction remain to be elucidated.

IV. Multiple Transcription Factors Participate in the Regulation of the TH Gene

We have begun to investigate the transcription factors that mediate the response of the TH gene to different signaling pathways, focusing on members of the CREB/ATF, Fos and Jun transcription factor families. We have isolated PC12 cell lines that are stably transfected with constructs expressing antisense RNA complementary to the 5′ region of CREB mRNA. CREB protein levels in these antisense-expressing cells are decreased by 80–90%. When these cell lines are transfected with TH-CAT constructs and treated with different inducing agents, the response of the TH gene promoter to cAMP or phorbol ester is dramatically inhibited. In contrast, the response to elevated Ca^{2+} is not reduced in the CREB-deficient cell lines. Hence, cAMP and phorbol ester, but not Ca^{2+}, require CREB for activating the TH gene.

Gel shift assays demonstrate that TH AP1 complex formation increases dramatically after treatment of PC12 cells with either cAMP-elevating agents, phorbol ester, or Ca^{2+}-elevating agents. Western blots demonstrate that almost all members of the Fos and Jun families are present in PC12 cell nuclei after treatment with these stimuli, and supershift gel shift assays show that they all can bind to the TH AP1 site. We have recently overexpressed Fos, Jun, and JunB in PC12 cells and shown that each of these factors activates the TH gene promoter. Hence, regulation at the TH AP1 site is extremely complex, involving multiple transactivating factors.

V. Conclusions

Environmental stimuli that activate the sympathetic nervous system induce adrenal TH by mechanisms that stimulate TH gene transcription rate. These mechanisms are initiated by the generation of multiple second messengers, leading to the activation and/or induction of multiple transcription factors that transactivate the CRE and AP1 sites in the proximal TH gene promoter. The participating transcription factors and the precise molecular steps in the signaling pathways remain to be elucidated. Similarly, increased cell–cell contact activates signal transduction pathways that stimulate TH gene transcription rate and that presumably participate in maintaining basal TH expression. The signaling pathways mediating this cell–cell contact response are distinct from those mediating the transsynaptic response, but more research is needed to elucidate these pathways.

Acknowledgments

The authors acknowledge the collaboration of Dr. Dona Chikaraishi on the transgenic mice studies. The work was supported by NIDA grant 05014 and Smokeless Tobacco Research Council grant 0481 to AWT. KMP was supported by the NIDA training grant 07232 and CAO by the Pharmacological Sciences training grant GM08427.

References

1. Kumer, S. C., and Vrana, K. E. (1996). Intricate regulation of tyrosine hydroxylase activity and gene expression. *J. Neurochem.* **67**, 443–462.
2. Fossom, L. H., Carlson, C. D., and Tank, A. W. (1991). Stimulation of tyrosine hydroxylase gene transcription rate by nicotine in rat adrenal medulla. *Mol. Pharmacol.* **40**, 193–202.
3. Fossom, L. H., Sterling, C. R., and Tank, A. W. (1991). Activation of tyrosine hydroxylase by nicotine in rat adrenal gland. *J. Neurochem.* **57**, 2070–2077.
4. Osterhout, C. A., Chikaraishi, D. M., and Tank, A. W. (1997). Induction of tyrosine hydroxylase protein on a transgene containing tyrosine hydroxylase 5′ flanking sequences by stress in mouse adrenal gland. *J. Neurochem.* **68**, 1071–1077.
5. Carlson, C. D., and Tank, A. W. (1994). Increased cell–cell contact stimulates the transcription rate of the tyrosine hydroxylase gene in rat pheochromocytoma PC18 cells. *J. Neurochem.* **62**, 844–853.
6. Fader, D., and Lewis, E. J. (1990). Interaction of cyclic AMP and cell-cell contact in the control of tyrosine hydroxylase mRNA. *Mol. Brain Res.* **8**, 25–29.

Tohru Yamakuni,* Toshibumi Yamamoto,† Masato Hoshino,‡ Hideko Yamamoto,§ Si-Young Song,* Mayuko Kunikata-Sumitomo,* and Shiro Konishi*

* Mitsubishi Kasei Institute of Life Sciences
Machida, Tokyo 194, Japan

† Laboratory of Molecular Recognition
Graduate School of Integrated Science
Yokohama City University, Yokohama 236, Japan

‡ Department of Physiological Chemistry
The Tokyo Metropolitan Institute of Medical Science
Tokyo 113, Japan

§ Department of Psychopharmacology
Tokyo Institute of Psychiatry
Tokyo 156, Japan

A New Regulatory Protein of Catecholamine Synthesizing-Enzyme Expression

To search for novel proteins that play crucial roles in dynamic cellular processes of rat cerebellar morphogenesis such as neuronal differentiation and migration, synaptogenesis, and synaptic rearrangement, we have analyzed developmental changes in soluble proteins of rat cerebellum using a two-dimensional high-performance liquid chromatography (HPLC) technique. In this analysis, we have found a protein designated V-1, whose expression reached maximal levels during early postnatal development. Direct protein sequencing and cDNA cloning revealed that V-1 is a novel protein that consists of 117 amino acids and contains 2.5 contiguous repeats of the cdc10/SWI6 motif, alternatively called the cell cycle motif or ankyrin repeat (1). We have also found that the V-1 gene is ubiquitously expressed in neurons of murine brains (2). These observations suggest that the V-1 protein plays a general role in the regulation of neuronal functions.

To explore the functional role of the V-1 protein, we first performed stable V-1–overexpression experiments using PC12 cells, a rat pheochromocytoma-derived cell line, which is well known as a model experimental system to study neuronal differentiation. To generate PC12 cell lines that overexpress the V-1 gene in a stable manner, we transfected PC12D cells with a human elongation factor 1 gene promotor-driven expression vector carrying the rat V-1 cDNA. Following neomycin selection, neomycin-resistant clones were screened by Western blot analysis with anti-V-1 antibody (3). Further assay of several positive clones resulted in the isolation of two V-1–overexpressing clones, V1-

Advances in Pharmacology, Volume 42

1054-3589/98 $25.00

46 and V1-69, whose V-1 expression levels are fivefold higher than that of the parent cell, which endogenously expresses the V-1 protein. On the other hand, there were no significant differences in V-1 expression levels among control clones transfected with the vector alone, C-7 and C-9. These stable clones were used for further phenotype analyses.

We initially tested whether there are any changes in neuron-like properties, including neurotransmitter biosynthesis, ion channel currents, and nerve growth factor–inducible neurite elongation and branching among these stable transfectants. By screening, we found an increase in dopamine content in the V-1–overexpressing clones compared with in the control clones. Further, we determined tyrosine hydroxase (TH) enzyme activities and repetitively quantified dopamine contents. These studies yielded the following results: TH enzyme activities in homogenates of the V-1–overexpressing clones were increased compared with those in the control clones (Table I). The contents of dopamine in the V-1–overexpressing clones were also remarkably higher than those of the control clones. The results of our biochemical and neurochemical assays indicated that changes in TH enzyme activity appear to be responsible for the increases in dopamine content observed in the two V-1–overexpressing clones. These results suggest that the V-1 protein regulates catecholamine biosynthesis via control of the expression of catecholamine biosynthesizing enzymes.

To test this notion, we examined the expression of mRNA for four kinds of catecholamine biosynthesizing enzymes: TH, AADC, DBH, and PNMT in the V-1–overexpressing clones and in the control clones by Northern analysis. Consistent with the result of the assays of TH enzyme activities, TH mRNA levels in the V-1–overexpressing clones were higher than those in the control clones. The levels of the AADC mRNA were upregulated in the V1-46 and V1-69 clones in comparison with those in the C-7 and C-9 clones where AADC mRNA levels were lower than those of the parent cell. Moreover, expression of mRNA encoding DBH, which is essential for (nor)adrenaline biosynthesis, was increased in the V-1–overexpressing clones compared with those in the control clones. By contrast, no PNMT mRNA expression was detected in either

TABLE I TH Enzyme Activities and Dopamine Contents in the V-1–Overexpressing Transfectants and the Vector Control Transfectants[a]

Clone	*TH enzyme activities (n mol/min/mg)*	*Dopamine contents (pmol/well)*
C-7	0.9	14
C-9	1.6	63
V1-46	3.7	499
V1-69	2.6	357

[a] Assays of TH enzyme activity (n = 6) and dopamine (n = 5) were performed. The protein content was 148 Mg/dish (the mean). The Lowry method was used for protein assays. For assays of TH enzyme activity and dopamine, 5×10^5 cells were plated on 35-mm dishes and cultured in the culture medium without neomycin. Cells were harvested 48 hr later. For determination of TH enzyme activity, cells were homogenated in 0.32 M sucrose. Cells were also extracted with 0.4 N perchloric acid and the dopamine contents assayed by HPLC-ECD.

the parent cell or the V-1–overexpressing clones. This result is consistent with the data reported previously, showing that there is no detectable PNMT enzyme activity in PC12 cells (4). The results of our Northern analysis suggest that the V-1 protein upregulates the coordinate expression of TH, AADC, and DBH mRNAs, which are required for the (nor)adrenergic phenotypes.

To our knowledge, this study is the first to suggest the molecular mechanism that controls the coordinate transcriptional upregulation of TH, AADC, and DBH genes. An unusual characteristic of the V-1 protein is that about 80% of it consists of 2.5 repeats of the cdc10/SWI6 motif. The V-1 protein lacks DNA binding domains, however. The cdc10/SWI6 motif has been demonstrated to be essential for protein–protein interactions with specific partner proteins that regulate functions, such as transcription (5). It is thus reasonable to suggest that upregulation of mRNA expression of three catecholamine biosynthesizing enzymes observed in the V-1–overexpressing clones results from the cooperative action of the V-1 protein with a partner protein that influences transcription factors or transcription regulators that regulate the TH, AADC, and DBH genes. This notion is supported by our immunohistochemical data showing that the V-1 protein is strongly expressed in noradrenergic and adrenergic cells. Our working hypothesis is that the novel protein V-1 may be involved in the control of catecholaminergic cell specification.

Acknowledgments

We wish to thank Dr. David Saffen for critically reading the manuscript and discussion. We would like to acknowledge the secretarial expertise provided by T. Hiratsuka. This work was supported partially by the Research Grant (5B-4) for Nervous and Mental Disorders from the National Ministry of Health and Welfare (T. Yamakuni).

References

1. Taoka, M., Isobe, T., Okuyama, T., Watanabe, M., Kondo, H., Yamakawa, Y., Ozawa, F., Hishinuma, F., Kubota, M., Minegishi, A., Song, S.-Y., and Yamakuni, T. (1994). Murine cerebellar neurons express a novel gene encoding a protein related to cell cycle control and cell fate determination proteins. *J. Biol. Chem.* **269,** 9946–9951.
2. Fujigasaki, H., Song, S.-Y., Kobayashi, T., and Yamakuni, T. (1996). Murine central neurons express a novel member of the cdc10/SWI6 motif-containing protein superfamily. *Mol. Brain Res.* **40,** 203–213.
3. Song, S.-Y., Asakai, R., Kenmotsu, N., Taoka, M., Isobe, T., and Yamakuni, T. (1996). Changes in the gene expression of a protein with the cdc10/SWI6 motif, V-1, during rat follicular development and corpus luteum formation. *Endocrinology* **137,** 1423–1428.
4. Greene, L. A., and Tischler, A. S. (1976). Establishment of a noradrenergic clonal line of rat adrenal pheochromocytoma cells which respond to nerve growth factor. *Proc. Natl. Acad. Sci. USA* **73,** 2424–2428.
5. Blank, V., Kourilsky, P., and Israel, A. (1992). NFk-B and related protein: Rel/dorsal homologies meet ankyrin-like repeats. *Trends Biochem. Sci.* **17,** 135–140.

T. H. Joh, J. H. Son, C. Tinti, B. Conti, S. J. Kim, and S. Cho

Laboratory of Molecular Neurobiology
Cornell University Medical College at The W. M. Burke Medical Research Institute
White Plains, New York 10605

Unique and Cell-Type-Specific Tyrosine Hydroxylase Gene Expression

Over the past 10 years, we have demonstrated unique tyrosine hydroxylase (TH) gene expression and regulation *in vivo*. Although some controversial results exist, two important DNA elements, activating protein-1 (AP1) and cyclic adenosine monophosphate (cAMP)–responsive element (CRE), have been extensively investigated for their activity. We have shown that CRE in the upstream sequence of the TH gene is essential for basal and cAMP-stimulated transcription. Moreover, AP1 is an important element for TH gene regulation. Cis-acting elements for tissue/cell-type–specific TH gene expression are identified to be localized within a 9.1-kb upstream sequence. However, some data suggest that the TH gene is not regulated in a uniform fashion in every cell type, but is regulated in a cell-type–specific manner. For instance, the TH gene in dopaminergic (DA) cells of the substantia nigra pars compacta (SNc) does not respond to cAMP, and these SNc DA cells do not express c-fos. This indicates that TH gene regulation depends on many factors which are cell-type–specific. In this chapter, we demonstrate data that show unique and cell-type–specific TH gene regulation.

Inducible cAMP early repressor (ICER) modulates TH gene expression in the adrenal medulla and PC12 cells but not in other catecholaminergic cells (1). The cyclic nucleotide response element–binding (CREB) protein and cyclic nucleotide response modulator (CREM) family have been known to participate in TH gene transcription through CRE in the immediate upstream sequence of the TH promoter. Members of the CREB/CREM/ATF (activating transcription factor) family of transcription factors either enhance or repress transcription after binding to the CREs of numerous genes. PC12 and rat adrenal gland–derived nuclear proteins retarded a TH-CRE oligonucleotide in gel mobility shift assays with virtually identical patterns. These differed substantially from patterns exhibited by extracts from locus ceruleus (LC) or from neuroblastoma (SK-N-BE(2)C) and LC-derived (CATHa) cell lines (Fig. 1). Forskolin stimulation of PC12 cells and reserpine treatment of rats increased, in nuclear extracts derived from cells and adrenal glands, respectively, the amount of a fast-moving CRE/protein complex that was supershifted by an anti-CREM antibody. Subsequent Western, Northern, and polymerase chain reaction analyses indicated

Advances in Pharmacology, Volume 42

1054-3589/98 $25.00

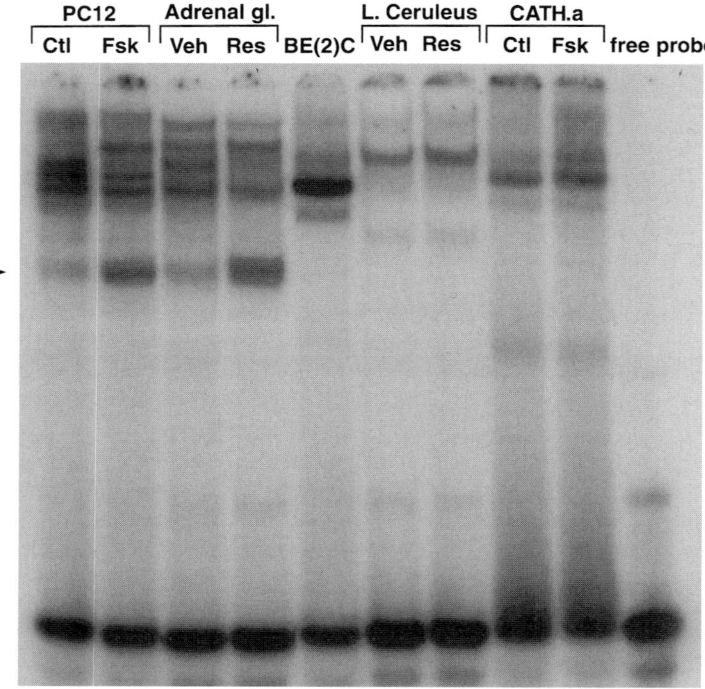

FIGURE 1 Gel mobility shift assay: identication of TH-CRE binding proteins and analysis of the effects of forskoline and reserpine. In PC12 cells and adrenal gland, the formation of a fast-running complex (*arrow*) was enhanced by the pharmacological treatments. In the last lane, the probe underwent electrophoresis in the absence of nuclear proteins. Ctl, control; Fsk, forskolin; Veh, vehicle; Res, reserpine. (Reprinted by permission from *JBC* Vol. 271, No. 41, pp. 25375–25381.)

that a specific member of the CREM family, the ICER, was strongly induced in both systems. Cotransfection of PC12 cells with TH2400CAT plasmid and the expression vector pCMV-ICER-Ib demonstrated that ICER efficiently represses transcriptional activity of the TH gene promoter. In addition, protein kinase A (PKA)-stimulated transcriptional activity of the promoter was effectively suppressed by ICER. These results suggest that ICER can modulate cAMP-stimulated transcription of the TH gene and provide a model accounting for rapid reversal of increased TH transcription following elevations in cAMP.

At least 9.1 kb upstream of the TH gene is necessary to express in a tissue-specific and in a developmentally correct manner in transgenic mice, and sufficient to control transcriptional induction of TH by reserpine or cold stress in the central nervous system. Various data suggest that a relatively small DNA sequence of the TH upstream is sufficient to express cell-type specificity of TH. Earlier reports demonstrated that 4.8 kb of the 5′ flanking region contained some tissue-specific element(s) determined by chloramphenicol acetyltransferase (CAT) assay using regional brain dissections. To delineate the DNA sequence necessary for tissue-specific expression of the rat TH gene, transgenic mice were produced containing a 0.15, 2.4, and 9.0 kb of 5′ flanking sequence fused to the lacZ reporter gene. Reporter gene expression in the transgenic animals was monitored by both X-gal histochemical staining and β-gal immunohistochemistry and was compared with TH mRNA and protein expression. Transgenic mice bearing 9.0 kb, but not the smaller constructs with either 2.4 kb or 0.15 kb of 5′ flanking sequence, fused to lacZ were able to direct a high level expression of β-gal at levels equivalent to the endogenous TH in central CA (catecholamine) cells, and to a lesser degree to the adrenal gland. Our study demonstrated that the crucial CA neuron-specific DNA element(s) reside between −9 kb and −2.4 kb of the 5′ flanking region of the rat TH gene. Because Chikaraishi's group has shown that the 4.8-kb upstream sequence can express strongly in noradrenergic cells of the LC but not in specific expression in DA cells of the SN (2), it can be assumed that the DNA segment between −9.1 kb and 4.8 kb may be an important part of the rat TH gene for cell-type–specific expression in DA cells of the SNc.

It was demonstrated in the past that TH expression appears permanently within cells destined to be CA-secreting during adult life, and transiently in several cell types that will not express TH in adulthood. We examined the early ontogeny of TH expression in transgenic mouse embryos by following the expression of a lacZ reporter, driven by the 9.1-kb 5′ upstream promoter of the rat TH gene. The first β-gal expression in the embryonic brain appeared within distinct regions, such as the ventral prosencephalon, the ventral and dorsolateral mesencephalon, and the rostral and caudal rhombencephalon. The level of β-gal expression in all of these tissues decreased at E13.5, but a distinct adult pattern of β-gal expression started to emerge in the SN and VTA (ventral tegmental area) in the central nervous system and the adrenal medulla in the periphery. Thus, our findings indicate that proximally 9.0 kb of the 5′ promoter region of the rat TH gene encodes sufficient information to direct development of the appropriate catecholaminergic lineage cells in the central and peripheral nervous system during embryogenesis.

The results given are consistent with our view that a 9.1 kb upstream region of the rat TH is necessary for tissue-specific expression of the rat TH gene in most CA cells in the central nervous system. However, we have not yet confirmed that it expresses in all peripheral CA cells. For instance, one out of nine lines of our transgenic mice expressed lacZ in sympathetic nerve cells of the superior cervical ganglia (SCG). This means that the 9.1 kb upstream region is not sufficient to express in all CA cells in the peripheral nervous system. Further studies are necessary to identify proper DNA segment(s) to express in all CA cell types in rat tissues. This raises the important question of whether the 9.1 kb upstream region is sufficient to control TH gene regulation by the pharmacological and physiological stimulation that causes TH gene regulation both *in vivo* and *in vitro*.

We extended our studies on specificity of transgene expression to demonstrate cell type-specific–functional regulation of lacZ expression using manipulations known to alter endogenous TH expression (3). Alterations in lacZ reporter expression should parallel changes in endogenous TH levels if the DNA elements mediating these functional changes of TH expression *in vivo* reside within the 9.0 kb of the TH promoter region. Densitometry and image analysis were used to quantify lacZ expression following acute reserpine administration, which upregulates endogenous TH. At 48 hr postinjection, analysis of OD values indicated a significant increase of X-gal staining in the LC and VTA but not in the SN or olfactory bulb of reserpine-treated transgenic animals. These data showed that the 9.1-kb sequence also mediates cell-type–specific transsynaptic regulation of reporter gene expression (3).

The TH gene is downregulated in noradrenergic cells of the SCG by ectopic expression of DA in target cells of the pineal gland (PG).

We recently demonstrated that transgenic animals ectopically expressing TH and thereby *de novo* DA production in the PG also exhibit a dramatic decrease in TH fiber density arising from the SCG and terminating in the PG (4). Our results indicated, however, that, contrary to expectation, the decrease in TH immunostained fibers is associated with normal innervation in the PG but lower levels of TH and norepinephrine transporter mRNA in the SCG. Test results using histology of the presynaptic SCG neurons are also abnormal, showing heterogeneous TH and norepinephrine transporter mRNA density and TH immunostaining. Although the mechanisms for this unique, altered TH expression have yet to be determined, we think that the altered target neurotransmitter phenotype produces both quantitative (amount) and qualitative (phenotype) retrograde changes in gene expression in presynaptic neurons. Whether this type of CA cellular response occurs in the central nervous system or is limited to the SCG is not known.

The results given here demonstrate that the TH gene is not regulated in a uniform fashion, but is regulated by tissue- or cell-type–specific conditions. Second messenger system, immediate early gene products, nuclear transcription factors, and target cell neurotransmitter phenotypes, among others, are important factors in determining cell-type specificity for TH gene regulation.

References

1. Tinti, C., Conti, B., Cubells, J. F., Kim, K. S., Baker, H., and Joh, T. H. (1996). Inducible cAMP early repressor can modulate tyrosine hydroxylase gene expression after stimulation of cAMP synthesis. *J. Biol. Chem.* **271,** 25375–25381.
2. Banerjee, S. A., Hoppe, P., Brilliant, M., and Chikaraishi, D. M. (1992). 5′ flanking sequences of the rat tyrosine hydroxylase gene target accurate tissue-specific, developmental, and transynaptic expression in transgenic mice. *J. Neurosci.* **12,** 4460–4467.
3. Min, N., Joh, T. H., Corp, E. S., Baker, H., Cubells, J. F., and Son, J. H. (1996). A transgenic mouse model to study transsynaptic regulation of tyrosine hydroxylase gene expression. *J. Neurochem.* **67,** 11–18.
4. Cho, S., Son, J. H., Park, D. H., Aoki, C., Song, X., Smith, G. P., and Joh, T. H. (1996). Reduced sympathetic innervation after alteration of target cell neurotransmitter phenotype in transgenic mice. *Proc. Natl. Acad. Sci. USA* **93,** 2862–2866.

D. M. Jacobowitz, K. R. Isaacs, and V. Cimini

Laboratory of Clinical Sciences
National Institute of Mental Health
Bethesda, Maryland 20892

Triple Colocalization of Tyrosine Hydroxylase, Calretinin, and Calbindin D-28k in the Periventricular-Hypophyseal Dopaminergic Neuronal System

Earlier studies have revealed that catecholamine inhibition of α-melanocyte-stimulating hormone (α-MSH) release is mediated by dopaminergic (DA) nerves in the intermediate lobe of the pituitary (ILP). Other studies have shown that the dopaminergic neurons terminating in the ILP originate in the periventricular nucleus of the hypothalamus (1). A recent observation revealed that tyrosine hydroxylase (TH) colocalized with calretinin (CR), a calcium-binding protein, in the rat ILP (2). In addition, TH also colocalized with calbindin-D28k (CB), another calcium-binding protein (CaBP) (unpublished results). This suggested to us that a triple colocalization of TH, CR, and CB existed in the fibers of the ILP.

The purpose of this study was to determine, using an immunohistochemical colocalization procedure, whether a triple colocalization of TH, CR, and CB existed in the DA fibers of the ILP. Furthermore, having demonstrated that such a triple colocalization exists, we set out to map the dopaminergic cells in the rat hypothalamus in a search for the origin of these pituitary fibers.

Four rats were fixed by perfusion with 200 ml of 10% neutral buffered formalin, and the brains and pituitaries were postfixed for 1–2 hr at 4°C. Sections were cut in a cryostat and were routinely processed for indirect immunofluorescence. For triple immunofluorescent labeling, goat anti-CR sera (1 : 2500), rabbit anti-CB sera (1 : 5000), and monoclonal antityrosine hydroxylase sera (1 : 2000) were diluted in phosphate-buffered saline (PBS) containing 1% normal donkey serum and 0.3 Triton X-100. The primary antibodies were visualized using donkey anti-goat Texas Red (1 : 100), donkey anti-rabbit fluorescein isothiocyanate (FITC, 1 : 100), and donkey anti-mouse AMCA (1 : 50). Double fluorescent immunostaining used rabbit anti-CR sera (1 : 2500) or rabbit anti-CB sera with monoclonal antibodies to TH with 1% normal goat serum and 0.3% Triton X-100 in PBS. Goat anti-rabbit FITC (1 : 300) and goat anti-mouse Texas Red (1 : 100) were used to reveal primary antibodies. Antibodies to dopamine-beta-hydroxylase (DBH) (1 : 1000) were used to confirm that the fibers in the ILP were not noradrenergic.

Specificity controls consisted of preadsorbing the CR antibodies with 0.1 μM recombinant CR using preimmune serum or no primary antibodies.

Rat brain sections were assigned to 1 of 13 rostral to caudal levels according to a standard atlas. DA nuclei A11 to A15v were identified by their TH-immunoreactive (TH-ir) cells and location in the brains in relation to standard landmarks. All TH-ir cells were classified as either TH-only, TH+CR, TH+CB, or TH+CR+CB. The percentage of DA cells in each category was calculated for each animal at each level.

In immunocytochemical preparations of rat pituitary glands, varicose dopaminergic nerve fibers formed a plexus enveloping the endocrine cells of the intermediate lobe (Fig. 1), and the lack of anti-DBH staining confirmed the DA nature of these fibers. Triple colocalization of TH, CR, and CB revealed that all three neurochemicals were present in these fibers in what appeared to be a 1 : 1 relationship between the CaBPs (CR, CB) and the TH-containing fibers. No immunoreactive cell bodies were observed in the ILP. Pituitary stalk section resulted in the disappearance of the triple colocalized fibers in the intermediate lobe.

We have identified four subpopulations of TH-ir cells: (1) TH-ir only, (2) TH+CR-ir cells, (3) TH+CB-ir cells, and (4) TH+CR+CB-ir cells. Triple colocalized cells (TH+CR+CB) were found in three subsets of DA perikarya

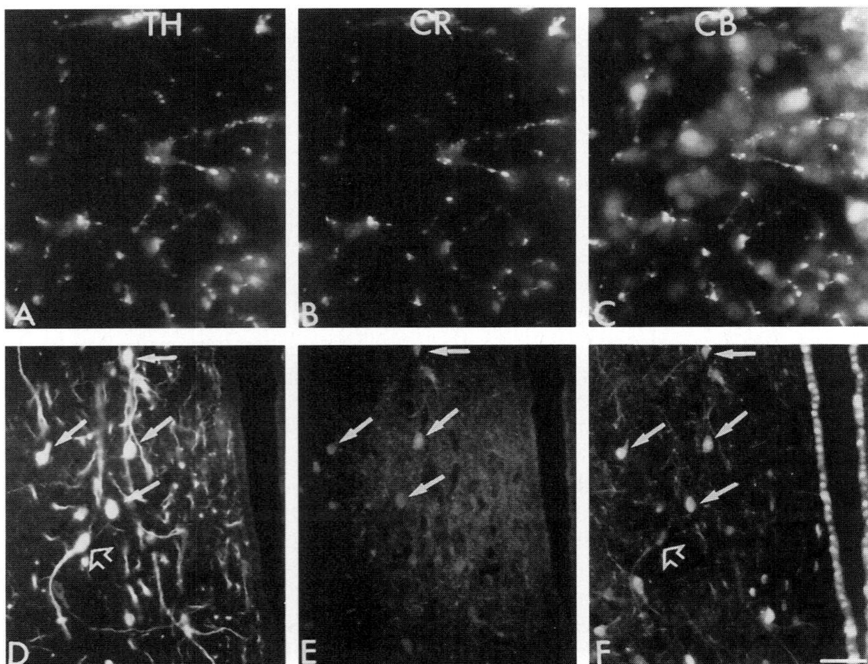

FIGURE 1 Triple colocalization of CR, TH, and CB in the intermediate lobe of the rat pituitary (*A–C*). Note the 1 : 1 correlation between the three antigens; all TH-ir fibers also include CR and CB. Magnification bar equals 25 μm. In the A14 nuclei (*D–F*), many of the TH-ir cells are also CR- and CB-ir (*closed arrows*), although some TH- and CB-ir cells can be identified (*open arrows*). Magnification bar equals 55 μm.

in the diencephalon of the brain (i.e., the A11, A14, and A15v cells) (see Fig. 1). No triple colocalized cells were noted in the A12 (arcuate nucleus) and A13 cell groups. In the A11 and A14 cell groups, the triple colocalized cells represent 3.1% and 3.3% of the total TH-ir population, respectively. In the A14 cell group, the TH+CR cells represent 24.3% of the total TH-ir population and the TH+CB cells represent 1.9%. The A15v group had a 12.8% colocalization rate in the TH+CR subpopulation, but only a 1% colocalization rate for the TH+CB cells. Cells in this nucleus were triple colocalized only 1% of the time.

This study revealed that the DA (TH-ir) fibers that innervate the ILP gland also contain two CaBPs: CR and CB. CR is a predominantly neuronal protein that shares 58% amino acid homology with CB. It has been suggested that CR and CB may act as intracellular calcium buffers within neurons, thereby serving a neuroprotective role. The fact that TH-ir nerves in the ILP and that A11 and A14 cell bodies in the hypothalamus are DA rather than noradrenergic has been amply documented. Smelik (3) first showed a DA innervation of the ILP, although he believed that the cell bodies were located in the arcuate nucleus (A12 cells). Björklund and Falck (4) reported that dopamine was the predominant catecholamine in the ILP, although in the cat, significant amounts of norepinephrine were also found. Anatomical, neurochemical, and neuroendocrinological evidence (1) indicates that dopamine terminals in the ILP originate from A14 cells in the hypothalamic periventricular nucleus (5). We have observed that a triple colocalization of TH, CR, and CB was contained in about 3.3% of the A14 DA cell bodies, and we suggest that these are the cells that innervate the ILP. Because of the triple colocalization within approximately 3.1% of the A11 cells, this system may also project to the ILP. In addition, a few cells (about 1%) of the A15v system (6) may also project to the ILP. The DA nature of the A15 cells has not been established.

The sites of innervation of the TH+CR and TH+CB cells remain to be studied. Rogers (7) studied TH-containing cells that colocalized with either CR or CB in the rat brain. He also reported that double colocalization (TH+CR and TH+CB) was present in the A11, A14, and A15v cell groups. Numerical differences between these two studies probably represent sampling differences. Rogers did not study triple colocalizations.

Many TH-ir cells show great heterogeneity in CaBP content. The presence of CaBP in a variety of neuronal subpopulations has repeatedly been shown to express neurochemical heterogeneity, the significance of which, however, continues to elude us. The functional significance of the CaBPs may very well be linked to our understanding of this property of heterogeneity within subsets of brain neurons, ganglia, and even nonneuronal CaBP-containing cells (testis, ovary, pituitary). In this regard, Rogers (7) suggested that the observed heterogeneity may be a reflection of neuronal activity.

Much work has supported the conclusion that catecholamines inhibit the release of α-MSH from the ILP. Furthermore, the inhibition of α-MSH release has been shown to be mediated by dopamine. Blockade of DA neurotransmission *in vivo* has been reported to stimulate α-MSH release, whereas DA receptor stimulation seems to inhibit α-MSH release *in vitro,* as well as *in vivo.*

Prior studies have shown that isolated melanotrophs prepared from the intermediate lobe responded briskly, with an increased output of α-MSH when challenged with high K^+, and that this response depended on extracellular calcium and was inhibited by a calcium channel blocker (8). Their results indicated that the activation of the voltage-dependent calcium channels allowed calcium to enter the cell in amounts sufficient to provide a strong stimulus to secretion. These results also suggested that the spontaneous secretory activity of these cells could be explained by the existence of some channels that were in the open state under basal conditions.

The significance of the presence of CR and CB in the hypothalamic-ILP DA system is unknown. It has been proposed that the presence of CaBPs in neurons in general may provide protection from calcium-induced neurotoxicity. The release of dopamine, which inhibits α-MSH secretion, would result in an increase in calcium in the extracellular milieu. The presence of two CaBPs, CR and CB, would provide extra buffering capacity to prevent neurotoxicity. This, however, remains a working hypothesis, and we do not begin to fathom the true function of these CaBPs.

References

1. Goudreau, J. L., Lindley, S. E., Lookingland, K. J., and Moore, K. E. (1992). Evidence that periventricular dopamine neurons innervate the intermediate lobe of the rat pituitary. *Neuroendocrinology* **56**, 100–105.
2. Cimini, V., Isaacs, K. R., and Jacobowitz, D. M. (1997). Calretinin in the rat pituitary: Colocalization with thyroid stimulating hormone. *Neuroendocrinology* **65**, 179–188.
3. Smelik, P. G. (1966). A dopaminergic innervation of the intermediate lobe of the pituitary? *Acta Physiol. Pharmacol. Néerl.* **14**, 1.
4. Björklund, A., and Falck, B. (1969). Pituitary monoamines of the cat with special reference to the presence of an unidentified monoamine-like substance in the adenohypophysis. *Z. Zellforsch.* **93**, 254–264.
5. Björklund, A., and Lindvall, O. (1984). Dopamine-containing systems of the CNS. *In* Handbook of Chemical Neuroanatomy, Vol. 2. (A. Björklund and T. Hökfelt, eds.) pp. 55–122. Elsevier, New York.
6. Hökfelt, T., Martensson, R., Björklund, A., Kleinau, S., and Goldstein, M. (1984). Distributional maps of tyrosine-hydroxylase-immunoreactive neurons in the rat brain. *In* Handbook of Chemical Neuroanatomy, Vol 2. (A. Björklund and T. Hökfelt, eds.) pp.277–379. Elsevier, New York.
7. Rogers, J. H. (1992). Immunohistochemical markers in rat brain: Colocalization of calretinin and calbindin-D28k with tyrosine hydroxylase. *Brain Res.* **587**, 203–210.
8. Tomiko, S. A., Taraskevich, P. S., and Douglas, W. W. (1981). Potassium-induced secretion of melanocyte-stimulating hormone from isolated pars intermedia cells signals participation of voltage-dependent calcium channels in stimulus-secretion coupling. *Neuroscience* **6**, 2259–2267.

Seymour Kaufman

Laboratory of Neurochemistry
National Institute of Mental Health
Bethesda, Maryland 20892

Genetic Disorders Involving Recycling and Formation of Tetrahydrobiopterin

The discovery in the late 1950s that the enzymatic conversion of phenylalanine to tyrosine in the liver is dependent on a previously unknown coenzyme and our subsequent demonstration in 1963 that the active factor is tetrahydrobiopterin (BH_4) (1) opened up a new field of enzymology: the pterin-dependent oxygenases. This work also paved the way for the finding reported independently in 1964 by my own group (2), and by Nagatsu *et al.* (3), that BH_4 is also the coenzyme for tyrosine hydroxylase. Later, it was shown that this pterin also functions in the same manner with tryptophan hydroxylase.

Although these enzyme studies made it highly likely that BH_4 is essential for catecholamine and serotonin synthesis *in vivo,* it was not until the description of variant forms of phenylketonuria (PKU) that are caused by blocks in the *de novo* synthesis or recycling of BH_4 that this link was conclusively established.

Figure 1 shows how BH_4 functions with the three aromatic amino acid hydroxylases and also outlines the biosynthetic pathway for BH_4, together with the known genetic blocks in the synthesis and recycling of BH_4. The aromatic hydroxylases catalyze oxidative reactions in which the amino acids are hydroxylated and BH_4 is converted to 4a-hydroxytetrahydrobiopterin (4a-OH-BH_4). Although this compound breaks down fairly rapidly nonenzymatically to quinonoid dihydrobiopterin (q-BH_2), the reaction is also catalyzed by an enzyme, 4a-carbinolamine dehydratase (reaction 7). The recycling of BH_4 is completed by the action of dihydropteridine reductase (DHPR)(reaction 8) (4).

As shown in Figure 1, BH_4 is synthesized *in vivo* from the purine nucleotide GTP. The first reaction, in the pathway (reaction 1), catalyzed by guanosine triphosphate (GTP) cyclohydrolase, converts the purine to a phosphorylated neopterin derivative. (Neopterin differs from biopterin in having a trihydroxypropyl side chain at position 6 of the pterin ring rather than the dihydroxypropyl side chain of biopterin). In the next reaction, catalyzed by 6-pyruvoyl-tetrahydropterin synthase, the neopterin intermediate is converted to 6-pyruvoyl-tetrahydropterin (reaction 2), which is converted to BH_4 by sepiapterin reductase (reaction 3).

Figure 1 also shows the metabolic blocks known to cause hyperphenylalaninemia (HPA). Classical PKU is caused by a block in the conversion of phenylalanine to tyrosine (reaction 6) caused by defects in phenylalanine hydroxylase.

Advances in Pharmacology, Volume 42

1054-3589/98 $25.00

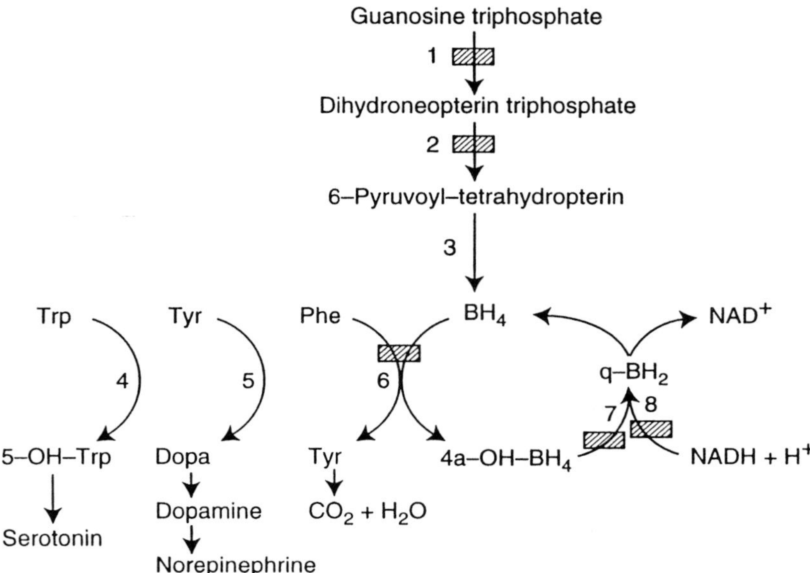

FIGURE 1 Scheme showing the reactions catalyzed by the three aromatic amino acid hydroxylases, as well as those involving the regeneration and the biosynthesis of tetrahydrobiopterin (BH_4). The position of the genetic metabolic blocks that lead to hyperphenylalaninemia are indicated in the cross-hatched boxes. See text for details. BH_4, tetrahydrobiopterin; Trp, tryptophan; Tyr, tyrosine; Phe, phenylalanine; NAD^+, nicotinamide-adenine dinucleotide; 4a-OH-BH_4, 4a-hydroxytetrahydrobioteri; q-BH_2, quinonoid-dihydrobiopterin; NADH, reduced form of nicotinamide-adenine-dinucleotide. (Reprinted from Frontiers and New Horizons in Amino Acid Research. 1992. 653–659. With kind permission from Elsevier Science-NL, Sara Burgerhartstraat 25, 1055 KV Amsterdam, The Netherlands.)

During the past 20 years, variant forms of PKU or HPA have been described due to genetic blocks in either the recycling or the *de novo* synthesis of BH_4. The first of these variants was traced to a deficiency of DHPR, a defect that leads to a failure in BH_4 recycling, with consequent decreases in the activity of the three hydroxylases. We and others reported that a deficiency of the dehydratase (reaction 7) is the cause of a mild form of HPA. In addition to biopterin, these last patients excrete 7-biopterin, whose formation is due to the rearrangement of the 4a-hydroxytetrahydropterin in the absence of the dehydratase. With respect to BH_4 synthesis, cases of HPA due to deficiencies of GTP cyclohydrolase (reaction 1) and of pyruvoyl–tetrahydropterin synthase (reaction 2) have been reported (5).

Except for dehydratase deficiency, all of the variants are characterized by developmental delay and progressive neurological deterioration with symptoms of parkinsonism, including hypokinesis, drooling, swallowing difficulty, truncal hypotonia, and increased limb tone accompanied by myoclonus or dystonic limb movements and very brisk tendon jerks (5, 6). If untreated, *these are lethal diseases,* in part because of the severe deficits in central neurotransmitters. The only variant that appears not to be associated with low levels of monoamine neurotransmitters is dehydratase deficiency, in all probability because, as mentioned, the reaction catalyzed by this enzyme can occur nonenzymatically.

The variants that affect BH_4 metabolism can be differentially diagnosed by analysis of the levels of urinary neopterin and biopterin. In GTP cyclohydrolase deficiency, levels of both neopterin and biopterin are low, as would be expected from the position of the metabolic block. A block in reaction 2 leads to a marked increase in levels of neopterin with low levels of biopterin. In DHPR deficiency, biopterin levels are higher than normal, but the pattern is somewhat variable.

Accurate early diagnosis of these defects in BH_4 metabolism is essential because treatment is maximally effective only when it is initiated as soon as possible after birth. Treatment is aimed at decreasing blood levels of phenylalanine by dietary restriction of the amino acid and normalizing low cerebrospinal fluid (CSF) levels of the neurotransmitters. In addition to a diet low in phenylalanine, DHPR deficiency is treated by the oral administration of the neurotransmitter precursors, L-dopa and 5-hydroxytryptophan, together with a decarboxylase inhibitor (carbidopa). Because DHPR has been implicated in maintaining normal brain levels of tetrahydrofolate, defects in this enzyme can lead to folate deficiency. When CSF folate levels are low, administration of a source of tetrahydrofolate (e.g., 5-formyltetrahydrofolate), together with the neurotransmitter precursors, is essential (5, 6). In addition to limiting phenylalanine intake, defects in BH_4 synthesis are treated by administration of a combination of L-dopa and 5-hydroxytryptophan and BH_4. Some, but not all, patients of this type have been successfully treated by administration of large doses of BH_4 alone. The response to treatment is variable. It seems likely that to be fully effective, treatment of BH_4 synthesis defects may have to be initiated prenatally (5).

Compared with classical PKU, which has an incidence of about 1 in 12,000 births, defects in BH_4 metabolism are much rarer, occurring in around 1 in 500,000 to 1 in 1 million births.

References

1. Kaufman, S. (1963). The structure of phenylalanine hydroxylation cofactor. *Proc. Natl. Acad. Sci. USA* **50,** 1085–1093.
2. Brenneman, A. R., and Kaufman, S. (1964). The role of tetrahydropteridines in the enzymatic conversion of tyrosine to 3,4-dihydroxyphenylalanine. *Biochem. Biophys. Res. Commun.* **17,** 177–183.
3. Nagatsu, T., Levitt, M., and Udenfriend, S. (1964). Tyrosine hydroxylase: The initial step in norepinephrine biosynthesis. *J. Biol. Chem.* **239,** 2910–2917.
4. Kaufman, S. (1993). The phenylalanine hydroxylating system. *In* Advances in Enzymology. (A. Meister, ed). pp 77–264. Wiley, New York.
5. Scriver, C. R., Eisensmith, R. C., Kaufman, S., and Woo, S. L. C. (1995). The hyperphenylalaninemias. *In* The Metabolic and Molecular Bases of Inherited Disease. (C. R. Scriver, A. L. Beaudet, W. S. Sly, and D. Valle, eds.), pp. 1015–1076. McGraw-Hill, New York.
6. Smith, I., and Brenton, D. P. Hyperphenylalaninaemias. *In* Inborn Metabolic Diseases—Diagnosis and Treatment. (J. Fernandes, J.-M. Saudubray, and G. VandenBerghe, eds.), pp. 147–160. Springer-Verlag, Berlin.

T. Nagatsu and H. Ichinose

Institute for Comprehensive Medical Science
School of Medicine
Fujita Health University
Toyoake, Aichi 470-11 Japan

Genetic Basis of Dominant Dystonia

Dominant dystonia was first described by Segawa *et al.* (1) in 1971, as hereditary progressive dystonia (HPD) with marked diurnal fluctuation. Because L-dopa can completely cure the symptom, HPD was also termed *dopa-responsive dystonia* (DRD) by Nygaard *et al.* (2) and Tanaka *et al.* (3) mapped the HPD/DRD gene to chromosome 14q. HPD/DRD is thought to be a striatal dopamine (DA) deficiency. DA is synthesized from tyrosine via dopa by tyrosine hydroxylase (TH) and aromatic amino acid decarboxylase. TH requires tetrahydrobiopterin (BH_4) as cofactor. The first rate-limiting enzyme for BH_4 synthesis is guanosine triphosphatase (GTP) cyclohydrolase I (GCH).

We cloned the GCH gene (4), mapped the gene to chromosome 14q22.1-22.2, and identified it as the causative gene for HPD/DRD in collaboration with Segawa and Tsuji (5, 6). The entire coding region including the splicing junction of the GCH gene, was sequenced using polymerase chain reaction (5, 6). GCH activity was measured in phytohemagglutinin-stimulated mononuclear blood cells (5). We found 6 independent mutations in 10 HPD/DRD families; 4 missense mutations due to single-base changes (Leu 79 Pro, Arg 88 Trp, Asp 134 Val, Gly 201 Glu), and 2 frameshift mutations due to a 2-base insertion and a 13-base deletion (3 ins GG, 511 del 13 bp). All HPD patients were heterozygous in terms of these mutations, with a mutated gene and a normal gene. The mutated enzyme expressed in *Escherichia coli* had no detectable GCH activity. None of these mutations was present on 108 chromosomes from 54 unrelated Japanese individuals. All patients with HPD/DRD showed very low (2–20% of normal values) GCH activities compared with normal individuals. Two unaffected male carriers showed higher GCH activities (about 40% of normal values), although they had the same genetic defects as patients. This is in contrast to recessive GCH deficiency, in which we found two missence mutations (Arg 184 His, Met 211 Ile). In recessive GCH deficiency, both alleles of the GCH gene were mutated, resulting in no detectable activity and severe neurological symptoms, probably due to hyperphenylalaninemia and deficiencies of catecholamine and serotonine. In 4 HPD/DRD families we could not find any mutations in all of the exons, including the splicing junctions of the GCH gene (5). Hirano *et al.* (7) reported a Japanese family with HPD/DRD, in which GT→CT substitution at the 5′ end of intron 1 caused both exon 2

Advances in Pharmacology, Volume 42

1054-3589/98 $25.00

skipping and frameshift mutation. Bandmann *et al.* (8) reported 5 missense mutations and 1 nonsense mutation in 9 British families with 33 affected family members and in 3 sporadic cases. No mutations were identified in 4 families and in 2 sporadic cases. Furukawa *et al.* (9) (reported in 2 Japanese HPD/DRD patients a nonsense mutation in exon 1 and a G-to-T transversion at the intron 1 exon 2 boundary of GCH gene causing exon 2 skipping and frameshift mutation. Furukawa et al. (9) also reported an autopsied case of HPD/DRD of English-Irish ancestry, in which the GCH gene revealed G-to-T transversion in exon 1, resulting in a nonsense mutation (Glu 65 Ter). In this HPD/DRD case, the levels of DA were reduced in substantia nigra and striatum, but pathological studies revealed normal numbers of hypopigmented substantia nigra neurons. Since the first discovery of the GCH mutation in HPD/DRD in 1994 (5), 16 different mutations have been reported (Table I).

Reduced GCH activity in mononuclear blood cells stimulated with phytohemagglutinin from HPD/DRD patients, less than 20% of the normal controls, agrees with the clinical findings that neopterin content in cerebrospinal fluid, which is thought to reflect the GCH activity in the brain, was less than 20% of normal levels in HPD/DRD patients. GCH gene mutations, reduced GCH activity in mononuclear blood cells, and reduced neopterin content in cerebrospinal fluid in HPD/DRD indicate that reduced DA content in the striatum is caused by mutated GCH, reduced GCH activity, low BH_4 content, and low TH activity.

However, several problems remain to be solved. The first concern is the molecular mechanism in which the chimeric GCH protein, composed of wild-type and mutant subunits, shows GCH activity less than 20% of normal values. GCH is a 300-kDa decamer composed of a homologous subunit of about 30 kDa. In missence mutations, the mechanism causing decreased GCH activity, not to 50% but less than 20% of the wild-type enzyme, could be explained by dominant negative effects between wild-type and mutant subunits. However, in frameshift mutations, the mutant subunit may not be able to interact with the wild-type subunit. One possibility is that although the defect in the GCH gene is essential, mutations in the regulatory noncoding domain of the GCH gene may be involved in the lowered expression of GCH in HPD/DRD patients. The second concern is the mechanism that only nigrostriatal DA neurons are affected in HPD/DRD. In recessive GCH deficiency without any GCH activity and BH_4 severe neurological symptoms are developed probably due to decreases of all BH_4-requiring monooxygenases, low TH (low catecholamines), low phenylalanine hydroxylase (hyperphenylalaninemia), and low tryptophan hydroxylase (low serotonin levels). HPD/DRD is caused by partial decrease in BH_4. This may suggest that the nigrostriatal DA neurons are highly susceptible to BH4 and DA deficiencies. In recessive DRD reported in Germany in 1995 by Lüdecke *et al.,* (10) a point mutation (Q381K, L205P) in the TH gene may cause the residual activity of about 15% of the corresponding wild-type human TH type 1. Thus, in dominant or recessive dystonia, a partial decrease in TH activity and DA in the nigrostriatal neurons may cause the symptoms of DRD. In Parkinson's disease, GCH activity, BH_4 content, TH activity, and DA content are all decreased to less than 20% of the normal values owing to the cell death of the nigrostriatal dopamine neurons by unknown mechanisms. The

TABLE I Mutations of GCH in HPD/DRD

Patient/family	*Exon/ intron*	*Base-pair change*		*Mutation*	*Reference no.*
(1) Sa	Exon 1	3 3 ATG GAG → ATG GG GAG	3insGG	Frameshift mutation	5
(2) —	Exon 1	GAG → TAG	Glu65Ter	Nonsense mutation	9
(3) Y	Exon 1	CTG → CCG	Leu79Pro	Missense mutation	6
(4) K	Exon 1	CGG → TGG	Arg88Trp	Missense mutation	5
(5) Mo	Exon 1	CGG → CCG	Arg88Pro	Missense mutation	8
(6) —	Exon 1	TCA → TAA	Ser114Ter	Nonsense mutation	9
(7) —	Intron 1	AG → AA	delExon2	Frameshift mutation	9
(8) —	Intron 2	GT → CT	delExon2	Frameshift mutation	7
(9) Su	Exon 2	GAC → GTC	Asp134Val	Missense mutation	5
(10) Ro	Exon 3	CAT → CCT	His153Pro	Missense mutation	8
(11) I	Exon 4	511 G ATT GTA GAA ATC TAT 511 → GA - - - - - - - - - - - - - - - T	511del13bp	Frameshift mutation	6
(12) N	Exon 5	GGA → GAA	Gly201Glu	Missense mutation	5
(13) Be	Exon 5	GGG → AGG	Gly203Arg	Missense mutation	8
(14) Ha	Exon 6	CGA → TGA	Arg216Ter	Nonsense mutation	8
(15) Hu	Exon 6	AAA → AGA	Lys224Arg	Missense mutation	8
(16) Sm	Exon 6	TTC → TCC	Phe234Ser	Missense mutation	8

parkinsonian symptoms may appear only after dopamine content is decreased to less than 20% of the normal values. These results support the hypothesis that the nigrostriatal DA neurons may be most sensitive to a partial (less than 20%) decrease in DA. The third concern is the mechanism of marked diurnal fluctuation; dystonia symptoms become aggravated toward the evening and are partially alleviated in the morning after sleep. Because a low level of GCH activity remains in HPD/DRD patients, they might continue to synthesize BH_4 at a low rate. However, the rate would not be high enough to supplement the consumption of BH_4 during the day, thus aggravating symptoms toward the evening. DA in the nigrostriatal neurons may be supplemented during sleep. The fourth concern is the mechanism of marked (4 : 1) female predominance in HPD/DRD. Male and female DA neurons may differ in vulnerability to adverse conditions of decreased TH activity. We observed that our normal control data for GCH activity in mononuclear blood cells showed much higher levels in males than in females. Higher GCH activity may protect males from the appearance of symptoms. In conclusion, dominant dystonia is caused by mutations of the GCH gene and is due to resultant partial striatal DA deficiency in the nigrostriatal DA neurons.

Acknowledgment

This work was supported by a Grant-in-Aid for Scientific Research from the Ministry of Education, Science, Culture and Sports of Japan.

References

1. Segawa, M., Ohmi, K., Itoh, S., Acyama, M., and Hayakawa, H. (1971). Childhood basal ganglia disease with remarkable response to L-dopa, hereditary basal ganglia disease with marked diurnal fluctuation. *Shinyo (Tokyo)* **24**, 667–672.
2. Nygaard, T. G., Wilhelmsen, K. C., Risch, N. J., Brown, D. L., Trugman, J. M., Gilliam, T. C., Fahn, S., and Week, D. E. (1993). Linkage mapping of dopa-responsive dystonia (DRD) to chromosome 14q. *Nature Genet.* **5**, 386–391.
3. Tanaka, H., Endo, K., Tsuji, S., Nygaard, T. G., Weeks, D. E., Nomura, Y., and Segawa, M. (1995). The gene for hereditary progressive dystonia with marked diurnal fluctuation maps to chromosome 14q. *Ann. Neurol.* **37**, 405–408.
4. Ichinose, H., Ohye, T., Segawa, M., Nomura, Y., Endo, K., Tanaka, H., Tsuji, S., Fujita, K., Ichinose, H. (1995). *J. Biol. Chem.* **270**, 10062–10071.
5. Ichinose, H., Ohye, T., Takahashi, E., Seki, N., Hori, T., Segawa, M., Nomura, Y., Endo, K., Tanaka, H., Tsuji, S., Fujita, K., and Nagatsu, T. (1994). Hereditary progressive dystonia with marked diurnal fluctuation caused by mutations in the GTP cyclohydrolase I gene. *Nat. Genet.* **8**, 236–242.
6. Ichinose, H., Ohye, T., Segawa, M., Nomura, Y., Endo, K., Tanaka, H., Tsuji, S., Fujita, K., and Nagatsu, T. (1995). GTP cyclohydrolase I gene in hereditary progressive dystonia with marked diurnal fluctuation. *Neurosci. Lett.* **196**, 5–8.
7. Hirano, M., Tamaru, Y., Nagai, Y., Ito, H., Imai, T., and Ueno, S. (1995). Exon skipping caused by a base substitution at a splice site in the GTP cyclohydrolase I gene in a Japanese family with hereditary progressive dystonia/dopa responsive dystonia. *Biochem. Biophys. Res. Commun.* **213**, 645–651.
8. Bandmann, O., Nygaard, T. G., Surtees, R., Marsden, C. D., Wood, N. W., and Harding A. E. (1996). Dopa-responsive dyatonia in British patients: New mutations of the GTP-cyclohydrolase I gene and evidence for genetic heterogeneity. *Hum. Mol. Genet.* **5**, 403–406.
9. Furukawa, Y., Shimadzu, M., Rajput, A. H., Shimizu, Y., Tagawa, T., Mori, H., Yokochi, M., Narabayashi, H., Hornykiewicz, O., Mizuno, Y., and Kish, S. J. (1996). GTP-cyclohydrolase I gene mutations in hereditary progressive and dopa-responsive dystonia. *Ann. Neurol.* **39**, 609–617.
10. Lüdecke, B., Dworniczak, B., and Bartholomé, K. (1995). A point mutation in the tyrosine hydroxylase gene associated with Segawa's syndrome. *Human Genet.* **95**, 123–125.

Klaus Bartholomé and Barbara Lüdecke

University Children's Hospital
44791 Bochum, Germany

Mutations in the Tyrosine Hydroxylase Gene Cause Various Forms of L-Dopa-Responsive Dystonia

Tyrosine hydroxylase (TH, EC 1.14.16.2) is the rate-limiting enzyme in the biosynthesis of dopamine. Because this enzyme is located in the brain and adrenal glands, the enzyme's activity cannot be measured directly. Many attempts have failed to correlate a specific disease with deficient TH levels. Because the reduced enzyme activity of TH causes low levels of L-dopa, the expected clinical symptoms should be similar to those seen in parkinsonism.

L-Dopa-responsive dystonia (DRD) is a disease possibly caused by a TH deficiency, as seen in a patient with DRD low levels of homovanillic acid as measured in the cerebrospinal fluid. There is a dramatic response on a low-dose therapy with L-dopa.

The clinical manifestation of DRD has a broad spectrum. Dominant as well as autosomal recessive traits of inheritance are known. The dominant form seems to be caused by a deficiency of GTP-cyclohydrolase I (1). In this form, many questions remain unanswered.

During the past 2 years, we have found various mutations in the TH gene causing DRD in the mild as well as in the severe form. The mild form of DRD is characterized by a dystonia of the lower extremities, diurnal fluctuations, and a good response to L-dopa therapy. The mean age at the onset of symptoms is about 4 years. Two siblings with this condition carried a point mutation in exon 11 of the TH gene (2). This mutation (Q381K) converts glutamine to lysine. In this family, a sister and the parents were heterozygotes without any clinical symptoms. This mutation was expressed by a coupled *in vitro* transcription–translation system (3). In this system, the activity of the mutant enzyme was about 15% of the wild-type form. This residual enzyme activity corresponds well to the mild form of the disease seen in these patients. It also may explain the diurnal fluctuation of the dystonia, because during night the stores for L-dopa can be refilled.

On the other side of the spectrum, there is the severe form of DRD caused by a point mutation in exon 5 (L205P) changing leucine into proline. Substitution of a leucine residue in an α-helical region by proline is expected to change the structure and catalytic function of the enzyme. The recombinant mutant enzyme revealed a residual activity of 0.3 to 16% of the wild-type TH in three

Advances in Pharmacology, Volume 42

1054-3589/98 $25.00

complementary expression systems. This patient with the severe form of DRD had the first symptoms at the age of 3 months. The main symptoms were brief jerky movements, later generalized rigidity, expressionless face, ptosis, and drooling. Analyses of the cerebrospinal fluid revealed a very low level of the dopamine metabolite homovanillic acid. There were no biochemical signs of a deficiency in the cofactor system. A moderate dose of L-dopa resulted in a marked improvement in hypokinesia and other parkinsonian symptoms. At the age of 3 years, the patient has mild motor and speech delay (4).

Another mutation in exon 7 cannot be correlated to a disease, because an A to G transition in codon 240 is a silent mutation. There also is a frequent sequence variant in the human TH gene in exon 2 (5), which might be an important finding for other clinical groups working in this field.

Studies on the fifth mutation located in exon 5 have not been completed. We found this mutation in only one allele. Because this patient suffers from the severe form of DRD with early onset of symptoms, we assume a compound heterozygote situation. Work is in progress to find the second mutation. Compound heterozygotes also are frequently found in other metabolic diseases, such as phenylketonuria. They contribute to the great variety of inborn errors of metabolism.

This chapter correlates various forms of DRD with the known molecular defect; many further mutations can be expected. Because DRD remains a little-known diagnosis and can be treated easily, more attention should be drawn to dystonias in adults and in children.

References

1. Ichinose, H., Ohye, T., Takahashi, E., Seki, N., Hori, T., Segawa, M., Nomura, Y., Endo, K., Tanaka, H., Tsuji, S., Fujita, K., and Nagatsu, T. (1994). Hereditary progressive dystonia with marked diurnal fluctuation caused by mutations in the GTP cyclohydrolase I gene. *Nat. Genet.* **8**, 236–242.
2. Lüdecke, B., Dworniczak, B., and Bartholomé, K. (1995). A point mutation in the tyrosine hydroxylase gene associated with Segawa's syndrome. *Hum. Genet.* **95**, 123–125.
3. Knappskog, P. M., Flatmark, T., Mallet, J., Lüdecke, B., and Bartholomé, K. (1995). Recessively inherited L-DOPA-responsive dystonia caused by a point mutation (Q381K) in the tyrosine hydroxylase gene. *Hum. Mol. Genet.* **4**, 1209–1212.
4. Lüdecke, B., Knappskog, P. M., Clayton, P. T., Surtees, R. A. H., Clelland, J. D., Heales, S. J. R., Brand, M. P., Bartholomé, K., and Flatmark, T. (1996). Recessively inherited L-DOPA-responsive parkinsonism in infancy caused by a point mutation (L205P) in the tyrosine hydroxylase gene. *Hum. Mol. Genet.* **5**, 1023–1028.
5. Lüdecke, B., and Bartholomé, K. (1995b). Frequent sequence variant in the human tyrosine hydroxylase gene. *Hum. Genet.* **95**, 716.

Rolando Meloni, Sylvie Dumas, and Jacques Mallet

Laboratoire de Génétique Moléculaire de la Neurotransmision et des Processus Neurodégénératifs
Bat C.E.R.V.I., CNRS
Hôpital La Pitié-Salpétrière
75013 Paris, France

Catecholamine Biosynthetic Enzyme Expression in Neurological and Psychiatric Disorders

The crucial rate-limiting role played by tyrosine hydroxylase (TH) in the catecholamine biosynthetic pathway could *a priori* implicate it in neurological or psychiatric diseases. TH is highly regulated at several levels by a diversity of mechanisms, and changes in TH activity can result from short-term regulation with activation of the enzyme by phosphorylation or feed-back inhibition by catecholamines and long-term regulation involving TH gene expression. Thus, the study of variations in TH gene regulation could yield new leads for investigating the etiopathogenesis of neuropsychiatric diseases.

The TH gene has been implicated in the genetics of one neurological disease, hereditary progressive dystonia (HPD), an L-dopa responsive form of dystonia in childhood. A transition C > T in the highly conserved exon 11 of TH has been shown to be responsible for HPD (1). Interestingly, the study of TH alternative splicing has led to the implication of TH in another neurological disease, progressive supranuclear palsy (PSP), a disease characterized by atrophy of several brain nuclei. The TH gene shows alternative splicing in different species, with the highest complexity in the splicing mechanism achieved in homo sapiens. We have previously demonstrated that alternative splicing of exons 1 and 2 of human TH generates four different mRNA species whose corresponding proteins present differently regulated activities. In a further study of an alternative splicing mechanism, we have shown that the diversity of TH mRNA is greater than previously described, finding three new species derived from alternative splicing of exon 3 in various human adrenal medulla samples. Interestingly, the proportion of the new TH mRNA was unusually higher than normal in the adrenal medulla of patients with PSP (2). Although this finding shows a novel regulatory potential for the alternative splicing of exon 3, its relevance to the pathogenesis of PSP remains to be established. However, the altered pattern of TH mRNA expression may contribute to the deregulation of TH activity observed in these patients, suggesting that PSP could also have a genetic etiology.

Several studies have, moreover, implicated TH in psychiatric diseases. In our laboratory we have investigated the role of the TH gene in bipolar disorders

Advances in Pharmacology, Volume 42

1054-3589/98 $25.00

(BPD) and schizophrenia (SZ), using molecular genetics techniques. Family, twin, and adoption studies provide evidence that BPD and SZ have a strong genetic component. However, these diseases do not show a classical mendelian mode of inheritance, suggesting that the phenotype is determined by an unknown number of genes (polygenicity) that may vary (heterogeneity) with variable penetrance and the intervention of environmental factors. Thus, the genetic complexity of BPD and SZ requires a composite approach using the classical parametric method of linkage assuming a major gene in very large families, nonparametric methods based on allele sharing in sibling pairs and affected pedigree member familial studies, and association analysis in case-control and case-parents studies (3).

Using a nonparametric approach, we have shown that the TH gene is associated both with BPD and SZ. Initially, we performed an association study on BPD on 100 patients and 65 controls. Our results showed significant association of BPD with a TaqI and a BglII restriction fragment length polymorphism (RFLP), two polymorphic biallelic genetic markers flanking the TH gene (4). To confirm this result in another population, we collected a sample of 64 BPD patients and 64 controls strictly matched for their geographical origin. In this study we used a more informative marker: the HUMTH01 microsatellite, a tetrameric repeated sequence located in the first intron of the TH gene. In our case-control population we found at this marker 5 different alleles (T6, T7, T8, T9, and T10, characterized by the repetition of the core motif TCAT from 6 to 10 times). We found that the T7–T10 genotype was overrepresented in the patient population (14 patients versus 4 controls), giving a significant association of this allele with BPD. Moreover, the patients with this genotype were characterized clinically by having familial history of affective disorders and/or delusive symptoms during manic or depressive episodes (5).

The sequencing of the different alleles showed that the T10 allele, the most frequent allele in our population, presents a sequence variant with a constant one base pair deletion in the fifth of its 10 repetitions. This deletion constitutes the imperfect repeat (TCAT)4 CAT (TCAT)5 (allele T10i), which is the most frequent allele in the Caucasian general population, whereas the corresponding perfect repeat, that is, its nondeleted (TCAT)10 sequence (allele T10p), is very rare (34.7% and 1.1%, respectively, in a sample of 186 Caucasians in the United States) (6). In the association study for BPD, we found only the T10i allele represented in both the patients and the controls.

In another association study we evaluated the possible involvement of the TH gene in SZ with particular attention to the imperfect and perfect repeat alleles of the sequence variant in the HUMTH01 microsatellite. The population we used for the association study consisted of 94 unrelated chronic schizophrenic patients and 145 unaffected controls from Normandy (northwestern France). The T10p allele, the (TCAT)10 perfect repeat form of the HUMTH01 microsatellite, was absent from the control French population but present in 5 of the 94 (5%) SZ patients, giving a highly significant association with SZ. This finding was replicated in another population of different ethnic origin: 44 unrelated chronic SZ patients and 44 unaffected controls from Sousse (eastern Tunisia). The allele frequencies in the Tunisian patients and control population were significantly different from those in the corresponding groups in the French population, due to differences in the frequencies of alleles other than the T10p allele. Nevertheless, consistent with the result obtained in the French sample, the T10p allele was present in 4 of 44 (9%) schizophrenic patients and in none of the controls. However, the T10p allele, in spite of having an higher frequency in this second SZ population, did not attain statistical significance compared with the control population, most probably because of the relatively small size of the sample used. Thus, the perfect repeat was rare and found only in the SZ patients, in both the French and the Tunisian populations (7). No variants other than the single base difference between the T10i and the T10p alleles were found either in the tetrarepeat or in its flanking sequences. It is at present unclear whether it is this HUMTH01 polymorphism or another polymorphism in disequilibrium with it that is functionally implicated in conferring vulnerability to SZ. However, we have found that the two sequence variants are able to bind nuclear proteins, hinting that this repeated sequence could play a role in the regulation of the TH gene.

The results we have obtained show that a composite approach using molecular biology and molecular genetics techniques is fruitful in establishing an eventual role for the TH gene in neurological and psychiatric diseases. The study of the alternative splicing mechanism has led to the discovery of new TH mRNA species that could be relevant in PSP, whereas the molecular genetic studies have established that a mutation in the TH gene is responsible for HPD and that variations in its sequence could be implicated in the genetics of more complex psychiatric diseases, such as BPD and SZ. Interestingly these variations could represent the molecular basis of a further form of TH gene regulation. Further studies in these domains may open new perspectives for the investigation of the pathogenesis of neuropsychiatric diseases.

References

1. Knappskog, M., Flatmark, T., Mallet, J., Lüdecke, B., and Bartholomé, K. (1995). Recessively inherited L-DOPA responsive dystonia caused by a point mutation (Q381K) in the tyrosine hydroxylase gene. *Hum. Mol. Genet.* **4**, 1209–1210.
2. Dumas, S., Le Hir, H., Bodeau-P/an, S., Hirsch, E., Thermes, C., and Mallet, J. (1996). New species of human tyrosine hydroxylase mRNA are produced in variable amounts in adrenal medulla and are overexpressed in progressive supranuclear palsy. *J. Neurochem.* **67**, 19–25.
3. Lander, E. S., and Schork, N. J. (1994). Genetic dissection of complex traits. *Science* **265**, 2037–2048.
4. Leboyer, M., Malafosse, A., Boularand, S., Campion, D., Gheysen, F., Samolyk, D., Henriksson, B., Denise, E., des Lauriers, A., L. A., Lepine, J. P., Zarifian, E., Clerget-Darpoux, F., and Mallet, J. (1990). Tyrosine hydroxylase polymorphisms associated with manic-depressive illness, *Lancet* **335**, 1219.
5. Meloni, R., Leboyer, M., Bellivier, F., Barbe, B., Samolyk, D., Allilaire, J. F., and Mallet, J. (1995). Association of manic-depressive illness with tyrosine hydroxylase microsatellite marker. *Lancet* **345**, 932.
6. Puers, C., Hammond, H. A., Jin, L., Caskey, T., and Schumm, J. W. (1993). Identification of repeat sequence heterogeneity at the polymorphic short tandem repeat locus HUMTH01

(AATG)n and reassignment of alleles in population analysis by using a locus-specific allelic ladder. *Am. J. Hum. Genet.* **53**, 953–958.
7. Meloni, R., Laurent, C., Campion, D., Ben, H. B., Thibaut, F., Dollfus, S., Petit, M., Samolyk, D., Martinez, M., Poirier, M. F., and Mallet, J. (1995). A rare allele of a microsatellite located in the tyrosine hydroxylase gene found in schizophrenic patients. *C. R. Acad. Sci. III* **318**, 803–809.

Esther L. Sabban and Bistra B. Nankova

Department of Biochemistry and Molecular Biology
New York Medical College
Valhalla, New York 10595

Multiple Pathways in Regulation of Dopamine β-Hydroxylase

Dopamine β-hydroxylase (DBH, EC 1.14.17.1) catalyzes the hydroxylation of dopamine to form norepinephrine. Aberrations in the noradrenergic pathway and/or DBH are associated with a variety of human disorders, including hypertension, hypotension, congestive heart failure, depression, idiopathic Parkinson's disease, and disorders related to copper deficiency.

DBH is localized in membrane-bound (77-kDa) and soluble (73-kDa) forms in neurosecretory vesicles of the noradrenergic neurons of the central and peripheral nervous systems and in the chromaffin granules of the adrenal medullary cells. Experiments from several groups reveal that both forms of DBH arise from a single translational product. The 77-kDa form has an uncleaved signal sequence (1, 2). Several treatments, such as exposure to cyclic adenosine monophosphate (cAMP) analogues or to nerve growth factor (NGF), have been shown to reduce the proportion of the 77-kDa form of DBH.

The expression of DBH is elevated *in vivo* in response to a variety of transsynaptic signals, hormones, growth factors, and stress. There is now a large body of experiments demonstrating that with prolonged conditions that stimulate catecholamine biosynthesis, the expression of DBH also is markedly elevated. This includes DBH mRNA levels, immunoreactive protein, and activity. DBH gene expression is activated by subset of conditions which elevate TH gene expression. In some situations they are activated concomitantly, whereas often activation of DBH gene expression requires more severe or more prolonged treatments than that of TH. This scenario occurs both *in vivo* as well as in tissue culture models.

Cultured cells (rat PC12 cells, human neuroblastoma cell lines, and bovine adrenal chromaffin cells) have been used to investigate the mechanisms of regulation of DBH gene expression. Although not as well examined as the

Advances in Pharmacology, Volume 42

1054-3589/98 $25.00

regulation of TH, it is clear that DBH gene expression in cultured cells is indeed markedly elevated by some of the treatments that activate TH gene expression: glucocorticoids, elevated cAMP, nicotine, short-term NGF, bradykinin, and phorbol esters. Only TH is induced by membrane depolarization, ionomycin, and increased cell density. Elevated cAMP and nicotine have a biphasic effect on DBH expression in PC12 cells. They elicit activation after 12–24 hr of treatment. However, after 2 days or more of continuous administration, DBH expression is reduced to well below control levels, that is, essentially turned off (3).

In vivo, DBH gene expression has been shown to be elevated by several important physiological and pharmacological stimuli. These include treatment of rats with reserpine, or nicotine or electrical preganglionic stimulation. Exposure to stress markedly activates DBH gene expression. We have shown previously that repeated immobilizations increase rat adrenal medullary DBH mRNA maximally to about 400–500% of control levels (4). Run on assays of transcription revealed elevations as compared to controls after even a single episode of stress, suggesting a transcriptionally mediated mechanism.

The rat DBH promoter sequence (3) contains a number of putative regulatory elements. There are several AP2-like sites. Previously it has been shown by others that the AP2 motif at position −136 to −115 binds AP2 transcription factor and regulates the basal levels of DBH gene expression in tissue culture cells. We examined binding to this element in extracts from adrenal medulla of controls and rats immobilized for 2 hr daily for 4 consecutive days. The results revealed that AP2 specific complexes were formed. However, the patterns were similar in both experimental groups.

The rat DBH-1 element is important for both basal expression and activation of DBH transcription by cAMP and phorbol esters (5). The sequence of this element is shown in Figure 1. This region contains two adjacent cAMP regulatory elements (CRE1, CRE2). Interestingly, there are several differences between the rat and human DBH promoters within the regions implicated in regulation by cAMP. The rat DBH CRE1 contains a core CRE/AP1 sequence identical to ENKCRE2 in the enkephalin gene. CRE2 overlaps with binding sites for the recently identified homeobox protein, arix, which appears to be important for tissue-specific expression in noradrenergic cells (6).

Our studies suggest that the DBH-1 promoter region also may be involved in activation of DBH gene expression in response to repeated immobilization stress. Electrophoretic mobility shift assays were carried out with DBH-1 element and nuclear extracts from adrenal medulla of controls or rats exposed to immobilization stress (see Fig. 1). At least two specific complexes were formed. The formation of the complex with lower mobility was greatly increased by repeated immobilization stress (7). This complex was recognized by a fos family–specific antibody, but cross-linking experiments indicated that the involvement of c-fos is unlikely (4). A replacement mutation of the DBH CRE1 (identical to ENKCRE2) prevented the binding. Inclusion of antisera to various transcription factors was used to identify the proteins involved. There was no detectable binding of CREB or ATF-1 immunoreactive proteins. However, the

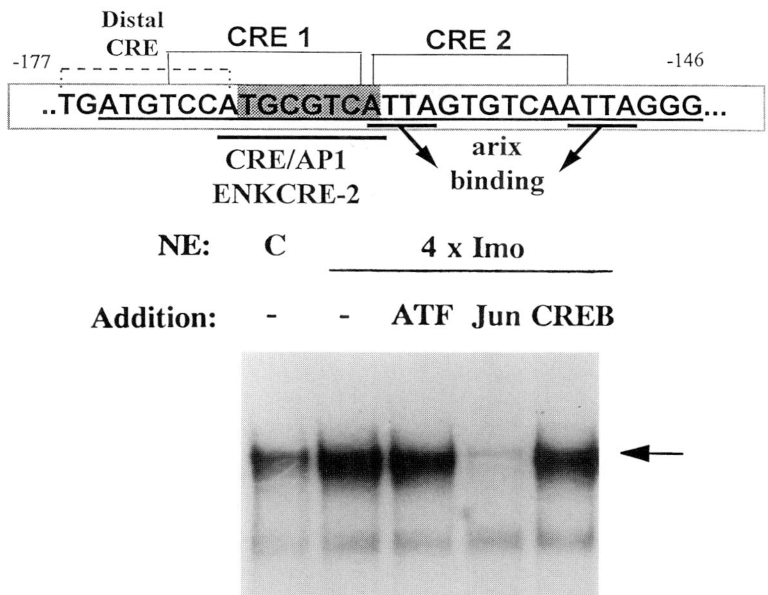

FIGURE 1 Electrophoretic mobility shift assay of the DNA-protein complexes formed with DBH-1 promoter element. A 30-bp oligonucleotide, containing rat DBH-1 enhancer (*shown on top*) was radiolabeled and incubated with 5-Mg nuclear extracts from adrenal medulla of control (C) or repeatedly immobilized animals (on 4 consecutive days for 2 hr each, 4 × Imo). The resulting DNA-protein complexes were resolved on 6% native polyacrylamide gel. In lanes 3 to 5, the nuclear extracts were preincubated with antibodies to ATF-1, Jun-family of transcription factors, or CREB. The arrow indicates the complex, increased by immobilization stress and competed by Jun-family–specific antisera.

slower mobility complex was completely competed with antisera to jun family members (see Fig. 1), suggesting involvement of AP1-like factors in stress-elicited activation of DBH transcription.

Using electrophoretic mobility shift assays, we have shown that the transcription factors are not identical to those reported to interact with its human analogue. For example, the human DBH promoter oligonucleotide analogous to DBH-1 has been shown by others to bind YY1. However, we found no detectable binding of YY1 in nuclear extracts from the adrenal medulla, although these transcription factors are abundant and bind a YY1 consensus oligonucleotide. Use of a longer oligonucleotide to include the distal sequences homologous (but not identical) to the human DBH CRE did not yield additional complexes. Different patterns of complexes were formed between rat DBH-1 and nuclear extracts from PC12 cells, compared with the adrenal medulla. These studies indicate that transcriptional activation is one of the mechanisms of regulation of DBH gene expression in response to long-term activation of the sympathoadrenal system.

Acknowledgments

Supported by NIH grants NS 28869 and 32166.

References

1. Feng, Z., Angeletti, R. H., Levin, B. E., and Sabban, E. L. (1992). Glycosylation and membrane insertion of newly synthesized rat dopamine β-hydroxylase in a cell free system without signal cleavage. *J. Biol. Chem.* **267,** 2108–2115.
2. Houhou, L., Lamouroux, A., Faucon Biquet, N., and Mallet, J. (1995). Expression of human dopamine β-hydroxylase in mammalian cells infected by recombinant vaccinia virus. *J. Biol. Chem.* **270,** 12601–12606.
3. McMahon, A., and Sabban, E. L. (1992). Regulation of expression of dopamine β-hydroxylase in PC12 cells by glucocorticoids and cyclic AMP analogues. *J. Neurochem.* **59,** 2040–2047.
4. Sabban, E. L., Hiremagalur, B., Nankova, B., and Kvetnansky, R. (1995). Molecular biology of stress-elicited induction of catecholamine biosynthetic enzymes. *Ann. N. Y. Acad. Sci.* **771,** 327–338.
5. Shaskus, J., Greco, D., Asnani, L. P., and Lewis, E. J. (1992). A bifunctional genetic regulatory element of the rat dopamine β-hydroxylase gene influences cell type specificity and second messenger–mediated transcription. *J. Biol. Chem.* **267,** 18821–18830.
6. Zellmer, E., Zhang, Z., Greco, D., Rhodes, J., Cassel, S., and Lewis, E. J. (1995). A homeodomain protein selectively expressed in noradrenergic tissue regulates transcription of neurotransmitter biosynthetic genes. *J. Neurosci.* **16,** 4102–4112.
7. Nankova, B., Devlin, D., Kvetnansky, R., Kopin, I. J., and Sabban, E. L. (1993). Repeated immobilization stress increases the binding of c-fos-like proteins to a rat dopamine β-hydroxylase promoter enhancer sequence. *J. Neurochem.* **61,** 776–779.

Steven A. Thomas and Richard D. Palmiter

Howard Hughes Medical Institute
Department of Biochemistry
University of Washington
Seattle, Washington 98195

Examining Adrenergic Roles in Development, Physiology, and Behavior through Targeted Disruption of the Mouse Dopamine β-Hydroxylase Gene

We have taken a novel approach to studying the adrenergic nervous system by genetically altering mice so that they do not produce norepinephrine (NE) and epinephrine (Epi). This was accomplished by targeted disruption of the dopamine β-hydroxylase (DBH) gene (1). An important consideration of this mutation is that dopamine (DA), the precursor of NE, is stored in and secreted from adrenergic terminals and cell bodies. Our goal is to understand how the nervous system develops and functions in the context of this neurotransmitter switch. Interestingly, there is a small minority of patients with autonomic failure that have now been documented to be DBH-deficient (2, 3). The major clinical findings in these patients are orthostatic hypotension and exercise intolerance. Given the existence of these patients, it came as a great surprise when we discovered that elimination of DBH in mice results in fetal lethality.

To disrupt the DBH locus, the proximal promoter and first exon were deleted. A sensitive RT-PCR assay confirmed the predicted absence of the DBH message in knockout fetuses. In addition, it showed that message for TH appears as early as E8.5, and that for DBH 1 day later. The developmental appearance of catecholamines parallels that for the mRNAs: DA first appears at E9.5, NE first appears at E10.5 and becomes maximal by E12.5, and Epi first appears around E14.5 and becomes maximal at birth.

To determine when the mutant fetuses die, pregnant females were sacrificed at specific times during gestation. All of the mutant fetuses examined were alive and appeared grossly normal at E10.5. However, by E11.5, 40% of the mutant fetuses had died. Even fewer survived to later stages, such that by birth only about 10% of the mutant fetuses were still viable. One explanation for the fact that some DBH −/− mice are born is that maternal catecholamines can cross the placenta in sufficient quantities to rescue some fetuses. DBH −/− females that survived to adulthood were mated to assess fetal survival in the absence of maternal NE and Epi. As before, almost all the mutant fetuses were

Advances in Pharmacology, Volume 42

1054-3589/98 $25.00

alive at E10.5, but about two-thirds had died 1 day later, and no mutant pups were found born to DBH −/− females. These results indicate that NE is essential for fetal survival and suggest that the DBH-deficient human patients may be the lucky few who have survived this condition.

Measurement of catecholamines at E11.5 verified the idea that catecholamines can cross the placenta. Small but detectable levels of NE and Epi were present in DBH −/− fetuses from heterozygous females. However, NE and Epi were below the limit of detection in DBH −/− fetuses from homozygous females. As expected for the precursor of NE, DA was present at high levels in the mutant fetuses. The fetal lethality is due to the absence of NE rather than excess DA, because mice lacking both NE and DA (tyrosine hydroxylase −/−) had a very similar phenotype (4).

To examine how fetal development might be altered, serial sagittal sections were cut from fetuses age E10.5 to E15.5 and stained with hematoxylin and eosin. In many but not all fetuses, subtle changes were observed in the cellular organization of the heart but not other tissues. In one E13.5 fetus, hemorrhaging in the liver was also observed. These observations suggest that fetal demise is due to cardiovascular failure.

To study adult physiology and behavior in the mutant mice, rescuing the DBH −/− fetuses became paramount. Pregnant females were treated with a precursor of NE whose conversion does not depend on DBH. Dihydroxyphenylserine (DOPS) was provided in the maternal drinking water, beginning at E8.5 and continuing until birth. A dose-dependent rescue was seen such that at 1 mg/ml of DOPS, the expected frequency of DBH −/− pups was born. When carbidopa, an inhibitor of aromatic amino acid decarboxylase and thus the conversion of DOPS into NE, was added with DOPS to the maternal drinking water, rescue was no longer observed.

Because catecholamines can cross the placenta, it was possible that derivatives of catecholamines that are agonists for specific adrenergic receptor subtypes might prove useful in defining the receptors required for fetal development. To test this, agonists were added to the maternal drinking water from E8.5 to E13.5, at which time fetal survival was assessed. At doses of 0.01 mg/ml and higher, the β-agonist isoproterenol (Iso) increased survival of the mutant fetuses to 90% (as compared to 50% without any drug). The inclusion of the α_1-agonist phenylephrine with Iso resulted in 100% survival at E13.5. Interestingly, phenylephrine alone did not enhance survival.

To determine whether β_1, β_2, or both adrenergic receptor subtypes are important during development, selective receptor antagonists were added to maternal drinking water containing 0.01 mg/ml Iso. The β_1-selective antagonist CGP 20712A reduced survival from 90 to 70% over a broad range of doses. When the β_2-selective antagonist ICI 118,551 was included with Iso and CGP 20712A, survival fell to 50%, the value observed when no drugs were added. These results suggest that both β-receptor subtypes are critical for fetal development.

Disruption of the DBH gene also revealed that NE is important for postnatal survival and the timing of the adolescent growth phase. About 40% of the mutant mice born die within several days of birth and another 20% die during the third, fourth, and fifth postnatal weeks. In addition, the mutant mice weigh about half of normal during adolescence. However, adults typically reach 80–90% of normal weight, and this difference is reflected in the weights of the major organs and other tissues, such as white fat. Brown fat is an interesting exception, however, weighing in at over twice that of normal. The increased mass is not due to an increase in cell number or RNA/cell. Grossly, brown fat is much paler than normal in the mutant mice, a reflection of its increased fat content as seen at the microscopic level (5).

Nonshivering thermogenesis in response to cold is largely achieved by the uncoupling of oxidative phosphorylation in brown fat, a tissue that has substantial sympathetic innervation. Uncoupling occurs by insertion of a proton channel into the inner membrane of mitochondria, generating heat as it dissipates the proton gradient. The uncoupling protein (UCP) is expressed exclusively in brown fat, and its abundance is regulated at the message level by β-adrenergic stimulation. To assess the capacity for heat production in the mutant mice, message for UCP was quantified by solution hybridization. At the thermoneutral temperature of 30°C, mice express about 15% of the mRNA for UCP that they do at room temperature. Message for UCP in mutant mice at room temperature was only 10% of normal, but full expression could be restored by administering DOPS, demonstrating that impaired expression was due to the physiological rather than the developmental absence of NE.

NE and Epi play an additional role in cold acclimatization because they also mediate vasoconstriction and piloerection. A clear prediction then is that adaptation to cold should be impaired in mutant mice. Indeed, DBH −/− mice only survive between 1 and 2 hr at 4°C, whereas controls survive indefinitely. Administration of DOPS prior to being placed at 4°C diminished the rate of heat loss in the mutant mice by over threefold. Even at room temperature, DBH −/− mice have a lower body temperature than do controls. A likely explanation for this difference was a reduction in metabolic rate due to the inactivity of brown fat. However, when oxygen consumption was measured by indirect calorimetry, basal metabolic rate (BMR) was found to be elevated—25% in the mutant mice. Food and water intake were also increased, allowing the mice to maintain a normal energy balance. Body temperature, BMR, and food and water intake all normalized after restoration of NE with DOPS.

The increase in BMR may represent compensation for greater heat loss in the mutant mice. To test this possibility, the mice were kept in a thermoneutral environment for 2 weeks. The increased ambient temperature normalized body temperature between the mutants and the controls. However, BMR and food and water intake remained elevated in the mutant mice. Because thyroid hormone is a major regulator of BMR, levels of this hormone were quantitated by radioimmunoassay. Total thyroxin and triiodothyronine, and free thyroxin did not differ between the mutant and control mice. These observations suggest that NE plays a fundamental role in determining BMR that is independent of thyroid hormone.

Reproduction in the DBH −/− mice is also impaired. The majority of pups born to mutant females does not survive beyond several days after birth, even though the pups are DBH +/−. Affected females are repeat offenders in that prior birthing experience does not improve neonatal survival. Because the pups are scattered within the bedding in about 50% of the cases, rather than neatly gathered in the nest, a deficit in maternal behavior was suspected. This was confirmed when young virgin females were examined for maternal behavior by scattering several pups within their cage. Control females almost always retrieved all three pups within several minutes during a second exposure the next day. The majority of DBH −/− females investigated but did not retrieve the pups.

To test whether the defect in maternal behavior is developmental or physiological, mutant females were injected with DOPS around the time of birth. Administration of DOPS the evening prior to and the morning after giving birth resulted in over 75% of mutant females that had previously abandoned their litters acting maternally and raising their litters. Remarkably, mutant females act maternally with subsequent litters even though NE is no longer present.

In conclusion, the genetic approach of disrupting the DBH gene to produce mice devoid of NE and Epi has proven to be interesting and useful. Unanticipated phenotypes have been observed at the developmental, physiological, and behavioral levels. The use of DOPS to restore NE in these animals is a powerful tool for examining whether observed deficits result from developmental or physiological requirements for NE. These mice should continue to be important for defining the mechanisms by which NE and Epi exert their effects.

References

1. Thomas, S. A., Matsumoto, A. M., and Palmiter, R. D. (1995). Noradrenaline is essential for mouse fetal development. *Nature* **374**, 643–646.
2. Craig, I., Porter, C., and Craig, S. (1992). Dopamine β-hydroxylase deficiency and other genetically determined causes of autonomic failure. B. The molecular genetics of dopamine β-hydroxylase. *In* Autonomic Failure: A Textbook of Clinical Disorders of the Autonomic Nervous System, 3rd edition. (C. J. Mathias and R. Bannister, eds.), pp. 721–758. Oxford University Press, New York.
3. Mathias, C. J., and Bannister, R. (1992). Dopamine β-hydroxylase deficiency and other genetically determined causes of autonomic failure. A. Clinical features, investigation, and management. *In* Autonomic Failure: A Textbook of Clinical Disorders of the Autonomic Nervous System, 3rd edition. (C. J. Mathias and R. Bannister, eds.), pp. 721–758. Oxford University Press, New York.
4. Zhou, Q.-Y., Quaife, C. J., and Palmiter, R. D. (1995). Targeted disruption of the tyrosine hydroxylase gene reveals that catecholamines are required for mouse fetal development. *Nature* **374**, 640–643.
5. Thomas, S. A., and Palmiter, R. D. (1997). Thermoregulatory and metabolic phenotypes of mice lacking noradrenaline and adrenaline. *Nature* **387**, 94–97.

David Robertson and Nannette Hale

Clinical Research Center
Vanderbilt University
Nashville, Tennessee 37232

Genetic Diseases of Hypotension

Norepinephrine and epinephrine are crucial determinants of minute-to-minute neural regulation of blood pressure and are also present at crucial central nervous system sites likely to be involved in a variety of behaviors. Norepinephrine and epinephrine thus seem so important to human beings that it seemed unlikely for many years that subjects without these catecholamines would survive the perinatal period and develop to adulthood.

This view has changed with recognition of a congenital syndrome of severe orthostatic hypotension, noradrenergic failure, and ptosis of the eyelids in two young adults (1, 2). The syndrome differs from familial dysautonomia and various other autonomic disorders seen in adults in that the defect can be localized to the noradrenergic and adrenergic tissues. There is virtual absence of norepinephrine, epinephrine, and their metabolites. However, there is greatly increased dopamine in plasma, cerebrospinal fluid, and urine (1–6).

Parents of patients with dopamine β-hydroxylase (DBH) deficiency have appeared normal, but there has usually been a history of spontaneous abortions and stillbirths in the mothers of affected patients. There has been delay in opening of the eyes in neonates and persistent ptosis of the eyelids in several patients. There have also been hypoglycemia, hypotension, and hypothermia in the prenatal period. It is possible that excessive central nervous system dopamine contributed to the hypothermia in these patients and may also have contributed to the recurrent vomiting observed in the first year of life in two patients. With severe hypoglycemia or hypotension, seizures have occasionally been seen in adults. The full expression of DBH deficiency, as far as is currently known, is shown in Table I (7, 8).

As children, DBH-deficient patients have had a markedly reduced ability to exercise, perhaps because of hypotension engendered by the physical exertion. Because of occasional syncope, anticonvulsive medications have been given in some patients, even though no abnormality was seen on the electroencephalogram. Symptoms have generally worsened in late adolescence, and by early adulthood, patients complain of profound orthostatic hypotension, especially early in the day and during hot weather or after alcohol ingestion. In addition to ptosis of the eyelids, there is reduced exercise tolerance, a tendency for nasal stuffiness to occur, especially in the supine posture. A male patient had appropriate erectal function, but retrograde rather than antegrade ejaculation. Presyncopal symptoms in these patients have included dizziness, blurred vision, dyspnea, nuchal discomfort, and occasionally chest pain. The mechanism of the chest pain is not understood but has been replicated in one patient by the administration of isoproterenol during the supine posture. It is, therefore,

Advances in Pharmacology, Volume 42

1054-3589/98 $25.00

TABLE I Characteristics of Dopamine β-Hydroxylase Deficiency[a]

Feature	*Frequency (%)*
Severe orthostatic hypotension	100
Impaired ejaculation (n = 2)	100
Plasma dopamine ⋙ plasma norepinephrine	100
Ptosis of the eyelids	67
Complicated perinatal course	67
Nocturia	67
Hypoprolactinemia	67
Hyperextensible/hyperflexible joints	50
High palate	50
Nasal stuffiness	50
Mild behavioral changes	33
Hypomagnesemia	33
Seizures (with hypotension)	33
Brachydactyly	33
Sluggish deep-tendon reflexes	33
Weak facial musculature	33
Hypotonic skeletal muscles	33
Raised blood urea nitrogen	33
Atrial fibrillation	16
T-wave abnormalities (ECG)	16
Hypoglycemia	16

[a] Data are taken from the first six published cases.
ECG, electrocardiogram.

possible that this chest pain is somehow related to excessive pumping action of the heart associated with the intact parasympathetic withdrawal accompanying upright posture.

Findings during the physical examination usually include a normal or low normal supine blood pressure and a normal heart rate but a standing blood pressure that is less than 80 mm Hg systolic. Heart rate rises on standing but appears to have an attenuated elevation given the very low blood pressure with upright posture. Pupils are somewhat small but respond to light and accommodation. Parasympatholytics dilate the eye appropriately.

Many specialized tests differentiate these patients from those with familial dysautonomia. Cholinergic sensitivity as assessed by conjunctival methacholine was normal, in that there was no response. Intradermal histamine evoked a typical flare reaction, whereas this does not occur in familial dysautonomia. These patients are further distinguished from familial dysautonomia in having normal tearing, intact corneal and deep tendon reflexes, normal sensory function, and normal senses of taste and smell. Moreover, no subjects thus far recognized have been of Ashkenazi Jewish extraction. Atrial fibrillation, quite resistant to therapy, developed in 1 patient at age 40. Findings of laboratory tests of autonomic maneuvers are shown in Table II.

TABLE II Autonomic Maneuvers in Dopamine β-Hydroxylase Deficiency[a]

Finding	*Frequency (%)*
Orthostatic hypotension >40 mm Hg systolic	100
Abnormal Valsalva maneuver	100
Sweating present	100
Sinus arrhythmia present	100
Atropine tachycarida >25 beats/min	100
Pressor clonidine response	100
Absent pressor tyramine response	100
Pressor efficacy of DOPS	100
Absent pressor isometric handgrip > 10 mm Hg	87
Absent cold pressor response > 10 mm Hg	87

DOPS, dihydroxyphenylserine

The DBH gene is now recognized to be located at 9q34 (9–12). Although the precise genetic cause of DBH deficiency has still not been reported, it is clear that there is no recognizable DBH enzyme in some of the patients, even when a polyclonal antibody against DBH is used to measure it (13).

Efforts to treat DBH deficiency have also led to surprising observations. Fludrocortisone at relatively high doses has successfully raised blood pressure with some benefit. Indomethacin also has been of modest benefit in raising blood pressure, but one patient had aggressive ideation while receiving this drug. The monoamine oxidase inhibitor tranylcypromine also produced paranoid thinking in one patient. There has been a reasonable pressor response to phenylpropanolamine (25 and 50 mg), perhaps because of the hypersensitive 1-adrenoreceptors in these patients. Kuchel *et al.* proposed that the excessive dopamine in these patients might be eliciting a depressor response (14). We tested this by administration of metyrosine (-methyl-p tyrosine). Normally this agent lowers blood pressure by reducing norepinephrine levels, but DBH-deficient patients have dopamine rather than norepinephrine in their neurons. We found that as the dose of metyrosine was gradually increased, plasma and urinary dopamine levels fell and there was a significant increase in blood pressure corresponding to this. Blood pressure actually rose into the hypertension range in the supine posture at the highest dose of metyrosine tested.

A more favorable long-term result has been achieved with L-dihydroxyphenylserine (L-DOPS) (15). This agent is a prodrug acted upon by endogenous dopa decarboxylase to yield norepinephrine. Patients have had remarkable resolution of their orthostatic hypotension on this agent. Presumably, their noradrenergic neurons are able to use this supply of norepinephrine effectively. Long-term experience with this drug indicates continued efficacy at the 250-mg or 500-mg t.i.d. regimen.

DBH deficiency and its successful treatment by DOPS has provided valuable lessons in catecholamine pharmacology (4, 19) and encourages us to hope that other autonomic disorders may one day also yield to genuinely effective therapeutic intervention. Indeed, other newly recognized genetic autonomic and catecholamine disorders are now being recognized, including tetrahydrobiopterin deficiency, dopa decarboxylase deficiency (20), Menkes disease (21), monoamine oxidase deficiency (22) and other disorders of dopamine metabolism (23).

Acknowledgments

Supported in part by the National Institutes of Health grants HL44589 and RR00095 (General Clinical Research Center), National Aeronautics and Space Administration grants NAG 9-563, NAGW 3873, and NCC 2-696, and a grant from the International Life Sciences Institute.

References

1. Robertson, D., Goldberg, M. R., Hollister, A. S., Onrot, J., Wiley, R., Thompson, J. G., and Robertson, R. M. (1986). Isolated failure of autonomic noradrenergic neurotransmission: Evidence for impaired beta-hydroxylation of dopamine. *N. Engl. J. Med.* **314,** 1494–1497.
2. Man in't Veld, A. J., Boomsma, F., Moleman, P., and Schalekamp, M. A. D. H. (1987). Congenital dopamine-β-hydroxylase deficiency: A novel orthostatic syndrome. *Lancet* **1,** 183–187.
3. Biaggioni, I., Goldstein, D. D., Atkinson, T., and Robertson, D. (1990). Dopamine-beta-hydroxylase deficiency in humans. *Neurology* **40,** 370–373.
4. Man in't Veld, A. J., Boomsma, F., van den Meiracker, A. H., and Schalekamp, M. A. D. H. (1987). Effect of an unnatural noradrenaline precursor on sympathetic control and orthostatic hypotension in dopamine-β-hydroxylase deficiency. *Lancet* **2,** 1172–1175.
5. Mathias, C. J., Bannister, R. B., Cortelli, P., Heslop, K., Polak, J. M., Raimbach, S., Springall, D. R., and Watson, L. (1990). Clinical, autonomic and therapeutic observations in two siblings with postural hypotension and sympathetic failure due to inability to synthesize noradrenaline from dopamine because of a deficiency of dopamine beta hydroxylase. *Q. J. Med.* **75,** 617–633.
6. Thompson, J. M., O'Callaghan, C. J., Kingwell, B. A., Lambert, G. W., Jennings, G. L., and Esler, M. D. (1995). Total norepinephrine spillover, muscle sympathetic nerve activity and heart-rate spectral analysis in a patient with dopamine-hydroxylase deficiency. *J. Auton. Nerv. Syst.* **55,** 198–206.
7. Robertson, D., Hollister, A. S., Biaggioni, I. (1990). Dopamine-beta-hydroxylase deficiency and cardiovascular control. *In* Hypertension: Pathophysiology, Management, and Diagnosis. (J. Laragh and B. M. Brenner, eds.), pp. 749–758. Raven Press, New York.
8. Robertson, D., Perry, S. E., Hollister, A. S., Robertson, R. M., and Biaggioni, I. Dopamine-β-hydroxylase deficiency: A genetic disorder of cardiovascular regulation. *Hypertension* **18,** 1–8.
9. Craig, S. P., Buckle, V. J., Lamouroux, A., Malet, J., and Craig, I. W. (1988). Localization of the human dopamine beta hydroxylase (DBH) gene to chromosome 9q34 *Cytogenet. Cell Genet.* **48,** 48–50.
10. Kobayashi, K., Morita, S., Mizuguchi, T., Sawada, H., Yamada, K., Nagatsu, I., Fugita, K., and Nagatsu, T. (1994). Functional and high level expression of human dopamine-hydroxylase in transgenic mice. *J. Biol. Chem.* **47,** 29725–29731.
11. Robertson, D., and Davis, T. (1995a). Recent advances in the treatment of orthostatic hypotension. *Neurology* **45 (Suppl 4),** S26–S32.

12. Robertson, D. Autonomic failure *In* Handbook of Autonomic Nervous System Dysfunction. (A. D. Korczyn, ed.), pp. 129–148. Dekker, New York.
13. O'Connor, D. T., Cervenka, J. H., Stone, R. A., Levine, G. L., Palmer, R. J., Franco-Bourland, R. E., Madrazo, I., Langlais, P. J., Robertson, D., and Biaggioni, I. (1994). Dopamine 8-hydroxylase immunoreactivity in human cerebrospinal fluid: Properties, relationship to central noradrenergic activity and variation in Parkinson's disease and congenital dopamine 8-hydroxylase deficiency. *Clin. Sci.* **86,** 149–158.
14. Kuchel, O., Debinski, W., Larochelle, P., Robertson, D., Hollister, A. S., Biaggioni, I., and Robertson, R. M. (1986). Isolated failure of autonomic noradrenergic transmission. (Letter). *N. Engl. J. Med.* **315,** 1357–1358.
15. Man in't Veld, A. J., Boomsma, F., van den Meiracker, A. H., and Schalekamp, M. A. D. H. (1987). Effect of an unnatural noradrenaline precursor on sympathetic control and orthostatic hypotension in dopamine-β-hydroxylase deficiency. *Lancet* **2,** 1172–1175.
16. Biaggioni, I., and Robertson, D. (1987). Endogenous restoration of norepinephrine by precursor therapy in dopamine-beta-hydroxylase deficiency. *Lancet* **2,** 1170–1172.
17. Tulen, J. H. M., Man in't Veld, A. J., Dzoljic, M. R., Melchese, K., and Moleman, P. (1990). D,L-threo-3,4-DOPS enhances rapid eye movement sleep in patients with congenital dopamine-hydroxylase deficiency. *J. Clin. Psychopharmacol.* **10,** 73–74.
18. Tulen, J. H. M., Man in't Veld, A. J., Dzoljic, M. R., Melchese, K., and Moleman, P. (1991). Sleeping with and without norepinephrine: Effects of metoclopramide and D,L-3,4-dihydroxyphenylserine on sleep in dopamine beta-hydroxylase deficiency. *Sleep* **14,** 32–38.
19. Gray, J. (1992). Criteria for selection and evaluation of good teaching cases. *J. Clin Pharmacol.* **32,** 779–797.
20. Hyland, K. (1996). Tetrahydrobiopterin deficiency and aromatic L-amino acid decarboxylase deficiency. *In* Primer on the Autonomic Nervous System. (D. Robertson, P. A. Low, and R. J. Polinsky, eds.), pp. 201–204. Academic Press, San Diego.
21. Hoeldtke, R. (1996). Menkes disease. *In* Primer on the Autonomic Nervous System. (D. Robertson, P. A. Low, and R. J. Polinsky, eds.), pp. 208–210. Academic Press, San Diego.
22. Breakefield, X. (1996). Monoamine oxidase deficiency states. *In* Primer on the Autonomic Nervous System. (D. Robertson, P. A. Low, and R. J. Polinsky, eds.), pp. 210–212. Academic Press, San Diego.
23. Kuchel, O. (1996). Disorders of catecholamine metabolism. *In* Primer on the Autonomic Nervous System. (D. Robertson, P. A. Low, and R. J. Polinsky, eds.), pp. 212–216. Academic Press, San Diego.

Stephen G. Kaler,*† Courtney S. Holmes,* and David S. Goldstein*

*Clinical Neuroscience Branch, NINDS
National Institutes of Health
Bethesda, Maryland 20892

†Children's Hospital National Medical Center
Washington, D.C. 20010

Dopamine β-Hydroxylase Deficiency Associated with Mutations in a Copper Transporter Gene

Menkes disease and occipital horn syndrome (OHS) are human X-linked recessive disorders of copper metabolism associated with a considerable spectrum of neurological symptoms, ranging from fatal infantile neurodegeneration (classic, severe type) to mild dysautonomia with few other symptoms (OHS) (1). Defects in a gene (ATP7A) encoding a highly conserved copper-transporting adenosine triphosphatase (ATPase) produce these clinical syndromes (2–4). Over the past 6 years, the National Institutes of Health has offered a treatment protocol (parenteral copper replacement) for infants and children with these disorders. In this chapter, we focus on what is arguably the most important aspect of the trial, characterization of a distinctive neurochemical pattern that has proven extremely useful as a rapid and reliable diagnostic marker for Menkes disease and OHS. This abnormal neurochemical pattern includes the plasma and/or cerebrospinal fluid (CSF) concentrations of five catechols (dihydroxyphenylalanine, dihydroxyphenylacetic acid, dopamine, norepinephrine, and dihydroxyphenylglycol) whose levels are directly influenced by the activity of dopamine β-hydroxylase (DBH), a copper-containing enzyme. These concentrations can be precisely quantitated by high-performance liquid chromatography (HPLC) with electrochemical detection.

As we previously documented (5), Menkes disease patients show a neurochemical pattern that is similar to that in patients with congenital absence of DBH: elevated levels of dihydroxyphenylalanine, dihydroxyphenylacetic acid, and dopamine and low levels of dihydroxyphenylglycol. As expected, ratios of dihydroxyphenylalanine to dihydroxyphenylglycol (DOPA : DHPG) and dihydroxyphenylacetic acid to dihydroxyphenylglycol (DOPAC : DHPG) were significantly elevated in comparison to normal controls. In comparison to DBH-absent individuals, data from our Menkes cohort indicated partial DBH deficiency. We have found this assay extremely reliable in distinguishing affected

Advances in Pharmacology, Volume 42

1054-3589/98 $25.00

from unaffected infants and children (Table I). It has particular utility for rapidly diagnosing or excluding the diagnosis in the newborn period, during which other biochemical markers are unreliable (6).

Presumably, the Menkes copper ATPase is normally required to incorporate copper into DBH apoenzyme, and mutations in ATP7A impair this process. Recent intracellular localization of the Menkes ATPase to the Golgi network (7) implies that addition of copper may occur during processing of DBH in the secretory pathway. To determine the molecular defects underlying partial DBH deficiency in our patients, we evaluated the ATP7A locus by reverse transcription–polymerase chain reaction (RT-PCR) and DNA sequencing. In 11 patients from four families with clinical and biochemical evidence of Menkes disease (6 individuals) or occipital horn syndrome (5 individuals), we identified four mutations in the copper-ATPase gene. Each of the mutations altered normal pre-mRNA splicing and led to exon skipping, although the clinically mild patients showed some proper splicing (approximately 20–35% of normal). Interestingly, in patients with relatively better neurologic outcomes, plasma and CSF DOPA:DHPG ratios tended to be closer to the normal ranges (Table II). We are continuing to characterize the neurochemical and molecular features of patients whose clinical syndromes are within the Menkes disease–OHS spectrum, as well as murine models of defective copper transport.

In summary, quantitation of plasma and/or CSF catechol levels provides a sensitive and specific indicator of partial DBH deficiency in humans with mutations in ATP7A. The assay has a particularly valuable application in the very early identification of affected infants, for whom diagnosis by clinical or alternative biochemical parameters is problematic, and who may benefit significantly from early recognition and treatment. Splice site mutations at the ATP7A locus, which reduce but do not eliminate proper splicing, are associated with milder clinical phenotypes, including typical OHS. Mouse models of Menkes disease and OHS possess defects in a highly homologous copper transport gene, and can be used to further explore the relationship between abnormal copper transport and DBH deficiency.

TABLE I Diagnosis or Exclusion of Menkes Disease and OHS by Plasma and/or CSF Catechol Assay, 1990–1996

Older infants/children/adults	
Affected	26
Unaffected	7
At-risk newborns	
Affected	8
Unaffected	6
Total evaluated	47

TABLE II ATP7A Mutations Associated with DBH Deficiency

Patient no.	*ATP7A mutation*	*Plasma DOPA : DHPH*	*CSF DOPA : DHPG*
1	IVS 8, AS, dup5	14.2	1.9
2	IVS 8, AS, dup5	8.2	1.9
3[a]	IVS 8, AS, dup5	6.4	0.8
4	Q724H	9.8	1.2
5	Q724H	13.2	3.6
6	Q724H	7.4	ND
7[b]	S833G	5.4	1.1
8	IVS 21, DS, +3 A → T	120.0	ND
9	IVS 21, DS, +3 A → T	41.0	ND
10	IVS 21, DS, +3 A → T	78.6	ND
11	IVS 21, DS, +3 A → T	18.9	1.4
Normal controls		1.4–3.3[c]	0.3–0.7[d]

[a] Successful response to early copper replacement therapy.
[b] Typical OHS.
[c] Pediatric controls from Kaler *et al.* (5) (number increased to 13).
[d] Adult controls from Kaler *et al.* (5) (n = 8).
ND, not done.

References

1. Kaler, S. G. (1994). Menkes disease. *Adv. Pediatr.* **41,** 263–304.
2. Vulpe, C., Levinson, B., Whitney, S., Packman, S., Gitschier, J. (1993). Isolation of a candidate gene for Menkes disease and evidence that it encodes a copper-transporting ATPase. *Nat. Genet.* **3,** 7–13.
3. Kaler, S. G., Gallo, L. K., Proud, V. K., Percy A. K., Mark, Y., Segal, N. A., Goldstein, D. S., Holmes, C. S., Gahl, W. A. (1984). Occipital horn syndrome and a mild Menkes phenotype associated with splice site mutations at the Menkes locus. *Nat. Genet.* **8,** 195–202.
4. Kaler, S. G. (1996). Menkes disease mutations and response to early copper histidine treatment. (Letter). *Nat. Genet.* **13,** 21–22.
5. Kaler, S. G., Goldstein, D. S., Holmes, C., Salerno, J., Gahl, W. A. (1993). Plasma and cerebrospinal fluid neurochemical pattern in Menkes disease. *Ann. Neurol.* **33,** 171–175.
6. Kaler, S. G., Gahl, W. A., Berry, S. A., Holmes, C. S., Goldstein, D. S. (1993). Predictive value of plasma catecholamine levels in neonatal detection of Menkes disease. *J. Inherited Metab. Dis.* **16,** 907–908.
7. Adam, A. N., Dierick, H. A., Escara-Wilke, J. F., and Glover, T. W. (1996). Localization of the Menkes copper transport protein to the Golgi complex. *Am. J. Hum. Genet.* **59,** Suppl, A244.

M. Mannelli,* R. Lanzillotti,* C. Pupilli,* L. Ianni,*
A. Natali,† and F. Bellini*

*Department of Clinical Pathophysiology
Endocrinology Unit
†Department of Urology
University of Florence
Florence, Italy

Glucocorticoid–Phenylethanolamine-*N*-methyltransferase Interactions in Humans

Phenylethanolamine-*N*-methyltransferase (PNMT) is the enzyme responsible for conversion of norepinephrine (NE) to epinephrine (Epi). It is highly concentrated in the adrenal medulla but has also been demonstrated, although to a much limited extent, in other human tissues, such as lung, heart, kidney, liver, spleen, and pancreas (1). Nonetheless, the adrenal medulla is the main source of plasma Epi. Adrenalectomized patients show a dramatic decrease in plasma Epi so that plasma Epi is generally considered a good index of adrenal medulla secretion. From *in vitro* and *in vivo* studies performed in several animal species, it is well known that in adrenal medulla PNMT activity is dependent on the high levels of glucocorticoids received from the cortex (2). *In vivo* data from studies performed in humans are scanty. In children with hypocortisolism due to adrenocotricotropic hormone (ACTH) deficiency, plasma Epi levels are lower (3).

To study the *in vivo* glucocorticoid–adrenal PNMT interaction in humans, we evaluated basal and stimulated plasma Epi in patients affected by hypocortisolism (HP) due to ACTH deficiency (n°7; 3 females, 4 males; mean + SD age = 33.8 + 11.7 yr), as well as in patients with endogenous hypercortisolism (4) (Cushing's syndrome [CS]) (n°9; 7 females, 2 males; mean + SD age = 41.5 + 18.7 yr) and in a normocortisolemic control group (CG1) (n°20; 13 females, 7 males; mean + SD age = 34.1 + 10.3 yr). Stimulation of Epi secretion was achieved by intravenous glucagon. Patients with HP were studied while receiving thyroid, glucocorticoid, and sexual steroid replacement therapy. To evaluate whether an increased cortisol delivery to the adrenal medulla may affect PNMT activity, we also measured (as an indirect *in vivo* index of PNMT activity) Epi and NE levels and their ratio in adrenal venous blood in 8 patients with CS (5 females, 3 males; mean + SD age = 39 + 11 yr) undergoing adrenalectomy (3 bilateral hyperplasia, 1 monolateral adenomatous hyperplasia, 3 adenomas,

Advances in Pharmacology, Volume 42

1054-3589/98 $25.00

1 carcinoma) and in a control group (CG2) of 12 patients (6 females, 6 males; mean + SD age = 53 + 12 yr) undergoing surgery for renal diseases. Adrenal venous blood was sampled by direct venipuncture at surgery. Plasma Epi and NE were assayed by a radioenzymatic assay using CAT-A-Kit (Amersham, Bucks, U.K.). The limit of assay sensitivity for Epi is 10 pg/ml. Lower results were given this value. Statistical analysis of the results was performed using Student's t-test for unpaired data (basal conditions) or two-way analysis of variance (ANOVA) and Student's t-test as appropriate (stimulated conditions). Results are reported as mean + SE. Epi and NE concentrations are reported as pg/ml.

In patients affected by HP, basal plasma Epi resulted lower than in CG1 (29.3 + 19.9 vs 11.9 + 2.6; $p < .02$), whereas basal plasma NE was unchanged (186.0 + 62.5 vs 172.8 + 87.5). In CG1, intravenous glucagon caused a significant increase in plasma Epi ($p < .001$) while in patients with HP, glucagon increased plasma Epi not significantly (Fig. 1). When comparing the response

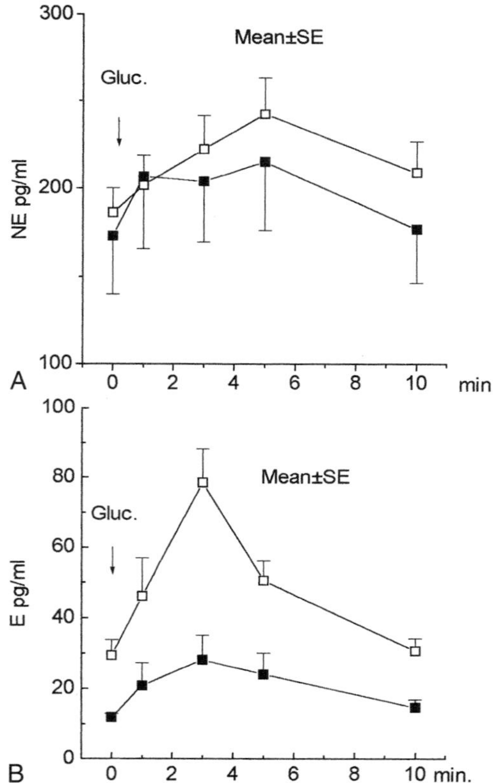

FIGURE 1 (A) Norepinephrine (NE) and (B) epinephrine (E) plasma levels before and after glucagon in patients with HP due to ACTH deficiency (*closed symbols*) and in normocortisolemic control subjects (*open symbols*).

to glucagon in the two groups by evaluating the differences between the basal values and the maximal Epi increases, the net Epi increase observed in HP patients was significantly lower than in the CG1 (18.6 + 19.9 vs 57.3 + 36.9; $p < .02$).

The ratio of Epi to NE in the adrenal vein of patients with CS did not produce different results (4) from that found in CG2 (4.61 + 2.1 vs 4.71 + 2.2). In patients with CS, basal plasma Epi results were not significantly different from those of CG1 (21.3 + 12.1 vs. 29.3 + 19.9) whereas plasma NE was found significantly decreased (119.6 + 65.5 vs 186.0 + 62.5; $p < .02$). In patients with CS, glucagon caused a net increase in plasma Epi, which was significantly lower than that observed in CG1 (19.4 + 11.9 vs 57.3 + 36.8; $p < .01$).

Our data also confirm that in humans, adrenal PNMT activity seems to depend on the normal cortisol supply to the medulla from the cortex. In fact, in adults affected by HP due to ACTH deficiency, the adrenal medulla secretion is impaired in basal as well as in stimulated conditions (see Fig.1). The deficiency of Epi secretion is very probably one of the most important factors causing an impaired response to hypoglycemia in these patients.

Although a normal glucocorticoid secretion seems necessary for normal Epi synthesis, a condition of chronic hypercortisolism, as present in CS, does not seem able to increase adrenal Epi synthesis further through an increased PNMT activity. In fact, the similar Epi/NE ratio found in the adrenal vein of patients with and without chronic hypercortisolism suggests that in humans, a normal cortisol secretion is sufficient to activate PNMT maximally. These results confirm the finding that in the adrenal medullas obtained at surgery from patients with CS, the PNMT activity resulted similar to that found in control human adrenal glands obtained at autopsy.

A possible role for surgical stress in modifying PNMT activity and therefore Epi/NE ratio in the adrenal vein of our control patients seems improbable. In fact, the Epi/NE ratio we found is very similar either to the Epi/NE content of normal human chromaffin tissue or to the Epi/NE ratio measured in the adrenal vein of normal volunteers during much less stressful conditions (adrenal catheterization) (5).

Therefore, in CS, the conversion of NE to Epi is not enhanced, and an increase in adrenal Epi production cannot account for the reduced glucose tolerance and the high blood pressure, which are very common in this pathological condition. On the contrary, during chronic hypercortisolism, the overall function of the adrenal medulla seems to be reduced, as suggested by the lower plasma Epi and the lower Epi response to glucagon. In addition, in CS the sympathetic system seems to be inhibited, as suggested by the lower NE plasma levels.

Because we have previously demonstrated (4) that in patients with CS the metabolic clearance rate of Epi is not altered, the decrease in basal and stimulated plasma catecholamines indicates that an overall reduction in sympathetic–adrenal activity is present during chronic endogenous hypercortisolism.

A possible explanation for this finding is that the sympathetic–adrenal system might be inhibited by the hypertensive and hyperglycemic effects of chronic hypercortisolism. The mechanisms through which a chronic cortisol excess causes these effects are only partly known, but it is worth noting that a possible mechanism might be the activation of extra-adrenal PNMT (1) and the increase in nonadrenal Epi synthesis. This increase might be sufficient to determine some biological effects, such as insulin resistance in the muscle and increased output in the heart, without affecting Epi plasma levels significantly. An important contribution to this hypothesis might come from the demonstration that nonadrenal PNMT can be activated by glucocorticoids at concentrations similar to those observed in the plasma of patients with CS.

In conclusion, our data confirm that in humans, similar to findings in many other animal species, an intact function of the adrenal cortex is necessary to permit Epi synthesis by the chromaffin cells. Because during chronic hypercortisolism, adrenal PNMT activity is not enhanced, the adrenal PNMT activation by cortisol seems to already be at a maximal level in normal conditions. On the contrary, during chronic hypercortisolism, the catecholamine release by the adrenal medulla and the sympathetic nerves seems to be reduced, probably as a compensatory response to some biological effects of cortisol. Part of these effects might be induced through the activation of nonadrenal PNMT.

Acknowledgments

We wish to thank Mrs. Nadia Misciglia for her excellent technical assistance.

References

1. Kennedy, B., Bigby, T. D., and Ziegler, M. G. (1995). Nonadrenal epinephrine-forming enzymes in humans. Characteristics, distribution, regulation and relationship to epinephrine levels. *J. Clin. Invest.* **95,** 2896–2902.
2. Pohorecky, L. A., and Wurtman, R. J. (1971). Adrenocortical control of epinephrine synthesis. *Pharmacol. Rev.* **23,** 1–35.
3. Rudman, D., Moffitt, S. D., Fernhoff, P. M., Blackston, R. D., and Faraj, B. A. (1981). Epinephrine deficiency in hypocorticotropic hypopituitary children. *J. Clin. Endocrinol. Metab.* **53,** 722–729.
4. Mannelli, M., Lanzillotti, R., Pupill, C., Ianni, L., Conti, A., and Serio, M. (1994). Adrenal medulla secretion in Cushing's syndrome. *J. Clin. Endocrinol. Metab.* **78,** 1331–1335.
5. Planz, G., and Planz, R. (1979). Dopamine-beta-hydroxylase, adrenaline, noradrenaline and dopamine in the venous blood of adrenal gland in man: A comparison with levels in the periphery of the circulation. *Experientia* **35,** 207–208.

M. J. Evinger

Department of Pediatrics
SUNY at Stony Brook
Stony Brook, New York 11794

Determinants of Phenylethanolamine-*N*-methyltransferase Expression

Expression of the phenylethanolamine-*N*-methyltransferase (PNMT) gene involves coordination of three major regulatory mechanisms: Hormonal, neurally mediated, and cell-specific determinants all contribute to the physiological expression of this enzyme (Fig. 1). Axelrod and Wurtman provided our first understanding of the influences governing epinephrine production in the 1960s. (1) These workers demonstrated that the decline in adrenal PNMT enzymatic activity following hypophysectomy in rats could be reversed by administration of the synthetic glucocorticoid, dexamethasone. It was more than two decades later that glucocorticoids were shown to increase steady-state levels of PNMT mRNA, primarily by enhancing the rate of PNMT gene transcription (2). Ross and coworkers (3) further established that this hormonal action is specifically mediated via a glucocorticoid-responsive element (GRE) encoded within the 5′ regulatory region of the rat PNMT gene.

Neurally mediated regulation of PNMT expression in adrenal chromaffin cells occurs predominantly through the influence of cholinergic stimuli on nicotinic and muscarinic receptors. A single neurotransmitter, acetylcholine, is capable of acting through two separate intracellular signaling pathways to stimulate transcription of the PNMT gene (4). Expression of transiently transfected rat PNMT promoter constructs in primary cultures of bovine chromaffin cells reveals that both nicotinic and muscarinic agonists can selectively and independently enhance expression from the PNMT gene. These effects can be blocked with antagonists for each receptor subtype. The use of 5′ nested deletion constructs has revealed separate responsive regions on the PNMT promoter for nicotinic and muscarinic stimuli. Thus, cholinergic-regulated expression of the PNMT gene involves not only different transduction pathways but also distinctive responsive regions of the gene.

Although the importance of cell-specific determinants has been recognized in a number of systems, only recently have we begun to characterize those sequences responsible for the highly restricted expression of PNMT. Not surprisingly, the PNMT gene likewise possesses information that directs the site of its expression. During migration of primitive neuroblasts from the neural crest, an early commitment is made by these cells to follow a neural, rather than nonneural, course of differentiation. Moreover, these sympathoadrenal

Advances in Pharmacology, Volume 42

1054-3589/98 $25.00

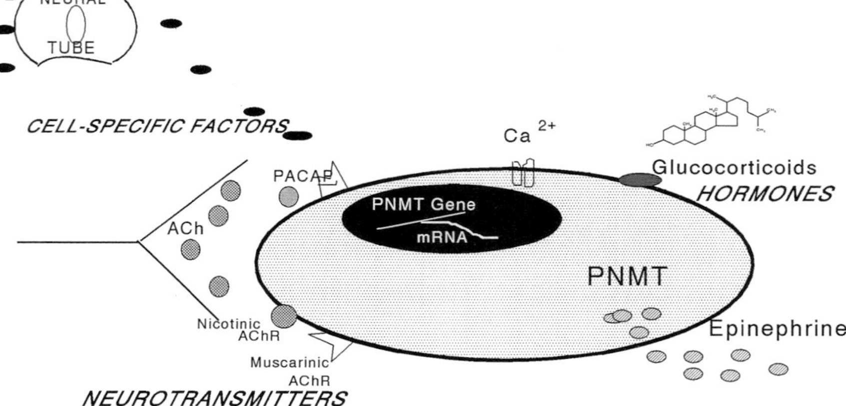

FIGURE 1 ACh, acetylcholine; PNMT, phenylethanolamine *N*-methyltransferase; AChR, acetylcholine receptor; PACAP, pituitary adenylate cyclase activating polypeptide; mRNA, messenger RNA.

precursors must also decide whether to pursue neuronal differentiation—giving rise to sympathetic ganglia—or to adopt a more endocrine phenotype—ultimately assuming residence in the adrenal medulla. One additional developmental decision results in either a noradrenergic or an adrenergic phenotype in the adrenal medulla. We sought to determine which portions of the PNMT gene encode information designating appropriate adrenal medullary expression.

Transient transfections of upstream PNMT promoter regions were conducted utilizing an array of pheochromocytoma, neuroblastoma, glioma, fibroblast, and myoblast continuous cell lines. Luciferase reporter gene expression was expressed relative to that observed with the −863 bp PNMT construct in primary bovine adrenal chromaffin cells. On comparison of the levels of expression of eight 5′ "nested deletion" PNMT promoter constructs, it was noted that deletions of three specific regions (designated as distal, mid, and proximal cell-specific elements) resulted in enhanced expression of the PNMT constructs. Relative levels of expression were dependent on the transfected cell type and on the length of the PNMT construct.

Deletion of the most distal element produces enhancement of reporter gene expression in nonadrenergic, nonchromaffin cell types. Specifically, deletion of this region results in increases ranging from 1.5- to fivefold as the transfected cells include related neuroblastoma stem cells, noradrenergic PC12 cells, and highly unrelated cells, for example, fibroblasts and glioma lines. Electrophoretic mobility shift analyses reveal a limited number of bands following incubation of nuclear extracts from these cell lines, with radiolabeled double-stranded oligonucleotides encoding this region. Currently, efforts are underway to characterize those factors responsible for this aspect of PNMT cell-specific expression.

A second group of cell-specific elements shares sequence similarities with previously characterized neuron specific silencers (5, 6). These neural silencer sequences encoded within regulatory regions of selected neural genes interact with specific nuclear proteins, thereby effectively keeping these genes turned off in nonneural tissues. REST (RE1-silencing transcription factor; neural restrictive silencer factor) is one nuclear silencing protein of the Kruppel family of transcription factors with demonstrated ability to repress transcription of the Na^+(II) channel (5) and SCG-10 genes (6).

The PNMT mid and proximal cell-specific regions each share approximately 70% sequence similarity with a derived consensus sequence for neural specific silencers. Deletion of the regions of the PNMT promoter between −647 and −572 bp and between −572 and −537 bp produces enhancement of reporter gene constructs transfected into nonneuronal cell types, for example, C6 glioma and L6 myoblast lines. In addition to these functional aspects, binding assays with oligonucleotides generate a highly reproducible DNA–protein complex, with altered electrophoretic mobility following incubation of nuclear extract proteins from all cells tested thus far with ^{32}P-labeled double-stranded oligonucleotides encoding either the mid or proximal cell-specific element. However, an additional band is generated when L6 nuclear extracts are incubated with either of the oligonucleotide sequences. The pattern is the same for each of these elements, although the intensity of autoradiographic bands is three- to fivefold greater with the oligo encoding the mid element.

To determine whether REST (5, 6) is a component of the complexes observed with the PNMT oligonucleotides, competition and supershift analyses have been performed using unlabeled sequences and antisera specific for neural silencing mechanisms characterized in other systems. When 100-fold molar excess of double-stranded oligonucleotides encoding the Na^+ (II) channel or SCG-10 RE1 sequence is incubated with L6 nuclear extract prior to addition of ^{32}P-labeled oligos encoding the PNMT mid or proximal cell-specific regions, competition is indicated by a decrease in the intensity of the faster migrating complex. Similarly, incubation of L6 nuclear extract with polyclonal antisera to REST (5) results in supershifting of the PNMT complex to a more slowly migrating form. Thus, REST appears to be one component of the L6 complexes resolved by electrophoretic mobility shift assays.

When L6 myoblast cells are transiently cotransfected with the −757 PNMT–luciferase construct plus an equimolar amount of a plasmid expressing a dominant negative form of REST, luciferase expression increases. As previously reported, expression of Na^+(II) RE1-CAT also increases (5). Conversely, when PC12 cells (expressing minimal endogenous REST) are transfected with a constitutively expressing wild-type REST plasmid, expression of cotransfected Na^+(II) CAT decreases approximately 10-fold (5). In contrast, reporter gene expression driven by the −757 PNMT promoter remains unchanged in PC12 cells and even increases approximately threefold in BE2 neuroblastoma cells. Thus, we must consider that protein factors binding to the PNMT promoter show certain activities distinct from those of REST on the Na^+(II) channel gene. Characterization of the factors binding to the PNMT neural cell-specific elements is currently in progress.

In summary, we have demonstrated that the major influences establishing the physiological level of PNMT expression can be classified as hormonal, neural, and cell-specific in nature. Extending our previous studies showing glucocorticoid regulation through a specific GRE on the PNMT gene, we have, in similar manner, shown that cholinergic stimulation of the chromaffin cell activates transcription of the PNMT gene through two separate intracellular pathways; furthermore, sequences conveying responsiveness to nicotinic and muscarinic stimuli map to distinct regions of the PNMT promoter.

Cell-specific determinants also play critical roles in specifying the highly restricted expression of the PNMT gene. PNMT mid and proximal cell-specific elements share sequence similarities with neural silencers in the type II Na^+ channel and SCG-10 genes. However, differences in gel shift patterns and in expression levels following cotransfection with REST-expressing plasmids suggest that PNMT silencing mechanisms are not identical to these neural silencers. Although REST can bind to these sequences, other data suggest the presence of additional factors acting at these sites. Resolving the identities of the proteins binding to these PNMT cell-specific elements should provide important information about the functions of these major determinants of PNMT expression.

Acknowledgments

We gratefully acknowledge Gail Mandel and her laboratory for generously providing REST antisera, REST expression plasmids, and Na+(II) RE1-CAT. This work was supported by NIH grant GM46588 to MJE.

References

1. Wurtman, R. J., and Axelrod, J. (1965). Adrenaline synthesis: control by the pituitary gland and adrenal glucocorticoids. *Science* **150**, 1464–1465.
2. Evinger, M. J., Towle, A. C., Park, D. H., Lee, P., and Joh, T. H. (1992). Glucocorticoids stimulate transcription of the rat phenylethanolamine *N*-methyltransferase (PNMT) in vivo and in vitro. *Cell. Mol. Neurobiol.* **12**, 193–215.
3. Ross, M. E., Evinger, M. J., Hyman, S. E., Carroll, J. M., Mucke, L., Comb, M., Reis, D. J., Joh, T. H., and Goodman, H. M. (1990). Identification of a functional glucocorticoid response element in the phenylethanolamine *N*-methyltransferase promoter using fusion genes introduced into chromaffin cells in primary culture. *J. Neurosci.* **10**, 520–530.
4. Evinger, M. J., Ernsberger, P., Raganathan, S., Joh, T. H., and Reis, D. J. (1994). A single transmitter regulates gene expression through two separate mechanisms: Cholinergic regulation of PNMT mRNA via nicotinic and muscarinic pathways. *J. Neurosci.* **14**, 2106–2116.
5. Chong, J. A., Tapia-Ramirez, J., Kim, S., Toledo-Aral, J. J., Zheng, Y., Boutros, M. C., Altshuller, Y. M., Frohman, M. A., Kraner, S. D., and Mandel, G. (1995) REST: A mammalian silencer protein that restricts sodium channel gene expression to neurons. *Cell* **80**, 949–957.
6. Schoenherr, C. J., and Anderson, D. J. (1995).The neuron-restrictive silencer factor (NRSF): A coordinate repressor of multiple neuron-specific genes. *Science* **267**, 1360–1363.

Dona L. Wong, Steven N. Ebert, and Kyoji Morita

Nancy Pritzker Laboratory
Department of Psychiatry and Behavioral Sciences
Stanford University School of Medicine
Stanford, California 94305

Neural Control of Phenylethanolamine-*N*-methyltransferase via Cholinergic Activation of Egr-1

Phenylethanolamine-*N*-methyltransferase (PNMT, EC 2.1.1.28), the final enzyme in the catecholamine biosynthetic pathway, converts norepinephrine to epinephrine, utilizing S-adenosylmethionine as the methyl donor and coenzyme. Like other enzymes involved in catecholamine biosynthesis, PNMT is both neurally and hormonally regulated. In the adrenal medulla, the major source of peripheral catecholamines, PNMT expression is neurally controlled via the splanchnic nerve, which provides cholinergic innervation to the medulla. Administration of the catecholamine reuptake inhibitor, reserpine, leads to reflex stimulation of the splanchnic nerve and a rise in PNMT enzymatic activity and mRNA. Thus, it appears that neural stimulation activates cholinergic receptors, which, in turn, activate the PNMT gene. However, the molecular mechanism for the neural activation of PNMT gene expression, beginning with cholinergic receptor activation through PNMT gene activation, remains to be elucidated.

We have demonstrated that the immediate early gene family member, Egr-1, is one transcription factor controlling PNMT gene expression (1). Within the proximal 863 bp of upstream promoter regulatory sequences of the PNMT gene are two potential Egr-1 binding elements at −45 and −165 bp upstream of the site of transcription initiation based on consensus sequence identity. If RS1 cells, derived from the PC12 cell line, are transiently transfected with a PNMT promoter–reporter gene construct, pRP863LUC, consisting of the proximal 863 bp of PNMT promoter–regulatory sequences linked to the firefly luciferase gene, and a functional Egr-1 expression construct, pCMVEgr-1, Egr-1 activates the PNMT promoter to stimulate luciferase expression as much as four- to fivefold. Gel mobility shift assays in the presence of an anti-Egr-1 antibody confirm that Egr-1 protein will complex with 21 bp oligonucleotides, encoding these Egr-1 binding sites. Transient transfections assays with site-directed mutant PNMT promoter–luciferase reporter gene constructs further show that mutation of the Egr-1 binding elements markedly reduces PNMT promoter and luciferase reporter gene activation by Egr-1.

Egr-1 appears to be an important activator of endogenous PNMT gene expression as well. Transfection of the RS1 cells with the Egr-1 expression

Advances in Pharmacology, Volume 42

1054-3589/98 $25.00

plasmid alone markedly increases PNMT mRNA as measured by RNase protection assay. In addition, reserpine administration to rats elevates both Egr-1 mRNA and PNMT mRNA in a manner consistent with Egr-1 serving as a transcriptional activator of the PNMT gene. Egr-1 mRNA quickly rises in response to reserpine, peaking at 30 min, followed by the subsequent rise in PNMT mRNA (peak at 6 hr). Finally, immunohistochemistry shows that reserpine induces Egr-1 protein expression in the chromaffin cells of the adrenal medulla, and further, that Egr-1 is primarily localized to the nuclei of these cells, again consistent with its participation in the transcriptional activation of PNMT.

To examine whether Egr-1 might mediate the neural activation of the PNMT gene, the effects of the cholinergic agonist, carbachol (CCh), on luciferase reporter gene expression and Egr-1 mRNA were examined (2). Carbachol induced a time-dependent rise in luciferase, with a peak 1.5-fold induction at 6 hr, after which luciferase activity declined and returned to basal levels by 36 hr. Maximum induction of luciferase was achieved at a concentration of 100 μM CCh. Carbachol increased Egr-1 mRNA levels as well, and the rise in Egr-1 mRNA preceded that of luciferase (30 min). The effects of cholinergic antagonists (10 μM) on CCh-induced stimulation of luciferase were also examined to determine the specific cholinergic receptor subtype responsible for PNMT promoter induction. Atropine, but not hexamethonium, blocked CCh activation of luciferase, suggesting that muscarinic receptors mediated PNMT promoter stimulation. To confirm this finding, we examined the effects of specific cholinergic agonists. Bethanecol (50 μM), a muscarinic agonist, maximally induced luciferase activity approximately 1.7-fold, whereas nicotine itself did not significantly alter reporter gene expression. Atropine also blocked CCh induction of Egr-1 mRNA, although hexamethanium had no effect whatsoever. Finally, bethanecol elevated Egr-1 RNA in a dose-dependent manner, inducing peak expression at 50 μM, whereas nicotine did not alter basal Egr-1 mRNA expression at all. Thus, *in vitro,* it would appear that activation of muscarinic cholinergic receptors stimulates Egr-1 mRNA expression and PNMT promoter activity in a manner consistent with Egr-1 mediating the neural control of PNMT gene expression.

In PC12 cells, from which the RS1 cells are derived, muscarinic receptors are the predominant cholinergic receptor subtype, and of that subtype, m1 and m4 receptors appear most abundant (3). However, in adrenal chromaffin cells, both muscarinic and nicotinic receptors are present. Using an *in vivo* paradigm, we therefore examined changes in endogenous Egr-1 and PNMT mRNA in the adrenal medulla of rats treated with either nicotine (2 mg/kg i.p.) or muscarine (0.1 mg/kg/i.p.). Both nicotine and muscarine elevated Egr-1 mRNA (~15-fold) at 30 min after administration. Basal levels of Egr-1 mRNA expression were restored by 8 hr. In contrast, PNMT mRNA showed no significant change in response to either nicotine or muscarine at 30 min, but was elevated twofold by both cholinergic agonists at 8 hr. Thus, *in vivo,* activation of muscarinic as well as nicotinic receptors appears to mediate neural regulation of the PNMT gene.

To provide further evidence for the role of Egr-1 in the neural control of PNMT gene expression, we examined the effects of CCh on PNMT promoter activity when the Egr-1 binding site was mutated. The wild-type PNMT promoter–luciferase reporter gene construct pRP863LUC or the Egr-1 mutant PNMT promoter–luciferase reporter gene construct pRPmutALUC was transfected into RS1 cells, followed by treatment with 100 μM CCh for 6 hr. Although full activation of luciferase reporter gene expression was observed with the wild-type construct, CCh was unable to induce luciferase activity when the Egr-1 site was mutated, indicating that an intact Egr-1 binding element is critical for CCh induction.

The role of Egr-1 in the neural activation of the PNMT gene was further examined *in vivo* as well. Metrazole, which activates the splanchnic nerve, and cocaine, which activates the hypothalamic–pituitary–adrenal axis (hormonal pathway), were administered acutely to rats at doses of 70 mg/kg s.c. and 15 mg/kg i.p., respectively. Metrazole elevated adrenal medullary Egr-1 mRNA 22-fold after 30 min and PNMT mRNA 3.2-fold after 8 hr of drug treatment. Consistent with its activation of hormonal pathways, cocaine elevated adrenal corticosterone levels nearly eightfold and also elevated PNMT mRNA levels 3.7-fold, but only slightly increased Egr-1 mRNA, with the temporal rise in corticosterone and Egr-1 mRNA being coincident. Hence, only in the case of neural activation by metrazole are quantitative and temporal changes in Egr-1 and PNMT mRNA consistent with Egr-1 acting as a transcriptional activator of PNMT gene expression (3).

Because phorbol esters have been shown to stimulate Egr-1 expression, we also examined whether phorbol 12-myristate-13-acetate (PMA) could activate the PNMT promoter (). PMA (80 nM maximally stimulated luciferase activity 4.5-fold at 6 hr in RS1 cells transfected with pRP863LUC alone. Activity remained constant up to 12 hr and then rapidly declined, so that by 36 hr, basal levels of luciferase were restored. Moreover, maximum induction of luciferase was achieved in the presence of 40 nm PMA. PMA also elevated Egr-1 mRNA expression. A 6.0-fold increase in Egr-1 mRNA was apparent at 30 min, and by 4 hr, Egr-1 mRNA had been restored to its low basal levels. We further demonstrated that PMA induction of PNMT requires synthesis of Egr-1 protein and its nuclear translocation. If RS1 cells are treated with 80 nM PMA for 1 hr, immunohistochemistry shows that there is an increase in Egr-1 protein expression in the nuclei of the RS1 cells. The rise in nuclear Egr-1 protein can be completely blocked by pretreatment of the RS1 cells with 50 μg/ml of cycloheximide, a protein synthesis inhibitor, prior to PMA treatment. Finally, PMA will not induce luciferase activity in RS1 cells transfected with the Egr-1 mutant PNMT promoter–reporter gene construct (pRPmutALUC).

Because PMA has been reported to be involved in the protein kinase C (PKC) signal transduction pathway, we next examined whether cholinergic receptor activation might stimulate PKC, leading to Egr-1 induction and an increase in PNMT gene transcription. RS1 cells transfected with the wild-type PNMT promoter–luciferase reporter gene construct were pretreated with chelerythrine (1–5 μM) or staurosporine (0–100 nM), followed by 100 μM CCh. Neither of these PKC inhibitors blocked luciferase induction by CCh. Because the action of PMA is not always related to PKC activation, we examined whether inhibitors of other protein kinases might block CCh induction of luciferase reporter gene expression in the RS1 cells transfected with pRP863LUC. The calcium–calmodulin protein kinase (CAM kinase) inhibitor, trifluoperazine (0–50 μM), and the protein kinase A (PKA) inhibitor, IBMX (0–100 μM), did not inhibit CCh induction of luciferase. In fact, in the case of IBMX, a significant activation of luciferase was apparent both in the presence and in the absence of CCh. PMA is also known to stimulate phosphotyrosine phosphatases (PTPases), which, in turn, control other regulatory proteins. Although orthovanadate, a PTPase inhibitor, does stimulate luciferase activity, the effect is nonspecific because there is no significant difference in luciferase expression elicited from the PNMT promoter–luciferase reporter gene construct and a promoterless luciferase–reporter gene construct. Moreover, orthovanadate did not alter CCh-induced luciferase expression seen with the wild-type construct.

The relationship between cholinergic activation of the PNMT gene and intracellular Ca^{2+} has also been examined, because Ca^{2+} entry into the cell is essential for catecholamine secretion. Although the calcium channel blocker D-600 (0–100 μM) did not inhibit CCh-stimulated luciferase expression in RS1 cells transfected with the PNMT promoter–reporter gene construct, thus demonstrating that the influx of extracellular Ca^{2+} is not required for the stimulatory action of CCh, mobilization of intracellular Ca^{2+} stores does appear to be critical. When RS1 cells transfected with the wild-type reporter gene construct were pretreated with the permeable intracellular Ca^{2+} chelator, BAPTA/AM, the rise in luciferase activity induced by CCh was fully suppressed. In contrast, the impermeant Ca^{2+} chelator BAPTA had no effect. In addition, thapsigargin (50 nM) also suppressed carbachol induction of the PNMT promoter, whereas ryanodine (0–20 μM) did not, thereby indicating that mobilization of thapsigargin-sensitive Ca^{+2} stores is required for CCh.

In summary, Egr-1 appears to be one transcription factor involved in the neural control of PNMT gene expression. However, cholinergic induction of the PNMT gene does not appear to occur through a mechanism associated with PKC, PKA, CAM kinase, or protein–tyrosine phosphorylation, although it does appear to depend on the mobilization of intracellular Ca^{2+}, which may be derived from inositol phosphate–sensitive intracellular stores. Because PKC and CAM kinases are not apparently involved in the cholinergic activation of the PNMT promoter, CCh and PMA may activate a novel pathway independent of protein phosphorylation. However, protein–tyrosine kinases have been shown to participate in Ca^{2+}-dependent signaling events, and we are investigating both of these alternative possibilities.

References

1. Ebert, S. N., Balt, S. L., Hunter, J. P. B., Gashler, A., Sukhatme, V., and Wong, D. L. (1994). Egr-1 activation of rat adrenal phenylethanolamine *N*-methyltransferase gene. *J. Biol. Chem.* **269**, 20885–20898.
2. Morita, K., and Wong, D. L. (1996). Role of Egr-1 in cholinergic stimulation of phanylethanolamine *N*-methyltransferase promoter. *J. Neurochem.* **67**, 1344–1351.

3. Morita, K., Ebert, S. N., and Wong, D. L. (1995). Role of transcription factor Egr-1 in phorbol ester-induced phenylethanolamine *N*-methyltransferase gene expression, *J. Biol. Chem.* 270, 11161–11167.
4. Vilaro, M. T., Wiederhold, K. H., Palacios, J. M., and Mengod, G. (1991). Muscarinic cholinergic receptors in the rat caudate-putamen and olfactory belong predominantly to the M4 class: In situ hybridization and receptor autoradiography evidence. *Neuroscience* 40, 159–167.

Harvey B. Pollard,*† Hung Caohuy,*† Allen P. Minton,‡ and Meera Srivastava*†

* Department of Anatomy and Cell Biology
Uniformed Services University School of Medicine, USUHS
Bethesda, Maryland 20814

† Laboratory of Cell Biology and Genetics, NIDDK
National Institutes of Health
Bethesda, Maryland 20892

‡ Laboratory of Biochemical Pharmacology, NIDDK
National Institutes of Health
Bethesda, Maryland 20892

Synexin (Annexin VII) Hypothesis for Ca^{2+}/GTP-Regulated Exocytosis

Guanosine triphosphate (GTP) and its nonhydrolyzable analogue, GTPγS, are known to promote Ca^{2+}-dependent exocytotic secretion from chromaffin and many other cell types by a mechanism thought to involve as yet unknown proteins in the GTPase superfamily (1–4). It has been presumed that these two ligands might act through the "fusion machine," proposed for constitutive and calcium-regulated vesicle trafficking (5–7). The current fusion machine hypothesis envisions a core complex formed between plasma membrane syntaxin and SNAP-25 and the synaptic vesicle protein synaptobrevin/VAMP (8), with vesicular synaptotagmin identified as a low-affinity calcium sensor, that interacts with regulatory syntaxin 1 (9–10). However, with the exception of syntaxin, which is lethal, knockout mutations in the mouse for these genes have had little or no effect on secretory processes. Furthermore, none of the proteins presently identified in the hypothetical fusion machine actually fuse membranes (7, 11–13). Therefore, the general opinion has been that the calcium

Advances in Pharmacology, Volume 42

1054-3589/98 $25.00

sensor has yet to be discovered, and that other GTP-binding proteins, as yet unidentified, might control the activity of the fusion complex (6, 12).

I. Action of Ca^{2+} and GTP at a Common Exocytotic Site

The site of GTP action in exocytosis has been thought to be closely associated with the site of calcium action at some common site (3, 14–16). The affinity of this site for Ca^{2+} has been shown to be in the 50- to 200-μM range, as estimated from electrophysiological studies with caged calcium in chromaffin cells (17–21), neurons (22–24), and digitonin-permeabilized chromaffin cells (25). Therefore, a good experimental goal would seem to be to search for a membrane fusion protein that is activated by both calcium and GTP. Parenthetically, Gomperts *et al.* (14) predicted that the protein in question would also be activated by protein kinase C.

II. Synexin as the Common Exocytotic Site for Ca^{2+} and GTP Action

Chromaffin and many other cell types contain a membrane fusion protein, synexin (annexin VII), associated with secretory vesicles and plasma membranes, which have an appropriate intrinsic K_d for Ca^{2+} of approximately 200 μM (26–28). In the earliest days of synexin research, when nanomolar levels of Ca^{2+} were thought to drive secretion, it was sometimes conventional to dismiss the synexin K_d level as "too high" to be relevant to secretory processes. However, in view of the current data already summarized here, the K_d would now appear to be "just right." More important for the question of how GTP acts to promote secretion, we have recently reported that synexin drives the calcium-dependent membrane fusion reaction in a manner further activated by GTP (29). The mechanism of fusion activation depends on the unique ability of synexin to bind and hydrolyze GTP in a calcium-dependent manner, both *in vitro* and *in vivo* in SLO (streptolysin-O)-permeabilized chromaffin cells. Inasmuch as previous immunolocalization studies place synexin at exocytotic sites in chromaffin cells, we have concluded that synexin may both detect and mediate the Ca^{2+}/GTP signal for exocytotic membrane fusion.

III. Synexin as a Ca^{2+} and GTP Switch for Membrane Fusion

We have, therefore, envisioned the mechanism of this process explicitly in terms of a G-protein-like molecular switch (30–33) in which calcium-conditional activation leads synexin into a GTP-bound fusogenic "on" state

(Fig. 1). Subsequently, inactivation to the "off" state occurs by GTP hydrolysis. In the off state (S^o), synexin slowly hydrolyzes GTP to GDP by an Mg^{2+}-dependent reaction. On elevation of the free [Ca^{2+}] into the 50- to 200-μM range, synexin converts to the on state (S^*) and binds GTP to form the $S^{**} \cdot Ca^{2+}/Mg^{2+} \cdot$ GTP complex. This process can be further potentiated by mastoparan, a peptide known to activate both heterotrimeric G-proteins (34) and exocytosis in chromaffin cells (35). Fusion then ensues until GTP is hydrolyzed and the calcium is reduced, leaving synexin again in the off state. This model is generally characteristic of the molecular switch mechanism identified for different members of the GTPase superfamily in which a conditional activator induces an active GTP-bound state, and for which inactivation occurs by GTP hydrolysis. In the case of synexin, the conditional activator seems to be calcium.

IV. Kinetic Evidence for a Role for Polymeric, GTP-Bound Synexin in Defining the Optimal Fusion Rate

In addition to binding of GTP and Ca^{2+}, a further identifying feature of the hypothetical "fusion scaffold" protein has been proposed to be self-association

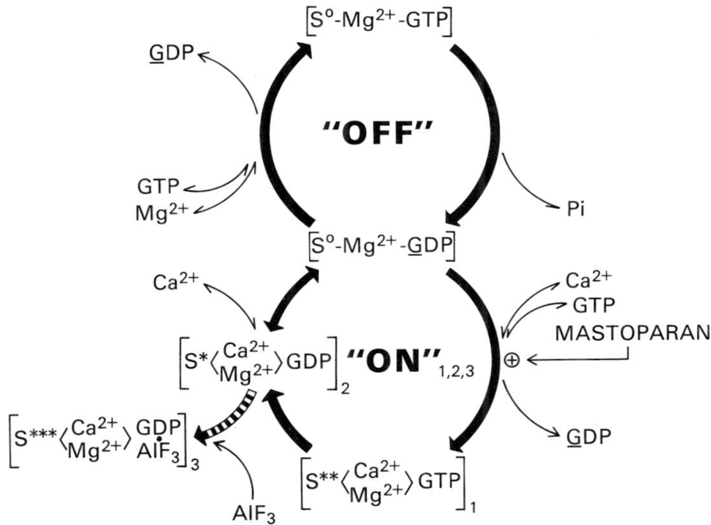

FIGURE 1 Model for synexin-driven membrane fusion regulated by Ca^{2+} and GTP. Under a resting, low-calcium condition, synexin exists in an "off" state (S^o-Mg^{2+}-GDP), which is formed by a process of constitutive Mg^{2+}-dependent hydrolysis of GTP. On elevation of Ca^{2+}, synexin binds GTP and assumes a stable "on" state, which can drive membrane fusion (S^{**}-Ca^{2+}/Mg^{2+}-GTP). This transition can be activated by mastoparan. On hydrolysis of the GTP to GDP, a transient (S^*-Ca^{2+}/Mg^{2+}-GDP) complex is formed. Then, with the reduction in the free Ca^{2+} concentration, the off state (S^o-Mg^{2+}-GDP) complex is reformed, and the cycle can recur. AlF_3 can activate the GDP-bound form of synexin, so long as elevated Ca^{2+} levels are also present. From Caohuy *et al.* (29).

(3). Indeed, it has been shown by light scattering and electron microscopy studies (36) that the Ca^{2+}-activated form of synexin has a polymeric structure. Consistently, the protein concentration dependence for many synexin-dependent reactions has consistently been sigmoidal, with Hill coefficients (n_H) of approximately 2. These data have been interpreted to indicate that the aggregation and fusion reactions depend on at least two synexin subunits forming a polymer, which can interact simultaneously with two target membranes (36).

As a further test of this concept, we compared GTP activation of synexin-driven membrane aggregation with binding of GTP to synexin. As shown in Figure 2, the nonlinear relationship can be modeled by a system in which two Ca^{2+}-activated synexin molecules on different membranes can either bind (1) or not bind (0) a GTP molecule. The rate of fusion can be represented by the equation,

$$\text{Rate} = R_{00} \cdot p_{00} + R_{01} \cdot p_{01} + R_{11} \cdot p_{11}$$

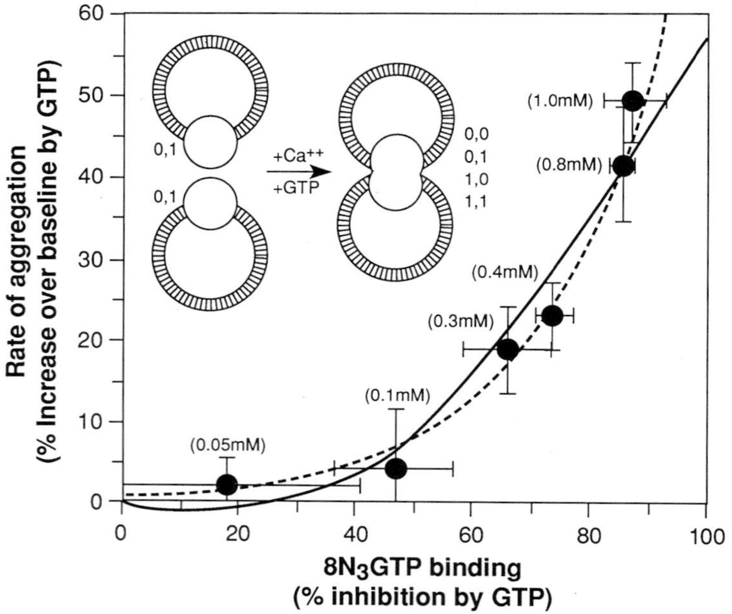

FIGURE 2 The relationship between the rates of the membrane aggregation or fusion by synexin and the inhibition of $8N_3$-γ-^{32}[P]-GTP binding caused by various concentrations of GTP. Values on the y-axis are the averages ± averages of the SD of the chromaffin granule aggregation and phosphatidylsevine (PS) liposome fusion assays from Caohuy *et al.* (29, Figs. 1A and 1B). The x-axis values are data taken from Caohuy *et al.* (29, Fig. 3). The solid line is a fit of an equation that models the system as if at least two calcium-activated synexin molecules must each bind GTP independently to drive membrane aggregation or fusion. The details are in the text. The dashed line is the fit of a single exponential to the data, $y = 0.7\ e^{0.0484x}$, $r^2 = 0.970$.

in which p is the probability of interaction of two synexin molecules, with the GTP either bound or not bound, and R is the fusion rate for each condition. If S is the fraction of synexin bound to GTP (29), the simplest binomial model gives the values of p as $p_{00} = (1 - S)^2$; $p_{01} = 2S(1 - S)$; and $p_{11} = S^2$. Assuming only the experimentally verified value of 1.0 for R_{00} when [GTP] = 0, the calculated value for R_{01} is 0.9 and for R_{11} is 1.6. The solid line in Figure 2 shows the fit of this equation to the experimental data. Thus, according to this simplest model, optimal fusion activity would indeed occur only if both synexins bound GTP. We conclude that whereas the reaction is undoubtedly more complex, the latter hypothesis is in general agreement with the shape and fit of the curve. Thus, at least two GTP-bound synexin molecules are necessary for an optimal fusion reaction.

V. Conclusion

These studies give strong support to the concept that the function of synexin in cells is to transduce the calcium signal for exocytosis by binding GTP, polymerizing, and directly mediating a membrane fusion process. It is true that synexin alone is able to mediate efficiently both membrane contact and fusion reactions *in vitro* (37, 38). However, the exocytosis process is likely to be much more complicated. Therefore, although the fusion reaction, per se, may be mediated by synexin or related molecules, *in vivo* it remains likely that additional contributions to specific docking processes may be derived from other protein components in the cell.

References

1. Ahnert-Hilger, G., Wegenhorst, U., Stecher, B., Spicher, K., Rosenthal, W., and Gratzl, M. (1992). Exocytosis from permeabilized bovine adrenal chromaffin cells is differently modulated by guanosine 5′[γ-thio]triphosphate and guanosine 5′-[βγ-imino]triphosphate. *Biochem. J.* **284,** 321–326.
2. Aridor, M., Rajmilevich, G., Beaven, M. A., and Sagi-Eisenberg, R. (1993). Activation of exocytosis by the heterotrimeric G protein G_{i3}. *Science* **262,** 1569–1572.
3. Monck, J. R., and Fernandez, J. M. (1994). The exocytotic fusion pore and neurotransmitter release. *Neuron* **12,** 707–716.
4. Neher, E. (1988). The influence of intracellular calcium concentration on degranulation of dialyzed mast cells from rat peritoneum. *J. Physiol. (Lond.)* **395,** 193–214.
5. Bennett, M. K., and Scheller, R. H. (1993). The molecular machinery for secretion is conserved from yeast to neurons *Proc. Natl. Acad. Sci. U.S.A.* **90,** 2559–2563.
6. Ferro-Novick, S., and Jahn, R. (1994). Vesicle fusion from yeast to man. *Nature* **370,** 191–193.
7. Rothman, J. E., and Orci, L. (1992). Molecular dissection of the secretory pathway. *Nature* **355,** 409–415.
8. Sudhof, T. C. (1995). The synaptic vesicle cycle: A cascade of protein-protein interactions. *Nature* **375,** 645–653.
9. Chapman, E. R., Hansen, P. I., An, S., and Jahn, R. (1995). Ca^{2+} regulates the interaction between synaptotagmin and syntaxin. *J. Biol. Chem.* **270,** 23667–23671.

10. Morgan, A., and Burgoyne, R. D. (1995). A role for soluble NSF attachment proteins (SNAPs) in regulated exocytosis in adrenal chromaffin cells. *EMBO J.* **14,** 232–239.
11. Morgan, A. (1995). Exocytosis. *Essays Biochem.* **30,** 77–95.
12. O'Connor, V., Augustine, G. J., and Betz, H. (1994). Synaptic vesicle exocytosis: Molecules and models. *Cell* **76,** 785–787.
13. White, J. (1992). Membrane fusion. *Science* **258,** 917–923.
14. Gomperts, B. D., Cockroft, S., Howell, T. W., Nüsse, O., and Tatham, P. E. R. (1987). The dual effector system for exocytosis in mast cells: Obligatory requirements for both Ca^{2+} and GTP. *Biosci. Rep.* **7,** 369–381.
15. Howell, T. W., Cockroft, S., and Gomperts, B. D. (1987). Essential synergy between Ca^{2+} and guanine nucleotides in exocytotic secretion from permeabilized rat mast cells. *J. Cell Biol.* **105,** 191–197.
16. Okano, K., Monck, J. R., and Fernandez, J. M. (1993). GTPγS stimulates exocytosis in patch-clamped rat melanotrophs. *Neuron* **11,** 165–172.
17. Augustine, G. J., and Neher, E. (1992). Calcium requirements for secretion in bovine chromaffin cells. *J. Physiol.* (*Lond.*) **450,** 247–271.
18. Heinemann, C., Chow, R. H., Neher, E., and Zucker, R. S. (1994). Kinetics of the secretory response in bovine chromaffin cells following flash photolysis of caged Ca^{2+}. *Biophys. J.* **67,** 2546–2557.
19. Neher, E., and Augustine, G. J. (1992). Calcium gradients and buffers in bovine chromaffin cells *J. Physiol.* **450,** 273–301.
20. Neher, E., and Zucker, R. S. (1993). Multiple calcium-dependent processes related to secretion in bovine chromaffin cells. *Neuron* **10,** 21–30.
21. von Rüden, L., and Neher, E. (1993). A Ca-dependent early step in the release of catecholamines from adrenal chromaffin cells. *Science* **262,** 1061–1065.
22. Heldelberger, R., Heinemann, C., Neher, E., and Mathews, G. (1994). Calcium dependence of the rate of exocytosis in a synaptic terminal *Nature* **371,** 513–515.
23. Rosenboom, H., and Lindau, M. (1994). Exo-endocytosis and closing of the fission pore during endocytosis in single pituitary nerve terminals internally perfused with high calcium concentrations *Proc. Natl. Acad. Sci. U.S.A.* **91,** 5267–527.
24. Thomas, P., Wong, J. G., Lee, A. K., and Almers, W. (1993). A low affinity Ca^{2+} receptor controls the final steps in peptide secretion from pituitary melanotraphs *Neuron* **11,** 93–104.
25. Bittner, M. A., and Holz, R. W. (1992). Kinetic analysis of secretion from permeabilized adrenal chromaffin cells reveals distinct components *J. Biol. Chem.* **267,** 16219–16225.
26. Creutz, C. E., Pazoles, C. J., and Pollard, H. B. (1978). Identification and purification of an adrenal medullary protein (synexin) that causes calcium dependent aggregation of isolated chromaffin granules. *J. Biol. Chem.* **253,** 2858–2866.
27. Hong, K., Duzgunez, N., Ekerdt, R., and Papahadjopoulos, D. (1982). Synexin facilitates fusion of specific phospholipid membranes at divalent cation concentrations found intracellularly. *Proc. Natl. Acad. Sci. U.S.A.* **79,** 4642–4644.
28. Raynal, P., and Pollard, H. B. (1994). Annexins: The problem of assessing the biological role for a gene family of multifunctional calcium- and phospholipid-binding proteins. *BBA Biomembranes* **1197,** 63–93.
29. Caohuy, H., Srivastava, M., and Pollard, H. B. (1996). Membrane fusion protein synexin (annexin VII) as a Ca^{2+}/GTP sensor in exocytotic secretion. *Proc. Natl. Acad. Sci. U.S.A.* **93,** 10797–10802.
30. Bourne, H. R., Sanders, D. A., and McCormick, F. (1990). The GTPase superfamily: A conserved switch for diverse cell functions. *Nature* **348,** 125–132.
31. Bourne, H. R., Sanders, D. A., and McCormick, F. (1991). The GTPase superfamily: conserved structure and molecular mechanism. *Nature* **349,** 117–127.
32. Gilman, A. G. (1987). G proteins: Transducers of receptor generated signals. *Annu. Rev. Biochem.* **56,** 615–649.

33. Ross, E. M. (1988). Receptor-G protein-effector: The design of a biochemical switchboard. *Concepts Biochem.* **65,** 937–942.
34. Higashijima, T., Burnier, J., and Ross, E. M. (1990). Regulation of G_i and G_o by mastoparan, related amphiphilic peptides and hydrophobic amines. *J. Biol. Chem.* **265,** 14176–14186.
35. Vitale, N., Mukai, H., Rouot, B., Thiersé, D., Aunis, D., and France-Bader, M. F. (1995). Exocytosis in chromaffin cells. *J. Biol.Chem.* **268,** 14715–14723.
36. Creutz, E. C., Pazoles, C. J., and Pollard, H. B. (1979). Self-association of synexin in the presence of calcium: Correlation with synexin-induced membrane fusion and examination of the structure of synexin aggregates. *J. Biol. Chem.* **254,** 553–558.
37. Nir, S., Stutzin, A., and Pollard, H. B. (1987). Effect of synexin on aggregation and fusion of chromaffin granule ghosts at pH6. *Biophys. Biochem. Acta.* **903,** 309–318.
38. Pollard, H. B., Rojas, E., Pastor, R. W., Rojas, E. M., Guy, H. R., and Burns, A. L. (1991). Synexin: Molecular mechanism of calcium-dependent membrane fusion and voltage-dependent calcium channel activity. Evidence in support of the "hydrophobic bridge hypothesis" for exocytotic membrane fusion. *Ann. N. Y. Acad. Sci.* **635,** 328–351.

Dieter Bruns and Reinhard Jahn

Howard Hughes Medical Institute
Yale University, School of Medicine
New Haven, Connecticut 06510

Monoamine Transmitter Release from Small Synaptic and Large Dense-Core Vesicles

Calcium-regulated transmitter release is the primary means of communication between neurons. The vesicle hypothesis of chemical transmission proposes that neurotransmitter is discharged from synaptic vesicles at specialized sites in the nerve terminal, the "active zones" (1). This hypothesis has formed the foundation of our current understanding of synaptic transmission, but the biophysical properties of transmitter release and the molecular structures involved in promoting the exocytotic step remain unknown. Most studies of neurotransmitter release utilize electrophysiological measurements of postsynaptic responses to infer presynaptic events. Furthermore, electrical measurements of the membrane capacitance have been used to follow directly the changes in cell–surface area associated with exocytosis of vesicles in a wide variety of cells, such as mast cells, chromaffin cells, and vertebrate neurons (2). Electrochemical methods have been refined by the development of small carbon fibers that can be used as amperometric sensors for the detection of transmitter release from single secretory cells (3, 4). Amperometry monitors the oxidation

Advances in Pharmacology, Volume 42

1054-3589/98 $25.00

or reduction of transmitter at the surface of a carbon fiber, enabling a direct and highly sensitive measurement of secretion. Many secreted products, for example, epinephrine, dopamine, or serotonin, are readily oxidizable. We have exploited this technique to study transmitter release from neurons at high time resolution (5). Retzius cells of the leech (*Hirudo medicinalis*) synthesize, store, and release serotonin as their only classical transmitter. Due to the large size of these cultured neurons (diameter of about 80 μm), the releasing cell surface can be directly approached with a carbon fiber. Furthermore, the fiber can be positioned over discrete regions of the cell, such as the soma or the axon stump, the latter being the preferential site of synapse formation when these neurons are cocultured with a chemoreceptive "follower" cell. In the first set of experiments, the tip of the fiber was manipulated to the distal end of the axon stump of a single Retzius cell, gently touching its membrane. On stimulation of action potentials in the Retzius cell, discrete spikelike oxidation currents were detected that followed the action potential with a short delay. These signals were seen only when the potential applied to the fiber was greater than about 200 mV (exceeding the oxidation potential for serotonin) and were reversibly blocked on superfusion of the cells with Cd^{2+} solution, demonstrating their dependence on calcium influx during the action potential. These findings suggest that the signals represent Ca^{2+}-triggered exocytosis of quantal packets of serotonin.

We characterized two types of signals, small and large oxidation currents. Small and large events differed with respect to their amplitude, charge, and kinetics, suggesting that they are caused by the exocytosis of distinctly sized vesicles. In close correlation, electron microscopy demonstrated that two types of vesicles, small synaptic vesicles (SSVs) and large dense-core vesicles (LDCVs), are present at these axonal recording sites. Our morphological data showed that SSVs are clustered, directly apposing the plasma membrane, whereas LDCVs often surround these clusters and are scattered throughout the cytoplasm. Small events occurred more rapidly and more frequently after single action potentials than large events. The different time-dependent coupling of small and large events to action potentials agrees with the spatial organization of SSVs and LDCVs, but the involvement of other molecular processes controlling the secretion rate of the different vesicle types cannot be excluded.

Recent evidence has shown that the membrane proteins synaptobrevin, SNAP-25, and syntaxin play a key role in exocytosis and are specifically targeted by clostridial proteases that are potent inhibitors of transmitter release (6). We cloned the leech homologue of the vesicle protein synaptobrevin and found it to be highly homologous to its mammalian isoform. The sensitivity of leech synaptobrevin to tetanus toxin cleavage is preserved. Both small and large events were completely abolished when Retzius cells were microinjected with tetanus-toxin light chain (TeNT-LC), corroborating our hypothesis that the identified oxidation signals reflect exocytotic fusion events.

At somatic recording sites, only large events were detected, and consistent with our hypothesis, only LDCVs were observed in the soma of these neurons. These findings strongly suggest that small events are due to the exocytosis of SSVs, whereas large events reflect transmitter release from LDCVs. Furthermore, these data demonstrate that the plasma membrane of the neuronal soma provides the molecular machinery to promote vesicle fusion. Interestingly, this finding is complementary to the immunohistochemical observations of several investigators showing that syntaxin and SNAP-25, which have been implicated to act as plasma membrane receptors for secretory organelles, are widely distributed in nonsynaptic areas of neuronal plasma membranes.

Having shown that transmitter release from SSVs and LDCVs can be distinguished, we next asked how many transmitter molecules are released per vesicle. To answer this question, the number of electrons transferred per oxidized serotonin molecule was determined. Using a specially designed experimental set-up we measured that at least 4 mol of electrons are transferred per mole of serotonin reacted. A single SSV thus releases, on average, 4700 serotonin molecules and the intravesicular concentration of serotonin is 270 mM. The calculation of the intravesicular transmitter concentration assumes that the stored molecules are freely diffusible. The amount of serotonin per SSV is approximately half the number of acetylcholine molecules in frog SSV, estimated by iontophoretic application of transmitter at the neuromuscular synapse. LDCVs release 15,000 to 300,000 serotonin molecules. Given the size heterogeneity of LDCVs in these neurons (range, from 55–155 nm in diameter), one might suggest that the transmitter is stored at similar intravesicular concentration.

Several investigators have noted that the initial stage of exocytosis in nonneuronal secretory cells is the formation of a fusion pore, which establishes an electrical and diffusional connection between the vesicle interior and the extracellular spaces. In these cells, the fusion pore expands slowly and delays the moment when vesicles and plasma membrane merge completely. Using amperometry, it has been shown that transmitters and hormones leak out well before fusion is complete. This phenomenon was observed as a small "foot" signal preceding the bulk exocytosis that is represented by the main spike of the amperometric signal. In the neurons used in our studies, about 30% of the large events were preceded by a similar foot signal that lasted, on average, 540 μs (Fig. 1). Furthermore, these foot signals were often initiated by a stepwise increase in the current amplitude, pointing to a rapid first opening of a preassembled fusion pore. These results suggest that the small size and the slow expansion of the fusion pore delays the release of transmitter from LDCVs in neurons. In contrast, a discernible preceding foot signal was not observed in the time course of small events. In fact, exocytosis of SSVs appeared to be initiated by a rapid upstroke of the transmitter signal, with a mean 50–90% rise time of 90 μs. The 10 to 90% rise time of very rapid events, which probably arise from vesicles opening closest to the detector, is only 60 μs, reaching the resolution limit of our measuring system. In line with our observations of transmitter release from LDCVs, one might suggest that the rapid upstroke of the transmitter signal reflects the initial opening of a similar fusion pore. The mean time constant of decay of small events was 360 μs, with a median value of 260 μs of the corresponding frequency distribution. Our measurements are limited by the rate of diffusion of transmitter to the surface of the carbon fiber, hence reflecting an upper limit for the speed of transmitter discharge. Therefore, whether the fusion pore that forms upon SSV exocytosis rapidly expands or does not dilate at all remains to be clarified. However, the size of SSVs keeps diffusion distances short and makes these vesicles well suited for rapid and discrete signaling between neurons.

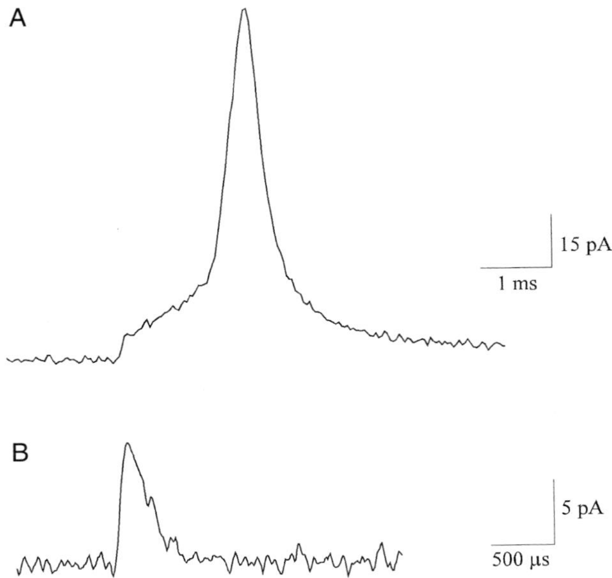

FIGURE 1 Kinetic characteristics of transmitter discharge from SSVs and LDCVs. (*A*) "Foot" signals often precede the main phase of transmitter discharge from LDCVs. The foot had a mean duration of 540 ± 40 μs and a mean amplitude at its maximum value of 10 ± 1 pA (n = 230). The integral of the foot signal reflects, on average, 4 ± 0.5% of the event's total integral. Half of the foot signals were initiated by a stepwise increase of the current signal, with an amplitude of 7.5 ± 0.7 pA (median, 4.5 pA) and a 50–90% rise time of 110 ± 10 μs (median, 60 μs; n = 110). (*B*) Exemplary recording of transmitter discharge from an SSV at high resolution. The vesicle released 4200 transmitter molecules, with a time constant of 160 μs. (Reprinted with permission from *Nature*, **377**, pp. 62–65, Figures 3B and 3E [1995].)

References

1. del Castillo, J., and Katz, B. (1956). Biophysical aspects of neuromuscular transmission. *Prog. Biophys. Biophy. Chem.* **6**, 121–170.
2. Matthews, G. (1996). Synaptic exocytosis and endocytosis: Capacitance measurements. *Curr. Opin. Neurobiol.* **6**, 358–364.
3. Chow, R. H., and von Rüden, L. (1995). Electrochemical detection of secretion from single cells. *In* Single Channel Recording, 2nd ed. (B. Sakmann and E. Neher, eds.), pp. 245–275. Plenum Press, New York.
4. Alvarez de Toledo, G., Fernandez-Chacoon, R., and Fernandez, J. M. (1993). Release of secretory products during transient vesicle fusion. *Nature* **363**, 554–558.
5. Bruns, D., and Jahn, R. (1995). Real-time measurement of transmitter release from single synaptic vesicles. *Nature* **377**, 62–65.
6. Ferro-Novick, S., and Jahn, R. (1994). Vesicle fusion from yeast to man. *Nature* **370**, 191–193.

Antonio G. García, Almudena Albillos, Maria F. Cano-Abad, Esther García-Palomero, M. Hernández-Guijo, Carlos J. Herrero, Richard B. Lomax, and Luis Gandía

Departamento de Farmacología
Facultad de Medicina
Universidad Autónoma de Madrid
Arzobispo Morcillo
28029-Madrid, Spain

Calcium Channels for Exocytosis in Chromaffin Cells

On biophysical grounds, the voltage-dependent Ca^{2+} channels of bovine adrenal medullary chromaffin cells were initially believed to consist of a homogeneous population (1). The subsequent availability of powerful pharmacological tools allowed the separation of various subcomponents in the whole-cell currents through neuronal Ca^{2+} channels. According to their pharmacological sensitivities, high-treshold Ca^{2+} channels have been classified as follows (2): L-type (blocked by 1,4-dihydropyridines), N-type (blocked by ω-conotoxin GVIA or by ω-conotoxin MVIIA), P-type (inhibited by nanomolar concentrations of ω-agatoxin IVA), and Q-type (blocked by micromolar concentrations of ω-agatoxin IVA or by ω-conotoxin MVIIC); R-type channels are resistant to all known toxins.

Earlier results from our and other laboratories led to the conclusion that bovine chromaffin cells expressed L-type Ca^{2+} channels (15–20%), N-type channels (30–40%), as well as P-type channels (30–40%). An R-type channel resistant to combinations of all blockers also seemed to be present. By using low and high concentrations of ω-agatoxin IVA, high concentrations of ω-conotoxin MVIIC, and low and higher concentrations of Ba^{2+} as charge carrier, we have recently come to the conclusion that the P-type component can be resolved into P- and Q-subtypes of Ca^{2+} channels. This has been achieved through the use of 20-nM and 2-μM concentrations of ω-agatoxin IVA, as well as a 3-μM concentration of ω-conotoxin MVIIC, in voltage-clamped bovine chromaffin cells superfused with low (2 mM) and high (10 mM) extracellular Ba^{2+} solution. We have found that the divalent cation concentration strongly modified the ability of ω-toxins to block specific Ca^{2+} channel subtypes (3); this was particularly relevant for ω-conotoxin MVIIC, whose blockade was markedly slowed down and decreased in the presence of 10-mM Ba^{2+}, as compared with 5- or 2-mM Ba^{2+} (Fig. 1). Accordingly, we revised our previous classification of Ca^{2+} channels and came to the conclusion that the P/Q component consisted mostly of Q-type Ca^{2+} channels (about 30–40%). The scarcity of P-channels has been recently corroborated with video-imaging analysis of

Advances in Pharmacology, Volume 42

1054-3589/98 $25.00

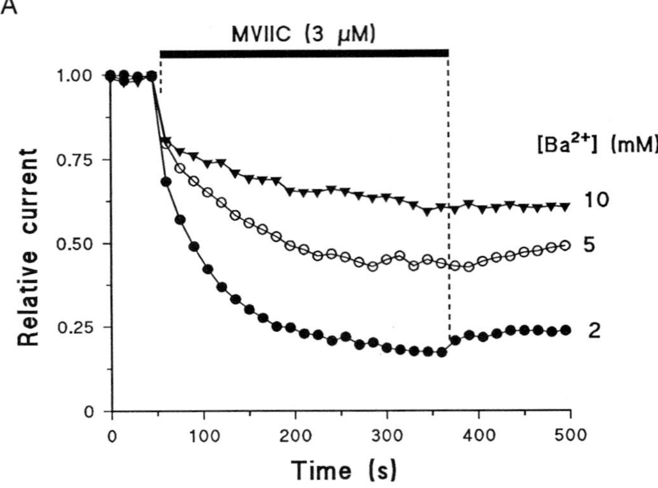

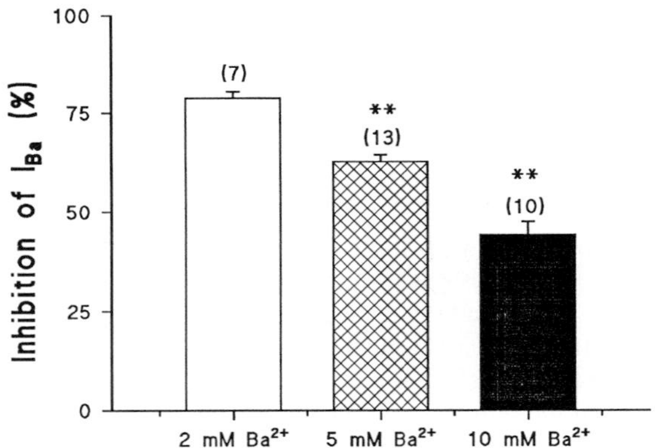

FIGURE 1 Distinct blocking effects of ω-conotoxin MVIIC of whole-cell Ba^{2+} currents through Ca^{2+} channels in voltage-clamped chromaffin cells superfused with an external solution containing different Ba^{2+} concentrations. Cells were voltage-clamped at −80 mV; test depolarizing pulses to 0 mV were applied at 10-sec intervals (3). In (A) the time-course of blockade is shown; in (B) means ± SEM of the inhibition of current by ω-conotoxin MVIIC (3 μM) is shown. In parentheses, number of cells. $^{**}P < .001$ with respect to the inhibition of current seen in 2-mM Ba^{2+}.

$[Ca^{2+}]_i$ transients in fura-2-loaded cells stimulated with brief pulses of high K^+. In this study, we found that L-type Ca^{2+} channels were more abundant in immunologically identified noradrenaline-containing cells, while Q-type Ca^{2+} channels predominated in adrenaline-containing cells (4).

Not all Ca^{2+} channel subtypes of bovine chromaffin cells participate equally in the control of the delivery of external Ca^{2+} to the secretory machinery to trigger a secretory response on cell depolarization. Thus catecholamine release triggered by brief pulses of high-K^+ solutions in superfused cells, measured online with an electrochemical detector, was partially inhibited by 3-μM furnidipine, by 2-μM ω-agatoxin IVA, or by 3-μM ω-conotoxin MVIIC. Combined furnidipine plus ω-agatoxin IVA (2 μM), or combined furnidipine plus ω-conotoxin MVIIC blocked completely the secretory signal. Neither ω-conotoxin GVIA (1 μM) nor a low concentration of ω-agatoxin IVA (20 nM) affected the release of catecholamines. This suggests that only L- and Q-type Ca^{2+} channels are involved in the control of exocytosis, and that in the gland *in situ,* L- and Q-type Ca^{2+} channels must be strategically located at exocytotic microdomains on the secretory surface of the chromaffin cell looking at the vessel lumen (5). Consistent with the relative density of channels in cell subtypes, L-type Ca^{2+} channels dominate secretion in noradrenaline cells, while Q-type channels control the secretory response in adrenaline cells (4).

For long time, no functional role was found for materials coreleased with adrenaline and noradrenaline during exocytosis (i.e., ATP and the opiate peptide methionine–enkephalin). Recent findings suggest that they may regulate the kinetics of Ca^{2+} channels in chromaffin cells in two ways: (1) slowing down of the current activation and (2) inhibition of the current amplitude (6, 7). Inhibition of I_{Ba} by adenosine triphosphate (ATP) or methionine–enkephalin was partially removed by strong depolarizing prepulses (facilitation of Ca^{2+} channels) (1). This voltage-dependent modulation occurred mainly on N/P/Q-types of Ca^{2+} channels, because it was partially or completely suppressed in the presence of ω-conotoxin GVIA or in the presence of ω-conotoxin MVIIC, respectively; however, it was insensitive to dihydropyridines. L-type Ca^{2+} channels are modulated through a voltage-independent pathway. The modulation of Ca^{2+} channels by exogenous ATP and methionine–enkephalin also takes place by endogenously released ATP and opiates, as the following experiments suggest (8): (1) A soluble vesicle lysate (SVL) obtained from purified bovine adrenal medulla chromaffin vesicles mimicks the effects of exogenous ATP and of methionine–enkephalin on whole-cell currents through Ca^{2+} channels. (2) Stop-flow of the superfusing extracellular solution also mimics those effects. (3) In the cell-attached configuration, 70% of the control patches (n = 28) and 80% of the patches containing ATP and methionine–enkephalin (in the presence of nifedipine) exhibited non-L-type ensemble currents with markedly delayed activation kinetics; they could be accelerated by step depolarizations. In contrast, 70% of the patches containing the opiate and ATP receptor antagonists naloxone and suramin showed Ca^{2+} channel activities with rapid activation and fast inactivation kinetics (9).

In conclusion, bovine chromaffin cells express a variety of voltage-dependent Ca^{2+} channels. L- and Q-type Ca^{2+} channels control the delivery of Ca^{2+} ions to the secretory machinery, thus controlling the rate and extent of secretion. In its turn, ATP and opiates coreleased with the catecholamines inhibit non-L-type Ca^{2+} channels in a voltage-dependent manner through a G-protein–mediated mechanism and L-type channels in a voltage-independent manner. This forms part of a feedback autoinhibitory mechanism. Thus, Ca^{2+} channels control secretion, and, in their turn, secreted products control the activity of Ca^{2+} channels.

Acknowledgments

Supported by grants from Fundación Ramón Areces and DGICYT, Spain.

References

1. Fenwick, E. M., Marty, A., and Neher, E. (1982). Sodium and calcium channels in bovine chromaffin cells. *J. Physiol.* **331,** 599–635.
2. Olivera, B. M., Miljanich, G., Ramachandran, J., and Adams, M. E. (1982). Calcium channel diversity and neurotransmitter release: The ω-conotoxins and ω-agatoxins. *Annu. Rev. Biochem.* **63,** 823–867.
3. Albillos, A., García, A. G., Olivera, B., and Gandía, L. (1996). Re-evaluation of the P/Q Ca^{2+} channel components of Ba^{2+} currents in bovine chromaffin cells superfused with low and high Ba^{2+} solutions. *Pflugers Arch. Eur. J. Physiol.* **432,** 1030–1038.
4. Lomax, R. B., Michelena, P., Núñez, L., García-Sancho, J., García, A. G., and Montiel, C. (1997). Different contributions of L- and Q-type Ca^{2+} channels to Ca^{2+} signals and secretion in chromaffin cell subtypes. *Am. J. Physiol.* **272,** 476–484.
5. López, M. G., Villarroya, M., Lara, B., Martínez-Sierra, R., García, A. G., and Gandía, L. (1994). Q- and L-type Ca^{2+} channels dominate the control of secretion in bovine chromaffin cells. *FEBS Lett.* **349,** 331–337.
6. Gandía, L., García, A. G., and Morad, M. (1993). ATP modulation of calcium channels in bovine chromaffin cells. *J. Physiol.* **470,** 55–72.
7. Albillos, A., Carbone, E., Gandía, L., García, A. G., and Pollo, A. (1996). Opioid inhibition of Ca^{2+} channel subtypes in bovine chromaffin cells: Selectivity of action and voltage-dependence. *Eur. J. Neurosci.* **8,** 1561–1570.
8. Albillos, A., Gandía, L., Michelena, P., Gilabert, J.-A., del Valle, M., Carbone, E., and García, A. G. (1996). The mechanism of calcium channel facilitation in bovine chromaffin cells. *J. Physiol.* **494,** 687–695.
9. Carabelli, V., Albillos, A., García, A. G., and Carbone, E. (1996). Calcium channel kinetics changes reveal localized opioids and ATP secretion in bovine chromaffin cells. *Soc. Neurosci. Abs.* (in press)

T. C. Cunnane and A. Smith

Department of Pharmacology
Oxford University
Oxford, United Kingdom

Characteristics of Transmitter Secretion from Individual Sympathetic Varicosities

In recent years it has become clear that noradrenaline and adenosine triphosphate (ATP) function as cotransmitters in a wide variety of autonomically innervated tissues (1). In suitable *in vitro* preparations (vas deferens, blood vessels), the characteristic features of neurotransmitter release at the level of the individual varicosity can be studied on an impulse-to-impulse basis. Electrophysiological recording techniques have been developed to study action potential propagation and neurotransmitter release, at the level of the varicosity, in postganglionic sympathetic nerve terminals. Several techniques have been employed, including differentiation of the rising phases of intracellularly recorded excitatory junction potentials (discrete events) and focal extracellular recording of the nerve terminal action potential and associated excitatory junction currents (NTIs, EJCs). A discussion of the methodology and the relative merits of these techniques can be found elsewhere (2, 3). Much of our data have come from studies on the sympathetic nerves innervating the guinea pig–isolated vas deferens; we will focus in the majority of this chapter on results obtained using this preparation. Briefly, studies have shown that (1) transmitter release from individual varicosities occurs intermittently following low-frequency nerve stimulation, (2) only a single quantum is normally secreted when the release mechanism of a varicosity is activated by the nerve action potential, (3) intermittence of transmitter release is not due to failure of nerve impulse propagation in the varicose nerve terminal network but rather to a low probability of release, and (4) the probability of transmitter release increases with stimulation frequency, that is, frequency-dependent facilitation. The mechanisms underlying frequency-dependent facilitation remain fairly poorly understood, but it is clear that calcium plays a pivotal role.

What is the nature of the calcium channels controlling transmitter release in postganglionic nerve terminals? At stimulation frequencies of 1 Hz or lower, transmitter release seems to depend solely on calcium entry into sympathetic nerve terminals through N-type calcium channels. However, at higher frequencies of stimulation, at least one other pharmacologically distinct calcium channel regulates transmitter release. When N-type calcium channels are blocked, a substantial component of release, termed *residual release,* is revealed at higher stimulation frequencies.

Advances in Pharmacology, Volume 42

1054-3589/98 $25.00

The characteristic features of residual release can be briefly summarized. First, the initial EJP in a train is abolished by ω-conotoxin GVIA (ω-CTX GVIA), suggesting that N-type calcium channels dominate release evoked by single pulses at low frequencies of stimulation. Second, during short trains of stimuli, there is a progressive increase in the amplitude of successive EJPs that summate to produce a large depolarization, which often leads to muscle contraction. Third, the magnitude of residual release depends not only on the frequency of nerve stimulation, but also on the train length. Fourth, residual release is insensitive to various combinations of selective calcium channel blockers but is abolished by a reduction in the extracellular calcium concentration or by low concentrations of cadmium. Residual release is, therefore, dependent on calcium entry through a pharmacologically distinct voltage-dependent calcium channel (4).

One surprising finding is that residual release is inhibited by ryanodine in a use- and time-dependent manner, suggesting that this pharmacologically distinct voltage-dependent calcium channel is located close to, or is in some way coupled to, ryanodine receptors (RyRs). Ryanodine receptors are involved in excitation–contraction coupling in cardiac and skeletal muscle but are not thought to play any role in action potential–evoked transmitter release. Nevertheless, immunocytochemical and autoradiographic investigations have revealed a widespread distribution of RyRs in the central nervous system, suggesting that intraneuronal calcium stores and calcium-induced calcium release (CICR) have important functions in neuronal calcium dynamics. In our experiments, ryanodine (10 μM) caused an inhibition of residual release in a use- and time-dependent manner consistent with its pharmacological mode of action at the single-channel level, that is, the channel needs to be activated before ryanodine can bind to the channel protein and lock it into a permanently open, reduced-conductance state. It was necessary to stimulate nerves regularly at intervals (e.g., 10 stimuli at 20 Hz delivered every 20 sec for 20–60 min) to reveal the inhibitory effect of ryanodine. Ryanodine did not decrease EJP amplitudes in previously *unstimulated* time-matched controls. The inhibitory effects of ryanodine were first observed after 20–30 min, and the maximal inhibition occurred after 60–90 min. Ryanodine inhibited all EJPs in a series of trains and did not appear to affect significantly the rate of development of facilitation. The amplitude and time course of individual spontaneous EJPs were unaffected by these procedures, showing that the site of action of ryanodine was prejunctional. These results indicate that ryanodine-sensitive calcium stores play an important role in transmitter release evoked by modest-to-high frequencies of stimulation. Similar results were obtained using the adenosine triphosphatase (ATPase) inhibitors thapsigargin (1–3 μM) and cyclopiazonic acid (30–100 μM). The inhibitory effect of ryanodine on residual release was potentiated by the application of caffeine (5 mM), which is also an agonist at the RyR (5).

One striking finding was that the potassium channel blocker 4-aminopyridine (4-AP) could overcome the inhibitory effect of ryanodine. This channel blocker prolongs the nerve terminal action potential, thereby allowing increased calcium influx (through calcium channels other than N-type because these have been irreversibly blocked by ω-CTX GVIA) into the nerve terminal to activate the release mechanism without the need for CICR. Ryanodine (10–30 M) had little effect on EJPs evoked at 1 Hz in the absence of N-type calcium channel blockade. Therefore, ryanodine-sensitive calcium stores do not appear to be involved in transmitter release mechanisms at low stimulation frequencies. The lack of effect of ryanodine on EJPs evoked by low stimulation frequencies further demonstrates that ryanodine has no effect on the sensitivity of the postjunctional membrane to released neurotransmitter. The inhibitory effect of ryanodine on residual release was also investigated using focal extracellular recording techniques. Ryanodine reduced the probability of occurrence of EJCs without any detectable change in the configuration of the nerve terminal impulse. Therefore, ryanodine appears to act at the depolarization-secretion coupling step and not by modifying impulse propagation in sympathetic nerve terminals. Ryanodine binds to the RyR only when the channel is open. The inhibition of residual release produced by ryanodine was shown to be use-dependent by the following experiment. Submaximal stimuli were used to excite only a small proportion of the nerve fibers. Control, fully facilitated EJPs (in the presence of ω-CTX GVIA and ryanodine >70 min) evoked by 5 submaximal stimuli at 10 Hz delivered every 2 sec at 20 V had amplitudes of 2.5 mV. When the stimulation voltage was increased from 20 to 30 V, EJP amplitudes reached 16mV. However, after continued stimulation at 30 V, EJP amplitudes progressively decreased and 60 min later were 3.0 mV. We interpret these results to mean that when the stimulus strength was increased, previously unstimulated nerve fibers were recruited that were unaffected by the presence of ryanodine; that is, ryanodine was unable to bind to its receptor in these nerves because they were previously unstimulated, but once the nerves were activated, ryanodine could access the channel and cause depletion of intraneuronal calcium stores. This finding is consistent with the idea that in previously unstimulated nerves, ryanodine was unable to bind to its receptor, but once the release mechanism was activated, ryanodine produced the expected inhibitory effects (5).

In summary, the critical determinant of neurotransmitter release is an increase in the intraneuronal calcium concentration following invasion of the nerve terminal by the action potential. This transient increase in the intraneuronal calcium level is thought to arise solely by calcium entering through clusters of voltage-dependent calcium channels located close to the sites of neurotransmitter release, that is, active zones. However, when N-type calcium channels are blocked by ω-conotoxin GVIA, a novel form of ryanodine-sensitive release is revealed. These findings suggest that CICR mechanisms play an important role in neurotransmitter release in sympathetic nerve terminals by amplifying the calcium entering through voltage-gated calcium channels at modest stimulation frequencies.

References

1. Burnstock, G. (1990). Cotransmission. The Fifth Heymans Lecture—Ghent, February 17, 1990. *Int. Pharmacodyn. Ther.* **304**, 7–33.
2. Brock, J. A., and Cunnane, T. C. (1988). Electrical activity at the sympathetic neuroeffector junctions in the guinea-pig vas deferens. *J. Physiol.* **399**, 607–632.

3. Cunnane, T. C., and Searl, T. (1994). Neurotransmitter release mechanisms in autonomic nerve terminals. *In* Molecular and Cellular Mechanisms of Neurotransmitter Release. (Stjärne, Lennart; Greengard, Paul; Grillner, and Tomas; Hökfelt, and David Ottoson, eds.), pp. 425–459.
4. Smith, A. B., and Cunnane, T. C. (1996). ω–Conotoxin GVIA-resistant neurotransmitter release in postganglionic sympathetic nerve terminals. *Neuroscience* 70, 817–824.
5. Smith, A. B., and Cunnane, T. C. (1996). Ryanodine-sensitive calcium stores involved in neurotransmitter release from sympathetic nerve terminals of the guinea pig. *J. Physiol* (in press).

M. R. Bennett

Neurobiology Laboratory
University of Sydney
Sydney, Australia

Neurotransmitter Release at Individual Sympathetic Varicosities, Boutons

I. The Secretosome Hypothesis

Synaptic vesicles undergo exocytosis in a manner that is regulated by a set of vesicle-associated proteins, at least one of which undergoes a calcium-triggered conformational change leading to exocytosis. It is most likely that this protein is synaptotagmin, which contains an N-terminus that spans the membrane of the synaptic vesicle and a C-terminus in close association with the presynaptic membrane–bound protein syntaxin, which is in turn complexed in many terminals with the N-type voltage-dependent calcium channel. Taken together, these proteins with the synaptic vesicle constitute a secretory unit that has been termed the *secretosome* (1).

A. Development of the Secretosome and of the Fourth-Power Relationship between Calcium and Transmitter Release at an Autonomic Synapse

The calyciform nerve terminal of the chick ciliary ganglion is an ideal preparation for analyzing the development of the molecular components that go to make up the secretory apparatus, in particular the secretosome. We

Advances in Pharmacology, Volume 42

1054-3589/98 $25.00

have described the expression of the vesicle-associated proteins synaptotagmin, syntaxin, and synapsin in calyciform nerve terminals at different stages of development together with the calcium dependence of transmission and of facilitation, augmentation, and posttetanic potentiation (PTP) (2). The results show that within a few hours of synapse formation within the ganglion (stage 34), both syntaxin and synapsins 1a and 11a are at high levels and can be localized to the nerve terminals, whereas the C2a domain of synaptotagmin 1 is at very low levels in the ganglion. At these early times, the cooperativity between calcium and transmitter release is lower than the fourfold that is found at the mature synapses. This fourfold cooperativity does not emerge until the level of the C2a domain of synaptotagmin increases to mature amounts. The results suggest a causal relationship between the appearance of the mature form of synaptotagmin and the fourfold cooperativity of the action of calcium on transmitter release observed at the synapse.

B. Nonuniform Probability of Transmitter Releases at Different Sympathetic Varicosities

It is possible to record the electrical signs of transmission at visualized varicosities of sympathetic neuromuscular junctions with loose-patch electrodes to show that adjacent varicosities on the same nerve terminal have different probabilities for secretion of transmitter (3). Antibodies to the N-terminus of synaptotagmin (SNAb) can be used to determine the extent of release from the varicosities, because when a vesicle opens to the extracellular space in the process of exocytosis, the N-terminus in the lumen of the vesicle is permanently labeled. This technique has been developed to determine the extent of exocytosis at different sympathetic varicosities of single terminals in the vas deferens during nerve stimulation. Varicosities that contain the same number of synaptic vesicles, as indicated by their similar staining with antibodies to the vesicle proteoglycan SV2, show very different levels of labeling with SNAb following nerve stimulation. The results are in agreement with those using electrophysiological techniques, namely that different varicosities in the same nerve terminal branch have very different probabilities for the secretion of the contents of a synaptic vesicle.

II. Calcium Influx at Different Sympathetic Varicosities and the Probability for Secretion

A description of the stochastic events that occur between the opening of an N-type calcium channel and exocytosis triggered by a conformational change in synaptotagmin has been given both for the case of all the calcium channels complexed in secretosomes and for the case in which there is a random distribution of both calcium channels and synaptic vesicles with their associated proteins (4). The results show that different numbers of vesicles are secreted on the opening of a calcium channel, depending on these spatial distributions of channels and vesicles. If secretosomes are distributed in a regular array at release sites and incorporate all of the classes of calcium channels that mediate secretion there, as they may at the active zones of somatic neuromuscular junctions and possibly in boutons, or if they are found in uniform density within release sites, then it might be expected that the probability for secretion at different release sites within a single nerve terminal will differ according to the size of the release site. There is evidence that this is the case, because larger active zones within a somatic neuromuscular junction possess greater probabilities for the secretion of a quantum. These observations suggest that the probability for secretion at a release site should be proportional to the calcium influx at the release site following an action potential, even given the stochastic nature of channel opening under these circumstances. The calcium level observed some 10 ms or so after a stimulus in a nerve terminal at a release site, at a time when the calcium is at uniform concentration throughout the terminal at a site, is proportional to the calcium entry (see Fig. 3 in Brain and Bennett [5]). With this in mind, the calcium changes in strings of varicosities following a single impulse to the sympathetic nerves were determined. There were variations in the calcium concentration observed within different varicosities a few milliseconds after an impulse, suggesting that there is nonuniformity in the calcium influx between the varicosities. Such observations are consistent with the secretosome hypothesis.

The excitatory junction potential at sympathetic neuromuscular junctions is elevated after a short high-frequency tetanus. This declines to control levels over two different exponential phases characterized by time constants of about 0.4 sec and 6.0 sec. These are the same time constants as those for the F2 facilitation and augmentation phases of elevated transmitter release following a short tetanus at motor–nerve terminals. The question arises as to whether the F2 facilitation and augmentation phases of transmitter release elevation at these nerve terminals are accompanied by similar phases of elevated terminal calcium in the sympathetic varicosities. Calcium indicators in individual varicosities showed two exponential phases of decline in intraterminal calcium following an impulse, with time constants of F2 facilitation (about 0.4 sec) and augmentation (about 6.0 sec). These time constants are the same as those of the electrically measured F2 facilitation and augmentation, suggesting that the rate of sequestering of the residual calcium in the varicosities is responsible for these phases of increased efficacy of transmission.

III. Calcium Influx at Different Sympathetic Boutons and the Probability for Secretion

According to the secretosome hypothesis, the probability for secretion at a release site should be proportional to the calcium influx at the release site following an action potential, as argued in the case of transmitter release at varicosities. The calcium changes in strings of boutons following a single impulse to the preganglionic sympathetic nerves of the superior cervical ganglion were determined after previously loading the terminals with calcium indicators. There were variations in the calcium concentration observed within different boutons a few milliseconds after an impulse, suggesting that there is nonuniformity in the calcium influx between the boutons. Such observations are consistent with the secretosome hypothesis.

The efficacy of synaptic transmission through the superior cervical ganglion is elevated following a single impulse or a short train of impulses and then declines to the control level over three different exponential phases, namely those of F1 facilitation, F2 facilitation, and augmentation, with time constants of about 50 ms, 0.5 sec, and 7.0 sec, respectively. A fourth phase of elevated synaptic potential amplitude occurs after a long high-frequency train—PTP—which has a time constant of about 2 min. The question arises as to whether there is a temporal relationship between the time course of these changes in synaptic efficacy after a tetanus and the decline in intraterminal calcium that would suggest that the latter gives rise to the former. Calcium changes in individual boutons of preganglionic nerve terminals of the superior cervical ganglion following trains of impulses were determined with calcium indicators in the boutons. The results showed that, as in sympathetic varicosities, there are three exponential phases of decline in the intraterminal calcium, with time constants of about 0.5 sec, 7.0 sec, and 2.0 min. Because these are similar to the time constants of decline of F2 facilitation, augmentation, and PTP observed with electrophysiological techniques, it is likely that the time course of residual calcium sequestration in the boutons determines the time course of these different phases of increased synaptic efficacy.

IV. Summary

The secretosome hypothesis has been studied at autonomic nerve terminals. Evidence is presented that there is a nonuniform secretion probability at different release sites of these terminals. This can be correlated with their influx of calcium ions following an impulse, as expected according to the secretosome hypothesis.

References

1. Bennett, M. R. (1996). Neuromuscular transmission at an active zone: The secretosome hypothesis. *J. Neurocytol.* **25**, 869–891.
2. Lin, Y. Q., Brain, K. L., Nichol, K. A., Morgan, J. J., and Bennett, M. R. (1996). Vesicle-associated proteins and calcium in nerve terminals of chick ciliary ganglia during development of facilitation. *J. Physiol.* **497**, 639–656.
3. Lavidis, N. A., and Bennett, M. R. (1993). Probabilistic secretion of quanta from visualized varicosities along single sympathetic nerve terminals. *J. Auton. Nerv. Syst.* **43**, 41–50.
4. Bennett, M. R., Gibson, W., and Robinson, J. (1995). Probabilistic secretion of quanta: Spontaneous release at active zones of varicosities, boutons and endplates. *Biophys. J.* **69**, 42–56.
5. Brain, K. L., and Bennett, M. R. (1995). Calcium in the nerve terminals of chick ciliary ganglia during facilitation, augmentation and potentiation. *J. Physiol.* **489**, 637–648.

Arun R. Wakade, Dennis Przywara, and Taruna D. Wakade

Department of Pharmacology
School of Medicine
Wayne State University
Detroit, Michigan 48201

Appropriate Target Cells Are Required for Maturation of Neurotransmitter Release Function of Sympathetic Neurons in Culture

There are two properties of sympathetic neurotransmitter release, originally described in mature neuroeffector organs, that form the basis of the present work. One is the enhancement of release by increased stimulation frequency. The other involves facilitation of release by pharmacological agents that increase action potential duration. An increase in the release of norepinephrine (NE) when an identical number of stimulating pulses are delivered at increasing frequencies was first observed in perfused spleen of the cat by Brown and Gillespie in 1957. A frequency-dependent increase in the release of [^{3}H]NE from preloaded tissues has also been demonstrated in rat vas deferens and chick heart. A classic agent used to increase neuronal action potential duration is tetraethylammonium (TEA), which selectively inhibits repolarizing K^+ currents. A 20- to 40-fold facilitation of [^{3}H]NE release by TEA has been reported in a variety of sympathetic neuroeffector organs, including, guinea pig and chick hearts, cat spleen, and rat salivary gland and vas deferens. Interestingly, maximum facilitation of [^{3}H]NE release by TEA occurs with a single stimulus in each of the sympathetic neuroeffector organs tested. The enhancement of release is inversely related to stimulation frequency.

The use of cell culture techniques has allowed extensive study of the Ca^{2+}-dependent regulation of [^{3}H]NE release at the cellular level. Culture conditions have been specifically designed to promote optimum survival of sympathetic neurons and eliminate nonneuronal cells in a defined medium. We have previously shown that chick sympathetic neurons in culture exhibit regulated, Ca^{2+}-dependent, exocytotic release of [^{3}H]NE when electrical or other depolarizing stimuli are administered. They express membrane receptors coupled to production of inositol 1,4,5-trisphosphate and diacylglycerol stimulation of protein kinase (PKC), as well as adenylate cyclase–coupled receptors that stimulate

Advances in Pharmacology, Volume 42

1054-3589/98 $25.00

cyclic adenosine monophosphate (cAMP) production. Agents that affect voltage-dependent Ca^{2+} entry or second-messenger levels can also affect [^{3}H]NE release. Thus, sympathetic neurons in culture appear to be a good model of neurons *in vivo*. However, sympathetic neurons in the body seek their target during development and continue to grow in an environment dominated by nonneuronal cells.

When we use cultures of nerve growth factor (NGF)–supported chick sympathetic neurons for detailed study of transmitter release properties, some of their responses are significantly different from their counterparts growing in the body (1). Compared with transmitter release from neuroeffector preparations, NGF-supported chick sympathetic neurons have higher basal and evoked release (1.5–2% of total [^{3}H]NE content); the release response to increasing stimulation frequency is negative rather than positive and is less sensitive to facilitation by TEA. These differences in the functional properties of sympathetic neurons growing under two different environments, and the fact that these neurons can be supported in culture by several agents (excess K^+, phorbol esters, or forskolin), suggest that multiple trophic inputs are essential for normal development of transmitter release properties either *in vivo* or *in vitro.* To address this issue, we monitored release of [^{3}H]NE, voltage-clamped Ca^{2+} currents and intracellular Ca^{2+} concentration ($[Ca^{2+}]_i$) in embryonic chick sympathetic neurons cultured alone, with target cells, and in the presence of various trophic and supportive agents.

The effects of different targets on release of [^{3}H]NE are summarized in Table I and described later along with the corresponding changes in Ca^{2+} handling that we observed under the various culture conditions. Chick embryo sympathetic neurons cultured alone exhibited maximal elevation of $[Ca^{2+}]_i$ and release of [^{3}H]NE when given 10 stimulating pulses at 1 Hz but not at 10 Hz, yielding a negative frequency-release response. Stimulation-evoked release was only slightly enhanced by the K^+ channel blocker TEA (10 mM). Increasing TEA to 100 mM caused no further enhancement of release. However, when sympathetic neurons were cocultured with cardiac cells of the chick embryo, electrically stimulated transmitter release was reduced by three- to fivefold. A similar reduction was found in the stimulated rise in $[Ca^{2+}]_i$ measured in neurites, the sites of transmitter release. Cocultured neurons had a positive stimulation frequency –[^{3}H]NE release response and a five- to sevenfold facilitation of release by 10 mM TEA. Voltage-clamped Ca^{2+} current density measured in cell bodies of neurons grown alone was 0.61 ± 0.13 pA/μm^2. Ca^{2+} current density decreased to 0.19 ± 0.03 pA/μm^2 in cocultured neurons ($p < .01$, n = 7), consistent with the observed reduction of $[Ca^{2+}]_i$ and [^{3}H]NE release. Sympathetic neurons supported by NGF plus ciliary neurotrophic factor (CNTF) and/or neurotrophin-3 (NT-3) did not exhibit the trophic changes observed in neurons cocultured with cardiac cells. To determine if the trophic effect of targets on [^{3}H]NE release properties was more than a peculiar property of avian peripheral sympathetic neurons, neonatal rat superior cervical ganglion (SCG) neurons were also tested. SCGs were relatively insensitive to TEA when cultured alone, but [^{3}H]NE release was greatly facilitated by TEA when tested in SCG neurons cocultured with rat neonatal cardiac myocytes.

TABLE I Characterization of Trophism

Culture Conditions	*1 Hz*	*1 Hz plus TEA*	*10 Hz*	*Ratio (10Hz/1Hz)*	*TEA/control (1 Hz)*
SN alone	0.87 ± 0.07	1.47 ± 0.12^{b}	0.75 ± 0.03	0.86	1.7
SN + CCs	0.28 ± 0.03^{a}	1.88 ± 0.09^{b}	0.58 ± 0.09^{b}	2.07^{c}	6.7^{c}
SN + hepatocytes	0.36 ± 0.09^{a}	1.6 ± 0.23^{b}	0.65 ± 0.22^{b}	1.80^{c}	4.8^{c}
SN + skeletal muscle	1.50 ± 0.15^{a}	n.d.	1.30 ± 0.11	0.86	n.d.
SN + sensory neurons	0.84 ± 0.11	1.63 ± 0.15^{b}	0.79 ± 0.01	0.94	1.94
SN + ciliary ganglion	1.41	n.d.	0.963	0.68	n.d.
SN + water-lysed CCs	0.41 ± 0.06^{a}	0.98 ± 0.16^{b}	$0.27 \pm 0.04^{a,b}$	0.65	2.39
SN + ethanol-fixed CCs	0.89 ± 0.02	n.d.	0.72 ± 0.06	0.80	n.d.
SN + CC membranes	0.25 ± 0.026^{a}	$0.49 \pm 0.03^{a,b}$	0.25 ± 0.06^{a}	1.0	1.96
SN + conditioned medium	0.62 ± 0.19^{a}	0.83 ± 0.15	0.41 ± 0.08^{a}	0.66	1.34

Values are fractional release of [^{3}H]NE ($\times 10^{-2}$) and the mean ± SEM of four to 18 experiments, except ciliary ganglia data (one observation). The value of the release ratio indicates a negative (10Hz/1Hz < 1) or positive (10Hz/1Hz > 1) frequency-release response.

[a] $p < .01$ (Student's *t*-test) compared with control (SN alone).

[b] $p < .01$ compared with 1Hz without TEA for the same culture condition.

[c] $p < .001$ compared with control.

n.d., not done; SN, sympathetic neurons; CC, cardiac cell.

Changes in Ca^{2+} handling and release properties were produced within 24 hr by sympathetic neuroeffector cells, such as cardiac cells and hepatocytes, but not by skeletal muscle cells, sensory neurons, ciliary ganglion neurons, or fibroblasts. Preparations of cardiac cell membranes, cardiac cells fixed in ethanol or lysed in water, were ineffective in altering transmitter release properties of sympathetic neurons. Using culture media conditioned by cardiac cells also did not affect the release properties. Furthermore, the changes in Ca^{2+} handling and release properties did not occur spontaneously in neurons grown alone for up to 6 days. The frequency and TEA responses of neurons grown with cardiac cells or hepatocytes are characteristic of responses seen in sympathetic neuroeffector organs.

In embryonic chick sympathetic neurons, both excess K^+ and phorbol ester activate PKC and support neuronal survival (4). Therefore, our ongoing studies are examining whether these agents affect Ca^{2+} handling and transmitter release properties in a manner similar to those of cardiac cells. Recent data show that compared with NGF, phorbol ester-supported neurons have significantly lower Ca^{2+} current density and exhibit decreased binding of [^{125}I] ω conotoxin GVIA (Wakade *et al.*, unpublished data). These results suggest that reduced expression of N-type Ca^{2+} channels underlies the trophic change in sympathetic transmitter release properties and supports the interpretation that the PKC signaling pathway is active in altering Ca^{2+} channels and transmitter release during normal development of sympathetic neurons. Finally, the similar changes in transmitter release properties produced by appropriate target cells and PKC stimulation raise the possibility that target cells, via an unidentified trophic signal, activate a neuronal PKC cascade that controls the development of transmitter release properties by altering Ca^{2+} channel gene expression. Regardless of the mechanisms involved, it is clear that multiple trophic inputs distinct from NGF, NT-3, or CNTF are required for normal physiological function of sympathetic neurons (2) and that physiological targets play a crucial role in development of normal transmitter release properties by controlling Ca^{2+} homeostasis in sympathetic neurons (3).

References

1. Wakade, A. R., and Wakade, T. D. (1988). Comparison of transmitter release properties of embryonic sympathetic neurons growing in vivo and in vitro. *Neuroscience* **27,** 1007–1019.
2. Przywara, D. A., Kulkarni, J. S., Wakade, T. D., and Wakade, A. R. (1996). Importance of protein kinase C for normal development of transmitter release properties in embryonic chick sympathetic neurons in culture. *Neuroscience* **72,** 815–820.
3. Wakade, A. R., Przywara, D. A., Bhave, S. V., Mashalkar, V., and Wakade, T. D. (1995). Cardiac cells control transmitter release and calcium homeostasis in sympathetic neurons cultured from embryonic chick. *J. Physiol.* (*Lond.*) **488,** 587–600.
4. Wakade, A. R., Wakade, T. D., Malhotra, R. K., and Bhave, S. V. (1988). Excess K^+ and phorbol ester activate protein kinase C and support the survival of chick sympathetic neurons in culture. *J. Neurochem.* **51,** 975–983.

Thomas C. Westfall, Laura A. McCullough, Lillian Vickery, Linda Naes, Chun Lian Yang, Song-Ping Han, Terry Egan, Xiaoli Chen, and Heather MacArthur

Department of Pharmacological and Physiological Science
Saint Louis University School of Medicine
St. Louis, Missouri 63104

Effects of Neuropeptide Y at Sympathetic Neuroeffector Junctions

It is now well accepted that sympathetic nerves express at least three neurotransmitters/neuromodulators, namely norepinephrine(NE), neuropeptide Y (NPY), and adenosine triphosphate (ATP). Evidence for the existence of these three cotransmitters has come from studies showing that they are located in sympathetic nerves and that they can be released under appropriate conditions. In addition, the application of each transmitter mimics a phase of sympathetic nerve stimulation and each phase can be blocked with appropriate antagonists (1).

The purpose of the present studies were severalfold: (1) to examine the prejunctional effects of NPY on cotransmitter release, (2) to examine the postjunctional effects of NPY on cotransmitter function, (3) to further examine a cotransmitter role for NPY, (4) to determine the mechanism of action of NPY on cotransmitter release, and (5) to characterize and determine the mechanism of action of NPY on catecholamine synthesis. Studies were carried out using two model systems. These included the perfused mesenteric arterial bed of the rat and the nerve growth factor (NGF)–differentiated pheochromocytoma (PC12) cell. Both preparations were set up and utilized as previously described (2, 3). The release of mediators were induced by periarterial nerve stimulation of the mesenteric arterial bed or depolarization with high K^+ (50 mM) of the PC12 cell. Samples were collected and release quantified by HPLC-EC detection of NE, high-perfromance liquid chromatography (HPLC)–fluorometric analysis of ATP, and radioimmunoassay (RIA) of NPY. Postjunctional responses in the perfused mesenteric arterial bed were quantified by continuously monitoring perfusion pressure.

Nerve stimulation of the perfused mesenteric arterial bed resulted in the release of NE, ATP, and NPY. The release of NE was inhibited by NPY and NPY analogues. IC_{50} values for the inhibition of NE release by NPY, PYY, NPY 13-36, and $Leu^{31}Pro^{34}NPY$ were 3.5, 20, 25, and 185 nM, respectively. This pattern is most consistent with the inhibitory effects being mediated by activation of NPY-Y_2 receptors. Postjunctionally, NPY and certain analogues

Advances in Pharmacology, Volume 42

1054-3589/98 $25.00

evoked a potentiation of the increase in perfusion pressure to nerve stimulation as well as a wide variety of vasoactive agents. NPY and the Y_1 selective agonist $Leu^{31}Pro^{34}NPY$ produced potentiation of the increase in perfusion pressure to phenylephrine, angiotensin II, and arginine vasopressin, as well as the cotransmitters, NE and ATP. The postjunctional potentiative action appears to be due to activation of NPY-Y_1 receptors since it was mimicked by $Leu^{31}Pro^{34}NPY$ but not the Y_2 selective agonist NPY13-36. In addition, it was antagonized by the Y_1 selective antagonist BIBP3226.

We investigated the possible role of NPY in the development or maintenance of hypertension in the spontaneous hypertensive rat (SHR). We observed that periarterial nerve stimulation of NE release was greater in the mesenteric arterial bed of 8- to 10-week-old SHRs compared with age-matched Wistar-Kyoto (WKY) or in DOCA salt hypertensive rats and their control. The ability of NPY to inhibit the evoked release of NE from the perfused mesenteric arterial bed of 8- to 10-week-old SHRs was attenuated compared with WKY rats while the postjunctional potentiation of the contractile effect induced by nerve stimulation or the exogenous administration of NE or ATP was enhanced. These results are consistent with alterations in the normal function of NPY playing a role in the development of hypertension in the SHR.

We next utilized the NGF-differentiated PC12 cell to study the modulation and mechanism for the NPY-induced inhibition of both catecholamine (CA) and NPY release, as well as CA synthesis. Differentiation of PC12 cells with NGF induces neurite outgrowth and expression of a sympathetic neuronal phenotype, which synthesizes, stores, and releases CA, NPY, and ATP. The PC12 cell expresses receptors for all three mediators and contains a variety of ion channels, second-messenger systems, and G-proteins. It was seen that the Y_2 selective agonist NPY13-36 and PYY13-36 produced a concentration-dependent inhibition of CA or NPYir release, respectively. The inhibition of release by the NPY analogues was blocked by pretreatment of the cells for 18 hr with pertussis toxin, suggesting the involvement of a guanosine triphosphate (GTP)-binding protein of the G_i or G_o subtype. The K^+-evoked release of CA and NPYir was also modulated by both ATP and adenosine. Changes in intracellular Ca^{2+} $[Ca^{2+}]_i$ was monitored by measuring Fura2-fluorescence emitted by a single cell in a monolayer in response to pulses of K^+. It was seen that the Y_2 agonist NPY13-36 produced a concentration-dependent attenuation of the K^+ evoked increase in $[Ca^{2+}]_i$ by decreasing the influx of extracellular Ca^{2+}. It was observed that both the L-type Ca^{2+} channel antagonist, nifedipine, and the N-type Ca^{2+} channel antagonist, ω-conotoxin GVIA attenuated the K^+-evoked release of CA and NPYir. However, it was observed that the Y_2 agonist could not produce a further inhibitory effect on CA or NPYir release in the presence of a maximum concentration of ω-conotoxin GVIA. All of these results are consistent with the Y_2-mediated inhibitory effects on transmitter release being due to inhibition of Ca^{2+} influx through voltage-gated Ca^{2+} channels. This was further investigated using an electrophysiological approach (discussion follows).

Because it is known that neurotransmitter synthesis as well as release can be modulated by auto- and heteroreceptors, we investigated the effects of NPY on basal and depolarization-stimulated catecholamine (CA) synthesis utilizing

NGF-differentiated PC12 cells (4). It was observed that NPY-inhibited depolarization stimulated CA synthesis as determined by *in situ* measurement of DOPA production in the presence of the decarboxylase inhibitor m-hydroxybenzylhydrazine (NSD-1015). The inhibition by NPY was concentration-dependent and was prevented by pretreatment with pertussis toxin in a similar way as for the inhibition of K^+-induced release of CA from these cells. This suggests the involvement of a GTP-binding protein of the G_i or G_o subtype. The NPY analogue [$Leu^{31}Pro^{34}$]NPY also caused inhibition of DOPA production but was less potent than NPY itself, while PYY and NPY13-36 had no significant effect. This pattern is most consistent with the involvement of the NPY-Y_3 receptor subtype.

We next carried out studies utilizing multiple selective Ca^{2+} channel and protein kinase agonists and antagonists to elucidate the mechanisms by which NPY modulates CA synthesis (5). The L-type Ca^{2+} channel blocker, nifedipine, inhibited the depolarization-induced stimulation of DOPA production by approximately 90% and attenuated the inhibitory effect of NPY. In contrast, the N-type Ca^{2+} channel blocker, ω-conotoxin GVIA, inhibited neither the stimulation of DOPA production nor the effect of NPY. Antagonism of Ca^{2+}/calmodulin-dependent protein kinase (CaM kinase) greatly inhibited the stimulation of DOPA production by depolarization and prevented the inhibitory effect of NPY, whereas alterations in the cyclic adenosine monophosphate (cAMP)–dependent protein kinase pathway modulated DOPA production but did not prevent the inhibitory effect of NPY. Stimulation of Ca^{2+}/phospholipid-dependent protein kinase (PKC) with phorbol-12-myristate 13-acetate (PMA) did not affect the basal rate of DOPA production in NGF-differentiated PC12 cells but did produce a concentration-dependent inhibition of depolarization-stimulated DOPA production. In addition, NPY did not produce further inhibition of DOPA production in the presence of PMA, and the inhibition by both PMA and NPY was attenuated by the specific PKC inhibitor chelerythrine. These results indicate that NPY inhibits Ca^{2+} influx through L-type voltage-gated channels, possibly through a PKC-mediated pathway, resulting in attenuation of the activation of CaM kinase and inhibition of depolarization-stimulated CA synthesis.

Other studies using whole-cell patch-clamp recording examined the effects of NPY on Ba^{2+} currents in the NGF-differentiated PC12 cells in order to demonstrate that activation of the different receptor subtypes results in differential Ca^{2+} channel modulation. NPY was found to inhibit the voltage-activated Ba^{2+} current in a reversible fashion with an EC_{50} of 13 nM. Experiments using selective NPY analogues revealed that most of this inhibition was mediated by the Y_2 and Y_3 receptor subtypes. It was observed that the Y_2 effect was confined to inhibition of N-type Ca^{2+} channels. NPY also inhibited L-type Ca^{2+} channels in these cells, as demonstrated both by occlusion of the effects of nifedipine and by reduction of BAY K 8644-enhanced tail currents. The inhibition of L-channels was prevented by the PKC inhibitors chelerythrine and PKC 19-36 and was mimicked by the PKC activator PMA. The results of these studies indicate that in NGF-differentiated PC12 cells, NPY acts via Y_2 receptors to inhibit N-type Ca^{2+} channels concomitant with CA and NPYir release and via

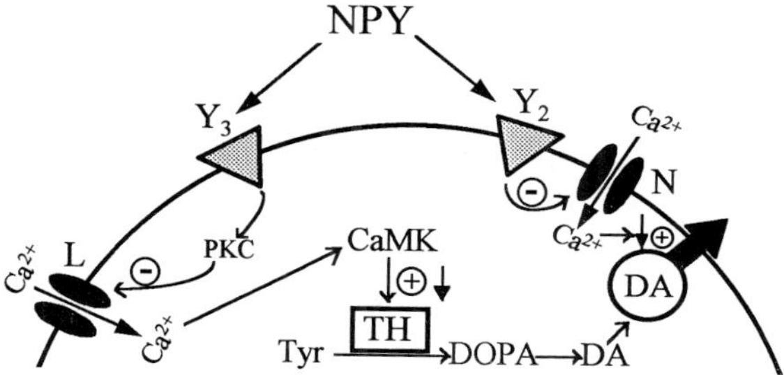

FIGURE 1 The proposed mechanism of differentiated regulation of catecholamine synthesis and release induced by NPY in NGF-differentiated PC12 cells. NPY, neuropeptide Y; DA, dopamine; TH, tyrosine hydroxylase; CaMK, calcium/calmodulin-dependent protein kinase; L, L-type Ca^{2+} channel; N, N-type Ca^{2+} channel. (Reproduced with permission from *Journal of Neurochemistry* 67:1090–1099, 1996.)

Y_3 receptors and a PKC-dependent pathway to inhibit L-type Ca^{2+} channels and CA synthesis.

In conclusion, NPY is a cotransmitter with CA and ATP in sympathetic neurons and contributes to the contractile effect of vascular smooth muscle following sympathetic nerve stimulation. NPY modulates the release of all three cotransmitters from sympathetic nerves via an action on prejunctional Y_2 receptors and potentiates the postjunctional contractile effect of nerve stimulation and vasoactive agents via an action on Y_1 receptors. In addition to its modulation of CA release, NPY also inhibits the depolarization-induced increase in CA synthesis via an action on Y_3 receptors. The Y_2-induced inhibition of CA and NPY release is mediated by an inhibition of Ca^{2+} influx through N-type Ca^{2+} channels. Finally the Y_3-induced inhibition of CA synthesis is mediated by an inhibition of L-type Ca^{2+} channels and a PKC-dependent pathway. Figure 1 represents a schema summarizing the idea that NPY-induced inhibition of CA synthesis and release are mediated through these distinct receptors and mechanisms. The observation that transmitter synthesis and release can be differentially modulated by NPY implies an additional level of control in the presynaptic regulation of sympathetic neurotransmission.

Acknowledgments

This work was supported by grants HL-26319, HL 35202, and 5-T32-GM-08306 from the National Institutes of Health and the American Heart Association–Missouri Affiliate.

References

1. Morris J. L., and Bibbins, I. C. (1992). *In* Autonomic Neuroeffector Mechanism. (J. Burnstock and C. H. V. Hoyle, eds.), pp. 33–119. Harwood Academic Publishers, Reading, U.K.
2. Chen, X., and Westfall, T. C. (1994). Modulation of intracellular calcium transients and dopamine release by neuropeptide Y in PC-12 cells. *Am. J. Physiol.* **266,** C784–C793.

3. Westfall, T. C., Carpentier, S., Chen, X., Beinfeld, M. C., Naes, L., and Meldrum, M. J. (1987). Prejunctional and postjunctional effects of neuropeptide Y at the noradrenergic neuroeffector junction of the perfused mesenteric arterial bed of the rat. *J. Cardiovasc. Pharmacol.* **10**, 716–722.
4. McCullough, L. A., and Westfall, T. C. (1995). Neuropeptide Y inhibits depolarization-stimulated catecholamine synthesis in rat pheochromocytoma cells. *Eur. J. Pharmacol.* **287**, 271–277.
5. McCullough, L. A., and Westfall, T. C. (1996). Mechanism of catecholamine synthesis inhibition by neuropeptide Y: Role of Ca^{2+} channels and protein kinases. *J. Neurochem.* **67**, 1090–1099.

David G. Nicholls

Neurosciences Institute
Department of Pharmacology
Ninewells Medical School
University of Dundee
Dundee DD1 9SY, United Kingdom

Strategies for Receptor Control of Neurotransmitter Release

The arrival of an action potential at a central nervous system varicosity does not automatically result in exocytosis. Indeed, both glutamatergic and catecholaminergic varicosities can display a quantal release probability of considerably less than unity. Because an action potential must pass through a varicosity to continue its propagation, this implies that there is an inherent unreliability in the coupling between the action potential and exocytosis. The high-power dependency of exocytosis on external Ca^{2+} concentration suggests that Ca^{2+} entry may be the prime probability-determining step in action potential–exocytosis coupling. Presynaptic receptors may either facilitate or inhibit release, and the evidence, obtained from studies with isolated nerve terminals (synaptosomes), will now be reviewed that the two classes of receptor may modulate exocytosis by respectively enhancing presynaptic action potentials and inhibiting Ca^{2+} channels. Experiments in our laboratory have focused on glutamate as a transmitter; however, it is evident that parallel interlocking signal transduction pathways may control the release of catecholamine neurotransmitters.

Transmitter exocytosis from synaptosomes may be evoked either by a "clamped" depolarization following KCl addition, or during spontaneous tetrodotoxin-sensitive activity initiated by the addition of dendrotoxin I (Dtx) or

Advances in Pharmacology, Volume 42

1054-3589/98 $25.00

4-aminopyridine (4AP) (1). These agents inhibit a presynaptic K^+ channel, which is responsible for stabilizing the membrane potential and preventing such firing. As long as precautions are taken to remove any tonic inhibition, for example, from leaked adenosine, protein kinase C (PKC) activity does not influence the rate or extent of KCl-evoked glutamate exocytosis. In contrast, 4AP-induced glutamate exocytosis is totally under the control of PKC: Addition of phorbol ester immediately prior to 4AP enhances glutamate exocytosis up to fivefold, while addition of the PKC inhibitor Ro 31-8220 virtually abolishes 4AP-induced release (2). Because the KCl-evoked results demonstrate that the phosphorylation state of PKC substrates does not influence exocytosis per se, some upstream locus must be sought. Fluorescent monitoring of the time- and population-average membrane potential reveals that phorbol esters greatly potentiate the 4AP-evoked depolarization of the synaptosomes, indicating that the action potentials are being prolonged or facilitated by the kinase (2). Dendrotoxin, which has been shown in electrophysiological studies at cell somata to inhibit A-type K^+ channels, acts like 4AP, while the phorbol ester–mediated facilitation of release can be mimicked by Ba^{2+} or clofilium. It is, therefore, necessary to conclude that glutamatergic nerve terminals possess, in addition to the constitutively open K^+ channel responsible for the resting polarized membrane potential, two further K^+ channels: one sensitive to Dtx and 4AP but not PKC that provides a negative feedback to prevent spontaneous firing of the terminals, and one that is insensitive to 4AP and Dtx but is inhibited by PKC, Ba^{2+}, or clofilium.

More physiologically relevant activation of the release-regulating PKC can be achieved by addition of the group 1 metabotropic glutamate receptor (mGluR) agonist 1S,3R-ACPD (3). The agonist generates a transient pulse of diacylglycerol (DAG), which decays within 30 sec. The combination of 1S,3R-ACPD and arachidonic acid (AA) precisely mimics the acute effects of phorbol esters on population depolarization, Ca^{2+} elevation, and glutamate exocytosis in the synaptosomal preparation. However, the receptor undergoes rapid homologous desensitization, which underlies the short duration of the DAG production (4). The desensitization is mediated by PKC and is delayed in the presence of the PKC inhibitor, staurosporine. The extreme sensitivity of the facilitatory receptor to desensitization may explain the difficulty in observing facilitatory effects of 1S,3R-ACPD in brain slice preparations; in addition, 1S,3R-ACPD alone is unable to enhance glutamate release unless low concentrations of AA are also present. AA does not influence DAG formation by 1S,3R-ACPD but does allow the agonist to enhance the phosphorylation state of two major presynaptic PKC substrates, MARCKS and GAP-43. It is concluded, therefore, that the locus of AA action is at the kinase itself (2), in accordance with data obtained with purified PKC isoforms demonstrating asynergistic activation of the kinase by DAG and AA. The potent regulation of 4AP-evoked glutamate exocytosis by this PKC-mediated pathway indicates that virtually every glutamatergic varicosity in the rat cerebral cortex must possess this facilitatory receptor.

The presence of inhibitory presynaptic receptors on glutamatergic terminals is well attested. Adenosine A1 agonists reduce synaptosomal glutamate exocytosis by 50–60%; significantly, they are equally effective against release evoked by KCl depolarization and by 4AP; thus, some mechanism other than action potential modulation must underlie this regulation. Indeed, no modulation of membrane potential can be detected on A1 agonist addition to polarized, KCl-depolarized, or 4AP-depolarized synaptosomes. The possibility of K^+ channel activation can, therefore, be eliminated. In contrast, the elevation in bulk cytoplasmic free Ca^{2+}, $[Ca^{2+}]_c$, is inhibited, supporting the consensus that such inhibitors act presynaptically by inhibiting release-coupled Ca^{2+} channels (5).

Synaptosomes prepared from young (3 wk) rats possess an additional inhibitory receptor, sensitive to L-AP4 and with a pharmacology consistent with a group 3 mGluR (4). This inhibitory mGluR coexists with the adenosine receptor and acts by the same inhibitory mechanism—in particular, the two receptor-mediated inhibitions are mutually occlusive.

The extensive regulation of release obtained with both the inhibitory and facilitatory receptors makes it highly likely that the two opposing mechanisms coexist on the same varicosities. It was, therefore, of importance to establish whether one was dominant under conditions where both were activated. Initial studies showed that adenosine A1 agonists failed to inhibit when release was facilitated in the presence of 4AP and phorbol ester. This was more than a mere "swamping" of the inhibitory response, because the adenosine inhibition

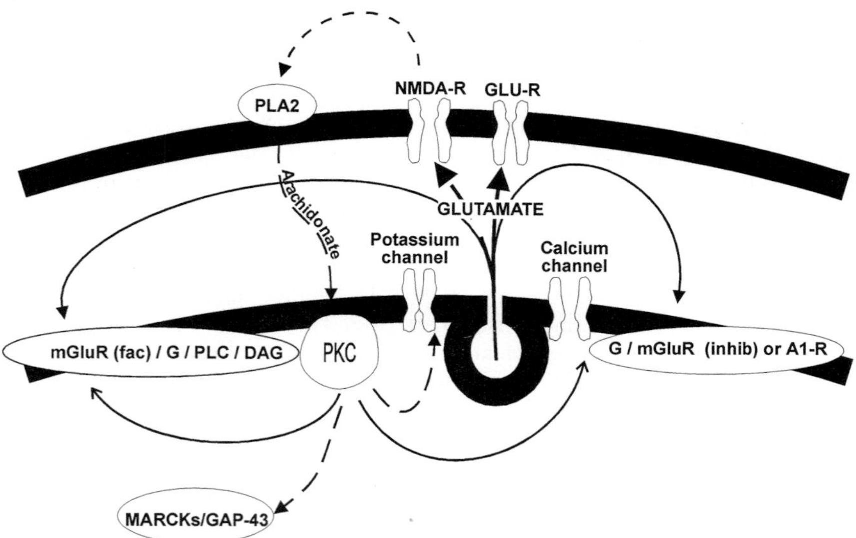

FIGURE 1 Interactions between presynaptic facilitatory and inhibitory receptors in glutamatergic terminals. The facilitatory mGluR is linked via phospholipase C to protein kinase C. The inhibitory mGluR and adenosine A1 receptors are coupled via a G-protein to the release-coupled Ca^{2+} channel. In the absence of AA, receptor-mediated PKC activation causes homologous desensitization of the facilitatory receptor and transiently suppresses the inhibitory pathway. In the presence of AA, perhaps generated postsynaptically, persistent phosphorylation of PKC substrates (GAP-43, MARCKS) occurs, together with a facilitation of glutamate release mediated by action potential enhancement.

of KCl-evoked release was also suppressed by the phorbol ester, even though no facilitation is observed in these conditions. Thus, PKC is acting at a locus distinct from that at which facilitation occurs. The interaction of the adenosine receptor with its G-protein is unaffected by PKC activation, because agonist-enhanced GTPγS binding to membrane preparations is still observed. No phosphorylation of G-proteins can be observed, and by elimination, the most likely locus for PKC-mediated suppression is at the site on the Ca^{2+} channel that governs G-protein interaction, although such phosphorylation would have to occur without altering the normal channel kinetics.

This suppression is not merely an artifact of prolonged phorbol ester exposure because it can be reproduced by 1S,3R-ACPD activation of PKC, even in the absence of AA. Such a differential requirement for AA for these two PKC-mediated pathways allows the role of the fatty acid to be investigated. In the absence of AA, suppression is as transient as the DAG peak, reversing within 60 sec; however, when 2 μM of AA is present, 1S,3R-ACPD-mediated suppression persists for at least 10 min. AA thus locks DAG-activated PKC in a constitutively active state, even after DAG has returned to basal (5).

These pathways are summarized in Figure 1. The physiological significance of the coordinated facilitation of release, suppression of inhibition, and GAP-43/MARCKS phosphorylation following PKC activation may be to maximize release into the synaptic cleft during the induction of synaptic plasticity to ensure that glutamate-evoked changes continue to completion, with the AA originating postsynaptically via phospholipase A_2 activation (5).

References

1. Tibbs, G. R., Barrie, A. P., Van-Mieghem, F., *et al.* (1989). Repetitive action potentials in isolated nerve terminals in the presence of 4-aminopyridine: Effects on cytosolic free Ca^{2+} and glutamate release. *J. Neurochem.* **53**, 1693–1699.
2. Coffey, E. T., Herrero, I., Sihra, T. S., *et al.* (1994). Glutamate exocytosis and MARKS phosphorylation are enhanced by a metabotropic glutamate receptor coupled to a protein kinase C synergistically activated by diacylglycerol and arachidonic acid. *J. Neurochem.* **63**, 1303–1310.
3. Herrero, I., Miras-Portugal, M. T., and Sánchez-Prieto, J. (1992). Positive feedback of glutamate exocytosis by metabotropic presynaptic receptor stimulation. *Nature* **360**, 163–166.
4. Vázquez, E., Budd, D., Herrero, I., *et al.* (1995). Co-existence and interaction between facilitatory and inhibitory metabotropic glutamate receptors and the inhibitory adenosine A1 receptor in cerebrocortical nerve terminals. *Neuropharmacology* **34**, 919–927.
5. Sánchez-Prieto, J., Budd, D. C., Herrero, I., *et al.* Presynaptic receptors and the control of glutamate exocytosis. *Trends Neurosci.* **19**, 235–239.

Lennart Stjärne

Department of Physiology and Pharmacology
Karolinska Institutet
S-17177 Stockholm, Sweden

Pattern of Adenosine Triphosphate and Norepinephrine Release and Clearance: Consequences for Neurotransmission

The biochemistry and pharmacology of sympathetic neurotransmission are relatively well understood. This chapter discusses its microphysiology (i.e., neurotransmission at the varicosity level). First, some general neurobiological principles for synaptic transmission are described (1, 2). Then, these insights are applied to sympathetic neuromuscular transmission mediated by adenosine triphosphate (ATP) and norepinephrine (NE) in a frequently employed vascular model, isolated rat tail artery (RTA).

I. General Principles

Nerve impulses release neurotransmitters in multimolecular packets of preset size, probably the contents of single vesicles (quanta). At the exit point, the transmitter concentration equals that in the vesicle, about 100 mM. This is the pressure head that drives diffusion of the released transmitter first to, then away from its various targets: intra-peri, or extrasynaptic receptors; reuptake transporters; degrading enzyme molecules; and so on (1). The peak concentration and time course of clearance at these targets help determine the fractional occupancy and degree of desensitization of the receptors and, ultimately, the amplitude and time course of the effector responses. The probability that the effector response will be "distorted" (e.g., by saturation of reuptake transporters or desensitization of postsynaptic receptors) is minimal when quanta are released from random sites, and maximal when they are repeatedly released from the same site, with short intervals (Fig. 1).

These principles apply as well when exocytosis itself is viewed as the nerve impulse–induced "response." No current method measures quantitatively the per-pulse transmitter exocytosis per se. What is measured is (e.g., electrical, electrochemical, or mechanical) signals (Fig. 1) that, at best, reflect the amplitude and time course of the rise and fall in the concentration of a released

Advances in Pharmacology, Volume 42

1054-3589/98 $25.00

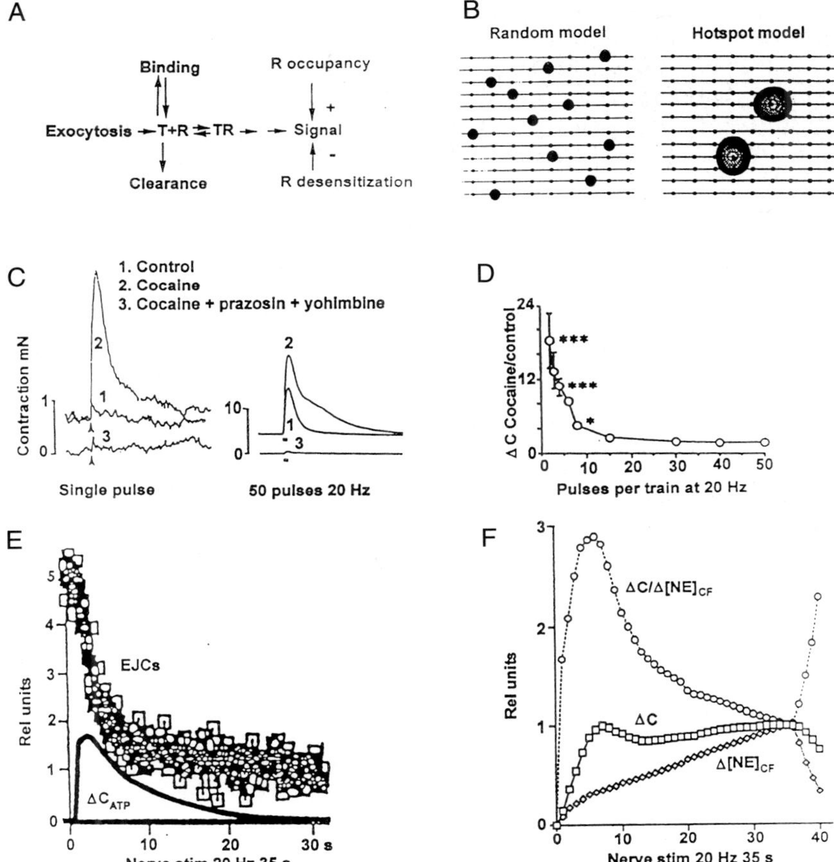

FIGURE 1 (*A*) No current method measures transmitter exocytosis per se. What is measured is signals (in this chapter, the EJC, $\Delta[\text{NE}]_{\text{CF}}$, and ATP- and NE-evoked contractile responses), which at best represent the concentration of the released transmitter at the receptors (i.e., exocytosis minus clearance). Even this does not hold true if the released transmitter saturates reuptake transporters or desensitizes the receptors. T, transmitter (ATP or NE); R, ATP (P2x)- and NE (CF electrode and or α)-receptors. (*B*) Two extreme, mutually exclusive models of the spatial pattern of release of single sympathetic transmitter quanta from >108 varicosities by a train of 11 nerve impulses. In both, the probability of monoquantal release of ATP and NE (p) from the average varicosity is .01. In the random model p in individual varicosities is uniform, in the hotspot model extremely nonuniform. For further explanations, see the text. (*C*) Contractile responses to a single pulse or to a train of 50 pulses at 20 Hz, in the absence of drug (1), after block of NE reuptake (2), and after further block of the NE-mediated component of the contraction to reveal the residual ATP-mediated contraction (3). (*D*) The ratio between the contractile response (ΔC) in the presence and absence of cocaine during stimulation with 1–50 pulses at 20 Hz. (*E*) Peak amplitude of EJCs (n = 6) and time course of the ATP-mediated contraction (Δ_{ATP}) during 700 pulses at 20 Hz. (*F*) Time course of $\Delta[\text{NE}]_{\text{CF}}$, of the (mainly) NE-mediated contraction (ΔC) and of the ratio $\Delta\text{C}/\Delta[\text{NE}]_{\text{CF}}$ during and after nerve stimulation with 700 pulses at 20 Hz. (*C, D* reprinted with permission from *Neuroscience* 60:1021, 1994. *E* reprinted with permission from *Acta Physiol. Scand.* 149:501–517, 1993; and from *Prog. Neurobiol.* 47:45–94, 1995.)

transmitter at its (biological or artificial) receptor (i.e., exocytosis minus clearance) (1, 4). Not even this is true if the kinetics of the signals are slower than those of the examined events, or when clearance is slowed by saturation of reuptake transporters, or if the receptors are desensitized by a high and prolonged receptor occupancy (1, 2). The error in estimates of exocytosis introduced by these factors depends on the pattern of quantal release; it is minimal in the random model and maximal in the hotspot model (Fig. 1B). An additional problem is that nerve impulses induce both quantal exocytosis and (an often much larger) molecular leakage of some substances (e.g., ATP), which occur both as transmitter and nontransmitter (4). To quantitatively estimate per-pulse transmitter exocytosis, one needs, therefore, a signal that is specific, sensitive, and linear to concentration; distinguishes between transmitter exocytosis and molecular leakage of the same substance; has sufficiently fast kinetics; and is mediated by receptors that are not desensitized by the released transmitter.

II. Sympathetic Neurotransmission in Rat Tail Artery

The discussion will be based on experiments in RTA under two extreme conditions: stimulation with single pulses or with high-frequency trains, that is, when distortion of the relationship between exocytosis and effector responses is minimal or maximal, respectively.

A. The Preparation

In RTA, sympathetic nerve terminals form a dense two-dimensional network at the adventitial surface of the tunica media (12–15 layers of smooth-muscle cells). Most of the varicosities form junctional contacts with muscle cells. The main electrical and mechanical transmitters are ATP and NE, respectively. The postjunctional effects of ATP are mediated via P2x-purinoceptors, those of NE via α_2 and α_1-adrenoceptors. As in all other preparations examined in this regard, each pulse in trains of 5–150 pulses at 1–20 Hz increases the fractional overflow of tritium-labeled NE by approximately 1/50,000 of the NE content of the tissue. This corresponds to exocytosis of the NE content of a single, small, dense-cored vesicle from 3% or less of the varicosities. Analysis by extracellular recording of the nerve terminal spike and spontaneous and action potential-evoked excitatory junction currents (EJCs) indicates that each nerve impulse probably releases a single ATP quantum from 3% or less of the varicosities. The intermittency of per-pulse (NE and ?) ATP exocytosis is not due to failure of the nerve impulse to invade the terminals but to a low probability of excitation–secretion coupling in most varicosities (4).

B. Methods

The experiments to be described here were made in preparations from adult male Sprague-Dawley rats (200–300 g). When the aim was to study ATP and NE release, the nerves in the proximal end of 2- to 3-cm lengths of the artery were stimulated via a suction electrode. The spontaneous EJCs recorded by extracellular electrodes (diameter, 100–150 μm) were used to detect exocytosis of single ATP quanta, and evoked EJCs were used as measures of the number of ATP quanta released per pulse. The electrochemical NE oxidation current at a carbon fiber (CF) electrode (diameter, 8 μm; length, 50–100 μm) was used to measure online the concentration of released NE at the CF electrode; the signal was termed $\Delta[NE]_{CF}$. The contractile responses of ring preparations to electrical field stimulation or to bath-applied agonists were recorded isometrically (4).

Advantages of EJCs are that they ignore nontransmitter ATP and selectively detect quanta of transmitter ATP, that their high spatial resolution allows them to be used to detect exocytosis of single ATP quanta from individual sites, and that their fast time course enables them to be used to monitor the per-pulse release of quanta at frequencies up to 20 Hz. A disadvantage is that P2x-purinoceptors in RTA become desensitized during prolonged exposure to high concentrations of ATP (3). A question arising is whether a decline in the EJC amplitude during a high-frequency train is due to reduced per-pulse exocytosis of ATP or to desensitization of P2x-receptors (4). Advantages of the electrochemical method are that the receptor, the CF electode, is not desensitized by NE and that the kinetics of the $\Delta[NE]_{CF}$ response itself (to a jet of exogenous NE) are faster than of $\Delta[NE]_{CF}$ caused by the nerve stimulation (4). The time course of the $\Delta[NE]_{CF}$ response thus probably reflects the rise and fall in the concentration of released NE at the electrode. It should be noted, however, that the slow time course of $\Delta[NE]_{CF}$ precludes its use to analyze per-pulse NE release at frequencies above 0.2 Hz.

C. Localization of Activated Postjunctional Receptors

The ATP quanta released by a single pulse trigger two P2x-purinoceptor–mediated responses: a fast EJC (time to peak, < 10 ms; duration, 50–100 ms) and a usually small contraction (time to peak, < 0.8 sec; total duration, about 2 sec). High concentrations (300–500 μM) of the competitive P2x antagonist, suramin, are needed to fully block these responses. The ATP concentration required to activate the receptors may thus be ≥ 300 μM. The activated low-affinity P2x-receptors are, therefore, likely to be intrajunctional. The time required for diffusion and/or enzymatic inactivation to reduce the local ATP level to below this threshold may be ≤ 1 ms (1). The time course of the EJCs may thus reflect the "off" rate of ATP from the receptor complex rather than the biologically effective lifetime of released ATP, as commonly assumed. The NE quanta released by a single pulse cause $\Delta[NE]_{CF}$ to peak in ≤ 200 ms, return to baseline in 1.5–4,5 sec, and trigger a small α_2-adrenoceptor–mediated depolarization (time to peak, >15 sec; duration, ≤ 40 sec) and two contractions, one via α_1-adrenoceptors (time to peak, 3–4 sec; total duration, 15 sec) and one (time to peak, 7–9 sec; total duration, 30 sec) via α_2-adrenoceptors. These responses are blocked by low concentrations (≤ 0.1 μM) of competitive α-antagonists. This plus the fact that $\Delta[NE]_{CF}$ does not exceed 0.5 μM NE, even during high-frequency trains, suggests that, similarly to the CF electrode, the

α-receptors driving the smooth-muscle responses may be located at a distance from the release sites (i.e. peri- or extrajunctionally) (4).

D. Roles of NE Clearance Vary with the Stimulation Parameters

The increase in the amplitude and duration of the NE-mediated contractions caused by addition of cocaine to block neuronal reuptake of NE varies with the length and frequency of stimulus trains. As shown in Fig. 1C, cocaine increased the small contractile response to a single pulse (0.1% of the neurogenic maximum) by 20-fold or more and the large response to 50 pulses at 20 Hz by twofold or less. Figure 1D shows that the enhancing effect of cocaine was initially inversely related to the train length, but all responses to more than 15 pulses were increased by about twofold. The small size of the control contraction caused by a single pulse (i.e., release of NE quanta from 1% of the varicosities) is, thus, not due to too little NE release but to "instant" reuptake of ($\geq$95% of?) the released NE, probably into the releasing varicosities. That the effect of cocaine declined with increasing train length to reach a constant level at 15 pulses or more suggests that rapidly repeated release saturates the NE reuptake transporters near, but not in the surround of active sites. The roles of NE clearance, thus, vary strongly with the impulse parameters (4).

F. Depression of Release or Desensitization of Receptors?

As shown in Figure 1E, the amplitude of the first few EJCs increased during trains of 700 pulses at 20 Hz, but later EJCs declined in size to near the noise level, and the ATP-mediated contractile response was also transient (4). The results imply that P2x-receptor channels ceased to open during ongoing stimulation. The question whether this was due to depression of ATP release (due to depletion of a small releasable pool) or to desensitization of P2x-receptors cannot be answered conclusively at present, but both alternatives require release according to the hotspot model. The finding that the electrical and contractile responses to exogenous ATP were the same before, during, and after 20-Hz trains shows that most P2x-receptors were intact, but it does not fully exclude the extremely local desensitization of P2x-receptors predicted by this model (4).

Similar analysis of the NE-mediated neurotransmission during 700 pulses at 20 Hz is shown in Figure 1F. It can be seen that $\Delta[\mathrm{NE}]_{\mathrm{CF}}$ grew progressively throughout the stimulation, that the NE-mediated contraction was biphasic, and that the contractile efficacy of released NE (as reflected in the ratio $\Delta\mathrm{C}/\Delta[\mathrm{NE}]_{\mathrm{CF}}\Delta\mathrm{C}$) during and after the stimulus train was triphasic (4). It is important to keep in mind that the $\Delta[\mathrm{NE}]_{\mathrm{CF}}$ and contractile responses to fixed concentrations of exogenous NE are essentially constant (4). The decline in the ratio $\Delta\mathrm{C}/\Delta[\mathrm{NE}]_{\mathrm{CF}}\Delta\mathrm{C}$ during, and the rebound after, nerve stimulation is, therefore, likely to be due to reversible desensitization restricted to the α-adrenoceptors near the NE hotspots that drive the contraction.

III. Conclusions

Both the parameters used as measures of the amount of the sympathetic cotransmitters ATP and NE released by nerve impulses, and the smooth muscle contractile responses depend in part on the pattern of exocytosis, and in part on postsecretory factors. The kinetics of the contractions are determined by those of transmitter–receptor complexes and/or postreceptor mechanisms. That RTA, which "normally" virtually ignores single pulses, contracts vigorously to high-frequency train is not due to facilitation of per-pulse release but probably to increasing saturation of local NE reuptake. The profound decline in the amplitude of the EJCs and the ATP-mediated contraction during high-frequency trains implies that P2x-receptor channels cease to open during ongoing stimulation; the effect may be due to depression of per-pulse ATP release, but it could be due, alternatively, to desensitization of P2x-receptors. Similarly, the decline in the contractile efficacy of released NE during, and the rebound after, high-frequency trains may be due to reversible desensitization of the α-adrenoceptors driving the contraction. All of these results seem to require a hotspot model of quantal release of ATP and NE.

Acknowledgment

This work was supported by the Swedish Medical Research Council (project B96-14X-03027-27A) and Karolinska Institutets Fonder. I thank the members of the group (in alphabetical order) Jian-Xin Bao, François Gonon, Mussie Msghina, and Eivor Stjärne.

References

1. Clements, J. D. (1996). Transmitter timecourse in the synaptic cleft: Its role in central synaptic function. *Trends Neurosci.* **19,** 163–171.
2. Jones, M. V., and Westbrook, G. L. (1996). The impact of receptor desensitization on fast synaptic transmission. *Trends Neurosci.* **19,** 96–101.
3. Evans, R. J., and Kennedy, C. (1994). Characterization of P2-purinoceptors in the smooth muscle of the rat tail artery: A comparison between contractile and electrophysiological responses. *Br. J. Pharmacol.* **113,** 853–860.
4. Stjärne, L., and Stjärne, E. (1995). Geometry, kinetics and plasticity of release and clearance of ATP and noradrenaline as sympathetic cotransmitters: Roles for the neurogenic contraction. *Prog. Neurobiol.* **47,** 45–94.

Ivar von Kügelgen, Jorge Gonçalves, Bernd Driessen, and Klaus Starke

Pharmakologisches Institut
Universität Freiburg
D-79104 Freiburg, Germany

Corelease of Noradrenaline and Adenosine Triphosphate from Sympathetic Neurones

I. Noradrenaline and Adenosine Triphosphate Corelease

Neurones in the peripheral and central nervous system use adenosine triphosphate (ATP) as an extracellular signaling molecule. The probably best-studied example for this transmitter role of ATP is noradrenaline–ATP cotransmission. The vesicles of the postganglionic sympathetic axons store noradrenaline and ATP. In many sympathetically innervated tissues, both noradrenaline and ATP are involved in the postjunctional responses to nerve action potentials (1–3). In agreement with this cotransmitter role of ATP, electrical stimulation of axons and depolarization by high K^+ concentrations elicit the release of ATP from sympathetically innervated tissues (1–3). This overflow constitutes one basic piece of evidence for the view that ATP is a transmitter in the postganglionic sympathetic nervous system. However, caution is necessary in the interpretation of some of these data, because in smooth-muscle tissues the overflow of ATP induced by sympathetic nerve stimulation originates only partly from nerve endings: To a large extent it comes from nonneural cells due to the postjunctional effects of the neurotransmitters (4–8). To overcome this problem, the postjunctional responses to released noradrenaline and ATP were blocked in some studies to abolish the nonneural release of ATP. In guinea pig vas deferens, the blockade of postjunctional responses by the α_1-adrenoceptor antagonist prazosin combined with the P2X-purinoceptor-antagonist suramin or the P2X-purinoceptor-desensitizing agent α,β-methylene-ATP greatly attenuated the evoked overflow of ATP; the remaining part of the overflow of ATP is likely to reflect neural release of ATP (6, 8, 9). Electrical stimulation or stimulation by high K^+ concentrations has also been shown to elicit an overflow of ATP from preparations with a low number of nonneural cells (e.g., from cultured chick postganglionic sympathetic neurones) (10). Figure 1 shows the overflow of [^{3}H]-noradrenaline and ATP from these cultured neurones induced by 100 electrical pulses applied at 10 Hz. Prazosin or suramin did not change the evoked overflow of ATP in agreement with a neural origin of ATP. The evoked overflow of ATP was abolished by tetrodotoxin as well as by omission

Advances in Pharmacology, Volume 42

1054-3589/98 $25.00

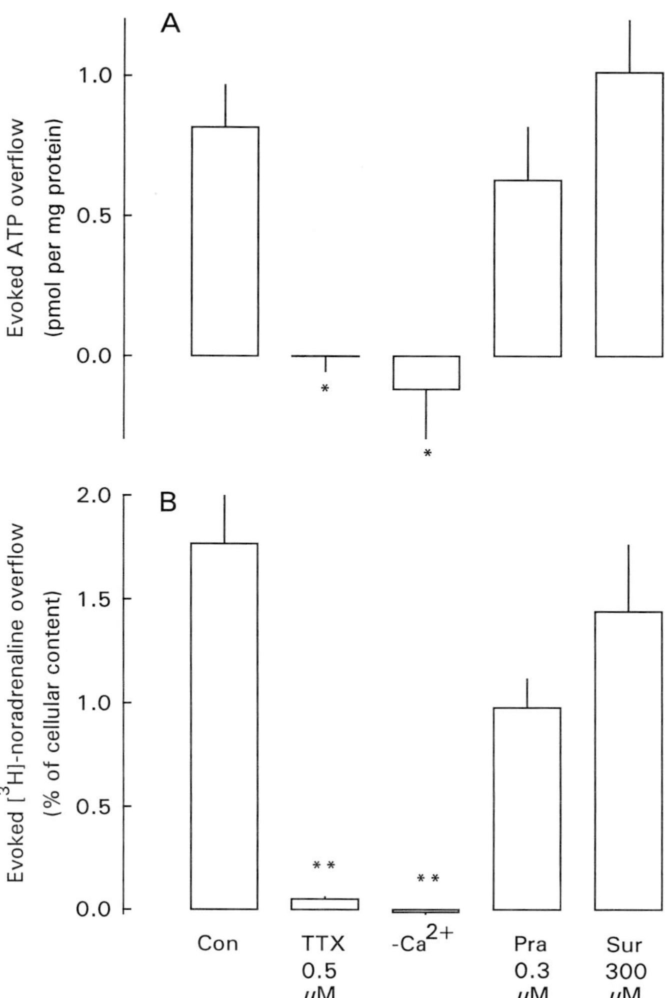

FIGURE 1 (*A*) Electrically evoked ATP and (*B*) tritium overflow from cultured chick sympathetic neurones and effects of tetrodotoxin (TTX), prazosin (Pra), suramin (Sur), and omission of calcium ($-Ca^{2+}$). Cultured chick sympathetic neurones were preincubated with [^{3}H]-noradrenaline and then superfused and stimulated electrically by 100 pulses applied at 10 Hz. Ordinates show the evoked overflow of ATP (luciferase technique; expressed as pmol/mg protein) and the evoked overflow of [^{3}H]-noradrenaline (expressed as a percentage of the cellular content). Drugs were added to the medium and calcium was omitted from the medium 24 min before electrical stimulation. Means ± SEM of six to 12 experiments. * and ** indicate significant differences from respective control (con; $p < .05$ and $p < .01$, respectively). Data from von Kügelgen *et al.* (10).

of calcium from the medium (Fig. 1), indicating that it was due to an action potential–evoked, Ca^{2+}-dependent exocytosis.

Transmitter release can also be elicited chemically by activation of presynaptic receptors. This is also true for the release of ATP from postganglionic sympathetic axons. In guinea pig vas deferens and in synaptosomes prepared from the myenteric plexus of the guinea pig ileum activation of presynaptic nicotinic and serotonin receptors induced not only the release of noradrenaline but also the release of ATP (6, 11, 12). Hence, the release of ATP from postganglionic sympathetic nerves resembles in several aspects the release of noradrenaline.

II. Modulation through Presynaptic Receptors

In addition to eliciting transmitter release (see preceding discussion), chemical signals can also modulate ongoing release from nerve terminals through activation of presynaptic receptors. This modulation through presynaptic receptors is a characteristic feature of the release of neurotransmitters (13). The first evidence for a presynaptic modulation of ATP release from sympathetic axons came from studies measuring effector responses. These studies showed, for example, that blockade of presynaptic α_2-adrenoceptors increased both the adrenergic and the purinergic components of stimulation-evoked responses, suggesting that the release of both cotransmitters is modulated by α_2-autoinhibition (2). Studies measuring the evoked overflow of ATP and noradrenaline from isolated tissues confirmed this view (14). In guinea pig vas deferens, α_2-adrenoceptor agonists such as UK-14304 decreased and the antagonist rauwolscine increased the evoked overflow of ATP and noradrenaline (8). In addition to α_2-adrenoceptors, a variety of presynaptic receptors that modulate the release of ATP from postganglionic sympathetic axons have now been identified; these receptors include β_2-adrenoceptors, adenosine A_1-receptors, prostaglandin E_2 receptors, as well as receptors for peptides such as neuropeptide Y and angiotensin II (14).

Postganglionic sympathetic axons also possess receptors for ATP itself. ATP and ATP analogues inhibit the release of noradrenaline by activation of presynaptic P2-purinoceptors in sympathetically innervated tissues, such as the mouse vas deferens (15), the rat iris (16), the rat heart atrium (17), and the rat tail artery (18). In these and several other peripheral tissues, the presynaptic P2-purinoceptors have also been shown to be activated by ATP released by nerve activity; the presynaptic P2-purinoceptors mediate a negative feedback in which cotransmitter ATP inhibits the subsequent release of noradrenaline (16–19). The effects of activation of presynaptic P2-purinoceptors on the release of ATP itself has not been studied directly by measuring the overflow of ATP (14), but the results of an electrophysiological study on guinea pig mesenteric arteries indicate a modulation of ATP release through presynaptic P2-purinoceptors (20).

III. Differential Control of Release

The modulation of ATP release by presynaptic receptors is in many cases quantitatively similar to the presynaptic modulation of noradrenaline release (14). However, there are also some examples for a differential modulation of the release of noradrenaline and ATP from the same tissue. Studies on guinea pig vas deferens indicated that the release of noradrenaline was affected by α_2-autoinhibition to a greater extent than the release of ATP (7, 8). Rauwolscine (10 μM), for example, increased the electrically evoked overflow of [^{3}H]-noradrenaline by 270% but the evoked overflow of ATP only by 100% (8). Similar results were obtained when the effects of blockade of presynaptic α_2-adrenoceptors on the evoked overflow of *endogenous* noradrenaline and *endogenous* ATP were compared in guinea pig vas deferens (21). Moreover, in synaptosomes prepared from the guinea pig myenteric plexus, clonidine reduced the evoked overflow of [^{3}H]-noradrenaline but not the evoked overflow of ATP (12). A striking example for differential modulation was recently observed in guinea pig vas deferens: activation of presynaptic β_2-adrenoceptors enhanced the overflow of [^{3}H]-noradrenaline but reduced the evoked overflow of ATP (22, 23). Adenosine A_1-receptors are a third example for differential modulation of cotransmitter release. In guinea pig vas deferens, adenosine and the A_1-selective analogue 2-chlorocyclopentyladenosine reduced the electrically evoked overflow of ATP more markedly than the overflow of [^{3}H]-noradrenaline (24), indicating that activation of adenosine A_1-receptors affects predominantly the release of ATP—the converse of the modulation through α_2-adrenoceptors (see preceding discussion). A change in the ratio of the evoked overflow of noradrenaline to the evoked overflow of ATP was also observed using different modes of stimulation (e.g., electrical stimulation vs stimulation by nicotine or different frequencies for applying electrical pulses), again indicating that the mechanisms controlling the release of noradrenaline differ from those controlling the release of ATP (6, 9).

The mechanisms of this differential control of cotransmitter release are yet unknown. The overflow of noradrenaline and ATP from guinea pig vas deferens was blocked by guanethidine or pretreatment of animals with 6-hydroxydopamine, indicating that the release of both was derived from postganglionic sympathetic axons (6, 25). There remain at least two possibilities. First, in the same axons there may be different populations of vesicles with a different cotransmitter mixture, and the probability of exocytosis from noradrenaline-rich and ATP-rich vesicles may be controlled through different mechanisms (e.g., distinct signal transduction systems); second, there may be different populations of axons, each population with a homogeneous population of vesicles, the axons differing in their vesicular noradrenaline/ATP ratio as well as in their control of the probability of exocytosis.

IV. Conclusion

In conclusion, postganglionic sympathetic axons release noradrenaline and ATP on nerve action potentials. The release of both cotransmitters is modulated through presynaptic receptors. In some cases, the mechanisms controlling the release of noradrenaline and ATP seem to differ. Postganglionic sympathetic axons possess, in addition to presynaptic α_2-adrenoceptors, presynaptic P2-purinoceptors mediating a negative feedback control by cotransmitter ATP of the subsequent release of noradrenaline.

References

1. Burnstock, G. (1990). Co-transmission. *Arch. Int. Pharmacodyn.* **304,** 7–33.
2. von Kügelgen, I., and Starke, K. (1991). Noradrenaline-ATP co-transmission in the sympathetic nervous system. *Trends Pharmacol. Sci.* **12,** 319–324.
3. Hoyle, C. H. V. (1992). Transmission: Purines. *In* Autonomic Neuroeffector Mechanisms. (G. Burnstock and C. H. V. Hoyle, eds.), pp. 367–407. Harwood, Chur.
4. Westfall, D. P., Sedaa, K., and Bjur, R. A. (1987). Release of endogenous ATP from rat caudal artery. *Blood Vessels* **24,** 125–127.
5. Sedaa, K. O., Bjur, R. A., Shinozuka, K., and Westfall, D. P. (1990). Nerve and drug-induced release of adenine nucleosides and nucleotides from rabbit aorta. *J. Pharmacol. Exp. Ther.* **252,** 1060–1067.
6. von Kügelgen, I., and Starke, K. (1991). Release of noradrenaline and ATP by electrical stimulation and nicotine in guinea-pig vas deferens. *Naunyn Schmiedebergs Arch. Pharmacol.* **344,** 419–429.
7. Sperlágh, B., and Vizi, E. S. (1992). Is the neuronal ATP release from guinea-pig vas deferens subject to α_2-adrenoceptor-mediated modulation? *Neuroscience* **51,** 203–209.
8. Driessen, B., von Kügelgen, I., and Starke, K. (1993). Neural ATP release and its α_2-adrenoceptor-mediated modulation in guinea-pig vas deferens. *Naunyn Schmiedebergs Arch. Pharmacol.* **348,** 358–366.
9. von Kügelgen, I., and Starke, K. (1994). Corelease of noradrenaline and ATP by brief pulse trains in guinea-pig vas deferens. *Naunyn Schmiedebergs Arch. Pharmacol.* **350,** 123–129.
10. von Kügelgen, I., Allgaier, C., Schobert, A., and Starke, K. (1994). Corelease of noradrenaline and ATP from cultured sympathetic neurons. *Neuroscience* **61,** 199–202.
11. Al-Humayyd, M., and White, T. D. (1985). 5-Hydroxytryptamine releases adenosine 5′-triphosphate from nerve varicosities isolated from myenteric plexus of guinea-pig ileum. *Br. J. Pharmacol.* **84,** 27–34.
12. Hammond, J. R., MacDonald, W. F., and White, T. D. (1988). Evoked secretion of [^{3}H]noradrenaline and ATP from nerve varicosities isolated from the myenteric plexus of the guinea pig ileum. *Can. J. Physiol. Pharmacol.* **66,** 369–375.
13. Starke, K., Göthert, M., and Kilbinger, H. (1989). Modulation of neurotransmitter release by presynaptic autoreceptors. *Physiol. Rev.* **69,** 864–989.
14. von Kügelgen, I. (1996). Modulation of neural ATP release through presynaptic receptors. *Semin. Neurosci.* **8,** 247–257.
15. von Kügelgen, I., Schöffel, E., and Starke, K. (1989). Inhibition by nucleotides acting at presynaptic P_2-receptors of sympathetic neuro-effector transmission in the mouse isolated vas deferens. *Naunyn Schmiedebergs Arch. Pharmacol.* **340,** 522–532.
16. Fuder, H., and Muth, U. (1993). ATP and endogenous agonists inhibit evoked [^{3}H]-noradrenaline release in rat iris via A_1 and P_{2Y}-like purinoceptors. *Naunyn Schmiedebergs Arch. Pharmacol.* **348,** 352–357.
17. von Kügelgen, I., Stoffel, D., and Starke, K. (1995). P_2-Purinoceptor-mediated inhibition of noradrenaline release in rat atria. *Br. J. Pharmacol.* **115,** 247–254.
18. Gonçalves, J., and Queiroz, G. (1996). Purinoceptor modulation of noradrenaline release in rat tail artery: Tonic modulation mediated by inhibitory P_{2Y}- and facilitatory A_{2A}-purinoceptors. *Br. J. Pharmacol.* **117,** 156–160.

19. von Kügelgen, I., Kurz, K., and Starke, K. (1993). Axon terminal P_2-purinoceptors in feedback control of sympathetic transmitter release. *Neuroscience* **56**, 263–267.
20. Fujioka, M., Cheung, D. W. (1987). Autoregulation of neuromuscular transmission in the guinea-pig saphenous artery. *Eur. J. Pharmacol.* **139**, 147–153.
21. Todorov, L., Bjur, R., and Westfall, D. P. (1995). Modulation of sympathetic cotransmission by endogenous ATP and NE. *FASEB J.* **9**, A371.
22. Driessen, B., Bültmann, R., Gonçalves, J., and Starke, K. (1996). Opposite modulation of noradrenaline and ATP release in guinea-pig vas deferens through prejunctional β-adrenoceptors: Evidence for the β_2 subtype. *Naunyn Schmiedebergs Arch. Pharmacol.* **353**, 564–571.
23. Gonçalves, J., Bültmann, R., and Driessen, B. (1996). Opposite modulation of cotransmitter release in guinea-pig vas deferens: Increase of noradrenaline and decrease of ATP release by activation of prejunctional β-adrenoceptors. *Naunyn Schmiedebergs Arch. Pharmacol.* **353**, 184–192.
24. Driessen, B., von Kügelgen, I., and Starke, K. (1994). P_1-Purinoceptor-mediated modulation of neural noradrenaline and ATP release in guinea-pig vas deferens. *Naunyn Schmiedebergs Arch. Pharmacol.* **350**, 42–48.
25. Kirkpatrick, K., and Burnstock, G. (1987). Sympathetic nerve-mediated release of ATP from the guinea-pig vas deferens is unaffected by reserpine. *Eur. J. Pharmacol.* **138**, 207–214.

Zofia Zukowska-Grojec

Department of Physiology and Biophysics
Georgetown University Medical Center
Washington, D.C. 20007

Neuropeptide Y: An Adrenergic Cotransmitter, Vasoconstrictor, and a Nerve-Derived Vascular Growth Factor

The sympathetic nerves possess, in addition to norepinephrine (NE), several nonadrenergic cotransmitters, the major ones being purines and neuropeptide Y (NPY). NPY is, in fact, the most abundant peptide in the central and the sympathoadrenomedullary nervous systems, and its presence and structure are highly conserved throughout the evolution; for example, human and shark NPY differ only by three amino acids (1). Together with peptide YY and pancreatic polypeptide, they form a family of peptides that subserve pleiotropic functions of neurotransmitters, neuromodulators, hormones, and paracrine regulators. In addition to neuronal sources, NPY is present extraneuronally in

Advances in Pharmacology, Volume 42

1054-3589/98 $25.00

platelets, lymphocytes, and endothelial cells. Elevated levels of immunoreactive NPY are found during conditions of intense and prolonged sympathetic nerve activation, such as stress, exhaustive exercise, hypertension, and myocardial infarction (1). The functions of NPY as a sympathetic cotransmitter and a neuromodulator in the cardiovascular system have been extensively studied (1) and reviewed elsewhere in the book. This review briefly summarizes our studies of vascular actions of NPY (Fig. 1).

NPY exerts multiple actions in blood vessels. Acutely, it causes vasoconstriction by activation of its specific Y1 receptors and potentiates the actions of other sympathetic and nonsympathetic vasoconstrictors (1). However, not all blood vessels are sensitive to NPY. While small resistance vessels in some vascular beds, such as cerebral, coronary, and splanchnic, are extremely sensitive to NPY, larger vessels, such as the aorta or the pulmonary arteries, are resistant to NPY-induced vasoconstriction (1). Recent discoveries of specific and Y1-selective receptor antagonists (2) corroborate these findings and indicate that endogenously released NPY does not contribute to resting vascular tone. Thus, in basal conditions, catecholamines appear to provide the primary neurogenic influence on blood vessels.

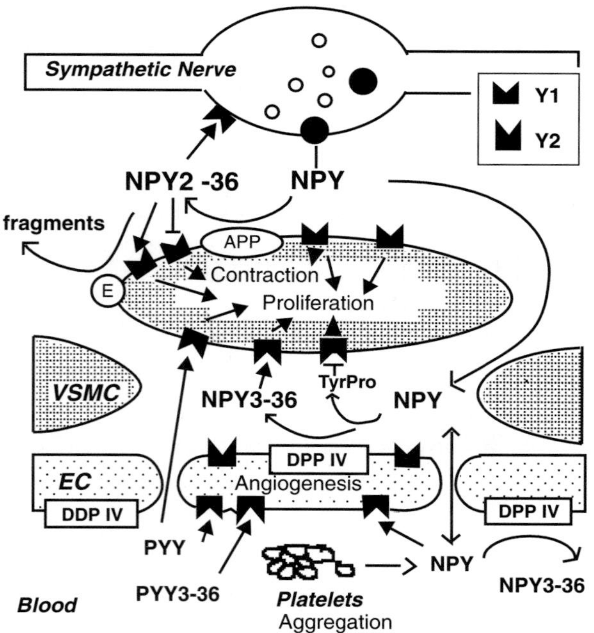

FIGURE 1 Vascular neural and paracrine NPY system. NPY is released from the sympathetic nerves, endothelium, and platelets and acts on Y1 receptors to cause vasoconstriction and on Y1 and Y2 receptors to stimulate VSMC proliferation and endothelial angiogenesis. DPPIV and APP inactivate NPY as a Y1 agonist and convert it into a Y2 agonist. Additionally, circulating PYY mimics the NPY-Y1 and -Y2 activities.

However, the situation changes during stress and in other states where the sympathetic nerve activity becomes highly activated. In rats subjected to immobilization stress, which markedly elevates plasma catecholamine levels, pressor responses are blocked by combined α- and β-adrenergic blockade, but increases in vascular resistance in the superior mesenteric bed are not (2). Fifty percent of the mesenteric vasoconstriction is resistant to total adrenergic blockade (with prazosin, yohimbine, and propranolol) and also remains unchanged by blockade of vasopressin and angiotensin II receptors (2). Recently, we have confirmed that the nonadrenergic (cold) stress-mediated component of the mesenteric vasoconstriction is NPY-dependent and mediated by NPY-Y1 receptors (2).

What determines increased vascular responsiveness to NPY during stress is not yet clear. It may simply be the result of greater NPY release and/or decreased peptide inactivation. While the former is true for many stress conditions, particularly when prolonged and exhaustive, the latter is not. In rats stressed by immobilization, NPY clearance appears to be increased in parallel to the increased release (1). Two ectopeptidases have been implicated in the process of NPY degradation (3). Dipeptidyl peptidase IV (DPPIV) (3), a serine protease present on endothelial cells and activated lymphocytes (where it is known as cd26), cleaves Tyr-Pro (3, 4) off of the NPY's N-terminus, and aminopeptidase P (APP) (3), present on vascular smooth-muscle cells (VSMCs), eliminates Tyr (4). The products of these enzyme reactions, C-terminal fragments of NPY devoid of Tyr (4), are no longer able to stimulate vasoconstrictive Y1 receptors but remain active at the Y2 receptors, which are believed to be nonvasoconstrictive. Thus, increased NPY degradation would presumably eliminate the vasoconstrictive activity of the peptide.

Vascular responsiveness to NPY is increased not only in stress, but also in other states with elevated plasma catecholamine levels (e.g., in SHR). In endotoxemia, the NPY-mediated responsiveness is not augmented, yet NPY is the only agonist to which vessels do not develop desensitization (1). We hypothesized that high circulating catecholamine levels sensitize blood vessel to NPY. In support of this notion, we have previously found that prolonged exposure of blood vessels to NE markedly upregulates vasopressor responses *in vivo* and *in vitro* and augments calcium-stimulatory (Ca^{2+} uptake into cultured VSMC) responses to NPY (1). This effect is long-lasting, resistant to adrenergic and calcium-channel blockers, and dependent on thapsigargin-sensitive intracellular calcium stores, indicating activation by NPY of a unique signaling pathway, distinct from that used by adrenergic agonists (1). The NE-induced sensitization to NPY is mediated by β_2-adrenergic (mimicked by isoproterenol) and not α_1-adrenergic receptors and, in part, is due to stimulation of cyclic adenosine monophosphate. It appears, therefore, that vasoconstrictor hyperresponsiveness to NPY in hyperadrenergic states is due to β-adrenergic "priming" of VSMCs; the exact nature of this cross-talk remains to be established.

The efficacy of NPY as a vasoconstrictor would then depend on several factors: the adrenergic status, the activity of NPY degrading enzymes, and the density the Y1 receptors. Although low adrenergic activity, activation of peptidases, and low Y1 receptor density may eliminate NPY's vasoconstrictor action, they do not abolish other vasoactive properties of the peptide. We

(1) have found, and others (4) have confirmed, that NPY stimulates VSMC proliferation independently of serum factors and with an efficacy similar to that of platelet-derived growth factor. The mitogenic effect of NPY is, in part, mimicked by the Y1- and, even more so, the Y2-receptor agonists, products of DPPIV and APP cleavage. Thus, these peptidases do not actually inactivate NPY but convert it to other biologically active forms.

Finally, we have found that NPY exerts similar growth-promoting activity on endothelial cells and *in vivo* stimulates angiogenesis (5). As with the mitogenic effect in VSMCs, the angiogenic action of NPY is, in part, mimicked by Y1 and Y2 agonists (5). Whether this indicates that the growth promoting effects of NPY require cooperation of both receptors or that they are mediated by a novel receptor that recognizes both types of agonists, remains to be determined. Interestingly, the growth-promoting activities of NPY occur at concentrations much lower (< nm) that those required for vasoconstriction (> nM), suggesting that NPY may be foremost the nerve-derived trophic factor, and only at high levels of sympathetic nerve activity is released in large enough quantities to be vasoconstrictive. Thus, physiologically, NPY may be the mediator that regulates blood vessel growth and remodeling during development, the long-suspected sympathetic trophic factor.

References

1. Zukowska-Grojec, Z., and Wahlestedt, C. (1993). Origin and actions of neuropeptide Y in the cardiovascular system. *In* The Biology of Neuropeptide Y and Related Peptides. (W. F. Colmers and C. Wahlestedt, eds.), pp. 315–338. Humana Press, Totowa, New Jersey.
2. Zukowska-Grojec, Z., Dayao, E. K., Karwatowska-Prokopczuk, E., Hauser, G. J., and Doods, H. N. (1996). Stress-induced mesenteric vasoconstriction in rats is mediated by neuropeptide Y Y1 receptors. *Am. J. Physiol.* , H796–H800.
3. Mentlein, R., Dahms, P., Grandt, D., and Kruger, R. (1993). Proteolytic processing of neuropeptide Y and peptide YY by dipeptidyl peptidase IV. *Regul. Peptides* **49**, 133–144.
4. Erlinge, D., Brunkwall, J., and Edvinsson, L. (1994). Neuropeptide Y stimulates proliferation of human vascular smooth muscle cells: Cooperation with noradrenaline and ATP. *Regul. Peptides* **50**, 259–265.
5. Zukowska-Grojec, Z., Karwatowska-Prokopczuk, E., Yeh, Y-Y., Chen, W-T., Rose, W., and Grant, D. (1995). Endothelial neuropeptide Y (NPY) system and angiogenesis. *Circulation* **192,** 3430.

Markus Haass
Department of Cardiology
University of Heidelberg
Heidelberg, Germany

Neuropeptide Y: A Cardiac Sympathetic Cotransmitter?

Neuropeptide Y (NPY: 36 amino acids; mol wt 4250) is colocalized with catecholamines in central and peripheral sympathetic nerves and the adrenal medulla (1, 2). On sympathetic activation, NPY has been shown in various organs (e.g., spleen, heart) to be coreleased with norepinephrine (NE) from peripheral postganglionic sympathetic nerve endings. As a neurotransmitter and neuromodulator, NPY affects the cardiovascular system both directly (e.g., as a potent vasoconstrictor and indirectly (e.g., by presynaptic inhibition of both vagal and sympathetic efferents) (1, 2). In addition NPY has been found to be mitogenic, stimulating proliferation of vascular smooth-muscle cells (2). This short chapter focuses on our attempts to further characterize the relationship of NPY to NE release under physiological and pathophysiological conditions, such as congestive heart failure and myocardial ischemia. First, results obtained in human beings are reported, followed by a summary of relevant experimental studies. The plasma concentrations of NPY were determined by a polyclonal radioimmunoassay and those of NE by a radioenzymatic assay (in human studies) and by high-performance liquid chromatography and electrochemical detection (in experimental studies), respectively (1, 2). All human studies were approved by the ethical committee of the University of Heidelberg.

I. Human Studies

Under resting conditions, the venous plasma concentrations of NPY were found to be 200 to 250 times smaller than the corresponding plasma concentrations of NE (Table I). During graded bicycle exercise, the plasma concentrations of both NPY and NE increased significantly; however, higher work loads (>100 watts) were required for NPY to increase in healthy volunteers (26 ± 1 years; n = 8). Similarly, the acute sympathetic activation achieved by a modified cold-pressor test (immersion of both legs up to the knees in a bucket of ice-cold water for 2 min) was not sufficient to significantly increase the plasma concentrations of NPY, although the plasma concentrations of NE more than doubled (see Table I). The plasma half-life of NPY, calculated from the plasma elimination kinetics after graded bicycle exercise, was approximately 10.3 min and, therefore, much longer than that of NE (1.3 min). The exact mechanism of the clearance of NPY from the plasma is not yet known (2).

Advances in Pharmacology, Volume 42

1054-3589/98 $25.00

TABLE I Venous Plasma Concentrations of Neuropeptide Y and Norepinephrine, Systolic and Diastolic Blood Pressure, and Heart Rate in Healthy Volunteers and in Patients with Acute and Chronic Congestive Heart Failure

	n	*NPY (pmol/liter)*	*NE (nmol/liter)*	*SBP (mm Hg)*	*DBP (mm Hg)*	*HR (beats/min)*
Healthy volunteers						
Rest (control)	8	4.0 ± 0.3	0.8 ± 0.1	117 ± 4	71 ± 3	62 ± 3
100 watts	8	4.0 ± 0.4	1.4 ± 0.2	133 ± 6	66 ± 3	116 ± 5[a]
200 watts	8	7.0 ± 1.2[a]	3.6 ± 0.7	144 ± 5[a]	70 ± 5	158 ± 8[a]
Rest (control)	8	4.5 ± 0.5	0.7 ± 0.1	116 ± 5	76 ± 5	74 ± 4
Cold-pressor test	8	4.1 ± 0.3	1.5 ± 0.2[a]	131 ± 4[a]	91 ± 3	75 ± 6
Congestive heart failure patients						
NYHA 0 (control)	16	4.3 ± 0.6	1.1 ± 0.1	118 ± 3	74 ± 3	69 ± 3
NYHA I-II	13	5.0 ± 0.3	1.0 ± 0.1	122 ± 6	75 ± 4	77 ± 5
NYHA III-IV	18	7.0 ± 0.8	3.0 ± 0.3[a]	103 ± 3[a]	68 ± 2	94 ± 5[a]
Acute pulmonary edema	9	24.8 ± 5.5	8.3 ± 1.8	149 ± 7[a]	81 ± 6	102 ± 4[a]

Healthy volunteers were evaluated at rest, during graded bicycle exercise, and during cold-pressor test.
[a] $p < .05$ versus respective control.
SBP, systolic blood pressure, DBP, diastolic blood pressure; HR, heart rate.

In chronic congestive heart failure patients (53 ± 2 years; n = 31), the venous plasma concentrations of NE increased with the severity of the disease, indicating progessive sympathetic activation (see Table I). However, the plasma concentrations of NPY only tended to be increased in severe congestive heart failure (NYHA III–IV) as compared with age-matched control patients (55 ± 2 years; n = 16) with normal left ventricular function. Although half of the patients with severe heart failure (and none of the NYHA I–II patients) had plasma concentrations of NPY above the normal range, plasma NPY may not be regarded as a sensitive marker of the severity of chronic congestive heart failure in individual patients. However, in acute heart failure patients (79 ± 2 years) with pulmonary edema, an approximately fivefold increase in the plasma concentrations of NPY was observed (see Table I). Although the plasma concentrations of NPY have been claimed to increase with age, the huge increase in acute cardiac decompensation (pulmonary edema) cannot be simply explained by the fact that older patients were affected.

It has been suggested that the stress-induced changes in the plasma concentrations of NPY are under inhibitory control from estrogens (2). However, the number of premenopausal female patients in each group (≤15%) was too small to test whether there were any sex-related differences. In conclusion, NPY and NE share some common features (e.g., release by intense acute sympathetic activation), but both transmitters are characterized by different release and elimination kinetics.

In human right atrial tissue samples, the concentrations of NPY amounted to 2.97 ± 0.86 pmol/g (n = 11), with the molar ratio of NPY to tissue NE concentrations being approximately 1 : 2000. To prove whether NPY was released from the human heart, the plasma concentrations of NPY were determined simultaneously in the coronary sinus and the radial artery in eight patients with congestive heart failure (56 ± 4 years; NYHA II–III). Measurements performed under resting conditions revealed no arteriovenous concentration difference of plasma NPY across the heart (5.0 ± 0.5 pmol/liter vs 5.0 ± 1.1 pmol/liter, respectively). However, during graded bicycle exercise, the coronary sinus concentrations of NPY increased more than the arterial concentrations in most of the patients, thus pointing to a net secretion of NPY during exercise. Similarly, Kaijser *et al.* (3) found a significant cardiac net secretion of NPY during graded bicycle exercise in combination with hypoxia in healthy volunteers, the combination being a very strong stimulus of the efferent sympathetic nervous system.

II. Experimental Studies

Because a detailed study of cardiac NPY release is not feasible in humans, the perfused guinea pig heart was chosen to further characterize cardiac NPY release in comparison to NE release. In this model, several different stimuli, such as electrical stimulation, potassium depolarization, and nicotine, all resulted in a concomitant overflow of both NPY and NE into the coronary sinus (1, 4). The molar ratio of the concentrations of NPY and NE in the coronary sinus was approximately 1 : 500. Pharmacological depletion of sympathetic storage vesicles (by pretreatment of the animals with syrosingopine) resulted in a significant reduction (30–40% of control) of the tissue concentrations of NPY in all four chambers of the heart and a marked attenuation of stimulation-evoked NPY release. These findings further support the sympathetic origin of cardiac NPY release. The stimulated corelease of both transmitters was shown to be calcium-dependent and thus exocytotic. It required calcium entry through voltage-dependent calcium channels (preferentially N-type calcium channels) and activation of protein kinase C. The corelease of NPY and NE evoked by electrical stimulation was similarly modulated by presynaptic inhibitory α_2-adrenoceptors, muscarinic M_2 acetylcholine receptors and adenosine A_1-receptors. However, blockade of the neuronal norepinephrine re-uptake carrier (uptake$_1$-carrier) by either desipramine or nisoxetine had a divergent effect on NPY and NE overflow from the perfused guinea pig heart. Whereas NE overflow was increased by blockade of the uptake$_1$-carrier, NPY overflow was attenuated. Additional experiments with simultaneous inhibition of presynaptic α_2-adrenoceptors revealed that the effect of uptake$_1$-blockade on NPY overflow was mainly due to an augmented negative feedback inhibition via α_2-adrenoceptors, which itself was mediated by an increase in NE in the synaptic cleft.

In additional experiments, the impact of single factors of myocardial ischemia on sympathetic transmitter release was investigated (4). Simulated cardiac energy depletion (anoxia and cyanide intoxication, both in combination with glucose withdrawal to prevent ongoing glycolysis) markedly potentiated the exocytotic corelease of NPY and NE. Hyperkalemia and acidosis, on the other hand, were shown to attenuate the evoked transmitter release. The effect of all the single factors of myocardial ischemia on sympathetic transmitter release was mediated by their impact on the intracellular calcium concentration, as shown by microfluorimetry (Fura-2) in bovine adrenal medullary chromaffin cells (4). The exocytotic corelease of NPY and NE from the guinea pig heart evoked by electrical stimulation was already suppressed after short periods ($\geq$2 min) of global myocardial ischemia (stop-flow). This inhibitory effect was rapidly (i.e., within 2 min) reversed after reperfusion (5). Longer periods of global myocardial ischemia ($>$10 min) resulted in a "spontaneous" release of both NPY and NE. This corelease of both transmitters was calcium-independent and inhibited by uptake$_1$-blocking agents, such as desipramine, nisoxetine, or nomifensine (5). These features are not compatible with an exocytotic release mechanism. So far, NPY release has been thought to occur solely through depolarization-evoked, calcium-dependent fusion of sympathetic storage vesicles with the cell membrane at the synapse (1). Accordingly, the calcium-independent nonexocytotic norepinephrine release evoked by tyramine has been shown by several groups not to be accompanied by NPY release (1). Therefore, the exact mechanism underlying the calcium-independent NPY release observed during ongoing myocardial ischemia needs to be further determined.

Acknowledgments

This research was supported by a grant from the German Research Foundation (DFG; Sonderforschungsbereich 320—Cardiac function and its regulation).

References

1. Haass, M., Richardt, G., Schömig, A. (1992). Relationship between neuropeptide Y and norepinephrine release from cardiac sympathetic nerves. *In* Stress: Neuroendocrine and Molecular Approaches. (R. Kvetnansky, R. McCarty, and J. Axelrod, eds.), pp. 211–218. Gordon and Breach Science, New York.
2. Zukowska-Grojec, Z. (1995). Neuropeptide Y—A novel sympathetic stress hormone and more. *Ann. N. Y. Acad. Sci.* **771,** 219–233.
3. Kaijser, L., Pernow, J., Berglund, B., Lundberg, J. M. (1990). Neuropeptide Y is released together with noradrenaline from the human heart during exercise and hypoxia. *Clin. Physiol.* **10,** 179–188.
4. Krüger, C., Haunstetter, A., Gerber, S., Serf, CH., Kaufmann, A., Kübler, W., Haass, M. (1995). Nicotine-induced norepinephrine release in guinea-pig heart, human atrium and bovine adrenal chromaffin cells: Modulation by single components of ischaemia. *J. Mol. Cell. Cardiol.* **27,** 793–803.
5. Haass, M., Krüger, C., Haunstetter, A. (1994). Dual effect of myocardial ischemia on the co-release of norepinephrine and neuropeptide Y from the guinea-pig heart. *Circulation* **90,** (Suppl 1), 1–370. (Abstract).

Elizabeth D. Abercrombie, Peter DeBoer, and Marten J. Heeringa

Center for Molecular and Behavioral Neuroscience
Rutgers University
Newark, New Jersey 07102

Biochemistry of Somatodendritic Dopamine Release in Substantia Nigra: An *in Vivo* Comparison with Striatal Dopamine Release

A noteworthy characteristic of nigrostriatal dopamine (DA) neurons is their ability to release transmitter not only at the nerve terminals in striatum but also from the somatodendritic region in substantia nigra. Dopaminergic neurons of substantia nigra pars compacta (SNc) possess an extensive dendritic arbor that can be divided on morphological grounds into two domains (1). First, there is a plexus of dopaminergic dendrites that ramify laterally and remain confined largely to the SNc; second, there typically exists a single, very lengthy dendrite that extends ventrally to ramify within the substantia nigra pars reticulata (SNr). DA released from these dendritic domains may play an important functional role in the autoregulation of the electrophysiological activity of the dopaminergic neurons and in the regulation of GABA release from striatonigral afferents to SNr, respectively (2, 3). Using *in vivo* microdialysis to monitor extracellular DA levels in SNr and in striatum, we have compared the characteristics of dendritic versus nerve terminal DA release to determine whether differences exist in the mechanisms regulating dopaminergic neurotransmission at these two sites (4).

In the first set of experiments, we sought to examine the extent to which dendritic DA appeared to be released via a classical impulse-dependent, exocytotic mechanism and whether evidence could be obtained for a two-compartment storage model for dendritic DA. These commonly accepted concepts regarding dopaminergic neurotransmission were developed based primarily on studies of DA release from the nerve terminal. Local application of 10^{-6} M tetrodotoxin via a microdialysis probe decreased extracellular DA to nondetectable levels in both SNr and striatum. A linear dose-response relation for extracellular DA was obtained in response to the DA releaser amphetamine (AMPH; 0.25, 2, 5, and 10 mg/kg i.p.) in both SNr and in striatum; at each dose, peak DA levels after AMPH administration were approximately eightfold lower in SNr relative to striatum. Reserpine administration (5 mg/kg i.p.), which disrupts vesicular storage of monoamines, significantly decreased the tissue level of DA in SNpr, SNc, and striatum. In these brain areas, reserpine-

Advances in Pharmacology, Volume 42

1054-3589/98 $25.00

induced reductions in tissue DA concentration occurred within 2 hr and persisted at 24 hr postdrug. *In vivo* microdialysis measurements revealed that reserpine administration decreased extracellular DA to nondetectable levels in SNr as well as in striatum with a time course similar to that observed for the reserpine-induced reductions in tissue DA levels. These data suggest that in SNr, as well as in striatum, DA release is impulse-dependent and there exists both a cytoplasmic pool of transmitter that is released by AMPH and a storage pool of transmitter that is reserpine-sensitive.

Thus, a two-compartment model consisting of a storage pool and a free cytoplasmic pool appears to apply equally to intracellular DA in both SNr and striatum. We used an approach aimed at further testing this model directly by using AMPH as a pharmacological tool to demonstrate the existence of the two DA compartments. Low doses of AMPH are thought to selectively induce release of DA from an extravesicular, cytoplasmic pool, whereas high doses of AMPH also appear to displace DA from the stored pool. In both structures, it was observed that reserpine treatment significantly attenuated the release of DA evoked by a high dose of AMPH (10 mg/kg i.p.) given 2 hr later. In contrast, DA efflux in response to a low dose of AMPH (2 mg/kg i.p.) was not altered by reserpine pretreatment, either in SNr or in striatum.

In a second set of experiments, we evaluated the relative role of DA autoreceptors in the regulation of extracellular DA concentrations in SNr and in striatum. First, we compared the dose-response effects on extracellular DA of the D2 agonist quinpirole (QUIN; 3, 30, 300, and 3000 μg/kg i.p.) and of the D2 antagonist haloperidol (HAL; 0.05, 0.5 mg/kg i.p.) Systemic administration of QUIN produced dose-related decreases in extracellular DA in both SNr and striatum; however, the threshold dose was lower (3 vs 30 μg/kg) and the maximal relative inhibition was greater (70% vs 40%) in striatum compared with SNr. Similarly, the increase in extracellular DA elicited by systemic administration of HAL was both more sensitive in terms of threshold dose (0.05 vs 0.5 mg/kg) and of larger magnitude in terms of maximal relative augmentation (100% vs 40%) in striatum versus SNr.

We next sought to determine the extent to which spontaneous release of DA was regulated by local endogenous dopaminergic tone at inhibitory autoreceptors in the two structures. Because increases in dendritic DA efflux in response to systemic administration of the D2 antagonist HAL may involve long-loop circuit effects as well as local antagonism of autoreceptor function in SNr, we determined the ability of HAL locally applied via the dialysis probe to alter extracellular DA in that structure as compared with striatum. In striatum, local application of 10^{-6} M HAL produced an increase in DA output of approximately 100%. In SNr, however, application of this concentration of HAL had no significant effect on DA efflux (Fig. 1). The lack of effect of local HAL application to increase extracellular DA level in SNr cannot be attributed to inadequate blockade of D2 receptors in that region because the same concentration of HAL was able to block the decrease in DA output produced by local application of the D2 agonist QUIN (see Fig. 1). Rather, we conclude that, unlike striatum where local autoreceptor tone appears to exist, in SNr there is little or no endogenous autoreceptor tone regulating spontaneous DA output. In support of this conclusion, we have observed that microinjection of HAL

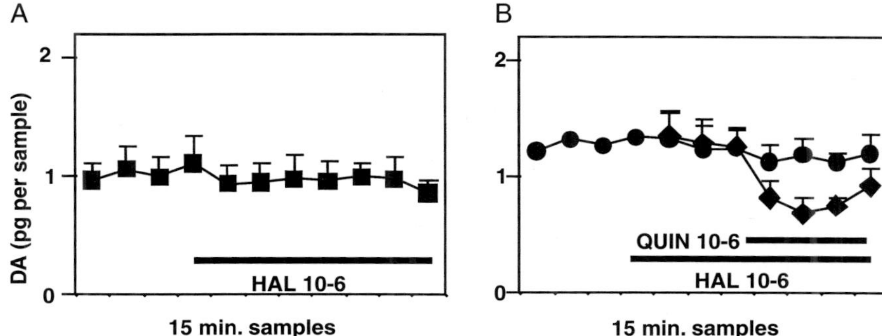

FIGURE 1 (A) Local application of the D2 antagonist HAL via the microdialysis probe did not affect the extracellular concentration of DA in SNr. (B) Local application of the D2 agonist QUIN (*diamonds*) significantly inhibited DA output in SNr, and this effect was blocked by coapplication of HAL (*circles*).

(50 μg) into striatum elicits approximately a 40% increase in DA efflux in SNr and that this increase is not additive with that produced by systemic administration of 0.5 mg/kg HAL.

In summary, it appears that the release of DA from dendrites within SNr, as from nerve terminals in striatum, occurs via a mechanism of impulse-dependent exocytosis. Our data suggest that DA in dendrites exists within both a free cytoplasmic compartment as well as in a reserpine-sensitive storage compartment and that the latter compartment serves as the substrate for the spontaneous release of dendritic DA. The storage compartment for DA in dendrites has characteristics that are virtually identical to the vesicular storage pool described in nerve terminals. The role of traditional vesicles as the physical substrate for DA storage and release in dendrites is controversial, however, because the visualization of dopaminergic vesicles in SNr is a rare occurrence (5). Dendritic DA release differs from striatal DA release in that it appears not to be regulated by autoreceptor-mediated inhibition under basal conditions. This latter result would at first seem to be discrepant with a recent report in which neuroleptic-induced increases in the electrophysiological activity of SNc dopaminergic neurons were attributed to local blockade of autoreceptors that are tonically activated by somatodendritic release of DA (6). These observations may suggest, however, that differences exist even within substantia nigra such that the release of DA occurring from the population of laterally extending DA dendrites within SNc is subject to different regulatory conditions, as compared with DA release occuring from the ventrally extending dendrites found within SNr.

References

1. Tepper, J. M., Sawyer, S. F., and Groves, P. M. (1987). Electrophysiologically identified nigral dopaminergic neurons intracellularly labeled with HRP: Light-microscopic analysis. *J. Neurosci.* 7, 2794–2806.

2. Bunney, B. S., Walters, J. R., Roth, R. H., and Aghajanian, G. (1973). Dopaminergic neurons: Effect of antipsychotic drugs and amphetamine on single cell activity. *J. Pharmacol. Exp. Ther.* **185,** 560–571.
3. Cameron, D. L., and Williams, J. T. (1993). Dopamine D1 receptors facilitate transmitter release. *Nature* **366,** 344–347.
4. Heeringa, M. J., and Abercrombie, E. D. (1995). Biochemistry of somatodendritic dopamine release in substantia nigra: An in vivo comparison with striatal dopamine release. *J. Neurochem.* **65,** 192–200.
5. Nirenberg, M. J., Chan, J., Liu, Y. J., Edwards, R. H., and Pickel, V. M. (1996). Ultrastructural localization of the vesicular monoamine transporter-2 in midbrain dopaminergic neurons: Potential sites for somatodendritic storage and release of dopamine. *J. Neurosci.* **16,** 4135–4145.
6. Pucak, M. L., and Grace, A. A. (1994). Evidence that systemically administered dopamine antagonists activate dopamine neuron firing primarily by blockade of somatodendritic autoreceptors. *J. Pharmacol. Exp. Ther.* **271,** 1181–1192.

B. H. C. Westerink,* W. J. Drijfhout,† M. vanGalen,*
Y. Kawahara,‡ and H. Kawahara‡

* University Center for Pharmacy
Groningen, The Netherlands

† Pharma Bio-Research International
Zuidlaren, The Netherlands

‡ Department of Dental Anesthesiology
Kyushu Dental College
Fukuoka, Japan

The Use of Dual-Probe Microdialysis for the Study of Catecholamine Release in the Brain and Pineal Gland

Dual-probe microdialysis is a variant of the multiple push–pull cannulae method that has been applied successfully to dopamine neurones in anaesthetized cats (1). The approach consists of implanting a microdialysis probe in the cell-body area where the projections originate and implanting a second probe in the terminal area of the same projections. Both probes can be used to sample released neurotransmitter. Drugs can be perfused via the probe

Advances in Pharmacology, Volume 42

1054-3589/98 $25.00

(retrograde microdialysis) into the cell-body area, and effects on neurotransmitter output are measured in the terminal areas of the same projection.

Drugs are often administered to small brain areas with help of injection cannulae. However, this method has certain disadvantages. The use of injection cannulae may cause locally high concentrations of injected drugs, and damage to the surrounding neuronal tissue has been demonstrated. Moreover, high concentrations of the injected drug may cause unwanted leakage to other brain areas. When drugs are applied to nerve tissue by retrograde microdialysis, the mechanical contact between the infused solution and brain tissue is minimized. Moreover, a regular concentration gradient of the infused drug is built up in the surrounding tissue. The maximal concentration of the drug in nerve tissue does not exceed the concentration in the infusion fluid. For interpretation of effects of the infused drugs, it is of great importance to know the diffusion speed of the compound out of the probe into brain tissue. We have recently developed a simple method to determine experimentally the diffusion speed of various centrally acting drugs (Westerink and deVries, in preparation). This method is not based on analytical chemistry of the infused drug, but on estimation of the effects that the infused drug causes on local neurotransmitter release. With this approach, only drugs can be studied that influence extracellular dopamine (DA) or acetylcholine in the striatum. In brief, two probes are implanted at a fixed distance (1 mm) in the striatum. The drug is infused in one of the probes. The diffusion speed of the infused drug can be followed in time by measuring DA or acetylcholine in the second probe. The studied drugs (agonists–antagonists of D_2 and muscarinic receptors, TTX, uptake blockers, etc.) displayed large (3 orders of magnitude) differences in diffusion speed.

Next, we present the use of dual-probe microdialysis for three different catacholaminergic neuronal systems:

1. The DA neurones that originate in the A9 and A10 and innervate the striatum, nucleus accumbens, and prefrontal cortex (PFC)
2. The norepinephrine (NE) neurones that originate in the locus ceruleus (LC) and innervate the PFC
3. Neurones located in the dorsomedial hypothalamus (DMH) that are transsynaptically connected to NE neurones that innervate the pineal gland

I. A9 and A10 Dopamine Neurones

A9 and A10 dopamine neurons innervate several forebrain structures, such as the striatum, nucleus accumbens, and PFC. With respect to the mechanism of action of certain centrally acting drugs (antipsychotics, drugs of abuse), the mesocortical DA neurons of the A10 are of particular interest. However, this neuronal pathway represents only some hundreds of neurones, which means that they are difficult to study with electrophysiological techniques. We found that dual-probe microdialysis is a technique suitable for studying selectively mesocortical DA neurons.

Infusions of receptor-specific compounds into the substantia nigra (SN) and ventral tegmental area (VTA) demonstrated that the three types of DA neurones all contain N-methyl-D-aspartate (NMDA), non-NMDA, D_2, cholinergic, $GABA_A$, and $GABA_B$ receptors (2–4). Infusion of antagonists revealed that a tonic inhibition—at the cell-body level—mediated by $GABA_A$ and D_2 receptors was detectable in A9 as well as A10 neurones. A small tonic excitation via NMDA and non-NMDA receptors could be established in mesocortical neurones.

There are several differences between the pharmacological responsiveness of the three DA systems to receptor stimulation or blockade:

1. A10 neurones differ from A9 neurones with respect to interactions with the $GABA_A$ receptor. Infusion of muscimol into the SN increased DA release in the ipsilateral striatum. However, when muscimol was infused into the VTA, extracellular DA release was inhibited in the nucleus accumbens as well as the PFC (Fig.1). A GABAergic interneuron that was hypothesized to act in the SN is apparently not necessary to be postulated in the VTA.
2. Mesocortical DA neurones are more sensitive to interaction with D_2 and cholinergic receptors in comparison with mesolimbic and nigrostriatal ones.
3. A10 neurones are more responsive to NMDA and non-NMDA agonists than are A9 neurones.

The dual-probe method also was used to analyze the neurochemistry of certain behavioral conditions. Stress (handling) and reward (feeding) was ap-

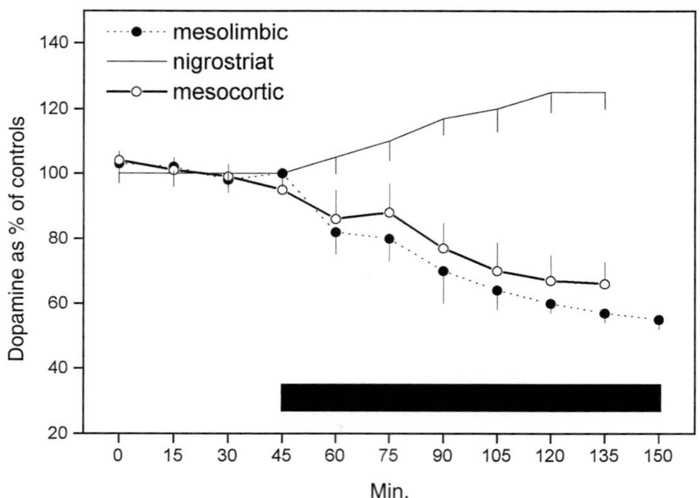

FIGURE 1 Dual-probe microdialysis of the A9 and A10 DA neurons. Muscimol was infused (indicated by a black bar) into the SN (10 μmol/liter) or VTA (20 μmol/liter), and extracellular DA was recorded in the ipsilateral striatum (*no symbols*), nucleus accumbens (*closed circles*), or PFC (*open circles*).

plied to animals during administration of receptor-specific compounds into the VTA. It appeared that glutamatergic afferents to the VTA, by acting via NMDA-receptors, mediate the stress- or reward-induced increase in DA release in the nucleus accumbens and PFC.

II. Norepinephrine Neurones in the Locus Ceruleus

Dual-probe microdialysis of the LC is performed by implanting one probe near the LC and a second probe into terminal areas such as the PFC. Infusion of clonidine (100 μmol/liter) into the LC for a period of 15 min induced a rapid and pronounced decrease in extracellular NE in the PFC (to about 20% of control), apparently via α_2-receptors localized on somatodendritic sites of the NE neurones. NE levels rapidly normalized after termination of clonidine infusion. Infusion of carbachol into the LC (100 μmol/liter) resulted in a clear stimulation of NE release in the PFC to about 200% of controls. It is concluded that with help of clonidine and carbachol, LC neurons can be reversibly inhibited and stimulated for a short period in freely moving animals. The dual-probe method is a useful approach for studying the pharmacology of afferents to the LC. Moreover, it also can be used to determine the participation of afferents to the LC during certain behavioral conditions (arousal, stress, transmission of sensory messages, etc.).

By analyzing simultaneously DA and NE in the PFC, we established a clear interaction between the two catecholaminergic systems.

III. Norepinephrine Neuron Innervation in the Pineal Gland

Microdialysis of the pineal gland was developed to study the release of melatonin. As expected, melatonin levels in the pineal gland displayed a strong diurnal rhythm. Because released melatonin is under the influence of sympathetic innervation, we developed a method to detect the minute (subpicogram) quantities of NE that were released by the sympathetic innervation of the pineal gland. For the first time, the actual NE release in the pineal gland was detectable (5). Extracellular NE measured in the pineal showed a remarkable rhythmicity with the nature of an on–off switch. Being switched on about 1.5 hr after lights off, it was switched off again about 1.5 hr before lights on, fully symmetrically to the light–dark cycle.

Dual-probe microdialysis of the suprachismatic nuclei (SCN)-hypothalamus-pineal axis was achieved by implanting one probe in the DMH and a second probe in the pineal gland. Infusion of muscimol (10 μmol/liter; during 40 min in the dark phase) into the DMH resulted in a rapid decline of extracellular NE in the pineal gland. NE levels dropped to 30% of controls and returned to normal values when infusion of muscimol was terminated. These results strongly suggest that the circadian clock located in the SCN utilizes a GABA-containing projection to transmit diurnal information to the innervation of the pineal gland.

References

1. Nieoullon, A., Cheramy, A., and Glowinski, A. (1977). Release of dopamine in vivo from cat substantia nigra. *Nature* **266**, 375–377.
2. Westerink, B. H. C., Santiago, M., and de Vries, J. B. (1992). The release of dopamine from nerve terminals and dendrites of nigrostriatal neurons induced by excitatory amino acids in the conscious rat. *Naunyn Schmiedebergs Arch. Pharmacol.* **345**, 523–529.
3. Santiago, M., and Westerink, B. H. C. (1992). The role of GABA-receptors in the control of nigrostriatal dopaminergic neurons: A microdialysis study in awake rats. *Eur. J. Pharmacol.* **219**, 175–181.
4. Westerink, B. H. C., Kwint, H. F., and deVries, J. B. (1996). The pharmacology of mesolimbic dopamine neurons: A dual-probe microdialysis study in the ventral tegmental area and nucleus accumbens of the rat brain. *J. Neurosci.* **16**, 2605–2611.
5. Drijfhout, W. J., van der Linde, A. G., Kooi, S. E., Grol, C. J., and Westerink, B. H. C. (1996). Norepinephrine release in the rat pineal gland: The input from the biological clock measured by microdialysis. *J. Neurochem.* **66**, 748–755.

F. Gonon and B. Bloch

CNRS UMR 5541
Université de Bordeaux 2
33076 Bordeaux, France

Kinetics and Geometry of the Excitatory Dopaminergic Transmission in the Rat Striatum *in Vivo*

It is generally assumed that dopamine exerts an inhibitory influence on the discharge activity of target neurons in the striatum and nucleus accumbens (NACC) mostly on the basis of many electrophysiological studies using local application of exogenous dopamine or of dopaminergic agents. However, excitatory effects also have been reported (1). The aim of our studies was to further

Advances in Pharmacology, Volume 42

1054-3589/98 $25.00

investigate the postsynaptic effects of endogenous dopamine when its release is manipulated at a distance, using stimulation of the impulse flow. The evoked dopamine release was monitored in every experiment using electrochemical techniques before electrophysiological recording of single target neurons. This approach allowed us to describe the kinetics of the dopaminergic transmission on a millisecond scale.

I. Effects of Dopamine Released by Chemical Stimulation of the Impulse Flow

Dopamine release in NACC can be reproducibly evoked by chemical stimulation of dopaminergic cell bodies using N-methyl-D-aspartate (NMDA) microinjection in the ventral tegmental area (VTA)(2). In fact, when NMDA (65 nl, 100 μM) was pressure-injected in the core of the VTA, the discharge activity of dopaminergic neurons was strongly enhanced for 3–4 min with a maximum 30 to 60 sec after injection (2). We showed by *in vivo* electrochemistry that such injections induced a sixfold increase in the extracellular dopamine level in NACC. Moreover, the kinetics of this increase closely paralleled that of the discharge activity of dopaminergic neurons (2). In a recent study we investigated the effects of these NMDA injections in VTA on the discharge activity of 182 single neurons in NACC (3). Among them, 142 were strongly excited: Their mean discharge rate was more than doubled during the 2 min that followed NMDA injection. Only nine of them were repeatedly inhibited. Moreover, the kinetics of these excitations closely paralleled those of the evoked DA overflow (Fig. 1).

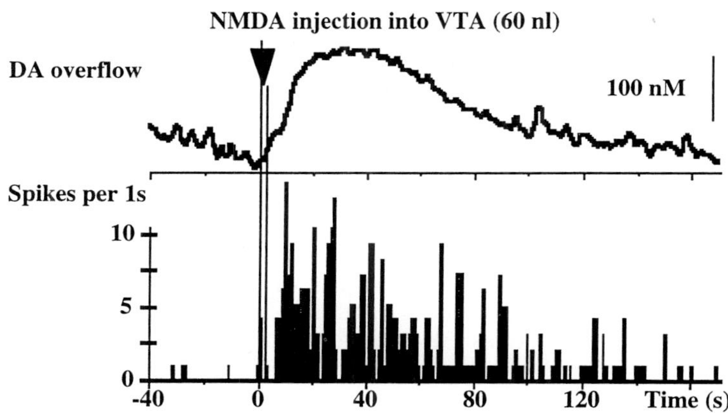

FIGURE 1 NMDA (65 nl, 100 μM) was pressure-injected in VTA with a glass micropipette (2–4). The overflow of dopamine in NACC was monitored with a carbon-fiber electrode (4). This electrode was then removed from the brain and calibrated *in vitro* for dopamine (4). Thereafter, the discharge activity of a single neuron in NACC was monitored extracellularly, and its response to the same NMDA injection was recorded (3).

We assume that these excitatory effects were actually due to the dopamine overflow evoked in NACC by NMDA injection in VTA. First, both neuronal excitation and evoked dopamine release exhibited the same time course (see Fig. 1) and the same geometry: When the NMDA pipette was raised 0.75 mm above the core of the VTA, both effects vanished (2, 3). Second, when dopaminergic neurons were previously lesioned with 6-hydroxydopamine, NMDA injections were no longer effective. Third, systemic administration of the D1 antagonist SCH 23390 (0.25 mg/kg s.c.) completely blocked the excitatory responses of all 15 neurons tested but did not affect the evoked dopamine overflow. Similar blockage was observed with local applications of the same antagonist (100 μM, 40 nl) in the vicinity of recorded neurons (3).

Our conclusion that dopamine is excitatory via D1 receptors does not imply that it is capable of triggering discharge activity on its own. We showed that the dopamine overflow evoked by NMDA injections in VTA strongly facilitated the neuronal responses evoked in NACC by electrical stimulation of the hippocampus (3). Thus, as already suggested (1), dopamine may instead facilitate the activity of target neurons receiving other excitatory inputs.

II. Kinetics of Dopamine Release and Elimination

In the rat, dopaminergic neurons exhibit two kinds of discharge activity: single spikes and bursts of two to six action potentials at 15 Hz. It is noteworthy that NMDA injection in VTA promotes the bursting activity of dopaminergic neurons (2). The dopamine overflow evoked by single-action potentials or by bursts was studied on a pulse-by-pulse basis (5). Electrical stimulations of the medial forebrain bundle (MFB), consisting of 1 to 4 pulses at 15 Hz were applied every 20 sec. The dopamine overflow evoked in the striatum by these stimulations was monitored with a carbon-fiber electrode combined with amperometry. The amplitude of this overflow was positively correlated with the number of pulses and corresponded to a maximal increase in extracellular dopamine of 0.5–1 μM. The main mechanism that explains why the dopamine overflow evoked by a burst far exceeds that observed with a single action potential is extracellular accumulation of the released dopamine as a result of overcoming of the dopamine re-uptake (5).

During its raising phase, the kinetics of the dopamine overflow is fast, a 50% increase of 20 ms. Considering the time for dopamine diffusion from the release sites to the electrode surface, it is likely that release is even faster than measured. This is consistent with the hypothesis that dopamine is released by exocytosis. However, dopamine clearance is significantly slower than release: After the last stimulation pulse, the extracellular dopamine decreased to baseline with a half-life of 50–100 ms (5).

III. Kinetics of the Postsynaptic Excitatory Response to Dopamine

In a first step, the previously described electrochemical approach was used to confirm in each experiment that electrical stimulations of the MFB were capable of triggering the expected dopamine overflow. Then, in the same animals, the discharge activity of single striatal neurons was monitored extracellularly, and the responses evoked by electrical stimulations consisting of 4 pulses at 15 Hz were recorded. Various kinds of inhibitions and excitations were observed, but only one was strongly attenuated by SCH 23390 (0.5 mg/kg s.c.): a delayed excitation that started 100 ms after the fourth pulse and lasted for 300–600 ms (6). Therefore, the dopamine overflow occurs 300 ms before this excitatory response. This delayed postsynaptic response is consistent with the fact that D1 dopamine receptors are coupled with G-protein.

IV. Importance of the Bursting Impulse Flow

The amplitudes of these delayed excitations were positively correlated with the number of stimulating pulses: They were barely detectable after single-pulse stimulations but more and more obvious from 2 to 4 pulses (6). Because the dopamine overflow is also correlated with the number of pulses, this further suggests that these excitations were actually caused by the dopamine overflow. It is noteworthy that dopaminergic neurons respond to appetitive stimuli by a burst of two to six action potentials. We have shown that this bursting impulse flow is more potent than single-action potentials in enhancing the extracellular dopamine level. Moreover, bursts are required to trigger postsynaptic excitatory responses mediated by dopamine acting on D1 receptors. All together, these data suggest that dopamine mediates, via the D1 receptors, a phasic excitatory transmission involved in motivation and reward.

V. Geometry of the Dopaminergic Transmission

As illustrated in Figure 1, there is a good correlation between the amplitude of the dopamine overflow and that of the excitatory response. This is also true concerning the effects of MFB electrical stimulations: Both overflow and excitation are positively correlated with the number of stimulation pulses. This suggests that these postsynaptic excitations are mediated by the extracellular dopamine, which is actually monitored by the carbon fiber electrode. Because the active part of this electrode is the cylindrical surface of one carbon fiber that is 8 μm in diameter and 250 μm long, it monitors the extracellular dopamine in the extrasynaptic space. Therefore, our data suggest that this dopaminergic transmission occurs outside synaptic clefts. This interpretation is in excellent agreement with morphological studies that showed that the D1 receptors are mostly located on dendritic spines outside symmetric contacts formed by dopaminergic terminals on these dendrites (7).

This conclusion does not imply that dopamine can act far away from its release sites. In fact, we have shown that dopamine elimination by re-uptake is fast, with a half-life of 50–100 ms in the striatum and the mesolimbic area (5). The maximal distance for dopamine diffusion before complete elimination is, thus, inferior to 15 μm (6). Therefore, this dopaminergic transmission should be considered perisynaptic.

References

1. Haracz, J. L., Tschanz, J. T., Wang, Z. R., White, I. M., and Rebec, G. V. (1993). Striatal single-unit responses to amphetamine and neuroleptics in freely moving rats. *Neurosci. Biobehav. Rev.* **17**, 1–12.
2. Suaud-Chagny, M. F., Chergui, K., Chouvet, G., and Gonon, F. (1992). Relationship between dopamine release in the rat nucleus accumbens and the discharge activity of dopaminergic neurons during local in vivo application of amino-acids in the ventral tegmental area. *Neuroscience* **49**, 63–72.
3. Gonon, F., and Sundstrom, L. (1996). Excitatory effects of dopamine released by impulse flow in the rat nucleus accumbens *in vivo*. *Neuroscience* **75**, 13–18.
4. Dugast, C., Suaud-Chagny, M. F., and Gonon, F. (1994). Continuous in vivo monitoring of evoked dopamine release in the rat nucleus accumbens by amperometry. *Neuroscience* **62**, 647–654.
5. Chergui, K., Suaud-Chagny, M. F., and Gonon, F. (1994). Nonlinear relationship between impulse flow, dopamine release and dopamine elimination in the rat brain in vivo. *Neuroscience* **62**, 641–645.
6. Gonon, F. (1997). Prolonged and extrasynaptic excitatory action of dopamine mediated by D1 receptors in the rat striatum *in vivo*. *J. Neurosci.* (in press).
7. Caillé, I., Dumartin, B., and Bloch, B. (1996). Ultrastructural localization of D1 dopamine receptor immunoreactivity in the rat striatonigral neurons and its relation with dopaminergic innervation. *Brain Res.* **730**, 17–31.

George E. Mickelson, Paul A. Garris, Melissa Bunin, and R. Mark Wightman

Department of Chemistry and Curriculum in Neurobiology
University of North Carolina at Chapel Hill
Chapel Hill, North Carolina 27599

In Vivo and *in Vitro* Assessment of Dopamine Uptake and Release

Carbon-fiber microelectrodes have made it possible to study the dynamic processes that regulate the concentration of dopamine in the brain. Their small size and rapid response enable neurotransmission to be studied on the millisecond time scale with micrometer spatial resolution. On this time scale, dopamine concentrations are controlled by two opposing processes, release from terminals, which occurs rapidly, and neuronal uptake that occurs on a slower time

Advances in Pharmacology, Volume 42

1054-3589/98 $25.00

scale. Using this technique, we have probed the question of whether dopamine normally exits the synapse to act as a volume neurotransmitter (1) rather than as a classical synaptic neurotransmitter. The prototypical neurotransmitter is acetylcholine at the neuromuscular junction. At that structure, the efflux of acetylcholine from the synapse following release is prevented by binding to receptors and degradation by acetylcholine esterase.

Dopamine synapses in the striatum are much smaller than at the neuromuscular junction, and unless efficient mechanisms exist for its retention, diffusional efflux from the synapse is likely. To test this hypothesis *in vivo,* efflux evoked by a four-pulse, 100-Hz train applied to dopamine fibers in the medial forebrain bundle was compared with that evoked by a single pulse. The duration of a 100-Hz train is too rapid for appreciable uptake to occur. The observed release was 4 times the amplitude of a single pulse, providing strong evidence that dopamine normally exits the synapse during impulse flow. Binding of dopamine to intrasynaptic receptors does not impede efflux from the synapse because the addition of D1 or D2 antagonists did not alter the maximal response of short trains. Similarly, the administration of a dopamine uptake inhibitor does not alter the maximal quantity of released dopamine, although its effects on the uptake rate were clearly apparent. Thus, the conclusion we drew from these experiments is that dopamine normally escapes the synaptic cleft, diffuses into the extrasynaptic fluid, and then acts with remote receptors. Its extracellular diffusion is restricted to a few micrometers because of interaction with the dopamine transporter. In support of this view, recent anatomical papers have shown that both dopamine receptors (2) and transporters (3) are extrasynaptic.

If the majority of the dopamine released by an impulse leaves the synapse, then the amount observed in the extracellular fluid should directly reflect the vesicular origin of released dopamine. The estimated amount of dopamine released per terminal per impulse from our single-pulse experiments was 1000 molecules, a value similar to that measured during release from a cultured neuron (4). While at the level of single cells, individual secretory events can be directly detected (4), this is not the case in intact tissue, where multiple terminals contribute to the measured response. In this work, we have further probed the quantal nature of dopamine release in brain slices prepared from the rat anterior caudate putamen. We examined dopamine release while the current intensity of the stimulation and the external Ca^{2+} concentration were varied.

Slices were prepared with a vibratome, placed in a Scottish-type chamber, and superfused with Krebs buffer, preheated to 34°C, at 1 ml/min, as described previously (5). Carbon-fiber microcylinder electrodes were inserted 75 μm below the surface of the slice, approximately 200–250 μm from the stimulation electrode pair. Extracellular dopamine was identified by background-subtracted, fast-scan cyclic voltammograms by its characteristic oxidation peak at 550 mV versus SSCE (6). The scan rate was 460 V/sec, and voltammograms were repeated at 10-ms intervals. Peak currents were converted to concentration using postcalibration in dopamine standard solutions. Release was elicited with a bipolar stimulation electrode placed on the surface of the slice. Biphasic constant-current pulses, 2 ms each phase, were employed.

Maximal evoked dopamine release in caudate slices was monitored with Ca^{2+} concentrations of 1.2 and 2.4 mM. The amplitude of the stimulation current of the two-pulse, 10-Hz trains was varied up to 200 μA. Results for both external Ca^{2+} concentrations from a single location in a slice are shown in Figure 1A, while Figure 1B shows the time course of release and uptake (upper panel) and the background-subtracted cyclic voltammogram (lower panel). For all of the experiments reported here, the voltammograms identified the detected substance as dopamine. With 2.4 mM of Ca^{2+}, no evoked release occurs with stimulation pulses less than 20 μA, but between 40 and 70 μA, an abrupt increase in release is seen that reaches a plateau. A second transition occurs at 80–100 μA. At stimulation currents above 150 μA, the maximal dopamine concentration increases in a linear fashion with the stimulation amplitude (data not shown). In contrast, with 1.2 mM of external Ca^{2+}, the response shows a single plateau over this range of stimulation amplitudes, and the ascent to the plateau has a lower slope than with 2.4 mM of Ca^{2+}. Note that the amplitude of this plateau is the same as that seen with 2.4 mM of Ca^{2+}. At both Ca^{2+} concentrations, the error in the measurements is much smaller than the amplitude of the observed plateaus.

Mean data from multiple slices showed a similar response (1.2 mM Ca^{2+}, n = 9 slices; 2.4 mM Ca^{2+}, n = 6 slices). The size of the error bars in the

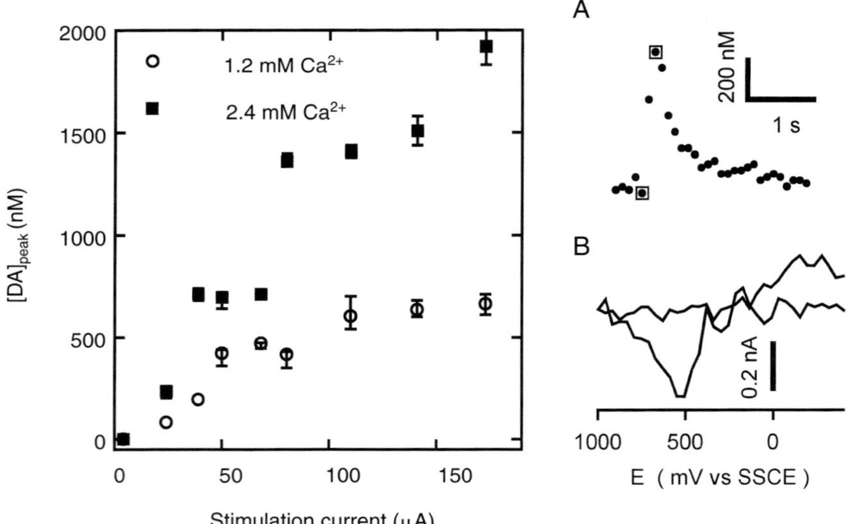

FIGURE 1 Maximal evoked dopamine concentration observed during local electrical stimulation measured at a single location in a slice as a function of stimulation current amplitude. The superfusion buffers contained either 1.2 or 2.4 mM of Ca^{2+}. The Krebs buffer consisted of (in millimoles): NaCl, 126; KCl, 2.5; NaH_2PO_4, 1.2; $CaCl_2$, 1.2 or 2.4; $MgCl_2$, 1.2; $NaHCO_3$, 25; glucose, 11; and HEPES, 20, with the pH adjusted to 7.4 with NaOH and saturated with 95% O_2/5% CO_2. The stimulus train consisted to two pulses at 10 Hz. Each point is the average of the concentration evoked by four to eight individual stimulations, with error bars indicating the SEM. The order of the stimulation amplitude was randomly determined. Circles: buffer contained 1.2 mM of $CaCl_2$; squares: 2.4 mM of $CaCl_2$. (A) [DA] versus time during and after a 40-μA stimulation; average of four stimulations. Stimulus on–off is indicated by the open squares. (B) Background-subtracted cyclic voltammogram taken at the peak of the [DA] versus time response above; also an averaged result from four stimulations.

pooled data was larger because of the different amplitude of evoked dopamine release as a consequence of the heterogeneity of dopamine distribution in the anterior caudate putamen. Nevertheless, two plateaus were apparent for the maximal evoked dopamine concentration with 2.4 mM of Ca^{2+}, with the first of equal amplitude to the single plateau observed with 1.2 mM of Ca^{2+} (between 100 and 150 μA).

The nonlinear relationship between dopamine release, stimulation current, and the external [Ca^{2+}] is exactly the relation anticipated for a quantized release probability associated with vesicles. Thus, according to this model, the quantized levels represent the population of terminals near the electrode shifting from 0 to 1 to 2 vesicles released per terminal as the amount of Ca^{2+} entry increases. The increase in released dopamine between plateaus has a more gradual slope with lower [Ca^{2+}] consistent with the fourth-order dependence of dopamine release on [Ca^{2+}] (7). Multiple quantal release events that occur for a prolonged time after depolarization have been previously observed at amacrine cells (8). Such events could explain the increase in evoked dopamine concentration caused by uptake inhibitors following a single pulse stimulation of high intensity in the caudate putamen of rat brain slices (CPu) (5). If the larger stimulation amplitude evokes multiple, sequential release events that continue for several hundred milliseconds, the overall sequence of release and uptake will be temporally intertwined and the maximal concentration will increase in the presence of uptake inhibitors.

Acknowledgment

Research was supported by the National Institutes of Health (NS-15841).

References

1. Garris, P. A., Ciolkowski, E. L., Pastore, P., and Wightman, R. M. (1994). Efflux of dopamine from the synaptic cleft in the nucleus accumbens of the rat brain. *J. Neurosci.* **14,** 6084–6093.
2. Sesack, S. R., Aoki, C., Pickel, V. M. (1994). Ultrastructural localization of D_2 receptor-like immunoreactivity in midbrain dopamine neurons and their striatal targets. *J. Neurosci.* **14,** 88–106.
3. Nirenberg, M. J., Vaughan, R. A., Uhl, G. R., Kuhar, M. J., and Pickel, V. M. (1996). The dopamine transporter is localized to dendritic and axonal plasma membranes of nigrostriatal dopaminergic neurons. *J. Neurosci.* **16,** 436–447.
4. Bruns, D., and Jahn, R. (1995). Real-time measurement of transmitter release from single synaptic vesicles. *Nature* **377,** 62–65.
5. Jones, S. F., Garris, P. A., and Wightman, R. M. (1995). Different effects of cocaine and nomifensine on dopamine uptake in the caudate-putamen and nucleus accumbens. *J. Pharmacol. Exp. Ther.* **274,** 396–403.
6. Millar, J., Stanford, J. A., Kruk, Z. L., Wightman, R. M. (1985). Electrochemical, pharmacological and electrophysiological evidence of rapid dopamine release and removal in the rat caudate nucleus following electrical stimulation of the median forebrain bundle. *Eur. J. Pharmacol.* **109,** 341–348.
7. Nachshen, D. A., and Sanchez-Armass, S. (1987). Co-operative action of calcium ions in dopamine release from rat brain synaptosomes. *J. Physiol.* **387,** 415–423.
8. Borges, S., Gleason, E., Tirelli, M., and Wilson, M. (1995). The kinetics of quantral transmitter release from retinal amacrine cells. *Proc. Natl. Acad. Sci., USA,* **92,** 6896–6900.

PART B

CATECHOLAMINE REUPTAKE AND STORAGE

Heinz Bönisch* and Lee Eiden†

*Institute of Pharmacology and Toxicology
University of Bonn
D-53113 Bonn, Germany

†Section on Molecular Neuroscience
Laboratory of Cellular and Molecular Regulation
National Institute of Mental Health
National Institutes of Health
Bethesda, Maryland 20892-4090

Overview

Catecholamines are poised for release from neurons and endocrine cells in a two-step process. First, catecholamines are synthesized from tyrosine, and then they are packaged into secretory granules or vesicles by an energy-dependent process. Packaging in secretory vesicles is critical to neurotransmission. This is amply demonstrated by recent knockouts of the gene encoding the vesicular monoamine transporter. In mice, homozygous knockout of the VMAT2 gene is lethal (1). In nematodes loss of VMAT completely abrogates behaviors ascribed to individual neurons secreting dopamine (DA) (Rand *et al.*, p. 940). Following secretion from neurons but not from endocrine cells, catecholamines are recaptured from the synapse by specific re-uptake mechanisms. Again, evidence from mice in which the DA membrane transporter is knocked out (Fumagalli *et al.*, p. 179) indicates that re-uptake is required

Advances in Pharmacology, Volume 42

1054-3589/98 $25.00

not only to terminate postsynaptic action (as indicated by a requirement for dopamine transporter (DAT) in the action of cocaine and amphetamine), but also to abet enzymatic synthesis from tyrosine in maintaining DA levels within neurons. The mild yet psychopharmacologically significant phenotype imparted to neurons by the presence of plasma membrane catecholamine transporters makes these transporters attractive therapeutic targets. The vesicular monoamine transporter has two very similar isoforms (VMAT1 and VMAT2) one of which (VMAT2) is used by all monoaminergic neurons (NE, DA, Epi, 5HT, and histamine), perhaps explaining why its absence in mice is lethal (Uhl *et al.*, p. 1024). The plasma membrane transporters (because they potentially must distinguish among amines in the extracellular space) are substrate-specific, hence their attractiveness as targets for therapeutics aimed at specific manipulation of dopaminergic, noradrenergic, or serotonergic systems.

Part B is divided into two portions, reflecting the two parallel processes, vesicular uptake and plasma membrane re-uptake, that contribute to neurotransmitter storage, secretion, and recycling in catecholaminergic neurons. Costorage of the peptide modulators and secretory granule proteins that precede catecholamines into the storage vesicle, accompany catecholamines on release to target cells and sometimes modulate catecholamine secretion itself is also a subject of this part.

I. Plasma Membrane Catecholamine Transporters

Fast inactivation of a neurotransmitter released into the synaptic cleft is a prerequisite for fine control over an effector sytstem. As a rule, termination of synaptic transmission occurs by re-uptake via specific neurotransmitter transporters located in the plasma membrane of the neuron from which the transmitter is released and, for some neurotransmitters, additionally by transport into neighboring nonneuronal cells (e.g., glial cells). After their transport into neurons, neurotransmitters are taken up by vesicular transporters into storage vesicles and/or degraded by metabolizing enzymes. This concept also holds true for the catecholamines DA, norepinephrine (NE), and epinephrine (Epi). However, the existence of a specific transporter for Epi in mammalians is still a matter of controvery (see following discussion). The transporters for DA and NE are important targets for psychomotor stimulants and tricyclic antidepressants, respectively.

A. Molecular Biology of Catecholamine Transporters

Since the first description of cloning of a neurotransmitter transporter (rat GABA transporter) by Guastella *et al.* (2), an increasing number of neurotransmitter transporters have been cloned (for review, see Amara and Kuhar [3] and Borowsky and Hoffman [4]). These transporters have been subdivided into at least three large families on the basis of sequence homology, ion dependence, and predicted topology. The plasma membrane transporters for DA, NE, serotonin, GABA, glycine, proline, taurine, and betaine belong to the family of Na^+- and Cl^--dependent transporters with 12 putative transmembrane domains (TMs) and a large extracellular domain between TM3 and TM4. The topology of this family has been confirmed by immunofluorescence microscopy studies (5). Plasma membrane transporters for glutamate (and related amino acids) form a family of Na^+-dependent transporters with eight to 10 potential transmembrane domains. Finally, a third family consists of H^+-dependent vesicular transporters (with 12 potential transmembrane domains) for monoamines and acetylcholine.

Together with the serotonin transporter, the catecholamine transporters for DA and NE (DAT and NET) form a subfamily of monoamine transporters within the Na^+- and Cl^--dependent transporter families. This subfamily shares a high degree of homology and also some overlap in substrate and inhibitor affinities. The human NET (hNET) was the first cloned monoamine transporter (6). Meanwhile the bovine (7) and recently also the rat (8) NET have been cloned. For the DAT, the same species homologues have been cloned (4). However, despite an extensive search for expression of an Epi transporter in rat brain regions expressing the Epi synthetizing enzyme phenylethanolamine-*N*-methyltransferase (PNMT), little or no DAT or NET or signals for a related transporter were detected by Larong *et al.* (9). This suggests that Epi-synthetizing neurons either do not express such a transporter or express a molecularly distinct transporter.

Now Blakely and Apparsundaram (p. 206) report the molecular cloning and characterization of an Epi transporter from bullfrog sympathetic ganglia, which, interestingly, is not expressed in the frog brain. This transporter (fET) is a new member of the family of Na^+- and Cl^--dependent transporters and exhibits 75% amino acid identity with the hNET, its closest relative in the gene family. The close relationship to the hNET was also evident from similar affinities at both transporters of a series of substrates and inhibitors. The fET shares with NETs high affinity for tricyclic antidepressants. Among the catecholamines, Epi is transported by the fET with highest maximum velocity of transport (Vmax), followed by NE and DA, whereas the Vmax rank order at the hNET is NE > DA > Epi. This clearly indicates the importance of Vmax as a parameter to characterize a transport system. This had already been emphasized earlier by Graefe and Bönisch (10), who had used the ratio Vmax/Km to characterize the "effectiveness" of substrate transport by the DAT and NET. Without considering Vmax, one might incorrectly conclude that DA is a "better" substrate than NE, because DA is known to exhibit a lower Km (or higher apparent "affinity") at the NET compared with NE. As noted by Blakely and Apparsundaram (p. 206), the close relationship between fET and NET makes it difficult to pharmacologically dicriminate fET or a related mammalian ET from NETs. However, with a new member in the family of catecholamine transporters, the additional knowledge of common structural elements might help to elucidate common residues in catecholamine transporters necessary for recognition and transport of catecholamines.

B. Substrate Transport and Ion Coupling

Structural requirements of compounds recognized as transported substrates by catecholamine transporters as well as the ion-dependence of transport have been extensively studied (for review, see, e.g., Iversen [11], Paton [12], and

Graefe and Bönisch [10]). However, the availability of transfected cells expressing a high number of a defined catecholamine transporter together with new techniques now allow deeper and more detailed insights. Thus, rotating-disk electrode voltametry (RDEV) is being used to follow rapid kinetics of transport of DA and NE in synaptosomal preparations and in transfected cells expressing a DAT or a NET (13, 14). Justice and colleagues (p. 191) used this technique to study, in the same assay, outward transport of DA induced by various phenylethylamine derivatives (relative to the uptake of DA) in LLC cells expressing the human NET. Outward transport is due to inward transport of the outward transport–inducing agent. They confirmed the importance of the ethylamine side chain for optimal transport and showed that the nitrogen-to-phenyl ring distance is critical for transport. They also confirmed methylphenidate as an inhibitor but not a substrate and found no stereoselectivity for d- and l-amphetamine in inducing outward transport of DA or NE. The latter observation is consistent with the result of affinity studies carried out by Blakely and Apparsundaram (p. 206), who showed that amphetamine displays stereoselective inhibition of DATs, whereas both stereoisomers had comparable affinity for NETs and fET. Furthermore, the results obtained by Justice and collegues (p. 191) confirm earlier studies studies in which amphetamine had been shown to be transported by the NET and that transport and not lipophilic diffusion is important for its NE "releasing" effect (15, 16).

Gu and colleagues (p. 175) used the polarized epithelial cells LLC-PK_1 or MDCK stably transfected with hNET or rDAT (or plasma membrane vesicles isolated from these cells) to study ion coupling of the transporters. In agreement with earlier studies (17, 18), they demonstrate the importance of transmembrane Na^+ and Cl^- gradients for catecholamine transport and intracellular accumulation by both transporters and conclude that both ions are cotransported with the substrate. However, in contrast to the NET, substrate transport by the DAT exhibits a sigmoidal Na^+ dependence, indicating that more than one Na^+ ion is cotransported. Furthermore, it is shown that K^+ has little effect on NET or DAT transport in the absence of a K^+ gradient, indicating that K^+ is not required for transport (or is countertransported as by the serotonin transporter). However, a K^+ gradient, by creating a negative membrane potential–stimulated amine transport, this indicates that substrate transport is accompanied by a net positive charge movement, presumably also is due to transport of the cationic form of the substrate. For DA transport by the DAT, even two positive charges are expected to be moved with each transport cycle. Thus, the membrane potential is expected to be an important regulatory factor for DA transport by the DAT.

Electrogenicity and a hyperpolarization-induced increase of DA transport had been demonstrated in *Xenopus* oocytes expressing hDAT (19), and alteration of DA transport in rat striatum by hyperpolarization of DA neurons due to D_2 DA receptor activation is in agreement with voltage dependence of DA transport (20). By electrochemically measuring clearances of DA and by patch-clamp monitoring changes in membrane potential in DA neurons of isolated rat brain slices containing substantia nigra pars compacta (SNc), Zahniser and coworkers (p. 195) show that membrane depolarization diminishes DAT activity in rat SNc. This model also seems to be useful for studying the effects of activation of presynaptic DA receptors on DAT.

Sodium-dependent and antidepressant-sensitive NE-induced currents in NET-transfected cells that exceeded the prediction of stoichiometry of ion cotransport has been demonstrated (21). This and a similar observation in other members of the Na^+- and Cl^--dependent transporters suggest that these transporters have channel modes of conductance (22, 23). Using whole-cell patch electrodes, DeFelice (p. 186) measured inward currents due to NE transport in stably transfected HEK293 cells expressing about 10^6 copies per cell of hNET protein. It is shown that both mean current and variance of fluctuations are larger than predicted from fixed-stoichiometry models of coupled transport. In addition, by measuring transient currents in the absence of external Na^+ (a condition that blocks NE transport or keeps the transporter in a "closed" state), DeFelice demonstrates a constitutive Na^+ leak through the transporter that can be blocked by choline, a cation presumably not able to enter the closed transporter. It is concluded that only one Na^+ and one NE^+ are required to open the transporter and that up to 300 Na^+ ions move through the transporter for each NE molecule.

C. Regulation of Catecholamine Transporters

While regulation of adrenergic and dopaminergic receptor function and expression has been studied extensively, little is known about acute modulation (apart from by ion gradients or the membrane potential) and regulation of cell surface expression of catecholamine transporters, or of expression of their mRNAs. Catecholamine transporters are endowed with N-linked glycosylation sites located in the second putative extracellular loop. A reduction of these sites (by site-directed mutagenesis) has been shown by Nguyen and Amara (24) and Melikian *et al.* (25) not to affect ligand recognition but to reduce cell surface expression of the NET protein and transport activity. Because NETs and DATs possess consensus sequences for phosphorylation by protein kinase C (PKC), phosphorylation of the transporters by activation of PKC is suggested to regulate these transporters. Kitayama *et al.* (26) examined the effect of PKC activating phorbol esters in COS cells expressing the rat DAT and found that PKC activation decreases DA transport. In COS cells transiently expressing the human or bovine NET (with one and three PKC sites, respectively) and in the human neuroblastoma cell line SK-N-SH, Bönisch and coworkers (p.) demonstrate a relatively fast reduction of NE transport by PKC-activating phorbol esters but not by an inactive phorbol ester. Transport reduction is due to a decrease in the Vmax; Km of NE transport is not influenced by activation of PKC. The reduction of Vmax is presumably a consequence of the phorbol ester–induced reduction in the number of transporters expressed in the plasma membrane, because activation of PKC (by phorbol esters) causes a decrease of Bmax of nisoxetine binding. Because similar results were also obtained in COS cells expressing a mutant hNET lacking a consensus sequence for PKC, direct phosphorylation of PKC sites of the transporter as an explanation of the effects of phorbol esters could be excluded. Amara and coworkers (p. 164) present similar results in MDCK cells stably expressing NET and DAT. They show that activation of PKC (by phorbol esters) causes a rapid decrease in Vmax of DA and NE transport (without change in their Km). Furthermore, they show by indirect immunofluorescence a loss of transporter protein staining at the cell surface with a concomitant increase in intracellular staining. These results indicate that activation of PKC (e.g., by the second messenger diacylglycerol) is indirectly involved in the regulation of cell surface expression of catecholamine transporters.

Increasing intracellular cyclic adenosine monophosphate (cAMP) in rat hypothalamic cells has been shown to enhance DA uptake (27). However, it is unknown whether this effect is due to phosphorylation of PKA sites present in the rDAT. On the other hand, the human and bovine NETs as well as the recently cloned PC12 cell rat NET (8) possess no consensus sequences for PKA. The chapter by Bönisch and coworkers (p. 183) reports effects of cAMP on NE transport in SK-N-SH and PC12 cells that express the hNET and rNET, respectively. The effect of cAMP seems to be tissue-specific, because cAMP exhibited no effect in SK-N-SH cells but decreased NE transport in PC12 cells.

In vitro experiments of cocultivation of substantia nigra neurons with striatal neurons show that target cells influence DAT mRNA expression and maintenance of the dopaminergic phenotype (28). Tyrosine hydroxylase (TH) and DAT are considered characteristic markers for dopaminergic neurons. Thus, Hoffman and coworkers (p. 202), by using *in situ* hybridization, have examined whether regulation of TH and DAT mRNA in the rat central nervous system (CNS) are coordinate. Hemisection of the median forebrain bundle causes a coordinate change in both DAT and TH mRNA. However, immobilization produces differential increases in the two mRNAs in different dopaminergic neurons. This suggests stimulus-specific regulation of TH and DAT mRNA expression.

D. Localization of Catecholamine Transporters

In the CNS, DATs are expressed only in dopaminergic cells. However, DAT protein has been shown to differ in the degree of glycosylation in the striatum and nucleus accumbens where DA uptake also was shown to differ (29, 30). Furthermore, the farther diffusion of DA in the prefrontal cortex (PFC) than in the striatum has been explained by lower expression of DAT in the PFC. These results indicate variation in the expression of DAT in the CNS. In contrast to DAT, NET expression is not restricted to the CNS or to noradrenergic cells. Uptake of NE by a NET also takes place in some nonneuronal cells, such as rat pulmonary endothelial cells (see Overview C, report by Bryan-Lluka and colleagues), fibroblast-like cells of the rabbit dental pulp (31), rabbit myometrial cells (32), and human placental syncytiotrophoblasts (33).

Gu and colleagues (p. 175) have examined the sorting properties of biogenic amine transporters (DAT, NET, and SERT) expressed in polarized epithelial cells. They were expected to be sorted to the apical plasma membrane. However, in LLC-PK_1 cells, all three transporters were sorted basolaterally; this was also observed for the NET and SERT in MDCK cells. Only DAT was sorted apically in these cells. These results question the hypothesis that neurons and epithelial cells use the same sorting and signal machinery.

Several groups reported on the localization of the DAT protein or mRNA in the rat CNS. Using *in situ* hybridization, immunohistochemistry, and electron microsopy, Kuhar and coworkers (p. 168) show highest mRNA levels in the substantia nigra and lowest levels in the caudal linear nucleus. DAT protein in midbrain neurons is distributed throughout the neuron, but outside the midbrain (e.g., in tuberoinfundibular neurons) it was prominent in the nerve terminal region, suggesting DAT recycling between intracellular and cell surface pools in these neurons. Surprisingly, at the electron microscopic level, DAT was found surrounding the synapse rather than in the "active" zone of the synapse, suggesting that beside re-uptake diffusion of DA is also of importance. Furthermore, Sesack *et al.* (p. 171) report a relative lack of DAT protein within most DA axon varicosities of the PFC, suggesting a greater paracrine role for DA in the PFC. Dependent on the brain region, often an uneven distribution of DAT mRNA or a lack of DAT mRNA relative to TH mRNA in dopaminergic cells of the rat CNS was observed by Hoffman *et al.* (p. 202). For example, the hypothalamus contains fewer DAT-positive cells than TH-positive cells in the A11, A12, and A14 cell groups and no DAT mRNA in the A15 cell groups.

E. Role of DAT in Physiology and as Pathophysiological Marker

DA in the CNS is involved in many physiological functions (e.g., movement control, cognition, affect, neuroendocrine secretion), and a dysfunction of DA neurotransmission plays an important role in the pathophysiology of disorders such as schizophrenia, Tourette's syndrome or Parkinson's disease. DAT is responsible for termination of dopaminergic neurotransmission. Drugs of abuse, such as cocaine or amphetamines, are known to affect the transport activity of DAT; thus, their long-term abuse changes the expression of DAT. Because DAT is a good marker for intact DA neurons, cocaine analogues which bind to DAT, can be used to image *in vivo* DAT by positron emission tomography (PET) or single photon emission computed tomography (SPECT) and thus the number of intact DA neurons in the human brain.

To examine the role of DAT in neurotransmission and in other physiological functions, Giros *et al.* (34) recently developed a mouse model in which the DAT gene was inactivated by homologous recombination. Mice lacking DAT are hyperactive and indifferent to cocaine and amphetamines. The chapter by this group (Fumagalli *et al.*, p. 179) explains the reason for the pronounced weight deficit in adult males and the lack of appropriate lactation in nursing mothers. These mice develop a marked hypopituitarism with a substantial decrease in the size of anterior and intermediate (but not posterior) lobe of the pituitary and a concommitant decrease in the density of D2 DA receptors. The changes in the content of prolactin and growth hormone are suggested to explain the aforementioned deficits.

Innis (p. 215), Wong *et al.* (p. 219), and Volkow *et al.* (p. 211) show the utility of PET and SPECT in imaging of DAT in primates and humans, including patients with Parkinson's disease and other neurodegenerative disorders, such as Lesch-Nyhan disease, Rett syndrome, Tourette's syndrome, and stimulant abuse.

II. Vesicular Transporters, Catecholamine Storage, and Polypeptide Costorage

Nils-Ake Hillarp reported 39 years ago that practically all of the catecholamines in homogenates of the adrenal medulla could be recovered from particles with a buoyant density slightly different from mitochondria (35). Energy-dependent catecholamine uptake and concentration into the "chromaffin granules" was demonstrated soon after, explaining how metabolically labile catecholamines could be accumulated in endocrine cells (see Winkler and Westhead [36]). About 10 years after the identification of the catecholamine storage organelle, costorage of the chromogranins in these same particles was described (see Winkler and Fischer-Colbrie [37]). Chromogranins were found to be coreleased with catecholamines from the adrenal medulla, explaining the process of neuroendocrine secretion (exocytosis). These basic facts have since evolved into a cellular picture of hormone storage and regulated secretion in chromaffin cells and noradrenergic neurons. Most recently, a molecular understanding of catecholamine uptake and storage, vesicular dynamics, and protein costorage in neurons and endocrine cells has been developed, with parallel progress in molecular analysis of catecholamine re-uptake (discussed previously). The progress in basic research has been rewarded by an enhanced ability to compass aspects of autonomic, neurodegenerative, and especially psychiatric illness within the molecular framework of catecholamine uptake, storage and protein costorage.

Four "themes" of catecholamine vesicular uptake and protein costorage are touched on in the contributions of Part B: (1) the molecular determinants of transport mediated by VMAT1 and VMAT2; (2) the regulation of synthesis of bioactive polypeptide hormones costored with catecholamines and their physiological functions, once released into the extracellular space; (3) the biogenesis of catecholaminergic secretory vesicles and protein targeting to vesicles, including the genesis of small dense-core vesicles (SDCVs) of noradrenergic neurons; and (4) imaging of catecholamine nerve terminals in human neurodegenerative and psychiatric disease with *in vivo* markers for both plasma membrane and vesicular amine transport proteins.

A. Molecular Biology of the Vesicular Monoamine Transporters

Progress in understanding the structure, function, and physiology of VMATs accelerated when the two isoforms of the VMAT were cloned in 1992 (38). Schuldiner and coworkers (p. 223) describe the likely phylogenetic origin of VMATs, which exchange catecholamines for protons across the vesicular membrane, in bacterial antibiotic resistance proteins called TEXANs: 4- or 8-transmembrane domain–containing *T*oxin *EX*porting *AN*tiporters (hence the name) that exchange a positively charged toxin molecule within the bacterium for an extracellular proton. The analogy between TEXANs and catecholamine transporters is empirical, and therefore, TEXANs provide an ideal model system for assessing minimal structural requirements for substrate recognition and transport—and conceptual—vesicular transporters may have evolved from toxin exporters, and therefore neurotransmitters may have evolved from toxins. The concept of the 'toxin neurotransmitter' in fact is in force as a possible explanation for idiopathic neurodegenerative disease, such as Parkinson's. Loss of function of vesicular transporters could result in, for example, increased cytoplasmic DA and its neurotoxic metabolites, possibly shortening the lifetime of dopaminergic cells in the brain. This concept is now directly testable with VMAT2 heterozygous knockout mice, and homozygous knockouts in *C. elegans* (see Uhl *et al.*, p. 1024 and Rand *et al.*, p. 940). In addition to work with TEXANs, Schuldiner's group has demonstrated the importance of histidine between TMs 10 and 11 of VMAT for proton-coupled substrate transport and aspartate residues in TMs 10 and 11 for steps in transport distal to substrate binding, as inferred by preservation of reserpine binding, but not serotonin transport, in these mutants. Schuldiner *et al.* discuss the importance of mutagenesis coupled with existing information on the effect on transport of specific chemical reagents to develop mechanistic models for catecholamine transport by VMATs. Jeff Erickson (p. 227) discusses the structural basis for functional differences in transport, substrate recognition, and inhibitor binding between VMAT1 and VMAT2, the endocrine and neuronal isoforms of VMAT. Both are expressed in catecholamine-containing chromaffin cells in various species, although in the rat and human, VMAT1 is expressed only in neuroendocrine cells, and VMAT2 is the only isoform expressed in neurons (39, 40). VMAT2 is expressed in a type of endocrine cell, the oxyntic mucosal histamine-secreting ECL cell of the stomach, consistent with the far greater affinity of histamine for VMAT2 compared with VMAT1. Using chimeras of VMAT1 and VMAT2, Erickson demonstrates that replacement of the middle transmembrane domains of hVMAT2 with those from VMAT1 results in a VMAT2-like chimera with respect to interaction with amphetamine and tetrabenazine, but a VMAT1-like affinity for histamine. This suggests that multiple domains of VMAT2 are required for interaction with tetrabenazine, while residues critical for histamine interaction are present in the domain encompassing the conserved aspartate-containing TM 6 (plus surrounding regions). Finn and colleagues (p. 232) discuss the critical importance of the aspartate residue in TM1 of VMAT2 for substrate recognition, but not reserpine binding, and point out the correspondence to a similarly critical aspartate residue in the first TMs of both the β-receptor and the plasma membrane DAT. Finn and coworkers also used production of VMAT1/2 chimeras and site-directed mutagenesis of VMAT1 and VMAT2 to assess domains of VMAT2 important for tetrabenazine binding. They report results consistent with multiple sites overlapping with substrate recognition domains for binding of tetrabenazine to VMAT2. Intriguingly, charged residues in TMDs 2 and 11 are individually important for serotonin transport but are not required if doubly mutated to eliminate charge altogether. One of these is the conserved aspartate in TMD 11, suggesting its role is in maintenance of transporter conformation rather than in transport function per se. Henry and his colleagues (p. 236) discuss the use of iodoketanserin binding to allow covalent labeling of regions of the transporter near those binding specific inhibitors of transport to be identified. Because VMAT2 rather than VMAT1 apears to be the major isoform expressed in bovine adrenal, it is perhaps significant that both VMAT1 and VMAT2 of cow, unlike human and rat, can bind tetrabenazine. The relative importance of aspartic acid residues in TMs 1, 6, 10, and 11 is underscored by additional functional mutagenesis studies, referenced in the contributions summarized here and the description elsewhere in this volume, of a recently cloned "primordial" VMAT from *C. elegans* (Rand *et al.*, p. 940), in which these residues are also conserved.

B. Synthesis, Processing, Targeting, and Function of Polypeptides Costored with Catecholamines

The regulation of synthesis and costorage of secretory proteins in catecholamine-containing cells and their roles in catecholamine secretion are described in a series of papers by Winkler *et al.* (p. 257), Mahata *et al.* (p. 260), and Eiden *et al.* (p.). Considering their ubiquitous expression in virtually all neurons and neuroendocrine cells and special abundance in sympathoadrenal catecholaminergic cells, remarkably little is firmly established about what the chromogranins do. Winkler and colleagues (p. 257) discuss the multiple roles of the chromogranins, possible actors in vesiculogenesis as well as prohormone precursors for biologically active peptides, in chromaffin cells and noradrenergic neurons, focusing in particular on secretoneurin, a biologically active peptide processed from secretogranin II by PCI/III cleavage in secretory granules. Mahata *et al.* (p. 260) have discovered that catestatin, a polypeptide derived from chromogranin A by proteolytic processing and presumably coreleased with catecholamines from catecholaminergic neurons and chromaffin cells, inhibits catecholamine release induced by nicotine via indirect action at the nicotinic cholinergic receptor. A potential autocrine role for a chromogranin A–derived peptide is thus added to the list of functions for chromogranin. Coregulation of neuropeptide gene expression and processing along with catecholamine biosynthesis are important in maintaining (in Winkler's phrase) a defined "secretory cocktail" within catecholamine/neuropeptide-containing cells. The coupling of neuropeptide gene transcription to calcium influx during secretion, so-called stimulus-secretion-synthesis coupling is a mechanism whereby this occurs (Eiden *et al.*, p. 264). Activation of second-messenger pathways operating *in vivo* and studied in chromaffin cells *in vitro* results in dramatic changes in the secretory cocktail that may be relevant to the role of neuroendocrine cells in responding to pathophysiological states or states of chronically enhanced neurosecretion. Several laboratories are now defining the cis- and trans-signaling determinants responsible for the acquisition of peptidergic phenotypes in catecholaminergic cells, using chromaffin and PC12 cells and cultured sympathetic neurons as model systems to identify cAMP-responsive element (CRE) and TPA-responsive element (TRE) gene sequences (also shared by catecholaminergic biosynthetic enzyme genes) on which depolarization, protein kinase, and extracellular response kinase (ERK) pathways act via trans-activators such as AP1, cAMP response element-binding protein (CREB), and others (41, 42).

C. Catecholamine Storage Vesicule Genesis

Current unanswered questions about the genesis of the vesicles into which catecholamines are packaged include the functional importance of lumenal structures, including small dense-core vesicles, that store catecholamines, and their spatial, biochemical, and catalytic properties relevant for exocytotic catecholamine release (neurotransmission) or pharmacological release of catecholamine stores. Nirenberg and colleagues (p. 240) localized VMAT2 by immune electron microscopy to "tubulovesicular" structures. Especially in dendrites of dopaminergic neurons, these might function as the morphological substrates for dendritic release of DA in substantia nigra. Dirkx *et al.* (p. 243) describes the localization of the insulin autoimmune antigen ICA 512, a receptor tyrosine phosphatase–like protein, to the membrane of neurosecretory granules. ICA 512 is likely to become an important marker for study of the differentiation of synaptic vesicle and secretory vesicle biogenesis in catecholamine-containing neuroendocrine cells. West *et al.* (p. 247) describe a system for tagging proteins bound for axons versus dendrites in hippocampal neurons, which allows production and "fate mapping" of chimeras of such proteins reintroduced into cells with dendritic and axonal polarity. Using the transferrin receptor as a model protein, West and colleagues (p. 247) provide evidence that axonal localization is probably the "default" condition for polarized neuronal proteins, while specific dendritic retention signals are required for routing to the latter compartment. Regarding targeting to specific organelles within nerve terminals, these workers describe employment of chimeras of SV2, a vesicular protein assumed to be a transporter but with unknown function, and the glucose transporter, which is targeted to neuronal plasma membrane, expressed in PC12 cells. The studies yielded surprisingly crisp results about the location of cytoplasmically oriented signals for endosomal retention and targeting of membrane-bound proteins to secretory vesicles that are likely to lead to further progress in vesicular targeting of additional molecules, including the vesicular amine transporters, which bear significant structural similarity to SV2. Tao-Cheng and Eiden (p. 250) discuss the use of immune electronmicroscopical localization of vesicular antigens, some introduced by retrovirally mediated gene transfer, to study how proteins are routed to secretory granules or synaptic vesicles in PC12 cells. Surprisingly, proteins transported down the axon by different mechanisms (constitutive vesicles or large dense-core vesicles [LDCVs]) can end up in the same place (small synaptic vesicles [SSVs]). VMAT2, a marker for both LDCVs and SDCVs in noradrenergic neurons, localizes only to LDCVs in PC12 cells, indicating that SDCVs are not merely "SSVs with catecholamines" but organelles with a unique biogenesis in noradrenergic neurons. The contribution of Bauerfeind *et al.* (p. 253) concludes this group of papers by discussing the recycling of synaptic vesicles, invoking a model in which dynamin, amphiphysin, and synaptojanin interactions, controlled by multiple phosphorylation and dephosphorylation events, lead to "clathrinization" and endocytosis of secretory vesicles. *Inactivation* of this pathway by the calcium-activated phosphatase calcineurin, and *activation* of exocytosis involving calcineurin, provide a potential molecular switch for coordinating exo- and endocytosis in neurons.

An intriguing ongoing story in catecholamine cell biology is the origin and function of SDCVs of noradrenergic nerve terminals, which are absent from catecholamine-containing endocrine cells. LDCVs are electron-opaque because they contain proteins like chromogranins. SDCVs are electron-opaque (although they contain no soluble proteins such as chromogranin A, or neuropeptides) because tissue treatment with permanganate or dichromate produces electron-dense deposits of β-hydroxylated catecholamines. Without treatment,

SDCVs look like SSVs under the electron microscope and have, therefore, been assumed to be essentially identical to SDCVs but for their catecholamines and the vesicular transporter that puts the catecholamines in them. The results of Tao-Cheng and Eiden (vide infra) suggest otherwise: VMAT2 is found in SDCVs in noradrenergic cells, yet does not target to SSVs in PC12 cells. How different are SDCVs, and how do they arise in noradrenergic cells? And what do they do? Bauerfeind and Huttner have hypothesized that SDCVs are an early endosome-derived "hybrid" of SDCVs and SSVs (43). Are SDCVs a secretory pool at all? De Potter and colleagues (44) argue that they are not: The ratios of neuropeptide Y (NPY) to catecholamines released from sheep noradrenergic nerves are constant across a broad range of stimulus intensities and equivalent to the neuropeptide-CA ratios of LDCVs. These are much higher than the ratios found in SDCVs, suggesting that SDCVs do not participate in exocytotic catecholamine release in peripheral noradrenergic nerves. The ratios of SDCVs to LDCVs vary widely among noradrenergic tissues and mammalian species, in part corresponding with relative tonic nerve activity. Thus, whatever their functional role, SDCV levels may in part reflect of the rate of neurotransmission at the nerve terminal.

D. *In Vivo* Imaging of Vesicular Transporters

Why might the functional status of SDCVs in neurotransmittion at noradrenergic nerve terminals be important? Correlation of the density of vesicular and plasma membrane uptake binding sites, with synaptic dynamics, underlies attempts to use *in vivo* imaging of synaptic markers for staging and understanding the pathogenesis of CNS neurodegenerative disease. Contributions by Volkow, Innis, and Wong on plasma membrane transporters (discussed earlier) discuss *in vivo* imaging of DAT in the human brain of patients with Parkinson's disease as well as neuropsychiatric disorders. Frey *et al.* (p. 269) report on imaging of the vesicular amine transporters VAChT and VMAT2, with labeled congeners of their respective inhibitors vesamicol and tetrabenazine, in Parkinson's and Alzheimer's disease patients. While methodological issues remain paramount in the interpretation of such studies, it seems clear that direct observation of the number of plasma membrane and vesicular transporter molecules present in different regions of the diseased brain through progression of disease is an achieved goal.

III. Future Directions

Although studies with chimeric DAT/NET transporters have helped to identify regions within these transporters involved in substrate or inhibitor binding, an exact knowledge of the ligand recognition sites is, however, still lacking. Knowledge of these sites could contribute to the development of new antidepressants. Studies in site-directed mutants of transporters and the use of photoaffinity ligands (which are not yet available for NET) are possible approaches for identifing these binding sites.

Whether the activity or expression of NET is regulated or modulated by activation of presynaptic receptors remains to be examined; this might eventually help to clarify the delay in the onset of antidepressants.

The promoter region of the NET is still unknown, and it is not yet understood how gene transcription of the DAT gene is driven. It remains to be shown whether variant promoter sequences might be involved in psychiatric disorders. However, the report by Lesch *et al.* (45), showing an association of anxiety-related traits with a polymorphism in the serotonin transporter gene regulatory region, raises the possibility that similar polymorphisms or genetic variants may also exist in DAT and NET transporter genes. Such variants, if they exist, could be important factors in the susceptibility to some disease processes. These may not be limited to psychiatric or neurological disorders. NET also participates importantly in inactivating NE released by sympathetic nerves, and impaired NE re-uptake has been demonstrated in certain circulatory disorders, such as congestive heart failure (46).

The development of knockout mice lacking the NET and DAT genes can be anticipated to not only further understanding of the physiological and pathological significance of DA and NE systems in the CNS, but also clarify the importance of these systems in the periphery.

It can be anticipated that SPECT and PET imaging of DATs will prove useful for diagnosing patients in early states of neurodegenerative disorders or to monitor progression of disease or the efficacy of neuroprotective agents.

For the vesicular transporters, what remains challenging in the clinical realm is correlating of *in vivo* imaging data with actual degree of patency of neurotransmission at affected synapses. Even without profound insight into their functional significance, however, monitoring the onset of neuronal degeneration in at-risk populations, may produce the lead time to begin potential neurotrophin and other preventive rather than restorative therapies. The importance of gene dosage (i.e., the surprising phenotypic penetrance of heterozygous null mutations of the VMAT and DAT loci) also points up the potential for identification of genetic abnormalities in vesicular as well as plasma membrane transporter genes involved in neuropsychiatric or neurodegenerative disease. These would be worthwhile clinical payoffs for the basic research elucidating the molecular methods underlying catecholamine re-uptake and storage.

References

1. Takahashi, N., Miner, L., Sora, I., Ujike, H., Revay, R., Kostic, V., Jackson-Lewis, L., Przedborski, S., and Uhl, G. (1997). VMAT2 knockout mice: Heterozygotes display reduced amphetamine-conditioned reward, enhanced amphetamine locomotion and enhanced MPTP toxicity. *Proc. Natl. Acad. Sci. USA,* in press.
2. Guastella, J., Nelson, N., Nelson, H., Czyk, L., Keynan, S., Miedel, M., Davidson, N., Lester, H., and Kanner, B. (1990). Cloning and expression of a rat brain GABA transporter. *Science* **249,** 1303–1306.
3. Amara, S. G., and Kuhar, M. J. (1993). Neurotransmitter transporters: Recent progress. *Annu. Rev. Neurosci.* **16,** 73–93.
4. Borowsky, B., and Hoffman, B. (1995). Neurotransmitter transporters: Molecular biology, function, and regulation. *Int. Rev. Neurobiol.* **38,** 139–199.

5. Brüss, M., Hammermann, R., Brimijoin, S., and Bönisch, H. (1995). Antipeptide antibodies confirm the topology of the human norepinephrine transporter. *J. Biol. Chem.* **270,** 9197–9291.
6. Pacholczyk, T., Blakely, R. D., and Amara, S. G. (1991). Expression cloning of a cocaine- and antidepressant-sensitive human noradrenaline transporter. *Nature* **350,** 350–354.
7. Lingen, B., Brüss, M., and Bönisch, H. (1994). Cloning and expression of the bovine sodium- and chloride-dependent noradrenaline transporter. *FEBS Lett.* **342,** 235–238.
8. Brüss, M., Pörzgen, P., Bryan-Lluka, L. J., and Bönisch, H. (1997). Cloning of the (uptake$_1$) noradrenaline transporter from PC12 cells. *Naunyn Schmiedebergs Arch. Pharmacol.* **355,** (suppl), R27.
9. Larong, D., Amara, S. G., and Simerly, R. B. (1993). Cell-type specific expression of catecholamine transporters in the rat brain. *J. Neurosci.* **14,** 4903–4914.
10. Graefe, K. H., and Bönisch, H. (1988). The transport of amines across the axonal membranes of noradrenergic and dopaminergic neurones. *In* Handbook of Experimental Pharmacology, 90/I. (U. Trendelenburg and N. Weiner eds.), pp. 193–245. Springer-Verlag, Berlin, Heidelberg, New York.
11. Iversen, L. L. (1967). The Uptake and Storage of Noradrenaline in Sympathetic Nerves. Cambridge University Press, Cambridge.
12. Paton, D. M. (1976). The Mechanism of Neuronal and Extraneuronal Transport of Catecholamines. Raven Press, New York.
13. Meiergerd, S. M., and Schenk, J. O. (1994). Striatal transporters for dopamine: Catechol structure-activity studies und susceptibility to chemical modification. *J. Neurochem.* **63,** 1683–1692.
14. Burnette, W. B., Bailey, M. D., Kukoyi, S., Blakely, R. D., Trowbridge, C. G., and Justice, J. B. (1996). Investigation of human norepinephrine transporter kinetics using rotating disk electrode voltametry. *Anal. Chem.* **68,** 2932–2938.
15. Bönisch, H. (1984). The transport of (+)amphetamine by the neuronal noradrenaline carrier. *Naunyn Schmiedebergs Arch. Pharmacol.* **327,** 267–272.
16. Wall, S. C., Gu, H. H., and Rudnick, G. (1995). Biogenic amine flux mediated by cloned transporters stably expressed in cultured cell lines: amphetamine specificity for inhibition and efflux. *Mol. Pharmacol.* **47,** 544–550.
17. Harder, R., and Bönisch, H. (1985). Effect of monovalent ions on the transport of noradrenaline across the plasma membrane of neuronal cells (PC 12 cells). *J. Neurochem.* **45,** 1154–1162.
18. Gu, H. H., Wall, S., and Rudnick, G. (1996). Ion coupling stoichiometry for the norepinephrine transporter in membrane vesicles from stably transfected cells. *J. Biol. Chem.* **271,** 6911–6916.
19. Sonders, M. S. (1996). Characterization of two steady-state conductances mediated by the electrogenic human dopamine transporter. *Biophys. J.* **70,** A98.
20. Cass, W. A., and Gerhardt, G. A. (1994). Direct evidence that D2 dopamine receptors can modulate dopamine uptake. *Neurosci. Lett.* **176,** 259–263.
21. Galli, A., DeFelice, L. J., Duke, B. J., Moore, K. R., and Blakely, R. D. (1995). Sodium-dependent norepinephrine-induced currents in norepinephrine-transporter-transfected HEK-293 cells blocked by cocaine and antidepressants. *J. Exp. Biol.* **198,** 2197–2212.
22. DeFelice, L. J., and Blakely, R. D. (1996). Pore models for transporters? *Biophys. J.* **70,** 579–580.
23. Sonders, M., and Amara, S. G. (1996). Channels in transporters. *Curr. Opin. Neurobiol.* **2,** 294–302.
24. Nguyen, T. T., and Amara, S. G. (1996). N-linked oligosaccharides are required for cell surface expression of the norepinephrine transporter but do not influence substrate or inhibitor recognition. *J. Neurochem.* **67,** 645–655.
25. Melikian, H. E., Ramamoorthy, S., Tate, C. G., and Blakely, R. D. (1996). Inability to N-glycosylate the human norepinephrine transporter reduces protein stability, surface

trafficking, and transport activity but not ligand recognition. *Mol. Pharmacol.* **50**, 266–276.
26. Kitayama, S., Dohi, T., and Uhl, R. G. (1994). Phorbol esters alter functions of the expressed dopamine transporter. *Eur. J. Pharmacol.* **268**, 115–119.
27. Kadowaki, K., Hirota, K., Koike, K., Ohmichi, M., Kiyama, H., Miyake, A., and Tanizawa, O. (1990). Adenosine 3′,5′-cyclic monophosphate enhances dopamine accumulation in rat hypothalamic cell cultures containing dopaminergic neurons. *Neuroendocrinology* **52**, 256–261.
28. Perrone-Capano, C., Tino, A., Amadoro, G., Pernas-Alonso, R., and DiPorzio, U. (1996). Dopamine transporter gene expression in rat mesencephalic dopaminergic neurons is increased by direct interaction with target striatal cells in vitro. *Mol. Brain Res.* **39**, 160–166.
29. Lew, R., Vaughan, R., Simantov, R., Wilson, A., and Kuhar, J. M. (1991). Dopamine transporters in the nucleus accumbens and the striatum have different apparent molecular weights. *Synapse* **8**, 152–153.
30. Jones, R. S., Garris, A. P., Kilts, D. C., and Wightman, M. R. (1995). Comparison of dopamine uptake in the basolateral amygdaloid nucleus, caudate-putamen, and nucleus accumbens of the rat. *J. Neurochem.* **64**, 2581–2589.
31. Marino, V., DeLaLande, I. S., Parker, D. A. S., Dally, J., and Wing, S. (1992). Extraneuronal uptake of noradrenaline in rabbit dental pulp: Evidence of identity with uptake$_1$. *Naunyn Schmiedebergs Arch. Pharmacol.* **346**, 166–172.
32. DeLaLande, I. S. V., Marino, V., Kennedy, J. A., Parker, D. A., and Seamark, R. F. (1991). Distribution of extraneuronal uptake$_1$ in reproductive tissues: Studies on cells in culture. *J. Neural. Transm. Suppl.* **34**, (suppl), 37–42.
33. Ramamoorthy, S., Prasad, P. D., Kulanthailev, P., Leibach, F. H., Blakely, R. D., and Ganapathy, V. (1993). Expression of a cocaine-sensitive norepinephrine transporter in the human placental syncytiotrophoblast. *Biochemistry* **32**, 1346–1353.
34. Giros, B., Jaber, M., Jones, S. R., Wightman, R. M., and Caron, M. G. (1996). Hyperlocomotion and indifference to cocaine and amphetamine in mice lacking the dopamine transporter. *Nature* **379**, 606–612.
35. Hillarp, N-A. (1958). Isolation and some biochemical properties of the catechol amine granules in the cow adrenal medulla. *Acta Physiol. Scand.* **43**, 82–96.
36. Winkler, H., and Westhead, E. (1980). The molecular organization of adrenal chromaffin granules. *Neuroscience* **5**, 1803–1823.
37. Winkler, H., and Fischer-Colbrie, R. (1992). The chromogranins A and B: The first 25 years and future perspectives. *Neuroscience* **49**, 497–528.
38. Schuldiner, S. (1994). A molecular glimpse of vesicular transporters. *J. Neurochem.* **62**, 2067–2078.
39. Peter, D., Liu, Y., Sternini, C., de Giorgio, R., Brecha, N., and Edwards, R. H. (1995). Differential expression of two vesicular monoamine transporters. *J. Neurosci.* **15**, 6179–6188.
40. Weihe, E., Schäfer, MK-H., Erickson, J. D., and Eiden, L. E. (1994). Localization of vesicular monoamine transporter isoforms (VMAT1 and VMAT2) to endocrine cells and neurons in rat. *J. Mol. Neurosci.* **5**, 149–164.
41. Mahata, S. K., Wu, H., Mahata, M., Minth-Worby, C., Parmer, R. J., and O'Connor, D. T. (1997). Acquisition of peptidergic phenotypes by catecholamine storage vesicles: Cis and trans signaling determinants. *Exp. Neurol.*, in press.
42. MacArthur, L., and Eiden, L. E. (1996). Neuropeptide genes: Targets of activity-dependent signal transduction. *Peptides* **17**, 721–728.
43. Bauerfeind, R., Jelinek, R., Hellwig, A., and Huttner, W. B. (1995). Neurosecretory vesicles can be hybrids of synaptic vesicles and secretory granules. *Proc. Natl. Acad. Sci. U.S.A.* **92**, 7342–7346.
44. De Potter, W. P., Partoens, P., Schoups, A., Llona, I., and Coen, E. P. (1997). Noradrenergic neurons release both noradrenaline and neuropeptide Y from a single pool: The large dense cored vesicles. *Synapse* **25**, 44–55.

45. Lesch, K. P., Bengel, D., Heils, A., Sabol, S. Z., Greenberg, B. D., Petri, S., Benjamin, J., Muller, C. R., Hamer, D. H., and Murphy, D. L. (1996). Association of anxiety-related traits with a polymorphism in the serotonin transporter gene regulatory region. *Science* **274**, 1527–1531.
46. Eisenhofer, G., Friberg, P., Rundqvist, B., Quyyumi, A. A., Lambert, G., Kaye, D. M., Kopin, I. J., Goldstein, D. S., and Esler, M. D. (1996). Cardiac sympathetic nerve function in congestive heart failure. *Circulation* **93**, 1667–1676.

S. G. Amara,* M. S. Sonders,* N. R. Zahniser,† S. L. Povlock,* and G. M. Daniels*

*Vollum Institute and Howard Hughes Medical Institute
Oregon Health Sciences University
Portland, Oregon 97201

†Department of Pharmacology
University of Colorado Health Sciences Center
Denver, Colorado 80262

Molecular Physiology and Regulation of Catecholamine Transporters

Catecholamine transporters found in the plasma membrane of neurons and possibly glial cells mediate the removal of neurotransmitter from the extracellular space, thus limiting the activation of receptors during neuronal signaling. Two distinct carrier subtypes, the dopamine and norepinephrine transporters (DAT and NET), are responsible for the reaccumulation of catcholamines into dopaminergic and noradrenergic neurons, respectively. These carriers, together with the serotonin transporter, are the primary targets for a wide variety of clinically important antidepressants, antihypertensives, stimulants, and stimulant drugs of abuse (1). We have been examining several novel functional properties and regulatory mechanisms that contribute to the impact DAT and NET have on catecholaminegic neurotransmission and that may be relevant to the actions of specific classes of drugs. Electrophysiological analyses of the currents associated with transporters have provided insights into the electrical properties and the pathways of ion permeation of the carrier proteins (2). Recent work from our laboratory has shown that dopamine uptake by human DAT (hDAT) is an electrogenic and voltage-dependent process. Although one might assume that the currents associated with these sodium-coupled cotrans-

Advances in Pharmacology, Volume 42

1054-3589/98 $25.00

porters would reflect the net charge of substrates and ions transported, electrophysiological investigations of neurotransmitter transporters demonstrate a much richer variety of electrical behaviors more akin to those of ion channels. Transporter-mediated currents not only provide tools for examining the mechanisms by which these pumps operate, but also suggest the possibility that transporters may also influence the signaling properties of excitable cells by means apart from the modulation of extracellular neurotransmitter concentrations. A second line of investigation has examined how intracellular signaling mechanisms regulate the number of carriers expressed at the cell surface, focusing on a particularly striking effect of protein kinase C (PKC) activation on the internalization and sequestration of NET and DAT proteins. Our overall goals are to relate protein structural features to novel functional properties and regulatory mechanisms and to establish the contributions of the DA and NE carriers to neuronal function.

I. Currents Associated with the Human Dopamine Transporter

We have used voltage clamp techniques in *Xenopus* oocytes expressing hDAT to analyze currents generated by the movement of substrates and ions and to determine how various drugs influence these currents (3). Dopamine uptake by DAT is electrogenic, giving rise to an inward transport-associated current. One of the unique advantages of the oocyte system is that the uptake of [^{3}H]DA can be measured while controlling the oocyte membrane potential by voltage clamp. Using this technique across a range of potentials, we have demonstrated that hyperpolarization increases the velocity of [^{3}H]DA uptake, as would be expected for an electrogenic carrier which couples the inward movement of DA and of net positive charge. This finding suggests one possible mechanism by which DA or drugs acting at DA autoreceptors could influence presynaptic DA uptake. Because activation of $D_{2/3}$ autoreceptors ordinarily increases an outward K^+ current in dopaminergic neurons, the resulting hyperpolarization would be predicted to increase the clearance of extracellular DA by DAT.

An unpredicted finding of these experiments, however, was that the charge flux (i.e., current) accompanying DA uptake exceeds what would have been expected based on the generally accepted coupling stoichiometry of 2 Na^+/1 Cl^-/1 DA^+. Correlation of [^{3}H]DA uptake velocity with net charge movement in voltage-clamped oocytes yields a mean charge to DA flux ratio ranging between 2.5+ (at −30 mV) and 11.5+ (at −120 mV) charges per DA molecule. One interpretation of the increase in net charge DA flux ratio with membrane hyperpolarization is that, aside from the stoichiometric translocation of DA with cosubstrate ions, hDAT also regulates a transmembrane current to which is activated by, but not stoichiometrically coupled to, DA translocation.

hDAT mediates another macroscopic current that can be easily distinguished from the transport-associated currents, namely, a constitutively active leak current. In contrast to the transport-associated current, the leak conductance is blocked both by hDAT substrates (such as DA, amphetamine, and

MPP$^+$) and by translocation inhibitors (such as cocaine, GBR12935, and methylphenidate). Because the transport-associated and leak currents have different voltage and ionic dependences, it is straightforward to determine from the current-voltage plot of an individual drug whether it acts as a substrate or as a nonsubstrate translocation inhibitor at hDAT. Because all hDAT substrates are also known to be Ca^{2+}-independent "releasers" of preloaded DA, this electrophysiological assay allows the discrimination of drugs that promote efflux from those that prevent uptake through hDAT—a distinction that can be complicated by the lipophilicity and multiple sites of action of many DAT ligands.

II. Regulation of Transporter Cell Surface Expression by PKC

Cell lines derived from a canine kidney cell line (MDCK cells) stably expressing catecholamine transporters have been used to explore how intracellular signaling mechanisms regulate the activity and number of carriers expressed at the cell surface. These studies have focused on the effects of PKC activation on the internalization and sequestration of DAT and NET proteins. Catecholamine transport velocity can be acutely regulated by activation of second messenger systems. In our studies, the activation of PKC by phorbol esters causes a rapid reduction of 20–50% in transport activity of both DAT and NET by decreasing the carrier V_{max} without altering the apparent substrate affinities. This decrease occurs rapidly, with maximal inhibition achieved within 20 minutes. No change in transport activity is observed when cells are treated with 4α-PMA, an inactive analogue of PMA. Furthermore, incubation of the cells for 16 hours with 100 nM PMA to deplete the endogenous PKC activity abolished PMA-induced inhibition of transport.

Examination of DAT and NET expressing cells by indirect immunofluoresence revealed a striking difference in protein distribution on activation of PKC. In control cells, transporter localization is characterized by intense staining at the cell surface (Fig. 1). In addition, there is a less intense perinuclear staining, presumably due to immature carrier protein transiting through the secretory pathway. In contrast, cells treated with PMA show a marked decrease in plasma membrane staining and a concomitant dramatic increase in intracellular staining. This staining appears as very intense punctate staining suggestive of a vesicular localization. Cell surface biotinylation was also utilized to assess the quantity of transporter protein found at the plasma membrane. The amount of biotinylated carrier protein was decreased significantly in PMA-treated cells when compared with control cells, confirming the results of immunofluorescence localization and providing further evidence that the PKC-mediated reduction in transport activity is due to a loss of transporter protein at the cell surface.

Thus far we have no evidence that any alterations in transport activity involve direct phosphorylation of either NET or DAT. Attempts were made to assess *in vivo* phosphorylation by treating cells expressing DAT or NET with PMA, followed by immunoprecipitation with transporter-specific antibodies.

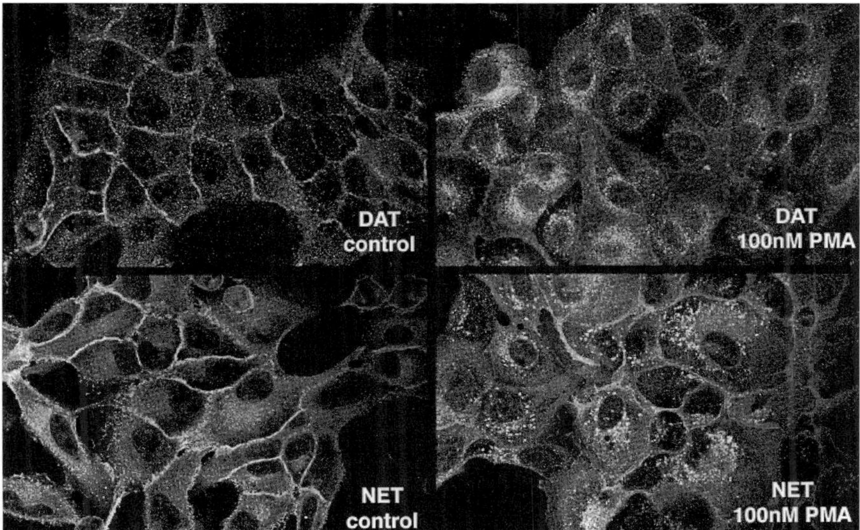

FIGURE 1 PMA causes a dramatic shift of carriers from the surface to intracellular compartments. The cellular localization of transporters was examined by indirect immunofluorescence with confocal laser scanning microscopy in control and treated cells. MDCK cells stably expressing DAT or NET were treated for 1 hr with either 100 nM PMA or vehicle control. Cells were then fixed, permeabilized, and incubated with anti-NET or anti-DAT antisera followed by incubation with a fluorescein-conjugated donkey antirabbit antibody. Control cells (*left panels, top and bottom*) showed intense immunofluorescent staining at the cell surface and membrane border, while cells treated with PMA (*right panels, top and bottom*) showed reduced surface staining and a dramatic increase in perinuclear staining.

Although [^{35}S]-labeled carriers could be precipitated in a parallel experiment, there was no evidence for a direct phosphorylation that correlated with changes in carrier activity. Furthermore, removal of some of the putative phosphorylation sites from DAT had no apparent influence on the V_{max} or K_m of the carrier.

Acknowledgments

These investigations have been supported by NIH grants DA07595 and DA04216 and the Howard Hughes Medical Institute.

References

1. Amara, S. G. (1995). Monoamine transporters: Basic biology with clinical implications. *Neuroscientist* 1, 259–267.

2. Sonders, M., and Amara, S. G. (1997). Channels in transporters. *Curr. Opin. Neurobiol.* 6, 294–302.
3. Sonders, M., Zahniser, N., Zhu, S-J., Kavanaugh, M. P., and Amara, S. G. Multiple ionic conductances of the human dopamine transporter: The actions of dopamine and psychostimulants. *J. Neurosci.* 17, 960–974.

M. J. Kuhar,* R. Vaughan,† G. Uhl,† C. Cerruti,† R. Revay,† C. Freed,† M. Nirenburg,‡ and V. Pickel‡

*Yerkes Regional Primate Research Center
Emory University
Atlanta, Georgia 30329

†Addiction Reseach Center
National Institute on Drug Abuse
Baltimore, Maryland 21224

‡Cornell University Medical College
New York, New York 10021

Localization of Dopamine Transporter Protein by Microscopic Histochemistry

The dopamine transporter (DAT) is an important protein at dopaminergic synapses. It is involved in the inactivation of and retrieval of released dopamine (DA), a target or "receptor" for psychostimulant drugs such as cocaine, a target for the therapeutic drug methylphenidate, a mediator of drug-induced neurotoxicity, and perhaps the only unique marker for dopaminergic neurons (1).

Following the cloning of a DAT cDNA by several groups, it has become feasible to map the distribution of mRNA for DAT by *in situ* hybridization, as well as to generate antibodies against DAT peptide fragments and to carry out immunohistochemistry. This is a brief summary of the findings of these studies in our laboratory. While there is another transporter for DA, the vesicular transporter, our focus has been the cell surface DAT.

In *in situ* hybridization experiments, we found DAT mRNA only in cell groups previously known to be dopaminergic. However, the amount of probe that bound differed from cell group to cell group, indicating that different levels of DAT mRNA were present. This in turn suggests that different cells make different quantities of DAT, something that was verified in immunohistochemi-

Advances in Pharmacology, Volume 42

1054-3589/98 $25.00

cal studies (discussion follows). In the midbrain ventral tegmentum, dopaminergic cell groups exhibiting probe binding from highest levels to lowest are as follows: substantia nigra zona compacta, parabrachial nucleus, the paranigral nucleus, the rostral linear nucleus, the interfascicular nucleus, and the caudal linear nucleus. There was perhaps a three- to fourfold difference in the average hybridization density from the highest cell densities in the substantia nigra to the lowest in the caudal linear nucleus. Within each nucleus there was a broad variation in densities over the individual cells. For example, in the substantia nigra, a few cells had almost 100 grains per 100 square microns, while several had as few as 30 grains per 100 square microns (2).

The observation that different dopaminergic cell groups express different quantities of DAT is intriguing and suggests that a functional consequence of this may be a basis for "volume" or "paracrine" transmission at dopaminergic synapses. The hypothesis would be that relatively low levels of DAT would be associated with greater probabilities of volume transmission, because there would be less chance for DA to be recaptured by the neuron.

In immunohistochemical studies (3, 4), the dopaminergic neurons of the midbrain had the highest levels of DAT immunoreactivity. DAT was clearly in cell bodies, dendrites, axons, and nerve terminals. The finding that DAT is distributed throughout the neuron was not completely surprising because DAT is obviously made in the cell body and transported to other parts of the neuron. However, outside the midbrain, such as in the tuberoinfundibular neurons, DAT staining was prominent in the nerve terminal region in the external layer of the median eminence, but very weak in the cell bodies in the arcuate nucleus. The latter results suggest that DAT is accumulated and stabilized in the nerve terminals of these neurons. Although the mechanism of this compartmentation or stabilization is unknown, the localization of DAT is consistent with the hypothesis of DAT recycling between intracellular and cell surface pools at the synapse.

These results with the light microscope suggest that DAT is found in high levels throughout some dopaminergic neurons, such as those in the midbrain, but is mainly sequestered in nerve terminal areas in other dopaminergic neurons. The functional consequence of such an extensive localization in midbrain neurons would be a careful intracellular sequestering of DA by these neurons. Also, compartmentation of DAT to nerve terminal regions suggests key role for DAT in nerve terminals, which, of course, is not surprising.

The electron microscopic localization of DAT was carried out in collaboration (5) and will be presented in detail in other chapters of this book. We will describe one finding here that may have some interesting functional consequences. At the synapse, DAT was found in the presynaptic plasma membrane, as expected, with some DAT in tubulovesicles inside the nerve terminal. However, there was little or no DAT in the synaptic thickening or the so-called active zone. Thus, DAT was really found *surrounding* the synapse rather than *in* the synapse. If this is an accurate depiction of the distribution of DAT levels, then DA is not taken up in the synapse but is taken up *after* it diffuses out of the synapse. This would suggest that diffusion of DA plays a bigger role in the inactivation of DA in the synaptic cleft than previously suspected. DA must diffuse out of the cleft before it is taken up. Much evidence indicates that re-uptake is critical for reducing neurotransmitter levels, but, according to this notion, diffusion is critical as well.

In summary, DAT is found only in dopaminergic cell in brain, but in varying levels. The lower levels of DAT may be associated with volume transmission. In some neurons, such as the arcuate nucleus, there appeared to be a build up of DAT in nerve terminal regions, suggesting a compartmentation or enrichment in nerve terminals. Finally, DAT is enriched in perisynaptic regions rather than in the synapse, suggesting that diffusion may be more important in cessation of transmission than was previously thought.

References

1. Kuhar, M. J. (1993). Otto Krayer Award Lecture: Neuotransmitter transporters as drug targets. *Pharmacologist* **35**, 28–33.
2. Cerruti, C., Pilotte, N. S., Uhl, G. R., and Kuhar, M. J. (1994). Reduction in dopamine transporter mRNA after cessation of repeated cocaine administration. *Mol. Brain Res.* **22**, 132–138.
3. Freed, C., Revay, R., Vaughan, R. A., Kriek, E., Grant, S., Uhl, G. R., and Kuhar, M. J. (1995). Dopamine transporter immunoreactivity in rat brain. *J. Comp. Neurol.* **359**, 340–349.
4. Revay, R., Vaughan, R., Grant, S., and Kuhar, M. J. (1996). Dopamine transporter immunohistochemistry in median eminence, amygdala, and other areas of the rat brain. *Synapse* **22**, 93–99.
5. Nirenberg, M. J., Vaughan, R. A., Uhl, G. R., Kuhar, M. J., and Pickel, V. M. (1996). The dopamine transporter is localized to dendritic and axonal plasma membranes of nigrostriatal dopaminergic neurons. *J. Neurosci.* **16**, 436–447.

Susan R. Sesack,* Valerie A. Hawrylak,*
Margaret A. Guido,* and Allan I. Levey†

*Departments of Neuroscience and Psychiatry
University of Pittsburgh
Pittsburgh, Pennsylvania 15260

†Department of Neurology
Emory University
Atlanta, Georiga 30322

Cellular and Subcellular Localization of the Dopamine Transporter in Rat Cortex

The dopamine transporter (DAT) plays a critical role in regulating the duration of dopamine's synaptic actions and the extent to which dopamine can diffuse in the extracellular space (1). Previous light microscopic immunocytochemical studies of the rat striatal neuropil have demonstrated a dense localization of DAT (2). Furthermore, ultrastructural subcellular studies indicate that most of the DAT protein in striatal dopamine axons is distributed at the periphery of synapses and at nonsynaptic membrane sites (3). This extensive distribution of the DAT protein is likely to greatly restrict the extracellular diffusion of dopamine in this brain region.

In rodents, the rostromedial cortex also receives a dopamine innervation, and behavioral studies indicate that this input is important for proper cognitive functioning. Separate dopamine systems derive from A10 or A9 dopamine neurons and terminate, respectively, in deep layers of the prelimbic or superficial layers of the anterior cingulate cortices. Several neurochemical observations suggest that extracellular dopamine in the prelimbic division, or prefrontal cortex (PFC), undergoes less regulation by DAT-mediated uptake, compared with the striatum. (1) Relative to total tissue content, extracellular dopamine levels are 20 times higher in the PFC than in the striatum. (2) In the PFC, levels of the extracellular metabolite, HVA, are higher than the intracellular metabolite, DOPAC, while the reverse is true in the striatum. (3) Extracellular dopamine diffuses farther in the PFC than in the striatum or other forebrain areas (4). These observations are typically explained on the basis of a reduced availability of DAT, secondary to a lower density of cortical dopamine axons. However, we wished to explore the hypothesis that the neurochemical profile of dopamine overflow and diffusion in the rat PFC is consistent with a restricted distribution of the DAT protein in individual dopamine axons. We tested this hypothesis by using a light and electron microscopic immunocytochemical approach.

Advances in Pharmacology, Volume 42

1054-3589/98 $25.00

Eleven adult male Sprague-Dawley rats were anesthetized, and 10 animals were perfused with 3.75% acrolein and 2% paraformaldehyde. The remaining animal was perfused with 4% paraformaldehye and 0.2% glutaraldehyde. The brains were removed, postfixed for 30 min, and sectioned at 50 μm on a vibratome. Sections were treated for 30 min with 1% sodium borohydride to improve antigenicity and reduce nonspecific labeling. Sections were incubated for 30 min in a blocking serum consisting of 1% bovine serum albumin and 3% normal goat serum. Steps taken to maximize immunostaining in sections for electron microscopy included the use of 0.04% Triton X-100 or freeze-thaw and, in several cases, the use of two night incubations in primary antibody. Sections for light microscopy included 0.4% Triton.

Two different primary antibodies directed against the N-terminus of the DAT protein were used: a rabbit polyclonal (1 : 100) and a rat monoclonal antibody (1 : 1000). Both were tested for specificity by Western blot analysis against cloned transporter (2) (Levey, personal communication). The immunocytochemical staining results with either antibody were comparable, but only the rat antibody was used in quantitative studies. Primary antibody incubation was for 15 hr at room temperature or 40 hr at 4°C. In a preadsorption control experiment, 100 μg/mL of fusion protein antigen was added to the rat primary antibody for 2 hr prior to use. To compare DAT immunoreactivity with another marker for dopamine terminals, adjacent sections from three animals were incubated in rabbit anti-tyrosine hydroxylase (TH) antiserum (1 : 1000). The secondary antibodies employed were biotinylated goat anti-rabbit (1 : 400) or donkey anti-rat (1 : 100), and avidin biotin peroxidase complex (Vectastain Elite) was used (1 : 200). Bound peroxidase was visualized by the addition of 3,3′-diaminobenzidine and H_2O_2. Sections for light microscopy were slide mounted, while sections for electron microscopy were osmicated, dehydrated, and plastic embedded.

The electron microscopic results from six rats with the best morphology and most robust staining with the rat antibody were quantified. One to two tissue sections per region per animal were examined, and the surface of the tissue was sampled at random until at least 40 DAT-immunoreactive processes were photographed. Processes were then numbered on the micrographs from upper left to lower right, and a random number generator was used to select 30 processes per region per animal (total of 180 immunoreactive processes per region for DAT and 90 observations for TH in the PFC). The processes were then traced using an image analysis system, which calculated their maximum diameter along the short axis. For processes containing DAT immunoreactivity in only one portion, only the immunoreactive area was traced. For processes with eccentric shape, such as longitudinal sections through both varicose and nonvaricose portions of an axon, the diameter represented the most varicose portion. The data were analyzed statistically by a two-way ANOVA, with main effects being either region and animal (for DAT) or marker and animal (for DAT vs TH in the PFC). The interaction between main effects was also examined in each case.

By light microscopy, DAT-immunoreactive fibers were abundantly expressed in the neuropil of the dorsolateral striatum beneath the corpus callosum. This immunostaining was absent from sections incubated in primary antibody preadsorbed with the fusion protein antigen. The anterior cingulate cortex also showed robust immunostaining for DAT, particularly in clusters of axons in layer III. However, the immediately adjacent prelimbic PFC showed sparse labeling for DAT that was dramatically lower than that seen in the anterior cingulate cortex. Furthermore, many of the axons that were present were difficult to visualize without differential interference contrast optics. This weak immunolabeling was observed despite the use of two night incubations in primary antibody that contained a high concentration of detergent to enhance penetration. Finally, this degree of immunostaining for DAT was considerably weaker than that seen with other markers for dopamine axons, such as TH antibodies. These light microscopic observations suggest that individual dopamine axons in the rat PFC are relatively lacking in the DAT protein.

By electron microscopic examination of the dorsolateral striatum, peroxidase immunoreactivity for DAT was abundantly expressed in axon varicosities, some of which made punctate symmetric synapses on spines or distal dendrites. In the superficial layers of the anterior cingulate cortex, immunoreactivity for DAT was also frequently localized to axon varicosities. In addition, DAT immunolabeling was observed in numerous small-diameter axons. In the deep layers of the prelimbic PFC, immunoreactivity for DAT was localized almost exclusively to small-diameter axons. In a few cases, varicose axons cut in a longitudinal plane showed DAT immunoreactivity only in the preterminal regions and not in varicose segments that formed synapses. These results were observed consistently in all animals, regardless of the fixative used, the primary antibody employed, the use of one- or two-night incubations in antibody, or the use of Triton detergent or freeze-thaw to enhance antibody penetration. As a positive control for the ability to detect proteins in dopamine axon varicosities, immunoreactivity for TH was assessed in adjacent sections of the PFC from three animals. TH labeling was seen in intervaricose axons, as well as in numerous varicosities, some of which formed synapses on spines and small dendrites.

The quantitative analysis of process diameter by region and marker produced the following means and standard deviations: DAT-PFC, 0.137 ± 0.049; DAT-CING, 0.180 ± 0.074; DAT-STR, 0.218 ± 0.084; TH-PFC, 0.215 ± 0.088. By ANOVA, there was a significant overall effect of region ($p < .0001$) and no significant effect of animal. However, a significant interaction between region and animal ($p < .012$) suggested that some of the region effect might be explained by animal differences. Post-hoc analyses using Tukey's studentized range test on all pair-wise comparisons with a simultaneous significance of $p < .05$ revealed a significant difference in diameter between the PFC and striatum in all six animals. For the anterior cingulate cortex, significant differences with the PFC were observed in three of the six animals, while significant differences with the striatum were seen in only two of the six animals. For the ANOVA comparing markers, there was an overall significant effect of marker ($p < .0001$) with no effect of animal and no interaction effect.

The combined results of light and electron microscopic immunocytochemical studies suggest that dopamine axon varicosities in the PFC express a relative paucity of immunoreactivity for the DAT protein. This finding is consistent with *in situ* hybridization studies showing lower mRNA levels for DAT in the A10 relative to the A9 dopamine cell groups (5). The fact that the dopamine inputs to the dorsolateral striatum and anterior cingulate cortex both originate from A9 dopamine neurons is consistent with their observed greater content of DAT protein. The data also suggest that DAT protein contained in dopamine axons in the PFC is localized at a distance from synaptic release sites. Although the exact subcellular distribution was not determined by immunogold labeling, the more sensitive avidin-biotin peroxidase method failed to reveal DAT protein within most dopamine axon varicosities in this region. However, it should be noted that intervaricose segments of axons can sometimes form synapses, in which case the DAT protein would be localized closer to the release site. Finally, an alternative interpretation of our data is that dopamine axon varicosities in the PFC express a biochemically modified DAT protein that has an altered antigenicity. Such a protein also might be expected to function differently from the striatal DAT. However, the fact that our findings were consistent with two different antibodies argues against this latter explanation of the data. Nevertheless, additional antibodies should be tested once they become available.

In conclusion, dopamine's high extracellular levels (relative to tissue content) and considerable diffusion in the PFC may be secondary to restricted localization and/or altered function of the DAT protein within individual axon varicosities. While these data appear to suggest a greater paracrine role for dopamine in the PFC, compared with the striatum and other forebrain regions, dopamine's actual sphere of influence may be limited by the low density of its receptors in the cortex.

Acknowledgment

Supported by U. S. Public Health Service grant MH50314.

References

1. Giros, B., Jaber, M., Jones, S. R., Wightman, R. M., and Caron, M. G. (1996). Hyperlocomotion and indifference to cocaine and amphetamine in mice lacking the dopamine transporter. *Nature* **379**, 606–612.
2. Ciliax, B. J., Heilman, C., Demchyshyn, L. L., Pristupa, Z. B., Ince, E., Hersch, S. M., Niznik, H. B., Land evey, A. I. (1995). The dopamine transporter: Immunocytochemical characterization and localization in brain. *J. Neurosci.* **15**, 1714–1723.
3. Nirenberg, M. J., Vaughan, R. A., Uhl, G. R., Kuhar, M. J., and Pickel, V. M. (1996). The dopamine transporter is localized to dendritic and axonal plasma membranes of nigrostriatal dopaminergic neurons. *J. Neurosci.* **16**, 436–447.
4. Garris, P., and Wightman, R. M. (1994). Different kinetics govern dopaminergic transmission in the amygdala, prefrontal cortex, and striatum: An *in vivo* voltammetric study. *J. Neurosci.* **14**, 442–450.
5. Shimada, S., Kitayama, S., Walther, D., and Uhl, G. (1992). Dopamine transporter mRNA: Dense expression in ventral midbrain neurons. *Mol. Brain. Res.* **13**, 359–362.

H. Gu,* M. J. Caplan,† and G. Rudnick*

*Department of Pharmacology
†Department of Cellular and Molecular Physiology
Yale University School of Medicine
New Haven, Connecticut 06510

Cloned Catecholamine Transporters Expressed in Polarized Epithelial Cells: Sorting, Drug Sensitivity, and Ion-Coupling Stoichiometry

I. Production of Biogenic Amine Transporter Cell Lines in LLC-PK_1 and MDCK

To facilitate investigation of neurotransmitter transport, we stably transfected two polarized epithelial cell lines with cDNAs encoding the rat GABA, dopamine (DA), and 5-HT transporters (rGAT-1, rDAT, and rSERT, respectively) and the human norepinephrine (NE) and DA transporters, (hNET and hDAT). LLC-PK_1 cells were initially chosen because preliminary experiments had demonstrated that they could be stably transfected with GAT-1. These experiments led to studies on the distribution of GAT isoforms in epithelial cells and neurons in culture. Another reason for choosing this cell line was that membrane vesicles from LLC-PK_1 cells had been shown to be useful for transport studies.

In total, we generated nine cell lines for this project. In the pig kidney cell line (LLC-PK_1), we transfected rGAT-1, hNET, rSERT, and rDAT. In Modin-Darby Canine Kidney (MDCK) cells, we transfected those same cDNAs and also hDAT. The transport characteristics and inhibitor sensitivities for the cell lines were consistent with what was known for each transporter *in vivo* and in other heterologous expression systems. The advantage of stably transfecting the three biogenic amine transporters into the same cellular background was that it allowed us to compare in detail the catalytic and pharmacological properties of these proteins. We found that NET and DAT share a very similar substrate specificity. They both act as catecholamine transporters with a preference for DA over NE. In both inhibitor sensitivity and ion dependence, however, NET was more similar to SERT than to DAT. Tricyclic antidepressants inhibited NET and SERT but not DAT. DAT was also the only biogenic amine transporter to show a sigmoidal Na^+ dependence for transport, suggesting that more than one Na^+ ion was cotransported together with DA.

Advances in Pharmacology, Volume 42

1054-3589/98 $25.00

II. GAT, NET SERT, and DAT Vesicles

A detailed understanding of biogenic amine transport mechanism requires knowing how many Na^+ and Cl^- ions are cotransported, and K^+ ions countertransported, with each molecule of substrate. This ion coupling stoichiometry is difficult to determine using rate measurements in intact cells. However, plasma membrane vesicles have been very useful for stoichiometry determination in many transport systems, including SERT and GAT-1. Homogenization of LLC-PK_1 cells was known to produce a preparation of membrane vesicles that retained the transport properties of the original cells, and this was found also for LLC-PK_1 and MDCK cells expressing neurotransmitter transporters. Dilution of vesicles equilibrated with high-K^+ medium into NaCl buffer generated Na^+, Cl^-, and K^+ gradients across the vesicle membrane. Under these conditions, the vesicles transiently accumulated the appropriate substrates in response to the applied gradients (1).

For each transporter, external Na^+ was required for transport, and accumulation was stimulated by a transmembrane Na^+ gradient (out>in). Likewise, external Cl^- ion must be present for each system to transport, and the transmembrane Cl^- gradient (out>in) also stimulated accumulation. These results support the prediction that both Na^+ and Cl^- are cotransported by all members of this neurotransmitter transporter family. The K^+ gradient (in>out) stimulated 5-HT accumulation by vesicles from LLC-SERT cells, and K^+ stimulated 5-HT transport in those vesicles even in the absence of a gradient. This result is in accordance with previous evidence that K^+ is countertransported by SERT. The presence of K^+ had little effect on NET, DAT, or GAT-1 transport in the absence of a K^+ gradient (1).

Because all of the membrane vesicles were prepared from the same parental cell line, each transporter could be exposed to the same driving forces, and the response to those forces (a function of the coupling stoichiometry) could be compared. From previous studies in plasma membrane vesicles from brain and platelet, the ion coupling stoichiometry for GAT-1 and SERT was well known. By imposing different ion gradients and comparing the effect on transport in vesicles from LLC-NET, LLC-SERT, and LLC-GAT, we were able to estimate the ion-coupling stoichiometry for NET. Because K^+ was not required for transport into LLC-NET MDCK-NET or MDCK-DAT vesicles, and K^+ did not stimulate transport in the absence of a K^+ gradient, we concluded that K^+ was not countertransported by NET or DAT. However, NET- and DAT-mediated transport in vesicles from MDCK cells was stimulated by a K^+ gradient (in>out) in the presence of valinomycin. We concluded that a diffusion potential (negative inside) generated by the K^+ gradient stimulated transport indirectly, indicating that net positive charge crossed the membrane in the same direction as substrate flux.

The KM for DA transport by NET and DAT was not pH dependent, suggesting that the cationic form of the substrate is transported. Taking all of these observations together and comparing the behavior of NET, GAT-1, and SERT in LLC-PK_1 vesicles, we concluded that a single Na^+ and one Cl^- are cotransported by NET with each molecule of catecholamine substrate (1). For

DAT, the results are consistent with a similar stoichiometry, or one in which more than one Na^+ ion is cotransported with DA. Future experiments will be directed toward evaluating the Na^+ stoichiometry.

III. Amphetamine Specificity of Catecholamine Transporters

Stably transfected LLC-PK_1 cells also were useful for characterizing the amphetamine specificity of SERT, NET, and DAT. In these cell lines, amphetamine and its derivatives inhibited biogenic amine influx and caused efflux of previously accumulated amine (2). These effects are likely to underlie amphetamine-induced biogenic amine release *in vivo*. By measuring the sensitivity of each cell line to a variety of amphetamine derivatives, we defined the specificity of each transporter. In agreement with the similar substrate sensitivity of LLC-NET and LLC-DAT cells, these two catecholamine transporters were also similar in their sensitivity to amphetamine derivatives. This finding reinforces the notion that amphetamine derivatives interact with plasma membrane biogenic amine transporters primarily as substrates. LLC-SERT cells, in contrast, demonstrated a distinctly different profile of amphetamine sensitivity consistent with their unique substrate specificity. Whereas amphetamine and methamphetamine were the most potent at inhibiting substrate influx into LLC-NET and LLC-DAT cells, *p*-chloroamphetamine and MDMA were the most potent inhibitors of uptake by LLC-SERT cells. We were able to classify compounds as either inhibitors or amine releasers by comparing the potency of transport inhibition with that for stimulation of amine efflux by various compounds. Methylphenidate, for example, was a potent inhibitor but did not cause efflux under conditions where amphetamine or MDMA stimulated rapid reversal of transport (2).

IV. Sorting in Epithelial Cells

We have examined the sorting properties of biogenic amine transporters expressed in polarized epithelial cells. Initially this project was designed to test the hypothesis that neurons and epithelial cells utilize the same sorting signals and sorting machinery to determine which plasma membrane domain will be the target for delivery of surface and secreted proteins. Although initial studies with GAT-1 appeared to bear out this hypothesis, more recent results with biogenic amine transporters argue against such a simple hypothesis (3). As these studies developed, it became apparent that relatively small differences in primary sequence between closely related members of the neurotransmitter transporter family could dramatically alter their distribution between apical and basolateral domains of MDCK cells.

In contrast with the GABA transporters, where multiple isoforms are found in the brain and the periphery, only one isoform each of the biogenic amine transporters SERT, NET, and DAT has been identified. We have used LLC-

PK_1 and MDCK cells, described previously, to examine the sorting properties of these three transporters. By immunocytochemistry, all three biogenic amine transporters were sorted basolaterally in LLC-PK_1 cells, in contrast with GAT-1 and GAT-3 but similar to GAT-2 and the betaine gaba transporter (BGT-1). In MDCK cells, transport, surface biotinylation, and immunofluorescence measurements all indicated that, although SERT and NET are found in the basolateral membrane, DAT is sorted apically. Thus, targeting of DAT is different in two kidney epithelial cell lines. Moreover, DAT and NET, which have approximately 70% sequence identity, are sorted to opposite sides of MDCK cells. Perhaps the most important finding from this work is that the simple notion that neurons and epithelial cells use the same sorting signals and machinery cannot be correct. According to the hypothesis of Dotti and Simons (4), each of the biogenic amine transporters, which were expected to be localized to axonal presynaptic terminals *in vivo,* should be sorted to the apical plasma membrane domain of cultured epithelial cells. Instead, we found that NET and SERT in MDCK cells, and all three biogenic amine transporters in LLC-PK_1 cells, were sorted basolaterally.

V. NET–DAT Chimeras

We have investigated the difference between NET and DAT that is responsible for their sorting to opposite sides of MDCK cells. Giros *et al.* (5) constructed chimeric NET–DAT transporters by dividing the sequence into four cassettes, which were linked in different combinations. Because many of these chimeras were functional for DA transport, we tested some of the active constructs by transfection into MDCK cells to identify the region responsible for sorting. By confocal immunofluorescence microscopy, cells expressing chimera A, which contained NET sequence only in its N-terminal portion (amino acids 1–129, including the first two putative transmembrane domains), showed a pattern of expression similar to that of wild-type NET. Transporter expression was restricted to the basolateral regions of the cell in both cases. In contrast, replacement of the C-terminal portion of DAT with NET sequence did not affect the apical distribution of DAT. Chimera F contained DAT sequence through amino acid 434 and NET sequence from that point (between predicted transmembrane helices 8 and 9) to the C-terminus. This chimera was predominantly found in the apical membrane, as was wild-type DAT. To further define the N-terminal sequence apparently responsible for basolateral sorting of NET, we generated a deletion mutant of DAT lacking the first 50 amino acid residues and a chimera in which the first 50 residues were replaced with NET sequence. The deletion mutant is sorted, like DAT, to the apical membrane, but the NET-DAT chimera is sorted basolaterally. These data suggest that NET contains a basolateral sorting signal within the N-terminal 50 residues.

References

1. Gu, H. H., Wall, S., and Rudnick, G. (1996). Ion coupling stoichiometry for the norepinephrine transporter in membrane vesicles from stably transfected cells. *J. Biol. Chem.* **271**, 6911–6916.

2. Wall, S. C., Gu, H. H., and Rudnick, G. (1995). Biogenic amine flux mediated by cloned transporters stably expressed in cultured cell lines: Amphetamine specificity for inhibition and efflux. *Mol. Pharmacol.* **47**, 544–550.
3. Gu, H. H., Ahn, J., Caplan, M. J., Blakely, R. D., Levey, A. I., and Ruduick, G. (1996). Cell-specific sorting of biogenic amine transporters expressed in epithelial cells. *J. Biol. Chem.* **271**, 18100–18106.
4. Dotti, C. G., and Simons, K. (1990). Polarized sorting of viral glycoproteins to axons and dendrites of hippocampal neurons in culture. *Cell* **62**, 63–72.
5. Giros, B., Wang, Y., Suter, S., Mcleskey, S., Pifl, C., Caron, M. (1994). Delineation of discrete domains for substrate, cocaine, and tricyclic antidepressant interactions using chimeric dopamine-norepinephrine transporters. *J. Biol. Chem.* **269**, 15985–15988.

Fabio Fumagalli,* Sara Jones,* Roger Bossé,*
Mohamed Jaber,* Bruno Giros,* Cristina Missale,†
R. Mark Wightman,‡ and Marc G. Caron*

*Howard Hughes Medical Institute Laboratories
Departments of Cell Biology and Medicine
Duke University Medical Center
Durham, North Carolina 27710

†Curriculum in Neurobiology and Department of Chemistry
University of North Carolina
Chapel Hill, North Carolina 27514

‡Division of Pharmacology
Department of Biomedical Sciences and Biotechnology
University of Brescia
Brescia, Italy

Inactivation of the Dopamine Transporter Reveals Essential Roles of Dopamine in the Control of Locomotion, Psychostimulant Response, and Pituitary Function

Dopamine (DA) is an important modulator of many physiological functions. In the central nervous system (CNS), DA is involved in the control of movement, cognition, and affect, as well as neuroendocrine secretion. In the

Advances in Pharmacology, Volume 42

1054-3589/98 $25.00

periphery, DA regulates pituitary and parathyroid hormone synthesis and secretion, in addition to modulating certain retinal, cardiovascular, and kidney functions. DA is also thought to influence gastrointestinal physiology (1, 2). In the CNS, the DA transporter (DAT) presumably plays an important role in terminating dopaminergic transmission by uptaking DA from extracellular spaces back into presynaptic terminals. The role of the DAT in the CNS has been largely inferred from the clinical and psychosocial effects of drugs such as antidepressants and psychostimulants. These drugs interfere with the function of DAT. However, the involvement of DAT in the peripheral as well as the hypothalamic–neuroendocrine actions of DA has remained more enigmatic. DAT belongs to the large family of Na^+- and Cl^--dependent transporters containing 12 putative transmembrane domains. Other members of this family include transporters for norepinephrine, serotonin, gamma aminobutyric acid, glycine, creatine, and betaine. In an attempt to develop an animal model in which the role of the DAT in the control of dopaminergic neurotransmission and in other physiological functions of DA could be directly examined, we have inactivated the mouse DAT gene by homologous recombination (3).

Homozygous DAT knockout (−/−) animals are viable. However, adult (−/−) DAT animals show a pronounced weight deficit (50–70% of normal animals). (−/−) DAT mothers have a markedly reduced ability to nurse. The most obvious behavioral characteristic of these mice is a five- sixfold increase in spontaneous locomotor activity when tested in an open field. The level of activity of the (−/−) DAT animals is nearly as high as that achieved by treatment of wild-type animals with psychostimulants. Indeed, treatments of (−/−) DAT mice with high doses of cocaine or amphetamine yielded no further overall increase in locomotor activity, indicating that the locomotor effects of these drugs is mediated by interactions with DAT. These findings suggest that, in the absence of DA uptake, (−/−) DAT animals are hyperdopaminergic. In an attempt to examine whether this hyperdopaminergic behavioral phenotype was reflected at the biochemical level, several parameters were examined. Messenger RNAs for genes that are normally under dopaminergic control, such as preproenkephalin and dynorphin, showed variations indicative of increased dopaminergic tone. Similarly mRNAs for pre- and postsynaptic D2 dopamine receptors, as well as striatal D1 receptors, were markedly (>50%) diminished. As assessed by autoradiography, D1 and D2 receptor numbers in the striatal and ventral midbrain areas were also decreased by more than 50%, presumably reflecting downregulation of these receptors in response to the elevated dopaminergic tone of these animals.

Even more interestingly, at the neurochemical level, these animals displayed properties reminiscent of what might be expected of parkinsonian animals. When striatal levels of DA were measured by high-performance liquid chromatography, they were found to be 5–10% of those of normal animals. Although at the gross morphological level the dopaminergic cell bodies of the substantia nigra–ventral tegmental area and the terminal fields of the striatum appeared essentially normal, *in situ* hybridization and Western blots for tyrosine hydroxylase (TH) in these areas revealed important adaptive changes. A marked decrease in the levels of TH protein was observed in these regions with no significant changes in TH mRNA, suggesting that posttranslational control mechanisms might be important in regulating TH activity in dopaminergic neurons (MJ, SJ, RB, BG, and MGC, in preparation).

Thus, in the absence of a functional DAT, the (−/−) DAT mice have undergone major adaptive changes in their biochemical and neurochemical mediators of dopaminergic transmission in an attempt to dampen the presumed increased DA signal. This situation is schematized in Figure 1. However, none of these changes appear sufficient to effectively quench the increased signal, because the (−/−) DAT animals still display spontaneous hyperlocomotion. This paradox may be explained when one considers that, as measured by fast cyclic voltammetry, DA persists at the synapse of (−/−.) DAT animals about 100–300 times longer than in normal animals (SJ, R. G, MJ, BG, RMW, and MGC, in preparation). These findings suggest that the DAT plays a most crucial role in maintaining neuronal dopaminergic tone.

Additional striking features of the DAT knockout mice include their weight deficit and lack of appropriate lactation of nursing mothers. These properties suggest that a hypothalamo-pituitary dysfunction might exist in these mice. Although low amounts of DAT mRNA are present in the cell bodies of the tuberoinfundibular system (arcuate nucleus and zona incerta), the functional role of DAT in this area of the brain has remained obscure (4). DA released from hypothalamic nuclei is known to modulate synthesis and secretion of prolactin from anterior pituitary lactotrophs. In rodents, hypothalamic DA also

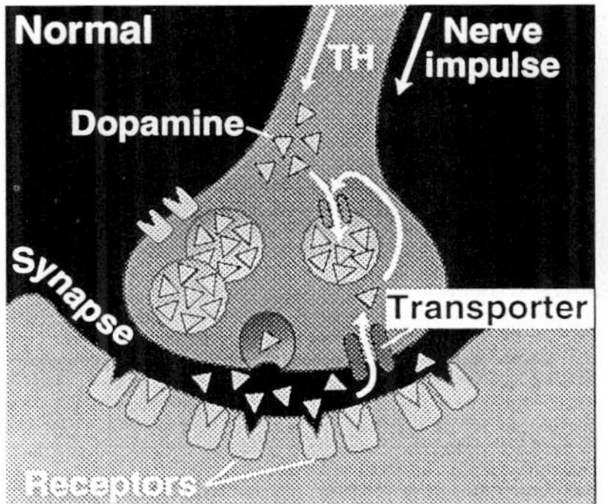

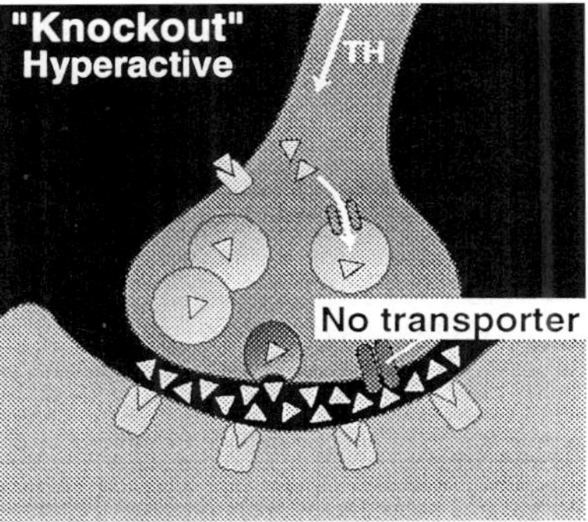

FIGURE 1 Adaptive changes in biochemical and neurochemical mediators of DA transmission in DAT knockout mice. Rectangles represent pre- and postsynaptic receptors; triangles depict stored and released DA; circles represent vesicles within presynaptic terminals; tyrosine hydroxylase, the rate-limiting enzyme of DA synthesis, is abbreviated TH, and the difference in size of the symbols reflects different amounts of TH. In the normal mice, after stimulation, DA is released from presynaptic terminals. Once in the synaptic cleft, DA interacts with pre- and postsynaptic receptors to elicit its effects and is then rapidly carried back into nerve terminals by the DAT, which regulates synaptic levels of DA. In the DAT knockout mice, due to the absence of transporter, DA is not cleared away and remains in the synapse 300 times longer than normal. Despite lower amounts of released DA (3), absence of DAT leads to increased functional DA, which presumably will occupy pre- and postsynaptic receptors, resulting in downregulation of receptor numbers, a decrease in DA content, and ultimately a decrease in TH. Modified from *Science* 271, 909, 1996.

modulates the secretion of α-melanocyte–stimulating hormone (α-MSH) from the intermediate lobe. These effects of DA are mediated via D2 DA receptors on anterior pituitary lactotrophs and α-MSH-secreting cells of the intermediate lobe (5).

Interestingly, (−/−) DAT mice show a marked hypopituitarism, with glands of knockout animals being about half the size of those of normal animals. (−/−) DAT mice show a marked decrease in the size of both the anterior and intermediate lobes of the pituitary, whereas the posterior lobe, which is known not to be under dopaminergic control, is unaffected. Autoradiographic studies reveal that the density of the D2 DA receptors is reduced by more than 75%, in both anterior and intermediate lobes. Using radioimmunoassays, the pituitary content of both prolactin and, unexpectedly, growth hormone was decreased by 85 and 70%, respectively in (−/−) DAT animals compared with wild-type littermates. The extent of the changes in the content of prolactin and growth hormone may explain both the growth and nursing deficits of these animals.

Taken together, these findings provide direct evidence for a role of DAT in the tuberoinfundibular DA system. Moreover, these results establish the importance of DA and its hypothalamic re-uptake mechanism in the function of the pituitary gland. Further elucidation of the mechanisms underlying this pituitary phenotype should reveal whether DA also plays a role in pituitary development.

References

1. Giros, B., and Caron, M. G. (1993). Molecular characterization of the dopamine transporter. *Trends Pharmacol. Sci.* **14**, 43–49.
2. Lackovic, Z., and Relja, M. (1983). Evidence for a widely distributed peripheral dopaminergic system. *Fed. Proc.* **42**, 3000–3004.
3. Giros, B., Jaber, M., Jones, S. R., Wightman, R. M., Caron, M. G. (1996). Hyperlocomotion and indifference to cocaine and amphetamine in mice lacking the dopamine transporter. *Nature* **379**, 606–612.
4. Meister, B., Elde, R. (1993). Dopamine transporter mRNA in neurons of the rat hypothalamus. *Neuroendocrinology* **58**, 388–395.
5. Gingrich, J. A., Caron, M. G. (1993). Recent advances in the molecular biology of dopamine receptors. *Annu. Rev. Neurosci.* **16**, 299–321.

H. Bönisch, R. Hammermann, and M. Brüss

Institute of Pharmacology and Toxicology
University of Bonn
D-53113 Bonn, Germany

Role of Protein Kinase C and Second Messengers in Regulation of the Norepinephrine Transporter

Re-uptake of released norepinephrine (NE) into presynaptic nerve terminals is responsible for the rapid termination of neurotransmission in noradrenergic synapses. Transport of NE by the neuronal NE transporter (NET) is absolutely dependent on extracellular Na^+ and Cl^- and is selectively inhibited by, for example, desipramine and nisoxetine, and Na^+ and Cl^--dependent binding of these inhibitors to the NET has been used to localize the tissue distribution of NET (1, 2). Aside from its physiological role, the NET in the central nervous system is the primary target for tricyclic antidepressants. However, in spite of its physiological and therapeutic importance, little is known about the involvement of second messengers in the regulation of NET expression and function.

cDNAs of a series of neurotransmitter transporters (NTTs) have been cloned and the NET has been shown to be a member of the superfamily of structurally related Na^+ and Cl^--dependent NTTs for monoamines (dopamine, serotonin, and NE) and certain amino acids, such as GABA and glycine; transporters of this family are structurally characterized by 12 transmembrane domains, intracellular N- and C-termini, and a large second extracellular loop (3). Cloning of the human and bovine NET cDNAs (4, 5) has revealed one common consensus sequence for phosphorylation by protein kinase C (in the following, termed *PKC site*) within the second intracellular loop and, for the bovine NET (bNET), two additional PKC sites at the C-terminal end of bNET. In this paper we provide evidence that stimulation of PKC by a phorbol ester causes downregulation of NE transport and that the second-messenger cyclic adenosine monophosphate (cAMP) may be involved in tissue-specific regulation of NETs.

To examine whether the PKC sites of NETs are involved in the regulation of NE transport, uptake of [^{3}H]NE was studied in cells expressing the hNET (with one PKC site), bNET (with three PKC sites), and a mutant hNET, $hNET_{S259A}$, in which (by site-directed mutagenesis) serine (at amino acid position 259 of hNET) was exchanged against alanine, resulting in the destruction of the single potential phosphorylation site for PKC of the hNET. For NE transport studies, COS-7 cells were transfected with the corresponding cDNA by means of the calcium phosphate method and used 48 hr posttransfection. The human

Advances in Pharmacology, Volume 42

1054-3589/98 $25.00

neuroblastoma cells, SKN-SH-SY5Y (SKN cells), which constitutively express the hNET, were also used in this study. Effects of second messengers on NE transport were studied in SKN cells and in the rNET-expressing rat PC12 pheochromocytoma cells.

To determine uptake of NE, cells were washed three times with Krebs-Ringer-Hepes (KRH) buffer containing 10 μM pargyline (to inhibit monoamine oxidase) and 10 μM U-0521 (to inhibit COMT) and preincubated for 30 min (at 37°C) in this buffer. Therafter, cells were incubated at 37°C for 2–5 min (as indicated) in KRH buffer containing 10 nM [^{3}H]NE (53.4 Ci/mmol, NEN) and washed three times with ice-cold KRH buffer to terminate uptake. Specific uptake was defined by subtracting the uptake in the presence of 10 μM nisoxetine from total uptake. When short-term effects of the compounds under study were determined, the compounds were present during incubation and preincubation of the cells. When cells during culture had already been exposed for 24 hr to these compounds, they were not further present during incubation or preincubation. For kinetic analysis, cells were incubated for 2 min in 10 nM [^{3}H]NE with 0.1–2.0 μM unlabeled NE. Binding of [^{3}H]nisoxetine to the NE transporter on intact cell membranes was determined by pretreating the cells are described for NE uptake. Cells were incubated with 0.5–20.0 nM [^{3}H]nisoxetine (83 Ci/mmol, Amersham) at 4°C for 3 hr and then washed three times with KRH buffer. Nonspecific binding was determined in the presence of desipramine (1 μM). At the end of uptake or binding experiments, cells were solubilized in 0.1% Triton X-100 to determine protein content and radioactivity.

Short (10 min) exposure of COS-7 cells expressing either the bNET, the hNET, or the mutant $hNET_{S259A}$ to the PKC-activating phorbol ester PMA (phorbol 12-myristate-13-acetate, 160 nM), the inactive phorbol ester 4αPD (4α-phorbol 12, 13-didecanoate, 160 nM), the PKC inhibitor staurosporine (0.5 μM), or the phosphatase inhibitor okadaic acid (1 μM) did not cause a significant change in [^{3}]NE uptake. When COS-7 cells expressing the various NETs were exposed for 30 min with the aforementioned compounds and with the same concentrations, [^{3}H]NE uptake by hNET, bNET, and $hNET_{S259A}$ was significantly reduced by PMA to 61.3%, 44.4%, and 71.9%, respectively, of the corresponding control values. On the other hand, uptake of [^{3}H]NE by all three transporters was not affected by 4αPD or staurosporin. Interestingly, uptake of [^{3}]NE by all three transporters was not different from controls if cells were exposed to both PMA and staurosporin; that is, the inhibitory effect of PMA was antagonized by the PKC inhibitor staurosporin. Incubation of COS-7 cells (expressing either hNET or bNET) with okadaic acid affected neither [^{3}H]NE uptake nor the inhibitory effect of PMA. Prolonged treatment with PMA (160 nM) for up to 24 hr of COS-7 cells expressing bNET, hNET, or $hNET_{S259A}$ resulted in a time-dependent reduction of [^{3}H]NE uptake, which, after about 3 hr, remained at a constant reduced level of 35-70% of corresponding controls. A similar PMA-induced reduction of [^{3}H]NE uptake was also observed in SKN cells. Under the same conditions, 4αPD had no influence on [^{3}H]NE uptake in either of these cells. The PMA-induced reduction of [^{3}H]NE uptake (after 24 hr) was due to a significant reduction in maximum transport velocity (Vmax) with no significant effect on apparent affinity (Km): Vmax was reduced to 29.5%, 31.8%, and 28.6% of control values in COS-7 cells expressing bNET, hNET, and $hNET_{S259A}$, respectively. In SKN cells, PMA caused a reduction of Vmax from 4.0 pmol/mg/min (control) to 1.9 pmol/mg/min (i.e., to 47.5% of the control value). [^{3}H]nisoxetine binding was reduced by 24 hr pretreatment of SKN cells with PMA, but it remained unaffected by pretreatment of the cells with 4αPD. Analyses of the binding data revealed no significant differences in Kd values from control cells (controls = 12.1 nM vs 8.8 nM after pretreatment with PMA. However, PMA pretreatment caused a significant reduction in Bmax of [^{3}H]nisoxetine binding from 529 fmol/mg (controls) to 287 fmol/mg (i.e., to 54.3% of the control value). Thus, the PMA-induced reduction of NE transport was a consequence of the reduction in the number of NET molecules expressed in the plasma membrane of the cells.

To study the effects of cAMP on human NET, 2-min uptake of [^{3}H]NE (10 nM) was determined in SKN cells after 15-min or 24-hr exposure of the cells to lipophilic cAMP analogues (8Br-cAMP or db-cAMP) or to adenylate cyclase–stimulating agents (cholera toxin, forskolin). In addition, 8Br-cGMP was included in this study. [^{3}H]NE uptake in SKN cells remained unchanged if cells were incubated for 15 min with 8Br-cAMP (1 or 10 mM), choleratoxin (100 mg/ml), forskolin (100 μM), or 8Br-cGMP (1 mM). These agents, as well as db-cAMP (1 mM), were also without effect on NE transport when present for 24 hr (i.e., during culture of the cells). A more physiological stimulation of adenylate cyclase through activation of the prostaglandin E_1 (PGE_1 receptor of SKN cells by 10 μM PGE_1 (present for 24 hr) was also without effect in [^{3}H]NE uptake, and NE transport also remained unchanged if degradation of cAMP was inhibited by IBMX (500 μM). Uptake of [^{3}H]NE under control conditions amounted to 755 fmol/mg protein. In addition, in COS-7 cells expressing the hNET, a 24-hr treatment of the cells with forskolin (100 μM), 8Br-cAMP (1 mM), or 8Br-cGMP (1 mM) also did not affect [^{3}H]NE uptake, which amounted in these cells to 2269 fmol/mg protein. Interestingly, 24-hr incubation of PC12 with forskolin (100 μM), cholera toxin (100 ng/ml), or db-cAMP (1 mM) strongly reduced [^{3}H]NE uptake to 30.5%, 40.6%, and 36.2%, respectively, of control values (100% = 451 fmol/mg protein). However, NE uptake remained unchanged when PC12 cells were incubated for 24 hr with db-cGMP (1 mM).

The results of this study indicate that human NET (and NET in COS-7 cells) is downregulated in a staurosporin-sensitive way by the activation of PKC. Because NE transport in a variant hNET, lacking a potential phosphorylation site for PKC, was also downregulated by the PKC-activating phorbol ester PMA, a direct phosphorylation of the transporter at the PKC site(s) of NETs can be excluded. The PMA-induced reduction in the density of expressed transporters (measured as a decrease in Bmax of nisoxetine binding, with a concomitant decrease in Vmax of NE transport) could be due to PKC-mediated changes in plasma membrane incorporation and/or redistribution of the NET protein. Similar effects of phorbol esters were also observed for other members of the family of Na^+/Cl^- dependent NTTs (3). The effect of cAMP on NE transport seems to be tissue-specific, because cAMP exhibited no effect in SKN cells but decreased NE transport in rat PC12 cells, whereas in rat midbrain neurons, cAMP had been shown to increase the expression of NET mRNA and protein.

The lack of effect of 8Br-cGMP on NE transport by the human and rat NET suggests a minor role of this second messenger in the regulation of NET. However, further studies with measurement of mRNA and protein expression of NET are necessary to obtain more direct information about the regulation of the NET at the transcriptional or translational level.

Acknowledgment

Supported by the Deutsche Forschungsgemeinschaft (SFB 400).

References

1. Graefe, K-H., and Bönisch, H. (1988). The transport of amines across the axonal membranes of noradrenergic and dopaminergic neurones. *In* "Handbook of Experimental Pharmacology, 90/I (U. Trendelenburg and N. Weiner, eds.), pp. 193–245. Springer-Verlag, Berlin Heidelberg, New York.
2. Bönisch, H., and Brüss, M. (1994). Catecholamine transporter of the plasma membrane. *Ann. N. Y. Acad. Sci.* **733**, 193–202.
3. Borowsky, B., and Hoffman, B. J. (1995). Neurotransmitter transporters: Molecular biology, function, and regulation. *Int. Rev. Neurobiol.* **38**, 139–199.
4. Pacholczyk, T., Blakely, R. D., and Amara, S. G. (1991). Expression cloning of a cocaine- and antidepressant-sensitive human noradrenaline transporter. *Nature* **350**, 350–354.
5. Lingen, B., Brüss, M., and Bönisch, (1994). Cloning and expression of the bovine sodium- and chloride-dependent noradrenaline transporter. *FEBS Lett.* **342**, 235–238.

Louis J. DeFelice and Aurelio Galli

Department of Pharmacology and Center for Molecular Neuroscience
Vanderbilt University School of Medicine
Nashville, Tennessee 37232

Electrophysiological Analysis of Transporter Function

Plasma membrane transporters selective for catecholamines efficiently clear these transmitters from the synapse, thereby regulating the spatial and temporal dimensions of synaptic transmission. For the past three decades, cocaine- and antidepressant-sensitive catecholamine transporters have been studied almost

Advances in Pharmacology, Volume 42

1054-3589/98 $25.00

exclusively by radiotracer flux analyses. Although useful in the definition of ionic and pharmacological sensitivities, these paradigms are limited in providing mechanistic explanations of transporter function, particularly the role of voltage in regulating catecholamine transport. Advances in the molecular biology of neurotransmitter transporters, and the heterologous expression of these transporters in transient and stable expression systems, provide a new route to study the mechanisms underlying catecholamine transport and the role of voltage in regulating transport. The human norepinephrine (NE) transporter (hNET) was the first catecholamine transporter to be cloned and expressed in nonneuronal cells. The sequence predicted from this clone encodes a 617 amino acid peptide bearing 12 hydrophobic regions of sufficient length to consider them transmembrane domains. Multiple homologues have since been identified using conserved sequences in hNET and the GABA transporter, GAT1, including serotonin, dopamine, and proline transporters (1).

From radioligand uptake assays, the translocation of NE via NETs should be electrogenic (2) with stoichiometry: Na/Cl/NE. Thus, we may expect the generation of current associated with NE transport and possibly a role of voltage in regulating transport. Several laboratories have described transporter-associated currents that exceed the predictions of stoichiometric transport (3–5). For hNET, we have documented the presence of NE-induced currents in transfected mammalian cells that exceed the predictions of fixed stoichiometric transport (6, 7). These currents are sensitive to transporter agonists and antagonists, and they depend on extracellular Na. Although it is evident that many neurotransmitter transporters do not maintain a strict ratio of transmitter flux to ionic current, the mechanistic relationship between transmitter transport and the transmitter-induced currents is unknown. This chapter focuses on a comparison of NE uptake, NE-induced macroscopic currents, current fluctuations, single-channel events, and transient pre-steady-state currents in a heterologous expression system. We propose a model of hNET that is compatible with these data and suggest a role for the NE-induced current in synapses.

The existence of a channel mode of conduction in a previously established fixed-stoichiometry transporter affects the models that we may evoke to explain transporter function. Carrier models require binding, translocation of the carrier, and unbinding of the transported species and are unlikely to explain the existence of a channel mode of conduction. Alternating access models, in which binding, occlusion, and unbinding occur, and in which transporters sequentially face outward and inward, never create an open pore. A modification of the alternating access model includes a pore with gates on either side of the membrane that are controlled by the substrate and the cotransported ions. Gates are not open at the same time, and fixed or, in some cases, variable stoichiometry again governs the transport of transmitter and coupled ions through an occluded pore. Single-file models, similar to those used to describe ion channel conduction, proposing that transmitters and cotransported ions hop from binding site to binding site through an open pore, have also been suggested as a mechanism for cotransport. The large currents that we observe restrict the applicability of such models to hNET. We and others have proposed models that combine the properties of transporters and ion channels. In transporter–channel models, in addition to coupling with ion gradients that activate and drive uptake, the transmitters induce currents in transporters. This latter capability is similar to a ligand-gated ion channel, where the transmitter acts as the ligand.

To investigate the mechanism of NE transport in hNET, we have used stably transfected HEK-293 (human embryonic kidney) cells studied under voltage clamp (6, 7). All experiments described here were done at 37°C. The bath solutions contained (in mM): 130 Na, 1.3 K, 1.5 Ca, 0.5 Mg, 5 dextrose, 10 HEPES; 7.35 pH, and 300 mosm. The anions are Cl except for 0.5 SO_4 and 1.3 PO_4. Whole-cell patch electrodes contain (in mM): 120 K, 0.1 Ca, 2 Mg, 1.1 EGTA, and 10 HEPES; 7.35 pH, 270 mosm, 0.1 uM free Ca. In some protocols we replaced Na with Li or choline. Transfected HEK-293 cells express approximately 10^6 copies/cell of hNET protein. These cells take up radiolabled NE, and hNET antagonists or NE competitors, such as guanethidine (GU), inhibit this uptake. NE or GU induces current in these transfected cells, and the induced currents saturate with substrate concentration. Concentrations of desipramine (DS) or cocaine that inhibit NE uptake also inhibit the substrate-induced current.

An analysis of substrate-induced current fluctuations provides an estimate of the unitary events underlying the induced current. At −120 mV and at saturating NE concentrations, the NE-induced inward current is I approximately 60 pA. At the same voltage, the NE-induced noise variance is σ^2 equals approximately 20 pA^2 (no capacity compensation, 2000 Hz bandwidth). The single-channel currents predicted from a ratio of the variance to the mean is, therefore, *i* equals approximately 0.3 pA. Both the mean current and the variance of the fluctuations are larger than predicted from fixed-stoichiometry models of coupled transport. The variance obeys an equation similar to induced current, namely: $\sigma^2 = \sigma^2_{max}$ [NE]/(K_m + [NE]), where at −120 mV and in 30 μM NE, $K_M = 0.6$ μM. GU-induced currents and current fluctuations follow a similar pattern. With 30 μM NE or GU in the pipette, inside-out patches from hNET-293 cells contain inward currents similar in size to those predicted from fluctuation analysis. Cell-detached patch currents are comparable to those seen in cell-attached patches, but the openings are more pronounced. For single-channel experiments, we selected ultra low-noise patches with more than 50 GΩ seal resistance. To minimize extraneous noise, we coated the electrodes close to the tip and brought patches near the surface of the bath. These inward currents are voltage dependent, and they are blocked by adding 2 μM DS to the bath. The mean conductance of these events were estimated from amplitude histograms obtained at different voltages (7). For inside-out patches voltage clamped between 0 and −80 mV, the GU-induced inward current varies linearly with voltage and has an extrapolated reversal potential near 0 mV and a conductance $\gamma = 3$ pS. These elementary currents explain the substrate-induced macroscopic currents and the whole-cell current fluctuations.

To further elucidate the mechanisms associated with NE transport and NE-induced currents, we have also studied transient pre-steady-state currents that arise from voltage steps. Because both parental and transfected cells have membrane-associated charge that could contribute to the transients, we subtract capacitance-uncompensated traces to eliminate background transients. Let Q(V) represent the area under the transporter-associated transient current. The data so far indicate that the pre-steady-state charge, Q(V), does not represent movement of the bound charge associated with NE/Na/Cl. In that case, the average cell we record from would have to contain well over 10^7 transporters on the surface. The total number of hNETs per cell measured by binding assays is 10^6. A possible explanation is that we are selecting cells with higher than average transporter densities: 10^6 transporters per cell represents the mean for a population of cells; assuming a standard deviation equal to the mean, we would expect 8% of the cells to have more than 10^7 transporters. Correlating the Q(V) with the NE-induced I(V) indicates an interdependence between the pre-steady-state and the induced current. Nevertheless, Q_{max} determined from a Boltzmann fit to a broad range of voltages predicts an unreasonable number of transporters. We conclude that Q must represent an additional charge not associated with NE/Na/Cl movement.

To investigate the source of Q, we studied the effect of DS on the transient. In addition to blocking NE uptake and NE-induced current, DS also blocks the transporter-associated transient currents. Because it is unlikely that NE/Na/Cl bound to the transporter is the source of Q, we considered the movement of substrates and ions into the closed transporter. To examine whether the substrate influences the transient, we measured Q in the presence and absence of NE or GU. With GU and DS both present, or with DS alone, Q is approximately the same size at all voltages. Thus, the transients are not associated with the presence of the substrate and probably not with the movement of charged residues within the transporter. This leaves the movement of cotransported ions as a possible source of Q. To investigate this possibility, we analyzed the transient currents in the presence and absence of external Na, using Li or choline to replace Na. Substituted LiCl for NaCl resulted in a decrease of the control current (no NE, 130 mM Na). This indicates that there is a constitutive Na leak through the transporter. However, with Li replacing Na, the transient remains essentially unchanged. Thus, Li, although it blocks both NE uptake and the NE-induced current, nevertheless evokes a pre-steady-state transient current similar to Na. To explore this further, recent unpublished experiments use choline as a replacement for Na. These data show that choline reduces the pre-steady-state transient with the same strength as DS. Thus, although choline and Li both block translocation of substrate and substrate-induced current, choline also blocks the transient current. Most likely, choline cannot enter the closed transporter. Although Li can enter the transporter, it cannot gate it; our model is that Li generates a transient current but inhibits the transport of substrate and the substrate-induced current.

We conclude that in solutions containing Na and NE, the pre-steady-state charge resulting from steps to negative voltages represents the movement of Na ions into the closed transporter. Once Na and NE bind to the transporter, they enable it to translocate substrate and Na. Saturation kinetics of uptake and of induced current indicates that only one Na and one NE are required to open the transporter. (We have not measured the saturation kinetics of Cl.) From fluctuation analysis and single-channel measurements, we infer that at −120 mV, approximately 300 ions translocate per NE-induced channel opening. This is near the ratio of NE-induced current to NE uptake as estimated from V_{max} data in similar cells. Thus, up to 300 Na ions move through the transporter for each NE molecule. These may be distinct molecular events, but they appear to be coupled events. In uptake assays, the voltage is not controlled and is unlikely to be as high as −120 mV. Therefore, we cannot exclude the possibility that, although only one NE gates the transporter, more than one NE molecule may move through the open pore. The consequence of this model is that electrical as well as chemical gradients may drive NE uptake, similar to the electrochemical gradients that drive ions through channels. The movement of a large number of Na ions through the open transporter would contribute to the depolarization of the presynaptic terminal, possibly suppressing uptake through an autoregulatory mechanism.

Acknowledgment

We wish to acknowledge the support of National Institutes of Health grant NS-34075.

References

1. Amara, S., and Kuhar, M. J. (1993). Neurotransmitter transporters: Recent progress. *Annu. Rev. Neurosci.* 16, 73–93.
2. Rudnick, G. (1996). Mechanisms of biogenic amine neurotransmitter transporters, *In Neurotransmitter Transporters: Structure, Function, and Regulation.* (M. E. A. Reith, ed.), pp. 73–100. Humana Press, Totowa, NJ.
3. Lester, H. A., Mager, S., Quick, M. W., and Corey, J. L. (1994). Permeation properties of neurotransmitter transporters. *Annu. Rev. Pharmacol. Toxicol.* **342,** 219–249.
4. DeFelice, L. J., and Blakely, R. D. (1996). Pore models for transporters? *Biophys. J.* **70,** 579–580.
5. Sonders, M., and Amara, S. G. (1996). Channels in transporters. *Curr. Opin. Neurobiol.* **2,** 294–302.
6. Galli, A., DeFelice, L. J., Duke, B. J., Moore, K. R., and Blakely, R. (1995). Sodium dependent norepinephrine-induced currents in norepinephrine-transporter-transfected HEK-293 cells blocked by cocaine and antidepressants. *J. Exp. Biol.* **198,** 2197–2212.
7. Galli, A., Blakely, R., and DeFelice, L. J. (1996). Norepinephrine transporters have channel modes of conduction. *Proc. Natl. Acad. Sci. U.S.A.* **93,** 8671–8676.

J. B. Justice,* K. S. Danek,* J. W. Kable,* E. L. Barker,†
and R. D. Blakely†

*Department of Chemistry
Emory University
Atlanta, Georgia 30322

†Department of Pharmacology
Vanderbilt University
Nashville, Tennessee 37235

Voltammetric Approaches to Kinetics and Mechanism of the Norepinephrine Transporter

Rotating-disk electrode voltammetry (RDEV) was used to follow the time course of uptake and efflux of catecholamines in cells expressing the human norepinephrine transporter (LLC-NET cells). The method has previously been applied to the study of the dopamine transporter (DAT) in tissue homogenates and synaptosomal preparations (1). It has also been used to determine kinetic parameters of dopamine (DA) and norepinephrine (NE) at the hNET in LLC-NET cells (2). Induced efflux experiments allow for kinetic analysis of substrates not oxidized at the applied potential and provide information on both inward and outward substrate transport in the same assay.

To conduct these experiments, cells are grown at 37°C until confluent, washed with 20 ml Krebs-Ringer-HEPES (KRH) buffer and harvested by scraping in two 2.5-ml volumes of room temperature KRH buffer. The harvested cells are transferred to a single glass test tube and centrifuged at 2500 rpm (1000 × g) for 2 min. The pellet is divided for use in four replicate transport assays. The pellet is resuspended in room-temperature KRH buffer and transferred to the electrochemical cell. The potential is applied (+0.45 V vs Ag/AgCl) and the current monitored for baseline acquisition. Following baseline, DA, NE, dihydroxybenzylamine (DHBA), or other substrate is added to bring the solution to the desired initial concentration, causing a sharp rise in the observed oxidation current. As uptake occurs, the decreasing current is recorded for 6 min as a new steady-state is approached. To induce efflux, 6 μl of unbuffered electrolyte as a control or 1, 10, or 100 μM of selected substrate is added to the suspension, and the efflux of DA is monitored. Data are expressed as the ratio of the initial rate of efflux to initial rate of uptake. This normalizes the data for the number of cells and level of transporter expression in each assay.

An example of the time course of the efflux of 1 μM DA by 10 μM *l*-amphetamine is represented in Figure 1A. The initial rate of uptake is obtained from a regression performed on the first 10 sec of data collected following the addition of DA. The time course data also provide the steady-state concentration

Advances in Pharmacology, Volume 42

1054-3589/98 $25.00

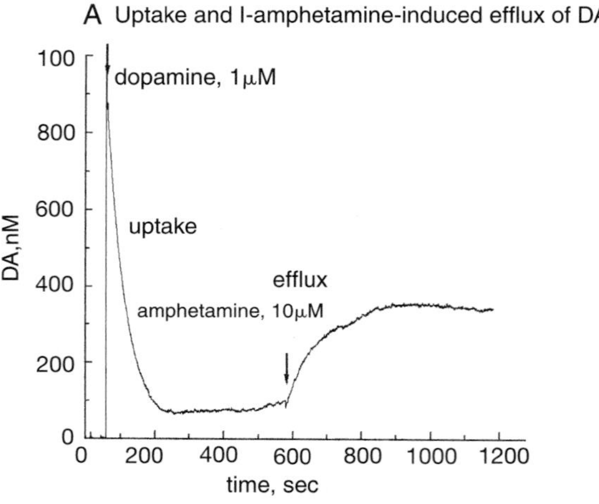

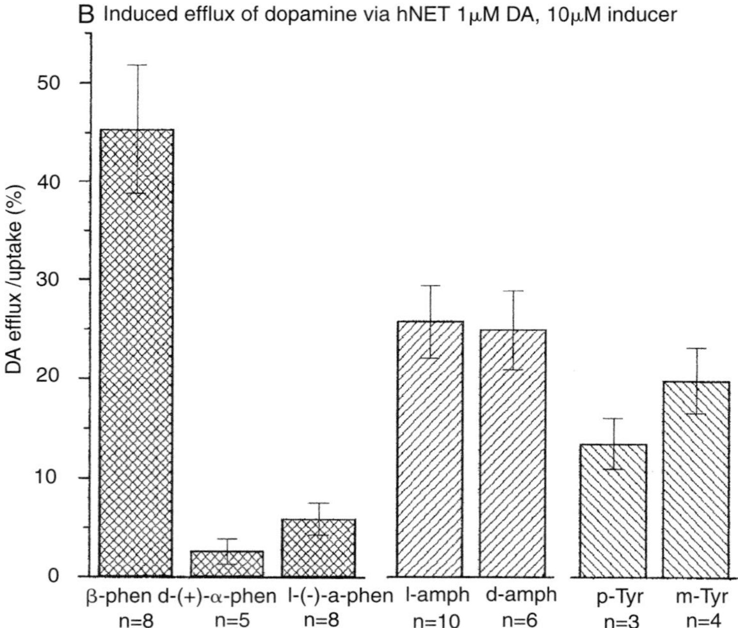

FIGURE 1 (*A*) Uptake of 1 μM DA by LLC-NET cells, followed by efflux induced by 10 μM *l*-amphetamine. (*B*) Ratio of induced efflux to uptake of DA for α and β-phenethylamines, *d*- and *l*-amphetamine, and m- and p-tyramine.

of DA, which in this example is about 80 nM. Following the addition of *l*-amphetamine, DA concentration increases due to net outward DA transport. The initial efflux rate is obtained from linear regression on the first 20 sec following addition of amphetamine or other substrate. On the addition of *l*-

amphetamine, there is a large amount of *l*-amphetamine on the outside of the cell relative to DA so an influx of *l*-amphetamine occurs. *l*-Amphetamine is transported inward and dissociates from the transporter, allowing DA to bind. The complex then reorients to the outside, causing an increase in DA in the medium. At first, there is little *l*-amphetamine relative to the high concentration of DA on the inside of the cells to compete for outward transport. However, as it accumulates intracellularly, it competes with DA for binding to the empty transporter, and DA efflux decreases. At the same time, DA concentration in the medium increases, causing an increase in the inward flux of DA. The result is seen as a maximum in DA concentration in the medium when the increasing inward and decreasing outward DA fluxes are equal. Beyond the maximum, net DA transport is inward as the system approaches a new steady-state.

Figure 1B depicts the results of the induced efflux of DA taken up from a 1-μM DA solution by the addition of 10 μM of various substrates. β-Phenylethylamine was found to induce efflux of DA at 45.4 $\pm$ 6.5% (mean $\pm$ SEM; n = 8) of the initial rate of DA uptake. The *d*- and *l*-α-phenylethylamines produced very little DA efflux (DA efflux rate/influx rate: 1.4 $\pm$ 0.3% [n = 4] and 6.0 $\pm$ 1.6% [n = 8], respectively). There was no statistical difference in the induced efflux rates of these two isomers (p = .134, n.s.). β-Phenylethylamine induced efflux of DHBA at only 13.5 $\pm$ 4.3% (n = 7) of its initial uptake rate. These results for DHBA versus DA and α versus β-phenylethylamines indicate that the nitrogen to phenyl ring distance on the substrate is critical for transport. In addition, benzylamine was found not to induce efflux of DA (efflux rate/influx rate <1%), suggesting that the methyl group on the α-carbon of the α-phenylethylamines is not what is causing the lower transport activity. Structures in which the side-chain amine group was extended (3-phenyl-1-propylamine), part of a ring (isoquinoline, 3-aminoquinoline), or replaced by $NHCONH_2$ (phenyl urea) did not induce efflux. These results, indicating the importance of the ethylamine side chain for optimal transport, agree with previous work performed in rat striatal tissue with the DAT (3).

β-Phenylethylamine (100 μM) was found to induce efflux of DA (1 μM) at 37.5 $\pm$ 4.6% (n = 4) of its initial rate uptake. This was the statistically same rate observed (p = .346, n.s.) as for 10 μM β-phenylethylamine, indicating that the transport rate is near maximal with 10 μM β-phenylethylamine.

From Figure 1B it is also evident that the presence of a hydroxy ring substituent, as in the tyramines, serves to slow transport relative to β-phenylethylamine at a 10μM concentration. The efflux-uptake ratio of DA following a 10-μM m-tyramine addition was 19.9 + 3.3% (n=4). p-Tyramine induced efflux of DA at a similar rate (13.5 $\pm$ 2.6%, n = 3, p =. 18, n.s.). This effect on the rate of transport is not as great as shortening the side-chain length. Other para substituents (at 10 μM) were also examined for relative rates of induced efflux of DA via hNET. The relative order was H > OH > NH_2 = F = CH_3 > Cl > Br > OCH_3 > NO_2. The correlation of induced efflux rate with volume of substituent was $r = -.95$. The results indicate that even bulky substituents at the para position allow transport.

Amphetamine is known to inhibit the uptake of norepinephrine across the neuronal membrane and participate in the transmitter's release through the transporter. Fisher and Cho (3) hypothesized that amphetamine-induced release was through an exchange diffusion mechanism. Due to the highly lipophilic nature of amphetamine, it is also capable of entering the cell by diffusion, complicating the analysis (5). RDEV allows for measurement of fast initial rates, limiting the effects of diffusion. On the addition of amphetamine (10 μM), the initial efflux rate of DA or NE obtained represents only transporter-mediated efflux. There was no difference found in the induced efflux of DA by *l*- versus *d*-amphetamine (efflux rate/influx rate = 25.8 ± 3.7% [n = 10] and 25.0 ± 4.0% [n = 6], respectively; p =. 88, n.s.). The induced efflux of NE by *l*-amphetamine was also the same as by *d*-amphetamine (efflux rate/influx rate = 39.4 ± 3.2% [n = 10] and 33.6 ± 1.7% [n = 6], respectively; p = .14, n.s.).

The effect of using methylphenidate as an effluxing agent was also examined. Methylphenidate, as with *d*-amphetamine, has been used to treat attention deficit disorder with hyperactivity (4). Its mode of action is believed to be similar to that of cocaine in that it inhibits the DA transporter (4). To test for neurotransmitter-releasing effects, methylphenidate was added as an effluxing agent following the uptake and accumulation of NE by the LLC-NET cells. No efflux of NE was observed at 10- or 100-μM additions of methylphenidate, in agreement with earlier work on hNET (6).

References

1. Meiergerd, S. M., and Schenk, J. O. (1994). Striatal transporter for dopamine: Catechol structure-activity studies and susceptibility to chemical modification. *J. Neurochem.* **63,** 1683–1692.
2. Burnette, W. B., Bailey, M. D., Kukoyi, S., Blakely, R. D., Trowbridge, C. G., and Justice, J. B. Jr. (1996). Investigation of human norepinephrine transporter kinetics using rotating disk electrode voltammetry. *Anal. Chem.* **68,** 2932–2938.
3. Fisher, J. F., and Cho, A. K. (1979). Chemical release of dopamine from striatal homogenates: Evidence for the exchange diffusion model. *J. Pharmacol. Exp. Ther.* **208,** 203–209.
4. Volkow, N. D., Ding, Y. S., Fowler, J. S., Wang, G. J., Logan, J., Gatley, J. S., Dewey S., Ashby, C., Liebermann, J., Hitzemann, R., and Wolf, A. P. (1995). Is methylphenidate like cocaine? *Arch. Gen. Psychiatry* **52,** 456–463.
5. Wong, T. W., Van Frank, R. M., and Horng, J. (1972). Accumulation of amphetamine and p-chloro-amphetamine into synaptosomes of rat brain. *J. Pharm. Pharmacol.* **24, 171–173.**
6. Wall, S. C., Gu, H., and Rudnick, G. (1995). Biogenic amine flux mediated by cloned transporters stably expressed in cultured cell lines: Amphetamine specificity for inhibition and efflux. *Mol. Pharmacol.* **47,** 544–550.

N. R. Zahniser,*‡ G. A. Gerhardt,*†‡ A. F. Hoffman,* and C. R. Lupica*‡

Departments of *Pharmacology, †Psychiatry, and ‡Neuroscience Program
University of Colorado Health Sciences Center
Denver, Colorado 80262

Voltage-Dependency of the Dopamine Transporter in Rat Brain

The neuronal dopamine transporter (DAT) is critical for terminating DA neurotransmission. Uptake of DA by DAT is dependent on cotransport of Na^+ and Cl^- ions. Results of biochemical studies from several labs (1) suggest that two Na^+ ions and one Cl^- ion are cotransported with each DA molecule. DA is also positively charged at physiological pH. Thus, at a minimum, there should be a net inward movement of two positive charges with each transport cycle, thereby generating an inward current. The suggestion that DAT is an electrogenic transporter also predicts that the translocation of DA, or DA uptake, should be voltage-dependent. A number of investigators (1) have previously shown that uptake of [^{3}H]DA into rat striatal preparations is diminished by agents (e.g., elevated KCl concentrations, veratridine, batrachotoxin, and ouabain) that induce membrane depolarization. However, in these preparations it is difficult to distinguish definitively the effects of agents such as these on membrane potential from their effects on the ionic gradients required for DA uptake.

In cells such as *Xenopus* oocytes, membrane potential can be manipulated independently of ionic gradients. Using *Xenopus* oocytes expressing the cloned human (h) DAT and the two-electrode voltage clamp technique, Sonders and colleagues (2) recently demonstrated that hDAT is an electrogenic transporter. DA ($K_{0.5} = 2.4\ \mu M$) induces an inward, transport-associated current at potentials below -20 mV. The magnitude of these transport-associated currents is voltage-dependent. Furthermore, hDAT uptake velocity is also dependent on membrane potential. Under voltage clamp, membrane hyperpolarization increases DA uptake velocity in hDAT-expressing oocytes, whereas depolarization diminishes it (~25% change/30 mV). This latter result is interesting for several reasons. First, it suggests a novel mechanism by which neuronal activity can transiently regulate DAT velocity. Second, it suggests that membrane hyperpolarization, by receptors such as presynaptic D2 DA autoreceptors, could also influence DAT activity. Indeed, several groups have demonstrated that D2 receptor ligands can modulate DA uptake measured in rat striatum both *in vivo* and *in vitro* (3).

In situ DAT activity in the brain can be monitored using electrochemical recording(s) to measure local changes in clearance of exogenously applied DA

Advances in Pharmacology, Volume 42

1054-3589/98 $25.00

(4). Thus, to address the question of whether DAT activity in the brain shows a voltage dependency similar to that expressed in the oocyte, we have begun to measure DA clearance in an *in vitro* rat brain slice preparation containing substantia nigra pars compacta (SNc). In addition, we have also used patch-clamp recording to monitor membrane potential in DA neurons in the same slice.

High-speed chronoamperometry was used to monitor clearance of exogenously applied DA in SNc in rat brain slices. Brain slices (400 μm) were superfused at a rate of 2 ml/min with artificial cerebrospinal fluid buffer. When finite amounts of DA (200 μM barrel concentration) were pressure-ejected at 5-min intervals from a micropipette positioned 250–300 μm from a Nafion-coated, single-carbon-fiber electrochemical electrode, reproducible DA signals were observed in SNc. Chronoamperometric measurements were made continuously at 5 Hz and averaged to 1 Hz using an IVEC-10 system. Ejections of DA resulted in average signal amplitudes of 0.52 ± 0.11 μM (mean $\pm$ SEM, N = 9). The mean time course of the signals, defined as the time for the signal to rise to a maximal value and to decay by 80% (T_{80}), was 84 ± 11 sec (N = 9). The clearance rate, which considers changes in amplitude as a function of time, was determined from the slope of the decay curve between the T_{20} and T_{60} time points. Under our experimental conditions, the basal clearance rate in SNc was 12.0 ± 2.1 nM/sec (N = 9). In comparison, in brain slices containing caudate-putamen (CPu), larger volumes of DA had to be ejected to achieve signals with equivalent amplitudes. This result is consistent with the higher DAT density in CPu versus SNc. However, signal time courses and clearance rates were similar in both brain regions. These results demonstrate that exogenous DA clearance, which we have previously shown to reflect DAT activity in CPu *in vivo* (3), exhibits similar characteristics in SNc and CPu in brain slices.

The SNc contains not only DAT, but also norepinephrine transporters (NETs) and serotonin transporters (SERTs), and evidence exists that DA can be taken up by NET and SERT in rat SN (4). However, bath application of desipramine (200 nM), a selective NET inhibitor, did not significantly alter the DA signals or clearance in SNc. Preliminary results with citalopram (500 nM), a selective SERT inhibitor, also suggested that it produced no effect on DA clearance. In contrast, cocaine inhibits DAT, NET, and SERT with approximately equal affinity. After 10 min of exposure to cocaine (50 μM), both DA signal amplitudes and T_{80} values were significantly increased from control values by 76% and 33%, respectively (ANOVA with Newman-Keuls post hoc comparison; N = 9). Because both signal amplitude and time course changed, clearance rates were not significantly altered (control, 12.0 ± 2.1 nM/sec; cocaine, 14.0 ± 2.8; N = 9) . During washout, the changes in T_{80} were more persistent than those in amplitude. Exposure to nomifensine (5 μM), an inhibitor with five- to 10-fold higher affinity for NET than DAT, produced effects similar to cocaine. Taken together, these results suggest that under our experimental conditions, DA clearance in SNc is primarily due to DA uptake by DAT. However, additional experiments with other selective inhibitors and in other brain regions are needed to establish conclusively the relationship between DA clearance and DAT activity in SNc.

As our first approach to answering the question of whether DAT in the central nervous system is voltage-dependent, we have used the patch-clamp method to monitor the change in the membrane potential of SNc neurons induced by elevating the concentration of KCl in the superfusion buffer. Neurons were patched in SNc in the hemisphere opposite to that where DA clearance was measured. Resting membrane potential was recorded every 30 sec for the duration of the experiment. Tetrodotoxin (TTX, 0.5 μM) was included in the superfusion buffer throughout these experiments in order to block voltage-gated sodium channels and to ensure that any effects of KCl were due to direct voltage changes and not due to changes in Na^+ gradients. Neurons were identified as being DA-like by their action potentials of relatively long duration. As expected, sequential increases in KCl concentrations from 10 to 40 mM depolarized the cell membrane. Whereas superfusion with 10 and 20 mM of KCl produced approximately 5 and 10 mV depolarizations, respectively, from baseline, 40 mM of KCl produced as much as a 40 mV depolarization. Furthermore, the magnitude of depolarization with 40 mM KCl was not stable.

Based on these results, the effect of 30 mM of KCl was determined on DA clearance in SNc at 2, 7, 12, and 17 minutes after elevating the KCl concentration in the superfusion buffer containing 0.5 μM of TTX. No endogenous DA release was detected when the KCl concentration was increased to 30 mM in the presence of TTX. In contrast, 30 mM of KCl rapidly and transiently increased the time course (T_{80}) of the signal from the exogenously applied DA but had little effect on the signal amplitude. The greatest changes were seen 2 min after increasing the KCl concentration and are reported here. Control amplitudes were 0.88 ± 0.13 μM (N = 3); amplitudes in the presence of 30 mM of KCl in the same slices were 0.90 ± 0.14. In contrast, the T_{80} parameter increased significantly (control, 120 ± 13 sec; 30 mM KCl, 199 ± 26; N = 3; ANOVA followed by Newman-Keuls). The DA clearance rate decreased nonsignificantly by 37% (control, 12.0 ± 2.5 nM/sec; 30 mM KCl, 7.6 ± 2.0; N = 3). These preliminary observations suggest that membrane depolarization, similar to its effect on the activity of hDAT expressed in oocytes, diminishes DAT activity in rat SNc.

Future studies will employ concomitant patch-clamp recording and DA clearance measurements in SNc slice preparations to test further the hypothesis that DAT activity in rat brain is voltage-dependent. Additional strategies will be used to hyperpolarize, as well as depolarize, membrane potential. hDAT-expressing oocytes will be used to confirm that the effects of these various strategies are due to changes in membrane potential, as opposed to changes in ionic gradients and/or intracellular signaling. The results of these studies should enhance our understanding of mechanisms by which DAT activity can be transiently regulated and may also provide insight into interactions between presynaptic receptors and DAT.

Acknowledgments

Supported by National Institutes of Health grants DA 04216, DA 00174, NS 09199, and DA 07725.

References

1. Krueger, B. K. (1990). Kinetics and block of dopamine uptake in synaptosomes from rat caudate nucleus. *J. Neurochem.* **55**, 260–267.
2. Sonders, M. S., Zhu, S-J., Zahniser, N. R., Kavanaugh, M. P., and Amara, S. G. (1997). Multiple ionic conductances of the human dopamine transporter: The actions of dopamine and psychostimulants. *J. Neurosci.* **17**, 960–974.
3. Cass, W. A., and Gerhardt, G. A. (1994). Direct evidence that D2 dopamine receptors can modulate dopamine uptake. *Neurosci. Lett.* **176**, 259–263.
4. Cass, W. A., Zahniser, N. R., Flach, K. A., and Gerhardt, G. A. (1993). Clearance of exogenous dopamine in rat dorsal striatum and nucleus accumbens: Role of metabolism and effects of locally applied uptake inhibitors. *J. Neurochem.* **61**, 2269–2278
5. Kelly, E., Jenner, P., and Marsden, C. D. (1985). Evidence that [^{3}H]dopamine is taken up and released from nondopaminergic nerve terminals in the rat substantia nigra *in vitro*. *J. Neurochem.* **45**, 137–144.

Emmanuel N. Pothos* and David Sulzer†

*Departments of Neurology and Psychiatry
Columbia University
New York, New York 10032

†Department of Neuroscience
New York State Psychiatric Institute
New York, New York 10032

Modulation of Quantal Dopamine Release by Psychostimulants

The effects of the psychostimulants cocaine and *d*-amphetamine (AMPH) resemble the effects of the dopamine (DA) precursor L-dihydroxyphenylalanine (L-dopa) in that they elevate extracellular DA. However, we find that while L-dopa increases the quantal size of DA release, psychostimulants reduce quantal size. Therefore, L-dopa potentiates stimulation-dependent release, whereas psychostimulants decrease stimulation-dependent release while simultaneously elevating stimulation-independent background levels via uptake blockade or promotion of reverse transport.

I. Electrochemical Methods

Because postsynaptic recordings are unsuitable for observing quantal release of slow-acting neurotransmitters like DA, we used amperometry (1) to

Advances in Pharmacology, Volume 42

1054-3589/98 $25.00

directly monitor vesicle release from PC12 cells (2) and midbrain DA neurons. A +700-mV potential was applied to a 5-μm carbon-fiber electrode placed on PC12 cell bodies or on DA neuron axonal varicosities (identified by immunostaining for the vesicular monoamine transporter; VMAT2). Following stimulation of exocytosis by secretagogues or current injection, spikes of faradic current resulting from DA oxidation estimate the number of molecules released (quantal size); quantal sizes in this report are based on an assumption of 2 e^- donated/molecule.

II. Variation of Quantal Size in Midbrain DA Neurons

In recordings from sister cultures of rat ventral tegmental area neurons, stimulation by 80 mM KCl and 10 nM α-latrotoxin produced apparent unitary events that lasted 100–200 μs (duration at half width = 81 ± 4 μs) and represented 1800 ± 100 molecules (mean ± SEM; n = 56 events; Fig. 1A). Some of these cultures were fixed and processed for tyrosine hydroxylase (TH) immunochemistry: Each process from which events were recorded was DAergic (12 of 12 neurons), although only 40% of the total neurons in these cultures are DAergic.

Following exposure to 20 μM L-dopa for 30 min, the quantal size increased to 5000 ± 700 molecules (n = 35 events; Fig. 1B). Because TH is the rate-limiting step in DA synthesis, this indicates that vesicles can be loaded by increasing cytosolic DA levels and may provide a functional basis for the control of TH activity by transcript regulation and phosphorylation. In the larger events, distinctions in shape such as the expression of a "foot" preceding the full spike, could be discerned (see lowest trace in Fig. 1B).

III. Variations in Quantal Size in PC12 Cells

To observe the effects of psychostimulants on quantal release in a tractable preparation where tens to hundreds of release events can be recorded from individual cells, we used the PC12 line. Average quantal sizes, presumably representing dense core vesicles, range from 100 to 800 K molecules, depending on culture conditions. As observed in the midbrain DA neurons, L-DOPA increased quantal size, apparently by elevating cytosolic substrate (3). Moreover, we found that lowered cytosolic DA reduced quantal size because the D2 agonist quinpirole at an exposure that inhibited TH activity by 50% also decreased quantal size by approximately 50% (E. N. Pothos, V. Davila, and D. Sulzer, in preparation).

We previously used the PC12 cell line to help elucidate the mechanism of action of AMPH. Analogous to findings in isolated chromaffin granules, AMPH (10 μM for 10 min) decreased quantal size by 50%. Because weak bases that are not transporter substrates also redistribute catecholamines from isolated granules and induce DA release in cultures, this appears to follow collapse of the vesicle electrochemical gradient that provides energy for transmitter

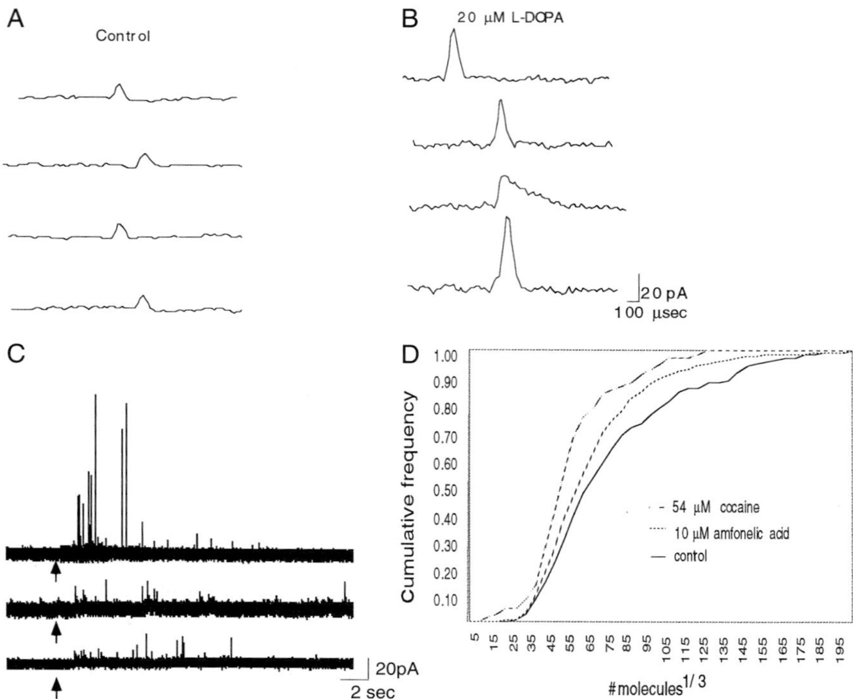

FIGURE 1 Examples of mechanisms by which the vesicle free energy gradient alters quantal size. (*A*) Representative evoked amperometric spikes observed at varicosities of ventral tegmental area DA axons in monolayer culture. (*B*) Following L-dopa, which elevates cytosolic DA levels and increases the driving potential for vesicle uptake, quantal events are larger, and variations in shape are observed that may reflect the expression of the fusion pore. (*C*) Psychostimulant drugs reduce quantal size. The upper trace shows amperometric spikes following stimulation of a control PC12 cell with 80 mM of KCl. The lower traces show spikes following incubation with 54 μM of cocaine for 30 min or 10 μM of amfonelic acid for 30 min, respectively. (*D*) Plots of the cumulative frequency of the quantal sizes indicate that following cocaine or amfonelic acid exposure, quantal size is decreased (p < .02, KS-Z = 1.56 for cocaine; $p < .01$, KS-Z = 1.84 for amfonelic acid; Kolmogorov-Smirnov two-sample test). The y axis expresses the cubed root of the quantal size, generally a normal distribution, as expected from the normal distribution of vesicle volumes (3).

accumulation. The resulting redistribution of DA to the cytosol may then initiate reverse transport across the DA transporter (DAT) (4).

To examine the effects of psychostimulants that are DAT blockers, we exposed PC12 cultures to 10 μM cocaine for 30 min, which reduced the average quantal size to 50% of control levels (n = 9 cells, 109 events for cocaine; 16 cells, 393 events for controls) or to 54 μM cocaine, which reduced quantal size to 27% of controls (n = 12 cells, 100 events; traces C and D of Fig. 1).

Potential mechanisms that may explain reduced quantal size following cocaine are: (1) a quinpirole-like activation of autoreceptors and resulting TH inhibition, (2) an AMPH-like collapse of the vesicle pH gradient by a weak base action, and (3) DAT blockade. We have ruled out autoreceptor activation as an explanation because the D2 antagonist sulpiride, used at exposures that block the effects of quinpirole, does not inhibit quantal size reduction by cocaine. We cannot rule out the possibility that collapse of the vesicle pH gradient contributes because a relatively small collapse of acidic gradients can be observed in microscopic observations using weak base vital dyes. However, we found that the DAT blocker amfonelic acid (10 μM for 30 min), which does not collapse pH gradients, reduced average quantal size to 39% of control levels (n = 14 cells, 661 events). This similar effect by a DAT blocker that is not a weak base suggests that the decreased quantal size by DAT blockade is due to a reduced cytosolic gradient.

In addition to manipulations of the vesicular free energy gradient, we have identified other mechanisms that can alter the quantal size of DA release. Quantal size can be regulated by expression of the VMAT (B. C. Sun, Y. Liu, R. Edwards, and D. Sulzer, in preparation), indicating that vesicular transport can be a rate-limiting step in the accumulation of transmitter. Moreover, in PC12 cells, release evoked by α-latrotoxin rather than elevated K^+ or nicotine results in a different mean quantal size, suggesting that the total vesicle contents are not generally released during exocytosis and that expression of the fusion pore can also be rate-limiting for DA flux from dense core granules (E. N. Pothos and D. Sulzer, in preparation).

In summary, psychostimulants may reduce the quantal size of DA release by altering the vesicular free energy used to provide for the accumulation of neurotransmitter. This is accomplished by altering the cytosolic–intravesicular concentration or the vesicle electrochemical gradient. While reduced quantal size does not underlie the abuse potential of drugs that elevate DA overflow, inhibition of local stimulation-dependent release may disturb associated behaviors, such as working memory or classical conditioning. Indeed, while both L-dopa and psychostimulants elevate overall extracellular levels, the results indicate that the "signal to background" of local stimulation–dependent DA input is far higher for L-dopa.

Acknowledgments

E. N. Pothos is an Aaron Diamond Foundation Fellow and recipient of a 1995 NARSAD Young Investigator Award. This research is supported in part by grants from the Aaron Diamond Foundation, NARSAD, the Parkinson's Disease Foundation and NIDA.

References

1. Leszczyszyn, D. J., Jankowski, J. A., Viveros, O. H., Diliberto, E. J. Jr., Near, J. A., and Wightman, R. M. (1991). Secretion of catecholamines from individual adrenal medullary chromaffin cells. *J. Neurochem.* 56, 1855–1863.

2. Chen, T. K., Luo, G., and Ewing, A. G. (1994). Amperometric monitoring of stimulated catecholamine release from rat pheochromocytoma (PC12) cells at the zeptomole level. *Anal. Chem.* **66**, 3031–3035.
3. Finnegan, J. M., Pihel, K., Cahill, P. S., Haung, L., Zerby, S. E., Ewing, A. G., Kennedy, R. T., and Wightman, R. M. (1996). Vesicular quantal size measured by amperometry at chromaffin, mast, pheochromocytoma, and pancreatic beta cells. *J. Neurosci.* **66**, 1914–1923.
4. Pothos, E., Desmond, M., and Sulzer, D. (1996). L-3,4-Dihydroxyphenylalanine increases the quantal size of exocytic dopamine release in vitro. *J. Neurochem.* **66**, 629–636.
5. Sulzer, D., Chen, T. K., Lau, Y. Y., Kristensen, H., Rayport, S., and Ewing, A. (1995). Amphetamine redistributes dopamine from synaptic vesicles to the cytosol and promotes reverse transport. *J. Neurosci.* **15**, 4102–4108.

Beth J. Hoffman,* Miklós Palkovits,*† Karel Paćak,‡§ Stefan R. Hansson,* and Éva Mezey‡

*Laboratory of Cell Biology
NIMH, NIH
Bethesda, Maryland 20892

†Laboratory of Neuromorphology
Semelweis University Medical School
Budapest 1094, Hungary

‡Clinical Neurosciences Branch
NINDS, NIH
Bethesda, Maryland 20892

§Department of Medicine
Washington Hospital Center
Washington, D.C. 20010

Regulation of Dopamine Transporter mRNA Levels in the Central Nervous System

Dysfunction of dopamine (DA) neural systems are thought to underlie neuropsychiatric disorders and psychostimulant abuse. Abnormalities of nigrostriatal dopaminergic neurons cause motor impairment associated with Parkinson's disease. Dysfunction of mesolimbic and mesocortical DA neurons have been implicated in schizophrenia and in drug addiction. The diencephalic DA neurons have been associated with adaptive, homeostatic responses to a variety of stressful stimuli through neuroendocrine regulation.

Advances in Pharmacology, Volume 42

1054-3589/98 $25.00

Re-uptake of DA through the plasma membrane DA transporter (DAT) is the primary mechanism for terminating the action of DA extracellularly. DAT and tyrosine hydroxylase (TH), the rate-limiting enzyme in catecholamine synthesis, are both considered characteristic markers for dopaminergic neurons in the central nervous system (CNS). In fact, levels of DAT as measured by radioligand binding are often used as an indicator of dopaminergic innervation. Consequently, DAT and TH would be expected to be colocalized, although not necessarily expressed at identical levels, in dopaminergic neurons. Under circumstances requiring an adaptive change in extracellular DA levels, TH activity and DAT activity might be expected to be regulated in opposing directions so as to increase DA synthesis and decrease re-uptake of DA from the extracellular space. However, elimination of DAT from early embryogenesis using transgenic animals (1) resulted in an increased DA half-life extracellularly but a concomitant downregulation of DA synthesis, most likely as a result of stimulation and desensitization of DA autoreceptors. Recent cloning of the DAT cDNAs (2) from several species has permitted a detailed analysis of DAT mRNA (3) and protein expression (4).

To address some of these issues, we have determined the cellular localization of DAT mRNA compared with TH mRNA using *in situ* hybridization histochemistry (ISHH) with cRNA probes. For synthesis of riboprobes, we have used polymerase chain reaction (PCR) to generate DAT-specific templates bearing both T7 (antisense) and T3 (sense) RNA polymerase promoters. As a control, two non-overlapping probes were synthesized from regions of the mRNA corresponding to transmembrane domains 7 through 10 and the carboxy-terminus/3′ untranslated, both regions that are less well conserved among the monoamine transporters. Both probes gave identical results, but the carboxy-terminal probe resulted in a higher signal-to-noise ratio. Further, we have determined the effect of unilateral transection of the median forebrain bundle (MFB) at the level between the diencephalon and the mesencephalon on DAT and TH mRNA levels by ISHH in control, sham-operated, and brain stem–transected animals. This transection severs all ascending fibers, including nigrostriatal, striatonigral, mesocortical, and mesolimbic pathways. Additionally, we have determined the effect of a single 3-hr immobilization stress on DAT and TH mRNA levels in dopaminergic neurons. Increased levels of plasma adrenocorticotropic hormone were used to confirm stressed status between controls (33–50 pg/ml) and immobilized animals (402–962 pg/ml).

We have identified four classes of ISHH patterns: (1) complete overlap of DAT with TH neurons, (2) partial overlap of DAT with TH-positive neurons, (3) no detectable DAT in TH-positive cells, and (4) no detectable TH in DAT-positive cells (Table I). These results are generally in good agreement with a previous study (3). Of particular note is the hypothalamus, which contains fewer DAT-positive cells than TH-positive cells in the A11, A12, and A14 cell groups and no DAT mRNA detected in the A15 cell group. Regions not previously noted contained DAT mRNA but no TH mRNA, including the ventral premamillary nucleus, the Barrington nucleus (at the level of the locus ceruleus, which was DAT mRNA–negative), and scattered cells of the pontine tegmentum.

Following immobilization stress, DAT mRNA expression increased in parallel with TH mRNA in the periventricular subdivision of the arcuate nucleus,

TABLE I Distribution of DAT mRNA-Positive Cells Compared with TH mRNA-Positive Cells

1. Complete overlap with TH-positive neurons
   - A8 cell group
     - Dorsolateral SN
     - Retrorubral field
     - Supraleminiscal cells
   - A9 cell group
     - Zona compacta and reticularis of SN
     - Cells ventral to nucleus ruber
   - A10 cell group
     - Ventral tegmental area
     - Tegmental decussation (A10c)
     - Ventral periaqueductal gray (A10dc)
   - A13 cell group
     - Zona incerta
2. Partial overlap with TH-positive neurons
   - A10 cell group (fewer DAT than Th cells)
     - Rostral linear nucleus
     - Cells over the interpeduncular nucleus
   - A11 cell group
     - Posterior hypothalamus (fewer DAT than TH cells)
     - Dorsal hypothalamus (no DAT)
     - Periaqueductal (no DAT)
   - A12 cell group
     - Ventrolateral subdivision (no DAT)
   - A14 cell group
     - Hypothalamic and preoptic periventricular cells
     - Paraventricular nucleus
3. No DAT in TH-positive cells
   - A2 cell group
     - Nucleus of the solitary tract
   - A15 cell group
     - Supraoptic nucleus
     - Scattered cells in the dorsomedial nucleus
4. No TH in DAT-positive cells
   - Ventral premamillary nucleus
   - Barrington nucleus
   - Scattered cells in the pontine tegmentum

SN, substantia nigra.

rostral A11 cell group, and the zona incerta (A13). Interestingly, the ventrolateral subdivision of the arcuate nucleus, which contains TH-positive, DAT-negative cells in control animals, remains DAT-negative despite an increase in TH mRNA after stress. These results suggest that while these cells do respond to immobilization stress, either (1) this group of cells is DAT-negative rather than expressing a low level of DAT mRNA or (2) DAT mRNA levels are unresponsive in these cells. In a subgroup of A10 cells, DAT mRNA increased with no visible change in TH mRNA. Moreover, DAT mRNA increased in the

Barrington nucleus, which projects to the spinal cord and scattered cells in the pontine tegmentum, cell groups that contain only DAT. Because there are clearly differences in the effects of different stressors (6), it will be important to examine these differences and levels of DAT mRNA in comparison to other catecholaminergic markers such as NET mRNA, vesicular monoamine transporter mRNAs, and synthetic enzymes.

Hemisection of the MFB either resulted in mRNAs for (1) decreases in both DAT and TH mRNA ipsilateral to the transection in zona incerta (A13), medial portion of the substantia nigra (A9) and ventral tegmental area (A10), or (2) a lack of any visible side difference for both TH and DAT mRNAs in the hypothalamic DA cell groups (A11, A12, A14, A15). These results suggest that maintenance of DAT and TH mRNA levels requires either nigrostriatal or striatonigral innervation (which contains GABA, substance P, dynorphin, and D1 DA receptors), or both, in some cell groups but not in others. Either the axotomy does not have the same effect for all cells groups or the descending fibers are important to maintain higher levels of mRNA. Further studies will be required to differentiate the critical pathway and specific innervation. Interestingly, *in vitro* studies of cocultivation of substantia nigra neurons with striatal neurons suggests that the target cells are important for the maintenance of the dopaminergic phenotype and DAT mRNA expression (6).

While there is considerable overlap between TH and DAT mRNA expressing cells, there are DA cells that express little or no DAT and DAT-positive cells that contain no TH mRNA. Lack of DAT mRNA does not imply a lack of DA synthesis. Therefore, DAT may not be a suitable marker as the sole criterion for identification of dopaminergic neurons. Certainly, these studies should be extended to correlating mRNA levels with protein levels as well as numbers of functional transporters at the plasma membrane for a better understanding of the relevant physiology. More interestingly, the presence of DAT mRNA in nondopaminergic neurons suggests the importance of tightly regulating DA overflow under certain physiological conditions at the level of target neurons. Additional studies will be needed to address the innervation patterns and where the DAT protein is present in these nondopaminergic neurons.

Regulation of TH and DAT mRNAs is not necessarily coordinate. At the mRNA level, DAT and TH are regulated in concert under certain conditions, such as innervation or denervation in particular cell groups, and independently in response to immobilization stress. This suggests that regulation of TH and DAT has some common regulatory factors that may be closely linked to cellular mechanisms specifying transmitter phenotype. However, there appear to be distinct regulatory factors that specify cell-specific and stimulus-specific responses. Thus, different regulatory factors appear to regulate DAT and TH mRNA levels in individual neurons.

References

1. Giros, B., Jaber, M., Jones, S. R., Wightman, R. M., and Caron, M. G. (1996). Hyperlocomotion and indifference to cocaine and amphetamine in mice lacking the dopamine transporter. *Nature* 379, 606–612.

2. Borowsky, B., and Hoffman, B. J. (1995). Neurotransmitter transporters: Molecular biology, function and regulation. *Int. Rev. Neurobiol.* **38,** 139–199.
3. Lorang, D., Amara, S. G., and Simerly, R. B. (1994). Cell-type-specific expression of catecholamine transporters in the rat brain. *J. Neurosci.* **14,** 4903–4914.
4. Freed, C., Revay, R., Vaughan, R. A., Kriek, E., Grant, S., Uhl, G. R., and Kuhar, M. J. (1995). Dopamine transporter immunoreactivity in rat brain. *J. Comp. Neurol.* **359,** 340–349.
5. Pacak, K., Palkovits, M., Kopin, I. J., and Goldstein, D. S. (1995). Stress-induced norepinephrine release in the hypothalamic paraventricular nucleus and pituitary-adrencortical and sympathoadrenal activity: *In vivo* microdialysis studies. *Front. Neuroendocrinol.* **16,** 89–150.
6. Perrone-Capano, C., Tino, A., Amadoro, G., Pernas-Alonso, R., and di Porzio, U. (1996). Dopamine transporter gene expression in rat mesencephalic dopaminergic neurons is increased by direct interaction with target striatal cells in vitro. *Mol. Brain Res.* **39,** 160–166.

Randy D. Blakely and Subramaniam Apparsundaram

Department of Pharmacology and Center for Molecular Neuroscience
Vanderbilt University Medical Center
Nashville, Tennessee 37232

Structural Diversity in the Catecholamine Transporter Gene Family: Molecular Cloning and Characterization of an L-Epinephrine Transporter from Bullfrog Sympathetic Ganglia

In contrast to the inactivation of acetylcholine at the neuromuscular junction by enzymatic hydrolysis, the actions of the catecholamine neurotransmitters dopamine (DA) and L-norepinephrine (NE) are terminated at central and peripheral synapses by rapid clearance, mediated by specific presynaptic transporters (1, 2). Molecular cloning studies have revealed that distinct gene products encode DA and NE transporters (DATs and NETs, respectively). Anatomical mapping studies reveal that dopaminergic neurons elaborate DATs, whereas noradrenergic neurons express NETs (3), though each carrier can efficiently transport both catecholamines. Moreover, kinetic studies indicate that, at satu-

Advances in Pharmacology, Volume 42

1054-3589/98 $25.00

rating substrate concentrations, DATs transport DA more efficiently than L-NE, with the reverse true for NETs. Thus, distinct gene products appear to have been evolved in concert with the structural diversification of the catecholamines themselves. Interestingly, neurons in the rodent brain, which express the enzyme phenylethanolamine-N-methyltransferase (PNMT) and thus are presumed to synthesize epinephrine (Epi) from NE, exhibit little or no expression of DAT or NET (3). These findings suggest that synapses formed by PNMT-positive neurons either do not require rapid clearance to carry out efficient chemical signaling or elaborate a molecularly distinct transporter. In this regard, Epi-synthesizing neurons in amphibians have been reported to express catecholamine transporters with a pharmacology similar to that displayed by NET, though a specific carrier has yet to be isolated and characterized. Thus, catecholamine uptake in the innervated frog heart is sensitive to antidepressants and cocaine. Unlike mammalian NETs, the cocaine-sensitive catecholamine uptake process in the innervated frog heart appears to preferentially accumulate Epi over NE. Radioligand binding studies have also identified the presence of high-affinity desipramine binding sites in frog heart with a sensitivity to antagonists like that of human NET (hNET) (4, 5). Structural analysis of such an Epi transporter (ET) might give additional insights into the requirements for the catecholamine binding pocket and/or the residues involved in antagonist binding as well as help establish whether an evolutionary path exists for optimization of Epi transport efficiency, in keeping with findings of transfer rates of DA and L-NE by DAT and NET.

Because of the relatively small number of PNMT-positive neurons in the rodent brain, we decided to utilize a more convenient source for cloning an Epi transporter. We isolated mRNA and prepared a cDNA library from sympathetic ganglia of the bullfrog, *Rana catesbiana*. Degenerate oligonucleotides (5′-CCGCTCGAGAA(C/T)GT(G/C)TGGCGG(C)TT(C/T)CC(A/G/C/T)TA-3′ and 5′-GCTCTAGAGCTG(A/G)GTIGC(A/G)GC(A/G)TC(A/G)A(T/G)CCA-3′), designed to match conserved sequences in TMD1 and TMD6 of NET and homologues (1), were first utilized to amplify potential transporter sequences from sympathetic RNA. One amplification product of 711 bp was found to exhibit a high degree of inferred amino acid sequence with hNET (1), and Northern blots showed the cognate mRNA to be highly enriched in sympathetic ganglia but not in peripheral nonneural tissues. This partial clone was used to design specific oligonucleotides to screen the cDNA library. Multiple overlapping clones were identified, and one encoded a 2514 bp cDNA frog ET (fET) possessing a complete open reading frame for a protein of 630 amino acids (Fig. 1A), which was suitable for functional characterization in transfected HeLa cells (1). The inferred sequence of fET exhibits 75% amino acid identity with hNET (1), its closest relative in the gene family. A more detailed structural analysis of this clone has been presented elsewhere and the full sequence has been deposited in the GenBank database (Accession # U72877). Here, we briefly describe features of its transport characteristics inferred from studies in transfected cells.

We expressed fET in HeLa cells using the vaccinia T7 expression system (1). To assess directly the efficiencies of substrate transport at fET versus hNET, we performed uptake assays with radiolabeled l-Epi, l-NE and DA in both

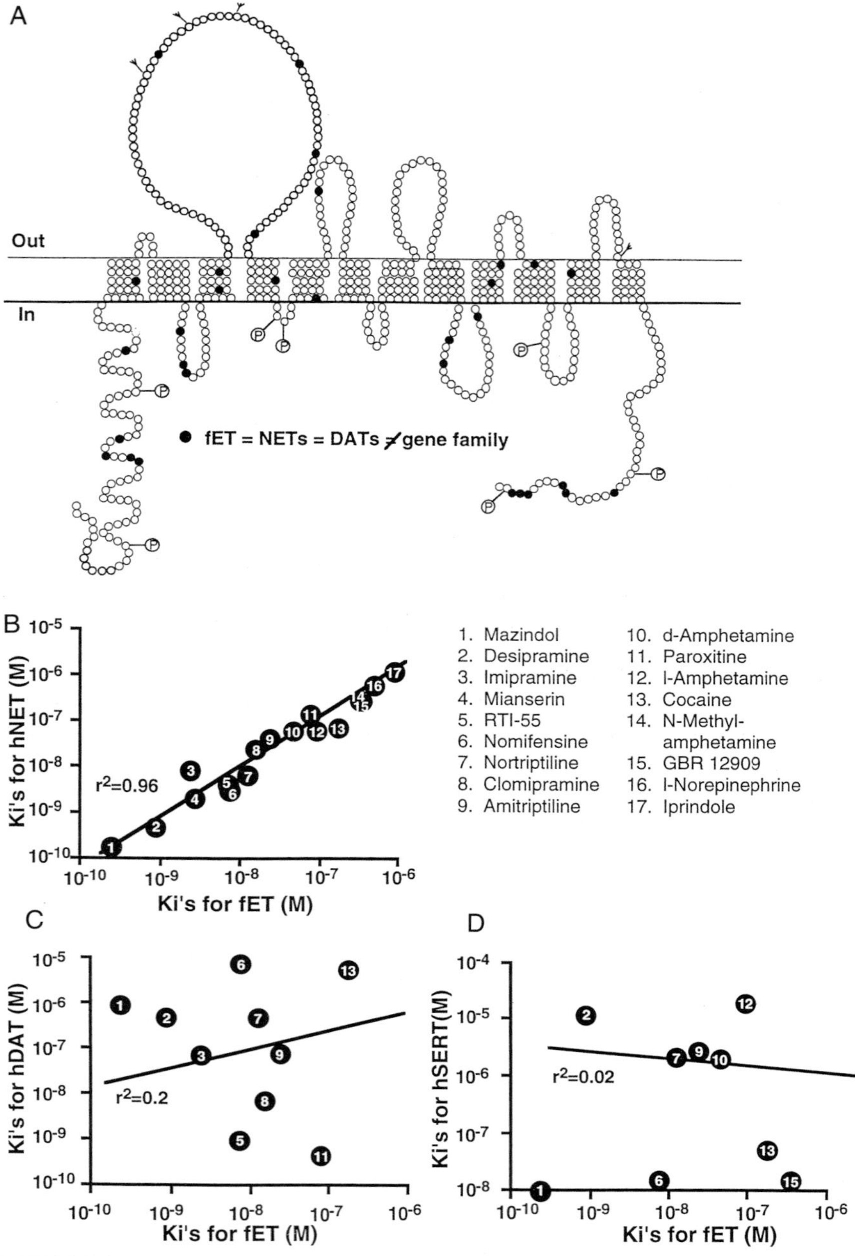

FIGURE 1 (*A*) Hydrophobicity-based model of the frog L-epinephrine transporter (fNET) and identification of catecholamine transporter–specific residues. (*B*) Correlation analysis of antagonist Ki for blocking [^{3}H]NE uptake by fET and hNET. (*C*) Correlation analysis of antagonist Ki for blocking [^{3}H]NE uptake by fET versus [^{3}H]DA transport by hDAT. (*D*) Correlation analysis of antagonist Ki for blocking [^{3}H]NE uptake by fET versus [^{3}H]5HT transport by hSERT.

fET and hNET transfected cells. These experiments confirmed saturability of substrate transport with apparent single-site kinetics for all substrates. Km values at fET (l-Epi, 1520 ± 152 nM; l-NE, 560 ± 92 nM; DA, 47 ± 9 nM) bore the same rank order relationship as observed for hNET (l-Epi, 2872 ± 24 nM; l-NE, 737 ± 77 nM; DA, 91 ± 6 nM). Moreover, we observed clear differences in maximal velocities (Vmax) of transport for different catecholamine substrates at fET and hNET, particularly evident as (1) an enhanced capacity for l-Epi transport and (2) a markedly reduced efficiency for DA transport at fET. The rank order of Vmax for catecholamines at fET was l-EPI>l-NE≫DA (1.0: 0.93: 0.13 relative to l-Epi), whereas at hNET, the order was l-NE>DA>l-Epi (1.0: 0.59: 0.53, relative to l-NE). In preliminary studies, we have also found NE- and Epi-induced currents in *Xenopus laevis* oocytes injected with fET cRNA.

We also performed parallel transfections with fET and hNET cDNAs and tested antagonists against the single substrate [^{3}H]-l-NE, because NE is a relatively equivalent substrate for both carriers. Absolute Ki values of all agents at fET were consistent with potencies observed for hNET, in keeping with its greater structural similarity to this carrier in the gene family. Indeed, correlation analyses of antagonist Ki at all three cloned human biogenic amine transporters demonstrate significant relationships only between antagonist Ki for fET and hNET (hNET vs fET, r^2 = .96; hDAT vs fET, r^2 = .20; hSERT vs fET, r^2 = .02, Fig. 1B–D). Cocaine exhibited submicromolar potency for blockade of uptake mediated by fET, in keeping with the activity of this nonspecific amine uptake inhibitor in the frog sympathetic nervous system.

Overall, we found 31 residues (Fig. 1A) that are conserved selectively in catecholamine transporters (fET = NETs = DATs ≠ gene family). These residues may define the ability of fET, NETs, and DATs to recognize and transport catecholamines, interact nonspecifically with antagonists, or undergo common modes of trafficking and regulation. Catecholamine transporters also may utilize residues for substrate recognition that are common to homologous 5HT transporters (SERTs). By analogy with models for the interaction of catecholamines with adrenoceptors, a negatively charged aspartate in transmembrane domain 1 (TM1) has been suggested to interact with the amine group of substrates, whereas Ser residues in TM7 have been proposed to interact with hydroxyl moieties of catecholamines (2). Although other models to explain these data are tenable, these residues are also conserved in SERTs, NETs, and fET, suggesting a common contribution in organization of the amine-binding pocket. More importantly, we have found fET to differ significantly from NET and DAT in Epi and DA translocation capacity at saturating substrate concentrations. Which residues have diverged to account for these functional distinctions? We find that 25 amino acids are conserved in NETs and DATs but not in fET (fET ≠ NETs = DATs). Finally, fET shares with NETs, but not with DATs, high affinity for tricyclic antidepressants and low affinity for DAT-selective inhibitors. In addition, amphetamine displays stereoselective inhibition of DATs, whereas both stereoisomers have comparable affinity for NETs and fET. fET and NETs share 53 residues not conserved in DATs. Additional studies, integrating recent efforts to map important residues by chimera studies, are warranted to determine the degree to which the distinctive substrate translocation efficiency and NET-like antagonist recognition are determined by these sites.

In summary, our identification of fET, an amphibian catecholamine transporter with enhanced efficiency for L-Epi transport, reveals an evolutionary path for divergence between NETs and ETs and raises the question of whether the two transporters coexist in a single species. We have been unable to amplify fET RNA from frog brain, suggesting that an additional NET-like transporter is responsible for NE clearance at bullfrog noradrenergic synapses in the bullfrog CNS. PNMT and L-Epi are also found in the mammalian brain. As mentioned, *in situ* hybridization studies indicate the absence of both DAT and NET mRNA in L-Epi-synthesizing neurons in the brainstem (3). Thus, if L-Epi is cleared at these sites by re-uptake, these terminals might elaborate a catecholamine transporter distinct from DAT and NET, like fET. The close relationship between fET and hNET antagonist sensitivities suggests that, if such a carrier exists in mammals, it will be difficult to discriminate from NETs with present pharmacologic agents.

References

1. Pacholczyk, T., Blakely, R. D., and Amara, S. G. (1991). Expression cloning of a cocaine- and antidepressant-sensitive human noradrenaline transporter. *Nature* **350**, 350–354.
2. Barker, E. L., and Blakely, R. D. (1995). Norepinephrine and serotonin transporters: Molecular targets of antidepressant drugs. *In* Psychopharmacology: The Fourth Generation of Progress. (F. E. Bloom, and D. J. Kupfer eds.), pp. 321–333. Raven Press, New York.
3. Lorang, D., Amara, S. G., and Simerly, R. B. (1993). Cell-type specific expression of catecholamine transporters in the rat brain. *J. Neurosci.* **14**, 4903–4914.
4. Pimoule, C., Schoemaker, H., and Langer, S. Z. (1987). [^{3}H] desipramine labels with high affinity: The neuronal transporter for adrenaline in the frog heart. *Eur. J. Pharmacol.* **137**, 277–280.
5. Schoemaker, H., Pimoule, C., de Oliceira, A-M., and Langer, S. Z. (1988). Radioligand binding to the neuronal adrenergic and noradrenergic transporter. *In* Progress in Catecholamine Research, Part A: Basic Aspects and Peripheral Mechanisms. (A. Dahlström, R. H. Belmaker, and M. Sandler, eds.), pp. 135–139. Liss Inc., New York, NY.
6. Apparsundaram, S., Moore, K. R., Malone, M. D., Hartzell, H. C., and Blakely, R. D. (1997). Molecular cloning and characterization of an L-epinephrine transporter from sympathetic ganglia of the bullfrog, rana catesbiana. *J. Neurosci.* **17**, 2691–2702.

N. D. Volkow,*† J. S. Fowler,† Y-S. Ding,† G-J. Wang,† and S. J. Gatley†

*Department of Psychiatry
State University of New York
Stony Brook, New York 11794

†Medical and Chemistry Departments
Brookhaven National Laboratory
Upton, New York 11973

Positron Emission Tomography Radioligands for Dopamine Transporters and Studies in Human and Nonhuman Primates

Interest in the Dopamine transporter (DAT) has been stimulated, in part, by the fact that DAT constitutes the main target site for the reinforcing properties of cocaine. Additionally, because transporters are localized on the presynaptic terminal, they serve as markers of DA neurons. Several radioligands have been examined for their suitability as positron emission tomography (PET) probes of the nerve terminal DAT (1). These include [^{11}C]nomifensine, [^{11}C]cocaine, [^{11}C] RTI-55, [^{11}C]WIN 35 428, [^{11}C]methylphenidate, and [^{11}C]*d-threo*-methylphenidate. F-18 GBR compounds have also been prepared as DAT radioligands.

These PET radioligands differ with respect to their affinities for DAT, their specific-to-nonspecific binding ratios, and their specificity for the DAT as well as their kinetics. For example, the cocaine analogues WIN 35 428 and RTI-55 have affinities for the DAT approximately 10 and 100 times higher than those of cocaine, respectively, and nomifensine and *d-threo*-methylphenidate have affinities intermediate between those of cocaine and WIN 35 428. Their times to maximum uptake vary by a factor of over 200, that is, between about 6 min for cocaine and over 1200 min for the high-affinity cocaine analogue RTI-55 (also known as β-CIT). The initial uptake in brain for these radioligands is high and corresponds for cocaine, methylphenidate, and *d-threo*-methylphenidate to approximately 7–10% of the injected dose.

The suitability of one DAT radioligand versus another is likely to vary as a function of the patients studied. For example, under conditions of decreased DAT availability and/or to examine regions with low DAT densities, ligands with very high affinities, such as [^{11}C]WIN 35428, may be desirable. However, for quantification in patients with decreased cerebral blood flow, ligands with

Advances in Pharmacology, Volume 42

1054-3589/98 $25.00

a relatively lower affinity for the DAT, such as [^{11}C]*d-threo*-methylphenidate may be more appropriate. Also, one tracer may be more suitable for certain experiments than for others. For example, when quantifying transporter occupancies for DAT blockers with fast association and dissociation rates, such as for cocaine, tracers with fast pharmacokinetics, such as [^{11}C]cocaine, may be advantageous.

These radioligands have been used to investigate both the normal human brain as well as the brain of neuropsychiatric patients. For example, studies using [^{11}C]nomifensine, [^{11}C]cocaine, [^{123}I]RTI-55, and [^{11}C]*d-threo*-methylphenidate have documented decreases in DAT with aging (1). The rate of DAT loss has been estimated to be approximately 6–7% per decade. These reductions are consistent with the age-related decrements in DAT and in DA cells documented in postmortem studies.

In patient populations, these radioligands have been used to assess DA cell degeneration in subjects with Parkinson's disease who show marked reductions in DAT when compared with healthy age-matched controls (1). Studies in cocaine abusers have shown that while there are increases in DAT shortly after withdrawal, there are decreases or no changes with protracted withdrawal. A preliminary study done in violent alcoholics showed significant elevations of DAT when compared with nonalcoholic subjects. In contrast, nonviolent alcoholics had lower or no changes in DAT levels when compared with age-matched controls (1).

These radioligands also have been used to measure DAT occupancies by drugs that inhibit the DAT and to assess if there is a relation between DAT blockade and the "high." For example, PET and [^{11}C]cocaine have been used to evaluate the relation between cocaine doses and DAT occupancies in the baboon brain (2). The time activity curves in striatum (ST) and in cerebellum (CB) and the time activity concentration of the nonmetabolized tracer in plasma were used to obtain distribution volumes (DV). The ratio of the DV in ST to that in CB (DV_{ST}/DV_{CB}), which corresponds to Bmax/Kd + 1 and is insensitive to changes in cerebral blood flow, was used as a index of DAT availability. DAT occupancies were calculated as percent change in Bmax/Kd from the paired baseline scan. Cocaine significantly inhibited [^{11}C]cocaine binding at all doses investigated. Percent change in the Bmax/Kd estimates for total striatal binding was dose related and ranged between 76 and 83% for a 1-mg/kg i.v. cocaine dose and 41 and 46% for the 0.1-mg/kg i.v. cocaine dose. The results differ from those reported in a single photon emission computed tomography (SPECT) study that used [^{123}I] RTI-55 as a DAT ligand (3). Administration of 20 and 40 mg of cocaine in humans led to reductions of only 6 and 17%. These low values were interpreted as reflecting the slow dissociation of [^{123}I]β-CIT from the transporters, making the estimates more representative of the plasma concentrations of cocaine at 60 min after its administration than of those achieved at peak concentration (4–6 min). DAT radioligands with very slow dissociation constants are likely to underestimate occupancies achieved by cocaine, the fast pharmacokinetics of which allow for only a very short period at peak concentration when it can compete maximally with the radioligand. The slower the dissociation rate of the radioligand, the larger is likely to be the underestimation. In this respect, the use of [^{11}C]cocaine as a ligand is ideal because its kinetics at tracer doses are similar to those at pharmacological doses. These results document that i.v. doses frequently abused by humans induce significant blockade of DAT-binding sites, and that even doses associated with mild behavioral effects blocked more than 50% of DAT.

PET and [11C]cocaine also have been used to evaluate the time course of the effects of RTI-55 on cocaine binding in baboon brain (4). [11C]Cocaine binding was measured prior to and 90 min, 24 hr, 4–5 days, and 11–13 days after RTI-55 (0.3 mg/kg i.v.). RTI-55 significantly inhibited [11C]cocaine binding at 90 min and 24 hr after administration. The half-life for the clearance of RTI-55 from DAT was estimated to be 2–3 days in the baboon brain. These results document long-lasting inhibition of cocaine binding by RTI-55 and corroborate that binding kinetics of RTI-55 in striatum observed in imaging studies with [123I]RTI-55 represent binding to DAT.

These radioligands also have been used to assess if there is a relation between DAT blockade by psychostimulant drugs and their behavioral effects in human subjects. This is illustrated by a study that evaluated the relation between methylphenidate-induced "high" and DAT inhibition (5). PET and [11C]*d-threo*-methylphenidate were used to estimate DAT occupancies at different times after methylphenidate (0.375 mg/kg i.v.). Methylphenidate pretreatment significantly reduced the binding of [11C]*d-threo*-methylphenidate at 7 and at 60 min after a single dose. The estimates for Bmax/Kd were reduced from a baseline value of (mean ± S.D.) 1.81 ± 0.3 to a values of 0.29 ± 0.2 at 7 minutes and of 0.36 ± 0.1 at 60 min after injection of methylphenidate. This corresponds to a DAT occupancy by methylphenidate of 84 ± 7% at 7 min and of 80 ± 7% at 60 min. Methylphenidate induced a "high" in most of the subjects. There was no relation between the level of DAT occupancy measured either at 7 or at 60 min and the subjective perception of the "high" (Fig. 1). Failure to observe a correlation between DAT and the "high" may reflect the fact that other variables are necessary for the "high", that an inadequate range of methylphenidate doses was used, that these were nonabusing subjects, and/or that there are differences between methylphenidate and cocaine

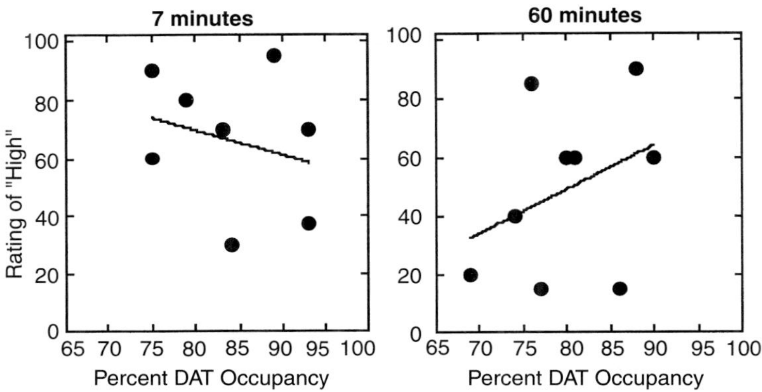

FIGURE 1 Relation between self-reports for methylphenidate-induced "high" and DAT occupancy at 7 min and at 60 min after its administration.

vis-à-vis their reinforcing properties. Recent studies were established in which a relation between DAT blockade and the cocaine-induced "high" in cocaine abusers was corroborated (6).

Acknowledgments

Research was the Department of Energy supported by (contract DE-ACO2-76CH00016) and (grants DA06891, DA09490-01, and DA06278).

References

1. Volkow, N. D., Fowler, J. S., Gatley, J. S., Logan, J., Wang, G-J., Ding, Y-S., and Dewey, S. L. (1996). Evaluation of the human brain dopamine system with PET. *J. Nucl. Med.* **37**, 1242–1256.
2. Volkow, N. D., Gatley, J. S., Fowler, J. S., Logan, J., Fischman, M., Gifford, A. N., Pappas, N., King, P., Ding, Y-S., and Wang, G-J. (1996). Cocaine doses equivalent to those abused by humans occupy most of the dopamine transporters. *Synapse* **24**, 339–402.
3. Malison, R. T., Best, S. E., Wallace, E. A., McCance, E., Laruelle, M., Zoghbi, S. S., Baldwin, R. M., Seibyl, J. S., Hoffer, P. B., Price, L. H., Kosten, T. R., and Innis, R. B. (1996). Euphorigenic doses of cocaine reduce [^{123}I]β-CIT SPECT measures of dopamine transporter availability in human cocaine addicts. *Psychopharmacology* **122**, 358–362.
4. Volkow, N. D., Gatley, J., Fowler, J. S., Logan, J., Dewey, S. L., Ding, Y-S., Pappas, N., King, P., Mac Gregor, R. R., Kuhar, M. J., Carroll, F. I., and Wolf, A. P. (1993). Long-lasting inhibition of in vivo cocaine binding to dopamine transporters by 3β-(4-iodophenyl)tropane-2-carboxylic acid methyl ester; RTI-55 or b-CIT. *Synapse* **19**, 206–211.
5. Volkow, N. D., Wang, G-J., Fowler, J. S., Gatley, S. J., Ding, Y-S., Logan, J., Dewey, S. L., Hitzemann, R., and Lieberman, J. (1996). Relationship between psychostimulant induced high and dopamine transporter occupancy. *Proc. Nat. Acad. Sci. U.S.A.* **93**, 10388–10392.
6. Volkow, N. D., Wang, G. J., Fischman, M. W., Foltin, R. W., Fowler, J. S., Abumrad, N. N., Vitkun, S., Logan, J., Gatley, S. J., Pappas, N., Tremann, H., Sheak, R. (1990). Relationship between subjective effects of cocaine and dopamine transporter occupancy. *Nature* **386**, 827–830.

Robert B. Innis

Departments of Psychiatry and Pharmacology
Yale University School of Medicine
West Haven, Connecticut 06516

West Haven Veterans Administration Medical Center
West Haven, Connecticut 06516

Single Photon Emission Computed Tomography Imaging of Dopaminergic Function: Presynaptic Transporter, Postsynaptic Receptor, and "Intrasynaptic" Transmitter

Parkinson's disease (PD) is a progressive, disabling neurodegenerative disorder characterized clinically by its major motor symptoms: tremor at rest, bradykinesia, rigidity, and postural instability. The disorder is characterized pathologically by the degeneration of dopaminergic neurons in the substantia nigra, resulting in an 80–99% reduction in striatal dopamine (DA) concentration. DA replacement with either the precursor L-dopa or direct DA receptor agonists effectively reverses the motor deficits of the disease early in its course. However, as the disease progresses, patients develop disabilities from a number of drug-induced side effects, progression of motor dysfunction (due to continued degeneration of DA nerve terminals), and an array of nonmotor, non-DA responsive symptoms.

We have developed a promising single photon emission computed tomography (SPECT) radiotracer—[^{123}I]β-CIT (2β-carbomethoxy-3β-(4-iodophenyl)tropane (also designated RTI-55)—that labels DA transporters (DATs), which are located on the terminals of DA neuronal projections from the substantia nigra to the striatum. Thus, this tracer is a marker for the neurons that degenerate in PD. This article reviews studies with [^{123}I]β-CIT, which suggest it will be a useful as a research and potential clinical imaging to measure dopaminergic terminal innervation of the striatum

Neuroreceptor imaging can be used not only to quantify the target receptor, but also to assess changes in the synaptic concentration of neurotransmitter following pharmacological challenges. We will review the development and validation of a SPECT paradigm designed to provide, in the same experiment, a measure of the baseline D2-binding potential (BP) and of the reduction in D2 BP induced by amphetamine, which is an indirect measure of an amphetamine-induced increase in synaptic DA concentration.

Advances in Pharmacology, Volume 42

1054-3589/98 $25.00

I. [^{123}I]β-CIT: Probe of Dopamine and Serotonin Transporters

Multiple saturable binding sites for [^{3}H]cocaine have been identified in the central nervous system, the majority of which are associated with DA, serotonin (5HT), and norepinephrine (NE) transporters. Some of the cocaine phenyltropane congeners are more potent than cocaine itself in behavioral stimulation, monoamine uptake inhibition, and affinity for monoamine transporters. For example, the radiolabeled phenyltropane fluoro derivative [^{123}I]β-carbomethoxy-3β-(4-fluorophenyl)tropane (WIN 35,428; also designated CFT) is used as a potent *in vitro* ligand for the DAT.

The search for an iodinated analogue of CFT led to the development of β-carbomethoxy-3β-(4-iodophenyl)tropane, designated as β-CIT in analogy to CFT and by the code number RTI-55. ^{125}I-labeled β-CIT has been used as an *in vitro* probe with homogenate binding studies [^{125}I]β-CIT has high and approximately equivalent affinities for DATs and 5HT transporters. However, the vast majority (i.e., >90%) of striatal binding is associated with the DAT, which is consistent with the relative density of DATs to 5HT transporters in striatum of greater than 10 : 1.

SPECT brain imaging with [^{123}I]β-CIT in nonhuman primates has shown that brain activity is concentrated in two areas: the striatum and the hypothalamic/midbrain region (1). Displacement studies demonstrated that striatal uptake was associated with DATs, because it was displaced by GBR 12909, a selective DA uptake inhibitor, but not by citalopram, a selective 5HT uptake inhibitor. The inverse was true in the hypothalamic/midbrain area, suggesting that the uptake in this area was associated primarily with 5HT transporters. Maprotiline, a selective NE uptake inhibitor, did not affect [^{123}I]β-CIT uptake in either striatum or hypothalamic/midbrain region. Infusion of L-dopa (50 mg/kg i.v.) failed to displace striatal [^{123}I]β-CIT binding, suggesting that the much lower clinical doses used for Parkinson's patients would not induce displacement of the tracer.

II. Studies in Parkinsonian Patient

SPECT imaging studies with [^{123}I]β-CIT at Yale University have demonstrated marked reduction of striatal uptake in patients with idiopathic PD compared with healthy subjects with a similar mean age. Our initial SPECT work in idiopathic PD patients and healthy controls has focused on quantitation and reproducibility of the striatal signal in healthy subjects and PD patients. These studies demonstrated the following: (1) striatal [^{123}I]β-CIT uptake is reduced in healthy controls as a function of age, (2) decreased [^{123}I]β-CIT uptake in PD patient is correlated with disease severity, (3) differences in striatal uptake completely separate controls and PD patients in a linear discriminant function analysis (2), and (4) in hemiparkinson patients, diminished striatal tracer uptake is most pronounced contralateral to the symptomatic side but also contralateral to the asymptomatic side, suggesting that [^{123}I]β-CIT SPECT is sensitive to changes in brain prior to the onset of clinical symptoms (3). In addition, we have shown a high degree of test–retest reproducibility of several [^{123}I]β-CIT

SPECT outcome measures in healthy subjects (2) and PD patients, supporting the feasibility of performing reliable SPECT imaging in serial patient studies.

III. Amphetamine Challenge to Measure Dopamine Release

Neuroreceptor imaging can be used not only to quantify the target receptor, but also to assess changes in the synaptic concentration of neurotransmitter following pharmacological challenges. We will review the development and validation of a SPECT paradigm designed to provide, in the same experiment, a measure of the baseline D2 BP and of the reduction in D2 BP induced by amphetamine, which is an indirect measure of an amphetamine-induced increase in synaptic DA concentration.

IV. Amphetamine Challenge and SPECT Imaging in Healthy Subjects

D-Amphetamine challenge studies in conjunction with SPECT imaging of the D2 receptor have been performed in human subjects (4). A bolus plus constant infusion administration schedule of the D2-receptor radiotracer [^{123}I]IBZM was used to obtain a stable baseline for reliable quantitation of the *d*-amphetamine effect. Eight healthy subjects first underwent a controlled experiment to demonstrate that stable levels of striatal and occipital activities could be maintained from 150 to 420 min during programmed infusion of the tracer. Next, seven subjects underwent the experiment with *d*-amphetamine. The experimental conditions were identical except that 0.3 mg/kg of amphetamine was injected intravenously at 240 min. Behavioral effects of *d*-amphetamine were measured by self-rating on the following analogue scales: euphoria, altertness, restlessness, and anxiety. The *d*-amphetamine injection induced a 15 $\pm$ 4% (mean $\pm$ SD) decrease in D2-receptor availability, measured as the specific-to-nonspecific equilibrium partition coefficient (V_3''). The *d*-amphetamine injection induced marked increases in euphoria, alertness, and restlessness scores. The intensity of these behavioral responses correlated with the decrease in D2 availability measured with SPECT. In contrast, the anxiety response was milder and not correlated with the decrease in D2 availability. These studies demonstrate the feasibility of using [^{123}I]IBZM programmed infusion and SPECT imaging to measure endogenous DA release after *d*-amphetamine challenge and to study brain neurochemical correlates of emotions.

V. Amphetamine/[^{123}I]IBZM in Schizophrenia

The amphetamine/[^{123}I]IBZM paradigm has recently been extended to a group of 15 drug-free schizophremic patients and 15 healthy subjects matched for age, sex, race, and parental socioeconomic status (5). The amphetamine-induced (0.3 mg/kg i.v.) displacement of equilibrium [^{123}I]IBZM striatal uptake was significantly greater ($p = .014$) in the schizophrenic patients (19.5 ± 4.1%) than in the healthy subjects (7.6 ± 2.1%). Plasma levels of *d*-amphetamine were not significantly different between the two groups. Increased displacement of [^{123}I]IBZM in the patients was correlated with the emergence or worsening of positive psychotic symptoms, such as hallucinations and delusions.

VI. Future Directions

Preliminary studies suggest that [^{123}I]β-CIT is a very promising agent, but additional studies will be needed to demonstrate both its utility and applicability in routine clinical settings. SPECT imaging of DATs may be used: (1) to diagnose patients in early stages of the disease, (2) to monitor the progression of the disease over time and to assess the efficacy of putative neuroprotective agents (such as antioxidants, free radical trappers, iron chelators, and trophic factors) to slow the progression of the disease, (3) to monitor the growth or rejection of fetal tissue transplanted into patients with PD, and (4) to distinguish the idiopathic disorder from other parkinsonian syndromes like striatonigral degeneration, which have more widespread pathology and are believed to have less severe loss of DA-containing neurons in substantia nigra. These potential clinical uses of SPECT imaging with [^{123}I]β-CIT will presumably require moderately large clinical trials to adequately measure their sensitivity, reproducibility, and clinical utility.

The application of [^{123}I]β-CIT in routine clinical studies will require the commercial preparation and distribution of a moderately short lived ($T_{1/2}$ = 13 hr) radiotracer. Assuming these logistical problems are addressed, a relatively simple SPECT imaging measurement will be required. For example, most clinical nuclear medicine departments could not perform compartmental modeling with a densely sampled arterial input function for measurement of the density of DATs. A relatively simple target-to-background ratio (like the ratio of striatal to occipital activity) would be relatively easy to implement in clinical settings. Additional studies are now in progress in our laboratory to assess the stability, reproducibility, and validity of this simple outcome measure to provide a value linear with the density of DATs in the striatum.

Acknowledgments

The studies reported from Yale University represent the dedicated efforts of many members of the Neurochemical Brain Imaging Program, and the Departments of Psychiatry, Diagnostic Radiology, and Neurology, and the Yale University School of Medicine.

References

1. Laruelle, M., Baldwin, R. M., Malison, R. T., *et al.* (1993). SPECT imaging of dopamine and serotonin transporters with [^{123}I]β-CIT: Pharmacological characterization of brain uptake in nonhuman primates. *Synapse* **13**, 295–309.
2. Seibyl, J. P., Marek, K. L., Quinlan, D., *et al.* (1995). Decreased single-photon emission computed tomographic [^{123}I]β-CIT striatal uptake correlates with symptom severity in idiopathic Parkinson's disease. *Ann. Neurol.* **38**, 589–598.

3. Marek, K. L., Seibly, J. P., Zoghbi, *et al.* (1996). [^{123}I]β-CIT SPECT imaging demonstrates bilateral loss of dopamine transporters in hemi-Parkinson's disease. *Neurology* **46**, 231–237.
4. Laruelle, M., Abi-Dargham, A., van Dyck, C. H., *et al.* (1995). SPECT imaging of striatal dopamine release after amphetamine challenge in humans. *J. Nucl. Med.* **36**, 1182–1190.
5. Laruelle, M., Abi-Dargham, A., van Dyck, C., *et al.* (1996). SPECT imaging of amphetamine-induced dopamine release in drug free schizophrenic subjects. *Proc. Natl. Acad. Sci. U.S.A.* **93**, 9235–9240.

D. F. Wong,* G. Ricaurte,* G. Gründer, R. Rothman,† S. Naidu,* H. Singer,* J. Harris,* F. Yokoi,* V. Villemagne,* S. Szymanski,* A. Gjedde,‡ and M. Kuhar§

*Department of Radiology
Johns Hopkins University
Baltimore, Maryland 21287

†National Institute of Drug Abuse
Baltimore, Maryland 21224

‡Aarhus University
8000 Aarhus, Denmark

§Neuroscience Branch
Emory University
Atlanta, Georgia 30322

Dopamine Transporter Changes in Neuropsychiatric Disorders

The dopamine transporter (DAT) protein is a member of Na^+-dependent membrane transporters with very high affinity and substrate specificity. Reuptake of dopamine (DA) into the presynaptic neuron by means of DAT is believed to be the primary mechanism for termination of dopaminergic neurotransmission. Moreover, DAT is the key target for psychostimulants like amphetamine or cocaine as well as for neurotoxins like 1-methyl-4-phenylpyridine (MPP^+). The highest concentrations of DATs are found in the basal ganglia, corresponding to the amount of DA nerve terminals in this brain region. Numerous studies demonstrating parallel losses of DA levels and DAT after lesions of nigrostriatal DA neurons suggest that the density of DAT is an excellent marker of the structural integrity of the dopaminergic system.

In vivo imaging of DAT has been performed mainly in patients with Parkinson's disease to demonstrate the marked loss of nigrostriatal DA nerve terminals in this disorder. However, positron emission tomography (PET) and single photon emission computed tomography (SPECT) imaging of DAT have been shown to be useful in various other neurodegenerative disorders, including

Advances in Pharmacology, Volume 42

1054-3589/98 $25.00

Lesch-Nyhan disease (LND), Rett syndrome, Tourette's syndrome, stimulant abuse, and progressive supranuclear palsy.

Using PET and [^{11}C]WIN 35 428 as the radiotracer (1), we have recently reported a marked 50–63% reduction of tracer binding to DATs in the caudate and a 64–75% reduction in the putamen of six patients with classic LND compared with 10 normal controls (2). The almost total absence of hypoxanthine-guanine phosphoribosyl transferase activity leads to hyperuricemia, choreoathetosis, dystonia, and compulsive self-injury, which characterize this devastating X-linked disease of infantile onset. A postmortem study of three LND patients and the results of animal studies had suggested the link between diminished DA function and self-injurious behavior in LND. The observed difference in the caudate-cerebellum ratio between the LND patients and the controls was even greater when we performed a partial volume correction of caudate time activity curves, which was necessary because volumetric magnetic resonance imaging studies had detected a 30% reduction in caudate volume in our patient sample. These results were confirmed soon thereafter in a PET study with [^{18}F]fluorodopa (3). The ratio of specific-to-nonspecific binding of this tracer as an index of dopa decarboxylase activity was lowered in various brain regions of 12 patients with LND compared with normal controls. The ratio was diminished to between 31% in the putamen and 57% in the ventral tegmental area.

Rett syndrome is a severe neurodevelopmental disease of females, characterized by microcephaly with cognitive retardation and autistic behavior, abnormalities of movement and muscle tone, and seizures. Postmortem studies suggest involvement of the basal forebrain cholinergic as well as nigrostriatal dopaminergic systems, with reduced melanin content and tyrosine hydroxylase activity in the substantia nigra.

We have studied 12 patients with Rett syndrome with a single injection of [^{11}C]WIN 35 428 (2). When compared with age- ($n = 11$) or age/gender-matched ($n = 5$) healthy controls, we found a significant reduction in binding potential (k_3/k_4) of up to 45% both in the caudate and in the putamen of the patients. Interestingly, while we found low to low-normal values for the receptor density (Bmax) of postsynaptic D2-like DA receptors in the caudate in 12 patients with Rett syndrome measured with [^{11}C]NMSP, Chiron *et al.* (4) have reported markedly increased specific binding of [^{123}I]iodolisuride, a SPECT tracer for the D2-like DA receptor, in 11 children with Rett syndrome. These authors concluded that the dopaminergic deficiency in Rett syndrome, which could be confirmed by our aforementioned findings *in vivo,* consecutively leads to an upregulation of postsynaptic receptors. The divergence of these findings is presently unexplained.

Tourette's syndrome is a disorder that is usually characterized by complex cognitive and behavioral features such as obsessions and compulsions, impulsivity, coprolalia, self-injurious behavior, and involuntary motor and vocal tics. An increase in numbers of DATs has been reported in Tourette's syndrome from postmortem studies (5). This finding was corroborated by SPECT studies using the ligand [^{123}I]β-CIT in a small number of patients (6). In five neuroleptic-free adult patients with Tourette's syndrome, striatal [^{123}I]β-CIT binding was a mean of 37% higher than in the age- and gender-matched healthy controls. However, our PET studies with [^{11}C]WIN 35 428 showed inconsistent results; although early findings pointed to increased numbers of DATs in patients with clinical Tourette's syndrome (7), this result is yet to be confirmed consistently in a larger patient sample. Moreover, imaging of presynaptic DA metabolism with [^{18}F]fluoro-dopa revealed no abnormality in presynaptic dopaminergic function in 10 patients with this disorder, compared with normal subjects (8). Because seven of these patients were treated with neuroleptic drugs, these results must be considered cautiously. Also, striatal [^{11}C]raclopride and [^{11}C]NMSP binding potentials, respectively, in PET studies (8, 9), as well as striatal uptake of [^{123}I]IBZM in SPECT investigations (10), were reported to be normal in patients with Tourette's syndrome, compared with normal subjects. It is unclear why some studies suggest a defect in dopaminergic (especially presynaptic) function and some do not. One explanation could be that the patient population is biologically heterogeneous.

It is now well established that the psychostimulant cocaine exhibits its psychotropic action via DAT. Volkow *et al.* (11) have recently shown that there is a direct relationship between the subjective effects of cocaine and DAT occupancy and that commonly abused cocaine doses block between 60–80% of DATs. It may be possible to develop cocaine antagonists for the treatment of cocaine addiction, that will prevent cocaine binding without blocking DAT. Preliminary baboon studies with cocaine antagonists are being carried out by our group to measure DAT occupancy as a prelude to treatment trials in humans. *Papio anubis* baboons underwent two [^{11}C]WIN 35 428 PET scans, one baseline (with saline) and a second with GBR 12909, either 1 mg/kg or 3 mg/kg, 90 min before the second tracer injection. When occupancy of the DAT was determined as percentage change in binding potential (k_3/k_4), we found a marked occupancy of DAT in the range of 50–70%.

A marked reduction of striatal DAT densities has been reported in methamphetamine treated animals (12). In baboon studies with methamphetamine treatment, we could demonstrate a dose-related reduction in DAT densities. These studies were followed by postmortem analysis of neurochemical parameters. In recent years, the synthetic amphetamine analogue methcathinone (2-methylamino-1-phenylpropanone,ephedrone, "Cat") has emerged as a recreational drug of abuse. In animals, methcathinone, like methamphetamine, produces toxic effects on brain DA and serotonin neurons. Moreover, our studies in human subjects with a history of methcathinone abuse suggest a DAT decline as well (13). We employed [^{11}C]WIN 35 428 imaging in five subjects who abused Cat. Preliminary analysis revealed a −26% and a −17% decrement in caudate and putamen DAT, respectively, compared with age-matched controls.

Of the three neurodevelopmental disorders studied, there seems to be a progression from markedly low to borderline elevated in LND, Rett syndrome, and Tourette's syndrome. These observations undoubtedly relate to degrees of involvement of the dopamine neuronal loss/transporter loss, and/or transporter upregulation in these young adults. This suggests that a dopaminergic involvement is justified in all of these disorders, and the rationale for the degree and direction has yet to be elucidated. These findings are illustrated for the putamen in Figure 1.

Studies involving methcathinone and methamphetamine abuse confirmed the prior studies that significant DAT loss can occur in these drug abuse paradigms. Baboon studies with a neurochemical correlate will determine the dose-response curve. The human studies with methcathinone document the dopaminergic damage in such substance abusers.

The preliminary studies in baboons demonstrating DAT occupancy are an important means of screening drugs capable of blocking cocaine's actions at the DAT and will be an important target of cocaine treatment strategies in the

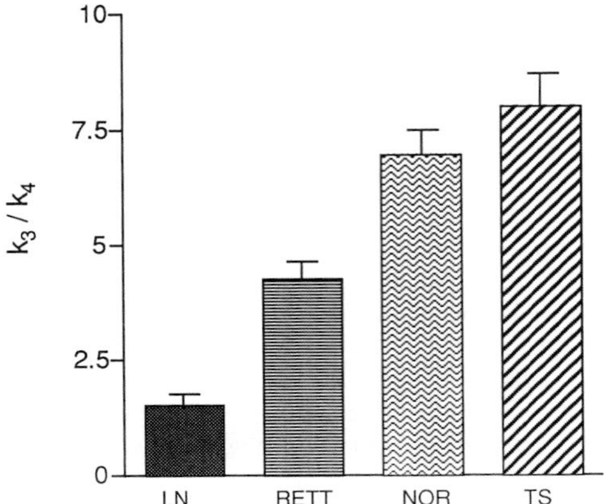

FIGURE 1 Mean ± SEM DAT binding of [^{11}C]WIN 35 428 for six LN, three RETT, 10 normals, and six Tourette subjects. DAT is significantly reduced for the LN compared with the other groups. LN, Lesch-Nyhan disease; RETT, Rett syndrome; NOR, normals; TS, Tourette's syndrome.

future. Thus, in analogy to receptor imaging studies that have been used to screen neuroleptics for antipsychotic action, it is likely that such imaging studies will be very useful in imaging DAT occupancy for such drug development.

Acknowledgments

Supported by NIH grants DA09482, HD24061, DA09487, MH42821, MD24448, Tourette Syndrome Assoc.

References

1. Wong, D. F., Yung B, Dannals, R. F., Shaya, E. K., Ravert, H. T., Chen, C. A., Chan, B., Folio, T., Scheffel, U., Ricaurte, G. A., Neumeyer, J. L., Wagner, H. N. Jr., and Kuhar, M. J. (1993). *In vivo* imaging of baboon and human dopamine transporters by positron emission tomography using [^{11}C]WIN 35,428. *Synapse* **15**, 130–142.
2. Wong, D. F., Harris, J.C., Naidu, S., Yokoi, F., Marenco, S., Dannals, R. F., Ravert, H. T., Yaster, M., Evans, A., Rousset, O., Bryan, R. N., Gjedde, A., Kuhar, M. J., and Breese, G. R. (1996). Dopamine transporters are markedly reduced in Lesch-Nyhan disease in vivo. *Proc. Natl. Acad. Sci. U.S.A.* **93**, 5539–5543.
3. Ernst, M., Zametkin, A. J., Matochik, J. A., Pascualvaca, D., Jons, P. H., Hardy, K., Hankerson, J. G., Doudet, D. J., and Cohen, R. M. (1996). Presynaptic dopaminergic deficits in Lesch-Nyhan disease. *N. Engl. J. Med.* **334**, 1568–1572.
4. Chiron, C., Bulteau, C., Loc'h, C., Raynaud, C., Garreau, B., Syrota, A., and Maziere, B. (1993). Dopaminergic D2 receptor SPECT imaging in Ratt syndrome: increase of specific binding in striatum. *J. Nucl. Med.* **34**, 1717–1721.
5. Singer, H. S., Hahn, I. H., and Moran, T. H. (1991). Abnormal dopamine uptake sites in postmortem striatum from patients with Tourette's syndrome. *Ann. Neurol.* **30**, 558–562.

6. Malison, R. T., McDougle, C. J., van Dyck, C. H., Scahill, L., Baldwin, R. M., Seibyl, J. P., Price, L. H., Leckman, J. F., and Innis, R. B. (1995). [123I]beta-CIT SPECT imaging of striatal dopamine transporter binding in Tourette's disorder. *Am. J. Psychiatry* **152**, 1359–1361.
7. Wong, D. F., Singer, H., Marenco, S., Brown, J., Yung, B., Yokoi, F., Chan, B., Matthews, W., Musachio, J., and Dannals, R. (1994). Dopamine transporter reuptake sites measured by [^{11}C]WIN 35,428 PET imaging are elevated in Tourette syndrome. *J. Nucl. Med.* **35**, 130P. (Abstract).
8. Turjanski, N., Sawle, G. V., Playford, E. D., Weeks, R., Lammertsma, A. A., Lees, A. J., and Brooks, D. J. (1994). PET studies of the presynaptic and postsynaptic dopaminergic system in Tourette's syndrome. *J. Neurol. Neurosurg. Psychiatry* **57**, 688–692.
9. Singer, H. S., Wong, D.F., Brown, J. E., Brandt, J., Krafft, L., Shaya, E., Dannals, R. F., and Wagner, H. N. Jr. (1992). Positron emission tomography evaluation of dopamine D-2 receptors in adults with Tourette syndrome. *Adv. Neurol.* **58**, 233–239.
10. George, M. S., Robertson, M. M., Costa, D. C., Ell, P. J., Trimble, M. R., Pilowsky, L., Verhoeff, N. P. (1994). Dopamine receptor availability in Tourette's syndrome. *Psychiatry Res.* **55**, 193–203.
11. Volkow, N. D., Wang, G. J., Fischman, M. W., Foltin, R. W., Fowler, J. S., Abumrad, N. N., Vitkun, S., Logan, J., Gatley, S. J., Pappas, N., Hitzemann, R., and Shea, C. E. (1997). Relationship between subjective effects of cocaine and dopamine transporter occupancy. *Nature* **386**, 827–830.
12. Seiden, L. S., Ricaurte, G. A. Neurotoxicity of methamphetamine and related drugs. *In* Psychopharmacology—A Generation of Progress. (Meltzer, H. Y., ed.), pp. 359–366. Raven Press, New York, NY.
13. Ricaurte, G., Wong, D. F., Szabo, Z., Yokoi, F., Scheffel, U., Matthews, W., Ravert, M., Dannals R., and Naidu, S. (1996). Reductions in brain dopamine and serotonin transporters detected in humans previously exposed to repeated high doses of methcathinone using PET. *Soc. Neurosci.* **22**, 1915 (Abstract)

Shimon Schuldiner, Sonia Steiner-Mordoch, and Rodrigo Yelin

Alexander Silberman Institute of Life Sciences
Hebrew University
Jerusalem, 91904 Israel

Molecular and Biochemical Studies of Rat Vesicular Monoamine Transporter

Classical neurotransmitters are stored in synaptic vesicles and storage organelles of secretory cells. Transport of the monoamines serotonin, dopamine, norepinephrine, epinephrine, and histamine into storage organelles in a variety of cells is catalyzed by vesicular monoamine transporters (VMATs). Accumulation of the neurotransmitter depends on the proton electrochemical gradient generated by the vesicular H^+–adenosine triphosphatase and involves the

Advances in Pharmacology, Volume 42

1054-3589/98 $25.00

VMAT-mediated exchange of two lumenal protons with one cytoplasmic amine (1–5).

A model of the mechanism of action of VMAT has been proposed based on a large body of biochemical data. In this model, the first step in the cycle is translocation of a single H^+, which generates the binding form of the transporter (6). The energy invested in the transporter by H^+ flux is released by ligand binding and is converted into vectorial movement of a substrate molecule across the membrane or directly into binding energy as measured with the high-affinity ligand [^{3}H]reserpine. In the case of a substrate, a second conformational change results in the ligand binding site being exposed to the vesicle interior, where the substrate can dissociate. The second H^+ in the cycle may be required to facilitate the conformational change or to allow for release of the positively charged substrate from the protein (6). Binding occurs also in the absence of a proton electrochemical gradient ($\Delta\mu_{H+}$) but many times slower (7, 8).

A clue for the molecular basis of some of these processes was obtained using diethyl pyrocarbonate (DEPC), a reagent relatively specific for His residues (9, 10). The inhibition by DEPC was specific for His groups because transport could be restored by hydroxylamine (9). DEPC inhibited transport but had no effect on binding of reserpine, indicating that the inhibition of transport was not due to a direct interaction with either of the known binding sites. Interestingly, however, the acceleration of reserpine binding by $\Delta\mu_{H+}$ was inhibited (9). The results suggested that either proton transport or a conformational change induced by proton transport was inhibited by DEPC. Practically identical results were obtained with phenylglyoxal, a reagent specific for Arg residues.

Cloning of VMAT (11, 12) made it feasible to try to identify the residue(s) modified by DEPC. Only one His (H419) is conserved in the VMATs from different species and in the two subtypes (5). Replacement of H419 with either Cys (H419C) or Arg (H419R) completely abolished transport as measured in permeabilized CV-1 cells transiently transformed with plasmids coding for the mutant proteins. Reserpine binding to the mutant proteins in the absence of $\Delta\mu_{H+}$ was at levels comparable to those detected in the wild type. However, $\Delta\mu_{H+}$ did not accelerate reserpine binding to either H419R or H419C proteins. These results suggested that His 419 is associated with H^+ translocation or in conformational changes occurring after substrate binding (13).

Aspartate (ASP) residues have been implicated in recognition of the cationic amino group of the ligand in β-adrenergic receptors (14, 15) and in the plasma membrane dopamine transporter (16, 17). Four Asp residues are fully conserved in putative transmembrane segments I, IV, VI, and IX of the various vesicular neurotransmitter transporters (VNTs) (5). In addition, biochemical evidence was available that N,N′-dicyclohexylcarbodiimide (DCC) inhibits VMAT-mediated transport (18–20). DCC reacts with a carboxyl residue, the availability of which is influenced by the occupancy of the tetrabenazine (TBZ) binding site (19). Reaction with this carboxyl residue inhibits not only overall transport activity but also TBZ and reserpine binding. As with all chemical modifiers, indirect effects, such as steric hindrance by the DCC moiety or indirect effect on the structure of the protein, cannot be ruled out at present. Mutagenesis studies of the roles of these Asp residues on VMATs should elucidate this point. Merickel and collaborators have replaced rat VMAT2 Asp 33 (equivalent to

D34 in rVMAT1) with Glu and Asn (21). While the D34E protein transports normally, the D34N is inactive. The D34N protein binds [^{3}H]reserpine, and binding is accelerated by $\Delta\mu_{H+}$. [^{3}H]Reserpine binding, however, was inhibited only by high concentrations of serotonin, suggesting impaired substrate recognition by this mutant (21).

Two other conserved Asp residues (D404 and D431) are located in putative transmembrane segments nearby H419 (segments X and XI). Both Asp residues seem to be important for activity, as hinted from their full conservation and their presence in the membrane domain. In addition, in both cases, (D404 and D431), even conservative replacements have dramatic effects on [^{3}H]serotonin transport activity (22). In the case of D431, complete inhibition is observed in D431E. The mutant protein D431E binds [^{3}H]reserpine and transports the first H^+, as judged by the ability of $\Delta\mu_{H+}$ to accelerate binding. Serine replacement (D431S) yields a protein practically indistinguishable from D431E, suggesting that a carboxyl moiety in residue 431 situated in a very critical location is important in the ability of VMAT to catalyze one of the steps beyond substrate binding and translocation of the first H^+. In addition, because [^{3}H]reserpine binding and coupling to $\Delta\mu_{H+}$ are normal in D431S as well as in D431E, it can be concluded that the overall structure of rVMAT1, as judged by this criterion, is not altered by removal of a negative charge in the membrane domain, a quite unanticipated finding. Surprisingly, the presence of Cys at position 431 and not the lack of a negative charge at that position (compare with D431S) has a deleterious effect on either expression or on protein stability (22).

Replacement of Asp 404 with Glu (D404E) yields a particularly interesting protein with a modified pH dependence of the transport reaction. Although there is a slight effect on [^{3}H]reserpine binding as well, the most dramatic effect is on [^{3}H]serotonin transport. The effect is complex, because it is not just a shift but also a sharpening of the optimum. We cannot, therefore, conclude at present whether the effect is due to a change in pKa of a single residue, but we can assume that it is on one of the steps beyond substrate recognition and transport of the first H^+, as measured in this report by [^{3}H]reserpine binding and its sensitivity to $\Delta\mu_{H+}$. A tempting speculation is that the shift of position of the carboxyl in D404E, as opposed to wild type, results in a change in its pKa because of the different interactions with other residues in the protein. In addition, because the last steps of transport include H^+, a direct involvement of D404 in this step can be speculated. Changes in pKa of carboxyl residues on shift of their position in proteins have been well documented in several cases (23).

VMAT1 and VMAT2 are two distinct and highly similar subtypes of VMAT. The two subtypes are coded by genes located in different chromosomes (24) differ in their tissue and subcellular distribution (25–27) and in their sensitivity to TBZ (28). TBZ is a potent inhibitor of VMAT activity that apparently binds to a site different than the substrate and reserpine binding sites. Surprisingly, modification of a residue (Asp 404) in VMAT1, which is conserved in both transporters, causes a change in the mutant's affinity to TBZ. The shift in position of the carboxyl moiety may indirectly induce a change in the conformation of the protein such that now it can recognize TBZ with higher affinity. The effect can also be due, as discussed earlier for the shift in the pH optimum, to a modification in the environment of the carboxyl moiety at position 404, which may be interacting directly with TBZ. Interestingly in this respect is the mode by which DCC inhibits VMAT-mediated transport (18–20). DCC reacts with a carboxyl residue, the availability of which is influenced by the occupancy of the TBZ binding site (19). Site-directed mutagenesis provides a tool to isolate mutants incapable of performing partial reactions and, therefore, provides strong support for mechanistic models.

References

1. Kanner, B. I., and Schuldiner, S. (1987). *CRC Crit. Rev. Biochem* **22,** 1–38.
2. Njus, D., Kelley, P. M., and Harnadek, G. J. (1986). *Biochim. Biophys. Acta* **853,** 237–265.
3. Johnson, R. (1988). *Physiol. Rev.* **68,** 232–307.
4. Schuldiner, S. (1994). *J. Neurochem.* **62,** 2067–2078.
5. Schuldiner, S., Shirvan, A., and Linial, M. (1995). *Physiol. Rev.* **75,** 369–392.
6. Rudnick, G., Steiner-Mordoch, S. S., Fishkes, H., Stern-Bach, Y., and Schuldiner, S. (1990). *Biochemistry* **29,** 603–608.
7. Weaver, J. A., and Deupree, J. D. (1982). *Eur. J. Pharmacol.* **80,** 437–438.
8. Scherman, D., and Henry, J-P. (1984). *Molec. Pharmacol.* **25,** 113–122.
9. Suchi, R., Stern-Bach, Y., and Schuldiner, S. (1992). *Biochemistry* **31,** 12500–12503.
10. Isambert, M., and Henry, J. (1981). *FEBS Lett.* **136,** 13–18.
11. Liu, Q-R., Lopez-Corcuera, B., Nelson, H., Mandiyan, S., and Nelson, N. (1992). *Proc. Natl. Acad. Sci. U.S.A.* **89,** 12145–12149.
12. Erickson, J., Eiden, L., and Hoffman, B. (1992). *Proc. Natl. Acad. Sci. U.S.A.* **89,** 10993–10997.
13. Shirvan, A., Laskar, O., Steiner-Mordoch, S., and Schuldiner, S. (1994). *FEBS Lett.* **356,** 145–150.
14. Strader, C., Sigal, I., Register, R., Candelore, M., Rands, E., and Dixon, R. (1987). *Proc. Natl. Acad. Sci. U.S.A.* **84,** 4384–4388.
15. Strader, C., Sigal, I., Candelore, M., Rands, E., Hill, W., and Dixon, R. (1988). *J. Biol. Chem.* **263,** 10267–10271.
16. Amara, S. G., and Kuhar, M. J. (1993) *Ann. Rev. Neurosci.* **16,** 73–93.
17. Kitayama, S., Shimada, S., Xu, H., Markham, L., Donovan, D., and Uhl, G. (1992). *Proc. Natl. Acad. Sci. U.S.A.* **89,** 7782–7785.
18. Schuldiner, S., Fishkes, H., and Kanner, B. I. (1978). *Proc. Natl. Acad. Sci. U.S.A.* **75,** 3713–3716.
19. Suchi, R., Stern-Bach, Y., Gabay, T., and Schuldiner, S. (1991). *Biochemistry* **30,** 6490–6494.
20. Gasnier, B., Scherman, D., and Henry, J. (1985). *Biochemistry* **24,** 1239–1244.
21. Merickel, A., Rosandich, P., Peter, D., and Edwards, R. (1995). *J. Biol. Chem.* **270,** 25798–25804.
22. Steiner Mordoch, S., Shirvan, A., and Schuldiner, S. (1996). *J. Biol. Chem.* **271,** 13048–13054.
23. Lanyi, J., Tittor, J., Varo, G., Krippahl, G., and Oesterhelt, D. (1992). *Biochim. Biophys. Acta.* **1099,** 102–110.
24. Peter, D., Finn, J. P., Klisak, I., Liu, Y. J., Kojis, T., Heinzmann, C., Roghani, A., Sparkes, R. S., and Edwards, R. H. (1993). *Genomics* **18,** 720–723.
25. Weihe, E., Schafer, M-H., Erickson, J., and Eiden, L. (1995). *J. Molec. Neurosci.* **5,** 149–164.

26. Nirenberg, M., Liu, Y., Peter, D., Edwards, R., and Pickel, V. (1995). *Proc. Natl. Acad. Sci. U.S.A.* **92**, 8773–8777.
27. Peter, D., Liu, Y., Sternini, C., de Giorgio, R., Brecha, N., and Edwards, R. (1995). *J. Neurosci.* **15**, 6179–6188.
28. Peter, D., Jimenez, J., Liu, Y. J., Kim, J., and Edwards, R. H. (1994). *J. Biol. Chem.* **269**, 7231–7237.

J. D. Erickson
Neuroscience Center and Department of Pharmacology
LSU School of Medicine
New Orleans, Louisiana 70112

A Chimeric Vesicular Monoamine Transporter Dissociates Sensitivity to Tetrabenazine and Unsubstituted Aromatic Amines

Active transport of biogenic amines into secretory organelles of neurons and neuroendocrine cells requires the presence of both a transmembrane H^+ electrochemical gradient, established and maintained by a vacuolar-type H^+–adenosine triphosphatase, and a reserpine-sensitive transporter molecule, which catalyzes the exchange of H^+ ions for amine substrates. Two isoforms of the human vesicular monoamine transporter (hVMAT1 and hVMAT2) have recently been cloned and shown to differ with respect to their tissue distribution, substrate affinity, and sensitivity toward various inhibitors, including neurotoxic and psychoactive compounds (1).

Functional chimeras have been constructed between hVMAT1 and hVMAT2 (60% identity) to identify domains responsible for the observed pharmacologic differences between these proteins. Tetrabenazine (TBZ) and two unsubstituted aromatic amines, amphetamine and histamine, were evaluated because of the marked differences in their interaction with hVMAT1 and hVMAT2. The ability of these compounds to inhibit or compete with [^{3}H]serotonin (5HT) for uptake by hVMAT1 or hVMAT2 was measured after expression in digitonin-permeabilized fibroblastic (CV-1) cells. The affinity of hVMAT2 and hVMAT1 for [^{3}H]SHT was similiar, exhibiting Km values of approximately 0.7 and 1.5 μM, respectively. Only the transport of [^{3}H]5HT mediated by hVMAT2 was inhibited by TBZ with an inhibition constant (IC_{50}) of approximately 50 nM. HVMAT1 activity was unaffected by 10 μM of TBZ.

Advances in Pharmacology, Volume 42

1054-3589/98 $25.00

HVMAT2 exhibited an apparent affinity (IC 50) of approximately 2 μM for amphetamine while hVMAT1 displayed 10- 20-fold lower affinity for this drug. The apparent affinity (IC 50) of hVMAT2 for histamine was also 10- to 20-fold greater than that of hVMAT1. However, histamine is a poor substrate for hVMAT2 relative to amphetamine and 5HT, exhibiting a Km for transport of approximately 200 μM (2).

Discrete domains of hVMAT1 and hVMAT2 are involved in the differential interaction of unsubstituted aromatic amines with [^{3}H]5HT vesicular uptake. A hVMAT1/hVMAT2 chimera (2/1/2@*Nar*I/*Bam*HI) has been constructed that dissociates the interaction of amphetamine from that of histamine (Fig. 1). The 2/1/2@*Nar*I/*Bam*HI chimera clearly displayed hVMAT2-like sensitivity toward amphetamine, whereas it exhibited hVMAT1-like sensitivity towards histamine. Thus, specific amino acid residues between Ile 265 in putative transmembrane domain (TMD) 6 and Tyr 418 in TMD 11 of hVMAT2 may be important for the increased potency of histamine to compete with [^{3}H]5HT for uptake, compared with hVMAT1. The lack of strong electron donating substituents on the aromatic rings of both histamine and amphetamine may play an important role in the decreased apparent affinity (10- to 20-fold) displayed by hVMAT1 compared with hVMAT2. While several TMDs are likely to contribute to substrate recognition, dissociation of the interaction of histamine from that of amphetamine in the 2/1/2 @ Nar/BamHI chimera suggests that the aromatic rings of these compounds interact with different domains of the proteins. On the other hand, the differences in the apparent affinities of both transporters for amphetamine and histamine (200-fold), may be due to a differential interaction of the aromatic rings of these compounds with amino acids in the same TMDs.

The potency of TBZ and the selective inhibition of [^{3}H]5HT uptake by VMAT2 by TBZ suggest that several domains unique to VMAT2 are important for this high-affinity interaction. The 2/1/2@*Nar*I/*Bam*HI chimera clearly displayed hVMAT2-like sensitivity toward TBZ, as was observed with amphetamine (see Fig. 1). TBZ was, however, approximately 5 times less potent to inhibit ^{3}H-5HT uptake by this chimera than was observed with hVMAT2. The relative affinity of TBZ for several additional hVMAT chimeras are shown in Table I. Extending the hVMAT1 sequences to the *Bst*BI site at the junction of TMD 12 and the cytoplasmic domain, as in the 2/1/2@*Nar*I/*Bst*BI chimera, resulted in dramatic reduction in the potency of TBZ, indicating that critical residues important for TBZ binding exist between putative TMD11 and TMD12. Furthermore, residues between Lys 20 and Phe 135 are also important for the interaction of TBZ with hVMAT2, because low-affinity inhibition was observed with the 2/1/2@*Nru*I/*Nhe*1 chimera. While amino acids between *Nru*I and *Nhe*I and between *Bam*HI and *Bst*BI may be required for the hVMAT2-like sensitivity toward TBZ observed with the 2/1/2@*Nar*I/*Bam*HI chimera these regions alone or together do not suffice to confer the high-affinity interaction characteristic of hVMAT2, suggesting that the region between *Nhe*I and *Nar*I is also important. Replacing the N-terminus or the C-terminus of hVMAT2 with hVMAT1 sequences (1/2@*Nru*I; 2/1@*Bst*BI) did not significantly reduce the ability of TBZ to inhibit [^{3}H]5HT uptake.

A recent analysis of functional chimeras between rat VMAT1 and VMAT2 indicated that the same domains affecting histamine recognition similarly influ-

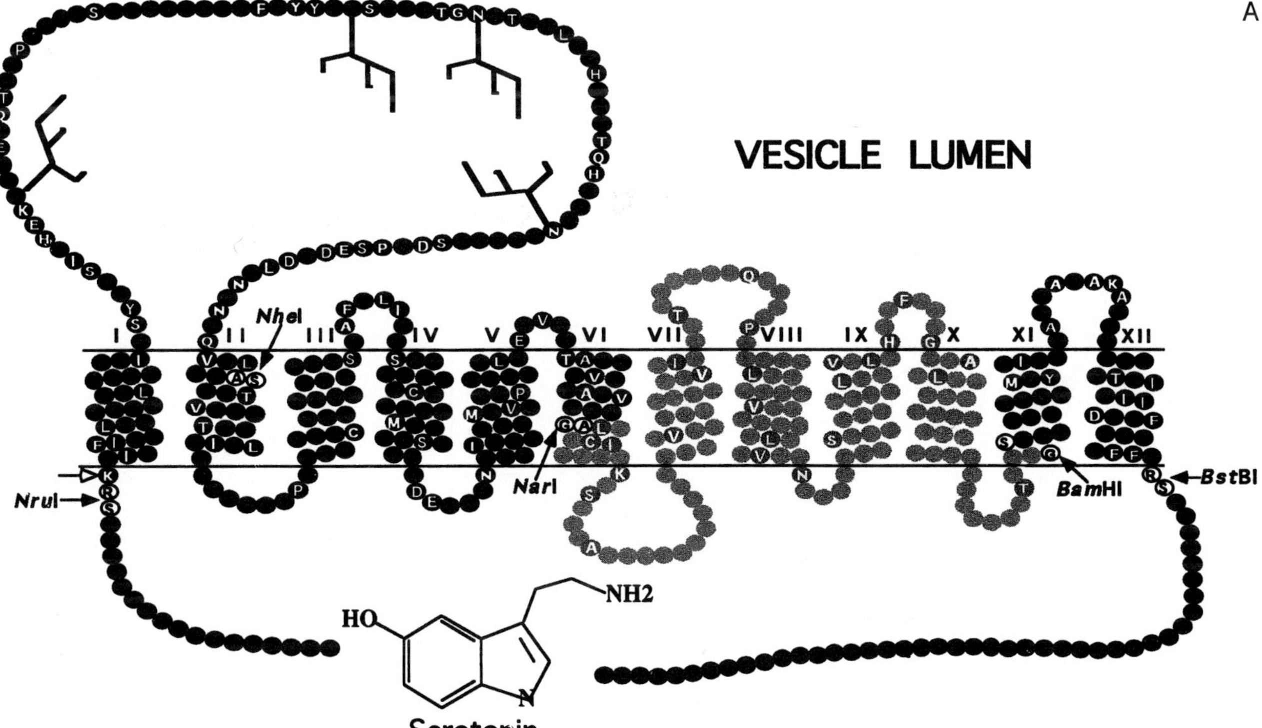

FIGURE 1 Chimeric vesicular transporter dissociates the differential sensitivity of hVMAT1 and hVMAT2 to amphetamine and histamine. (*Upper panel*) Predicted secondary structure of hVMAT1/hVMAT2 chimera (2/1/2@*Nar*I/*Bam*HI). Functional chimeras between hVMAT1 and hVMAT2 were constructed following introduction of unique restriction sites by site-directed mutagenesis at positions that encode conserved amino acids. Amino acids indicated in black circles represent residues conserved among rat, bovine, and human VMAT2. Amino

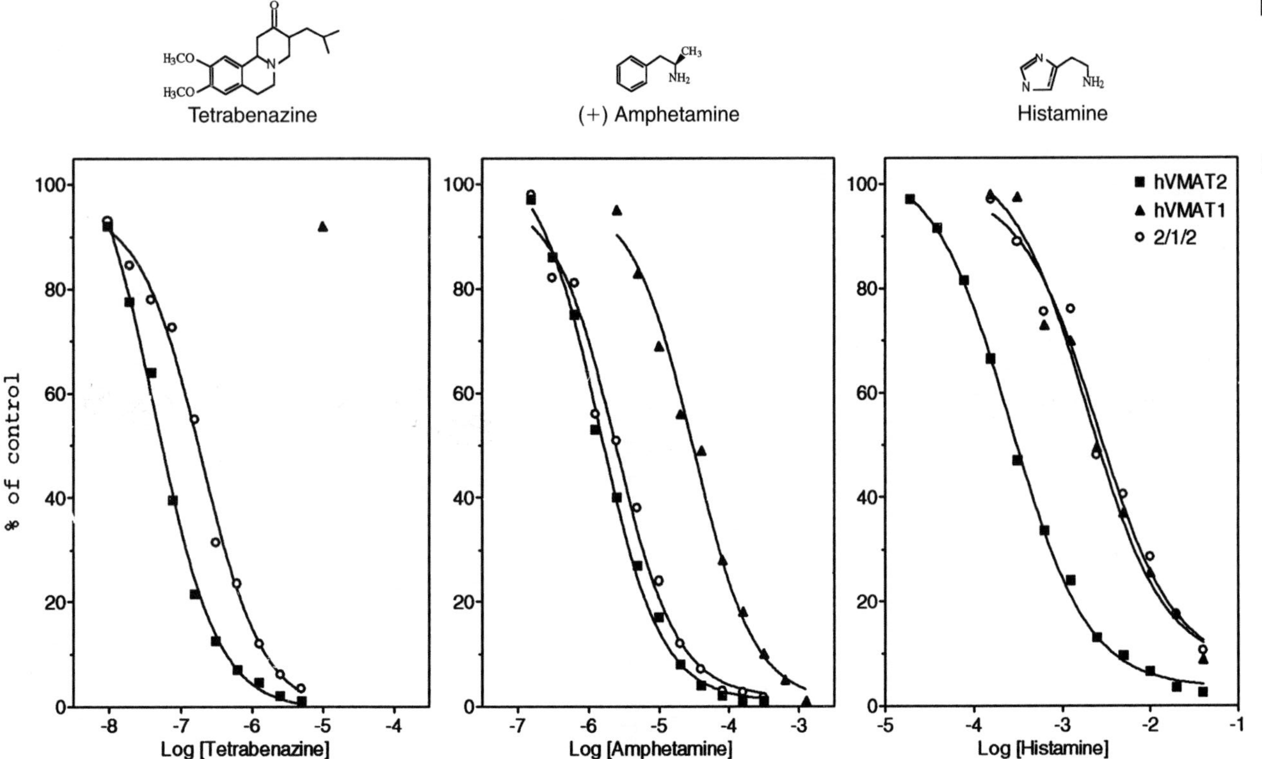

FIGURE 1 Continued acids indicated in grey circles represent residues conserved among rat and human VMAT1. White arrow indicates amino acid photolabeled by [^{125}I]AZIK (see text). (*Lower panel*) Inhibition of ^{3}H-5HT uptake in digitonin-permeabilized fibroblasts expressing hVMAT1, hVMAT2, and 2/1/2@*Nar*I/*Bam*HI chimera by tetrabenazine, amphetamine, and histamine.

TABLE I IC_{50} Values of hVMAT1/hVMAT2 Chimeras for TBZ Inhibition

hVMAT2	~50 nM
2/1/2@NarI/BamHI	~250 nM
2/1/2@NarI/BstBI	~5 μM
2/1/2@NruI/NheI	~5 μM
hVMAT1	>20 μM
2/1@NheI	>20 μM
2/1/2@NheI/BamHI	>20 μM
1/2/1@Nru/NheI	>20 μM

enced sensitity to TBZ (3). Two putative domains of rat VMAT2, TMD5 through TMD8 and TMD9 through TMD12, were each required for increased substrate affinity and sensitivity to TBZ, but neither domain alone sufficed to confer VMAT2-like interactions.

Photolabeling of VMAT2 with 7-azido-8-[^{125}I]iodoketanserin ([^{125}I]AZIK) is inhibited by TBZ and has been a useful probe to identify at least one domain of VMAT2 that may be important for this interaction. We suggested that the [^{125}I]AZIK photolabeled domain of VMAT2 was located within the first putative eight TM domains based on the fact that a TBZ-sensitive photolabeled VMAT2 proteolytic cleavage product was glycosylated (4). Recently, radiomicrosequencing of [^{125}I]AZIK-labeled rat VMAT2 indicated derivatization of Lys 20 at the putative junction between the N-terminus and TMD1 (5). While the Lys 20 is conserved in rat, bovine, and human VMAT2, it is also present in rat VMAT1. Interestingly, human VMAT1 has a Gln at this position (Gln 20) that may contribute to the fact that TBZ does not interact with human VMAT1 at all compared with only a 100-fold difference in potency of TBZ between rat VMAT1 and VMAT2.

The specific interactions of substrates, inhibitors, neurotoxins, and psychoactive substances with VMAT1 and VMAT2 may facilitate a basic understanding of the molecular mechanisms of vesicular H^+/monoamine antiport. Conserved Asp residues are predicted to lie within putative TMD I, VI, X, and XI. It is likely that these putative TMDs, and perhaps others, participate in the formation of a gated pore through which protonated substrates and H^+ ions are exchanged. Future chimeric and site-directed mutagenic work may reveal which domains of the VMATs compose this pore, how it is gated, and the specific amino acids involved in the H^+/substrate translocation pathways.

References

1. Erickson, J. D., Schäfer, M. K-H., Bonner, T. I., Eiden, L. E., and Weihe, E. (1996). Distinct pharmacological properties and distribution in neurons and endocrine cells of two isoforms of the human vesicular monoamine transporter. *Proc. Natl. Acad. Sci. U.S.A.* **93**, 5166–5171.
2. Erickson, J. D., Eiden, L. E., Schäfer, M. K-H., and Weihe, E. (1995). Reserpine- and tetrabenazine-sensitive transport of ^{3}H-histamine by the neuronal isoform of the vesicular monoamine transporter. *J. Mol. Neurosci.* **6**, 277–287.

3. Peter, D., Vu, T., and Edwards, R. H. (1996). Chimeric vesicular monoamine transporters identify structural domains that influence substrate affinity and sensitivity to tetrabenazine. *J. Biol. Chem.* **271**, 2979–2986.
4. Erickson, J. D., Eiden, L. E., and Hoffman, B. J. (1992). Expression cloning of a reserpine-sensitive vesicular monoamine transporter. *Proc. Natl. Acad. Sci. U.S.A.* **89**, 10993–10997.
5. Sievert, M. D., and Ruoho, A. E. (1996). Identification of drug binding sites on the synaptic vesicle monoamine translocator. *FASEB J.* **10**, A1233 (#1350).

J. P. Finn, A. Merickel, D. Peter, and R. H. Edwards

Departments of Neurology and Physiology
UCSF School of Medicine
San Francisco, California 94143

Ligand Recognition by the Vesicular Monoamine Transporters

Vesicular neurotransmitter transport differs from re-uptake across the plasma membrane in several important respects. Plasma membrane transport uses the Na^+ gradient across the plasma membrane as the driving force for transport, and vesicular transport uses the proton electrochemical gradient across the vesicle membrane (1). In the case of monoamines, cocaine and antidepressants inhibit plasma membrane transport, whereas reserpine and tetrabenazine inhibit vesicular transport. In addition, molecular cloning has found that the proteins responsible for plasma membrane transport have no sequence similarity to the vesicular transporters (2, 3). Molecular cloning has also identified three distinct proteins responsible for plasma membrane monoamine transport. Although functional analysis indicates that each can recognize multiple monoamines as substrates, selective expression within dopamine, norepinephrine, and serotonin cell populations of the brain suggests that each probably transports a single substrate *in vivo*. In contrast, molecular cloning has identified two vesicular monoamine transporters (VMATs), one expressed in the adrenal medulla and other neuroendocrine tissues (VMAT1) and the other in the brain (VMAT2) (4, 5). Indeed, VMAT2 occurs in multiple central monoamine cell groups, indicating that it packages dopamine, norepinephrine, and serotonin *in vivo* (6, 7). Functional analysis also indicates that VMAT1 recognizes multiple monoamines as substrates, but unlike VMAT2, VMAT1 does not recognize or transport histamine (8–10). Interestingly, both VMATs

Advances in Pharmacology, Volume 42

1054-3589/98 $25.00

protect against the parkinsonian toxin MPP^+, apparently by sequestering it inside secretory vesicles and away from its primary site of action in mitochondria (11, 12). The sequence of VMATs also shows sequence similarity to a class of bacterial antibiotic resistance proteins, indicating evolution from ancient detoxification systems (11). Thus, VMATs recognize a wide range of substrates that reflect their role in signaling and neural protection.

Classical pharmacologic studies have provided information and tools that help to explain how VMATs recognize such a wide range of substrates. The inhibitor reserpine binds almost irreversibly to both VMATs, and monoamines inhibit reserpine binding with potencies similar to their apparent affinity as substrates (13, 14), suggesting that reserpine binds to the site of substrate recognition. Furthermore, the imposition of a pH gradient across the vesicle membrane accelerates reserpine binding, providing a measure of coupling to the driving force and suggesting that the translocation of a proton out of the vesicle reorients the substrate recognition site to the cytoplasmic face of the membrane. In contrast to reserpine, the inhibitor tetrabenazine binds reversibly to VMAT2 but not VMAT1 (10), accounting for the ability of this drug to deplete central monoamine stores more effectively than peripheral stores. In addition, substrates inhibit tetrabenazine binding only at very high concentrations, suggesting that tetrabenazine does not bind at the site of substrate recognition. The presence of a pH gradient also does not influence tetrabenazine binding. Despite these observations suggesting that reserpine and tetrabenazine bind at distinct sites, tetrabenazine inhibits reserpine binding. To explain these findings, investigators have invoked the existence of two distinct VMAT conformations that bind exclusively to one of these inhibitors (15). Alternatively, reserpine and tetrabenazine may bind to distinct but overlapping sites. To understand how VMATs recognize substrates, we have used a variety of approaches.

Although VMATs have no sequence similarity to the plasma membrane transporters or receptors for monoamines, they do contain individual residues implicated in substrate recognition by these other proteins. To assess the role of a transmembrane aspartate implicated in substrate recognition by the β-adrenergic receptor (16) and the plasma membrane dopamine transporter (17), we have mutagenized the aspartate in transmembrane domain (TMD) 1 of VMAT2 to asparagine and found that the mutant has no transport activity (18). Importantly, the mutant protein remains coupled to the driving force as measured by reserpine binding. However, serotonin no longer inhibits reserpine binding to the mutant, indicating a selective role for this residue in substrate recognition. A set of serines in TMD3 of VMAT2 appear to have a similar role, as suggested by previous study of the β-adrenergic receptor and dopamine transporter. Presumably, the amino group of the substrate interacts with the aspartate, and the hydroxyl groups with the serines (19).

In addition to those residues required for substrate recognition, the comparison of VMAT1 and VMAT2 indicates that other residues may modulate the recognition of ligand. VMAT2 has an apparent affinity for most substrates approximately threefold higher than VMAT1. In addition, only VMAT2 recognizes histamine, and tetrabenazine inhibits only VMAT2. To identify the domains responsible for these differences, we constructed and analyzed a series of VMAT1/VMAT2 and VMAT2/VMAT1 chimeras (20). Surprisingly, the same two major domains contribute to the high-affinity interactions of VMAT2 with serotonin, histamine, and tetrabenazine, despite the apparent differences between interaction with substrate and tetrabenazine. In particular, these two domains, one spanning TMD5–8 and the other TMD9–12, are both required for these high-affinity interactions and so appear to cooperate. Additional domains, such as the cytoplasmic N-terminus of VMAT2, also contribute to the high-affinity interactions. Nonetheless, additional regions selectively modulate the interaction with serotonin, consistent with its properties as a high-affinity substrate. Thus, multiple domains appear to influence the interaction with ligands.

In addition to the analysis of chimeras, several observations suggest that histamine interacts differently with VMAT2 than other substrates. Unlike other monoamines, histamine does not inhibit reserpine binding (9). Indeed, reserpine does not inhibit histamine transport with the same high potency that it inhibits transport of other monoamines. However, histamine inhibits tetrabenazine binding (A.M., R.H.E., unpublished observations), raising the possibility that tetrabenazine binds at the site of histamine recognition. Rather than binding to an entirely distinct site or conformation, tetrabenazine may, therefore, bind to a site that overlaps with the site for recognition of typical substrates. Thus, it is of considerable interest to determine whether the same residues responsible for the recognition of histamine by VMAT2 but not VMAT1 also account for the selective sensitivity of VMAT2 to tetrabenazine. Indeed, preliminary observations indicate that of the three residues responsible for the differences between VMAT1 and VMAT2 in the domain spanning TMD9–12, two account for the differences in interaction with both histamine and tetrabenazine, and one accounts for the differences in interaction with tetrabenazine but not histamine (J.P.F., R.H.E., in preparation). The results, thus, indicate that the sites for interaction with histamine and tetrabenazine overlap but are not identical. However, real understanding of how VMATs recognize and translocate substrates requires more information about structure.

Charge pairs stabilize interactions between the transmembrane domains of many polytopic membrane proteins, and VMATs also have several charged residues predicted to reside in TMDs. However, the current model for transmembrane topology indicates only one basic residue, Lys 139 in TMD2. To determine whether this residue forms a charge pair with aspartates in TMDs 6, 10, or 11, we have neutralized each of these residues and found that only Lys 139, Asp 400 in TMD10, and Asp 427 in TMD11 are required for transport activity (A.M., R.H.E., submitted). We then constructed double mutants and found that the combination of K139A and D427N restored transport function to virtually normal levels. Thus, these two residues appear to form a charge pair, providing some of the first tertiary structural information about VMATs. However, the double mutant does show some differences from the wild-type protein. In particular, it has a significantly reduced apparent affinity for substrate, suggesting that the charge pair, although not required for transport activity, helps to stabilize the substrate recognition site. Interestingly, reversal of the charges at Lys 139 and Asp 427 does not confer transport activity. Nonetheless, this double mutant remains coupled to the driving force, as measured by reserpine binding, but serotonin does not inhibit reserpine binding, again indicating a selective defect in substrate recognition.

In summary, we have identified residues that have a variety of roles in substrate recognition by VMATs. Analogous to other membrane proteins that interact with monoamines, VMATs contain an aspartate in TMD1 and serines in TMD3 that are required for transport activity. Additional residues scattered throughout the protein modulate the interaction with substrate. Finally, a charge pair between residues in TMDs 2 and 11 appears to provide a structural framework for substrate recognition.

Acknowledgments

This work was supported by grants from the National Institute of Mental Health (to R.H.E.) and by fellowships from the NIMH (to D.P.) and the National Institute of Health (to J.P. F. and A.M.).

References

1. Kanner, B. I., and Schuldiner, S. (1987). Mechanisms of storage and transport of neurotransmitters. *CRC Crit. Rev. Biochem.* **22,** 1–38.
2. Amara, S. G., and Kuhar, M. J. (1993). Neurotransmitter transporters: Recent progress. *Annu. Rev. Neurosci.* **16,** 73–93.
3. Edwards, R. H. (1992). The transport of neurotransmitters into synaptic vesicles. *Curr. Opin. Neurobiol.* **2,** 586–594.
4. Erickson, J. D., Eiden, L. E., and Hoffman, B. J. (1992). Expression cloning of a reserpine-sensitive vesicular monoamine transporter. *Proc. Natl. Acad. Sci. U.S.A.* **89,** 10993–10997.
5. Liu, Y., Peter, D., Roghani, A., Schuldiner, S., Prive, G. G., Eisenberg, D., Brecha, N., and Edwards, R. H. (1992). A cDNA that supresses MPP+ toxicity encodes a vesicular amine transporter. *Cell* **70,** 539–551.
6. Peter, D., Liu, Y., Sternini, C., de Giorgio, R., Brecha, N., and Edwards, R. H. (1995). Differential expression of two vesicular monoamine transporters. *J. Neurosci.* **15,** 6179–6188.
7. Weihe, E., Schafer, M. K., Erickson, J. D., and Eiden, L. E. (1994). Localization of vesicular monoamine transporter isoforms (VMAT1 and VMAT2) to endocrine cells and neurons in rat. *J. Mol. Neurosci.* **5,** 149–164.
8. Erickson, J. D., Schafer, M. K-H., Bonner, T. I., Eiden, L. E., and Weihe, E. (1996). Distinct pharmacological properties and distribution in neurons and endocrine cells of two isoforms of the human vesicular monoamine transporter. *Proc. Natl. Acad. Sci. U.S.A.* **93,** 5166–5171.
9. Merickel, A., and Edwards, R. H. (1995). Transport of histamine by vesicular monoamine transporter-2. *Neuropharmacology* **34,** 1543–1547.
10. Peter, D., Jimenez, J., Liu, Y., Kim, J., and Edwards, R. H. (1994). The chromaffin granule and synaptic vesicle amine transporters differ in substrate recognition and sensitivity to inhibitors. *J. Biol. Chem.* **269,** 7231–7237.
11. Liu, Y., Roghani, A., and Edwards, R. H. (1992). Gene transfer of a reserpine-sensitive mechanism of resistance to MPP+. *Proc. Natl. Acad. Sci. U.S.A.* **89,** 9074–9078.
12. Reinhard, J. F. Jr., Diliberto, E. J. Jr., Viveros, O. H., and Daniels, A. J. (1987). Subcellular compartmentalization of 1-methyl-4-phenylpyridinium with catecholamines in adrenal

medullary chromaffin vesicles may explain the lack of toxicity to adrenal chromaffin cells. *Proc. Natl. Acad. Sci. U.S.A.* **84,** 8160–8164
13. Schuldiner, S., Liu, Y., and Edwards, R. H. (1993). Reserpine binding to a vesicular amine transporter expressed in Chinese hamster ovary fibroblasts. *J. Biol. Chem.* **268,** 29–34.
14. Weaver, J. H., and Deupree, J. D. (1982). Conditions required for reserpine binding to the catecholamine transporter on chromaffin granule ghosts. *Eur. J. Pharmacol.* **80,** 437–438.
15. Darchen, F., Scherman, E., and Henry, J. P. (1989). Reserpine binding to chromaffin granules suggests the existence of two conformations of the monoamine transporter. *Biochemistry* **28,** 1692–1697.
16. Strader, C. D., Sigal, I. S., Register, R. B., Candelore, M. R., Rands, E., and Dixon, R. A. F. (1987). Identification of residues required for ligand binding to the beta-adrenergic receptor. *Proc. Natl. Acad. Sci. U.S.A.* **84,** 4384–4388
17. Kitayama, S., Shimada, S., Xu, S., Markham, L., Donovan, D. M., and Uhl, G. R. (1992). Dopamine transporter site-directed mutations differentially alter substrate transport and cocaine binding. *Proc. Natl. Acad. Sci. U.S.A.* **89,** 7782–7785.
18. Merickel, A., Rosandich, P., Peter, D., and Edwards, R. H. (1995). Identification of residues involved in substrate recognition by a vesicular monoamine transporter. *J. Biol. Chem.* **270,** 25798–25804.
19. Strader, C. D., Candelore, M. R., Hill, W. S., Sigal, I. S., and Dixon, R. A. F. (1989). Identification of two serine residues involved in agonist activation of the beta-adrenergic receptor. *J. Biol. Chem.* **264,** 13572–13578.
20. Peter, D., Vu, T., and Edwards, R. H. (1996). Chimeric vesicular monoamine transporters identify structural domains that influence substrate affinity and sensitivity to tetrabenazine. *J. Biol. Chem.* **271,** 2979–2986.

J. P. Henry, C. Sagné, D. Botton, M. F. Isambert, and B. Gasnier

Service de Neurobiologie Physico-Chimique
CNRS-UPR 9071
Institut de Biologie Physico-Chimique
5005 Paris, France

Molecular Pharmacology of the Vesicular Monoamine Transporter

The function of neurotransmitter vesicular transporters is to concentrate the cytosolic neurotransmitter into secretory vesicles, using as an energy source the proton electrochemical gradient generated by the adenosine triphosphate (ATP)-dependent H^+ pump present in the vesicle membrane. These transporters

Advances in Pharmacology, Volume 42

1054-3589/98 $25.00

are present not only on synaptic vesicles derived from the endosomal pathway, but also on vesicles derived from the trans Golgi network, such as the large dense-core vesicles or the adrenal medulla chromaffin granules. The transporter from the latter ones, the chromaffin granule vesicular monoamine transporter (VMAT), which has a broad specificity toward catecholamines but also serotonin and histamine, has been the object of many studies, specially because chromaffin granules are easy to purify in large quantities. However, another interesting aspect of VMAT is the existence of an attractive pharmacology (1) comprising substrates, such as metaiodobenzyl-guanidine or methyl-4-phenylpyridinium (MPP^+), and inhibitors, such as reserpine (RES), tetrabenazine (TBZ), or ketanserin (KET). The two substrates are also substrates of the monoamine plasma membrane transporters DAT, NET, and SERT, whereas the inhibitors are more specific, with the exception of KET, which is also an antagonist of HT2 receptors. The most powerful inhibitor is RES (Ki in the subnanomolar concentration range); TBZ and KET inhibit ATP-dependent noradrenaline uptake with IC_{50} (concentration giving a 50% inhibition) of 3 and 50 nM, respectively. The characteristics of the corresponding binding sites have been determined, leading to the conclusion that TBZ and KET shared some common site, but that high-affinity RES binding occurred on another site. This last conclusion was supported by two types of evidence:

1. The kinetics of RES binding were dependent on $\Delta\mu_{H}{}^{+}$ generation, whereas those of TBZ and KET were insensitive.
2. RES was displaced from its binding site by concentrations of substrates lower by two orders of magnitude than those required to displace the other ligands.

This pharmacology allowed several interesting developments. First, the inhibitors prove to be useful markers of monoamine storage vesicles. For instance, TBZ has been developed as a positron emission tomography radioligand for brain imaging by the group of M. Kilbourn. Second, the purification of the VMAT from bovine chromaffin granules has been possible only through its ligand-binding activity. In our laboratory, we developed the covalent labeling of VMAT by the technique of photoaffinity labeling, using 7-azido-8-iodoketanserin (AZIK), a KET derivative. The labeled protein has been purified in a denatured state, an approach convenient for obtaining sequence data. An original purification procedure has been set up, which can be applied to other labeled intrinsic membrane proteins (2). Third, in view of the existence of two ligand binding sites, we proposed a model for $\Delta\mu_{H}{}^{+}$-dependent monoamine translocation. The high-affinity RES binding site was considered a substrate charge site directed toward the cytoplasmic compartment, whereas the TBZ binding site would be a discharge site oriented toward the vesicle matrix. Each site would be associated with a conformation of the protein, and the conformational change would require the proton electrochemical gradient (1).

How have our views on the pharmacology of VMAT been changed by the cloning of the corresponding cDNA? A first surprise has been the existence of two related but distinct genes encoding VMAT. The signification of this finding is still unclear. In rats, VMAT1 is expressed in chromaffin cells and in endocrine cells from intestine, stomach, and the sympathetic nervous system. On the other hand, VMAT2 is expressed in neurons in the sympathetic and central nervous systems. VMAT2 is also present in some endocrine cells, such as a small population of adrenal medulla chromaffin cells or the histaminergic cells in the oxyntic mucosa of the stomach (3, 4). There are large species differences, and, for instance, in bovine chromaffin cells, VMAT2 is the major isoform expressed (5), as shown by analysis of the N-terminal sequence of the protein isolated from chromaffin granule membrane.

Analysis of the pharmacology of VMAT2 expressed in COS cells indicated a reasonable agreement with the data obtained on bovine chromaffin granule vesicles, for the inhibition of ATP-dependent noradrenaline uptake and for [^{3}H]TBZOH binding. The [^{3}H]TBZOH equilibrium dissociation constant was 6.7 nM and 3.0 nM, for VMAT2 and chromaffin granule membrane, respectively. The pharmacology of bovine VMAT1 was tested on COS cells transfected in the same conditions. Consistent with results obtained on rat and human VMAT1, inhibition of [^{3}H]5HT uptake indicated a threefold decrease of the relative affinity of the substrates (5HT, dopamine, noradrenaline, and adrenaline) for VMAT1 compared with VMAT2 . The affinity for the inhibitors RES and KET was unchanged, but TBZ was a 10-fold less potent inhibitor of VMAT1 than of VMAT2. The low TBZ affinity of VMAT1was confirmed by the fact that no saturation isotherm could be obtained with [^{3}H]TBZOH in the concentration range used for VMAT2.

These results deserve two comments. First, the low affinity of bovine, rat, and human VMAT1 for TBZ has been observed after expression in heterologous cells, which are neither endocrine nor neuronal. We tested [^{3}H]TBZOH binding on PC12 cells. These cells have been shown to express VMAT1. A saturable binding could be demonstrated, characterized by a K_D of 2.2 nM and a Bmax of 127 fmol/mg of protein. This value, which is 10 times lower than that observed on chromaffin cells, is consistent with the lower secretory granule content of PC12 cells. This result is also consistent with previous data from our laboratory, indicating an affinity in the nanomolar concentration range for [^{3}H]TBZOH binding in rat adrenal medulla. Though VMAT2 is also expressed in rat chromaffin cells, the Bmax to catecholamine ratio was similar to that obtained with bovine chromaffin granules, thus showing that the high-affinity binding sites were not minor components. The hypothesis that the low affinity for TBZ of transfected VMAT1 originates in its expression in heterologous cells will have to be investigated in more detail.

A second surprising fact is the different behavior of TBZ and KET. Previous work on bovine chromaffin granule VMAT1, presumably VMAT2, showed clearly competitive binding of TBZ and KET (1). The results observed with $VMAT_1$ indicate that this isoform has a different TBZ/KET binding site.

Recently, we took advantage of the covalent labeling of VMAT by the photoactivable KET derivative, AZIK, to get some information on the localization of the KET binding site. Bovine chromaffin granule membranes were photolabeled with AZIK, solubilized by detergents, and VMAT was purified. After digestion with the endoproteases V8 or Lys C, which cleave peptide bonds after, respectively, acidic or lysine residues, and analysis by monodimensional electrophoresis, the radioactivity was only found in a 7-kDa peptide, which was considered to be associated with the KET binding site. The N-terminal sequence of this peptide was identical to that of bovine VMAT2, suggesting strongly the involvement of the N-terminal part of the protein in the KET binding site. From the sequence of VMAT2, cleavage by Lys C was anticipated to occur at the level of Lys 55. To test this hypothesis, a K55E mutant of bVMAT2 was constructed and expressed in COS cells. The mutant was active, and it could be labeled by AZIK. However, proteolysis by Lys C did not generate the 7-kDa–labeled peptide, and the radioactivity was redistributed in higher molecular weight products. It can, thus, be proposed that the 2–55 segment of the polypeptidic chain, which in the current topological model of VMAT corresponds to the cytosolic N-terminus and the first transmembrane segment, participates in the KET/TBZ binding site.

The results obtained on the inhibition of 5HT uptake by recombinant VMAT1 and VMAT2 indicated that the KET binding site of VMAT1 had a low affinity for TBZ. The possible involvement of the first transmembrane segment in high-affinity TBZ binding was tested by preparing a construct in which the 2–55 segment of VMAT1 was substituted for that of VMAT2. Expression of this construct in COS cells gave a low level of TBZ binding. However, Scatchard analysis showed a low Bmax, perhaps indicating an impairment at the expression level and a K_D value similar to that of VMAT2. A similar result has been described for rat chimeras: The replacement of VMAT2 with VMAT1 from the N-terminus to residue 38, after the first transmembrane segment, did not alter the TBZ sensitivity of [^{3}H]5HT uptake, and larger replacements were required to decrease this sensitivity, indicating that, in recombinant VMAT1, the low sensitivity to TBZ cannot be ascribed to a limited domain but is likely to involve a change in the conformation of the protein.

Acknowledgments

This work was supported by the Centre National de la Recherche Scientifique by the BIOMED programme of the European commity (contract BMH 1 CT 93110).

References

1. Henry, J.P., and Scherman, D. (1989). Radioligands of the vesicular monoamine transporter and their use as markers of the monoamine storage vesicles. *Biochem. Pharmacol.* **38,** 2395–2404.
2. Sagné, C., Isambert, M. F., Henry, J. P., and Gasnier, B. (1996). SDS resistant aggregation of membrane proteins application to the purification of the vesicular monoamine transporters. *Biochem. J.* **316,** 825–831.
3. Peter, D., Liu, Y., Sternini, C., de Giorgio, R., Brecha, N., and Edwards, R. H. (1995). Differential expression of two vesicular monoamine transporters. *J. Neurosci.* **15,** 6179–6188.
4. Weihe, E., Schäfer, M. K. H., Erickson, J. D., and Eiden, L. E. (1994). Localization of vesicular monoamine transporter isoforms to endocrine cells and neurons in rats. *J. Mol. Neurosci.* **5,** 149–164.
5. Krecji, E., Gasnier, B., Botton, D., Isambert, M. F., Sagné, C., Gagnon, J., Massoulié, J., and Henry, J. P. (1993). Expression and regulation of the bovine vesicule monoamine transporter gene. *FEBS Lett.* **335,** 27– 32.

Melissa J. Nirenberg,* June Chan,* Yongjian Liu,† Robert H. Edwards,† and Virginia M. Pickel*

*Division of Neurobiology
Department of Neurology and Neuroscience
Cornell University Medical College
New York, New York 10021

†Departments of Neurology and Physiology
University of California, San Francisco
San Francisco, California 94143

Ultrastructural Localization of the Vesicular Monoamine Transporter 2 in Mesolimbic and Nigrostriatal Dopaminergic Neurons

In neurons, the vesicular monoamine transporter 2 (VMAT2) plays a critical role in the active, reserpine-sensitive uptake of monoamines into vesicles and other subcellular organelles (1). VMAT2 also can transport and sequester parkinsonism-inducing neurotoxins such as 1-methyl-4-phenylpyridinium (MPP^+) and, thus, also may be involved in neuroprotection. We have examined the ultrastructural localization of VMAT2 within two distinct populations of midbrain dopaminergic neurons: the mesolimbic dopaminergic neurons, which are relatively resistant to parkinsonism-inducing neurotoxins, and the nigrostriatal dopaminergic neurons, which are selectively vulnerable to both idiopathic and drug-induced parkinsonism (2). The neurons in these regions are known to release dopamine from their somata and/or dendrites in the midbrain, as well as from their axonal processes in the striatum.

We used immunocytochemistry with tissue derived from adult, male Sprague-Dawley rats to localize a well-characterized rabbit polyclonal antiserum directed against rat VMAT2 (3–5). The tissue sections were sampled from the following regions: (1) the medial substantia nigra (SN) and paranigral–parabrachial subnuclei of the ventral tegmental area (VTA), which contain the cell bodies and dendrites of nigrostriatal and mesolimbic dopaminergic neurons, respectively; (2) the dorsolateral striatum at the rostrocaudal level of the crossing of the anterior commissure, which contains nigrostriatal dopaminergic terminals; and (3) the nucleus accumbens (NAC) (dorsal core and medial shell), which contain mesolimbic dopaminergic terminals. Some of the tissue was dually labeled for VMAT2 and a commercially available mouse monoclonal

Advances in Pharmacology, Volume 42

1054-3589/98 $25.00

antiserum directed against the catecholamine-synthesizing enzyme tyrosine hydroxylase (TH; Incstar). In most cases, we used the immunogold-silver method for immunolabeling of VMAT2 and the avidin-biotin complex immunoperoxidase method for immunolabeling of TH, but in some, we reversed these markers.

In the SN and VTA, VMAT2 was prominently localized to TH-immunoreactive perikarya and dendrites (4). Within the TH-immunoreactive perikarya, VMAT2 was localized to synthetic organelles, particularly the lateral saccules of Golgi (Fig. 1A), as well as to electron-lucent tubulovesicular organelles that resembled smooth endoplasmic reticulum (SER). In dendrites, the VMAT2 immunolabeling was almost exclusively localized to cytoplasmic vesicles and tubulovesicles, many of which resembled SER and almost all of which were larger (>70 nm in diameter) than typical small synaptic vesicles (SSVs) (Fig. 1B). Quantitative analysis showed that there were significantly higher levels of VMAT2 within TH-immunoreactive dopaminergic dendrites in the

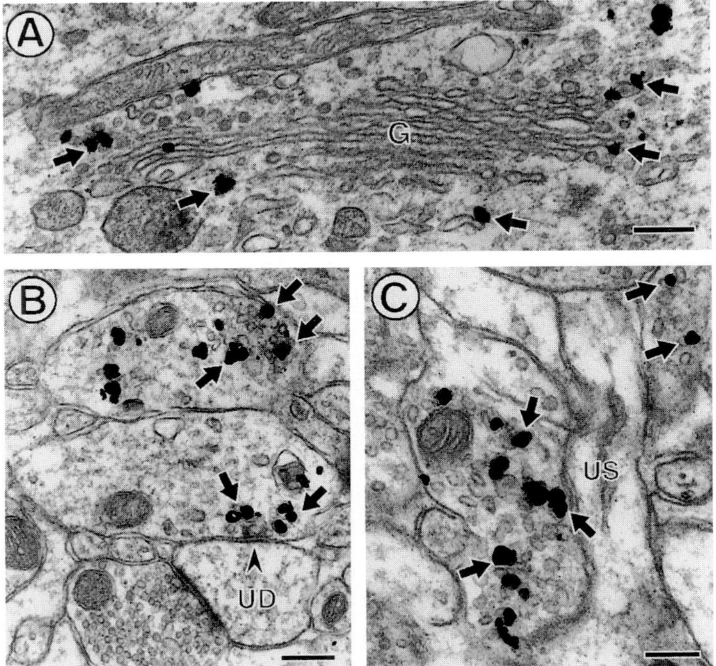

FIGURE 1 Electron micrographs showing immunogold-silver labeling for VMAT2. (*A*) Immunogold labeling for VMAT2 (*arrows*) observed along the lateral saccules of the Golgi apparatus (G) in a neuronal cell body in the SN. (*B*) VMAT2 immunolabeling (*arrows*) is localized to electron-lucent vesicles and tubulovesicles within two apposed dendrites in the VTA. The VMAT2 labeling in the lower dendrite is clustered near a dendrodendritic synapse (*arrowhead*) onto another unlabeled dendrite (UD). (*C*) Immunogold labeling for VMAT2 (*arrows*) is localized to SSVs within two axon terminals in the dorsolateral striatum. Both of the labeled terminals are apposed to an unlabeled dendritic spine (US). Bars, 0.25 μm.

VTA than in the SN. VMAT2-containing dendrites in the VTA were often apposed to and occasionally formed dendrodendritic synapses onto other unlabeled or VMAT2-containing dendrites (Fig. 1B). In both the SN and VTA, VMAT2 was only rarely localized to perikaryal and dendritic organelles that resembled typical SSVs or dense core vesicles (DCVs). In contrast, in axon terminals in the SN and VTA, almost all of the immunogold labeling for VMAT2 was localized to the membranes of typical SSVs and DCVs. Most of these terminals lacked detectable TH immunoreactivity and, thus, are presumably not catecholaminergic.

In both the dorsolateral striatum and the NAc, VMAT2 was extensively localized to axonal processes, within which it was localized both to SSVs (Fig. 1C) and to DCVs (5). The VMAT2-containing processes were often dually labeled for TH, consistent with the fact that most catecholaminergic axons in the striatum are dopaminergic. The dually labeled terminals often had the typical morphological characteristics of striatal dopaminergic terminals, including the presence of numerous tightly packed SSVs, the relative absence of DCVs, and the formation of small, symmetric synapses mostly onto dendrites and dendritic spines (Fig. 1C). Within these presumably dopaminergic processes, immunogold labeling for VMAT2 was localized to many but not all of the SSVs. In striatal axons with the morphological features of noradrenergic (TH-containing) or serotonergic (TH-lacking) terminals, VMAT2 was localized to many of the SSVs, some of the tubulovesicles, and almost all of the DCVs that were present (5). As we have previously observed in monoaminergic neurons in the solitary tract nuclei (3), almost all of the DCVs that were present in monoaminergic (VMAT2-containing) terminals were labeled with one or more gold particles for VMAT2.

In summary, we have shown that VMAT2 is prominently localized to tubulovesicular organelles in midbrain dopaminergic (TH-immunoreactive) perikarya and dendrites. In contrast, within catecholaminergic and noncatecholaminergic axonal processes in the midbrain and striatum, VMAT2 is almost exclusively localized to typical SSVs and DCVs. Within the presumptive dopaminergic terminals in the striatum, VMAT2 is almost exclusively localized to SSVs; whereas in other monoaminergic terminals seen in this and previous studies (3), VMAT2 is also extensively localized to DCVs. Quantitative comparison of midbrain dopaminergic dendrites showed that there are higher levels of VMAT2 in the VTA than in the SN, which may underlie the greater vulnerability of nigrostriatal dopaminergic neurons to neurotoxic insult (3). Together, these observations show that the subcellular localization of VMAT2 and accordingly the probable sites of storage and regulated release of monoamines differ with respect both to the transmitter type (e.g., dopaminergic vs nondopaminergic neuron) and the subcellular compartment (e.g., dendrites vs axon terminals). These findings have important implications regarding normal monoaminergic neurotransmission, as well as the differential susceptibilities of monoaminergic neurons to neurodegenerative disease.

Acknowledgments

V.M.P. receives salary support from the National Institute of Mental Health (NIMH; grant MH00078) and research support from the NIMH (grant MH40342) and the National

Institute on Drug Abuse (NIDA; grant DA04600). M.J.N. is supported by NIMH grant MH40342. R.H.E. is supported by the NIMH (MH50712) and by a National Alliance for Research on Schizophrenia and Depression Established Investigator Award.

References

1. Edwards, R. H. (1992). The transport of neurotransmitters into synaptic vesicles. *Curr. Opin. Neurobiol.* **2,** 586–594.
2. Roth, R. H., and Elsworth, J. D. (1995). Biochemical pharmacology of midbrain dopamine neurons. *In* Psychopharmacology: The Fourth Generation of Progress. (F. E. Bloom and D. J. Kupfer, eds.), pp. 227–243. Raven Press, New York.
3. Nirenberg, M. J., Liu, Y., Peter, D., Edwards, R. H., Pickel, V. M. (1995). The vesicular monoamine transporter 2 is present in small synaptic vesicles and preferentially localizes to large dense core vesicles in rat solitary tract nuclei. *Proc. Natl. Acad. Sci. U.S.A.* **92,** 8773–8777.
4. Nirenberg, M. J., Chan, J., Liu, Y., Edwards, R. H., Pickel, V. M. (1996). Ultrastructural localization of the vesicular monoamine transporter 2 in midbrain dopaminergic neurons: Potential sites for somatodendritic storage and release of dopamine. *J. Neurosci.* **16,** 4135–4145.
5. Nirenberg, M. J., Chan, J., Liu, Y., Edwards, R. H., and Pickel, V. M. (1997). The vesicular monoamine transporter-2: Immunogold localization in striatal axons and terminals. *Synapse* **26**(2), 194–198.

R. Dirkx, Jr.* J-M. Hermel,* D. U. Rabin,† and M. Solimena*

*Department of Internal Medicine
Section of Endocrinology
Yale University School of Medicine
New Haven, Connecticut 06520

†Biopharmaceutical Division
Bayer Corporation
West Haven, Connecticut 06516

ICA 512, a Receptor Tyrosine Phosphatase-Like Protein, Is Concentrated in Neurosecretory Granule Membranes

The islet cell autoantigen (ICA) 512 (also known as IA/2) is a recently identified autoantigen of insulin-dependent diabetes mellitus. The open reading frame of human ICA 512 cDNA encodes for a protein of 979 amino acids

Advances in Pharmacology, Volume 42

1054-3589/98 $25.00

with a predicted molecular weight of 106 kDa (1, 2). Relevant features of its primary sequence include a signal peptide and two putative N-glycosylation sites in the putative extracellular domain (amino acids 1–575), a single transmembrane domain (amino acids 576–600), and a cytoplasmic domain (amino acids 601–979) which is homologous to receptor protein tyrosine phosphatases (RPTPs). Interestingly, ICA 512 contains only one protein tyrosine phosphatase (PTP) "core domain" (amino acids 907–917), rather than two, as usually found in receptor PTPs (RPTPs). Moreover, ICA 512 expressed in bacteria did not display PTP activity when tested with several common PTP substrates. Thus, the function of ICA 512 is still unknown. By Northern blot, ICA 512 transcripts were detected in pancreatic islets, brain, and pituitary, suggesting an enrichment of ICA 512 in neuroendocrine tissues.

To begin addressing the structure and function of ICA 512, we investigated its tissue distribution and intracellular localization using rabbit antibodies directed against either its recombinant extracellular domain (amino acids 389–575, 92-18 antibodies) or recombinant cytoplasmic domain (amino acids 643–979, 89-59 antibodies) (3). Confocal microscopy on rat tissues demonstrated that ICA 512 is expressed in virtually all neuroendocrine cells, including neurons of the autonomic nervous system; chromaffin cells of the adrenal medulla; α-, β-, and δ-pancreatic islet cells; cells in the anterior and intermediate pituitary, as well as neurons of the autonomic nervous system and of the hypothalamus and the amygdala in the brain. The strongest ICA 512 immunoreactivity was found in the posterior pituitary, which contains the nerve endings of neurons located in the hypothalamus and the highest concentration of neurosecretory granules in the entire body.

In *r*at *in*sulinoma (RIN m5F) cells, an *in vitro* model of pancreatic β-cells, ICA 512 was colocalized with insulin but not with markers of constitutively exocytosing membranes (transferrin receptor), lysosomes (lgp 120) or synaptic-like microvesicles (synaptophysin). No ICA 512 immunoreactivity was detected at the cell surface of nonpermeabilized RIN m5F cells in resting conditions, suggesting that ICA 512 is not resident at the plasma membrane of β-cells. A prominent staining of the plasma membrane region, however, was observed following the incubation of living RIN m5F cells with the 92-18 antibodies in conditions stimulating regulated exocytosis (55 mM KCl in Krebs-Ringer HEPES buffer) for 10 min. This data demonstrated that, on depolarization, ICA 512 is exposed at the cell surface and becomes accessible to antibodies directed against its extracellular domain, further suggesting its enrichment in secretory granules.

By immunoelectron microscopy on ultrathin cryosections of rat posterior pituitary, a prominent immunogold labeling of neurosecretory granules, but not synaptic vesicles (the other major class of regulated secretory vesicles), was observed on the sections incubated with the 89-59 and 92-18 antibodies. Specificity of these stainings was supported by the lack of gold particles on other structures, including the plasma membrane and mitochondria of the neurosecretosomes as well as the cytoplasm and nuclei of surrounding pituicites. Virtually no gold particles were detected on sections immunostained with the 89-59 and 92-18 preimmune sera. This conclusively demonstrated that ICA 512 is localized on secretory granules.

To establish the fate of ICA 512 after its exposure at the cell surface, RIN m5F cells were incubated with the 92-18 antibody in stimulating buffer for 1 hr and then were either immediately fixed and permeabilized or allowed to recover for 14 hr before fixation and permeabilization. Internalized 92-18 antibodies were then visualized with rhodamine-conjugated goat anti-rabbit IgG. In RIN 5mF cells fixed immediately after incubation, internalized 92-18 antibodies were detected in the region of the plasma membrane, throughout the cytoplasm, and in a perinuclear region. This staining partially overlapped with that of transferrin receptor, suggesting the internalization of ICA 512 in endosomal structures. After 14 hr of recovery, 92-18 antibodies colocalized for the most part with insulin in the Golgi complex region as well as at the cell periphery. At this time point, however, there was only a limited overlap between internalized 92-18 antibodies and transferrin receptor. This suggested that internalized ICA 512 recycles to the Golgi complex where it is sorted into newly formed secretory granules and, thus, participates in repeated rounds of exo-endocytosis of secretory granules (Fig. 1).

By Western blotting on postnuclear supernatants of bovine posterior pituitary homogenates, both anti-ICA 512 antibodies recognized a broad band with an apparent electrophoretic mobility of 70 kDa. Upon subcellular fractionation of bovine posterior pituitary, this 70 kDa protein was exclusively recovered in high speed (100,000 $\times$ *g*) pellets and could only be extracted from membranes with detergents (2% Triton X-100), but not with high salt (1 M NaCl) or high pH (0.1 M Na_2C0_3, pH 11.5) treatments, as expected for an intrinsic membrane protein. Following sucrose density gradient fractionation of bovine posterior pituitary homogenates, the peak of the 70 kDa protein was present in dense fractions (1.7–1.8 M sucrose) and colocalized with the peak of the secretory granule marker secretogranin II. Conversely, the 70 kDa protein and the synaptic vesicle marker synaptophysin migrated at different densities, consistent with our immunocytochemical findings indicating that ICA 512 is not associated with synaptic vesicles. Thus, the biochemical and immunological properties of the 70 kDa protein recognized by anti-ICA 512 antibodies as well as its intracellular distribution strictly matched those of ICA 512.

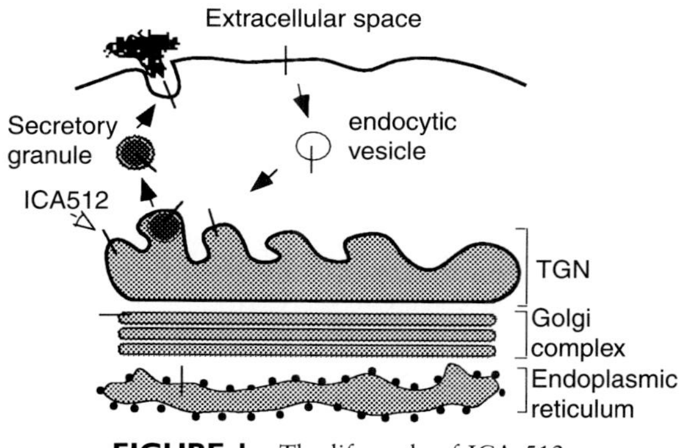

FIGURE 1 The life cycle of ICA 512.

Thirteen picomoles of the 70 kDa protein affinity purified from bovine posterior pituitary with 89-59 antibodies were processed for N-terminal microsequencing. The partial amino acid sequence, which included 14 amino acids, was 79% identical and 93% similar to the predicted sequences of human and mouse ICA 512, starting from residue 449. In both the human and mouse ICA 512 sequences, serine 449 is preceded by two lysines at positions 447 and 448. This di-lysine motif is a typical consensus sequence for convertases acting on secretory granule proteins, suggesting that a proteolytic cleavage between residues 448 and 449 is involved in the generation of the 70 kDa transmembrane fragment of ICA 512. This ICA 512 transmembrane fragment (amino acids 449–979) has a predicted molecular weight of 60 kDa. The apparent slower mobility of this fragment by SDS-gel electrophoresis may result from its putative N-glycosylation at asparagines 506 and 524 and its acidic isoelectric point (pI). Two-dimensional gel electrophoresis showed that the 70 kDa protein has a pI of 6.0, a value that fits with the predicted pI of the transmembrane fragment of ICA 512 generated after cleavage between amino acids 448 and 449. The predicted pI of full-length ICA 512, on the other hand, is 6.7. Taken together, this data demonstrated that the 70 kDa protein recognized by anti-ICA 512 antibodies corresponds to a cleaved fragment of ICA 512. Lack of an ICA 512 form larger than 70 kDa in the neurosecretosomes of the bovine posterior pituitary suggests that its processing occurs somewhere between the endoplasmic reticulum and mature secretory granules *before* it arrives in the neurosecretosomes.

In conclusion, we have shown that ICA 512 is a specific membrane marker of neurosecretory granules. To our knowledge, ICA 512 is the first member of the RPTP family found to be resident in an intracellular compartment. More importantly, our data suggests that ICA 512 and, thus, tyrosine phosphorylation–dephosphorylation plays a role in the life-cycle of neurosecretory granules. In view of these considerations, it is likely that the elucidation of ICA 512's function will generate new insights into the connection between signal transduction and vesicular trafficking in neuroendocrine cells, including catecholamine-secreting cells.

Acknowledgments

This work was partially supported by a Juvenile Diabetes Foundation International Career Development Award and NIH pilot grant DK-45735 to M.S.

References

1. Rabin, D. U., Pleasic, S. M., Shapiro, J. A., Yoo-Warren, H., Oles, J., Hicks, J. M., Goldstein, D. E., and Rae, P. M. (1994). Islet cell antigen 512 is a diabetes-specific islet autoantigen related to protein tyrosine phosphatases. *J. Immunol.* **152,** 3183–3188.
2. Lan, M. S., Lu, J., Goto, Y., and Notkins, A. L. (1994). Molecular cloning and identification of a receptor-type protein tyrosine phosphatase, IA-2, from human insulinoma. *DNA Cell Biol.* **13,** 505–514.
3. Solimena, M., Dirkx, R., Hermel, J-M., Pleasic-Williams S., Shapiro, J. A., Caron L., and Rabin, D. U. (1996). ICA 512, an autoantigen of insulin-dependent diabetes mellitus is an ubiquitous intrinsic membrane protein of neurosecretory granules. *EMBO J.* **15,** 2102–2114.

A. E. West,* C. Provoda,* R. L. Neve,† and K. M. Buckley*

*Department of Neurobiology
Harvard Medical School
Boston, Massachusetts 02115

†Molecular Neurogenetic Laboratory
McLean Hospital
Belmont, Massachusetts 02178

Protein Targeting in Neurons and Endocrine Cells

Neurons are highly polarized cells with domains that subserve distinct functions. The somatodendritic region of the neuron is specialized to receive incoming signals from other neurons at postsynaptic sites containing high concentrations of neurotransmitter receptors. The axon, in contrast, is designed to send outgoing signals by exocytosis of neurotransmitter-filled synaptic vesicles at the presynaptic membrane. This functional polarity is reflected in the polarized distribution of a number of proteins, both membrane and cytoskeletal, but the underlying molecular mechanism for this polarity is poorly understood. Targeting to each domain of the neuron may require a specific signal; alternatively, targeting to only one domain may be specific, while localization to the other may occur by default.

In addition to sorting to the proper domain, proteins in neurons must also be sorted to the correct subcellular organelle. For example, synaptic vesicle proteins synthesized in the endoplasmic reticulum (ER) and posttranslationally modified in the Golgi must first be targeted to the axon and then to the nerve terminal. A number of different studies taken together suggest that synaptic vesicle proteins are finally assembled into mature synaptic vesicles at the synapse; consequently, synaptic vesicle proteins may contain multiple targeting signals to reach their final destination.

We have been investigating the molecular basis for targeting of membrane proteins in neurons and to synaptic vesicles in two different systems. Using dissociated cultures of rat embryonic hippocampal neurons, we are examining the mechanism for sorting of proteins to dendrites or axons. In the pheochromocytoma cell line PC12, we are determining the sequences that are necessary to target an exogenous protein to synaptic vesicles.

To determine the molecular basis for targeting of membrane proteins in neurons, we have looked at the distribution of two type II membrane proteins with different distributions in neurons. The transferrin receptor, a protein that recycles between the plasma membrane and endosomes, is restricted entirely to the cell body and dendrites of hippocampal neurons in culture. In contrast, the synaptic vesicle protein synaptobrevin is concentrated in the axon and

Advances in Pharmacology, Volume 42

1054-3589/98 $25.00

nerve terminals, and as the neurons mature in culture, the protein becomes progressively restricted to synaptic contacts. We began by constructing a series of chimeras between the transferrin receptor and synaptobrevin. We used a cDNA for human transferrin receptor (hTfR), so that the expressed protein could be easily distinguished from the endogenous rat protein. The chimeric cDNAs were synthesized by standard techniques using the polymerase chain reaction (PCR) and expressed in neurons using the herpes virus amplicon system (1). Rat hippocampal neurons were infected with recombinant virus on day 7 *in vitro* and fixed 24 hr after infection to determine the location of the expressed proteins by indirect immunofluorescence. Control experiments determined that viral infection alone produced no change in the targeting of endogenous proteins; similarly, overexpression of the wild-type hTfR did not change its exclusive somatodendritic localization.

When the cytoplasmic domain of synaptobrevin was added on to the entire hTfR sequence (Add) or when the cytoplasmic domain of the hTfR was replaced with that of synaptobrevin, the chimeric protein showed a marked shift in localization. Each of the chimeric proteins could be easily detected in the axons as well as the cell bodies and dendrites of infected neurons, in contrast to the wild-type hTfR, which is exclusively somatodendritic. To determine if the difference in distribution could be attributed to a defined sequence in synaptobrevin, we made a series of deletions, 10 amino acids each, throughout the entire cytoplasmic domain of the Add chimera, because this chimera had a more robust axonal staining pattern than the replacement chimera. None of the deletions was sufficient to cause the protein distribution to revert to that of the wild-type transferrin receptor (i.e., exclusively somatodendritic). These results suggest that the signal for axonal targeting in synaptobrevin is diffuse and not contained within any of the 10 amino acid deletions. We next examined the targeting of the chimera to synaptic vesicles by fractionation of cell homogenates in glycerol velocity gradients. We found that although expression of the chimera in the axon overlaps with synaptophysin localization by indirect immunofluorescence, the chimera is not targeted to synaptic vesicles. However, the chimera in which amino acids 61–70 of the synaptobrevin sequence were deleted was detected in the synaptic vesicle peak, suggesting that this deletion can enhance targeting to synaptic vesicles. A similar result was shown by Grote *et al.* (2) for this deletion in an epitope-tagged synaptobrevin cDNA expressed in PC 12 cells. Taken together, these results suggest that targeting to synaptic vesicles is mediated by two separate signals: one that directs the protein to the axon and a second signal that allows entry into synaptic vesicles. It also demonstrates that the final stage of sorting to synaptic vesicles in neurons is mediated by similar if not identical processes as in PC12 cells.

To investigate the basis for dendritic targeting of membrane proteins, we took the cDNA for the hTfR and made a series of deletions through its cytoplasmic domain. The distribution of the expressed proteins revealed that any mutation of the N-terminal 18 amino acids or throughout the region known to affect endocytosis in other cell types (3) changes the distribution of the protein such that it is now present in the axon. The overlap between the dendritic targeting signal in transferrin receptor and the signal for endocytosis was apparent in that all versions of the protein that were able to enter the axon also appear to have more prominent plasma membrane localization, compared with the wild-type receptor.

We conclude from these results that the sequences that mediate endocytosis in the transferrin receptor overlap with those that are responsible for dendritic localization, suggesting that the actual sorting event may be mediated by an intracellular compartment, such as a sorting endosome. Transferrin receptor that does not return to the correct intracellular compartment due to a defect in endocytosis is no longer restricted to the cell body and dendrites of the neuron.

To determine if sequences can be identified in synaptic vesicle proteins that mediate targeting to synaptic vesicles, we have made chimeras between cDNAs encoding the synaptic vesicle protein SV2 and the glucose transporter GLUT1. Both proteins are part of the same family of transmembrane transporters; each is predicted to contain 12 membrane-spanning domains and have some sequence identity, particularly in the first half of each protein. The SV2 protein is found in endosomes, large dense-core vesicles, and synaptic vesicles in PC12 cells, while GLUT1 is localized predominantly to the plasma membrane and recycles at a low rate through endosomes. cDNAs for the chimeras were synthesized by PCR to include a myc-epitope tag, subcloned into the pcDNA3 expression vector, and transfected into PC12 cells by electroporation. The subcellular distribution of the expressed proteins was determined 2 days after electroporation by fractionation of a postnuclear supernatant in glycerol velocity gradients, followed by Western blotting with antibodies specific for the epitope tag. Only one chimera was detected in the peak synaptic vesicle fractions. The chimera targeted to synaptic vesicles consists of GLUT1, in which the cytoplasmic loop between TM6 and TM7 is replaced by the cytoplasmic loop of SV2 from TM 6 to TM7. This sequence contains approximately 80 amino acids, and we are currently making a series of deletions of 10 amino acids throughout to determine the targeting sequence in more detail. We are also analyzing the distribution of the chimeras on sucrose gradients designed to separate large dense-core vesicles from endosomes, synaptic vesicles, and other membranes to determine if the same targeting information is used for both types of regulated secretory vesicles.

References

1. Lim, F., Hartley, D., Starr, P., Lang, P., Song, S., Yu, L., Wang, Y., and Geller, A. I. (1996). Generation of high-titer defective HSV-1 vectors using an IE 2 deletion mutant and quantitative study of expression in cultured cortical cells. *Biotechniques* **20**, 460–469.
2. Grote, E., Hao, J. C., Bennet, M. K., and Kelly, R. B. (1995). A targeting signal in VAMP regulating transport to synaptic vesicles. *Cell* **81**, 581–589.
3. Trowbridge, I. S., Collawn, J. F., and Hopkins, C. R. (1993). Signal-dependent membrane protein trafficking in the endocytic pathway. *Annu. Rev. Cell Biol.* **9**, 129–161.

J. H. Tao-Cheng,* and L. E. Eiden†

*Laboratory of Neurobiology
NINDS, NIH
Bethesda, Maryland 20892

†Section on Molecular Neuroscience Laboratory of Cell Biology
NIMH, NIH
Bethesda, Maryland 20892

The Vesicular Monoamine Transporter VMAT2 and Vesicular Acetylcholine Transporter VAChT Are Sorted to Separate Vesicle Populations in PC12 Cells

PC12 cells, a rat pheochromocytoma-derived cell line, contain two distinct types of regulated secretory vesicles: large dense core vesicles (LDCVs) and small clear synaptic vesicles (SSVs). By subcellular fractionation studies, these two populations of vesicles contain and take up catecholamine and acetylcholine, respectively (1). To further characterize the biogenesis of different types of secretory vesicles and their components in intact PC12 cells, the subcellular distribution of several secretory vesicle–associated proteins were localized by electron microscopy immunocytochemistry (EM ICC).

Vesicular transporters of transmitters are excellent markers for functional neuroanatomy and for potential identification of vesicular contents at the subcellular level (2). In the present study, the distribution of two vesicular monoamine transporters (VMAT1 and VMAT2), a vesicular acetylcholine transporter (VAChT), a transmembrane transporter (SV2, substrate yet unknown), and two other proteins were examined.

VMAT1 is endogenous in PC12 cells, and it is abundant on LDCVs (Fig. 1A). This location of VMAT1 is consistent with the fact that LDCVs take up norepinephrine in PC12 cells. However, some VMAT1 staining is also on SSVs (Fig. 1A inset) with unknown function. Because SSVs do not take up norepinephrine in PC12 cells, it is possible that these VMAT1 located on SSVs are functionally inactive.

By contrast, VMAT2 is not endogenous in PC12 cells. In the present study, VMAT2 was expressed in PC12 cells under a constitutive viral promoter to see if it would be sorted to secretory vesicles (2). All PC12 cells in the present study were grown for 13–17 days on eight-well plastic chamber slides coated with matrigel and in the presence of nerve growth factor. The cells were pro-

Advances in Pharmacology, Volume 42

1054-3589/98 $25.00

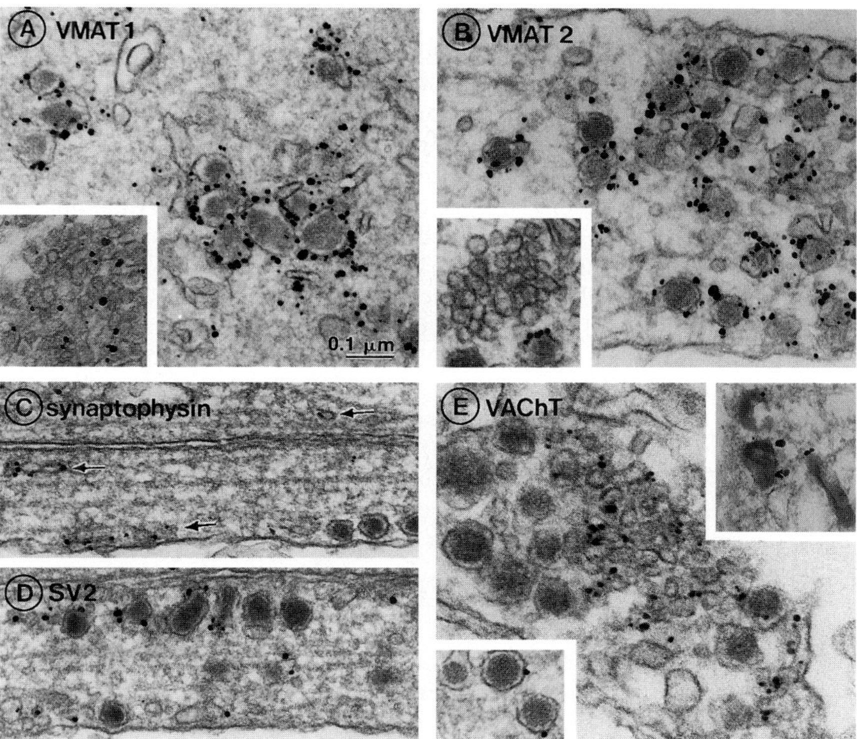

FIGURE 1 Immunocytochemical electron microscopy in PC12 cells. (*A*) VMAT1 is primarily located on LDCVs. Some staining is also present on SSVs (*inset*). (*B*) VMAT2 is present on LDCVs but not SSVs (*inset*) of VMAT2-transfected PC12 cells. (*C*) Synaptophysin is prominent on constitutive vesicles (*arrows*) in the neurites. (*D*) SV2 is prominent on LDCVs in the neurites (*E*) VAChT is preferentially located on SSVs. Some LDCVs are positive (*lower inset*). The HRP-filled early endosomes (*upper inset*) are labeled by VAChT.

cessed for pre-embedding EM ICC with a small gold probe plus silver enhancement (3).

In the transfected PC12 cells, VMAT2 is preferentially localized on the LDCVs and not on the SSVs (Fig. 1B). The lack of a prominent population of VMAT2-positive synaptic vesicles in these PC12 cells is in contrast to the distribution of endogenous VAMT2 in central and peripheral adrenergic neurons. In these adrenergic neurons *in vivo,* VMAT2 is localized on LDCVs and on a distinct population of catecholamine-containing synaptic vesicles, which are termed *small dense-core vesicles* (SDCVs). It should be noted that, unlike the SDCVs in adrenergic neurons, the SSVs in the transfected PC12 cells do not contain a dense core after permanganate or dichromate fixation. Thus, these SSVs are clearly different from the SDCVs in that the SSVs lack both VMAT2 and a dense core.

One possible explanation for the absence of VMAT2 from the SSVs in these transfected PC12 cells is that additional factors may be required for the sorting of VMAT2 into these SSVs. Alternatively, It is possible that SDCVs in noradrenergic neurons are biosynthetically distinct from the SSVs in PC12 cells, and that VMAT2 expression alone is not sufficient for the genesis of a separate population of SDCVs in these PC12 cells.

Subcellular distributions of three other SSV components, synaptophysin, SV2, and VAChT, were also analyzed. Synaptophysin (found in SSV membranes) and SV2 (found in SSV and LDCV membranes) appear to be transported separately from the cell body to reach the neuritic terminals. The bulk of synaptophysin (Fig. 1C) is carried by constitutive vesicles (4), while the bulk of SV2 (Fig. 1D) is carried by LDCVs. The two proteins are added to the plasma membrane via fusion of the respective vesicles during exocytosis. The two proteins are then recycled from the plasma membrane into early endosomes and sorted into SSVs.

The VAChT is preferentially localized on SSVs (Fig. 1E), compared with LDCVs or constitutive vesicles. Presence of VAChT on LDCVs (lower inset of Fig. 1E) suggests that this compartment may be the vesicle type carrying VAChT to the nerve terminals in PC12 cells. Due to the low staining intensities of VAChT except on SSVs, transport of VAChT via constitutive vesicles cannot be definitely ruled out. In any event, the prominent appearance of VAChT on SSVs probably represents a significant concentration of VAChT, through selective sorting at the early endosomes, from one of these compartments. The upper inset of Figure 1E shows an early endosome filled with endocytosed horseradish peroxidase (HRP) and labeled with VAChT. Such early endosomes are also positive for synaptophysin and SV2.

Chromogranin A, an acidic glycoprotein expressed in neuronal and endocrine cells, is clearly localized in the core of the LDCV and absent from SSV. The silver grains of chromogranin A staining are distinctly separated from the LDCV membranes that are positive for SV2 and VMAT1. Chromogranin A is packaged into LDCV at the Golgi and carried from the cell body by LDCVs.

Proteins destined for cholinergic SSVs may arrive via LDCVs or constitutive vesicles and recycle through the early endosomes, where they are concentrated for inclusion in the SSV membrane. Proteins in this category include SV2, VAChT, and synaptophysin. VMAT2, the neuronal VMAT in LDCVs and SDCVs of noradrenergic neurons, is not targeted to cholinergic SSVs when expressed in PC12 cells. Thus, SDCVs of noradrenergic neurons may represent a distinct population of synaptic vesicles that is biosynthetically separate from cholinergic SSVs in PC12 cells.

Acknowledgments

We thank Virginia Tanner-Crocker for expert technical assistance in EM ICC, and Dr. Erik Schweitzer for SV2 antibody.

References

1. Bauerfeind, R., Regnier-Vigouroux, A., Flatmark, T., and Hüttner, W. B. (1993). Selective storage of ACh, but not catecholamines, in neuroendocrine synaptic-like microvesicles of early endosomal origin. *Neuron* **11**, 105–121.
2. Weihe, E., Tao-Cheng, J-H., Schäfer, MK-H., Erickson, J. D., and Eiden, L. E. (1996). Visualization of the vesicular acetylcholine transporter in cholinergic nerve terminals and its targeting to a specific population of small synaptic vesicles. *Proc. Natl. Acad. Sci. U.S.A.* **93**, 3547–3552.
3. Tanner, V., Ploug, T., and Tao-Cheng, J-H. (1996). Subcellular localization of SV2 and other secretory vesicle components in PC12 cells by an efficient method of pre-embeddingEM immunocytochemistry for cell cultures. *J. Histochem. Cytochem.* **44**, 1481–1488.
4. Regnier-Vigouroux, A., Tooze, S. A., and Hüttner, W. B. (1991). Newly synthesized synaptophysin is transported to synaptic-like microvesicles via constitutive secretory vesicles and the plasma membrane. *EMBO* **10**, 3589–3601.

Rudolf Bauerfeind, Carol David, Detlev Grabs, Peter S. McPherson, Yasuo Nemoto, Vladimir I. Slepnev, Kohji Takei, and Pietro De Camilli

Department of Cell Biology and HHMI
Yale University School of Medicine
New Haven, Connecticut 06510

Recycling of Synaptic Vesicles

Fast synaptic transmission is mediated by synaptic vesicle exocytosis. Rapidly after exocytosis, the synaptic vesicle membranes are internalized and reutilized for the formation of new synaptic vesicles. A widely accepted model describes the synaptic vesicle cycle as a modification of the receptor-mediated recycling pathway present in all cells. This pathway, by which cell surface receptors, like transferrin or low-density lipoprotein receptors, are internalized and recycled back to the surface, involves two distinct vesicular transport steps: (1) clathrin-mediated budding from the plasma membrane and fusion with early endosomes and (2) budding from endosomes of vesicles destined to the plasma membrane. This model is consistent with the increase in number of clathrin-coated vesicles and in number and size of endosome-like tubulovesicular structures observed in nerve terminals after massive stimulation. However, we have recently suggested that these tubulovesicular structures are formed by bulk endocytosis and are, therefore, homologous to the plasma membrane (1). This indicates that synaptic vesicles reform by a single clathrin coat–mediated budding step that occurs in parallel from the cell surface and from these internal membranes.

Advances in Pharmacology, Volume 42

1054-3589/98 $25.00

The involvement of clathrin in synaptic vesicle reformation suggested by ultrastructural studies is supported by immunocytochemical, biochemical, and genetic evidence. Clathrin is highly concentrated in nerve terminals. The membrane composition of synaptic vesicles and of clathrin-coated vesicles purified from synaptosomes is very similar. The clathrin adaptor complex AP2, a component of the plasma membrane–derived clathrin coat, binds to synaptotagmin. The paralytic *shibire Drosophila* mutants, whose nerve terminals are depleted of synaptic vesicles at the restrictive temperature due to selective impairment of synaptic vesicle endocytosis, harbor mutations in dynamin, a guanosine triphosphatase (GTPase) implicated in clathrin-mediated endocytosis (2).

The three major components of the coat are clathrin (heavy and light chain), AP2, and the protein AP180/AP3. Assembly of the clathrin coat starts with the binding of AP2 to the plasma membrane, which appears to involve multiple interactions, including direct binding of its α subunit to the cytoplasmic domain of synaptotagmin. The AP2 lattice acts as a template for the recruitment of clathrin triskelia, which assemble to form a hexameric clathrin coat. A rearrangement of the clathrin lattice results in the progressive bending of the membrane to form a clathrin-coated pit.

Fission of clathrin-coated pits to form free vesicles requires dynamin. Dynamin was originally described as a microtubule-binding protein, but its function in synaptic vesicle recycling was revealed when the similarity to the protein encoded by the *Drosophila* gene *shibire* was discovered. The temperature-sensitive mutant *shibire* has normal neuromuscular transmission at permissive temperature but becomes paralyzed with increasing temperature due to a reversible block in synaptic vesicle endocytosis. There is no effect on exocytosis, with the result that the terminals become rapidly depleted of synaptic vesicles.

Dominant negative mutants of dynamin, when ectopically expressed, also block the fission of clathrin-coated pits to generate free vesicles, a reaction that is dependent on GTP hydrolysis *in vitro*. Morphological studies of lysed synaptosomes that were treated with cytosol, ATP, and GTPγS (3) provided the first mechanistic insights into the role of dynamin in clathrin-coated vesicle formation. Under these conditions numerous clathrin-coated buds were observed connected to the plasma membrane by long tubular membrane invaginations. Most importantly, these tubules were surrounded by regularly spaced electron-dense dynamin rings. These structures were remarkably similar to structures formed by purified dynamin around microtubules and to the collars seen at the neck of invaginated plasma membrane vesicles in the *Drosophila* mutant *shibire*. In a separate study, it was shown that purified dynamin itself can spontaneously assemble to form open rings and stacks of rings of similar dimensions. Dynamin may oligomerize around microtubules *in vitro,* because microtubules represent a template of the appropriate size for dynamin assembly. However, the physiological assembly site for dynamin is likely to be the stalk of a clathrin-coated pit.

Based on these findings and on results of dynamin transfection studies, the following scenario for the function of dynamin in a late step in clathrin-coated vesicle budding can be hypothesized. Once a narrow vesicle stack is formed, dynamin is required to drive vesicle fission. Direct or indirect interaction of dynamin with components of the clathrin coat may be responsible for concentrating dynamin at the coated pit. Here, elementary dynamin units (most likely tetramers) oligomerize into an open ring (or lock–washer structure) at the neck of the invaginated pit. At this stage, a conformational change of the ring that correlates with GTP hydrolysis leads to the fission reaction. Alternatively GTP-dynamin may recruit or activate other proteins that drive the fission reaction. Only one, or very few, dynamin rings are formed under physiological conditions. In contrast, stabilization of the GTP-bound conformation of dynamin by GTPγS prevents disassembly of the dynamin ring and allows the progressive addition of dynamin units to its open ends. As a result, dynamin assembles into a spiral, which leads to the elongation of the vesicle neck (3).

In addition to the basic constituents of the clathrin coat (clathrin, AP2, and AP180) and to dynamin, there is evidence that several other factors participate in clathrin-mediated endocytosis of synaptic vesicles by assisting in coat recruitment and assembly, regulating dynamin GTPase activity, and mediating interactions of the clathrin coat with the cytoskeletal matrix.

The formation of a clathrin coat appears to generally correlate with the recruitment of src homology 3 (SH3) domain–containing proteins. SH3 domain–containing proteins are often part of signal transduction pathways and in cytoskeletal matrices. SH3 domains bind to proline-rich sequences and can recruit proteins containing the sequences to sites of clathrin-coated vesicle formation. An abundant SH3-containing protein of the nerve terminal is amphiphysin, the SH3 domain of which binds to the proline-rich carboxy-terminus of dynamin *in vitro* (4). In addition, coprecipitation and colocalization studies suggest an interaction of the two proteins *in situ*. A different region of amphiphysin binds the earlike appendage domain of AP2 (4). Thus, amphiphysin may cooperate with other mechanisms in recruiting dynamin in proximity of clathrin coats and in generating a local high concentration of dynamin that facilitates the self-assembly of dynamin rings when a vesicle stack is formed. In addition, or alternatively, amphiphysin may modulate the GTP hydrolysis rate of dynamin, because binding of SH3 domain–containing proteins to dynamin was found to enhance its GTPase activity. Dynamin and amphiphysin undergo parallel dephosphorylation in response to nerve terminal stimulation that is consistent with the hypothesis that they may participate in a same function. Inibitor experiments indicate that the Ca^{2+}-calmodulin–dependent protein phosphatase calcineurin is responsible for this dephosphorylation, providing an attractive link between the induction of neurotransmitter release and the regulation of synaptic vesicle endocytosis and recycling.

A role of amphiphysin in endocytosis is supported by genetic studies in yeast. Portions of amphiphysin are homologous to Rvs 167 and Rvs 161, two yeast proteins whose mutations produce alterations in the peripheral cytoskeleton and defective endocytosis.

Until recently, most studies on synaptic vesicle recycling have focused on the role of proteins. There is increasing evidence, however, for a role of lipids and of phosphoinositides in particular, in synaptic vesicle endocytosis (5). Affinity chromatography studies on inositol polyphosphates have led to the demonstration that the clathrin adaptor AP2, AP180, and the synaptic vesicle protein synaptotagmin, which binds AP2 and, therefore, has a putative role in endocytosis, bind the phosphorylated inositol ring of PtdIns metabolites. Additionally, a binding of the pleckstrin homology domain of dynamin to both InsP3 and PtdIns(4,5)P2 has been demonstrated. It was also shown that manipulation of synaptotagmin binding to inositol polyphosphates in the squid giant axon affects synaptic vesicle endocytosis. A possible role of phosphoinositides in synaptic vesicle endocytosis is converging with evidence obtained from other systems, indicating an important role of these lipids in coat-mediated budding reactions.

Strong support for a physiological role of phosphoinositides in synaptic vesicle recycling has come from the identification in the nerve terminal of the phosphoinositide 5-phosphatase synaptojanin (6), an enzyme that removes the phosphate on the 5′ position of the inositol ring. This protein was shown by subcellular fractionation and by immunofluorescence to be colocalized with dynamin and to undergo stimulation-dependent calcineurin-mediated dephosphorylation in parallel with dynamin and amphiphysin. Synaptojanin has a proline-rich C-terminus that binds amphiphysin and Grb 2. It may, therefore, be recruited at the clathrin coat in parallel with dynamin. The N-terminal region of synaptojanin is similar to the cytoplasmic domain of the yeast protein Sac 1. Although the function of Sac 1 remains unclear, genetic studies have linked it to the function of the peripheral actin cytoskeleton, to vesicular traffic, and to phosphatidylinositol metabolism, corroborating the hypothesis that synaptojanin may function in synaptic vesicle endocytosis via a function in PtdIns metabolism. An attractive model for synaptojanin's function in synaptic vesicle endocytosis is that cleavage of PtdIns(4,5)P2 localized on the membrane of the vesicle bud is part of a cycle of biosynthesis and degradation of specific phosphoinositides, which correlates with the exoendocytotic cycle of the vesicles.

The phosphorylated inositol ring could represent a versatile molecular module that can be modified by phosphorylation and dephosphorylation and, therefore, used to bind, recruit, and activate peripheral proteins during the vesicle cycle. These proteins may be the coat proteins themselves, proteins of the actin cytoskeleton (many of which are known to bind phosphoinositides [5]), or other regulatory proteins. Alternatively a change in the phosphoinositide composition may affect the property of the lipid bilayer.

References

1. Takei, K., Mundigl, O., Daniell, L., and De Camilli, P. (1996). The synaptic vesicle cycle: A single vesicle budding step involving clathrin and dynamin. *J. Cell Biol.* **133**, 1237–1250.
2. Bauerfeind, R., David, C., Galli, T., McPherson, P. S., Takei, K., and De Camilli, P. (1995). Molecular mechanisms in synaptic vesicle endocytosis. *Cold Spring Harb. Symp. Quant. Biol.* **60**, 397–404.
3. Takei, K., McPherson, P. S., Schmid, S. L., and De Camilli, P. (1995). Tubular membrane invaginations coated by dynamin rings are induced by GTP-gamma S in nerve terminals. *Nature* **374**, 186–190.
4. David, C., McPherson, P. S., Mundigl, O., and De Camilli, P. (1996). A role of amphiphysin

in synaptic vesicle endocytosis suggested by its binding to dynamin in nerve terminals. *Proc. Natl. Acad. Sci. U.S.A.* **93,** 331–335.
5. De Camilli, P., Emr, S. D., McPherson, P. S., and Novick, P. (1996). Phosphoinositides as regulators in membrane traffic. *Science* **271,** 1533–1539.
6. McPherson, P. S., Garcia, E. P., Slepnev, V. I., David, C., Zhang, X. M., Grabs, D., Sossin, W. S., Bauerfeind, R., Nemoto, Y., and De Camilli, P. (1996). A presynaptic inositol-5-phosphatase. *Nature* **379,** 353–357.

H. Winkler, A. Laslop, B. Leitner, and C. Weiss

Department of Pharmacology
University of Innsbruck
A-6020 Innsbruck, Austria

The Secretory Cocktail of Adrenergic Large Dense-Core Vesicles: The Functional Role of the Chromogranins

Our knowledge on the composition of the secretory cocktail released from adrenergic large dense-core vesicles (LDCVs) is mainly based on studies of adrenal LDCVs (i.e., the chromaffin granules) (1). For LDCVs of adrenergic neurons, the best-investigated ones are those of bovine splenic nerves (2), whereas for those present in the adrenergic terminals, data are more limited.

In chromaffin granules of adrenal medulla, the major peptide components of the secretory cocktail are the chromogranins A and B (3). Their relative concentration varies from species to species. In addition, further peptides are present as minor components: secretogranin II, 7B2 and enkephalin precursors of intermediate sizes, and other neuropeptides. In adrenal medulla, the endoproteolytic processing of the chromogranins but also of the enkephalin precursor is rather limited, although chromaffin granules contain significant concentration of the prohormone convertases PC1 and PC2. These proteases, as we and others (4) have shown, are involved not only in the processing of neuropeptides but also in that of chromogranin B and secretogranin II. In stably transfected PC12 cells, both PC1 and PC2 can convert secretogranin II to secretoneurin, and chromogranin B to PE-11 (representing the chromogranin B sequence 552–562). Several studies have indicated that if chromaffin cell cultures are treated

Advances in Pharmacology, Volume 42

1054-3589/98 $25.00

with reserpine the proteolytic processing of the enkephalin precursor and also that of chromogranin A is increased (see Wolkersdorfer *et al.* [5] for references). We have now extended these observations: Reserpine not only increases the processing of these components, but also that of chromogranin B and secretogranin II. Furthermore, another drug, α-methyl-para-tyrosine, which, like reserpine, depletes the catecholamines in chromaffin granules, also increased proteolysis. Finally, we showed that catecholamines can inhibit the endoproteases at millimolar concentrations. Thus, in chromaffin granules with the extremely high concentration of catecholamines (400 mM), endogenous proteolysis of chromogranins and neuropeptides is inhibited. This offers an explanation why, in adrenal medulla, chromogranin processing is limited, whereas in other endocrine tissues and in brain, proteolysis of chromogranins is much more pronounced.

LDCVs in the adrenergic nerve also contain chromogranins A and B and secretogranin II, but processing seems more complete. Thus, whereas the intact precursor peptides are the major species in adrenal LDCVs, in nerves (e.g., bovine splenic nerve), intermediate and smaller peptides predominate. This is shown in Figure 1 for chromogranin A and secretogranin II. Apparently, in the nerve, the free peptides GE-25 (derived from chromogranin A), PE-11 (derived from chromogranin B; not shown), and secretoneurin (derived from secretogranin II) are dominant or represent at least a significant proportion

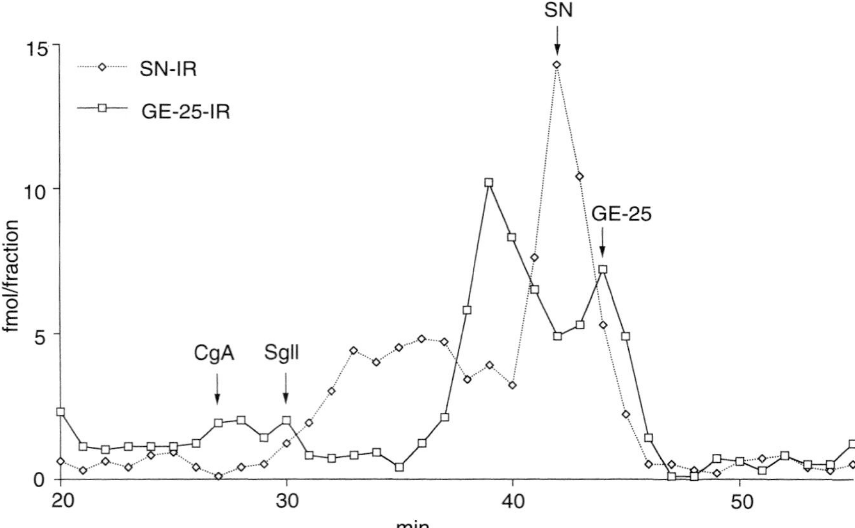

FIGURE I Processing of chromogranins in bovine splenic nerve. Boiled extracts from bovine splenic nerve were subjected to molecular sieve chromatography followed by radioimmunoassay. The antisera against GE-25 (chromogranin A: 367–391) and secretoneurin (secretogranin II: 154–186) also recognize the precursor molecules (chromogranin A: CgA; and secretogranin II: SgII). Most of the immunoreactivity is found in intermediate and free peptides. SN, secretoneurin.

of the immunoreactivity. The precursor peptides are practically absent. For secretogranin II, this has also been demonstrated for the vas deferens, where only the free peptide secretoneurin appears present. There are two possible explanations:

1. The neuronal LDCVs, when compared with adrenal LDCVs, contain more prohormone convertases responsible for processing. In fact, in neuronal LDCVs, the PC1 and PC2 ratio to secretogranin II was more than 5 times higher when compared with adrenal LDCVs (6).
2. Neuronal LDCVs slowly accumulate catecholamines during axonal transport (2): Even LDCVs in terminals probably have a lower catecholamine concentration (0.12–0.2 M) compared with adrenal LDCVs (0.4 M). Thus, the critical concentration of catecholamines inhibiting proteolysis may not be reached before processing of the chromogranins is complete.

Previous discussion on the function of the chromogranins was influenced by the fact that in the anterior pituitary, in the parathyroid gland, and in the adrenal medulla, their proteolytic processing is rather limited. This seemed to favor the idea that they might function as intact molecules. This may still be the right assumption for the adrenal, where the chromogranins represent the major peptide present in the soluble content of chromaffin granules. Thus a "filling" of the granules with these peptides may be necessary to stabilize these relatively large vesicles for catecholamine storage. However, in the other neuroendocrine tissues, especially in neurons, the proteolytic processing of the chromogranins is extensive. Thus, small peptides are formed, and this is consistent with a function of these peptides after secretion. Indeed, peptides derived from chromogranin A (parastatin, vasostatin, and pancreastatin), from chromogranin B (secretolytin), and from secretogranin II (secretoneurin) have been shown to be functionally active molecules. At present, receptors for these peptides are in the process of being defined. When this occurs, it will provide the final proof that the chromogranins are proneuropeptides.

References

1. Winkler, H., Apps, D. K., and Fischer-Colbrie, R. (1986). The molecular function of adrenal chromaffin granules: Established facts and unresolved topics. *Neuroscience* **18**, 261–290.
2. Klein, R. L. (1982). Chemical composition of the large noradrenergic vesicles. *In* Neurotransmitter Vesicles: Composition, Structure and Function. R. L. Klein, H. Lagercrantz, and H. Zimmermann (eds.), pp. 133–173. Academic Press, New York.
3. Winkler, H., and Fischer-Colbrie, R. (1992). The chromogranins A and B: The first 25 years and future perspectives. *Neuroscience* **49**, 497–528.
4. Dittié, A. S., and Tooze, S. A. (1995). Characterization of the endopeptidase PC2 activity towards secretogranin II in stably transfected PC12 cells. *Biochem. J.* **310**, 777–787.
5. Wolkersdorfer, M., Laslop, A., Lazure, C., Fischer-Colbrie, R., and Winkler, H. (1996). Processing of chromogranins in chromaffin cell culture: Effects of reserpine and α-methyl-p-tyrosine. *Biochem. J.* **316**, 953–958.
6. Egger, C., Kirchmair, R., Hogue-Angeletti, R., Fischer-Colbrie, R., and Winkler, H. (1993). Different degrees of processing of secretogranin II in large dense core vesicles of bovine adrenal medulla and sympathetic axons correlate with their content of soluble PC1 and PC2. *Neurosci. Lett.* **159**, 199–201.

Sushil K. Mahata,* Manjula Mahata,* Seung Hyun Yoo,† Laurent Taupenot,* Hongjiang Wu,* Vanita R. Aroda,* Carolyn V. Livsey,* Joseph P. Taulane,‡ Murray Goodman,‡ Robert J. Parmer,* and Daniel T. O'Connor*

*Departments of Medicine and ‡Chemistry
and Center for Molecular Genetics
University of California and Veterans Administration Medical Center
San Diego, California 92161

†Laboratory of Neurochemistry
NIDCD, NIH
Bethesda, Maryland 20892

A Novel, Catecholamine Release-Inhibitory Peptide from Chromogranin A: Autocrine Control of Nicotinic Cholinergic-Stimulated Exocytosis

When nicotinic cholinergic stimulation triggers catecholamine release from chromaffin cells or noradrenergic axons, the process is exocytotic (all-or-none) corelease of vesicle constituents: catecholamines, neuropeptides, and acidic chromogranins, the major component being chromogranin A. The ultimate functional role(s) of chromogranin A has remained elusive, but evidence implicates chromogranin A as a pro-hormone (1).

Proteolysis of chromogranin A takes place both within secretory granules and extracellularly, giving rise to several smaller, biologically active peptides, such as pancreastatin (which inhibits insulin and parathyroid hormone release), β-granin or vasostatin (which inhibits parathyroid hormone release and relaxes vascular smooth muscle), and parastatin (which inhibits parathyroid hormone release). Thus, we and others have also searched for chromogranin A fragments that might modulate catecholamine release. In 1988, Simon et al. (2) reported that trypsin-digestion of chromogranin A generated a peptide that inhibits release of catecholamines from bovine chromaffin cells, but the identity of the responsible peptide has remained elusive.

I. Methods

Catecholamine secretion from PC12 and bovine chromaffin cells was monitored in cells whose chromaffin granules were prelabeled with $[^3H]$-L-

Advances in Pharmacology, Volume 42

1054-3589/98 $25.00

norepinephrine. For neuronal differentiation, PC12 cells were split to approximately 50% confluence and treated with nerve growth factor (2.5S form, 100 ng/ml). The medium was changed every other day, with nerve growth factor addition to the new medium. After 5 days of treatment, neurite-bearing cells were used for secretion studies. $^{22}Na^+$ and $^{45}Ca^{2+}$ cellular uptake studies were done in response to secretagogues in PC12 cells.

II. Secretion–Inhibitory Domain within Chromogranin A

In search of the specific secretion–inhibitory domain within chromogranin A, we synthesized 15 peptides (range, 19–25 residues) spanning 336 amino acids, or 78% of the length (431 amino acids) of the bovine chromogranin A mature protein, and tested their efficacy on nicotine-induced norepinephrine secretion from rat pheochromocytoma (PC12) cells, screening each peptide at a 10-μM concentration. Only bovine chromogranin A 344–364 (RSMRLSFRARGYGFRGPGLQL) profoundly decreased nicotine-induced secretion of norepinephrine ($IC_{50} \approx 200$ nM). We named this peptide "catestatin," because it inhibits catecholamine release. This region, bounded on its N-terminus by a furin recognition site (RXXR) (3) and on its C-terminus by a dibasic site (RR), was extensively processed within chromaffin vesicles *in vivo*.

We tested the effects of synthetic bovine (chromogranin A 344–364; RSMRLSFRARGYGFRGPGLQL), human (chromogranin A 352–372; SSMKLSFRARAYGFRGPGPQL), or rat (chromogranin A 367–387; RSMKLSPRARAYGFRDPGPQL) catestatin on nicotine-induced norepinephrine release from PC12 cells, and found that each species' form of catestatin inhibited nicotine-induced norepinephrine release, though the bovine form showed the greatest potency. We also tested the effects of bovine catestatin on nicotine-induced norepinephrine release from bovine chromaffin cells, as well as neurite-bearing PC12 cells; the catecholamine-release inhibitory activity of catestatin was confirmed in each.

III. Mechanism of Secretion Inhibition

We tested several secretagogues, which act at stages in the secretory signal transduction pathway later than the nicotinic cholinergic receptor: membrane depolarization (55 mM KCl) to open voltage–gated calcium channels, an alkaline earth ($BaCl_2$, 2 mM; whose effects require participation of calcium channels), a calcium ionophore (A23187, 1 μM), or alkalinization of the chromaffin vesicle core (chloroquine, 1 mM), and found that catestatin suppressed norepinephrine release only when triggered by nicotine. A pertussis toxin experiment suggested that inhibitory G-proteins were not involved in catestatin signaling.

We also explored whether catestatin specifically disrupts nicotinic cationic signaling pathways. Catestatin blocked nicotine-induced uptake of $^{22}Na^+$ into PC12 cells in dose-dependent fashion, and this blockade paralleled that of catecholamine release (for the $^{22}Na^+$ effect, $IC_{50} \approx 250$ nM). Catestatin abolished

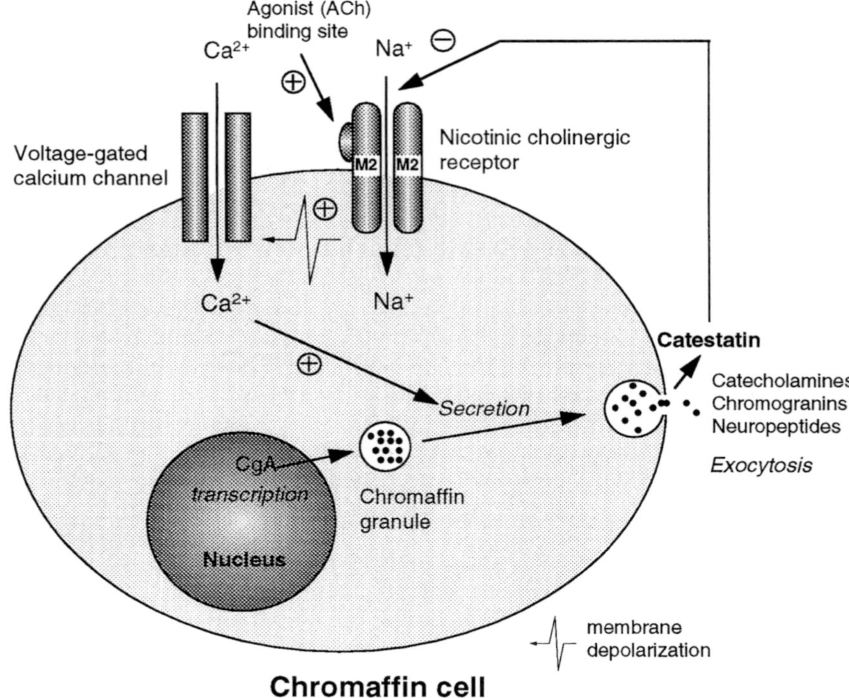

FIGURE 1 The action of the chromogranin A fragment catestatin on catecholamine secretion from chromaffin cells. The physiologic secretagogue for chromaffin cells is acetylcholine, acting at neuronal-type nicotinic cholinergic receptors. Such receptors are extracellular ligand-gated cation channels, permitting cytosolic influx of sodium, which depolarizes the cell membrane, permitting cytosolic influx of calcium through voltage-gated calcium channels. Antagonists at nicotinic receptors may be either competitive (with the agonist binding site) or noncompetitive (often cation channel blockers). CgA, chromogranin A; Ach, acetylcholine; M2, cation channel domain of nicotinic cholinergic receptor; (+, stimulation; (−), inhibition.

nicotine-induced uptake of $^{45}Ca^{2+}$ in PC12 cells and strongly inhibited the nicotine-induced rise in cytosolic calcium in individual bovine chromaffin cells, imaged by the fluorescent dye Indo-1.

Exposure of PC12 cells to a spectrum of nicotine doses (10–1000 μM), alone or with catestatin (10 μM), revealed that nicotine never overcame catestatin's inhibition of norepinephrine release, even at the highest nicotine dosage. Thus, catestatin is likely to be a noncompetitive nicotinic cholinergic antagonist (Fig. 1).

Prior exposure of chromaffin cells to nicotine causes desensitization of catecholamine release upon rechallenge with nicotine. In dose-dependent fashion (from 0.1-10 μM), catestatin protected PC12 cells against nicotinic desensitization.

IV. Structure and Function

To explore the minimal catestatin domain exerting catecholamine secretion–inhibitory effect, we synthesized several (N-terminal, C-terminal, and bidirectional) truncated catestatin peptides. Catecholamine-release inhibitory mapped toward the N-terminus of catestatin, in a region of 12 or fewer amino acids.

The circular dichroism spectrum of human catestatin in the far ultraviolet revealed no structure (i.e., all random coil) in pure H_2O, but inclusion of as little as 25% trifluoroethanol (an organic solvent of relatively low dielectric constant) provoked substantial (47–63%) β-pleated sheet conformation.

A hydrophobic moment plot predicted that the catestatin region of chromogranin A is likely to be a β-pleated sheet, and β-sheet structure was also suggested by homology modeling.

The catestatin region in bovine chromogranin A is bounded on the N-terminus by a furin recognition site (3), and at the C-terminus by a dibasic site, RR. The RR site is conserved across all mammalian species, though the RXXR motif is perfectly conserved only in bovine chromogranin A. Monobasic (R) sites are also abundant in this region. Several laboratories have documented proteolytic cleavage in the catestatin region of chromogranin A (4, 5).

V. Conclusions

We identified a novel chromogranin A peptide catestatin (bovine chromogranin A 344–364), which is a potent ($IC_{50} \approx 200$ nM) inhibitor of catecholamine secretion. This domain's primary structure is conserved across mammalian species, is flanked by proteolytic cleavage sites, and is processed from chromogranin A in chromaffin granules. We characterized the inhibitory effects of this peptide on secretion, specifically as a noncompetitive nicotinic cholinergic antagonist, and established its effects on nicotinic signal transduction, as well as protection against prior nicotinic desensitization. As an endogenous nicotinic antagonist, catestatin may provide a novel homeostatic, negative-feedback mechanism for antagonism of both stimulation and desensitization of catecholamine release (Fig. 1). Studies of this peptide may yield new tools to intervene diagnostically or therapeutically in nicotine-regulated catecholamine release. Finally, catestatin's protection against nicotinic desensitization may be advantageous to an organism during circumstances of prolonged stress, perhaps guarding against premature termination of secretory responses to physiologic nicotinic stimulation.

Acknowledgments

Supported by the National Institutes of Health, the Department of Veterans Affairs, and the American Heart Association.

References

1. Winkler, H., and Fischer-Colbrie, R. (1992). The chromogranins A and B: The first 25 years and future perspectives. *Neuroscience* **49**, 497–528.
2. Simon, J-P., Bader, M-F., and Aunis, D. (1988). Secretion from chromaffin cells is controlled by chromogranin A-derived peptides. *Proc. Natl. Acad. Sci. U.S.A.* **85**, 1712–1716.
3. Molloy, S. S., Bresnahan, P. A., Klimpel, K., Leppla, L., and Thomas, G. (1992). Human furin is a calcium-dependent serine endoprotease that recognizes the sequence Arg-X-X-Arg and efficiently cleaves anthrax toxin protective antigen. *J. Biol. Chem.* **267**, 16396–16402.
4. Sigafoos, J., Chestnut, W.G., Merrill, B. M., Taylor, L. C., Diliberto, E. J. Jr., and Viveros, O. H. (1993). Novel peptides from adrenomedullary chromaffin vesicles. *J. Anat.* **183**, 253–264.
5. Kirchmair, R., Leitner, B., Fischer-Colbrie, R., Marksteiner, J., Hogue-Angeletti, R., and Winkler, H. (1995). Large variations in the proteolytic formation of a chromogranin A-derived peptide (GE-25) in neuroendocrine tissues. *Biochem. J.* **310**, 331–336.

Lee E. Eiden, Youssef Anouar, Chang-Mei Hsu, Linda MacArthur, and Sung Ho Hahm

Section on Molecular Neuroscience
Laboratory of Cellular and Molecular Regulation
NIMH, NIH
Bethesda, Maryland 20892

Transcription Regulation Coupled to Calcium and Protein Kinase Signaling Systems through TRE- and CRE-Like Sequences in Neuropeptide Genes

Depolarization and calcium influx stimulates the secretion of catecholamines, neuropeptides, and secretory proteins such as chromogranin A from chromaffin cells. Calcium entry during enhanced secretory activity also triggers a compensatory biosynthesis of neuropeptides, or 'stimulus-secretion-synthesis coupling' (1). Neuropeptide secretion and biosynthesis in chromaffin cells is modulated by second messengers in addition to calcium. In contrast to depolarization, adenylate cyclase stimulation increases long-term but not acute secre-

Advances in Pharmacology, Volume 42

1054-3589/98 $25.00

tion and also upregulates neuropeptide biosynthesis (Fig. 1A, B). Stimulation of protein kinase C, after a latency period, causes an augmentation of potassium-induced secretion, followed by inhibition at longer times of exposure (Fig. 1A). Phorbol esters cause a profound increase in biosynthesis of the neuropeptides galanin and vasoactive intestinal polypeptide (VIP) and a modest increase in enkephalin biosynthesis in acutely cultured chromaffin cells (Fig. 1B).

The effects of second-messenger stimulation on biosynthesis of different neuropeptide genes can also be distinguished on the basis of requirement for new protein synthesis. Stimulation of enkephalin mRNA production by potassium depolarization (i.e., stimulus-secretion-synthesis coupling) is largely independent of new protein synthesis. In contrast, upregulation of enkephalin mRNA levels following treatment with forskolin and phorbol ester is abolished by inhibition of new protein synthesis with cycloheximide (CHX), whereas phorbol ester–stimulated galanin gene transcription is not blocked by CHX, indicating that it is mediated through preexisting rather than induced trans-activating factor(s) (Anouar, Hsu, and Eiden, in preparation; Fig. 1C).

Several of the cis-active elements on the enkephalin, galanin, and VIP genes responsible for basal and regulated expression in chromaffin and neuroblastoma cells have been characterized. Galanin gene expression, and stimulation by both cyclic adenosine monophosphate (cAMP) and phorbol ester, in chromaffin cells depends on an element, the GTRE, that binds both cyclic response element-binding (CREB)- and AP1-like proteins in gelshift assays (Anouar, Hsu and Eiden, in preparation; [2]). The enkephalin promoter uses an element, ENKCRE-2, that binds to CREB and not AP1 proteins in central nervous system tissues (3), but binds mainly to the AP1 proteins c-jun and junD and to a novel fos-immunoreactive protein complex, in addition to CREB, in chromaffin cell nuclear extracts (4). The fos-like protein-containing complex is present in nuclear extracts from quiescent chromaffin cells. Therefore, it, along with CREB, is a candidate for the preexisting trans-acting protein(s) that mediate stimulus-secretion-synthesis coupling in chromaffin cells.

In contrast to galanin and enkephalin, VIP gene regulation by protein kinases A and C appears to be modular; that is, it depends on two separate cis-active elements in the VIP gene. Contrary to previous findings in nonneuronal cells transfected with a VIP reporter gene (5), the VIP proximal CRE mediates only the effects of cAMP and is required for basal gene expression in neuroblastoma cells (6). A newly identified upstream consensus TRE mediates the effects of protein kinase C on VIP gene transcription.

The chromogranin A gene, which contains a consensus CRE in its proximal promoter, is stimulated only modestly by forksolin and in a negative manner by phorbol ester in chromaffin and insulinoma cells (7). The chromogranin A CRE element has been reported to be required for neuroendocrine-specific regulation of this gene (8).

Candidate cis- and trans-acting factors that mediate differential transcriptional responses of each gene to calcium, protein kinase A, and protein kinase C, are summarized in Table I. Cis- and trans-acting elements are employed in a highly cell- and gene-specific manner to regulate constitutive expression, coupling of gene transcription to secretion, and stimulation by protein kinase signaling pathways in neuroendocrine cells. Cell-specific utilization of individual

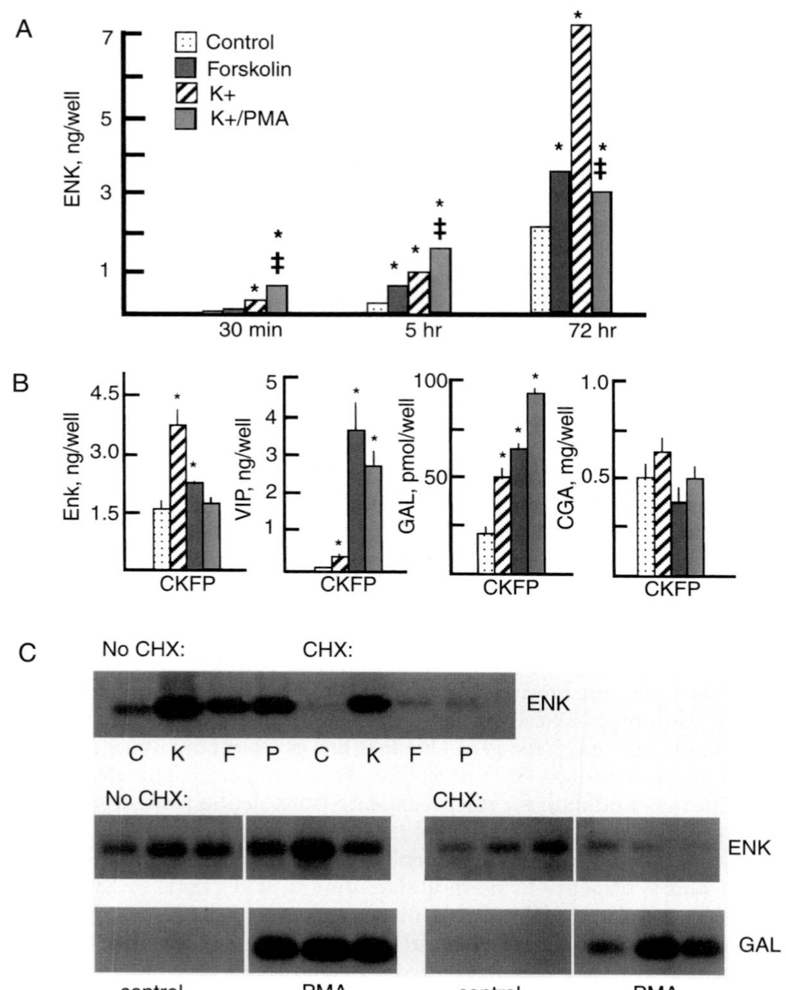

FIGURE 1 Regulation of neuropeptide secretion and biosynthesis via calcium, protein kinase A, and protein kinase C signaling pathways. (*A*) Enkephalin secretion from bovine chromaffin cells in primary culture treated with 40 mM of KCl, 25 μM of forskolin, or KCl plus 0.1 μM of PMA for the times indicated. *, $p < .05$, Student's *t*-test. (*B*) Differential effect on cellular peptide or protein levels measured by radioimmunoassay following stimulation for 72 hr with KCl, forskolin, or PMA at the concentrations noted in (*A*). Data are from references 9, 10, and unpublished observations. (*C*) *Upper:* differential block of enkephalin (ENK) mRNA induction following treatment with KCl, PMA, or forskolin by cycloheximide (CHX; 0.5 μg/ml). CHX was added to the bovine chromaffin cell cultures 30 min before addition of drugs, and cells were harvested for measurement of mRNA levels by Northern blot hybridization 18 hr later. *Lower:* In contrast to blockade by CHX of PMA-induced enkephalin (ENK) mRNA elevation, galanin mRNA (GAL) induction by PMA is unaffected by pretreatment with CHX. Data in C, Anovar, Hsu, and Eiden, in preparation.

TABLE I Summary of Putative/Candidate cis- and trans-Acting Factors Mediating ENK, GAL, VIP, and CGA Gene Regulation by Second-Messenger Signaling Pathways in Neuroendocrine Cells

Neurosecretory protein gene	*Sequence and location*	*Candidate trans-acting factor(s)*	*Proposed function*	*References*
ENK	ENK TSE, 5′ distal	None identified	Cell-specific expression	11
	ENKCRE2, proximal (TGCGGTCA)	c-jun, junD, novel fra; CREB	stimulation by Ca^{2+}, PKA, PKC	4, 12
GAL	GTRE, proximal (TGACGCGG)	Constitutive AP1 like protein(s); CREB	Contributory to cell-specific expression; stimulation by PKA and PKC; Ca^{2+}?	2; Anouar, Hsu, and Eiden, unpublished
VIP	VIP TSE, 4 kb 5′	None identified	Cell-specific expression; stimulation by PKC	6, 13; Hahm and Eiden, unpublished
	VIP TRE, 1.3 kb 5′ (TGACTCA)	None identified		
	VIP CRE, proximal (<u>CGTCA</u>TACTG<u>TGACG</u>)	CREB?	Stimulation by PKA; cell-specific expression	
CGA	CGA CRE, proximal (TGACGTCA)	CREB; neuroendocrine cell–specific CREB-binding proteins?	Cell-specific expression; PMA downregulation?	7, 8

Bovine (GAL, ENK, CGA) or human (VIP) gene sequences are shown. Proximal sequences are located within the first 100 bases upstream of the start of transcription of the gene. ENK, enkephalin; GAL, galanin; CGA, chromogranin; ENK TSE, eukephalin tissue-specifier element; PKA, protein kinase A; PKC, protein kinase C.

cis-active elements on each neuroepeptide gene (e.g., lack of phorbol ester stimulation through the VIP and CGA CREs in neuroblastoma and chromaffin cells) may reflect neuroendocrine cell-specific occupancy of these elements by constitutive factors that mask further stimulation or even result in a relative decrease in transcription when protein kinase pathways are stimulated. These mechanisms of differential regulation of neuroepeptide genes within catecholamine-storing tissues may serve to tailor the mixture of neuropeptides cosecreted with catecholamines and other classical neurotransmitters from neuroendocrine cells to diverse physiological conditions.

References

1. MacArthur, L., and Eiden, L. E. (1996). Neuropeptide genes: Targets of activity-dependent signal transduction. *Peptides* **17**, 721–728.
2. Anouar, Y., MacArthur, L., Cohen, J., Iacangelo, A. L., and Eiden, L. E. (1994). Identification of a TPA-responsive element mediating preferential transactivation of the galanin gene promoter in chromaffin cells. *J. Biol. Chem.* **269**, 6823–6831.
3. Konradi, C., Kobierski, L. A., Nguyen, T. V., Heckers, S., and Hyman, S. E. (1993). The cAMP-response–element-binding protein interacts, but Fos protein does not interact, with the proenkephalin enhancer in rat striatum. *Proc. Natl. Acad. Sci. U. S.A.* **90**, 7005–7009.
4. MacArthur, L. (1996). AP-1 related proteins bind to the enkephalin CRE-2 element in adrenal chromaffin cells. *J. Neurochem.* **67**, 2256–2264.
5. Fink, J. S., Verhave, M., Walton, K., Mandel, G., and Goodman, R. H. (1991). Cyclic AMP- and phorbol ester-induced transcriptional activation are mediated by the same enhancer element in the human vasoactive intestinal peptide gene. *J. Biol. Chem.* **266**, 3882–3887.
6. Hahm, S.H., and Eiden, L. E. (1996). Tissue-specific expression of the vasoactive intestinal peptide gene requires both an upstream tissue specifier element and the 5′ proximal cAMP-responsive element. *J. Neurochem.* **67**, 1872–1881.
7. Iacangelo, A. L., Grimes, M., and Eiden, L. E. (1991). The bovine chromogranin A gene: Structural basis for hormone regulation and generation of biologically active peptides. *Mol. Endocrinol.* **5**, 1651–1660.
8. Wu, H., Rozansky, D. J., Webster, N. J. G., and O'Connor, D. T. (1994). Cell type-specific gene expression in the neuroendocrine system. A neuroendocrine-specific regulatory element in the promoter of chromogranin A, a ubiquitous secretory granule core protein. *J. Clin. Invest.* **94**, 118–129.
9. Waschek, J. A., Pruss, R. M., Siegel, R. E., Eiden, L. E., Bader, M-F., and Aunis, D. (1987). Regulation of enkephalin, VIP and chromogranin A biosynthesis in actively secreting chromaffin cells: Multiple strategies for multiple peptides. *Ann. N. Y. Acad. Sci.* **493**, 308–323.
10. Rökaeus, A., Pruss, R. M., and Eiden, L. E. (1990). Galanin gene expression in chromaffin cells is controlled by calcium and protein kinase signalling pathways. *Endocrinology* **127**, 3096–3102.
11. Kaplitt, M. G., Kwong, A. D., Kleopoulos, S. P., Mobbs, C. V., Rabkin, S. D., and Pfaff, D. W. (1994). Preproenkephalin promoter yields region-specific and long-term expression in adult brain after direct in vivo gene transfer via a defective herpes simplex viral vector. *Proc. Natl. Acad. Sci. U.S.A.* **91**, 8979–8983.
12. Van Nguyen, Kobierski, L., Comb, M., and Hyman, S. E. (1990). The effect of depolarization on expression of the human proenkephalin gene is synergistic with cAMP and dependent upon a cAMP-inducible enhancer. *J. Neurosci.* **10**, 2825–2833.
13. Waschek, J. A., Hsu, C-M., and Eiden, L. E. (1988). Lineage-specific regulation of the vasoactive intestinal peptide gene in neuroblastoma cells is conferred by 5.2 kilobases of 5′-flanking sequence. *Proc. Natl. Acad. Sci. U.S.A.* **85**, 9547–9551.

Kirk A. Frey,*†‡ Donald M. Wieland,* and Michael R. Kilbourn*

Departments of *Internal Medicine (Nuclear Medicine) and †Neurology and
‡The Mental Health Research Institute
The University of Michigan
Ann Arbor, Michigan 48109

Imaging of Monoaminergic and Cholinergic Vesicular Transporters in the Brain

Studies of chemically defined neurons may offer important neurochemical perspectives on normal aging and on age-associated neurodegenerative diseases. Conditions including Parkinson's disease (PD) and Alzheimer's disease (AD), for example, share age-associated increasing incidences and may be defined by pathologic losses of neurons and synapses in specific brain regions, or involving select chemical neuronal classes. The pathophysiologies of most neurodegenerative disorders are unknown, despite detailed descriptions of the ultimate neuropathologic effects of the disease processes. Postmortem analyses, however, may be confounded by the intrusions of secondary, disease-compensatory or treatment-induced changes. Thus, *in vivo* investigations conducted early in the course of neurodegenerative diseases offer unique opportunities to determine the initial and most specific neurochemical alterations. Furthermore, *in vivo* neurochemical measures may have the potential to distinguish disease-modifying from symptomatic therapies. In this instance, objective, quantitative markers of neuronal or synaptic integrity, targeting the involved cell types and brain regions, may offer considerable advantage over clinical examinations or other indirect measures in tracking the course and progression of neuropathology. With these goals and applications in mind, our laboratories have focused on the development of *in vivo* imaging markers of the dopaminergic and cholinergic neurons and synapses of the human brain, employing positron emission tomography (PET) or single photon emission tomography (SPECT) of radiolabeled neurochemical tracers.

Among the possible biochemical markers of specific neurons and synapses, including neurotransmitter synthetic enzymes, plasma membrane uptake transporters, and neurotransmitter receptors, a quantitative relationship between neuron or synapse integrity and the level of marker expression in macroscopic tissue samples is difficult to establish. The activities and expressed protein levels of these markers are well recognized to undergo compensatory regulation in response to therapeutic drugs and in response to disease conditions. Recent molecular and pharmacologic advances now permit the study of an additional class of neuronal markers, the distinct vesicular neurotransmitter transporters

Advances in Pharmacology, Volume 42

1054-3589/98 $25.00

in monoaminergic and cholinergic neurons. Although neurotransmitter release from synaptic vesicles is a regulated process, it has been suggested that a major aspect of this regulation may reside in the partitioning of vesicles between active (cycling) and reserve (bound) pools in nerve terminals (1). We hypothesized that vesicular markers not distinguishing these states might serve as quantitative indicators of synaptic density, with minimal or no influence of regulatory changes. Measures of the vesicular neurotransmitter transporters may, thus, provide unique insights into conditions that alter synaptic regulation versus neuronal integrity.

We synthesized both unlabeled and radioligand series targeting the neuronal vesicular monoamine transporter type-2 (VMAT2) or the vesicular acetylcholine transporter (VAChT). Ligand-binding specificities were established *in vitro* in direct binding and competition assays and after selective lesions of dopaminergic or cholinergic projection pathways. The possible regulation of VMAT2 and of VAChT by drugs that alter synaptic transmission were also studied. Single photon- or positron-emitting ligand analogues were then developed for *in vivo* SPECT and PET imaging of VMAT2 and VAChT.

In vitro studies identified specific VMAT2 binding either with [^{3}H]methoxytetrabenazine (2) or with stereochemically resolved (+)-[^{3}H]dihydrotetrabenazine (DTBZ). Binding of both ligands was saturable, reversible, and of high affinity (K_ds 4 nM and 1.5 nM, respectively) to an apparently homogeneous population of binding sites on intact, slide-mounted brain sections. The brain regional pattern of binding was consistent with the known distributions of dopaminergic, noradrenergic, and serotonergic neurons and their projections. After 6-hydroxydopamine lesions, linear correlation of striatal VMAT2 binding and nigral tyrosine hydroxylase–immunoreactive neuron density was observed over a wide range of lesion severity (2). Furthermore, VMAT2 binding was not affected by 2-wk pretreatment of animals with a variety of drugs leading to regulatory changes in dopamine receptors or in the synaptic membrane dopamine transporter (3). *In vivo* studies in rodents revealed cerebral biodistributions of the tracers paralleling the *in vitro* pattern of specific VMAT2 binding and additionally confirmed lack of interfering effects of nonvesicular ligand, dopaminergic drugs. Chromatographic analysis after the injection of DTBZ revealed only unchanged tracer in brain, permitting use of total tissue tracer levels in subsequent tracer kinetic analyses.

Human imaging experiments employing PET and [^{11}C]DTBZ revealed separable effects of tracer delivery (blood-to-brain transport) and VMAT2 binding on the cerebral time courses of activity after intravenous injection (4). We detected significant age-associated losses of striatal VMAT2 binding in normal subjects, corresponding to linear 0.8% per year losses of VMAT2 from the putamen. Patients with PD demonstrated substantial reductions in VMAT2 in the putamen and more minor losses in the caudate nucleus in the early stages. Patients with severe, advanced PD demonstrated near-complete losses of VMAT2 binding throughout the striatum. Patients with multiple-systems atrophy and some patients with sporadic olivopontocerebellar atrophy also demonstrate reduced striatal VMAT2 binding, supporting prior hypotheses of phenotypic overlap between these two conditions.

Initial *in vitro* analyses of the VAChT, based on [^{3}H]vesamicol binding, revealed multiple sites of interaction for the ligand. Specific VAChT binding was identified for benzovesamicol-derivative ligands, including methylaminobenzovesamicol (MABV) and 5-iodobenzovesamicol (IBVM). Similarly to the VMAT2, VAChT expression was not regulated by drug treatments affecting ACh turnover and release. After fornix lesions, virtually complete loss of hippocampal VAChT binding was observed in *in vitro* VAChT binding assays. Intravenous injections of [^{125}I]IBVM led to initial uptake in proportion to cerebral blood flow, but delayed retention in a pattern consistent with the *in vitro* distribution of VAChT, including heavy labeling of cholinergic nuclei and projections (5). Fimbrial lesion reduced *in vivo* retention of [^{125}I]IBVM to background levels, confirming the specificity of IBVM biodistribution.

Human brain imaging of VAChT *in vivo* was developed on the basis of SPECT employing [^{123}I]IBVM. After intravenous injections, tracer entered the brain readily and continued to rise over 12–18 hr in the striatum. The brain regional tracer distribution at 24 hr postinjection paralleled the VAChT distribution observed in *in vitro* assays. Tracer kinetic analyses confirmed close correlation of *in vivo* [^{123}I]IBVM binding to VAChT and the levels of delayed tracer retention, permitting simplified experimental determinations of immediate tracer uptake and 24-hr delayed retention in subsequent clinical studies of aging and dementing disorders. In contrast to striatal VMAT2, cerebral VAChT binding was not significantly affected by aging. Unexpectedly, however, VAChT binding showed relatively minor losses in the cerebral cortex of patients with AD (6), despite extensive prior reports of reduced cortical presynaptic cholinergic markers such as choline acetyltransferase (CAT) and acetylcholinesterase activities. There are apparent distinctions between AD patients with presenile versus senile onset of symptoms, the former demonstrating apparently greater VAChT reductions. In addition, PD patients with dementia reveal reduced cortical VAChT, in keeping with postmortem findings of concomitant AD changes in approximately half of demented PD patients.

We have confirmed the SPECT VAChT findings in postmortem *in vitro* assays of AD (7). Samples of temporal neocortex reveal nonsignificant, approximately 15% losses of VAChT binding, yet have over 40% losses of CAT activity in parallel assays. Both VAChT and CAT are significantly reduced in the hippocampal formation of AD, but again, the VAT reduction is less than half of the magnitude of CAT change. These findings suggest the possibility of persistent cholinergic nerve terminals in AD cortex, but with reduced or absent CAT expression. Surviving cholinergic terminals may, thus, provide an attractive target substrate for the possible treatment of the cholinergic lesion in AD with neurotrophic factors that might restore CAT and other obligatory transmitter synthetic activities.

In summary, vesicular neurotransmitter transporters appear to provide measures of presynaptic cholinergic and monoaminergic neuronal integrity, unaffected by use- or drug-induced regulatory changes. The roles of cholinergic and monoaminergic neurons in neurodegenerative diseases and their responses to potential disease-modifying therapies can now be specifically studied *in vivo* with PET and SPECT measures.

Acknowledgments

The studies reported in this work were supported in part by U.S. Public Health Service grants AG08671 and MH47611 from the National Institute of Aging and the National Institute of Mental Health, respectively.

References

1. Greengard, P., Valtorta, F., Czernik, A. J., and Benfenati, F. (1993). Synaptic vesicle phosphoproteins and regulation of synaptic function. *Science* **259,** 780.
2. Vander Borght, T. M., Sima, A. A. F., Kilbourn, M. R., Desmond, T. J., Kuhl, D. E., and Frey, K. A. (1995). [^{3}H]Methoxytetrabenazine: A high specific activity ligand for estimating monoaminergic neuronal integrity. *Neuroscience* **68,** 995–962.
3. Vander Borght, T. M., Kilbourn, M. R., Desmond, T. J., Kuhl, D. E., and Frey, K. A. The vesicular monoamine transporter is not regulated by dopaminergic drug treatments. *Eur. J. Pharmacol.* **294,** 577–583.
4. Frey, K. A., Koeppe, R. A., Kilbourn, M. R., Vander Borght, T. M., Albin, R. L., Gilman, S., and Kuhl, D. E. (1996). Presynaptic monoaminergic vesicles in Parkinson's disease and normal aging. *Ann. Neurol.* **40,** 873–884.
5. Jung, Y-W., Frey, K. A., Mulholland, G. K., del Rosario, R., Sherman, P. S., Raffel, D. M., Van Dort, M. E., Kuhl, D. E., and Wieland, D. M. (1996). Vesamicol receptor mapping of brain cholinergic neurons with radioiodine labeled positional isomers of benzovesamicol. *J. Med. Chem.* **39,** 3331–3342.
6. Kuhl, D. E., Minoshima, S., Fessler, J. A., Frey, K. A., Foster, N. L., Ficaro, E. P., Wieland, D. M., and Koeppe, R. A. (1996). In vivo mapping of cholinergic terminals in normal aging, Alzheimer's disease and Parkinson's disease. *Ann. Neurol.* **40,** 399–410.
7. Murman, D. L., Desmond, T. J., Higgins, D. S., Hermanowicz, N. S., Penney, J. B. Jr., and Frey, K. A. (1994). Autoradiographic quantification of the vesamicol receptor in Alzheimer's disease. *Neurology* **44,** (Suppl. 2), A389.

PART C

CATECHOLAMINE METABOLISM

From Molecular Understanding to Clinical Diagnosis and Treatment

Alan A. Boulton* and Graeme Eisenhofer†

*Neuropsychiatry Research Unit
Department of Psychiatry
University of Saskatchewan
Saskatoon, Saskatchewan
Canada, S7N 5E4

†Clinical Neuroscience Branch
National Institute of Neurological Disorders and Stroke
National Institutes of Health
Bethesda, Maryland 20892

Overview

Catecholamines are metabolized by multiple pathways involving oxidative deamination catalyzed by monoamine oxidase (MAO), *O*-methylation by catechol *O*-methyltransferase (COMT) and conjugation by sulfotransferases or glucuronidases. Aldehyde reductase and aldehyde dehydrogenase are other enzymes that participate in sequence with MAO in the production of respective glycol and acid deaminated metabolites. Additionally, alcohol dehydrogenase contributes to formation of homovanillic acid and vanillylmandelic acid (VMA), the final end-products of catecholamine metabolism. This multiplicity of metabolic enzymes ensures redundancy in the mechanisms of catecholamine inactivation and also leads to a wide array of metabolites (Fig. 1).

Advances in Pharmacology, Volume 42

1054-3589/98 $25.00

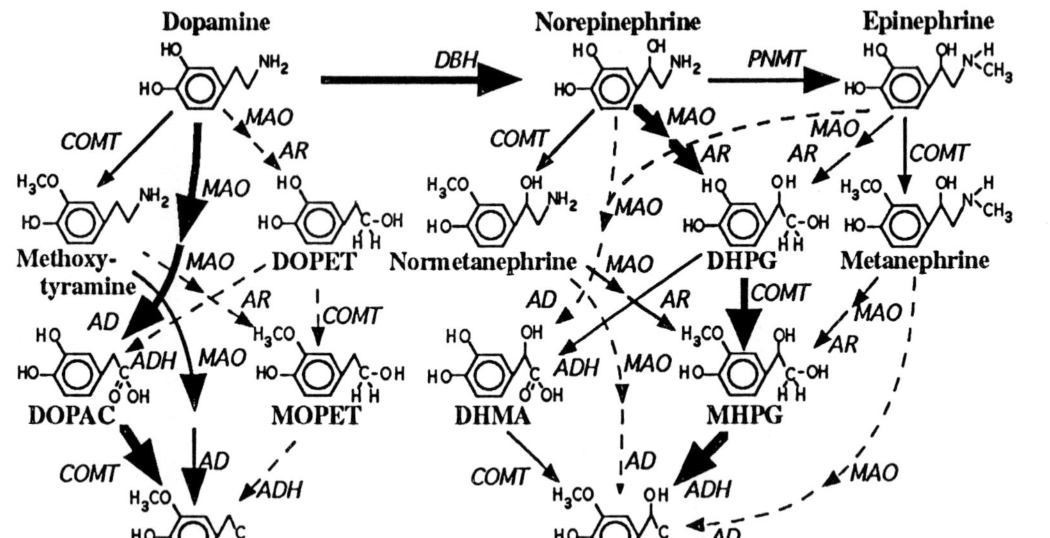

FIGURE 1 Pathways of catecholamine metabolism. The most active pathways are shown by the more solid arrows and the least active by the dashed arrows. The enzyme responsible for each step is shown at the head of each arrow. The pathways of sulfate and glucuronide conjugation are not shown because each of the parent amines and metabolites may, to some extent, be conjugated (see later in this section). DBH, dopamine β-hydroxylase; PNMT, phenylethanolamine-N-methyltransferase; COMT, catechol *O*-methyltransferase; MAO, monoamine oxidase; AR, aldehyde reductase; AD, aldehyde dehydrogenase; ADH, alcohol dehydrogenase; DOPET, dihydroxyphenylethanol; DHPG, dihydroxyphenylglycol; DOPAC, dihydroxyphenylacetic acid; MOPET, methoxyhydroxyphenylethanol; DHMA, dihydroxymandelic acid; MHPG, methoxyhydroxyphenylglycol; HVA, homovanillic acid; VMA, vanillylmandelic acid.

The pathways of metabolism depend on compartmentalization of the various catabolic enzymes among different cells and tissues. MAO is the sole metabolizing enzyme present in catecholaminergic neurons. Thus, the deaminated metabolite of norepinephrine, dihydroxyphenylglycol (DHPG), provides a marker of intraneuronal metabolism (1). This has indicated that metabolism of norepinephrine within neurons is mainly from transmitter leaking from vesicles (2); the contribution of transmitter recaptured by the membrane transporter is small by comparison (Fig. 2). Because COMT is confined to extraneuronal tissues, the O-methylated metabolites of catecholamines provide markers of extraneuronal transport and metabolizing systems. This has indicated that the contribution of extraneuronal uptake and metabolism to peripheral norepinephrine turnover is minor (3).

Differences in metabolism among tissues and organs must also be considered. VMA, the end-product of norepinephrine and epinephrine metabolism and the predominant metabolite excreted in urine, is produced almost exclusively in the liver, mainly from the deaminated metabolites DHPG and 3-methoxy-4-hydroxyphenylglycol (4). Over 80% of the latter metabolite is derived from the DHPG initially produced in neurons (5). Thus, intraneuronal metabolism is the main determinant of norepinephrine turnover; most of

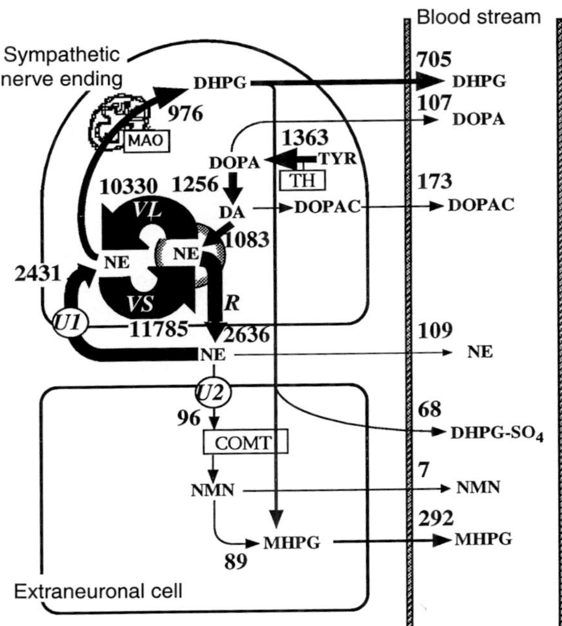

FIGURE 2 The processes of synthesis, release (R), neuronal reuptake (U1), extraneuronal uptake (U2), vesicular leakage (VL) and sequestration (VS), metabolism and turnover of NE in sympathetic nerve endings in relation to extraneuronal tissues and the bloodstream of the human heart. Numbers at the head of each arrow represent rates of each process in pmol/min. Data from Eisenhofer *et al.* (2). TYR, tyrosine; other abbreviations as in Figure 1.

this is dependent on leakage of transmitter from vesicular stores with the contribution from re-uptake only becoming important with increases in exocytotic release (2). The contribution of vesicular leakage to metabolism and turnover acts as a gearing mechanism so that only proportionally small percentage increases in transmitter synthesis are required to compensate for large increases in exocytotic release.

These considerations indicate how an understanding of catecholamine metabolism can provide insight into the various functions of catecholaminergic systems covered in other sections of this volume. Moreover, as shown later in this section, an understanding of catecholamine metabolism can help interpret, diagnose, and treat disorders of central and peripheral catecholaminergic systems.

To the reductionalist and linear thinker, investigations of catecholamine metabolism may appear to offer little in the way of "hot" areas of study. However, as shown later in this section, catecholamine metabolism does have its hot areas of investigation. Perhaps more importantly, and as illustrated in this section, studies of catecholamine metabolism provide an ideal contextural framework for bridging molecular biologic, genetic, neurochemical, psychological, and clinical areas of interest. This should represent an important goal for advancing future knowledge about catecholamine systems in health and disease.

I. Oxidative Deamination

MAO has a venerable history, having been discovered in 1910 and characterized by Hare, Blashko, Zeller, and Quastel, among others. The separation of MAO into two isozymes, A and B, was first suggested by Gorkin in 1963 and demonstrated by Johnston and Squires, respectively, in 1968 on the basis of sensitivity to clorgyline and later to preferential substrates and other selective inhibitors (Table I). The archetypical inhibitors are clorgyline for MAO-A and deprenyl for MAO-B. These isozymes are found in all mammalian species and most other species, although fish seem to possess a sort of hybrid MAO. Some cells or tissues can exhibit a single isozyme so that only MAO-B is found, for example, in platelets and lymphocytes and MAO-A in skin fibroblasts and placenta. In humans, MAO-B is particularly plentiful in the brain and liver, while MAO-A is highest in liver, lung, and intestine. A more interesting heterogeneity relates to compartmentalization, so that MAO-A is found in the cell bodies of all catecholaminergic neurons, while MAO-B is present only in serotoninergic neurons. MAO-B, however, is also present in all glia and astrocytes. Because the preferred substrate for MAO-B seems to be β-phenylethylamine, an endogenous amine present in the brain in only tiny quantities, it is valid to ask whether the major substrate for MAO-B has yet been identified. It is a fact, however, that when MAO-A is compromised, MAO-B will oxidatively deaminate those substrates that are usually oxidized by MAO-A (see references 6–13 for extensive documentation).

TABLE I Substrates and Inhibitors for MAO-A and MAO-B

MAO-A	*MAO-B*	*MAO-A/MAO-B (nonselective)*
Substrates		
5-hydroxytryptamine	2-Phenylethylamine	Adrenaline[a]
	MPTP	Noradrenaline[a]
	Aliphatic amines	Dopamine[a]
	N′-methylhistamine	Tyramine(s)
		Kynuramine
		Tryptamine
Inhibitors		
Clorgyline	Deprenyl	Tranylcypromine
Moclobemide[b]	Ro 19-6327[b]	Isocarboxazid
Harmine[b]	MDL 72 145	Phenelzine
Harmaline[b]	MDL 72 974	Iproniazid
Ro-41-1049	Pargyline	Nialamide
MDL 72 394	Ro 16-6491	
Brofaromine[b]	Rasagiline (TVP-1012)	
Lilly 51641	Aliphatic N-methyl propargylamines	
Tetrahydro-β-carboline[b]	Lilly 54761	
Cimoxatone[b]	Caroxazone[b]	
Toloxatone[b]	MD-780236[b]	
Befloxatone[b]	Milacemide[b]	
BW 1370 U 87[b]		
Amiflamine[b]		

[a] The catecholamines appear to be selectively metabolized *in vivo* by MAO-A (see Lenders *et al.*, p. 297).
[b] Selective, reversible inhibitors.

Both isozymes use the same flavin cofactor (FAD), and binding to the cofactor occurs on the cysteine residue of a pentapeptide sequence (ser-gly-gly-cys-tyr). The isozymes, however, are structurally distinct and unique proteins of similar molecular weight, as revealed by immunological studies, partial proteolysis and peptide mapping, and photodependent inactivation of the active sites. Both isozymes are found on the outer membrane of mitochondria and arise from different genes but with a common ancestry located on the X chromosome (Xp11·23). As outlined by Chen and Shih in this volume (p. 292), the promoter regions on these genes possess different transcription binding sites. The three-dimensional structure of MAO, however, remains to be elucidated.

MAO-A and MAO-B are expressed during development at different times; MAO-B correlates with glial proliferation and differentiation, while MAO-A correlates more with neuronal development. From the preceding discussion then, it is permissible to ask what is MAO-B's role, why is it so ubiquitously distributed, why is it present in serotoninergic but not catecholaminergic cell bodies, and what are the identities of all its substrates?

The earlier MAO inhibitors (MAOIs) of both MAO-A and MAO-B were irreversible, nonselective inhibitors such as tranylcypromine (Parnate), phenelzine (Nardil), and isocarboxazid (Marplan). They were introduced in the 1950s as antidepressants; while these compounds were clinically effective, they fell into disrepute because of the so-called cheese effect. That is, they caused hypertensive crises after patients ingested foods containing amines such as p-tyramine. After virtually disappearing from clinical practice MAOIs are now being successfully reintroduced as selective, reversible, MAO-A inhibitors. These MAO-A inhibitors, such as moclobemide and brofaromine, are effective antidepressants, and they lack the cheese effect; they are also being introduced as treatments for panic attacks and agoraphobia, social phobias, and bulimia nervosa (13).

MAO-B inhibitors have not been particularly successful in the treatment of psychiatric disorders, but they have been used more interestingly, as we shall see, to treat some neurodegenerative conditions, such as Parkinson's and Alzheimer's disorders and even aging (in animals). The first clinical trial of (−)deprenyl in Parkinson's disease was by Birkmayer *et al.* in the early 1970s. They showed both symptomatic relief as well as a prolongation of life (14). This was confirmed in 1989 in double blind trials although these trials—the so-called DATATOP study—have been criticized on methodological grounds. In Alzheimer's disease, several trials have shown the benefit of (-)deprenyl administered in relatively low doses (10 mg/day); in particular, it improves cognition, particularly episodic memory and learning. In animals (rats), there are also claims that (−)deprenyl can improve spatial memory and enhance cognition, although whether this is from the deprenyl or one of its metabolites (amphetamine) is not clear. A different MAO-B irreversible inhibitor, 2-pentyl N-methyl propargylamine, was without effect (13). The effects of (-)deprenyl on life span in rodents are controversial, with results varying from a dramatic increase in life expectancy in old rats to a small increase in average life span but not maximal life span in more recent studies (13). As Ruehl *et al.* convincingly show later in this section (p. 316), however, old dogs exhibiting cognitive dysfunction are a good model for human age-related cognitive decline, dementia, and Alzheimer's; lasting improvement followed from chronic (−)deprenyl (0.5 mg/kg/day) ingestion. In the veterinary area (−)deprenyl has now been successfully introduced as a treatment for Cushing's disease and cognitive decline (aging) in companion animals (pet dogs).

Considerable insight into the importance of MAO and its role in metabolism has followed from the discovery of some humans lacking MAO-A, MAO-B, or MAO-AB. Lenders *et al.* (p. 297) describe that the significance of MAO-B deletion seems somewhat inconsequential, causing only an increase in β-phenylethylamine excretion and no apparent disturbance in behavior. Deletion of MAO-A, on the other hand, is associated with considerable changes in the neurochemical profile as well as behavioral changes, such as borderline mental retardation along with impaired impulse control and stress-induced aggression. If both MAO-A and MAO-B are deleted, then this is a very severe condition accompanied by deletion of the closely associated Norrie gene and producing severe mental retardation. Surprisingly, however, even this severe deletion does not quickly end life, because the few subjects so far studied survive until young adulthood. What such studies are showing is that MAO-A seems to be more important than MAO-B. As more patients are found, their behavioral and metabolic profiles will be identified, and these, coupled with profiles obtained after enzyme inhibition studies and transgenic models in animals, will shed further light on the role(s) of MAO. In a transgenic mouse lacking MAO-A, Cases *et al.* (15) have shown that brain 5-hydroxytryptamine (5HT) metabolism is substantially changed (i.e., brain levels of 5HT markedly increased in young animals) and that this is accompanied by behavioral abnormalities (i.e., increased offensive aggression in adult animals) and cytoarchitectural changes in the somatosensory cortex. Ingestion of parachlorophenylalanine (i.e., a tryptophan hydroxylase inhibitor) reversed all the behavioral effects of the gene deletion.

The dynamic imaging studies, described here by Fowler and colleagues (p. 304), that use positron emission tomography and deuterated ^{11}C-labeled deprenyl ([^{11}C]L-deprenyl-D2) as ligand have confirmed that MAO-B increases with age, although the extent of this increase is less than postmortem studies had indicated. What is not yet known is whether this age-related increase is caused by increasing numbers of glia and astrocytes. The further application of this method has shown that there is a 40% decrease in MAO-B activity in the brains of cigarette smokers (16). Such a reduction is clinically significant and correlates with the reduction in platelet MAO-B seen in smokers (17). Because smoking is supposed to protect to some extent against Parkinson's disease and Alzheimer's (18), and schizophrenic and severely depressed patients are thought to self-medicate by smoking (19, 20), the role of MAO-B and its inhibition and neuroprotection in neuropsychiatric disorders is raised yet again. As shown by Berlin and colleagues (21), reversible MAO inhibitors (moclobemide, for example) may be useful in programs of smoking cessation. Of course, as suggested by the work of Oreland *et al.* (p 301), it is also possible that inhibition of MAO-B by tobacco smoke is related to changes in nicotine-induced dopaminergic neurotransmission. Another possibility is the condensation of biogenic amines with constituents of smoke or other endogenous metabolites to produce alkaloids such as the β-carbolines and isoquinolines (17, 22). Naoi *et al.* (23) have shown that R-N-methylsalsolinol is present in cerebrospinal fluid and is a dopaminergic neurotoxin capable of producing a syndrome resembling Parkinson's disease in rats.

(−)Deprenyl is effective as a monotherapy and as an adjunct treatment in Parkinson's disease; several expected as well as new mechanisms of action have been described (Table II). These include:

1. MAO blockade, which in primates and after chronic treatment inhibits both MAO-B and MAO-A and leads to an increase in the levels of dopamine and other amines. This alleviates some of the symptoms of parkinsonism. Of course if L-dopa is also being ingested, inhibition of astroglial MAO-B activity will also increase dopamine levels (11, 13).
2. Blockade of dopamine uptake, thus enhancing neurotransmission. Although D-deprenyl and its metabolites D-amphetamine and D-methamphetamine are more potent blockers of dopamine and norepinephrine uptake than L-deprenyl, the latter is still a potent

TABLE II Mechanisms of Action of (−)Deprenyl and Some Other MAO-B Inhibitors

Expected and known	*Unexpected and new*
Blockade of DA metabolism	Gene regulation
Blockade of DA uptake	Trophic effects
Amphetaminergic effects	Prevention of apoptosis
Potentiation of DA neurotransmission	Cell cycle effects
	Cell-cell interaction effects

DA, dopamine.

blocker and at much lower doses than is needed by other MAO-B blockers, such as the aliphatic N-methyl propargylamines (24).

3. Because both enantiomers of deprenyl are metabolized to amphetamine and methamphetamine, it has been argued that (−)deprenyl's beneficial effects may stem from amphetamine's well-known alerting and releasing properties. At the most commonly used human dose (10 mg/day), however, this seems unlikely in the case of the L(−)enantiomer.
4. Induction of superoxide dismutase and catalase, which would reduce oxidative stress, reactive oxygen species, and free radicals. This suggestion seems more doubtful because studies have failed to confirm earlier claims of upregulation (25).
5. Potentiation of dopaminergic and noradrenergic neurotransmission (26). An undisputed effect of MAO-B blockade is to increase brain β-phenylethylamine levels, it is now established that β-phenylethylamine is a neuromodulator and amplifies the synaptic actions of dopamine.

The other actions of (−)deprenyl, the aliphatic *N*-methylpropargylamines (27, 28), rasagiline (29), and some other compounds were unexpected. One of the first to be recognized was the upregulation of the mRNA of the biogenic amine synthetic enzyme aromatic L-amino acid decarboxylase (30). MAO-A inhibitors, reversible and irreversible, were without effect, as were the (+) enantiomers of these compounds. This was the first indication that (−)deprenyl possessed actions unrelated to MAO-B inhibition, because only the (−) form was active and PC12 cells do not contain MAO-B. Upregulation of superoxide dismutase-1 (SOD-1) (31) has also now been demonstrated, while the glial marker GFAP (32) and the low-affinity nerve growth factor receptor (p75) (26) are downregulated.

(−)Deprenyl has now been shown to be an antiapoptotic agent and to effect neuronal rescue in a variety of *in vitro* and *in vivo* models (see Table III with its associated references, as well Paterson and Tatton, p. 312 and Boulton *et al.*, p. 308). While these antiapoptotic actions are not fully understood, several details are becoming known; in PC12 cells, for example, the action of (−)deprenyl is dependent on gene transcription and protein synthesis. While apoptotic death after trophic withdrawal in PC12 cells is independent of protein synthesis, rescue by (−)deprenyl requires protein synthesis, and the antiapop-

TABLE III Experimental Paradigms in which (−)Deprenyl Exerts a MAO-B Independent Effect

In vivo (ref. no.)	*In vitro (ref. no.)*
Facial motor neuron axotomy (34, 35)	PC 12 cells (47)
Ischemia (36)	Cerebellar granule cells (48)
Gliosis (37)	Fibroblasts (49)
Aging (13)	Primary neuronal cells (50–52)
Kainate toxicity (38)	Gene expression (30–33)
NMDA antagonists (39)	Cell-cell interaction (53)
MPTP toxicity (40, 41)	Mitochondrial depolarization
DSP-4 toxicity (42–44)	Reactive astrocytes (54)
AF64A toxicity (45)	Process growth (55)
Optic nerve crush (46)	

NMDA, N-methyl-D-aspartate; MPTP, N-methyl-4-phenyl-1,2,3,6-tetrahydropyridine.

totic gene bcl-2 is increased and the proapoptotic gene bax is decreased (see Paterson and Tatton, p. 312). (−)Deprenyl also prevents the loss of mitochondrial function in these cells.

In cerebellar granule cells, apoptosis can be induced by lowering the K^+ concentration or by exposing the cells to cytosine arabinoside (ara C) (48). (−)Deprenyl and some other MAO-B inhibitors prevent the apoptotic death caused by ara C but not that caused by low K^+. This indicates an involvement of the tumor suppresser gene p53 because low K^+ apoptosis is p53 independent, while ara C–induced death is p53 dependent.

As described in the papers in this volume by Paterson and Tatton (p. 312) and Boulton and colleagues (p. 308), the upshot of all this is that (−)deprenyl and some other MAO-B inhibitors are turning out to be antiapoptotic drugs and neural rescue agents. They rescue in animal models of Parkinson's disease (i.e., MPTP model), amyotrophic lateral sclerosis (FMN model), stroke (unilateral global cerebral ischemia), and perhaps Alzheimer/epilepsy (kainate model). The *in vitro* paradigms are allowing a more detailed assessment of their mechanisms of action, including the loss of function in the cell and which proteins in the apoptotic cascade are involved, as well as involvement in the cell cycle. These developments seem to point to new clinical therapies for a variety of neurodegenerative conditions in which apoptotic death is involved. When the target(s) for these MAO-B inhibitors have been identified, structure–activity studies can follow along with the assessment of other rescue compounds.

II. O-Methylation and Conjugation

A. Catechol-O-Methyltransferase

O-methylation of catechol compounds, preferentially at the 3-hydroxy position, is catalyzed by catechol *O*-methyltransferase (COMT) with S-adenosyl-L-methionine as the methyl donor. Apart from the O-methylation of catecholamines and their deaminated metabolites, COMT is also involved in the metabolism of catecholestrogens, compounds implicated in carcinogenesis (Overview G).

Development of drugs to inhibit the enzyme as potential adjuncts for the treatment of Parkinson's disease has provided considerable impetus for recent advances in understanding the role of COMT in the catecholamine metabolism. Not only has this generated useful pharmacological tools to selectively block the enzyme, but the associated cloning of the human enzyme (55, 56) has led to important advances on several other fronts (57): (1) regulation of COMT gene expression (58), (2) enzyme structure and function (59, 60), (3) cellular and subcellular distribution (61–63), and (4) genotypic variation and influences on catecholamine metabolism (64–69).

The human COMT gene maps to chromosome 22q11.2 (70). Its transcription is regulated by two promoters that are responsible for two 1.5-kb and 1.3-kb mRNA species corresponding to two forms of COMT: (1) membrane bound COMT and (2) soluble COMT (56, 57). The latter form can also be formed from the 1.5-kb mRNA species that is responsible for the membrane-bound enzyme.

The chapter by Vidgren (p. 328) provides information about the three-dimensional structure of COMT derived from x-ray crystallography studies. This information shows how Mg^{2+} functions at the active site to help ionize the catechol hydroxy groups and how a lysine residue participates in the exchange of a proton for the methyl group from S-adenosyl-L-methionine. The findings provide a logical basis for design of new and selective inhibitors of the enzyme.

Polymorphisms associated with a thermolabile low-activity COMT and a thermostable high-activity COMT have been described (64–66). Several studies have suggested possible links between this polymorphism and neurological and psychiatric disorders (65–69). The possibility of an association of these polymorphisms with certain aspects of personality and behavior is strengthened by a study in patients with velocardiofacial syndrome, a disorder with deletions on chromosome 22q11 (69). Patients with this deletion and the low-activity variant of COMT appear more prone to some of the psychiatric disturbances of the disorder than patients with the high-activity allele.

It has also been suggested that COMT polymorphisms may influence response to COMT inhibitors as adjuncts in the treatment of Parkinson's disease (57, 66). As argued by Männistö (p. 324), the "second-generation" COMT inhibitors need not cross the blood brain barrier to increase the effectiveness of L-dopa in the treatment of Parkinson's disease. The preclinical data indicate that use of these drugs depends largely on inhibition of the peripheral O-methylation of the oral L-dopa and subsequent increased delivery to the brain.

The studies described by Nutt, (p. 331), comparing entacapone (primarily a peripheral COMT inhibitor) and tolcapone (both a central and peripheral COMT inhibitor), show that both drugs are equally effective in increasing the duration of the clinical response to L-dopa. Reduction in 3-O-methyldopa in patients does not materially affect L-dopa penetration across the blood brain barrier. It remains to be established whether inhibition of COMT in the brain provides additional therapeutic effects to those of peripheral inhibition.

B. Glucuronidases

Conjugation of catecholamines and their deaminated and O-methylated metabolites at the ring hydroxy positions with either a sulfate or a glucuronide residue leads to production of more easily excreted end-products. As described by Burchell and colleagues (p. 335), at least 20 different human cDNAs for UDP-glucuronosyltransferases (UGTs) have been identified and classified into two subfamilies according to sequence identities. Unlike the single-copy UGT-1 gene subfamily, the UGT-2 subfamily of isozymes appears to be encoded by independent genes. Wide species differences exist in the importance of glucuronidation for metabolism of catecholamines. The pathway does not appear important for catecholamine metabolism in humans, but as described by Tephly *et al.* (p. 343), it is important for metabolism of a wide variety of other endogenous compounds and drugs.

C. Sulphotransferases

In humans, catecholamines and their deaminated and O-methylated metabolites are sulfoconjugated, but the extent of sulfoconjugation varies considerably from compound to compound (Table IV). Dopamine and the O-methylated metanephrines are extensively sulfoconjugated, whereas norepinephrine and epinephrine are only moderately conjugated, and the metabolic end-products, homovanillic acid and VMA, are excreted almost entirely in the free form. As outlined by Coughtrie (p. 339), the thermolabile enzyme responsible for conjugation of catecholamines is a member of the phenolsulfotransferase (PST) family, known as monoamine-sulfating PST (M-PST). Although the various sulfotransferases are present in a wide variety of tissues, M-PST is mainly expressed in the small intestine, but hardly at all in the liver (71). This is consistent with other data showing substantial production of sulfoconjugates

TABLE IV Concentrations of Free and Sulfate-conjugated Catecholamines and Metabolites in Normal Human Plasma

Compound	*Free*	*Conjugated*	*% Conjugated*
Dopamine	0.08 ± 0.01	12.50 ± 2.00	99
Norepinephrine	1.38 ± 0.08	4.88 ± 0.43	78
Epinephrine	0.35 ± 0.11	0.70 ± 0.22	67
DOPAC	6.9 ± 0.03	3.3 ± 0.04	32
DHPG	5.53 ± 0.20	6.09 ± 0.33	52
Methoxytyramine	0.03 ± 0.01	2.8 ± 0.02	94
Normetanephrine	0.25 ± 0.02	9.13 ± 0.95	97
Metanephrine	0.20 ± 0.02	3.9 ± 0.47	95
MHPG	14.34 ± 0.54	25.55 ± 2.73	74
HVA	44.94 ± 2.34	0	0
VMA	33.65 ± 2.33	0	0

Data from Eisenhofer *et al. (2, 4).*

Results represent mean ± SEM concentrations. For abbreviations, see Figure 1.

by mesenteric organs, but not by the liver (4). Considerable production of dopamine sulfate by gastrointestinal tissues reflects the substantial amount of dopamine produced in these organs and is also consistent with other findings showing up to 100-fold increases in plasma dopamine sulfate after meals (Eisenhofer, unpublished results).

Three PST genes (including that for M-PST) have been localized to chromosome 16p12.1-p11.2 and were, at one stage, considered candidate genes for Batten disease (72–74). Polymorphisms of the PST locus have been identified (75). Considering the presence of M-PST in gastrointestinal tissues and the major source of sulfoconjugates from these tissues, it might be expected that genetic variants of the enzyme would be associated with disturbances in the digestive tract. Because M-PST is also in platelets, other possibilities include an involvement in migraine (76).

III. Catecholamine Metabolizing Systems

Compared with neuronal catecholamine transport and metabolizing systems, relatively little is known about the extraneuronal transport and metabolizing systems or the importance of these systems in terminating the actions of centrally and peripherally released catecholamines. Gründemann and coworkers (p. 346) describe one transporter, OCT1, initially cloned from rat kidney cDNA (77) that may also function in the liver as a transporter of organic cations, including catecholamines and serotonin. Furthermore, Martel (p. 350) describes experiments using isolated hepatocytes and perfused rat liver, indicating that the catecholamine transporter in rat liver is distinct from the classical neuronal (uptake-1) transporter and extraneuronal (uptake-2) transporters. Two transporters are suggested to be involved: the organic cation transporter OCT1 (78) and a P-glycoprotein transporter.

Removal of catecholamines in lungs is by a desipramine-sensitive process (79, 80) distinct from the classical uptake-2 process described by Iversen (81). In the report by Bryan-Lluka and colleagues (p. 353), pulmonary endothelial cells are described as having a transporter with the same properties as the neuronal norepinephrine transporter. However, in contrast to neurons, where deamination is the sole pathway of metabolism, in endothelial cells metabolism of catecholamines is by both O-methylation and deamination, with the former pathway predominating at low substrate concentrations.

Research about extraneuronal catecholamine uptake and metabolizing systems has been hindered by lack of appropriate pharmacological agents to block the membrane transport process. Identification of certain cyanine analogues as relatively selective blockers of this process has provided useful new tools for the study of extraneuronal uptake and metabolizing systems (82). Use of one of these compounds, disprocynium 24, by Graefe and colleagues (p. 360) to block extraneuronal uptake in anesthetized rabbits establishes the function and importance of this process for removal of circulating catecholamines. Whereas neuronal uptake predominates over extraneuronal uptake for removal of catecholamines before entry into the circulation, extraneuronal mechanisms are more important after entry of catecholamines into the circulation (3, 83, 84). This conclusion is consistent with the series arrangement of neuronal and extraneuronal transporters proposed by Trendelenburg to operate between sites of release and the circulation (85, 86). However, consideration also must be given to the importance of the liver and kidneys, which together remove over 70% of circulating catecholamines (87). As pointed out by Graefe *et al.*, (p. 360), blockade of renal and hepatic organic cation transporters in addition to uptake-2 are likely responsible for much of the changes in the disposition of catecholamines observed after administration to disprocynium 24 to experimental animals.

While neuronal transport and metabolizing systems predominate over extraneuronal systems for inactivation of norepinephrine released by sympathetic nerves, the relative roles of these systems in the brain have not been established. Schömig and coworkers (p. 356) present evidence based on uptake of the neurotoxin, 1-methyl-4-phenylpyridinium, that glial cells possess a monoamine transporter sensitive to selective blockers of extraneuronal uptake. This work not only raises the possibility that the extraneuronal transporter may be involved in the mechanism of action of some neurotoxins, but also suggests a potentially important role of glia in terminating the actions of catecholamines within the brain (88).

Catecholaminergic systems within the brain are considerably more difficult to study than those in the periphery, particularly when it comes to their investigation in humans. Past clinical studies of brain catecholaminergic systems relied heavily on measurements of catecholamine metabolites in plasma and urine. However, the brain contributes typically less than 10% to plasma levels of any one catecholamine metabolite (89, 90). Using gradients in metabolite levels from inflowing arterial to outflowing internal jugular plasma, Lambert and colleagues (p. 364) show that turnover of norepinephrine in the brain often parallels that in sympathetic nerves. This methodology provides a much more precise method of assessing neurotransmitter turnover in the human brain than was previously possible. The results raise questions about the validity of previous suggestions that central nervous system (CNS) noradrenergic systems have a generally inhibitory influence on the sympathetic nervous system.

In contrast to catecholamines released by neuronal systems, which exert their actions close to their sites of release, catecholamines released by the adrenal medulla exert their actions at distant sites. Thus, catecholamines released by nerves require inactivation close to their sites of release, whereas those released by the adrenals do not. Despite this, Tyce and colleagues (p. 370) show, using perfused dog adrenals, that considerable metabolism of catecholamines takes place within the adrenal gland. The pattern of metabolite production indicates that both MAO and COMT participate in adrenal catecholamine metabolism. In line with this, Eisenhofer and Lenders (p. 374) show considerable production of O-methylated catecholamine metabolites by human adrenals. In particular, over 90% of plasma metanephrine is derived from metabolism of epinephrine within the adrenals and very little from metabolism after release of epinephrine into the circulation (87). Substantial production of metanephrines within chromaffin tissue provides an explanation of why plasma metanephrines provide such a sensitive test for diagnosis of pheochromocytoma (91).

IV. Future Directions

Studies of patients with selective deficiencies of the MAO-B isozyme have shown that MAO-B is unimportant for catecholamine metabolism *in vivo* and raise the question about what function the enzyme has. This also has implications for the mechanism of action of MAO-B inhibitors as neuroprotective agents in the treatment of various neurodegenerative disorders; already, accumulated data indicate that the mode of action of MAO-B inhibitors is unrelated to inhibition of catecholamine metabolism. Future attempts to establish the mechanism of neuroprotection of MAO-B inhibitors will prove useful to understanding and developing new and more effective treatments for neurodegenerative disease processes.

The wealth of recent advances at molecular levels concerning *O*-methylation and conjugation has brought not only increased understanding of these enzymes, but also a new host of tools to explore their functional significance. For example, identification of the various members of the PST gene family has made it possible to identify which tissues express the isozyme, M-PST, which is responsible for catecholamine metabolism. This not only helps identify the source of catecholamine sulfoconjugates, but will also help elucidate the functional significance of M-PST in physiological systems and in clinical disorders involving catecholamine systems.

Cloning and identification of extraneuronal catecholamine transporters has lagged behind comparable research on neuronal membrane transporters. Perhaps this is partly because in the periphery these transporters do not appear as important as neuronal transporters in terminating the actions of neuronally released catecholamines. In the brain, however, the disposition of released catecholamines may be different than that in the periphery; thus, future advances in extraneuronal transporter and metabolizing systems could have tremendous importance for understanding the biological basis of psychiatric and neurological disorders as well as in the design of novel drugs for treatment of these conditions.

Progress in the genetic mapping of transporter proteins and metabolizing enzymes and identification of associated genetic variants and polymorphisms can be expected to have important implications for phenotyping of catecholamine systems. Genetic variants or polymorphisms in COMT and sulfoconjugating enzymes have been identified. Also, the existence of variants in MAO isozymes with more mild effects on deamination than those already identified seems likely. It remains to be established what influence these variants have on individual variations in catecholamine systems and determinants of personality and behavior or susceptibility to disease.

The use of transcerebral differences in plasma catecholamine metabolites offers tremendous scope for future advances in our understanding of CNS catecholamine systems in neurological and psychiatric disorders. This requires cooperation among investigators from various disciplines (e.g., psychiatrists, neurologists, and cardiologists for catheterizations), but if this can be achieved, the combination with CNS scanning techniques would provide enormous potential for clinical investigation of CNS function.

The advances described in this section illustrate how the area of catecholamine metabolism offers an ideal context for bridging basic and clinical areas of research. It can be anticipated that future investigation of catecholamine metabolism will continue to provide the conceptual mortar for cementing together integrative understanding of catecholamine systems in health and disease.

Acknowledgments

Alan A. Boulton was supported by Saskatchewan Health and Novartis Canada Ltd.

References

1. Goldstein, D. S., Eisenhofer, G., Stull, R., Folio, C. J., Keiser, H. R., and Kopin, I. J. (1988). Plasma dihydroxyphenylglycol and the intraneuronal disposition of norepinephrine in humans. *J. Clin. Invest.* **81,** 213–220.
2. Eisenhofer, G., Friberg, P., Rundqvist, B., Quyyumi, A. A., Lambert, G., Kaye, D. M., Kopin, I. J., Goldstein, D. S., and Esler, M. D. (1996). Cardiac sympathetic nerve function in congestive heart failure. *Circulation* **93,** 1667–1676.
3. Eisenhofer, G. (1994). Plasma normetanephrine for examination of extraneuronal uptake and metabolism of norepinephrine in rats. *Naunyn Schmiedebergs Arch. Pharmacol.* **349,** 259–269.
4. Eisenhofer, G., Åneman, A., Hooper, D., Rundqvist, B., and Friberg, P. (1996). Mesenteric organ production, hepatic metabolism, and renal elimination of norepinephrine and its metabolites in humans. *J. Neurochem.* **66,** 1565–1573.
5. Eisenhofer, G., Pecorella, W., Pacak, K., Hooper, D., Kopin, I.J., and Goldstein, D. S. (1994). The neuronal and extraneuronal origins of 3-methoxy-4-hydroxyphenylglycol. *J. Auton. Nerv. Syst.* **50,** 93–107.
6. Amine Oxidases and Their Impact on Neurobiology. (1990). (P. Riederer and M. B. H. Youdim, eds.). *J. Neural Transm. Suppl.* **32,** 1–491.
7. Amine Oxidases: Function and Dysfunction. (1992). (K. F. Tipton, M. B. H. Youdim, C. J. Barwell, B. A. Callingham, and G. A. Lyles, eds.), *J. Neural Transm. Suppl.* **41,** 1–457.
8. Inhibitors of Monoamine Oxidase B: Pharmacology and Clinical Use in Neurodegenerative Disorders. (1993). (I. Szelenyi, ed.), pp. 1–360. Birkhauser Verlag, Berlin.
9. Clinical Advances in Monoamine Oxidase Therapies. (1994). Ed (S. H. Kennedy, ed.), pp. 1–302. American Psychiatric Press, New York, NY.
10. Monoamine Oxidase Inhibitors in Neurological Diseases. (1994). (A. Lieberman, C. W. Olanow, M. B. H. Youdim, and K. Tipton, eds., pp. 1–371. Marcel Dekker, New York, NY.
11. Current Neurochemical and Pharmacological Aspects of Biogenic Amines. (1995). (P. H. Yu, K. F. Tipton, and A. A. Boulton, eds.), pp. 1–358. Elsevier, Amsterdam.
12. Berry, M. D., Juorio, A. V., and Paterson, I. A. (1994). The functional role of monoamine oxidase A and B in the mammalian central neuron system. *Prog. Neurobiol.* **42,** 375–391.
13. Berry, M. D., Juorio, A. V., and Paterson, I. A. (1994). Possible mechanisms of action of (−) deprenyl and other MAO-B inhibitors in some neurological and psychiatric disorders. *Prog. Neurobiol.* **44,** 141–161.
14. Youdim, M. B. H., and Riederer, P. (1997). Understanding Parkinson's disease. *Sci. Am.* **276,** 52–59.
15. Cases, O., Seif, I., Grimsby, J., Gaspar, P., Chen, K., Pourmin, S., Müller, U., Aguet, M., Babinett, C., Shih, J. C. and De Maeyer, E. (1995). Aggressive behaviour and altered

amounts of brain serotonin and norepinephrine in mice lacking MAO-A. *Science* **268,** 1763–1766.
16. Fowler, J. S., Volkow, N. P., Wang, G. J., Pappas, N., Logan, J., MacGregor, R. R., Alexoff, D., Shea, C. Schlyer, O., Wolf, A. P., Warner, D., Zezulkova, I., and Cilento, R. (1996). Inhibition of MAO-B in the brains of smokers. *Nature* **379,** 733–735.
17. Yu, P. H., and Boulton, A. A. (1987). Irreversible inhibition of monoamine oxidase by some components of cigarette smoke. *Life Sci.* **41,** 675–682.
18. Morens, D. M., Grandinetti, A., Reed, D., White, L. R., and Ross, G. W. (1995). Cigarette smoking and protection from Parkinson's disease: False association or etiologic clue? *Neurology* **45,** 1041–1051.
19. Glassman, A. H. (1993). Cigarette smoking: Implications for psychiatric illness. *Am. J. Psychiatry.* **150,** 546–553.
20. de Leon, J. (1996). Smoking and vulnerability for schizophrenia. *Schizophr. Bull.* **22,** 405–409.
21. Berlin, I., Said, S., Spreux-Varoquaux, O., and Launay, J. M. (1997). Monoamine oxidases and smoking. Potential therapeutic efficacy of a reversible monoamine oxidase inhibitor in smoking cessation. *Exp. Neurol.* (in press)
22. Durden, D. A., Davis, B. A., Yu, P. H., and Boulton, A. A. (1988). Formation of cyanomethyl-derivatives of biogenic amines from cigarette smoke: Their synthesis and identification. *In* Trace Amines: Comparative and Clinical Neurobiology. (A. A. Boulton, A. V. Juorio, and R. G. H. Downer, eds.), pp. 359–368. Humana Press, Toyota, New Jersey.
23. Naoi, M., Maruyama, W., and Dostert, P. (1997). An endogenous dopamine-derived N-methyl-(R)-salsolinol and Parkinsons disease: Animal model, biosynthesis and peripheral marker. *Exp. Neurol.* (in press)
24. Fang, J., and Yu, P. H. (1994). Effects of L-deprenyl, its structural analogues and some MAO inhibitors on dopamine uptake. *Neuropharmacology* **33,** 763–768.
25. Lai, C. T., Zuo, D-M., and Yu, P. H. (1994). Is brain superoxide dismutase activity increased following chronic treatment with L-deprenyl? *J. Neural Transm.* Suppl. **41,** 221–229.
26. Paterson, I. A., Juorio, A. V., and Boulton, A. A. (1990). 2-Phenylethylamine: A modulator of catecholamine transmission in the mammalian central nervous system? *J. Neurochem.* **55,** 1827–1837.
27. Li, X-M., Juorio, A. V., and Boulton, A. A. (1995). Some new mechanisms underlying the actions of (−)deprenyl: Possible relevance to neurodegeneration. *Prog. Brain Res.* **106,** 99–112.
28. Yu, P. H., Davis, B. A., Zhang, X., Zuo, D-M., Fang, J., Lai, C. T., Li, X-M., Paterson, I.A., and Boulton, A. A. (1995). Neurochemical, neuroprotective and neurorescue effects of aliphatic N-methylpropargylamines; new MAO-B inhibitors without amphetamine-like properties. *Prog. Brain Res.* **106,** 113–131.
29. Lamensdorf, I., and Finberg, F. M. (1997). Chronic MAO-B inhibition with (−)deprenyl or rasagiline (TVP-1012) elevates striatal dopamine release. *Exp. Neurol.* (in press).
30. Li, X-M., Juorio, A. V., Paterson, I. A., Zhu, M. Y., and Boulton, A. A. (1992). Specific, irreversible, MAO-B inhibitors stimulate gene expression of aromatic L-amino acid decarboxylase in PC12 cells. *J. Neurochem.* **59,** 2324–2327.
31. Li, X-M., Qi, J., Juorio, A. V., and Boulton, A. A. (1995). (−)Deprenyl enhances NGF-induced increase in superoxide dismutase mRNA in PC12 cells. *J. Neurochem.* **65,** (Suppl), S103D.l
32. Li, X-M., Qi, J., Juorio, A. V., and Boulton, A. A. (1993). Reduction in GFAP mRNA abundance induced by (−)deprenyl and other MAO-B inhibitors in C6 glioma cells. *J. Neurochem.* **61,** 1573–1576.
33. Li, X-M., Konradi, C. Qi, J., Juorio, A. V., and Boulton, A. A. (1996). (−)Deprenyl down-regulates the low affinity NGF receptor (LNGFR, p75) in PC 12 cells. *J. Neurochem.* **66,** (Suppl), S11A.

34. Salo, P. T., and Tatton, W. G. (1992). Deprenyl reduces the death of motoneurones caused by axotomy. *J. Neurosci. Res.* **31,** 394–400.
35. Ansari, K. S., Yu, P. H., Kruck, T. X., and Tatton, W. G. (1993). Death of axotomised immature rat facial motor neurones: Stereospecific rescue by R(−)deprenyl independently of MAO inhibition. *J. Neurosci.* **13,** 4042–4053.
36. Paterson, I. A., Barber, A. J., Gelowitz, D. L., and Voll, C. L. (1997). (−)Deprenyl reduces delayed neuronal death of hippocampal pyramidal cells. *Neurosci. Biobehav. Rev.* **21,** 181–186.
37. Biagini, G., Zoli, M., Fuxe, K., and Agnati, L. F. (1993). L-Deprenyl increases GFAP immunoreactivity selectively in activated astrocytes in rat brain. *Neuroreport* **4,** 955–958.
38. Gelowitz, D. I., and Paterson, I. A. (1995). Deprenyl enhances functional recovery following kainic acid-induced neuronal death. *Soc. Neurosci. Abs.* **21,** 3269.
39. Zhang, X., Boulton, A. A., Zuo, D-M., and Yu, P. H. (1996). MK-801 induces neuronal death in rat retrosplenial cortex: Prevention by cycloheximide and R(−)-2-Hexyl-N-methyl propargylamine. *J. Neurosci. Res.* **46,** 82–89.
40. Tatton, W. G., and Greenwood, C. E. (1991). Rescue of dying neurons: A new action for deprenyl in MPTP parkinsonism. *J. Neurosci. Res.* **30,** 666–672.
41. Yu, P. H., Davis, B. A., Durden, D. A., Barber, A., Terleckyj, I., Nicklas, W. G., and Boulton, A. A. (1994). Neurochemical and neuroprotective effects of some aliphatic propargylamines: New selective nonamphetamine-like MAO-B inhibitors. *J. Neurochem.* **62,** 697–704.
42. Bertocci, B., Gill, G., and DaPrada, M. (1988). Prevention of the DSP-4-induced noradrenergic neurotoxicity by irreversible, not by reversible, MAO-B inhibitors. *Pharmacol. Res. Commun.* **20,** 131–132.
43. Finnegan, K. T., Skratt, J. S., Irwin, I., Delanney, L. E., and Langston, J-W. (1990). Protection against DSP-4-induced neurotoxicity by deprenyl is not related to its inhibition of MAO-B. *Eur. J. Pharmacol.* **184,** 119–126.
44. Yu, P. H., Davis, B. A., Fang, J., and Boulton, A. A. (1994). Neuroprotective effects of some MAO-B inhibitors against DSP-4-induced noradrenaline depletion in the mouse hippocampus. *J. Neurochem.* **63,** 1820–1828.
45. Ricci, A., Mancini, M., Strocchi, P., Bongrani, S., and Bronzetti, E. (1992). Deficits in cholinergic neurotransmission markers induced by ethylcholine mustard aziridinium (AF64A) in the rat hippocampus: Sensitivity to treatment with the MAO-B inhibitor L-deprenyl. *Drugs Exp. Clin. Res.* **18,** 163–171.
46. Buys, Y. M., Trope, G. M., and Tatton, W. G. (1995). (−)Deprenyl increases the survival of rat retinal ganglion cells after optic nerve crush. *Curr. Eye Res.* **14,** 119–126.
47. Tatton, W. G., Ju, W. Y. L., Holland, D. P., Tai, C., and Kwan, M. (1994). (−)Deprenyl reduces PC12 cell apoptosis by inducing new protein synthesis. *J. Neurochem.* **63,** 1572–1575.
48. Paterson, I. A., Warrington, R., and Boulton, A. A. (1997). R-Deprenyl and R-2-heptylmethylpropargylamine prevent p53-dependent but not p53-indepdendent apoptosis. ISN/ASN Meeting Abstract, Boston.
49. Skibo, G., Ahmed, I., Yu, P. H., Boulton, A. A., and Federoff, S. (1992). L-Deprenyl, a MAO-B inhibitor, acts on the astroglia cell cycle and the G1/G0 boundary. Abstracts of the Annual Meeting of the Society of Cellular Biologists.
50. Roy, E., and Bedard, P. J. (1993). Deprenyl increases survial of rat foetal nigral neurones in culture. *Neuroreport* **4,** 1183–1186.
51. Mytilineou, C., Radcliffe, P. M., Leonardi, E. K., Werner, P., and Olanow, W. C. (1997). L-Deprenyl protects mesencephalic dopamine neurons from glutamate receptor-mediated toxicity in vitro. *J. Neurochem.* **68,** 33–39.
52. Mytilineou, C., Radcliffe, P.,M., and Olanow, W. C. (1997). L-(−)-Desmethylselegiline, a metabolite of selegiline(L-(−)-deprenyl) protects mesencephalic dopamine neurons from excitotoxicity in vitro. *J. Neurochem.* **68,** 434–436.

53. Koutsilieri, E., O'Callaghan, J. F. X., Chen, T-S., Riederer, P., and Rausch, W-D. (1994). Selegiline enhances survival and neurite outgrowth of MPP+−treated dopaminergic neurons. *Eur. J. Pharmacol.* **269,** R3–R4.
54. Seniuk, N. A., Henderson, J. T., Tatton, W. G., and Roder, J. C. (1994). Increased CNTF gene expression in process-bearing astrocytes following injury is augmented by L(−)Deprenyl. *J. Neurosci. Res.* **37,** 278–286.
55. Lundström, K., Salminen, M., Jalanko, A., Savolainen, R., and Ulmanen, I. (1991). Cloning and characterization of human placental catechol-O-methyltransferase. *DNA Cell Biol.* **10,** 181–189.
56. Tenhunen, J., Salminen, M., Lundstrom, K., Kiviluoto, T., Savolainen, R., and Ulmanen, I. (1994). Genomic organization of the human catechol O-methyltransferase gene and its expression from two distinct promoters. *Eur. J. Biochem.* **223,** 1049–1059.
57. Lundstrom, K., Tenhunen, J., Tilgmann, C., Karhunen, T., Panula, P., and Ulmanen, I. (1995). Cloning, expression and structure of catechol-O-methyltransferase. *Biochim. Biophys. Acta* **1251,** 1–10.
58. Tenhunen, J. (1996). Characterization of the rat catechol-O-methyltransferase gene proximal promoter: Identification of a nuclear protein-DNA interaction that contributes to the tissue-specific regulation. *DNA Cell Biol.* **15,** 461–473.
59. Vidgren, J., Svensson, L. A., and Liljas, A. (1994). Crystal structure of catechol O-methyltransferase. *Nature* **368,** 354–358.
60. Lotta, T., Vidgren, J., Tilgmann, C., Ulmanen, I., Melen, K., Julkunen, I., and Taskinen, J. (1995). Kinetics of human soluble and membrane-bound catechol-O-methyltransferase: A revised mechanism and description of the thermolabile variant of the enzyme. *Biochemistry* **34,** 4202–4210.
61. Karhunen, T., Tilgmann, C., Ulmanen, I., Julkunen, I., and Panula, P. (1994). Distribution of catechol-O-methyltransferase enzyme in rat tissues. *J. Histochem. Cytochem.* **42,** 1079–1090.
62. Kastner, A., Anglade, P., Bounaix, C., Damier, P., Javoy-Agid, F., Bromet, N., Agid, Y., and Hirsch, E. C. (1994). Immunohistochemical study of catechol-O-methyltransferase in the human mesostriatal system. *Neuroscience* **62,** 449–457.
63. Karhunen, T., Tilgmann, C., Ulmanen, I., and Panula, P. (1995). Catechol-O-methyltransferase (COMT) in rat brain: Immunoelectron microscopic study with an antiserum against rat recombinant COMT protein. *Neurosci. Lett.* **187,** 57–60.
64. Boudikova, B., Szumlanski, C., Maidak, B., and Weinshilboum, R. (1990). Human liver catechol-O-methyltransferase pharmacogenetics. *Clin. Pharmacol. Ther.* **48,** 381–389.
65. Saito, T., Yu, Y. M., Szumlanski, C. L., and Weinshilboum, R. M. (1996). Human catechol-O-methyltransferase pharmacogenetics: Description of a functional polymorphism and its potential application to neuropsychiatric disorders. *Pharmacogenetics* **6,** 243–250.
66. Syvänen, A-C., Tilgmann, C., Rinne, J., and Ulmanen, I. (1997). Genetic polymorphism of catechol-O-methyltransferase (COMT): Correlation of genotype with individual variation of S-COMT activity and comparison of the allele frequencies in the normal population and Parkinsonian patients in Finland. *Pharmacogenetics* **6,** (in press)
67. Gutierrez, B., Bertranpetit, J., Guillamat, R., Valles, V., Arranz, M. J., Kerwin, R., and Fananas, L. (1997). Association analysis of the catechol O-methyltransferase gene and bipolar affective disorder. *Am. J. Psychiatry* **154,** 113–115.
68. Sham, P. C., Vallada, H., Xie, T., Tang, X., Murray, R. M., Liu, X., and Collier, D. A. (1996). Preferential transmission of the high activity allele of COMT in schizophrenia. *Psychiatr. Genet.* **6,** 131–133.
69. Lachman, H. M., Morrow, B., Shprintzen, R., Veit, S., Parsia, S. S., Faedda, G., Goldberg, R., Kucherlapati, R., and Papolos, D. F. (1996). Association of codon 108/158 catechol-O-methyltransferase gene polymorphism with the psychiatric manifestations of velo-cardio-facial syndrome. *Am. J. Med. Genet.* **67,** 468–472.

70. Winqvist, R., Lundstrom, K., Salminen, M., Laatikainen, M., and Ulmanen, I. (1992). The human catechol-O-methyltransferase (COMT) gene maps to band q11.2 of chromosome 22 and shows a frequent RFLP with BglI. *Cytogenet. Cell. Genet.* **59,** 253–257.
71. Rubin, G. L., Sharp, S., Jones, A. L., Glatt, H., Mills, J. A., and Coughtrie, M. W. H. (1996). Design, production and characterization of antibodies discriminating between the phenol- and monoamine-sulphating forms of human phenol sulphotransferase. *Xenobiotica* **26,** 1113–1119.
72. Dooley, T. P., and Huang, Z. (1996). Genomic organization and DNA sequences of two human phenol sulfotransferase genes (STP1 and STP2) on the short arm of chromosome 16. *Biochem. Biophys. Res. Commun.* **228,** 134–140.
73. Dooley, T. P., Probst, P., Obermoeller, R. D., Siciliano, M. J., Doggett, N. A., Callen, D. F., Mitchison, H. M., and Mole, S. E. (1995). Phenol sulfotransferases: Candidate genes for Batten disease. *Am. J. Med. Genet.* **57,** 327–332.
74. Taschner, P. E., de Vos, N., Thompson, A. D., Callen, D. F., Doggett, N., Mole, S. E., Dooley, T. P., Barth, P. G., and Breuning, M. H. (1995). Chromosome 16 microdeletion in a patient with juvenile neuronal ceroid lipofuscinosis (Batten disease). *Am. J. Hum. Genet.* **56,** 663–668.
75. Henkel, R. D., Galindo, L. V., and Dooley, T. P. (1995). Detection of a HindIII restriction fragment length polymorphism in the human phenol sulfotransferase (STP) locus. *Hum. Genet.* **95,** 245–246.
76. Jones, A. L., Roberts, R. C., Colvin, D. W., Rubin, G. L., and Coughtrie, M. W. (1995). Reduced platelet phenolsulphotransferase activity towards dopamine and 5-hydroxytryptamine in migraine. *Eur. J. Clin. Pharmacol.* **49,** 109–114.
77. Gründemann, D., Gorboulev, V., Gambaryan, S., Vehyl, M., and Koepsell, H. (1994). Drug excretion mediated by a new prototype of polyspecific transporter. *Nature* **372,** 549–552.
78. Martel, F., Vetter, T., Russ, H., Grundemann, D., Azevedo, I., Koepsell, H., and Schomig, E. (1996). Transport of small organic cations in the rat liver. The role of the organic cation transporter OCT1. *Naunyn Schmiedebergs Arch. Pharmacol.* **354,** 320–326.
79. Bryan-Lluka, L. J., Westwood, N. N., and O'Donnell, S. R. (1992). Vascular uptake of catecholamines in perfused lungs of the rat occurs by the same process as uptake$_1$ in noradrenergic neurons. *Naunyn Schmiedebergs Arch. Pharmacol.* **345,** 319–326.
80. Eisenhofer, G., Smolich, J. J., and Esler, M. D. (1992). Different desipramine-sensitive pulmonary removals of plasma epinephrine and norepinephrine in dogs. *Am. J. Physiol.* **262,** L360–L365.
81. Iversen, L. L. (1965). The uptake of catecholamines at high perfusion concentrations in the rat isolated heart: A novel catecholamine uptake process. *Br. J. Pharmacol.* **25,** 18–33.
82. Russ, H., Engel, W., and Schömig, E. (1993). Isocyanines and pseudoisocyanines as a novel class of potent noradrenaline transport inhibitors: Synthesis, detection, and biological activity. *J. Med. Chem.* **36,** 4208–4213.
83. Friedgen, B., Wolfel, R., Russ, H., Schomig, E., and Graefe, K. H. (1996). The role of extraneuronal amine transport systems for the removal of extracellular catecholamines in the rabbit. *Naunyn Schmiedebergs Arch. Pharmacol.* **354,** 275–286.
84. Eisenhofer, G., McCarty, R., Pacak, K., Russ, H., and Schomig, E. (1996). Disprocynium24, a novel inhibitor of the extraneuronal monoamine transporter, has potent effects on the inactivation of circulating noradrenaline and adrenaline in conscious rat. *Naunyn Schmiedebergs Arch. Pharmacol.* **354,** 287–294.
85. Trendelenburg, U. (1986). The metabolizing system involved in the inactivation of catecholamines. *Naunyn Schmiedebergs Arch. Pharmacol.* **332,** 201–207.
86. Trendelenburg, U. (1990). The interaction of transport mechanisms and intracellular enzymes in metabolizing systems. *J. Neural Transm.* Suppl. **32,** 3–18.
87. Eisenhofer, G., Rundquist, B., Aneman, A., Friberg, P., Dakak, N., Kopin, I. J., Jacobs, M. C., and Lenders, J. W. (1995). Regional release and removal of catecholamines and extraneuronal metabolism to metanephrines. *J. Clin. Endocrinol. Metab.* **80,** 3009–3017.

88. Russ, H., Staust, K., Martel, F., Gliese, M., and Schomig, E. (1996). The extraneuronal transporter for monoamine transmitters exists in cells derived from human central nervous system glia. *Eur. J. Neurosci.* **8**, 1256–1264.
89. Lambert, G. W., Eisenhofer, G., Cox, H. S., Horne, M., Kalff, V., Kelly, M., Jennings, G. L., and Esler, M. D. (1991). Direct determination of homovanillic acid release from the human brain, an indicator of central dopaminergic activity. *Life Sci.* **49**, 1061–1072.
90. Lambert, G. W., Kaye, D. M., Vaz, M., Cox, H. S., Turner, A. G., Jennings, G. L., and Esler, M. D. (1995). Regional origins of 3-methoxy-4-hydroxyphenylglycol in plasma: Effects of chronic sympathetic nervous activation and denervation, and acute reflex sympathetic stimulation. *J. Auton. Nerv. Syst.* **55**, 169–178.
91. Lenders, J. W., Keiser, H. R., Goldstein, D. S., Willemsen, J. J., Friberg, P., Jacobs, M. C., Kloppenborg, P. W., Thien, T., and Eisenhofer, G. (1995). Plasma metanephrines in the diagnosis of pheochromocytoma. *Ann. Intern. Med.* **123**, 101–109.

Kevin Chen and Jean Chen Shih

Department of Molecular Pharmacology and Toxicology
School of Pharmacy
University of Southern California
Los Angeles, California 90033

Monoamine Oxidase A and B: Structure, Function, and Behavior

Monoamine oxidase (MAO) A and B (amine:oxygen oxidoreductase [deaminating], EC 1.4.3.4) are key isoenzymes that degrade biogenic and dietary monoamines (1). They also oxidize xenobiotics, including the parkinsonism-producing neurotoxin MPTP (1-methyl-4-phenyl-1,2,3,6-tetrahydropyridine). These isoenzymes are integral proteins of the outer mitochondrial membrane and are distinguished by differences in substrate preferences, inhibitor specificities, tissue and cell distribution, and immunological properties. MAO-A preferentially oxidizes serotonin, norepinephrine, and epinephrine and is inactivated by the irreversible inhibitor clorgyline. MAO-B oxidizes phenylethylamine and benzylamine and is inactivated by the irreversible inhibitors pargyline and deprenyl; dopamine, tyramine, and tryptamine are oxidized by both forms. Placental tissue contains predominantly MAO-A, whereas platelets and lympho-

Advances in Pharmacology, Volume 42

1054-3589/98 $25.00

cytes express only MAO-B. Most human tissues express both forms of the enzyme.

I. The Cloning of Human Liver MAO-A and MAO-B Has Unequivocally Demonstrated That the Two Forms of the Enzyme Are Made of Different Polypeptides and Are Coded for by Different Genes

The application of recombinant DNA techniques has provided exciting and insightful information concerning the structure and function of MAO-A and MAO-B. Two oligonucleotide probes designed from the amino acid sequences of purified human liver MAO-B fragments were used to screen a human liver cDNA library. A full-length cDNA clone (2.5 kb) encoding a subunit of human MAO-B was isolated. Subsequently, an oligonucleotide probe designed from a peptide fragment of human placental MAO-A, unique for MAO-A, was used in conjunction with an internal fragment of MAO-B cDNA to screen the same cDNA library. A full-length cDNA clone (2.1 kb) encoding human MAO-A was isolated (2). These are the first full-length cDNAs for MAO-A and MAO-B to be cloned. Comparison of the deduced amino acid sequences of the human liver MAO-A and MAO-B shows approximately a 70% identity. Expression of functional enzymes by transient transfection of the cDNAs provides unequivocal evidence that the different catalytic activities of MAO-A and MAO-B reside in their primary amino acid sequences. This work has answered a question that has puzzled scientists since the discovery of these two forms of the enzyme in 1968, namely, are they different proteins? Or the same protein differentially modified by carbohydrates or lipids?

II. MAO-A and MAO-B Are Located on the X-Chromosome, Made of 15 Exons with Identical Intron–Exon Organization, Suggesting They Are Derived from a Common Ancestral Gene

Using the MAO-A and MAO-B cDNA probes, we have mapped the chomosomal locations of both genes. They are located on the X-chromosome in humans and mice. In humans, they are closely located between bands Xp11.23 and Xp22.1 and are deleted in patients with Norrie's disease, a rare X-linked recessive neurologic disorder characterized by blindness, hearing loss, and mental retardation. Both genes were isolated from X-chromosome–specific libraries by utilizing MAO-A and MAO-B cDNA specific fragments. They were found to consist of 15 exons and exhibited identical exon–intron organization. Exon 12 codes for the covalent flavin adenine dinucleotide (FAD)–binding site and is the most conserved exon, sharing 93.9% amino acid identity between MAO-A and MAO-B. These results suggest that MAO-A and MAO-B are derived from duplication of a common ancestral gene (3).

III. The Organization of MAO-A and MAO-B Promotor Are Distinctly Different, Even Though They Share 60% Sequence Identity

The promoter regions of human MAO-A and MAO-B genes have been characterized using a series of 5′ flanking sequences linked to a human growth hormone reporter gene and transfected into both human and animal cells. The maximal promoter activity for MAO-A was found in a 0.14-kb PvuII/DraII fragment (A 0.14) and in a 0.15-kb PstI/NaeI fragment (B 0.15). Both fragments are GC-rich, containing potential Sp1-binding sites, and share approximately 60% sequence identity. However, the organization of the transcription elements is distinctly different between the two (4). MAO-A 0.14 fragment lacks a TATA box, consists of three Sp1 elements, and exhibits bidirectional promoter activity. MAO-B promoter fragment B 0.15 consists of two clusters of overlapping SP1 sites separated by CACCC element. The different promoter organization of MAO-A and MAO-B genes provides the basis for their different tissue- and cell-specific expression.

IV. There Are Two Cysteines in MAO-A and Three Cysteines in MAO-B That Are Important for Their Catalytic Activity

There are nine cysteine residues in the deduced amino acid sequences of both MAO-A and MAO-B in human liver. Their role in MAO-A and MAO-B catalytic activities was studied by site-directed mutagenesis. The wild-type and mutant cDNAs were transiently transfected into COS cells and assayed for MAO-A and MAO-B catalytic activity. The catalytic activities of seven MAO-A mutants and in six MAO-B mutants were similar to those of the wild-type enzymes, proving that these cysteines are not necessary for enzymatic activity. However, substitution of MAO-A Cys 374 and 406 and MAO-B Cys 156, 365, and 397 with serine resulted in complete loss of catalytic activity, which was not due to unsuccessful transfection, as indicated by Western blot analysis. The loss of catalytic activity in the MAO-A Cys 406 and MAO-B Cys 397 mutants is probably due to the prevention of covalent binding of the enzyme to the necessary FAD cofactor. The loss of catalytic activity of MAO-A Cys 374 and MAO-B Cys 156 and 365 suggests that these cysteines are important for catalytic activity, but whether they are involved in forming the active site or are required to maintain the appropriate conformation of MAO-A and MAO-B remains a future study.

V. The C-Terminus of Only MAO-B Is Critical for Maintaining Its Catalytic Activity

To determine which regions of the MAOs confer the substrate and inhibitor selectivities, we have constructed chimeric MAO enzymes by reciprocally exchanging the corresponding N-terminals and C-terminals of MAO-A and MAO-B and then determining the catalytic properties of these chimeric enzymes. The chimerics were transfected into mammalian COS cells, and MAO-A and MAO-B catalytic activities were determined using 5HT or PEA as a substrate. Our results show that switching the N-terminals has no effect on either MAO-A or MAO-B catalytic activity. Interestingly, when MAO-B C-terminal residues 393–520 were replaced with MAO-A C-terminal residues 402–527, catalytic activity was not detectable. The lack of catalytic activity was not due to non-expression of the enzyme as shown by Western analysis. When the C-terminus of MAO-A was switched with MAO-B, little effect was observed on MAO-A catalytic activity. These results suggests that the C-terminal of MAO-B but not MAO-A is critical for maintaining the enzyme in an active form.

VI. Replacement of MAO-A Amino Acids 161–375 with the Corresponding Region of MAO-B Results in a Change of Catalytic Properties That Resembles Those of MAO-B

To determine the role of the internal regions on the substrate and inhibitor selectivities, two chimerics were made by exchanging the segments of amino acids 161–375 of MAO-A with the corresponding regions of MAO-B (amino acids 152–366). Our results show that replacement of MAO-A amino acids 161–375 by the corresponding region of MAO-B, termed AB161-375A, converted MAO-A catalytic properties to ones typical of MAO-B. Serotonin (5HT), a preferred substrate for MAO-A, was not oxidized by AB161-375A nor by wild-type MAO-B. Furthermore, AB161-375 A was more sensitive to the MAO-B–specific inhibitor deprenyl than to the MAO-A–specific inhibitor clorgyline. However, the reciprocal chimera in which a MAO-B segment was replaced with the corresponding region of MAO-A, termed BA152-366B, lacked catalytic activity. The lack of catalytic activity was not due to aberrant expression but rather to an inactive protein, as demonstrated by Western blot analysis. These results demonstrate that MAO-B amino acids 152–366 contain a domain that confers substrate and inhibitor selectivity.

VII. MAO-A–Deficient Mice Exhibit Increased Serotonin Levels and Aggressive Behavior

In collaboration with Drs. Isabelle Seif and Edward De Maeyer at the Centre National de la Recherche Scientifique, Institut Curie, Orsay, France, we have discovered a line of transgenic mice in which an interferon (INF-β) transgene was integrated into the location of MAO-A gene, resulting in MAO-A–deficient mice (5). Interestingly, serotonin concentrations in MAO-A–deficient pup brains were increased up to ninefold, compared with wild-type. However, in adult brains, the serotonin levels were increased only twofold when compared with wild-type due to the appearance of MAO-B, which was not present in the pups. In pup and adult brains, norepinephrine concentrations were reversed by the serotonin synthesis inhibitor parachlorophenylalanine. Adults manifested a distinct behavioral syndrome, including enhanced aggression in males. These mice are a valuable model for studying the role of monoamine neurotransmitters in aggressive behavior.

Acknowledgments

This work was supported by grant R01 MH37020, R37 MH39085 (Merit Award), and Research Scientist Award K05 MH00796, from the National Institute of Mental Health. Support from the Boyd and Elsie Welin Professorship is also appreciated.

References

1. Shih, J. C. (1990). Molecular basis of human MAO A and B, an invited review article. *Neuropsychopharmacology* **4,** 1–7.
2. Bach, A. W. J., Lan, N. C., Johnson, D. L., Abell, C. W., Bemkenek, M. E., Kwan, S. W., Seeburg, P. H., and Shih, J. C. (1988). CDNA cloning of human liver MAO A and B: Molecular basis of differences in enzymatic properties. *Proc. Natl. Acad. Sci. U.S.A.* **85,** 4934–4938.
3. Grimsby, J., Chen, K., Wang, L. J., Lan, N. C., and Shih, J. C. (1991). Human MAO A and B genes exhibit identical exon itron organization. *Proc. Natl. Acad. Sci. U.S.A.* **88,** 3637–3641.
4. Zhu, Q-S., Grimsby, J., Chen, K., and Shih, J. C. (1992). Promoter organization of human MAO A and B genes. *J. Neurosci.* **12,** 4437–4446.
5. Cases, O., Seif, I., Grimsby, J., Gasper, P., Chen, K., Pournin, S., Muller, U., Aguet, M., Babinet, C., Shih, J. C., and De Maeyer, E. (1995). Aggressive behavior and altered amounts of brain serotonin and norepinephrine in mice lacking MAO A. *Science* **268,** 1763–1766.

Jacques W. M. Lenders,* Han G. Brunner,*
Dennis L. Murphy,† and Graeme Eisenhofer‡

*Department of Internal Medicine and Department of Human Genetics
St. Radboud University Hospital
6525 GA Nijmegen, The Netherlands

†Laboratory of Clinical Neuroscience/NIMH
‡Clinical Neuroscience Branch/NINDS
National Institutes of Health
Bethesda, Maryland 20892

Genetic Deficiencies of Monoamine Oxidase Enzymes: A Key to Understanding the Function of the Enzymes in Humans

Monoamine oxidase (MAO) plays a pivotal role in the oxidative deamination of catecholamines, serotonin, and the trace amines, phenylethylamine and tyramine. Pharmacological inhibition of MAO in animals and humans has profound neurophysiological effects by modulating neurotransmitter function. Consequently, it has been assumed that genetically determined variations of MAO activity might be associated with behavioral and mood disorders. Although the different substrate and inhibitor specificities of the two isoenzymes, MAO-A and MAO-B, are well documented (1, 2), their relative roles for the metabolism of catecholamines *in vivo* are less established.

The metabolism of catecholamines by MAO *in vivo* in humans can be examined in several ways. First, it is possible to block MAO pharmacologically by specific inhibitors. The major limitation of this approach is that there are no inhibitors of MAO that are absolutely specific for one isoenzyme over the other. A second approach is to examine the metabolic fate of radiolabeled catecholamines and their metabolites. However, the use of radiotracers in humans is limited. A third possibility is to examine the catecholamines and their metabolites in subjects who have a genetically determined absence of one or both MAO isoenzymes. Knowledge of the specific patterns of metabolites associated with deficiency of a specific (iso-) enzyme might offer diagnostic possibilities for detection of patients with genetic disorders of biogenic amine metabolism associated with psychiatric disturbances.

We have described several subjects with genetically determined deficiencies of MAO-A and MAO-B and a combined deficiency of MAO-A and MAO-B

Advances in Pharmacology, Volume 42

1054-3589/98 $25.00

(MAO-AB) (3–6). In the subjects with the selective deficiency of MAO-A, the clinical phenotype is characterized by borderline mental retardation in combination with impaired impulse control and stress-induced aggression. Tyramine sensitivity was increased in these subjects because the pressor dose of tyramine was less than 1/10 of that required in subjects with depression but comparable to that in patients treated with selective MAO-A inhibitors. Plasma concentrations of free and sulfate-conjugated normetanephrine (NMN) and sulfate-conjugated metanephrine (MN) and the 5-hydroxytryptamine (5HT) content of platelets were increased considerably in these subjects, whereas plasma concentrations of catecholamines were within the normal range. Plasma concentrations of free MN were only slightly increased above normal, compared with the more highly elevated plasma concentrations of free NMN and sulfate-conjugated MN. Studies in rats have also shown that inhibition of MAO resulted in larger elevations of plasma NMN than of plasma MN (7). As demonstrated previously, the contribution of the adrenal glands to the plasma concentration of NMN is much less than their contribution to the plasma concentration of MN (40% vs 90%) (8), indicating that deamination does not play an important role in the degradation of catecholamines within the adrenal glands. It has not been resolved whether the greater increases in sulfate-conjugated than free MN, induced by MAO inhibition, have to be attributed to a reduced sulfate-conjugation of MN formed from epinephrine (Epi) outside the adrenals or to a reduced extra-adrenal deamination of sulfate-conjugated MN. Plasma concentrations of the deaminated metabolites dihydroxyphenylglycol (DHPG), methoxyhydroxyphenylglycol (MHPG), vanillylmandelic acid (VMA), dihydroxyphenylacetic acid (DOPAC), and homovanillic acid (HVA) are strongly reduced in the MAO-A–deficient subjects. A pattern similar to that in plasma was seen in urine: increased excretion of 5HT, NMN, and 3-methoxytyramine and decreased excretion of MHPG, VMA, and HVA. Excretion of all trace amines, however, was within normal limits, indicating that MAO-A is not essential for their metabolism in the presence of MAO-B.

In contrast to the findings in MAO-A–deficient subjects, MAO-B deficiency is associated with neither abnormal behavior nor mental retardation. The neurochemical alterations in these subjects were also much less severe than in the MAO-A deficient subjects. Plasma catecholamine levels and the 5HT content of platelets were normal, as were also the O-methylated and deaminated metabolites of the catecholamines. The sole abnormal finding in these subjects was an elevated urinary excretion of phenylethylamine with a normal excretion of tyramine. Together with the normal phenylethylamine excretion in the MAO-A–deficient subjects, our data suggest that it is unlikely that phenylethylamine is involved in abnormal behavior.

The subjects with the combined deficiency of MAO-A and MAO-B had Norrie's disease and were severely mentally retarded. Tyramine sensitivity was even more pronounced than in the selective MAO-A–deficient subjects. Despite the complete absence of MAO-A and MAO-B, the plasma norepinephrine (NE) and Epi levels were normal while the plasma dopamine level was slightly increased. This does suggest that dopamine is a substrate for both isoenzymes, although MAO-A seems to be more important than MAO-B. Despite the normal plasma catecholamine concentrations, the MAO-AB–deficient subjects showed a qualitatively similar but somewhat more pronounced pattern of metabolic alterations than the selective MAO-A–deficient subjects: the O-methylated metabolites were elevated similarly, and the deaminated metabolites were more depressed. These findings from the MAO-A–, MAO-B–, and MAO-AB–deficient subjects indicate that MAO-A is considerably more important than MAO-B for metabolism of NE, Epi, and their O-methylated amine metabolites *in vivo*. In addition, it is apparent that the two isoenzymes do not have equal complementary capacities for the deamination of catecholamines when one of the two isoenzymes is lacking. These findings agree also with those in rats (7), in which pharmacological inhibition of MAO-B failed to affect either the deaminated or O-methylated amine metabolites of NE and Epi, whereas inhibition of MAO-A resulted in similar directional changes to those observed here. Normal urinary and platelet levels of 5HT in MAO-B–deficient subjects and substantial increases in selective MAO-A– and combined MAO-AB–deficient subjects indicate that MAO-B is less important than MAO-A for the *in vivo* metabolism of 5HT. Phenylethylamine excretion was much more elevated in the MAO-AB–deficient than in the MAO-B–deficient subjects, indicating that MAO-A can partially substitute for MAO-B in the deamination of this trace amine. Tyramine excretion was severely increased in subjects with the combined MAO-AB deficiency in contrast to subjects with selective MAO-A or MAO-B deficiency, confirming that tyramine is a substrate for both isoenzymes and also indicating that one isoenzyme can fully compensate for the absence of the other.

The principal metabolite from intraneuronal deamination of NE is DHPG, and in humans 98% of the plasma DHPG is derived from intraneuronally deaminated NE (8). In the patients lacking MAO-A and MAO-AB, plasma DHPG levels were severely reduced, in contrast to the normal DHPG levels in those lacking MAO-B. This observation confirms that MAO-A and not MAO-B is responsible for the intraneuronal deamination of NE. As distinct from the intraneuronal origin of DHPG, NMN is mainly derived from extraneuronal metabolism; as a consequence, plasma NMN levels were strongly increased in the subjects with the MAO-A and MAO-AB deficiencies. The ratios of the plasma sulfate–conjugated NMN to DHPG, representing the relative contributions of extra- versus intraneuronal metabolism of NE, were within normal limits in the MAO-B–deficient subjects elevated by about 100-fold in the MAO-A–deficient subjects and by more than several thousand-fold in all five subjects with the combined MAO-AB deficiency. For comparison, subjects who received deprenyl in a dose inhibiting both MAO-A and MAO-B demonstrated a mean increase of this ratio of about 20-fold, which is substantially less than in subjects lacking MAO-AB.

On the basis of the data from the subjects with the genetically determined MAO deficiencies, there are two arguments that challenge the hypothesis that low MAO-B and abnormal behavior are causally related (9, 10). First, the subjects with the MAO-B deficiency, though blind and hearing impaired, were individuals with completely normal behavior. Second, subjects with the MAO-A and MAO-AB deficiencies exhibited the most severe alterations of the metabolism of catecholamines and 5HT, while these abnormalities were absent in the MAO-B–deficient subjects. It remains to be determined, however, which of the metabolic alterations, caused by the absence of MAO-A, is responsible for the behavioral disorders. Cases and coworkers (11) reported markedly elevated 5HT concentrations in the brains of transgenic mice with a partial deletion of the MAO-A gene. These elevated 5HT concentrations could be reversed by postnatal administration of a 5HT synthesis inhibitor. A potential role for 5HT in the development of the pathological behavior in these mice was further supported by the finding that the elevated brain 5HT levels were related to histological changes in the cortex of these mice. Another recent study also pointed to a role of 5HT, showing structural changes in the serotoninergic neurons of rats that were exposed to MAO-inhibitors during gestation (12).

In conclusion, comparison of the selective deficiencies of MAO-A and MAO-B and the combined deficiency of MAO-AB provides a unique insight into the metabolic roles of the MAO isoenzymes in humans *in vivo*. In contrast to the selective deficiency of MAO-A and the combined deficiency of MAO-AB, selective deficiency of MAO-B does not lead to disturbances in behavior or to severe neurochemical derangements. It remains to be established whether the metabolic consequences of the lack of these enzymes contribute to the different clinical phenotypes observed in patients with deficiencies of either or both isoenzymes.

References

1. Weyler, W., Hsu, Y-P. P., and Breakefield, X. O. (1990). Biochemistry and genetics of monoamine-oxidase. *Pharmacol. Ther.* **47,** 391–417.
2. Berry, M. D., Juorio, A. V., and Paterson, I. A. (1994). The functional role of monoamine oxidases A and B in the mammalian central nervous system. *Prog. Neurobiol.* **42,** 375–391.
3. Brunner, H. G., Nelen, M., Breakefield, X. O., Ropers, H. H., and van Oost, B. A. (1993). Abnormal behavior associated with a point mutation in the structural gene for monoamine oxidase A. *Science* **262,** 578–580.
4. Brunner, H. G., Nelen, M. R., van Zandvoort, P., Abeling, N. G. G. M., van Gennip, A. H., Wolters, E. C., Kuiper, M. A., Ropers, H. H., and van Oost, B. A. (1993). X-linked borderline mental retardation with prominent behavioral disturbance: Phenotype, genetic localization, and evidence for disturbed monoamine metabolism. *Am. J. Hum. Genet.* **52,** 1032–1039.
5. Murphy, D. L., Sims, K. B., Karoum, F., de la Chapelle, A., Norio, R., Sankila, E-M., and Breakefield, X. O. (1990). Marked amine and amine metabolite changes in Norrie disease patients with an X-chromosomal deletion affecting monoamine oxidase. *J. Neurochem.* **54,** 242–247.
6. Lenders, J. W. M., Eisenhofer, G., Abeling, N. G. G. M., Berger, W., Murphy, D. L., Konings, C. H., Bleeker Wagemakers, L. M., Kopin, I. J., Karoum, F., Gennip, A. H., and Brunner, H. G. (1996). Specific genetic deficiencies of the A and B isoenzymes of monoamine oxidase are characterized by distinct neurochemical and clinical phenotypes. *J. Clin. Invest.* **97,** 1–10.
7. Eisenhofer, G., and Finberg, J. P. M. (1994). Different metabolism of norepinephrine and epinephrine by catechol-O-methyltransferase and monoamine oxidase in rats. *J. Pharmacol. Exp. Ther.* **268,** 1242–1251.
8. Eisenhofer, G., Friberg, P., Pacak, K., Goldstein, D. S., Murphy, D. L., Tsigos, C., Quyyumi, A. A., Brunner, H. G., and Lenders, J. W. M. (1995). Plasma metadrenalines: Do they

provide useful information about sympatho-adrenal function and catecholamine metabolism? *Clin. Sci.* **88**, 533–542.
9. von Knorring, L., Oreland, L., and Winblad, B. (1984). Personality traits related to monoamine oxidase activity in platelets. *Psychiatr. Res.* **12**, 11–26.
10. Devor, E. J., Cloninger, C. R., Hoffman, P. L., and Tabakoff, B. (1993). Association of monoamine oxidase (MAO) activity with alcoholism and alcoholic subtypes. *Am. J. Med. Genet.* **48**, 209–213.
11. Cases, O., Seif, I., Grimsby, J., Gaspar, P., Chen, K., Pournin, S., Müller, U., Aguet, M., Babinet, C., Shih, J. C., and De Maeyer, E. (1995). Aggressive behavior and altered amounts of brain serotonin and norepinephrine in mice lacking MAO-A. *Science* **268**, 1763–1766.
12. Whitaker-Azmitia, P. M., Zhang, X., and Clarke, C. (1994). Effects of gestational exposure to monoamine oxidase inhibitors in rats: Preliminary behavioral and neurochemical studies. *Neuropsychopharmacology* **11**, 125–132.

Lars Oreland, Jonas Ekblom, Håkan Garpenstrand, and Jarmila Hallman

Department of Medical Pharmacology
Uppsala University
S-75124 Uppsala, Sweden

Biological Markers, with Special Regard to Platelet Monoamine Oxidase (trbc-MAO), for Personality and Personality Disorders

Monoamine oxidase (E.C. 1.4.3.4) activity in platelets (trbc-MAO) has for 20 years been known to correlate with personality traits such as sensation seeking and impulsiveness, as a result of the pioneering work of Buchsbaum, Belmaker, and Murphy (1, 2). This paradigm was based on the observation that the low trbc-MAO subjects were socially more active, but that they also had more psychiatric disorders and had relatives who were more psychiatrically disturbed than the relatives of the high trbc-MAO subjects. The trbc-MAO activity is characterized by a considerable variability between individuals, whereas, the intraindividual variation of activity over time is low. The only known ways of affecting trbc-MAO are short-term physical activity, administration of adrenaline or lithium, vitamine B-12 deficiency (increased activity),

Advances in Pharmacology, Volume 42

1054-3589/98 $25.00

smoking (discussion follows), and MAO inhibiting drugs (reduced activity). The strong genetic influence on trbc-MAO activity, which forms a cornerstone for the notion of this enzyme activity being a genetic marker, has been confirmed in several twin studies. Thus, in a recently published study, the heritability was estimated to be 0.75, both in males and in females (1, 2).

The correlation between trbc-MAO and extraversion- and impulsiveness-related personality traits has been confirmed in numerous studies using a variety of personality inventories. Moreover, a correlation between trbc-MAO and outcome scores in computerized neuropsychological tests, such as reaction time, many failed inhibitions, and maze checking time, has been demonstrated (1, 2). Furthermore, correlations between behavioral traits in monkeys, likely to be related to the personality traits mentioned previously, have been shown to correlate with trbc-MAO, as well as trbc-MAO and behavior in new-born babies. A confounding factor, which might influence the results with trbc-MAO, is smoking, because there is accumulating evidence that smoking lowers the trbc-MAO activity. The results with monkeys, newborn babies, and the neurospycholgical tests (in which no smokers participated), however, show that smoking cannot explain the correlations found.

I. Alcoholism and Drug Abuse

At about the same time as the appearance of the works showing associations between trbc-MAO and personality, the first reports on low trbc-MAO in connection with alcohol abuse were published. This observation agreed well with the high-risk paradigm of Buchsbaum and coworkers. The finding of reduced trbc-MAO activity in alcoholics was later further explored, and it could be shown that the low activity was present ony in type-2 alcoholics but not in type-1 alcoholics. Type-2 alcoholics are characterized by a strong genetic load for abuse, early age at the debut of alcohol intake, often mixed abuse, and often social complications. This form occurs only in males. Type-1 alcoholics are characterized by low genetic load and late debut of the abuse, as well as by less frequently mixed abuse or social complications. It could also be shown that type-2 alcoholics have exactly the temperament associated with low trbc-MAO. From these observations, it seems very likely that at least a major part of the genetic disposition for abuse occurring in type-2 alcoholics is the personality traits reflected by low trbc-MAO (1, 2).

II. Criminality

Low trbc-MAO has been found in violent criminal offenders as well as in consecutive clients undergoing forensic psychiatric examination. In a recent series of investigations, it was shown that low trbc-MAO has a high predictive value for continued criminal activity in boys convicted of crime. The predictive value was particularly high if combined with scores on Hare's psychopathy check list (PCL). As a matter of fact, both low trbc-MAO and scores on the

PCL alone had quite weak predicitve values. It should be noted that trbc-MAO has, in several studies, been found not to correlate with psychopathy. This notion is supported by results of a recent study showing that type-2 alcoholics differed significantly from type-1 alcoholics with more antisocial personality disorder, diagnosed according to DSM-III R. However, although the type-2 alcoholics had lower trbc-MAO activity than the type-1 alcoholics, there was no correlation between trbc-MAO and antisocial personality (2).

III. Possible Molecular Mechanisms

Possible mechanisms underlying the connection between trbc-MAO and vital parts of the temperament can be summarized as follows (1, 2):

1. Trbc-MAO (which is only of the B form) is correlatd to brain MAO-B. This, however, does not seem to be the case. Direct studies have failed to demonstrate such a correlation, and, furthermore, complete blockade of brain MAO-B does not seem to induce any significant changes in temperament.
2. The findings with regard to trbc-MAO are not limited to this enzyme and are shared with other mitochondrial enzymes. This would mean that the number or total mass of mitochondria would be linked to temperament. MAO is bound to the outer mitochondrial membrane, and in order to explore this hypothesis, we have studied the correlation between MAO and some inner membrane enzymes in the platelet mitochondria. Indeed, significant correlations could be found between MAO and cytochrome oxidase as well as with cytochrome isocitrate dehydrogenase. If a correlation would prevail also in brain tissue, individuals with low trbc-MAO activity might also have low energy supply in neurons. It could be speculated further that different types of neurons might have a different response toward such a shortage and that, for example, seronergic neurons might function at a lower level.
3. Most likely, however, seems to be the hypothesis that trbc-MAO is a genetic marker for, for example the size or capacity of the central serotonergic system. There is much evidence that especially serotonergic functions are linked to trbc-MAO activity. They are described elsewhere but can be summarized here as showing a correlation between trbc-MAO activity and CSF levels of 5-HIAA and that personality disorders such as type 2 alcoholism and violence are linked both to low levels of CSF 5-HIAA and to low trbc-MAO activity. Because no gene alleles have been reported to be linked to different trbc-MAO activities, it would seem likely that the common link between trbc-MAO and central serotonergic function might be on the transcriptional level. Thus, it can be speculated that transcription factors regulating trbc-MAO expression also regulate the expression of some component of the serotonergic system (or the whole system).

It is obvious that not only the central (serotonergic) function, which is reflected by trbc-MAO, is of importance for the temperament. As a continuation of our studies on biological markers for personality, we have started a program for mapping a variety of likely genetic markers with the aim of creating genetic marker "profiles" for various temperaments. Recently, two papers were published on a weak correlation between long dopamine D4 receptor gene alleles, due to the number of repeats in the third exon, and novelty seeking (3). Obviously, there are differences between the Cloninger concept, novelty seeking, and the Zuckerman concept, sensation seeking, but there should be some overlap. In 30 psychiatric patients with various diagnoses (mainly depressed and psychotic), we could not find any correlation between trbc-MAO activity and distribution of D4R gene allele distribution. Neither was there any correlation with the occurrence of different gene alleles for tyrosine hydroxylase, which has been shown to be linked to manic depressive disorder and schizophrenia (4).

References

1. Oreland, L., and Hallman, J. (1995). The correlation between platelet MAO activity and personality—a short review of findings and a discussion on possible mechanisms. *Prog. Brain Res.* **106**, 77–84.
2. von Knorring, L., and Oreland, L. (1996). Platelet MAO in type 1/2 alcoholics. *Alcohol. Clin. Exp. Res.* **20**, 224A–230A.
3. Ebstein, R. P., Novick, O., Umansky, R., Priel, B., Osher, Y., Blaine, D., Bennett, E., Nemanov, L., Katz, M., and Belmaker, R. (1996). Dopamine D4 receptor (D4DR) exon III polymorphism associated with the human personality trait of novelty seeking. *Nature Genetics.* **12**, 78–80.
4. Mallet, J., Meloni, R., and Laurent, C. (1994). Catecholamine metabolism and psychiatric or behavioral disorders. *Curr. Opin. Genet. Dev.* **4**, 419–426.

J. S. Fowler, N. D. Volkow, G. J. Wang, N. Pappas, C. Shea, R. R. MacGregor, and J. Logan

Chemistry and Medical Departments
Brookhaven National Laboratory
Upton, New York 11973

Visualization of Monoamine Oxidase in Human Brain

Monoamine oxidase (MAO; EC: 1.4.3.4) is a flavin-containing enzyme that exists in two subtypes, MAO-A and MAO-B. MAO-A and MAO-B are different gene products, and they also differ in their substrate and inhibitor

Advances in Pharmacology, Volume 42

1054-3589/98 $25.00

selectivities and their cellular localizations. In human brain, MAO-B predominates (B:A = 4:1) and is largely compartmentalized in cell bodies of serotonergic neurons and in glia. Many studies of human brain MAO-B postmortem report that MAO-B increases with age and in neurodegenerative disease (1). This is consistent with investigations showing that the number of glial cells increases with age in the normal human brain (2) and in neurodegenerative disease.

As an initial step in the investigation of the feasibility of detecting and tracking neurodegenerative processes in the living human brain, we measured brain MAO-B in normal healthy subjects ($n = 21$; age range, 23–86 yr; 9 women and 12 men; nonsmokers). The studies followed the guidelines of the Human Subjects Research Committee at Brookhaven National Laboratory, and subjects gave informed consent after the procedures had been explained to them. We used positron emission tomography (PET) and deuterium-substituted [^{11}C]L-deprenyl ([^{11}C]L-deprenyl-D2) (3). MAO-B was assessed using a model term $\lambda\kappa_3$, which is a function of MAO-B activity. A blood-to-brain influx constant (K_1), which is related to brain blood flow, was also calculated. Regions of interest were occipital cortex, frontal cortex, cingulate gyrus, parietal cortex, temporal cortex, pons, thalamus, basal ganglia, cerebellum, and global regions.

The regional distribution of MAO B was highest in the basal ganglia and the thalamus, with intermediate levels in the frontal cortex and cingulate gyrus and lowest levels in the parietal and temporal cortices and cerebellum. The model term $\lambda\kappa3$ showed a significant increase with age ($p < .004$) in all brain regions examined except for the cingulate gyrus (with a trend for the parietal cortex). The results of correlation analysis for the global region is shown in Figure 1A. The same patterns remained when the correlation analysis was performed separately for men and women.

[^{11}C]L-deprenyl-D2 has tracer characteristics that allow a plasma-to-brain transfer constant, a model term that is related to blood flow, to be extracted from dynamic PET data. In contrast to $\lambda\kappa_3$, which increased with age, K_1 significantly ($p < .01$) decreased in all brain regions except for the pons and the cerebellum. The highest correlation coefficients were in the cingulate gyrus, the frontal cortex, the temporal cortex, and the parietal cortex, consistent with other studies. Individual data for K_1 for the global region versus age is shown in Figure 1B.

This study confirms several postmortem studies reporting increases in brain MAO-B with age, though the rate of increase is lower than most studies. The whole brain and the cortical regions and the basal ganglia, thalamus, pons, and cerebellum showed an average increase of 7.1 ± 1.3% per decade. The frontal cortex showed a rate of increase of 5.7% per decade, which is similar to that reported by Fowler and coworkers (4) but far lower than the increase of 51% per decade reported in a recent postmortem study (5). The only brain region where we observed no increase with age is the cingulate gyrus. Though the increases with age are statistically significant, it is noteworthy that there is also a large variability among subjects in the same age range, as can be seen from Figure 1A. The factors that account for the difference in magnitude between this PET study and postmortem studies (and to differences between postmortem studies) and to the large intersubject variability are not known.

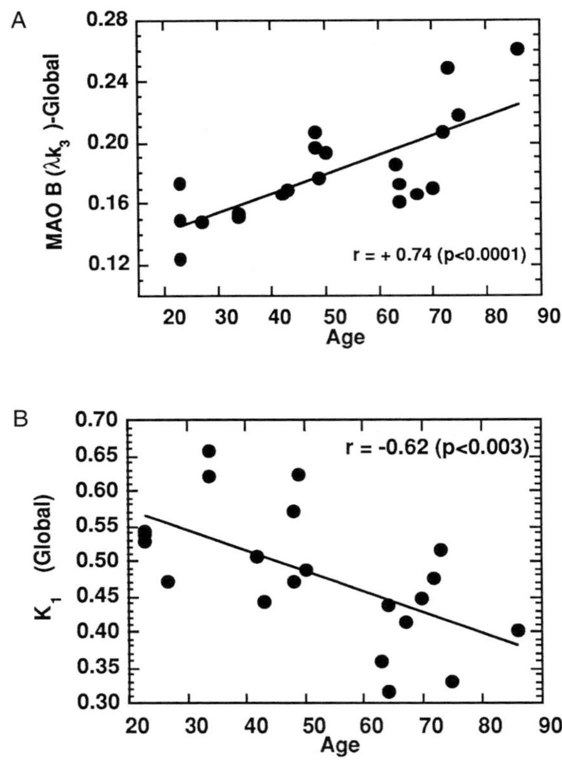

FIGURE 1 (*A*) Individual values of MAO-B activity (represented by $\lambda\kappa_3$) versus age, (*B*) individual values of plasma-to-brain transfer constant (K_1), which is related to blood flow versus age. (Reprinted with permission from *Neurobiology of Aging*, Vol. 18, No. 4, 1997.)

However, it is likely that differences in subjects contribute to differences between different studies. In this regard, the difficulty in distinguishing mild dementia from normality in postmortem studies of normal aging has been noted and may have been a factor (2). Smoking status was not controlled in the postmortem studies and may have accounted for some of the differences, based on the report that smokers have reduced brain MAO-B (6). It would be interesting and important to examine this issue retrospectively.

In summary, we have observed that brain MAO-B increases with age in healthy normal subjects who show typical patterns of age-related decreases in blood flow. However, the increases we observed are generally smaller than those reported for most postmortem studies, and there is also a relatively large variability in MAO-B even within relatively small age ranges in this group of normal, healthy subjects. Thus, the extent to which increases in brain MAO B reflect age-related increases in glial cells or whether there are other variables contributing to the results of this PET study and to the range of results reported in the literature requires further investigation.

Acknowledgments

This research was carried out at Brookhaven National Laboratory under contract DE-AC02-76CH00016 with the U.S. Department of Energy and was supported by its Office of Health and Environmental Research and also by the National Institutes of Health (NS 15638 and NS 15380). The authors are grateful to David Schlyer, Robert Carciello, Richard Ferrieri, Donald Warner, David Alexoff, Payton King, Noelwah Netusil, Thomas Martin, Darrin Jenkins, Christopher Wong, Carol Redvanly, and Alfred Wolf for their advice and assistance in performing these studies. They are also grateful to the subjects who volunteered for these studies.

References

1. Strolin Benedetti, M., and Dostert, P. (1989). Monoamine oxidase, brain ageing and degenerative diseases. *Biochem. Pharmacol.* **38,** 555–561.
2. Terry, R. D., DeTeresa, R., and Hansen, L. A. (1987). Neocortical cell counts in normal human adult aging. *Ann. Neurol.* **21,** 530–539.
3. Fowler, J. S., Wang, G-J., Logan, J., Xie, S., Volkow, N. D., MacGregor, R. R., Schlyer, D., Pappas, N., Alexoff, D., Patlak, C., and Wolf, A. P. (1995). Selective reduction of radiotracer trapping in brain by deuterium substitution: Comparison of [^{11}C]L-deprenyl and [^{11}C]L-deprenyl-D2 for MAO B mapping in human brain. *J. Nucl. Med.* **36,** 1255–1262.
4. Fowler, C. J., Wiberg, A., Oreland, L., Marcusson, J., and Winblad, B. (1980). The effect of age on the activity and molecular properties of human brain monoamine oxidase. *J. Neural Transm.* **49,** 1–20.
5. Sastre, M., and Garcia-Sevilla, J. A. (1993). Opposite age-dependent changes of a_2-adrenoreceptors and nonadrenoreceptor [^{3}H]idazoxan binding sites (I_2-imidazoline sites) in the human brain: Strong correlation of I_2 with monoamine oxidase B sites. *J. Neurochem.* **61,** 881–889.
6. Fowler, J. S., Volkow, N. D., and Wang, G-J., Pappas, N., Logan, J., MacGregor, R., Alexoff, D., Shea, C., Schlyer, D., Wolf, A. P., Warner, D., Zezulkova, I., and Cilento R. (1996). Inhibition of monoamine oxidase B in the brains of smokers. *Nature* **379,** 733–736.

Alan A. Boulton, Peter H. Yu, Bruce A. Davis, I. Alick Paterson, Xi-Min Li, Augusto V. Juorio, David A. Durden, and Lillian E. Dyck

Neuropsychiatry Research Unit
University of Saskatchewan
Saskatoon, Saskatchewan
Canada, S7N 5E4

Aliphatic *N*-Methylpropargylamines: Monoamine Oxidase-B Inhibitors and Antiapoptotic Drugs

Deprenyl is the archetypical monoamine oxidase (MAO)-B inhibitor. It has been shown to offer symptomatic relief in the treatment of Parkinson's patients, and it has also been claimed to be neuroprotective. Possible disadvantages are its amphetaminergic metabolites and dopamine uptake–blocking properties. In addition, since its neurorescue effects were discovered somewhat serendipitously, it is quite possible that other MAO-B inhibitors or even different classes of compounds could be better rescue agents.

In 1989, Yu *et al.* (1) showed that the aliphatic amines were excellent substrates for MAO-B. By adding an *N*-methyl and propargyl blocking group, a series of compounds—the aliphatic *N*-methylpropargylamines—were discovered and shown to be potent, selective, irreversible MAO-B inhibitors (1). Structure activity studies in which changes in the nature of the groups attached to the nitrogen atom, including the effects of chain length substitutions at either the α-carbon atom or on the terminal carbon atom or unsaturation or remote branching revealed that:

1. MAO activity is eliminated if the methyl group on the nitrogen atom is changed.
2. Optimum potency and selectivity are achieved by the blocking group being either propargyl or 2-butyn-1-yl.
3. Four to seven carbon branched-chain compounds are as potent as, or more potent than, deprenyl as MAO inhibitors *in vivo. In vitro,* they are somewhat less effective than deprenyl.
4. *In vivo* and *in vitro* selectivity of the 1- and 2-hexyl analogues is much superior to deprenyl.
5. The presence of unsaturation or branching (at other than the α-carbon) has no effect on potency or selectivity; the presence of both, however, in some cases produced a marked increase in selectivity.

Advances in Pharmacology, Volume 42

1054-3589/98 $25.00

6. Substituting with hydroxyl or carbethoxy groups destroys MAO inhibiting activity.

The aliphatic *N*-methylpropargylamines possess an asymmetric carbon atom and so are optically active. The R(-)form is about 20 times more active than the S(-)form, although the S forms are still potent MAO-B inhibitors. They do not block the uptake of dopamine into rat striatal slices, whereas both enantiomers of deprenyl do block uptake, with the S form being much more potent than the R.

When (R)-N-2-heptyl-*N*-methylpropargylamine[(-)2HMP] is administered to rats (10mg/kg), both subcutaneously and orally, it penetrates into the brain as well as the peripheral organs. The brain-plasma ratio is quite high and its pharmacokinetic profile is better than is the case for (−)deprenyl. See Table I for tissue levels and comparative brain-plasma ratios.

In addition to being an MAO-B inhibitor, deprenyl has been shown to prolong life in rodents, exert an effect on the cell cycle, up- and downregulate certain genes, rescue dying substantia nigra neurons after MPTP treatment but in a way that does not involve MAO-B (i.e., (−)deprenyl given 3 days after MPTP injections), and rescue dying neurons in the immature facial axotomy model and at a concentration at which MAO-B inhibition could not be occurring. (−)Deprenyl has also been shown to exhibit trophic-like effects on reactive astrocytes and to activate astrocytes after striatal lesions. It stimulates neurite outgrowth in cocultured dopaminergic neurons after treatment with methyl-4-phenylpiridinium (MPP^+), and it increases tyrosine hydroxylase levels in mesencephalic cultures.

TABLE I Biodistribution of (−)2HMP

				Brain-plasma ratio	
Time	*Brain*	*Liver*	*Plasma*	*(−)2HMP*	*(−)deprenyl*
Subcutaneous					
5	4426 ± 437	1023 ± 231	1125 ± 409	3.9	
15	6204 ± 912	2279 ± 363	443 ± 82	14.0	
30	3709 ± 633	2755 ± 306	328 ± 32	11.3	
60	1214 ± 78	1468 ± 128	144 ± 13	8.4	
120	416 ± 163	758 ± 259	58 ± 24	7.2	
240	204 ± 58	519 ± 113	38 ± 15	5.4	
Oral					
5	50 ± 27	1420 ± 844	240 ± 200	0.2	—
15	85 ± 52	5134 ± 1633	34 ± 11	2.5	0.2
30	109 ± 66	1625 ± 583	33 ± 28	3.3	0.2
60	26 ± 16	2375 ± 982	3.1 ± 1.2	8.4	0.3

Time in minutes, tissue levels in ng/g. (−)2HMP, 10 mg/kg, s.c. and oral, n = 4, values mean ± SEM. Samples, 1 g or 1 ml, homogenized in 0.1 N $HClO_4$ in the presence of (±)2HMP-d_8, centrifuged, extracted 3× with ethyl acetate:hexane (1:1 v/v). Remaining aqueous solution made alkaline and after derivatization with pentafluorobenzoyl chloride (PFB) and extraction, separated by gas chromatography (GC), and then quantified by mass spectrometry dectron impact (MS-EI) using the fragment ions 96 and 102.

Deprenyl also blocks apoptotic death after serum (or nerve growth factor [NGF]) withdrawal in PC12 cell cultures, and it rescues rat retinal ganglion cells after crushing and rat hippocampal cells after a variety of insults (ischemia, kainate, DSP-4, etc.). In many of these examples, the rescue is via an antiapoptotic mechanism, because in the absence of (−)deprenyl, the morphological and biochemical characteristic features of programmed cell death have been demonstrated. When (−)deprenyl acts as an antiapoptotic agent, there is an absolute stereochemical requirement, with only the *R*(−)*enantiomer* being active.

We have investigated (3) the effects of several of the aliphatic *N*-methylpropargylamines (usually (-)2HMP) in both *in vitro* and *in vivo* models in which the neuronal death is apoptotic. These models include trophic withdrawal in PC12 cells (4), toxin (cytosine arabinoside)-induced death of cerebellar granule cells as well as unilateral global cerebral hypoxia–ischemia in the rat (5), and the effects of kainate on HSP_{70} as a marker of hippocampal neuronal cell death in the rat (6). In some gene expression studies (7), we have assessed the regulation of the mRNAs for aromatic L-amino acid decarboxylase (AADC) and SOD-1, which are upregulated by (−)deprenyl and the aliphatic *N*-methylpropargylamines, while GFAP and the low-affinity NGF receptor (p75) are downregulated. Because with the exception of GFAP, which was assessed in C6 glioma cells, the other mRNAs were assessed in PC12 cells, which do not contain MAO-B, we again can conclude an action for the MAO-B inhibitory drugs that does not involve MAO-B enzyme activity. A summary of our findings for the aliphatic N-methylpropargylamines in these various *in vitro* and *in vivo models* is given in Table II.

TABLE II Summary of Responses of Aliphatic *N*-Methylpropargylamines in Several *in Vitro* and *in Vivo* Models

			Gene expression					
	PC12	*Cerebellar granules*	*AADC*	*GFAP*	*SOD-1*	*LNGFR*	*Ischemia*	*Kainate (HSP_{70})*
(+) 2HMP	X	X	X			X	X	√
(−) 2HMP	√	√	↑			↓	√	√
(±) 2HxMP	√		↑				√	√
(+) 2HxMP			X			X		X
(−) 2HxMP			↑	↓	↑	↓		√
(±) 2PeMP	√						√	
(+) 2PeMP			X			X		X
(−) 2PeMP	√		↑			↓		√
(+) 2BuMP						X	X	X
(−) 2BuMP						↓	√	X
2PrMP							√	
1HxMP							√	
1PrMP							X	
(+) Deprenyl	X		X			X	X	X
(−) Deprenyl	√		↑	↓	↑	↓	√	√

X denotes no effect; √ denotes rescue; ↑ denotes upregulation; ↓ denotes downregulation; (−) is the active enantiomer; (+), the inactive enantiomer; and (±), a racemic mixture

In summary, then, our recent studies have demonstrated new mechanisms of action for most MAO-B inhibitors, which raise the intriguing question of what is MAO-B really doing? The inhibitors are antiapoptotic at very low concentrations, although their site of action in the apoptotic cascade is not yet clear. The possibility that some of these MAO-B inhibitor drugs will find a use in the treatment of various neurodegenerative disorders and stroke seems quite promising.

Acknowledgments

We thank Saskatchewan Health and CIBA-GEIGY Canada for continuing financial support.

References

1. Yu, P. H., Davis, B. A., and Boulton, A. A. (1992). Aliphatic propargylamines: Potent, selective, irreversible monoamine oxidase B inhibitors. *J. Med. Chem.* **35,** 3705–3713.
2. Yu, P. H., Davis, B. A., and Boulton, A. A. (1993). Effect of structural modification of alkyl N-propargylamines on the selective inhibition of monoamine oxidase B activity. *Biochem. Pharmacol.* **46,** 753–757.
3. Yu, P. H., Davis, B. A., Zhang, X., Zuo, D. M., Fang, J., Lai, C. T., Li, X-M., Paterson, I. A., and Boulton, A. A. (1995). Neuroprotective and neurorescue effects of aliphatic N-methylpropargylamines: New MAO-B inhibitors without amphetamine-like properties. *Prog. Brain Res.* **106,** 113–121.
4. Tatton, W. G., Ju, W. Y. L., Holland, D. P., Tai, C., and Kwan, M. (1994). (−)Deprenyl reduces PC12 cell apoptosis by inducing new protein synthesis. *J. Neurochem.* **63,** 1572–1575.
5. Paterson, I. A., Barber, A. J., Gelowitz, D. L., and Voll, C. C. (1996). (−)Deprenyl reduces delayed neuronal death of hippocampal pyramidal cells. *Neurosci. Biobehav. Res.* (in press)
6. Zhang, X., Boulton, A. A., and Yu, P. H. (1996). Expression of HSP_{70} and limbic seizure-induced neuronal death in the rat brain. *Eur. J. Neurosci.* **8,** 1432–1440.
7. Li, X-M. Juorio, A. V., and Boulton, A. A. (1995). Some new mechanisms underlying the actions of (−)deprenyl: Possible relevance to neurodegeneration. *Prog. Brain Res.* **106,** 99–112.

I. A. Paterson* and W. G. Tatton†

*Neuropsychiatry Research Unit
University of Saskatchewan
Saskatoon, Saskatchewan
Canada S7N 5E4

†Institute for Neuroscience
Dalhousie University
Halifax, Nova Scotia
Canada B3H 4H7

Antiapoptotic Actions of Monoamine Oxidase B Inhibitors

R-Deprenyl is an irreversible inhibitor of monoamine oxidase (MAO) with selectivity for MAO-B and, as such, is thought to influence neuronal survival by preventing necrosis induced by oxidative radicals resulting from monoamine metabolism. There is now, however, accumulating evidence that R-deprenyl and certain other MAO-B inhibitors can prevent apoptotic death of neurones by a novel mechanism. The finding that R-deprenyl could prevent (MPTP)–induced death of dopaminergic neurones, even when R-deprenyl was given 3 days after MPTP when all MPTP has been metabolized and cleared, suggested an action of R-deprenyl that did not depend on MAO-B inhibition (1). This was followed by studies on the survival of facial motor neurones following facial nerve axotomy in 14-day-old rats (2, 3) that showed that R-deprenyl could prevent apoptosis of the cholinergic motor neurones at low doses, which did not inhibit MAO-B, but that S-deprenyl was ineffective, even at high doses.

We have used a modified Levine preparation (4) to produce unilateral global cerebral ischemia in conscious rats to study the effects of R-deprenyl and aliphatic propargylamines (5) on ischemia-induced apoptosis in the rat hippocampus. This model produces delayed neuronal death in the CA1 and other subfields of the hippocampus, which involves apoptosis (demonstrated by TUNEL staining and nuclear staining with bis-benzamide and propridium iodide) 24–72 hr after ischemia. Administration of R-deprenyl (1.0–0.01 μmol/kg, s.c., once daily) reduced the delayed neuronal death in the CA1. The drug was effective when the first dose was given at the time of ischemia or up to 8 hr after the ischemic episode, and it had to be given throughout the period of neuronal death. In contrast, S-deprenyl was ineffective, even at a dose of 10 μmol/kg. A variety of branched-chain and straight-chain aliphatic propargylamines were examined in the hypoxia–ischemia model. A range of branched-chain compounds, 2-heptyl to 2-propyl, were effective in reducing delayed neuronal death, but where a chiral center existed, the R-isomers were effective but the S-isomers were ineffective. A dose-response study with R-2-heptyl methylpropargylamine (R-2HMP) found that it had an ED_{50} between 1 and

Advances in Pharmacology, Volume 42

1054-3589/98 $25.00

10 nmol/kg. A straight-chain compound (1-hexyl methylpropargylamine) was also effective, demonstrating that the chiral center in the branched-chain compounds is not required for activity.

Another model of apoptosis in hippocampal neurones *in vivo* is the delayed neuronal death following kainic acid–induced seizures. We have found that neuronal death following seizure reaches a maximum after 3 days, with no further death at later times. Using TUNEL staining, we have observed apoptotic neurones 18 hr after stage 3–5 seizures but have not found apoptotic cells following milder seizures. Apoptosis is seen in the CA3 field and less consistently in the CA1 and CA4. The delayed neuronal death can be prevented by R-deprenyl and R-2HMP, administered at the end of seizure activity. Using the mouse, we found that R-deprenyl and R-2HMP (but not S-deprenyl and S-2HMP) could prevent delayed neuronal death following kainate-induced seizures. In the rat, R-deprenyl (1.0–0.1 μmol/kg, s.c.) prevented delayed neuronal death in the CA3 and sometimes in the CA1. It would appear that the neurones spared by R-deprenyl are functional. In a simple behavioral test examining the habituation of locomotor activity on repeated exposure to an open field, kainate-treated animals lose the habituation seen in control rats. It was found that treatment with R-deprenyl following kainate-induced seizures reversed the effects of the seizures (6). Neurones rescued by R-deprenyl can survive after the withdrawal of R-deprenyl treatment. Termination of R-deprenyl administration 4 days after seizures resulted in a loss of the rescued neurones after 7 days, but with 14 days treatment, the rescued neurones survived at least 14 days after the last R-deprenyl injection.

The ability to prevent delayed neuronal death is not a function of MAO inhibition. There were potent MAO-B inhibitors (e.g., MDL 72 974, RO 16-6491) that did not prevent delayed neuronal death, and the S-isomers of deprenyl and 2HMP are also inhibitors of MAO (though less selective than the R-isomers). Conversely, some of the compounds that prevent delayed neuronal death are very poor MAO inhibitors. Studies with *in vitro* models have revealed novel mechanisms of action of R-deprenyl.

The mechanisms by which R-deprenyl prevents apoptosis in PC12 cells have been studied extensively. In this model, protein synthesis–independent apoptosis is induced in differentiated PC12 cells by withdrawal of trophic factor support (nerve growth factor and serum). Using protein synthesis and transcription inhibitors, it was shown that the antiapoptotic action of R-deprenyl is dependent on gene transcription and protein synthesis in the first few hours following trophic withdrawal (7). Using differential display polymerase chain reaction (PCR), it was found that the expression of most genes is reduced 6 hr after withdrawal of trophic support but that a few genes showed increased expression. Treatment with R-deprenyl following trophic withdrawal not only prevented the loss of many bands in the differential-display PCR, but also induced new bands, suggesting that R-deprenyl is selectively turning on gene expression. New protein synthesis by PC12 cells was greatly reduced by trophic withdrawal. R-Deprenyl reversed the loss of new proteins in the nuclear and mitochondrial fractions of the cells but not in the cytoplasmic or plasma membrane fractions, again demonstrating selective changes in gene expression. A number of genes suspected to be altered during apoptosis have been examined using reverse-transcription PCR and Western blotting. The expression of the antiapoptotic gene bcl-2 was reduced by trophic withdrawal, as were the genes for superoxide dismutase, while the expression of the proapoptotic gene bax was increased. These changes in gene expression were reversed by R-deprenyl treatment. The expression of genes for structural proteins such as neurofilament-light was also reduced by trophic withdrawal but was not reversed by R-deprenyl treatment.

Loss of mitochondrial function is an early event preceding apoptosis (8), and loss of mitochondrial integrity appears to be an important trigger of apoptosis (9). Studies with fluorescent dyes, which estimate mitochondrial membrane potential, using epifluorescent and confocal microscopes, showed that there was a loss of mitochondrial function in preapoptotic PC12 cells, which is prevented by R-deprenyl. This is consistent with the changes in gene expression observed in these cells, and these data suggest that R-deprenyl activates a select set of genes to maintain mitochondrial function and prevent the induction of apoptosis.

Experiments with cultured cerebellar granule cells suggest that R-deprenyl and R-2HMP may prevent only some forms of apoptosis. Cultures of cerebellar granule cells obtained from 6- to 8-day-old Wistar rat pups were grown on glass in 35-mm Petri dishes for 3 days and then used for experiments. Cultures of cerebellar granule cells can be induced into apoptosis by lowering the concentration of K^+ in the culture medium or by adding of a high concentration of cytosine arabinoside (ara C). Apoptosis was induced either by adding ara C to the medium (final concentration of 100 μM) or by changing the cultures into a low K^+ medium. Twenty-microliter aliquots of drug solutions (antiapoptotic drugs, drug vehicles) were added to the medium of the cultures, and 24 hr later the cultures were fixed and stained with bis-benzamide. Normal and apoptotic nuclei were counted to a total of 90–120 cells per culture. R-Deprenyl and R-2HMP completely blocked completely apoptosis induced by ara C with an EC_{50} of about 10^{-8} M, while S-deprenyl and S-2HMP were ineffective. In contrast, while a similar degree of apoptosis could be induced by switching the cultures to a low K^+ medium, neither R-Deprenyl nor R-2HMP could prevent the apoptosis. These two methods of inducing apoptosis differ in their mechanisms, the low $[K^+]$ induction is P53-independent, whereas the ara C induction is P53-dependent (10). The prevention of P53-dependent apoptosis by these compounds, while failing to prevent P53-independent apoptosis, suggests that these compounds may work specifically in models of P53-dependent apoptosis.

In conclusion, the MAO-B inhibitors R-deprenyl and R-2HMP can prevent neuronal apoptosis both *in vivo* and *in vitro*. *In vivo*, this can result in permanent sparing of neurones and lead to functional improvements. The antiapoptotic action of these compounds, however, is independent of their MAO-B inhibitory properties. The mechanism appears to involve selective changes in gene expression to preserve mitochondrial integrity and prevent the induction of P53-dependent apoptosis.

Acknowledgments

This work was supported by Novartis Canada and Saskatchewan Health.

References

1. Tatton, W. G., and Greenwood, C. E. (1991). Rescue of dying neurons: A new action for deprenyl in MPTP parkinsonism. *J. Neurosci. Res.* **30,** 666–672.
2. Salo, P. T., and Tatton, W. G. (1992). Deprenyl reduces the death of motorneurons caused by axotomy. *J. Neurosci. Res.* **31,** 394–400.
3. Ansari, K. S., Yu, P. H., Kruck, T. P., and Tatton, W. G. (1993). Rescue of axotomised immature rat facial motorneurons by R(−)deprenyl: Stereospecificity and independence from monoamine oxidase inhibition. *J. Neurosci.* **13,** 4042–4053.
4. Paterson, I. A., Barber, A. J., Gelowitz, D. L., and Voll, C.L. (1997). (−)Deprenyl reduces delayed neuronal death of hippocampal pyramidal cells. *Neurosci. Biobehav. Rev.* (in press)
5. Boulton, A. A., *et al.* (1997). Aliphatic N–methylpropargylamines: MAO-B inhibitors and antiapoptotic drugs. *Adva. Pharmacol.* **42,** (in press)
6. Gelowitz, D. L., and Paterson, I. A. (1995). Deprenyl enhances functional recovery following kainic acid-induced neuronal death. *Soc. Neurosci. Abs.* **21,** 326.9.
7. Tatton, W. G., Ju, W. Y., Holland, D. P., Tai, C., and Kwan, M. (1994). (−)Deprenyl reduces PC12 cell apoptosis by inducing new protein synthesis. *J. Neurochem.* **63,** 1572–1575.
8. Zamzami, N., *et al.* (1995). Reduction in mitochondrial potential constitutes an early irreversible step of programmed lymphocyte death *in vivo*. *J. Exp. Med.* **181,** 1661–1672.
9. Lui, X., Kim, C. N., Yang, J., Jemmerson, R., and Wang, X. (1996). Induction of apoptotic program in cell-free extracts: Requirement for dATP and Cytochrome C. *Cell* **86,** 147–157.
10. Enokido, Y., Araki, T., Aizawa, S., and Hatanaka, H. (1996). P53 involves cytosine arabinoside-induced apoptosis in cultured cerebellar granule neurons. *Neurosci. Lett.* 203, 1–4.

W. W. Ruehl,* J. Neilson,† B. Hart,† E. Head,‡, D. S. Bruyette,§ and B. J. Cummings‖

*Deprenyl Animal Health, Inc.
Overland Park, Kansas 66210

†School of Veterinary Medicine
University of California
Davis, California 95616

‡Institute for Brain Aging and Dementia
University of California
Irvine, California 92717

§West Los Angeles Animal Hospital
Los Angeles, California 90025

‖Harvard Medical School
Mailman Research Center
Belmont, Massachusetts 02178

Therapeutic Actions of L-Deprenyl in Dogs: A Model of Human Brain Aging

Severe cognitive dysfunction (CD) or dementia is an important clinical syndrome that increases in prevalence from 1 to 3% in people aged 65–70 years to as high as 47% in the very elderly (>85 years); the prevalence of age-related cognitive decline (ARCD) is even greater. Practicing veterinarians have long been aware of geriatric behavioral changes in pet dogs, often described by the pet owners as "normal aging" or "senility." In this chapter, we summarize the potential of the canine as a model of human brain aging, ARCD, dementia, and Alzheimer's disease (AD), as well as discuss the response of dogs with CD to therapy with the monoamine oxidase (MAO) inhibitor L-deprenyl HCl (selegiline HCl, Anipryl).

To make the diagnosis of dementia in either a human or canine patient, there must be multiple cognitive or behavioral problems that cause a significant decline from a previous level of social or occupational functioning (1). In people, the multiple deficits include memory impairment plus one or more of the following: language disturbance; impaired ability to carry out motor tasks despite intact motor function; failure to recognize or identify familiar objects despite intact sensory function; and decreased ability to perform complex tasks, planning, organizing, or abstract thought. Behavioral deficits also occur in dogs in the areas of decreased social interactions, changes in activity and sleep patterns, getting lost or becoming disoriented, and housetraining problems (1) (discussion follows). Despite recent advances in brain imaging techniques, confirmation of AD as the cause of the dementia usually requires postmortem histopathology; thus, for most

Advances in Pharmacology, Volume 42

1054-3589/98 $25.00

medical and veterinary clinicians, the antemortem diagnosis of "probable AD" remains one of exclusion, based in large part on history.

Animal models have been utilized in the continually increasing research on aging and associated cognitive dysfunction at levels from the molecular to the epidemiological. However, progress has been hampered by the lack of a completely adequate animal model. The majority of studies on the effects of aging on learning and memory have been conducted in rodents and primates, each of which has specific advantages and limitations. It is not clear that results from rat studies can be extrapolated to other species such as dogs, cats, humans, or other primates. Primates are difficult to handle and expensive to maintain, compared with canines or rodents. In addition, primate tissue can be as difficult to obtain as human tissue, and pathological changes are not usually observed until a monkey is at least 20 years old.

We have been investigating the elderly canine as a model for ARCD, dementia, and AD. The dog is an excellent model in many respects (1, 2). Dogs have moderate life spans of approximately 12–20 years, depending on the breed and are considered by veterinarians to be elderly at approximately 7.5–12.0 years, depending on size. Importantly, pet dogs live with people and share common environmental (and often nutritional) risk factors for aging and age-associated disorders. In contrast to primates, dogs are easier to obtain and handle and can be specially bred for research, thus obtaining a genetically homogeneous population. Their nutrition and husbandry are consistent and controlled. Furthermore, dogs are highly motivated to perform consistently on tests of cognitive function when using a food reward, so deprivation paradigms are not prerequisites.

How prevalent are geriatric behavior problems in dogs? To assess the prevalence of age-related behavior changes in elderly dogs, some of us (JN, BH, WWR) are conducting a survey of the owners of dogs that are more than 10 years old. Dogs are excluded if they have medical conditions or are receiving medication that might affect their behavior. To date, 66 dogs have been evaluated, and preliminary results suggest that age-related behavioral changes include: decreased activity and playfulness (67%), housetraining problems (30%), changes in social interactions (23%), changes in sleep patterns (23%), and disorientation (14%). Although the study is ongoing, these preliminary results along with observations from a clinical trial (1) suggest that elderly dogs often exhibit behavioral or cognitive problems indicative of CD. In some geriatric dogs, the behavioral changes may be sufficiently severe to disrupt the dog's function; such dogs fulfill a canine equivalent of the DSM IV definition for dementia.

There are more than 60 diseases or conditions that cause dementia in people, the most common of which are AD and vascular dementia. There are strong similarities between human and canine with respect to age-related neuropathology. Canines experience Alzheimer's-like neuropathology within the same brain regions as humans, including formation of β-amyloid plaques, age-related cerebral vascular changes, thickening of the meninges, dilation of the ventricles, and age-related reactive gliosis. At least some of the pathology present within aged canine brains may be genetically linked. Recently, some of us studied the pathology of elderly dog brains, focusing on plaque morphology and patterns of amyloid deposition (2). There are subtle differences between dog and human with respect to plaque morphology and β-amyloid processing, despite complete sequence homology of the protein.

Since Pavlov's pioneering studies on classical conditioning at the beginning of this century, canines have been evaluated in a variety of cognitive tests. Studies by members of our group confirmed that dogs, like other species, suffer age-dependent cognitive deterioration (2). Additional work has established that CD on some tasks but not others correlates with increasing age. A greater variance in individual cognitive performance occurs with increasing age, as previously reported for aged nonhuman primates and humans. Such dysfunction correlates even more strongly with the amount of β-amyloid present in the brain.

Neurochemical and pharmacological considerations are also important when discussing pathology and pathogenesis. The enzyme MAO-B catalyzes dopamine breakdown with the secondary production of free radicals. In a variety of species, brain MAOB activity is higher in the aged than in the young, and MAOB activities can be very high in patients with neurodegenerative disorders such as parkinsonism and AD (3). A variety of neurotransmitter abnormalities, including dopaminergic imbalances, have been described in Alzheimer's patients. Dopaminergic depletion in the prefrontal cortex has been correlated to CD in monkeys.

Cognitive deficits have been demonstrated in people with an imbalance of the hypothalamic-pituitary-adrenal (HPA) axis. Dysregulation of the HPA axis in both humans and dogs is associated with hypothalamic dopamine depletion and behavioral or cognitive signs (4). Dogs with pituitary-dependent hyperadrenocorticism (PDH, Cushing's disease) also exhibit behavioral or cognitive signs, such as lethargy and decreased interaction with their owners. Further work is needed to determine the extent to which dopamine depletion, cortisol excess, and/or HPA axis dysregulation contribute to the CD of these patients.

The pharmaceutical, L-deprenyl, by inhibiting MAO-B and facilitating dopamine release, should theoretically help restore dopaminergic balance in the prefrontal cortex and other brain regions of an individual with CD and, thus, reverse or slow the progression of clinical signs. Furthermore, L-deprenyl is known to decrease the production of free radicals, enhance clearance of free radicals, and exert protective and "rescue" effects on damaged neurons (5). Numerous clinical trials of L-deprenyl in demented people have demonstrated encouraging results (3, 6).

Head *et al.,* (7) evaluated the effects of L-deprenyl on cognitive performance of laboratory-housed canines as a model system for treatment of CD. Administration of L-deprenyl was found to have a beneficial effect on spatial short-term memory in aged dogs but not in young ones (7).

Bruyette *et al.,* (4) conducted clinical trials of L-deprenyl in pet dogs with PDH. The diagnosis was confirmed, based on presence of clinical signs, laboratory abnormalities, dexamethasone suppression, and other endocrine tests, as needed. Dogs were treated with 1–2 mg/kg of L-deprenyl orally once daily and monitored for 6 mos. Therapy with L-deprenyl resulted in a return towards normal of the dexamethasone suppression test results ($p < .001$), as well as improvement ($p < .05$) at various time points in all common clinical parameters.

Based on the rationale and findings noted here, a clinical trial in pet dogs was conducted to evaluate the safety and efficacy of L-deprenyl for treatment of canine CD and to evaluate the utility of elderly pet dogs with CD as a model system (1). Dogs were nominated on the presence of one or more geriatric problems with housetraining; interest in food; activity or attention to environment, people, or other animals; awareness or orientation to surroundings; ability to recognize familiar places or people; ability to recognize commands or when called by name; hearing; navigation of stairs; tolerance to being alone; compulsive behavior; circling; tremors; sleep–wake cycle; inappropriate vocalization; and stiffness or weakness. Dogs were excluded if they exhibited evidence of a concurrent, general medical condition. Sixty-nine dogs fulfilled the enrollment criteria and were administered 0.5-mg/kg L-deprenyl tablets once daily in open-label fashion; subjects were re-evaluated monthly during the 3-mo study. Dogs were 7–19 years of age (mean, 14 years), weighed 3.6–36.0 kg (mean, 14 kg), and at enrollment exhibited multiple signs of CD. After 1 mo of L-deprenyl therapy, the population was improved in all 15 parameters evaluated ($p < .05$ for 13 parameters). Benefits were generally maintained at months 2–3, with the percentages of individuals who improved with respect to global function at each month being 77%, 76%, and 78%, respectively.

In conclusion, information from the studies noted in this chapter and other sources establishes that elderly dogs with cognitive dysfunction provide an excellent spontaneously occurring animal model of ARCD, dementia, AD, and other neurodegenerative disorders. Some similarities include: (1) occurrence of clinically relevant behavioral and cognitive problems; (2) common prevalence, which likely increases with age; (3) deterioration of performance on formal cognitive tests with increasing age; (4) Alzheimer's-like pathology, increasing with age; and (5) response to L-deprenyl therapy. The population of pet dogs who exhibit CD may also offer an additional model system for epidemiology and pathology studies, development of diagnostics, and the evaluation of cognitive enhancer drugs.

References

1. Ruehl, W. W., Bruyette, D. S., DePaoli, A., Cotman, C. W., Head, E., Milgram, N. W., and Cummings, B. J. (1995). Canine cognitive dysfunction as a model for human age-related cognitive decline, dementia and Alzheimer's disease: Clinical presentation, cognitive testing, pathology and response to l-deprenyl therapy. *Prog. Brain Res.* **106,** 217–225.
2. Cummings, B. J., Head, E., Ruehl, W., Milgram, N. W., and Cotman, C. W. (1996). The canine as an animal model of human aging and dementia. *Neurobiol. Aging* **17,** 259–268.
3. Tariot, P. N., Schneider, L. S., Patel, S. V., and Goldstein, B. (1993). Alzheimer's disease and l-deprenyl: Rationales and findings. *In* Inhibitors of Monoamine Oxidase B. (I. Szelenyi, ed.), pp. 301–317. Birkhauser Verlag, Basel.
4. Bruyette, D. S., Ruehl, W. W., and Smidberg, T. L. (1995). Canine pituitary-dependent hyperadrenocorticism: A spontaneous animal model for neurodegenerative disorders and their treatment with l-deprenyl. *Prog. Brain Res.* **106,** 207–215.
5. Tatton, W. G., and Greenwood, C. E. (1991). Rescue of dying neurons: A new action for deprenyl in MPTP Parkinsonism. *J. Neurosci. Res.* **30,** 666–672.
6. Sano, M., Ernesto, C., Thomas, R. G., Klauber, M. R., Schafer, K., Grundman, M., Woodbury, P., Growdon, J., Cotman, C. W., Pfeiffer, E., Schneider, L. S., and Thal, L. J. (1997). A controlled trial of selegiline, alphatocopherol, or both as treatment for Alzheimer's disease. *N. Engl. J. Med.* **336,** 1216–1222.
7. Head, E., Hartley, J., Kameka, A. M., Mehta, R., Ivy, G. O., Ruehl, W. W., and Milgram, N. W. (1996). The effects of l-deprenyl on spatial short term memory in young and aged dogs. *Prog. Neuropsychopharmacol. Biol. Psychiatr.* **20,** 515–530.

M. Gassen, B. Pinchasi, and M. B. H. Youdim

Department of Pharmacology
Eve Topf and National Parkinson's Foundation Centers
Bruce Rappaport Family Research Institute
Faculty of Medicine, Technion
Haifa 31096, Israel

Apomorphine Is a Potent Radical Scavenger and Protects Cultured Pheochromocytoma Cells from 6-OHDA and H_2O_2-Induced Cell Death

The mixed-type dopamine D1–D2-receptor agonist apomorphine has had an impressive career as a replacement for L-dopa in the therapy for Parkinson's disease in the late stage of the disease (1). Earlier attempts to introduce apomorphine as a dopamine replacement had failed due to its extremely short duration of action and serious gastrointestinal side effects. Presently, the pharmacokinetic problems have been overcome with more sophisticated drug delivery devices, such as portable infusion pumps or insulin pens. Coadministration of the peripherally acting dopamine receptor antagonist domperidone prevents the undesired noncentral effects of apomorphine, which has become the medication of choice for patients suffering from on–off fluctuations after years of L-dopa administration.

However, the discovery that catecholamines can be cytotoxic has raised the question of the long-term effects of the treatment in Parkinson's disease with drugs like L-dopa or apomorphine on the progression of the disease. Catecholaminergic drugs can act as reducing agents and iron chelators. They can also interfere with cellular redox reactions and promote the formation of free radicals, which are thought to be important mediators of neurodegeneration. On the other hand, catecholamines can also act as radical scavengers and may, thus, slow down the progression of neurodegenerative diseases. In this study we demonstrated, that apomorphine can act as a potent free radical scavenger *in vitro* and in cell culture.

Free oxygen–derived radicals, namely superoxide ($O_2^{\cdot -}$) and hydroxyl (OH^-) radicals, are extremely reactive species and react with proteins, DNA, and, most importantly with polyunsaturated fatty acids, which are part of biological membranes. A major source of free radicals is cellular respiration (e.g., by the mitochondrial respiratory chain). Due to the short half-life of radical species, they will cause damage mainly close to the site where they are formed. For this reason, free radical biochemistry can be easily studied in

Advances in Pharmacology, Volume 42

1054-3589/98 $25.00

isolated mitochondria. We examined the effect of dopamine and apomorphine on the formation of thiobarbituric acid reactive substances (TBARS) from radical-induced lipid and DNA oxidation. Incubation of rat brain mitochondria with ascorbic acid (50 μM) and $FeSO_4$ (1–10 μM) leads to a 10-fold increase of TBARS formation as measured after 2 hr. This effect can be almost completely abolished by addition of submicromolar concentrations of apomorphine (2). The effective concentrations depend on the concentration of Fe^{2+} in the system and on the overall concentration of TBARS. Oxidation of apomorphine was monitoredby photometric determination of the colored oxidation products at $\lambda = 619$ nm and found to be accelerated in the presence of mitochondria, reflecting the protection of mitochondrial lipids and DNA. Kinetic experiments revealed a negative correlation between apomorphine oxidation and TBARS formation. The former reaction occurs at a high rate during early incubation, completely suppressing TBARS. When apomorphine oxidation slows down later on, increasing amounts of TBARS are generated in the system.

Dopamine also shows antioxidant properties in the rat brain mitochondrial system, although it was not quite as effective (IC 50 = 3.45 μM for 2.5 μM Fe^{2+}; 6.59 μM for 5.0 μM Fe^{2+}). On the basis of these data, iron chelation by dopamine may be a major contribution to the observed inhibition of TBARS formation. This can be ruled out for apomorphine, because this agent provides almost complete inhibition of TBARS formation at concentrations much lower than the Fe^{2+} concentration. Apomorphine also protects against oxidation of proteins. Free radicals induce cysteine–cysteine and tyrosine–tyrosine cross-links and react with proline, arginine, lysine, and threonine, leading to the formation of new keto and aldehyde functionalities (3). To detect these, we labeled the carbonyl groups with the specific reagent 2,4-dinitrophenylhydrazine. We found that forcing conditions (250 μM Fe^{2+}, 15 mM ascorbic acid) were needed to induce a threefold increase of protein carbonyls in the mitochondrial proteins. This effect could be reduced by 50% in the presence of 100 μM of apomorphine.

A key question that remains concerns the balance between catecholamine toxicity and possible beneficial effects due to antioxidation. We looked at this problem in PC12 cell culture, a well-established system to study apoptotic and necrotic cell death (4). Oxidative stress can be induced by various agents, like H_2O_2, organic hydroperoxides, or 6-hydroxydopamine.

We treated PC12 cells with H_2O_2 and 6-hydroxydopamine and observed cell death in a concentration-dependent manner within 24 hr. (Fig. 1) There was no significant difference of the sensitivity between cells that were grown in medium containing 15% serum (one-third fetal calf serum, two-thirds horse serum) and those that had been differentiated for 6 days with an additional 100 μg/ml 7S-NGF (5), if all the nerve growth factor had been washed out prior to the experiment. Although it takes 24 hr to observe the maximum cell death, only 2 hr exposure to the toxic agent is sufficient to induce the full damage. The viability of the cells has been alternatively measured by counting cells after trypane blue exclusion or by measurement of metabolic conversion of 3-[4,5-dimethylthiazol-2-yl]-2,5-diphenyltetrazolium bromide (MTT) into a colored product, which can be dissolved in isopropanol and quantified photometrically.

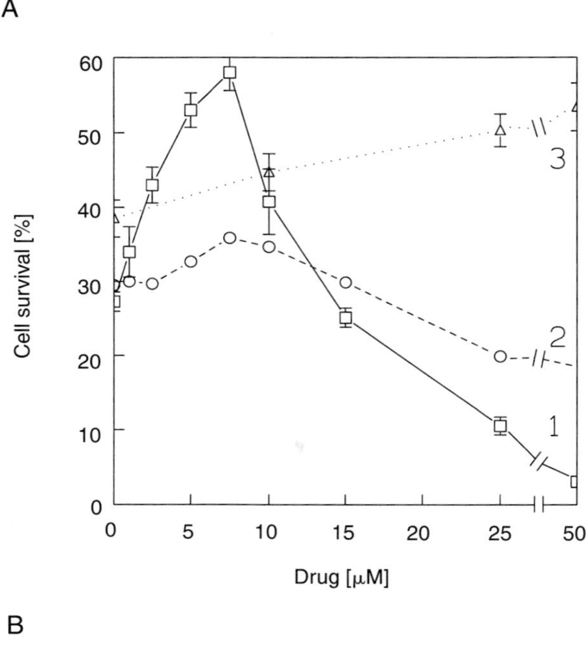

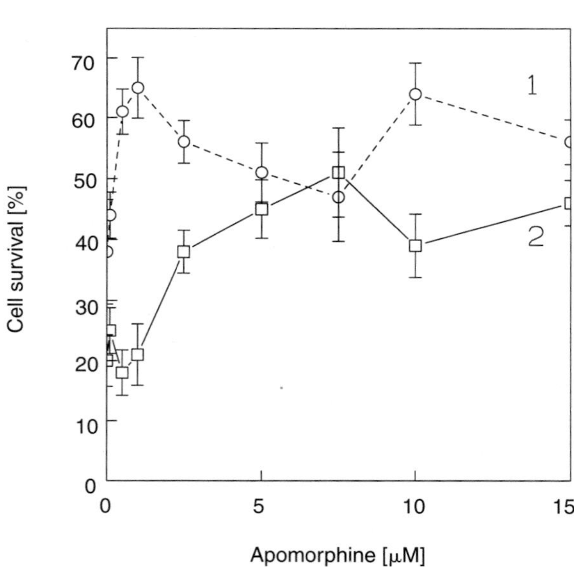

FIGURE 1 Antioxidant effects of apomorphine (*A*) against 600 μM H_2O_2 (1: apomorphine added simultaneously with H_2O_2; 2: 1 hr after H_2O_2; dopamine added simultaneously) and (*B*) against 6-hydroxydopamine (1: 150 μM; 2: 200 μM).

Both methods produced equivalent results, and we used the faster MTT-procedure for routine experiments.

Exact EC 50 values were obtained: 400 μM for H_2O_2 and 150 μM for 6-hydroxydopamine were necessary to kill 50% of the cultured cells. In this system, dopamine and apomorphine were tested for their ability to protect PC12 cells from the oxidative insults. At the same time, the toxicity can be monitored to obtain information about the therapeutic window of the agents. Despite both being catecholamines, apomorphine and dopamine differ markedly in their toxicity and in their potency to protect the cells against H_2O_2. Apomorphine is by far more efficient as an antioxidant: only 5 μM improve the rate of survival from 50% to 80% in the presence of 400 μM of H_2O_2 and from 30% to 58% with 600 μM of H_2O_2. The same concentration of dopamine does not provide any significant protection, since as much as 125 μM of dopamine are required to obtain the same effect. However, the toxicity of apomorphine is much higher, with ED 50 = 50 μM, than that of dopamine, which, unlike 6-hydroxydopamine, does not lead to any significant cell degeneration at concentrations below 250 μM. Any protection against H_2O_2 by apomorphine depended on the presence of the drug during the insult. Preincubation with apomorphine and washout prior to H_2O_2 addition or addition of H_2O_2 1 hr after the toxin did not improve the survival as compared with controls treated only with the oxidant (Fig. 1a).

Apomorphine but not dopamine was able to provide some protection against 6-hydroxydopamine insults. The survival rate after 150 μM of 6-hydroxydopamine was improved from 50 to 65% with only 1 μM of apomorphine; at 200 μM of 6-hydroxydopamine, an improvement from 30 to 48% was obtained with 5 μM of apomorphine. This is the first example of one catecholamine attenuating the toxicity of another (Fig. 1b).

Because the dopamine prodrug L-dopa and apomorphine are widely used in the treatment of Parkinson's disease, it is of high interest to further investigate the influence of these agents on the biochemical processes involved in the progression of neurodegeneration. We found that apomorphine is a representative of a catecholamine with pronounced antioxidant effects, although problems might arise due to only a small therapeutic window. However, we consider this study also as an incentive for the design of novel, less toxic catecholaminergic dopamine receptor agonists with antioxidant properties.

Acknowledgments

We wish to to thank ASTA Medica (Frankfurt/Main, Germany), the Golding Parkinson Fund (Technion, Haifa, Israel), and the National Parkinson Foundation (USA). M.G. thanks the Minerva Foundation (Heidelberg, Germany) for his postdoctoral fellowship.

References

1. Gancher, S. T., Nutt, J. G., and Woodward, W. R. (1995). Apomorphine infusional therapy in Parkinson's disease: Clinical utility and lack of tolerance. *Mov. Disord.* **10,** 37–43.

2. Gassen, M., Glinka, Y., Pinchasi, B., and Youdim, M. B. H. (1996). Apomorphine is a highly potent free radical scavenger in rat brain mitochondrial fraction. *Eur. J. Pharmacol.* **308**, 219–226.
3. Stadtman, E. R. (1993). Oxidation of free amino acids and amino acid residues in proteins by radiolysis and by metal-catalyzed reactions. *Annu. Rev. Biochem.* **62**, 797–821.
4. Vimard, F., Nouvelot, A., and Duval, D. (1996). Cytotoxic effects of an oxidative stress on neuronal-like pheochromocytoma cells. *Biochem. Pharmacol.* **51**, 1389–1395.
5. Greene, L. A., Aletta, J. M., Rukenstein, A., and Green, S. H. (1987). PC12 pheocytoma cells, culture, nerve growth factor treatment and experimental exploidation. *Methods Enzymol.* **147**, 207–216.

Pekka T. Männistö

Department of Pharmacology and Toxicology
University of Kuopio
FIN-70211 Kuopio, Finland

Catechol O-Methyltransferase: Characterization of the Protein, Its Gene, and the Preclinical Pharmacology of COMT Inhibitors

Catechol O-methyltransferase (COMT) is a widespread enzyme that catalyzes the transfer of the methyl group of S-adenosyl-L-methionine to one of the phenolic group of the catechol substrate in the presence of Mg^{2+} and following a sequential ordered mechanism. High COMT activity is found particularly in the liver, kidney, and gut wall (1).

A single COMT gene codes two separate enzymes, the soluble (S-COMT) and membrane-bound (MB-COMT) forms. S-COMT contains 221 amino acids. MB-COMT has an additional amino terminal extension of 43 (rat) or 50 (human) amino acids. The hydrophobic 17 and 24 amino acid residues in rat and humans, respectively, form an alfahelical transmembrane domain that serves as a membrane anchor. Otherwise, the two proteins are similar. Synthesis of recombinant S-COMT in *E. coli* and MB-COMT in the insect cells using baculovirus vectors has helped in clarifying the biochemistry, physiology, and pharmacology of COMT (1, 2).

Advances in Pharmacology, Volume 42

1054-3589/98 $25.00

At the moment, it is thought that MB-COMT is partially responsible of the termination of dopaminergic and noradrenergic synaptic neurotransmission. S-COMT, on the other hand, is a high-capacity enzyme mainly responsible for the elimination of biologically active or toxic, particularly exogenous catechols and some hydroxylated metabolites. Accordingly, MB-COMT is a dominating isoenzyme in the human brain.

Early COMT inhibitors, like gallates, tropolone, and U-0521 (3′,4′-dihydroxy-2-methyl-propiophenone) have IC 50 and Ki values in micromolar range or higher. Owing to unfavorable pharmacokinetics their efficacy *in vivo* has been low. Moreover, most of them lack selectivity and are quite toxic.

Second-generation COMT inhibitors that are very potent, highly selective, and orally active were independently developed in three laboratories in the late 1980s. Despite CGP 28014, nitrocatechol is the key structure of the most of these molecules. We have divided the current COMT inhibitors into three groups: (1) mainly peripherally acting nitrocatechol-type compounds (entacapone, nitecapone), (2) broad-spectrum nitrocatechols having activity both in the periphery and the brain (tolcapone, Ro 41-0960), and (3) atypical compounds and pyridine derivatives (CGP 28014, 3-hydroxy-4-pyridone and its derivatives), some of which are not COMT inhibitors *in vitro* but may instead inhibit catechol O-methylation by some other mechanism (3).

The common features of the newest compounds are excellent potency, high selectivity, low toxicity, and activity through oral administration. Their biochemical properties have been fairly well characterized. Most of these compounds do not affect any other enzymes studied. However, 1,2-dimethyl-3-hydroxypyridin-4-one is also an effective iron chelator that inhibits tyrosine and tryptophan hydroxylases, and thereby the formation of catecholamines and 5-hydroxytryptamine (1–3).

The main use of COMT inhibitors will be as adjunct (or additional adjunct) in the therapy of Parkinson's disease. The standard therapy of Parkinson's disease is oral L-dopa given with a dopa decarboxylase (DDC) inhibitor (e.g., carbidopa and benserazide), which does not reach the brain. When the peripheral DDC is inhibited, the concentration of 3-OMD in plasma is many times that of dopa. Because the half-life of 3-OMD is about 15 hr, compared with about 1 hr for dopa, the concentration of 3-OMD remains particularly high during chronic therapy, especially if new slow-release L-dopa preparations are used. Although the opinions about the real nature and actions of 3-OMD vary, it certainly is not a useful metabolite but rather has a potential to be clinically harmful (1–4).

COMT inhibitors improve the brain entry of L-dopa and decrease 3-OMD formation in the peripheral tissues. The dose of L-dopa could be decreased, compared with the present combination therapy. The dose interval of L-dopa could be prolonged. COMT inhibitors should also decrease fluctuations of the dopamine formation.

A summary of some preclinical evidence supporting the suggested use is given here. In *in vivo* studies, nitrocatechol-type inhibitors prevent the formation of COMT-derived metabolites at doses of 3–30 mg/kg. The duration of COMT inhibition in various tissues has varied from less than 60 min to 12 hr, depending on the dose and the compound. Generally, tolcapone and Ro 41-0960 are more potent and longer acting than entacapone and nitecapone (2).

All types of COMT inhibitors potentiate the L-dopa–induced turning behavior of rats having unilateral nigral lesions. This has generally been kept as a reliable rat model of Parkinson's disease. It is noteworthy that the peripherally acting compound (entacapone) is practically as effective as the broad-spectrum compound (tolcapone). This suggests that the majority of the beneficial action is peripheral in origin, evidently through enhanced bioavailability of L-dopa. Also, an atypical compound, CGP 28014, is equally active as the real COMT inhibitors (5). It is also noteworthy that when given alone, none of the COMT inhibitors significantly affects the turning behavior (Fig. 1).

In the case of entacapone and tolcapone, but not with CGP 28014, enhanced extracellular dopamine levels can be demonstrated in L-dopa/carbidopa-treated rats, using a striatal microdialysis technique. However, when given without L-dopa, extracellular dopamine levels are not increased in spite of elevated dihydroxyphenylacetic acid and reduced homovanillic acid (HVA) levels (2).

Analyzed from brain homogenates, all COMT inhibitors work effectively at 3–30 mg/kg as inhibitors of 3-OMD (mostly peripheral action), HVA, and 3-MT formation (mostly central actions) (3).

There are possibilities of drug interactions with COMT inhibitors. Both widely used inhibitors of dopa decarboxylase, benserazide and carbidopa, are substrates and weak inhibitors of COMT. The affinity of benserazide for COMT is higher than that of carbidopa. COMT inhibition would, therefore, increase particularly the plasma levels and brain penetration of benserazide. When it is coadministered with tolcapone to *Rhesus* monkeys, a dose-dependent decrease of the brain dopamine formation takes place. High doses of nitrocatechols may influence the absorption of carbidopa and increase the peripheral formation of dopamine. Because several adrenergic drugs having a catechol structure are

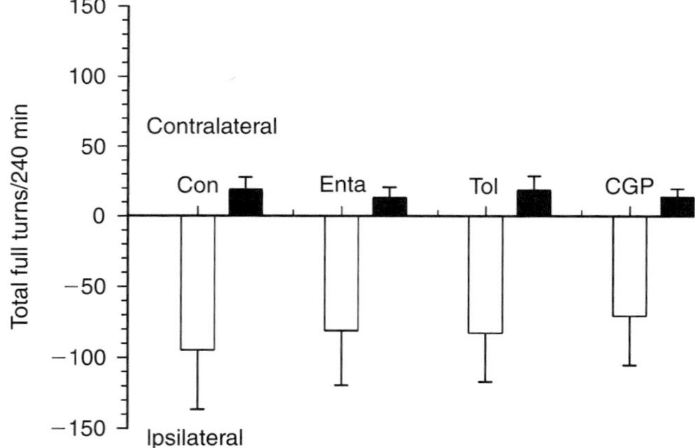

FIGURE 1 Effects of entacapone (Enta), tolcapone (Tol), and CGP 28014 at 30-mg/kg intraperitoneal doses in the rat model of Parkinson's disease. Both contralateral and ipsilateral circlings were analyzed for 240 min. Mean ± SEM, n = 6–8.

COMT substrates it is possible to prolong or even potentiate in some cases their actions by COMT inhibitors. Such drugs include bronchodilating compounds (adrenaline, isoprenaline, rimiterol), dopamine agonists (dobutamide, fenoldopam, apomorphine), antihypertensive drugs (α-methyldopa) and a β-blocking drug (nadolol) (1).

There is also a possibility for potentiating interactions with endogenous catecholamines during stress and exercise and adverse drug interactions with, for example, exogenous noradrenaline and the drugs mentioned previously. However, interaction studies in animals have not supported this conclusion. Evidently, the capacity of S-COMT in the peripheral tissues is so high that only a minor general COMT inhibition can ever been achieved.

Estrogens are easily hydroxylated to catecholestrogens that serve as COMT substrates. Both COMT activity and estrogen concentrations are high in breast cancer tissue. Estrogens stimulate tumor growth and catecholestrogens inhibit it. COMT inhibition would then inhibit cancer growth. The consequences of the use of COMT inhibitors to prevent the metabolism of catecholestrogens need further study. No clinical data are available at the moment.

The time course of the COMT activity in the conceptus also favors the view that catecholestrogens have a role in conceptus-maternal signaling during the establishment of pregnancy. Estrogen-induced implantation of the fertilized ovum may be mediated through formation of catecholestrogens that stimulate the synthesis and/or release of prostaglandins. Interference in this delicate system by COMT inhibitors would be worth studying as a novel contraceptive mechanism. No clinical data support this idea so far (1).

We want to point out, however, that during the first trimester, COMT is likely to protect the placenta and the developing embryo from activated hydroxylated compounds formed by aryl hydrocarbon and estrogen hydroxylases (1). Direct effects of COMT inhibitors in the brain are also evident. Even in Parkinson's disease there also is, besides the loss of nigrostriatal dopamine, a diffuse depletion of other dopaminergic and noradrenergic innervation in the brain. Central COMT inhibtion would be beneficial in restoring noradrenaline and dopamine levels and in improving the symptoms (decreased cognition, etc.) caused by the general lack of these transmitters (1, 2). I want to stress again, however, that centrally acting COMT inhibitors do not work better than the peripheral COMT inhibitors in improving the efficacy of L-dopa on Parkinson's disease.

Noradrenaline and dopamine deficiency in the synaptic cleft also depicts some forms of depression. The remedy of this deficit by the use of tricyclic uptake inhibitors and monoamine oxidase inhibitors has become an established therapy for depression. In analogy, the COMT inhibitors reaching the brain would decrease the metabolism of noradrenaline and dopamine and be beneficial in depression, as has been demonstrated in animal studies (1, 2).

References

1. Männistö, P. T., Ulmanen, I., Lundström, K., Taskinen, J., Tenhunen, J., Tilgmann, C., and Kaakkola, S. (1992). Characteristics of catechol O-methyltransferase (COMT) and properties of selective COMT inhibitors. *Prog. Drug. Res.* **39**, 291–350.

2. Kaakkola, S., Gordin, A., and Männistö, P. T. (1994). General properties and clinical possibilities of new selective inhibitors of catechol O-methyltransferase (COMT). *Gen. Pharmacol.* **25**, 813–824.
3. Männistö, P. T., Tuomainen, P., and Tuominen, R. K. (1992). Different in vivo properties of three new inhibitors of catechol O-methyltransferase in the rat. *Br. J. Pharmacol.* **105**, 569–574.
4. Männistö, P. T., and Kaakkola, S. (1989). New selective COMT inhibitors: Useful adjuncts for Parkinson's disease? *Trends Pharmacol. Sci.* **10**, 54–56.
5. Törnwall, M., and Männistö P. T. (1993). Effects of three types of catechol O-methylation inhibitors on L-3,4-dihydroxyphenylalanine-induced circling behaviour in rats. *Eur. J. Pharmacol.* **250**, 77–84.

Jukka Vidgren
Orion Pharma
FIN-02101 Espoo, Finland

X-Ray Crystallography of Catechol O-Methyltransferase: Perspectives for Target-Based Drug Development

Catechol O-methyltransferase (COMT) plays an important physiological role in the metabolism of catecholamines and a large number of other substances containing the catechol structure. COMT also inactivates L-dopa, which is used as a drug in the treatment of Parkinson's disease. COMT catalyzes the transfer of the methyl group from the coenzyme S-adenosyl-L-methionine (AdoMet) to one of the phenolic hydroxyl groups of a catechol or substituted catechols. The presence of magnesium ions is required for the catalysis. The reaction products are O-methylated catechol and S-adenosyl-L-homocysteine (AdoHcy). COMT enzyme is common in almost all mammalian tissues. The highest activities have been found in liver and kidney.

During the last decade there was a remarkable interest in COMT. Basic biochemical and molecular biology research has given detailed insights into the function and role of the enzyme (1). The solution of the atomic structure has allowed the analysis of the molecular reaction mechanism and the structure-based design of potent inhibitors (2) .

COMT occurs in two distinct forms, soluble and membrane-bound, the latter having a 50 amino-acids-long membrane anchor region. Enzyme kinetic studies have shown that the membrane-bound COMT has about 10 times lower

Advances in Pharmacology, Volume 42

1054-3589/98 $25.00

Km value than soluble COMT, but at saturating substrate concentrations, the catalytic number of both enzyme forms is similar. At low concentrations, catecholamines are methylated more efficiently by membrane-bound COMT (3).

The crystal structure of rat soluble COMT has been determined to high resolution. COMT has a single domain α/β-folded structure, in which eight α-helices are arranged around the central mixed β-sheet. The active site of COMT consists of the AdoMet-binding domain and the catalytic site. The catalytic site is a rather simple environment formed by a few amino acids that are important for substrate binding and catalysis of the methylation reaction, and by the metal ion (see Fig. 1) (2). The magnesium ion plays a crucial role for the catalytic activity of COMT. The Mg^{2+} bound to the enzyme makes the hydroxyl groups of the catechol substrate more easily ionizable. In a vicinity of one hydroxyl from the bound substrate, there is a lysine residue, which accepts the proton of that moiety, and subsequently the methyl group is transferred from the AdoMet to the hydroxyl. Lysine acts as a catalytic base in a general base-catalyzed nucleophilic reaction (4, 5).

Human, rat, and pig enzymes have a high degree of sequence homology, containing about 80% identical amino acids. The most significant difference between the catalytic sites of these enzymes lies in residue 38, the hydrophobic tryptophan in rat and human COMT and the polar arginine in pig COMT. Kinetic data show that the Km values of common substrates for rat and human COMT are very similar, whereas pig COMT shows considerably higher Km

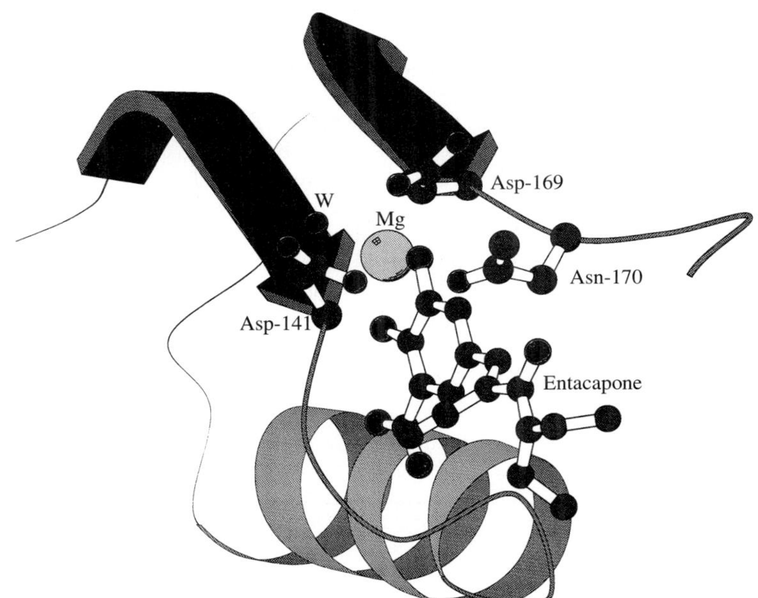

FIGURE 1 Binding of Mg^{2+} ion and the tight-binding inhibitor entacapone in COMT. W represents a water molecule.

values for catechol substrates. The same tendency is apparent for competitive inhibitors represented by the lower Ki values for rat and human COMT.

In the 1980s the new therapeutic principle of using COMT inhibitors as adjuvants in the L-dopa therapy of Parkinson's disease started an active period in the development of COMT inhibitors as potent drug molecules. Nitrocatechol derivatives have been shown to be the most active inhibitors of catechol O-methyltransferase. Entacapone (OR-611) and tolcapone (Ro-40-7592) have been extensively studied in the treatment of Parkinson's disease. Both compounds are very potent and selective tight-binding inhibitors of human COMT with Ki values of 0.3 nM. They differ mainly in their pharmacokinetic properties: Entacapone acts peripherally, while tolcapone inhibits COMT both peripherally and centrally. Although nitrocatechols possess the basic catechol structure and bind very well into the active site of COMT, they are not methylated. The electron-withdrawing nitro groups strongly stabilize the ionized catechol–COMT complex, and the energy barrier for the methylation is too high. This is the reason why these compounds act as inhibitors but not as substrates (4, 5).

Quantitative structure activity relationship studies of substituted catechols indicated the importance of the acidity of the hydroxyl groups long before the crystal structure was determined. The results led to the conclusion that the lipophilicity of the side chain could play a significant role in the binding. Later when the crystal structure was determined, it became clear that the shallow groove in the active site of COMT allows the long side chains of inhibitors to reach the surface of the enzyme. The crystal structure of COMT in complex with an inhibitor, coenzyme, and magnesium showed unambiguously the atomic arrangements of the ligands with the COMT residues. The structure proved that the magnesium ion has an essential role in the binding of substrates and in the methyl transfer. Further, the structure indicated that, ideally, one of the ortho-positions of catechol hydroxyls is substituted with a nitro group, while the other ortho-position should be unsubstituted. The two remaining positions in the catechol ring can be substituted rather freely, for example, to modify the physicochemical properties of the compounds.

The best inhibitors of COMT are nitrocatechols, so that one of their hydroxyl groups is ionized under physiological conditions. Ionized molecules penetrate poorly the blood brain barrier. In clinical trials it has been shown that the beneficial effect on the L-dopa metabolism in the treatment of Parkinson's disease is reached with a peripheral COMT inhibitor such as entacapone. The three-dimensional structure of COMT with the simple active site and the role of Mg^{2+} ion limits the possibilities for designing noncatechol-type inhibitors.

References

1. Salminen, M., Lundström K., Tilgmann, C., Savolainen, R., Kalkkinen, N., and Ulmanen, I. (1990). Molecular cloning and characterization of rat liver catechol-O-methyltransferase. *Gene* **93**, 241–247.
2. Vidgren, J., Svensson, L. A., and Liljas, A. (1994). Crystal structure of catechol O-methyltransferase. *Nature* **368**, 354–358.

3. Lotta, T., Vidgren, J., Tilgmann, C., Ulmanen, I., Melen, K., Julkunen, I., and Taskinen, J. (1995). Kinetics of human soluble and membrane-bound catechol O-methyltransferase: A revised mechanism and description of the thermolabile variant of the enzyme. *Biochemistry* **34**, 4202–4210
4. Vidgren, J., and Ovaska, M. (1997). Structural aspects in the inhibitor design of catechol O-methyltransferase. *In* Structure Based Drug Design. P. Veerapandian, ed.), pp. 343–363. Marcell Decker, New York, NY.
5. Ovaska, M. (1997). The mechanism of catalysis and inhibition of catechol O-methyltransferase. (submitted)

John G. Nutt
Departments of Neurology and Physiology and Pharmacology
Oregon Health Sciences University
Portland, Oregon 97201

Catechol O-Methyltransferase Inhibition and the Treatment of Parkinson's Disease

Catechol O-methyltransferase (COMT) plays an important role in the metabolism of the dopamine precursor, L-dopa, and of dopamine. As such, inhibitors of COMT could alter the metabolism of L-dopa and dopamine to therapeutic advantage by a number of mechanisms in L-dopa–treated patients with Parkinson's disease (PD). The introduction of entacapone and tolcapone, reversible inhibitors of COMT that are well absorbed and well tolerated, has permitted investigation of the clinical importance of COMT in Parkinson's disease. Entacapone is primarily a peripheral inhibitor of COMT; tolcapone, on the other hand, penetrates the blood brain barrier and inhibits brain COMT as well as peripheral COMT (1).

I. L-Dopa Absorption

COMT is in high concentration in the gut and liver and may contribute to first-pass metabolism of levodopa. Inhibition of gut and liver COMT would then be expected to increase the bioavailability of orally administered L-dopa. Pharmacokinetically, this should be manifest by an increase in peak concentration of plasma L-dopa and a decrease in the time-to-peak concentration. Neither entacapone nor tolcapone increases the peak concentration of levodopa or

Advances in Pharmacology, Volume 42

1054-3589/98 $25.00

shortens the time-to-peak concentration in subjects with PD (2, 3). These results are in contrast to the effect of inhibition of first-pass decarboxylation of L-dopa; the aromatic amino acid decarboxylase inhibitors, carbidopa and benserazide, markedly increase the peak plasma L-dopa concentrations. These observations argue that first-pass O-methylation of L-dopa is not quantitatively important. To the extent that dyskinesia and, perhaps, some mental effects are related to peak concentrations of L-dopa, this lack of effect on peak concentrations by COMT inhibitors may be beneficial.

Higher doses of entacapone and tolcapone delay the absorption of L-dopa by uncertain mechanisms. Despite increasing COMT inhibition with very high doses of COMT inhibitors, the area under the curve (AUC) of levodopa does not increase because of the delayed absorption (4). Thus, doses of COMT inhibitors chosen for clinical use will have to balance the extent to which COMT is inhibited with the extent to which L-dopa absorption is delayed.

II. L-Dopa Elimination

COMT is in high concentrations in many other peripheral tissues. Tolcapone and entacapone can be shown to reversibly inhibit RBC COMT by greater than 80% in humans (4, 5). Inhibition of peripheral tissue COMT could alter the elimination of L-dopa. Pharmacokinetically, this should be manifest by an increase in elimination half-life and in AUC. The elimination half-life of levodopa is variably increased by COMT inhibition, but the AUC of L-dopa is robustly increased (2–5).

III. Antiparkinsonian Effects of COMT Inhibition

The slowed elimination and increased AUC of L-dopa translate into an increase in the duration of clinical response to single doses of L-dopa in patients with PD and a fluctuating (on–off) response. The maximum antiparkinsonian effect is not altered, and dyskinesia and other dopaminergic adverse effects are similar to those with L-dopa alone (2, 3). Positron emission tomography (PET) studies indicate that the delivery of L-dopa to brain is increased, while the entry of 3-O-methyldopa (3-OMD)is reduced, roughly in proportion to the changes in the peripheral pharmacokinetics of L-dopa induced by COMT inhibition.

The effect of COMT inhibition is more robust during repetitive dosing with the COMT inhibitors and L-dopa; the average plasma concentrations of L-dopa are increased, daily L-dopa requirements are reduced, and the amount of "on" time is increased. Because tolcapone and entacapone are reversible inhibitors of COMT with a short plasma half-life and most patients have a drug-free period of hours during the night, there is no accumulation of L-dopa from day to day (3).

The changes in clinical response induced by tolcapone and entacapone are roughly proportional to the changes in peripheral L-dopa pharmacokinetics.

Thus, the central inhibition of COMT by tolcapone is of uncertain importance at this point in time.

Clinical studies clearly indicate that COMT inhibition augments the antiparkinsonian actions of L-dopa. The important issue is to determine if this augmentation offers clinical benefits that cannot be obtained by increasing the dose of L-dopa or using controlled-release preparations of L-dopa. Two potential differences are: (1) peak concentrations are not increased by COMT inhibition as they would be with administering larger L-dopa doses and (2) absorption of COMT inhibitors and L-dopa may be more predictable than absorption of controlled-release preparations.

IV. L-Dopa Transport into Brain

3-OMD, the product of methylation of L-dopa, has long been of interest in the treatment of PD. It was first speculated to be beneficial, serving as a slow release pool of L-dopa. Subsequently, it was suggested to be detrimental and the cause of a poor response to L-dopa in patients with PD. The mechanism for this adverse effect is believed to be by competition with L-dopa for transport by the large neutral amino acid (LNAA) transporter at the blood-brain barrier. It can be demonstrated in animals and humans that high concentrations of 3-OMD will inhibit the entry of L-dopa into brain and block its pharmacological effects. The issue is whether 3-OMD in the clinical setting is in concentrations to significantly interfere with L-dopa transport. Because concentrations of plasma 3-OMD range from about 20 to 100 nmol/ml and the concentrations of the other plasma LNAAs (largely leucine, isoleucine, phenylalanine, tyrosine, valine, and tryptophan) total 400 to 700 nmol/ml, it would seem that twofold changes in concentrations of 3-OMD are unlikely to markedly influence the total of LNAAs competing with L-dopa for transport at the blood-brain barrier.

Acute COMT inhibition by entacapone or tolcapone markedly reduces the concentrations of COMT formed after single doses of L-dopa in a dose dependent manner (4, 5). During repetitive dosing with entacapone in PD patients on long-term L-dopa therapy, 3-OMD decreased by about 60% (3). The clinical response to a single dose of L-dopa, however, did not change with the reduction of plasma 3-OMD. This observation is consistent with the reasoning that 3-OMD, in the concentrations occurring during long-term L-dopa treatment, does not materially affect L-dopa penetration of the blood-brain barrier. PET studies also find that varying the concentration of 3-OMD within the range seen clinically does not alter the influx of [^{18}]F-DOPA also suggesting that L-dopa transport is not changed by COMT inhibition.

V. Inhibition of Brain COMT

Both entacapone and tolcapone increase extracellular brain dopamine concentrations after L-dopa. Only tolcapone reduces formation of homovanillic acid in brain, indicative of tolcapone's central inhibition of COMT (1). Inhibition of *brain* COMT could increase availability of L-dopa for decarboxylation to dopamine and reduce the catabolism of dopamine by O-methylation to 3-O-methyltyramine and thereby increase the pharmacological response to L-dopa. The unanswered question is whether central inhibition of COMT adds to the augmentation produced by increasing the amount of L-dopa delivered to brain by reducing its peripheral O-methylation.

Another potential benefit of inhibition of brain COMT is sparing of S-adenosylmethionine, the methyl donor for O-methylation reactions. Low cerebrospinal fluid S-adenosylmethionine concentrations have been linked to depression and dementia. Both depression and dementia are common accompaniments of PD, and perhaps sparing S-adenosylmethionine would reduce the prevalence of these problems.

References

1. Kaakkola, S., and Wurtman, R. (1993). Effects of catechol-O-methyltransferase inhibitors and L-3,4-dihydroxyphenylalanine with or without carbidopa on extracellular dopamine in rat striatum. *J. Neurochem* **60**, 137–144.
2. Roberts, J. W., Cora-Locatelli, G., Bravi, D., Amantea, M. A., Mouradian, M. M., and Chase, T. N. (1993). Catechol-O-methyltransferase (COMT) inhibitor tolcapone prolongs levodopa/carbidopa action in parkinsonian patients. *Neurology* **43**, 2685–2688.
3. Nutt, J. G., Woodward, W. R., Beckner, R. M., Stone, C. K., Berggren, K., Carter, J. H., Gancher, S. T., Hammerstad, J. P., and Gordin, A. (1994). Effect of peripheral catechol-O-methyltransferase (COMT) inhibition on the pharmacokinetics and pharmacodynamics of levodopa in parkinsonian patients. *Neurology* **44**, 913–919.
4. Dingemanse, J., Jorga, K. M., Zurcher, G., Schmitt, M., Sedek, G., Da Prada, M., and van Brummelen, P. (1995). Pharmacokinetic-pharmacodynamic interaction between the COMT inhibitor tolcapone and single-dose levodopa. *Br. J. Clin. Pharmacol.* **40**, 253–262.
5. Keranen, T., Gordin, A., Harjola, V., Karlsson, M., Korpela, K., Pentikainen, P. J., Rita, H., Seppala, L., and Wikberg, T. (1993). The effect of catechol-O-methyltransferase inhibition by entacapone, on the pharmacokinetics and metabolism of levodopa in healthy volunteers. *Clin. Neuropharmacol.* **16**, 145–156.

B. Burchell, C. H. Brierley, G. Monaghan, and D. J. Clarke

Department of Molecular and Cellular Pathology
Ninewells Hospital and Medical School
Dundee, DD1 9SY, Scotland

The Structure and Function of the UDP-Glucuronosyltransferase Gene Family

The UDP-glucuronosyltransferases (UGTs) have evolved to catalyze the glucuronidation of potentially hazardous endogenous compounds such as catecholamines and bilirubin and environmental chemicals (1). The UGT gene family is a major participant in the chemical defense required by plant and animal environmental warfare. The role of glucuronidation may not be confined to chemical defense, because rat and bovine olfactory epithelia have been shown to contain specific UGT isoforms responsible for the glucuronidation of odorants catalyzing signal termination and excretion of stimulants (1).

A vast array of chemicals are encountered in the modern environment, where the chemical revolution has outpaced biological evolution of humans leading to serious unpredictable toxicities. Even some conjugates remain biologically active (1). Nonetheless, the majority of chemicals are effectively glucuronidated by a large family of enzymes. Individual UGTs are differentially regulated in tissues and during development by hormones and xenobiotics, providing a different panel of enzymes for organ-specific function.

I. UGT1 Subfamily

At least 20 different human UGT cDNAs have been identified and classified into two subfamilies based on sequence identities.[1] The UGT1 subfamily proteins share an identical C-terminal coding sequence, whereas the N-terminal 246 amino acids were as little as 38% similar.

Southern blot analysis indicated that the region encoding the conserved 3′ half of four separate human UGT1 cDNAs (the common domain) was a single copy in the human genome, suggesting a role for alternative splicing in the synthesis of different isoforms. In support of this, the common domain and the isoform-specific 5′ half of the four UGT1 cDNAs colocalized to chromosome 2 at 2q.37 (2). Owens and Ritter (3) described the existence of a gene complex by the isolation of overlapping cosmid clones containing six alternative substrate-determining first exons upstream of the four exons that make up the common domain. The human liver cDNA clone HP4 has the same common domain se-

Advances in Pharmacology, Volume 42

1054-3589/98 $25.00

quence as other UGT1 cDNAs but contained a novel substrate-determining exon that was termed UGT1A8 and was not among those already genomically cloned. This suggested that the UGT1 gene locus was larger than had previously been described, presumably extending further upstream. Human genomic Southern blotting indicated the presence of multiple sequences homologous to the 5′ portion of HP4 (2). Further work has shown that the human UGT1 gene is a single copy gene consisting of four common exons and more than 13 variable exons that span more than 200 kb of the human genome (C. Brierley, B. Burchell, unpublished work and Fig. 1). Genetic defects have been observed in the UGT1 gene associated with hyperbilirubinemia. Defects in the coding regions of the gene cause loss of single or many isoforms. Defects in regulatory regions are implicated in polymorphic variation in drug and endobiotic glucuronidation.

II. UGT2 Subfamily

The UGT2 subfamily has been subdivided into the UGT2A (olfactory-specific isoforms) and the UGT2B (steroid/bile and specific isoforms). To date, 10 distinct UGT2B cDNAs have been isolated.

Sequence differences that occur along the entire length of the human UGT2B cDNAs (unlike UGT1 cDNAs) strongly suggest that the UGT2B isoenzymes are encoded by independent genes.Indeed, the UGT2B4 gene has been determined to have five introns (1.2–4.9 kb) and six exons in a gene structure of 17.5 kb (see Fig. 1)

Oligonucleotide primers specific for three of the human UGT2B cDNAs (UGT2B4, UGT2B8, UGT2B15) were used to map the gene(s) to chromosome 4 utilizing polymerase chain reaction and a panel of human rodent somatic cell hybrid lines. A yeast artificial chromsome DNA of 195 kb was isolated using UGT2B4 and also was observed to contain genes encoding UGT2B9 and UGT2B15. This genomic DNA containing three UGT genes was localized to 4q13 by FISH analysis (4).

The UGT2B subfamily has probably evolved by whole gene duplication and divergencies rather than exon duplication, as observed in UGT1 subfamily. The subfamilies expanded in different ways on different chromosomes, but similar exon–intron structures encode a protein family with similar functional architecture despite the amino acid sequence diversity (see Fig. 1).

III. Substrate Specificity of Cloned–Expressed Human Liver UGTs

Glucuronidation of drug molecules containing a wide range of acceptor groups has been reported, including phenols (e.g., propofol, paracetamol, naloxone), alcohols (e.g., chloramphenicol, codeine, oxazepam), aliphatic amines (e.g., ciclopiroxalimine, lamotrigine, amitriptyline), acidic carbon atoms (e.g., feprazone, phenylbutazone, sulphinpyrazone), and carboxylic acids (e.g., na-

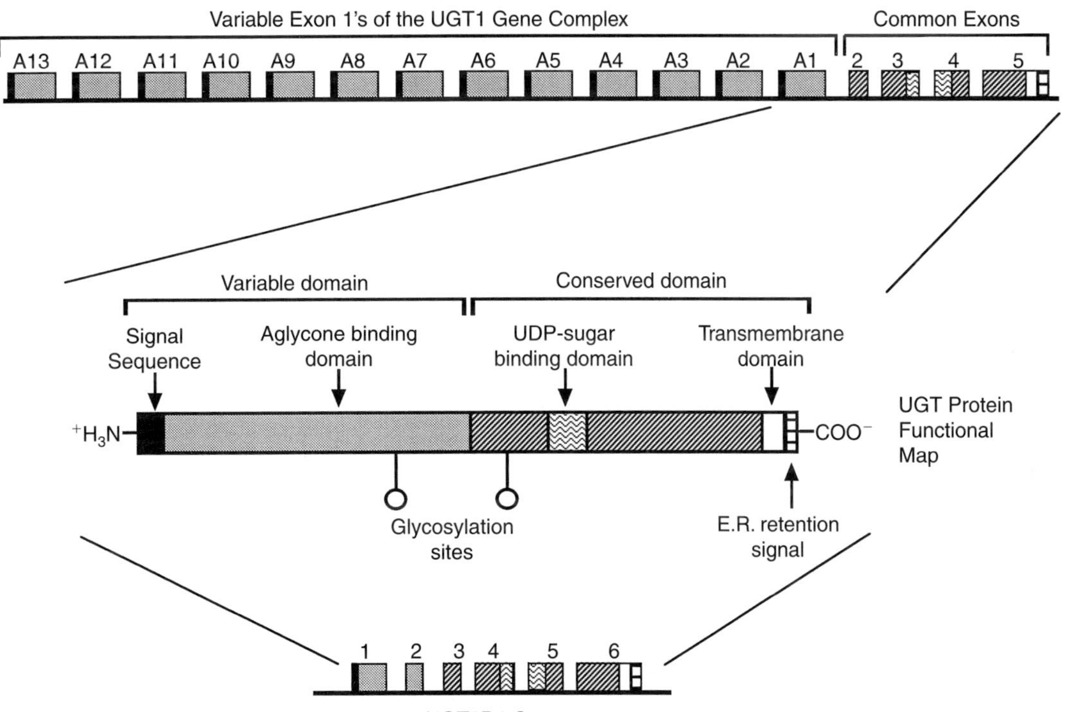

FIGURE 1 The relationship of exon organization of UGT genes to a UGT protein architectural map. The different shaded areas relate the domains of the exons to the functional protein map.

proxen, zomepirac, ketoprofen). This indicates the variability of acceptor groups that can be conjugated to glucuronic acid in humans (1).

Heterologous expression of UGT1 family members has shown that they catalyze the glucuronidation of bilirubin, phenols, amines and carboxylic acids (5). UGT1A6 exhibited a limited substrate specificity for planar phenolic compounds, whereas UGT1A8 was more promiscuous in acceptance of nonplanar phenols, anthraquinones, flavones, aliphatic alcohols, aromatic carboxylic acids, steroids, and many drugs of varied structure.

A cloned human bilirubin UGT (UGT1A1) accepted a wide diversity of compounds such as phenols, anthraquinones, flavones, and steroids. Seven substrates were glucuronidated at a comparable or higher rates than bilirubin. Octyl gallate was glucuronidated at the highest rate among all substrates tested. The UGT1A1 isoform exhibited stereospecificity toward simple nonplanar phenols and estrogens.

UGT1A4 has been shown to catalyze the *N*-glucuronidation of drugs and xenobiotics. Subfamily 2 contains at least five UGTs catalyzing steroid or bile acid glucuronidation; UGT2B4 and UGT2B7 (steroid/bile acid UGTs) also catalyzed the glucuronidation of some xenobiotics (5).

A group of 10% of the whole population was determined to be poor glucuronidators of (S) oxazepam and suggested a genetic relationship to the UGT2B7 isoform, indicating how polymorphic variation of UGT genes affects the response to clinically used drugs. All of these UGTs have the potential to glucuronidate neurotransmitters, and brain UGTs require characterization.

Acknowledgment

We thank the Wellcome Trust for their support of this research program.

References

1. Clarke, D. J., and Burchell, B. (1994). The uridine diphosphate glucuronosyltransferase multigene family: Function and regulation. *Handb. Exp. Pharmacol.* **112**, 3–43.
2. Brierley, C. H., and Burchell, B. (1993). Human UDP-glucuronosyltransferase chemical defence, jaundice and gene therapy. *Bioessays* **15**, 749–754.
3. Owens, I. S., and Ritter, J. K. (1992). The novel bilriubin/phenol UDP-glucuronosyltransferase UGT 1 gene locus: Implication for multiple nonhemolytic familial hyperbilirubinaemia phenotypes. *Pharmacogenetics* **2**, 93–108.
4. Monaghan, G., Clarke, D. J., Povey, S., Gee See, C., Boxer, M., and Burchell, B. (1994). Isolation of human YAC contig encompassing a cluster of UGT 2 genes and its regional localization to chromosome 4q13. *Genomics* **23**, 496–499.
5. Burchell, B., Brierley, C. H., and Rance, D. (1995). Specificity of human UDP-glucuronosyltransferases and xenobiotic glucuronidation. *Life Sci.* **57**, 1819–1831.

Michael W. H. Coughtrie

Department of Molecular and Cellular Pathology
University of Dundee
Ninewells Hospital and Medical School
Dundee DD1 9SY, Scotland

Catecholamine Sulfation in Health and Disease

The sulfation reaction catalyzed by the cytosolic sulfotransferases (STs) plays an obviously key, but still somewhat ill-defined, role in the biological function of numerous important and potent endogenous compounds. The list includes steroids, bile acids, iodothyronines, catecholamines, and other bioactive amines, such as 5-hydroxytryptamine. This function must, of course, be considered along with the role of sulfation in the detoxication of xenobiotics and in steroid hormone biosynthesis (1, 2). The sulfation system should be viewed as a whole, because it comprises not only the ST enzymes but also a number of other components, including the arylsulfatases, which are capable of hydrolyzing the sulfate conjugates of many compounds; the bifunctional sulfate activating enzyme (ATP sulfurylase–APS kinase), which synthesizes the sulfation reaction cosubstrate PAPS (3′ phosphoadenosine 5′-phosphosulfate) from ATP and inorganic sulfate; and the transport mechanisms, which import and export the polar sulfate conjugates into and out of cells. Our understanding of the complex interrelations between these various components and their individual contributions to "sulfation" is still in its infancy, but it is clear that genetically and/or environmentally determined levels of each will critically shape an individual's overall sulfation capacity.

To date we do not know the complete extent of the human ST[a] enzyme family. A number of isoforms, all coded by distinct but related genes, have been purified and their cDNAs and genes cloned and sequenced. On the basis of the similarity of their derived amino acid sequences and their substrate preferences, these different forms of ST can be classified into one of two subfamilies. There is a single hydroxysteroid sulfotransferase (HST), which will accept numerous steroids (including dehydroepiandrosterone, testosterone, pregnenolone, cortisol), bile acids, cholesterol, and xenobiotic alcohols. The other subfamily, called phenolsulfotransferase (PST), comprises three closely related enzymes ($\geq$ 93% amino acid sequence identity) often called P-PST (or TS-PST),

[a] A new nomenclature system for the cytosolic sulfotransferases is under development, which will group the enzymes purely based on their derived amino acid sequences. The nomenclature used here is that used until now by our laboratory (and others).

Advances in Pharmacology, Volume 42

1054-3589/98 $25.00

ST1A2, and M-PST (or TL-PST), which sulfate phenolic xenobiotics, iodothyronines, catecholamines, *N*-hydroxyarylamines, and arylhydroxamic acids. A single estrogen ST, which conjugates endogenous and xenobiotic estrogens, also belongs to the PST subfamily. Allelic variants of most of these STs exist within the human population, some of which appear to confer altered substrate and/or PAPS affinity on the enzymes.

The human ST that appears to be principally responsible for the conjugation of catecholamines and other bioactive amines is M-PST (for monoamine-sulfating phenolsulfotransferase), also known as TL-PST (for thermolabile phenolsulfotransferase, in recognition of its instability at high temperatures compared with P-PST). The enzyme has been purified and the cDNA isolated from a number of human tissues, including liver, platelets, and a breast cancer cell line, by different laboratories. The derived amino acid sequence indicates a protein of subunit molecular weight, 34,196 daltons, that appears to exist as a dimer, at least *in vitro*. The gene coding for M-PST (currently called *STM*) resides on chromosome 16, close to those for P-PST and ST1A2. Heterologous expression of M-PST in mammalian cells (e.g., stable expression in V79 cells) or in bacteria has permitted the examination of M-PST's substrate preferences (3). In addition to dopamine, tyramine, 5-hydroxytryptamine, epinephrine, and norepinephrine, the enzyme also sulfates numerous xenobiotics such as vanillin, 1-naphthol, acetaminophen and α-methyldopa. In our hands, highlevel expression of M-PST in *E. coli* followed by purification using ion exchange and affinity chromatography can yield up to 10 mg/liter of highly purified enzyme suitable for structural analysis (unpublished data from this laboratory).

Recently, we have prepared antipeptide antibodies that can selectively distinguish between P-PST/ST1A2 and M-PST and have begun to use these to assess the tissue distribution of M-PST, because this may give us some clues as to the function of this enzyme. On carrying out such studies, it is immediately obvious that M-PST is expressed at extremely low levels in human liver, whereas expression appears highest in the small intestine. Tissues as diverse as the brain, platelets, skin, and endometrium all express the enzyme at levels much higher than liver, and various human cell lines, including breast cancer and embryonic lung epithelium, also display appreciable levels of M-PST. The implications of this for the sulfation of catecholamines are significant but perhaps not fully appreciated. In the circulation, catecholamines are present principally (between approximately 60 and 90%) as their sulfate conjugates, and the lack of a significant role for the liver in the generation of catecholamine sulfates is supported by *in vivo* work (4). If the enzyme responsible for the formation of catecholamine sulfates (i.e., M-PST) is barely expressed in the liver, it raises the interesting question of where these conjugates actually come from. The major site of expression of M-PST seems to be the small intestinal mucosa, and it is, of course, possible that this is the major contributing source of circulating conjugated catecholamines. Further investigation of this is clearly indicated.

The selectivity of the tissues in which M-PST is expressed may suggest some clues as to the enzyme's function. For instance, the high level of expression in the small intestinal mucosa may indicate an evolutionary role in detoxication of dietary xenobiotics, in addition to any function it may have in supplying conjugated catecholamines for transport to other tissues. Indeed, certain dietary chemicals are known to be substrates for this enzyme. It is, however, the potential role of catecholamine sulfation in the brain that has aroused considerable interest over the years, because obviously the ability to modulate amine neurotransmitter function in the brain through sulfation would put the M-PST enzyme in the same league as monoamine oxidase (MAO) and catechol O-methyltransferase (COMT), which are usually considered to be the major pathways of amine–catechol neurotransmitter metabolism. There are two interesting observations relevant to M-PST in the brain. First, the Km displayed by M-PST toward catecholamines such as dopamine is significantly lower than that observed for MAO, suggesting that catecholamines may be preferentially metabolized by sulfation in cells expressing both ST and MAO, at least at low substrate concentrations. Second, expression of PST(s) has been demonstrated by immunohistochemistry to be principally neuronal, using an antibody admittedly unable to distinguish the various members of the PST subfamily. These observations led us to wonder whether sulfation of catecholamines and other amine neurotransmitters, such as 5-hydroxytryptamine, may be involved in neurological conditions where such compounds are implicated. We, therefore, assessed sulfation of dopamine and 5-hydroxytryptamine in platelets from a large population of migraineurs and a similar control population (5). We found that the population of migraineurs showed a 20–30% reduction in the sulfation of these two neurotransmitters, illustrated by the population distribution in Figure 1, which shows a shift toward the left among the migraine sufferers. The implications for this reduction are unclear, but coupled with the fact that numerous dietary compounds are known to be potent inhibitors of intestinal

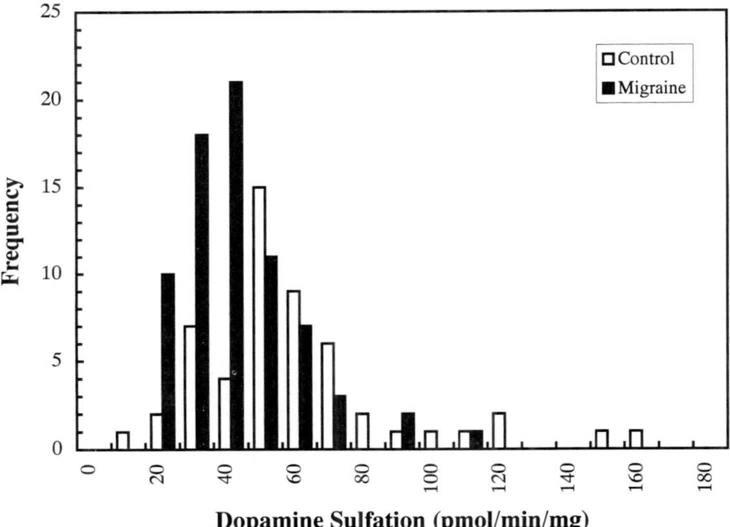

FIGURE 1 Population distribution of platelet M-PST sulfotransferase activity (dopamine as substrate) in groups of migraineurs and control individuals

M-PST, it is interesting to speculate that a combination of reduced sulfation of neurotransmitters and inhibition of intestinal M-PST may contribute to an altered homeostasis of biogenic amines in migraineurs.

In conclusion, sulfation of biogenic amines is certainly an important biological process, and as we learn more of the enzyme responsible for these reactions, as well as the other components of the sulfation system involved, we are sure to understand more of its function in the normal human, as well as the possible consequences of its disruption for the disease process.

Acknowledgments

Work on M-PST and catecholamine sulfation in my laboratory has been supported by the Migraine Trust, the Medical Research Council, the Biotechnology and Biological Sciences Research Council, the Commission of the European Communities, and Glaxo Wellcome.

References

1. Weinshilboum, R., and Otterness, D. (1994). Sulfotransferase enzymes. *In* Conjugation-Deconjugation Reactions in Drug Metabolism and Toxicity. (F. C. Kauffman, ed.), pp. 45–78. Springer-Verlag, Berlin.
2. Coughtrie, M. W. H. (1996). Sulphation catalysed by the human cytosolic sulphotransferases—Chemical defence or molecular terrorism? *Hum. Exp. Toxicol.* **15**, 547–555.
3. Jones, A. L., Hagen, M., Coughtrie, M. W. H., Roberts, R. C., and Glatt, H. (1995). Human platelet phenolsulfotransferases—cDNA cloning, stable expression in V79 cells and identification of a novel allelic variant of the phenol-sulfating form. *Biochem. Biophys. Res. Commun.* **208**, 855–862.
4. Tyce, G. M., Van Dyke, R. A., Rettke, S. R., Atchinson, S. R., Wesner, R. H., Dickson, E. R., and Krom, R. A. (1987). Human liver and conjugation of catecholamines. *J. Lab. Clin. Med.* **109**, 532–537.
5. Jones, A. L., Roberts, R. C., Colvin, D. W., Rubin, G. L., and Coughtrie, M. W. H. (1995). Reduced platelet phenolsulphotransferase acitvity towards dopamine and 5-hydroxytryptamine in migraine. *Eur. J. Clin. Pharmacol.* **49**, 109–114.

T. R. Tephly, M. D. Green, B. L. Coffman, C. King, Z. Cheng, and G. Rios

Department of Pharmacology
The University of Iowa
Iowa City, Iowa 52242

Metabolism of Endobiotics and Xenobiotics by UDP-Glucuronosyltransferase

All organisms are exposed to vast numbers of chemical compounds that are potentially toxic were it not for metabolic mechanisms available to the organism to moderate their effects. Chemical substances may be xenobiotics, such as drugs, or compounds presented to the organism from environmental or dietary sources. The elimination of many such compounds and their detoxification involves several different types of metabolic reactions. One of the most important is conjugation through glucuronic acid catalyzed by UDP-glucuronosyltransferases (UGTs). In addition to xenobiotics, many endobiotics, such as steroids or bilirubin, are glucuronidated. Formation of glucuronides from xeno- and endobiotics generally results in the formation of products that are more hydrophilic and more readily excreted by the kidney or liver.

UGTs have been cloned from several different species and, at this time, over 30 different UGT gene products have been described. A UGT nomenclature has been established to assist in classifying these proteins, and two UGT gene families have been described based on evolutionary divergence (1). The UGT1 family consists of a number of UGTs that result from alternate splicing of multiple exon 1s with common exons 2–5. The UGT2A subfamily contains gene products that are specifically expressed in olfactory epithelium, whereas the UGT2B subfamily consists of numerous enzymes that, in general, demonstrate substrate specificity for endogenous steroids and bile acids as well as many xenobiotics (1).

Our laboratory has purified a number of UGTs to homogeneity from liver of several species, including humans. Recently, we have identified a number of UGTs from both family 1 and 2 that are important for the metabolism of xenobiotics and endobiotics, such as steroids and estrogen catechols. This chapter discusses the metabolism of important amines, opioids, and endobiotics, such as steroids and their catechol derivatives.

I. Glucuronidation of Amines

Many isoforms of UGT from both family 1 and 2 catalyze the glucuronidation of primary and secondary amines. For example, UGT2B2 (3α-

Advances in Pharmacology, Volume 42

1054-3589/98 $25.00

hydroxysteroid UGT) was purified and found to specifically react with the 3α-hydroxy position of androsterone, but it was surprising to find that it also catalyzed the N-glucuronidation of α-naphthylamine and 4-aminobiphenyl. UGT2B3 and UGT1.6 also catalyze the glucuronidation of aromatic primary amines.

In humans, many important therapeutic agents, such as the antihistamines, antipsychotics, and tricyclic antidepressants, are tertiary amines. Many of these tertiary amines, in humans, are converted to and excreted as quaternary ammonium-linked glucuronides. Stably expressed human UGT1.4 protein has recently been discovered to catalyze the N-glucuronidation of a wide variety of tertiary amines (2). To date, only human UGT1.4 has been identified to catalyze the formation of quaternary ammonium-linked glucuronides from tertiary amines such as amitriptyline, imipramine, trifluoperazine, and clozapine (2). This isoform also actively catalyzes primary and secondary amine substrates. We have been amazed at the diversity of substrates for UGT1.4; this protein is also the only isoform of UGT recognized to catalyze the O-glucuronidation of progestin compounds and the O-glucuronidation of a large number of sapogenins, plant steroids that have been used as precurors for the synthetic production of endogenous steroidal compounds. Indeed, the efficiency of glucuronidation for sapogenins is the highest of all known substrates for UGT1.4. The very low Km values and high turnover rates for sapogenins such as hecogenin, diosgenin, tigogenin are similar to the efficiency of glucuronidation for progestins and certain androgens, such as 5α-androstane-3α,17β-diol. Because sapogenins are ubiquitous in nature, they pose a potential drug-drug interaction. UGT1.4, however, is not an important catalyst for the glucuronidation of estrogen catechols.

II. Glucuronidation of Opioid Compounds

In 1986, we reported the first purification of an enzyme that catalyzed the glucuronidation of morphine. More recently, Coffman *et al.* (3) demonstrated the purification of two rat liver morphine UGTs, one that appeared to have a number of features similar to rat UGT2B1 and another that had an N-terminal amino acid sequence similar to human UGT1.1. UGT2B1 and UGT1.1 have been demonstrated to catalyze the glucuronidation of opioid substances. UGT2B1 has high activity toward many opioid substrates, whereas UGT1.1 catalyzes the glucuronidation of one particular type of opioid substrate, the oripavine-type opioid compounds, such as buprenorphine and diprenorphine (4). UGT1.1 has also been shown to be the major bilirubin isoform.

Recently, we have demonstrated the stable expression of two UGTs from human liver that catalyze the glucuronidation of opioids, UGT1.1 and UGT2B7. Similar to results found for rat liver UGT1.1, buprenorphine is a major substrate, whereas other opioids, such as naltrexone and nalorphine, are substrates with poorer reactivity with this expressed protein. Morphine glucuronidation rates are also quite low with UGT1.1. However, UGT2B7 catalyzes the glucuronidation of opioid substrates such as morphine, nalorphine, and buprenor-

TABLE I Catechol Estrogen Glucuronidation by Stably Expressed Human UGTs

Substrate	*UGT1.1b*	*UGT1.4*	*UGT2B7*	*UGT2B15*
2-Hydroxyestrone	479[a]	ND	6.0	3.0
4-Hydroxyestrone	ND	ND	1403	14.0
2-Hydroxyestradiol	449	8.0	11.0	0
4-Hydroxyestradiol	ND	6.0	67	1.3
2-Hydroxyestriol	22	5.0	73	3.0

[a] Rates expressed as pmol/min/mg protein.
ND, none detected

phine at very high rates. Furthermore, the glucuronidation of morphine at the 3-OH position and the 6-OH position has been demonstrated for the first time, and codeine, which posseses only a 6-OH moiety, has been shown to be glucuronidated by UGT2B7. Another major feature of glucuronidation catalyzed by UGT1.1 and UGT2B7 is that each reacts with estrogen catechols.

III. Glucuronidation of Catechol Estrogens by UGTs

Table I shows the glucuronidation of estrogen catechols by human UGTs. UGT1.1 and UGT2B7 have been discussed with respect to their activity toward opioid compounds. Furthermore, UGT1.1 has been recognized for some time as the major bilirubin isoform of human liver, and its absence in liver results in the hyperbilirubinemia of Crigler-Najjar disease. UGT1.1 shows very high activity toward 2-hydroxyestrogens as well as high efficiency toward bilirubin glucuronidation (see Table I), whereas UGT2B7 and UGT2B15 show higher activities toward 4-hydroxyestrogenic catechols. However, at this time we have not been able to demonstrate the formation of catechol amine glucuronides by any of these UGT isoforms.

Because of the reactivity of UGT1.1 with many xenobiotic compounds and endobiotics, such as bilirubin and estrogen catechols, there is a potential drug–endobiotic interaction that may occur under certain conditions. For example, buprenorphine is under investigation as a potential compound for the treatment of opioid abuse as well as the abuse of cocaine. Its reactivity with UGT1.1 is as high as that observed for bilirubin. It is possible that under certain conditions of hepatic disease, where there is a compromise of the activity of UGT1.1, buprenorphine and bilirubin may interact with UGT1.1 such that a hyperbilirubinemia may occur. It is also possible that catechol estrogen glucuronidation may be compromised where drugs used in the therapy of pain may interact to force a decrease in the rate of glucuronidation with catechol estrogens.

Acknowledgment

This research was supported by National Institutes of Health grant GM26221.

References

1. Burchell, B., Nebert, D. W., Nelson, D. R., Bock, K. W., Iyanagi, T., Jansen, P. L., Lancet, D., Mulder, G. J., Chowdhury, J. R., Siest, G., Tephly, T. R., and Mackenzie, P. I. (1991). The UDP-glucuronosyltransferase gene superfamily: Suggested nomenclature based on evolutionary divergence. *DNA Cell Biol.* **10**, 487–494.
2. Green, M. D., and Tephly, T. R. (1996). Glucuronidation of amines and hydroxylated xenobiotics and endobiotics catalyzed by expressed human UGT1.4 protein. *Drug Metab. Dispos.* **24**, 356–363.
3. Coffman, B. L., Rios, G. R., and Tephly, T. R. (1996). Purification and properties of two rat liver phenobarbital-inducible UDP-glucuronosyltransferases that catalyzed the glucuronidation of opioids. *Drug Metab. Dispos.* **24**, 329–333.
4. King, C., Green, M. D., Rios, G. R., Coffman, B. L., Owens, I. S., Bishop, W. P., and Tephly, T. R. (1996). The glucuronidation of exogenous and endogenous compounds by stably expressed rat and human UDP-glucuronosyltransferase 1.1. *Arch. Biochem. Biophys.* **232**, 92–100.

Dirk Gründemann, Tilo Breidert, Folker Spitzenberger, and Edgar Schömig

Department of Pharmacology
University of Heidelberg
69120 Heidelberg, Germany

Molecular Structure of the Carrier Responsible for Hepatic Uptake of Catecholamines

The liver is the principal organ for the inactivation of circulating catecholamines. In the first step, the catecholamines are carried across the sinusoidal membrane into hepatocytes by a saturable mechanism. So far, this transport mechanism has not been identified on a molecular level. To see if only a single type of membrane-inserted transport protein is responsible for hepatic uptake of catecholamines and to examine transport mechanism, tissue distribution, and physiological and pathophysiological importance of that carrier, it would be immensely helpful to uncover the primary structure of the hepatic catecholamine carrier.

OCT1, a transporter for organic cations from the renal proximal tubule, was cloned from rat kidney by functional expression cloning with *Xenopus laevis* oocytes (1). OCT1 consists of 556 amino acids. Like many other

Advances in Pharmacology, Volume 42

1054-3589/98 $25.00

membrane-inserted transporters, it is predicted to have 12 transmembrane segments. In support of this topology, three putative N-glycosylation sites are contained in a large extracellular loop between transmembrane segments 1 and 2, and potential phosphorylation sites are located intracellularly.

By functional expression, both in oocytes and in 293 cells, a cell line derived from human embryonic kidney, OCT1 has been shown to transport radiolabeled substrates like tetraethylammonium (TEA) and 1-methyl-4-phenylpyridinium (MPP^+). OCT1-mediated transport can be inhibited by an array of organic cations like, for instance, N^1-methylnicotinamide, mepiperphenidol, procainamide, quinine, and tetrapentylammonium. The most potent inhibitors of OCT1 known so far are decynium22 and cyanine863, with Ki values of 400 and 100 nmol/liter, respectively (1).

Originally, OCT1 was cloned from a rat kidney cDNA library. Based on functional data from heterologous expression, it was assigned to the basolateral membrane of proximal tubular epithelial cells. However, it was soon learned that large amounts of OCT1 mRNA could also be found in liver (1).

Meanwhile, it has been established that OCT1 is functionally expressed in hepatocytes. When the Ki values determined for MPP^+ uptake into cultured hepatocytes were compared with the respective Ki values of 293 cells transiently transfected with an eucaryotic expression vector that carried the OCT1 cDNA, a highly significant correlation was found. The slope of the regression line was near unity (2). This strongly suggests identity of transport mechanisms. OCT1 probably is responsible for the so-called type I hepatic uptake of small organic cations at the sinusoidal membrane (2).

From the data at hand, OCT1 qualified as a candidate for the hepatic catecholamine carrier. It was the aim of the present study to investigate whether OCT1 actually accepts monoamine transmitters as substrates.

To express OCT1 by stable transfection of 293 cells, the cDNA of OCT1 was inserted into the eucaryotic expression vector pcDNA3. The resulting plasmid, pcDNA3OCT1, was introduced by a cationic liposome technique into 293 cells, which were grown on Petri dishes. Under selection by the aminoglycoside G418, only those cells were supposed to survive that stably integrated the plasmid into their genome.

To verify success of stable transfection, the G418-resistant cells were assayed for transcription of OCT1 cDNA by reverse transcription–polymerase chain reaction and Northern analysis. With both methods, OCT1 cRNA was detected in 293 cells transfected with pcDNA3OCT1 (293_{OCT1} cells), but in neither nontransfected cells (293_{WT} cells), nor cells stably transfected with pcDNA3. To demonstrate functional expression of OCT1, initial rates of uptake of radiolabeled substrates were measured. Nonspecific uptake was defined as uptake in the presence of 10 μM cyanine863. Compared with 293_{WT} cells, 293_{OCT1} cells showed markedly increased specific uptake rates for TEA (11.8 ± 2.2 vs 0.06 ± 0.08 pmol min^{-1} mg $protein^{-1}$, 2 μmol/liter ^{14}C-TEA) and MPP^+ (1.26 ± 0.13 vs 0.04 ± 0.002 pmol min^{-1} mg $protein^{-1}$, 0.1 μmol/liter ^3H-MPP^+). 293 cells stably transfected with pcDNA3 gave results similar to 293_{WT} cells. Finally, a pharmacological fingerprint of the MPP^+ uptake mechanism in 293_{OCT1} cells was taken. For that purpose, Ki values for five known inhibitors of OCT1 were determined. When these

Ki values were compared with the Ki values for OCT1 transiently expressed in *Xenopus* oocytes (1), a highly significant correlation was found (slope = 0.98 ± 0.06; r = 0.994, p < .001). It was, thus, confirmed that OCT1 was functionally expressed in 293_{OCT1} cells.

With an array of radiolabeled substrates, it was then examined which substrates beyond TEA and MPP^+ OCT1 actually is able to transport. Not surprisingly, specific uptake of alanine, glutamate, 3-*O*-methylglucose, and vincristine was not influenced by expression of OCT1 (Fig. 1). Also, choline and cimetidine, which have been proposed in the literature as substrates for renal organic cation transport mechanisms, were not transported by OCT1. Interestingly, however, OCT1 was found to transport dopamine, noradrenaline, and 5HT. The factors of stimulation of specific uptake by expression of OCT1 were 30 for MPP^+, 21 for 5HT, 17 for dopamine, and 11 for noradrenaline. Uptake of histamine, another biogenic amine, was unchanged relative to controls.

OCT1-mediated uptake of dopamine, noradrenaline, and 5HT was analyzed in more detail. For each substrate, from the time course of uptake into 293_{OCT1} cells, the kinetic constant for inwardly directed transport (k_{in}) was calculated (2). Notably, the k_{in}, expressed as clearance, for 5HT, dopamine, and noradrenaline was not much different from that for MPP^+ (3.1 ± 0.2, 6.9 ± 0.8, and 1.2 ± 0.1 vs 15.6 ± 1.8 μ/liter min^{-1} mg $protein^{-1}$; the respective k_{in} values for 293_{WT} cells were 0.5 ± 0.1, 0.9 ± 0.4, 0.2 ± 0.1, and 1.2 ± 0.1 μ/liter min^{-1} mg $protein^{-1}$).

Uptake of MPP^+, noradrenaline, dopamine, and 5HT was inhibited by cyanine863 with similar Ki values (0.09 [95% c.i. 0.067, 0.12], 0.14 [0.10,

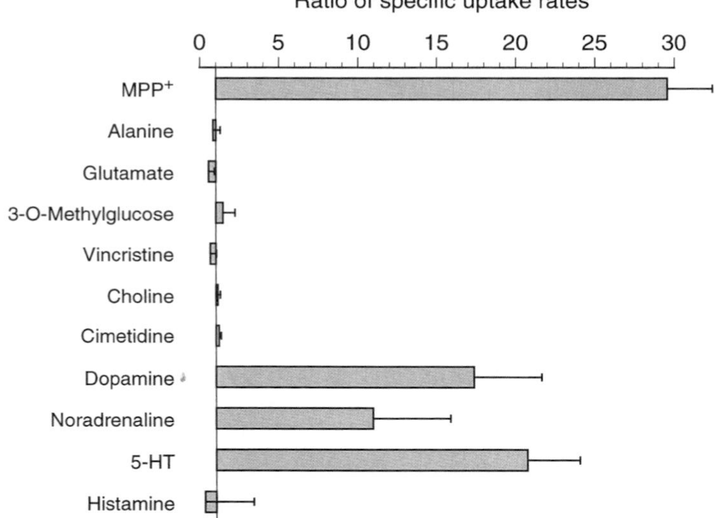

FIGURE 1 Transport activities induced by stable expression of OCT1. Shown is the ratio of specific uptake ± SEM of various radiolabeled compounds into 293_{OCT1} cells and 293_{WT} cells.

0.18], 0.25 [0.22, 0.29], and 0.27 [0.21, 0.35] μmol/liter). This confirms that a single transporter (i. e. OCT1) was indeed responsible for uptake of the monoamine transmitters.

Uptake into 293_{OCT1} cells of noradrenaline was resistant to inhibitors of the neuronal noradrenaline transporter (100 nM desipramine), the vesicular monoamine transporter (10 nM reserpine), and the extraneuronal monoamine transporter (1 μM corticosterone). Citalopram (300 nM), an inhibitor of neuronal 5HT uptake, and cocaine (1 μM), an inhibitor of neuronal dopamine uptake, also failed to affect the uptake into 293_{OCT1} cells of the respective substrates.

Finally, Ki values for the inhibition of MPP^+ uptake by 5HT, dopamine, and noradrenaline were determined as 0.65 (95% c.i. 0.49, 0.86), 1.1 (0.86, 1.4), and 2.8 (2.1, 3.7) mmol/liter, respectively.

This study demonstrates that OCT1 actually translocates 5HT and the catecholamines dopamine and noradrenaline. Velocity of transport was analyzed on the basis of clearances, which are normalized transport rates at nonsaturating substrate concentrations. Compared with MPP^+, which is an excellent substrate for OCT1 and other catecholamine transporters, the monoamines are transported in substantial amounts, with the order dopamine > 5HT > noradrenaline. In fact, dopamine was taken up as avidly as TEA, the archetypical organic cation transporter substrate. Uptake was specific in that histamine, another biogenic amine, was not translocated. With Ki values around 1 mmol/liter, OCT1 has only modest affinity for these substrates. This agrees with the notion, however, that good substrates do not have very low Ki values.

In conclusion, we have demonstrated that OCT1 avidly translocates 5HT and the catecholamines dopamine and noradrenaline. From previous work, it is known that OCT1 is functionally expressed in hepatocytes. Together, these findings suggest that OCT1 is responsible for hepatic uptake of catecholamines and other biogenic amines. For the final piece of evidence, it will be necessary to show that uptake of catecholamines into the perfused liver can be inhibited by prototypical inhibitors of OCT1.

Acknowledgments

The authors thank Anke Ripperger for skillful technical assistance. This work was supported by Deutsche Forschungsgemeinschaft (SFB 176).

References

1. Gründemann, D., Gorboulev, V., Gambaryan, S., Veyhl, M., and Koepsell, H., (1994). Drug excretion mediated by a new prototype of polyspecific transporter. *Nature* **372,** 549–552.
2. Martel, F., Vetter, T., Russ, H., Gründemann, D., Azevedo, I., Koepsell, H., and Schömig, E. (1996). Transport of small organic cations in the rat liver: The role of the organic cation transporter OCT1. *Naunyn Schmiedebergs Arch. Pharmacol.* **354,** 320–326.

Fátima Martel

Department of Biochemistry
Institute of Pharmacology and Therapeutics
Faculty of Medicine
4200 Porto, Portugal

Catecholamine Uptake and Metabolism in the Liver

It has been known for several decades that the liver plays a major role in the inactivation of circulating catecholamines through an enormous capacity for removal and metabolic degradation of these amines. The capacity of the liver to metabolically inactivate catecholamines is in agreement with the presence of huge amounts of the catecholamine-metabolizing enzymes (monoamine oxidase [MAO], catechol *O*-methyltransferase [COMT], aldehyde dehydrogenase and reductase, and conjugating enzymes) in this organ. The removal and metabolism of catecholamines were studied using three different experimental models: rat freshly isolated hepatocytes, liver slices, and perfused liver.

The removal and metabolism of [^{3}H]noradrenaline ([^{3}H]NA) and [^{3}H]-adrenaline ([^{3}H]AD) were studied using isolated hepatocytes. The removal of both amines from the medium was quantitatively similar and mainly due to metabolism, which represented more than 95% of total removal, for both amines. There was no accumulation of [^{3}H]amines in the cells, as the cell–medium ratio was about unity. For [^{3}H]AD, *O*-methylation predominated, because *O*-methylated and deaminated metabolites ([^{3}H]OMDA) and [^{3}H]MN were the most abundant metabolites. For [^{3}H]NA, deamination predominated: [^{3}H]OMDA and [^{3}H]DOMA were the most abundant metabolites. In experiments with [^{3}H]NA, simultaneous inhibition of MAO and COMT originated a strong reduction in the formation of all [^{3}H]metabolites, but surprisingly there was no increase in the amount of [^{3}H]NA in cells (1).

The removal and metabolism of [^{3}H]NA were also studied using liver slices. The degree and pattern of metabolism of this amine were similar to those of the isolated cells. Likewise, there was no accumulation of [^{3}H]NA in the tissue. Moreover, the results of simultaneous MAO and COMT inhibition (namely the lack of increase of tissue content in [^{3}H]NA) were also very similar to those obtained with isolated hepatocytes (1).

Finally, the removal and metabolism of [^{3}H]NA were studied in the perfused liver (being the amine infused through the portal vein in a nonrecirculating system). Removal of [^{3}H]NA by the perfused liver corresponded to 25% of the amount of [^{3}H]NA infused through the portal vein. Of the [^{3}H]NA removed by the liver, almost 90% was metabolized, the most abundant metabolites being present in the [^{3}H]OMDA fraction, followed by [^{3}H]NMN, [^{3}H]DOPEG, and [^{3}H]DOMA. Simultaneous inhibition of MAO and COMT produced a

Advances in Pharmacology, Volume 42

1054-3589/98 $25.00

significant reduction in the formation of all [^{3}H]metabolites, with the exception of [^{3}H]OMDA. Once again, this treatment did not increase the tissue content in [^{3}H]NA (2). The difference in the pattern of metabolism of [^{3}H]NA and in the effects of simultaneous inhibition of MAO and COMT observed between perfusion and incubation experiments (using either isolated hepatocytes or liver slices) is probably related to differences in the experimental models used.

In conclusion, it was shown that: (1) the rat liver possesses a huge capacity to remove and metabolize AD and NA, being AD preferentially *O*-methylated and NA preferentially deaminated; (2) there is no accumulation of catecholamines in the liver, even after simultaneous inhibition of MAO and COMT, possibly due to the existence of a very active extrusion system; (3) the pattern of metabolism of catecholamines is dependent on the experimental model studied.

Despite it being known for a long time that the liver has an important role in the removal of circulating catecholamines, until recently few studies were conducted on this subject. Consequently, the mechanism(s) responsible for the hepatic uptake of these compounds remained poorly characterized.

Using rat isolated hepatocytes and liver slices, we found that the uptake$_1$ inhibitor desipramine (1 μM) and the uptake$_2$ inhibitor corticosterone (40 μM) produced similar discrete effects on the removal and metabolism of either [^{3}H]NA or [^{3}H]AD. Both compounds slightly decreased (to 60–70% of control) the formation of [^{3}H]OMDA and of [^{3}H]NMN or [^{3}H]MN (1). Moreover, corticosterone (40 μM) did not affect removal and metabolism of [^{3}H]NA by the perfused rat liver (2).

When isolated hepatocytes were incubated with either [^{3}H]AD or [^{3}H]NA under conditions of simultaneous inhibition of MAO and COMT, desipramine (1 μM) did not change the cell content of [^{3}H]AD or [^{3}H]NA, and corticosterone (40 μM) produced only a small decrease in the cell content of [^{3}H]AD. Probenecid (100 μM), an inhibitor of the renal transport of organic anions, was devoid of effect on the uptake of both amines, and cyanine 863 (10 μM), an inhibitor of the renal transport of organic cations, had only a partial inhibitory effect on the uptake of [^{3}H]AD. Moroever, bilirubin (200 μM) caused a significant increase (50%) in the amount of [^{3}H]AD and [^{3}H]NA present in the cells, possibly because bilirubin and catecholamines compete for the same extrusion transport system (3). Taken together, these results suggest that the mechanism responsible for the hepatic uptake of catecholamines is neither uptake$_1$ nor uptake$_2$.

More recently, the characteristics of the hepatic uptake of catecholamines have been studied using the compound [^{3}H]1-methyl-4-phenylpyridinium ([^{3}H-MPP$^+$). Several facts suggest that MPP$^+$ and catecholamines are taken up by hepatocytes through the same transport system. First, [^{3}H]MPP$^+$ is efficiently taken up and accumulated by rat isolated hepatocytes and by rat cultured hepatocytes, being the factor of accumulation of this compound in these cells superior to 50 (4, 5). Second, uptake of [^{3}H]MPP$^+$ by isolated hepatocytes is sensitive to catecholamines. The type of inhibition exerted by (−)-AD, which is the most potent catecholamine in inhibiting [^{3}H]MPP$^+$ uptake, is competitive (4). Third, in cultured hepatocytes, [^{3}H]MPP$^+$, similarly to catecholamines, is taken up by a carrier-mediated mechanism distinct from uptake$_1$ and uptake$_2$ (5). Recently, it was found that the type-I hepatic transporter of organic cations seems to be involved in the uptake of [^{3}H]MPP$^+$ by cultured hepatocytes and isolated hepatocytes (5, 6).

Interestingly enough, some of the type-I transport substrates are also substrates of P-glycoprotein. Moreover, other substrates–inhibitors of P-glycoprotein (amiloride, daunomycin, vinblastin, cyclosporin A, bilirubin, and progesterone) produced marked reductions in the uptake of [^{3}H]MPP$^+$ by isolated hepatocytes. These results suggested the involvement of P-glycoprotein in the hepatic uptake of [^{3}H]MPP$^+$ (6). Preliminary results obtained with isolated hepatocytes incubated with [^{3}H]AD suggest that P-glycoprotein is also involved in the uptake of catecholamines.

P-glycoprotein, the multidrug resistance gene product, is located in the canalicular membrane of hepatocytes, where it is known to mediate the excretion of several compounds, mainly cations. So, how does one explain that a transporter known to remove drugs from cells is taking up MPP$^+$ and catecholamines? This question is currently under study in our laboratory, and we hope that it can be answered in the near future.

In conclusion, for the hepatic uptake of catecholamines: (1) the uptake systems for catecholamines uptake$_1$ and uptake$_2$ are not involved; (2) the type-I hepatic transporter of organic cations seems to play a significant role; and (3) a P-glycoprotein transporter also seems to be involved, at least in freshly isolated rat hepatocytes.

Acknowledgment

Research work supported by JNICT (PRAXIS/2/2.1/SAU/1251/95).

References

1. Martel, F., Azevedo, I., and Osswald, W. (1993). Uptake and metabolism of ^{3}H-adrenaline and ^{3}H-noradrenaline by isolated hepatocytes and liver slices of the rat. *Naunyn Schmiedebergs Arch. Pharmacol.* **348**, 450–457.
2. Martel, F., and Azevedo, I. (1995). The fate of ^{3}H-(−)-noradrenaline in the perfused rat liver. *J. Auton. Pharmacol.* **15**, 309–319.
3. Martel, F., Azevedo, I., and Osswald, W. (1994). Uptake of ^{3}H-catecholamines by rat liver cells occurs mainly through a system which is distinct from uptake$_1$ or uptake$_2$. *Naunyn Schmiedebergs Arch. Pharmacol.* **350**, 130–135.
4. Martel, F., Martins, M. J., and Azevedo, I. (1996). Inward transport of ^{3}H-MPP$^+$ in freshly isolated rat hepatcytes: Evidence for interaction with catecholamines. *Naunyn Schmiedebergs Arch. Pharmacol.* **354**, 305–311.
5. Martel, F., Vetter, T., Russ, H., Gründemann, D., Azevedo, I., Koepsell, H., and Schömig, E. (1996). Transport of small organic cations in the rat liver. The role of the organic cation transporter OCT1. *Naunyn Schmiedebergs Arch. Pharmacol.* **354**, 320–326.
6. Martel, F., Martins, M. J., Hipólito-Reis, C., and Azevedo, I. (1996). Inward transport of ^{3}H-MPP$^+$ in isolated rat hepatocytes: Putative involvement of a P-glycoprotein transporter. *Br. J. Pharmacol.* **119**, 1519–1524.

Lesley J. Bryan-Lluka,* Kristy M. James,* Heinz Bönisch,† Peter Pörzgen,† Karen S. Guice,‡ and Keith T. Oldham‡

*Department of Physiology and Pharmacology
The University of Queensland
Brisbane, Qld 4072, Australia

†Department of Pharmacology and Toxicology
University of Bonn
D-53113 Bonn, Germany

‡Department of Surgery
Duke University Medical Center
Durham, North Carolina 27710

Catecholamine Uptake and Metabolism in Rat Lungs

The pulmonary circulation is an important site for clearance of circulating catecholamines in various species, including human, rat, rabbit, and dog lungs, mainly by uptake and metabolism in endothelial cells of the lung microvasculature (1). Although there were early reports that pulmonary catecholamine uptake occurs by a transporter with the properties of both extraneuronal uptake$_2$ and neuronal uptake$_1$ (1), we provided pharmacological evidence from studies in intact perfused lungs of rats that the transporter in the pulmonary endothelial cells has the same properties as the neuronal norepinephrine transporter (NET), uptake$_1$ (2).

The subsequent fate of the catecholamines is metabolism by catechol O-methyltransferase (COMT) and monoamine oxidase (MAO) in the cells. Extensive studies on the kinetics of metabolism of norepinephrine and dopamine in perfused lungs of rats have shown that O-methylation is the predominant metabolic fate of both amines at low concentrations (3), but COMT becomes saturated at much lower perfusion concentrations than does MAO (4). Half-saturation of O-methylation occurred when rat lungs were perfused with 9.8 nM norepinephrine, but 100-fold higher amine concentrations were required to saturate deamination and the NET in the lungs (4). Determination of rate constants for deamination in lungs perfused with a nonsaturating amine concentration (1 nM) showed that about 80% of the deamination of both norepinephrine and dopamine occurs by MAO-A and the remaining 20% by MAO-B, with no detectable metabolism by semicarbazide-sensitive amine oxidases (3). Despite the similarities between the pulmonary metabolic profiles of norepinephrine and dopamine in the lungs, a major difference is that the rate constants for both O-methylation and deamination of dopamine are more than an order of magnitude greater than those for norepinephrine (4). The values for norepinephrine in the lungs are of the same order of magnitude as

Advances in Pharmacology, Volume 42

1054-3589/98 $25.00

the rate constants determined for both amines for O-methylation in myocardial cells and for deamination in noradrenergic neurones and myocardial cells of rat hearts (see values in ref. 4). The reason for the markedly higher activities of both COMT and MAO for metabolism of dopamine than norepinephrine in the lungs is not yet clear. Nevertheless, the result does account for the failure to detect any pulmonary clearance of dopamine in early studies in which only accumulation of total radioactivity in the lungs, and not the appearance of metabolites in the venous effluent, was measured (1). This led to the erroneous conclusion that dopamine was not a substrate for uptake in the lungs (1).

An alternative fate of catecholamines in the lungs is efflux from the cells (5), and this aspect was further investigated in this study. Specific pathogen-free rats were anesthetized and the lungs were isolated and perfused at 10 ml/min with Krebs solution (MAO and COMT inhibited) at 37°C, initially for 15 min and then with the addition of 1 nM [^{3}H]norepinephrine for a loading period of 10 min. The efflux of [^{3}H]norepinephrine was then measured from venous effluent samples collected from a cannula in the left atrium at 1-min intervals during perfusion with Krebs solution for 21 min. The efflux of norepinephrine was expressed as the fractional rate of loss (FRL) of the amine from the lungs, because this is a measure of the instantaneous rate constant for efflux. The rate constant for efflux of [^{3}H]norepinephrine was 0.0163 min^{-1} (corresponding to a half-time of 42 min), reached steady-state (and hence a constant FRL) from about the 10th min of efflux and was increased when the Na^+ concentration in the Krebs solution was reduced from 143 mM to 25 mM (5). The addition of the NET inhibitors, nisoxetine or imipramine, from the 16th min to the end of efflux caused a small and gradual increase in the FRL of [^{3}H]norepinephrine, while the substrates, dopamine and adrenaline, caused a rapid and marked increase in the FRL (5). These results correspond with those described previously for efflux of [^{3}H]norepinephrine from noradrenergic neurones and suggest that both the NET and passive diffusion contribute to efflux of norepinephrine from the pulmonary endothelial cells; this proposal was investigated later in the study. Additional experiments showed that 1 μM of dopamine included from the 16th min of efflux caused a rapid increase in the FRL of [^{3}H]norepinephrine, reaching a peak effect after 2 min of 0.0285 $\pm$ 0.0031 min^{-1} (n = 4) that was 69% inhibited ($p < .01$) by the NET inhibitor, desipramine (1 μM) but not by the 5-hydroxytryptamine (5HT) transporter inhibitor, citalopram (1 μM). Experiments in which dopamine at concentrations of 100–600 nM or 100 or 300 μM (maximal effects) were added from the 16th min of efflux (n = 3–6) showed that 50% of the maximal dopamine-induced increase in FRL of [^{3}H]norepinephrine occurred at 273 nM of dopamine (95% confidence limits: 182–363 nM; n = 24). This value corresponds very closely to the Km of dopamine uptake in rat lungs of 246 nM (2) and shows that the effects of substrates on catecholamine efflux from rat lungs are mainly determined by the affinity of the substrate for the NET.

A further series of experiments was carried out to determine the contributions of the NET and diffusion to the total efflux of norepinephrine and the extent of re-uptake in the rat lungs. In these experiments, the previously described method was modified in that MAO was only partially inhibited, and efflux of both [^{3}H]norepinephrine and its deaminated metabolite, [^{3}H]3′,4′-

dihydroxyphenylglycol ([^{3}H]DOPEG) were measured. The FRL values were determined before and after the addition of Krebs solution (controls, n = 4), 0.35 μM of desipramine (n = 5), or 1.5 μM of desipramine (n = 6), to achieve 0, 88% and 97% inhibition of NET, respectively, for the last 5 min of the 25-min efflux period. Calculations from the values determined for the FRL of [^{3}H]norepinephrine and [^{3}H]DOPEG under these experimental conditions showed that the NET contributed 81% and diffusion 19% to the total efflux of norepinephrine from the rat lungs, with 90% of this norepinephrine being subject to re-uptake by the NET. Hence, the norepinephrine efflux measured under control conditions represents only 10% of the total norepinephrine leaving the cells by both the NET and diffusion.

To examine the properties of the NET in the lungs in more detail, experiments were carried out on rat lung microvascular endothelial cells in culture. Specific pathogen-free rats were euthanased and peripheral lung sections were minced under sterile conditions. The cells were cultured for three passages under selective growth conditions and then characterized as pure cultures of endothelial cells. Most experiments were carried out on cells from passages 6–12. Four days after subculturing, medium was replaced by Krebs-Hepes buffer with or without 10 μM of desipramine (MAO and COMT inhibited) for 15 min at 37°C, and then norepinephrine uptake was measured by the addition of 10 nM of [^{3}H]norepinephrine for a further 2 min. After washing with ice-cold buffer and lysis of the cells, their [^{3}H] and protein contents were determined. There was very little norepinephrine uptake in cells from passages 6–10 (45.5 ± 6.2 fmol/mg protein, n = 6), and desipramine had no effect (34.6 ± 8.5 fmol/mg protein, n = 6, $p > .05$, paired *t*-test). In contrast, in parallel experiments on the endothelial cells with 10 nM of [^{3}H]5HT, the uptake of 5HT was 1980 ± 57 fmol/mg protein (n = 5) and was 97% inhibited by 10 μM of citalopram (n = 5, $p < .001$, paired *t*-test). Hence, the pulmonary microvascular endothelial cells in culture express the 5HT transporter but not the NET. Further experiments were carried out to determine whether NET expression decreases with subculturing and whether a range of agents could stimulate NET expression in the cells. However, there was no desipramine-sensitive uptake of norepinephrine (n = 3; $p > .05$) in the endothelial cells at passages 1–3 nor in cells (passages 6–12) in which 1 μM of angiotensin II, 1 μM of captopril, 15 mM of KCl, 10 μM of norepinephrine, 1 mM of dibutyryl cAMP, or 0.1 or 25 μM of dexamethasone was added to the medium for the last 48 hr of culture. Hence, there was no evidence of NET expression in the rat pulmonary microvascular endothelial cells under any of these culture conditions. In the final series of experiments, RNA was extracted with Trizol (Gibco) from endothelial cells cultured under standard conditions and used for reverse transcription and polymerase chain reaction with primers determined from the known part of the rat NET sequence. There was a detectable, but small, amount of rat NET cDNA formed from the RNA isolated from the endothelial cells at passages 4 and 9, and the sequence was identical to that of the corresponding fragment of the known rat NET sequence.

In conclusion, in intact rat lungs, catecholamine uptake by the pulmonary NET, which appears histologically to be in the microvascular endothelial cells, is followed by metabolism by COMT and/or MAO or efflux from the cells, mainly by the NET. However, rat pulmonary microvascular endothelial cells in culture do not show expression of the NET under various culture conditions. The reason for the discrepancy between results from intact lungs and cultured pulmonary microvascular endothelial cells is not yet known.

References

1. Nicholas, T. E., Strum, J. M., Angelo, L. S., and Junod, A. J. (1974). Site and mechanism of uptake of ^{3}H-l-norepinephrine by isolated perfused rat lungs. *Circ. Res.* **35**, 670–680.
2. Bryan-Lluka, L. J., Westwood, N. N., and O'Donnell, S. R. (1992). Vascular uptake of catecholamines in perfused lungs of the rat occurs by the same process as uptake$_1$ in noradrenergic neurones. *Naunyn Schmiedebergs Arch. Pharmacol.* **345**, 319–326.
3. Scarcella, D. L., and Bryan-Lluka, L. J. (1995). A kinetic investigation of the pulmonary metabolism of dopamine in rats shows marked differences compared with noradrenaline. *Naunyn Schmiedebergs Arch. Pharmacol.* **351**, 491–499.
4. Bryan-Lluka, L. J. (1995). Evidence for saturation of catechol-O-methyltransferase by low concentrations of noradrenaline in perfused lungs of rats. *Naunyn Schmiedebergs Arch. Pharmacol.* **351**, 408–416.
5. Westwood, N. N., Scarcella, D. L., and Bryan-Lluka, L. J. (1996). Evidence for uptake$_1$-mediated efflux of catecholamines from pulmonary endothelial cells of perfused lungs of rats. *Naunyn Schmiedebergs Arch. Pharmacol.* **353**, 528–535.

Edgar Schömig, Hermann Russ, Kerstin Staudt, Fátima Martel, Martin Gliese, and Dirk Gründemann

Department of Pharmacology
University of Heidelberg
69120 Heidelberg, Germany

The Extraneuronal Monoamine Transporter Exists in Human Central Nervous System Glia

The inactivation of released transmitters is essential for synaptic signal transmission. From various studies on sympathetically innervated peripheral organs, it is well known that both the neuronal type of noradrenaline transporter (uptake$_1$) and the extraneuronal monoamine transporter (uptake$_2$) are important for the inactivation of released catecholamines. The neuronal noradrenaline transporter is driven by the transmembrane Na^+ and Cl^- gradients and is

Advances in Pharmacology, Volume 42

1054-3589/98 $25.00

sensitive to desipramine-like antidepressants. By contrast, the extraneuronal monoamine transporter operates independently of the Na^+ gradient. It is, at least partially, driven by the membrane potential and is sensitive to corticosterone.

In the central nervous system (CNS), the close proximity between catecholaminergic neurons and glia cells raises the question whether the extraneuronal monoamine transporter contributes to the inactivation of centrally released catecholamines. There is strong evidence in support of the view that glia cells accumulate monoamines by a saturable transport system; the underlying mechanism, however, is still under discussion (1, 2).

Two developments opened the possibility to readdress this question: (1) The isocyanines and pseudoisocyanines represent a novel class of selective and highly potent inhibitors of the extraneuronal monoamine transporter (3). (2) The neurotoxin 1-methyl-4-phenylpyridinium (MPP^+) turned out to be an excellent substrate of the extraneuronal monoamine transporter that is superior to noradrenaline for *in vitro* studies (4).

The uptake of tritiated MPP^+ was investigated in isolated incubated rat cerebral cortex slices as well as in primary cultures of human astrocytes and various human gliomas. Cerebral cortex slices of 10 mg were prepared from rats pretreated with reserpine in order to deplete intracellular catecholamine stores and to avoid vesicular storage of $[^3H]MPP^+$. The tissue slices accumulated $[^3H]MPP^+$, both in the absence and presence of 10 μmol/liter of cocaine. Cocaine was used to inhibit the neuronal catecholamine carriers. Both in the absence and in the presence of cocaine, the isocyanine 1-methyl-1′-isopropyl-2,4′-cyanine (10 μmol/liter) inhibited $[^3H]MPP^+$ uptake by 69% and 66%, respectively ($n = 4$, $p < .05$).

The uptake of $[^3H]MPP^+$ was determined also on primary cultures of human astrocytes and various human gliomas. All tested tissue cultures showed 1-methyl-1′-isopropyl-2,4′-cyanine–sensitive accumulation of $[^3H]MPP^+$. Specific (i.e., 1-methyl-1′isopropyl′2,4’-cyanine–sensitive) accumulation was most pronounced in human astrocytes and in the human glioma cell line HTZ146. Because the supply of human astrocytes is limited, MPP^+ accumulation was characterized in HTZ146 human glioma cells in more detail. Initial rates of specific $[^3H]MPP^+$ transport were saturable, the Km and Vmax being 23 μmol/liter (95% confidence limits 19, 28) and 120 $\pm$ 10 pmol/(min · mg protein) ($n = 4$). The results of the Scatchard analysis were compatible with a single mechanism being involved in specific $[^3H]MPP^+$ transport. To characterize $[^3H]MPP^+$ uptake in HTZ146 cells, initial rates of specific $[^3H]MPP^+$ transport were determined in the absence and presence of various compounds. The resulting inhibitory potencies were compared with the known inhibitory potencies of these compounds for the inhibition of the extraneuronal monoamine transporter (Fig.1).

Three key findings support the view that MPP^+ transport in HTZ146 cells is due to the extraneuronal monoamine transporter: (1) Specific MPP^+ transport in HTZ146 cells was saturable, and the half-saturating concentration exactly fits the known Km of MPP^+ transport in Caki-1 cells via the extraneuronal monoamine transporter. (2) All tested inhibitors of the extraneuronal monoamine transporter—which cover a wide variety of chemical structures—affected MPP^+ transport in HTZ146 cells. (3) Most importantly, there is a highly

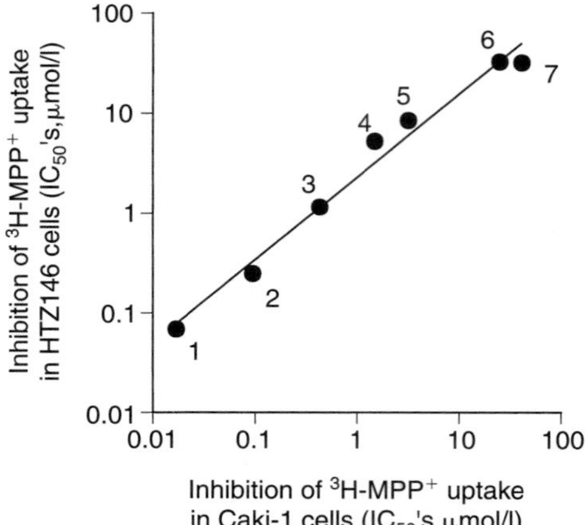

FIGURE 1 Correlation between the inhibitory potencies of various compounds for the inhibition of [^{3}H]MPP$^+$ uptake in Caki-1 cells and HTZ146 cells. Each number represents one of the tested compounds (1, decynium 22; 2, corticosterone; 3, cyanine 863; 4, O-methylisoprenaline; 5, quinine; 6, clonidine; 7, quinidine). There is a highly significant correlation (r = .991, N = 7, p < .001). The slope of the regression line is 0.83 ± 0.06. Data are taken from Russ *et al.* (5).

significant positive correlation between the IC50's for the inhibition of MPP$^+$ transport in HTZ146 cells and for the inhibition of the extraneuronal monoamine transporter measured as MPP$^+$ transport in Caki-1 cells. The slope of the corresponding regression line nearly equals unity (see Fig. 1).

These findings demonstrate that the extraneuronal monoamine transporter exists in cells that stem from human CNS glia. The physiological role of the extraneuronal monoamine transporter in the CNS may be characterized by a second line of defense. It inactivates that fraction of monoamines that escapes neuronal re-uptake and thus prevents uncontrolled spreading of the signal.

Another series of experiments was performed to test whether MPP$^+$ is able to make use of the extraneuronal monoamine transporter in reverse of its normal transport direction. For this purpose, HTZ146 cells were preloaded for 20 min with 100 nmol/liter [^{3}H]MPP$^+$. After this loading period, monophasic efflux of [^{3}H]MPP$^+$ was measured for 25 min. In the absence of an inhibitor, the rate constant of efflux was 0.0230 ± 0.0010 min^{-1} (n = 6). In other words, 2.3% of intracellular [^{3}H]MPP$^+$ left the HTZ146 cells per minute. The pseudoisocyanine decynium 22 (1.4 μmol/liter) and corticosterone (5 μmol/liter)—at concentrations amounting to about 20 times of the known Ki for the inhibition of the extraneuronal monoamine transporter—reduced the rate constants of efflux to 0.0035 ± 0.0005 min^{-1} and 0.0054 ± 0.0010 min^{-1}, respectively (n = 6).

This finding raises the possibility that the extraneuronal monoamine transporter in the CNS is involved in the mechanism of 1-methyl-4-phenyl-1,2,3,6-tetrahydropyridine (MPTP) neurotoxicity. MPP$^+$ is formed from the protoxin MPTP by monoamine oxidase B, which within the CNS is located predominantly inside glia cells. To exert its toxic effect on dopaminergic neurons, the organic cation MPP^+ has to leave the glia cells to cross the extracellular space and to enter dopaminergic neurons. The present experiments demonstrate that nonspecific diffusion of MPP^+ out of glia cells is negligible. When the extraneuronal monoamine transporter was blocked by decynium 22, only 0.35% of intracellular MPP^+ left the glia cells per minute. On the omission of an inhibitor, the efflux of MPP^+ from HTZ176 cells increased more than sixfold. In case of a marked outwardly directed concentration gradient, MPP^+ makes use of the extraneuronal monoamine transporter to leave the glia cells. Blockade of the extraneuronal monoamine transporter prevents MPP^+ from leaving the glia cells.

Acknowledgments

We thank Professor Dr. Ulrich Bogdahn for his help with primary cultures of astrocytes and HTZ cells. We are grateful to Mrs. Michaela Hoffmann and Mrs. Brigitte Dürner for technical assistance. This study was supported by the Sonderforschungsbereich 176 of the Deutsche Forschungsgemeinschaft.

References

1. Kimelberg, H., and Pelton, E. (1983). High affinity uptake of [3H]norepinephrine by primary astrocyte cultures and its inhibition by tricyclic antidepressants. *J. Neurochem.* **40**, 1265–1270.
2. Paterson, L. A., and Hertz, L. (1989). Sodium-independent transport of noradrenaline in mouse and rat astrocytes in primary culture. *J. Neurosci. Res* **23**, 71–77.
3. Russ, H., Sonna, J., Keppler, K., Baunach, S., and Schömig, E. (1993). Cyanine-related compounds: A novel class of potent inhibitors of extraneuronal noradrenaline transport. *Naunyn Schmiedebergs Arch. Pharmacol.* **348**, 458–465.
4. Russ, H., Gliese, M., Sonna, J., and Schömig, E. (1992). The extraneuronal transport mechanism for noradrenaline (uptake2) avidly transports 1-methyl-4-phenyl-pyridinium (MPP+). *Naunyn Schmiedebergs Arch. Pharmacol.* **346**, 158–165.
5. Russ, H., Staudt, K., Martel, F., Gliese, M., and Schömig, E. (1996). The extraneuronal transporter for monoamine transmitters exists in cells derived from human central nervous system glia. *Eur. J. Neurosci.* **8**, 1256–1264.

Karl-Heinz Graefe,* Bernd Friedgen,* Reinhard Wölfel,*
and Edgar Schömig†

*Department of Pharmacology and Toxicology
University of Würzburg
D-97078 Würzburg, Germany

†Department of Pharmacology
University of Heidelberg
D-69120 Heidelberg, Germany

Removal of Circulating Catecholamines by Extraneuronal Amine Transport Systems

The main enzymes involved in the catabolism of catecholamines, namely monoamine oxidase (MAO) and catecholamine *O*-methyltransferase (COMT), are intracellular enzymes. Therefore, cellular uptake processes are primarily responsible for the removal of catecholamines from the excellular fluid. Two uptake processes for catecholamines are well known: uptake$_1$ (which brings about uptake into noradrenergic neurons and capillary endothelial cells of the pulmonary circulation) and uptake$_2$ (which mediates transport into myocardial cells, vascular as well as nonvascular smooth muscle cells, and glia cells). While the contribution by uptake$_1$ to the removal of circulating noradrenaline (NA), adrenaline (A), and dopamine (DA) has been quantified (1), very little is known about the part played by uptake$_2$. In an attempt to determine the role of uptake$_2$ in the clearance of extracellular catecholamines, 1,1′diisopropyl-2,4′-cyanine iodide (disprocynium 24) was used as a tool in the rabbit. Disprocynium 24 is a novel agent that blocks uptake$_2$ (without affecting uptake$_1$) with high potency (i.e., with an IC 50 of 14 nM) (2).

Anesthetized rabbits were infused with trace amounts of ^{3}H-labeled NA, A, and DA, and heart rate, mean arterial blood pressure, the cardiac output of plasma (CO_p), and the plasma clearance of the infused [^{3}H]catecholamines (Cl_{tot}; as determined from steady-state [^{3}H]catecholamine concentrations in arterial plasma) were measured both before and during treatment with disprocynium 24 (270 nmol/kg i.v. followed by i.v. infusion of 80 nmol/[kg min]) or vehicle (DMSO in saline). As the values of Cl_{tot} increased linearly with increasing CO_p, they were expressed as a percentage of CO_p (which on average amounted to 110 ml/[kg min]). Two groups of animals were studied: group I (no further treatment) and group II (MAO and COMT inhibited by pretreatment with clorgyline [5 mg/kg i.v.] plus pargyline [20 mg/kg i.v.] and treatment with tolcapone [3 mg/kg i.v. followed by 1.5 mg/kg i.v. given every 30 min], respectively). Group I included seven vehicle controls and 10 animals given disprocynium 24, whereas group II comprised seven controls and nine disprocynium

Advances in Pharmacology, Volume 42

1054-3589/98 $25.00

24–treated animals. In a second series of experiments, the effect of disprocynium 24 (same dose as in first series) on the plasma clearance due to renal excretion (Cl_u) of endogenous NA, A, and DA as well as infused [^{3}H]NA, [^{3}H]A, and [^{3}H]DA was determined for 60-min periods of urine collection (Cl_u = renal excretion rate/plasma concentration). Again, these experiments were carried out in animals with no further treatment (group III) and in animals in which MAO and COMT were inhibited as described previously (group IV). Each of the two latter groups comprised 13 vehicle-treated and 13 disprocynium 24–treated animals.

In the experiments involving animals of groups I and II, the treatment with disprocynium 24 resulted in a steady-state plasma concentration of the drug of about 640 nM. When compared with vehicle controls, disprocynium 24 did not alter heart rate and mean arterial pressure but increased CO_p by 21% ($p < .001$). Hence, disprocynium 24 decreased the calculated total peripheral resistance. These hemodynamic drug effects were very much the same in groups I and II. Baseline values of Cl_{tot} for [^{3}H]NA, [^{3}H]A, and [^{3}H]DA were 86, 83, and 157% of CO_p, respectively, in group I and 66, 72, and 100% of CO_p, respectively, in group II. Hence, the combined inhibition of MAO and COMT reduced Cl_{tot} of [^{3}H]NA, [^{3}H]A, and [^{3}H]DA by 24, 14, and 36% ($p < .01$ for each), respectively. Disprocynium 24 significantly ($p < .05$) reduced Cl_{tot} of [^{3}H]NA, [^{3}H]A, and [^{3}H]DA in groups I (by 33, 31, and 22%, respectively) and II (by 29, 22, and 17%, respectively).

Uptake$_2$ has the capacity to function as an irreversible site of loss for (and mediates a steady net removal of) catecholamines only when "sink" mechanisms are operative subsequent to inward transport (e.g., amine metabolism through MAO and COMT activity). In many tissues, net removal through uptake$_2$ quickly approaches zero when MAO and COMT are inhibited, because inward transport then gives rise to intracellular amine accumulation followed by amine efflux out of the cells (3). Therefore, the inhibitory effect of disprocynium 24 on [^{3}H]catecholamine clearances was expected to be much less pronounced after MAO und COMT inhibition (group II) than under normal conditions (group I). But contrary to expectation, the percent inhibition of the [^{3}H]catecholamine clearances produced by disprocynium 24 was similar in groups I and II (see preceding discussion). Moreover, when the disprocynium 24–sensitive component of [^{3}H]NA, [^{3}H]A, and [^{3}H]DA clearance was determined by subtracting the value of clearance in the presence of disprocynium 24 from that observed prior to disprocynium 24 treatment, it was found that, after MAO and COMT inhibition, there was still a statistically significant, disprocynium 24–sensitive removal of [^{3}H]NA, [^{3}H]A, and [^{3}H]DA. Hence, disprocynium 24 must have reduced [^{3}H]catecholamine clearances not only by blocking uptake$_2$, but also by blocking related extraneuronal amine transporters that are capable of mediating net removal of circulating catecholamines even after MAO and COMT inhibition. These amine transporters may bring about net removal of catecholamines as a result of uptake followed by excretion or, simply, as a result of catecholamine excretion (which may take place not only in the kidneys, but also in the liver, intestine, and glandular tissues).

Cyanine analogues closely related to disprocynium 24 are known inhibitors of the tubular secretion of organic cations, including catecholamines in the kidney (4). Indeed, disprocynium 24 was recently shown to inhibit (IC 50 110 nM) an organic cation transporter (OCT1) isolated from the rat kidney (5). Moreover, OCT1 is also functionally expressed in rat hepatocytes (5). Hence, in addition to the present results, there is evidence that part of the inhibitory effect of disprocynium 24 on the plasma clearance of [^{3}H]catecholamines may well be due to inhibiton of organic cation transporters related to, but not identical with, uptake$_2$. To examine whether disprocynium 24 inhibits organic cation transport *in vivo,* the experiments involving animals of groups III and IV were carried out. In these experiments, the renal excretion of catecholamines was measured in rabbits treated with either disprocynium 24 or vehicle. As mentioned previously, part of the renal catecholamine excretion is due to tubular secretion and, hence, the result of organic cation transport (4).

In vehicle-treated animals, Cl_u of endogenous NA, A, and DA was 5.2, 7.2, and 153.6 ml/(kg min), respectively, in group III and 7.0, 10.4, and 134.3 ml/(kg min), respectively, in group IV. Obviously, MAO and COMT inhibition (as in group IV) increased Cl_u of endogenous NA (by 35%; $p < .05$) and A (by 44%; $p < .05$) but had no effect on Cl_u of endogenous DA. Similar Cl_u values were obtained for infused [^{3}H]NA and [^{3}H]A but not for infused [^{3}H]DA. In fact, Cl_u of [^{3}H]DA (4.9 ml/[kg min] in group III and 15.4 ml/[kg min] in group IV; $p < .01$) was considerably smaller than Cl_u of DA. This difference is readily explained by the fact that most of the urinary DA was derived from DA formed in the tubular epithelium of the kidney. In support of this contention, it was found that the fraction of urinary DA derived from the circulation (which is given by the specific activity of [^{3}H]DA in urine expressed as a fraction of the specific activity of [^{3}H]DA in plasma) was 2.3% in group III and 7.8% in group IV. Hence, the renal contribution to the urinary DA excretion amounted to about 98% of the total in group III and 92% of the total in group IV ($p < .01$ for the group difference). Similar considerations for NA indicated that one-third to one-fourth of the urinary NA (32% in group III and 24% in group IV) was derived from renal sources of NA formation (e.g., sympathetic nerves). In contrast to DA and NA, all the A in urine was derived from the circulation, because the ratio of "specific activity of [^{3}H]A in urine/specific activity of [^{3}H]A in plasma" did not differ from unity.

Disprocynium 24 significantly ($p < .01$) reduced Cl_u of endogenous NA, A, and DA by 73, 72, and 90%, respectively in group III and by 72, 72, and 80%, respectively, in group IV; the corresponding inhibition figures for the Cl_u of circulating [^{3}H]NA, [^{3}H]A, and [^{3}H]DA were 49, 62, and 61%, respectively, in group III and 64, 69, and 66%, respectively, in group IV. These results indicate that the degree of inhibition of Cl_u by disprocynium 24 was more pronounced for endogenous than for infused ^{3}H-labeled catecholamines. This was especially true for DA and NA and less so for A, indicating that disprocynium 24 preferentially inhibited the urinary excretion of those components of DA and NA that are derived from renal sources of amine formation. Accordingly, the fractions of urinary DA and NA that originated from plasma increased in response to disprocynium 24: The plasma component of urinary DA increased from 2.3 to 6.8% in group III ($p < .01$) and 7.8 to 13.3% in group IV ($p < .05$), whereas the plasma component of urinary NA increased from 68 to 106%

in group III ($p < .01$) and 76 to 95% in group IV. Disprocynium 24 did not alter the ratio "specific activity in urine/specific activity in plasma" for [^{3}H]A.

In conclusion, disprocynium 24 reduces the clearance of circulating catecholamines even after the combined inhibition of MAO and COMT. This was taken to indicate that the drug interferes with the removal of circulating catecholamines by blocking not only uptake$_2$ but also related organic cation transporters. In support of this conclusion, it was found that disprocynium 24 likewise inhibited the renal excretion of catecholamines. Because this effect of disprocynium 24 was more pronounced for urinary catecholamines that originated from the kidney than for those derived from the circulation, it can be inferred that disprocynium 24 inhibits the tubular secretion of catecholamines and, hence, organic cation transport in the kidney.

Acknowledgments

This work was supported by the Deutsche Forschungsgemeinschaft (grants 490/8 and SFB 176 TP A13) and the Senator Kurt und Inge Schuster-Stiftung, Würzburg, Germany.

References

1. Friedgen, B., Halbrügge, T., Graefe, K-H. (1994). Roles of uptake$_1$ and catechol-O-methyltransferase in removal of circulating catecholamines in the rabbit. *Am. J. Physiol.* **267**, (*Endocrinol. Metab.* **30**), E814–E821.
2. Russ, H., Sonna, J., Keppler, K., Baunach, S., and Schömig, E. (1993). Cyanine-related compounds: A novel class of potent inhibitors of extraneuronal noradrenaline transport. *Naunyn Schmiedebergs Arch. Pharmacol.* **348,** 458–465.
3. Trendelenburg, U. (1988). The extraneuronal uptake and metabolism of catecholamines. *In* Trendelenburg U, Weiner N (eds) Catecholamines I, Handbook of Experimental Pharmacology, vol. 90/I. (U. Trendelenburg and N. Weiner, eds.), pp. 279–319. Springer, Berlin.
4. Rennick, B. R. (1981). Renal tubule transport of organic cations. *Am. J. Physiol.* **240,** (*Renal Fluid Electrolyte Physiol.* **9**), F83–F89.
5. Martel, F., Vetter, T., Russ, H., Gründemann, D., Azevedo, I., Koepsell, H., and Schömig, E. (1996). Transport of small organic cations in the rat liver; the role of the organic cation transporter OCT1. *Naunyn Schmiedebergs Arch. Pharmacol.* **354,** 320–326.

G. W. Lambert, D. M. Kaye, J. M. Thompson, A. G. Turner, C. Ferrier, H. S. Cox, M. Vaz, D. Wilkinson, I. T. Meredith, G. L. Jennings, and M. D. Esler

Human Autonomic Function Laboratory and Alfred and Baker Medical Unit
Baker Medical Research Institute
Prahran Vic 3181, Australia

Catecholamine Metabolites in Internal Jugular Plasma: A Window into the Human Brain

Following von Euler's characterization of norepinephrine (NE) as the sympathetic neurotransmitter and the subsequent demonstration of a direct relationship between rates of sympathetic nerve firing and neurotransmitter release, the potential to use NE washout, or spillover to plasma, as an index of sympathetic nervous function was clear. Maas and colleagues (1) extended this reasoning to central nervous system neurons and demonstrated the utility of using direct internal jugular vein blood sampling techniques in the assessment of central nervous system neuronal activity, by demonstrating a reduction in 3-methoxy-4-hydroxyphenylglycol (MHPG) jugular venous overflow from the brain of stump-tailed monkeys following clonidine administration. The clinical application of such methodology, however, is a relatively recent development (2–4).

Using percutaneously placed catheters in either the right, left, or both internal jugular veins, combined with cerebral blood flow scans to differentiate between cortical and subcortical venous drainage of the brain, we have examined brain catecholaminergic processes in over 200 subjects, comprising healthy volunteers and patients with essential hypertension, heart failure, and pure autonomic failure. A variety of pharmacological interventions have also been examined. Venoarterial plasma concentration differences and internal jugular blood or plasma flow was used, according to the Fick Principle, to quantify the amount of neurotransmitter stemming from the brain. To assess the possible involvement of brain catecholaminergic neuronal systems in the modulation of sympathetic outflow, sympathetic nervous function was determined concurrently via norepinephrine isotope dilution methodology and microneurographic nerve recording.

In the healthy subjects examined, positive venoarterial plasma concentration increments of the catecholamine precursor, dihydroxyphenylalanine (DOPA); NE and its metabolites, dihydroxyphenylglycol (DHPG) and MHPG; and the dopamine metabolite, homovanillic acid (HVA) were evident in the internal jugular venous effluent (Fig. 1). There was a net extraction of tritium-

Advances in Pharmacology, Volume 42

1054-3589/98 $25.00

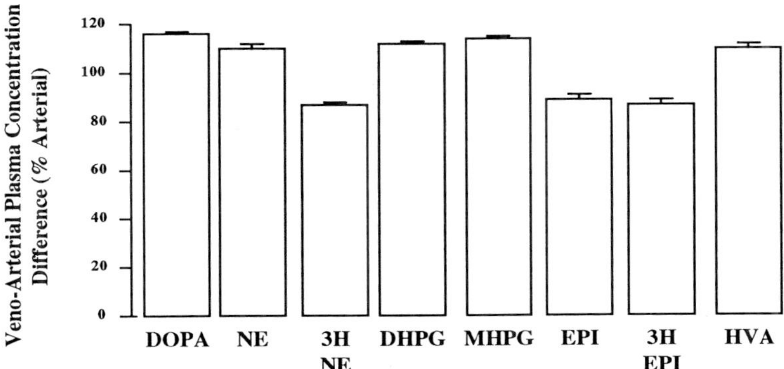

FIGURE 1 Internal jugular–arterial plasma concentration gradients of a variety of neurochemicals.

labeled [^{3}H]NE and epinephrine (Epi) and endogenous Epi across the brain. There was no detectable release of either dopamine or dihydroxyphenylacetic acid from the brain.

The cerebrovascular circulation, however, is subject to a rich sympathetic innervation, and, hence, the actual source of these neurochemicals washing into the cerebral effluent may be open to some conjecture. To elucidate the origin of the overflow of NE and its metabolites into the internal jugular vein, we studied patients with pure autonomic failure in whom there was biochemical evidence of almost complete postganglionic sympathetic denervation. The spillover of NE to plasma for both the body as a whole (3.37 ± 0.24 vs 0.74 ± 0.18 nmol/min, $p < .001$) and the heart (153 ± 26 vs 15 ± 7 pmol/min, $p < .001$) was, as expected, substantially reduced in these patients, yet the estimated central nervous system turnover of NE, based on the combined internal jugular venous overflows of NE and its lipophilic metabolites DHPG and MHPG, was no different than that of the healthy, age-matched subjects (0.87 ± 0.29 vs 0.74 ± 0.20 nmol/min). This observation supports the contention that the internal jugular venous NE and metabolites we measured emanate from central noradrenergic neurons and not from cerebrovascular sympathetic nerves.

In a subset of the healthy individuals (n = 13), resting muscle sympathetic nervous activity was recorded in parallel with estimates of central nervous system NE turnover. Muscle sympathetic nerve impulses occurred at irregular bursts in synchrony with the cardiac rhythm, with the mean burst frequency being 19 ± 3 bursts/min. Linear regression analysis of the cerebral NE turnover data, irrespective of the internal jugular vein sampled, revealed a significant positive relationship between the estimated central nervous NE turnover and the level of muscle sympathetic nervous activity ($y = 6.0x + 13.8$; $r = .64$, $p = .02$). Pertinent to this finding is our previous demonstration of a dependence of sympathetic outflow for the body as a whole, and to the skeletal muscle vascular bed and the kidneys in particular, on central nervous system NE turnover following intravenous infusion of desipramine (5).

To further investigate the relationship between brain NE, sympathetic nervous activity, and blood pressure regulation, we also examined brain catecholamine turnover in response to blood pressure reduction elicited via intravenous administration of the ganglion-blocking drug, trimethaphan, and the centrally acting antihypertensive agent, clonidine. Both systolic and diastolic blood pressures were significantly reduced 60 min following initiation of trimethaphan administration ($p < .05$). In parallel with this blood pressure reduction, the rate of spillover of NE for the body as a whole fell by approximately 30% (4.40 ± 0.87 vs 2.95 ± 0.38 nmol/min, $p < .05$). The concomitant reductions in blood pressure and total body NE spillover to plasma were accompanied by an over fivefold increase in the estimated turnover of NE in the brain (0.50 ± 0.50 vs 2.59 ± 1.08 nmol/liter, $p < .05$). Following clonidine administration, systolic and mean arterial blood pressures were also significantly reduced ($p < .05$). Clonidine resulted in greater than 50% reduction in the rate of spillover of NE to plasma for the body as a whole (2.85 ± 0.62 vs 1.24 ± 0.17 nmol/min, $p < .05$). The overflow of MHPG into the internal jugular vein was reduced by 33% following clonidine treatment ($p < .05$).

The substantial increase in brain NE turnover following trimethaphan presumably results from a compensatory response in sympathoexcitatory forebrain noradrenergic neurons in the face of interruption of sympathetic neural traffic and reduction in arterial blood pressure, while the reduction in central nervous system NE turnover in response to intravenous clonidine administration underlies the blood pressure–lowering action of the drug. Taken together with our previous observations documenting a link between central nervous system noradrenergic neuronal activity and the cardiac sympathoexcitation seen in patients with congestive heart failure (3, 4) or the elevated renal sympathetic activity in patients with essential hypertension (2), clearly the old idea of brain noradrenergic neurons subserving a predominantly sympathoinhibitory role is no longer tenable.

References

1. Maas, J. W., Hattox, S. E., Landis, D. H., and Roth, R. H. (1977). A direct method for studying 3-methoxy-4-hydroxyphenethyleneglycol (MHPG) production by brain in awake animals. *Eur. J. Pharmacol.* **46,** 221–228.
2. Ferrier, C., Jennings, G. L., Eisenhofer, G., Lambert, G., Cox, H., Kalff, V., Kelly, M., and Esler, M. D. (1993). Evidence for increased noradrenaline release from subcortical brain regions in essential hypertension. *J. Hypertens.* **11,** 1217–1227.
3. Lambert, G. W., Kaye, D. M., Lefkovits, J., Jennings, G. L., Turner, A. G., Cox, H. S., and Esler, M. D. (1995). Increased central nervous system monoamine neurotransmitter turnover and its association with sympathetic nervous activity in treated heart failure patients. *Circulation* **92,** 1813–1818.
4. Kaye, D. M., Lambert, G. W., Lefkovits, J., Morris, M., Jennings, G., and Esler, M. D. (1994). Neurochemical evidence of cardiac sympathetic activation and increased central nervous system norepinephrine turnover in congestive heart failure. *J. Am. Coll. Cardiol.* **23,** 570–578.
5. Esler, M. D., Wallin, B. G., Dorward, P. K., Eisenhofer, G., Westerman, R., Meredith, I., Lambert, G., Cox, H. S., and Jennings, G. (1991). Effects of desipramine on sympathetic nerve firing and norepinephrine spillover to plasma in humans. *Am. J. Physiol.* **260,** R817–R823.

Mika Scheinin,* Ari Illi,† Markku Koulu,*
and Pirjo Ojala-Karlsson*

*Department of Pharmacology and Clinical Pharmacology
University of Turku
FIN-20520 Turku, Finland

†Orion Research Center
FIN-02101 Espoo, Finland

Norepinephrine Metabolites in Plasma as Indicators of Pharmacological Inhibition of Monoamine Oxidase and Catechol O-Methyltransferase

The formation of 3,4-dihydroxyphenylglycol (DHPG) from norepinephrine (NE) is dependent on deamination of NE by intraneuronal monoamine oxidase type A (MAO-A) and reduction of the resulting aldehyde. Another main metabolite of NE, 3-methoxy-4-hydroxyphenylglycol (MHPG), is formed by methylation of DHPG by catechol O-methyltransferase (COMT) and by deamination and reduction of the COMT-dependent methylated metabolite of NE, normetanephrine (NM) (1–3). Metabolite concentrations in blood plasma are determined by the rates of their influx into the circulation and by their rates of removal, either by excretion or metabolism. Acute alterations of neuronal NE release have relatively minor effects on the concentrations of DHPG and MHPG in plasma (4), which suggests that measurement of DHPG and MHPG levels may give reliable information on the activity of the catabolic enzymes acting on NE. We report here the results from studies investigating the usefulness of plasma DHPG and MHPG concentrations as indicators of pharmacological inhibition of MAO-A and COMT activity *in vivo* in human subjects. Such indicators are needed in the clinical development of new inhibitors of MAO and COMT.

All study subjects were healthy male volunteers. All studies were carried out after appropriate informed consent and Investigation Review Board (IRB) approval. Concentrations of NE, DHPG, and MHPG in plasma were measured using high-performance liquid chromatography (HPLC) with coulometric electrochemical detection. Single doses (50–300 mg) of moclobemide (Roche) were used to reversibly inhibit MAO-A. Selegiline (deprenyl; Orion-Pharma) was used in some experiments to produce long-lasting inhibition of MAO-B; 10-mg doses were administered either once or repeatedly over 4 days. These

Advances in Pharmacology, Volume 42

1054-3589/98 $25.00

doses are used clinically in the treatment of depression (moclobemide) and Parkinson's disease (selegiline). Reversible inhibition of COMT was achieved by single or repeated doses of entacapone (up to 800 mg t.i.d. for 1 week) or nitecapone (up to 100 mg t.i.d. 1 week)(Orion-Pharma), two novel nitrocatechol derivatives. The former is currently in clinical trials as adjuvant treatment in Parkinson's disease; it acts by inhibiting the O-methylation of levodopa, thus increasing its bioavailability. Both drugs have poor access into the central nervous system (CNS) and inhibit COMT activity mainly in peripheral tissues.

Moclobemide induced dose-related, rapid (half-life, about 30 min) decreases in plasma DHPG (up to 80% after 300 mg) and slower, up to 50%, decreases in plasma MHPG. Plasma DHPG was, thus, a more sensitive indicator of acute MAO-A inhibition than plasma MHPG. The DHPG decrease was specific for MAO-A inhibition, because neither single nor repeated doses of selegiline (causing nearly 100% inhibition of MAO-B activity in blood platelets) had appreciable effects on plasma DHPG levels. Platelet MAO-B activity was reduced by about 25% after moclobemide 300 mg. Plasma NE and epinephrine levels were unchanged after moclobemide and selegiline. Moclobemide significantly reduced the urinary excretion of the deaminated catecholamine metabolites 3-methoxy-4-hydroxymandelic acid (up to 44%), and homovanillic acid (up to 40%) and increased the excretion of the nondeaminated, methylated metabolites NM (up to 114%), metanephrine (up to 50%), and 3-methoxytyramine (up to 180%). The excretion of 5-hydroxyindoleacetic acid, the main deaminated metabolite of 5-hydroxytryptamine, was not influenced by single doses of moclobemide.

Inhibition of COMT was studied both in resting subjects and in combination with marked stimulation of the sympathetic nervous system with submaximal bicycle exercise lasting 12 min. Entacapone increased plasma DHPG levels (up to threefold after 800 mg) similarly after single and repeated doses (t.i.d. for 7 days) and decreased MHPG in plasma more markedly after repeated administration (up to 50% after 800 mg t.i.d.). These doses inhibit soluble COMT in erythrocytes by up to 60–80%. Again, DHPG appeared to be a more sensitive indicator of pharmacological enzyme inhibition than MHPG, presumably because of its faster turnover and less complicated pathways of formation and elimination compared with MHPG. Nitecapone had effects similar to those of entacapone.

Inhibition of COMT by entacapone or nitecapone had no effects on the plasma concentrations of unconjugated NE and epinephrine, neither at rest nor during physical exercise, which elevated plasma NE approximately sixfold (to about 12 nmol/liter) and epinephrine approximately 10-fold (to about 1 nmol/liter). The plasma concentrations of catecholamine conjugates were slightly elevated by repeated administration of entacapone and nitecapone, but not by single doses. Methylated metabolites in urine were also measured in some studies. DHPG in plasma was a clearly more sensitive indicator of altered catecholamine metabolism than the methylated metabolite concentrations in a 6-hr urine fraction (NM, metanephrine, 3-methoxytyramine, 3-methoxy-4-hydroxymandelic acid, homovanillic acid). The expected reductions in the plasma concentrations of NM and metanephrine could not be documented after administration of the COMT inhibitors; resting plasma levels of these metabolites were already at the lower limit of quantitation of the employed HPLC assay, and no further reductions could be reliably demonstrated.

When single doses of moclobemide (150 mg) and entacapone (200 mg) were given together, plasma DHPG levels were intermediate between those after either drug alone. The drug combination reduced plasma MHPG concentrations similarly as moclobemide alone, indicating that MHPG levels in plasma are mainly determined by MAO-A activity. In the CNS, a significant proportion of MHPG may be derived from NM by extraneuronal (MAO-B–dependent) deamination and aldehyde reduction; in peripheral tissues and in circulating plasma, no appreciable amounts of MHPG seem to be derived from MAO-B–dependent deamination of NM. This agrees with earlier findings in rats (5).

In conclusion, the concentration of DHPG in plasma appears to be a sensitive and specific indicator of pharmacological inhibition of MAO-A activity, as already suggested by others (2, 6). We suggest that DHPG measurements could be used in early clinical studies to determine the dosage and dose interval of new inhibitors of MAO-A, especially because no equally simple and convenient alternatives are available. Whether DHPG responses could be used to monitor drug intake and dose requirements and/or to predict therapeutic efficacy in depressed patients receiving MAO-A inhibitors remains to be determined. Pharmacological inhibition of COMT is also accurately reflected in increased DHPG and decreased MHPG concentrations in plasma. Determination of these metabolites may, thus, be helpful in early clinical studies on new COMT inhibitors.

Acknowledgments

The authors are grateful to the coinvestigators of the original studies, J. Akkila, A. Gordin, S. Heinävaara, S. Kaakkola, A. Kaarttinen, A. Kallio, J. Kallio, S. Karhuvaara, T. Keränen, K. Pyykkö, S. Sundberg, J. Vuorinen, and R. H. Zimmer, and to R. Pohjola and T. Pulska for technical assistance. The studies were supported financially by Orion-Farmos Pharmaceuticals, Finland, and F. Hoffman–La Roche & Co., Switzerland.

References

1. Kopin, I. J. (1985). Catecholamine metabolism: Basic aspects and clinical significance. *Pharmacol. Rev.* **37**, 333–364.
2. Izzo, J. L., Thompson, D. A., and Horwitz, D. (1985). Plasma dihydroxyphenylglycol (DHPG) in the *in vivo* assessment of human neuronal norepinephrine metabolism. *Life Sci.* **37**, 1033–1038.
3. Goldstein, D. S., Eisenhofer, G., Stull, R., Folio, C. J., Keiser, H. R., and Kopin I. J. (1988). Plasma dihydroxyphenylglycol and the intraneuronal disposition of norepinephrine in humans. *J. Clin. Invest.* **81**, 213–220.
4. Scheinin, M., Karhuvaara, S., Ojala-Karlsson, P., Kallio, A., and Koulu, M. (1991). Plasma 3,4-dihydroxyphenylglycol (DHPG) and 3-methoxy-4-hydroxyphenylglycol (MHPG) are insensitive indicators of alpha-2-adrenoceptor mediated regulation of norepinephrine release in healthy volunteers. *Life Sci.* **49**, 75–84.

5. Eisenhofer, G., Pecorella, W., Pacak, K., Hooper, D., Kopin, I. J., and Goldstein, D. S. (1994). The neuronal and extraneuronal origins of plasma 3-methoxy-4-hydroxyphenylglycol in rats. J. Auton. Nerv. Syst. 50, 93–107.
6. Brown, M. J., and Monks, N. J. (1984). Plasma dihydroxyphenylglycol concentration: A simple, specific and sensitive index of monoamine oxidase A activity in man. *In* Monoamine Oxidase in Disease. (K. F. Tipton, P. Dostert and and M. S. Benedetti, eds.), pp. 559–560. Academic Press, London.

G. M. Tyce,* S. L. Chritton,* R. D. Barnes,† L. E. Ward,† L. W. Hunter,† and D. K. Rorie†

Departments of *Physiology and Biophysics and †Anesthesiology
Mayo Clinic/Foundation
Rochester, Minnesota 55905

The Adrenal Gland as a Source of Dihydroxyphenylalanine and Catecholamine Metabolites

Considerable quantities of dihydroxyphenylalanine (DOPA), the precursor of catecholamines, and of the catecholamine metabolites are present in plasma of humans and experimental animals. Several of these compounds have been considered to be indices of activity of peripheral sympathetic nerves. However, the highest concentrations of catecholamines in the body are present in the adrenal gland, although little attention has been paid to this organ as a source of DOPA and of the catecholamine metabolites.

In these studies, the isolated perfused dog adrenal gland was used to examine the following:

1. The characteristics of the releases of DOPA, catecholamines, and their metabolites and in some experiments, the releases of the neuropeptides metenkephalin (Met-Enk) and neuropeptide Y (NPY). Met-Enk and NPY are costored with the catecholamines in chromaffin cells and are released exocytotically with catecholamines.
2. The effects of cocaine on basal and evoked releases of catecholamines and their metabolites. Cocaine inhibits reuptake of released norepinephrine (NE) in sympathetic neurons.
3. The effects of nitric oxide (NO) on releases of catecholamines from the adrenal gland. NO inhibits release of NE in some blood vessels.

Advances in Pharmacology, Volume 42

1054-3589/98 $25.00

Isolated dog adrenal glands were perfused *ex situ* with oxygenated Krebs-Ringer solution at 37°C (1). After a 60-min stabilization, perfusates were collected during (1) a 10-min basal period, (2) a 2-min stimulation with carbachol (3×10^{-5} M) or with the nicotinic agonist 1,1-dimethyl-4-phenylpiperazinum (DMPP, 3×10^{-6} M or 5×10^{-5} M), (3) an 8-min poststimulation period, and (4) a 30-min stabilization period. This sequence was repeated two or three times (i.e., with three stimulations, S_1, S_2, and S_3). In some studies, cocaine (10^{-5} M) was added before S_2, and in other studies, a nitric oxide synthase (NOS) inhibitor, N^G-monomethyl-L-arginine (L-NMMA, 3×10^{-4} M), was added before S_3.

Epinephrine (Epi), NE, dopamine (DA), DOPA, metanephrine (MN), nometanephrine (NM), 3-methoxytyramine (3MT), 3,4-dihydroxyphenylglycol (DOPEG), 3,4-dihydroxyphenylacetic acid (DOPAC), and 3-methoxy,4-hydroxyphenylglycol (MHPG) were quantified in perfusates by high-performance liquid chromatography with electrochemical detection (1). The neuropeptides were determined by radioimmunoassay (2). Net evoked effluxes were calculated by subtracting prior basal efflux from total evoked efflux. To determine the effect of NOS inhibition on evoked release, the ratios of net evoked effluxes in S_3 were expressed as a percentage of those in S_2. Comparisons were made of S_3/S_2 in the presence versus the absence of L-NMMA during S_3 (3).

During the first 60 min of stabilization of each gland, the concentrations of the catecholamines and of the neuropeptides declined exponentially. Substantial amounts of the metabolites MN, NM, DOPEG, DOPAC, and MHPG were also present in perfusates, and the concentrations of these metabolites did not change in the initial 60 min of stabilization. DOPA was also detectable in perfusates, and its efflux declined exponentially in a manner similar to the catecholamines and the neuropeptides, suggesting a similar origin in the chromaffin cells.

After 60 min, the mean ($\pm$SEM) overflows of Epi, NE, DA, DOPA, DOPEG, DOPAC, MN, NM, and MHPG were 4500, 680, 57, 18, 380, 36, 230, 93, and 19 pmoles/min, respectively (Table I). The effluxes of Met-Enk and NPY were 2.4 and 0.01 pmoles/min, respectively. Noticeably, vanilmandelic acid, homovanillic acid, 3MT, 3-*O*-methyldopa, 3,4-dihydroxymandelic acid, 3,4-dihydroxyphenylethanol, and 3-methoxy-4-hydroxyphenylethanol were not detected in superfusates.

Carbachol evoked releases one- to fourfold above basal effluxes of Epi, NE, DA, NPY, Met-Enk, and, surprisingly, DOPA, but had no significant effects on the effluxes of the metabolites. In Ca^{2+}-free Krebs-Ringer, carbachol did not evoke releases of the catecholamines, the neuropeptides, or DOPA.

Cocaine had no effect on the basal or evoked effluxes of the catecholamines, DOPA, the neuropeptides, or of any of the metabolites, including DOPEG. In one experiment, saponin was introduced, and its addition increased the effluxes of DOPA as well as of the catecholamines and the neuropeptides, but it had no effect on effluxes of the metabolites.

When L-NMMA was added to the perfusate, there was a 25% increase in the basal effluxes of NE, Epi, and DA within 10 min. These effects of L-NMMA were not evident in the presence of arginine (10^{-3} M) or when Ca^{2+} was absent from the perfusate.

TABLE I Basal Effluxes of Catecholamines and Metabolites in Dog Adrenal Gland, Mesenteric Artery, and Portal Vein

Analyte	*Adrenal gland (pmole/min)**	*Mesenteric artery (fmole/min)*	*Portal vein (fmole/min)*
E	4,500 ± 2000	ND	ND
NE	680 ± 200	68 ± 11	90 ± 11
DA	57 ± 10	8 ± 5	35 ± 13
DOPA	18 ± 1	49 ± 14	30 ± 9
DOPEG	380 ± 100	678 ± 52	409 ± 34
DOPAC	36 ± 10	ND	ND
MN	230 ± 80	ND	ND
NM	93 ± 60	+	+
MHPG	19 ± 4	+	+

Data as mean ± SE of 5 to 9 determinations.
* Data from Chritton *et al.* (2).
+, not detectable under basal conditions but can be detected during nerve stimulation; ND, not detectable.

The effects of L-NMMA on evoked releases of catecholamines were dependent on the concentration of DMPP used to stimulate release. At a low dose of DMPP (3×10^{-6} M), L-NMMA increased releases of NE and DA, whereas at a high dose (5×10^{-5}M), L-NMMA inhibited releases of Epi and DA (3).

From these studies it can be concluded that the adrenal medulla is potentially a source of DOPA in plasma. However, the adrenal gland produced much less DOPA relative to NE than is found in blood vessels (Table I) (4). DOPA was released from the adrenal gland by carbachol in a Ca^{2+}-dependent manner, and its initial washout was similar to that of the catecholamines and the neuropeptides. The time course of DOPA release was the same as that of the catecholamines. In addition, DOPA increased in perfusates similarly to the neuropeptides and the catecholamines after application of saponin. All of these data suggest that DOPA was localized in storage vesicles in the chromaffin cells and was released from these vesicles exocytotically. An alternate explanation of the Ca^{2+} dependence of DOPA release may be that the activity of tyrosine hydroxylase is regulated by the cytoplasmic Ca^{2+} concentration, which is increased during exocytotic release.

DOPEG was the catecholamine metabolite present in highest concentrations in perfusates (see Table I). The amounts of DOPEG in perfusates were much less relative to the precursors (NE and Epi) in the adrenal glands than in superfusates from canine blood vessels (see Table I) (4). In blood vessels, we have proposed that under basal conditions, DOPEG in superfusates originates by the action of monoamine oxidase (MAO) on NE subsequent to translocation of NE from vesicles into cytoplasm. This conclusion is reached because basal efflux of DOPEG is unchanged in the presence of inhibitors of neuronal uptake of released NE (4). During evoked release of NE from blood vessels, the concentration of DOPEG in superfusates increases, and this increase is blocked by inhibitors of NE reuptake. Thus, in peripheral sympathetic nerves,

DOPEG is formed by oxidative deamination of NE subsequent to two distinct processes: translocation of NE from vesicles into neuroplasm and reuptake of NE after release. However, in the adrenal gland, cocaine had no effect on DOPEG or on NE overflow either under basal conditions or during evoked release, indicating that there is no reuptake process for NE in this organ. This would be expected in view of the endocrine function of the Epi and NE released from the adrenal gland. Thus, in the adrenal gland, DOPEG overflow probably originates solely after translocation of NE from storage vesicles into neuroplasm, and its formation can be taken as an index of this process.

The lower ratios of DOPEG to precursor catecholamines in the adrenal gland compared with blood vessels probably reflect differences in the vesicular transporters in these two tissues or to differences in MAO activity.

O-Methylated metabolites were abundant in perfusates of the adrenal gland. Catechol *O*-methyltransferase (COMT) is probably not expressed in chromaffin cells, so NE and Epi would not access this enzyme until after release and reuptake into nonchromaffin cells. Despite this more circuitous access of the catecholamines to COMT than to MAO, the total amounts of *O*-methylated metabolites were quite similar to those of DOPEG.

Endogenously produced NO appeared to inhibit basal releases of NE, Epi, and DA from adrenal glands. Basal release has previously received little attention, but it has been recognized that nonexocytotic release from sympathetic nerves can be substantial (5). The importance of this basal release may relate to the maintenance of tone in target organs. The enzyme NOS is present in chromaffin cells, in ganglion cells, in the network of nerve fibers of the intrinsic plexus, and in endothelial cells of blood vessels in the adrenal gland.

NO had little effect on effluxes of DOPEG or of DOPAC during basal or evoked releases, indicating that vesicular retention of catecholamines was not affected. The effects of NO on evoked release from the adrenal gland depended on the intensity of stimulation, and further studies are required to elucidate the mechanisms involved.

References

1. Chritton, S. L., Dousa, M. K., Yaksh, T. L., and Tyce, G. M. (1991). Nicotinic- and muscarinic-evoked release of canine adrenal catecholamines and peptides. *Am. J. Physiol.* **260**, R589–599.
2. Chritton, S. L., Chinnow, S. L., Grabau, C., Dousa, M. K., Lucas, D., Roddy, D., Yaksh, T. L., and Tyce, G. M. (1996). Adrenomedullary secretion of DOPA, catecholamines, catechol metabolites and neuropeptides. *J. Neurochem.* (submitted)
3. Barnes, R. D., Ward, L. E., Tyce, G. M., and Rorie, D. K. (1997). Role of nitric oxide in modulation of evoked catecholamine efflux from canine adrenal gland. Intern Anesth Res Soc 71st Clinical and Scientific Congress, San Francisco, March 14–18, 1997. (abstract).
4. Tyce, G. M., Hunter, L. W., Ward, L. E., and Rorie, D. K. (1995). Effluxes of 3,4-dihydroxyphenylalanine, 3,4-dihydroxyphenylglycol, and norepinephrine from four blood vessels during basal conditions and during nerve stimulation. *J. Neurochem.* **64,** 833–841.
5. Ward, L. E., Hunter, L. W., Grabau, C. E., Tyce, G. M., and Rorie, D. K. (1996). Nitric oxide reduces basal efflux of catecholamines from perfused dog adrenal glands. *J. Auton. Nerv. Syst.* **61**(3), 235–242.

Graeme Eisenhofer* and Jacques W. M. Lenders†

*Clinical Neuroscience Branch
National Institute of Neurological Disorders and Stroke
National Institutes of Health
Bethesda, Maryland 20892

†Department of Internal Medicine
St. Radboud University Hospital
Nijmegen, 6525 GA The Netherlands

Clues to the Diagnosis of Pheochromocytoma from the Differential Tissue Metabolism of Catecholamines

Pheochromocytoma is a tumor of chromaffin cells most commonly arising from the adrenal medulla and usually presenting as hypertension. The clinician is usually alerted to the presence of the tumor if the patient does not respond to antihypertensive therapy or complains of symptoms such as flushing, attacks of anxiety, headaches, sweatiness, or palpitations. These symptoms reflect the effects of high circulating levels of catecholamines secreted by a tumor. Measurements of catecholamines in urine or plasma, or of catecholamine metabolites in urine, typically provide the standard initial test in the diagnosis of the tumor.

Understanding the utility and limitations of measurements of urine or plasma levels of catecholamines and their metabolites for diagnosis of pheochromocytoma requires an understanding of catecholamine metabolism. This is not restricted to knowledge of the sources of catecholamines and the pathways of their metabolism, but more importantly requires an appreciation of how catecholamines are metabolized differentially within neuronal and extraneuronal compartments, among tissues and organs and before and after their entry into the bloodstream.

I. Differential Metabolism of Catecholamines

Only a small amount of norepinephrine released from nerves escapes neuronal re-uptake to be metabolized by extraneuronal tissues. Thus, most norepinephrine is deaminated initially within sympathetic nerves to 3,4-dihydroxyphenylglycol (DHPG), and most of this is formed from transmitter leaking from storage vesicles, not from metabolism of recaptured transmitter after release (1). The DHPG formed in nerves is metabolized to 3-methoxy-4-

Advances in Pharmacology, Volume 42

1054-3589/98 $25.00

hydroxyphenylglycol (MHPG) in extraneuronal tissues (2); both metabolites are avidly extracted by the liver to form vanillylmandelic acid (VMA), the major end-product of norepinephrine and epinephrine metabolism (3). Because most MHPG and VMA are formed originally from DHPG produced in nerves (2, 3)—not from catecholamines metabolized extraneuronally before and after their release into the bloodstream—none of these metabolites is a particularly good marker for pheochromocytoma.

In contrast to these metabolites, the metanephrines (normetanephrine and metanephrine) are formed exclusively in extraneuronal tissues from O-methylation of norepinephrine and epinephrine (4, 5). In contrast to intraneuronal metabolism, a substantial amount of extraneuronal metabolism is dependent on removal of circulating catecholamines (2). Because pheochromocytomas secrete catecholamines directly into the bloodstream, production of metanephrines from circulating catecholamines provides one reason why these metabolites should be better markers for the tumor than DHPG, MHPG, and VMA.

However, a far more important reason for the use of metanephrines as reliable markers for a pheochromocytoma is their considerable synthesis within chromaffin tissue. At least 90% of metanephrine and up to 40% of normetanephrine are formed from epinephrine and norepinephrine within the adrenals before their release into plasma (4, 5). Concentrations of metanephrines in tumor tissue, of over 3 orders of magnitude higher than circulating concentrations in the same patients, also indicate their substantial production in pheochromocytoma tissue.

II. Metanephrines Versus Catecholamines

A problem with use of plasma or urinary catecholamines for diagnosis of pheochromocytoma is that some tumors are quiescent or encapsulated and may not secrete large amounts of catecholamines; other tumors appear to secrete catecholamines episodically. Thus, plasma levels of catecholamines are normal in some patients with pheochromocytoma (Fig. 1), and the presence of a pheochromocytoma cannot be reliably excluded using measurements of plasma or urinary catecholamine concentrations. In contrast, plasma metanephrines (either normetanephrine or metanephrine or both) show much more consistent increases above normal than plasma catecholamines (see Fig. 1) and appear to reliably exclude the presence of a pheochromocytoma (6). Additionally, plasma metanephrines show much larger increases above normal than plasma catecholamines.

Why are plasma metanephrines better than catecholamines for diagnosis of pheochromocytoma? The answer is simple. Many tumors may be relatively quiescent and not actively secrete catecholamines, whereas all tumors appear to actively metabolize catecholamines to the O-methylated metanephrine derivatives.

Measurements of plasma metanephrines can also offer other advantages over catecholamines for diagnosis of pheochromocytoma. Whereas there is no relationship of tumor size to plasma levels of norepinephrine, the size of tumors

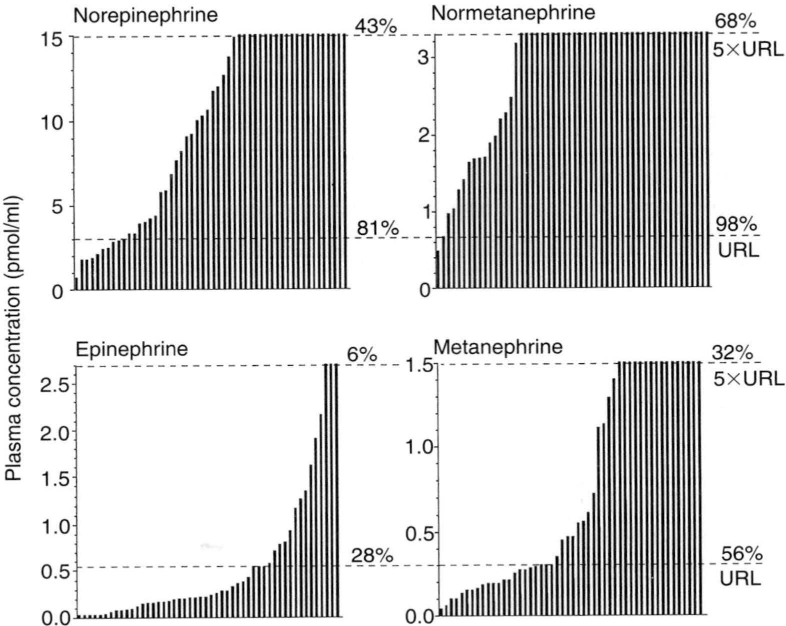

FIGURE 1 Comparison of plasma concentrations of norepinephrine (*upper left*), normetanephrine (*upper right*), epinephrine (*lower left*), and metanephrine (*lower right*) in 52 patients with pheochromocytoma. Concentrations for each patient are shown by the vertical bars. The lower dashed horizontal lines for each panel depict the upper reference limits (URL) determined from a large population of normal subjects and hypertensive patients. The upper dashed lines indicate concentrations 5 times the upper reference limits (5 × URL). Nearly 20% of pheochromocytoma patients had plasma concentrations of norepinephrine within the normal range, whereas only one patient had a normal plasma concentration of normetanephrine; however, this patient had an elevated plasma metanephrine. Thus, even in this patient, an elevated metanephrine was consistent with the presence of a tumor. Many more patients showed elevated plasma concentrations of metanephrine than of epinephrine (56% vs 25%). Although plasma metanephrine concentrations are normal in 44% of patients, they nevertheless provide much better supporting evidence for a tumor than does epinephrine (72% normal). Patients with pheochromocytoma show much more severe elevations in plasma metanephrines than in catecholamines; 66% had concentrations of normetanephrine over fivefold the upper reference limits, compared with 43% for norepinephrine; for metanephrine and epinephrine, this difference was even more pronounced (32% vs 6%). Thus, plasma metanephrines not only reliably exclude a pheochromocytoma, but also provide a more definitive positive diagnosis for the presence of the tumor than do catecholamines.

is positively correlated with plasma normetanephrine concentrations. Thus, the level of increase above normal in plasma normetanephrine can provide an indication of the size of a tumor, whereas plasma levels of norepinephrine provide no information about tumor size (6). Another source of information that can be provided by the metanephrines is the adrenal versus extra-adrenal location of a tumor. Patients having tumors with an adrenal location show much larger increases in plasma concentrations of metanephrine relative to normetanephrine than patients with extra-adrenal tumors (6). In particular, a high plasma metanephrine level combined with a metanephrine-to-normetanephrine ratio above 0.2 suggests the presence of an adrenal tumor. The reason for the difference in adrenal and extra-adrenal tumors is the proximity of adrenal tumors to the influence of steroids, which increase the expression of phenylethanolamine-*N*-methyltransferase, the enzyme required to convert norepinephrine to epinephrine.

III. Conclusions

Plasma metanephrines are formed extraneuronally and to a large extent within chromaffin tissues (e.g., adrenal medulla and pheochromocytomas). Therefore, these metabolites are better markers for a pheochromocytoma than are other catecholamine metabolites, which are derived mainly from metabolism of norepinephrine within sympathetic nerves.

Pheochromocytomas may be quiescent and not secrete catecholamines, whereas they always appear to actively metabolize catecholamines to the free metanephrines. Thus, plasma measurements of free metanephrines offer a superior and more reliable test over plasma catecholamines for diagnosis of pheochromocytoma.

Normal plasma metanephrines reliably exclude a pheochromocytoma, whereas normal catecholamines do not. Thus, where excluded, no other tests are necessary. This means that measurements of plasma metanephrines avoid a missed diagnosis and minimize the need to run multiple diagnostic tests to exclude the presence of a tumor. Plasma metanephrines also can provide useful information about the size and location of the tumor.

References

1. Goldstein, D. S., Eisenhofer, G., Stull, R., Folio, C. J., Keiser, H. R., and Kopin, I. J. (1988). Plasma dihydroxyphenylglycol and the intraneuronal disposition of norepinephrine in humans. *J. Clin. Invest.* **81,** 213–220.
2. Eisenhofer, G., Pecorella, W., Pacak, K., Hooper, D., Kopin, I. J., and Goldstein, D. S. (1994). The neuronal and extraneuronal origins of plasma 3-methoxy-4-hydroxyphenylglycol in rats. *J. Auton. Nerv. Syst.* **50,** 93–107.
3. Eisenhofer, G., Åneman, A., Hooper, D., and Friberg, P. (1996). Mesenteric organ production, hepatic metabolism, and renal elimination of norepinephrine and its metabolites in humans. *J. Neurochem.* **66,** 1565–1573.
4. Eisenhofer, G., Friberg, P., Pacak, K., Goldstein, D. S., Murphy, D. L., Tsigos, C., Quyyumi, A. A., Brunner, H. G., and Lenders, J. W. M. (1995). Plasma metanephrines: Do they provide useful information about sympatho-adrenal function and catecholamine metabolism? *Clin. Sci.* **88,** 533–542.
5. Eisenhofer, G., Rundqvist, B., Åneman, A., Friberg, P., Dakak, N. A., Kopin, I. J., Jacobs, M-C., and Lenders, J. W. M. (1995). Regional release and removal of catecholamines and extraneuronal metabolism to metanephrines. *J. Clin. Endocrinol. Metab.* **80,** 3009–3017.
6. Lenders, J. W. M., Keiser, H. R., Goldstein, D. S., Willemsen, J. J., Friberg, P., Jacobs, M-C., Kloppenborg, P. W. C., Thien, T., and Eisenhofer, G. (1995). Plasma metanephrines in the diagnosis of pheochromocytoma. *Ann. Intern. Med.* **123,** 101–109.

PART D

CATECHOLAMINE RECEPTORS AND SIGNAL TRANSDUCTION

David S. Goldstein

Clinical Neuroscience Branch
National Institute of Neurological Disorders and Stroke
National Institutes of Health
Bethesda, Maryland 20892-1424

Overview

Receptors in the brain and periphery mediate the physiological actions of catecholamines. The myriad different effects exerted by only three endogenous catecholamines—norepinephrine (NE), epinephrine (Epi), and dopamine (DA)—in different organs depend on numerous types and subtypes of catecholamine receptors and intracellular mechanisms.

Recent research in the area of catecholamine receptors has focused on: (1) Structure, classification, and tissue localization of adrenoceptor subtypes: (2) Intracellular mechanisms influenced by catecholamine receptor occupation, including negative feedback loops responsible for receptor desensitization; (3) pharmacology of neurological, psychiatric, and cardiovascular drugs acting at catecholamine receptors; (4) Roles of catecholamine receptor subtypes in behavioral and physiological processes; and (5) Catecholamine receptors in pathophysiological states.

I. Structure, Classification, and Tissue Localization of Catecholamine Receptor Subtypes

Distinctions based on genetic sequences for identifying types and subtypes of catecholamine receptors have outpaced those based on pharmacological

Advances in Pharmacology, Volume 42

1054-3589/98 $25.00

techniques such as ligand binding. As a result, the functional importance of many of these distinctions remains unclear. The following table summarizes classification schemas for these subtypes, lists the associated G-proteins and intracellular messengers that couple catecholamine receptor occupation with cellular functional changes, and notes some of the specific agonists and antagonists referred to in this review.

Schwinn (p. 390) has surveyed expression of α_1-adrenoceptor subtypes in various organs of humans. The α_{1A} subtype predominates in human heart, liver,

Subtype	*G-Protein*	*Effector*	*Specific Agonist*	*Specific Antagonist*
α_1	G_q/G_{11}	Phospholipase C-β	Phenylephrine	Prazosin
α_{1A}	G_q/G_{11}	Phospholipase C-β		Tamsulosin
α_{1B}	G_q/G_{11}	Phospholipase C-β		WB4101 (low affinity)
α_{1D}	G_q/G_{11}	Phospholipase C-β		
α_2	G_i, G_o	Inh. Adenyl cyclase**	BHT 920	Yohimbine
α_{2A}	G_i, G_o	Inh. Adenyl cyclase**	Oxymetazoline	BRL 44408
α_{2B}	G_i, G_o	Inh. Adenyl cyclase**		Prazosin***; ARC 239
α_{2C}	G_i, G_o	Inh. Adenyl cyclase**		Prazosin***; ARC 239

(*) In humans.
(**) plus other mechanisms (see text)
(***) Prazosin also blocks α_1-adrenoceptors nonselectively.

Subtype	*G-Protein*	*Effector*	*Specific Agonist*	*Specific Antagonist*
β	G_s	Adenyl cyclase	Isoproterenol	Propranolol
β_1	G_s	Adenyl cyclase	Dobutamine	Metoprolol
β_2	G_s	Adenyl cyclase	Terbutaline	ICI118551
β_3	G_s	Adenyl cyclase	BRL 37344; CL316243; CGP 12177	
D1-like	G_s	Adenyl cyclase	SKF 38393	SCH-23390
D1 (D1A)	G_s	Adenyl cyclase		
D5 (D1B)	G_s	Adenyl cyclase		
D1C*	G_s	Adenyl cyclase		
D1D*	G_s	Adenyl cyclase		
D2-like			Sulpiride	NCQ-298
D2 long (D2A)**	G_1, G_o	Inh. Adenyl cyclase		Raclopride
D2 short (D2B)**	G_1, G_o	Inh. Adenyl cyclase		
D3	G_i, G_o(?)	Inh. Adenyl cyclase (?)	Quinpirole	Nafadotride
D4	G_i, G_o	Inh. Adenyl cyclase		Clozapine; NGD-94

* Not found in mammals, including humans.
** Based on length of third cytoplasmic loop.

brain, and prostate. Transcriptional regulation appears to underlie tissue-specific expression of this subtype. The author speculates that drugs influencing this regulation in the prostate gland may prove useful in the treatment of benign prostatic hypertrophy.

In humans, α_1-adrenoceptors play relatively minor roles in mediating catecholamine effects in the heart, and kidneys play an important role in regulation of vascular tone. Michel and coworkers (p. 394) have described a radioreceptor assay to identify the relative sizes of populations of vascular α_1-adrenoceptor subtypes in humans. In this technique, subjects receive a single high dose of a subtype-selective antagonist, and the plasma is applied to rat-1 fibroblasts stably expressing one of the subtypes. From the extent of inhibition of binding in the radioreceptor assay, one can estimate the amount of binding by that subtype *in vivo*. The authors have not yet identified the subtypes mediating vasoconstriction in humans.

Adrenoceptor occupation presumably results in conformational changes that initiate the signaling process. Perez and coworkers (p. 398) have proposed a "salt-bridge" hypothesis to explain structural changes corresponding to receptor activation in α_{1B}-adrenoceptors. Disruption of a salt-bridge between transmembrane loops 3 and 7 could activate the receptor by releasing the receptor from a conformational restraint. The protonated amine on Epi may compete with the positive charge at position K 331 for the negative aspartic acid residue at position 125.

Different species express different subtypes of adrenoceptors. Niznik and coworkers (p. 404) summarized evidence for the existence of four distinct subtypes of D1 receptors, including cloning and isolation of the genes for all four subtypes in distinct species—crocodile and chicken. Homologues of D1C and D1D receptor subtypes, however, have not yet been identified in humans.

Relatively little is known about interplay among adrenoceptor subtypes. As noted by Schwartz and coworkers (p. 408), the D3 receptor seems to be coexpressed with the D1 receptor but not with the D2 receptor in several central neural tissues.

Binding sites for hydrophilic ligands such as catecholamines occur in water-accessible crevices that extend from the extracellular surface of the receptor to within the mainly hydrophobic membrane domains of the receptor. Javitch (p. 412) mapped the binding-site crevice of the D2 receptor, using the substituted-cysteine accessibility method. The third transmembrane segment (M3) appears to form an α-helix, with one side facing the binding-site crevice; M5 and M7 do not.

II. Intracellular Mechanisms

Many studies have focused on refining understanding of mechanisms of adrenoceptor desensitization in response to pharmacological doses of agonists. No study has measured actual *in vivo* concentrations of agonist at the receptors, and little remains known about mechanisms of denervation supersensitivity.

Other studies have considered intracellular mechanisms elicited by agonist occupation of specific adrenoceptor subtypes; mechanisms of termination of cellular action on removal of agonist; the stoichiometry of receptors, G-protein subunits, and adenyl cyclase in determining transductional effectiveness; and mechanisms of effects of adrenoceptor occupation on cellular growth and differentiation.

As pointed out in the review by Lefkowitz and coworkers (p. 416), adrenoceptor desensitization reflects the effects of several processes, including phosphorylation of the receptor, sequestration (agonist-induced dissociation of the receptor from the cell membrane), inactivation of intracellular messengers, and decreased synthesis of receptor protein.

Rapid desensitization appears to result from receptor phosphorylation, which uncouples the receptor from its G-protein. At least in the case of β_2-adrenoceptors, receptor phosphorylation can occur by the actions of two types of kinases. One type is second messenger–dependent kinases such as protein kinase A (PKA) and protein kinase C (PKC). Because different types of agonists and receptors can use the same second messengers, kinase-induced phosphorylation can provide a basis for "cross-talk" in regulation of different receptors simultaneously. A second type of kinase phosphorylating receptors is the G-protein–coupled receptor kinases (GRKs), which phosphorylate only agonist-occupied receptors. GRKs operate indirectly, in that the phosphorylated receptor binds cytosolic proteins, called arrestins, and the arrestins sterically block the G-protein activation that normally would follow agonist occupation of the receptor. Of the six mammalian cDNA sequences for GRKs isolated so far, all code kinases that phosphorylate intramembrane domains of the receptor. Probably the most extensively studied GRK is GRK2 (β-adrenoceptor kinase 1, βARK1).

Resensitization occurs rapidly on removal of the agonist, by the actions of phosphatases that catalyze dephosphorylation of the receptor. The phosphatases are latent, membrane-associated members of the PP-2A family. One, designated the G-protein–coupled receptor phosphatase (GRP), has the ability to dephosphorylate not only the βARK-phosphorylated β_2-adrenoceptor but also the βARK-phosphorylated α_{2C}-adrenoceptor, although GRP cannot catalyze dephosphorylation of the PKA-phosphorylated β_2-adrenoceptor.

Resensitization by dephosphorylation appears to require sequestration of the receptor from the membrane, into a cytoplasmic population of vesicles. Decreasing the vesicular pH probably alters the receptor conformation in such a way as to associate the receptor with the phosphatase.

Five reports have dealt with various aspects of β-adrenoceptor desensitization and resensitization. Ferguson and coworkers (p. 420) have focused on the roles of GRK-mediated phosphorylation and arrestin binding on sequestration of β_2-adrenoceptors. Mutation of tyrosine to alanine at position 326 in the M7 transmembrane domain yielded a sequestration-defective mutant that exhibited desensitization but not resensitization. Sequestration was found to depend on arrestin binding, and GRK-mediated phosphorylation, while not absolutely required, enhanced the receptor's affinity for arrestin.

Karoor and Malbon (p. 425) have described effects of insulin, which is a growth factor that possesses intrinsic tyrosine kinase (TK) activity, and of insulin-like growth factor I (IGF-I) on phosphorylation of the β-adrenoceptor. Insulin-induced phosphorylation of tyrosyl residues in the C-terminal region of the receptor uncoupled the receptor from the stimulatory G-protein, possibly accounting for catecholamine counterregulatory effects of insulin. This site of action contrasts with the third transmembrane loop site of action of PKA and PKC and with the second transmembrane loop site of action of IGF-I.

Goodman and coworkers (p. 429) have explored whether arrestin-dependent sequestration of the β_2-adrenoceptor depends on an interaction with clathrin, the main structural protein of coated pits in vesicular membranes containing sequestered β_2-adrenoceptors. β-Arrestin bound reversibly and with high affinity to clathrin, and β-adrenoceptor stimulation with isoproterenol led to punctate colocalization of clathrin and β-arrestin.

Most studies of β-adrenoceptor desensitization have focused on the β_2 subtype. Bouvier and Rousseau (p. 433) have studied desensitization of the β_1 and β_3 subtypes. The β_3 subtype resists both agonist-induced uncoupling and sequestration, and the β_1 subtype exhibits fewer of these processes than does the β_2 subtype. The findings in the β_3 subtype appear to result at least partly from absence of phosphorylation sites for PKA or βARK in the third cytoplasmic loop (i_3) and carboxy-terminal tail of the receptor. Studies of chimeras indicated that domains within these regions and in the i_2 cytoplasmic loop contribute to uncoupling but do not appear to enable sequestration, implying that the molecular mechanisms of the two processes differ. β_3-Adrenoceptors also differ from β_2-adrenoceptors in long-term regulation, with slower and less extensive down-regulation and without agonist-induced endocytosis or receptor degradation contributing to the loss of binding sites. In general, findings in β_1-adrenoceptors were intermediate between those in the other two subtypes.

Liggett (p. 438) has summarized structural determinants of rapid desensitization of α_2-adrenoceptors expressed in CHW-1102 fibroblasts. Exposure to Epi reduced α_{2A}-adrenoceptor function, as measured by inhibition, via G_i, of forskolin-stimulated cyclic adenosine monophosphate (cAMP) levels or adenyl cyclase activity. The desensitization appeared to result from receptor phosphorylation by a GRK, with the target for the phosphorylation a series of four serine residues in the third transmembrane loop. In contrast, α_{2C}-adrenoceptors failed to exhibit either phosphorylation or desensitization, and α_{2B}-adrenoceptors exhibited only partial desensitization.

As pointed out in the review by Burris and coworkers (p. 443), exposure to both antagonists and agonists at dopamine receptors can increase the densities of the receptors in mammalian cells transfected with DNA that encode dopamine receptor subtypes. For instance, in HEF-293 expressing transfected D2L receptors, sulpiride, a D2 antagonist, and quinpirole, a D2 agonist, increase the density of the D2L receptors. Burris and coworkers report analogous findings for D3 receptors. Because agonists and antagonists with opposite effects on the activity of adenyl cyclase can produce the same effect, alterations in coupling of the receptors to G-proteins seem unlikely to mediate directly the changes in receptor density.

Sibley and coworkers (p. 447) have summarized recent findings about mechanisms of rapid agonist-induced desensitization of D1 receptors. As for β-adrenoceptors, this desensitization depends on receptor phosporylation via the PKA pathway. Mutant Chinese hamster ovary (CHO) cells with deficient PKA activity had decreased dopamine- or cAMP analog–induced desensitization of D1 receptors, as well as decreased downregulation of membrane D1 receptor binding. In C6 glioma cells expressing either wild-type or mutant D1 receptors with four specific amino acid substitutions at potential PKA sites, the mutant cells exhibited decreased and delayed desensitization. The carboxy terminus of the D1 receptor contains consensus sequences for phosphorylation by GRKs and PKA, whereas the carboxy terminus of the D2 receptor does not contain phosphorylation sites. Based on studies of chimeric D1/D2 receptors transfected into CHO or C6 cells, phosphorylation sites in the carboxy terminus do not appear to be required for agonist-induced desensitization of D1 receptors.

Insel and coworkers (p. 451) have studied intracellular mechanisms of phospholipase activation after agonist stimulation of α_1-adrenoceptors. In MDCK-D1 cells, epithelial cells derived from the distal tubule-collecting duct region of the dog kidney, agonist occupation of α_{1b}-adrenoceptors appears to activate multiple phospholipases, including phospholipase C (PLC) and phospholipase A_2 (PLA_2). The authors hypothesize that PLC activation begins a cascade of diacylglycerol (DAG) formation and PKC activation, leading to activation of mitogen-activated protein kinase (MAPK) and then phosphorylation of cytosolic PLA_2, eventually leading to arachidonic acid release.

β-Adrenoceptor signaling has long been thought to depend exclusively on coupling between the receptor and G_s to activate adenyl cyclase. In an oral presentation based on a submitted abstract, however, Xiao and coworkers provided evidence for coupling of the β_2-adrenoceptor to a pertussis toxin-sensitive (i.e., non-G_s) protein. In isolated heart cells from TG4 transgenic mice overexpressing the β_2-adrenoceptor, the baseline amplitude of contraction exceeded that in wild-type mice, indicating spontaneous activation of the β_2-adrenoceptor in the absence of agonist. Moreover, the β_2-adrenoceptor agonist, zinterol, augmented contraction in pertussis toxin-treated but not in untreated cells, indicating the presence of an inhibitory pertussis toxin-sensitive G-protein.

High circulating levels of catecholamines would be expected, if anything, to downregulate β-adrenoceptors. In an oral presentation based on a submitted abstract, Christensen and coworkers presented evidence that, on the contrary, lymphocyte mRNA for β_2-adrenoceptors increased threefold in exercising humans.

In rat striatum, D1-like receptors may interact with both G_s protein, activating adenyl cyclase, and with G_q protein, activating PLC. Friedman and coworkers orally presented findings summarized in a submitted abstract about intracellular mechanisms consequent to activation of D1-like receptors. Mice with knockout of D1A receptors lacked DA-induced stimulation of cAMP formation in the frontal cortex but had normal DA-induced phosphoinositide hydrolysis. Thus, within the family of D1-like receptors, those coupled via G_q protein appear to differ from those coupled via G_s protein.

All receptors in the D2-like receptor family are susceptible to pertussis toxin inactivation, indicating linkage to $G_{i/o}$ proteins. In the presence of pertussis toxin, the α-subunits of the G-protein undergo ADP-ribosylation at a cysteine residue near the C-terminus, preventing coupling of the receptor with the G-protein. Huff and coworkers (p. 454) have compared signal transduction pathways modulated by subtypes of D2-like receptors. In transfected CHO 10001 cells, D2, D3, and D4 receptors all inhibit forskolin-induced cAMP accumulation; increase extracellular acidification via an amiloride-sensitive Na^+-H^+ exchanger (NHE-1); and stimulate cell division, probably from activation of the MAPK signaling cascade. D2 and D4 receptors also potentiate adenosine triphosphate-induced arachidonic acid release, via a PKC-dependent pathway. G_{i2} coupling accounted for inhibition of cAMP formation modulated by D2A receptors but not for Na^+-H^+ exchange or arachidonic acid release. At least two G-proteins, G_{i2} and G_o, therefore, appear to mediate signal transduction consequent to occupation of D2 receptors.

Hydrolysis of bound guanosine triphosphate (GTP) terminates the signal initiated by receptor occupation. *In vivo,* GTPase-activating proteins (GAPs) accelerate this hydroylsis. One such GAP is PLC-β, generated as a G-protein-regulated effector. Ross and coworkers (p. 458) have put forth the notion that when an effector, such as PLC-β, also functions as a GAP, then termination of receptor–second messenger coupling may occur by hydrolysis of the bound GTP, without dissociation of the G-protein from the receptor. Maintenance of a portion of the G-protein in a form bound to GTP would help explain steady-state activation of the G_q-protein. Ross and coworkers also report discovery of anew GAP for the G_z-protein, a member of the G_i family. The physiological effector remains unidentified.

Milligan and coworkers (p. 462) have studied the stoichiometry of $G_{s\alpha}$-coupled β_2-receptors, $G_{s\alpha}$, and type II adenyl cyclase in NG-108-15 neuroblastoma x glioma hybrid cells. Levels of expression of adenylyl cyclase appear to constitute the limiting element in transductional effectiveness.

The mitogen-activated protein kinase (ERK1/2) pathway appears to constitute a point of convergence for several signals that regulate cellular growth and differentiation. Because receptors that couple with heterotrimeric G-proteins can participate in cell proliferation and transformation, Luttrell and coworkers (p. 466) have studied mechanisms by which α_{1B}- and α_{2A}-adrenoceptors stimulate ERK1/2 activity in transiently transfected COS-7 and CHO cells: Whereas α_{2A}-adrenoceptors signal via the pertussis toxin–sensitive family of G_q/11 proteins, α_{1B}-adrenoceptors signal via the pertussis toxin-insensitive family of G_q/11 proteins. Two mechanisms for ERK1/2 activation correspond to these different types of G-protein families, one dependent on Ras and tyrosine phosphorylation and mediated by $G_{\beta\chi}$ subunits from pertussis toxin–sensitive G-proteins and the other independent of Ras and dependent on PKC, mediated by G_α subunits from pertussis toxin–insensitive G-proteins.

III. Pharmacology

Kobilka and Gether (p. 470) have considered structural bases for different types of drug effects at adrenoceptors, including true agonism, where the drug activates the receptor fully; partial agonism, where the drug elicits only a fractional response; true antagonism, where the drug exerts no direct effect on basal receptor activity; and inverse agonism (or negative antagonism), where the drug decreases basal receptor activity. The investigators used nitrobenzdioxazol iodoacetamine (IANBD) fluorescence spectroscopy to note graded changes in receptor configuration in response to drugs representative of these four classes, attributable to polarity changes in the environment of cysteine residues in transmembrane regions 3 and 6 of the receptor. The different classes of drugs may, therefore, stabilize different receptor conformations.

Antipsychotic drugs thought to act as DA receptor antagonists may actually act as inverse agonists; that is, they may suppress agonist-independent basal activity of the receptors. In an oral presentation based on a submitted abstract, Strange and Hall reported results of studies of cAMP production in CHO cells stably expressing the human D2 (short) receptor. Constitutive activity of the receptors inhibited adenyl cyclase in these cells, and several antipsychotic drugs, including chlorpromazine, clozapine, domperidone, and haloperidol, reversed this basal inhibition.

Drugs active at receptors for α_2-adrenoceptors can also bind to imidazoline–guanidinium receptive sites. As noted in an update by Lanier and coworkers (p. 474), ligands at the I_1 imidazoline binding site include clonidine, and ligands at the I_2 binding site include guanabenz and idazoxan. The family of I_2 binding proteins includes the two isoforms of monoamine oxidase (MAO). Agmatine constitutes the only known bioactive endogenous substance that binds to imidazoline receptors. Intracellular effector mechanisms associated with imidazoline–guanidinium receptive sites remain unknown.

Drugs that block D2 receptors can be used effectively to treat psychosis; however, D2 antagonists also can produce parkinsonian side effects. Antipsychotics that do not elicit extrapyramidal effects are called "atypical neuroleptics." Seeman and coworkers (p. 478) have summarized current hypotheses to explain the absence of extrapyramidal effects by atypical neuroleptics. Some drugs have relatively low affinity for D2 receptors, so that high endogenous dopamine levels in the striata may displace the drugs; alternatively, atypical neuroleptics may block both D2 and D4 receptors, the latter being sparse in human striata. Roth and coworkers (p. 482) have emphasized that atypical neuroleptics bind to receptor subtypes for serotonin (5HT). All tested atypical neuroleptics had higher affinity for $5HT_{2A}$ than to D2 receptors.

Van Tol (p. 486) has reported that Val 194 to Gly 194 polymorphism in M5 of the D4 receptor eliminates high-affinity binding of DA and clozapine, and expression of the polymorphism in CHO-K1 cells attenuates the extent of inhibition of forskolin-stimulated cAMP levels, compared with CHO-K1 cells expressing the wild-type receptor. The polymorphism, therefore, appears to reduce the ability of the receptor to fold into the high-affinity active state.

Tallman (p. 490) has described a new drug, NGF 94-1, which appears to act as an antagonist specifically at D4 receptors. Human hippocampal, hypothalamic, and several cortical regions exhibited intense binding of [^{3}H]NGF 94-1, whereas the caudate–putamen did not. The clinical utility of the drug remains unknown. Animal models to test potential antipsychotic efficacy may be irrelevant here, because the models relate mainly to functions of D2 receptors.

IV. Catecholamine Receptors in Physiology and Behavior

The three subtypes of α_2-adrenoceptor all couple via pertussis toxin–sensitive G_i and G_o proteins to inhibit adenyl cylcase, activate receptor-operated membrane K^+ channels, and inhibit voltage-sensitive Ca^{2+} channels. MacMillan and coworkers (p. 493) have used a "hit and run" gene targeting approach in mouse embryonic stem cells to produce mice with an aspartate-to-asparagine mutation of the α_{2A}-adrenoceptor at the 79 position. The α_{2A}-adrenoceptor appears responsible for centrally mediated hypotensive reponses to α_2-adrenoceptor agonists (in contrast with suggestions that these agonists work by way of interactions with imidazoline receptors) and for sedative, anesthetic-sparing, and analgesic effects of α_2-adrenoceptor agonists. Consistent with the latter, in a presentation based on a submitted abstract, Mizobe and coworkers reported that administration of an antisense oligodeoxynucleotide-inhibiting expression of α_{2A}-adrenoceptors attenuated dexmedetomidine-induced hypnotic and analgesic responses in rats, supporting a role of α_{2A}-adrenoceptors in these processes. The α_{2A}-adrenoceptor appears responsible for hypertensive reponses immediately after administration of α_2-adrenoceptor agonists.

Lafontan and coworkers (p. 496) have noted that α_{2A}-adrenoceptors may also play a role in fat-cell metabolism. Human fat cells possess more α_2-adrenoceptors than they do β_1- or β_2-adrenoceptors. Moreover, in contrast with β-adrenoceptors, α_{2A}-adrenoceptors in human fat cells failed to exhibit desensitization.

Rohrer and coworkers (p. 499) have used a gene knockout to evaluate the roles of subtypes of β-adrenoceptors in fetal development and adult cardiac physiology. Only about 10% of the expected number of β_1-adrenoceptor knockout mice were recovered at weaning, indicating the importance of this receptor subtype for normal fetal development. In contrast, knockout of the β_2-adrenoceptor did not affect survival to birth. Animals with β_1-adrenoceptor knockout also had no cardiac chronotropic or inotropic response to isoproterenol.

Koch and coworkers (p. 502) have summarized effects of manipulations of various components of the myocardial β-adrenoceptor system, from studies of lines of transgenic mice with altered receptor signaling. Mice with increased expression specifically of myocardial β_2-adrenoceptors (by using the murine α-myosin heavy-chain gene promoter) had increased myocardial adenyl cyclase activity and increased myocardial contractility at baseline, similar to values obtained for isoproterenol-stimulated activity in nontransgenic animals. Peptides derived from the $G_{\beta c}$-binding domain of βARK1 inhibit βARK1 activity *in vitro,* preventing βARK1-induced phosphorylation and desensitization of the receptor. Transgenic mice expressing this inhibitor specifically in the heart had increased left ventricular contractility and augmented responses to isoproterenol. The former finding indicates that βARK1 exerts a tonic inhibitory effect on cardiac β-adrenoceptors, even in the absence of agonist. Conversely, transgenic mice with overexpression of βARK1 had decreased basal adenyl cyclase activity and decreased cardiac contractile responses to isoproterenol. Transgenic mice with cardiac-specific overexpression of GRK5 also had decreased basal adenyl cyclase activity and had decreased cardiac adenyl cyclase responses to isoproterenol. Thus, at least two GRKs appear to modulate β-adrenoceptor signaling in the mouse heart. In contrast, animals overexpressing βARK1 had attenuated angiotensin II–induced increases in left ventricular contractility, whereas those overexpressing GRK5 had normal responses to angiotensin II, indicating that βARK1 but not GRK5 can modulate responsiveness to occupation of angiotensin II receptors in the mouse heart. The investigators have hypothesized that gene transfer approaches based on increasing expression of β_2-adrenoceptors or of a βARK1 inhibitor may provide a novel therapy for cardiovascular disease states such as heart failure.

One might predict from this that chronic administration of a β-adrenoceptor antagonist would decrease expression of GRKs in the heart. In a summary of recently published data, Hammond and coworkers (p. 507) have noted that chronic β_1-adrenoceptor blockade using bisprolol reduced left ventricular βARK activity and increased by twofold β-adrenoceptor–dependent stimulation of ventricular adenyl cyclase activity.

About 6–8% of humans have a polymorphism of the β_2-adrenoceptor, with a Thr-to-Ile switch at position 164. In a presentation based on a submitted abstract, Green and coworkers studied the hearts of transgenic mice expressing either type of receptor. The Thr-to-Ile polymorphism was associated with decreased resting heart rate, indices of contracility, and isoproterenol-stimulated adenyl cyclase activity.

As noted by Strosberg (p. 511), fat cells such as in brown adipose tissue possess high concentrations of β_3-adrenoceptors. A single residue substitution (Trp to Arg) at position 64 has been associated with a tendency toward obesity in humans. Whether this substitution affects caloric balance and lipolysis in animals remains unknown.

Waddington and coworkers (p. 514) have analyzed "ethogram" data about behaviors of D1A knockout mice. The affected animals had increased locomotion and grooming but decreased sniffing, rearing, and sifting, inconsistent with simple changes such as hyperactivity.

No physiological, useful, selective drugs for subtypes of D2-like receptors have been developed. Antisense oligonucleotides bind to complementary mRNA, stopping translation of the mRNA. The antisense approach provides an alternative means to evaluate the presence and functions of the coded proteins. Creese and Tepper (p. 517) have used this approach and obtained evidence for the presence of both D2 and D3 somatodendritic autoreceptors in the rat substantia nigra. D2 and D3 "knockdown" appeared to affect spontaneous contralateral rotation additively, consistent with increased DA release in the ipsilateral striatum due to decreased inhibitory modulation by the autoreceptors, despite an absence of changes in baseline spontaneous firing.

One may also study the roles of subtypes of DA receptors by generating mutant mice lacking the receptors. Saiardi and coworkers (p. 521) have established a line of mice lacking the D2 receptor, via homologous recombination. D2-deficient mice had reduced, slow movements and increased pituitary prolactin and pro-opiomelanocortin mRNA. In contrast, D1-deficient mice had hyperlocomotion. The investigators have proposed that the D2-deficient mice may constitute an animal model of Parkinson's disease. In a presentation based on a submitted abstract, Steiner and coworkers reported that transgenic mice lacking D3 receptors had increased locomotion and rearing. From behavioral tests the authors inferred that transgenic animals lacking D3 receptors had decreased anxiety, rather than "hyperactivity."

V. Pathophysiological States

Whereas many reports have noted pathophysiological or psychopathological effects of altered expression of adrenoceptor subtypes, as described in the previous section, relatively few recent studies have attempted to explain disease states in terms of abnormal adrenoceptor function.

Vascular tissue from spontaneously hypertensive rats (SHRs) have increased α_1-adrenoceptor–induced phosphoinositide metabolism. In a presentation based on a submitted abstract, Wu and de Champlain reported that vascular smooth muscle cells from SHRs had increased basal inositol phosphate (IP) formation and larger increments in IP formation in response to phenylephrine than did cells from WKY normotensive rats. Vascular smooth muscle cells from SHRs also had decreased isoproterenol-induced accumulation of cAMP. These alterations would bias SHRs toward increased adrenoceptor-mediated vasoconstriction in SHRs.

Genetically obese mice or mice made obese by high-fat feeding have decreased adipocyte expression of all three subtypes of β-adrenoceptors. In a presentation based on a submitted abstract, Collins and coworkers noted that administration of a selective β_3-adrenoceptor agonist, CL316243, prevented diet-induced obesity in A/J mice. Administration of leptin increased catecholamine biosynthesis, assessed by the AMPT tyrosine hydroxylase inhibitor method, by 100% in interscapular brown adipose tissue. The authors speculated that both leptin and β_3-agonism increase brown adipocyte thermogenesis.

During increased dietary salt intake, DA acting at D1-like receptors exerts a natriuretic effect, and SHRs have decreased natriuretic responses to D1 agonists, despite normal numbers of D1-like, D1A, and D1B receptors on proximal tubule cells. SHRs appear to lack high-affinity binding of agonists at D1-like receptors, and they have decreased proximal tubular cAMP and IP production in response to D1-like agonists, despite normal responses to forskolin and GTP and its analogues. Jose and coworkers (p. 525) have put forward tissue-specific uncoupling of renal proximal tubular D1-like receptors as an explanation for these findings. The presumed uncoupling seems to sort genetically with hypertension. Recent experiments in immortalized proximal tubular cells confirmed attenuated responses to D1-like receptor agonists in SHRs, possibly related to D1A receptor sequestration.

VI. Future Directions

Future research about catecholamine receptors would benefit from more attention to feedback loops that operate not only intracellularly but also systemically. At the intracellular level, compensatory activation of alternative effectors and effector sharing by different receptor subtypes merit increased consideration. At the systemic level, future work should consider the identification and mechanisms of action of catecholamine receptors in the brain and periphery that modulate release of catecholamines from nerve terminals and consequently exert effects on post- or extra-synaptic catecholamine receptors.

Intense interest about intracellular mechanisms underlying receptor desensitization and sequestration in response to pharmacological doses of ligands contrasts with lack of interest about mechanisms underling denervation supersensitivity, a clinically relevant issue.

Use of recently introduced genetic techniques, such as transgenic mice and antisense oligonucleotides, will become routine in studies of roles of adrenoceptor subtypes in physiological and psychological phenomena and development of drugs acting at adrenoceptor subtypes. Application of these techniques would be especially helpful in elucidating mechanisms of action of α_2-adrenoceptor agonists that also bind to imidazoline receptors.

Catecholamines not only bind to membrane receptors but also undergo well-known uptake into cells, via transporters. Whether and to what extent cytoplasmic catecholamines interact with sequestered receptors remain unknown.

Future clinical research will probably include methods to visualize catecholamine receptors in the brain and periphery, using positron-emitting ligands for adrenoceptor subtypes, including assessments of release of endogenous ligands, as assessed by *in vivo* displacement of positron-emitting ligands.

Debra A. Schwinn and Madan M. Kwatra

Departments of Anesthesiology, Pharmacology, and Surgery
Duke University Medical Center
Durham, North Carolina 27710

Expression and Regulation of α_1-Adrenergic Receptors in Human Tissues

α_1-Adrenergic receptors (α_1ARs) are members of the larger family of G-protein–coupled receptors that bind the endogenous catecholamines epinephrine and norepinephrine. α_1ARs are important in many physiologic processes, such as myocardial inotropy, hypertrophy, arrhythmias, vasoconstriction, and prostate disease. These receptors couple predominantly via Gq proteins to activation of phospholipase C-β, resulting in hydrolysis of membrane phospholipids, production of IP_3, mobilization of intracellular calcium, and ultimately myocardial and smooth-muscle contraction. cDNAs encoding three α_1ARs (α_{1a}, α_{1b}, α_{1d}) have been cloned from several species; a fourth subtype (α_{1L}) has been described pharmacologically (see review for detailed analysis of nomenclature changes in the α_1AR field [1]). To study regulation of α_1ARs in human tissues, our laboratory began by cloning human homologues for each α_1AR subtype, stably expressed each in rat-1 cells, and characterized their pharmacology in detail (Table I) (2). In addition to subtype selectivity for various ligands, we have demonstrated that all three human α_1AR subtypes couple to phosphoinositide hydrolysis (a $\gg$ b $>$ d) in a pertussis toxin–insensitive manner (see Table I) (2). To determine which α_1AR subtypes are likely to be important in specific

Advances in Pharmacology, Volume 42

1054-3589/98 $25.00

TABLE I Characteristics of Cloned α_1AR Subtypes

	α_{1a}	α_{1b}	α_{1d}
Protein size (amino acids) (1)	466	515	560
Number of potential glycosylation sites	2	4	2
Human chromosome location (1)	8	5	20
Chloroethylclonidine inactivation (2)	+/−	+++	++
Subtype selective ligand (2)	(+)niguldipine 5-methylurapidil RS-17053 Oxymetazoline SNAP5272	?Spiperone	BMY7378 Norepinephrine
PI hydrolysis (fold over basal) (2)	×15.5	×4.6	×1.5
Human tissue distribution (4) (mRNA expression ≥++)[a]	Liver Heart Cerebellum Cortex Prostate	Spleen Kidney Heart Fetal brain	Cerebral cortex

[a] Many α_1AR subtypes are expressed at low levels in many tissues. See original reference for details (4). Numbers inside parentheses references.
PI, phosphoinositide.

human diseases, we next characterized the distribution of each subtype in various tissues.

α_1AR subtype tissue distribution is unique and varies with the species studied (rat, rabbit, mouse, dog, vs human). For example, while the α_{1b} is present in rat liver, only the α_{1a} subtype is expressed in native human liver (3, 4). Although individual α_1AR subtypes coexist in many human tissues, Northern analysis and RNase protection assays demonstrate that mRNA encoding α_{1a}ARs predominates in human heart, liver, brain, and prostate (see Table I); in addition, new data from our laboratory suggest an important role of α_{1a}ARs in human vasculature, particularly resistance vessels. Because our laboratory is focused on understanding mechanisms underlying benign prostatic hypertrophy (BPH), and α_1ARs are known to be important in the dynamic component (smooth-muscle contraction) of BPH, we were interested in elucidating the α_1AR subtype present in prostate smooth muscle. In human prostate, mRNA encoding the α_{1a}AR subtype represents 75% of the total α_1AR mRNA, with mRNA encoding α_{1d} and α_{1b} subtypes present to a lesser extent (5). Distribution of α_1AR subtypes can further be localized using *in situ* hybridization, studies to specific cells. Using *in situ* hybridization, we demonstrated that α_{1a}AR mRNA is the only subtype present in human prostate smooth muscle (suggesting a role in BPH), while α_{1b} mRNA and α_{1d} mRNA are present in prostate epithelial cells. Confirmation of our findings at a protein level has now been presented by four independent laboratories utilizing human prostate smooth-muscle contraction assays combined with various subtype-selective ligands. Many pharmaceutical companies have developed α_{1a}AR-selective antagonists for the therapy of BPH based on these data. Hence, knowledge regarding the distribution of α_1AR subtypes is important in both designing therapeutic targets for treating disease as well as designing experiments aimed at elucidating mechanisms underlying disease processes.

In addition to subtype distribution data, understanding regulation of α_1ARs (both at an mRNA and protein level) is important in elucidating mechanisms underlying various human diseases. To understand what controls expression of α_{1a}ARs in BPH, we hypothesize that transcriptional regulation is important. α_1AR subtypes have recently been shown to be transcriptionally regulated in myocardial hypertrophy, with α_{1a} mRNA increasing and α_{1b} and α_{1d} mRNA decreasing with norepinephrine in rat neonatal myocytes (6). We recently cloned 6 kb of the 5′ untranslated region (5′UTR) of the human α_{1a}AR gene from a human genomic library, identified the transcriptional start site and putative promoter, and elucidated regions important in basal transcription activity (manuscript submitted). Human cell lines not expressing α_{1a}ARs endogenously exhibit low transcriptional activity, suggesting cell-specific expression of α_{1a}ARs. Elucidating mechanisms underlying prostate-specific modulation of α_1ARs is the goal of these studies.

To understand the function and regulation of prostate α_1ARs at a protein level, we initiated studies to biochemically characterize these receptors. Because photoaffinity labeling has been claimed to be different between α_1AR subtypes (absent in the α_{1d}), we first examined this biochemical property using cloned α_1ARs. α_1ARs (bovine α_{1a}, hamster α_{1b}, rat α_{1d}) were stably expressed in rat-1 cells at a density of 1–2 pmol/mg membrane protein, and photolabeled using a ^{125}I-labeled azido analogue of prazosin. Sodium dodecyl sulfate–polyacrylamide gel electrophoresis (SDS-PAGE) and autoradiography of photolabeled membranes revealed a single major band in the case of each receptor that was not present when the membranes were photolabeled in the presence of excess unlabeled prazosin (Fig. 1). The photolabeled receptor displays the following apparent molecular weights on SDS-PAGE: α_{1a} 60 kD, α_{1b} 80 kD, α_{1d} 70 kD. Although the α_{1b}AR has 45 fewer amino acids than the α_{1d}AR, the higher molecular weight observed for the α_{1b} can be explained by the fact that it has four potential glycosylation sites compared with two on the α_{1d}. Clearly, photoaffinity labeling with azidoprazosin occurs for all three α_1AR subtypes and can be used to identify the residues involved in ligand binding. Further studies of the human α_{1a}AR protein have begun in our laboratory using the Sf9 cells–baculovirus system. This expression system not only is suitable for overexpressing G-protein–coupled receptors, but also appears to be a good system for studying receptor–G protein interactions. This is because Sf9 cells naturally express three main G-proteins (G_o, $G_{q/11}$, G_s) and additional G-proteins can easily be introduced. Using photoaffinity labeling with [^{32}P]azidoanilido guanosine triphosphate followed by immunoprecipitation with antibodies specific for various Gα-subunits, we have found that stimulation of human α_{1a}AR in Sf9 cells activates $G_{q/11}$ but not G_s or G_o (manuscript in preparation). Further characterization of receptor–G-protein interactions in Sf9 cells and human prostate should help us better define the G proteins involved in α_{1a}AR function in this tissue.

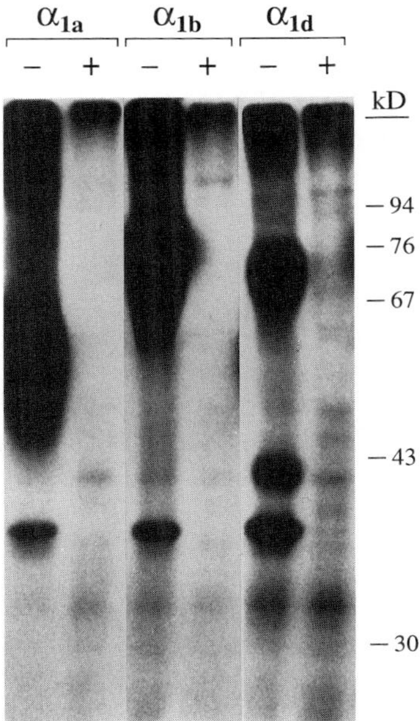

FIGURE 1 Photoaffinity labeling of α_1AR subtypes.

In summary, α_1ARs are important in many human diseases, with the α_{1a}AR subtype playing a particularly important role in urinary symptoms associated with BPH. Characterization of prostate α_1ARs, including tissue-specific regulation, should facilitate our understanding of mechanisms underlying this common disease of elderly men.

References

1. Michel, M. C., Kenny, B., and Schwinn, D. A. (1995). Classification of α_1-adrenoceptor subtypes. *Nauyn Schmiedebergs Arch. Pharmacol.* **351**, 1–10.
2. Schwinn, D. A., Johnston, G. I., Page, S. O., Mosley, M. J., Wilson, K. H., Worman, N. P., Campbell, S., Fidock, M. D., Furness, L. M., Parry-Smith, D. J., Peter, B., and Bailey, D. S. (1995). Cloning and pharmacological characterization of human *alpha-1* adrenergic receptors: Sequence corrections and direct comparison with other species homologues. *J. Pharmacol. Exp. Ther.* **272**, 134–142.
3. Price, D. T., Chari, R. S., Berkowitz, D. E., Meyers, W. C., and Schwinn, D. A. (1994). Expression of α_1-adrenergic receptor subtype mRNA in rat tissues and human SK-N-MC neuronal cells: Implications for α_1-adrenergic receptor subtype classification. *Mol. Pharmacol.* **46**, 221–226.
4. Price, D. T., Lefkowitz, R. J., Caron, M. G., Berkowitz, D., and Schwinn, D. A. (1994). Localization of mRNA for three distinct α_1-adrenergic receptor subtypes in human tissues: Implications for human α-adrenergic physiology. *Mol. Pharmacol.* **45**, 171–175.

5. Price, D. T., Schwinn, D. A., Lomasney, J. W., Allen, L. F., Caron, M. G., and Lefkowitz, R. J. (1993). Dentification, quantification, and localization of mRNA for three distinct alpha$_1$ adrenergic receptor subtypes in human prostate. *J. Urol.* **150**, 546–551.
6. Rokosh, D. G., Stewart, A. F. R., Chang, K. C., Bailey, B. A., Karliner, J. S., Camacho, S. A., Long, C. S., and Simpson, P. C. (1996). α_1-adrenergic receptor subtype mRNAs are differentially regulated by α_1-adrenergic and other hypertrophic stimuli in cardiac myocytes in culture and *in vivo*. *J. Biol. Chem.* **271**, 5839–5843.

Martin C. Michel,* Katsunari Taguchi,*
Rafael S. Schäfers,* Timothy J. Williams,†
David E. Clarke,† and Anthony P. D. W. Ford†

*Department of Medicine
University of Essen
45122 Essen, Germany

†Neurobiology Unit
Roche Bioscience
Palo Alto, California 94303

α_1-Adrenoceptor Subtypes in the Human Cardiovascular and Urogenital Systems

Three subtypes of α_1-adrenoceptors have been defined in pharmacological and molecular terms and are designated α_{1A}, α_{1B}, and α_{1D}. However, some data are not fully explained by these classifications, indicating the possible existence of additional subtypes. In particular, the low apparent affinity of prazosin in some functional assays has led to the proposal of an α_1-adrenoceptor subtype with low prazosin affinity, which has been designated α_{1L}; whether this truly represents an additional subtype or is the effect of specific assay conditions has remained controversial (see following discussion). Because another presentation in this section (Schwinn) extensively covers the distribution of mRNA for the three α_1-adrenoceptor subtypes in human tissues, this article focuses on radioligand binding and functional data.

α_1-Adrenoceptors mediate numerous effects in the cardiovascular and urogenital systems. In contrast to, for example, rats, α_1-adrenoceptors do not play a major role in the human heart or kidney. Thus, in the human heart, only few α_1-adrenoceptors exist. While they can couple to inotropic responses, they do so only much weaker than β-adrenoceptors. They also seem to play only a

Advances in Pharmacology, Volume 42

1054-3589/98 $25.00

minor role in the cardiac conduction system but can affect cardiac function profoundly by regulating coronary blood flow. Similarly, in the human kidney, only few, if any, α_1-adrenoceptors have been detected, and they do not appear to play an important role in the regulation of renovascular resistance, renin release, and tubular function.

In the human vasculature, α_1-adrenoceptors play an important role in the regulation of vascular tone. While α_2-adrenoceptors can also couple to human vascular smooth-muscle contraction, α_1-adrenoceptors appear to be more important in most vascular beds. Thus, the blood pressure–elevating effects of systemically infused noradrenaline (10–160 ng/kg/min) in healthy volunteers are almost completely blunted by pretreatment with 2 mg of the α_1-adrenoceptor antagonist, doxazosin, while pretreatment with 15 mg of the α_2-adrenoceptor antagonist, yohimbine, has only minor inhibitory effects (1).

Only very limited information is available regarding the α_1-adrenoceptor subtypes mediating vasoconstriction in humans. Animal studies have suggested that within a given vessel type (e.g., aorta), each of the three subtypes may mediate contraction in a species-dependent manner. Moreover, within a species, (e.g., rat), each of the three subtypes can mediate contraction, depending on the vessel bed under investigation. An emerging picture from rat studies is that contraction of small resistance vessels is mainly mediated by the α_{1A}-adrenoceptor, while α_{1B}- and α_{1D}-adrenoceptors are more important in large conductance and capacitance vessels.

This heterogeneity between and within species clearly necessitates direct studies on the α_1-adrenoceptor subtypes mediating vasoconstriction in humans. Two main approaches for this can be envisioned. First, it is possible to study human blood vessels *in vitro*. Such studies should preferentially investigate small resistance vessels. Moreover, the examination of multiple vascular beds will be necessary to gain a representative insight. An alternative approach is the use of a radioreceptor assay. This technique has successfully been applied to studies of β-adrenoceptor subtypes in humans (2). For the radioreceptor assay approach, subjects are treated with a single high dose of a subtype-selective antagonist. At various time points thereafter, blood is withdrawn and the corresponding plasma used in a radioreceptor assay with homogeneous populations of the receptor subtypes of interest. From the inhibition in the radioreceptor assay, the occupancy of each subtype at any given time point can be calculated. This assay has three main advantages. First, the assay autocorrects for plasma protein binding of each drug. Second, the assay takes into account all drug metabolites to the degree to which they participate in receptor occupancy. Third, the assay allows for correlation of occupancy data with functional data (e.g., peripheral resistance), independent of the pharmacokinetic properties of the drug under investigation.

We have applied this technique to the question of which α_1-adrenoceptor subtype mediates human vasoconstriction *in vivo*. For this purpose, we have treated healthy subjects with placebo or a single dose of 0.4 mg of tamsulosin or 5 mg of terazosin in a single blind crossover manner. Before and 1, 3, 5, 7, 10, and 23.5 hr following drug intake, plasma samples were obtained for the radioreceptor assay. At each of these time points, the cardiovascular effects of the α_1-adrenoceptor agonist, phenylephrine, were determined. For the radioreceptor assay, cloned human α_1-adrenoceptor subtypes that were stably expressed in rat-1 fibroblasts were used. These experiments revealed that the α_{1A}-selective antagonist, tamsulosin, also behaves in an α_{1A}-selective manner *in vivo* in humans. Surprisingly, terazosin, which is not subtype-selective *in vitro,* also behaved as a subtype-selective agent *in vivo* with greater occupancy of α_{1A}- and α_{1D}-adrenoceptors, compared with α_{1B}-adrenoceptors. The analysis of the hemodynamic data is currently under way. Therefore, a definitive answer cannot be given yet with regard to the α_1-adrenoceptor subtype mediating vasoconstriction in humans *in vivo,* but the radioreceptor assay approach appears to be a promising tool for answering this question.

In the lower urinary tract, α_1-adrenoceptors couple to smooth-muscle contraction in the bladder neck, the urethra, and, in males, the prostate, while α_1-adrenoceptors do not appear important in the regulation of bladder detrusor activity under physiological conditions. Thus, α_1-adrenoceptor stimulation yields increased resistance to urinary flow, which can contribute to the symptoms of patients with benign prostatic obstruction. Detailed studies on α_1-adrenoceptor subtypes have been presented only for the prostate. However, because bladder neck, urethra, and prostate function as a unit, it is widely assumed that a similar situation exists in bladder neck and urethra. Within the human prostate, about 90% of the α_1-adrenoceptors are found on the stromal cells and only 10% on other cell types. In radioligand binding studies, approximately two-thirds of the prostatic α_1-adrenoceptors belong to the α_{1A}-subtype, with the remaining receptors most likely representing α_{1B}-adrenoceptors (3). In several studies, antagonist potencies for inhibition of human prostate contraction *in vitro* correlated well with those at cloned α_{1A}-adrenoceptors but not with those at cloned α_{1B}- or α_{1D}-adrenoceptors. Such data were obtained both with exogenous agonist (e.g., phenylephrine) or with endogenously released noradrenaline upon field stimulation. These observations indicate that in human prostate, α_{1A}-adrenoceptors not only dominate at the protein level, but also are most important functionally. The functional role of nonstromal α_1-adrenoceptors in the prostate and of the prostatic non-α_{1A}-adrenoceptors remains to be established.

In contrast to these data, more recent studies have found that some α_{1A}-selective antagonists have considerably lower antagonist potency in human prostate (4, 5). These include RS 17053, prazosin, S-niguldipine, and SB 216469. On the other hand, in the same studies, some drugs (e.g., tamsulosin and indoramin), have similar apparent affinities at cloned α_{1A}-adrenoceptors and in functional experiments with human prostate. This raises the possibility that the receptor subtype mediating contraction of human prostate is similar to but distinct from the α_{1A}-adrenoceptor and that it resembles the putative α_{1L}-adrenoceptor. However, we favor an alternative hypothesis. Recently, we have determined apparent antagonistic potencies of eight α_1-adrenoceptor ligands at cloned human α_{1A}-adrenoceptors stably expressed in Chinese hamster ovary cells under four conditions (6). Radioligand binding studies were performed on membranes in Tris/EDTA buffer at 20°C, on membranes in Ham's medium at 37°C, and on intact cells in Ham's medium at 37°C; in parallel inhibition of agonist, stimulated inositol phosphate formation was studied in intact cells in Ham's medium at 37°C. In the inositol phosphate experiments, drugs such as prazosin, RS 17053, and S-niguldipine displayed affinities that resembled those for antagonism of human prostate contraction. In contrast, these agents had high potency in radioligand binding experiments with membranes at 20°C. However, low potencies similar to the ones in the inositol phosphate experiments were seen when radioligand binding was performed at 37°C with intact cells. These experiments demonstrate that a single cloned receptor can have different apparent affinities for some antagonists, depending on the assay conditions. We have not observed similar differences between assay conditions for α_{1B}- or α_{1D}-adrenoceptors. Thus, we propose that the putative α_{1L}-adrenoceptor does not represent a distinct entity but rather a functional state of the α_{1A}-adrenoceptor.

In this context, three major questions remain: First, it has to be elucidated why some antagonists yield lower apparent affinities at cloned α_{1A}-adrenoceptors in intact cells than with cell membranes. Second, it is unclear why this occurs for some drugs but not for others. Third, it should be studied why this is seen with the aforementioned drugs in functional tests on some models of α_{1A}-adrenoceptors (e.g., human prostate, small mesenteric arteries) but not in others (e.g., rat perfused kidney, rat vas deferens). Finally, it would be interesting to know why this occurs with α_{1A}-adrenoceptors but not with α_{1B}- or α_{1D}-adrenoceptors.

In summary, α_1-adrenoceptors are most important in the human cardiovascular system in the vasulature. The subtype(s) mediating vasoconstriction in humans have not yet been firmly identified, but the radioreceptor assay appears to be a promising approach. In the human lower urinary tract, α_1-adrenoceptors mediate smooth-muscle contraction of bladder neck, urethra, and prostate. In the human prostate, the α_{1A}-adrenoceptor dominates at the protein level and probably is also most important for contraction. Low apparent affinities of some antagonists in functional tests with human prostate do not necessarily contradict the α_{1A}-adrenoceptor hypothesis but may relate to assay conditions. Why some drugs have much lower apparent affinities at α_{1A}-adrenoceptors when assayed in intact cells than with membranes remains to be determined.

Acknowledgments

Work in the authors' laboratories was supported in part by the Deutsche Forschungsgemeinschaft, Yamanouchi, and Boehringer Ingelheim.

References

1. Schäfers, R. F., Poller, U., Pönicke, K., Geissler, M., Daul, A. E., Michel, M. C., and Brodde, O.-E. (1996). Adrenergic and muscarinergic blockade and the cardiovascular effects of exogenously administered and endogenously released noradrenaline in humans. *Naunyn Schmiedebergs Arch. Pharmacol.* **335**, 239–249.
2. Brodde, O-E., Daul, A., Wellstein, A., Palm, D., Michel, M. C., and Beckeringh, J. J. (1988). Differentiation of β_1- and β_2-adrenoceptor-mediated effects in humans. *Am. J. Physiol.* **254**, H199–H206.

3. Michel, M. C., Grübbel, B., Taguchi, K., Verfürth, F., Otto, T., and Kröpfl, D. (1996). Drugs for treatment of benign prostatic hyperplasia: Affinity comparison at cloned α_1-adrenoceptor subtypes and in human prostate. *J. Auton. Pharmacol.* **16**, 21–28.
4. Ford, A. P. D. W., Arredondo, N. F., Blue, D. R., Bouhaus, D. W., Jasper, J. R., Kava, M. S., Lesnick, J., Pfister, J. R., Shiely, I. A., Vimont, R. L., Williams, T. J., McNeal, J. E., Stamey, T. A., and Clarke, D. E. (1996). RS-17053 (N-[2-cyclopropylmethoxy-phenoxy)-ethyl]-5-chloro-α,α-dimethyl-1H-indole-3-ethanamine hydrochloride), a selective α_{1A}-adrenoceptor antagonist, displays low affinity for functional α_{1A}-adrenoceptors in human prostate: Implications for adrenoceptor classification. *Mol. Pharmacol.* **49**, 209–215.
5. Chess-Williams, R., Chapple, C. R., Verfürth, F., Noble, A. J., Couldwell, C. J., and Michel, M. C. (1996). SB 216469, an antagonist which discriminates between the α_{1A}-adrenoceptor and the human prostatic α_1-adrenoceptor. *Br. J. Pharmacol.* **119**, 1093–1100.
6. Williams, T. J., Clarke, D. E., and Ford, A. P. D. W. (1996). Whole-cell radioligand binding assay reveals α_{1L}-adrenoceptor (AR) antagonist profile for the human cloned α_{1A}-AR in Chinese hamster ovary (CHO-K1) cells. *Br. J. Pharmacol.* **119**, Suppl., 359 P.

Dianne M. Perez, John Hwa, Ming-Ming Zhao, and James Porter

Department of Molecular Cardiology
Cleveland Clinic Research Institute
Cleveland, Ohio 44195

Molecular Mechanisms of Ligand Binding and Activation in α_1-Adrenergic Receptors

Three α_1-adrenergic receptor (α_1AR) subtypes have been cloned and pharmacologically characterized: α_{1a}-, α_{1b}-, and α_{1d}AR. The α_{1a}-subtype has 10 to 100-fold higher affinity for a number of agonists and antagonists when compared with the α_{1b}-subtype. However, there are currently no useful subtype-selective drugs readily available that can fully discriminate between the subtypes. To rationally design selective drugs, an understanding of subtype differences in the ligand binding pockets would be invaluable. We hypothesize that only a few nonconserved residues are critical for the ligand-binding specificity of each α_1AR subtype.

Nothing is known about the agonist-dependent molecular mechanisms of α_1AR stimulation. The activation mechanism for a related G-protein–coupled receptor, rhodopsin, has been described (1). The ligand isomerization of retinal when exposed to light breaks a constraining salt-bridge between transmembranes (TMs) 3 and 7, allowing the rhodopsin receptor to adopt an active

Advances in Pharmacology, Volume 42

1054-3589/98 $25.00

conformation that can now signal through transducin. A "constraining factor" has also been postulated for the α_{1b}AR subtype, holding the receptor in a basal configuration until bound by a receptor agonist (2). However, no molecular evidence for this α_{1b}AR constraining factor has been presented. In this chapter, the activational mechanism was also explored by testing the hypothesis that α_1ARs conserved the activational mechanism of rhodopsin in which a salt-bridge between TM 3 and TM 7 is disrupted.

I. Agonist Selectivity

Eight point mutations in converting α_{1b}AR to α_{1a}AR were analyzed by their ability to bind a panel of agonists. The A204V mutation had a five- to ten-fold increased binding affinity over WT (wild-type) α_{1b}AR for oxymetazoline, cirazoline, and methoxamine, a ligand-binding profile consistent with α_{1a}AR (Table I). These affinities were unchanged with GppNHp. The L314M mutant also demonstrated a significant increase in its affinity for these three agonists. No changes in antagonist pharmacology were observed.

TABLE I Agonist Binding Profiles of WT and Mutant α_1ARs in Converting α_{1b} to α_{1a}AR

	Agonists				
Mutants	*Oxymetazoline*	*Cirazoline*	*Methoxamine*	*Epinephrine*	*Phenylephrine*
Single					
WT α_{1b}	6.05 ± 0.05	5.64 ± 0.11	3.16 ± 0.06	5.29 ± 0.04	4.69 ± 0.06
S95T/F96S	5.54 ± 0.09	5.72 ± 0.03	3.26 ± 0.07	5.30 ± 0.23	4.64 ± 0.13
T174L	6.25 ± 0.12	5.76 ± 0.12	3.12 ± 0.03	5.47 ± 0.10	4.59 ± 0.07
L182F	5.94 ± 0.11	5.81 ± 0.17	3.16 ± 0.05	5.30 ± 0.02	4.78 ± 0.03
A204V	**6.90 ± 0.04**	**6.10 ± 0.09**	**3.96 ± 0.06**	**5.77 ± 0.03**	**5.55 ± 0.05**
S208A	6.18 ± 0.10	5.82 ± 0.06	3.16 ± 0.01	5.16 ± 0.10	4.46 ± 0.09
A313V	6.18 ± 0.01	5.87 ± 0.07	3.39 ± 0.07	5.49 ± 0.09	4.96 ± 0.02
L314M	**6.34 ± 0.03**	**6.20 ± 0.02**	**3.76 ± 0.05**	5.50 ± 0.04	4.81 ± 0.05
WTα_{1a}	**7.34 ± 0.10**	**6.38 ± 0.06**	**4.43 ± 0.12**	5.35 ± 0.08	4.96 ± 0.04
Combinations					
S95T/F96S/ L182F/S208A	5.89 ± 0.27	5.62 ± 0.03	3.21 ± 0.01	5.38 ± 0.19	4.47 ± 0.01
A204V/L314M	**7.27 ± 0.06**	**6.49 ± 0.04**	**4.57 ± 0.05**	5.43 ± 0.04	4.90 ± 0.05
Reverse					
V185A/M292L	5.91 ± 0.08	5.99 ± 0.04	3.79 ± 0.01	5.41 ± 0.05	4.85 ± 0.05

Single and combinational mutants are α_{1b}ARs that are mutagenized to corresponding α_{1a}-residues. Reverse mutant is an α_{1a} with corresponding mutagenized α_{1b}-residues. Competition binding studies were used to determine the K_i values (−LOG ± SE) of adrenergic agonists using [^{125}I]HEAT as the radioligand and membranes prepared from COS-1 cells expressing the receptor constructs. All competition binding isotherms were best fit to a single-site model. Bold numbers represent significant changes in affinity from the WT α_{1b} ($p < .001$).

Combination mutations were made to assess if the individual mutations were additive in their affects on changing the binding profiles. The combination (S95T/F96S, L182F, and S208A) involved mutations that by themselves had no major effect on ligand binding and were not different than that of WT α_{1b}AR. When the two mutants that alone showed a significant change toward the α_{1a}AR (L314M, A204V) were combined, the agonist-binding profile of this double mutant was now similar to that of WT α_{1a}AR. We also performed the equivalent reverse mutation in α_{1a}AR to confirm that the L314M and A204V mutations were specific. In combination, they imparted α_{1b}AR pharmacology.

II. Catechol Binding

Strader *et al.* (3) proposed that hydrogen bond interactions involving the hydroxyl groups on the phenyl ring are important for ligand binding to β_2AR. It was proposed that S 204 in TM 5 of the β_2AR forms a hydrogen bond with the meta-hydroxyl group while S 207 forms a hydrogen bond with the para-hydroxyl group. In support of this model, S to A mutants had a 30-fold decreased affinity for catechol agonists, with each serine contributing about 50% to the activation of the receptor. We have explored similiar mutations in the α_1AR to assess catechol binding.

Replacement of either S 188 or S 192 in TM 5 of the α_{1a}AR with an A did not significantly reduce the binding affinity for any of the agonists compared with WT (Table II). In fact, the binding affinity for phenylephrine was significantly *increased* (sevenfold) with the S192A mutant. These results are quite distinct from the β_2-paradigm in which either serine mutation was able to reduce agonist-binding affinity. To confirm a hydrogen bond interaction, the double mutant, S188/192A, was created and found to decrease the binding affinity by 25- to 100-fold for various agonists, consistent with a decrease in binding energy of $\Delta\Delta G = +3-5$ kcal/mol and equivalent to a disruption of a single hydrogen bond. Because either serine residue is sufficient in itself in maintaining the WT binding affinity, but the free energy values indicate only

TABLE II Agonist Binding Profiles of WT α_{1a}AR and Serine Mutations

	α_{1a}AR	S188A	S192A	S188A/S192A
Epinephrine (μM)	3.3 ± 0.4	2.6 ± 0.2	3.2 ± 0.7	398 ± 82[a]
Phenylephrine (μM)	6.2 ± 1.5	12.6 ± 1.6	0.9 ± 0.1[a]	479 ± 170[a]
Synephrine (μM)	52.5 ± 2.4	104.7 ± 11.4	28.2 ± 3.1	324 ± 133[a]
Norepinephrine (μM)	4.6 ± 0.3	9.8 ± 1.3	7.4 ± 1.1	115 ± 50[a]
K_D [^{125}I]HEAT (pM)	81	50	282	375

Competition binding studies were used to determine K_i values (μM ± SE) of membranes prepared from COS-1 cells expressing the receptor constructs. Epinephrine contains two hydroxyls on the phenyl ring, while phenylephrine and synephrine contain only the meta- or para-hydroxyls, respectively. All binding isotherms were best fit to a single-site model.
[a] Significant differences from the WT receptor ($p < .001$).

one hydrogen bond is formed, the data suggest that both serines contribute a weak hydrogen bond to the agonist. However, it seems that the meta-hydroxyl interaction with S 188 is the strongest, because with the WT receptor, phenylephrine has essentially the same binding affinity as epinephrine. In addition, based on affinity differences between phenylephrine and S 188 versus S 192, and assuming an initially direct relationship between affinity and distance, the meta-hydroxyl would be closer to S 188 than S 192. The activation requirements for α_1-ARs appear similiar to their binding interactions. At equal receptor numbers of 0.3 pmoles/mg protein, only Ser 188 plays a major role in receptor activation, contributing 70–90% of the wild-type response. On the other hand, the effect of S 192 on receptor activation is minimal.

To account for a stronger meta-hydroxyl interaction and a weaker para-hydroxyl effect, the catecholamine would dock in a planar orientation in the ligand-binding pocket as opposed to the skewed orientation in the βAR. Because there are three residues between the two serines in the helix in α_1ARs as opposed to the two residues between the βAR serines, the orientation of the ring is also different in these two receptor subtypes. The meta-S 188 interaction in α_1AR is closest to TM 4, while the meta-S 204 interaction in βAR is closer to TM 6, resulting in the catechol ring being rotated about 120° in α_1AR. Hence, it appears that the catechol docks in a unique manner in α_1ARs as compared with the β-paradigm.

III. Antagonist Selectivity

To explore the determinants of antagonist binding, we first constructed a chimera that consists of α_{1b}AR with TM 5 replaced with the corresponding region of α_{1a}AR. A portion of the second extracellular and third intracellular loops was also replaced due to the convenience of restriction sites. The chimera had a nine- to 29-fold increased binding affinity for the α_{1a}-selective antagonists, phentolamine and WB 4101. Another α_{1a}-selective antagonist, (+) niguldipine, showed no significant changes from an α_{1b}AR pharmacology.

The chimera contains eight different amino acids between α_{1b}- and α_{1a}ARs that are not due to species variations. However, only three point mutations located in the extracellular loop were responsible for the selectivity. These mutations (G196Q, V197I, T198N) individually increased in their binding affinity for both phentolamine (two- to threefold) and WB4101 (three- to fivefold) toward that of α_{1a}AR. These point mutations were then combined into a single receptor. The triple mutation displayed affinities similiar to both the $\alpha_{1b/a}$ chimera and the WT α_{1a}AR for both phentolamine and WB4101. We also performed the equivalent reverse mutation in α_{1a}AR to confirm the specificity. The results showed that the triple mutation in α_{1a} (Q177G, I178V, N179T) completely reversed the binding affinity for phentolamine and WB 4101 to that of α_{1b}AR. It appears likely that these three residues are indeed located in the extracellular loop and are not part of the TM domain; thus, antagonists may not bind deep within a pocket as agonists.

IV. Mechanism of α_1-Adrenergic Receptor Activation

Previous work has identified an interhelical salt-bridge holding the rhodopsin receptor in a basal conformational structure until activated by light (1). Analogous to rhodopsin, this potential salt-bridge between K 331 in TM 7 and D 125 in TM 3 could restrict α_{1b}AR to a basal conformation until bound by an agonist. Abolishing this ionic bond would allow α_{1b}AR to adopt an active conformation that would have properties of a constitutively active receptor. To test this hypothesis, site-directed mutagenesis eliminated this charged amino acid pair by mutating the K 331 to an A or a E.

If these mutant α_{1b}ARs are adopting an active protein conformation, then a higher AR agonist-binding affinity would be observed. Competition binding experiments demonstrated a significantly higher epinephrine-binding affinity for the mutant K331A compared with the WT α_{1b}AR. A calculated epinephrine Ki of 329 $\pm$ 53 nM for the K331A mutation was sixfold greater than the 1860 $\pm$ 227 nM Ki value determined for the WT α_{1b}AR. This high-affinity binding of epinephrine was not altered by the presence of 0.1 mM Gpp(NH)p. Significant increases in the norepinephrine- and methoxamine-binding affinity were also calculated for this same K331A mutation over the WT α_{1b}AR as well as for the K331E mutation. However, for both K331 mutants, no binding-affinity differences for α_1AR antagonists were noted.

As would be predicted for a constitutively active receptor, the potency of epinephrine to generate soluble [^{3}H]inositol phosphates was significantly greater for the K331A mutant versus the WT α_{1b}AR. The EC_{50} was 381 $\pm$ 101 nM for the mutant K331A versus 1849 $\pm$ 537 nM for the WT receptor. These experiments were performed at similiar receptor densities (0.1 pmol/mg). Significant leftward shifts of the epinephrine EC_{50} were also calculated for the K331E mutation (391 $\pm$ 89 nM) when compared with the WT receptor. The amount of basal IP_3 in the absence of agonist was then determined at various levels of α_{1b}AR density. The slope of the linear regression lines is an indication for the amount of basal IP_3 produced per α_{1b}AR. The amount of basal IP_3 produced for the K331A (14.0 $\pm$ 3.9 pmol/fmol) and the K331E (10.2 $\pm$ 0.2 pmol/fmol) mutations was significantly greater when compared with the WT receptor (3.1 $\pm$ 0.6 pmol/fmol).

If the salt-bridge hypothesis is correct, eliminating the negative charge at position 125 should also generate constitutively active α_{1b}ARs. To investigate this possibility, the D 125 of the WT α_{1b}AR was change to an A or a K. The binding properties for these D 125 mutations showed no significant changes in affinity for antagonists when compared with the WT receptor. However, there was a significantly lowered epinephrine affinity (threefold) for these D 125 mutant receptors when compared with the WT α_{1b}AR. This decrease in the epinephrine-binding affinity is likely due to the elimination of the conserved negative charged at position 125, shown in the βAR system to be responsible for docking with the protonated amine of epinephrine. Because this D 125 is the counterion necessary for receptor docking with the protonated amine of epinephrine, determining AR-agonist affinity and potency changes are no longer valid parameters for assessing the constitutively active properties of these D

125 mutations. However, the amount of basal IP_3 generated in the absence of agonists was quantified at various levels of α_{1b}AR density. A significantly increased quantity of basal IP_3 was produced by the D125A (23.8 $\pm$ 5.6 pmol/fmol) and D125K mutations (53.9 $\pm$ 3.7 pmol/fmol) when compared with the WT α_{1b}AR (3.1 $\pm$ 0.6 pmol/fmol). In summary, this constitutively active property of these D 125 mutations strongly suggests the existence of an D 125–K 331 salt-bridge in the WT α_{1b}AR. Docking of epinephrine initiates a competition between the protonated amine of the ligand and the positive charge of K 331 for the negative aspartic acid counterion. We speculate that competition for the D 125 disrupts the salt-bridge, allowing a translation movement of TM 3 toward the protonated amine of the ligand.

Acknowledgments

This work was supported in part by the National Institutes of Health grant RO1HL52544 (DMP) and an educational grant from Glaxo Wellcome, Inc. This work was performed during the tenure of an Established Investigatorship (DMP) from the American Heart Association.

References

1. Robinson, P. R., Cohen, G. B., Zhukovsky, E. A., and Oprian, D. D. (1992). Constitutively active mutants of rhodopsin. *Neuron* **9,** 719–725.
2. Kjelsberg, M. A., Cotecchia, S., Ostrowski, J., Caron, M. G., and Lefkowitz, R. J. (1992). Constitutive activation of the α_{1b}-adrenergic receptor by all amino acid substitutions at a single site. *J. Biol. Chem.* **267,** 1430–1433.
3. Strader, C. D., Candelore, M. R., Hill, W. S., Sigal, I. S., and Dixon, R. A. F., (1989). Identification of two serine residues involved in agonist activation of the β-adrenergic receptor. *J. Biol. Chem.* **264,** 13572–13578.

**Hyman B. Niznik,*§† Fang Liu,*§ Kim S. Sugamori,*§
Bruno Cardinaud,‡ and Phillipe Vernier‡**

Departments of *Pharmacology and †Psychiatry
University of Toronto
Toronto, Ontario, Canada

‡Institut Alfred Fessard
CNRS Gif-Sur-Yvette France

§The Molecular Neurobiology Laboratory
Clarke Institute of Psychiatry
Toronto, Ontario Canada M5T 1R8

Expansion of the Dopamine D1 Receptor Gene Family: Defining Molecular, Pharmacological, and Functional Criteria for D1A, D1B, D1C, and D1D Receptors

Dopamine D1 receptors have been operationally defined on biochemical and pharmacological criteria by their ability to promote adenylate cyclase activity and bind agonists and antagonists of the benzazepine and benzonapthazine class of compounds with high affinity. Molecular biological studies have revealed that native D1 receptors are comprised of two receptor subtypes, termed D1/D1A and D1B/D5, each of which when expressed in various cell lines stimulates the activity of adenylate cyclase. These receptors are, however, distinguishable on the basis of their primary structure, chromosomal localization, mRNA, and protein distribution as well as in their expressed pharmacological profiles. Of all the compounds tested, only a few unique distinguishing pharmacological features between these receptors are evident, however. Thus, the endogenous neurotransmitter dopamine and nonselective dopaminergic agonist ADTN display up to 10-fold higher affinity for D5/D1B receptors, while nonselective antagonists, such as butaclamol and spiperone, exhibit much higher affinities for D1/D1A than D5/D1B receptors. In contrast, the benzazepines do not discriminate between the various receptor subtypes and may be considered only as generic markers of the D1 receptor family. While most receptor genes have been characterized only in mammals, D1-like receptor gene diversity has been examined in a small set of other vertebrate species, particularly *Xenopus laevis* and *Gallus domesticus*. Vertebrate frog and chicken D1A and D1B receptors display overall sequence identities of approximately 80% and 65%, respectively, to mammalian D1/D1A-D5/D1B receptors and in agreement with their amino acid similarities exhibit pharmacological profiles for dopaminergic ago-

Advances in Pharmacology, Volume 42

1054-3589/98 $25.00

nists and antagonists virtually identical to their mammalian counterparts. Similarly, at a functional level, while vertebrate D1A and D1B receptors stimulate cyclic adenosine monophosphate (cAMP) accumulation, vertebrate D1B receptors exhibit constitutive adenylate cyclase activity that is inhibited by the inverse agonists, butaclamol and flupentixol, a defining characteristic of mammalian D5/D1B receptors.

Despite the existence of two distinct dopamine D1 receptors, biochemical, pharmacological, and behavioral evidence suggests the existence of additional D1-like receptor subtypes. Thus, D1 receptor stimulation in both brain and periphery has been shown to activate phospholipase C, stimulate K^+ efflux, and inhibit Na^+/H^+ exchange, all independent of adenylate cyclase activity. Moreover, behavioral studies have differentiated agonist-specific D1-like receptor-mediated behaviors from the activity of D1-stimulated adenylate cyclase. It is unclear whether these D1-like receptor-mediated responses can be accounted for solely by the presence of D1A and D1B receptors.

Direct support for the multiplicity and expansion of the dopamine D1 receptor gene family was initially provided in the vertebrate species, *Xenopus laevis* (1) and *Gallus domesticus* (2). Thus, in addition to D1A and D1B receptor genes, these species express a third D1 receptor subtype, termed D1C and D1D, respectively. D1-like receptor sequences not classified as either D1A or D1B have also been described in fish. The *Xen* D1C and chicken D1D receptors display overall amino acid and nucleotide sequence identity of approximately 52% with each other and with both mammalian D1A and D1B receptors. While D1C and D1D receptors stimulate the activity of adenylate cyclase, these receptors are each distinguishable from D1/D1A and D5/D1B receptor genes and from each other on the basis of their unique mRNA distribution and pharmacological profiles. In particular, the frog D1C receptor displays affinities for some discriminating compounds, such as DA, 6,7-ADTN, haloperidol, and spiperone with a rank order of potency D1A > D1C > D1B essentially intermediate to D1/D1A-D5/D1B receptor subtypes. For all other ligands tested, the D1C receptor exhibits affinities equal to or slightly higher than D1/D1A receptors. The D1D receptor, in contrast, exhibits affinities for discriminating agents ADTN and dopamine identical to the D5/D1B receptor subfamily while binding other dopaminergic agents, such as SKF-38393, apomorphine, pergolide, and lisuride, with affinities 10-fold higher than any other cloned mammalian or vertebrate D1A/D1B receptor subtype. With regard to antagonists, haloperidol exhibits considerably poorer affinity (>2 μM) for the D1D receptor.

Although *Xenopus* D1C and chicken D1D receptors significantly differ from either vertebrate or mammalian D1A and D1B receptors on the basis of their amino acid sequence and distinct pharmacological profiles, the functional and evolutionary relationship between these receptors is unclear and raises a number of intriguing questions. First, it is unknown whether *Xen* D1C and chicken D1D receptors are truly reflective of distinct D1 receptor subtypes or if their presence is merely associated with and restricted to these particular species. Second, confusion in receptor classification may occur when species-specific sequence differences between orthologous receptors (species homologues) give rise to distinct pharmacological profiles as recognized by specific ligands, or alternatively, when paralogous receptors (true subtypes) are not distinguished by different ligands (3). Examples of such heterodox behavior are observed in the D1 receptor family where flupentixol displays approximately sixfold higher affinity for the chicken than *Xenopus* D1B receptor. As such, it is unclear whether *Xen* and chicken D1C and D1D receptors are merely orthologous receptor subtypes, because we have not yet established, at both the molecular and pharmacological/functional levels, those attributes that define the existence of dopamine D1C and D1D receptors as true D1 receptor subtypes throughout the vertebrate phylum. To address these issues, we have undertaken to clone the entire complement of D1-like receptor genes from a number of early and late diverging vertebrate species. The assignment of each cloned receptor to defined D1 receptor subtype was achieved by combining molecular phylogeny, pharmacological, and functional approaches.

While the study is far from being complete, we have cloned from various cDNA and genomic libraries the genes encoding members of the vertebrate D1 receptor family. Molecular phylogenetic analysis (Vernier, P. *et al.*) of the available sequences from numerous vertebrate species clearly defines four D1 receptor subtypes, termed D1A, D1B, D1C, and D1D. Moreover, all four D1 receptor genes have been cloned and isolated from two distinct species, crocodile and chicken , clearly defining their paralogous nature. At present, only full-length sequences of the eel D1 receptor family (D1A, D1B, and D1C) have been extensively characterized, suggesting the existence of three distinct and well-established vertebrate D1 receptor subtypes arising prior to the evolutionary divergence of fish and tetrapods. The eel does not appear to contain the D1D receptor gene. At both the pharmacological and functional levels (cAMP formation, constitutive activity), eel D1A, D1B, and D1C receptors exhibit characteristics clearly attributable to mammalian or vertebrate receptor counterparts, suggesting that the particular molecular and pharmacological signatures distinguishing these receptor subtypes have been conserved throughout evolution (4). Work is ongoing to express and characterize all available cloned D1C receptors in order to establish the validity of these discriminating criteria.

As a prelude to this work, we have begun to search for subtype-specific D1 receptor ligands and identify specific amino acid and structural motifs that give rise to the distinct pharmacological and functional characteristics of multiple vertebrate D1-like receptors. In particular, we analyzed the ligand-binding profiles, affinity, and functional activity of 13 novel NOVO Nordisk (NNC) compounds at mammalian–vertebrate D1/D1A and D5/D1B as well as vertebrate D1C/D1D dopamine receptors transiently expressed in COS-7 cells (K. Sugamori and H. B. Niznik, unpublished). Of all the compounds tested, only one NNC-compound displayed preferential selectively for vertebrate D1C receptors exhibiting an estimated affinity ($\sim$0.5 nM) 10- to 20-fold higher than all other mammalian–vertebrate D1-like receptors. Functionally, the molecule fully inhibits D1C-mediated stimulation of adenylate cyclase activity, while inhibiting cAMP accumulation with somewhat lowered efficacy ($\sim$65%) at both D1-D1A and D5/D1B receptors. At vertebrate D1B receptors, NNC acts as a partial agonist stimulating adenylate cylase activity by approximately 35% relative to full agonists dopamine and SKF 82596 (10 μM), an effect that is blocked by selective D1 receptor antagonists SCH-23390 and NNC-01-0010. Surprisingly, mammalian D5 receptors do not, however, conserve the ability to recognize NNC as a partial agonist. To define some of the structural motifs that may regulate the pharmacological and functional actions by NNC, a series of carboxyl tail (CT) receptor chimeras were constructed. Substitution of the carboxyl terminus of vertebrate D1B receptors with CT sequences of the D1A receptor reduced the ability of NNC to activate cAMP. In marked contrast, vertebrate D1B/D1cCT receptor mutants displayed a marked enhancement in the ability of NNC to maximally stimulate cAMP accumulation (~70% relative to 10 μM dopamine) as well as increased responsiveness to the partial agonist SKF 38393. Substitution of D1A and D1C CTs with sequences encoded by the D1B receptor did not, however, produce receptors with functional characteristics significantly different than wild type. Taken together, these data clearly suggest that despite lacking the canonical N-methyl residue in the R-3 position, this compound is a fairly selective and potent D1C receptor antagonist and that sequence-specific motifs within the carboxyl terminal tail of D1-like receptors may modulate receptor activation of adenylate cyclase activity by partial agonists. Moreover, in line with the molecular phylogenic data described earlier, the differential pharmacological actions of NNC establishes, at least in vertebrates, that the D1C receptor constitutes a true D1 receptor subclass. With regard to the D1D receptors, it appears that a wide variety of Burroughs Wellcome compounds express selective and preferential high affinity (~0.9 nM) for this receptor subtype (A. Hamadanizadeh and H. B. Niznik, unpublished). While work is still ongoing, most of the compounds tested are antagonists at all D1-like receptors.

In summary, these data, when taken together, clearly suggest that D1A, D1B, D1C, and D1D receptors exhibit molecular, pharmacological, and functional attributes that unambiguously allow for their distinct classification in the vertebrate phylum as true D1 receptor subtypes. While mammalian homologues of D1C and D1D receptors have yet to be found and may indeed have been lost during the course of mammalian evolutionary divergence, the widespread appearance of multiple D1 receptor subtypes throughout the vertebrate phylum would suggest some gain or loss of function in mammals associated with these particular receptor subtypes. Elucidating these possible molecular and functional correlates may allow for a better understanding of mammalian dopamine D1-like receptor-mediated events, particularly those associated with the psychostimulant properties of drugs of addiction, such as cocaine, and in the maintenance of cognitive–emotional states in both health and disease.

References

1. Sugamori, K., Demchyshyn, L., Chung, M., and Niznik, H. B. (1994). D1A, D1B and D1C dopamine receptors from *Xenopus laveis*. *Proc. Natl. Acad. Sci. U.S.A.* **91,** 10536–10540.
2. Demchyshyn, L., Sugamori, K., Lee, F., Hamadanizedeh, S., and Niznik, H. B. (1995). The dopamine D1D receptor: Cloning and characterization of three functional and pharmacologically distinct avian dopamine D1 receptors. *J. Biol. Chem.* **270,** 4045–4012.

3. Vernier, P., Cardinaud, B., and Vincent, J. D. (1995). Evolution of the catecholaime receptor family. *TIPS*. **16**, 375–381.
4. Cardinaud, B., Sugamori, K. S., Coudouel, S., Vincent, J. D., Niznik, N. B., and Vernier, P. (1996). Early emergence of three dopamine D1 receptor subtypes in vertebrates: Molecular phylogenetic, pharmacological and functional criteria defining D1a, D1b and D1c receptors in european eel *anguilla anguilla*. *J. Biol. Chem.* **272**, 2778–2787.

J-C. Schwartz,* S. Ridray,† R. Bordet,† J. Diaz,† and P. Sokoloff*

*Unité de Neurobiologie et Pharmacologie (U. 109) Inserm
Centre Paul Broca
75014 Paris, France

†Laboratoire de Physiologie
Faculté de Pharmacie
Paris, France

D1/D3 Receptor Relationships in Brain Coexpression, Coactivation, and Coregulation

The identification of the dopamine (DA) D3 receptor (D3R), a D2-like receptor (1), has represented a challenge: to which extent its functions can be differentiated from those of the much more abundant D2R? The strategy we have followed to meet this challenge has consisted of (1) identification or designing selective D3R ligands, testing them on transfected cell lines; (2) characterizing the phenotype of D3R-expressing neurons in brain; and (3) assessing the responses mediated by the D3R and the regulation of its expression in brain.

Using these approaches, we show that, in contrast to D2R, highly segregated from D1R, D3R is coexpressed with D1R, with which interactions are shown at the level of neurochemical responses and regulation of gene expression.

I. Coexpression of D1R and D3R in Ventral Striatum and Functional Interactions

We had previously identified medium-size spiny neurotensin neurons in the shell part of nucleus accumbens as D3R-expressing neurons and suggested

Advances in Pharmacology, Volume 42

1054-3589/98 $25.00

that they respond in an opposite manner when compared with neurotensin neurons harboring D2R (2). This proposal was based on data obtained with a variety of neuroleptics with mixed D2R/D3R antagonist properties, which, on acute administration, all diminish neurotensin mRNA levels in D3R-expressing neurotensin neurons (in ventromedial shell), whereas they induce opposite effects on D2R-expressing neurotensin neurons (in cone). We have now confirmed that the former effect is, as assumed, selectively related to D3R blockade, because it is reproduced by administration of nafadotride in a D3R-selective dosage. In addition, the haloperidol-induced decrease in neurotensin mRNA in ventromedial shell is prevented on coadministration of D0-897, a novel highly selective and partial D3R agonist.

D3R and D1R are highly expressed in the islands of Calleja, a structure of poorly defined physiological role, receiving a sparse dopaminergic innervation from the mesencephalon (A9 or A10 neuronal groups?) and from which D2R is absent. Double hybridization studies showed D1R and D3R mRNAs to be coexpressed in a high percentage (~80%) of substance P granule cells, the major neuronal population of the islands.

The opposite roles of the two receptor subtypes are shown by studies in which the effects of agonist or antagonist administration were evaluated on the expression of c-fos. Whereas SKF 38393, a D1R agonist, enhances it markedly, quinpirole, a D2R/D3R agonist, has an opposite effect; in addition, endogenous dopamine must have a tonic effect on this system, because c-fos expression is decreased and increased on treatment by SCH 23390 and nafadotride, respectively (Table I).

In the ventromedial shell of nucleus accumbens, a large proportion of D3R-harboring neurons were found to coexpress D1R (63%) and substance P (71%) mRNAs. The two receptor subtypes seem to be involved in a synergistic manner in the control of the neuropeptide gene expression. In rats pretreated with reserpine (to avoid interference of changes in DA release), treatment with SKF 38390 slightly enhanced substance P mRNA levels, a change that was markedly

TABLE I Interactive Responses Mediated by D1 and D3 Receptors in Islands of Calleja and Ventromedial Shell of Nucleus Accumbens

Treatment	*Change in mRNAs (% of controls)*
	c-fos mRNAs in island of Calleja
SKF 38393 (10 mg/kg)	+271 ± 62**
Quinpirole (1 mg/kg)	−75 ± 8*
SKF 38393 + quinpirole	−19 ± 30 (NS)
SCH 23390 (0.5 mg/kg)	−75 ± 5*
Nafadotride (1 mg/kg)	+222 ± 76*
	Substance P mRNAs in accumbens shell
SKF 38393 (10 mg/kg)	+34 ± 8**
Quinpirole (1mg/kg)	−9 ± 6 (NS)
SKF 38393 + quinpirole	+190 ± 53**
SKF 38393 + quinpirole + nafadotride (1 mg/kg)	+63 ± 22*

NS, not significant. *$P < 0.05$; **$P < 0.01$

potentiated by cotreatment with quinpirole (see Table I). It can be concluded, therefore, that although D1R and D3R are largely coexpressed, their costimulation may result in either opposite or synergistic responses according to the neuronal type and kind of response considered.

II. D1R-Mediated Ectopic Induction of D3R in Denervated Striatum

The expression of the D3R in neurons of the ventral striatum is highly dependent on the dopaminergic innervation. Thus, after 6-OHDA–induced denervation, D3R mRNA levels and [^{3}H]7-OH-DPAT binding site number are drastically reduced (by over 50%) in nucleus accumbens (3). Considering the upregulation of D2R under such circumstance, the change seems paradoxical. The depletion is not consequent to DA (or DA cotransmitter) deprivation because it is not reproduced by chronic administration of D1R and D2R antagonists, reserpine, or cholecystokinin and neurotensin antagonists. In contrast, a D3R downregulation follows administration of GABA agonists or colchicine, suggesting that accumbic D3R expression is under the positive control of a trophic factor released by DA neurons in an activity-dependent manner.

By treating hemiparkinsonian rats with L-dopa under conditions leading to a sensitization of their turning behavior, we observed that D3R mRNA levels in ipsilateral nucleus accumbens progressively became enhanced over the control level. This change was accompanied unexpectedly by the appearance of the D3R signal as well as, after a few days, of D3R-binding sites in the ipsilateral dorsal striatum, an area where this receptor is normally not expressed (Fig. 1).

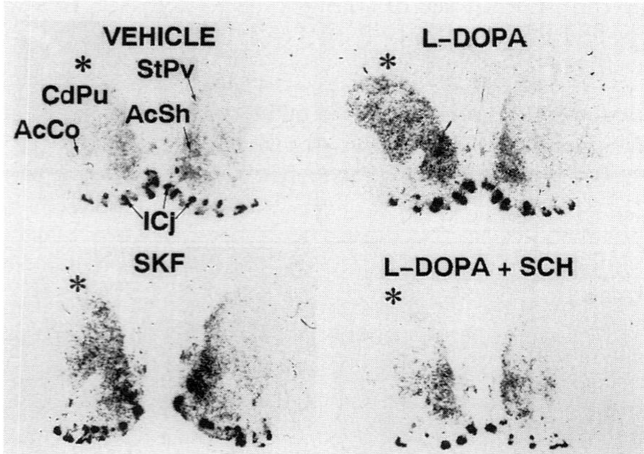

FIGURE 1 *In situ* hybridization of D3R mRNAs in frontal brain sections of rats unilaterally lesionned by 6-OHDA (*left*) and treated for 5 days with vehicle L-dopa (60 mg/kg, i.p. b.i.d.) alone or together with SCH 23390 (0.5 mg/kg) or with SKF 38393 (10 mg/kg, i.p., b.i.d.). StPv, periventricular striatum; CdPu, caudate putamen; AcSh, nucleus accumbens core; AcCo, nucleus accumbens core; ICj, islands of Calleja.

The ectopic induction by L-dopa is apparently consequent to D1R stimulation, being prevented by coadministration of SCH 23390 and mimicked by administration of SKF 38393. D3R induction in the denervated striatum seems to be responsible for the sensitization to L-dopa observed in the same animals. In support, the two processes develop during and decay on interruption of L-dopa with strictly parallel time courses. In addition, the L-dopa–induced turning behavior of rats previously sensitized by repeated L-dopa is inhibited on administration of nafadotride in low dosage, compatible with a selective D3R blockade. Because nafadotride has no such effect on turning behavior in nonsensitized hemiparkinsonian rats, this strongly suggests that ectopic D3R appearance and behavioral sensitization, a so far poorly understood process, are causally related.

The process might have therapeutic relevance because sensitization of the motor behavior in rats might reflect either beneficial or detrimental effects of chronic dopatherapy in parkinsonian subjects (i.e., the progressive resumption of complex motor behaviors or development of dyskinesias or psychiatric disturbances).

References

1. Sokoloff, P., and Schwartz, J. C. (1995). Novel dopamine receptors half a decade later. *Trends Pharmacol. Sci.* **16,** 270–275.
2. Diaz, J., Levesque, D., Lammers, C. H., Griffon, N., Martres, M. P., Schwartz, J. C., and Sokoloff, P. (1995). Phenotypical characterization of neurons expressing the dopamine D3 receptor in the rat brain. *Neuroscience* **65,** 731–745.
3. Leveque, D., Martres, M. P., Diaz, J., Griffon, N., Lammers, C. H., Sokoloff, P., and Schwartz, J. C. (1995). A paradoxical regulation of dopamine D3 receptor expression suggests the involvement of an anterograde factor from dopamine neurons. *Proc. Natl. Acad. Sci. U.S.A.* **92,** 1719–1723.

Jonathan A. Javitch

Center for Molecular Recognition
Departments of Psychiatry and Pharmacology
College of Physicians and Surgeons
Columbia University and New York State Psychiatric Institute
New York, New York 10032

Mapping the Binding-Site Crevice of the D2 Receptor

The dopamine receptors, like the homologous receptors for biogenic amines, bind neurotransmitters present in the extracellular medium and couple this binding to the activation of intracellular G-proteins. The binding sites of these receptors are formed among their seven, mostly hydrophobic, membrane-spanning segments and are accessible to charged, water-soluble agonists, like dopamine. Thus, each of these binding sites is contained within a water-accessible crevice, the binding-site crevice, extending from the extracellular surface of the receptor into the plane of the membrane. The surface of this crevice is formed by residues that contact specific agonists and/or antagonists and other residues that may affect binding indirectly.

In the homologous β2-adrenergic receptor, residues that contribute to binding have been identified in membrane-spanning segments, M3, M5, M6, and M7: Mutations of Asp 113, Ser 204, Ser 207, and Phe 290 altered binding and Trp 330 was affinity-labeled by an antagonist derivative. These five residues are identically conserved in all catecholamine receptors. In the dopamine D2 receptor, mutation of the residues that align with the first three, Asp 114, Ser 194, and Ser 197, also altered the binding of dopamine agonists and antagonists. Completely conserved residues, however, cannot account for the profound differences in binding specificities among the catecholamine receptors. Additional residues must contribute to binding, either directly or indirectly.

To identify the residues that form the surface of the binding-site crevice in the human D2 receptor, we have used the substituted-cysteine accessibility method (SCAM) (1–5). Consecutive residues in the membrane-spanning segments are mutated to cysteine, one at a time, and the mutant receptors are expressed in heterologous cells. If ligand binding to a cysteine-substitution mutant is near-normal, we assume that the structure of the mutant receptor, especially around the binding site, is similar to that of wild type and that the substituted cysteine lies in an orientation similar to that of the wild-type residue. In the membrane-spanning segments, the sulfhydryl of a cysteine can face either into the binding-site crevice, into the interior of the protein, or into the lipid bilayer; sulfhydryls facing into the binding-site crevice should react much faster with hydrophilic, lipophobic, and sulfhydryl-specific reagents. For such reagents, we use derivatives of methanethiosulfonate

Advances in Pharmacology, Volume 42

1054-3589/98 $25.00

(MTS): positively charged MTSethylammonium (MTSEA) and MTSethyltrimethylammonium (MTSET), and negatively charged MTSethylsulfonate (MTSES). These reagents are about the same size as dopamine, with maximum dimensions of approximately 10 Å by 6 Å. They form mixed disulfides with the cysteine sulfhydryl, covalently linking SCH_2CH_2X, where X is NH_3^+, $N(CH_3)_3^+$, or SO_3^-. We use two criteria for identifying an engineered-cysteine as forming the surface of the binding-site crevice: (1) The reaction with an MTS reagent alters binding irreversibly and (2) this reaction is retarded by the presence of agonists or antagonists.

A distinction between our approach and typical mutagenesis experiments is that we do not rely on the functional effects of a given mutation. The interpretation of the effects of typical mutagenesis experiments requires one to assume that functional changes, such as changes in binding affinity, caused by a mutation are only due to local effects at the site of the mutation and not due to nonlocal effects of the mutation on protein structure. The validity of this assumption is rarely proven for individual mutations. By contrast, it is unlikely that the protein segments lining the binding site would be grossly distorted in a dopamine-receptor mutant with near-normal binding properties. Thus, the engineered cysteine side-chain and the native side-chain are likely in close to the same position in the three-dimensional structure. We probe whether the engineered cysteine is accessible to our highly water soluble reagents, thereby determining whether it is on a water-accessible surface of the protein. An additional advantage of our approach is that we can determine whether a residue lines the binding-site crevice, even if mutation of the residue produces no functional change in the properties of the receptor.

We previously found that antagonist binding to wild-type D2 receptor was irreversibly inhibited by MTSEA and MTSET and that Cys 118, in the third membrane-spanning segment (M3), was responsible for this sensitivity (2). Therefore, we used the mutant in which Cys 118 was replaced by serine (C118S), which is insensitive to MTS reagents, as the starting point for further mutation. In our initial application of the substituted-cysteine accessibility method to the D2 receptor (3), we found that 10 of 23 residues tested in the M3 segment were exposed in the binding-site crevice (Fig. 1A). From the pattern of exposure, we inferred that M3 forms an α-helix, one side of which faces the binding-site crevice (Fig. 1B).

Unlike the positively charged MTSEA and MTSET, the negatively charged MTSES did not react with cysteines substituted for residues more cytoplasmic than Val 111. This is consistent with a negative electrostatic potential in the binding-site crevice, in part due to the negative charge of Asp 114. However, MTSES reacted with the mutants D108C, I109C, F110C, and V111C. At I109C, F110C, and V111C, 10 mM MTSES inhibited binding nearly as much as 1 mM MTSET. Because 1 mM MTSET and 10 mM MTSES are equireactive with simple thiols in solution, the rates of reaction of MTSET and MTSES with the residues located near the extracellular end of M3 are similar; this indicates that the electrostatic potential near these residues is not as negative as it is below Val 111.

Thirteen of 24 residues tested in the M5 segment of the D2 receptor were exposed in the binding-site crevice (4). Of the 13 exposed residues, 10 were

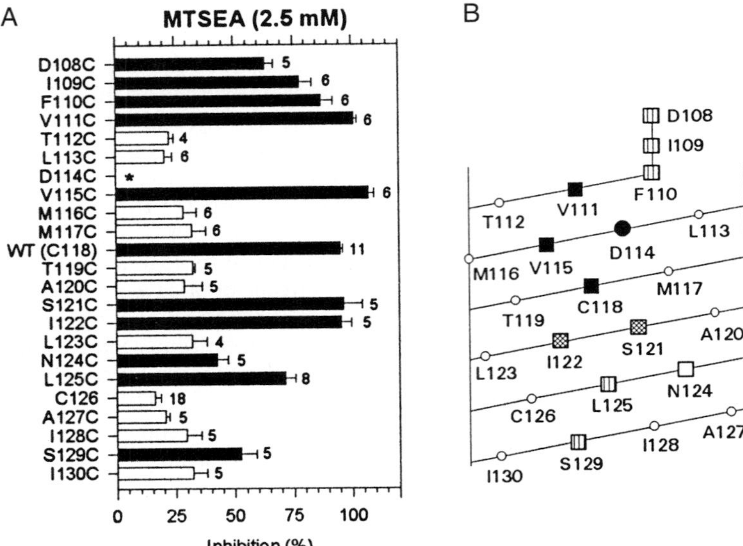

FIGURE 1 (*A*) The inhibition of specific [^{3}H]YM-09151-2 binding to intact cells transiently transfected with wild-type or mutant D2 receptors resulting from a 2-min application of 2.5 mM of MTSEA. The means and SEM are shown. The number of independent experiments for each mutant is shown next to the bars. Solid bars indicate mutants for which inhibition was significantly different ($p < .05$) than C118S by one-way ANOVA. In C118S, Cys 126 is the only cysteine present in M3, but it is insensitive to the MTS reagents. Cys 126 is, thus, present in all the mutants. WT, wild-type; *, no detectable binding. (*B*) Helical net representations of the residues in and flanking the M3 segment of the dopamine D2 receptor, summarizing the effects of MTSEA on [^{3}H]YM-09151-2 binding. Reactive residues are represented by squares, where the fill indicates the range of the second-order rate constants in $M^{-1}s^{-1}$ for reaction with MTSEA: solid squares, $k \geq 20$; hatched squares, $20 > k \geq 10$; striped squares, $10 > k \geq 3$; open square, $3 > k > 1$. Small open circles indicate that MTSEA had no effect on binding. The solid circle indicates no binding after cysteine substitution. D108 and I109 are represented outside of the α-helix in the loop from M2. (Reprinted with permission from *Neuron*, vol. 14, 825–836, 1995.)

consecutive. This pattern of exposure is inconsistent with the expectation that M5, like M3, forms a fixed α-helix, one side of which is exposed in the binding-site crevice. The exposed region of M5, which contains the serines likely to bind agonist, might loop out into the lumen of the binding-site crevice and be completely accessible to water and, thus, to MTSEA. Alternatively, the exposed region of M5 might be embedded in the membrane and also in contact with other membrane-spanning segments. At any instant, only a limited set of residues might be exposed in the binding-site crevice; however, M5 might move rapidly to expose different sets of residues.

Nine of 26 residues tested in the M7 segment reacted with the MTS reagents and were protected from reaction by the antagonist sulpiride (5). Again, the overall pattern of exposure is not consistent with a simple secondary structure of either α-helix or β-strand. M7 contains the extremely highly conserved residues Asn-Pro in the middle of the putative membrane-spanning segment. In soluble proteins, these residues have been observed to introduce kinks and twists in α-helices. The pattern of exposure of the cysteine substitution mutants to MTSEA can be explained if M7 is a kinked and twisted α-helix.

References

1. Akabas, M. H., Stauffer, D. A., Xu, M., and Karlin, A. (1992). Acetylcholine receptor channel structure probed in cysteine-substitution mutants. *Science* **258,** 307–310.
2. Javitch, J. A., Li, X., Kaback, J., and Karlin, A. (1994). A cysteine residue in the third membrane-spanning segment of the human dopamine D2 receptor is exposed in the binding-site crevice. *Proc. Natl. Acad. Sci. U.S.A.* **91,** 10355–10359.
3. Javitch, J. A., Fu, D., Chen, J., and Karlin, A. (1995). Mapping the binding-site crevice of the dopamine D2 dopamine receptor by the substituted-cysteine accessibility method. *Neuron* **14,** 825–831.
4. Javitch, J. A., Fu, D., and Chen, J. (1995). Residues in the fifth membrane-spanning segment of the dopamine D2 receptor exposed in the binding-site crevice. *Biochemistry* **34,** 16433–16439.
5. Fu, D., Ballesteros, J. A., Weinstein, H., Chen, J., and Javitch, J. A. (1996). Residues in the seventh membrane-spanning segment of the dopamine D2 receptor accessible in the binding-site crevice. *Biochemistry* **35,** 11278–11285.

Robert J. Lefkowitz, Julie Pitcher, Kathleen Krueger, and Yehia Daaka

Howard Hughes Medical Institute
Departments of Medicine and Biochemistry
Duke University Medical Center
Durham, North Carolina 27710

Mechanisms of β-Adrenergic Receptor Desensitization and Resensitization

A regulatory feature shared by many of the members of the superfamily of G protein–coupled receptors is that of desensitization. In response to prolonged or repeated agonist exposure, dampening of the signal transduction process is observed. Desensitization represents the summation of several different processes, including receptor phosphorylation, receptor sequestration (defined as the agonist-induced translocation of receptor away from the plasma membrane), enhanced degradation of intracellular messengers, and degradation of receptor protein. Rapid receptor desensitization, however, appears to be mediated by uncoupling of the receptor from its respective G protein, a consequence of receptor phosphorylation. In the case of the β_2-adrenergic receptor (β_2AR), phosphorylation by two distinct classes of serine–threonine kinases leads to receptor desensitization. The first class, the second-messenger–dependent kinases—cyclic adenosine monophosphate (cAMP)–dependent protein kinase (PKA) and protein kinase C—phosphorylate and directly uncouple β_2AR from G_s. Because these kinases phosphorylate receptors in an agonist-independent manner, this process permits cross-talk between receptor families (1).

The second class of enzymes, the G protein–coupled receptor kinases (GRKs), play a highly specialized role in receptor desensitization because only agonist-occupied receptors serve as substrates for these enzymes (1, 2). GRK phosphorylation of receptors does not directly inhibit the receptor–G protein interaction. Rather the GRK-phosphorylated receptor serves as a binding site for certain cytosolic proteins, members of the arrestin family. Binding of arrestin proteins sterically blocks β_2AR-mediated G protein activation. Thus, in the case of GRKs, the very signal that promotes activation of the G protein and the effector (i.e., ligand binding) also promotes the desensitization of that specific receptor. Once uncoupled from the G protein, receptor function can be restored only by receptor dephosphorylation, a process termed *resensitization*. The phosphatases and regulatory mechanisms involved in this resensitization process have only recently begun to be elucidated (discussion follows).

The cDNAs for six mammalian GRKs have been isolated. One feature common to all of the members of the GRK family is that they phosphorylate

Advances in Pharmacology, Volume 42

1054-3589/98 $25.00

membrane-incorporated receptor substrates. Efficient receptor phosphorylation thus requires membrane association of the kinase. Although the underlying mechanisms of membrane localization have been most extensively studied for RK and βARK, a common theme appears to be emerging among all the members of this family. The variable carboxyl-terminal domains of these enzymes appear to contain the structural information required for their membrane localization. These kinases, however, employ widely divergent mechanisms for affecting this membrane association.

The βARK-1–mediated phosphorylation of purified reconstituted β_2AR has previously been shown to be markedly enhanced in the presence of the $\beta\gamma$ subunits of heterotrimeric G proteins (G$\beta\gamma$). G$\beta\gamma$ addition does not directly activate the kinase but by virtue of the γ subunit, which is isoprenylated, promotes membrane association of the enzyme. Mapping of the G$\beta\gamma$-binding site of βARK defined a 125 amino acid residue domain, the distal end of which is located 19 residues from the carboxyl-terminus of the enzyme. Interestingly, the experimentally defined G$\beta\gamma$-binding domain maps to the carboxyl-terminus of the pleckstrin homology (PH) domain of this enzyme. PH domains are approximately 100 amino acid residue regions of protein homology found in numerous proteins involved in signal transduction processes.

The mapping of the G$\beta\gamma$ binding domain of βARK to the carboxyl-terminus of its constituent PH domain suggests one potential function for this domain: G$\beta\gamma$-mediated membrane association. Indeed, fusion proteins encompassing the PH domains of several other proteins, including those of phospholipase Cγ, oxysterol-binding protein, and guanine nucleotide–releasing factor, have been shown to bind to G$\beta\gamma$ subunits, although with varying affinities.

If the carboxyl-terminus of the PH domain of βARK binds G$\beta\gamma$, what role does the amino-terminus of this domain play? Following the observation that PH domains and lipid-binding molecules share structural similarities, it was demonstrated that fusion proteins encompassing the PH domains from several proteins, including βARK, bound specifically to phosphatidylinositol 4,5-bisphosphate (PIP_2). Furthermore, the amino-terminus of the PH domain was implicated as being the important site of lipid interaction. To assess any potential role for PIP_2 binding to the PH domain of βARK, purified β_2AR was reconstituted into vesicles composed of purified phosphatidylcholine (PC) in either the presence or absence of PIP_2. Somewhat surprisingly, in marked contrast to previous studies in which β_2AR was reconstituted in a heterogeneous lipid environment, when reconstituted in 100% PC no G$\beta\gamma$-mediated enhancement of βARK activity was observed. Indeed in a PC environment, the β_2AR represents a very poor substrate for this kinase in either the presence or absence of G$\beta\gamma$. Significant βARK-mediated β_2AR phosphorylation is observed only in the presence of both PIP_2 and G$\beta\gamma$, the presence of both PH domain–binding ligands being required for the effective membrane localization of the kinase. Because βARK-mediated phosphorylation of soluble substrates is unaffected in the presence of PIP_2 and/or G$\beta\gamma$, membrane targeting of the kinase appears to account for the enhanced receptor phosphorylation observed in the presence of these two PH domain–binding ligands (3).

The extent and duration of receptor desensitization depends not only on the activity of the GRKs, but also on the activity of the phosphatases responsible for reversing these phosphorylation events. Restoration of receptor function, resensitization, has been shown to occur rapidly on removal of agonist. Biochemical characterization of the phosphatase(s) responsible for the dephosphorylation of βARK-phosphorylated purified reconstituted β_2AR has revealed several unique features of this enzyme (4). Utilizing bovine brain homogenates as a source of phosphatase activity, the enzyme has been demonstrated to be (1) latent and (2) membrane associated. The membrane localization of this phosphatase places it in close proximity to its receptor substrate, and its latency suggests it represents a site of strict regulatory control. Significant dephosphorylation of βARK-phosphorylated β_2AR is observed only in the presence of activators of the PP-2A class of serine–threonine phosphatases; that is, activity is observed only after freezing and thawing phosphatase-containing fractions under reducing conditions or, alternatively, when assays are performed in the presence of protamine. That this enzyme is in fact a member of the PP-2A family was confirmed on the basis of its metal ion independence, insensitivity to the specific PP-1 inhibitor, I-2, and extreme sensitivity to the serine–threonine phosphatase inhibitor, okadaic acid (IC 50 = 0.1 nM). Following activation, this membrane-associated form of PP-2A dephosphorylates not only βARK-phosphorylated β_2AR, but also βARK-phosphorylated α_2C_2-adrenergic receptor and RK-phosphorylated rhodopsin. In stark contrast, PKA-phosphorylated β_2AR is not a substrate for this enzyme. In light of the properties of this enzyme, distinguishing it from previously characterized forms of PP-2A, coupled with its ability to dephosphorylate a number of GRK-phosphorylated receptor substrates, the enzyme is termed the G protein–coupled receptor phosphatase (GRP).

One critical question arising from the biochemical characterization of the GRP is the mechanism by which this latent phosphatase catalyzes dephosphorylation of GRK-phosphorylated receptor substrates *in vivo*. Resensitization of the β_2AR appears to require receptor sequestration, because sequestration-deficient mutants of this receptor or inhibitors of this sequestration process have been demonstrated to block receptor resensitization. These observations, coupled with the finding that vesicle-associated β_2ARs are less phosphorylated than their plasma membrane counterparts, have led to the proposal that receptor dephosphorylation occurs during sequestration into a vesicle population. Recent immunocytochemistry studies colocalize sequestered β_2AR with transferrin receptors, markers for endosomol vesicles. Because one of the characteristics of the endosomal pathway is vesicle acidification, could acidification be the signal promoting receptor dephosphorylation (5)? Indeed, *in vitro* GRP-mediated dephosphorylation of βARK-phosphorylated β_2AR is stimulated at acidic pH, with maximal dephosphorylation occurring near pH 5.0. In contrast, for model substrates such as PKA-phosphorylated casein more dephosphorylation occurs at pH 7.0 than at pH 5.0.

That acidification plays an important role in modulating the phosphorylation state of the β_2AR in cells was demonstrated by incubation of cells with ammonium chloride, an agent utilized to raise the pH of endocytic vesicles. Following exposure of β_2AR-transfected cells to isoproterenol, resulting in the agonist-specific phosphorylation of this receptor, ammonium chloride treatment dramatically inhibited receptor dephosphorylation. As much as 80% of the

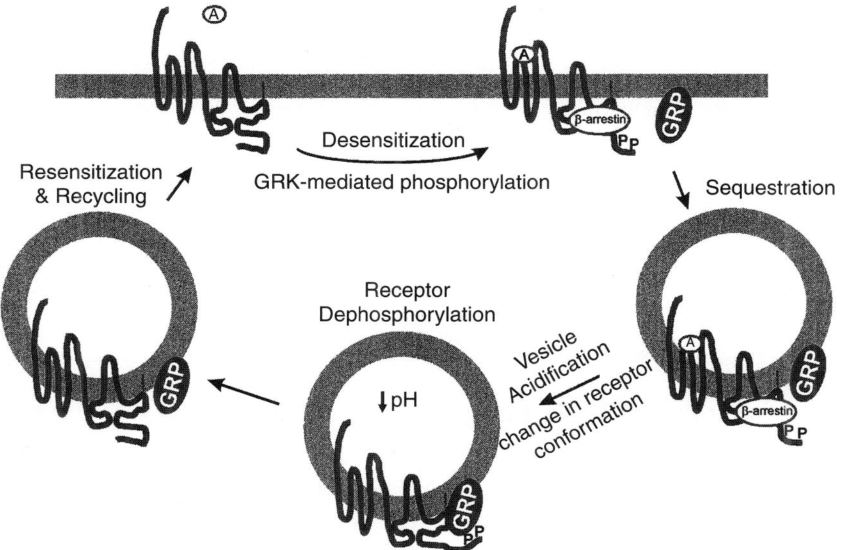

FIGURE 1 The role of sequestration in β_2AR resensitization. A, agonist; PP, sites of βARK phosphorylation on the carboxyl-terminus of the receptor (5). (Kreuger *et al.* Reprinted with permission from *The Journal of Biological Chemistry,* Vol. 272, p. 8, 1997.)

receptor dephosphorylation is inhibited after a 20-min incubation with ammonium chloride.

Moreover, it can be demonstrated that in vesicle but not plasma membrane fractions of cells, agonist stimulates an association of the PP-2A catalytic subunit with the receptor (as assessed by coimmunoprecipitation). This agonist-promoted association is essentially abrogated by ammonium chloride treatment of the cells.

Our current concepts of the role of receptor sequestration in receptor resensitization are summarized in Figure 1. Receptor phosphorylation is followed by β-arrestin binding which, as recently shown, leads to receptor internalization into clathrin-coated vesicles. As the pH drops within the vesicles, a conformational change in the receptor is induced that is transmitted across the vesicle membrane to its cytoplasmic surface where the phosphorylated domains of the receptor and the phosphatase are disposed. The conformationally altered receptor then associates with the phosphatase and is dephosphorylated. It subsequently recycles to the cell surface. Dephosphorylation of the GRK-phosphorylated β_2AR, like GRK-mediated phosphorylation, would, thus, appear to be highly dependent on receptor conformation.

References

1. Inglese, J., Freedman, N.J., Koch, W., and Lefkowitz, R. J. (1993). Structure and mechanism of the G protein-coupled receptor kinases. *J. Biol. Chem.* **268**, 23735–23738.

2. Premont, R. T., Inglese, J., and Lefkowitz, R. J. (1995). Protein kinases that phosphorylate activated G protein-coupled receptors. *FASEB J.* **9,** 176–182.
3. Pitcher, J. A., Touhara, K., Payne, E. S., and Lefkowitz, R. J. (1995). Pleckstrin homology domain-mediated membrane association and activation of the β-adrenergic receptor kinase requires coordinate interaction with G$\beta\gamma$ subunits and lipid. *J. Biol. Chem.* **270,** 11707–11710.
4. Pitcher, J. A., Payne, E. S., Csortos, C., DePaoli-Roach, A. A., and Lefkowitz, R. J. (1995). The G protein-coupled receptor phosphatase: A protein phosphatase type 2A with a distinct subcellular distribution and substrate specificity. *Proc. Natl. Acad. Sci. U.S.A.* **92,** 8343–8347.
5. Krueger, K. K., Daaka, Y., Pitcher, J. A., and Lefkowitz, R. J. (1997). The role of sequestration in G protein-coupled receptor resensitization: Regulation of β_2-adrenergic receptor dephosphorylation by vesicular acidification. *J. Biol. Chem.* **272,** 5–8.

Stephen S. G. Ferguson, Jie Zhang, Larry S. Barak, and Marc G. Caron

Howard Hughes Medical Institute and
Departments of Cell Biology and Medicine
Duke University Medical Center
Durham, North Carolina, 27710

Role of β-Arrestins in the Intracellular Trafficking of G-Protein–Coupled Receptors

G-protein–coupled receptors (GRs) transduce extracellular signals to the interior of cells via their interactions with heterotrimeric guanine nucleotide–binding regulatory proteins (G-proteins). G-protein α- and $\beta\gamma$-subunits in turn modulate the activity of a diverse variety of effector systems, which ultimately mediate the biological response to GR activation. Agonist activation of GRs is counteracted by intrinsic cellular mechanisms that turn off or dampen the agonist-generated signal, a phenomenon referred to as desensitization. Several processes potentially contribute to GR desensitization and include: receptor G-protein uncoupling, due to receptor phosphorylation; receptor sequestration, the internalization of plasma membrane–associated receptors; and downregulation, the loss of the total cellular complement of receptors due to reduced mRNA and protein synthesis, as well as increased lysosomal degradation.

The regulation GR signaling requires a coordinated and delicate balance between the processes governing receptor activation, desensitization, and resensitization. In the case of the β_2-adrenergic receptor (β_2AR), the most rapid

Advances in Pharmacology, Volume 42

1054-3589/98 $25.00

means of receptor desensitization is phosphorylation, which is achieved within seconds of agonist activation by either second-messenger–dependent protein kinases (e.g., cyclic adenosine monophophate [cAMP]–dependent protein kinase, PKA) or GR kinases (GRKs), which specifically phosphorylate the agonist-activated form of the receptor. GRK-mediated phosphorylation of agonist-activated β_2ARs promotes the binding of arrestin proteins (e.g., β-arrestin), which, when bound, further uncoupled the receptor (1).

While the mechanisms underlying receptor activation and desensitization have been fairly well characterized, until recently, the molecular mechanism(s) by which GR responsiveness can be re-established have remained unknown, except that receptor sequestration and dephosphorylation are likely involved in this process (2). However, recent studies now indicate that the same molecular intermediates required for receptor desensitization, GRK-mediated phosphorylation and β-arrestin binding, likely play an important role in receptor resensitization by serving as the signals and molecular determinants initiating β_2AR sequestration.

β_2AR sequestion occurs within minutes of agonist stimulation and is proposed to be the mechanism by which this receptor is resensitized. Agonist-dependent β_2AR sequestration involves the internalization of cell surface receptors to endosomes, where they are presumably dephosphorylated and recycled back to the plasma membrane as competent signaling receptors. The idea that sequestration plays a role in β_2AR resensitization originates from three experimental observations: (1) Following agonist stimulation, receptors isolated from the "light vesicular" ("sequestered") membrane pools exhibit lower stoichiometry of phosphorylation than receptors isolated from the plasma membrane. (2) Pharmacological treatments, which nonspecifically block receptor endocytosis, such as 0.45 M sucrose or concanavilin A, block receptor sequestration and resensitization without affecting desensitization. (3) Sequestration-defective receptor mutants that desensitize in response to agonist stimulation but do not resensitize have been described (2). It should be noted, however, that in the absence of specific inhibitors of β_2AR endocytosis, there is no direct evidence to suggest that sequestration is absolutely required for normal dephosphorylation and resensitization of wild-type β_2ARs. The ability to directly test the role of receptor sequestration in the re-establishment of normal receptor responsiveness has been hampered by the fact that the molecular mechanisms specifically directing β_2ARs for endocytosis has eluded characterization. However, the characterization of a sequestration-defective β_2AR mutant now suggests an important role for GRK-mediated phosphorylation and β-arrestin binding in the sequestration phenomenon.

The mutation of tyrosine residue 326 to an alanine in the highly conserved $NP(X)_{2,3}Y$ motif, found in the seventh transmembrane domain of the β_2AR, resulted in a sequestration-defective β_2AR mutant (Y326A), which maximally stimulated adenylyl cyclase and desensitized but did not resensitize. This mutant was also found to be impaired in its ability to serve as a substrate for GRK- but not PKA-mediated phosphorylation (3). GRK2-mediated rescue of Y326A mutant sequestration was dependent on intact sites for GRK- but not PKA-mediated phosphorylation (3). However, wild-type β_2ARs lacking putative sites for GRK-mediated phosphorylation still sequestered in HEK 293 cells, albeit poorly, in response to agonist stimulation. In addition, overexpression of a dominant-negative GRK2 (K220M) construct partially impaired both the phosphorylation and sequestration of wild-type β_2ARs (3).

Because β_2ARs lacking putative sites for GRK-mediated phosphorylation could sequester in response to agonist stimulation, it was clear that GRK-mediated phosphorylation was not an absolute requirement for sequestration. Rather, it suggested that phosphorylation either stabilized the conformation of the receptor required for sequestration or promoted the interaction of the receptor with some other cellular element(s) that could promote receptor sequestration, even in the absence of phosphorylation. Indeed, using the sequestration-defective β_2AR mutant (Y326A), we demonstrated that the essential element required for β_2AR sequestration was β-arrestin, the same molecular intermediate required for receptor desensitization (4). When overexpressed in HEK 293 cells, β-arrestins exhibited the capacity to rescue the sequestration of the Y326A mutant. In addition, β-arrestins facilitated the sequestration of receptor mutants lacking putative sites for GRK-mediated phosphorylation, clearly demonstrating that β_2AR sequestration can proceed in the absence of phosphorylation (4). However, GRK-mediated phosphorylation of the β_2AR, while not absolutely required for receptor sequestration, increased the affinity of the receptor for β-arrestin proteins, because β-arrestin–dependent rescue of Y326A sequestration was potentiated by the coexpression of low levels of GRK2 in HEK 293 cells (4). A role of β-arrestin in β_2AR sequestration was also demonstrated using a β-arrestin1 mutant (V53D), which specifically impaired receptor sequestration but not receptor desensitization and phosphorylation (4). Moreover, GRK phosphorylation itself was not sufficient to rescue Y326A mutant sequestration, because overexpression of GRK2 with the β-arrestin1–V53D mutant resulted in rescued phosphorylation but not sequestration (4).

These experiments clearly indicate that β-arrestins play a dual role in the regulation of β_2AR responsiveness: They not only bind and uncouple the agonist-activated GRK-phosphorylated β_2AR from its G protein, but also participate in directing the receptor for agonist-promoted sequestration. β-arrestins appear to act as adaptor-like proteins in β_2AR trafficking and either serve to recruit other cellular proteins that participate in the mobilization of receptors to endocytic organelles or execute this function themselves.

To better characterize the endocytic pathways utilized by the GR and to assess the specificity of the trafficking function of β-arrestin, we examined the effects of both β-arrestin and dynaminI on the agonist-promoted sequestration of two prototypic receptors, the β_2AR and the angiotensin II type 1A receptor (AT_{1A}R) (5). DynaminI is a large guanosine triphosphatase (GTPase), which is required for the pinching off of clathrin-coated vesicles from the plasma membrane. Dynamin mutants deficient in their GTPase activity cause clathrin-coated pits to accumulate at the plasma membrane and in doing so, prevent the clathrin-dependent endocytosis of plasma membrane–associated receptor proteins to endosomes. Overexpression of a GTP binding-defective dynamin mutant (K44A) in HEK 293 cells almost completely blocked the internalization of the β_2AR but not AT_{1A}R (5). This indicated that the internalization of these two receptor types was mediated by distinct endocytic mechanisms. Overexpression of the sequestration-specific dominant-negative β-arrestin–V53D mutant, like dynaminI-K44A, impaired β_2AR sequestration but had no effect on the ability of the AT_{1A}R to internalize in response to agonist stimulation (5). However, overexpression of the native β-arrestin protein increased the extent of AT_{1A}R sequestration by 50% (5). This β-arrestin–mediated increment in AT_{1A}R internalization could be blocked by coexpression of the dominant-negative dynamin mutant. In addition, AT_{1A}Rs but not β_2ARs sequestered well in COS7 cells (5). This impairment in COS7 cell β_2AR sequestration could be overcome by supplementing these cells with β-arrestin proteins (5). However, like that observed for the AT_{1A}R in both HEK 293 and COS7 cells, overexpression of dynaminI-K44A blocked the β-arrestin–dependent reversal of the β_2AR sequestration impairment (5).

These experiments indicate that, at least for the cell types tested, β_2AR sequesters by a clathrin-coated vesicle endocytic pathway, and the lack of dynamin-dependence of AT_{1A}R internalization suggests that GR internalization can occur via distinct pathways, even in the same cell type. The finding that β-arrestin overexpression can mobilize additional AT_{1A}Rs for dynamin-dependent endocytosis and can increase the dynamin-dependent sequestration of the β_2AR in COS7 cells indicates that β-arrestins likely serve as cellular trafficking molecules by specifically targeting GRs to clathrin-coated vesicles and suggests that there is plasticity in the choice of endocytic pathway utilized by GRs. This choice is dependent on not only receptor-specific determinants, but also the cellular milieu in which the receptor is expressed.

Taken together, the data presented here suggest a model for the role of β-arrestins in the regulation of β_2AR responsiveness (Fig. 1). Agonist activation of β_2AR leads to receptor–G-protein–coupling and the initiation of the biological

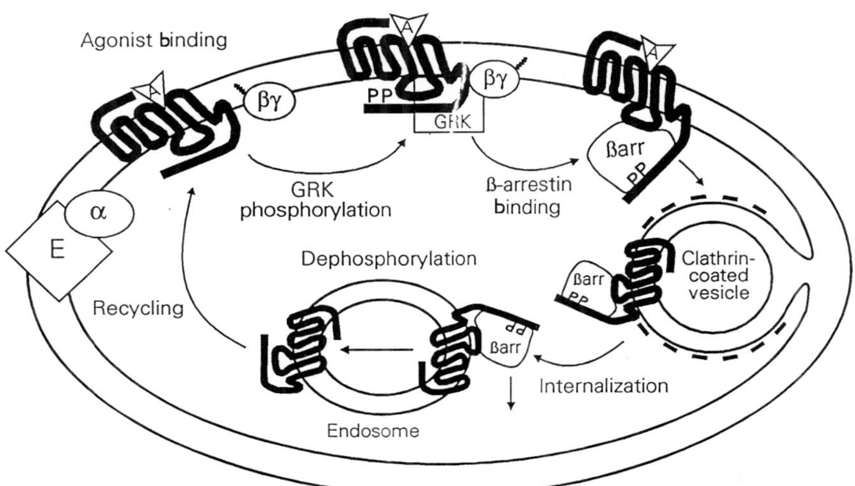

FIGURE 1 GR desensitization and resensitization following agonist activation. A, agonist, α, β, and γ, subunits of heterotrimeric G protein; E, effector enzyme; Barr, β-arrestin; PP, phosphate groups. See text for details.

response to receptor activation, which is followed by receptor desensitization as the consequence of receptor phosphorylation by second messenger–dependent protein kinases and GRKs. Agonist activation and GRK-mediated phosphorylation of the β_2AR promotes the binding of β-arrestin to the receptor, which not only serves to further uncouple the receptor from its G protein, but also mediates the mobilization of β_2ARs for dynamin-dependent clathrin-coated vesicle-mediated endocytosis. β_2ARs mobilized to endosomes are dephosphorylated, sorted, and subsequently recycled to the plasma membrane as functional receptors by mechanisms yet to be delineated. The identification of other components involved in β_2AR endocytosis, in particular those that associate with β-arrestins, will help provide a clearer understanding of the molecular mechanisms involved in receptor-mediated endocytosis, and thereby contribute to the regulation of GR responsiveness.

References

1. Lefkowitz, R. J. (1993). G protein-coupled receptor kinases. *Cell* **74**, 409–412.
2. Ferguson, S. S. G., Barak, L. S., Zhang, J., and Caron, M. G. (1996). G protein-coupled receptor regulation: Role of G protein-coupled receptor kinases and arrestins. *Can. J. Physiol. Pharmacol.* **74**, 1095–1110.
3. Ferguson, S. S. G., Ménard, L., Barak, L. S., Colapietro, A-M., Koch, W. J., and Caron, M. G. (1995). Role of phosphorylation in agonist-promoted β_2-adrenergic receptor sequestration. Rescue of a sequestration-defective mutant receptor by βARK1. *J. Biol. Chem.* **270**, 24782–24789.
4. Ferguson, S. S. G., Downey, W. E. III, Colapietro, A-M., Barak, L. S., Ménard, L., and Caron, M. G. (1996). Role of β-arrestin in mediating agonist-promoted G protein-coupled receptor internalization. *Science* **271**, 363–366.
5. Zhang, J., Ferguson, S. S. G., Barak, L. S., Ménard, L., and Caron, M. G. (1996). Dynamin and β-arrestin reveal distinct mechanisms for G protein-coupled receptor internalization. *J. Biol. Chem.* **271**, 18302–18305.

Vijaya Karoor and Craig C. Malbon

Department of Molecular Pharmacology
Diabetes and Metabolic Diseases Research Center
School of Medicine
SUNY Stony Brook
Stony Brook, New York 11794

G-Protein–Linked Receptors as Substrates for Tyrosine Kinases: Cross-Talk in Signaling

Growth factor receptors with intrinsic tyrosine kinase activity (TKR) and G-protein–linked receptors (GLRs) constitute the proximal elements of the dominant signaling pathways controlling cell proliferation, differentiation, and metabolism (1). Cross-talk and integration of signaling between GLRs and TKRs has been the focus of intense research. In 1992, it was shown that insulin treatment of DDT-1 hamster vas deferens smooth-muscle cells stimulated increased phosphotyrosine content of a GLR, the β_2-adrenergic receptor (β_2AR), as defined by phosphoamino acid analysis of β_2AR isolated from metabolically labeled cells (2). This was the first insight into a truly new dimension in cross-talk in cellular signaling at the most proximal point—a GLR acting as a substrate for protein phosphorylation by a TKR. The counterregulatory effects of insulin on catecholamine action, in particular the actions mediated by β_2AR, are well known, and biochemical evidence had revealed a primary locus of insulin action at the level of the GLR and not the more distal elements of G-protein–linked signaling (2).

Analysis of protein phosphorylation of low-abundance membrane proteins is a formidable task and requires the application of methods of protein chemistry operating in concert with metabolic labeling *in vivo* to define the sites of covalent modification. DDT-1 smooth-muscle cells, which express approximately 30,000 β_2AR per cell and display insulin-sensitive responses, were adopted as a model system. The first strategy developed to approach this task was to use synthetic peptides as probes of insulin receptor–catalyzed phosphorylation *in vitro* and, second, to use tryptic fragments of the synthetic peptides labeled *in vitro* as markers with which to follow the peptides derived from proteolytic digestion of β_2AR metabolically labeled and phosphorylated *in vivo* (3). Preferred substrate peptides for phosphorylation by insulin-stimulated, recombinant insulin receptor (rIR) were easily identified, providing a focus for more-detailed protein chemistry. The radiolabeled tryptic peptides provide valuable markers with which to identify sites of phosphorylation of β_2AR using reverse-phase high-performance liquid chromatography (HPLC), high-voltage electrophoresis–thin layer chromatography, and other separation procedures.

Advances in Pharmacology, Volume 42

1054-3589/98 $25.00

The dominant sites of protein phosphorylation of β_2AR *in vivo* in cells stimulated by insulin were confined to the C-terminal region of the receptor, although tyrosyl residues can be found in both the first and second (but not third) intracellular loops of this heptihelical membrane protein. The implications of these data are quite far-reaching, because the phosphorylation of β_2AR resulted in apparent uncoupling of the receptor from the stimulatory G-protein–controlling adenylylcyclase G_s and the phosphorylation was localized to the C-terminal tail of the β_2AR and not to the third intracellular loop that too clearly is involved in G-protein coupling. The biochemical data revealed the C-terminus to play an obligatory and important role in G-protein coupling. Thus, sites of protein phosphorylation associated with agonist-induced desensitization by protein kinase A and C in intracellular loop 3 were topologically distinct from the domain of the β_2AR acting as a substrate for insulin-stimulated phosphorylation *in vivo,* the C-terminal tail (3).

Mapping of the sites of insulin-stimulated phosphorylation in the C-terminal tail of the β_2AR revealed three tyrosyl residues of major import. Tyr 350, Tyr 354 and Tyr 364 are residues of β_2AR that are heavily phosphorylated in response to insulin. These sites were identified by both reverse-phase HPLC and two-dimensional peptide mapping studies (3). The peptide sequence harboring the Tyr 364 residue includes a recognition motif for a growth factor receptor tyrosine kinase. More revealing, phosphorylation of Tyr 350 creates a sequence motif (TyrP-Gly-Asn-Gly) with similarity to motifs known to interact with sem5 Src homology 2 (SH2) domains. This observation is seminal, because it reveals the ability of GLR to interact with a growing family of "adaptor" molecules well known in the TKR signaling literature, once the GLR has been phosphorylated on key tyrosyl residues by a TKR, or perhaps by a nonreceptor tyrosine kinase. In a simple but speculative model, insulin could express its counterregulatory effects on the β_2AR through phosphorylation of Tyr 350, which would allow for binding of the newly created SH2 domain by molecules like Grb2 and for steric blockade of β_2AR interaction with G_s (3).

Although the results from the peptide chemistry are exciting, it is crucial to be able to explore any suspected linkage of phosphorylation with the functional aspects of a GLR. Mutagenesis of the tyrosyl residues implicated in receptor G-protein coupling provided the avenue for this effort. The Tyr-to-Phe substitutions were performed and the mutant receptor expressed in cells largely deficient in β_2AR. In cells expressing the wild-type β_2AR, insulin counterregulates catecholamine-stimulated accumulation of intracellular cyclic adenosine monophosphate (cAMP). Mutagenesis of the Tyr350/Tyr354 residues attenuated sharply the ability of insulin to inhibit cAMP accumulation in response to the β_2AR agonist isoproterenol (3). Mutation of the Tyr 364 had a similar effect, attenuating the counterregulatory effects of insulin in the cells expressing these mutant receptors (3). In combination with results of the metabolic labeling *in vivo,* followed by fragmentation and peptide mapping of the β_2AR, the mutagenesis illuminates a central role of Tyr 350, Tyr 354 and Tyr 364 in insulin signaling. Analysis of the data in the GeneBank revealed many GLRs (that harbor tyrosyl residues embedded in one of the recognition motifs highlighted previously) as possible candidates for phosphorylation in response to stimulation by growth factors.

The linkage suspected by the study with synthetic peptides is that the GLR, in this case the β_2AR, is itself a substrate for phosphorylation by the TKR directly, not just an endpoint for a phosphorylation cascade initiated by a growth factor–like insulin. To establish whether there was a protein–protein interaction between the insulin receptor and β_2AR, reconstitution studies employing recombinant, resolved proteins *in vitro* are required. For the current purpose, rβ_2ARs expressed in baculovirus-infected Sf9 cells are used in combination with human rIR expressed in Chinese hamster ovary T cells. Purified rIR was able to phosphorylate the rβ_2AR in an insulin-sensitive manner. Insulin-like growth factor I (IGF-I) receptor also displayed the ability to phosphorylate the rβ_2AR *in vitro,* whereas PDGF receptor activation failed to demonstrate this capability (4). The ability of the reconstitution assays to demonstrate receptor–receptor interaction established that a GLR could act as a substrate for direct phosphorylation by a TKR, like the insulin or IGF-I receptors.

The ability of IGF-I both to counterregulate the stimulation of cAMP accumulation in response to isoproterenol and to stimulate phosphorylation of rβ_2AR in an *in vitro,* reconstituted system prompted an exploration of the site(s) of the β_2AR phosphorylated *in vivo* in response to IGF-I stimulation. Were the sites on GLR that are phosphorylated by the insulin receptor and IGF-I receptor the same, different, or overlapping? Recently, we have exploited the power of matrix-assisted laser desorption–ionization time-of-flight (MALDI TOF) mass spectrometry to analyze phosphopeptides derived from β_2AR isolated from metabolically labeled cells challenged with IGF-I (5). In combination with HPLC and two-dimensional peptide mapping of tryptic digests of the receptor, MALDI TOF mass spectrometry revealed the prominent sites of phosphorylation of the β_2AR in response to IGF-I to be Tyr 132/Tyr 141 residues (Table I). Interestingly, these sites are found in intracellular loop 2 of this heptihelical receptor, spatially distinct from those phosphorylated by the insulin receptor tyrosine kinase (Tyr 350, Tyr 354, and Tyr 364) and by the β_2AR kinase in the C-terminal region of the molecule.

The realization that phosphorylation of a GLR on specific tyrosyl residues can create recognition site(s) for binding of adaptor molecules and other signal-

TABLE I Identification by MALDI TOF Mass Spectrometry of Tryptic Phosphopeptides of the β_2AR Identifies Tyr 141 as the Site Phosphorylated in Response to IGF-I *in Vivo* as Compared with Tyr 350/Tyr 354 and Tyr 364 Phosphorylated in Response to Insulin

Major residues	*+ IGF-1*	*+ Insulin*
Tyr 141	919.4 (914)	
Tyr 141 + 2 Na+	964.0 (960)	
Tyr 350		1372.5 (1380)
Tyr 350/Tyr 354 + matrix + 1 Na^+		1700.6 (1704)
Tyr 364 + matrix + 2 Na^+		2186.0 (2181)

This table displays the mass values and identity of the tyrosyl residue–containing peptides of the β_2AR that compose the peaks identified in the MALDI TOF mass spectrum. Values in parentheses denote the expected molecular weight of the fragment.

ing elements was amplified by the demonstration of IGF-I–catalyzed phosphorylation of Tyr 132 and Tyr 141 residues of β_2AR. Phosphorylation of either Tyr 132 or Tyr 141 with flanking sequences YXXI and YXXL, respectively, creates a recognition domain for the binding of *Shc,* a key element in signaling by both TRK (insulin) and nonreceptor tyrosine kinases (e.g., *Src*). The extent to which GLRs not only employ adaptor molecules and other signaling elements with SH2 domains, and so on, in their functional repertoire, but also alter the signaling of interacting pathways that share these same elements remains to be elucidated. Clearly TKRs exert a subset of their actions via phosphorylation of GLR. It is interesting that both insulin and IGF-I counterregulate catecholamine action, but do so via targeted phosphorylation of sites on topologically distinct intracellular domains of GLRs. Phosphorylation of tyrosyl residues in either the second intracellular loop (via IGF-I) or the C-terminal tail (via insulin) sharply attenuates signaling from the β_2AR to a common element, the stimulatory G-protein of adenylyl cyclase G_s. To what extent phosphorylation alone, as opposed to creation of a recognition domain for an adapter molecule like G_{rb2} or *Shc* through phosphorylation, can explain the effects of growth factors on signaling via GLR remains an open question.

Acknowledgment

This work was supported by U.S. Public Health Service grant DK25410 from the National Institute of Diabetes, Digestive, and Kidney Disorders (NIDDK), National Institutes of Health.

References

1. Karoor, V., Shih, M., Tholanikunnel, B., and Malbon, C. C. (1996). Regulating expression and function of G-protein-linked receptors. *Prog. Neurobiol.* **48,** 555–568.
2. Hadcock, J. R., Port, J. D., Gelman, M. S., and Malbon, C. C. (1992). Cross-talk between tyrosine kinase and G-protein-linked receptors: Phosphorylation of β_2-adrenergic receptors in response to insulin. *J. Biol. Chem.* **267,** 26017–26022.
3. Karoor, V., Baltensperger, K., Paul, H., Czech, M. P., and Malbon, C. C. (1995). Phosphorylation of tyrosyl residues 350/354 of the β-adrenergic receptor is obligatory for counterregulatory effects of insulin. *J. Biol. Chem.* **270,** 25305–25308.
4. Baltensperger, K., Karoor, H., Paul, H., Ruoho, A., Czech, M. P., and Malbon, C. C. (1996). The β-adrenergic receptor is a substrate for the insulin receptor tyrosine kinase. *J. Biol. Chem.* **271,** 1061–1064.
5. Karoor, V., and Malbon, C. C. (1996). IGF-I stimulates phosphorylation of the β-adrenergic receptor *in vivo* on sites distinct from those phosphorylated in response to insulin. *J. Biol. Chem.* **271,** 29347–29352.

Oscar B. Goodman, Jr., Jason G. Krupnick, Francesca Santini, Vsevolod V. Gurevich, Raymond B. Penn, Alison W. Gagnon, James H. Keen, and Jeffrey L. Benovic

Department of Biochemistry and Molecular Pharmacology
Kimmel Cancer Institute
Thomas Jefferson University
Philadelphia, Pennsylvania 19107

Role of Arrestins in G-Protein–Coupled Receptor Endocytosis

One of the more pervasive of biological phenomena is the ability of a system to modulate its responsiveness in the presence of a continuous stimulus. This process, often termed *desensitization,* has been extensively characterized for hormonal signaling involving the β_2-adrenergic receptor (β_2AR) (1). β_2AR activation by catecholamines initiates a cascade of events that culminate in the cyclic adenosine monophosphate–dependent phosphorylation of multiple cell-specific target proteins. Within seconds to minutes after activation by agonist, β_2AR becomes phosphorylated by the β-adrenergic receptor kinase (βARK). β_2AR phosphorylation by βARK promotes the binding of another protein, termed *β-arrestin,* to the receptor, which effectively uncouples the β_2AR from the stimulatory G-protein and attenuates signaling. β_2AR uncoupling is rapidly followed by a loss, or sequestration, of cell surface β_2ARs into an intracellular compartment distinct from the plasma membrane. Recent studies suggest that β_2AR internalization may be important for receptor resensitization, via a process that involves dephosphorylation and recycling of the receptor back to the plasma membrane.

While the mechanisms involved in mediating G-protein–coupled receptor (GR) internalization remain poorly defined, recent studies support a role for receptor phosphorylation in this process. Overexpression of βARK in COS-7 or BHK-21 cells facilitates agonist-induced sequestration of the m2 muscarinic cholinergic receptor (m2AChR), while overexpression of a dominant-negative βARK reduces agonist-induced internalization. Similar studies have demonstrated that overexpression of βARK in HEK 293 cells rescues the sequestration of a β_2AR mutant (β_2AR-Y326A) that is defective in its ability to be sequestered, and that overexpression of dominant-negative βARK inhibits β_2AR sequestration. Interestingly, recent findings have revealed that overexpression of either β-arrestin or β-arrestin 2 (also termed *arrestin 3*) in HEK 293 cells is sufficient to restore the ability of β_2AR-Y326A to sequester (2).

Advances in Pharmacology, Volume 42

1054-3589/98 $25.00

There is general agreement that GRs are physically internalized into cells in an agonist-dependent manner, and that this process may occur by both clathrin- and non-clathrin-mediated processes. For β_2AR in particular, the available evidence supports receptor internalization predominantly through clathrin-coated pits (3). Based on the studies described, we hypothesized that arrestins, which bind directly to activated phosphorylated receptors (4), play a pivotal role in β_2AR internalization via their interaction with some component of the clathrin-coated pit. To explore this possibility, we examined whether arrestins interact with clathrin, the major structural protein of coated pits. In initial studies, *in vitro* translated radiolabeled β-arrestin and arrestin 3 were found to bind specifically to clathrin cages, while visual arrestin showed no appreciable binding (5). To determine whether β-arrestin and arrestin 3 interaction with clathrin occured directly, we assessed the binding of purified recombinant arrestins to clathrin in several assembled forms. β-Arrestin and arrestin 3, expressed in *E. coli* and purified to near homogeneity, bound specifically to pure clathrin cages, cross-linked clathrin cages, or intact clathrin-coated vesicles. Purified β-arrestin and arrestin 3 binding to clathrin cages occurred with high affinity, exhibiting half-maximal binding at 10 nM and 64 nM of clathrin, respectively. Moreover, arrestin 3 binding to clathrin was saturable with about 3–4 mol arrestin bound per mole of sedimented clathrin trimer. This stoichiometry is consistent with the triskelion shape of clathrin, composed of three heavy chains and three light chains. Collectively, these results and the established interaction of arrestins with the β_2AR provide compelling biochemical evidence for clathrin–arrestin interaction in the cell and its role in the β_2AR internalization process.

The ability of β-arrestin and arrestin 3 to bind both GRs and clathrin with high affinity suggests that nonvisual arrestins likely function as adaptor proteins to promote receptor localization in clathrin-coated pits. To address this question in intact cell studies, we assessed whether β_2ARs are localized in clathrin-coated pits in an agonist- and β-arrestin–dependent manner (5). COS-1 cells cotransfected with β-arrestin and Flag-tagged β_2AR were used in these studies. We visualized cell surface β_2ARs, without the background of any internalized receptors, by initially labeling fixed but unpermeabilized cells using anti-Flag antibodies. Clathrin-coated pits and vesicles were subsequently detected in the same cells following cell permeabilization using a specific anti-clathrin antibody. In the absence of agonist, an extremely fine-to-diffuse signal was obtained for β_2ARs on the cell surface, while a characteristic punctate pattern of clathrin-coated pits at the plasma membrane and greater intensity of Golgi clathrin in the perinuclear region were also observed. However, the distribution of the β_2ARs became punctate and largely coincident with that of plasma membrane clathrin-coated pits following a 10-min incubation with isoproterenol. Each of the β_2AR dots appeared coincident with a clathrin dot, demonstrating that a substantial proportion of the surface β_2ARs are localized in clathrin-coated pits in an agonist-dependent manner.

Because β-arrestin is thought to associate with the agonist-occupied phosphorylated β_2AR, we also investigated whether β-arrestin and β_2AR were colocalized. COS-1 cells coexpressing β-arrestin and Flag-tagged β_2AR were incubated in the presence or absence of isoproterenol for 10 min. Cell surface receptors were then labeled in intact cells, using anti-Flag antibodies, and total β-arrestin was detected using a C-terminal peptide antibody following cell permeabilization. While β-arrestin was diffusely distributed before addition of agonist, a distinct punctate pattern with substantial coincidence with surface β_2ARs was observed following agonist treatment. This demonstrates surface β_2ARs and β-arrestin colocalize in an agonist-dependent manner.

We next investigated whether β-arrestin is indeed localized in clathrin-coated pits. In these studies, COS-1 cells transfected with β-arrestin and Flag-tagged β_2AR were incubated in the presence or absence of isoproterenol for 10 min. Because coated pits contain both clathrin and the assembly protein AP-2, the cells were then double-stained for either β-arrestin and clathrin or β-arrestin and AP-2. Clathrin and AP-2 exhibit a distinctive punctate pattern that was not affected by the presence of isoproterenol. Following agonist treatment, β-arrestin was found in a punctate pattern that was essentially coincident with AP-2, a marker for plasma membrane coated pits, and with non-Golgi clathrin. Given the colocalization of surface β_2AR, β-arrestin, and plasma membrane clathrin and AP-2 following agonist challenge, we conclude that the β-arrestin–clathrin interaction demonstrated *in vitro* also occurs in intact cells in the presence of an activated receptor.

Much evidence has accumulated over the past decade implicating arrestins in the desensitization of GRs. Arrestins bind to the phosphorylated form of agonist-activated receptors and rapidly uncouple receptor G-protein interactions (1). Current evidence suggests that arrestin binding to GRs also promotes receptor internalization. Our studies suggest that in the absence of receptor activation, β-arrestin resides predominantly in the cytosol. Receptor activation promotes phosphorylation by a GR kinase (GRK), enhancing β-arrestin binding to the phosphorylated agonist-activated receptor, thereby uncoupling the receptor from further G-protein activation. The desensitized β_2AR, with β-arrestin bound to it, is now targeted for internalization, and on encountering a coated pit, β-arrestin interacts with the assembled clathrin lattice, resulting in selective endocytosis of the receptor. Because the affinity of β-arrestin binding to clathrin is weaker than the very low nanomolar affinity of β-arrestin binding to β_2AR, we favor a model in which β-arrestin initially binds to β_2AR before interacting with clathrin (Fig. 1).

It is likely that many other GR signaling systems utilize a similar mechanism of arrestin-promoted internalization of agonist-activated phosphorylated receptors through clathrin-coated pits. In addition to β_2AR, a number of other GRs have been shown to be phosphorylated in an agonist-dependent manner by GRKs. Moreover, the m2AChR, and β_1- and α_{1B}-adrenergic receptors have also been demonstrated to interact with arrestins. Interestingly, the domains of GRs shown to be critical for internalization overlap with the domains likely involved in interaction with GRK and arrestin. For example, mutagenesis of the C-terminal and/or third intracellular loop domains of the β_1-, β_2-, and α_{1B}-adrenergic, m1 and m3 muscarinic cholinergic, neurotensin, angiotensin IA and II, and parathyroid hormone receptors has been shown to diminish agonist-promoted sequestration. These same receptor domains have been implicated in GR interaction with GRKs and arrestins. Thus, these domains may be critical for receptor internalization because they modulate the ability of the receptor

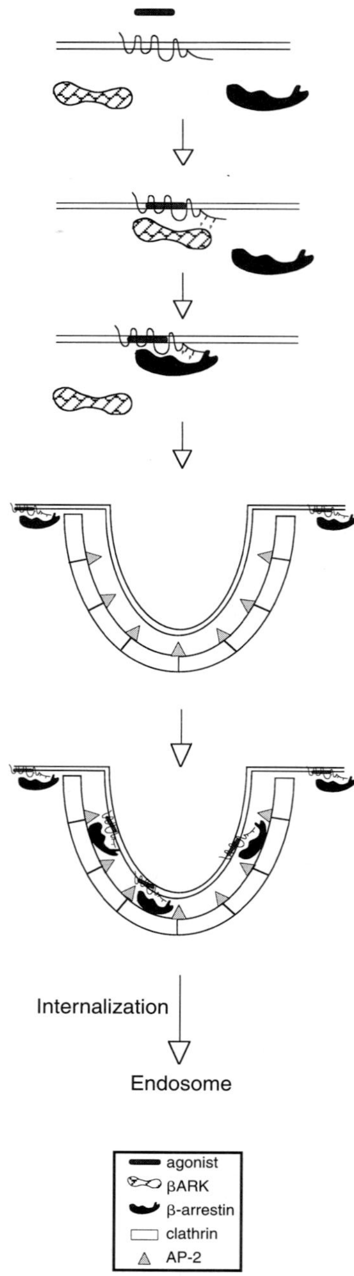

FIGURE 1 The agonist- and β-arrestin–dependent association of the β_2AR with clathrin-coated pits.

to interact with GRKs and/or arrestins. These findings suggest that agonist-dependent receptor phosphorylation, arrestin binding, and subsequent endocytosis may serve as a general mechanism for regulating many GRs.

References

1. Sterne-Marr, R., and Benovic, J. L. (1995). Regulation of G protein-coupled receptors by receptor kinases and arrestins. *Vitamins and Hormones* **51**, 193–234.
2. Ferguson, S. S. G., Downey, W. E., III, Colapietro, A-M., Barak, L. S., Menard, L., and Caron, M. G. (1996). Role of β-arrestin in mediating agonist-promoted G protein-coupled receptor internalization. *Science* **271**, 363–365.
3. von Zastrow, M., and Kobilka, B. K. (1994). Antagonist-dependent and independent steps in the mechanism of adrenergic receptor internalization. *J. Biol. Chem.* **269**, 18448–18452.
4. Gurevich, V. V., Dion, S. B., Onorato, J. J., Ptasienski, J., Kim, C. M., Sterne-Marr, R., Hosey, M. M., and Benovic, J. L. (1995). Arrestin interactions with G protein-coupled receptors: Direct binding studies of wild type and mutant arrestins with rhodopsin, β_2-adrenergic, and m2 muscarinic cholinergic receptors. *J. Biol. Chem.* **270**, 720–731.
5. Goodman, O. B., Jr., Krupnick, J. G., Santini, F., Gurevich, V. V., Penn, R. B., Gagnon, A. W., Keen, J. H., and Benovic, J. L. (1996) β-Arrestin acts as a clathrin adaptor in endocytosis of the β_2-adrenergic receptor. *Nature* **383**, 447–450.

Michel Bouvier and Guy Rousseau

Département de Biochimie et Groupe de Recherche sur le
Système Nerveux Autonome
Université de Montréal
Montréal, Quebec, Canada H3C 3J7

Subtype-Specific Regulation of the β-Adrenergic Receptors

Numerous studies over the past two decades have provided a wealth of information on the biochemical events leading to the transfer of information through β-adrenergic receptors (βARs). One of the most fascinating features of these signaling systems is their high degree of plasticity. This permits the cell to adapt to its environment and may play a major role in the sorting and integration of the information detected at the receptor levels. βARs, which belong to the family of G-protein–coupled receptors, have been shown to be the target of various regulatory processes. In particular, negative regulation and desensitization in response to sustained activation have been well documented.

Advances in Pharmacology, Volume 42

1054-3589/98 $25.00

Changes in the responsiveness of βARs are believed to contribute to the phenomenon of habituation and withdrawal syndrome associated with the use of β-adrenergic agonists and antagonists, respectively. Because βARs represent important pharmacological targets in the treatment of hypertension, angina pectoris, arrhythmias, heart failure, and obesity, analysis of the mechanisms leading to alterations of βAR responsiveness is essential, because it may lead to the identification of new means to normalize the receptor function.

I. β-Adrenergic Receptors and Their Signaling Pathways

βARs can be divided into three subtypes (β_1, β_2, β_3) that are coupled via G_s to the stimulation of the adenylyl cyclase, which is believed to mediate most of the biological actions of the three βARs through the activation of cyclic adenosine monophosphate (cAMP)–dependent protein kinase (PKA). However, it has also been proposed that $G\alpha_s$, following βAR stimulation, directly activates L-type calcium channel. It has also been suggested that the Na^+/H^+ antiporter is under the regulatory influence of β_2AR via its interaction with $G\alpha_{13}$. Whether the two other βAR isoforms can also interact with and activate $G\alpha_{13}$ remains to be determined. Potential coupling of the βARs to the mitogen activated protein kinase (MAPK) signaling pathway through the $\beta\gamma$ subunits has also recently been evoked.

The three βARs have different pattern of tissue expression. Although β_1AR and β_2AR are both expressed in the vast majority of tissues in mammals, their ratio of expression varies enormously. For example, in the heart, 70–80% of βAR belong to the β_1 subtype, whereas β_2AR largely predominates in the vasculature and the lungs. β_3AR shows a more restrictive expression pattern, being found almost exclusively in adipose tissue. However, functional expression of this subtype in human heart was recently suggested. Interestingly, this heart β_3AR was found to inhibit cardiac contractility. This contrasts with the well-characterized positive chronotropic and inotropic actions of β_1AR and β_2AR and raises questions about the signaling pathways involved in this subtype-selective response. Indeed, the physiological actions of β_1AR and β_2AR stimulation in the heart are classically believed to result from the cAMP-dependent phosphorylation of myocardial proteins involved in the positive contractile responses, such as L-type calcium channels and phospholambans. Clearly, alternative or complementary signaling pathway(s) will need to be invoked to explain the inhibitory action of heart β_3AR. Recent evidence also suggests subtype-specific signaling between β_1AR and β_2AR. Indeed, although convincing data show that, as β_1AR, β_2AR can promote positive chronotropic and inotropic responses in human ventricle, discrepancies between the efficacy of β_2AR to stimulate adenylyl cyclase and to modulate contractile function have raised questions regarding the precise relation between cAMP production and cardiac regulation. For example, in human heart, β_2AR was found to be responsible for most of the β-adrenergic–stimulated adenylyl cyclase activity while representing only 20% of the total βAR population and contributing only modestly to the inotropic and chronotropic responses. These observations can be partially explained by recent results obtained by us and others showing that human β_2AR couples more efficiently to the adenylyl cyclase than its β_1 counterpart (1–3). However, they also implicitly suggest that cAMP-independent βAR regulation of cardiac contractile function exists.

II. Molecular Basis of β-Adrenergic Receptor Regulation

One important regulatory mechanism of receptor function is known as agonist-induced desensitization and is characterized by the fact that the intensity of a response mediated by the receptor wanes over time despite the continuous presence of the stimulus. This has been best characterized for β_2AR. It is generally accepted that at least three distinct processes are involved: functional uncoupling, sequestration, and downregulation of the receptor. Uncoupling corresponds to a decreased receptor-mediated activation of the adenylyl cyclase with no change in receptor number or distribution. Sequestration consists of a cellular redistribution of β_2AR from the cell surface to an intracellular vesicular fraction. Downregulation refers to a loss of receptor sites and results in a decrease in total β_2AR number.

Phosphorylation of β_2AR has been shown to play an important role in rapid uncoupling. Two distinct protein kinases, the cAMP-dependent protein kinase (PKA) and the β-adrenergic receptor kinase (βARK), have been implicated. Using site-directed mutagenesis, we and others identified many of the phosphorylation sites involved. βARK-mediated phosphorylation of serine and threonine residues in the distal portion of the carboxyl terminus of the receptor has been shown to promote the association of the protein β-arrestin with β_2AR, thus inhibiting functional coupling of the receptor to G_s. In contrast, phosphorylation of β_2AR by PKA, does not favor the interaction of the receptor with β-arrestin, and the mechanisms by which it promotes functional uncoupling remain unknown. However, the two consensus sequences for PKA phosphorylation, one located in the third cytoplasmic loop and the other in the proximal portion of the carboxyl terminus of the receptor, are believed to play distinct roles. Our recent work suggests that in addition to phosphorylation, another postranslational modification, palmitoylation of its Cys 341, plays an important role in dictating the ability of β_2AR to couple to G_s. In fact, these studies have shown that palmitoylation modulates receptor function by controlling the accessibility of its carboxyl terminus to regulatory kinases.

Two distinct mechanisms contribute to the regulation of cell surface β_2AR concentration upon agonist stimulation. The first one is the rapid sequestration, which has been proposed to be a recycling pathway allowing uncoupled phosphorylated receptors to be internalized, dephosphorylated, and returned to the cell surface in an active conformation. Recent work suggested that binding of β-arrestin to the receptor plays a central role in sequestration. However, the identity of the receptor residues involved in the interaction with β-arrestin and in the promotion of agonist-induced sequestration remains elusive. Indeed, several studies have shown that mutation of βARK and PKA phosphorylation sites do not abolish sequestration, demonstrating that phosphorylation is not essential for sequestration. The second process regulating receptor density is the downregulation of receptor number, which occurs following longer-term stimulation and involves mechanisms acting both at the level of gene expression and at the level of intracellular receptor processing following internalization. Our studies and several others demonstrated that cAMP-mediated destabilization of β_2AR mRNA contributes to the agonist-induced downregulation of the receptor. Agonist-induced internalization and degradation of the receptor is also believed to contribute to the process. In this respect, we have identified potential molecular determinants of this intracellular processing. Mutation of Tyr 350 and Tyr 354 of β_2AR greatly impairs its ability to undergo agonist-induced downregulation. Similar motifs involving tyrosine residues have also been shown to play an important role in agonist-promoted endocytosis of other receptors, including the G-protein–coupled M2-muscarinic receptor.

III. Molecular Determinants of Subtype-Specific Regulation

As mentioned, most studies addressing βAR regulation were carried out in cell lines expressing the β_2 subtype. To a large extent, it was assumed that the same regulatory processes would apply similarly to the β_1 and β_3 subtypes. The results of our recent studies (see next paragraph) clearly challenge this notion. Indeed, we and others found β_3AR to be largely resistant to rapid desensitization resulting from functional uncoupling, while β_1AR was found to desensitize to a significantly lesser extent than β_2AR.

In all cell types studied, human β_3AR was found to be completely resistant to rapid agonist-promoted uncoupling and sequestration, while in the same surrogate cells, treatment with agonist caused rapid uncoupling and sequestration of β_2AR (4, 5). Because both uncoupling and sequestration involve regulatory processes taking place at the protein level, we reasoned that differences between the coding sequence of the two receptors were most likely responsible for their distinctive desensitization patterns. A comparison of their primary sequences reveals that none of the potential phosphorylation sites for PKA and βARK, located in the third cytoplasmic loop (i_3) and the carboxyl tail of β_2AR are conserved in β_3AR. To test the hypothesis that the lack of these phosphorylation sites is responsible for the absence of rapid desensitization, the i_3 and carboxy tail of β_3AR were substituted with those of β_2AR. This chimeric receptor was identified as β_3-3322, the number indicating the origin of the first, second, and third cytoplasmic loop and of the carboxyl tail. The introduction of these two β_2AR domains within the overall structure of β_3AR restores the occurrence of rapid agonist uncoupling but not the sequestration (4). However, the restoration of the uncoupling phenotype provided by β_2AR i_3 and the carboxyl tail was only partial. Indeed, a 15-min prestimulation with the agonist isoproterenol induces a 60% desensitization of wild-type β_2AR, while a maximum desensitization of only 35% is observed for β_2-3322. Therefore, additional chimeras, in which all the intracellular domains of β_3AR were substituted in various combinations for their β_2 counterparts, were constructed. The results obtained with 15 such chimeras revealed that the carboxyl tail, i_3, and i_2, added individually all partially restored the uncoupling phenotype and that their effects were additive (5). In fact, the β_3-3222 chimera was the only one to undergo a rapid desensitization almost identical to that of β_2AR. However, this combination did not restore agonist-promoted sequestration, suggesting that the molecular determinants of uncoupling and sequestration are different. An important finding of these studies is that in addition to the well-characterized phosphorylation sites of β_2AR i_3 and carobxyl tail, additional motif located in the receptor's i_2 also contributes to its agonist-promoted uncoupling.

The longer-term agonist-induced downregulation patterns of β_2AR and β_3AR were also found to differ. In all cells studied, downregulation of β_3AR was slower and never reached the levels observed for β_2AR (6). In fact, we found that for β_3AR, only regulation of the mRNA levels and not agonist-induced endocytosis and degradation of the receptor is involved in the long-term loss of binding sites. This contrasts with the additive contribution of these two processes for β_2AR downregulation. An interesting observation made in the course of these studies is that different extents of β_3AR downregulation can be induced by agonist treatment in different cell types. For instance, virtually no β_3AR downregulation could be observed in chinese hamster fibroblasts (CHW) cells, while a loss of more than 50% of the receptor was seen in mouse/ fibroblasts (LtK). This may reflect cell-specific differences in the efficacy of the mechanism regulating mRNA levels.

Although less marked than with β_3AR, significant differences between β_1AR and β_2AR regulation patterns were also observed (3, 7, 8). The results can be summarized by saying that the uncoupling, sequestration, and downregulation patterns of β_1AR are intermediary between those of β_2 and β_3. Sustained agonist stimulation led to β_1AR uncoupling but to a significantly lesser extent than that observed for β_2AR. Sequence analysis reveals that the PKA site located in the β_2AR carboxyl tail is not conserved in β_1AR. To determine whether this difference in the receptor carboxyl tail domain is responsible for the distinct uncoupling phenotype, a chimeric β_1/β_2AR was constructed. In this chimera, the carboxyl tail of β_1AR was substituted for that of β_2AR (β_1-1112). This substitution entirely restores a β_2-like agonist-promoted uncoupling, arguing that the lack of a PKA site in its carboxyl tail is responsible for the slower and more limited uncoupling pattern characteristic of β_1AR.

References

1. Levy, F. O., Zhu, X., Kaumann, A. J., and Birnbaumer, L. (1993). Efficacy of β_1-adrenergic receptors is lower than that of β_2-adrenergic receptors. *Proc. Natl. Acad. Sci. U.S.A.* **90**, 10797–10802.
2. Green, S. A., Holt, B. D., and Liggett, S. B. (1992). β_1 and β_2-adrenergic receptors display subtype-selective coupling to Gs. *Mol. Pharmacol.* **41**, 889–893.
3. Rousseau, G., Nantel, F., and Bouvier, M. (1996). Distinct receptor domains determine subtype-specific coupling and desensitization phenotypes for human β_1- and β_2-adrenergic receptors. *Mol. Pharmacol.* **49**, 752–760.

4. Nantel, F., Bonin, H., Emorine, L. J., Zilberfarb, V., Strosberg, A. D., Bouvier, M., and Marullo, S. (1993). The human β_3-adrenergic receptor is resistant to short term agonist-promoted desensitization. *Mol. Pharmacol.* **43,** 548–555.
5. Jockers, R., Da Silva, A., Strosberg, A. D., Bouvier, M., and Marullo, S. (1996). New molecular and structural determinants involved in β_2-adrenergic receptor desensitization and sequestration. *J. Biol. Chem.* **271,** 9355–9362.
6. Nantel, F., Marullo, S., Krief, S., Strosberg, A. D., and Bouvier, M. (1994). Cell specific downregulation of the β_3-adrenergic receptor. *J. Biol. Chem.* **269,** 13148–13155.
7. Suzuki, T., NGuyenne, T., Nantel, F., Bonin, H., Valiquette, M., Frielle, T., and Bouvier, M. (1992). Distinct regulation of β_1- and β_2-adrenergic receptors in Chinese hamster fibroblasts. *Mol. Pharmacol.* **41,** 542–548.
8. Zhou, X. M., Pak, M., Wang, Z., and Fishman, P. H. (1995). Differences in desensitization between human β_1- and β_2-adrenergic receptors stably expressed in transfected hamster cells. *Cell. Signal.* **7,** 207–217.

Stephen B. Liggett

Departments of Medicine and Pharmacology
University of Cincinnati College of Medicine
Cincinnati, Ohio 45267

Structural Determinants of α_2-Adrenergic Receptor Regulation

Desensitization is a common biological phenomenon observed in the signaling of G-protein–linked receptors, whereby the intensity of the cellular response wanes over time despite continuous exposure to agonist. Desensitization is critical for maintenance of homeostasis under normal physiologic conditions, may contribute to or act to compensate during pathologic states, and may limit the therapeutic effectiveness of administered agonists. While some general mechanisms appear to be common in G-protein–coupled receptor desensitization, studies have suggested that individual receptors or classes of receptors have evolved specific mechanisms for the process. The work herein describes the mechanisms responsible for short-term (minutes) agonist-promoted desensitization of the α_2-adrenergic receptors (α_2ARs).

Three human α_2AR subtypes have been identified and are classified pharmacologically as α_{2A}, α_{2B}, and α_{2C}. Sometimes these are referred to as α_2C10, α_2C2, and α_2C4, respectively, based on their human chromosomal localization. Our initial studies utilized the α_2C10 receptor recombinantly expressed in Chinese hamster fibroblasts (CHW-1102 cells). Agonist occupancy of the receptor in

Advances in Pharmacology, Volume 42

1054-3589/98 $25.00

these cells resulted in inhibition of forskolin-stimulated cyclic adenosine monophosphate (cAMP) levels (whole cell assays) or adenylyl cyclase activities (membrane assays) via coupling to G_i. After pre-exposure of cells for 5–30 min with the agonist epinephrine, followed by extensive washing, α_2C10 function was found to be reduced as manifested by an approximate fivefold shift in the EC50 for epinephrine-mediated inhibition (1). We found that this desensitization was maximal with saturating concentrations of agonist, was unaffected by blockade of receptor internalization, was blocked by the βAR kinase inhibitor heparin but not by PKA inhibitors, and was associated with receptor phosphorylation. A mutated receptor consisting of a deletion of a relatively large portion of the third intracellular loop that is rich in serine and threonine residues (Del257-332, Fig. 1A) was found to function normally but failed to undergo short-term agonist-promoted desensitization. Confirming the causal relationship between receptor phosphorylation and desensitization, this mutated receptor also failed to phosphorylate (1). These characteristics are all consistent with phosphorylation of the α_2C10 receptor by a G-protein–coupled receptor kinase (GRK), such as βARK, during agonist desensitization. Indeed, in a recombinant phospholipid vesicle system, βARK had been previously shown to phosphorylate purified receptor. In contrast, the mutated receptor that failed to phosphorylate nevertheless underwent desensitization after more prolonged (24-hr) agonist exposure. In both CHW (1) and CHO (2) cells, one of the major mechanisms involved in *long-term* desensitization was a decrease in the cellular content of G_i. In addition, in CHO cells, α_2C10 expression decreases by 30–40% (termed *downregulation*) during long-term agonist exposure, further contributing to the desensitization observed with these cells under these conditions (2). Thus, the mechanisms responsible for short- versus long-term agonist-promoted desensitization of α_2C10 are distinct and apparently not interdependent.

In additional studies, we expressed all three α_2AR subtypes in CHO cells and assessed agonist-promoted desensitization and phosphorylation (2). As before, the α_2C10 receptor underwent desensitization and phosphorylation. Interestingly, the α_2C4 receptor (human α_{2C}-subtype) displayed little desensitization or phosphorylation, despite high concentrations of agonist and exposure times of up to 1 hr. The α_2C2 receptor (human α_{2B}-subtype), on the other hand, did display agonist-promoted desensitization and phosphorylation, although the extent of desensitization was about one-half as much as α_2C10. To further understand these differences in desensitization between the subtypes, we have investigated the specific requirements for α_2C10 desensitization by phosphorylation using site-directed mutagenesis and recombinant expression (3). The mutants constructed are shown in Figure 1A. These included a smaller deletion mutation of the third loop, which was initially used to point toward this region as containing all the agonist-promoted phosphorylation sites, and variations on the EESSSS sequence, as shown. These constructs were cotransfected with βARK into COS-7 cells, and whole cell receptor phosphorylation studies were carried out in the absence and presence of 10 μM of the agonist UK14304 for 15 min. As shown in Figure 1B, substitution of all four serines with alanines resulted in no significant receptor phosphorylation. Successive elimination of each serine resulted in an approximate 25% decrease in phosphorylation, indicating that all four serines of wild-type α_2C10 are phosphorylated by βARK

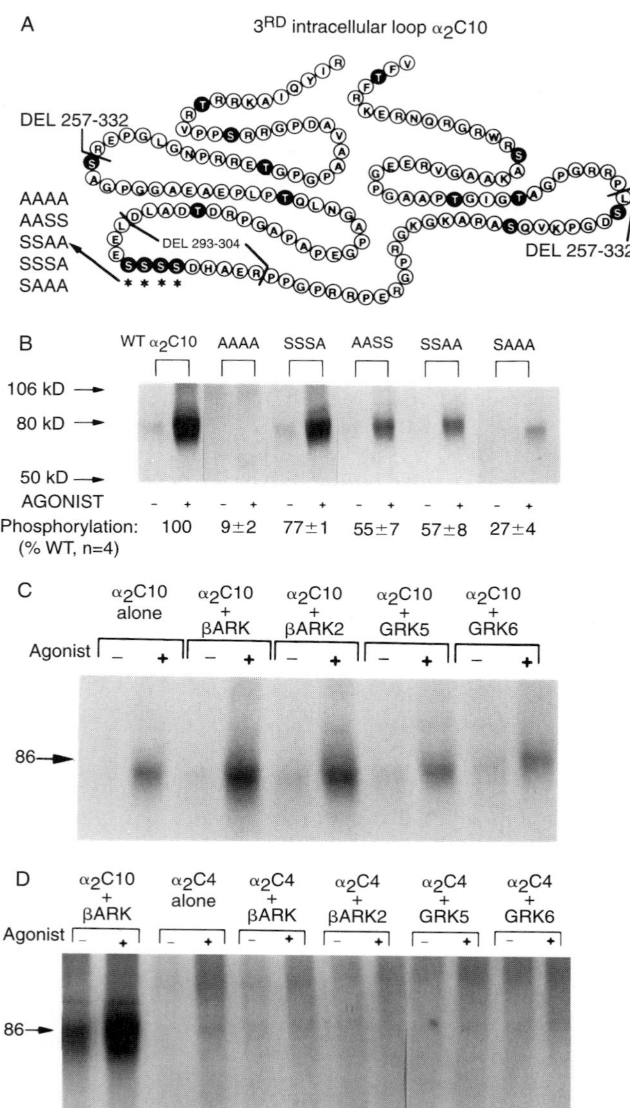

FIGURE 1 Agonist-promoted phosphorylation of α_2ARs. (*A*) The mutations discussed in the text. (*B*) The results of whole cell agonist-promoted phosphorylation of the various mutated receptors, coexpressed with βARK in COS cells. (Shown is a representative experiment with mean ± SE for four independent experiments). (*C, D*) Representative experiments of whole cell phosphorylation studies carried out with the α_2AR subtypes α_2C10 and α_2C4 in the presence of coexpression of the indicated GRKs in COS-7 cells. Modified from Eason *et al.* (3) and Jewell-Motz and Liggett (4). (Reprinted with permission from *JBC* **271,** 18082–18087, 1996.)

during agonist-promoted desensitization (3). Interestingly, in functional studies carried out in permanently transfected CHO cells, all of the mutated receptors failed to display short-term agonist-promoted desensitization. This suggests that a very precise conformation of the phosphorylated third loop is required for the binding of β-arrestin, which leads to receptor uncoupling. Thus, "partial" phosphorylation does not result in less desensitization but, in fact, none at all.

While these concepts with α_2C10 have been developed primarily around βARK being the GRK that phosphorylates this receptor, little is known about the substrate specificities of the various GRK isoforms for phosphorylation of α_2ARs. To further address this (4), COS-7 cells were cotransfected with α_2C10 or α_2C4, and either βARK (GRK2), βARK2 (GRK3), GRK5, or GRK6. Agonist-promoted phosphorylation in the whole cell setting was then assessed. The assumptions of these experiments, which are detailed elsewhere (4), are that these receptors were highly overexpressed, to the point that there was an insufficient amount of endogenous kinases to provide for phosphorylation of all receptors. Thus, the increase in phosphorylation that occurs with concomitant overexpression of a given GRK was interpreted as indicating that the transfected receptor was a substrate for that kinase. The amounts of cDNAs transfected were adjusted so that the same levels of expression of α_2ARs (as assessed by radioligand binding) and GRKs (as assessed by Western blots) were maintained. For α_2C10, βARK and βARK2 both enhanced agonist-promoted phosphorylation, while GRK5 and GRK6 did not (Fig. 1C). For α_2C4, no kinase was found to induce phosphorylation (Fig. 1D). To confirm these results in functional studies (4), the two receptor subtypes were overexpressed in HEK 293 cells, without or with coexpression of selected GRKs. Under these conditions, α_2C10 displayed no agonist-promoted desensitization in the absence of coexpression of a GRK. With coexpression of βARK, desensitization was observed which was manifested as a rightward shift in the dose-response curve for agonist-mediated inhibition of cAMP and a decrease in the extent of inhibition. In contrast, and in confirmation of the phosphorylation studies, coexpression of GRK6 had no effect on desensitization. Also consistent with the phosphorylation studies, α_2C4 desensitization could not be induced by coexpression of any GRK.

Two possible scenarios are plausible regarding the lack of desensitization–phosphorylation of α_2C4. One is that this receptor simply lacks the specific sites for GRK-mediated phosphorylation. Indeed, α_2C4 does not have an EESSSS sequence in the analogous region to α_2C10. However, there are several similar sequences in other locations within the third intracellular loop of α_2C4 that would appear to be reasonable GRK phosphorylation sites. This gives rise to the notion that perhaps the third-loop conformation that is induced or stabilized by agonist binding to α_2C4 is not favorable for GRK phosphorylation. To consider this latter possibility, we have constructed third-loop chimeric α_2C10/α_2C4 receptors. In preliminary experiments, we have found that an α_2C10 receptor with an α_2C4 third intracellular loop *does* undergo agonist-promoted phosphorylation. This supports the concept that the α_2C4 loop does contain GRK phosphorylation sites, but that in the context of the wild-type receptor, the binding of agonist does not achieve a favorable change in the conformation of the loop for GRK recognition or phosphorylation. Nevertheless, agonist binding to α_2C4 *is* sufficient to promote third-loop coupling to G_i. Thus, the information transferred by agonist interaction with the transmembrane-binding sites of α_2C4, which results in a conformational change in the third intracellular loop, is highly specific. So, despite the fact that both receptors functionally couple to G_i, the agonist-activated forms (R*) of α_2C10 and α_2C4 differ with respect to other agonist-promoted properties, such as GRK phosphorylation. (Interestingly, α_2C4 also does not sequester, which is another agonist-promoted function [2]). Thus R* might be considered as representative of several active states (R^{*1}, R^{*2}, R^{*3}, etc.). Further studies using complementary approaches are underway, but to date the evidence is supportive of this concept with these α_2AR subtypes. Another implication from these studies is the concept that synthetic α_2AR agonists that result in different third-loop conformations could, in essence, transmit different signals supporting some R* subsets but not others. Such agonists that do not evoke GRK phosphorylation, then, might be preferable, because one component of desensitization would be ablated.

Taken together, these studies have established a molecular basis for short-term agonist-promoted desensitization of α_2ARs, particularly the α_2C10-subtype. Such desensitization appears to be due entirely to phosphorylation of the receptor in the model systems examined to date. Phosphorylation and desensitization is receptor subtype–specific, being observed with α_2C10 and α_2C2 but not α_2C4. In addition, such desensitization is highly dependent on the expression of specific GRK isoforms, which is one factor in cell-type specificity of α_2AR desensitization. Finally, the conformation of the third intracellular loop, as induced by agonist binding to the transmembrane segments, also appears to be an important determinant of agonist-promoted desensitization.

References

1. Liggett, S. B., Ostrowski, J., Chesnut, L. C., Kurose, H., Raymond, J. R., Caron, M. G., and Lefkowitz, R. J. (1992). Sites in the third intracellular loop of the α_{2A}-adrenergic receptor confer short term agonist-promoted desensitization. Evidence for a receptor kinase-mediated mechanism. *J. Biol. Chem.* **267**, 4740–4746.
2. Eason, M. G., and Liggett, S. B. (1992). Subtype-selective desensitization of α_2-adrenergic receptors: Different mechanisms control short and long term agonist-promoted desensitization of α_2C10, α_2C4 and α_2C2. *J. Biol. Chem.* **267**, 25473–25479.
3. Eason, M. G., Moreira, S. P., and Liggett, S. B. (1995). Four consecutive serines in the third intracellular loop are the sites for βARK-mediated phosphorylation and desensitization of the α_{2A}-adrenergic receptor. *J. Biol. Chem.* **270**, 4681–4688.
4. Jewell-Motz, E. A., and Liggett, S. B. (1996). G-protein-coupled receptor kinase specificity for phosphorylation and desensitization of α_2-adrenergic receptor subtypes. *J. Biol. Chem.* **271**, 18082–18087.

Kevin D. Burris, Susanne M. Fausing, and Perry B. Molinoff
CNS Drug Discovery
Bristol–Myers Squibb Pharmaceutical Research Institute
Wallingford, Connecticut 06492

Regulation of D2 and D3 Receptors in Transfected Cells by Agonists and Antagonists

Changes the density and properties of receptors following exposure to agonists and antagonists is a phenomenon that occurs for many types of receptors that couple to guanine nucleotide–binding proteins (G-proteins). Exposure to agonists has most often been associated with a decrease in the density and sensitivity of receptors, whereas exposure to antagonists *in vivo* induces the opposite compensatory responses.

Dopamine D2 receptors in the central nervous system have been implicated in the regulation of motor function, as well as numerous psychobiological processes, including cognition, affect, and reward. The therapeutic actions of many drugs currently used to treat Parkinson's disease and schizophrenia are believed to involve activation or inhibition, respectively, of one or more subtypes of dopamine receptors.

In multiple studies with animals exposed to antagonists of dopamine receptors, an increase in the density of D2 receptors has been observed. Consistent effects of agonists on receptors have not been observed, although exposure of laboratory rats to L-dopa has been reported to result in supersensitivity of the behavioral effects of dopamine receptor agonists (1).

Investigation of the molecular mechanisms that underlie the regulation of D2 receptors in the brain has been hampered by the existence of multiple subtypes of D2-like receptors (i.e., D2L, D2S, D3, D4), which are difficult to distinguish given the lack of selective ligands. Studies of the mechanisms whereby agonists and antagonists regulate the density and functional properties of receptors have often utilized mammalian cells transfected with DNA that encode one or more subtypes of receptor. Cell lines grown in culture offer the advantage of a homogenous population of receptors, better control of the receptor microenvironment, and easier biochemical and genetic manipulation.

Exposure of HEK 293 cells expressing transfected D2L receptors (HEK-D2L cells) to sulpiride, an antagonist at D2 receptors, results in a time- and concentration-dependent increase in the density of receptors measured with the antagonist [^{125}I]NCQ-298 (2, 3). The ability of antagonists to stimulate upregulation of D2L receptors appears to be a general property of D2 receptor antagonists, because a variety of structurally distinct antagonists, including

1054-3589/98 $25.00

haloperidol, clozapine, and epidepride, also stimulate an increase in the density of receptors (2). Antagonists at D1 receptors (e.g., SCH-23390) have no effect on the density of receptors. Surprisingly, exposure of HEK-D2L cells to agonists (e.g., dopamine, quinpirole, and propylnorapomorphine) also results in an increase in the density of receptors (2). Exposure of cells to agonists and antagonists does not result in an increase in the levels of transfected $\beta 2$ receptors. This suggests that the effects of agonists and antagonists are not the result of a general increase in protein synthesis or activation of the promoter used to drive expression of transfected D2 and $\beta 2$ receptors (2, 3).

A variety of mechanisms that lead to an increase in receptor synthesis and/or decreased receptor degradation could mediate the increase in the density of receptors stimulated by agonists and antagonists. Pretreatment of HEK-D2L cells with cycloheximide, an inhibitor of protein synthesis, blocks the increase in density stimulated by agonists and antagonists (3, 4).

Furthermore, examination of the recovery of D2L receptors following inactivation by 1-ethoxycarbonyl-2-ethoxy-1,2-dihydroquinoline (EEDQ) suggests that an increase in the rate of synthesis of receptors, rather than a decrease in the rate of degradation, underlies the increase in receptor density stimulated by either agonists or antagonists (3). The apparent increase in receptor synthesis does not appear to require an increase in the synthesis of mRNA because actinomycin D does not antagonize the ability of agonists and antagonists to stimulate an increase in the density of receptors (3).

Taken together, these observations suggest that the increase in the density of receptors may involve an increase in translation of mRNA that encodes for D2L receptors or a posttranslation modification of receptors that facilitates recruitment of existing receptors into cell membranes.

D2L receptors in HEK 293 cells are coupled to inhibition of adenylyl cyclase activity through pertussis toxin–sensitive G-proteins (3, 4). The agonist [^{125}I]-7-OH-PIPAT labels a G-protein–coupled state of D2L receptors. In contrast to the increase in the density of binding sites for the antagonist [^{125}I]NCQ 298, exposure of HEK-D2L cells to agonists like quinpirole results in a time- and concentration-dependent decrease in the density of binding sites for [^{125}I]-7-OH-PIPAT (4). Furthermore, exposure of cells to quinpirole results in desensitization of D2L receptor–mediated inhibition of forskolin-stimulated cyclic adenosine monophosphate (cAMP) accumulation. In contrast to the effects of agonists, exposure of HEK-D2L cells to the antagonist sulpiride results in a time-and concentration-dependent increase in the density of binding sites for [^{125}I]-7-OH-PIPAT. However, the increase in the density of receptors coupled to G-proteins is not associated with an increase in the potency or efficacy with which quinpirole inhibits cAMP accumulation stimulated by forskolin (4). Quinpirole stimulates an increase in the density of D2L receptors in cells exposed to pertussis toxin at concentrations that ribosylate G-proteins and eliminate agonist-mediated inhibition of forskolin-stimulated cAMP accumulation (3, 4). The inability of pertussis toxin to block the effect of quinpirole suggests that the mechanism whereby agonists stimulate an increase in the density of D2 receptors does not involve a G_i/G_o-linked signal transduction pathway. Furthermore, the observation that both agonists and antagonists stimulate an increase in the density of receptors, though they have opposing effects on the activity of adenylyl cyclase, suggests that activation of D2 receptors and alterations in the coupling of receptors to G-proteins do not directly mediate the increase in receptor density.

D3 receptors are structurally and pharmacologically similar to D2 receptors. As seen with D2 receptors, exposure of cells that express transfected D3 receptors to agonists and antagonists results in an increase in the density of receptors (5) (Fig. 1, inset). The mechanisms whereby agonists and antagonists stimulate an increase in the density of D3 receptors are unknown. Cycloheximide blocks the increase in the density of D3 receptors that occurs following exposure of cells to the agonist NPA, suggesting that new protein synthesis is required for drug-induced upregulation of D3 receptors (5). However, exposure

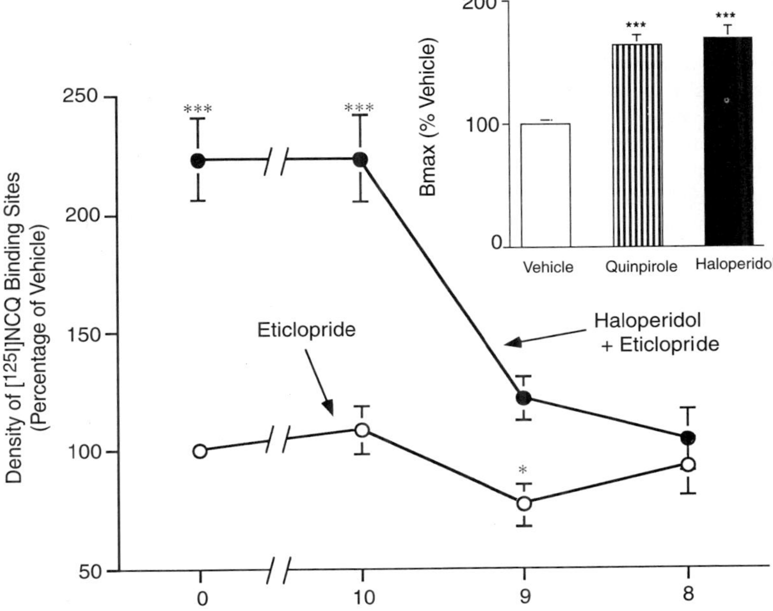

FIGURE 1 Effect of (-)eticlopride on the increase in the density of D3 receptors stimulated by haloperidol. HEK-D3 cells were exposed for 18 hours to increasing concentrations of (-)eticlopride in the absence and presence of 10 nM haloperidol. Bmax values for the binding of [^{125}I]-NCQ-298 were determined from linear transformation of saturation binding data. Results are expressed as a percentage of the average Bmax obtained from vehicle-treated cells. Results shown are the mean ± S.E.M. of seven to thirteen separate tissue preparations assayed in triplicate. The Bmax for vehicle-treated cells was 442 ± 30 fmol/mg protein, n = 13. [^{125}I]-NCQ-298 bound with a Kd of 192 ± 11 pM, n = 13 (vehicle), 154 ± 13 pM, n = 8 (exposure to haloperidol) and 215 ± 8 pM, n = 8 (exposure to 1 nM (-)eticlopride) to D3 receptors on membranes prepared from HEK-D3 cells. *P < 0.05 and ***P < 0.001 compared to vehicle using Student's t-test. *Inset,* Effects of haloperidol and quinpirole on the density of D3 receptors. HEK-D3 cells were exposed for 18 hours to vehicle, 10 nM haloperidol and 5 μM quinpirole. Results are expressed as a percentage of the average Bmax obtained from vehicle-treated cells. Results shown are the mean ± S.E.M. of ten to twenty separate tissue preparations asayed in triplicate. ***P < 0.001 compared to vehicle using Student's t-test.

of cells to NPA does not stimulate a significant increase in levels of mRNA encoding D3 receptors. As with D2 receptors, pertussis toxin does not block the ability of agonists to stimulate an increase in the density of D3 receptors (5).

In contrast to effects seen with cells that express D2 receptors, some antagonists do not stimulate an increase in the density of D3 receptors. In HEK-D3 cells, haloperidol but not eticlopride stimulated an increase in the density of D3 receptors (see Fig. 1). The lack of effect of eticlopride was not due to failure of eticlopride to interact with D3 receptors because eticlopride blocked the increase in the density of receptors stimulated by haloperidol (see Fig. 1). Two other benzamides, sulpiride and epidepride, did not stimulate an increase in the density of D3 receptors in transfected C6 glioma cells (5). However, epidepride blocked the increase in the density of receptors stimulated by quinpirole. The observation that some antagonists stimulate, whereas other antagonists block, an increase in the density of D3 receptors demonstrates that antagonists vary, perhaps through induction of different receptor conformations, in their interaction with and ability to modulate D3 receptors. The mechanisms whereby differences in receptor conformations lead to an increase in receptor density may involve posttranslational modification of receptors. For example, phosphorylation by members of multiple families of protein kinases has been implicated in the regulation of many subtypes of G-protein–coupled receptors. Preliminary evidence using activators and inhibitors of protein kinase C (PKC) suggests involvement of PKC in the regulation of D3 receptors by antagonists. Exposure of HEK-D3 cells to either haloperidol or phorbol-12-myristate-13-acetate (PMA), an activator of PKC, resulted in an increase in the density of D3 receptors. The increase in the density of receptors following exposure to PMA or haloperidol was attenuated by exposure of cells to sulpiride and two different inhibitors of PKC, bisindolylmaleimide I and calphostin C. The effects of these drugs on D3 receptors in the brain remain to be determined.

In conclusion, exposure to agonists and antagonists stimulates an increase in the density of D2 and D3 receptors. Understanding the mechanisms whereby agonists and antagonists regulate D2 and D3 receptors may help to explain the emergence of dystonias associated with exposure to dopamine receptor agonists and dyskinesias that follow long-term administration of dopamine receptor antagonists.

References

1. Jenner, P., and Marsden, C. D. (1987). Chronic pharmacological manipulation of dopamine receptors in brain. *Neuropharmacology* **26**, 931–940.
2. Filtz, T. M., Artymyshyn, R. P., Guan, W., and Molinoff, P. B. (1993). Paradoxical regulation of dopamine receptors in transfected 293 cells. *Mol. Pharmacol.* **44**, 371–379.
3. Filtz, T. M., Guan, W., Artymyshyn, R. P., Pacheco, M., Ford, C., and Molinoff, P. B. (1994). Mechanisms of up-regulation of D2L dopamine receptors by agonists and antagonists in transfected HEK-293 cells. *J. Pharmacol. Exp. Ther.* **271**, 1574–1582.
4. Boundy, V. A., Pacheco, M. A., Guan, W., and Molinoff, P. B. (1995). Agonists and antagonists differentially regulate the high affinity state of the D2L receptor in HEK 293 cells. *Mol. Pharmacol.* **48**, 956–964.
5. Cox. B. A., Rosser, M. P., Kozlowski, M. R., Duwe, K. M., Neve, R. L., and Neve, K. A. (1995). Regulation and functional characterization of a rat recombinant dopamine D3 receptor. *Synapse* **21**, 1–9.

David R. Sibley, Ana L. M. Ventura, Dong Jiang, and Chun Mak

Molecular Neuropharmacology Section
Experimental Therapeutics Branch
National Institute of Neurological Disorders and Stroke
National Institutes of Health
Bethesda, Maryland 20892

Regulation of the D1 Dopamine Receptor through cAMP-Mediated Pathways

D1 dopamine receptors are known to be dynamic entities that are subject to various forms of regulation (1). Perhaps the best studied of these is that of agonist-induced regulation that has been shown to involve both alterations in the expression of as well as the functionality of the receptor protein. Although the mechanisms underlying agonist-induced regulation have not been completely elucidated, evidence has accumulated implicating two major pathways, both of which involve protein phosphorylation (1). The first of these is the G-protein receptor kinase pathway, while the second is the cyclic adenosine monophosphate (cAMP)–dependent protein kinase (PKA) pathway. The latter pathway represents a typical negative feedback loop and has been studied extensively in our laboratory. This chapter briefly describes some of our more recent findings in this area.

As an initial approach to investigate the role of the PKA in the desensitization and downregulation of the D1 dopamine receptor, we stably expressed the rat cDNA for this receptor in mutant Chinese hamster ovary (CHO) cell lines deficient in PKA activity (2). The 10260 mutant CHO cell line has been characterized as expressing less than 10% of type I and type II PKA activities relative to the parental 10001 CHO cell line. The 10248 mutant CHO line lacks type II PKA activity but expresses low levels of type I PKA. The transfected parental and mutant cell lines were found to express between 1 and 3 pmol/mg protein of D1 receptor–binding activity (Bmax) as determined using [^{3}H]SCH 23390. All of the cell lines exhibited identical affinities for the radioligand. All three cell lines also demonstrated similar levels of dopamine-stimulated adenylyl cyclase activity with similar EC50 values for dopamine for producing this response. These results indicate that the mutant PKA isozymes have no effect on the expression of the D1 receptor or on its ability to couple to cAMP generation.

Pretreatment of all three CHO cells with dopamine (10^{-5} M for 18 hr) resulted in desensitization of the dopamine-stimulated adenylyl cyclase response. However, the maximum degree of desensitization varied among the three cell lines. In the parental cell line, dopamine stimulation of cAMP genera-

Advances in Pharmacology, Volume 42

1054-3589/98 $25.00

tion was nearly abolished subsequent to the treatment. In the 10260 cell line, the maximum extent of cAMP generation induced by dopamine was reduced by about 75%, whereas in the 10248 line, this activity was reduced only by about 50%.

Dopamine pretreatment also promoted, in a time- and dose-dependent fashion, greater than 90% downregulation of D1 receptor–binding activity in the parental cell line, but only about a 50% decrease in binding in the 10260 line and only a 25% decrease in the 10248 line. This reduction was due to decreased levels of receptor expression (Bmax). There was no change in affinity for the radioligand observed in any of these treatments.

Because dopamine treatment may be activating more than one regulatory pathway, we wished to assess the effects of selectively stimulating only the PKA-mediated pathway. To do this, we used treatment paradigms that increased intracellular levels of cAMP without activating the D1 receptor. Similar regulatory phenomena were observed on treatment of the cells with CPT-cAMP and 8-Bromo-cAMP, membrane-permeable analogues of cAMP. Treatment of the cells with CPT-cAMP (1 mM for 18 hr) decreased the maximum receptor binding activity (Bmax) in the parental cell line by about 50%; however, the decrease was smaller for the 10260 mutant line (about 25%) and not significant ($< 5\%$) with the mutant 10248 cells. Similar to the dopamine treatments, there was no change in the affinity of the radioligand. Similar results were obtained by treating the cells with forskolin, which indirectly elevates intracellular cAMP levels. Treatment with forskolin (10^{-6} M for 18 hr) downregulated the D1 receptors by about 50% in the parental cells, by about 25% in the 10260 mutant cells, and by about 10% in the 10248 cell line. It should be noted that the extent of receptor downregulation induced by the cAMP analogues or forskolin is less than that induced by dopamine. This suggests that more than one pathway exists for agonistinduced down-regulation of the D1 receptor (only one of which is mediated by cAMP).

Treatment of the cells with CPT-cAMP (1 mM for 18 hr) was also found to promote a functional desensitization of the D1 receptor. In the parental cells, the maximum extent of dopamine-stimulated cAMP accumulation was reduced by about 60%; in the 10260 cell line, this activity was reduced by about 40%; whereas in the 10248 line, the activity was down by about 15%. These results quantitatively agree well with the extent of receptor downregulation observed in these treatments. It should also be noted that the extent of desensitization induced by the cAMP analogues is less than that induced by dopamine, again suggesting that dopamine is inducing multiple pathways (cAMP- and non-cAMP-mediated) for regulating D1 receptor function and expression.

In summary, our studies with the mutant CHO cell lines have suggested the following conclusions. First, cAMP-dependent PKA appears to play a role in agonist-induced desensitization and downregulation of the D1 receptor in CHO cells. Second, because the attenuation of receptor desensitization–downregulation was greater in the 10248 cells than in the 10260 line, PKA II may play a greater role than PKA I in regulating D1 receptor function.

Because the effects of PKA in promoting regulation of the D1 receptor could be occurring at multiple levels, including phosphorylation of the receptor protein or phosphorylation of other components regulating receptor expression and function, we wished to directly assess the effect of direct phosphorylation of the receptor protein. To investigate this possibility, we used site-directed mutagenesis techniques to alter each of the four potential PKA sites to determine the effects on agonist-induced regulation. Using polymerase chain reaction–based methods, we created a mutant D1 receptor with the following amino acid substitutions: Thr 135 → Val 135, Ser 229 → Ala 229, Thr 268 → Val 268, and Ser 380 → Ala 380. Residue 135 is found within the second intracellular loop of the receptor, residues 229 and 268 are present in the third intracellular loop, whereas residue 380 is located in the carboxyl terminus of the receptor. For this series of experiments, we transfected the wild-type and mutant receptor into C6 glioma cells.

Characterization of the wild-type and mutant receptors stably expressed in C6 glioma cells suggested that both receptors are expressed at similar levels. There was no difference in the affinity of [^{3}H]SCH 23390 for either of the receptors. Dopamine stimulation of cAMP accumulation was also evaluated in the C6 cell lines. There was no difference in either the efficacy or potency of dopamine for producing this response for either of the receptors. Thus, the site-specific mutations have no effect on receptor expression, antagonist or agonist affinities, or the functional coupling with G_s to promote cAMP generation.

Initial experiments indicated that dopamine preincubation of the C6 cells results in a similar downregulation of both wild-type and mutant receptors. This effect was found to be primarily due to a reduction in the maximum binding capacity (Bmax) rather than a change in receptor affinity. Dopamine treatment was found to induce a maximal 75% receptor downregulation at about 6 hr with a half-life of about 1 hr in an identical fashion and time course for both wild-type and mutant receptors.

In contrast, dopamine-induced desensitization is attenuated in the mutant receptor. After 40 min of dopamine treatment, the wild-type receptor exhibits about a 75% reduction in the maximum cAMP response induced by dopamine, whereas the mutant receptor is desensitized by only about 10%. Detailed time course studies showed that the rate of agonist-induced desensitization is much quicker for the wild-type receptor in comparison to the mutant receptor. For the wild-type receptor, the half-life for desensitization is about 20 min, with maximum desensitization taking place by about 60 min. In contrast, the desensitization half-life for the mutant receptor is about 90 min with the maximum response occurring by 3 hr.

In summary, our studies with the PKA mutant receptor have suggested the following conclusions. First, PKA-mediated phosphorylation of the D1 receptor protein does not appear to be important for agonist-induced receptor downregulation. Reconciliation of these results with those of the CHO mutant cells would suggest that the effect of PKA on receptor downregulation occurs through the phosphorylation of another protein involved in either the expression or downregulation of the receptor. Second, phosphorylation of the D1 receptor by PKA on one or more sites appears to accelerate agonist-induced desensitization of the cAMP response. Future studies will involve mutating each of the PKA sites individually to determine the functional roles of each.

In a final series of experiments, we wished to assess the role of the long carboxyl (COOH) terminus of the D1 receptor in agonist-induced regulation. The long carboxyl terminus contains consensus sequences for phosphorylation by G-protein–coupled receptor protein kinases and also contains one of the PKA phosphorylation sites. In contrast, the D2 dopamine receptor has a short carboxyl terminus with no apparent phosphorylation sites. To assess the potential role of the D1 receptor carboxyl terminus in desensitization, we constructed a D1/D2 chimeric receptor in which the D1 receptor carboxyl terminus was replaced with that of the D2 receptor. The cDNAs of the chimeric D1/D2 and wild-type receptors were stably transfected into CHO and C6 cells. Cyclic AMP levels and [^{3}H]SCH 23390 binding assays were used as measures of receptor function and expression, respectively. Both the D1 wild-type and D1/D2 chimera receptors exhibited similar levels of expression and produced similar cAMP responses in both CHO and C6 cells. This suggests that the truncation of the D1 receptor COOH terminus has no effect on receptor expression, antagonist or agonist affinities, or the functional coupling with G_s to promote cAMP generation.

We initially examined the ability of dopamine to promote downregulation of the wild-type and D1/D2 chimeric receptors in CHO cells. Preincubation with dopamine promoted a loss in [^{3}H]SCH 23390 binding to the wild-type receptor with a half-life of 6 hr and a maximum loss in binding of about 80% by 18 hr. In contrast, the D1/D2 receptor downregulated more rapidly on dopamine exposure, demonstrating a half-time of about 2 hr and a maximum loss in binding by 6 hr. There was no difference between the two receptors with respect to the maximum loss in binding activity.

We also examined agonist-induced desensitization of the cAMP response for both receptor constructs. Incubation of C6 glioma cells expressing the wild-type receptor with dopamine resulted in a reduced maximal cAMP response. The half-life for this effect was about 30 min, with maximal desensitization taking place by 2 hr. In contrast, the chimeric receptor desensitized more rapidly, exhibiting a half-life of about 15 min with maximum desensitization occurring by 45 minutes. There was no difference between the two receptors with respect to the maximum desensitization observed.

Our conclusions from these preliminary experiments are the following. First, truncation of the COOH terminus of the D1 receptor appears to accelerate agonist-induced desensitization and downregulation of the receptor. Second, structural elements within the truncated COOH terminus are not required for agonist-induced D1 receptor desensitization and downregulation to occur. Future experiments will involve delineating each of the specific regions on the D1 receptor protein that is important for each of the various forms of agonist-induced regulation.

References

1. Sibley, D. R., and Neve, K. A. (1996). Regulation of dopamine receptor function and expression. *In* The Dopamine Receptors. (K. A. Neve, and R. L. Neve, ed.), pp. 383–424. Humana Press, Totowa, New Jersey.
2. Singh, T. J., Hochman, J., Verna, R., Chapman, M., Abraham, I., Pastan, I. H., and Gottesman, M. M. (1985). Characterization of a cAMP-resistant chinese hamster ovary cell mutant containing both wild-type and mutant species of type I regulatory subunit of cAMP-dependent protein kinase. *J. Biol. Chem.* **260**, 13927–13933.

Paul A. Insel, Maria A. Balboa, Naoki Mochizuki, Steven R. Post, Kazushi Urasawa, and Mingxhao Xing

Departments of Pharmacology and Medicine
University of California, San Diego
La Jolla, California 92093

Mechanisms for Activation of Multiple Effectors by α_1-Adrenergic Receptors

α_1-Adrenergic receptors (α_1ARs) mediate a wide variety of responses in target tissues, although the precise manner by which such actions occur is poorly understood. Studies of functional activity *in vivo* are complicated by indirect actions, such as alterations in blood flow, and compensatory changes, such as those caused in response to increases in blood pressure. Tissue culture cells offer alternative model systems in which the extracellular milieu can be more precisely defined and variables can be experimentally controlled. With this in mind, our laboratory has used the Madin Darby canine kidney (MDCK) cell line as a model to explore signaling mechanisms and regulation of α_1ARs.

MDCK cells, originally isolated over 30 years ago, are an epithelial cell line derived from the distal tubule–collecting duct region of the dog kidney. We chose to study these cells because we found that they co-expressed both α_1ARs and β_2ARs and thus reasoned that these cells would be a good model to contrast the properties of these two receptors when expressed in the same cells (1). Because there is considerable heterogeneity among MDCK cells, we isolated clonal lines of parental cells and focused subsequent studies on a particular line, MDCK-D1, which expresses a readily detectable number of adrenergic receptors (> 10,000/cell).

Our approach has involved a systematic study of α_1ARs using radioligand binding techniques and efforts to identify the GTP-binding protein(s) with which the receptors link and to characterize the signal transduction pathways regulated by these receptors. Other studies had shown that activation of αARs hyperpolarizes MDCK cells by activation of potassium channels. In initial studies we showed that MDCK cells possess α_1ARs, that these receptors were downregulated by exposure of the cells to agonists, and that agonists differentially regulated α_1- and α_2ARs. We also found that binding of agonists to these receptors was modulated by guanine nucleotides in a Mg^{2+}-dependent manner and that this effect, as well as α_1-adrenergic responses, were insensitive to pertussis toxin treatment. From these results, we concluded that α_1ARs in MDCK cells link to a G-protein distinct from known pertussis toxin–sensitive G-proteins. This conclusion is supported by preliminary studies in which we have used cDNA anti-sense to $G_{q/11}$ to show that inhibiting

Advances in Pharmacology, Volume 42

1054-3589/98 $25.00

expression of these G proteins decreased the ability of α_1ARs to activate arachidonic acid (AA) release but did not alter response to calcium ionophore, a phorbol ester, or to certain other classes of receptor agonists (e.g., ATP and bradykinin) (2).

A major thrust of our work has been devoted to defining the mechanisms by which α_1-agonists promote phospholipid hydrolysis in MDCK-D1 cells. Initial studies indicated that α_1ARs in MDCK cells, as in many other cells, promote phosphoinositide (PI) hydrolysis (by one or more types of phospholipase C [PLC]), and in MDCK cells, the agonists promote a prominent release of AA and AA metabolites, in particular, prostaglandin E_2 (PGE_2). In attempting to determine the relationship between agonist-promoted PI hydrolysis and AA release, we used a variety of approaches (including kinetic analyses, assessment of calcium dependence, quantitative analysis of AA-containing phospholipids, use of neomycin [which inhibited poly PI hydrolysis without altering PGE_2 formation], and response of cells grown under depolarizing conditions) that led us to conclude that the release of AA and AA metabolites and the PI hydrolysis were parallel and independent events promoted by α_1ARs. We inferred that the receptors were activating in parallel PLC and phospholipase (PLA_2) (3). Moreover, we found that α_1AR activation promoted a rapid and transient activation of protein kinase C (PKC) activity that was temporally linked to the ability of α_1ARs to promote a PLC-mediated hydrolysis of phosphatidylcholine (PC) to choline phosphate and diacylglycerol (DAG) (4). Other data indicate that the cells show a rapid activation of phospholipase D in response to α_1ARs. Thus, α_1ARs of MDCK-D1 cells regulate multiple phospholipases.

Much of our recent effort has been devoted to defining the detailed mechanisms by which the receptors regulate these phospholipases, in particular PLA_2. We have recently (5) described a variety of approaches to define the mechanism for AA release in MDCK-D1 cells in response to α_1-adrenergic agonists. A key aspect of these studies was our ability to detect enhanced PLA_2 activity in cell lysates from agonist-treated cells. Several results were consistent with the conclusion that this enhanced PLA_2 activity was secondary to activation of the 85-kDa cytosolic PLA_2 ($cPLA_2$). These results included: calcium-dependence of enzyme activity; inhibition of AA release in whole cells and cell lysates by the inhibitor $AACOCF_3$; and most directly, immunoprecipitation of enhanced enzyme activity with $cPLA_2$ antisera. We found that the activated PLA_2 in cell lysates promoted by α_1-adrenergic agonists could be reversed by treatment of the lysates with a phosphatase and that α_1-agonists could activate the 42-kDa mitogen-activated protein (MAP) kinase. The activation of both MAP kinase and $cPLA_2$ (and AA release in cells) was dependent on PKC and α_1-receptor–promoted AA release was inhibited by the MAP kinase cascade inhibitor PD098059. Taken together, these data suggest that the ability of α_1-agonists to activate PKC is "upstream" of MAP kinase and that MAP kinase is responsible for the phosphorylation and activation of $cPLA_2$.

Figure 1 provides a scheme that summarizes many of our results. Although this scheme reveals some answers to the questions that we set out to address, many questions remain. For example, we are not sure whether PKC itself may regulate $cPLA_2$ and other phospholipases independent of MAP kinase, nor do

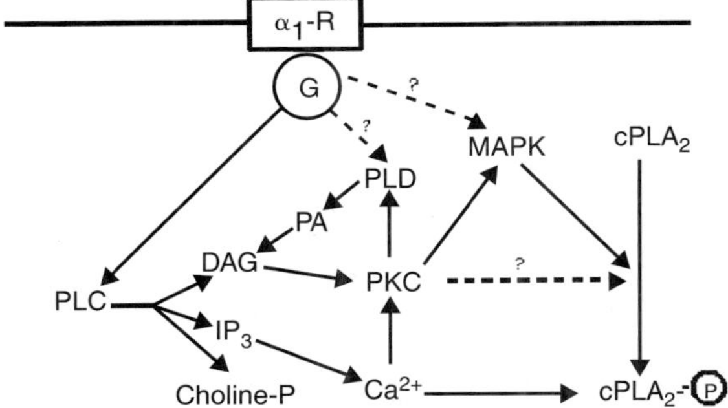

FIGURE 1 Phospholipase activation by α_{1b}-adrenergic receptors in MDCK-D1 cells. Interaction of α_{1b}-adrenergic receptors (α_1-R) on the cell membrane with one or more G-proteins (G) activates one or more forms of phospholipase C (PLC), resulting in production of diacylglycerol (DAG), inositol triphosphate (IP_3), and choline phosphate (Choline-P). α_1-Adrenergic receptors also activate phospholipase D (PLD), producing phosphatidic acid (PA), which is in turn converted to DAG. IP_3 elevates intracellular Ca^{2+}, which, together with DAG, activates protein kinase C (PKC), resulting in activation of PLD and MAP kinase (MAPK). MAP kinase, in turn, phosphorylates cytosolic phospholipase A_2 ($cPLA_2$) and consequently converts the lipase into an activated form ($cPLA_2$-P) in the presence of Ca^{2+}. α_1-Adrenergic activation of PLD and MAPK may also occur through other undefined pathways (as shown by the broken arrows).

we know which isozymes of PKC are involved. We believe, though, that the MDCK-D1 cell system has provided and will continue to provide a uniquely useful model to define pathways for α_1-receptor action.

References

1. Meier, K. E., Snavely, M. D., Brown, S. L., Brown, J. H., and Insel, P. A. (1983). Alpha$_1$- and beta$_2$-adrenergic receptor expression in the MDCK renal epithelial cell line. *J. Cell Biol.* **97**, 405–415.
2. Insel, P. A., Urasawa, K., and Mochizuki, N. (1995). Use of antisense to assess α_1-adrenergic receptor coupling to G proteins. *Pharmacol. Commun.* **6**, 73–77.
3. Slivka, S. R., and Insel, P. A. (1987). Alpha$_1$-adrenergic receptor-mediated phosphoinositide hydrolysis and prostaglandin E_2 formation in MDCK cells: Possible parallel activation of phospholipase C and phospholipase A_2. *J. Biol. Chem.* **262**, 4200–4207.
4. Slivka, S. R., Meier, K. E., and Insel, P. A. (1988). Alpha$_1$-adrenergic receptors promote phosphatidylcholine hydrolysis in MDCK-D_1-cells: A mechanism for rapid activation of protein kinase C. *J. Biol. Chem.* **263**, 11242–12246.
5. Xing, M., and Insel, P. A. (1996). Protein kinase C-dependent activation of cytosolic phospholipase A_2 and MAP kinase by alpha$_1$-adrenergic receptors in MDCK cells. *J. Clin. Invest* **97**, 1302–1310.

Rita M. Huff, Chris L. Chio, Mary E. Lajiness, and Linda V. Goodman

CNS Research
Pharmacia & Upjohn
Kalamazoo, Michigan 49080

Signal Transduction Pathways Modulated by D2-Like Dopamine Receptors

Dopamine receptors regulate ion channels to affect the excitability of neurons and turn on second-messenger systems to activate protein kinase cascades and transcription factors for long-term effects on gene expression. Characterization of the early events triggered by activation of dopamine receptors at a molecular level can contribute to an understanding of how changes in ion channel activation, protein phosphorylation, and alterations in the phenotype of the cells can occur.

The D2-like subfamily of dopamine receptors is a target for drugs used to treat Parkinson's disease and schizophrenia. This subfamily includes D2A, D2B, D3, and D4 receptors. D2A and D2B receptors are isoforms encoded by a single gene, which is differentially spliced to include (D2A) or exclude (D2B) 29 amino acids within the third intracytoplasmic loop. The D4 receptor is highly polymorphic in humans with more than 25 known alleles, which encode proteins with differences in composition of the third intracytoplasmic loops. All of the D2-like receptors are members of the G-protein–linked superfamily of receptors, and in native tissues most if not all D2-like receptor signaling is blocked by pertussis toxin, implicating linkage to the $G_{i/o}$ proteins. The α subunits of these G-proteins are adenosine diphosphate (ADP)-ribosylated on a cysteine residue near the C-terminus in the presence of pertussis toxin, which prevents receptor G-protein coupling. Generally, inhibition of adenylyl cyclase is mediated by G_i-linked receptors, and usually additional signaling events are also altered independently of adenylyl cyclase. Among the biochemical events initiated by D_2-like dopamine receptors in primary cultures of striatal cells are inhibition of vasoactive intestinal peptide–stimulated cyclic adenosine monophosphate (cAMP) accumulation and potentiation of arachidonic acid release stimulated by adenosine triphosphate (ATP) (1).

Studies of D2, D3, and D4 receptor–mediated signaling in our laboratory have been conducted primarily with transfected Chinese hamster ovary (CHO) 10001 cells. D2, D3, and D4 receptors inhibit forskolin-stimulated cAMP accumulation. D2 and D4 receptors potentiate ATP-stimulated arachidonic acid release through a protein kinase C (PKC)–dependent pathway, although the mechanism of PKC activation is independent of phospholipase C (PLC)

Advances in Pharmacology, Volume 42

1054-3589/98 $25.00

activation. D2, D3, and D4 receptors stimulate the rate of extracellular acidification through activation of an amiloride-sensitive Na^+/H^+ exchanger, NHE-1. D2, D3, and D4 receptors stimulate CHO cell division. This response is independent of changes in cAMP, arachidonic acid release, and Na^+/H^+ exchange. Stimulated mitogenesis is, however, sensitive to tyrosine kinase inhibition and most likely results from activation of the mitogen-activated protein kinase (MAPK) signaling cascade. D2 receptor activation results in a rapid stimulation of phosphorylation and activation of the MAPKs erk1 and erk2. Downregulation of PKC only reduces D2 stimulation of MAPK by half. Thus, D2 receptor activation of MAPK occurs through at least two signaling pathways, one of which includes PKC.

To determine if these multiple signaling pathways are a result of more than one kind of G-protein activation and to determine the G-proteins required for each signaling event, we used the strategy originally introduced by Taussig *et al.* (2) to determine the G-protein α subunits important for ion channel modulation by different receptors. Mutation of the codon for the cysteine residue, which is ADP-ribosylated by pertussis toxin to a codon for a serine residue results in a $G\alpha$-protein that can still couple to receptors but is insensitive to the effects of pertussis toxin. We carried out site-directed mutagenesis of human $G\alpha_{i2}$ and human $G\alpha_{i3}$. Ron Taussig (University of Michigan) kindly provided us with mutated rat $G\alpha_{oA}$ and mutated rat $G\alpha_{i1}$. Each of the mutated α-subunit cDNAs were transfected individually into CHO cells already expressing D2 receptors or cotransfected with D2 receptors. Cell lines were selected based on Northern analysis for the presence of transfected cDNA, Western blot analysis for elevated levels of the transfected G-protein α subunit, and receptor-binding analysis for D2 receptor expression. We selected several cell lines by these criteria, including R1, which expresses pertussis toxin–resistant G_{i2}, and G_o7, which expresses pertussis toxin resistant G_{oA}. Each of the cell lines had similar levels of D2A receptor expression, and in the absence of pertussis toxin, they had D2 receptor–mediated signaling characteristics equivalent to the parental cell line, L6. Following treatment with pertussis toxin (i.e., conditions in which the D2 receptors could only activate the pertussis toxin–resistant $G\alpha$ subunits), the D2-mediated signaling characteristics varied depending on the type of expressed mutated $G\alpha$ subunits.

In membranes prepared from pertussis toxin–treated R1 cells, D2 receptors stimulated [^{35}S]-GTPγS binding nearly to the same extent as in L6 cells. The dose response curves for dopamine and quinpirole stimulated-guanine nucleotide binding were nearly superimposable. In pertussis toxin–treated R1 cells, the dopamine dose-response curve for inhibition of cAMP was superimposable with the dose-response curve for dopamine in untreated L6 cells. Therefore, pertussis toxin–insensitive $G\alpha_{i2}$ could "rescue" D2-mediated cAMP inhibition. Pertussis toxin–insensitive $G\alpha_{i2}$ did not rescue D2-mediated potentiation of arachidonic acid release or D2-mediated increases in Na^+/H^+ exchange. The D2-mediated increase in MAPK activity in pertussis toxin–treated R1 cells was approximately half of the effect in the parental L6 cells. However, unlike the response in L6 cells, D2-mediated MAPK stimulation was insensitive to downregulation of PKC. Thus, we have learned that D2A receptors can couple to G_{i2} for inhibition of cAMP accumulation and for the PKC-independent stimulation of MAPK activity. G_{i2} is not involved in D2-mediated arachidonic acid release or Na^+/H^+ exchange.

All D2-mediated signaling was abolished in cells expressing mutated $G\alpha_{i3}$ and $G\alpha_{i1}$ after pertussis toxin treatment. From these results, it would appear that $G\alpha_{i3}$ and $G\alpha_{i1}$ are not coupled to D2 receptors, but we cannot rule out that lack of response was due to insufficient expression of mutated protein.

Dopamine added to the cells that express mutated $G\alpha_o$, G_o7 cells, had a modest ability to inhibit forskolin-stimulated cAMP accumulation after pertussis toxin treatment. In these pertussis toxin–treated cells, dopamine-potentiated ATP-stimulated arachidonic acid release was rescued. As mentioned earlier, this signaling event is downstream from D2-stimulated activation of PKC. So it was anticipated that D2 receptors would stimulate MAPK activity in pertussis toxin–treated G_o7 cells and that unlike in the R1 cells, the response would be abolished by PKC downregulation. This was found to be true. Thus, mutated $G\alpha_o$ rescued the D2-mediated responses that are dependent on PKC: potentiation of arachidonic acid release and PKC-dependent stimulation of MAPK activity.

Figure 1 summarizes the predicted D_2-mediated signaling pathways in CHO cells based on the rescue of D2-mediated signaling in cells that express mutated pertussis toxin–resistant $G\alpha$ subunits. The results have clearly demonstrated that D2 receptors couple to at least two and probably more G-proteins because changes in Na^+/H^+ exchange were not rescued by either $G\alpha_{i2}$ or $G\alpha_o$. Additionally, the results have indicated that even the relatively proximal signaling response of cAMP inhibition can result from D2 receptors coupling to more than one type of G-protein. With these cells we have developed systems in which

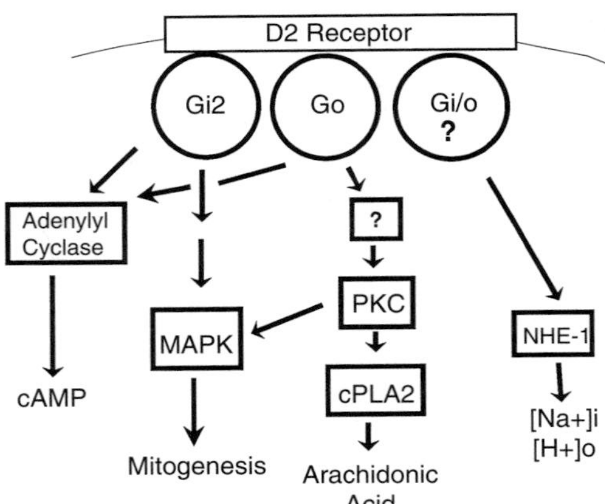

FIGURE 1 D2 Dopamine receptor signaling pathways in CHO cells. D2 receptors interact with at least two types of pertussis toxin sensitive G proteins to elicit intracellular signaling events. D2 receptors inhibit adenylyl cyclase and stimulate MAPK activity via G_{i2}. D2 receptors stimulate PKC resulting in potentiation of arachidonic acid release and stimulation of MAPK activity via G_0. The G protein which mediates D2-stimulated NHE-1 dependent sodium and hydrogen exchange is a third type of pertussis toxin sensitive G protein which has not yet been defined.

D2 coupling to a single G-protein type can be studied in isolation, and these could be invaluable for determining the potential for agonists to differentiate among signaling events.

Because the CHO cells are not neuronal, it is important to determine the relevance of the signaling events studied in CHO cells to neuronal functioning. We have now been able to relate a signaling outcome to a particular G-protein activation. As discussed, it has been demonstrated that D2 receptors inhibit cAMP accumulation and potentiate arachidonic acid release in primary cultures of striatal cells as well as in CHO cells. Although we have determined that D2 receptors can stimulate MAPK activity in dividing cells, it remains to be seen if this response happens in differentiated neurons. D2-like receptor activation has been shown to increase neurite extension and arborization in cultured rat embryonic frontal cortical neurons (3), and this response could result from D2 MAPK signaling. The capacity for D2 receptors to activate MAPK would imply a role for dopamine in synaptic remodeling. These potential novel aspects of dopaminergic neurotransmission should be considered as important aspects of Parkinson's disease and antipsychotic therapy.

References

1. Schinelli, S., Paolillo, M., and Corona, G. L. (1994). Opposing actions of D_1- and D_2 receptors on arachidonic acid release and cyclic AMP production in striatal neurons. *J. Neurochem.* **62**, 944–949.
2. Taussig, R., Sanchez, S., Rifo, M., Gilman, A. G., and Belardett, F. (1992). Inhibition of the omega-conotoxin sensitive calcium current by distinct G proteins. *Neuron* **8**, 799–802.
3. Todd, R. D. (1992). Neural development is regulated by classical neurotransmitters: Dopamine D2 receptor enhances neurite outgrowth. *Biol. Psychiatry* **31**, 794–807.

Elliott M. Ross, Jun Wang, Yaping Tu, and Gloria H. Biddlecome

Department of Pharmacology
University of Texas Southwestern Medical Center
Dallas, Texas 75235

Guanosine Triphosphatase–Activating Proteins for Heterotrimeric G-Proteins

Catecholamine receptors use selected heterotrimeric G-proteins to amplify their signals and convey them to intracellular effector proteins. Typically, β-adrenergic receptors activate adenylyl cyclase via G_s, α_1-adrenergic receptors activate phospholipase C-β (PLC-β) via G_q/G_{11}, and α_2-adrenergic receptors inhibit adenylyl cyclase via the three G_i's. Individual dopamine receptor isoforms are similarly selective among these G-proteins, and alternative pathways using G_{12} or G_{13} are also well documented.

G-proteins convey information by traversing a cycle of controlled GTP binding and hydrolysis. Receptors initiate a signal by promoting the binding of GTP to the G-protein α subunit, thereby activating the G-protein and causing it to activate its effector. Activation is terminated when bound GTP is hydrolyzed. Hydrolysis of Gα-bound GTP is an unusually and anomalously slow enzymatic reaction. Although the rate of deactivation of adenylyl cyclase is generally about equal to the rate of hydrolysis of G_s-bound GTP, physiological rates of termination of most G-protein signals in cells are much faster than the measured rate of hydrolysis of GTP by the relevant isolated G-protein. In such cases, hydrolysis of bound GTP is accelerated by GTPase–activating proteins (GAPs) (1, 2).

G-protein GAPs perform one or more important regulatory functions. The first GAPs for heterotrimeric G-proteins to be identified were the G-protein–regulated effectors PLC-β and the cyclic GMP phosphodiesterase of photoreceptor cells (3, 4). The GAP activity of effectors is thought to enhance the temporal resolution of signaling, as discussed earlier, or to increase the selectivity of the G-protein among different receptors (see ref. 2 and following text). The recently recognized RGS proteins, whose prototypes are involved in desensitization or antagonism of G-protein signaling, are also G-protein GAPs. Their primary function is assumed to be attenuation of signaling, although the regulation of RGS proteins is only now being investigated (1). GAPs for small monomeric GTP-binding proteins such as p21*ras*, whose GAPs were the first identified, also are negative signaling elements. Lastly, GAPs for some of the monomeric GTP-binding proteins involved in organelle trafficking are thought to terminate G-protein–mediated assembly, fusion, or transit functions, al-

Advances in Pharmacology, Volume 42

1054-3589/98 $25.00

though this role has not been conclusively demonstrated. We describe here the activities of two GAPs for heterotrimeric G-proteins, one an effector and the other a newly identified GAP whose physiological role is still unknown.

I. G_q GAP Activity and Activation of PLC-β

The G_q GAP activity of the PLC-β family was first described according to the ability of PLC-β1 to accelerate agonist-stimulated, steady-state GTPase activity in reconstituted unilamellar phospholipid vesicles that contained purified m1 muscarinic cholinergic receptor and G_q. Similar preliminary data have been obtained with α_{1C}-adrenergic receptors as well (T. Nguyen and E. M. Ross, unpublished). Kinetic evidence indicated that PLC increased the rate of hydrolysis of G_q-bound GTP up to 100-fold, yielding a deactivation lifetime as short as 500 ms (3, 5). Accelerated hydrolysis accounted well for physiologically fast rates of signal termination upon removal of agonist. However, such fast deactivation combined with previously measured rates of GTP binding were inconsistent with appreciable activation of G_q at steady state. More recent studies have largely resolved this discrepancy and have pointed to a novel association of receptor with G_q and PLC-β1(5).

In cases where effector is also a GAP, agonist initiates formation of a relatively stable complex of receptor, G_q, and PLC-β that remains intact during multiple turnovers of the GTPase cycle. Conventional schemes for receptor-stimulated GTPase cycling generally have assumed that receptor dissociates from GTP-activated G-protein and that reassociation of receptor and G-protein is a major rate-limiting event in the cycle. In the case of G_q and PLC-β1, GAP activity causes rapid hydrolysis of bound GTP, such that receptor does not dissociate during the lifetime of the activated G_q-GTP species and can catalyze GDP/GTP exchange immediately upon GTP hydrolysis. Such a complex allows maintenance of a substantial portion of the G_q to exist in the GTP-bound form at steady state despite extremely rapid GTP hydrolysis.

Several observations are consistent with the existence of such a complex. First, receptor-catalyzed nucleotide exchange on G_q is slow when initiated by addition of agonist. The initial rate of receptor-catalyzed [α-^{32}P]GTP binding displays a rate constant of about 0.3 min^{-1}, which is actually slower than the turnover rate of the entire steady-state GTPase cycle. This value is also much slower than the equilibrium exchange rate for GDP, which is greater than 25 min^{-1} (5). The approach to rapid GDP/GTP exchange upon addition of agonist—formation of the complex—is also slow, as is the decay of the complex upon addition of antagonist (rate constants of roughly 0.5–1.0 min^{-1}). While these kinetic data argue for a complex of proteins that binds and hydrolyzes GTP quickly, its composition is still inferential, and we are currently developing procedures to measure association of these proteins in the reconstituted vesicles.

In addition to providing enhanced speed, formation of a receptor-G_q-PLC complex that catalyzes rapid GTP hydrolysis can also enhance the functional specificity of G_q-receptor recognition. If signaling to PLC-β requires formation of the proposed complex, then the only receptors that will be able to activate PLC efficiently will be those that bind G_q tightly enough to remain associated during the phase of the GTPase cycle in which G_q is bound to GTP. The selectivity of receptor–G-protein recognition is not absolute, and most receptors will activate most G-proteins if they are overexpressed in cells or reconstituted into phospholipid vesicles at high concentrations (2). Preliminary data indicate that a β_1-adrenergic receptor that can catalyze GTPγS binding to G_q with reasonable efficiency in reconstituted vesicles but that does not stimulate PLC signaling in cells also fails to initiate cellular signaling because it cannot adequately maintain fast GDP/GTP exchange during the steady-state GTPase cycle.

II. A GAP for G_z

The GAP activity of effectors potentially allows discovery of new effectors according to their GAP activity toward the G-protein of interest. We recently identified a new GAP that is highly selective for G_z, a member of the G_i family that is expressed in brain, retina, platelets, and other tissues active in secretion (6, 7). Although G_z can mediate inhibition of adenylyl cyclase and respond to the α_2-adrenergic and other typically G_i-coupled receptors, its physiological effector is unknown. G_z is also unique among the G_i family in its extremely slow rate of GTP hydrolysis—the half-life of G_z-bound GTP is 7 min at 30°C—and a GAP is required for reasonable rates of G_z-mediated signaling (6).

G_z GAP activity, which we assay according to acceleration of the hydrolysis of $G\alpha_z$-bound [γ-^{32}P]GTP, is most abundant in brain, retina, spleen, and platelets. These are the tissues that also express $G\alpha_z$. G_z GAP was purified about 50,000-fold from bovine brain, its richest source. It is a tightly membrane bound protein and can be solubilized only with warm detergent. Several further steps were required to disaggregate the GAP and allow its purification. We do not know the cellular membrane to which it is bound, but its hydrophobic behavior strongly suggests that it is a membrane-spanning protein. Highly purified preparations of cerebral G_z GAP contain multiple proteins in the 22- to 28-kDa range that all display GAP activity. It is possible that the smaller proteins are proteolytic products of the largest.

G_z GAP binds $G\alpha_z$-GTP or its analogues ($G\alpha_z$-GTPγS, $G\alpha_z$-GDP/AlF_4) with 1- to 2-nM affinity and stimulates the rate of hydrolysis of $G\alpha_z$-bound GTP over 200-fold. The GAP can, therefore, interact productively with G_z at its normally low cellular concentration and accelerate hydrolysis enough to allow G_z to convey signals with physiologically reasonable kinetics. We estimate from purification data that G_z GAP is expressed at lower concentrations than is $G\alpha_z$, although this may be true only for a subset of cells in the brain. Regardless, because the GAP binds only weakly to G_z-GDP ($K_d > 100$ nM), it dissociates after hydrolysis is complete, and a single GAP molecule can, thus, act catalytically to promote deactivation of multiple molecules of G_z-GTP. Consistent with this behavior, G_z GAP amplifies agonist-stimulated, steady-state GTPase activity after it has been coreconstituted into phospholipid vesicles with m2 muscarinic receptor and trimeric G_z (7).

While its contribution to the speed of signal termination is clearly important, other physiological functions of the G_z GAP remain to be discovered. Its possible role as a G_z effector will probably remain obscure until its cDNA is cloned. Its GAP activity is regulated by several factors, the best studied of which are the G-protein $\beta\gamma$ subunits. G_z GAP displays its greatest activity toward $G\alpha_z$-GTP, and the addition of $\beta\gamma$ subunits markedly inhibits deactivation by forming heterotrimeric $G\alpha_z\beta\gamma$-GTP, which binds less tightly to the GAP and is, thus, a far worse target for the GAP effect than is the resolved α subunit. Because G_z is expressed at low levels, it will be sensitive primarily to $\beta\gamma$ subunits released from more abundant proteins, thus allowing other signaling pathways to inhibit the GAP and thereby either prolong G_z signaling, potentiate it, or both.

Acknowledgments

These studies were supported by grants from the National Institutes of Health (R37GM30355, T32GM07062), the R. A. Welch Foundation (I-0982), and Cadus Pharmaceutical Corp. We thank Karen Chapman and Jimmy Woodson for excellent technical assistance in this work.

References

1. Dohlman, H. G., and Thorner, J. (1997). RGS proteins and signaling by heterotrimeric G proteins. *J. Biol. Chem.* **272,** 3871–3874.
2. Ross, E. M., (1995). G protein GAP's: Regulation of speed, amplitude and signaling selectivity. *Recent Prog. Horm. Res.* **50,** 207–221.
3. Berstein, G., Blank, J. L., Jhon, D-Y., Exton, J. H., Rhee, S. G., and Ross, E. M. (1992). Phospholipase C-β1 is a GTPase-activating protein for $G_{q/11}$, its physiologic regulator. *Cell* **70,** 411–418.
4. Arshavsky, V. Y., and Bownds, M. D. (1992). Regulation of deactivation of photoreceptor G-protein by its target enzyme and cGMP. *Nature* **357,** 416–417.
5. Biddlecome, G. H., Berstein, G., and Ross, E. M. (1996). Regulation of phospholipase C-β1 by G_q and m1 muscarinic cholinergic receptor. Steady-state balance of receptor-mediated activation and GAP-promoted deactivation. *J. Biol. Chem.* **271,** 7999–8007.
6. Casey, P. J., Fong, H. K. W., Simon, M. I., and Gilman, A. G. (1990). G_z, a guanine nucleotide-binding protein with unique biochemical properties. *J. Biol. Chem.* **265,** 2383–2390.
7. Wang, J., Tu, Y., Woodson, J., Song, X., and Ross, E. M. (1997). A GTPase-activating protein for the G protein $G\alpha_Z$: Identification, purification and mechanism of action. *J. Biol. Chem.* **272,** 5732–5740.

G. Milligan, I. Mullaney, G. D. Kim, and D. MacEwan

Molecular Pharmacology Group
Division of Biochemistry and Molecular Biology
Institute of Biomedical and Life Sciences
University of Glasgow
Glasgow G12 8QQ, Scotland

Regulation of the Stoichiometry of Protein Components of the Stimulatory Adenylyl Cyclase Cascade

Despite enormous efforts to measure regulation of second-messenger production, very little is known about the absolute or relative levels of expression, or the cellular distribution, of the individual polypeptides that comprise distinct signaling cassettes and how alterations in these ratios might alter the effectiveness of signaling processes. Rapid progress has, however, been made in methods to transgenically alter levels of signal-transducing polypetides in mice, often in a tissue-specific manner. These studies can be invisaged to presage the use of such strategies in human gene therapy, and, thus, it is clearly of vital importance that simpler systems be developed in parallel to allow examination of how transgenic modification of the flux through signal transduction cascades might be modified by regulation of expression of levels of receptors, G-proteins, and effector enzymes. Furthermore, such systems would be particularly amenable to test many aspects of classical pharmacological receptor theory as to how ligand potency and efficacy might alter with designed manipulation of signaling-cassette stoichiometry. Although many such studies can be performed using transient transfection approaches, there has been a tendency in such work to fail to make a significant effort to control and analyze the levels of expression of the transfected polypeptides beyond that they are "overexpressed," and there are clear concerns in such studies about the appropriateness of targetting and protein folding of the expressed proteins. Based on both the endogenous expression of a considerable variety of receptors that regulate adenylyl cyclase activity and the elegant studies of Neubig and colleagues on potential compartmentalization of receptors that all mediate inhibition of adenylyl cyclase in these cells, we selected the neuroblastoma × glioma hybrid cell line, NG108-15, as the parental host for a series of studies in which we have systematically altered levels of each of a $G_s\alpha$-coupled receptor, $G_s\alpha$, and adenylyl cyclase. To do so, NG108-15 cells were stably transfected to express the wild-type or a constitutively activated mutant (CAM) variant of the β_2-adrenoceptor, to overexpress the long isoform of $G_s\alpha$ or to express type II adenylyl cyclase. Combinations

Advances in Pharmacology, Volume 42

1054-3589/98 $25.00

of these were generated by consecutive transfections with selections based on geneticin sulphate and hygromycin B. Levels of receptors were assessed in ligand-binding studies, G_s by semiquantitative immunoblotting, adenylyl cyclase by guanine nucleotide stimulated binding of [^{3}H]forskolin, and the regulation of the effectiveness of the signaling cascade using conventional adenylyl cyclase assays.

Initial experiments on parental NG108-15 cells demonstrated levels of adenylyl cyclase to represent quantitatively the least highly expressed element of the stimulatory adenylyl cyclase cascade ($\sim 2 \times 10^4$ copies per cell), as has also recently been demonstrated in isolated cardiac myocytes. NG108-15 cells express good levels of an IP prostanoid receptor, quantitatively undefined levels of an A_2 adenosine receptor, and low levels of a receptor for secretin, all of which stimulate adenylyl cyclase activity. None of these receptors is entirely suitable for detailed studies on signaling efficiency, due either to limited pharmacologies (IP prostanoid, secretin) or to concerns derived from production and release from the cells of an endogenous agonist (A_2 adenosine). As such, we isolated cell lines stably expressing varying levels of human β_2-adrenoceptor (1, 2). Higher levels of expression of the β_2-adrenoceptor resulted in higher basal adenylyl cyclase activity, an effect that we and others have also observed following stable transfection of the related NCB20 cell line. However, higher levels of the β_2-adrenoceptor did not result in greater maximal stimulation of adenylyl cyclase in response to isoprenaline but only a shift to the left of the concentration-effect curve, a set of observations consistent with the idea of adenylyl cyclase as the quantitatively limiting element of the cascade. Using a combination of different clones expressing differing levels of the receptor and an alkylating antagonist to eliminate access of agonist to varying proportions of these receptors, we were able to demonstrate that a full agonist such as isoprenaline required occupancy of only some 3000 receptors per cell (50 fmol/mg) to half-maximally activate the entire adenylyl cyclase population in these cells (3). As anticipated from pharmacological receptor theory, the intrinsic activity of partial agonists was also greater in cells expressing higher levels of the receptor. This could be measured both by conventional adenylyl cyclase assays and in whole cell assays in which agonist-induced formation of the activated $G_s\alpha$-adenylyl cyclase complex was detected as a target for the high-affinity binding of [^{3}H]forskolin (3). Given that $G_s\alpha$ is expressed at some 1×10^6 copies per cell in parental NG108-15 cells, further expression might not be anticipated to increase the effectiveness of the stimulatory adenylyl cyclase cascade.

To assess this directly, NG108-15 cells were transfected with a HA-epitope-tagged version of a long isoform of $G_s\alpha$ (4). In clone BST 15, stable expression of this construct was detected by immunoblotting with the tag-specific antibody. Serendipitously, the tagged polypeptide, although containing the same number of amino acids as the wild-type protein, migrated more slowly through SDS-PAGE, thus allowing an antiserum to the C-terminus of $G_s\alpha$ to detect both endogenous and the introduced form of the G-protein concurrently (4). In clone BST 15, total $G_s\alpha$ levels were increased from 1×10^6 copies per cell to 2×10^6 copies per cell. However, although the IP prostanoid receptor could interact with both the endogenous and introduced G-protein (4), and the epitope-tagged protein was capable of activating adenylyl cyclase, as judged by the increased capacity of cholate extracts of BST 15 cell membranes to reconstitute adenylyl cyclase activity to membranes of S49 cyc^- cells, ligands at each of the IP prostanoid receptor, the A_2 adenosine receptor, and the secretin receptor showed no significant differences in either their maximal ability or potency to stimulate adenylyl cyclase activity.

Based on reverse transcriptase–polymerase chain reaction (RT-PCR) analyses, parental NG108-15 cells express a variety of adenylyl cyclase isoforms, including types IV, V, and VI. However, they do not express detectable levels of type II adenylyl cyclase mRNA (5). As such, cells already transfected to express the β_2-adrenoceptor were further transfected, and clones were isolated that stably expressed this form. Based on maximal guanine nucleotide–stimulated binding of [^{3}H]forskolin, total cellular levels of adenylyl cyclase were increased by up to eightfold. The most obvious initial feature of adenylyl cyclase activity in the type II–expressing clones was that the cells displayed markedly increased basal adenylyl cyclase activity (5). Addition of maximally effective concentrations of isoprenaline also resulted in substantially higher adenylyl cyclase activity in these clones compared with those without type II adenylyl cyclase. However, no alteration in potency of isoprenaline was observed between the clones and, indeed, perhaps surprisingly, there was no alteration in the potency or the measured intrinsic activity of β_2-adrenoceptor partial agonists between the adenylyl cyclase–overexpressing clones and their parental cells (5). Furthermore, agonists at the A_2 adenosine and secretin receptors, which are unable to produce the same maximal stimulation of adenylyl cyclase activity as full agonists at the IP prostanoid or β_2-adrenoceptors in the parental cells, also displayed no alterations in their ability to activate adenylyl cyclase in the adenylyl cyclase type II–expressing cells when compared with isoprenaline or iloprost. They could cause activation of the introduced adenylyl cyclase, however, as the maximal adenylyl cyclase activity produced by these ligands was much greater than in the parental cells.

Following stable transfection of a CAM form of the β_2-adrenoceptor into NG108-15 cells, as anticipated, basal adenylyl cyclase activity was substantially higher in these cells than in clones expressing similar levels of the wild-type receptor. Isoprenaline was able to cause further activation but not to a higher level than could be achieved by the wild-type receptor, again indicating that the levels of expression of adenylyl cyclase are the limiting elements for transductional effectiveness. As such, if simply presented as "fold-stimulation" over basal activity, then in this system the CAM β_2-adrenoceptor appears to be producing a smaller activation than the wild-type receptor. Addition of the β_2-adrenoceptor inverse agonists betaxolol and sotalol caused a substantial inhibition of the CAM receptor–induced basal adenylyl cyclase activity, and the sustained presence of these compounds in the growth medium of the CAM β_2-adrenoceptor–expressing NG108-15 cells resulted in a substantial upregulation of the receptor. These effects were not observed with all β_2-adrenoceptor blockers. Neutral antagonists that were unable to cause a substantial alteration in basal adenylyl cyclase activity in these membranes also failed to cause upregulation of the receptor and were able to block the effects of the inverse agonists when coincubations were performed.

These studies have provided new insights into the regulation of the stimulatory adenylyl cyclase cascade. Decisions about which component of a signal transduction cascade to alter in transgenic studies designed to improve its efficiency should thus include whether the central aim is to increase the potency of an agonist drug or to increase the maximal capacity and output from the effector.

References

1. Adie, E. J., and Milligan, G. (1994). Regulation of basal adenylate cyclase activity in neuroblastoma × glioma hybrid, NG108-15, cells transfected to express the human β_2-adrenoceptor: Evidence for empty receptor stimulation of the adenylate cyclase cascade. *Biochem. J.* **303,** 803–808.
2. Adie, E. J., and Milligan, G. (1994). Agonist regulation of cellular G_s α-subunit levels in neuroblastoma × glioma hybrid NG108-15 cells transfected to express different levels of the human β_2-adrenoceptor. *Biochem. J.* **300,** 709–715.
3. MacEwan, D. J., Kim, G. D., and Milligan, G. (1995). Analysis of the role of receptor number in defining the intrinsic activity and potency of partial agonists in neuroblastoma × glioma hybrid NG108-15 cells transfected to express different levels of the human β_2-adrenoceptor. *Mol. Pharmacol.* **48,** 316–325.
4. Mullaney, I., and Milligan, G. (1994). Equivalent regulation of wild type and an epitope-tagged variant of $G_s\alpha$ by the IP prostanoid receptor following expression in neuroblastoma × glioma hybrid, NG108-15 cells. *FEBS Lett.* **353,** 231–234.
5. MacEwan, D. J., Kim, G. D., and Milligan, G. (1996). Agonist regulation of adenylate cyclase activity in neuroblastoma × glioma hybrid NG108-15 cells transfected to coexpress adenylate cyclase type II and the β_2-adrenoceptor. Evidence that adenylate cyclase is the limiting component for receptor-mediated stimulation of adenylyl cyclase activity. *Biochem. J.* **318,** 1033–1039.

Louis M. Luttrell, Tim van Biesen, Brian E. Hawes, Gregory J. Della Rocca, Deirdre K. Luttrell, and Robert J. Lefkowitz

The Howard Hughes Medical Institute and the Departments of Medicine and Biochemistry
Duke University Medical Center
Durham, North Carolina 27710

Regulation of Mitogen-Activated Protein Kinase Pathways by Catecholamine Receptors

Growing evidence suggests that receptors that couple to heterotrimeric G-proteins (GRs) participate in the regulation of cell proliferation in both physiological and pathophysiological states, and in cellular transformation in some, mostly neuroendocrine, human tumors. Indeed, the oncogenic potential of these receptors can be demonstrated *in vitro,* where stable expression of a constitutively activated mutant of the α1B-adrenergic receptor (AR) in Rat1 or NIH3T3 fibroblasts induces agonist-independent focus formation and tumor growth in nude mice (1). Until recently, however, little was known about the mechanisms of mitogenic signal transduction employed by these receptors.

GRs that mediate cellular responses to a variety of humoral, endothelium-, or platelet-derived substances have been shown to rapidly stimulate the mitogen-activated protein kinase (ERK1/2) pathway, a major point of convergence for signals regulating cell growth and differentiation. The best understood pathway of ERK1/2 activation is that mediated by growth factor receptors that possess intrinsic ligand-stimulated tyrosine kinase activity (RTK), such as the receptor for epidermal growth factor (EGF). Here, ligand binding leads to receptor tyrosine phosphorylation and binding of cytoplasmic Src-homology 2 (SH2) domain–containing adaptor proteins, such as Shc and Grb2, which recognize specific phosphotyrosine residues on the receptor. Activation of the low-molecular-weight G-protein Ras follows SH2 domain–mediated recruitment of the Ras-guanine nucleotide exchange factor mSos to the plasma membrane. Ras activation initiates a highly conserved protein phosphorylation cascade, involving the serine–threonine kinase Raf, the threonine–tyrosine MEK kinases, and ultimately ERK1/2.

To begin to understand the mechanisms whereby GRs mediate growth regulatory signals, we have studied the mechanisms whereby several GRs, including the α_1B AR and α_2A AR, stimulate ERK1/2 activity in transiently transfected COS-7 and Chinese hamster ovary (CHO) cell model systems. In these cells, clear heterogeneity exists between the mechanisms of ERK1/2

Advances in Pharmacology, Volume 42

1054-3589/98 $25.00

activation employed by receptors that signal via pertussis toxin–sensitive G_i family proteins, such as α_2A AR, and receptors that signal via pertussis toxin-insensitive $G_{q/11}$ family proteins, such as α_1B AR (Fig. 1) (2). ERK1/2 activation via α2A AR is sensitive to dominant negative mutants of Ras (N17Ras) and Raf (ΔNRaf), but is unaffected by downregulation of protein kinase C (PKC). This pathway is inhibited by coexpression of a polypeptide derived from the carboxyl terminus of GRK2 (GRK2ct), which functions as an intracellular sequestrant of free $G\beta\gamma$ subunits, and by genistein and herbimycin A, two pharmacological inhibitors of tyrosine protein kinases. Indeed, transient expression of $G\beta\gamma$ subunits, but not constitutively activated mutant $G_{i\alpha}$ subunits, is sufficient to induce Ras-dependent ERK1/2 activation in these cells. In contrast, ERK1/2 activation via α_{1B} AR is inhibited by PKC downregulation and ΔNRaf expression, but insensitive to N17Ras and GRK2ct expression or exposure to genistein and herbimycin A. Thus, at least two mechanisms exist for GR-mediated ERK1/2 activation in these cells: one a Ras-dependent pathway, dependent on tyrosine phosphorylation and mediated predominantly via $G\beta\gamma$ subunits derived from pertussis toxin–sensitive G-proteins; the other, a Ras-independent pathway, dependent on PKC and mediated via $G\alpha$ subunits derived from pertussis toxin–insensitive G-proteins.

The mechanism of $G\beta\gamma$ subunit–dependent activation of ERK1/2 mediated by G_i-coupled receptors bears striking similarity to that employed by classical RTKs, such as the EGF receptor (3). Stimulation of either endogenous G_i-coupled lysophosphatidic acid (LPA) receptors or transiently expressed α_2A AR in COS-7 cells results in a rapid three- to fivefold increase in tyrosine phosphorylation of the Shc adaptor protein and association of Shc with the SH2/SH3 domain–containing docking protein, Grb2. Ras guanine nucleotide exchange factor activity, reflecting recruitment of Grb2-mSos complexes, can be detected in Shc immunoprecipitates following G_i-coupled receptor stimulation. Expression of a dominant negative mutant of mSos, which forms catalytically inactive Grb2-mSos complexes, inhibits both G_i-coupled receptor– and EGF receptor–mediated ERK1/2 activation. These results indicate that the $G\beta\gamma$ subunit–mediated pathway, like the RTK pathway, requires phosphotyrosine-dependent recruitment of Ras guanine nucleotide exchange factors to achieve Ras activation.

GR-mediated increases in tyrosine phosphorylation of several proteins, including the EGF receptor and $p185^{neu}$, the platelet-derived growth factor receptor, the insulin-like growth factor I receptor, Shc, insulin receptor substrate-1, and focal adhesion kinase (FAK), have been described in various cell types. In COS-7 cells, stimulation of G_i-coupled receptors or overexpression of $G\beta\gamma$ subunits results in the association of Shc with tyrosine phosphoproteins of approximately 130 and 180 kDa, as well as Grb2. The 180-kDa Shc-associated tyrosine phosphoprotein band contains both the EGF receptor and $p185^{neu}$. G_i-coupled receptor stimulation results in a three- to fivefold increase in EGF receptor tyrosine phosphorylation, which mirrors an agonist-induced association between Shc and EGF receptor. Thus, Ras-dependent ERK1/2 activation in these cells correlates with formation of a multiprotein complex that contains EGF receptor, Shc, Grb2, and mSos. This suggests that EGF receptor and possibly other tyrosine phosphoproteins, such as FAK, may serve as scaf-

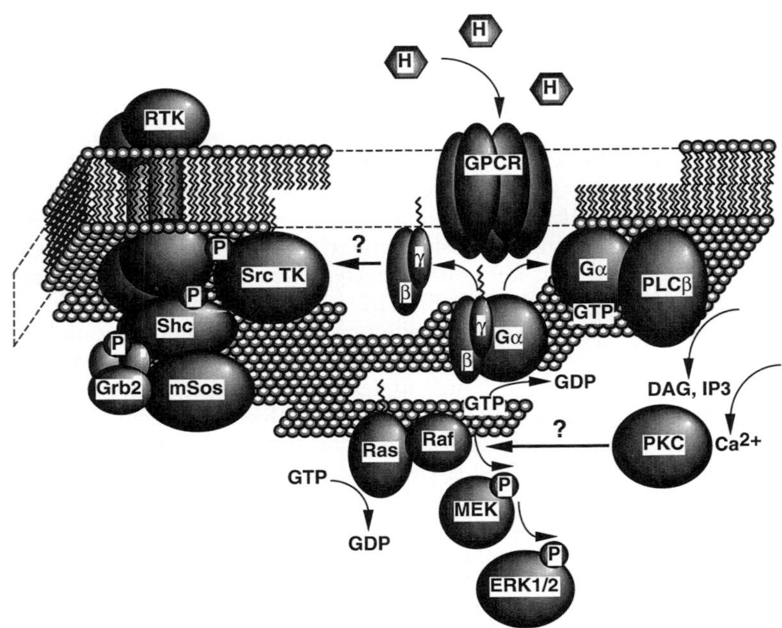

FIGURE 1 ERK1/2 activation via G_i-coupled and G_q-coupled receptors in COS-7 cells. Ras-independent ERK1/2 activation via $G_{q/11}$-coupled receptors is mediated by $G\alpha$ subunits and dependent upon PKC. The pathway of Ras-dependent ERK1/2 activation via G_i-coupled receptors is mediated by $G\beta\gamma$ subunits and requires activation of Src family tyrosine protein kinases via an unknown mechanism. Activation of the nonreceptor tyrosine kinase leads to the phosphotyrosine-dependent assembly of a membrane-associated Ras activation complex that contains the EGF receptor, the adaptor proteins Shc and GRB 2, and the Ras guanine nucleotide exchange factor, mSos. The two pathways converge at the level of Raf and proceed to ERK1/2 via the Raf, MEK, ERK kinase cascade. GPCR, G-protein–coupled receptor; P, phosphotyrosine; PLCβ, phospholipase-Cβ; GTP, guanosine triphosphate; GDP, guanosine diphosphate; DAG, diacylglycerol.

folding proteins for the assembly of a mitogenic signaling complex at the plasma membrane that resembles that formed following RTK activation.

Several lines of evidence suggest that the ubiquitous Src family nonreceptor tyrosine kinases play an essential role in the $G\beta\gamma$ subunit–dependent activation of ERK1/2 (4). *In vitro* kinase assays performed on Shc immunoprecipitates following LPA stimulation demonstrate rapid, transient recruitment of tyrosine kinase activity into Shc immune complexes. This recruitment of tyrosine kinase activity is pertussis toxin–sensitive and mimicked by cellular expression of $G\beta\gamma$ subunits. Protein immunoblots reveal a transient association of c-Src with the Shc-containing signaling complex, which coincides with the increases in Shc-associated tyrosine kinase activity and Shc tyrosine phosphorylation. LPA stimulates binding of EGF receptor to a GST fusion protein containing the c-Src SH2 domain, suggesting that EGF receptor, Shc, and c-Src coexist in a signaling complex. LPA stimulation or expression of $G\beta\gamma$ subunits results in c-Src activation as assessed by increased c-Src autophosphorylation. Overexpression of wild-type or constitutively active mutant c-Src, but not kinase inactive mutant c-Src, leads to increased tyrosine kinase activity in Shc immunoprecipitates, increased Shc tyrosine phosphorylation, Shc/Grb2 complex formation, and ERK1/2 activation.

Inhibition of endogenous Src family kinase activity by cellular expression of a kinase-inactive mutant c-Src inhibits $G\beta\gamma$ subunit– and G_i-coupled receptor–mediated phosphorylation of both EGF receptor and Shc. Expression of Csk, which inactivates Src family kinases by phosphorylating the regulatory C-terminal tyrosine residue (Y530), has the same effect. The G_i-coupled receptor-mediated increase in EGF receptor phosphorylation does not reflect increased EGF receptor autophosphorylation, assayed using an autophosphorylation-specific EGF receptor monoclonal antibody. Together, these observations suggest that $G\beta\gamma$ subunit–mediated activation of Src family nonreceptor tyrosine kinases, rather than the intrinsic kinase activity of the RTK, accounts for the G_i-coupled receptor–mediated tyrosine phosphorylation events that direct recruitment of the Shc and Grb2 adaptor proteins to the membrane. Coexpression of Csk, or a dominant negative mutant of SHPTP2, a phosphotyrosine phosphatase that activates Src family kinases by dephosphorylating Y530, inhibits $G\beta\gamma$ subunit– and G_i-coupled receptor–mediated ERK1/2 activation.

The mechanism whereby effectors of activated GRs stimulate Src family kinases is unknown. However, stimulation of phosphatidylinositol 3-kinase (PI3K) may play a role in Ras-dependent mitogen-activated protein kinase (MAPK) activation in some cells (5). G_i-coupled receptor– and $G\beta\gamma$ subunit–mediated MAPK activation in COS-7 and CHO cells is sensitive to the PI3K inhibitors wortmannin and LY294002 and to expression of a dominant negative form of the p85 regulatory subunit of PI3K. Interestingly, MAPK activation induced by expression of constitutively active mutant c-Src, mSos, and constitutively activated mutants of Ras and MEK is wortmannin-insensitive, suggesting that the PI3K-dependent step in the pathway may lie upstream of Src kinase activation. Understanding the mechanisms whereby GRs regulate tyrosine protein phosphorylation and the basis for cross-talk between GR receptor and receptor tyrosine kinase signaling pathways may ultimately provide strategies for selective activation or inhibition of cellular proliferation.

References

1. Allen, L. F., Lefkowitz, R. J., Caron, M. G., and Cotecchia, S. (1991). G-protein-coupled receptor genes as protooncogenes: Constitutively activating mutation of the α1B-adrenergic receptor enhances mitogenesis and tumorigenicity. *Proc. Natl. Acad. Sci. U.S.A.* **88,** 11354–11358.
2. Hawes, B. E., van Biesen, T., Koch, W. J., Luttrell, L. M., and Lefkowitz, R. J. (1995). Distinct pathways of Gi- and Gq-mediated mitogen activated protein kinase activation. *J. Biol. Chem.* **270,** 17148–17153.
3. van Biesen, T., Hawes, B. E., Luttrell, D. K., Krueger, K. M., Touhara, K., Porfiri, E., Sakaue, M., Luttrell, L. M., and Lefkowitz, R. J. (1995). Receptor-tyrosine-kinase- and $G\beta\gamma$-mediated MAP kinase activation by a common signalling pathway. *Nature* **376,** 781–784.

4. Luttrell, L. M., Hawes, B. E., van Biesen, T., Luttrell, D. K., Lansing, T. J., and Lefkowitz, R. J. (1996). Role of c-Src in G protein-coupled receptor- and Gβγ subunit-mediated activation of mitogen activated protein kinases. *J. Biol. Chem.* **271**, 19443–19450.
5. Hawes, B. E., Luttrell, L. M., van Biesen, T., and Lefkowitz, R. J. (1996). Phosphatidylinositol 3-kinase is an early intermediate in the Gβγ-mediated mitogen activated protein kinase signaling pathway. *J. Biol. Chem.* **271**, 12133–12136.

Brian K. Kobilka and Ulrik Gether

Division of Cardiovascular Medicine
Howard Hughes Medical Institute
Stanford University Medical School
Stanford, California 94305

Examination of Ligand-Induced Conformational Changes in the β_2-Adrenergic Receptor by Fluorescence Spectroscopy

Drugs acting as G-protein–coupled receptors are traditionally classified in biological assays as either true agonists, which fully activate the receptor, as partial agonists, which elicit a fractional response, or as antagonists, which block the response induced by agonists. In addition, recent data have suggested that antagonists should be subclassified into at least two categories: true (neutral) antagonists, which have no effect on basal receptor activity, and inverse agonists (also referred to as negative antagonists), which inhibit basal receptor activity occurring in the absence of agonist (1–3). A rather detailed understanding of the ligand-binding site for many compounds has been obtained through mutagenesis studies; however, very little is known about the actual molecular events and structural changes that occur in the receptor following ligand binding and during transmission of a signal across the membrane. Therefore, the structural basis for the biological classification of drug action remains unknown.

A classical model for receptor action proposes that receptors are simple bimodal switches with an unliganded "off" configuration and an agonist-dependent "on" configuration. This view has been challenged by the finding that several G-protein–coupled receptors have a significant level of basal activity in the absence of agonist (1, 3). This basal activity can be enhanced by selective mutations (4). These observations have been explained by a two-state model

1054-3589/98 $25.00

where the receptors are proposed to exist in a dynamic equilibrium between an inactive state (R) and an active state (R*) (R <=> R*) (4). According to the model, agonists are believed to preferentially bind to and stabilize R* and thereby strongly shift the equilibrium to the right, whereas partial agonists are considered less effective in stabilizing R*. Conversely, inverse agonists (negative antagonists) are believed to preferentially bind to and stabilize R and shift the equilibrium to the left, whereas true antagonists block agonist binding but do not alter the equlibrium between R and R*. However, the functional consequence of binding different classes of ligands could also be explained in a model predicting the existence of several ligand-induced conformational states, where agonists, partial agonists, neutral antagonists, and inverse agonists may stabilize distinct receptor conformations.

To date, the conformational state of G-protein–coupled receptors has only been inferred indirectly from the activity of the G-protein or the effector enzyme modulated by the G-protein. We have used fluorescence spectroscopy to directly monitor structural changes in the β_2-adrenergic receptor (5). The emission from many fluorescent molecules is strongly dependent on the polarity of the environment in which they are located. Thus, incorporation of fluorescent labels into proteins can be used as sensitive indicators of conformational changes and protein–protein interactions that cause changes in polarity of the environment surrounding the probe. Nitrobenzdioxazol iodoacetamide (IANBD) is a highly fluorescent, cysteine-selective reagent. The fluorescence intensity of IANBD increases as the polarity of the solvent decreases and is more than 10-fold stronger in n-butanol and n-hexane than in aqueous buffer. Labeling of the β_2-adrenergic receptor purified from SF-9 insect cells with IANBD revealed a strong fluorescence signal with an emission maximum at 523 nm. The blueshift in the emission maximum indicates that the modified cysteine(s) are located in an environment that, on the average, is of lower polarity than n-butanol but higher than n-hexane. This would likely involve labeling of one or more of the five cysteine residues that are located in the transmembrane, hydrophobic core of the receptor. The fluorescent labeling did not affect the binding properties of the receptor. We examined several β-adrenergic receptor compounds and compared the effect of these ligands on the fluorescence emission of IANBD-labeled β_2 receptor with their effect on adenylyl cyclase activity in membranes from SF-9 cells expressing β_2 receptors.

The compounds tested altered the fluorescence of IANBD-labeled β_2 receptor and the activity of adenylyl cyclase in membranes with a similar rank order of efficacy. The full agonists, epinephrine and isoproterenol, producing the largest increase in adenylyl cyclase activity, also caused the largest decrease in fluorescence (5–6%), whereas the relatively strong partial agonist, salbutamol, decreased fluorescence about 2.6% in agreement with a lower intrinsic activity in the adenylyl cyclase assay. In contrast, propranolol and ICI 118,551, which inhibit basal receptor activity, caused the largest increase in fluorescence. Alprenolol and pindolol, which in our biological assay exhibit either neutral antagonism or weak negative antagonism, induced a smaller but reproducible increase in fluorescence.

These studies represent the first direct examination of conformational changes in a G-protein–coupled receptor following ligand binding and the first attempt to correlate the functional properties of different receptor ligands with their effect on the conformational state of a purified, solubilized G-protein–coupled receptor. The changes in fluorescence of IANBD-labeled β_2 receptor most likely represent changes in the polarity of the environment surrounding one or more labeled cysteines in the hydrophobic core of the protein. To identify the specific cysteine(s) responsible for ligand-induced changes in fluorescence, we created a series of β_2-receptor mutants having a limited number of reactive cysteines. These mutants were expressed in Sf9 cells, purified, and labeled with IANBD. Agonist-induced changes in fluorescence were observed only in mutants retaining cysteine 125 in transmembrane 3 and cysteine 285 in transmembrane 6. These results indicate that transmembranes 3 and 6 are involved in the conformational changes that accompany agonist activation of the receptor.

To gain further insight into the mechanism of receptor activation, we used this technique for directly monitoring receptor structure to investigate a constitutively active receptor. Samama and coworkers previously showed that a discrete change in the carboxyl terminal part of the third intracellular loop leads to constitutive activation of the β_2-adrenergic receptor (constitutively activated mutant, CAM) (6). We expressed an epitope-tagged version of this mutant in SF-9 insect cells to obtain the large quantities of receptor needed for purification and subsequent structural characterization. The functional properties of CAM in insect cell membranes were similar to those observed for CAM in membranes from transfected mammalian cells.

The expression of functional CAM in SF-9 insect cells was considerably lower than for the wild type (3.4 ± 1.3 pmol/mg protein 8.7 ± 1.8 pmol/mg protein, mean ± SE, n = 4). However, incubation of the cells with either an agonist or an inverse agonist (also referred to as negative antagonist) during the 48-hr infection markedly increased the expression of CAM. This surprising lack of correlation between the increase in expression and the pharmacological properties of the added ligands strongly argues that the increased expression of CAM in the presence of ligand is due to biochemical stabilization of an inherently unstable protein. Observations during solubilization and purification of CAM further demonstrated that constitutive activation confers inherent instability to the receptor. Yields during all steps of the purification procedure were much lower for CAM than for the wild type (data not shown). Moreover, CAM denatured 4 times faster than the wild-type β_2 receptor at 37°C.

Purified CAM and wild-type receptor were labeled with IANBD. Time course analysis revealed that stimulation of IANBD-labeled CAM with the full agonist, isoproterenol, and the partial agonist, salbutamol, elicited substantially greater decreases in fluorescence emission than in the IANBD-labeled wild-type (WT) receptor. In addition, the ratio of the salbutamol response relative to the isoproterenol response increased from 0.40 in the WT to 0.76 in CAM. The changes were fully reversible by antagonist for both receptors.

Taken together, our findings delineate two novel properties of a constitutively activated receptor: structural instability and an exaggerated conformational response to drug binding. These findings, in particular the larger changes in fluorescence observed with CAM in response to agonists, are unexpected. If we assume that the CAM mutation simply causes the unliganded receptor to adopt a conformation mimicking the agonist-bound form of the WT receptor, then the effect of agonist binding might have been expected to produce little additional change in fluorescence. One likely explanation for this apparent discrepancy could be that constitutive activation of the β_2 receptor confers a higher degree of conformational flexibility to the receptor protein due to the disruption of stabilizing conformational constraints. This higher degree of conformational flexibility may allow CAM to more readily undergo transitions between the inactive state (R) and the active state (R*) in response to ligand binding, thus leading to larger fluorescence changes and structural instability. Thus, our results support the hypothesis that constitutive receptor activity is determined by the ability of the receptor to overcome the energy difference between the R and R* state and not by mutation-induced conformational change making R structurally similar to R*. The dramatic increase in basal and agonist-induced activity of CAM as compared with the WT can be explained by a smaller energy difference between the R and R* states. These studies demonstrate the value of constitutively active mutants as tools for understanding the molecular processes involved in activation of receptor proteins.

References

1. Schutz, W., and Freissmuth, M. (1992). Reverse intrinsic activity of antagonists on G protein–coupled receptors. *Trends Pharmacol. Sci.* **13**, 376–80.
2. Samama, P., Pei, G., Costa, T., Cotecchia, S., and Lefkowitz, R. J. (1994). Negative antagonists promote an inactive conformation of the beta 2-adrenergic receptor. *Mol. Pharmacol.* **4**, 390–394.
3. Chidiac, P., Hebert, T. E., Valiquette, M., Dennis, M., and Bouvier, M. (1994). Inverse agonist activity of beta-adrenergic antagonists. *Mol. Pharmacol.* **45**, 490–499.
4. Lefkowitz, R. J., Cotecchia, S., Samama, P., and Costa, T. (1993). Constitutive activity of receptors coupled to guanine nucleotide regulatory proteins. *Trends Pharmacol. Sci.* **14**, 303–307.
5. Gether, U., Lin, S., and Kobilka, B. K. (1995). Fluorescent labeling of purified beta 2 adrenergic receptor. Evidence for ligand-specific conformational changes. *J. Biol. Chem.* **270**, 28268–28275.
6. Samama, P., Cotecchia, S., Costa, T., and Lefkowitz, R. J. (1993). A mutation-induced activated state of the beta 2-adrenergic receptor. Extending the ternary complex model. *J. Biol. Chem.* **268**, 4625–4236.

Stephen M. Lanier,* Rita Raddatz,* and Angelo Parini†

*Department of Cell and Molecular Pharmacology and Experimental Therapeutics
Medical University of South Carolina
Charleston, South Carolina 29425

†Pharmacologie Moleculaire et Physiopathologie Renale
INSERM 4388
Institut Louis Bugnard
Toulouse, France

Relationship between α_2-Adrenergic Receptors and Imidazoline/Guanidinium Receptive Sites

Various compounds with an imidazoline or guanidinium moiety elicit a number of pharmacological effects on metabolism, secretion, ion transport, wakefulness, intraocular pressure dynamics, and cardiovascular–cerebrovascular function. These compounds include the α_1-adrenergic receptor (AR) agonist/α_2-AR antagonist cirazoline, the α_2-AR antagonist idazoxan, the α_2-AR agonist guanabenz, the ion transport inhibitor amiloride, and other structurally related ligands (Table I) (1). Although such ligands interact with known receptor systems, some of their functional effects are pharmacologically ill-defined, such as their centrally mediated effects on blood pressure and the ability of certain imidazoline derivatives to augment glucose-induced insulin secretion from pancreatic β cells. Indeed several studies indicate that these molecules interact with distinct membrane-bound proteins variously termed imidazoline/guanidinium receptive sites (IGRS), imidazoline binding proteins, I_1 and I_2 receptors, imidazoline receptors, nonadrenergic imidazoline binding sites, or imidazole receptors. These sites share the common property of not recognizing endogenous agonists for known monoamine receptors, but subgroups of these binding sites differ in their ligand recognition properties in various tissues.

The synthetic ligands recognized by imidazoline binding proteins elicit several cell responses; however, the signalling pathways involved and the relative importance of these proteins in these events are unclear. Such entities may represent an actual receptor in the target cell directly linked to various cell-signalling proteins, or these binding sites may be involved in the metabolism or transport of such ligands across the cell membrane. Although these proteins do not recognize endogenous agonists for known receptor-signalling systems, some members of this family of imidazoline binding proteins do interact with bioactive endogenous substances such as agmatine and/or as yet unidentified clonidine displacing substances.

Advances in Pharmacology, Volume 42

1054-3589/98 $25.00

TABLE I Properties of Imidazoline Binding Proteins

Imidazolines — Guanidiniums

cirazoline clonidine guanabenz amiloride

Imidazoline binding proteins	*Imidazoline binding site* I_1	I_2
Previous names	Imidazole binding site Imidazoline receptor	Imidazoline-guanidinium receptive site Nonadrenergic idazoxan binding site Idazoxan (I) receptor
Ligands	Clonidine Moxonidine Cirazoline Rilmenidine	Cirazoline Amiloride derivatives Guanabenz Idazoxan
Subcellular localization	Plasma membrane	Plasma membrane (?), mitochondria
Signal transduction mechanisms	Unknown; possible regulation of diacylglycerol formation	Unknown; possible regulation of monoamine oxidase
Subtype-selective radioligands	[^{3}H]clonidine [^{3}H]*p*-aminoclonidine [^{125}I]iodoclonidine	[^{3}H]idazoxan [^{3}H]BFI [^{125}I]AMIPI [^{125}I]AZIPI

[^{125}I]AMIPI, 2-(3-amino-4-iodophenoxyl)methyl imidazoline; [^{125}I]AZIPI, 2-(3-azido-4-iodophenoxy)methyl imidazoline; [^{3}]BFI, 2-(2-benzofuranyl)-2-imidazoline.

IGRS were initially defined by their ligand recognition properties, particularly their high affinity for imidazolines, such as cirazoline and idazoxan, and compounds with a guanidinium moiety, such as guanabenz and amiloride (1). An apparently related entity (imidazole binding site) was initially identified by Ernsberger *et al.* in radioligand-binding studies with the imidazoline [^{3}H]*p*-aminoclonidine and more recently with [^{125}I]clonidine and [^{3}H]moxonidine (2). This site exhibits nanomolar affinity for cirazoline and idazoxan but differs from IGRS in its distribution and expression level and by its lower affinity for guanidinium ligands. The differences in the ligand recognition properties of these two types of binding sites have resulted in the subgrouping of imidazoline binding sites as I_1 and I_2, with the latter most clearly related to IGRS as originally defined. The radioligands [^{3}H]clonidine, [^{125}I]clonidine, and [^{3}H]*p*-aminoclonidine are thus used to identify I_1-binding sites, whereas [^{3}H]idazoxan, [^{125}I]AMIPI, or [^{3}H]BFI are the ligands of choice for I_2- binding sites (IGRS). [^{3}H]idazoxan and [^{3}H]BFI may be selective for a subgroup of I_2-binding sites.

Cirazoline appears to interact with all members of the family of imidazoline binding proteins and, thus, the potential utility of the functionalized radioiodinated cirazoline derivatives as probes for structural characterization and localization of these entities (3). These probes include a radioiodinated probe for reversible binding([^{125}I]AMIPI) and a radioiodinated photoaffinity adduct ([^{125}I]AZIPI) to covalently label the ligand-binding subunit of imidazoline-binding proteins. Molecular confirmation of the heterogeneity within the family of imidazoline-binding proteins was recently provided using such probes, indicating that there are actually a number of binding sites for ligands of this class that differ in their ligand recognition properties, their apparent M_r, and their subcellular/tissue distribution. The demonstration that there is a family of distinct proteins that recognizes imidazoline and guanidinium molecules with varying affinities may explain the difficulty in precisely defining the pharmacology of this group of compounds.

Members of the imidazoline binding protein family are suggested to be G-protein–coupled, heptahelical receptors, monoamine oxidase isoforms, or potassium channels. The structural characterization of the I_1 group of imidazoline-binding proteins has been limited by their relatively low density, their restricted distribution, and the lack of functionalized ligands for use as molecular probes. In contrast, the relatively higher density of I_2 imidazoline-binding sites and the availability of high-affinity molecular probes have resulted in the identification of the ligand-binding subunit for these proteins. Unexpectedly and of particular significance, data indicate that two members of the I_2 family of imidazoline binding proteins are identical to the A and B isoforms of monoamine oxidase (MAO) (4, 5). The imidazoline binding domain on MAO-B is distinct from the enzyme-active site that recognizes the mechanism-based inhibitors such as pargyline and deprenyl, and it is not equally accessible in all tissues.

The demonstration that some of the subtypes of imidazoline-binding proteins in humans are indeed MAO isoforms is particularly exciting. MAO isoforms are therapeutic targets in the management of mood disorders and are also of particular interest in the etiology and treatment of neurodegenerative diseases, such as Parkinson's and Alzheimer's disease. Both diseases will become more prominent over the next 10 years as the percentage of the population over 65 years of age increases significantly. MAO-B is of interest in these diseases because of its ability to metabolize both neurotransmitters and xenobiotics and also because of the generation of damaging free radicals by the enzyme after substrate hydrolysis. Although side effects are often a complicating factor, inhibitors of enzyme activity clearly have utility in the management of certain mood disorders and the early stages of Parkinson's disease. The demonstration that the enzyme exhibits a unique domain that recognizes a broad group of pharmacologically active compounds and that such a site may also recognize endogenous substances assumes immediate significance. Two important questions arise: First, what are the functional consequences of occupying the imidazoline binding domain on substrate hydrolysis/selectivity? Second, how does one explain the variable accessibility of the imidazoline binding domain on MAO-B in different tissues?

Identification of additional imidazoline binding proteins is a goal of several laboratories. Many of the functional effects of imidazoline/guanidinium ligands suggested to involve distinct imidazoline binding proteins are difficult to explain based on the known functions of MAO. Thus, the identification of two types imidazoline binding proteins as MAO suggests a potentially novel therapeutic target for these ligands and at the same time energizes ongoing efforts to identify other members of the protein family.

Acknowledgments

The authors express their appreciation to Drs. V. Bakthavachalam and J. L. Neumeyer of Research Biochemicals for providing 2-[3-aminophenoxy]methyl imidazoline used as a precursor to AZIPI. This work was supported by National Institutes of Health (NIH) grants RO1-NS24821 (S. M. L.) and R44 GM46605 (S. M. L., J. L. N.), the Council for Tobacco Research grant 2235 (S. M. L.), and a NATO Award for International Scientific Exchange (S. M. L., A. P.); R. R. was supported in part by NIH Training to Improve Cardiovascular Drug Therapy grant 5-T32-HL07260-18 and an Individual National Research Service Award (F32NS10332).

References

1. Parini, A., Gargalidis Moudanos, C., Pizzinat, N., and Lanier, S. M. (1996). The elusive family of imidazoline binding proteins. *Trends Pharmacol. Commun.* **17**, 13–16.
2. Ernsberger, P., Meeley, M. P., Mann, J. J., and Reis, D.J. (1987). Clonidine binds to imidazole sites as well as alpha$_2$-adrenoceptors in the ventrolateral medulla. *Eur. J. Pharmacol.* **134**, 1–13.
3. Lanier, S. M., Ivkovic, B., Neumeyer, J., and Bakthavachalam, V. (1992). Identification of the ligand binding subunit of the imidazoline/guanidinium receptive site (IGRS). *J. Biol. Chem.* **268**, 16047–16051.
4. Tesson, F., Limon-Boulez, I., Urban, P., Puype, M., Vandckerckhove, J., Coupry, I., Pompon, D., and Parini, A. (1995). Localization of I$_2$-imidazoline binding sites on monoamine oxidases. *J. Biol. Chem.* **270**, 9856–9861.
5. Raddatz, R., Parini, A., and Lanier, S. M. (1995). Identification of imidazoline/guanidinium receptive sites as distinct domains on the enzyme monoamine oxidase: Cell-type specific access to binding domains. *J. Biol. Chem.* **270**, 15269–15270.

P. Seeman,* R. Corbett,† and H. H. M. Van Tol‡

*Pharmacology Department
University of Toronto
Toronto, Canada M5S 1A8

†Biological Research Department
Hoechst-Roussel Pharmaceuticals Inc.
Somerville, New Jersey 08876

‡Molecular Neurobiology Laboratory
Clarke Institute of Psychiatry
Toronto, Canada M5T 1R8

Dopamine D4 Receptors May Alleviate Antipsychotic-Induced Parkinsonism

I. Atypical Antipsychotics

The blockade of dopamine D2 receptors alleviates psychosis but also produces parkinsonism (1). Antipsychotic drugs that elicit few extrapyramidal signs of parkinsonism are termed *atypical neuroleptics.* What is the receptor basis for this atypical action? Current theories are that atypical neuroleptics

1. With low affinity for D2 may be readily displaced by endogenous dopamine.
2. May block D2 and muscarinic receptors.
3. May block D2 and serotonin-2A receptors (2).
4. May block D2 and D4 receptors.

II. Drug Inhibition Constant Depends on Radioligand

To determine why atypical antipsychotic drugs elicit low levels of parkinsonism, it is necessary to have accurate inhibition constants or dissociation constants (K values) for antipsychotic drugs at dopamine and serotonin receptors. However, despite standard experimental conditions internationally, it is known that the K of a particular neuroleptic may vary considerably between laboratories, particularly when different radioligands are used.

This dependence of the neuroleptic K on the radioligand was studied in more detail (3). It was found that the neuroleptic K depended on the tissue-buffer partition coefficient of the radioligand. For example, clozapine at D2 revealed a K of 390 nM using [^{3}H]nemonapride, 186 nM using [^{3}H]spiperone, and 83 nM using [^{3}H]raclopride. Haloperidol, as another example, had a K of

Advances in Pharmacology, Volume 42

1054-3589/98 $25.00

9.6 nM at D2 using [^{3}H]nemonapride, 2.7 nM using [^{3}H]spiperone, and 0.67 nM using [^{3}H]raclopride. These neuroleptic K values were related to the tissue-buffer partition coefficients of the radioligands (5 for [^{3}H]raclopride; 13.5 for [^{3}H]spiperone; 18.5 for [^{3}H]nemonapride) (4).

Hence, to obtain the real K of a neuroleptic, it was first necessary to obtain the neuroleptic K values using different radioligands of different solubility in the membrane, and then to extrapolate the data to low or "zero" ligand solubility. The extrapolated value represented the radioligand-independent K of the neuroleptic. These values were obtained for dopamine D2 and D4 receptors and for serotonin-2A receptors. The extrapolated value for an antipsychotic agreed with the K obtained directly using the radioactive form of the same antipsychotic drug. For example, clozapine revealed an extrapolated radioligand-independent value of 1.6 nM at D4, agreeing with that directly measured with [^{3}H]clozapine at D4. Such a match also occurred for [^{3}H]chlorpromazine, [^{3}H]haloperidol and [^{3}H]sertindole.

III. Loose and Tight Binding of Antipsychotics

Some atypical neuroleptics (remoxipride, clozapine, perlapine, seroquel, and melperone) had low affinity (loose binding) for the D2 receptor with radioligand-independent K values of 30–90 nM. Such low affinity values make these latter five drugs readily displaceable by high levels of endogenous dopamine in the caudate-putamen. The classical or typical neuroleptics, however, have radioligand-independent values of 0.3–5 nM at D2 (tight binding), making them less displaceable by endogenous dopamine.

IV. Drug Selectivity Is Ligand-Dependent

The receptor selectivity of a drug depends on the radioligands used. For example, olanzapine has a radioligand-independent K of 2 nM at D4. This value is statistically significantly lower than that of 3.7 nM for D2 and lower than that of 5.8 nM for the serotonin-2A receptor. However, if only the data for olanzapine using [^{3}H]spiperone were considered, then olanzapine would be viewed as preferring the serotonin-2A receptor. The receptor selectivity of haloperidol also depends on the ligand used.

V. Clozapine Occupancy of D2 Receptors

A clinical application of the ligand-dependency principle is the resolution of different positron tomography findings in the proportion of D2 receptors occupied by clozapine in humans. For example, clozapine occupies 48% of the D2 receptors in patients when measured with [^{11}C]raclopride (see Farde *et al.*, 1994, and Nordström *et al.*, 1994, in ref. 4), but between 0% and 22% when measured with [^{18}F]methylspiperone (Karbe *et al.*, 1991) or [^{18}F]fluoroethylspiperone (Louwerens *et al.*, 1993). By graphing the percentage of D2 occupied by clozapine versus the tissue-buffer partition of the radioligand, it is possible to extrapolate to zero partition, as is done with the *in vitro* data. Thus, the percentage of D2 occupied by clozapine is extrapolated to approximately 85%, if using a radioligand of zero partition (e.g., dopamine). Under clinical conditions, therefore, in the absence of any radioligand, clozapine occupies high levels of D2 in neuroleptic-treated patients. Thus, the different receptor hypotheses for the clinically atypical neuroleptics may now be examined using the values for the radioligand-independent dissociation constants.

A. Neuroleptics Displaceable by Endogenous Dopamine

Many atypical neuroleptics have a low affinity for D2 and, thus, may be readily displaced by high endogenous concentrations of dopamine in the caudate-putamen. This group includes remoxipride, clozapine, perlapine, seroquel, and melperone. The high values of 30–90 nM for the radioligand-independent K values of these atypical drugs indicate that they are loosely attached to the dopamine D2 receptors and may, therefore, be readily displaced by endogenous dopamine. The principle of displacement of a neuroleptic by endogenous dopamine has been shown for [^{3}H]raclopride, [^{11}C]raclopride, [^{3}H]spiperone, [^{3}H]methylspiperone, [^{18}F]N-methylspiperone, and [^{123}I]iodobenzamide.

Hence, loose neuroleptics would occupy more dopamine receptors in brain regions having low dopamine output (limbic regions, hypothalamus, and prefrontal cortex) but would occupy fewer dopamine receptors in regions having high dopamine output (caudate-putamen) as a result of the neuroleptic competition with endogenous dopamine. Hence, the fraction of dopamine receptors that are blocked in the caudate-putamen would be less than the fraction blocked in the nonstriatal regions, with corresponding fewer extrapyramidal signs, compared with the "tight" typical neuroleptics.

B. Combined Block of Dopamine D2 Receptors and Muscarinic Receptors

A second small group of two atypical neuroleptics, clozapine and thioridazine, block both D2 and muscarinic receptors. Clozapine causes no catalepsy in rats (below 160 mg/kg), but isoclozapine elicits catalepsy at 3 mg/kg. However, because both clozapine and isoclozapine have the same dissociation constant of approximately 10 nM at the muscarinic receptor, the absence of parkinsonism with clozapine cannot be explained by its anticholinergic action.

C. Block of Dopamine D2 Receptors and Serotonin-2A Receptors

Atypical neuroleptics may block both D2 and serotonin-2A receptors. Serotonin receptor block increases the release of dopamine, as measured by the fall in [^{11}C]raclopride binding to D2. Hence, the increased release of endogenous dopamine displaces some of the neuroleptic from D2, alleviating the parkinsonism caused by D2 block. Although this mechanism is reasonable, in reality there is little direct evidence to support the notion that the block of serotonin-2A alleviates extrapyramidal signs. For example, ritanserin does not alleviate haloperidol-induced dystonia in monkeys, unlike clozapine, which effectively reverses this extrapyramidal syndrome (D. Casey) (5).

Using the radioligand-independent K values, there is no relation between the ratio of the neuroleptic dissociation constants for these two receptors (D2/5HT-2A) and the dose that elicits catalepsy in rats (4). This ratio reflects the selectivity of the neuroleptic for the serotonin-2A receptor over the D2 receptor. For example, clozapine and isoclozapine have almost identical selectivity for the serotonin-2A receptor (compared with D2). Nevertheless, isoclozapine elicits catalepsy at about 3 mg/kg, while clozapine does not produce catalepsy at 100 mg/kg.

D. Block of Dopamine D2 and D4 Receptors

Atypical neuroleptics may block both D2 and D4. Omitting the antipsychotics with very low affinity for D2, a relation was found (for 20 drugs) between the antipsychotic doses for rat catalepsy and the D2/D4 ratio of the radioligand-independent K values (4).

An important feature in support of this idea is that clozapine and isoclozapine are considerably different in their values for the D2/D4 ratio of radioligand-independent dissociation constants, but they are in excellent relation to their different cataleptic potencies. This is in contrast to their identical values for the D2/5HT2A ratios of radioligand-independent dissociation constants.

Recent findings with D4-selective [^{3}H]ligands indicated little or no detectable amounts of true D4 dopamine receptors in either human control or schizophrenia striata. This means that the existence of the elevated D4-like sites in schizophrenia, although not representing genuine D4 receptors, may actually represent altered features of D2 or D2-like receptors, possibly a change in the balance of D2 monomers and dimers at the cell surface membrane.

VI. Conclusion

Atypical neuroleptics fall into two groups, those that have a low affinity for dopamine D2 receptors and those that are relatively selective for dopamine D4 receptors.

References

1. Seeman, P., Chau-Wong, M., Tedesco, J., and Wong, K. (1975). Brain receptors for antipsychotic drugs and dopamine: Direct binding assays. *Proc. Natl. Acad. Sci. U.S.A.* **72,** 4376–4380.
2. Meltzer, H. Y. (1995). The role of serotonin in schizophrenia and the place of serotonin-dopamine antagonist antipsychotics. *J. Clin. Psychopharmacol.* **15,** (Suppl 1), 2S–3S.

3. Seeman, P., and Van Tol, H. H. M. (1995). Deriving the therapeutic concentrations for clozapine and haloperidol: The apparent dissociation constant of a neuroleptic at the dopamine D_2 or D_4 receptor varies with the affinity of the competing radioligand. *Eur. J. Pharmacol.* **291,** 59–66.
4. Seeman, P., Corbett, R., Nam, D., and Van Tol, H. H. M. (1996). Dopamine and serotonin receptors: Amino acid sequences, and clinical role in neuroleptic Parkinsonism. *Jpn. J. Pharmacol.* **71,** 187–204.
5. Casey, D. (1995). The nonhuman primate model: Focus on dopamine D2 and serotonin mechanisms. *In* Schizophrenia, Alfred Benson Symposium 28, edited by Fog, R., Gerlach, J., and Hemmingsen, R., Munksgaard, Copenhagen, pp. 287–297.

Bryan L. Roth,* H. Y. Meltzer,† and Naseem Khan*

*Departments of Psychiatry, Neurosciences and Biochemistry
Case Western Reserve University Medical School
Cleveland, Ohio 44106

†Departments of Psychiatry and Pharmacology
Vanderbilt University Medical School
Nashville, Tennessee 37232

Binding of Typical and Atypical Antipsychotic Drugs to Multiple Neurotransmitter Receptors

Schizophrenia is a serious mental illness that afflicts nearly 1% of the world's population and for which there are no definitive treatments. For the past 40 years, typical antipsychotic drugs, exemplified by haloperidol and chlorpromazine have been the mainstays of treatment for schizophrenia. Because these drugs appear to act mainly via blockade of central D2 dopamine (DA) receptors, the hypothesis that DA is primarily involved in the pathogenesis of schizophrenia has gained wide acceptance. During the past 5–10 years, atypical antipsychotic drugs, exemplified by clozapine, have been developed. These drugs do not have many of the toxic side effects of typical antipsychotic drugs such as tardive dyskinesia, parkinsonism, dystonias, akathisia, and elevation of serum prolactin levels (1).

Clozapine remains the prototype of atypical antipsychotic drugs, and no currently available atypical antipsychotic drug appears to have the spectrum of efficacy of clozapine. A large number of putative atypical antipsychotic drugs have been developed that include risperidone, olanzepine, sertindole, ziprasidone, seroquel, melperone, and others (2).

Advances in Pharmacology, Volume 42

1054-3589/98 $25.00

A number of neurochemical hypotheses have been advanced to explain the unique efficacy of clozapine. These include a preferential blockade of mesolimbic D2 receptors, selective antagonism of D4 dopamine receptors, inhibition of α_1-adrenergic receptors, and a balanced blockade of 5HT2A serotonin and D2 dopamine receptors (S2/D2 hypotheses) (2). We have noticed, however, that many atypical antipsychotic drugs bind to a large number of neurotransmitter receptors, including multiple DA (D1, D2, D3, D4) and 5HT (5HT2A, 5HT2C, 5HT6, 5HT7) receptors. In this study, we investigated the spectrum of drug binding of clinically available atypical antipsychotic drugs to multiple DA and 5HT receptors and compared their binding spectrums with typical antipsychotic drugs. We discovered that atypical antipsychotic drugs are, in general, characterized by low D2 dopamine receptor affinity and relatively high affinities for various 5HT receptors (5HT2A, 5HT2C, 5HT6, 5HT7). Portions of this data set have been previously reported (3–5).

For these studies, receptor cDNAs were transiently transfected into COS-7 cells, as previously detailed (3), and 72 hr later, membranes were prepared for radioligand binding assays. For D2 and D4 dopamine receptor binding, [^{3}H]spiperone was used; for 5HT2A receptors, [^{3}H]ketanserin was used; for 5HT2C receptors, [^{3}H]mesulergine was used; and for 5HT6 and 5HT7 receptors, [^{3}H]lysergic acid diethylamide was the radioligand. Typically, specific binding represented more than 90% of total binding, and no more than 10% of the total radioligand was bound. Additionally, membrane protein concentrations were typically 10–20 μg/ml so that interference from differential solubility of radioligands in the lipid phase was avoided. Sources of drugs are as previously detailed (3–5), with the exception of sertindole, which was a gift from Lundbeck (Denmark), and ziprasidone (Pfizer Central Research, Groton, CT).

Figure 1 shows the log of the D2 affinity ratios for several atypical (A) and typical (B) antipsychotic drugs. Data have in most cases been previously reported by this lab (3–5). For a few compounds (ziprasidone, seroquel, zotepine, sertindole, olanzepine), new binding assays were performed for the following receptors: D4 dopamine, 5HT2A, 5HT2C, 5HT6, and 5HT7 serotonin. D2, D4, 5HT2A, and 5HT2C receptor values for ziprasidone were obtained from Seeger *et al.* (6), and D2 and values for sertindole were obtained from Perregaard *et al.* (7).

From Figure 1 it is evident that, in general, typical antipsychotic drugs are characterized as having higher affinities for the D2 receptors relative to other receptors. Additionally, typical antipsychotic drugs had relatively higher affinities for D2 receptors compared with atypical antipsychotic drugs. The average D2 dopamine affinity for typical antipsychotic drugs was 1.12 $\pm$ 1.73 nM, while for atypical antipsychotic drugs the average D2 receptor affinity was 22.3 $\pm$ 2.13 ($p < .01$). In terms of D4 affinities, typical antipsychotic drugs had affinities of 10 $\pm$ 1 nM compared with 16 $\pm$ 2 nM ($p > .05$). Thus, with this group of compounds, no differences were seen comparing typical and atypical antipsychotic drugs in terms of D4 dopamine receptor affinities.

In terms of D2 affinity ratios, the following distinguished atypical antipsychotic drugs from typical antipsychotic drugs: D2/5HT2A ($p < .0002$); D2/5HT2C ($p < .0008$); D2/5HT6 ($p < .01$); and D2/5HT7 ($p < .009$) and D2/D4 affinity ratios ($p < .03$). These results likely reflect the generally low affinity

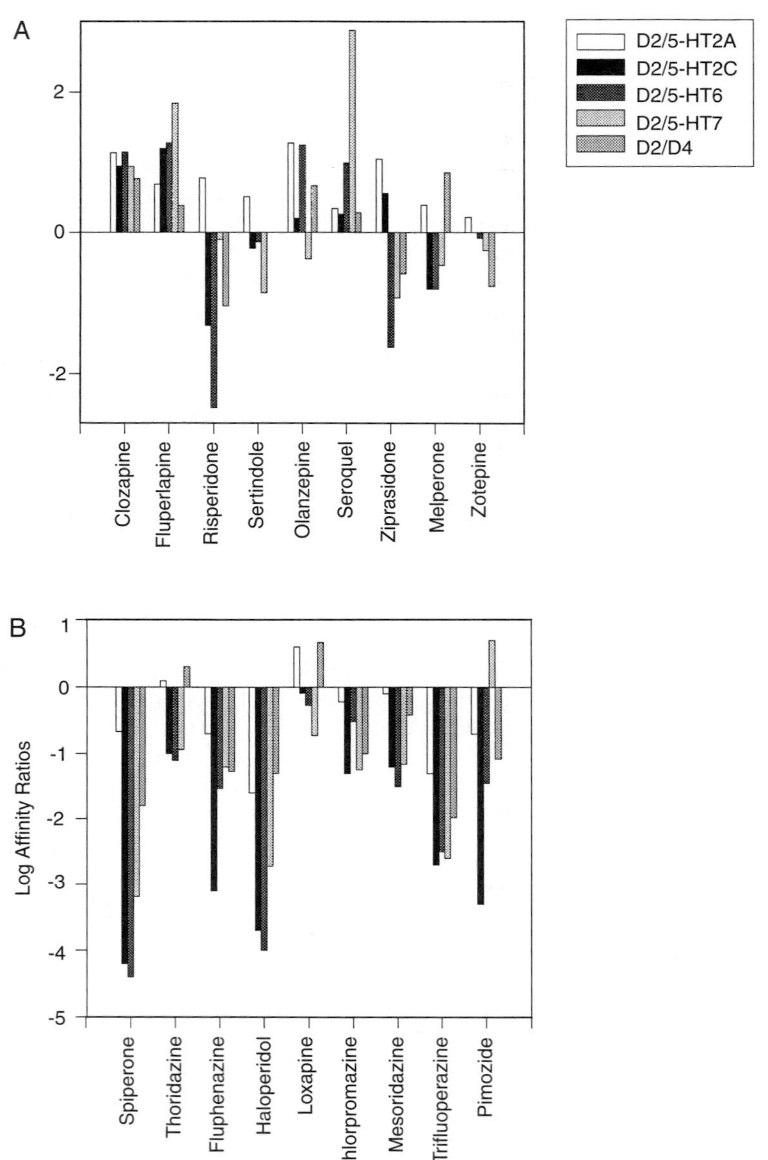

FIGURE 1 D2-dopamine receptor affinity ratios with various (*A*) atypical and (*B*) typical antipsychotic drugs. Shown are the log values of the affinity ratios of D2 versus 5HT2A, 5HT2C, 5HT6, 5HT7, and D4 receptors.

atypical antipsychotic drugs have for D2 dopamine receptors and their relatively higher affinities for receptors.

The fact that atypical antipsychotic drugs, as a group, have relatively high affinity for a number of different receptors suggests that ascribing the unique effects of any one member of this group to a particular receptor is likely to be unproductive. It is more likely that the unique effects of clozapine, for instance, are due to a spectrum of binding to several receptors.

All of the tested atypical antipsychotic drugs are characterized by having generally higher affinties for 5HT2A as compared with D2 dopamine receptors. This group of compounds has also been characterized clinically as having a relatively low incidence of extrapyramidal side effects (EPSs). One can predict that using this measure of atypicality (low EPS) as a criterion for drug development will yield compounds with relatively low EPS liability. It is equally clear, however, that this one measure alone does not describe all of the unique advantages of clozapine (e.g., efficacy in treatment-resistant schizophrenia, lack of tardive dyskinesia), and it is quite likely that other receptors will need to be targeted to obtain these unique properties.

Acknowledgments

This work was supported in part by GM52213, KO2 MH01366, and a grant from the Scottish Rite Schizophrenia Research Foundation to B. L. R. H. Y. M. was supported by MH41684.

References

1. Roth, B. L. (1994). Multiple serotonin receptors: Clinical and experimental aspects. *Ann. Clin. Psychiatry* **6,** 67–78.
2. Fatemi, S. H., Meltzer, H. Y., and Roth, B. L. Atypical antipsychotic drugs: Clinical and preclinical features. *In* Handbook of Experimental Pharmacology. (J. Csernansky, ed.) (in press)
3. Roth, B. L., Ciaranello, R. D., and Meltzer, H. Y. (1992). Binding of typical and atypical antipsychotic drugs to transiently expressed 5-HT1c receptors. *J. Pharmacol. Exp. Ther.* **260,** 1361–1366.
4. Roth, B. L., Craigo, S. C., Choudhary, M. S., Uluer, A., Monsma, F J. Jr., Shen, Y., Meltzer, H Y., and Sibley, D R. (1994). Binding of typical and atypical antipsychotic agents to 5-hydroxytryptamine (5-HT6) and 5-hydroxytryptamine7 (5-HT7) receptors. *J. Pharmacol. Exp. Ther.* **268,** 1403–1410.
5. Roth, B. L., Khan, N., Sibley, D. R., and Meltzer, H. Y. (1995). D4 dopamine receptor affinity does not distinguish between typical and atypical antipsychotic drugs. *Psychopharmacology* **120, 265–368.**
6. Seeger, T. G., Seymour, P. A., Schmidt, A. W., Zorn, S. H. Schulz, D. W., Lebel, L. A., McLean, S ., Guanowsky, V., Howard, H. R., Lowe, J. A., and Heym, J. (1995). Ziprasidone (CP-88,059): A new antipsychotic with combined dopamine and serotonin receptor antagonist activity. *J. Pharmacol. Exp. Ther.* **225,** 101–113.
7. Perregaard, J., Arnt, J., Bogeso, P., Hyttel, J., and Sanchez, C. (1992). Noncataleptogenic, centrally acting dopamine D-2 and serotonin 5-HT2 antagonists with a series of 3-substituted 1-(4-fluorophenyl)-1H-indoles. *J. Med. Chem.* **35,** 1092–1101.

Hubert H. M. Van Tol
Molecular Neurobiology Laboratory
Clarke Institute of Psychiatry
Toronto, Canada M5T 1R8

Structural and Functional Characteristics of the Dopamine D4 Receptor

The dopamine D4 receptor has structural, functional, and pharmacological characteristics closely related to the dopamine D2 receptor family. Both genomic structure (i.e., intron–exon organization) and primary sequence show highest similarity to the D2 receptor. Like the D2 receptor, D4 receptor activation will block adenylyl cyclase activity and stimulate G protein activated inwardly rectifying potassium channel (GIRK) opening. The rank order in affinity for the predominant endogenous amines is dopamine > epinephrine ≥ norepinephrine ≫ serotonin. The difference in affinity between dopamine and (nor)epinephrine is about 10-fold (Fig. 1). This raises the possibility that under certain physiological conditions, the D4 receptor could also be a target for (nor)epinephrine. Various characteristic D2 receptor ligands also bind to the D4 receptor with an equally high affinity. Conversely, there are also some ligands that strongly differ in affinity for the two receptors, as, for example, the antagonists, raclopride and S-sulpiride, and the agonist, bromocryptine, which have a much reduced affinity for the D4 receptor as compared with D2. Clozapine is an example of a ligand that has an increased affinity for the D4 receptor as compared with the D2 receptor (1). This observation has resulted in the hypothesis that some of clozapine's atypical features as an antipsychotic are mediated through blockade of D4 receptors (1).

The physiological functions mediated by the dopamine D4 receptor remain to be defined. Genetic and molecular biological analysis of polymorphisms and mutations occurring within the D4 receptor can contribute to the elucidation of the functional roles of the receptor. Several mutations and polymorphisms change the D4 receptor structure (2, 3). These include (1) an insertion–deletion of a four amino acid sequence immediately upstream from transmembrane 1 (TM1), (2) a frame-shift mutation in TM2, (3) a single nucleotide substitution that converts Val 194 into Gly 194, and (4) a highly polymorphic 48-bp tandem repeat in the third cytoplasmic loop. In addition, it has been observed that translation initiation of the D4 receptor can occur within the first TM region. We have investigated several structural variations of the receptor in more detail to gain further insight into the structure–function relationship of this receptor (4–9).

The Val 194 to Gly 194 polymorphism (8) is located immediately upstream from the Ser residues in TM5 that have been implicated in catecholamine

Advances in Pharmacology, Volume 42

1054-3589/98 $25.00

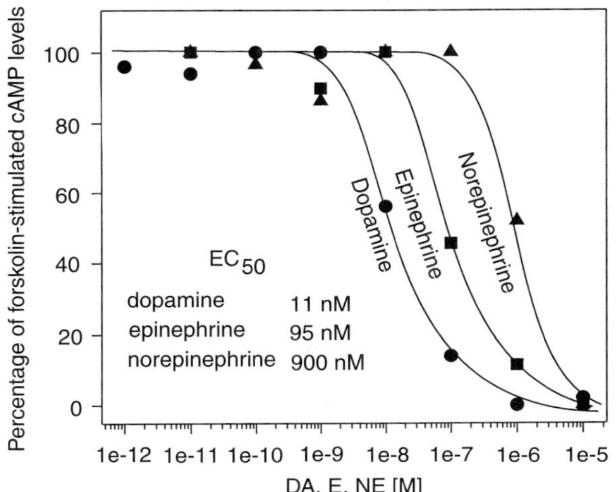

FIGURE 1 Functional activation of D4.2 receptors by dopamine and (nor)epinephrine. cAMP, cyclic adenosine monophosphate; DA, dopamine; E, epinephrine; NE, norepinephrine.

binding. The change of Val 194 into Gly 194 eliminates the high-affinity binding state for dopamine to the D4 receptor. Inclusion of Gpp[NH]p in the competiton binding assay does not further decrease the affinity of the mutant receptor for dopamine. When expressed in Chinese hamster ovary K1 (CHO-K1) cells, this mutant D4 receptor, unlike the wild-type D4 receptor, does not reduce forskolin-stimulated cAMP levels. From these experimental observations it appears that the mutant D4 receptor has a reduced ability of folding into the active high-affinity state and, therefore, is not functionally active. The Val 194 to Gly 194 change also resulted in the complete loss of the receptor's affinity for the atypical neuroleptics clozapine and olanzapine. The affinity of this receptor for haloperidol is only 20-fold reduced, while its affinity for chlorpromazine and spiperone was unaffected. The change of the Val 194 to a Gly 194 is likely to modify the overall structure of this region and possibly TM5 but not the entire structure of the receptor because low-affinity dopamine binding and spiperone, chlorpromazine, and haloperidol binding are minimally affected by this mutation. This suggests that structural determinants for the binding of clozapine and olanzapine are determined by the proper organization of TM5. Interestingly, an individual has been identified that was homozygous for this mutation and, therefore, could be considered as a functional knockout mutation for the D4 receptor. This individual did not have any obvious phenotype that correlated with this mutation.

One of the most remarkable polymorphisms in the human dopamine D4 receptor is a variable number of 48-bp tandem repeat (VNTR) in the putative third cytoplasmic loop (4, 5). We have identified individuals with 2–10 tandem repeat units (called D4.2, D4.3, D4.3, etc.). For the individual units that form the repeat, 19 different sequence variants have been identified. These different units can be found at various positions and frequencies within the VNTR. To date, we have identified 27 polymorphic variants of the D4 receptor (4–6). At the amino acid level, this has resulted in the identification of at least 20 different polymorphic forms of the D4 receptor with respect to this sequence. Unlike the unstable trinucleotide repeat polymorphisms that underlie various genetic disorders and confer genetic anticipation, this VNTR is transmitted in a normal mendelian pattern. This polymorphism appears to be primate-specific and has not been observed in rodents (6). No humans have yet been identified with less than the $\alpha\zeta$ repeat, and the α and ζ amino acid sequences are encoding in every individual analyzed thus far the first and last 16 amino acids. This conservation within the polymorphic region suggests a functional role for this region.

Genetic linkage and association studies using this particular polymorphism have been employed to obtain further insight into possible roles for this receptor in health and disease. Such studies suggest that the D4 locus is not a major contributor to schizophrenia and bipolar disorder (3). However, it has recently been reported that there is a positive association between the repeat polymorphism and novelty seeking and attention deficit hyperactivity disorder (10–12). Whether the repeat polymorphism is directly reponsible for these traits and disorders remains to be elucidated. By examining whether and how the repeat polymorphism alters the the functional characteristics of this receptor, we may obtain better insight into the relation between this receptor and the associated traits and/or disorders.

Pharmacological analysis has not resulted in the identification of major differences between various repeat variants. Deletion of the repeat sequence did not alter the pharmacological profile of the receptor and did not modify high-affinity dopamine binding and Gpp[NH]p sensitivity of the receptor. Nevertheless, small differences in pharmacological properties have been observed between variants with respect to their sensitivity to Na^+ or ionic strength and affinity for clozapine (4, 6). The differences are rather small and are unlikely to be responsible for clozapine responsiveness in antipsychotic treatment. This is supported by the report that there is no genetic association between clozapine response and the polymorphic repeat sequence (13).

The repeat sequence in the D4 receptor is located at about the same position as the alternatively spliced exon sequence of the D2 receptor. The two D2 receptor splice variants show some small differences in signal transduction and coupling to G-proteins. We investigated whether D4 repeat variants also differed in their functional coupling. CHO-K1 cells stably expressing different repeat variants of the D4 receptor could inhibit forskolin-stimulated cAMP production. Careful examination of the D4.2, D4.4, and D4.7 showed that the D4.7 variant displayed a two- to threefold lower potency for dopamine than did the D4.2 and D4.4 receptors (EC 50 $\sim$ 40 nM vs $\sim$15 nM) (7). Although such small variations are unlikely to be of importance for pharmacotherapy mediated through the D4 receptor, it could be of significance for biological functions mediated through this receptor.

Another level at which the repeat polymorphism could introduce functional variation is through differences in expression, possibly through altered translation effciency or RNA stability. We attempted to address the latter possibility by incorporating different repeat sequence variants into the 3′UTR of the luciferase reporter gene, driven from the MT-1 promoter. Transient transfection of such constructs into the somatomammotrophic GH4C1 cell line suggested that the reporter with the $\alpha\zeta$ in the 3′UTR was expressed at about twofold higher levels than the reporter gene with the $\alpha\beta\eta\varepsilon\beta\varepsilon\zeta$ sequence in the 3′UTR (unpublished). Whether *in vivo* differences in expression levels of different D4 variants occur is still unknown. Transgenic approaches will undoubtedly give further insight into the biological role of this receptor. However, if the repeat variation, which appears to be unique for primates, represents a gain of function for this receptor, additional approaches to gene knockout will be required. In this respect, our observation that the repeat sequence and immediate surrounding sequences contain putative SH3 binding domains is particularly intriguing. By using the yeast two-hybrid system and *in vitro* binding assays, we have obtained evidence that this region can interact with the SH3-SH2 adaptor protein Nck. However, mapping of the SH3-binding region indicates that the Nck-binding domain lies immediately upstream from the α-repeat unit. The suggests that the repeat is not essential for this interaction either. The possibilty that other, currently unidentified SH3-binding domains interact with this region cannot be excluded.

References

1. Van Tol, H. H. M., Bunzow, J. R., Guan, H-C., Sunahara, R. K., Seeman, P., Niznik, H. B., and Civelli, O. (1991). Cloning of a human dopamine D4 receptor gene with high affinity for the antipsychotic clozapine. *Nature* **350**, 610–614.
2. Seeman, P., and Van Tol, H. H. M. (1994). Dopamine receptors. *Trends Pharmacol. Sci.* **15**, 264–270.
3. Van Tol, H. H. M., and Seeman, P. (1995). The dopamine D4 receptor: A novel site for antipsychotic action. *Clin. Neuropharmacol.* **18**, (Suppl 1), S143–S153.
4. Van Tol, H. H. M., Wu, C. M., Guan, H-C., Ohara, K., Bunzow, J. R., Civelli, O., Kennedy, J., Seeman, P., Niznik, H. B., and Jovanovic, V. (1992). Multiple dopamine D4 receptor variants in the human population. *Nature* **358**, 149–152.
5. Lichter, J. B., Barr, C. L., Kennedy, J. L., Van Tol, H. H. M., Kidd, K. K., and Livak, K. J. (1993). A hypervariable segment in the human dopamine receptor D4 (DRD4) gene. *Hum. Mol. Genet.* **2**, 767–773.
6. Asghari, V., Schoots, O., Van Kats, S., Ohara, K., Jovanovic, V., Guan, H-C., Bunzow, J. R., Petronis, A., and Van Tol, H. H. M. (1994). The dopamine D4 receptor repeat: Analysis of different native and mutant forms of the human and rat gene. *Mol. Pharmacol.* **46**, 364–373.
7. Asghar, V., Sanyal, S., Buchwaldt, S., Paterson, A., Jovanovic, V., and Van Tol, H. H. M. (1995). Modulation of intracellular cyclic AMP levels by different human dopamine D4 receptor variants. *J. Neurochem.* **65**, 1157–1165.
8. Liu, I. S. C., Seeman, P., Sanyal, S., Ulpian, C., Rodgers-Johnson, P. E. B., Serjeant, G. R., and Van Tol, H. H. M. (1996). Dopamine D4 receptor variant in Africans, D4Valine194Glycine, is insensitive to dopamine and clozapine. Report of a homozygous individual. *Am. J. Med. Genet.* **61**, 277–282.
9. Schoots, O., Sanyal, S., Guan, H-C., Jovanovic, V., and Van Tol, H. H. M. (1996). Cryptic initiation at the human D4 receptor reveals a functional role for the amino-terminus. *Biochemistry* **35**, 3614–3618.
10. Benjamin, J., Li, L., Patterson, C., Greenberg, B. D., Murphy, D. L., and Hamer, D. H. (1996). Population and familial association between the D4 dopamine receptor gene and measures of novelty seeking. *Nat. Genet.* **12**, 81–84.

11. Ebstein, R. P., Novick, O., Umansky, R., Priel, B., Osher, Y., Blaine, D., Bennett, E. R., Nemanov, L., Katz, M., and Belmaker, R. H. (1996). Dopamine D4 receptor (D4DR) exon III polymorphism associated with the human personality trait of novelty seeking. *Nat. Genet.* **12,** 78–80.
12. LaHoste, G. J., Swanson, J. M., Wigal, S. B., Glabe, C., Wigal, T., King, N., and Kennedy, J. L. (1996). Dopamine D4 receptor gene polymorphism in association with attention deficit hyperactivity disorder. *Mol. Psychiatry* **1,** 121–124.
13. Shaikh, S., Collier, D., Kerwin, R. W., Pilowsky, L. S., Gill, M., Xu, X-M., and Thornton, A. (1993). Dopamine D4 subtypes and response to clozapine. *Lancet* **341,** 116.

John F. Tallman
Neurogen Corporation
Branford, Connecticut 06405

NGD 94-1: A Specific Dopamine-4–Receptor Antagonist

While the causes of schizophrenia remain obscure, many of the clinically important drugs for the treatment of schizophrenia interact with the receptors for dopamine. Until recently, there were thought to be two dopamine receptor subtypes (D1 and D2) that were distinguished by their linkage to different second messengers and by the pharmacology of their antagonists. These subtypes were thought to mediate the neurotransmitter functions of dopamine and the therapeutic activity of the neuroleptics. Their cloning and sequencing confirmed the differences of these two subtypes and their localizations. Molecular biological techniques have identified several additional members of the dopamine receptor family, based on their sequence homology to the original D1 and D2 subtypes. At present, five subtypes in the dopamine family and splice variants of these types exist (1, 2).

Although most neuroleptics possess clinical utility consistent with their D2 (and to some extent their D1) antagonist properties, they generally have neuroleptic properties typical of the whole class. In contrast, clozapine possesses unique clinical effectiveness against positive and negative symptoms and usefulness in treatment-refractory schizophrenic patients; this has been demonstrated clinically over a 20-year period. In addition, clozapine's lack of extrapyramidal side effects at therapeutic levels still are unique when compared with even the latest generation of antipsychotics. Several hypotheses about the possible mode of action of clozapine have been advanced. These include balanced activities at multiple biogenic amine and other receptors, unique *in vivo* effects on different groups of neurons, and specific effects on individual receptor subtypes for

Advances in Pharmacology, Volume 42

1054-3589/98 $25.00

dopamine and serotonin. Among the five distinct dopamine subtypes, the dopamine-4 (D4) receptor has been of particular interest because it is a high-affinity site of action of the atypical neuroleptic clozapine and many other neuroleptics (3). To examine the D4-receptor hypothesis, we created a series of highly specific blockers of the D4 receptor.

NGD 94-1 (2-phenyl-4(5)-[4-(2-pyrimidinyl)-piperazin-1-yl)-methyl]-imidazole dimaleate) is a highly selective agent at the D4 receptor. Its affinity for D4 receptors is 3 nM, while its affinity at D1, D2, D3, and D5 receptors is greater than 2 μM, NGD 94-1 has no significant affinity for a wide variety of monoamine or other neurotransmitter receptor or modulatory sites. The only other significant receptor activity is found at 5HT1A receptors with an affinity of 180 nM. The nonspecific agonist, quinpirole, is a specific D4 agonist and inhibits forskolin-stimulated adenylate cyclase activity in a cell line stably expressing only D4 receptors. NGD 94-1 inhibits this agonist effect of quinpirole without possessing significant agonist activity of its own. Thus, NGD 94-1 appears to be a D4 receptor antagonist in this system. In direct binding experiments, [^{3}H]NGD 94-1, prepared by the reductive tritiation of a halogenated precursor, binds to a single set of receptors in transformed cells expressing the human D4 receptor. The pharmacological binding profile of [^{3}H]NGD 94-1 is consistent with expected D4 pharmacology.

Using this ligand in normal tissue with naturally occurring receptors, low levels of specific [^{3}H]NGD 94-1 binding have been detected in human brain homogenates and autoradiographically visualized in rodent brain. In rodent brain, localization to hippocampal and some cortical regions is noted. [^{3}H]NGD 94-1 binding appears to have a highly localized distribution that is distinct from either D2- or D3-receptor localizations. Prominent [^{3}H]NGD 94-1 binding is localized in human hippocampal, hypothalmic, and several cortical regions; binding is not prominent in striatal areas in rodent and caudate–putamen in humans. This binding profile is consistent with the localization of D4 receptors, as demonstrated immunocytochemically by antibodies directed against unique protein sequences of this receptor (4). In addition, immunocytochemical localization indicates that the cells in cerebral cortex, hippocampus, and other regions containing D4 receptors in primates are in part GABAergic in nature. Additional labeling in cortical pyramidal cells was noted, and the immunocytochemical localization was consistent with the hypothesis that disinhibition of excitatory transmission may play a part in the action of clozapine. By inference, NGD 94-1's action behaviorally may have a similar basis.

In rats, NGD 94-1 is inactive in many of the traditional behavioral models that investigators consider predictive of antipsychotic activity. This reflects in no small part the dependence of many of these models on D2-receptor blockade. NGD 94-1 does not produce catalepsy and does not block amphetamine- or apomorphine-induced stereotypy. In these models, it is similar to clozapine and may be contrasted with the typical neuroleptic haloperidol. NGD 94-1 does not reduce spontaneous locomotor activity or rearing and does not inhibit the acoustic startle response; thus, in contrast to clozapine, NGD 94-1 is not sedating. In the rodent prepulse inhibition model, NGD 94-1 is effective in partly reversing the disruptive effects of apomorphine; both haloperidol and clozapine are active in this model. NGD 94-1 also presents improvement in performance retention in a passive avoidance task in rodents and improves retention in a spatial water maze task; both of these models indicate effects on learning and memory that are consistent with the hippocampal localization of the D4 receptors. No effects of NGD 94-1 were noted in a proconvulsant rodent model; this is in contrast to clozapine at higher doses, which shows clinically important proconvulsant effects. No acute effects on prolactin levels in the rat were noted; prolactin elevation is a D2-receptor–mediated effect.

In summary, NGD 94-1 possesses a preclinical profile that is at once similar to clozapine and also quite different from clozapine. The clinical utility of D4 antagonists, such as NGD 94-1, remains unknown. While it is possible that they may possess the full activities of a clozapine, it is more likely that, because of their unique specificity, they will possess a different profile and perhaps greater, but narrower, specificity for subsets of symptoms. In this regard, the recently demonstrated linkage between the D4-receptor gene and novelty-seeking temperament traits may underlie some aspects of the actions of D4 antagonists (5). Thus, it is quite possible that the D4-receptor blockers will be useful in addressing the negative symptoms or cognitive deficits that are a feature of schizophrenia and that are not currently addressed effectively by existing medications.

Acknowledgments

The work summarized here represents the collaborative efforts of many of my colleagues at Neurogen Corporation who are involved in our dopamine discovery program.

References

1. Seeman, P. (1995). Dopamine receptors and psychosis. Sci. Am. **2,** 28–37.
2. Civelli, O. (1995). Molecular biology of the dopamine receptor subtypes. *In* Psychopharmacology: The Fourth Generation of Progress. (F. E. Bloom and D. J. Kupfer, eds.), pp. 155–161. Raven Press, New York.
3. Van Tol, H. J. M., Bunzow, J. R., Guan, H., Sunahara, R. K., Seeman, P., Niznik, H. B., and Civelli, O. (1991). Cloning of the gene for a human dopamine D4 receptor with high affinity for the antipsychotic clozapine. *Nature* **350,** 610–614.
4. Mrzljak, L., Bergson, C., Pappy, M., Huff, R., Levenson, R., and Goldman-Rakic, P. S. (1996). Localization of dopamine D4 receptors in GABAergic neurons of the primate brain. *Nature* **381,** 245–248.
5. Benjamin, J., Lin, L., Patterson, C., Greenberg, B. D., Murphy, D. L., and Hamer, D. H. (1996). Population and familial association between the D4 dopamine receptor gene and measures of novelty seeking. *Nat. Genet.* **12,** 81–84.

L. B. MacMillan,* P. P. Lakhlani,* L. Hein,† M. Piascik,‡ T. Z. Guo,§ D. Lovinger,‖ M. Maze,§ and L. E. Limbird*

*Department of Pharmacology
Vanderbilt University
Nashville, Tennessee 37232

†Departments of Medicine, and Molecular and Cellular Physiology
Stanford University
Stanford, California 94305

‡Department of Pharmacology, Vascular Biology Research Group
University of Kentucky
Lexington, Kentucky 40536

§Department of Anesthesia
Stanford University
Stanford California 94305
Veterans Affairs, Palo Alto Health Care System
Palo Alto, California 94304

‖Department of Molecular Physiology and Biophysics
Vanderbilt University
Nashville, Tennessee 37232

In Vivo Mutation of the α_{2A}-Adrenergic Receptor by Homologous Recombination Reveals the Role of This Receptor Subtype in Multiple Physiological Processes

The endogenous catecholamines norepinephrine and epinephrine elicit physiological responses involving virtually every organ system. They function both as circulating hormones and as neurotransmitters in the central and peripheral nervous systems. Norepinephrine and epinephrine elicit their responses by interacting with adrenergic receptors (also referred to as adrenoceptors), a family of nine currently identified members of three major groups: α_1, α_2, and β (1). Adrenergic receptors belong to the superfamily of G-protein–coupled receptors and share the common features of an extracellular N-terminus, seven α-helical proposed transmembrane domains, and an intracellular C-terminus (2).

α_2-Adrenergic receptors (α_2ARs) are broadly associated with inhibitory actions of norepinephrine and epinephrine. Central nervous system α_2ARs inhibit neuronal firing and neurotransmitter release. Studies with drugs designed as α_2AR agonists and antagonists have demonstrated that α_2ARs participate in a wide range of central nervous system activities, including central regulation of blood pressure, sedation and analgesia, control of affect, and modulation of pituitary hormone release. In the periphery, α_2ARs inhibit neurotransmitter

Advances in Pharmacology, Volume 42

1054-3589/98 $25.00

release from peripheral nervous system neuronal terminals, inhibit insulin secretion by pancreatic β cells, and participate in the regulation of water and electrolyte balance in the kidney. Platelet α_2ARs activate aggregation, and α_2ARs localized to arteriolar and venous smooth muscle elicit a contractile response. These examples of α_2AR activity are not comprehensive but emphasize the broad spectrum of physiological functions that include α_2AR participation (3).

The α_2ARs that elicit these varied responses represent a structural and functional family of receptor subtypes defined by pharmacological measurements and molecular cloning. There are three α_2AR subtypes (α_{2A}-, α_{2B}-, and α_{2C}AR), and these subtypes all couple via pertussis toxin–sensitive G_i and G_o proteins to inhibition of adenylyl cyclase, activation of receptor-operated K^+ currents, and inhibition of voltage-sensitive Ca^{2+} currents.

Our intention in beginning these studies was to establish a mouse line that expressed a mutant α_{2A}AR receptor with selectively altered signal transduction capabilities. Studies from our laboratory had demonstrated that mutation of the highly conserved aspartate residue at position 79 in the α_{2A}AR to asparagine (D79N) resulted in a receptor that was uncoupled from a single signaling pathway. When expressed in anterior pituitary AtT20 cells, the D79N α_{2A}AR was able to couple to the inhibition of voltage-gated Ca^{2+} channels and of cyclic adenosine monophosphate (cAMP) production but was unable to couple to the activation of K^+ currents (4). We expected that expression of this D79N α_{2A}AR instead of the wild-type α_{2A}AR *in vivo* would reveal those physiological functions of the α_{2A}AR that relied on receptor-mediated K^+ current activation.

We successfully used the two-step "hit and run" gene-targeting approach in mouse embryonic stem cells (5) to establish a mouse line with this D79N α_{2A}AR mutation. Unexpectedly, we found that the density of functional D79N α_{2A}AR, evaluated using radioligand binding assays on membrane preparations from mouse brain, was dramatically reduced relative to wild-type α_{2A}AR density. This reduction in D79N α_{2A}AR density was not a consequence of altered transcription of the mutant gene because α_{2A}AR RNA levels were not different in wild-type and D79N mice. In addition, electrophysiological studies in locus ceruleus neurons from the mutant mice revealed that the D79N α_{2A}AR was not able to activate K^+ currents and that this mutant receptor was also unable to inhibit Ca^{2+} currents (6). Together, the reduced receptor density and perturbed coupling of the *in vivo* D79N α_{2A}AR may mean that the D79N mice represent a functional knockout of the α_{2A}AR subtype. Although these data indicate that functional deficits in the D79N mice cannot be interpreted in terms of α_{2A}AR signaling, such deficits can reveal the specific functional roles of the α_{2A}AR subtype.

We examined a variety of α_2AR-mediated functions in the D79N mice and demonstrated that the α_{2A}AR subtype is critically important to several prominent functions associated with α_2AR activity. Our studies provide evidence that the α_{2A}AR subtype mediates the decrease in blood pressure following α_2AR agonist administration. This centrally mediated hypotensive response has been the rationale for the use of α_2AR agonists as antihypertensive agents (3) but has been increasingly attributed to the interaction of these drugs with "imidazoline receptors" (7–9). Our findings clearly demonstrate the importance of the α_{2A}AR subtype to this therapeutic effect (10). In addition, we demonstrated that the α_{2A}AR subtype mediates the sedative, anesthetic-sparing and analgesic responses to α_2AR agonist administration (6). These α_2AR-mediated functions are exploited in clinical anesthesiology (11).

One goal in the generation of new pharmacologic agents is to increase specificity and reduce unwanted side effects. It has been presumed that elucidation of the roles of individual α_2AR subtypes would provide a rationale for the development of exclusively subtype-specific drugs that would elicit only a select subset of the many α_2AR-elicited functions. Our studies reveal that the α_{2A}AR subtype has a role central to several of the clinically desirable effects of α_2AR agonists, suggesting that these functions cannot be separated by administering subtype-specific drugs. Null mutation of the α_{2B}AR subtype has demonstrated that this subtype mediates the increase in blood pressure immediately following α_2AR agonist administration (12). Thus, drugs that are selective for the α_{2A}AR subtype relative to the α_{2B}AR subtype may hold promise as improved antihypertensive agents. Sedative side effects might be eliminated if high-affinity partial agonists are developed, especially if higher receptor occupancy is needed to evoke sedation in contrast to hypotension. The possibility that the α_{2A}AR subtype utilizes different signal transduction pathways to achieve one function versus another remains, but elucidation of any differences that might be exploited therapeutically will require novel experimental strategies.

References

1. Bylund, D. B., Eikenberg, D. C., Hieble, J. P., Langer, S. Z., Lefkowitz, R. J., Minneman, K. P., Molinoff, P. B., Ruffolo, R. R. Jr., and Trendelenburg, U. (1994). International union of pharmacology nomenclature of adrenoceptors. *Pharmacol. Rev.* **46,** 121–136.
2. Strader, C. D., Fong, T. M., Tota, M. R., Underwood, D., and Dixon, R. A. F. (1994). Structure and function of G protein-coupled receptors. *Annu. Rev. Biochem.* **63,** 101–132.
3. Ruffolo, R. R. Jr., Nichols, A. J., Stadel, J. M., and Hieble, J. P. (1993). Pharmacologic and therapeutic applications of α_2-adrenoceptor subtypes. *Annu. Rev. Pharmacol. Toxicol.* **32,** 243–279.
4. Surprenant, A., Horstman, D. A., Akbarali, H., and Limbird, L. E. (1992). A point mutation of the α_2-adrenoceptor that blocks coupling to potassium but not calcium currents. *Science* **257,** 977–980.
5. Hasty, P., Ramirez-Solis, R., Krumlauf, R., and Bradley, A. (1991). Introduction of a subtle mutation into the *Hox-2.6* locus in embryonic stem cells. *Nature* **350,** 243–246.
6. Lakhlani, P. P., MacMillan, L. B., Guo, T. Z., McCool, B. A., Lovinger, D. M., Maze, M., and Limbird, L. E. (1997). Substitution of a mutant α_{2A}-adrenergic receptor via 'hit and run' gene targeting reveals the role of this subtype in sedative, analgesic, and anesthetic-sparing responses *in vivo. Proc. Natl. Acad. Sci. USA.* (in press).
7. Ernsberger, P., Giuliano, R., Willette, R. N., and Reis, D. J. (1990). Role of imidazoline receptors in the vasodepressor response to clonidine analogs in the rostral ventrolateral medulla. *J. Pharmacol. Exp. Ther.* **253,** 408–418.
8. Tibirica, E., Feldman, J., Mermet, C., Gonon, F., and Bousquet, P. (1991). An imidazoline-specific mechanism for the hypotensive effect of clonidine: A study with yohimbine and idazoxan. *J. Pharmacol. Exp. Ther.* **256,** 606–613.
9. Bousquet, P., Feldman, J., Tibirica, E., Bricca, G., Greney, H., Dontenwill, M., Stutzmann, J., and Belcourt, A. (1992). Imidazoline receptors: A new concept in central regulation of the arterial blood pressure. *Am. J. Hypertens.* **5,** 47S–50S.

10. MacMillan, L. B., Hein, L., Smith, M. S., Piascik, M. T., and Limbird, L. E. (1996). Central hypotensive effects of the α_{2a}-adrenergic receptor subtype. *Science* **273**, 801–803.
11. Hayashi, Y., and Maze, M. (1993). Alpha$_2$ adrenoceptor agonists and anaesthesia. *Br. J. Anaesth.* **71**, 108–118.
12. Link, R. E., Desai, K., Hein, L., Stevens, M. S., Chruscinski, A., Bernstein, D., Barsh, G. S., and Kobilka, B. K. (1996). Cardiovascular regulation in α_2-adrenoceptor subtypes B and C knockout mice. *Science* **273**, 803–806.

M. Lafontan, S. Betuing, J-S. Saulnier-Blache, P. Valet, A. Bouloumié, C. Carpéné, J. Galitzky, and M. Berlan

Unité INSERM 317
Institut Louis Bugnard
Faculté de Médecine
Université Paul Sabatier
CHR Rangueil
31403 Toulouse Cedex 4, France

Regulation of Fat-Cell Function by α_2-Adrenergic Receptors

Normal or pathological development of white adipose tissue results from the combined regulation of hypertrophic and hyperplasic phases. Hyperplasia corresponds to the recruitment of new adipocytes and depends on the number of adipose precursors, while hypertrophy is linked to triglyceride storage and is dependent on lipogenic and lipolytic processes. Catecholamines play a major role among the hormones and neurotransmitters controlling the function of the cells composing white adipose tissue (e.g., fat-cell precursors, mature adipocytes, and vessels). The control of fat-cell function by the catecholamines involves at least five different adrenoceptor subtypes: three β-adrenoceptors (βARs) (e.g., β_1-, β_2-, and β_3-), one α_{2A}AR, and one α_{1B}AR, the importance of which differs according to cell differentiation stage and animal species. Some essential features concerning fat-cell α_2AR will be considered here; an extended review has been published (1). In fat cells, α_2ARs are coupled to adenylyl cyclase through G_i proteins (G_{i1}, G_{i2}, and G_{i3} proteins have been identified in fat cells). An important short-term metabolic action of catecholamines in fat cells is related to the control of the changes in cyclic adenosine monophosphate (cAMP) levels promoted by βAR-mediated stimulation versus α_2AR-mediated inhibition of adenylyl cyclase activity. Intracellular cAMP changes alone are sufficient to meet the requirements for stimulation of cAMP-dependent protein kinase and subsequent lipolysis activation by hormone-sensitive lipase. Selective α_2AR stimulation promotes adenylyl cyclase inhibition, reduction of intracellular cAMP

Advances in Pharmacology, Volume 42

1054-3589/98 $25.00

levels, and antilipolysis. The α_2-antagonist, RX821002, is a potent antagonist and the most appropriate ligand, under its tritiated form, for α_2AR identification in fat cells from humans and other mammals.

When considering the role of α_2-AR in mature adipocytes, there are minor species-specific variations in the pharmacology of fat-cell α_2AR that appear to be species variants of the α_{2A} subtype (encoded by the α_2-C10 gene) existing in human fat cells. The importance of the fat-cell α_2AR varies according to the species, the sex, the age, and the degree of obesity in humans and mammalian models. Striking differences exist in the disribution of fat cell α_{2A}AR according to the anatomical location of the fat deposits. In most human fat cells, α_2AR numerically predominate over β_1/β_2AR. Preferential recruitment of the α_2AR, at lower catecholamine concentrations than for the β_1/β_2AR, has been described for epinephrine and norepinephrine. Overexpression of α_2AR in human fat cells is always associated with weakened lipolytic responses to epinephrine and norepinephrine. α_2AR-dependent antilipolytic effects are weaker, and α_2AR number is lower in omental than in subcutaneous adipocytes in women. In obese men, reduction in the lipolytic response of subcuntaneous fat cells to ephinephrine involves changes in the functional balance between α_2- and βAR; reduced lipolytic response to catecholamines in obesity is related to greater α_2-adrenergic component. A recent study in 3T3-F442A adipocytes transfected with the human α_2-C10 gene and expressing a level of α_{2A}AR similar to that found in human fat cells, has clearly shown that the lipolytic response initiated by epinephrine in such adipocytes is reduced when compared with that of isoproterenol and mimicks the response that is observed in human fat cells (unpublished results). Desensitization and downregulation of adrenergic receptors is tissue- and subtype-selective; the fat-cell α_{2A}AR appears quite refractory to desensitization in contrast with β_1/β_2AR.

Important species-specific differences exist in fat-cell α_2AR distribution; weaker α_2-adrenergic effects are observed in various species (hamster, dog, rat, and rabbit). In a given species, heterogeneous distribution of β- and α_2AR has been described in fat cells originating from different anatomical sites, as in humans. The reduction of the lipolytic response to epinephrine in the aging animal is in part linked to an increased α_2-adrenergic responsiveness (and reduced β_3-adrenergic responsiveness). The enhancement of α_2-adrenergic responsiveness is generally associated with a concomitant increase in the α_2AR number in fat cells. The main factor involved in the genesis of the α_2-adrenergic responsiveness and rise in α_2AR number is the increase of fat-cell size rather than aging, as previously suspected. Dramatic fat-cell size reduction, promoted by various physiological experiments (e.g., fasting, prolonged caloric restriction, cold exposure, lactation of female hamsters, adrenalectomized hamsters, and streptozotocin-induced diabetic animals) was always associated with a strong decrease of the number of α_2-binding sites and a parallel reduction of α_2-adrenergic responsiveness.

Castration and short-day photoperiod exposure in the hamster, which promotes a "physiological" castration, reduced the α_2-adrenergic responsiveness and the number of α_2ARs in white fat cells. Administration of testosterone promoted a rapid and dose-dependent upregulation of fat-cell α_2AR in castrated hamsters or animals in short-day photoperiod; the effect was not observed with other sex steroids. Upregulation of fat-cell α_2AR was also induced by testosterone administration in both young and adult male hamsters adapted to long-day photoperiods, as well as in females. A specific transcriptional regulation of the adipocyte α_2AR gene by androgens was suspected, because increases in α_2AR mRNA levels coincided with the increase in α_2AR number (2). The effect of testosterone on α_2AR is tissue-specific and testosterone-dependent, and upregulation of α_2AR is limited to adipocytes. Moreover, the upregulating effect of testosterone on α_2AR expression was also observed in hamster fat-cell precursors cultured *in vitro* in the presence of testosterone (3).

α_2AR also exist in fat-cell precursors, and they are linked to transducing systems playing a role independent of cAMP production. Actions on preadipocyte growth have been described in rat fat-cell precursors, BFC-1β preadipocytes, and 3T3-F442A preadipocytes stably transfected with the human α_2-C10 gene, with expression levels of α_{2A}ARs ranging within physiological levels. In rat white fat-cell precursors, shown to express α_2- and βAR in serum-deprived medium, α_2AR stimulation led to the phosphorylation of p42 and p44 mitogen-activated protein kinases and an increase in cell number (4). Additionally, α_2-adrenergic stimulation, promoted by various α_2-agonists (UK14304, clonidine, epinephrine, dexmedetomidine), restored the spreading of cells previously retracted by serum withdrawal. This is associated with cell membrane ruffling, formation of lamelipodia and filipodia, appearance of focal adhesion plaques, and induction of actin stress fibers, as well as tyrosine phosphorylation of pp125 focal adhesion kinase. This effect is mediated by G_i proteins (it is pertussis toxin–sensitive) and is protein kinase C–independent. It is blocked by pretreatment of the cells with dihydrocytochalasin B (a blocker of actin polymerization), genistein (a tyrosine kinase inhibitor), and agents that increase cell cAMP content (forskolin, bromo-cyclic AMP, and isobutylmethylxanthine) (5).

These observations suggest that catecholamines, via α_2AR-dependent pathways, are able to regulate adhesion and actin cytoskeleton activity in white fat-cell precursors by interfering with the integrin-mediated pathway. In addition to their contribution to the regulation of lipolysis, α_2AR could also be an important component in the control of white adipose tissue hyperplasia by the sympathetic nervous system.

References

1. Lafontan, M., and Berlan, M. (1995). Fat cell α_2-adrenoceptors: The regulation of fat cell function and lipolysis. *Endocr. Rev.* **16**, 716–738.
2. Saulnier-Blache, J. S., Bouloumié, A., Valet, P., Devedjian, J-C., and Lafontan, M. (1992). Androgenic regulation of adipocyte alpha2-adrenoceptor expression in male and female Syrian hamsters: Proposed transcriptional mechanism. *Endocrinology* **130**, 316–327.
3. Bouloumié, A., Valet, P., Daviaud, D., Lafontan, M., and Saulnier-Blache, J. S. (1994). *In vivo* up-regulation of adipocyte α_2-adrenoceptors by androgens is the consequence of a direct action on fat cells. *Am. J. Physiol.* **267**, 926–931.
4. Bouloumié, A., Planat, V., Devedjian, J-C., Valet, P., Saulnier-Blache, J-S. Record, M., and Lafontan, M. (1994). Alpha2-adrenergic stimulation promotes preadipocyte proliferation: Involvement of mitogen-activated protein kinases. *J. Biol. Chem.* **269**, 30254–30259.
5. Bétuing, S., Daviaud, D., Valet, P., Bouloumié, A., Lafontan, M., and Saulnier-Blache, J. S. (1996). α2-Adrenoceptor stimulation promotes actin polymerization and focal adhesion in 3T3F442A and BFC-1β preadipocytes. *Endocrinology* **37**, 5220–5229.

D. K. Rohrer,* D. Bernstein,* A. Chruscinski,† K. H. Desai,* E. Schauble,* and B. K. Kobilka†

*Department of Pediatrics
†Howard Hughes Medical Institute
Stanford University
Stanford, California 94305

The Developmental and Physiological Consequences of Disrupting Genes Encoding β_1 and β_2 Adrenoceptors

β-Adrenergic receptors (βARs) are important modulators of many physiological processes, including cardiac chronotropic and inotropic states, vascular and smooth-muscle tone, lipid and carbohydrate metabolism, and behavior. Three βAR subtypes have been defined by molecular cloning (β_1,β_2,β_3). The role that individual βAR subtypes play in specific physiological processes has been traditionally defined using subtype-specific ligands. Unfortunately, many of these ligands either possess poor selectivity or are used at concentrations *in vivo* that can lead to occupation of nonspecific subtypes. The advent of gene disruption techniques in the mouse now enables one to selectively delete or alter cloned genes as a means of identifying the specific function(s) of their gene products. To better understand βAR subtype-specific functions in the context of either the whole animal or isolated organs and cells, we have disrupted the genes encoding both β_1- and β_2ARs.

Gene targeting vectors containing β_1AR or β_2AR gene sequences interrupted or partially replaced by a bacterial neomycin resistance gene cassette were flanked by a viral thymidine kinase gene cassette. A standard positive–negative selection strategy (G418 + gancyclovir) was used to isolate R1 embryonic stem (ES) cells, having undergone homologous recombination at β_1AR or β_2AR loci. Chimeric mice harboring these engineered ES cells were mated to wild-type mice to ascertain transmission of the disrupted alleles. Mice heterozygous for disrupted β_1AR or β_2AR genes were then intercrossed to generate homozygous β_1AR or β_2AR knockouts. Mendelian expectations would predict that such crosses would yield 25% wild-type, 50% heterozygous βAR knockouts, and 25% homozygous βAR knockouts. β_1AR heterozygote intercrosses performed in this way produce litters that are selectively deficient in homozygous β_1AR knockouts (1). When performed on a mixed-strain background (129Sv, C57B1/6J, and DBA2/J), approximately 30% of the expected number of homozygous β_1AR knockouts are recovered at weaning. When performed on a congenic

1054-3589/98 $25.00

background (129Sv only), approximately 10% of the expected number of homozygous β_1AR knockouts are recovered. Current evidence suggests that the β_1AR homozygous knockouts die in utero between days 10.5 and 18.5 of gestation (1). Homozygous β_1AR knockouts that do survive appear normal and are fertile, and interestingly, normal litter sizes can be recovered from homozygous β_1AR knockout intercrosses from the mixed background. While surviving β_1AR knockouts on the congenic 129Sv strain are viable, their fertility is greatly reduced and intercrosses between homozygous β_1AR knockouts rarely produce surviving knockout pups. These results suggest that β_1AR is critical for proper development and that strain-specific modifiers exist in either the C57B1/6J or DBA/2J mice that can rescue lethality on the 129Sv background. The specific cause of death in β_1AR knockouts is unknown. Preliminary experiments performed on β_2AR knockouts suggest that disruption of the β_2AR gene has no effect on mouse viability or survival.

The pharmacological profile of β_1AR and β_2AR knockouts is consistent with the loss of these specific receptor subtypes. When the nonspecific βAR antagonist [^{125}I]cyanopindolol is used in competition binding studies, excess unlabeled CGP 20712A (β_1AR-specific antagonist) reveals a selective loss of specific high-affinity sites in β_1AR knockout heart or lung membranes (1), while excess unlabeled ICI 118,551 (β_2AR-specific antagonist) reveals a selective loss of specific high-affinity sites in β_2AR knockout lung membranes. These results provide further evidence that the β_1AR and β_2AR gene disruptions have abolished βAR protein expression.

The physiological impact of knocking out either β_1- or β_2ARs has been studied in instrumented animals and in isolated tissues (1). For *in vivo* analysis of awake mice, animals were anesthetized with methoxyflurane, and stretched PE-10 tubing was inserted into the left carotid artery for blood pressure measurement and drug delivery. After recovery from anesthesia, baseline heart rate and blood pressure were determined (Crystal Biotech Dataflow software, Hopkington, MA), and the responses to various pharmacological agents were tested. While baseline blood pressure and heart rate were not significantly different between wild-type and β_1AR knockouts, isoproterenol had virtually no effect on heart rate in β_1AR knockouts, while causing a robust stimulation in wild-type mice. This lack of chronotropic stimulation was confirmed both in anesthetized mice monitored by ECG and in isolated spontaneously beating atria (Table I). Inotropic stimulation of wild-type or β_1AR knockout cardiac function by isoproterenol was tested *in vitro* using paced, right ventricular strips and an isometric force transducer. These experiments clearly showed that right ventricles from β_1AR-deficient mice are completely unaffected by isoproterenol, while wild-type ventricles are strongly stimulated (wild-type, 8.9-fold stimulation in twitch amplitude; β_1AR knockout, 0-fold stimulation, at 1 μmol/liter of isoproterenol; $p < .005$) (1). However, inotropic reserve as measured by forskolin stimulation is equivalent between wild-type and β_1AR-deficient ventricular preparations (4.9-fold vs 3.9-fold increase in twitch amplitude, respectively; p = NS). Early experiments examining the physiological responses of β_2AR knockouts suggest that the chronotropic response to isoproterenol is intact.

While our initial studies appeared to show that β_1AR knockouts were unable to increase heart rate in response to isoproterenol, technical changes

TABLE I Summary of Cardiovascular Measurements

	In Vivo					*In Vitro*
	Basal		*Isoproterenol Stimulated*			
	Mean BP	*HR*	*% Δ BP*	*% ΔHR*	*% Δ HR (anesth.)*	*% Δ (atrial rate)*
+/+	101.1 ± 9.5 (6)	520.6 ±24.8 (6)	−21.9 ±3.1 (6)	+18.6[a] ±5.9 (6)	+40.3[b] ±9.4 (4)	+58.1[b] ±11.7 (6)
−/−	91.9 ±5.3 (5)	542.6 ±28.9 (5)	−23.5 ±4.9 (5)	+1.8 ±0.5 (5)	−0.7 ±3.1 (5)	−2.6 ±2.9 (6)

All values are mean ± SEM; number of independent measurements is given inside parentheses.
[a] Significance at $p < .05$ for student's t-test comparing +/+ and −/− mice.
[b] Significance at $p < .01$ for student's t-test comparing +/+ and −/− mice. BP, mean blood pressure (mm Hg); HR, heart rate.

made in the mode of surgical anesthesia have led to a slight revision of our earlier findings. Isoflurane is currently being used as a surgical anesthetic for catheter placement, and animals instrumented in this way frequently show tachycardic responses to isoproterenol, exercise, or agents that induce a baroreflex response (e.g., sodium nitroprusside). Animals surgically treated in this way display lower basal heart rate values, though these baseline values do not appear to be significantly lower than wild types. Interestingly, the maximum heart rate achievable in isoflurane-instrumented mice is very close to the basal rate of methoxyflurane-instrumented mice, which explains their inability to be further stimulated by isoproterenol. These increases in heart rate are not βAR related. Wild-type mice, however, retain their ability to stimulate heart rate well beyond β_1AR knockouts, whether surgerized under methoxyflurane or isoflurane.

The creation of mice lacking either β_1- or β_2ARs allows us to test many of the long-held assumptions regarding subtype-specific functions of these receptors. Through simple breeding experiments, it should also be possible to create mice lacking both β_1- and β_2ARs. The development of all three animal types (β_1AR knockout, β_2AR knockout, and combination β_1/β_2AR knockout) and the concepts that emerge from their study should be valuable tools for furthering our understanding of βAR function and pharmacology.

References

1. Rohrer, D. K., Desai, K. H., Jasper, J. R., Stevens, M. E., Regula, D. P., Barsh, G. S., Bernstein, D., and Kobilka, B. K. (1996). Targeted disruption of the mouse β_1-adrenergic receptor gene: Developmental cardiovascular effects. *Proc. Natl. Acad. Sci. U.S.A.* **93**, 7375–7380.

Walter J. Koch,* Robert J. Lefkowitz,†‡
Carmelo A. Milano,* Shahab A. Akhter,*
and Howard A. Rockman‖

Departments of *Surgery, †Medicine, and ‡Biochemistry
Howard Hughes Medical Institute
Duke University Medical Center
Durham, North Carolina 27710

‖Department of Medicine/Cardiology
University of California
San Diego, California 92093

Myocardial Overexpression of Adrenergic Receptors and Receptor Kinases

With the successful insertion of foreign genes into the mouse genome, important *in vivo* transgenic models have emerged in several venues of biomedical research. We have recently developed several lines of transgenic mice in which β-adrenergic receptor (βAR) signaling has been altered. Manipulation of various components of the myocardial βAR system has increased our understanding of cardiovascular diseases, such as heart failure, in which adrenergic signaling plays a critical role.

Probably the most important receptors involved in beat-to-beat cardiac regulation are the adrenergic receptors. Myocardial βARs, for example, mediate increases in heart rate and contractility in response to increases in sympathetic neurotransmission. βARs in cardiac muscle include both the β_1 and β_2 subtypes, and in the human heart, as with most mammals, the β_1AR is the predominant subtype, approaching 75–80% of total βARs (1). Classically, β_1- and β_2ARs selectively couple to the adenylyl cyclase stimulatory G-protein, G_s, which triggers the catalysis of cyclic adenosine monophosphate (cAMP) formation and activation of cAMP-dependent protein kinase, which targets and phosphorylates several myocardial proteins involved in the positive chronotropic and inotropic response.

The regulation of myocardial βARs, like most G-protein–coupled receptors, involves desensitization mechanisms, which are characterized by a rapid loss of receptor responsiveness occurring through phosphorylation of receptors (2). Agonist-specific or homologous desensitization is initiated by a family of kinases, which contains at least six members, referred to as G-protein–coupled receptor kinases (GRKs) (2). The GRKs most abundantly expressed in the heart are βARK1 (GRK2) and GRK5.

Unique mechanisms for the cellular regulation of GRKs have been elucidated (2). Like most GRKs, βARK1 is a cytosolic enzyme that must translocate

Advances in Pharmacology, Volume 42

1054-3589/98 $25.00

to the membrane in order to phosphorylate its activated receptor substrate. The mechanism for translocation of βARK1 involves the physical interaction between the kinase and the membrane-bound $\beta\gamma$ subunits of G-proteins (G$\beta\gamma$). Peptides derived from the G$\beta\gamma$-binding domain of βARK have been shown to act as *in vitro* βARK inhibitors by competing for G$\beta\gamma$ and preventing translocation (2). In contrast to βARK, GRK5 does not undergo agonist-dependent translocation, but rather is constitutively membrane-bound (2) and, therefore, G$\beta\gamma$-independent. The precise *in vivo* roles for βARK1 and GRK5 in myocardial adrenergic signaling are not completely understood; however, their roles may be even more critical during pathophysiological conditions.

Changes in βAR signaling occur in several cardiac disorders, and the most well characterized alteration occurs in chronic congestive heart failure, where there is a loss of βAR density by approximately 50% (1). High levels of catecholamines during heart failure may also serve to desensitize the remaining βARs, probably through a GRK-mediated mechanism. In fact, βARK1 levels have been shown to be markedly elevated in tissue samples taken from the left ventricle (LV) of failing human hearts. This is consistent with, and may contribute to, the functional uncoupling of cardiac βARs seen in this condition.

I. Transgenic Manipulation of Myocardial Adrenergic Signaling

Despite significant advances in the understanding of βAR signaling in heart failure, it remains to be determined whether changes in βAR function can directly promote deterioration of heart function or are just secondary phenomena resulting as a consequence of enhanced local and systemic circulating catecholamines. Transgenic technology has made it possible to target adrenergic signaling components directly to the heart. The targeted overexpression of these molecules provides a powerful approach to understand how molecular alterations, which are known to occur in a disease, can modify the physiological phenotype. Several of these transgenic models are discussed here and are outlined in Table I.

A. Myocardial Overexpression of β_2ARs

To determine whether increasing the number of receptors would lead to greater G-protein coupling, transgenic mice were generated by overexpressing the human β_2AR (3). The murine α-myosin heavy chain (α-MyHC) gene promoter was utilized to specifically target β_2AR to the myocardium. Several lines of mice were generated, and one line had unexpectedly robust expression approaching 40 pmol/mg membrane protein (3). Surprisingly, cardiac βAR signaling in these animals was maximal even in the absence of exogenous agonist, as assessed by the measurement of several biochemical and physiological parameters. Baseline membrane adenylyl cyclase activity from transgenic hearts was increased twofold over baseline activity and equaled isoproterenol stimulated activity of a control nontransgenic mouse heart.

TABLE I Various Transgenic Mouse Models That Have Been Generated with Alterations in the βAR System

Mouse	*Transgene*	*Biochemistry*	*Physiological Phenotype*	*Principle*
TG4 (3)	αMyHC–human β2AR	↑ 200 fold βAR density ↑ AC activity	↑ Contractility ↑ Myocardial relaxation ↑ Heart rate	βAR can couple to G-proteins and stimulate AC in absence of agonist Supports *in vivo* existence of inverse agonists
TGβARK12 (4, 5)	αMyHC–bovine βARK1	↓ High-affinity coupling of βARs ↓ AC activity	↓ Inotropic and chronotropic response to βAR stimulation ↓ Inotropic response to AngII	*In vivo,* β1- and β2ARs and AT_1 receptors are targets for βARK1-induced desensitization ↑ Expression of βARK in heart failure contributes to blunted catecholamine responsiveness
TGMini27 (4)	αMyHC–carboxyl terminal 194 aa of βARK1	Peptide inhibitor competes for Gβγ binding to βARK1 ↓ Agonist-dependent GRK phosphorylation of rhodopsin and β2AR *in vitro*	↑ Basal contractility ↑ Myocardial relaxation Preserved isoproterenol responsiveness	↓ Desensitization of βARs with βARK inhibition βARK1 is a critical *in vivo* modulator of cardiac function
TGGRK5-45 (5)	α*MyHC*–bovine GRK5	↓ AC activity basally and in response to isoproterenol	↓ Basal contractility and inotropic responses to βAR stimulation No change in response to AngII	*In vivo,* β1- and β2ARs are targets for GRK5-induced desensitization, while AT_1 receptors are not targets for GRK5 action

Number inside parentheses are reference numbers.
αMyHC, αmyosin heavy chain; AC, adenylyl cyclase; βAR, β-adrenergic receptor; βARK, β-adrenergic receptor kinase; GRK, G-protein–coupled receptor kinase; AngII, angiotensin II; AT_1, angiotensin II type 1 receptor; aa, amino acid.

To assess the cardiac phenotype in the intact animal, *in vivo* contractile function was measured using a 2-Fr high-fidelity micromanometer catheter inserted into the LV. Overexpression of human β_2AR resulted in marked enhancement of LV contractility as assessed by the maximal first derivative of LV pressure, dP/dtmax, compared with negative littermate controls. LV relaxation, as assessed by peak negative dP/dt, and the time constant of isovolumic pressure decay (Tau) was also enhanced in the transgenic mice (3). With infusion of isoproterenol, no increase in dP/dtmax or Tau occurred, indicating that, at this extraordinary level of overexpression, myocardial βAR signaling and cardiac function are at a maximum.

B. Transgenic Manipulation of Myocardial GRK Activity

With the recent finding that βARK1 levels are elevated in chronic heart failure, which may contribute to diminished βAR signaling, βARK1 and other myocardial GRKs have become important targets of study. Inhibition of GRK-mediated βAR phosphorylation and desensitization is a potentially novel way to increase myocardial signaling. βARK1 activity can be inhibited *in vitro* by peptides derived from its G$\beta\gamma$-binding domain (2). Therefore, we have created transgenic mice with cardiac-specific targeting of the G$\beta\gamma$-binding domain of βARK1. In addition, mice were generated with cardiac-targeted overexpression of βARK1 (4). Interestingly, myocardial signaling and function were reciprocally altered in these two types of transgenic mice.

Mice overexpressing the βARK inhibitor in their hearts had increased basal *in vivo* LV contractility and enhanced responses to administered isoproterenol (4). These results indicate that βARK may exert a tonic inhibitory effect on myocardial β_1ARs even in the absence of agonist. Thus, signaling is increased when βARK is inhibited, which can lead to enhanced cardiac function (4). In transgenic mice with three- to fivefold βARK overexpression specifically in the myocardium, there was a significant attenuation of basal adenylyl cyclase activity that appeared to be mediated by an alteration in the β_1AR complex, because fewer receptors were in the high-affinity "coupled" state (4). In agreement with these biochemical data, *in vivo* hemodynamic measurements in anesthetized mice revealed significant attenuation of isoproterenol-induced LV dP/dtmax (4). This result suggests that myocardial β_1ARs are *in vivo* substrates for this GRK and strengthens the hypothesis that myocardial βARK activity is critical to cardiac physiology and pathophysiology. Thus, this information, coupled with the phenotype of the βARK-inhibitor mice, suggest that βARK1 inhibition may represent a novel therapeutic approach to enhance signaling and improve myocardial function.

Recently, transgenic mice have been generated with α-MyHC–targeted overexpression of GRK5. In a particular line of transgenic mice with approximate 30-fold overexpression of GRK5, myocardial adenylyl cyclase activity was significantly blunted under baseline conditions and also in response to isoproterenol (5). *In vivo* βAR cardiac contractility was also severely crippled in GRK5-overexpressing animals compared with nontransgenic littermate controls. (5). Thus, GRK5 appears to be capable, when overexpressed *in vivo,* of desensitizing and uncoupling myocardial βARs, which suggests that GRK5, like βARK1, may be critical to normal and compromised heart function.

With the generation of the α-MyHC-GRK transgenic mice, it is possible to directly compare responses in these animals to the α-MyHC–βARK1 mice. Because signaling through G-protein–coupled receptors other than βARs is also important for myocardial regulation, we have utilized βARK1 and GRK5 transgenic mice to study signaling through cardiac angiotensin II (AngII) receptors. AngII has been shown to be involved in the regulation of cardiac growth and myocardial inotropy (5). *In vivo* cardiac responses to AngII were studied in βARK1- and GRK5-overexpressing mice, and, interestingly, the data revealed specific differences between these two GRKs. LV dP/dtmax responses to AngII were significantly attenuated in βARK1-overexpressing animals, while GRK5 overexpressors had normal LV responses to AngII, indicating that myocardial AngII receptors are not regulated by GRK5-mediated desensitization (5). These data demonstrate the powerful nature of these mice as model systems to study *in vivo* substrate specificity of the GRKs and suggest that βARK1 and GRK5 may play distinct roles in the normal regulation of cardiac physiology.

II. Future Directions

Understanding cardiac contractility is essential for advances in the treatment of cardiac diseases, particularly heart failure. These transgenic mice have advanced the understanding of signaling through adrenergic receptors in the heart, under both normal and pathophysiological conditions, and represent important experimental models. Monitoring the physiological phenotype in these models will provide an extremely powerful approach to the understanding of disease processes where normal regulatory mechanisms have failed. Furthermore, transgenic mice overexpressing β_2ARs and a βARK inhibitor suggest that gene transfer may represent an exciting novel therapeutic approach for cardiovascular disease.

References

1. Brodde, O. (1993). Beta-adrenoceptors in cardiac disease. *Pharmacol. Ther.* **60,** 405–430.
2. Inglese, J., Freedman, N. J., Koch, W. J., and Lefkowitz, R. J. (1993). Structure and mechanism of the G protein-coupled receptor kinases. *J. Biol. Chem.* **268,** 23735–23738.
3. Milano, C. A., and Allen, L. F., Rockman, H. A., Dolber, P. C., McMinn, T. R., Chien, K. R., Johnson, T. D., Bond, R. A., and Lefkowitz, R. J. (1994). Enhanced myocardial function in transgenic mice overexpression the β_2-adrenergic receptor. *Science* **264,** 582–586.
4. Koch, W. J., Rockman, H. A., Samama, P., Hamilton, R., Bond, R. A., Milano, C. A., and Lefkowitz, R. J. (1995). Reciprocally altered cardiac function in transgenic mice overexpressing the β-adrenergic receptor kinase or a βARK inhibitor. *Science* **268,** 1350–1353.
5. Rockman, H. A., Choi, D-J., Rahman, N. U., Akhter, S. A., Lefkowitz, R. J., and Koch, W. J. (1996). Receptor specific *in vivo* desensitization by the G protein-coupled receptor kinase-5 in transgenic mice. *Proc. Natl. Acad. Sci. U.S.A.* **93,** 9954–9959.

H. Kirk Hammond, Peipei Ping, and Paul A. Insel

Veterans Administration Medical Center
San Diego and University of California–San Diego
La Jolla, California 92161

Cardiac G-Protein Receptor Kinase Activity: Effect of a β-Adrenergic Receptor Antagonist

The data reported in this document have been previously published (1). We have attempted to summarize the key features of this publication but have decided not to include detailed methodology, which can be found in the parent publication. Selected figures and tables from this publication have been reprinted, largely without alteration, from the original publication.

In this report, we will present a summary of the data that test the hypothesis that β-adrenergic receptor (βAR) activation affects G-protein receptor kinase (GRK) activity in the heart (1).

I. Effect of Bisoprolol on Cardiac G-Protein Receptor Kinase Activity

A. Background

βAR antagonists are used in the treatment of angina pectoris, hypertension, and heart failure. As agents that interfere with βAR stimulation, it is reasonable to propose that these agents may affect the expression and interaction of myocardial signal-transducing elements in the βAR-responsive adenylyl cyclase pathway. Clinical and experimental data suggest that chronic administration of βAR antagonists affects cell surface βAR number and physiological responsiveness in the heart (2). However, the mechanisms by which elements of the βAR–adenylyl cyclase pathway are altered by such treatment and the nature of these alterations remain to be firmly established. An element that has received little previous attention but that is likely to provide important insights into mechanisms for altered transmembrane signaling in the setting of reduced βAR activation is GRK activity. GRK participates in receptor desensitization by catalyzing phosphorylation of agonist-occupied receptors in concert with β-arrestin (3).

It has been reported that GRK activity is increased in failing left ventricles obtained from humans with heart failure (4, 5). Based on these studies and others alluded to, we have asked whether GRK activity may be altered in a

Advances in Pharmacology, Volume 42

1054-3589/98 $25.00

TABLE I LV Adenylyl Cyclase Activity After Bisoprolol Treatment

	Control	*Bisoprolol*	*p*
BASAL	29 ± 5	33 ± 6	NS
ISO + GTP	56 ± 5	113 ± 10	.0001
GTPγS	224 ± 24	222 ± 46	NS
AIF	267 ± 32	263 ± 42	NS
FORSK	349 ± 57	357 ± 46	NS

Data represent cAMP-produced in pmol/mg/min ± 1 SD and are net values (basal subtracted). p = control vs bisoprolol (unpaired, two-tailed t-test). N = 6 for each group.
ISO, 10 μM isoproterenol; GTP, 100 μM guanosine 5′-triphosphate; GTPγS, 100 μM guanosine 5′-0-(3-thiotriphosphate); AIF, 100 μM aluminum fluoride; FORSK, 100 μM forskolin.

manner dependent on the extent of βAR agonist activation. We, therefore, treated pigs for 35 days with the β_1AR antagonist bisoprolol (6) to determine the effects of reduced βAR activation on cardiac GRK activity. We tested the hypothesis that, in left ventricular myocardium, reduced βAR agonist activation would be associated with reduced βAR kinase activity.

B. Data: See Tables I and II and Figures 1 and 2

C. Summary of Results

To determine whether βAR agonist activation influences GRK activity in the heart, we examined the effects of chronic β_1AR antagonist treatment (bisoprolol, 0.2 mg/kg/d i.v., 35 days) on GRK activity and GRK2 protein content. Two novel alterations in cardiac adrenergic signaling associated with chronic reduction in βAR agonist activation were found. First, in the left ventricle, there was a twofold increase in βAR-dependent stimulation of ade-

TABLE II LV GRK Activity After Bisoprolol

	Control	*Bisoprolol*	*p*
Cytosol	23 ± 4	17 ± 4	.45
Membrane	15 ± 2	9 ± 2	.002
Total	37 ± 7	26 ± 4	.02
% in cytosol	60 ± 4	64 ± 4	.15

Values are mean ± 1 SD and represent ^{32}P-incorporated (fmol/mg/min), n = 5 in both groups. p values: Control vs bisoprolol (unpaired, two-tailed); n = 5 for each group.

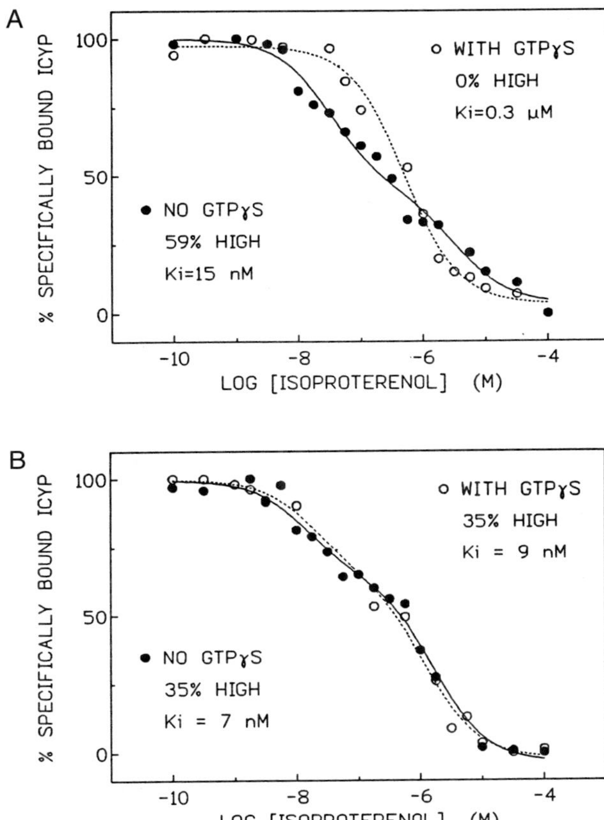

FIGURE 1 Competitive binding assays, performed to assess the proportion of left ventricular βAR displaying high- and low-affinity binding for (-)-isoproterenol. (A) Left ventricular membranes from a control animal; 59% of the receptors displayed high-affinity binding, and all of these were converted to low-affinity binding with the addition of GTPγS. (B) Left ventricular membranes from a bisoprolol-treated animal; 35% of the receptors displayed high-affinity binding, and these were resistant to upcoupling with GTPγS. Data from six control and six bisoprolol-treated animals are shown in Figure 2. (Reproduced from *The Journal of Clinical Investigation,* 1995, vol. 95, 1271–1280 by copyright permission of the American Society for Clinical Investigation.)

nylyl cyclase and a persistent high-affinity state of βAR. Second, there was a reduction in left ventricular βAR kinase activity, suggesting a previously unrecognized association between the degree of adrenergic activation and myocardial βAR kinase expression. The heart appears to adapt in response to chronic βAR receptor antagonist administration in a manner that would be expected to offset reduced agonist stimulation. The mechanisms for achieving this extend beyond βAR upregulation and include alterations in βAR/G_s interaction and myocardial GRK activity.

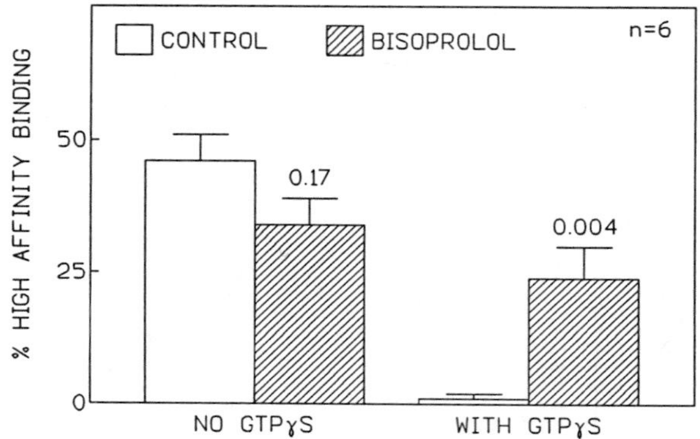

FIGURE 2 Competitive binding assays, performed to assess the proportion of left ventricular βARs displaying high- and low-affinity binding for (-)-isoproterenol. Left ventricular membranes from control and bisoprolol-treated animals displayed similar proportions of receptors showing high-affinity binding in the absence of GTPγS, but left ventricular membranes from bisoprolol-treated animals were resistant to upcoupling with GTPγS. Bars represent mean values; error bars represent 1 SEM. Numbers above bars denote *p*-values (unpaired, two-tailed). N = 6 for both groups.

References

1. Ping, P., Gelzer-Bell, R., Roth, D. A., Kiel, D., Insel, P. A., and Hammond, H. K. (1995). Reduced β-adrenergic receptor activation decreases G-protein expression and β-adrenergic receptor kinase activity in porcine heart. *J. Clin. Invest.* **95**, 1271–1280.
2. Brodde, O-E. (1991). β_1- and β_2-adrenoreceptors in the human heart: Properties, function, and alterations in chronic heart failure. *Pharmacol. Rev.* **43**, 203–242.
3. Benovic, J. L., Mayor, F., Staniszewski, C., Lefkowitz, R., and Caron, M. G. (1987). Purification and characterization of the β-adrenergic receptor kinase. *J. Biol. Chem.* **262**, 9026–9032.
4. Ungerer, M., Bohm, M., Elce, J. S., Erdmann, E., and Lohse, M. J. (1993). Altered expression of β-adrenergic receptor kinase (βARK) and β_1-adrenergic receptors in the failing human heart. *Circulation* **87**, 454–461.
5. Ungerer, M., Parruti, G., Bohm, M., Puzicha, M., DeBlasi, A., Erdmann, E., and Lohse, M. J. (1994). Expression of β-arrestins and β-adrenergic receptor kinases in the failing human heart. *Circ. Res.* **74**, 206–213.
6. Leopold, G., Ungethum, W., Pabst, J., Simane, Z., Buhring, K. U., and Wiemann, H. (1986). Pharmacodynamic profile of bisoprolol, a new β_1-selective adrenoceptor antagonist. *Br. J. Clin. Pharmacol.* **22**, 293–300.

A. Donny Strosberg

Institut Cochin de Génétique Moléculaire
Laboratoire d'Immuno-Pharmacologie Moléculaire
Université de Paris VII
CNRS 75014 Paris, France

Structure and Function of the β_3 Adrenoreceptor

Three adrenoreceptors are coupled positively to adenylyl cyclase: β_1AR, predominant in heart; β_2AR, the major subtype in lung and uterus; and β_3AR, mostly expressed in fat tissues (1). The cloning, sequencing, and expression in model systems of the newly discovered β_3AR has resulted in an extensive functional characterization (for recent reviews, see Strosberg and Piétri [2] and Strosberg [3]). Ligand binding and adenylyl cyclase activation studies helped define a pharmacological profile that is quite distinct from that of β_1- and β_2AR but strongly reminiscent of that of "atypical" adrenergic responses described in earlier studies (2, 4).

The relatively low affinity of β_3AR for noradrenaline compared with β_1- and β_2AR led to the suggestion that β_1- and β_2AR mediate the effects of circulating catecholamines, whereas β_3AR regulates the response to much higher local concentrations present, for example, in the sympathetic cleft during sympathetic stimulation. This would be the case in the densely innervated brown adipose tissue where β_3AR is abundantly expressed. Analyses done with a variety of compounds resulted in the definition of the following properties of β_3AR: (1) high potency of a novel class of compounds, initially described as potent activators of lipolysis and thermogenesis in white and brown adipose tissues; (2) partial agonistic activities of several β_1/β_2AR antagonists reflecting intrinsic sympathomimetic activities in heart tissue; (3) atypically low affinities and potencies for conventional β-antagonists; and (4) atypically low stereoselectivity index for reference agonist and antagonist enantiomers as compared with those reported for classical β_1- and β_2ARs.

β_3ARs have now been cloned, sequenced, and expressed for six different species: human, bovine, canine, murine, rat, and guinea pig. Partial sequence data are also available for monkey and hamster.

Most of these receptors share the same characteristic properties of β_3AR, although obvious species-specific differences have been described explaining in part why β_3 agonists, which were quite effective in stimulating lipolysis, thermogenesis, and resulting weight loss in rodents or dogs, failed to have similar effects in human trials.

β_3AR plays an important role in regulating energy metabolism and thermogenesis: Long-term treatment of rodents and dogs with β_3 agonists leads to weight reduction and reappearance of brown adipose tissue. The functional

Advances in Pharmacology, Volume 42

1054-3589/98 $25.00

expression of β_3AR in white omental fat was well documented by Arner's group (5, 6), which confirmed the earlier detection of β_3 mRNA in human white and brown adipose tissues by Krief *et al.* (7). Reappearance of β_3AR-rich brown fat in humans has been well documented for Finnish outdoor workers (8) and for pheochromocytoma patients with catecholamine-secreting adrenal tumors.

More recently, a single-residue substitution (Trp64Arg) was found to be correlated with a higher incidence of morbid obesity in Pima Indians (9) and in some Japanese patients (10) and with an increased dynamic capacity to add on weight (11) and develop non insulin-dependent diabetes mellitus in Western morbidity obese patients (12). This association has now been confirmed in 16 out of 20 recent surveys performed on thousands of individuals from a variety of origins. While not all studies reported positive results, most described linkage between the presence of the Arg 64 form and some of the symptoms of the metabolic syndrome related to obesity, diabetes, or both.

The development of β_3-selective agonists with high efficacy on the human receptor has benefitted considerably from the use of recombinant receptor expressed in model cell systems, which allowed detailed binding and adenylyl cyclase studies. The development of human immortalized brown adipocytes (13) now also permits the study of lipolysis and thermogenesis in response to β_3 agonists. These cells should, thus, allow the identification of new, human selective compounds for the treatment of obesity and diabetes.

Acknowledgments

Support for our work comes mostly from the Centre National de la Recherche Scientifique, the Institut National de la Santé et de la Recherche Médicale, the University of Paris VII, and the Ministry for Research and Technology. We are also grateful for help from the Fondation pour la Recherche Médicale Française.

References

1. Strosberg, A. D. (1993). Structure, function and regulation of adrenergic receptors. *Protein Sci.* **12**, 1198–1209.
2. Strosberg, A. D., and Piétri-Rouxel, F. (1996). Function and regulation of the β_3-adrenoceptor. *Trends Pharmacol. Sci.* **17**, 373–381.
3. Strosberg, A. D. (1997). Structure and function of the β_3-adrenergic receptor. *Annu. Rev. Pharmacol. Toxicol.* **37**, 421–450.
4. Emorine, L. J., Blin, N., and Strosberg, A. D. (1994). The human β_3-adrenoceptor: The search for a physiological function. *Trends Pharmacol. Sci.* **15**, 3–7.
5. Lönnqvist, F., Krief, S., Strosberg, A. D., Nyberg, B., Emorine, L. J., and Arner, P. (1993). Evidence for a functional β_3-adrenergic receptor in man. *Br. J. Pharmacol.* **110**, 929–936.
6. Enocksson, S., Shimizu, M., Lönnqvist, F., Nordenström, J., and Arner, P. (1995). Demonstration of an *in vivo* functional β_3-adrenoceptor in man. *J. Clin. Invest.* **95**, 2239–2245.
7. Krief, S., Lönnqvist, F., Raimbault, S., Baude, B., Arner, P., Strosberg, A. D., Ricquier, D., and Emorine, L. J. (1993). Tissue distribution of β_3-adrenergic receptor mRNA in man. *J. Clin. Invest.* **91**, 344–349.

8. Huttunen, P., Hirvonen, J., and Kinnula, V. (1981). The occurrence of brown adipose tissue in outdoor workers. *Eur. J. Appl. Physiol.* **46,** 339–345.
9. Walston, J., Silver, K., Bogardus, C., Knowler, W. C., Celi, F. S., Austin, S., Manning, B., Strosberg, A. D., Stern, M. P., Raben, N., Sorkin, J. D., Roth, J., and Shuldiner, A. R. (1995). Time of onset of non-insulin-dependent diabetes mellitus and genetic variation of the β_3-adrenergic receptor gene. *N. Engl. J. Med.* **333,** 343–347.
10. Kadowaki, H., Yasuda, K., Iwamoto, K., Otabe, S., Shimokawa, K., Silver, K., Walston, J., Yoshinaga, H., Yamada, N., Saito, Y., Hagura, R., Hakanuma, Y., Shuldiner, A. R., Yazaki, Y., and Kadowaki, T. (1995). A mutation in the β_3-adrenergic receptor gene is associated with obesity and hyperinsulinemia in Japanese subjects. *Biochem. Biophys. Res. Comm.* **215,** 555–560.
11. Clément, K., Vaisse, C., Manning, B., Basdevant, A., Guy-Grand, B., Shuldiner, A. R., Froguel, P., and Strosberg, A. D. (1995). Genetic variation in the β_3-adrenergic receptor and an increased capacity to gain weight in patients with morbid obesity. *N. Engl. J. Med.* **333,** 352–354.
12. Widen, E., Lehto, M., Kanninen, T., Walston, J., Shuldiner, A. R., and Groop, L. C. (1995). Association of a polymorphism in the β_3-adrenergic-receptor gene with features of the insulin resistance syndrome in virus. *N. Engl. J. Med.* **333,** 348–351.
13. Zilberfarb, V., Pietri-Rouxel, F., Jockers, R., Krief, S., De Louis, C., Issad, T., and Strosberg, A. D. (1997). Conversion of human immortalized brown pre-adipocytes into adipocytes leads to expression of functional β_3-adrenergic receptors. *J. Cell. Sci.* **110,** 801–807.

J. L. Waddington,* A. M. Deveney,* J. Clifford,* O. Tighe,† D. T. Croke,† D. R. Sibley,‡ and J. Drago§

Departments of *Clinical Pharmacology and †Biochemistry
Royal College of Surgeons in Ireland
Dublin 2, Ireland

‡National Institute of Neurological Disorders and Stroke, NIH
Bethesda, Maryland 20892

§Department of Anatomy
Monash University
Clayton, Victoria 3168 Australia

Behavioral Analysis of Multiple D1-Like Dopamine Receptor Subtypes: New Agents and Studies in Transgenic Mice with D1A Receptor Knockout

Molecular biology has revealed the existence of a broad family of dopamine (DA) D1-like receptor subtypes (D1A/D1, D1B/D5, D1C, D1D). This family plays an important role in the regulation of psychomotor behavior, often by way of both cooperative–synergistic and oppositional interactions with their D2-like (D2L/S, D3, D4) counterparts, but similarities or differences in the role(s) of individual family members are poorly understood, primarily due to the absence of agents able to distinguish selectively between members within each family. Furthermore, there is an expanding body of behavioral and other functional evidence that additional D1-like subtypes may exist (1). While the identification of selective agents is fundamental to resolving these issues, an alternative approach of targeted gene deletion has been applied to generate transgenic mice having knockout of individual DA receptor subtypes. However, the generation of D1A knockout animals independently in two laboratories (2, 3) has led to some conflicting results; for example, in terms of spontaneous behavior, such animals have been noted to show a decreased level of rearing but a normal level of locomotion (2), or to show hyperactivity in terms of photobeam interruptions (3).

In recent studies, new D1-like antagonists, the thienoazepine LY 270411 (4) and the benzoquinoxaline SDZ PSD 958, have been compared with BW 737C and SCH 23390 for their effects on behavioral responses to the selective D1-like agonist A 68930 versus selective D2-like agonist RU 24213. Grooming responses to A 68930 were readily blocked by each of LY 270411, SDZ PSD

Advances in Pharmacology, Volume 42

1054-3589/98 $25.00

958, BW 737C, and SCH 23390; however, the vacuous chewing response to A 68930 was blocked only by BW 737C. Typical sniffing and locomotor responses to RU 24213 were attenuated by SDZ PSD 958, BW 737C, and SCH 23390 in accordance with their regulation via cooperative–synergistic D1-like–D2-like interactions (1), but not by LY 270411; furthermore, myoclonic jerking to RU 24213 was released by SDZ PSD 958, BW 737C, and SCH 23390 in accordance with its regulation via oppositional D1-like–D2-like interactions (1), but not by LY 270411. These findings elaborate the notion that grooming induced by D1-like agonism is blocked by all known chemical classes of D1-like antagonist, while vacuous chewing is blocked only by isoquinoline D1-like antagonism; this suggests further that these behaviors are mediated via pharmacologically distinct subtypes of D1-like receptor. Furthermore, LY 270411 appears unique in its activity to readily block D1-like agonist-induced grooming without influencing behavioral responses to D2-like agonism; thus the site-mediating prototypical D1-like agonist-induced behavior may be dissociable pharmacologically from D1-like sites participating in functional interactions with D2-like receptors.

Among young adult homozygous D1A knockout mice (2), genotyped using a polymerase chain reaction technique, we found body weight to be reduced relative to wild-type controls from litters of the same generational age (black, female, ~ 8 wk: 19 ± 3 vs 13 ± 5 g, n = 9–11; −34%, $p < .02$), as noted previously (2, 3), in the absence of gross neurological deficit. The spontaneous behavior of these animals was evaluated by direct visual observation using an ethological approach similar to that which we have applied to rats (1, 4, 5); thus, we derived their "ethogram" in terms of simultaneous assessment of all manifested behaviors in their repertoire. Relative to wild-types, D1A knockout was associated with reductions in sniffing ($p < .01$), rearing free ($p < .01$), sifting ($p < .01$), and chewing ($p < .05$) of cage bedding and fecal pellets; and increases in locomotion ($p < .01$), grooming ($p < .01$), and intense grooming ($p < .01$); there were no significant alterations in total rearing, rearing against cage walls, rearing from a sitting position, climbing, or eating, and a low baseline level of vacuous chewing was unaltered. There were very few correlations between body weight and individual behaviors in either wild-type or knockout animals, none of which could account for the differences identified.

This preliminary profile suggests abnormalities of spontaneous behavior in D1A knockout mice; the resultant ethogram could not be encapsulated as indicating either hypo- or hyperactivity, because it involved significant shifts between multiple individual elements of behavior. Such a profile may both describe more completely the behavioral phenotype of D1A knockout and account for apparent contradictions in the existing literature (2, 3). In terms of prototypical D1-like–mediated behaviors, an increased level of grooming in association with D1A knockout may appear paradoxical. Whether this might reflect the involvement of other D1-like receptors, compensatory processes subsequent to developmental absence of D1A receptors, or other mechanisms, perhaps related to the complex genetic background of these animals, remains to be determined. It should not be overlooked that any (over)compensation to developmental absence of D1A receptors that might be evident at the level of tonic function may not endure in the face of phasic challenge with agonist and antagonist drugs. Classical pharmacological and knockout techniques are complementary in seeking to resolve the behavioral roles of multiple D1-like receptors.

At the classical pharmacological level, the anomalous benzazepine agent SK&F 83959 shows high affinity and selectivity for D1-like over D2-like receptors, fails to stimulate adenylyl cyclase (AC) and readily inhibits the stimulation of AC induced by DA, thus showing all the defining characteristics of a selective D1-like antagonist such as SCH 23390. However, SK&F 83959 readily induces prominent grooming, the most widely accepted behavioral model of D1-like receptor stimulation, and vacuous chewing, a more controversial model thereof; furthermore, SCH 23390 readily blocks SK&F 83959-induced grooming but not SK&F 83959-induced vacuous chewing, and indeed SCH 23390 also induces vacuous chewing when given alone. These studies (5) suggest that grooming and vacuous chewing have pharmacologically distinct substrates that appear to involve D1-like receptors linked to transduction mechanisms additional to or other than AC.

We have described the isochroman D1-like agonist A 68930 to share with the prototypical benzazepine D1-like agonist analogues of SK&F 38393 an action to induce prominent grooming but to be much more active than the benzazepines (with the exception of the anomalous agent SK&F 83959) in inducing vacuous chewing; furthermore, while SCH 23390 and its isoquinoline counterpart, the D1-like antagonist BW 737C, each readily blocked these grooming responses, only BW 737C but not SCH 23390 blocked the vacuous chewing response to A 68930 (1). On classical pharmacological grounds, these findings would suggest that grooming is mediated via a D1-like receptor that recognizes all known chemical classes of D1-like ligands, while vacuous chewing is mediated via a D1-like subtype that recognizes preferentially the isochromans, the isoquinolines, and those benzazepines which act as antagonists at the level of AC.

Acknowledgments

These studies are supported by the Royal College of Surgeons in Ireland and Forbairt; we thank Abbott, Lilly, Roussel-UCLAF, Sandoz, Schering, SmithKline Beecham, and Wellcome Foundation for making compounds available to us.

References

1. Waddington, J. L., Daly, S. A., Downes, R. P., Deveney, A. M., McCauley, P. G., and O'Boyle, K. M. (1995). Behavioural pharmacology of D1-like dopamine receptors: Further subtyping, new pharmacological probes and interactions with D2-like receptors. *Prog. Neuropsychopharmacol. Biol. Psychiatry* **19,** 811–831.
2. Drago, J., Gerfen, C. R., Lachowicz, J. E., Steiner, H., Hollon, T. R., Love, P. E., Ooi, G. T., Grinsberg, A., Lee, E. J., Huang, S. P., Bartlett, P. F., Jose, P. A., Sibley, D. R., and Westphal, H. (1994). Altered striatal function in a mutant mouse lacking D1A dopamine receptors. *Proc. Natl. Acad. Sci. U.S.A.* **91,** 12564–12568.

3. Xu, M., Moratalla, R., Gold, L. H., Hiroi, N., Koob, G. F., Graybiel, A. M., and Tonegawa, S. (1994). Dopamine D1 receptor mutant mice are deficient in striatal expression of dynorphin and in dopamine-mediated behavioural responses. *Cell* **79**, 729–742.
4. Deveney, A. M., and Waddington, J. L. Evidence for dopamine 'D1-like' receptor subtypes in the behavioural effects of two new selective antagonists, LY 270411 and BW 737C. *Eur. J. Pharmacol.* **317**, 175–181.
5. Deveney, A. M., and Waddington, J. L. (1995). Pharmacological characterization of behavioural responses to SK&F 83959 in relation to 'D1-like' dopamine receptors not linked to adenylyl cyclase. *Br. J. Pharmacol.* **116**, 2120–2126.

Ian Creese and J. M. Tepper

Center for Molecular and Behavioral Neuroscience
Aidekman Research Center
Rutgers, The State University of New Jersey
Newark, New Jersey 07102

Antisense Knockdown of Brain Dopamine Receptors

Dopamine (DA) neurons express receptors for DA in both their somatodendritic and axon terminal regions. Stimulation of somatodendritic autoreceptors inhibits the spontaneous activity of DA neurons, whereas stimulation of the axon terminal autoreceptors reduces the excitability of DA axon terminals and inhibits DA synthesis and release. DA receptors were originally classified as D1 or D2 receptors based on their differing affinities for various ligands and linkage to intracellular signaling pathways, and, within this classification, both autoreceptors were identified as D2 receptors. However, recent advances in molecular biology have demonstrated that there are two *families* of DA receptors: the D1-class family consists of the D_1 and D_5 receptors and the D2-class family of the D_2, D_3, and D_4 receptors. Although the different D2 receptor subtypes are differentially distributed throughout the central nervous system, both D_2 and D_3 mRNA have been identified in the substantia nigra (SN), and it has recently been argued that at least some DA autoreceptors may, in fact, be D_3 receptors. No selective drugs are available that can discriminate, at least in a physiologically useful manner, among members within the D2-class family. Thus, the precise subtype(s) of the somatodendritic and terminal autoreceptors remain unknown.

Antisense knockdown refers to the ability of specifically designed short sequences of antisense oligodeoxynucleotides (AON) to bind to their complementary mRNA and stop translation, thereby preventing the production of the protein that the mRNA codes for. Because it is so highly specific, it can be

Advances in Pharmacology, Volume 42

1054-3589/98 $25.00

used to probe the functions of receptor subtypes for which specific and selective agonists or antagonists do not exist. We previously reported the behavioral consequences following *in vivo* antisense knockdown of the D_2 receptor subtype in rat brain (1). In this report we describe the electrophysiological consequences following antisense knockdown of dopamine D_2 and/or D_3 receptors in DA SN neurons.

Male Sprague-Dawley rats weighing approximately 200 g were anesthetized with ketamine–xylazine, placed in a stereotaxic apparatus, and implanted with a 28-g stainless steel guide cannula at a 20-degree angle. Twenty-four hours later, a 33-g injection cannula, 1 mm longer than the guide, was filled with the appropriate substance and inserted into the guide cannula so that its tip was 500 μm dorsal to the SN pars compacta. The cannula was joined to a length of Teflon tubing connected through a fluid swivel to a microsyringe pump and saline, D_2 random, D_3 random, D_2, D_3, or D_2 + D_3 AON (10–20 μg/μl) was infused continuously at 0.1 μl/hr for 3–7 days. The D_2 AON was a 19-mer S-oligodeoxynucleotide complementary to codons 2–8 of the D_2 receptor mRNA with sequence 5′-AGGACAGGTTCAGTGGATC-3′, and the D_3 AON, also directed against codons 2–8, had the sequence 5′-TTATCTGGCTCAGAGGTGC-3′. D_2 and D_3 random oligodeoxynucleotide controls consisted of the same bases in a randomized, mismatched order. The AONs were synthesized by Oligos Inc. (Wilsonville, OR).

Three to 7 days after the start of the infusion, rats were anesthetized with urethane, the left femoral vein or a lateral tail vein was cannulated, and each rat was placed into a stereotaxic frame. A bipolar stimulating electrode was inserted into the ipsilateral neostriatum, and extracellular recordings of antidromically identified SN DA neurons were obtained. The firing pattern of each neuron was classified as pacemaker, random, or bursty on the basis of the neuron's autocorrelation histogram. The threshold current for each neuron was defined as the minimum stimulating current that evoked antidromic responses from neostriatum to 100% of the stimulus deliveries (2). To obtain an estimate of the excitability of the somatodendritic region of the DA neurons (3), the proportion of striatal-evoked antidromic responses consisting of the full initial segment-somatodendritic spike was counted while each neuron was stimulated at threshold. Following the establishment of a stable baseline firing rate, a dose of apomorphine HCl that was double the previous dose was injected intravenously every 2 min, starting with either 1 or 2 μg/kg. This was continued until complete inhibition of spontaneous activity was obtained, a cumulative dose of 2048 μg/kg was reached, or the cell was lost.

Neither D_2, D_3, nor D_2 + D_3 AON infusion had any effect on the mean spontaneous firing rate, the coefficient of variation of the interspike intervals, or the distribution of firing patterns of nigrostriatal neurons. Neurons recorded ipsilateral to D_2, D_3, or D_2 + D_3 AON infusions exhibited significantly lower thresholds for antidromic responding from the ipsilateral neostriatum. Previous experiments have shown that this threshold is modulated by terminal autoreceptors. Local striatal infusion of D2-family antagonists *in vivo* reduces the threshold, indicating that the terminal autoreceptors are stimulated by endogenous DA under physiological conditions (2). The results with D_2 and D_3 AONs exactly mimic the effects of administration of DA antagonists, which suggests that there exist both D_2 and D_3 autoreceptors on the axon terminals of nigrostriatal neurons.

Treatment with D_2, D_3, or $D_2 + D_3$ AON also increased the proportion of antidromic spikes consisting of the full initial segment-somatodendritic spike. D_2, D_3, and combined $D_2 + D_3$ AON infusions also produced a significant shift to the right in the apomorphine dose-response curve. Whereas control cells exhibited an ED_{50} of 10 μg/kg, D_2 or D_3 antisense treatment increased the ED_{50} to 475 μg/kg and 341 μg/kg, respectively. Simultaneous administration of D_2 and D_3 AONs produced an ED_{50} of 620 μg/kg. These data suggest that the majority of SN DA neurons may possess both D_2 and D_3 somatodendritic autoreceptors. The fact that the D_3 AON was just as effective as the D_2 AON is somewhat surprising, given that there is only a low level of D_3 mRNA present in the mesencephalon and that D_3 mRNA cannot be detected in many SN neurons (4).

However, the difficulty in detecting D_3 mRNA and/or binding in the midbrain may simply reflect the relative overabundance of D_2 mRNA and protein relative to that of D3. Furthermore, a study employing transfection of D_2 and D_3 receptors into a mesencephalic DA clonal line showed that D_3 receptors were more than twice as potent at inhibiting DA release than were D_2 receptors, despite the fact that B_{max} of the D_2 receptors was three times greater than that of the D_3 receptors (5), perhaps indicating that the receptor coupling mechanism(s) are more efficient in D_3 receptors. Whether this is also true *in vivo* remains to be determined. The fact that both D_2 and D_3 AONs blocked autoreceptor-mediated inhibition of firing more or less equally suggests the possibility that the normal electrophysiological response attributed to somatodendritic autoreceptors stimulation may require coactivation of both D_2 and D_3 receptors.

Administration of the D_2 AON significantly reduced D_2 receptor ([^{3}H]spiperone) binding in SN from a low of 32.4% to a high of 76.1%. The mean decrease in binding was 48.9 ± 3.3%. There was no significant decrease in [^{3}H]7-OH-DPAT binding following D_2 AON treatment. Conversely, D_3 AON did not significantly decrease [^{3}H]spiperone binding in SN (−4.3 ± 2.8%) but did decrease [^{3}H]7-OH-DPAT binding by 44.6 ± 5.8%. Combined treatment with D_2 and D_3 AONs decreased [^{3}H]spiperone binding by 56.2 + 4.7% and [^{3}H]7-OH-DPAT binding by 49.8 + 6.2%. In contrast, there was no change in D1-class ([^{3}H]SCH 23390) binding in SN after any AON infusion. Tyrosine hydroxylase immunostaining and Nissl staining after administration of D_2 AON failed to reveal any evidence of a nonspecific toxic effect of the AON on DA neurons.

Both D_2 and D_3 AON–treated rats exhibited a modest contralateral postural deviation during rest and intermittent or continuous spontaneous contralateral rotations at a relatively low rate. Both the contralateral postural deviation and the spontaneous contralateral rotation were evident by 24 hr after the start of the infusion and became maximal after 3 days of infusion. The combined $D_2 + D_3$ AON–treated group also differed significantly from either the D_2 AON– or the D_3 AON–treated groups, suggesting an additivity in the effects of D_2 and D_3 receptor knockdown on spontaneous rotation. These data suggest an increased DA release in the ipsilateral striatum as a result of D_2 and/or D_3 terminal autoreceptor knockdown.

It is interesting to note that although there were clear effects of both D_2 and D_3 AON treatment on the apomorphine dose-response relation, on the terminal excitability and on the proportion of antidromic responses consisting of the initial segment and somadendritic components, there was no effect of any AON treatment on the baseline spontaneous firing rate or pattern. The proportion of antidromic spikes consisting of the full spike is a measure of the level of excitability of the dendritic regions of the cell, which is related to the local membrane potential. We have shown previously that this parameter can vary independently of the firing rate of the DA neuron (3). The increased proportion of full-spike antidromic responses coupled with the lack of change in the baseline firing rate after D_2 or D_3 AON treatment supports our previous suggestion that DA somatodendritic autoreceptors are effectively stimulated by endogenous DA under normal physiological conditions, but that the endogenous activation of these receptors does not normally inhibit the firing of these neurons as a whole, but rather modulates the excitability of certain restricted dendritic regions (3).

The data presented here demonstrate that DA neurons possess functional D_2 and D_3 autoreceptors at both their axon terminal and somatodendritic regions. Both types of terminal autoreceptors modulate terminal excitability and presumably the release of DA from the nerve terminals and/or its synthesis, whereas somatodendritic D_2 and D_3 autoreceptors play a role in modulating the excitability of local dendritic regions.

Acknowledgments

We thank Drs. P. Celada, L. Martin, A. Ougazzal, B. C. Sun, and M. Zhang for their contributions to this research, which was supported by a Johnson & Johnson Discovery Research Fund, a Hoechst-Celanese Innovative Research Award and MH 52450 and MH-52383.

References

1. Zhang, M., and Creese, I. (1993). Antisense oligodeoxynucleotide reduces brain dopamine D2 receptors: Behavioral correlates. *Neurosci. Lett.* **161,** 223–226.
2. Tepper, J. M., Gariano, R. F., and Groves, P. M. (1987). The neurophysiology of dopamine nerve terminal autoreceptors. *In* Neurophysiology of Dopaminergic Systems—Current Status and Clinical Perspectives. (L. A. Chiodo and A. S. Freeman, eds.), pp. 93–127. Lakeshore Publishing, Grosse Point, MI.
3. Trent, F., and Tepper, J. M. (1991). Dorsal raphé stimulation modifies striatal-evoked antidromic invasion of nigral dopaminergic neurons *in vivo*. *Exp. Brain Res.* **84,** 620–630.
4. Diaz, J., Lévesque, D., Lammers, C. H., Griffon, N., Martres, M. P., Schwartz, J. C., and Sokoloff, P. (1995). Phenotypical characterization of neurons expressing the dopamine D3 receptor in the rat brain. *Neuroscience* **6,** 731–745.
5. Tang, L., Todd, R. D., and O'Malley, K. L. (1994). Dopamine D2 and D3 receptors inhibit dopamine release. *J. Pharmacol. Exp. Ther.* **270,** 475–479.

Adolfo Saiardi, Tarek Abdel Samad, Roberto Picetti, Yuri Bozzi, Ja-Hyun Baik, and Emiliana Borrelli

Institut de Génétique et de Biologie Moléculaire et Cellulaire
1, rue L. Fries
67404 Illkirch Cedex
Strasbourg, France

The Physiological Role of Dopamine D2 Receptors

Dopamine participates in the control of many physiological functions, such as the initiation and coordination of movements, motivated behaviors, and the synthesis and release of pituitary hormones (e.g., the well-documented inhibition of prolactin secretion). This occurs through the interaction with membrane receptors that belong to the family of seven transmembrane domain G-protein–coupled receptors. Five different dopamine receptors have been identified (1). A general subdivision into two groups has been made: the D1-like receptors, comprising D1 and D5; and the D2-like receptors, comprising D2, D3, and D4 receptors. Pharmacologically, while it is possible to discriminate between D1-like and D2-like receptors, it is more difficult to differentiate between members of each subfamily.

We have been interested in the study of the physiological activity of D2 receptors that are abundantly expressed in the striatum, substantia nigra, and hypophysis. Two isoforms of this receptor have been described that differ by their coupling characteristics to G-proteins *in vitro* (2). To study the physiology of the D2 receptor *in vivo,* we have generated mutant mice lacking a functional D2 receptor gene by homologous recombination.

We successfully established a line of D2R-null mice using conventional procedures (3). A normal mendelian segregation of the mutated genotypes was observed. A reduced fertility was observed in D2R-deficient mice together with a slight reduction of body weight and temperature.

[^{3}H]Spiperone-binding analysis revealed a severe reduction of binding sites in the heterozygous D2R-null mice (Bmax = 135 ± 19 fmol/mg protein) compared with the WT littermates (Bmax = 290 ± 20 fmol/mg protein), despite similar affinities for the ligand (Kd = 17.6 ± 3.0 and 25.5 ± 1.7 pM, respectively). A complete absence of D2 binding sites was demonstrated in the homozygous D2R-null mouse.

We then analyzed whether the expression of the other members of the D2 and D1 subfamily were altered in D2R-null mice. *In situ* binding analyses using [^{125}I]iodosulpride, an antagonist with high affinity for D2 and D3 receptors, and [^{3}H]spiperone, an antagonist with high affinity for D2 and D4 receptors, showed that the expression of D3 and D4 receptors is not affected in the D2R-

Advances in Pharmacology, Volume 42

1054-3589/98 $25.00

deficient background (Fig. 1, top). Similarly, *in vitro* binding analysis using a D1 receptor–specific ligand, [^{3}H]SCH 23390, revealed no significant differences either in the affinity for the ligand or in the D1 receptor number of sites between WT and D2R-deficient mice. These combined data demonstrate that the absence of D2 receptors does not affect the expression of other members of the dopamine receptor family.

We next addressed the functional consequence of the lack of D2 receptors in the basal ganglia and pituitary gland at the level of gene expression of peptides and proteins that are either localized in the same regions or whose expression is regulated by D2 receptors. We performed *in situ* hybridization of coronal brain sections from homozygous and control mice, using probes

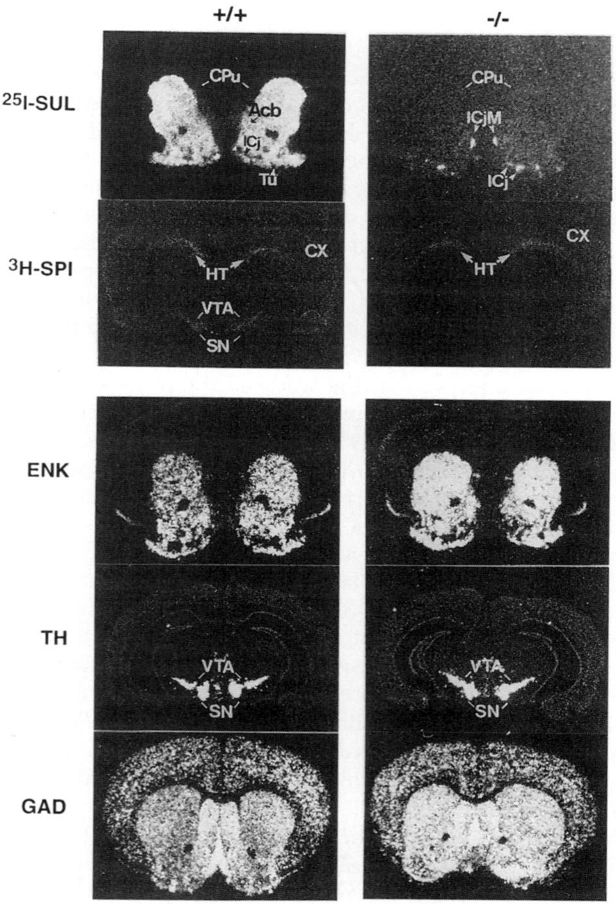

FIGURE 1 Expression of receptors and striatal markers in WT (+/+) and D2R-null (−/−) mice.

corresponding to the coding sequences of the enkephalin (ENK), substance P (SP), dynorphin (DYN), glutamic acid decarboxylase (GAD), and tyrosine hydroxylase (TH) genes. The first three peptides were highly expressed by striatal neurons, and their level of expression was influenced by the activity of D1 and D2 receptors. Interestingly, the expression of ENK mRNA was increased by 40% in the striatum of D2R-deficient mice, as compared with control animals (see Fig. 1, bottom). The level of expression of DYN and SP appeared not to be significantly altered in knockout mice.

We also analyzed the level of expression of two metabolic enzymes (GAD and TH) involved in the synthesis of γ-aminobutyric acid (GABA) and dopamine as a parameter to study whether the synthesis of these neurotransmitters was modified in homozygous D2R-deficient mice. TH expression was unchanged, also suggesting that in D2R-deficient mice the endogenous dopamine levels are not altered (see Fig. 1, bottom). These data were also confirmed by the analysis at the protein level. Strikingly, our *in situ* analysis revealed an increase of GAD expression in the cortex and also in the striatum of knockout versus normal mice (see Fig. 1, bottom).

A comparative behavioral study was performed analyzing the locomotor abilities of WT, heterozygous, and homozygous D2R-null mice. These studies clearly revealed a motor impairment in homozygous mice characterized by reduced and slow movements, sprawled hind legs, and backward movements. Homozygous D2-deficient mice showed a marked reduction of locomotion (74%) in the open field test, as compared with WT animals. Notably, in this test, homozygous mice did not show any vertical behavior (i.e., rearing). The number of rearings was also significantly affected in heterozygous mice (64%), while locomotion was reduced by 35%. D2R-deficient animals were also less coordinated, and the spontaneous movements were reduced.

D2 receptors are also specifically expressed by different cell types of the anterior pituitary gland. A well-documented function of D2Rs in this gland is the control of prolactin synthesis and release. D2Rs are also expressed on melanotropes in the intermediate lobe. We thus tested the expression of most pituitary hormones in the D2R deficient mice. The gonadotropin (follicle-stimulating and leuteinizing hormones) and growth hormone levels were not affected. In contrast, a threefold increase of prolactin and proopiohelano cortin (POMC) mRNA was observed. In addition, the size of the gland was greatly enlarged. The increase of prolactin levels was also correlated with a decrease of gonadal size in D2R-null mice. The histological aspect of these organs was normal in both sexes; however, the ovaries apparently contained a reduced number of follicles. Similarly, the tubular epithelium of knockout testis was also reduced. These data implicate for the first time a possible role for prolactin in gonad development.

D2 and D1 receptors are the most abundant dopaminergic receptors expressed in the brain. The parallel activation by dopamine of these two receptors in the basal ganglia leads to the activation of two pathways whose synergistic action controls coordinated movements. Unexpectedly, the elimination of D1 receptor expression by homologous recombination, which would be predicted to affect the locomotion-enhancing direct pathway, leads instead of hyperlocomotion in homozygous mice (4, 5). These results contradict pharmacological studies in which the use of D1 receptor antagonists results in a cataleptic response of treated animals. In contrast, D2R-deficient animals display a phenotype that broadly resembles D2 antagonist treatment, indicating that the effect observed by antagonist treatment is due to the blocking of D2 receptors and not to other members of the dopamine receptor family. The knockout of the D2R gene has evidenced the pleiotropic role of this receptor in the control of physiological functions. The most evident phenotypes are a serious locomotor impairment and abnormal hormonal function. Interestingly, in the D2R knockout animals, there is an increase in ENK mRNA expression, which could be responsible for the decreased movement of these animals. It is interesting to note that the phenotype exhibited by D2R-deficient mice, at the level of expression of striatal specific peptides, resembles the findings obtained by treatment of 6-OHDA–lesioned animals (6). This might indicate that the primary target of dopamine depletion in this model is the D2 receptor.

An unexpected result is that the cortical and striatal expression of GAD is increased. This feature may represent a compensatory mechanism in the brain of D2R-deficient mice. Finally, we propose that D2-deficient mice may serve as an animal model for Parkinson's disease. The genetic inactivation of the D2R gene apparently has physiological consequences analogous to the degeneration of dopaminergic fibers encountered in Parkinson's disease. The behavior of these mice strikingly resembles the lack of spontaneous movements, akinesy, abnormal gait, and posture characteristic of individuals affected by this disease. Thus, these animals represent a powerful tool, not only to test the *in vivo* effect of dopaminergic ligands specific for other dopaminergic receptors, but also for the study and design of drugs to alleviate the symptoms of Parkinson's disease.

References

1. Gingrich, J. A., and Caron, M. G. (1993). Recent advances in the molecular biology of dopamine receptors. *Annu. Rev. Neurosci.* **16**, 299–321.
2. Guiramand, J., Montmayeur, J. P., Ceraline, J., Bhatia, M., and Borrelli, E. (1995). Alternative splicing of the dopamine D2 receptor directs specificity of coupling to G-proteins. *J. Biol. Chem.* **270**, 7354–7358.
3. Baik, J. H., Picetti, R., Saiardi, A., Thiriet, G., Dierich, A., Depaulis, A., Le Meur, M., and Borrelli, E. (1995). Parkinsonian-like locomotor impairment in mice lacking dopamine D2 receptors. *Nature* **377**, 424–428.
4. Xu, M., Moratalla, R., Gold, L. H., Hiroi, N., Koob, G. E., Graybiel, A. M., and Tonegawa S. (1994). Dopamine D1 receptor mutant mice are deficient in striatal expression of dynorphin and in dopamine-mediated behavioral responses. *Cell* **79**, 729–742.
5. Drago, J., Gerfen, C. R., Lachowicz, J. E., Steiner, H., Hollon, T. R., Love, P. E., Ooi, G. T., Grinberg, A., Lee, E. J., Huang, S. P., Bartlett, P. F., Jose, P. A., Sibley, D. R., and Westphal, H. (1994). Altered striatal function in a mutant mouse lacking D1A dopamine receptors. *Proc. Natl. Acad. Sci. U.S.A.* **91**, 12564–12568.
6. Gerfen, C. R., Engber, T. M., Mahan, L. C., Susel, Z., Chase, T. N., Monsma, F. J. Jr., and Sibley, D. R. (1990). D1 and D2 dopamine receptor-regulated gene expression of striatonigral and striatopallidal neurons [see comments]. *Science* **250**, 1429–1432.

Pedro A. Jose,[*] Gilbert M. Eisner,[†] and Robin A. Felder[‡]

[*]Department of Pediatrics
Georgetown University Medical Center
Washington, D.C. 20007

[†]Department of Medicine
Georgetown University Medical Center
Washington, D.C. 20007

[‡]Department of Pathology
University of Virginia Health Sciences Center
Charlottesville, Virginia 22908

Regulation of D1 Receptor Function in Spontaneous Hypertension

Hypertension affects 15–20% of the adult population and is a major risk factor for stroke, myocardial infarction, and renal failure (1). Although multiple genes may interact to initiate hypertension, the kidney is a primary locus for the genesis and maintenance of hypertension (2) by its ability to influence sodium balance and produce hormones that can increase or decrease blood pressure. Indeed, renal transplantation studies in humans and in animals with genetic hypertension have shown the importance of the kidney in the genesis of hypertension. Thus, the transplantation of a kidney from a normotensive rat to a bilaterally nephrectomized, genetically hypertensive rat normalizes blood pressure; the converse procedure increases blood pressure in the recipient rat. In humans, recipients of a kidney transplant from a donor without a family history of hypertension require less antihypertensive treatment than recipients of a kidney from a donor with a family history of hypertension (3). There are now well-described examples of genetically transmitted mutations that impact on sodium transport by the kidney and produce hypertension (1).

Genetic abnormalities involving the regulation of electrolyte transport may be responsible for a considerable proportion of essential hypertension. From 20 to 70% of subjects with essential hypertension and 25% of subjects with normal blood pressure are salt-sensitive; that is, blood pressure increases with increased sodium chloride intake. Most of the salt transport in the kidney occurs in the proximal tubule, and several studies have reported abnormalities in the renal regulation of sodium transport at this site in human (4) as well as in animal models of genetic hypertension. The production of dopamine by renal proximal tubules and its effect on sodium transport in this nephron segment is regulated by sodium intake (5). During moderately increased sodium intake (5–10% of body weight), dopamine, via D1-like receptors, acts as an intrarenal natriuretic factor. This renal paracrine–autocrine effect is lost during states of

Advances in Pharmacology, Volume 42

1054-3589/98 $25.00

negative sodium balance. We, therefore, surmised that a defect in the regulation of dopamine and its effect on renal sodium transport may be important in the pathogenesis of spontaneous hypertension. Decreased renal production of dopamine has been reported in human essential hypertension, especially in those individuals with the salt-sensitive type. Those patients have also been reported to have an exaggerated natriuretic response to exogenous dopamine, although this natriuretic response may be related exclusively to the renal hemodynamic effect of D1 receptors. Other studies suggest that the natriuretic effect of endogenous dopamine is impaired in some hypertensive human subjects.

The resistance of rats with genetic hypertension to the natriuretic effect of dopamine and D1 agonists suggests a defect in or beyond the renal D1 receptor. However, the spontaneously hypertensive rat (SHR) and its normotensive control, the Wistar-Kyoto (WKY) rat, do not differ in D1-like receptor density in renal proximal tubules or in the expression of the two D1-like receptors cloned in mammals, the D1A and D1B receptors. While the affinity to antagonists of the renal D1-like receptor is similar in WKY rats and SHRs, the affinity to agonists is markedly different. High- and low-affinity binding sites are present in renal proximal tubules of WKY rat, whereas only a low-affinity binding site is present in the SHR. As with D1-like receptors in the brain, occupation of the renal proximal tubular D1-like receptor by agonists increases cyclic adenosine monophosphate (cAMP) and inositol phosphate production in normotensive rats; in kidneys of hypertensive rats, the response is markedly attenuated. The decreased ability of dopamine and D1 agonists to increase cAMP production in renal tubules of SHRs is not due to a defect in the responsiveness of G-proteins or effector enzymes. For example, forskolin, guanosine triphosphate (GTP), and its analogues increase cAMP production in renal proximal tubules to a similar extent in WKY rats and SHRs. Taken together, these observations indicate that the dopamine D1-like receptor is *uncoupled* from the G_s protein–adenylyl cyclase enzyme complex in proximal tubules and renal resistance vessels of genetically hypertensive rats. Moreover, there is receptor specificity, because parathyroid hormone increases cAMP production to a similar degree in proximal convoluted tubules of WKY rats and SHRs. Uncoupling of the D1-like receptor from its effector enzyme complex is nephron segment– and organ-specific. Thus, the ability of D1 agonists to increase cAMP production in cortical collecting ducts and in brain striatum is similar in normotensive and hypertensive rats. The phenotypic differences are primary and not secondary to the development of high blood pressure, because the uncoupling of the D1-like receptor from the G-protein–effector enzyme complex antedates the hypertension.

The uncoupling of the D1-like receptor from its effector enzyme complex in renal proximal tubules is not an isolated finding in one model of genetic hypertension. In another model of genetic hypertension, the Dahl salt-sensitive (DSS) rat, the same defect is present and is also observed prior to development of hypertension. While forskolin increases adenylyl cyclase activity to a similar extent in DSS and Dahl salt-resistant (DSRs) rats (rats that do not increase their blood pressure in response to increase NaCl intake), the stimulatory of fenoldopam, a D1 agonist, is greater in DSR rats than in DSS rats. In the DSS rat, the uncoupling is noted not only in proximal tubules but also in the thick ascending limb of Henle, the nephron segment that is apparently responsible for the sodium chloride retention in this model of genetic hypertension. The uncoupling of the D1-like receptor from its effector enzyme complex is associated with a failure of dopamine and D1 agonists to inhibit renal proximal tubular luminal Na^+/H^+ exchanger (NHE) and Na^+/K^+ adenosine triphosphatase (ATPase) activity. This is *not* due to any inherent abnormality in the NHE or the sodium pump. Thus, cAMP and GTPγS inhibit renal proximal tubular luminal NHE activity to a similar extent in WKY rats and SHRs. In addition, phorbol esters, presumably via activation of protein kinase C, inhibit renal proximal tubular Na^+/K^+ ATPase activity to a similar degree in these rat strains.

Many physiologic, biochemical, and molecular differences between normotensive and genetically hypertensive strains of rats have been shown to be caused by genetic drift and random fixation of alleles at loci that are not involved in blood pressure regulation. Commercially available SHRs are both hypertensive and hyperactive. In strains of rats in which the hyperactive phenotype has been bred from the hypertensive phenotype, we found that the uncoupling of the D1-like receptor from its effector enzyme complex in renal proximal tubules is associated with hypertension and not with hyperactivity. Furthermore, in a segregating population of F2 rats bred from female WKY rats and male SHRs, D1 agonists inhibit renal proximal tubular luminal NHE *in vitro* and increase sodium excretion *in vivo* only in rats with normal systolic blood pressures (<140 mm Hg) (6).

Chronic blockade of D1 receptors induces the development of hypertension in genetically normotensive rats, and inhibition of renal dopamine synthesis accelerates the development of hypertension in the SHR. Disruption of the D1A receptor gene (which is expressed in renal proximal tubules) in mice increases systolic blood pressure and results in diastolic hypertension (6). However, the defect in spontaneous hypertension is not due to a mutation of the D1A receptor gene. The mechanism of the uncoupling of the renal D1-like receptor from its effector proteins remains to be determined. Although, the uncoupling is akin to homologous desensitization (7), it is not due to agonist-induced desensitization because it is noted as early as 3 wk of age, when renal dopamine production is similar in SHR and its normotensive control, the WKY rat. Moreover, altering renal dopamine production does not affect sodium transport in the SHR. Immortalized renal proximal tubular cells from WKY rats, which do not produce dopamine in culture, retain their ability to increase cAMP and inositol phosphate production in response to D1 agonists. Immortalized proximal tubular cells from SHR, with D1A receptor density similar to those found in WKY rat, continue to have an attenuated response to D1 agonists (unpublished observations). Three mechanisms of desensitization have been described for G-protein–linked receptors, phosphorylation, sequestration, and downregulation. Downregulation is unlikely because renal proximal tubular receptor density is not decreased in spontaneous hypertension. However, preliminary evidence suggests that the D1A receptor is sequestered in immortalized renal proximal tubular cells from SHRs, the mechanism of which remains to be determined. We conclude that an abnormal regulation of the renal D1A receptor plays an important role in the pathogenesis of genetic hypertension, but the genetic defect remains to be discovered. It may, however, prove to be important in some forms of human hypertension.

References

1. Lifton, R. P. (1996). Molecular genetics of human blood pressure variation. *Science* **272,** 676–680.
2. Guyton, A. C., Coleman, T. G., Cowley, A. W. Jr., Scheel, K. W., Manning, R. D. Jr., and Norman, R. A. Jr. (1972). Arterial pressure regulation: Overriding dominance of the kidneys in long-term regulation and in hypertension. *Am. J. Med.* **52,** 584–594.
3. Cusi, D., and Bianchi, G., (1996). Renal mechanisms of genetic hypertension: From the molecular level to the intact organism. *Kidney Int.* **49,** 1754–1759.
4. Weder, A. B. (1983). Red-cell lithium-sodium countertransport and renal lithium clearance in hypertension. *N. Engl. J. Med.* **314,** 198–201.
5. Jose, P. A., Eisner, G. M., Drago, J., Carey, R. M., and Felder, R. A. (1996). Dopamine receptor signaling defects in spontaneous hypertension. *Am. J. Hypertens.* **9,** 400–405.
6. Albrecht, F. E., Drago, J., Felder, R. A., Printz, M. R., Eisner, G. M., Robillard, J. E., Sibley, D. R., Westphal, H., and Jose, P. A. (1996). Role of the D1A dopamine receptor in the pathogenesis of genetic hypertension. *J. Clin. Invest.* **97,** 2283–2288.
7. Lefkowitz, R. J., Cotecchia, S., Kjelsberg, M. A., Pitcher, J., Koch, W. J., Inglese, J., and Caron, M. G. (1993). Adrenergic receptors: Recent insights into their mechanism of activation and desensitization. *Adv. Second Messenger Phosphoprotein Res.* **28,** 1–9.

PART E

CATECHOLAMINES IN THE PERIPHERY

David S. Goldstein

Clinical Neuroscience Branch
National Institute of Neurological Disorders and Stroke
National Institutes of Health
Bethesda, MD 20892-1424

Overview

Catecholamines outside the brain act as effectors for a wide variety of homeostatic systems that regulate function of the heart and blood vessels, metabolism, and possibly immunity. Recent research related to catecholamines in the periphery has dealt with (1) assessment of peripheral catecholaminergic function, (2) catecholamines and stress, (3) catecholamines and pain, (4) catecholamines and neuroimmunology, (5) adrenomedullary secretion and cosecretion, (6) neurocardiology, (7) catecholamines and metabolism, and (8) roles of catecholamines in the brain in regulation of the cardiovascular system.

I. Assessment of Peripheral Catecholaminergic Function

Because the sympathetic nervous system consists of networks of filaments and terminals enmeshing blood vessels and pervading most organ parenchyma, assessing sympathetic "activity" in the body as a whole or in particular organs has posed a perennial challenge to clinical investigators. The main biochemical techniques for evaluating sympathoneural activity have depended on assays of plasma levels of the sympathetic neurotransmitter norepinephrine (NE). Rela-

Advances in Pharmacology, Volume 42

1054-3589/98 $25.00

tionships among sympathetic nerve traffic, NE release from nerve terminals, NE re-uptake, and entry of NE into regional venous plasma (spillover) remain controversial. Christensen and Knudsen (p. 540) have reported that application of lower-body negative pressure (LBNP) failed to increase the plasma appearance rate of NE, whereas skeletal muscle sympathetic traffic recorded using peroneal nerve microneurography increased in this setting, leading to skepticism about the validity of NE spillover measurements in indicating sympathetic "activity."

The investigators also have reported a close association between lymphocyte concentrations of catecholamines and of cyclic adenosine monophosphate (cAMP). The usefulness of lymphocyte NE concentrations in evaluating sympathoneural function remains unknown.

Based on mathematical analyses of different models relating the fate of infused [^{3}H]NE to sympathoneural release and re-uptake of endogenous NE, Kopin has demonstrated that the same general formula can apply to competing models (I. J. Kopin, unpublished observations). Because attenuation of neuronal reuptake of NE increases the ratio of spillover to release into the interstitial fluid in patients with a decreased complement of cardiac sympathetic terminals, regional NE spillover measurements underestimate the extent of loss of innervation.

Assessments of levels of NE in myocardial interstitial fluid, sampled using microdialysis, provide a potentially important, novel means to evaluate sympathetic function *in vivo*. Mertes (p. 544) found increased myocardial interstitial NE concentrations in the setting of brain death or coronary occlusion, even when plasma NE levels increased by relatively little. Occlusion of the left anterior descending coronary artery increased dialysate NE concentrations in the ischemic but not in the nonischemic region. In an oral presentation based on a submitted abstract, Akiyama and Yamazaki have confirmed that coronary occlusion increased microdialysate NE levels in the ischemic but not in the nonischemic region, with only partial inhibition by ω-conotoxin. Yamazaki has noted that local perfusion of classical neuropharmacological agents via a cardiac microdialysis probe produced expected neurochemical responses: increased microdialysate NE with tyramine, desipramine, or stellate stimulation, and decreased microdialysate dihydroxyphenylglycol (DHPG) with reserpine.

Sympathetic microneurography has refined understanding of homogeneous and heterogeneous patterns of sympathetic nerve firing. As noted in the review by Rea (p. 548), burst discharges occur simultaneously in multiple nerves (e.g., both legs), and plasma NE levels correlate positively with directly recorded skeletal sympathetic traffic, indicating a common central organ of sympathetic outflows; however, peroneal sympathetic fibers mediating cutaneous vasoconstriction and sweating undergo differential regulation from those mediating vasoconstriction in skeletal muscle. Because procainamide decreased and quinidine increased skeletal sympathetic traffic, administration of antiarrhythmic drugs in the same class can produce disparate effects on skeletal sympathoneural outflows.

Positron emission tomographic (PET) scanning provides a novel means to evaluate regional sympathetic innervation. In a presentation based on a submitted abstract, Raffel and coworkers have described results using three ^{11}C-labeled analogues of NE, [^{11}C]meta-hydroxyephedrine ([^{11}C]HED), [^{11}C]epinephrine ([^{11}C]Epi), and [^{11}C]phenylephrine ([^{11}C]PHEN). [^{11}C]HED undergoes rapid transport into sympathetic terminals and within the terminals into sympathoneural vesicles. Due to its lipophilicity, the drug has a relatively high rate of diffusion across vesicular and neuronal membranes, and the drug does not undergo metabolism by monoamine oxidase (MAO), which figures prominently in metabolism of endogenous NE. Vesicular uptake with relatively little leakage protects [^{11}C]Epi from metabolism by MAO. [^{11}C]PHEN leaks from vesicles more slowly than does [^{11}C]HED but more rapidly than does [^{11}C]Epi and undergoes metabolism by MAO. All three tracers can reveal abnormal sympathetic innervation in patients with disorders such as heart failure and diabetic autonomic neuropathy.

II. Catecholamines and Stress

Recent research has forced reconsideration of concepts about stress as a medical scientific idea. Continuing the line of thinking in Selye's theory, which defines stress as the nonspecific response of the body to any demand, Chrousos (p. 552) has summarized evidence consistent with the existence of a stress syndrome elicited on activation of a stress system. The central components of the system would include the parvicellular corticotropin-releasing hormone (CRH) and arginine vasopressin (AVP) neurons in the paraventricular nucleus (PVN) of the hypothalamus, CRH neurons in the paragigantocellular and barabrachial nuclei of the medulla and locus ceruleus (LC), and noradrenergic groups in the LC and medullary regions. Effector limbs of the system would be the hypothalamo-pituitary-adrenocortical (HPA) axis, the sympathoadrenal system, and the parasympathetic nervous system. Chrousos lists many disease states associated with increased or decreased activity of the HPA axis.

In contrast, Kvetnansky *et al.* and Pacak *et al.* (p. 556 and p. 561) have emphasized stressor specificity of neuroendocrine responses. In conscious rats, exposure to cold evoked selective sympathoneural activation, as indicated by plasma NE responses, whereas insulin-induced glucoprivation evoked selective adrenomedullary activation, as indicated by plasma Epi responses. Other stressors—especially immobilization—activated both the sympathoneural and adrenomedullary systems. Expression of mRNA for tyrosine hydroxylase (TH) in the adrenal gland increased similarly for the three stressors; however, adrenal denervation abolished the increment in TH mRNA in rats exposed to cold or undergoing insulin injection, whereas denervation did not affect the increment in TH mRNA expression in undergoing immobilization. Whereas immobilization and insulin elicited marked increases in plasma adrenocorticotropic hormone (ACTH) levels, c-fos expression in the PVN and LC increased much more with the former than with the latter stressor. The findings, therefore, demonstrated differences among stressors in pathways for expression of multiple genes.

Increased TH mRNA expression in response to immobilization depends on transcriptional mediation. Sabban and coworkers (p. 564) have used c-fos knockout mice to discover that immobilization-induced increases in transcription of TH mRNA do not require c-fos.

Treatment of patients with anxiety often includes benzodiazepines. In an oral presentation, Tulen and coworkers reported that administration of alprazolam attenuated plasma Epi responses during performance of the color-word test, a laboratory mental stressor, without affecting plasma NE responses to standing. The results support the view that the beneficial effect of alprazolam does not come at the cost of inhibition of orthostatic reflexes.

III. Catecholamines and Pain

Although, according to traditional views, the sympathetic nervous system constitutes an exclusively efferent effector system, recent findings have supported a role of the sympathetic nervous system in modulation of sensory information. Raja (p. 567) has described a model for sympathetically maintained pain in humans, where nociceptors normally unresponsive to sympathetic stimulation would become responsive to catecholamines after nerve injury. $Alpha_1$ adrenoceptors or α_{2B} adrenoceptors may mediate increased pain sensitivity in the hyperalgesic skin. Alternatively, the dorsal root ganglion may constitute the site of pathophysiological interaction between sympathetic efferent and somatic afferent traffic.

Palkovits and coworkers (p. 572) have applied immunohistochemical, *in situ* hybridization histochemical, tract-tracing, and *in vivo* microdialysis techniques to examine central pathways activated by acute pain induced by subcutaneous injection of formalin into the hindpaw of rats. This stressor increased plasma ACTH levels by four- to fivefold and produced large increases in microdialysate NE concentrations in the PVN. The nociceptive signals increased c-fos immunoreactivity ipsilaterally in marginal zone neurons of the dorsal horn of the spinal cord. More rostrally, c-fos immunoreactivity was noted bilaterally, in A1, A2, and A6 TH-containing catecholaminergic neurons and in the PVN, even in the setting of hemisection between the spinal cord and medulla oblongata. These results suggested the existence of axon collaterals that cross over in the spinal cord or lower medulla, confirmed by tract-tracing. Hemisection between the medulla and hypothalamus decreased formalin-induced NE responses ipsilaterally in the PVN, regardless of the side of the injection.

In the dorsal horn of the spinal cord, NE released from fibers descending from the brain stem gates access of nociceptive information by both pre- and postsynaptic actions on terminals of primary afferent fibers. Millan (p. 575) studied subtypes of α adrenoceptors that might mediate this antinociceptive effect. Administration of the α_{2A}-selective agonist S 18616 exerted antinociceptive and, at higher doses, sedative–hypnotic actions in mice. In an oral presentation, Zhang and coworkers reported evidence from microdialysis that in the dorsal horn of awake rats, depolarization releases NE and that the released NE undergoes neuronal reuptake.

Catecholamines and opioid peptides act synergistically to decrease pain when infused into the spinal subarachnoid space. Implantation of chromaffin cells, which secrete both types of compound, may offer a novel means to relieve intractable pain. Sagen (p. 579) has reported that in the rat, spinal subarachnoid implantation of adrenal medullary tissue, chromaffin cell suspensions, or encapsulated chromaffin cells all markedly reduced pain behavior in several models.

Tolerance did not develop. A phase I clinical trial of adrenal medullary allografts is under way in patients with intractable pain.

IV. Catecholamines and Neuroimmunology

In contrast with the well-known role of the HPA axis, the role of catecholaminergic systems in modulating immune responsiveness remains poorly understood. Several lines of evidence suggest such a role; however, studies have disagreed about what that role might be. Felten and coworkers (p. 583) have summarized research to date in this field. Lymphoid organs receive postganglionic sympathetic innervation, with TH-containing terminals contacting not only vascular and trabecular smooth-muscle cells, but also lymphocytes and macrophages. 6-Hydroxydopamine (6-OHDA)–induced neurotoxic destruction of sympathetic terminals produces several effects on lymphocyte numbers and function, including reductions of cell-mediated responses, T-cell proliferation, and delayed hypersensitivity. From studies of lymph node and splenic immunocytes of BALB/c mice given 6-OHDA, the authors obtained evidence for complex, organ-specific effects of chemical sympathectomy and for interactions between NE and production of cytokines.

The rapidity of acute stress-related lymphocytosis and of alterations in populations of natural killer (NK) and T-suppressor cells has led to speculation that the sympathetic nervous system may contribute to these phenomena. Mills and coworkers (p. 587) have used flow cytometry to study lymphocyte subpopulations (mature T lymphocytes, CD3/Anti-leu 4; T-helper lymphocytes, CD4/Anti-leu 3a; T-suppressor lymphocytes, CD8/Anti-leu 2a; B lymphocytes, CD19/Anti-leu 12; and NK, CD2-/CD16 + CD56) in subjects during exercise or exposure to a psychological stressor (speaking impromptu to a video camera). The speech task increased NK and T-suppressor cells and decreased T-helper and B cells. Exercise produced a lymphocytosis, with increased NK cells, T-suppressor cells, and T-helper cells. Plasma catecholamine levels at baseline, numbers of lymphocyte β_2 adrenoceptors, and the extent of catecholamergic activation correlated with the extent of redistribution of lymphocyte subsets.

Development and differentiation of sympathetic neurons requires nerve growth factor (NGF), the biological effects of which arise from occupation of two types of receptors: a low-affinity p75 transmembrane glycoprotein and a high-affinity tyrosine kinase transmembrane protein (TKA). As pointed out in the review by Aloe (p. 591), NGF exerts several generally proinflammatory effects on lymphocyte populations, and NGF levels increase in models of inflammatory autoimmune diseases, leading to speculation that NGF, perhaps in association with tumor necrosis factor-α, contributes to autoimmune pathologies.

V. Adrenomedullary Secretion and CoSecretion

Recent research has led to revision of the classical concept that adrenomedullary secretion results from splanchnic nerve traffic via increased occupation by acetylcholine of nicotinic receptors on chromaffin cells. Because effective cholinergic blockade abolishes acetylcholine-induced but not splanchnic stimulation–induced adrenomedullary secretion, other transmitters appear to affect chromaffin cell function. Of these, as noted in the review by Wakade (p. 595), pituitary adenylyl cyclase–activating polypeptide (PACAP) has a potency about 500 times greater than acetylcholine in evoking catecholamine secretion. In an oral presentation, Geng and coworkers reported that local infusion of PACAP into the adrenal gland of anesthetized dogs evoked dose-dependent catecholamine release. In perfused adrenal glands, electrical neural stimulation increased perfusate PACAP and vasoactive intestinal peptide (VIP) concentrations by two- to threefold. The cell bodies for PACAP and VIP release are located in dorsal root and nodose ganglia, indicating a sensory role for these peptides. Both PACAP and VIP selectively increase the ratio of secreted Epi to NE, and both activate adrenal TH, in contrast with acetylcholine, which does not. Because splanchnic stimulation also activates adrenal TH, these findings indicate that PACAP and VIP constitute neurotransmitter secretagogues that are independent of acetylcholine.

The vasoactive peptide adrenomedullin (AM), first isolated from human pheochromocytoma cells, exerts potent and long-duration hypotensive effects. Nishikimi (p. 599) has measured immunoreactive AM in arterial or regional venous plasma from patients with cardiovascular diseases or pheochomocytoma. Despite high levels of expression of AM mRNA in adrenal gland, lung, kidney, and heart, no important regional arteriovenous increments in AM concentrations were noted, although the pulmonary artery concentration exceeded the aortic concentration. Plasma AM levels increased as a function of New York Heart Association classification in heart failure patients, as a function of serum creatinine in renal failure patients, and as a function of World Health Organization classification in hypertensive patients. AM concentrations also correlated strongly positively with NE concentrations. Because the plasma clearance of AM has not yet been studied, the finding of only small arteriovenous increments in AM levels across organs does not necessarily imply other sites of production.

Chromaffin cells possess high-affinity binding sites for strychnine, a specific antagonist for the inhibitory central neurotransmitter glycine. The 48-kDa α subunit of the glycine receptor is homologous with the nicotine receptor family. Strychnine inhibits nicotine-induced catecholamine secretion from cultured bovine adrenomedullary cells. Yadid and coworkers (p. 604) have reported that strychnine displaces acetylcholine from membrane-binding sites, consistent with an interaction between glycine and acetylcholine in mediating adrenomedullary secretion.

Release of NE can occur by nonexocytotic means in the setting of energy depletion. In an oral presentation, Haass and coworkers provided evidence that in bovine adrenomedullary cells preloaded with [^{3}H]NE, nonexocytotic release occurs when the intracellular Na^+ concentration exceeds 10 mmol/liter.

VI. Neurocardiology

Peripheral catecholaminergic systems probably play a role in many neurocardiologic disorders, either secondarily as homeostatic effectors or primarily as etiologic factors. The hemodynamic hallmark of neurocardiogenic syncope is systemic vasodilation, with or without bradycardia. The vasodilation results from sudden, marked decreases in sympathoneural outflows. In contrast, plasma Epi levels generally increase in this setting. Lenders and coworkers (p. 607) have reported that people susceptible to neurocardiogenic syncope had relatively smaller forearm vasoconstrictor responses, smaller increments in NE spillover, and larger increments in plasma Epi levels during orthostasis than did people without such susceptibility. Sudden, central resetting of baroreflexes may produce the sympathoinhibition. Alternatively, the sympathoinhibition may result from increased activity of cardiac afferent inhibitory C-fibers.

Congestive heart failure constitutes the leading cause of hospitalization in the elderly and accounts for a substantial proportion of overall cardiovascular morbidity and mortality. Direct toxic effects of catecholamines on the heart, catecholamine-induced stimulation of the renin-angiotensin-aldosterone system, catecholamine-related enhancement of myocardial hypertrophy, and increased cardiac work during sympathoadrenal activation may combine to explain the worsened prognosis in patients with heart failure who have generalized sympathoadrenal activation. Carvedilol acts as a β-adrenoceptor blocker, α_1-adrenoceptor blocker, and antioxidant. Feuerstein and Ruffolo (p. 611) have reported remarkable beneficial effects of carvedilol from clinical trials. In particular, in a large, multicenter trial in the United States, carvedilol treatment decreased all-cause mortality by 65% in patients with heart failure. The authors suggest that the superiority of carvedilol over other β-blockers may result from additional suppression of oxygen radical formation, inhibition of inflammatory cell infiltration, blockade of apoptosis of myocardial cells, or α_1-adrenoceptor blockade–induced vasodilation.

In heart failure, cardiac NE spillover increases markedly and is related to prognosis. As noted in a review by Vaz and coworkers (p. 630), patients with early heart failure seem to have selectively increased cardiac NE spillover. Young patients with essential hypertension often have increased cardiac, renal, and skeletal muscle sympathoneural outflows, as indicated by measurements of directly recorded peroneal sympathetic traffic or of regional NE spillovers.

Patients with congestive heart failure have exercise intolerance. In an oral presentation, Runqvist and coworkers reported that during supine bicycle exercise, patients with mild-to-moderate heart failure had smaller increases in cardiac NE spillover than did healthy subjects but had normal increases in renal and whole body NE spillover. The authors suggested that cardioselective attenuation of NE release during exercise may help explain the exercise intolerance.

Goldstein and coworkers (p. 615) have summarized results of 6-[^{18}F]fluorodopamine PET scanning in patients with dysautonomias. The classification of dysautonomias has been confusing and the pathophysiology obscure. Goldstein and coworkers used thoracic 6-[^{18}F]fluorodopamine PET scanning and assessments of cardiac NE kinetics to examine myocardial sympathetic innervation and function in patients with dysautonomias. Patients with pure autonomic failure (sympathetic neurocirculatory failure without central neurodegeneration) or sympathetic failure and L-dopa/carbidopa-responsive parkinsonism had no myocardial 6-[^{18}F]fluorodopamine-derived radioactivity. Patients with the Shy-Drager syndrome (sympathetic neurocirculatory failure and L-dopa/carbidopa-unresponsive central neurodegeneration) had increased radioactivity (analogous to ganglion-blocked volunteers), and patients with multiple-system atrophy without sympathetic neurocirculatory failure had normal myocardial 6-[^{18}F]fluorodopamine-derived radioactivity. All five patients without detectable myocardial 6-[^{18}F]fluorodopamine-derived radioactivity had little or no cardiac NE spillover. The results indicate that the Shy-Drager syndrome, parkinsonism with sympathetic neurocirculatory failure, and multiple-system atrophy without sympathetic neurocirculatory failure constitute pathophysiologically distinct clinical entities.

In an oral presentation based on a submitted abstract, Ieda and coworkers provided supplementary neurochemical evidence for the view that the Shy-Drager syndrome differs from parkinsonism with sympathetic neurocirculatory failure. Plasma arginine vasopressin levels increased during head-up tilt in patients with autonomic failure in the setting of Parkinson's disease but not in patients with the Shy-Drager syndrome.

VII. Catecholamines and Metabolism

As patients with insulin-dependent diabetes mellitus (IDDM) become insulin-deficient over years, they not only lose their ability to suppress insulin release in response to hypoglycemia, they also lose their ability to secrete glucagon. IDDM patients, therefore, depend importantly on Epi as a glucose counterregulatory agent. Patients with absent responses of both Epi and glucagon have at least 25 times the risk of suffering episodes of severe, iatrogenic, insulin-induced hypoglycemia during intensive treatment than patients with present Epi and absent glucagon responses. Because awareness of hypoglycemia depends importantly on perceiving physiological changes elicited by high circulating Epi concentrations, patients with IDDM and defective glucose counterregulation also have a high frequency of hypoglycemia unawareness. Since recent antecedent hypoglycemia attenuates Epi responses to subsequent hypoglycemia, Cryer (p. 620) has put forward the hypothesis that hypoglycemia-associated autonomic failure reflects a functional disorder that differs from classical diabetic autonomic neuropathy.

The neurohumoral response to hypoglycemia appears to depend on glucose sensors in the ventromedial hypothalamus (VMH). As noted in the review by Maggs and Sherwin (p. 622), local perfusion of 2-deoxyglucose in the VMH elicits marked increases in plasma catecholamine levels and increases serum glucose levels, and VMH lesions attenuate catecholamine responses to hypoglycemia. Moreover, in a microdialysis study in which glucose was administered locally in the VMH, hypoglycemic animals had blunted catecholamine responses.

From the pattern of plasma catecholamine responses to hypoglycemia, researchers have inferred that glucoprivation evokes a mainly if not purely adrenomedullary response, with little if any overall sympathoneural activation. Maggs and Sherwin have also reported results of studies about effects of hypoglycemia on skeletal muscle and adipose tissue microdialysate levels of catecholamines in humans. In adipose tissue, NE constituted about 35% and in skeletal muscle about 50% of the total catecholamine increment, indicating that hypoglycemia augments release of NE from local sympathetic terminals.

Sweating occurs as a prominent sign of autonomic activation during hypoglycemia. Maggs and Sherwin have proposed that catecholamine-induced sweating and regional blood flow changes decrease core temperature, which may protect metabolically active organs such as the brain. The clinical importance of hypoglycemia-induced decreases in body temperature remains unknown.

Most clinical studies of counterregulatory hormonal responses to hypoglycemia have used experimental paradigms involving rapid decreases in blood glucose levels. Bolli (p. 627) has reported that more slowly developing hypoglycemia elicits a different neurohormonal response pattern, in which growth hormone and cortisol appear to play more prominent roles in glucose counterregulation. Bolli has confirmed the importance of catecholamines in glucose counterregulation by studying subjects treated with combined α- and β-adrenoceptor blockade; the experiments had to be interrupted early because of severe hypoglycemia despite increased secretion of glucagon, growth hormone, and cortisol. Catecholamine-induced hyperglycemia may occur partly indirectly via stimulation of lipolysis, because blockade of lipolysis attenuates alanine-induced gluconeogenesis.

The thermic effect of food (TEF) accounts for about 10% of daily energy expenditure. The TEF has obligatory and facultative components, and the sympathetic nervous system, via β adrenoceptors, appears to contribute importantly to the facultative component of TEF. Vaz and coworkers (p. 630) have reported about a doubling of forearm NE spillover after ingestion of a 750-kcal mixed meal, confirming previous microneurographic studies indicating substantial increases in skeletal sympathoneural traffic after glucose ingestion. The kidneys and skeletal muscle accounted for over 50% of the postprandial increment in spillover of NE into arterial plasma.

Administration of agonists or antagonists of adrenoceptor subtypes via a microdialysis probe enables evaluation of mechanisms of catecholamine-induced lipolysis *in vivo* in humans. In nonobese, healthy subjects with a microdialysis probe placed in abdominal, subcutaneous adipose tissue, Hagström-Toft (p. 634) has reported that agonists at all three β-receptor subtypes increase microdialysate glycerol levels. From studies of α- or β-adrenoceptor blockers, Hagström-Toft concludes that α adrenoceptors inhibit and β adrenoceptors contribute to lipolytic responses to hypoglycemia. Local β-adrenoceptor stimulation blocked the antilipolytic effect of insulin.

Because insulin produces vasodilation, sympathoneural activation in response to insulin theoretically could result from a baroreflex mechanism; however, in an oral presentation, Tack and coworkers reported different time courses between gradual insulin-induced vasodilation and rapid arterial NE responses in a euglycemic clamp study.

VIII. Catecholamines in the Brain and Regulation of the Cardiovascular System

Cells clustered in the rostral ventrolateral medulla (RVLM) constitute the main source of descending pathways to the sympathetic preganglionic neurons in the spinal intermediolateral columns. C1 cells in this region contain phenylethanolamine-*N*-methyltransferase and so presumably synthesize Epi. Guyenet and coworkers (p. 638) have studied electrophysiological properties of C1 neurons in neonatal rats. In coronal slices slightly caudal to the facial nucleus, RVLM cells identified after retrograde labeling with fluorescent tracers injected into the thoracic spinal cord contained about 65% C1 cells. Guyenet and coworkers report that most C1 cells fired slowly, spontaneously, and irregularly at rest, probably from a persistent sodium current that leads to slow, irregular fluctuations of membrane potential, whereas active cells fired more rapidly and regularly. Because application of various receptor antagonists did not influence firing, the firing results from intrinsic processes, not synaptic inputs. Several lines of evidence led to the conclusion that C1 cells possess somatodendritic α_2 adrenoceptors that probably contribute to inhibition of C1 cell firing via presynaptic control of transmitter release. C1 cells also possess stimulatory AT-1 receptors, with angiotensin II depolarizing C1 cells and increasing their firing rates; and they possess NK-1 receptors, with substance P (probably derived from the medullary raphe) also exciting the cells. In contrast to LC cells, C1 cells may not have a high-affinity amine transporter, they may use glutamate as the principal neurotransmitter, and they seem to play a role especially in vasoconstrictor responses mediated by sympathetic outflows.

The nucleus of the solitary tract (NTS) constitutes the main site of termination of baroreflex afferent fibers. Catecholamines or opioids in the NTS may modulate both cardiorespiratory and nociceptive reflexes. Pickel and coworkers (p. 642) have examined the immunocytochemical localization of the vesicular monoamine transporter-2 (VMAT2), responsible for reserpine-sensitive monoamine uptake in neurons; the β_2 adrenoceptor, implicated in presynaptic modulation of transmitter release; and the μ-opioid receptor (μOR), which participates in vagal reflexes. In axon terminals, VMAT2 seemed to be associated with large dense-core vesicles distant from synaptic sites. Both neurons and glia in the NTS possessed β_2 adrenoceptors. Some of the neurons also contained TH immunoreactivity, indicating the ability to synthesize catecholamines. μORs were localized in regions of the NTS receiving vagal afferents, indicating a role of μORs in presynaptic modulation of glutamate, a major transmitter released from vagal afferents.

Agmatine, a cationic amine formed by the decarboxylation of arginine via arginine decarboxylase, constitutes the first identified endogenous ligand at both α_2 adrenoceptors and imidazoline-binding sites. Reis and Regunathan (p. 645) have proposed that agmatine may function as a neurotransmitter. Brain and chromaffin cells contain agmatine, which can be detected in dense-core vesicles in the NTS. Radiolabeled agmatine undergoes release by depolarization or (in adrenal cells) nicotine. Among several pharmacological effects, agmatine dosedependently increases catecholamine release by adrenal chromaffin cells and enhances morphine-induced analgesia.

Recent studies summarized by Esler and coworkers (p. 650) applied a unique method, based on internal jugular venous sampling, [^{3}H]NE infusion, cerebral blood flow scanning, and neurochemical assays, to estimate jugular overflows of endogenous NE and its metabolites. In healthy men, central NE turnover (from the combined overflows of NE, methoxyhydroxyphenylglycol

[MHPG], and DHPG) correlated strongly positively with skeletal muscle sympathetic traffic, and in patients with pure autonomic failure, rates of jugular overflow of NE and its metabolites were not reduced, supporting the proposal that jugular NE overflow reflects release of NE in the brain and that in the forebrain NE release generally exerts a stimulatory effect on sympathoneural outflows. Heart failure patients had about a tripling of jugular NE turnover, which the authors suggest might underlie the markedly increased sympathoneural outflows in these patients.

IX. Conclusions and Future Directions

Studies using microdialysis, specific activity of [^{3}H]normetanephrine, and NE concentrations in lymph should enable estimates of NE concentrations in interstitial fluid and, thereby, estimates of release of NE from sympathetic nerves. Clinical use of PET scanning to evaluate regional sympathetic innervation and function will be expanded to patients with a variety of neurocardiological disorders, including orthostatic tachycardia syndrome, neurocardiogenic syncope, reflex sympathetic dystrophy, and congestive heart failure. Central neural pathways mediating responses to different stressors should be compared to determine stressor-specific mechanisms of neuroendocrine activation. The roles of NE, adrenoceptors, and NGF in neuropathic pain merit study both in animal models and in patients with pain syndromes. More convincing, consistent evidence is required to understand the role of catecholaminergic systems in components of immune responses. Increased use of microdialysis in adipose tissue and skeletal muscle should enhance understanding of catecholaminergic systems in patients with a variety of metabolic disorders, including IDDM and obesity. Elucidation of the physiological roles of agmatine, the first endogenous clonidine-displacing substance to be identified, will require development of means to block its production or effects.

Niels Juel Christensen and Jens H. Knudsen

Department of Endocrinology
Herlev Hospital
University of Copenhagen
DK-2730 Herlev, Denmark

Peripheral Catecholaminergic Function Evaluated by Norepinephrine Measurements in Plasma, Extracellular Fluid, and Lymphocytes, from Nerve Recordings and Cellular Responses

I. Plasma Norepinephrine and Microneurography

Sympathetic nervous system activity is somewhat differentiated, and the release of norepinephrine (NE) should, therefore, be measured in specific organs or tissues. The forearm venous plasma NE, which is often used as an index of sympathetic activity, mainly reflects muscle sympathetic nerve activity.

It has been shown in many studies that there is a close correlation between forearm venous plasma NE and muscle sympathetic nerve activity, as measured by microneurography. This has been observed not only at rest, but also during mental stress, hypoglycemia, and after raising and lowering arterial blood pressure, as well as during several pathophysiological states. Furthermore, both parameters increase with age (1).

This relationship between forearm venous plasma NE and muscle sympathetic nerve activity may be explained by several factors: (1) Skeletal muscle constitutes a major tissue; (2) approximately 50% of NE in forearm venous blood is derived locally from forearm tissue, mainly muscle, and (3) NE from the gastrointestinal tract is largely eliminated in the liver. An advantage of the microneurographic technique is that muscle sympathetic nerve activity can be recorded at short intervals, whereas with plasma NE measurements, steady-state values can be obtained only at 5- to 10-min intervals.

II. Regional Release of NE

Organ release of NE can be calculated from the venous-arterial difference of NE corrected for the extraction of NE across the organ and from the blood

Advances in Pharmacology, Volume 42

1054-3589/98 $25.00

flow to the organ. Studies in animals have indicated that the overflow is proportional to the impulse activity in sympathetic nerves to the organ, although this may not be true during all situations. Savard *et al.* (2) showed that the overflow of NE during one-legged exercise in a group of normal subjects was greater from the exercising leg than from the resting leg, suggesting that sympathetic nerve activity was greater in the exercising leg than in the resting leg. This problem cannot be examined by microneurography because one of the limitations of this technique is that the leg must be at rest during the examination or it may perform only very light exercise. It is possible, however, that the greater overflow from the exercising leg does not reflect a difference in sympathetic activity between the two sites, but it could be due to the greater blood flow in the exercising leg or reflect presynaptic regulation. Clearly, one cannot always be sure that the overflow of NE reflects differences in sympathetic activity. The exercising leg, however, contributes more to the increase in circulating catecholamines than does the resting leg.

III. The Clearance Concept

The secretion rate of plasma epinephrine can be measured as the product of the clearance and the endogenous arterial concentration of epinephrine. Changes in plasma epinephrine concentration during exercise have been shown largely to reflect changes in the secretion rate, whereas the clearance rate decreased only slightly at maximal work load.

The clearance concept cannot, however, be readily applied to NE. We can assume that tracer epinephrine infused intravenously will mix adequately with endogenous epinephrine in the blood and the specific activity of epinephrine will be the same in all parts of the organism. That is, it is possible to calculate only one secretion rate. This is not the case for tracer NE and endogenous NE. Tracer NE in plasma will mix with a certain fraction of NE released at nerve endings. The specific activity will, however, decrease towards the synapse. For the plasma appearance rate to be a valid index of NE release, tracer NE must mix with a constant fraction of NE released at nerve endings (Table I). It is difficult to know if this assumption is valid, but it is unlikely to be so during experiments, in which the capillary surface exchange area is decreased.

A number of studies indicate that there is a discrepancy between muscle sympathetic nerve activity, as measured by microneurography, and the plasma appearance rate of NE during moderate lower-body negative pressure

TABLE I The Clearance Concept

1. The clearance concept cannot be readily applied to NE.
2. The plasma appearance rate of NE is a reliable index only of sympathetic activity (a relative index), provided the tracer mixes with a constant fraction of NE released at nerve endings.
3. This assumption (constant error[a]) is unlikely to be fulfilled during several experimental conditions.

[a] The error is constant if the ratio of specific activity in plasma and at breakdown sites is constant.

to ÷ 15 mm Hg: Muscle sympathetic nerve activity increased approximately 55% and plasma NE increased 40%, but the plasma appearance rate did not change and the clearance decreased approximately 20% (3–6; unpublished results). This discrepancy between the muscle sympathetic nerve activity and the plasma appearance rate is most likely due to the problems with the clearance concept—plasma NE concentration and tracer NE during lower-body negative pressure did not increase or decrease, respectively, to the extent they should if mixing had been adequate (see Table I).

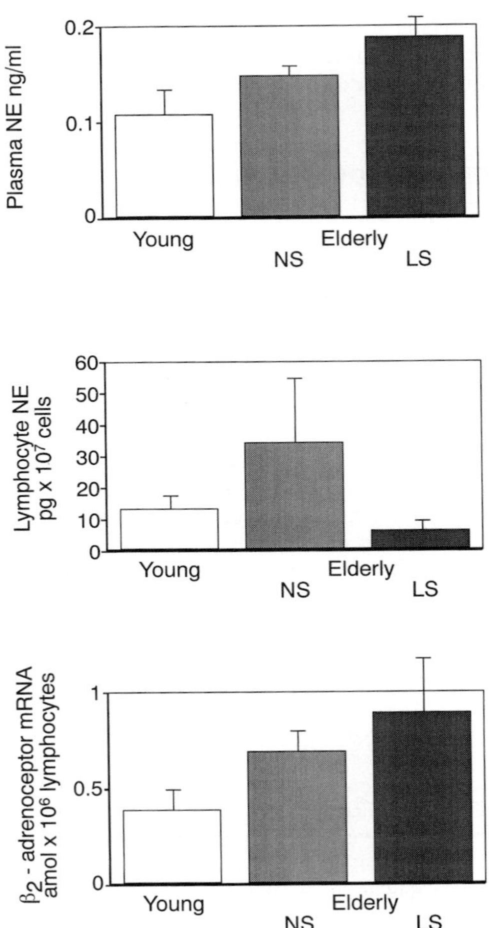

FIGURE 1 Plasma NE (ng/ml), lymphocyte NE (pg/10^7 lymphocytes), and β_2-adrenoceptor mRNA (amol/10^6 lymphocytes) in a group of young subjects and elderly nonsmokers and long-term smokers. NS, nonsmokers; LT, long-term smokers.

IV. Microdialysis

Microdialysis combined with a sensitive radioenzymatic technique may be used to study the extracellular fluid concentration of NE. This may be of particular significance in adipose tissue, in which sympathetic activity may be difficult to study with other techniques. We have observed that the NE concentration is substantially lower in extracellular fluid than in plasma, but the values tend to be correlated. The advantage of the microdialysis technique, however, is that it can be used to study effects of drugs on NE release in a small region, for example, in muscle tissue, without changing the plasma concentration of catecholamines and hemodynamics.

V. Lymphocyte Catecholamines

We have recently shown that lymphocytes contain catecholamines (7). The catecholamines may be derived from internalization of β_2 adrenoceptors because there was a close correlation between cyclic adenosine monophosphate and the level of catecholamines in lymphocytes. Further studies of lymphocyte NE have demonstrated a marked difference in the NE content of lymphocytes from elderly smokers as compared with nonsmokers. It is well known that plasma NE increases with age. This phenomen is most pronounced in long-term smokers but also may be observed in nonsmokers. In lymphocytes, NE concentration also was increased in elderly nonsmokers but was markedly reduced in smokers. Further studies have shown that β_2-adrenoceptor mRNA concentration is not reduced in long-term smokers, suggesting that the decrease in NE reflects a defective membrane function, especially of NK-cells, in long-term smokers (Fig. 1).

References

1. Christensen, N. J. (1991). The biochemical assessment of sympathoadrenal activity in man. *Clin. Auton. Res.* **1**, 167–172.
2. Savard, G. K., Richter, E. A., Strange, S., Kiens, B., Christensten, N. J., and Saltin, B. (1989). Norepinephrine spillover from skeletal muscle during exercise in humans: Role of muscle mass. *Am. J. Physiol.* **257**, H1812–H1818.
3. Rea, R. F., Hamdan, M., Clary, M. P., Randels, M. J., Dayton, P. J., and Strauss, R. G. (1991). Comparison of muscle sympathetic responses to hemorrhage and lower body negative pressure in humans. *J. Appl. Physiol.* **70**, 1401–1405.
4. Baily, R. G., Leuenberger, U., Leaman, G., Silber, D., and Sinoway, L. I. (1991). Norepinephrine kinetics and cardiac output during nonhypotensive lower body negative pressure. *Am. J. Physiol.* **260**, H1708–H1712.
5. Jacobsen, T. N., Morgan, B. J., Scherrer, U., Vissing, S. F., Lange, R. A., Johnson, N., Ring, W. S., Rahko, P. S., Hanson, P., and Victor, R. G. (1993). Relative contributions of cardiopulmonary and sinoaortic baroreflexes in causing sympathetic activation in the human skeletal muscle circulation during orthostatic stress. *Circ. Res.* **73**, 367–378.

6. Jacobs, M-C., Goldstein, D. S., Willemsen, J. J., Smits, P., Thien, T., and Lenders, J. W. M. (1996). Differential effects of low- and high-intensity lower body negative pressure on noradrenaline and adrenaline kinetics in humans. *Clin. Sci.* 90, 337–343.
7. Knudsen, J. H., Christensen, N. J., and Bratholm, P. (1996). Lymphocyte norepinephrine and epinephrine, but not plasma catecholamines predict lymphocyte cAMP production. *Life Sci.* 59, 639–647.

P. M. Mertes
Laboratory of Cell Biology
CHU de Brabois
54500 Vandoeuvre-les-Nancy, France

Cardiac Microdialysis

The idea of using microdialysis to sample extracellular fluid in the brain is more than 30 years old. The principle of microdialysis is to mimic the passive function of a capillary blood vessel by perfusing a tubular dialysis membrane implanted into a tissue. This allows for the recovery of a substance present in the interstitial fluid and absent from the perfusate or for administering substances into a tissue. The concentration of compounds in the dialysate, determined with appropriate analytical techniques, reflects the composition of extracellular fluid. The ultimate sensitivity and time resolution of the technique is determined by the analytical technique used.

Performing cardiac microdialysis requires specific considerations. Because of myocardial contraction, probes must be as flexible as possible, and special attention should be paid to the different connecting parts. Linear and concentric probes are easy to build when using thin, hollow fibers ranging from 2 to 300 μm and to introduce into the heart using a transfixient approach. The length of the dialysis membrane can be increased to 20 mm, allowing sampling of a larger tissue volume when compared with the 1- to 2-mm-length probes used in brain microdialysis.

Data obtained with microdialysis are usually presented as dialysate concentrations, and interstitial concentrations are estimated on the basis of *in vitro* recovery of the probe. However, several factors can affect recovery. It is dependent on perfusion speed, dialysis membrane surface area, and composition. Calculation of "true" interstitial concentrations based on *in vitro* recovery leads to significant underestimation due to several factors. First of all, the transport of substances in tissue is considerably slower compared with that in aqueous solutions. Another important factor is that mass transport through the mem-

Advances in Pharmacology, Volume 42

1054-3589/98 $25.00

brane can be affected by changes in heart rate and intracardiac pressure. The potential consequences of a change in heart rate or contractility appear to be complex. Probe recovery will increase or decrease with increasing heart rate, depending on the positive or negative pressure gradient between interstitium and the dialysis chamber. Different practical approaches have been proposed to assess *in vivo* recovery rates of microdialysis probes. These methods are usually based on the use of modifications of probe perfusion speed or on the addition of an internal standard to the perfusing solution. This allow us to determined *in vivo* norepinephrine (NE) recovery in pig hearts, using a 10-mm membrane (1). The *in vivo* calibration method used was based on the "difference method." Briefly, the probe is perfused with increasing concentrations of NE at a constant flow rate, and a linear relationship is established between NE concentration in the perfusate and the net outflow of NE towards interstitium. By using regression analysis, the mean relative recovery rate of NE is obtained from the slope of the line. In this case, it was of 34%, which should be compared with the 41% recovery rate determined *in vitro* for a 2-μ1/min perfusion speed.

Cardiac microdialysis has been applied in the study of interstitial fluid concentrations of various compounds. One of the first results, reported by Van Wylen *et al.*, concerned the evolution of interstitial purine metabolites following prolonged occlusion of the left anterior descending (LAD) coronary artery in dogs (2). A dramatic increase in dialysate adenosine concentrations was observed after 15 min of ischemia, but there was a decrease towards preischemia levels by 45 min of ischemia. A similar approach was used by Wikerström *et al.* in the preconditioned porcine heart (3). The preconditioning consisted of four 10-min ischemic periods. The index ischemia that followed lasted for 40 min. Lactate production in preconditioned hearts was less than in nonpreconditioned hearts during the index ischemia. Index ischemia resulted in a 12-fold adenosine increase in nonpreconditioned animals, compared with a limited 2.5-fold increase in preconditioned pigs. Therefore, microdialysis allowed them to demonstrate a powerful inhibition of adenosine nucleotide catabolic process due to ischemia by preconditioning.

Another proposed application of cardiac microdialysis was to monitor hydroxyl radical generation during ischemia by perfusing a probe with a solution containing salicylic acid at a rate of 0.5 mmol/μl/min (4). Then, the hydroxyl radical ·OH, which is extremely reactive, will react with salicylate to generate 2,3- and 2,5-dihydroxybenzoic acid, detected in dialysates using high-performance liquid chromatography–EC. Using this approach, Obata *et al.* (4) were able to monitor hydroxyl radical generation during ischemia in rat and dog hearts.

Cardiac microdialysis also has been used for *in vivo* peptide and catecholamine monitoring. Initially, we studied the consequences of brain death, which leads to major hemodynamic changes, usually referred to as an "autonomic storm" associated with histological damage to the heart, on cardiac neuroendocrine integration. We studied the evolution of endothelin 1 (ET-1) and angiontensin II (AngII). We demonstrated that ET-1 interstitial levels were considerably higher than those observed in plasma, supporting the hypothesis that ET-1 may act mainly as a paracrine factor rather than as a circulating hormone. Concerning AngII, the most interesting results we observed were significantly higher AngII dialysate levels compared with plasma levels during the control period. To our knowledge, this was the first *in vivo* demonstration of local myocardial AngII production. Some interesting results also were obtained when NE and neuropeptide Y (NPY) interstitial levels were monitored. Brain death induction leads to a dramatic increase in interstitial NE concentrations, lasting for almost 1 hr. To the contrary, no significant increase in plasma levels was observed during this period. This highlights the interest of cardiac microdialysis, which allowed us to demonstrate the onset of direct NE release from cardiac sympathetic nerve endings. Similar results were obtained for NPY interstitial concentrations. A sustained increase in NPY levels was observed in both ventricles, while circulating levels were almost undetectable. Cardiac microdialysis also was used to assess the consequences of myocardial infarction on cardiac NE and NPY (5). A slight increase in NPY levels was observed in both ischemic and nonischemic regions, suggesting moderate sympathetic activation. NE interstitial concentrations exhibited a completely different profile, with relatively stable NE concentration in the nonischemic area compared with a strong increase in the ischemic region (Fig. 1).

In our opinion, this increase was not explained by a simple reduction in vascular or metabolic clearance. It could be a consequence of a depletion in cardiac sympathetic nerve ending adenosine triphosphate content leading to a local nonexocytotic metabolic NE release, as suggested by Schömig *et al.* (6) in Langendorff perfused heart subjected to ischemia.

In conclusion, we can state that cardiac microdialysis represents a simple application of the classical technique developed for the brain. It is easily performed with minor adaptations. One should keep in mind, however, that this is an invasive technique requiring specific *in vivo* determination of probe recovery to assess "true" interstitial concentration of a compound. Moreover, espe-

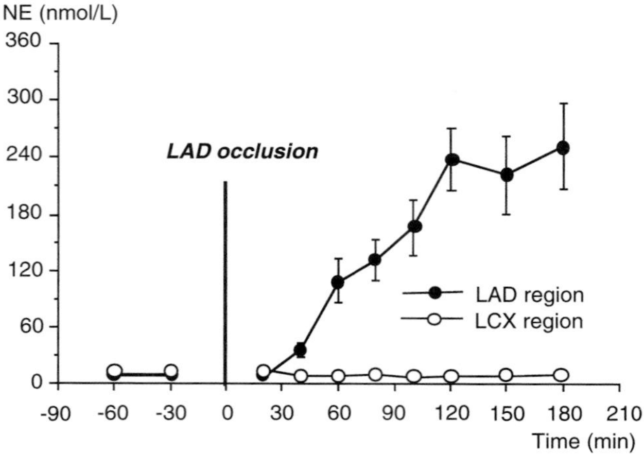

FIGURE 1 Changes in dialysate NE concentrations (mean ± SEM) obtained from ischemic (LAD) and nonischemic territories (LCX) following LAD coronary artery occlusion in pigs (n = 10). (From Ref. 5, with permission.)

cially when studying intercellular communication, if dialysate concentration of a substance could reflect the interstitial concentration of this product, a precise evaluation of metabolic and vascular clearance is required to allow for interpretation of results in terms of substance release.

References

1. Mertes, P., Carteaux, J., Jaboin, Y., Pinelli, G., El-Abbassi, K., Dopff, C., Atkinson, J., Villemot, J., Burlet, C., and Boulangé M. (1994). Estimation of myocardial interstitial norepinephrine release after brain death using cardiac microdialysis. *Transplantation* **57**, 371–377.
2. Van Wylen, D., Willis, J., Sodhi, J., Weiss, R., Lasley, R., and Mentzer, R. (1990). Cardiac microdialysis to estimate interstitial adenosine and coronary blood flow. *Am. J. Physiol.* **258**, H1642–H1649.
3. Wikström, B., Ronquist, G., and Waldenström, A. (1995). Dynamics of myocardial metabolism in the preconditioned porcine heart studied using continuous microdialysis. *Eur. Heart J.* **16**, 563–569.
4. Obata, T., Hosokawa, H., and Yamanaka, (1994). In vivo monitoring of norepinephrine and ·OH generation on myocardial ischemic injury by dialysis technique. *Am. J. Physiol.* **266**, H903–H908.
5. Mertes, P., El-Abbassi, K., Jaboin, Y., Michel, C., Beck, B., Pinelli, G., Carteaux, J., Villemot, J., and Burlet, C. (1996). Consequences of coronary occlusion on changes in regional interstitial myocardial neuropeptide Y and norepinephrine concentrations. *J. Mol. Cell. Cardiol.* **28**, 1995–2004.
6. Schömig, A., Haas, M., Richard, G. (1991). Catecholamine release and arrhytmias in acute myocardial ischaemia. *Eur. Heart J.* **12** (Sup. F), 38–57.

Robert F. Rea

Mayo Clinic and Foundation
Rochester, Minnesota 55905

Sympathetic Microneurography and Neurocirculatory Function: Studies of Ventricular Arrhythmias in Humans

I. Sympathetic Microneurography

It has long been appreciated that the sympathetic nervous system plays a crucial role in human cardiovascular control mechanisms, both in health and in disease. Probing the sympathetic nervous system in humans has been difficult, however, due to limitations of available techniques. The development of microneurographic techniques for the recording of sympathetic traffic in humans represents an important advance in this area.

In the course of experimental microelectrode recordings of peripheral nerve traffic in humans, investigators in Sweden in the late 1960s uncovered rhythmic nerve signals that were inconsistent with proprioceptive, somatomotor efferent, or sensory afferent activity. Subsequent experiments indicated that these rhythmic nerve signals were bursts of efferent sympathetic traffic destined for muscle vascular beds (muscle sympathetic nerve activity [MSNA]) (1). The findings supporting this conclusion were:

1. Elimination of burst activity with lidocaine block proximal but not distal to the recording site, indicating efferent propagation.
2. Lack of association of bursts with muscle activity or motor unit firing.
3. Simultaneity of burst discharges in multiple nerves (e.g., both legs), suggesting a common central origin of activity.
4. Reversible abolition of burst activity with ganglionic blocking agents, such as trimethaphan.
5. Correlation of burst frequency with plasma levels of norepinephrine within subjects under different conditions and between subjects.

Additional experiments uncovered a second type of efferent signal that was consistent with sympathetic traffic destined for cutaneous vascular beds mediating cutaneous vasoconstriction and sweating (skin sympathetic nerve activity [SSNA]) (2). The behavior of MSNA and SSNA was found to be regulated in very different ways. Most importantly MSNA is tightly governed by baroreceptor reflexes, both arterial and cardiopulmonary. This is evidenced by tight coupling or entrainment of bursts of MSNA to the cardiac rhythm;

Advances in Pharmacology, Volume 42

1054-3589/98 $25.00

with each arterial pulse wave, afferent baroreceptor discharges inhibit MSNA, and with each diastole, MSNA is disinhibited. Spectral analysis of MSNA recordings shows prominent peaks at the heart rate frequency. Unlike MSNA, SSNA is powerfully affected by brief emotional and thermal stimuli. Arousal stimuli, such as a loud noise, promptly result in bursts of SSNA but do not affect MSNA. These arousal responses are associated with changes in cutaneous electrodermal activity, suggesting that SSNA may include signals destined for sweat glands. SSNA includes cutaneous vasoconstrictor signals as well as that shown in recordings during body cooling. Figure 1 shows some of the differences between MSNA and SSNA.

II. Application of Sympathetic Microneurography in Studies of Ventricular Arrhythmias in Humans

An interaction between the sympathetic nervous system and ventricular arrhythmias in humans has been suggested by several clinical and experimental observations. First, heightened sympathetic activity has been documented, both by catecholamine measurements and by sympathetic microneurography, in patients with congestive heart failure, a group with a high incidence of ventricular arrhythmias. In addition, plasma norepinephrine levels correlate directly with overall mortality in congestive heart failure. Second, β-adrenergic blocking agents have been shown to reduce sudden death and ventricular arrhythmias after myocardial infarction. Third, in experimental animals, cardiac sympathetic stimulation lowers the ventricular fibrillation threshold and may precipitate ventricular fibrillation in the setting of acute myocardial ischemia (3).

The interaction between arrhythmias and sympathetic activity is, however, bidirectional. Isolated premature ventricular beats provoke large bursts of

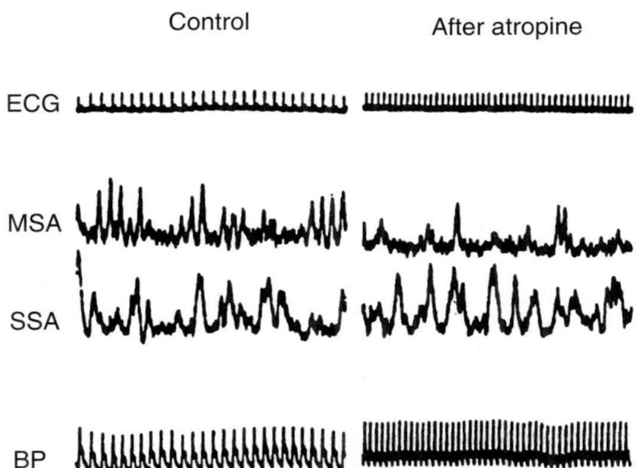

FIGURE 1 Simultaneous recording of MSNA and SSNA before and after atropine. Increased arterial pressure after atropine reflexly inhibits MSNA but not SSNA. From Fagius, J., Wallin, B. G., Sundlöf, G., Nerhed, C., and Englesson, S. *Brain* 108:423–438, 1985.

MSNA, probably as a result of transient declines in diastolic arterial pressure (4). The net effect of ventricular ectopy on sympathetic outflow is likely neutral because elevated arerial pressure with the postextrasystolic beat inhibits MSNA. When background MSNA is augmented by drug-induced hypotension, the increment in MSNA with isolated ectopic beats is diminished (5).

Sustained ventricular arrhythmias or ventricular tachypacing reflexly increases MSNA via an arterial baroreceptor reflex. In this situation, the role of the sympathetic nervous system is protective; without reflex sympathoexcitation, hypotension resulting from diminished cardiac output would be unopposed and catastrophic declines of arterial pressure would result. In patients with sympathetic failure, even modestly rapid tachyarrhythmias are poorly tolerated because of the lack of this compensatory mechanism (author's unpublished observation).

Evidence from experimental animals suggests that ventricular arrhythmias may provoke opposing inputs from arterial and cardiopulmonary baroreceptors to the brain stem. As outlined, tachycardia-induced hypotension unloads arterial baroreceptors and stimulates sympathetic activity. During ventricular tachycardia, however, intraventricular end-diastolic pressure and dimension increase stimulating cardiac mechanoreceptor activity that inhibits sympathetic outflow. The net effect is unclear, and published studies are conflicting. Some human-based studies have suggested, albeit indirectly, a reflexly mediated inhibition of sympathetic outflow and augmented vagal activity at the time of hemodynamic collapse during previously well tolerated tachyarrhythmias.

Antiarrhythmic drugs are widely used to suppress ventricular arrhythmias. Many of these agents block ion channel activities that are ubiquitous. It is not surprising, therefore, that these agents have a multiplicity of side effects, many of which are neural (central or peripheral).

Procainamide is a versatile drug, used for both atrial and ventricular arrhythmias. When given intravenously, it commonly causes hypotension. In a study of normal human volunteers, standard intravenous loading doses of procainamide caused mild-to-moderate declines in arterial and central venous pressures, peripheral vasodilation, and, interestingly, significant *decreases* in MSNA. Similar degrees of arterial and central hypotension produced with intravenous nitroprusside, however, cause marked sympathetic excitation. This untoward neural effect of procainamide is likely due to inhibition of ganglionic transmission, based on prior animal studies. Additionally, procainamide is related structurally to procaine, a known ganglionic blocker (6).

Quinidine, like procainamide, is a sodium channel blocker and shares the same Vaughan-Williams antiarrhythmic drug class (1A). Also like procainamide, when given intravenously, it may cause significant hypotension. In a study of normal human volunteers, standard intravenous loading doses of quinidine caused declines of central venous and arterial pressures and peripheral vasodilation similar to those provoked with procainamide, as outlined. Contrary to procainamide, MSNA increased markedly. Increases in MSNA were similar to those seen with equihypotensive doses of nitroprusside (7).

Lidocaine inhibited sympathetic activity in studies of experimental animals, and this was thought to be one of its antiarrhythmic mechanisms. In studies of patients with electrically inducible ventricular tachycardia, standard intravenous doses of lidocaine reduced the level of sympathetic excitation seen during ventricular tachycardia. However, lidocaine slowed the rate of the induced arrhythmia significantly, and the associated arterial pressure was higher. These data suggested that lidocaine has little direct effect on MSNA during ventricular arrhythmias but, rather, modulates MSNA via an effect on the arrhythmia rate and associated arterial pressure (8).

These studies illustrate the application of microneurography to an important problem in clinical medicine.

References

1. Delius, W., Hagbarth, K-E., Hongell, A., and Wallin, B. G. (1972). General characteristics of sympathetic activity in human muscle nerves. *Acta Physiol. Scand.* **84,** 65–81.
2. Bini, G., Hagbarth, K-E., Hynninen, P., and Wallin, B. G. (1980). Thermoregulatory and rhythm-generating mechanisms governing the sudomotor and vasoconstrictor outflow in human cutaneous nerves. *J. Physiol.* (*Lond.*) **306,** 537–552.
3. Meredith, I. T., Broughton, A., Jennings, G. L., and Esler, M. D. (1991). Evidence of a selective increase in cardiac sympathetic activity in patients with sustained ventricular arrhythmias. *N. Engl. J. Med.* **325,** 618–624.
4. Welch, W. J., Smith, M. L., Rea, R. F., Bauernfeind, R. A., and Eckberg, D. L. (1989). Enhancement of sympathetic nerve activity by single premature ventricular beats in humans. *J. Am. Coll. Cardiol.* **13,** 69–75.
5. Smith, M. L., Ellenbogen, K. A., and Eckberg, D. L. (1995). Baseline arterial pressure affects sympathoexcitatory responses to ventricular premature beats. *Am. J. Physiol.* **269,** H153–H159.
6. Rea, R. F., Hamdan, M., Schomer, S. J., and Geraets, D. R. (1991). Inhibitory effects of procainamide on sympathetic nerve activity in humans. *Circ. Res.* **69,** 501–508.
7. Mariano, D. J., Schomer, S. J., Rea, R. F. (1992). Effects of quinidine on vascular resistance and sympathetic nerve activity in humans. *J. Am. Coll. Cardiol.* **20,** 1411–1416.
8. Ellenbogen, K. A., Smith, M. L., Beightol, L. A., and Eckberg, D. L. (1992). Influence of lidocaine on human muscle sympathetic nerve activity during programmed electrical stimulation and ventricular tachycardia. *Am. Heart J.* **124,** 891–897.

George P. Chrousos

Section on Pediatric Endocrinology
Developmental Endocrinology Branch
Pediatric Endocrinology Training Program
National Institutes of Health
Bethesda, Maryland 20892

Stress as a Medical and Scientific Idea and Its Implications

I. The Stress Syndrome

Life exists by maintaining a complex dynamic equilibrium, or *homeostasis,* that is constantly challenged by intrinsic or extrinsic adverse forces or *stressors. Stress* is, thus, defined as a state of threatened homeostasis, which is re-established by a complex repertoire of physiological and behavioral adaptive responses of the organism. Hormones participate in a crucial fashion in the coordination of both basal and threatened homeostasis. Stress and related concepts, such as homeostasis and stressors, can be traced as far back as written science and medicine (1). However, it is Walter Cannon and Hans Selye who are responsible for the clear articulation of these concepts in the first half of this century.

The stress response is subserved by a complex system located in both the central nervous system (CNS) and the periphery (2). This system receives and integrates a great diversity of neurosensory (visual, auditory, somatosensory, nociceptive, visceral), blood-borne, and limbic–cortical signals, which arrive through distinct pathways. Activation of the stress system leads to behavioral and physical changes that are remarkably consistent in their qualitative presentation (Table I). These changes are normally adaptive and improve the chances of the individual for survival.

Behavioral adaptation includes increased arousal, alertness, and viligance; improved cognition; and focused attention, as well as euphoria and dysphoria. It also includes enhanced analgesia and elevations in core temperatures, along with concurrent inhibition of vegetative functions, such as appetite, feeding, and reproductive function. Concomitantly, physical adaptation occurs principally to promote an adaptive redirection of energy. Thus, oxygen and nutrients are shunted to the CNS and the stressed body site(s), where they are needed the most. Increases in cardiovascular tone (heart rate, cardiac ejection fraction, arterial blood pressure), respiratory rate, and intermediate metabolism (gluconeogenesis, lipolysis), all work in concert to promote availability of vital substrates. Detoxification functions to rid the organism of unnecessary metabolic products from the stress-related changes in metabolism are activated, while digestive function and growth, reproduction, and immunity are inhibited.

Advances in Pharmacology, Volume 42

1054-3589/98 $25.00

TABLE I Behavioral and Physical Adaptation during Acute Stress

Behavioral Adaptation
Adaptive redirection of behavior
Increased arousal and alertness
Increased cognition, vigilance, and focused attention
Euphoria or dysphoria
Heightened analgesia
Increased temperature
Suppression of appetite and feeding behavior
Suppression of reproductive behavior
Containment of the stress response
Physical Adaptation
Adaptive redirection of energy
Oxygen and nutrients directed to the CNS and stressed body site(s)
Altered cardiovascular tone, increased blood pressure and heart rate
Increased respiratory rate
Increased gluconeogenesis and lipolysis
Detoxification from toxic products
Inhibition of growth and reproductive systems
Inhibition of digestion-increased colonic motility
Containment of the inflammatory–immune response
Containment of the stress response

Adapted from Chrousos, G. P., and Gold, P. W. (1992). The concepts of stress and stress system disorders. *JAMA* 267, 1244.

The organism also activates restraining forces during stress, which prevent an overresponse from both central and peripheral components of the stress system. These forces are essential for successful adaptation. If they fail to timely contain the various elements of the stress response, the "adaptive" changes may turn excessive, prolonged, and maladaptive and may, thus, contribute to development of pathology.

Often, stress is of a magnitude and nature that allows the perception of control by the individual. As such, stress can be pleasant and rewarding. Seeking of novelty stress by an individual is related to this phenomenon and is pivotal for emotional and intellectual growth and development. It is of note that activation of the stress system occurs during both feeding and sexual activity, sine qua non functions for survival of self and species.

II. The Stress System

The central components of the stress system are located in the hypothalamus and the brain stem and include the parvicellular corticotropin-releasing hormone (CRH) and arginine vasopressin (AVP) neurons of the paraventricular nuclei (PVN) of the hypothalamus, and the CRH neurons of the paragigantocellular and parabranchial nuclei of the medulla, as well as the locus ceruleus (LC) and other mostly noradrenergic cell groups of the medulla and pons

(LC–noradrenergic sympathetic system) (2, 3). The peripheral limbs of the stress system are the hypothalamic-pituitary adrenal (HPA) axis, together with the efferent sympathetic–adrenomedullary system and components of the parasympathetic system.

There has been an exponential increase in knowledge regarding the interactions among the components of the stress system and between the stress system and other brain elements involved in the regulation of emotion, cognitive function, and behavior, as well as with the axes responsible for reproduction, growth, and immunity. Thus, CRH and the LC–noradrenergic system stimulate arousal and attention, as well as the mesocorticolimbic dopaminergic system, which is involved in anticipatory and reward phenomena. CRH inhibits appetite and activates thermogenesis via the catecholaminergic system. Also, reciprocal interactions exist between the amygdala and the hippocampus and the stress system, which stimulates these elements and is regulated by them.

CRH plays an important role in inhibiting gonadotropin-releasing hormone secretion during stress, while via somatostatin it also inhibits growth hormone (GH), thyrotropin-releasing hormone, and thyrotropin (TSH) secretion, suppressing, thus, reproduction, growth, and thyroid function (3). Interestingly, all three of these functions receive and depend on positive catecholaminergic input. The end-hormones of the hypothalamic-pituitary-adrenal (HPA) axis, glucocorticoids, on the other hand, have multiple roles. They simultaneously inhibit the CRH and LC–noradrenergic systems and stimulate the mesocorticolimbic dopaminergic system and the CRH peptidergic central nucleus of the amygdala. In addition, they directly inhibit pituitary gonadotropin, GH and TSH secretion, render the target tissues of sex steroids and growth factors resistant to these substances, and suppress the 5′ deiodinase, which converts the relatively inactive tetraiodothyronine (T_4) to triiodothyronine (T_3), contributing further to the suppression of reproductive, growth, and thyroid functions. They also have direct as well as insulin-mediated effects on adipose tissue, chronically promoting visceral adiposity. In addition, mainly through glucocorticoids, the HPA axis inhibits the immune–inflammatory reaction at multiple levels, including the secretion and action of cytokines and lipid and nonlipid mediators of inflammation (4). Interestingly, however, during stress, the combined action of glucocorticoids and catecholamines results in suppression of interleukin (IL-12) and stimulation of IL-10, and, hence, a shift from T-helper 1 to T-helper 2 activity (i.e., from cellular to humoral immunity) (5). This may be adaptive acutely but may render the chronically stressed individual vulnerable to certain viral or bacterial illnesses and tumors.

III. Clinical Implications

This new knowledge has allowed association of stress system dysfunction characterized by sustained hyperactivity or hypoactivity to various pathophysiologic states that cut across the traditional boundaries of medical disciplines. These include a range of psychiatric, endocrine, and inflammatory disorders and/or susceptibility to such disorders (Table II). Of special interest is atypical–

TABLE II States Associated with Altered HPA Axis Activity and Altered Regulation or Dysregulation of Behavioral and/or Peripheral Adaptation

Increased HPA axis	*Decreased HPA axis*
Chronic stress	Adrenal insufficiency
Melancholic depression	Atypical–seasonal depression
Anorexia nervosa	Chronic fatigue syndrome
Malnutrition	Fibromyalgia
Obsessive-compulsive disorder	Hypothyroidism
Panic disorder	Nicotine withdrawal
Excessive exercise (obligate athleticism)	Postglucocorticoid therapy
Chronic active alcoholism	Post–Cushing syndrome cure
Alcohol and narcotic withdrawal	Postpartum period
Diabetes mellitus	Postchronic stress
Central obesity (metabolic syndrome X)	Rheumatoid arthritis
Childhood sexual abuse	
Phychosocial short stature	
Attachment disorder of infancy	
"Functional" gastrointestinal disease	
Hyperthyroidism	
Premenstrual tension syndrome	
Pregnancy (last trimester)	

Updated from Chrousos, G. P., and Gold, P. W. (1992). The concepts of stress and the stress system disorders. *JAMA* **267**, 1244.

seasonal depression, which is associated with a hypofunctional stress system with weight gain in the dark winter months and/or autoimmune manifestations, and melancholic depression, which is associated with a hyperfunctional stress system and immunosuppression.

We hope that knowledge from apparently disparate fields of science and medicine, integrated into a working theoretical framework around the concept of the stress system, will allow generation and testing of new hypotheses on the pathophysiology and diagnosis of and therapy for a variety of human illnesses, reflecting systematic alterations in the principal effectors of the stress response.

References

1. Chrousos, G. P., Loriaux, D. L., and Gold, P. W. (1988). The concept of stress and its historical development. *Adv. Exp. Med. Biol.* **245**, 3–7.
2. Chrousos, G. P., and Gold, P. W. (1992). The concepts of stress system disorders: Overview of behavioral and physical homeostasis. *J.A.M.A.* **267**, 1244–1252.
3. Chrousos, G. P. (1996). Organization and integration of the endocrine system. *In* Pediatric Endocrinology. (M. A. Sperling, ed.), pp. 1–14. W. B. Saunders, Philadelphia.

4. Chrousos, G. P. (1995). The hypothalamic-pituitary-adrenal axis and immune-mediated inflammation. *N. Engl. J. Med.* **332**, 1351–1362.
5. Elenkov, I., Papanicolaou, D., Wilder, C. R., and Chrousos, G. P. (1996). Modulatory effects of glucocorticoids and catecholamines on human interleukin-12 and interleukin-10 production: Clinical implications. *Proc. Assoc. Am. Phys.* **108**, 374–381.

R. Kvetňanský,*§ K. Pacák,†§ E. L. Sabban,‡ I. J. Kopin,§ and D. S. Goldstein§

*Institute of Experimental Endocrinology
Slovak Academy of Sciences
833 06 Bratislava, Slovak Republic

†Department of Internal Medicine
Washington Hospital Center
Washington D.C. 20010

‡Department of Biochemistry and Molecular Biology
New York Medical College
Valhalla, New York 10595

§Clinical Neuroscience Branch
National Institute of Neurological Disorders and Stroke
National Institutes of Health
Bethesda, Maryland 20892

Stressor Specificity of Peripheral Catecholaminergic Activation

Selye defined *stress* as the nonspecific response of the body to any demand imposed on it. In 1975, Mason, in contrast, described *stress* as a multihormonal response pattern organized in a rather specific or selective manner, depending on the particular stimulus. By now, the weight of evidence supports the view that different neuroendocrine systems are activated specifically by different stressors. Some studies have demonstrated that even components of the same system might be activated differentially. For instance, the two branches of the sympathoadrenal system (SAS)—the adrenal medulla and the sympathetic nerves—can be activated independently by different stressors (1, 2). The studies have been based on determinations of plasma epinephrine (Epi) and norepinephrine (NE) levels. Measurements of plasma levels of the catecholamine (CA) precursor dihydroxyphenylalanine (DOPA) and of catecholamine metabolites give better information about mechanism of sympathoadrenal activation.

The aim of the present study was to examine activation of the two components of the SAS by simultaneous measuring the adrenomedullary (AM) and

1054-3589/98 $25.00

sympathoneural responses during exposure to different intensities of various stressors (14 situations). Plasma levels of NE, Epi, dopamine (DA), their precursor DOPA, and the metabolites dihydroxyphenylglycol (DHPG) and dihydroxyphenylacetic acid (DOPAC) were assessed, as well as gene expression of the CA biosynthetic enzyme tyrosine hydroxylase (TH) in the adrenal medulla.

Conscious cannulated male Sprague-Dawley rats (Taconic Farm) weighing 300–350 g were exposed to one of the following stressors at various intensities:

- *Saline i.v.* (SAL i.v.). Saline was injected via an indwelling catheter without handling the animals. This group was considered the control group.
- *Handling plus saline s.c.* (SAL s.c.). Saline was injected subcutaneously after handling the animals to position for injection.
- *Repeated handling* for 14 days (3 times daily for 1 min). The last handling before blood collection was performed by the same or different persons.
- *Insulin-induced hypoglycemia* (INS). Porcine regular insulin (Eli Lilly) was injected i.v. (0.1; 1.0; 3.0 IU/kg). Blood was collected at 0, 15, 45, 75, and 105 min after insulin injection.
- *Formalin-induced pain and tissue damage* (FORM). Formalin (1% or [illegible]%) was injected s.c. into the right leg. Intervals of blood collection were the same as after insulin injection, with an additional sample at 135 min.
- *Nonhypotensive and hypotensive hemorrhage* (HEM). The animals bled to 10% or 25% of estimated blood volume. Blood was collected as after formalin.
- *Cold exposure* (COLD). The animals were exposed for 3 hr to 4°C or −3°C. After baseline blood collection at room temperature, animals were carefully transferred together with their home cages into a cold chamber, and blood was collected at 15, 45, 105, and 165 min of cold. After 3 hr, the cooling system in the cold chamber was switched off and the doors opened, with room temperature attained in the chamber after less than 10 min. Blood collection continued 15, 30, and 90 min after the cooling system was turned off.
- *Immobilization* (IMMOB). The rat's limbs were taped to a metal frame with hypoallergic tape. Blood was collected in 15-min intervals up to 120 min.
- *Ether.* The animals were exposed for 3 min to ether vapor. Blood was collected immediately and 15 and 30 min later.

The majority of these stressors activated AM or sympathoneural or both components of SAS to markedly different extents. A single handling of rats increased both plasma Epi and NE levels: after repeated handling, no rise of Epi levels occurred, but the normal increase of NE levels was observed. Handling also significantly increased plasma levels of DOPA and DHPG. In repeatedly handled rats, where the last handling was done by a different person, the Epi response was not reduced, and the NE response was even potentiated.

Immobilization activated both the AM and sympathoneural systems, inducing many-fold increases in plasma Epi, NE, DA, DHPG, DOPAC, and DOPA levels. Insulin-induced hypoglycemia produced dose-dependent AM activation (plasma Epi levels reached extreme values over 5000 pg/ml—more than 60-fold increases). In contrast, plasma NE and DHPG levels increased by a maximum of about 2.5-fold. High insulin doses also induced two- to fourfold increases in plasma DA, DOPAC, and DOPA levels.

1% formalin administration did not significantly elevate values for any of the measured parameters, compared with values in the group with i.p. saline injection. Four percent formalin activated the AM system (plasma Epi levels increased by 12-fold), whereas, judging from plasma NE and DOPA changes, the sympathoneural system was not significantly activated.

Hemorrhage activated the AM system (plasma Epi levels were increased by fivefold), but almost no changes were found in plasma NE, DHPG, DOPA, DA, or DOPAC levels, indicating that the sympathoneural system was not activated, even by 25% hemorrhage, which induced hypotension.

An opposite picture was seen in cold-exposed rats. The AM system was not noticeably affected by cold (small Epi elevation in short intervals was instead affected by transfer of animals into the cold chamber), whereas the sympathoneural system was markedly activated by cold at both temperatures (several-fold increases in plasma levels of NE, DHPG, DOPA, DA, and DOPAC). Inhalation of ether vapors activated both the AM and sympathoneural systems.

As summarized in Figure 1, the greatest activation of the AM system was during insulin, immobilization, and 4% formalin administration, whereas the greatest sympathoneural system activation was during cold and immobilization. Plasma DHPG followed, in general, the NE changes, except for the levels in animals treated with 4% formalin and partly also by 3 IU of insulin, which were disproportionately large versus NE levels, suggesting an increase in NE turnover. Changes in plasma DOPA levels were very similar to those in plasma NE. Hemorrhage elicited small AM and almost no sympathoneural activation, whereas immobilization elicited extensive activation of both systems.

We have also studied effects of different stressors on expression of gene coding for TH, the rate-limiting enzyme in CA biosynthesis (3). Concentrations of TH mRNA in the adrenal medulla were measured by Northern blot. Three hours after immobilization for 2 hr, TH mRNA levels were increased 4.8-fold; 5 hr after insulin injection (5 IU/kg), the levels rose 3.8-fold; and 6 hr after the cold exposure, TH mRNA levels were elevated threefold.

Studies on the mechanism of regulation of TH gene expression showed that adrenal denervation, by cutting the splanchnic nerve, prevented the increases in AM TH mRNA levels of rats exposed to cold or insulin. Immobilization-induced increases in TH mRNA levels, however, were not affected by adrenal denervation. Thus, cold and insulin regulate TH gene expression in the adrenal medulla by neuronal pathways, whereas immobilization regulates this process by an as yet unidentified nonneuronal mechanism. Thus, these data demonstrate that even such a complicated basic process as gene expression could be specifically affected by different stressors.

In summary, the majority of stressors studied activated to different extents the two components of the sympathoadrenal system, as indicated by plasma levels of catechols and by increases in mRNA levels of TH in the adrenal

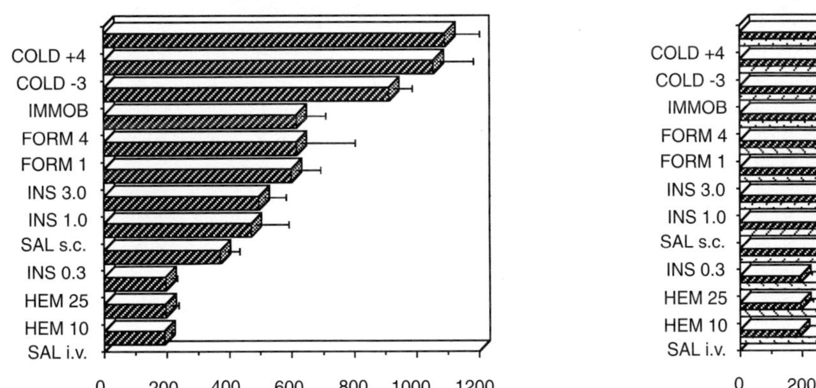

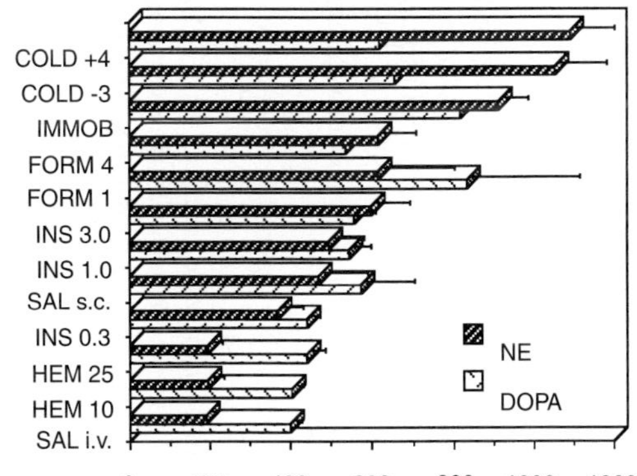

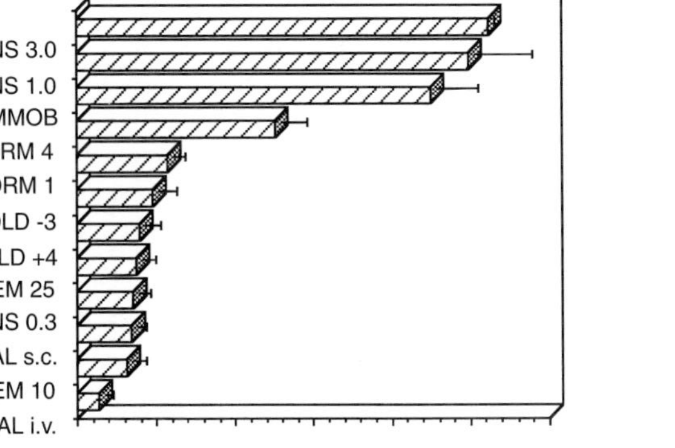

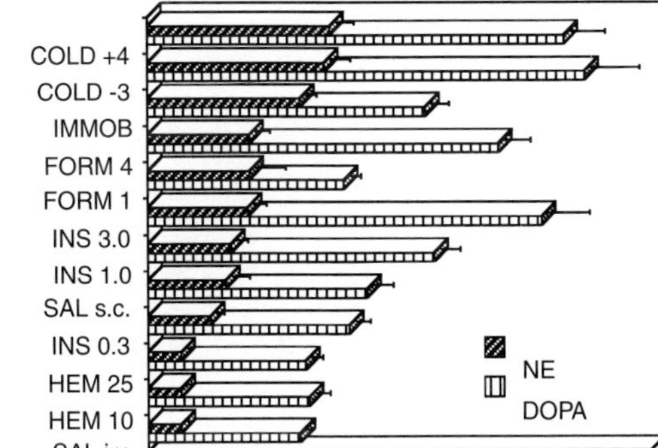

FIGURE 1 Levels of plasma NE, Epi, DOPA and DHPG in rats exposed to different intensities of various stressors. The values (means ± SEM, n = 6–8) are expressed in picograms per milliliter of plasma and represent maximal responses to stressors.

medulla. Thus, stress-induced sympathoadrenal system activation is not a nonspecific process.

Repeated handling (14 days) reduced plasma Epi levels, while NE levels stayed increased to the same extent as after the first exposure. When the last handling was performed by a different person, the reduction in Epi levels was no longer apparent and the NE response was even potentiated, consistent with the phenomenon of dishabituation, which also argues against Selye's doctrine of nonspecificity.

Exposure to cold evoked disproportionately large plasma NE, DHPG, and DOPAC responses; insulin and formalin, large plasma Epi and DHPG responses; and hemorrhage, small Epi and almost no NE responses, whereas immobilization elicited large increases in plasma levels of all of these compounds. The different stressors evoked markedly different responses of the two components of the SAS.

Immobilization, insulin, and cold regulate AM TH gene expression by different mechanisms.

In conclusion, the observed dissociation in responses of plasma Epi, NE, precursor, and metabolite levels, as well as TH gene expression during various stressors demonstrate that different pathways control the responses of the sympathoneural and AM systems. Our findings, therefore, are inconsistent with Selye's doctrine of nonspecificity of the stress response and suggest that each stressor has a neurochemical signature with distinct central and peripheral mechanisms. The majority of the central mechanisms remains unknown.

Acknowledgments

This study was supported in part by National Institute of Neurological Disorders and Stroke grant NS-32166, the Slovak Agency VEGA grants 95/5305/043 and 95/5305/272 and the U.S.–Slovak Project 93024.

References

1. Young, J. B., Rosa, R. M., and Landsberg, L. (1984). Dissociation of sympathetic nervous system and adrenal medullary responses. *Am. J. Physiol.* **247**, E35–E40.
2. Kopin, I. J. (1995). Definitions of stress and sympathetic neuronal responses. *Ann. N. Y. Acad. Sci.* **771**, 19–30.
3. Kvetňanský, R., Nankova, B., Hiremagalur, B., Viskupic, E., Vietor, I., Rusnak, M., McMahon, A., Kopin, I. J., and Sabban, E. L. (1996). Induction of adrenal tyrosine hydroxylase mRNA by single immobilization stress occurs even after splanchnic transection and in the presence of cholinergic antagonists. *J. Neurochem.* **66**, 138–146.

Karel Pacák,*† Judit S. Baffi,‡ Richard Kvetňanský,§ David S. Goldstein,† and Miklós Palkovits‡

*Department of Medicine
†Clinical Neuroscience Branch
National Institute of Neurological Disorders and Stroke
Department of Medicine
Washington Hospital Center
Washington, D.C. 20010

‡Laboratory of Cell Biology
National Institute of Mental Health
National Institutes of Health
Bethesda, Maryland 20892

§Institute of Experimental Endocrinology
Slovak Academy of Sciences
Bratislava, Slovak Republic

Stressor-Specific Activation of Catecholaminergic Systems: Implications for Stress-Related Hypothalamic-Pituitary-Adrenocortical Responses

The presence of noradrenergic terminals that synapse on corticotropin-releasing hormone (CRH) cells in the paraventricular nucleus (PVN) of the hypothalamus suggests a relationship between hypothalamic-pituitary-adrenocortical (HPA) function and central noradrenergic activity (1). Norepinephrine (NE) is thought to be a potent stimulator of CRH neurons. Surgical or chemical transection of the ascending ventral noradrenergic bundle markedly decreases tissue and extracellular basal concentrations of NE, immunoreactive CRH-41, and CRH messenger RNA (mRNA) in the ipsilateral PVN (2, 3). Coronal hemisection of the brain stem between the locus ceruleus and the rostral portion of the medulla oblongata also markedly decreases release of NE and expression of CRH mRNA in the PVN in rats exposed to immobilization stress (2).

Responses of central catecholaminergic systems as well as the HPA axis vary during exposure to different stressors. Extracellular NE levels in the PVN increase markedly with immobilization (IMMO) or with formalin (FORM)-induced pain and relatively little with insulin (INS)-induced hypoglycemia, cold (COLD) stress, or hemorrhage (HEM) (4). Levels of 3,4-dihydroxyphenylacetic

1054-3589/98 $25.00

acid, a metabolite of dopamine, increase in the locus ceruleus with IMMO or ether but are unchanged during INS-induced hypoglycemia (5). These and other findings have led to the hypothesis of the existence of stressor-specific central noradrenergic pathways participating in regulation of the HPA axis.

In the present study, conscious rats were exposed to one of several stressors—HEM, i.v. INS, s.c. FORM, COLD, or IMMO. Fos immunoreactivity of the immediate early gene *c-fos* was used to investigate changes in the activity of brain stem neurons. The results were correlated with previous findings about stress-induced central noradrenergic activation in the PVN using *in vivo* microdialysis and simultaneous measurements of plasma adrenocorticotropic hormone (ACTH) levels.

All experiments described in the present study were approved by the National Institute of Neurological Disorders and Stroke Animal Care and Use Committee. Adult male Wistar-Kyoto rats weighing 250 ± 10 g were used. After acclimatization in the laboratory, conscious rats were exposed to one of several stressors: HEM (25% of estimated blood volume), i.v. INS (3.0 IU/kg), s.c. FORM (4%), COLD (3°C), IMMO (2 hr), or handling and s.c. injection of physiological saline (SAL). After the acute experiment, rats were killed by ether overdose and perfused intracardially with a fixative solution (4% paraformaldehyde, 0.19% picric acid in 0.1 M phosphate buffer, ph 7.35) and postfixed in the same solution at 4°C overnight. Serial coronal sections of 40-μm thickness were cut through the lower brain stem and the hypothalamus with a freezing microtome at −20°C. Combined *c-fos* and tyrosine hydroxylase (TH) or *c-fos* and CRH immunostaining were performed using double immunoperoxidase methods.

IMMO induced a strong *c-fos* immunoreactivity in the entire ventrolateral medulla (including A1 and A5 noradrenergic neurons and C1 adrenergic neurons), the locus ceruleus, and A7 catecholaminergic cells. In the dorsomedial medulla, about 25–35% of cells were TH-positive. Simultaneously, a fairly high number of TH-negative cells, mainly in the nucleus of the solitary tract, showed strong *c-fos* activity. FORM-induced pain resulted in marked *c-fos* activation in the locus ceruleus and less activation in the A1 and A5 noradrenergic cell groups, with practically no activation in A2 noradrenergic cells. Exposure to COLD, INS, or HEM exerted only minor effects on *c-fos* activity in all brain stem catecholaminergic neurons (Table I). In the PVN, IMMO induced marked activation of *c-fos* in the parvicellular but not the magnocellular region. Similarly, exposure to FORM induced strong activation of *c-fos* in PVN CRH-containing cells. COLD exposure or HEM produced only moderate activation of CRH-containing cells in the PVH. HEM was only the stressor that activated *c-fos* in PVN magnocellular neurons. The smallest effect on *c-fos* in PVN CRH cells was seen after INS-induced hypoglycemia (see Table I).

The present results and our previous data illustrate the marked heterogeneity of both central and peripheral mechanisms of neuroendocrine responses. Noradrenergic activation in the PVN and lower brain stem seems to be stressor-specific.

Consistent with the present findings, our previous data showed that IMMO and FORM substantially increased levels of the PVN NE and plasma ACTH and corticosterone levels, with strongly positive correlations between PVN NE

TABLE I Effects of Various Stressors on *c–fos* Immunostaining in Brain Stem Catecholaminergic Cell Groups and the PVN of the Hypothalamus, Extracellular Concentrations of NE in the PVN, and Plasma Corticotropin (ACTH) Levels in Rats

		IMMO	*FORM 4%*	*COLD*	*HEM*	*INS*
PVN	NE	+++	++	+	±	+
Plasma	NE	+++	+++	+++	±	++
	ACTH	+++	++	±	+	++
Brainstem						
A1		+++	++	+	±	−
A2	*c-fos*	++	±	±	±	+
A6		+++	+++	+	−	−
PVN	*c-fos*	+++	+++	++	++	+

A1, A2, A6, brain stem catecholaminergic groups; +++, high level activity; ++, moderate level activity; +, low level activity; ±, just detectable; −, undetectable.

and plasma ACTH responses (see Table I). In contrast to IMMO, FORM did not induce significant *c-fos* activation in TH-positive A2 cells, suggesting that FORM-induced NE release in the PVN is derived mainly from A1 and A6 areas. Our previous studies showed that IMMO or FORM markedly increase NE release in the PVN and that NE released in the PVN is stimulatory to CRH neurons (4). The present data indicating that both stressors markedly increased *c-fos* activation in the PVN CRH neurons are consistent with a stimulatory role of NE on PVN CRH neurons.

INS-induced hypoglycemia resulted in only very moderate responses in *c-fos* immunostaining of CRH neurons in the parvicellular PVN. Moreover, there was only minor *c-fos* activation in A1, A2, and A6 catecholaminergic cells. This was consistent with our recent data showing small extracellular PVN NE responses measured by *in vivo* microdialysis (4). The findings suggest that hypoglycemia-induced ACTH release may not be controlled directly by CRH neurons in the PVN.

HEM or COLD exerted only minor effect on *c-fos* activation in A1, A2, and A6 catecholaminergic cells. In PVN CRH cells, there was slightly more *c-fos* activation in response to these stressors than in response to INS. As noted previously, only small increases in extracellular PVN NE and plasma ACTH occur in rats exposed to these stressors, with responses of NE PVN and plasma ACTH not related at all (4). Thus, PVN NE probably contributes little to HEM- or COLD-induced activation of the HPA axis.

In conclusion, the present results indicate that the magnitude of activation of *c-fos* in brain catecholaminergic regions and of PVN CRH neurons varies widely across stressors. The findings support the notion of stressor-specificity of responses of central catecholaminergic systems and the HPA axis and indicate that different central pathways regulate HPA reactivity. Future studies in this area should focus on mapping such central pathways and neurotransmitters involved in stressor-specific activation of the HPA axis.

References

1. Liposits, Z. S., Phelix, C., and Paull, W. K. (1986). Electron microscopy analysis of tyrosine hydroxylase, dopamine-β-hydroxylase and phenylethanolamine-N-methyltransferase immunoreactive innervation of the hypothalamic paraventricular nucleus in the rat. *Histochemistry* 84, 105–120.
2. Pacak, K., Palkovits, M., Makino, S., Kopin, I. J., and Goldstein, D. S. (1996). Brainstem hemisection decreases corticotropin-releasing hormone mRNA in the paraventricular nucleus but not in the central amygdaloid nucleus. *J. Neuroendocrinol.* 8, 543–551.
3. Sawchenko, P. E. (1988). Effects of catecholamine-depleting medullary knife cuts on corticotropin-releasing factor and vasopressin immunoreactivity in the hypothalamus of normal and steroid manipulated rats. *Neuroendocrinology* 48, 459–470.
4. Pacak, K., Palkovits, M., Kvetnansky, R., Yadid, G., Kopin, J. J., and Goldstein, D. S. (1995). Effects of various stressors on in vivo norepinephrine release in the hypothalamic paraventricular nucleus and on the pituitary-adrenocortical axis. *Ann. N. Y. Acad. Sci.* 771, 115–130.
5. Gaillet, S., Lachuer, J., Malaval, F., Assenmacher, and Szafarczyk, A. (1991). The involvement of noradrenergic ascending pathways in the stress-induced activation ACTH and corticosterone is dependent on the nature of stressors. *Exp. Brain Res.* 87, 173–180.

E. L. Sabban,* B. B. Nankova,* L. I. Serova,* B. Hiremagalur,* M. Rusnak,† E. Saez,‡ B. Spiegelman,‡ and R. Kvetňanský†

*Department of Biochemistry and Molecular Biology
New York Medical College
Valhalla, New York 10595

†Institute of Experimental Endocrinology
Slovak Academy of Science
Bratislava, Slovakia

‡Dana-Farber Cancer Institute
Boston, Massachusetts 02115

Regulation of Gene Expression of Catecholamine Biosynthetic Enzymes by Stress

The catecholamines are important mediators of the stress response. Activation of the sympathoadrenal system is among the early responses to stress. Prolonged or repeated stress is associated with development of cardiovascular

Advances in Pharmacology, Volume 42

1054-3589/98 $25.00

disorders, such as hypertension and myocardial infarction; mental disorders, such as depression; as well as gastrointestinal disorders and increased susceptibility to immune diseases and cancer. These effects of repeated stress are likely to be associated with alterations in gene expression. We have examined the effect of immobilization (IMMO) stress on the expression of the genes encoding some catecholamine biosynthetic enzymes, as well as GTPcyclohydrolase I (GTPCH), the rate-limiting enzyme in the tetrahydrobiopterin biosynthetic pathway. This was investigated in the adrenal medulla (AM), a major source of plasma epinephrine; the sympathetic ganglia (SG), a major source of plasma norepinephrine; and the locus ceruleus (LC), the major noradrenergic nucleus in the central nervous system that modulates the response of the hypothalamus and other brain areas to stress.

All of these genes were found to be activated by IMMO (1). However, there are important differences in the minimal time of IMMO required for the induction of these genes and in the factors mediating the stress response. For example, 5 min of single exposure to stress was sufficient to elicit a maximal increase in adrenal phenylethanolamine-N-methytransferase (PNMT) mRNA levels. However, to achieve maximal effect on the other genes of interest, longer times were required (greater than 30 min for tyrosine hydroxylase [TH] and repeated immobilizations for dopamine β-hydroxylase [DBH]) (1). In addition, an intact hypothalamic-pituitary-adrenal axis (HPA) was found to be required for induction of PNMT and GTPCH mRNAs. In contrast, hypophysectomy did not prevent the elevation of adrenal medullary TH mRNA by a single IMMO stress (2).

Tissue-specific differences in the regulation of these enzymes were also found. Thus, maximal (although transient) elevations of TH mRNA were observed in the AM with a single IMMO, while repeated stress was required for maximal elevations of TH mRNA in the SG. Moreover, exogenous administration of adrenocorticotropic hormone (ACTH) (s.c., 4 IU/rat/day for 7 days) alone mimicked the stress response in SG without affecting gene expression in the AM. Exposure of ACTH-treated animals to IMMO triggered different changes in TH mRNA levels in ganglia (no further increases of already elevated relative abundance) as compared with the AM (attenuated response to stress).

The effect of stress on TH gene expression in AM was found to be transcriptionally mediated. Pretreatment of rats with actinomycin D alone had no effect on adrenal TH mRNA levels, whereas it prevented the rise in TH mRNA levels in response to single IMMO (2). These results are further confirmed by using a nuclear run-on assay. The transcriptional activation of TH gene expression by a single IMMO stress correlates with the kinetics of c-fos induction. In addition, splanchnic denervation did not prevent the induction of TH or the elevation of c-fos (1, 3). Increased binding of nuclear proteins, including c-fos to the TH AP1 site was observed with IMMO (1, 2). These findings suggest that c-fos is an important candidate for mediating the transcriptional effect on TH gene expression triggered by stress.

In addition several investigators have shown that c-fos expression is markedly activated by stress in various tissues, including many central nuclei. Induction of c-fos has been used as a marker for stress-elicited activation of gene expression. To investigate whether the induction of c-fos by stress is required for

TABLE I Comparison of the Relative Abundance of TH mRNA in Adrenals of Control and Repeatedly Immobilized c-fos Knockout Mice

Genotype/sex	*Control*	*Immobilization*
Males		
Wild type (+/+)	1.0 ± 0.31	3.6 ± 0.53[a]
Heterozygous (+/−)	1.1 ± 0.51	2.5 ± 0.70[b]
Homozygous (−/−)	0.8 ± 0.35	8.9 ± 3.16[a]
Females		
Wild type (+/+)	1.0 ± 0.34	2.2 ± 0.45[b]
Heterozygous (+/−)	0.8 ± 0.34	2.8 ± 0.65[a]
Homozygous (−/−)	0.8 ± 0.13	2.5 ± 0.27[a]

Values represent the mean ± SEM.
[a] < .01.
[b] < .05.

stress-elicited changes in the expression of the genes involved in catecholamine biosynthesis, c-fos knockout mice (4) were used. Three genotypes (+/+ wild type, +/− heterozygous, and −/− homozygous) of six to eight per group were exposed to three daily IMMOs of 2 hr each. The animals were matched for age and litter as much as possible. The estrous cycles of the females were also taken into account.

Total RNA was isolated from adrenals of each animal, and the relative levels of TH mRNA were examined by Northern blots. The same filters were hybridized to probe for 18S rRNA. The results are shown in Table I.

In the AM, the basal levels of TH mRNA were similar in all genotypes and in males and females. The rise in TH mRNA in response to IMMO stress was about threefold in male and twofold in female (+/+) mice. Surprisingly, the exposure of heterozygous and homozygous mice to stress elicited at least as large an elevation as that observed in wild-type mice.

These results indicate that either c-fos is not required for the induction TH gene expression in AM or that alternative factors are involved. Studies on TH gene expression in cultured cells have previously identified two major promoter elements, an AP1 site and a cyclic AMP/calcium regulatory element (CRE/CaRE), each of which can independently activate TH transcription (5). The induction of c-fos is not the only change in transcription factors observed in the AM following IMMO. For example, there is increased phosphorylation of CRE-binding protein (CREB) within 5 min of immobilization. Therefore, activation of a pathway leading to utilization of the CRE/CaRE may circumvent the usage of the AP1 site. Alternatively, other members of the fos family of immediate early genes may compensate for lack of c-fos expression in the knockout animals. The results indicate that there are tissue-specific and enzyme-specific pathways in stress-elicited activation of genes for various enzymes involved in catecholamine biosynthesis.

Acknowledgments

Supported by National Institutes of Health grants NS 28869 and 32166.

References

1. Sabban, E. L., Hiremagalur, B., Nankova, B., and Kvetnansky, R. (1995). Molecular biology of stress-elicited induction of catecholamine biosynthetic enzymes. *Ann. N. Y. Acad. Sci.* **771**, 327–338.
2. Nankova, B., Kvetnansky, R., McMahon, A., Viskupic, E., Hiremagalur, B., Frankle, G., Fukuhara, K., Kopin, I. J., and Sabban, E. L. (1994). Induction of tyrosine hydroxylase gene expression by a nonneuronal nonpituitary-mediated mechanism in immobilization stress. *Proc. Natl. Acad. Sci. U.S.A.* **91**, 5937–5941.
3. Kvetnansky, R., Nankova, B., Hiremagalur, B., Viskupic, E., Vietor, I., Rusnak, M., McMahon, A., Kopin, I., and Sabban, E. (1996). Induction of adrenal tyrosine hydroxylase mRNA by single immobilization stress occurs even after splanchnic transection and in the presence of cholinergic antagonists, *J. Neurochem.* **66**, 138–146.
4. Johnson, R. S., Spiegelman, B. M., and Papaioannou, V. (1992). Pleiotropic effects of a null mutation in the c-fos proto-oncogene. *Cell* **71**, 577–586.
5. Sabban, E. L., (1996). Synthesis of dopamine and its regulation. *In* CNC Neurotransmitters and Neuromodulators Dopamine. (T. W. Stone, ed.), pp. 1–20. CRC Press, Boca Raton, FL.

Srinivasa N. Raja

Department of Anesthesiology and Critical Care Medicine
Division of Pain Medicine
Johns Hopkins Hospital
Baltimore, Maryland 21287-5354

Peripheral Modulatory Effects of Catecholamines in Inflammatory and Neuropathic Pain

The sympathetic nervous system (SNS) is a complex network of neurons and fibers in the peripheral and central nervous systems. Sympathetic efferent fibers innervate every organ in the body and control diverse involuntary functions. In the periphery, the SNS has been traditionally characterized as an effector organ. However, it is being recognized that the SNS may be also involved in the modulation of sensory information. Basic science and clinical

Advances in Pharmacology, Volume 42

1054-3589/98 $25.00

studies indicate that the SNS may play a role in pathological states associated with pain and hyperalgesia, such as inflammation and neuropathic pain (1).

I. Sympathetic Efferents in Normal Tissue

Earlier studies in experimental animals suggested that the SNS does not appear to be involved in signaling pain from healthy tissue (2). Physiological studies have demonstrated that the response properties of cold receptors, muscle spindles, low-threshold mechanoreceptors, hair follicles, tooth pulp afferents, and paccininan corpuscles may be altered by sympathetic stimulation. However, neurophysiological studies have failed to demonstrate an excitatory effect of sympathetic stimulation on thinly myelinated (Aδ) or unmyelinated (C) nociceptive afferent fibers innervating normal skin. In contrast to these studies, recent psychophysical studies in our laboratory provide pharmacological evidence that noradrenaline may influence pain sensation in normal human subjects. Studies indicate that norepinephrine (NE) produces hyperalgesia to heat stimuli when administered into the skin of normal subjects. How can the apparent discrepancy between the neurophysiological studies in the past and the more recent psychophysical studies be explained? The earlier animal studies utilized a mechanical search stimuli to identify nociceptors. More recent studies have recognized the presence of "silent" nociceptors in skin that are mechanically insensitive and hence may have been missed by the aforementioned search techniques (3).

II. Role of Sympathetics in Inflamed Tissues

Several lines of evidence suggest that sympathetics play a role in nociceptor responses from inflamed tissues. A-fiber nociceptors can be activated by sympathetic stimulation (SS) following sensitization by a heat injury. This activation is blocked by phentolamine, suggesting an α-adrenoceptor–mediated mechanism. Similarly, discharges of C-fiber nonciceptors innervating inflamed skin can be enhanced by noradrenaline and SS. Cutaneous C-fiber nociceptors in rats that were sensitized by the injection of a mixture of inflammatory mediators into the receptive field responded to SS and local arterial injection of NE. This activation was blocked by phentolamine, suggesting an α-adrenoceptor–mediated mechanism. Electrophysiological recordings from Freund's adjuvant-induced arthritic rats have shown that 50% of the C-fiber nociceptors respond to either SS or NE. The sympathetic activation of C-fibers was blocked or attenuated by the α_2-adrenergic antagonist, yohimbine, but less with the α_1-adrenergic antagonist, prazosin. These studies suggest that SS augments the activity of the nociceptors in inflamed tissue via an α-adrenoceptor mechanism.

Levine and coworkers propose an indirect mechanism for the role of sympathetics in inflammatory pain. They suggest that the site of action of NE is not on the primary afferents but instead is on the sympathetic postganglionic terminals (SPGTs). Topical application of chloroform to the hindpaw of rats results in hyperalgesia to mechanical stimuli that is enhanced by NE. The hyperalgesia is abolished by sympathectomy and yohimbine but not by prazosin. The authors postulate that NE released from SPGT activates autoreceptors (presynaptic α_2 receptors) on the SPGT in the same terminals, resulting in the release of prostaglandins, which may be the final mediators of the hyperalgesia.

In a recent psychophysical study, Drummond investigated the effects of NE on cutaneous hyperalgesia in humans, using topical application of capsaicin as a model to induce hyperalgesia (5). Capsaicin is known to sensitize C-fibers to heat and lower the pain threshold to mechanical stimul. The decrease in heat pain threshold (hyperalgesia) following capsaicin administration was prolonged at the site where NE was iontophoresed but not at the saline site, suggesting that NE increased heat hyperalgesia in skin treated with topical capsaicin. These effects of NE in humans are consistent with the neurophysiological observations in animals that NE and sympathoneural stimulation increase nociceptor discharges in inflamed skin. In behavioral experiments in rat, Kinnman and Levine also observed that the cutaneous hyperalgesia induced by intradermal capsaicin injection was dependant on sympathetic innervation (6). This hyperalgesia was abolished by intradermal injections of prazosin, suggesting a role of α_1 adrenoceptors.

III. Role of Sympathetics in Neuropathic Pain

Pain in certain patients is dependent on sympathetic activity in the painful area. Anesthetizing or surgically interrupting the sympathetic chain reduces or eliminates the pain in these patients. This pain state, which may be an aspect of reflex sympathetic dystrophy and causalgia, is termed *sympathetically maintained pain* (SMP). SMP may be a feature of a wide variety of disorders, including acute shingles, amputation neuromas, metabolic painful neuropathies, traumatic nerve injury, and soft-tissue injury.

Intact nerves are normally incapable of initiating impulses. However, ectopic activity develops in neuromas after nerve injury. Moreover, cell bodies and neuroma fibers develop chemosensitivity to adrenergic agonists after nerve injury. The adrenergic sensitivity disappears when axoplasmic transport is blocked in the axons near the neuroma site. Such ectopic chemosensitivity is thought to contribute to the pain in the clinical syndrome of causalgia. The response of fibers in experimental neuromas to close arterial injection of epinephrine or NE is blocked by the α-adrenergic antagonist, phentolamine, but not by the β-adrenergic antagonist, propranolol, indicating that α but not β adrenoceptors play a role in the pain of neuromas.

Although peripheral nerve injury and subsequent neuroma formation have been associated with chronic pain and dysesthesia, often precluding the use of prosthetic devices in amputees, few studies have attempted to understand the mechanisms of pain from neuromas in humans. Chabal and coworkers reported that perineuronal epinephrine, not saline, caused an intense increase in pain in subjects studied with neuroma pain following peripheral nerve injury (7).

Patients often commented that the appendage was "on fire." In preliminary studies, we have observed a dose-dependent increase in pain after perineuronal administration of NE in patients with postamputation stump pain. These observations indicate that neuromas in humans are also chemosensitive.

IV. Peripheral Sympathetic–Somatic Coupling after Partial Nerve Injury

Several animal models of neuropathic pain resulting from partial injury to peripheral nerves (e.g., sciatic nerve, lumbar spinal nerves) have been developed in the last few years. The animals exhibit varying degrees of spontaneous pain behavior and hyperalgesia to mechanical and thermal stimuli. Studies in these animal models indicate that the pain behavior is reversed after surgical and chemical sympathectomy, and hence are representative of SMP. The behavioral signs of hyperalgesia was rekindled by injecting NE intradermally into the sympathectomized paw. These observations suggest that sympathectomy suppresses but does not eliminate the mechanisms of pain in these animals. In a number of models tested, evidence supports the hypothesis that one of the sites of coupling between the SNS and nociceptors is in the periphery. Injuries to the auricular nerve (loose ligature, partial section, and stretching of the nerve; rabbit ear *in vitro* preparation) resulted in catechol sensitization in about 20% of the C-fiber nociceptors. Notably, the NE response of c-fiber nociceptors (CMHs) was blocked by anesthetizing the receptive field with lidocaine. In brief, nociceptors that are not normally responsive to SS become responsive to catecholamines after nerve injury. Anatomical and neurophysiological studies indicate that the dorsal root ganglion (DRG) is another potential site of interaction of sympathetic efferent and somatic afferent fibers following peripheral nerve injury.

Several lines of evidence support the hypothesis that in SMP α adrenoceptors develop in peripheral tissues (e.g., at the cutaneous terminals of nociceptors or at the site of nerve injury) such that the release of NE from the sympathetic terminals activates the nociceptors and leads to the sensation of pain (8). NE injected into the skin rekindles the former pain in patients who were relieved by sympathetic blockade but not in those who did not benefit from sympathetic blockade. In addition, intravenous phentolamine (an α-adrenergic blocker) but not propranolol relieves SMP. Thus, α adrenoceptors not β receptors appear to be the culprit. Drummond and associates (9) recently demonstrated, using quantitative autoradilography, that α_1 adrenoceptors were present in the epidermis and dermal papillae of normal individuals and in the hyperalgesic and normal skin of patients with reflex sympathetic dystrophy. The mean density of α_1 adrenoceptors was greater in the hyperalgesic skin of patients with reflex sympathetic dystrophy than in the skin of normal individuals. Based on this experimental evidence and clinical data from our studies, we described a model for the mechanism of pain in SMP in humans. The model hypothesizes that in certain instances, nerve injury may lead to the development of adrenergic sensitivity of nociceptors and that neuropathic pain may be attenuated in this subset of patients by sympatholytic interventions.

V. Pharmacology of Animal Models of SMP

In common with studies in patients, animal models for SMP suggest involvement of α adrenoceptors. The reports are, however, conflicting in regard to whether the α_1 or the α_2 adrenoceptors are involved. For example, in rat neuromas, phenylephrine was found to stimulate responsive nerve endings, suggesting α_1 involvement. In contrast, after nerve injury in the rabbit ear, the C-nociceptor responses were blocked by yohimbine and rauwolscine, indicating α_2 involvement. In a behavioral study in rats, selective adrenergic antagonists were used to determine if the hyperalgesic effects of NE in peripheral tissues were mediated via distinct α_2-adrenoceptor subtypes. NE-induced hyperalgesia was blocked by α_{2B} antagonists (imiloxan and SKF 104856). Differences observed in peripheral α_2-adrenoceptor mechanisms may arise from differences in subtypes, second-messenger coupling, species variations, or a combination of these factors. Future studies defining the adrenoceptors subtypes involved in the pathological sympathetic–somatic interactions leading to chronic pain may lead to more novel and specific therapies for certain chronic pain states.

Acknowledgment

Supported in part by National Institutes of Health grant NS 26363.

References

1. Raja, S. N. (1995). Role of the sympathetic nervous system in acute pain and inflammation. *Ann. Med.* 27, 241–246.
2. Jänig, W., and Koltzenburg, M. (1992). Possible ways of sympathetic-afferent interactions. *In* Reflex Sympathetic Dystrophy: Pathophysiological Mechanisms and Clinical Implications. (W. Jänig and R. F. Schmidt, eds.), pp. 213–243. VCH, New York.
3. Meyer, R. A., Davis, K. D., Cohen, R. H., Treede, R-D., and Campbell, J. N. (1991). Mechanically insensitive afferents (MIAs) in cutaneous nerves of monkey. *Brain Res.* **561**, 252–261.
4. Levine, J. D., Taiwo, Y. O., Collins, S. D., and Tam, J. K. (1986). Noradrenaline hyperalgesia is mediated through interaction with sympathetic postganglionic neurone terminals rather than activation of primary afferent nociceptors. *Nature* **323**, 158–169.
5. Drummond, P. D. (1996). Independent effects of ischemia and noradrenaline on thermal hyperalgesia in capsaicin-treated, skin. *Pain* **67**, 23–27.
6. Kinman, E., and Levine, J. D. (1995). Involvement of the sympathetic postganglionic neuron in capsaicin-induced secondary hyperalgesia in the rat. *Neuroscience* **65**, 283–291.
7. Chabel, C., Jacobson, L., Russell, L. C., Burchiel, K. J. (1992). Pain responses to perineuronal injection of normal saline, epinephrine, and lidocaine in humans. *Pain* **49**, 9–12.
8. Campbell, J. N., Meyer, R. A., and Raja, S. N. (1992). Is nociceptor activation by alpha-1 adrenergic receptors the culprit in sympathetically maintained pain? *Am. Pain Soc.* **1**, 3–11.
9. Drummond, P. D., Skipworth, S., and Finch, P. M. (1996). α_1-Adrenoceptors in normal and hyperalgesic human skin. *Clin. Sci.* **91**, 73–77.

M. Palkovits,*† J. Baffi,* Z. E. Tóth,* and K. Pacák‡

*Laboratory of Neuromorphology
Semmelweis University Medical School
Budapest 1094, Hungary

†Laboratory of Cell Biology
National Institute of Mental Health
National Institutes of Health
Bethesda, Maryland 20892

‡Department of Medicine
Washington Hospital Center
Washington, D.C. 20010

Brain Catecholamine Systems in Stress

Corticotropin-releasing hormones (CRHs) producing neurons in the hypothalamic paraventricular nucleus (PVN) are responsible for the activation of the pituitary-adrenal axis and influence the activity of the sympathoadrenal system. In addition to CRH, vasopressin- and oxytocin-containing neurons in the PVN also have a significant role in the organization of the stress response. CRH neurons, which project to the median eminence and influence adrenocorticotropic hormone (ACTH) release from the anterior pituitary, are concentrated in the medial parvocellular subdivision of the PVN, while those that project to the medullary and spinal cord autonomic centers occupy the caudal mediocellular subdivision of the nucleus. Vasopressin and oxytocin are synthesized by magnocellular neurons in the PVN, but vasopressin is also present in some parvicellular, mainly CR-containing neurons, and they can be activated by physiological or pathological alterations in the pituitary-adrenal axis. Neurons in the PVN receive adrenergic and noradrenergic inputs from the caudal ventrolateral medulla (A1 and C1 cell groups), the dorsomedial medulla (A2 and C2 cell groups), and the locus ceruleus (A6 cell group). In addition to these, descending fibers from the rostral lateral medullary noradrenergic neurons (A5 cell group) also participate in the control of the sympathoadrenal activity (i.e., in the stress response).

In the present study, immunohistochemical (for tyrosine hydroxylase [TH] and proto-oncogene *c–fos*), *in situ* hybridization histochemistry, tract-tracing, and *in vivo* microdialysis techniques (collection of the extracellular fluid in the PVN) were used to investigate the participation of the central catecholaminergic (CA) systems in the organization of responses to an acute neurogenic (pain) stress: subcutaneous injection of 0.5 ml of 4% formalin into the hind paw of rats. This single injection produces a local pain that develops in the first 5 min after injection and results in four- to fivefold increase in corticosterone and ACTH levels in the plasma and a similar increase in norepinephrine levels in

Advances in Pharmacology, Volume 42

1054-3589/98 $25.00

the PVN (3). Formalin also induced a rapid expression of *c-fos* and CRH mRNA in the PVN.

After formalin injection, pain (nociceptive) signals are carried by the peripheral nerve fibers to the lumbar spinal cord, where *c-fos* immunopositivity is seen in the marginal zone neurons of the dorsal horn. Until that level, the stress signal is *ipsilateral* to the injection. In higher levels, however, the stress-induced *c-fos* expression is *bilateral:* where *c-fos*-positive neurons were present bilaterally in the A1, A2, and A6 TH-positive CA neurons, as well as in the PVN (5); the formalin-induced CRH mRNA upregulation was also bilateral. Thus, the nociceptive axons (or axon collaterals) from the marginal zone of the spinal cord to brain stem CA and PVN neurons should cross over in or below the medulla oblongata. The possible crossover of the nociceptive spinal afferents to the medullary CA neurons and the crossover of the ascending CA neurons to the PVN were investigated after spinal and medullary surgical hemisections.

Hemisection between the spinal cord and the medulla oblongata did not influence formalin-induced elevation of plasma corticosterone and ACTH levels, and strong *c-fos* immunoreactivity was seen in the medullary CA cell groups and in the PVN *bilaterally,* as in the sham-operated animals. It is more likely that the ascending nociceptive neurons have axon collaterals that crossover at the spinal cord or within the lower medulla oblongata. The existence of such collaterals has been demonstrated by tract-tracing techniques: After unilateral injections of retrograde tracers into the ventrolateral medulla (A1 cell group), labeled neurons were seen in marginal neurons of the dorsal horn in both sides of the spinal cord (1). In addition to these, the existence of neuronal interconnections between the two A1 cell groups has also been demonstrated (1). Thus, both types of signal transfer of unilateral nociceptive signals to bilateral CA neurons in the medulla are likely (Fig. 1). Direct, probably similar interconnections exist between spinal nociceptive neurons and the A2 cell group in the nucleus of the solitary tract (2). Indeed, a fairly high number of fibers crossover in the commissural portion of the nucleus of the solitary tract.

Hemisection between the medullary CA cell groups and the PVN (unilateral knife cut at the pons–medulla oblongata border) decreased CRH immunoreactivity and CRH gene expression in the PVN ipsilateral to the transection but remained unchanged contralateral to the knife cut (4). The density of adrenergic and noradrenergic neuronal networks in the PVN was also reduced substantially 2 wk after surgery, but only on the side of the hemisection. This type of surgery attenuated formalin-induced elevation of norepinephrine levels in the PVN ipsilateral to the cut, regardless of whether formalin was injected into the ipsi- or contralateral hindleg. These data indicate that noradrenergic afferents to the PVN from the A1 and A2 CA neurons ascend mainly on the ipsilateral side. In addition to these, noradrenergic fibers to the PVN also arise in the locus ceruleus and join the ascending A1 and A2 fibers in the ventral noradrenergic bundle. A small percentage of ascending CA fibers may cross over into the hypothalamus at the level of the PVN, as is indicated by tract-tracing observations. Fibers labeled by an anterograde tracer, *Phaseolus vulgaris* leukoagglutinin, injected into the A1 region unilaterally, can be followed up

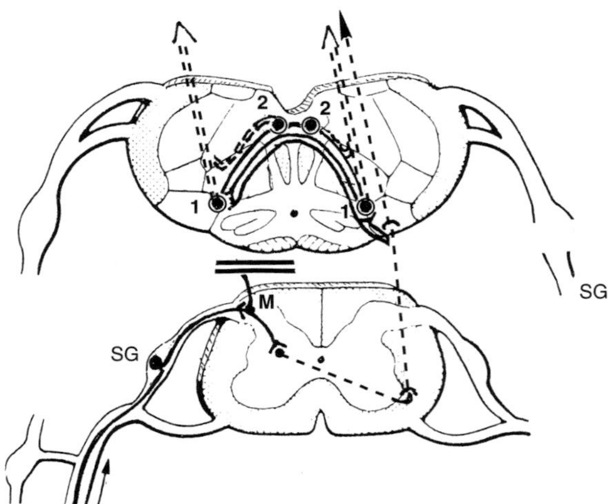

FIGURE 1 Variations for possible terminations of nociceptive axon collaterals on brain stem CA neurons, as well as interconnections between A1 and A2 neurons on the two sides (*single arrows*). CA fibers are indicated by double arrows. 1, A1 cells in the ventrolateral medulla; 2, A2 cells in the nucleus of the solitary tract; M, neurons in the marginal zone of the dorsal horn; SG, spinal ganglion.

to the PVN mainly ipsilaterally, but few of the labeled fibers cross over the top of third ventricle and terminate on PVN neurons contralateral to the injection.

Although, unilateral pontomedullary hemisection reduced TH and CRH immunoreactivity and CRH gene expression in the PVN, it did not influence the high plasma corticosterone and ACTH levels in response to formalin stress. It should be mentioned that acute immobilization stress in hemisected animals was able to induce CRH expression (4). These findings suggest that in brain stem–hemisected animals, the remaining intact ascending pathways are sufficient to allow normal, or almost normal, activation of the hypothalamo-pituitary-adrenal axis during stress.

The neuronal interconnection between the PVN and brain stem CA neurons is bidirectional: PVN neurons have descending bilateral projections to A1, A2, A5, and A6 TH-positive cells. After a microinjection of anterograde tracer in the PVN, labeled fibers can be followed to the lower brain stem ipsilateral to the side of the injection. Fibers crossover segmentally at the supramamillary level, in the pontine tegmentum at the level of the locus ceruleus, and in the medulla inside the commissural portion of the nucleus of the solitary tract and innervate CA neurons contralateral to the injection.

Inputs to the preganglionic neurons in the intermediolateral cell column of the spinal cord (sympathoadrenal output) arise from the ipsilateral A5 neurons (noradrenergic afferents) and from the PVN (bilateral peptidergic afferents).

References

1. Lima, D., Mendes-Ribeiro, J. A., and Coimbra, A. (1991). The spino-latero-reticular system of the rat: Projections from the superficial dorsal horn and structural characterization of marginal neurons involved. *Neuroscience* **45**, 137–152.
2. Menétrey, D., and Basbaum, A. I. (1987). Spinal and trigeminal projections to the nucleus of the solitary tract: A possible substrate for somatovisceral and viscerovisceral reflex activation. *J. Comp. Neurol.* **255**, 439–450.
3. Pacak, K., Palkovits, M., Kopin, I. J., and Goldstein, D. S. (1995). Stress-induced norepinephrine release in the hypothalamic paraventricular nucleus and pituitary-adrenocortical and sympatho-adrenal activity: In vivo microdialysis study. *Front. Neuroendocrinol.* **16**, 89–150.
4. Pacak, K., Palkovits, M., Makino, S., Kopin, I. J., and Goldstein, D. S. (1996). Brainstem hemisection decreases corticotropin-releasing hormone mRNA in the paraventricular nucleus but not in the central amygdaloid nucleus. *J. Neuroendocrinol.* **8**, 543–551.
5. Palkovits, M., Baffi, J. S., and Dvori, S. (1995). Neuronal organization of stress response: Pain-induced c-fos expression in brain stem catecholaminergic cell groups. *Ann. N.Y. Acad. Sci.* **771**, 313–326.

Mark J. Millan

Department of Psychopharmacology
Institut de Recherches Servier
78290 Croissy-sur-Seine (Paris), France

α_2-Adrenergic Mechanisms of Analgesia: Strategies for Improving Their Therapeutic Window and Identification of the Novel, Potent α_{2A}-Adrenergic Receptor Agonist, S 18616

Adrenergic pathways descending from the brain stem the dorsal horn (DH) of the spinal cord control the access of nociceptive information to the brain (1, 2). The antinociceptive actions of noradrenaline (NAD) in the DH are exerted both presynaptically on terminals of primary afferent fibers (PAFs), thereby inhib-

Advances in Pharmacology, Volume 42

1054-3589/98 $25.00

iting the release of substance P and glutamate, and postsynaptically on projection neurones targetted by PAFs (3). The discovery of multiple α-adrenergic receptors (α_2ARs) (α_{2A}, α_{2B}, α_{2C}, and α_{2D}/rα_{2A}, orthologous to hα_{2A}) raises the question as to their respective roles. α_{2A}ARs predominate in the DH and play a major role in mediating antinociception (2, 4). First, preferential α_{2A} agonists, such as guanfacine, guanabenz, and oxymetazoline, elicit antinociception. Second, their actions are blocked by the α_{2A}AR-selective antagonist, BRL 44408, but not by the $\alpha_{2B/2C}$AR antagonists, prazosin, ARC 239, and BRL 41992. Third, α_{2A}AR-targetted gene knockdown (antisense) and gene knockout (null mutation) strategies have confirmed that the antinociceptive actions of α_2AR agonists are exerted via α_{2A}ARs (M. Maze and L. E. Limbird, personal communication). Indeed, clonidine and dexmedetomidine (DMT), which elicit antinociception in both rodents and humans, display high affinity at α_{2A}ARs. These represent, thus, an attractive development target, and we have identified a spiro-oxazoline, S 18616 ([7,8](2-chlorobenzo)-2-amino-1-aza-3-oxa-[4,5]spirodeca-1,7-diene) possessing high efficacy and extraordinary potency at these sites (Table I).

S 18616 displays a modest preference for α_{2A}AR versus α_{2B}- and α_{2C}ARs. It also shows a marked preference for α_{2A}AR versus α_1ARs and I_2 receptors, while its preference for α_{2A}ARs versus I_1 receptors is also more pronounced than for clonidine. It is active at very low doses upon s.c. (not shown) or oral (see Table I) administration to mice, with a duration of action of 12–16 hr. S 18616 may show superior separation of antinociceptive to motor actions as compared with DMT and clonidine (see Table I), although this requires confirmation. S 18616 does exert sedative properties at higher doses, and the hypnotic–sedative actions of α_2AR agonists reflect the engagement of dendritic α_{2A}AR autoreceptors on noradrenergic neurones of the locus ceruleus (1, 2). Further, α_{2A}ARs in the brain stem and intermediolateral cell column of the spinal cord mediate the sympathoinhibitory actions of α_{2A}AR agonists (4). Thus, activation of α_{2A}ARs underlies not only the analgesic, but also the hypnotic–sedative and hypotensive actions of α_{2A}AR agonists. Moreover, reflecting an inhibition of ascending noradrenergic neurones, they similarly mediate the anxiolytic actions of α_2AR agonists. These roles open several therapeutic avenues for α_{2A}AR-selective agonists, including their use in the perioperative environment for the reduction of anesthesia requirements and (owing to their

TABLE I *In Vitro* and *In Vivo* Activity of S18616

	← *K_{iS} (nM)* →							← *ID_{50}* →		*Ratio*
	α_{2A}	*α_{2B}*	*hα_{2A}*	*hα_{2C}*	*α_1*	*I_1*	*I_2*	*FPL*	*RR*	*RR/FPL*
S 18616	*0.17*	*1.9*	*0.9*	*6.5*	*79*	*4.1*	*72*	*1.2*	*81*	*67.5*
DMT	2.1	3.8	12.9	26	355	198	>1000	9.4	80	8.5
Clonidine	8.4	27.0	73	444	562	29	>1000	56	350	6.3

Affinities are for native receptors, except hα_{2A}- and hα_{2C}ARs (cloned, human) transfected into an Sf9 baccilovirus. For methods, see Millan *et al.* (2) and Renouard *et al.* (5). 1994.
I, imidazoline; ID_{50}, inhibitory dose 50 (μg/kg, p.o.); FPL, formalin paw-lick; RR, rotarod.

hemostabilizing and sympatholytic actions) myocardioprotection (1, 6). However, the induction of hypotension and sedation by α_{2A}AR agonists raises the question as to whether satisfactory analgesia can be acheived in the absence of intolerable side effects in conscious subjects. This question can be answered only in the clinic with selective, potent, and high-efficacy drugs such as S 18616. Nonetheless, one may also envisage several alternative strategies for improving the therapeutic window of α_2AR agonists.

First, partial agonists may permit the separation of analgesic actions from side effects. This approach presupposes that the efficacy required for induction of analgesia is less than that for induction of hypotension or sedation. However, α_{2A}AR autoreceptors mediating motor actions are unlikely to be less sensitive than their postsynaptic counterparts mediating antinociception, and a reduction in agonist efficacy risks limiting maximal analgesia. Further, clonidine itself is a partial agonist. This approach has, thus, yet to prove successful. Second, drugs with varying degrees of efficacy may show differential rates of tolerance development, and the analgesic actions of α_2AR agonists may adapt more slowly than their secondary actions. Interestingly, there *are* indications that motor actions adapt more rapidly than analgesic actions (notably, with S 18616), but hypotensive effects are unlikely to disappear. Further, this is not relevant to acute, first-dose administration. Third, one might, conceivably, "build into" an α_{2A}AR agonist a component of activity palliating its undesirable effects. This would be analogous to antipsychotic agents possessing serotoninergic properties to moderate the extrapyramidal syndrome provoked by dopamine D_2-receptor blockade. Fourth, it may be possible to incorporate into a drug two synergistic mechanisms of analgesic activity. For example, a dual α_2AR agonist plus μ-opioidergic mechanism of action might allow for the optimization of analgesic actions while minimizing the side effects associated with stimulation of α_2AR and μ-opioid receptors, respectively (1). One possible example of such a drug is the weak opioid, tramadolol, which possesses adrenergic properties. Fifth, this strategy does not permit the titration of each of the components of activity separately, and a related approach would be to use an α_2AR agonist in association with a further analgesic agent in order to similarly augment the therapeutic window. Indeed, systemic or spinal coadministration of α_2AR agonists with μ-opioids or local anesthetics affords robust pain relief in the absence of severe side limits (6). Further, there is experimental evidence that α_2AR agonists elicit additive or synergistic antinociception (in the absence of intensified side effects) with agonists at muscarinic and serotonin$_{1B}$ receptors and with antagonists at the N-methyl-D-aspartate receptor (1, 3). Sixth, the preceding approach incorporates the strategy of local (intrathecal or epidural) administration of α_2AR agonists into the DH. Nevertheless, spinal application is not of universal utility and does not necessarily obviate all side effects of α_2AR agonists. Thus, actions of α_2AR agonists in the intermediolateral cell column modulate sympathetic outflow, while the occurrence of bradycardia and hypotension at high doses of spinal clonidine in humans may reflect redistribution (via the cerebrospinal fluid or circulation) to the brain stem (4, 6). Indeed, apart from motor effects in the ventral horn, sedation has been seen in some clinical studies with clonidine likewise indicating that the drug may access higher centers. Bolus administration and/or computerized control of injection rates may limit such problems (6). Further, an evaluation of the actions of lipophobic drugs (to minimize redistribution) would be of interest. Seventh, desipramine and other antidepressants inhibiting the re-uptake of NAD are effective analgesic agents, for example, against neuropathic pain, and a component of their activity may be exerted spinally (3). These drugs mimic the physiological release of NAD, which presumably acts with all available adrenergic receptor types (possibly including some as yet unknown). This might, arguably, be the most appropriate mode for obtaining analgesia in a manner comparable to that which occurs "naturally." Eighth, an opposite approach comprises the development of drugs interacting selectively with specific α_2AR subtypes and, as described previously, α_{2A}ARs constitute one possible target (2). However, α_{2B}ARs are found in the thalamus and in human DH, while α_{2C}ARs are found in both rat and human DH (albeit at low densities), in PAFs, and in sympathetic ganglia (4). Thus, they also may play a role in modulating nociception. In addition, an additional non-α_{2A}AR, possibly α_{2C}, also may contribute to antinociception at the spinal level in rats (2). Finally, the possible importance of α_1ARs should not be neglected (4).

In conclusion, there exist many potential strategies for an improvement in the therapeutic index of α_2AR-mediated analgesics. Of these, the development of α_2AR subtype-selective ligands remains particularly attractive. Unfortunately, to date, virtually all clinical data have been acquired with clonidine, which is a partial agonist of modest potency and only limited selectivity. There is, thus, an urgent need for novel ligands. In this regard, S 18616 should prove of use in the further exploration of the potential antinociceptive roles of adrenergic mechanisms, and its therapeutic properties will be of interest to evaluate.

Acknowledgments

A. Cordi, J-M. Lacoste, S. Girardon, K. Bervoets, and C. Dacquet are thanked for collaboration.

References

1. Hayashi, Y., and Maze, M. (1993). Alpha$_2$ adrenoceptor agonists and anaesthesia. *Br. J. Anaesth.* 71, 108–118.
2. Millan, M. J., Bervoets, K., Rivet, J-M., Widdowson, P., Renouard, A., Le Marouille-Girardon, S., and Gobert, A. (1994). Multiple alpha$_2$-adrenergic receptor subtypes. II. Evidence for a role of rat Rα_{2A}-ARs in the control of nociception, motor behaviour and hippocampal synthesis of noradrenaline. *J. Pharmacol. Exp. Ther.* **270**, 958–972.
3. Besson, J-M., and Guilbaud, G., eds. (1992). Towards the Use of Noradrenergic Agonists for the Treatment of Pain. pp. 233. Elsevier, Amsterdam.
4. Nicholas, A. P., Pieribone, V. A., and Hökfelt, T. (1996). The distribution and significance of CNS adrenoceptors examined with *in situ* hybridization. *Trends Pharmacol. Sci.* **17**, 245–255.

5. Renouard, A., Widdowson, P. S., and Millan, M. J. (1994). Multiple alpha$_2$-adrenergic receptor subtypes. I. Comparison of [^{3}H]RX821002-labelled rat Rα_{2A}-adrenergic receptors in cerebral cortex to human Hα_{2A}-adrenergic receptors and other populations of α_{2A}-adrenergic subtypes *J. Pharmacol. Exp. Ther.* **270**, 946–957.
6. Eisenach, J. C. (1994). Alpha-2 agonists and analgesia. *Exp. Opin. Invest. Drugs* **3**, 1005–1010.

Jacqueline Sagen

CytoTherapeutics, Inc.
Providence, Rhode Island 02906

Cellular Transplantation for Intractable Pain

Implantation of cells directly into the central nervous system (CNS) offers a novel means of providing sustained local delivery of pharmacologically active substances for the long-term management of chronic disorders. Furthermore, the use of cellular implant therapies allows for the delivery and utilization of agents that have short biological half-lives or barriers to CNS penetration. Work in our laboratory over the past 10 years has suggested that cellular transplantation may be a powerful approach for the alleviation of chronic pain. For these studies, adrenal medullary chromaffin were chosen as donor sources, because these cells produce and secrete catecholamines and opioid peptides, agents that reduce pain when administered directly into the spinal subarachnoid space and synergize to produce potent analgesia. In addition to these agents, chromaffin cells have been reported to secrete a variety of neurotrophic factors, cytokines, and other neuropeptides, as well as ascorbate and heme-containing proteins, which may aid in the restoration of spinal cord function in chronic pain syndromes.

Donor sources for these studies have included primarily either adrenal medullary tissue allografts or isolated xenogeneic chromaffin cells, although cell lines that can be genetically engineered have been utilized and are a likely avenue for future applications. For preclinical rodent studies, allogeneic tissues can be obtained from adult donors of the same strain by microdissection of adrenal medullary tissue from the adrenal cortex. Previous findings have shown that adrenal medullary tissue can be either transplanted following dissection or maintained in tissue culture for at least 30 days prior to implantation without apparent decrement in antinociceptive potency. The latter approach has been utilized in initial clinical studies, with adrenal medullary tissue derived from human organ donors (1). However, because the practical application of adrenal

Advances in Pharmacology, Volume 42

1054-3589/98 $25.00

medullary implantation in clinical pain management is limited by the availability of allogeneic human donor tissue, the future success of neural transplantation programs depends on the identification of alternate donor sources. Thus, xenogeneic chromaffin cells harvested from sources such as bovine adrenal glands may be an alternative. Our laboratory has utilized cellular isolation procedures to achieve relatively pure preparations of bovine chromaffin cells. Using this approach, we have achieved successful long-term survival of isolated bovine chromaffin cells implanted in the rat CNS following a short course with cyclosporin A (2). Another approach in xenotransplantation is the use of semipermeable polymer membranes to encapsulate xenogeneic cells and limit contact with the host immune system, while allowing diffusion of neuroactive substances and nutrients.

In preclinical rodent models, adrenal medullary tissue, cell suspensions, or encapsulated cells are implanted in the rat spinal subarachnoid space via a laminectomy at the L1-L2 level. Various pain models have been utilized in these studies, including acute analgesiometric tests, a chronic constriction nerve injury model for neuropathic pain, a complete Freund's adjuvant model for inflammatory arthritic pain, and the formalin test for phasic–tonic pain. The results of these studies indicated that implantation of chromaffin cells in the spinal subarachnoid space markedly reduced pain behaviors. Both acute and chronic pain responses were attenuated, although acute pain responses appeared to require stimulation of chromaffin cell surface nicotinic receptors, presumably to increase the release of catecholamines and opioid peptides necessary for producing analgesia to these more intense noxious stimuli. In contrast, alleviation of chronic pain responses did not require nicotine, suggesting that basal release from the implanted cells may be sufficient to attenuate chronic pain. The ability to secrete both catecholamines and opioid peptides is maintained in these implanted cells, as indicated by increased spinal cerebrospinal fluid (CSF) levels of both in animals with adrenal medullary transplants for at least 6 mo postimplantation follow-up. One of the possible limitations in pain management is the potential for tolerance development, because pain-reducing neuroactive substances are presumably released from the cells on a continual basis. While this issue has not been addressed in detail, studies using chronic pain models have generally revealed that there is no decrement in the analgesic potency of these implants over the time course of the pain symptomology. In addition, studies in our laboratory have revealed that there is not apparent cross-tolerance with systemic morphine injections in adrenal medullary–implanted animals. Rather, these implants appear to potentiate morphine, which could be due to the release of other agents from the implanted cells that could be additive or synergistic in their pain-reducing capabilities. For a review of these preclinical studies, see Czech and Sagen (3).

As a result of the success of the preclinical work, limited clinical trials have been initiated at several centers. At the University of Illinois at Chicago, five patients with intractable pain secondary to nonresectable cancerous lesions were implanted with adrenal medullary allografts from human donor glands. Adrenal medullary tissue dissected from two adrenal glands was maintained in explant culture for approximately 7 days, checked for chromaffin cell viability, and implanted in the spinal lumbar cistern via lumbar puncture. Follow-up included pain scoring (visual analogue scale [VAS]), analgesic consumption, and CSF sampling. In this uncontrolled study, four of the patients reported significant pain reduction and concomitant reduction in analgesic consumption. Pain reappeared in one of these patients and the rest remained pain-free, two for nearly 1 year postimplantation. Details of this study are reported in Winnie *et al.* (1). A protocol similar to the one followed in Chicago was conducted by Lazorthes *et al.* (4). The clinical trial involved eight patients suffering from intractable pain due to cancer. Consenting patients enrolled in the study had received inadequate pain control from oral morphine and were, thus, receiving opioids via implanted intrathecal pumps to maintain sufficient pain control prior to adrenal medullary implantation. Adrenal medullary tissue was prepared and implanted as described previously. A multidisciplinary pain evaluation demonstrated progressively decreased pain scores in six of the patients. Concomitant opioid analgesic intake was decreased in three of the patients and stabilized in three other patients. At another center, one patient with intractable pain was implanted with adrenal medullary tissue with "striking results" (Dr. R. Drucker-Colín, Universidad Nacional Autonoma de Mexico, personal communication). This patient showed gradual reduction in somatic pain VAS scores (from 10, most severe prior to implantation), reaching zero by approximately 1 mo posttransplantation and remaining at this level until death at 3 mo. There was a concomitant reduction in analgesic intake. A summary of these clinical studies is shown in Fig. 1.

Phase I clinical trials have been conducted at the University of Lausanne, Switzerland, to assess safety and preliminary efficacy utilizing encapsulated xenogeneic chromaffin cells from bovine adrenal glands (5). Approximately 2 million cells were loaded into the capsules, based on a linear scaling from animal results. Cell-loaded devices were tested for catecholamine output prior to implantation. In a preliminary report, three patients with terminal cancer pain were included. Of these, two markedly reduced opiate analgesic intake following implantation. CSF catecholamine levels were increased in two of the patients, and microscopic examination of retrieved devices revealed good cell viability and positive immunocytochemical staining for tyrosine hydroxylase. In a second report, seven patients with severe chronic pain inadequately managed with conventional therapies were enrolled. Of these, four patients who were originally receiving epidural morphine at the time of the implant decreased their analgesic usage during the study, with either a modest improvement or no worsening in pain ratings. Three of the other patients demonstrated improvements in McGill Pain ratings and two showed improved VAS scores. All devices were recovered after implant periods of 41 to 176 days. Postretrieval histology revealed viable chromaffin cells with positive immunostaining for tyrosine hydroxylase and Met-enkephalin in six of seven devices analyzed. A similar phase I clinical trial is underway in the United States and has been completely enrolled. Although results have not been analyzed in detail thus far, preliminary assessments suggest promising outcomes (personal communication).

In summary, results of these preclinical and initial clinical studies suggest that cellular implantation may be a means of delivering therapeutic pain-

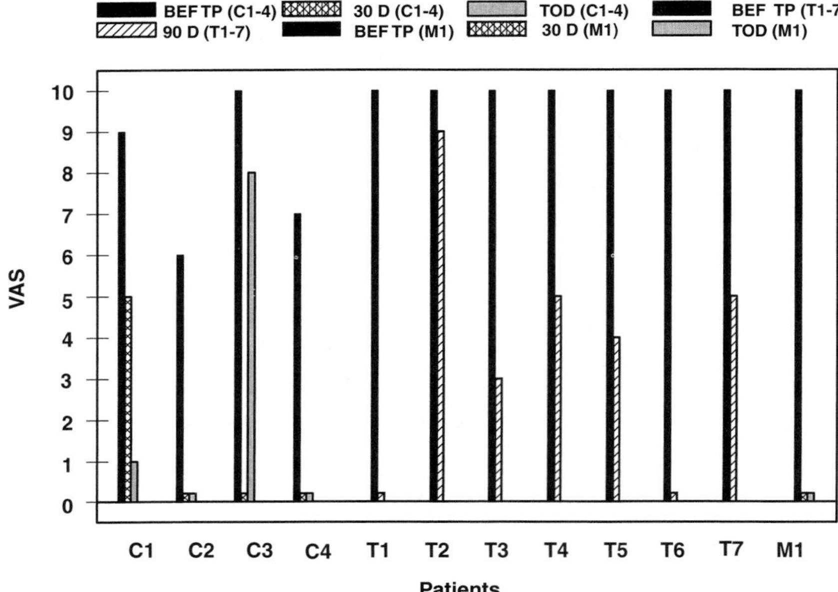

FIGURE 1 Pain scores as assessed by the VAS in cancer pain patients receiving adrenal medullary allografts at three centers: University of Illinois at Chicago (patients C1–C4), University Paul Sabatier, Toulouse (patients T1–T7), and Universidad Nacional Autonoma de Mexico (Patient M1). Pain scores are shown prior to implantation (BEF TP), 30 or 90 days postimplantation (30 D, 90 D), and just prior to time of death (TOD). (Reprinted from *Fetal Transplantation in Neurological Diseases,* T. B. Freeman and H. Widner, eds. Humana Press, in press.)

reducing agents, such as catecholamines and opioid peptides, on a long-term basis for the alleviation of chronic pain.

References

1. Winnie, A., Pappas, G. D., Gupta, T. K., Wang, H., Ortega, J., and Sagen, I. (1993). Subarachnoid adrenal medullary transplants for terminal cancer pain. *Anesthesiology* **79,** 644–653.
2. Ortega, J., Sagen, J., and Pappas, G. D. (1992). Shortterm immunosuppression enhances long-term survival of bovine chromaffin cell xenografts in rat CNS. *Cell Transplant.* **1,** 33–41.
3. Czech, K. A., and Sagen, J. (1995). Update on cellular transplantation into the CNS as a novel therapy for chronic pain. *Prog. Neurobiol.* **46,** 507–529.
4. Lazorthes, Y., Bès, J. C., Sagen, J., Tafani, M., Tkaczuk, J., Sallerin, B., Nahri, I., Verdié, J. C., Ohayon, E., Caratero, C., and Pappas, G. D. (1996). Transplantation of human chromaffin cells for intractable cancer pain control. *Acta Neurochir.* (*Wien*) **64,** 97–100.
5. Aebischer, P., Buscher, E., Joseph, J. M., Favre, J., de Tribolet, N., Lysaght, M., Rudnick, S., and Goddard, M. (1994). Transplantation in humans of encapsulated xenogeneic cells without immunosuppression: A preliminary report. *Transplantation* **58,** 1–3.

S. Y. Felten,*† K. S. Madden,* D. L. Bellinger,* B. Kruszewska,† J. A. Moynihan,†‡ and D. L. Felten*†

Center for Psychoneuroimmunology Research
Departments of *Neurobiology and
Anatomy, †Psychiatry, and ‡Microbiology and Immunology
University of Rochester Medical Center
Rochester, New York 14642

The Role of the Sympathetic Nervous System in the Modulation of Immune Responses

I. Neural-Immune Interactions

The importance of a role for the sympathetic nervous system in modulation of immune responses is suggested by a long history of studies examining the effects of stress and psychosocial factors on health and disease (1). Although the hypothalamo-pituitary-adrenal axis and glucocorticoids are clearly important in mediating effects on the immune system that can influence disease susceptibility and health outcome, it has become increasingly clear that other systems, including the sympathetic nervous system, are involved.

II. Role of the Sympathetic Nervous System

A. Sympathetic Innervation of Lymphoid Organs

Numerous anatomic studies (2) have shown that lymphoid organs, both primary and secondary, are innervated by norepinephrine (NE)-containing postganglionic sympathetic nerve fibers, as well as a variety of peptidergic fibers. At the light microscopic level, much of the innervation appears to be associated with the vasculature in these organs, although there are always fibers that appear to have no anatomic relationship to blood vessels. Immunocytochemistry of the rodent spleen at the ultrastructural level, using antibodies for tyrosine hydroxylase, has revealed that, in addition to abundant blood vessel and trabecular smooth-muscle innervation, there is direct contact between tyrosine hydroxylase–positive nerve terminals and both lymphocytes and macrophages.

In rodents, sympathetic innervation of lymphoid organs varies with the age of the animal. Innervation of the spleen decreases dramatically from 18 to 27 mo of age in the rat, in parallel with decreases in splenic T-cell compartments. In contrast, innervation of the thymus remains relatively constant as the thymus involutes with age, resulting in an increase in the density of innervation.

Advances in Pharmacology, Volume 42

1054-3589/98 $25.00

The presence of sympathetic nerve terminals near, or in direct contact with, lymphocytes in lymphoid organs, including thymus, spleen, lymph nodes, and bone marrow, and the presence of adrenergic receptors on lymphocytes suggest that sympathetic innervation and the transmitter NE may be important in the modulation of immune responses.

B. Sympathectomy

A number of studies from our laboratories and others have demonstrated alterations in lymphocyte function and immune responses after acute (1–14 days) sympathetic denervation using the neurotoxin 6-hydroxydopamine (6-OHDA) (3). These changes include reduced cell-mediated responses, including reduced T-cell proliferation, reduced delayed-type hypersensitivity, and changes in production of antibodies to T-dependent antigens. Taken as a whole, the literature suggests that 6-OHDA sympathectomy reduces many cell-mediated responses of the immune system, with the implication that NE may be important in providing optimal T-cell–mediated immune responses.

Antibody responses to T-dependent antigens have varied depending on the age, species, and strain of animal; type of antigen; antibody isotype; and technique for measuring. A number of studies have been done in our laboratories addressing some of these sources of variability.

III. Effects of Sympathectomy in Nonimmune Mice

These studies (4) were done to assess basic properties of lymph node and splenic immunocytes from BALB/c mice given 6-OHDA (100 mg/kg body weight). Compared with mice given vehicle (0.1% ascorbate in sterile saline) or pretreated with desmethylimipramine (10 mg/kg given 30 min prior to 6-OHDA), lymphocytes from lymph nodes from 6-OHDA–treated mice showed increased baseline proliferation most likely due to an increased proliferation of B lymphocytes. In general, measures of cell-mediated responses (% T cells, Con A–induced proliferation) were decreased, while measures of humoral responses (% B cell, lipopolysaccharide [LPS]-induced proliferation, immunoglobulin G [IgG] production and interferon-γ [IFN-γ] production) were increased. In contrast, lymphocytes recovered from the spleen showed a transient increase in proliferation, decreased T cells, decreased proliferation in response to mitogens (Con A and LPS), decreased cytokine production (interleukin-2 [IL-2] and IFN-γ), and no change in immunoglobulins. These results suggest that responses to 6-OHDA sympathectomy may be organ-specific.

IV. Effects of Sympathectomy on Cytokine Production and Antibody Response in Two Mouse Strains

One important source of variability may be the predisposition of various rodent strains to mount cell-mediated (Th1) or humorally mediated (Th2) responses after challenge with antigen. This predisposition is thought to be due to differences in the kind of T-helper cells produced in response to an antigen challenge. In turn, T-helper cell (Th) type is determined by the balance of cytokines produced.

For these experiments (5), two mouse strains were selected on the basis of their difference in Th response type: C57Bl/6J mice produce cytokines (IL-2 and IFN-μ) that drive Th1 responses and BALB/cJ mice produce cytokines (IL-4) that drive Th2 responses. Each strain was treated with 6-OHDA at 6–8 wk of age, depleting splenic NE, and then challenged with the antigen keyhole limpet hemocyanin (KLH). C57Bl/6J mice received 250 mg/kg of 6-OHDA in ascorbate vehicle, while BALB/cJ mice received 100 mg/kg of 6-OHDA. Previous studies showed that C57Bl/6J mice are more resistant to the effects of 6-OHDA, requiring more to produce the 95% level of splenic NE reduction that is the criterion level of denervation for all of our studies.

Because our previous studies had generally shown decreases in cell-mediated responses and increases in humoral responses, we predicted that TH1 cytokines would be decreased and TH2 cytokines increased, leading to increased antibody production. Denervation significantly increased KLH-induced *in vitro* proliferation in both strains of mice. Surprisingly, both IL-2 and IL-4 production in splenocytes from both strains was also increased. However, only the C57Bl/6J mice showed increased antibody responses as measured by serum titers of KLH-specific IgM, IgG, IgG1, and IgG2a. These findings suggest that NE modulates cytokine production in lymphocytes, but that the *in vivo* effects of that modulation may depend on strain, perhaps based on how powerfully driven by cytokines the Th2 response is. It is possible that increases in Th2 cytokines in mice that are already predisposed to Th2 responses (e.g., BALB/cJ mice) make less difference in antibody production than they do for mice normally predisposed to Th1 responses. In this instance, the cytokine response to denervation was similar, with both strains showing increased cytokine production, suggests that one role of NE may be to modulate (in this case, tonically inhibit) the production of signal molecules of the immune system.

V. Effects of Sympathectomy in Aged Rats

Additional studies comparing the effects of 6-OHDA sympathetic denervation on immune responses in adult and aged F344 rats show that aged rats, previously shown to have decreased splenic innervation and NE content, have decreased cell-mediated immune responses compared with young adult rats. Our studies (6) examined the effects of further depletion of NE in aged rats, using 6-OHDA–induced sympathectomy in 3-mo- and 17-mo-old F344 rats. Treatment with 6-OHDA increased KLH-specific antibody responses and mitogen-induced proliferation in both aged and young adult rats, with a much larger effect on aged rats. IL-2 production was not different, suggesting that other cytokines may be involved in this change in response.

The marked immunological effects of sympathectomy in aged rats may indicate a greater sensitivity to acute changes in sympathetic tone. The source of this sensitivity may be related to compensatory responses of the aging animal to gradual loss of sympathetic innervation in peripheral lymphoid organs.

VI. Cytokine Feedback

Not only does NE affect cytokine production, but it is also likely that some cytokines also regulate NE release both via central nervous system sympathetic responses to circulating cytokines and by interactions with local noradrenergic nerve terminals innervating lymphoid organs and other peripheral target sites where inflammation or immune responses take place. Recent preliminary evidence from our laboratories suggests that i.p. injections of IL-2 (50 μg/rat) into 3-mo-old F344 rats decrease splenic NE content. Immunocytochemical staining using an antibody to IL-2 receptor reveals little evidence for IL-2 receptor protein in sympathetic ganglia innervating the spleen. However, within 4 hr of IL-2 injection (50 μg), an increase in staining is seen. By 24 hr after injection, staining for IL-2R is robust in sympathetic cell bodies in the celiac-superior mesenteric plexus. The superior cervical ganglion from the same rats showed no evidence of IL-2R staining, suggesting that there may be some specificity of the response for sympathetics innervating in spleen. However, the route of administration (i.p.) of IL-2 may account for this difference.

Studies of the effects of sympathectomy on immune responses invariably suggest that NE and the sympathetic nervous system have important roles in modulation of immune responses. However, reports, including those from our laboratories, often do not agree on what that role may be. The aforementioned studies suggest that in this complicated interaction, there may be a number of sources of variability. In our hands, responses have been shown to vary by lymphoid organ studied, species and strain of experimental animal, and age. Certainly, a number of other possibilities exist, including sex, animal housing and diet, handling, and other environmental stimuli. Furthermore, all these variables are likely to interact. Despite these problems, several general findings seem to be repeated, suggesting that an important role of NE is the modulation of cytokine production. Evidence that cytokines may, in turn, have feedback regulation of sympathetics completes a hypothesized model system consistent with our understanding of other neural and neuroendocrine regulatory systems.

Acknowledgments

This work was supported by grants MH45681, MH18822, MH42076, the Whitehall foundation, and the Markey Charitable Trust.

References

1. Kiecolt-Glaser, J. K., and Gaser, R. (1991). Stress and immune function in human. *In* Psychoneuroimmunology, 2nd ed. (R. Ader, D. L. Felten, and N. Cohen, eds.), pp. 849–867. Academic Press, New York.
2. Felten, S. Y., and Felten, D. L. (1991). The innervation of lymphoid organs. *In* Psychoneuroimmunology. (R. Ader, D. L. Felten, and N. Cohen, eds.), pp. 27–69. Academic Press, San Diego.
3. Madden, K. S., Sanders, V. M., and Felten, D. L. (1995). Catecholamine influences and sympathetic neural modulation of immune responsiveness. *Annu. Rev. Pharmacol. Toxicol.* **35**, 417–448.
4. Madden, K. S., Felten, S. Y., Felten, D. L., Hardy, C. A., and Livnat S. (1994). Sympathetic nervous system modulation of the immune system. II. Induction of lymphocyte proliferation and migration in vivo by chemical sympathectomy. *J. Neuroimmunol.* **49**, 67–75.

5. Kruszewska, B., Felten, S. Y., and Moynihan, J. A., (1995). Alterations in cytokine and antibody production following chemical sympathectomy in two strains of mice. *J. Immunol.* 155, 4613–4620.
6. Madden, K. S., Felten, S. Y., Felten, D. L., and Bellinger, D. L. (1995). Sympathetic nervous system–immune system interactions in young and old Fischer 344 rats. *Ann. N. Y. Acad. Sci.* 771, 523–534.

Paul J. Mills, Michael G. Ziegler, Jalees Rehman, and Alan S. Maisel

University of California at San Diego and
The Veterans Affairs Medical Center
La Jolla, California 92093

Catecholamines, Catecholamine Receptors, Cell Adhesion Molecules, and Acute Stressor-Related Changes in Cellular Immunity

Studies repeatedly demonstrate that the immune system is highly responsive to acute stress. Typical changes include a lymphocytosis, with marked changes in natural killer and T-suppressor cells and more modest effects on T-helper cells (1, 2). The rapidity of immune responsivity to acute stress has led investigators to speculate that the sympathetic nervous system may, in part, regulate the phenomenon. Several lines of evidence support such an assumption. Studies using pharmacological blockade and infusion of adrenergic agonists support the β-adrenergic receptor as a major underlying mechanism (1, 2).

Cellular adhesion molecules (CAMs) constitute a family of specific cell surface receptors crucial to leukocyte trafficking out of the circulation (3). Adhesion is accomplished through a cascade of events involving different families of CAMs, including the selectins, the integrins, and the immunoglobulin-related CAMs (ICAMs). Chemoattractants on the endothelial surface bind to receptors on the surface of the leukocyte. These G-protein–mediated receptors then signal integrins on the leukocyte surface (previously in an inactive state) to bind to the immunoglobulins on the surface of the endothelium. On lymphocytes, for example, the integrin lymphocyte function–associated antigen 1 (LFA-1) binds to the intercellular adhesion molecule ICAM-1. LFA-1–ICAM-1 cou-

Advances in Pharmacology, Volume 42

1054-3589/98 $25.00

pling then results in the strong adhesion of the leukocyte to the endothelial wall and eventual transendothelial migration.

We present data from two separate studies that were designed to determine (1) β_2-adrenergic receptor and catecholamine involvement in lymphocyte subset redistribution following acute sympathetic activation and (2) whether CAM expression on lymphocytes might influence this phenomenon. Subjects were studied following both a psychologic and an exercise stressor. The psychologic stressor (N = 110) required presenting an impromptu speech in front of a video camera. For the exercise (N = 12), subjects exercised until exhaustion (approximately 15–18 min) on a treadmill, according to the Bruce protocol. For the exercise study, six of the volunteers then received the nonselective β-blocker propranolol (40 mg t.i.d.) and six the β_1-selective blocker metoprolol (50 mg b.i.d.). After 1 wk of treatment, subjects repeated the treadmill protocol, as described.

Flow cytometry (FACScan, Becton Dickinson, San Jose, CA) using Simu-SET software with CD45 gating was used to determine lymphocyte subsets (mature T lymphocyte, CD3/Anti-leu, 4; T-helper lymphocyte, CD4/Anti-leu 3a; T-suppressor lymphocyte, CD8/Anti-leu 2a; B lymphocytes, CD19/Anti-leu 12; natural killer, CD3-/CD16 + CD56) and CAMs (one member of each of the three reviewed CAM families was chosen for study; L-selectin, CD62L; the integrin LFA-1, CD11a; I-CAM1, CD54).

Lymphocyte β_2-adrenergic receptor sensitivity was determined in whole cells by quantifying cyclic adenosine monophosphate accumulation following incubation with 10 μM of isoproterenol (4). Lymphocyte β_2-adrenergic receptor density was determined in membranes by using radioligand binding with [^{125}I]iodopindolol at six concentrations from 10 to 320 pM (4). β-Adrenergic receptor density (Bmax) and the dissociation constant (Kd) were calculated using a nonlinear regression receptor-binding program (GraphPad Software, San Diego, CA). Plasma catecholamines were determined by using radioenzymatic assay (5).

The data analyses were directed at determining main effects of the tasks on the immune endpoints (repeated measures ANOVA) and toward determining relationships between the adrenergic and immune measures (multiple linear regression). For the regression analysis (the psychologic stress), the dependent variables were the speaking task cell subsets that showed significant alterations following the task. The independent variables included the respective baseline subset type, baseline catecholamines (norepinephrine and epinephrine), β-receptor sensitivity and density, the change in catecholamines (speaking task minus baseline values), and general demographic information such as age, race, blood pressure, and gender. General linear hypothesis testing was utilized to determine the significance of predictor variables or a combination thereof. Variables with significance at $p < .05$ were maintained in the model; those with a $p > .05$ were dropped from the model.

I. Study I

The speech task resulted in significant increases over baseline in natural killer cells ($p < .001$), T-suppressor cells ($p = .01$), and plasma norepinephrine and epinephrine ($p < .001$), and decreases in T-helper cells ($p < .01$) and B cells ($p < .001$).

The regression indicated that, in general, baseline and poststress adrenergic measures accounted for a significant portion of the variance in poststress immune values (all $p < .001$). For example, the stress-induced increase in natural killer cells was best predicted by a combination of the increase in plasma norepinephrine and a greater β_2-receptor sensitivity ($r^2 = .26, p < .001$). Similarly, the stress-induced decrease in T-helper cells was best predicted by a combination of the baseline norepinephrine level, the increase in norepinephrine, and the β_2-receptor sensitivity ($r^2 = .44, p < .001$). The pattern of relationships varied somewhat across the subsets examined. In some instances, such as CD56 cells, the baseline norepinephrine approached significance but did not enter in the model. For some, such as T cells and B cells, β_2-receptor measures did not significantly enter the model.

II. Study 2

Dynamic exercise led to an expected lymphocytosis, with increases in natural killer cells, T-helper cells, and T-suppressor cells ($p < .001$). At baseline, 74% of total lymphocytes expressed the CAM L-selectin, while 98% of total lymphocytes expressed the CAM LFA-1. Dynamic exercise led to a 1.5-fold increase in total lymphocytes expressing L-selectin ($p < .001$) and a twofold increase in total lymphocytes expressing LFA-1 ($p < .001$). While there was no overall change in the distribution of mixed lymphocytes expressing LFA-1 with exercise, there was a 15% decrease in distribution of cells expressing L-selectin ($p < .001$). ICAM-1 (CD54) was expressed on approximately 67% of total mixed lymphocytes, and despite an approximate twofold exercise-induced increase in these cells ($p < .001$), there was no change in their overall distribution following exercise.

T-suppressor cell responses to exercise could be differentiated based on whether the cell expressed L-selectin. At baseline, only 50% of T-suppressor cells expressed L-selectin, and exercise led to a 1.5-fold increase in these cell numbers ($p < .001$) but no change in their overall distribution (i.e., percentage of total lymphocytes). In contrast, those T-suppressor cells not expressing L-selectin showed a threefold increase in cell numbers ($p < .001$) and a significant 50% increase in overall distribution ($p < .001$). Approximately 86% of T-helper cells expressed L-selectin. T-helper cells expressing and not expressing L-selectin behaved essentially identically in terms of exercise effects, with both increasing in number and decreasing in overall distribution ($p < .001$) in response to exercise.

Propranolol yielded a significant 7.2% increase in the distribution of total lymphocytes expressing L-selectin ($p < .02$). Both propranolol and metoprolol led to a significant blunting in the number of total lymphocytes expressing LFA-1 following exercise ($p < .01$). While β blockade had no exercise-attenuating effect on T-suppressor cells expressing L-selectin, those T-suppressor cells not expressing L-selectin showed a significant effect of β blockade, with propranolol (but not metoprolol) decreasing the exercise-induced increase in number ($p = .015$) and distribution ($p = .040$). β Blockade had no effect on the number of T-helper cells expressing L-selectin but inhibited the exercise-induced decrease in their distribution ($p = .01$) β Blockade had no effect on T-helper cells not expressing L-selectin.

These studies sought to identify adrenergic and CAM involvement in lymphocytosis following acute immune activation. The studies found that an individual's initial sympathetic state (baseline catecholamines and β_2-adrenergic receptors) and the magnitude of sympathetic (catecholamine) activation were important factors underlying lymphocyte subset redistribution. The findings also suggest that experimental stress results in differential expression of CAMs on lymphocytes and that CAM expression may influence β-adrenergic effects. For example, the presence or absence of L-selectin on T-suppressor cells determined their postexercise distribution in circulation, with those expressing L-selectin remaining unchanged and those not expressing L-selectin showing a marked redistribution. In addition, only those T-suppressor cells not expressing L-selectin showed a significant attenuating effect of propranolol. These data suggest that L-selectin expression on T-suppressor cells inhibits the effects of sympathetic activation and β blockade. These effects were harder to differentiate in T-helper cells because a majority of these cells expressed L-selectin. Given that, compared with cell types such as the T helper, T-suppressor cells express a higher density and sensitivity of β_2-adrenergic receptors, it may be of interest to examine β-receptor expression in T-suppressor cells both expressing and not expressing L-selectin. The findings that β-adrenergic antagonists may modulate the expression of CAMs may be of therapeutic significance and suggest a direction for further study on mechanisms underlying the potential pathogenic effects of sympathetic arousal on immune function.

Acknowledgments

This work was supported by grants MO1-RR00827 and HL47074 from the National Institutes of Health. The authors are grateful to the Immunogenetics Laboratory at the Veterans Affairs Medical Center for determination of the cellular adhesion molecules and lymphocyte subsets.

References

1. Maisel, A. S., Harris, T., Reardon, A., and Martin, M. C. (1990). β-adrenergic receptors in lymphocyte subsets after exercise: Alterations in normal individuals and patients with congestive heart failure. *Circulation* **82,** 2003–2010.
2. Murray, D. R., Irwin, M., Rearden, A., Ziegler, M., Motulsky, H., and Maisel, A. (1992). Sympathetic and immune interactions during dynamic exercise: Mediation via β2-adrenergic dependent mechanism. *Circulation* **86,** 203–213.
3. Springer, T. (1990). Adhesion receptors of the immune system. *Nature* **346,** 425–434.
4. Mills, P. J., Dimsdale, J. E., Ziegler, M. G., Hauger, R., Nelesen, R. A., and Brown, M. (1993). Sympathetic alterations following sodium restriction and short-term captopril. *J. Am. Coll. Cardiol.* **21,** 177–181.
5. Kennedy, B., and Ziegler, M. G. (1990). A more sensitive radioenzymatic assay for catecholamines. *Life Sci.* **47,** 2143–2153.

Luigi Aloe

Institute of Neurobiology
Consiglio Nazionale delle Ricerche
I-00137 Rome, Italy

Nerve Growth Factor and Autoimmune Diseases: Role of Tumor Necrosis Factor-α?

Nerve growth factor (NGF) is a well-characterized neurotrophic protein that plays an essential role in the development and differentiation of peripheral sensory and sympathetic neurons (1). The biological effects of NGF are mediated by its binding with specific cell receptors, which exist in two distinct forms: a low-affinity receptor (p75), a transmembrane glycoprotein, and a high-affinity receptor (TrKA), transmembrane tyrosine kinase protein, the product of the trk-1 protooncogene (2). Both receptors are known to be localized in NGF-responsive cells. The essential role played by NGF was first demonstrated by injecting NGF antibody into developing rodents. These studies showed that removal of circulating NGF via administration of specific NGF antibodies results in death of peripheral sympathetic neurons (1).

Several lines of evidence published in recent years indicate that NGF is a retrogradely transported trophic molecule for the magnocellular basal forebrain cholinergic neurons (BFCNs), which project from the septum and the basal nucleus to the hippocampus (HI) and the cerebral cortex, respectively (1, 3). Impairment of these neurons is involved in the pathophysiology of normal aging and most probably of dementia of the Alzheimer type as well. The data, obtained on many experimental animal models, showing that administration of exogenous NGF significantly reduces neurodegenerative events of BFCN support this hypothesis. Some controversy, however, exists about the cellular source and mechanisms of synthesis and release of NGF in the central nervous system (CNS).

Although originally discovered for its effects on differentiation, survival, neurite outgrowth, and neurotransmitter expression in the peripheral and central nervous systems, NGF is now known to affect cells of the immune and endocrine systems in a variety of ways. NGF influences B- and T-lymphocyte proliferation, is an autocrine survival factor for memory B lymphocytes, stimulates immunoglobulin production, and promotes human hemopoietic colony growth factor and differentiation (1, 3–5). NGF also appears to participate in the control of specific neuroendocrine functions. For example, the hypothalamic content of NGF increases following stressful events, while NGF stimulates the pituitary-adrenocortical axis, enhancing the secretion of adrenocorticotropic hormone and the concentration of plasma glucocorticoids (1, 3). Moreover, genes encoding NGF are expressed in the developing female rat hypothalamus,

Advances in Pharmacology, Volume 42

1054-3589/98 $25.00

while fetal rats exposed to anti-NGF antibodies displayed marked neuroendocrine deficits after birth (3). Based on these and other findings, it was hypothesized that NGF plays a role in maintaining a balanced interplay between the nervous, endocrine, and immune systems (4). Because autoimmune diseases are characterized by an abnormal activation of the immune system and by a deregulation of the endocrine and nervous systems, we sought to determine whether NGF is involved in these pathologies.

Studies reported from our laboratory indicate that NGF increases in several inflammatory autoimmune diseases (3). A high NGF level has been found in patients with systemic sclerosis, lupus erythematosus, and multiple sclerosis (MS), and in rodents affected by experimental allergic encephalomyelitis (EAE), an animal model of inflammatory disorder that is considered to closely resemble the human disease MS, and in rats with rheumatoid arthritis (see Fig. 1). We have also shown that pretreatment with NGF antibody reduces or prevents the development of arthritis induced by carrageenan, suggesting a functional role of NGF in this type of peripheral inflammation (3). Similar results were obtained with arthritic transgenic mice (Tg-m) expressing high levels of tumor necrosis factor-α (TNF-α) in knee joints. Following is a brief summarization of our results on EAE and Tg-m.

EAE was induced in a susceptible Lewis rat strain by injecting guinea pig spinal cord homogenate. The pathology is characterized by paralysis of limbs and inflammation within the thalamic, brain stem, and spinal cord tissues and associated with lymphocyte infiltration and gliosis (3). Rats affected by EAE express high levels of NGF protein in the thalamus and brain stem (350 ± 100 vs 540 ± 13pg/g of tissue, $p < .05$). The increase of NGF is associated with the presence of NGF-immunopositive astrocytes and oligodendrocytes in the

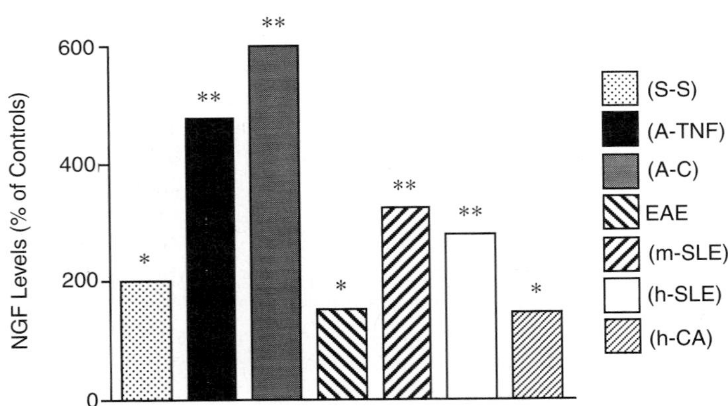

FIGURE 1 Levels of NGF in various inflammatory diseases expressed as percent of their controls: skin of humans affected by systemic sclerosis (S-S); arthritic TNF transgenic mice joint (A-TNF); synovinum of rats with arthritis induced by carrageenan (A-C); brain stem of rats with experimental allergic encephalomyelitis (EAE); mice with systemic lupus erythematosus (m-SLE); humans suffering systemic lupus erythematosus (h-SLE); humans suffering chronic arthritis (hCA): $^{*}p < .05$, $^{**}p < .01$.

thalamic microglia (Fig. 2) and brain stem regions. Interestingly, altered NGF levels also were found in the cerebrospinal fluid of MS patients (3). Because both EAE and MS are characterized by an overexpression of TNF, while *in vitro* studies indicated that TNF-α stimulates NGF production, it is likely that also *in vivo* TNF-α participates in the regulation of NGF synthesis.

To further verify this hypothesis, we have used Tg-m expressing high levels of TNF-α in the CNS. These mice, which spontaneously develop a chronic inflammatory demyelinating disease after the third postnatal week, display

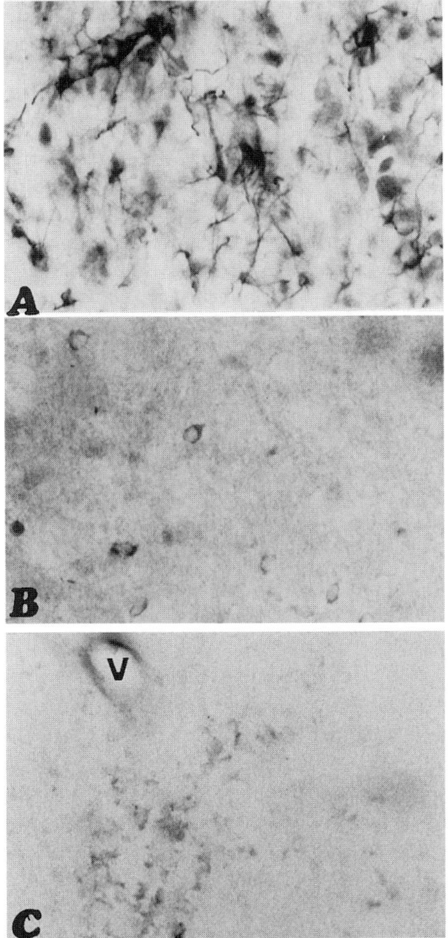

FIGURE 2 Immunohistochemical localization of NGF in (*A*) astrocyte, (*B*) oligodendrocytes, and (*C*) microglia in the thalamus of a rat 7 days after the induction of EAE with injection of guinea pig myelin. NGF positivity was also observed around blood vessel (v). No NGF immunoreactivity was found in glial cells in control rat brains or in brain of rats without clinical signs of the disease. Magnification, × 210.

progressively neurological symptoms ranging from mild tremors and ataxia to severe imbalances, seizures, and limb paresis (6).

Immunoenzymatic studies showed that the elevated presence of TNF-α in the brain of Tg-m differentially affected the constitutive presence of NGF. As compared with control mice, the levels of NGF significantly decreased in the HI (2220 $\pm$ 190 vs 1220 $\pm$ 43, $p < .001$), increased in the hypothalamus (210 $\pm$ 38 vs 1200 $\pm$ 860, $p < .01$) and brain stem (420 $\pm$ 32 vs 616 $\pm$ 45, $p < .01$), and remained unchanged in the cortex (1340 $\pm$ 230 vs 1320 $\pm$ 280). Why TNF-α differentially affects brain NGF levels is at present unknown. One possibility is that the local concentration of TNF-α is produced in different amounts within the various brain regions and that the effect of TNF on brain NGF levels is time- and dose-dependent. Thus, the local concentration of TNF in the hypothalamus and brain stem is able to stimulate NGF production, whereas in the HI, the level of TNF is too elevated, causing neurotoxic effect. The fact that TNF-α is expressed only a small area of the enthorinal cortex (6) could explain why the level of NGF in the cortex of Tg-m remains within baseline range.

To summarize, and also to answer the question posed in the title of this article, TNF-α should be considered as a positive and a negative mediator of NGF synthesis in inflammatory disease. Because both NGF and TNF are known to exert a wide range of effects in the nervous, endocrine, and immune systems, it is possible that an NGF/TNF-α interplay might provide useful insights for understanding mechanisms involved in autoimmune inflammatory diseases. The functional significance of NGF in autoimmune-based pathologies is currently under study.

Acknowledgments

This study was supported by project Sclerosi Multipla from the Istituto Superiore di Sanità, Roma, Italy. The author thanks R. Aloe for manuscript preparation.

References

1. Levi-Montalcini, R. (1987). The nerve growth factor 35 years later. *Science* **237,** 1154–1162.
2. Meakin, S. O., and Shoother, E. M. (1992). The nerve growth factor family of receptors. *Trends Neurosci,* **15,** 323–331.
3. Aloe, L., Skaper, S., Leon, A., and Levi-Montalcini, R. (1994). Nerve growth factor and autoimmune diseases. *Autoimmunity* **19,** 141–150.
4. Levi-Montalcini, R., Aloe, L., and Alleva, E. (1990). A role for nerve growth factor in nervous, endocrine and immune systems. *Prog. Neuroendocrinimmunol.* **3,** 1–10.
5. Torcia, M., Bracci-Laudiero, L., Lucibello, M., Nencioni, L., Labardi, D., Rubartelli, A., Cozzolino, F., and Aloe, L. (1996). Nerve growth factor is an autocrine survival factor for memory B lymphocytes. *Cell* **85,** 1–20.
6. Probert, L., Akassoglou, K., Pasparakis, M., Kontogeorgos, G., and Kollias, G. (1995). Spontaneous inflammatory demyelinating diseases in transgenic mice showing CNS-specific expression of TNF-α. *Proc. Natl. Acad. Sci. U.S.A.* **92,** 11294–11298.

Arun R. Wakade

Department of Pharmacology
School of Medicine
Wayne State University
Detroit, Michigan 48201

Multiple Transmitter Control of Catecholamine Secretion in Rat Adrenal Medulla

Secretion of adrenal medullary hormones is primarily controlled by the activity of splanchnic neurons innervating the chromaffin cells of this endocrine gland. Since 1934 (1) it has been accepted that these neurons are of cholinergic nature. Excitation of splanchnic neurons during exercise and stress and so on, releases acetylcholine (ACh), which then activates the cholinergic receptors of the chromaffin cells to stimulate secretion of catecholamines (CAs) and other biochemical functions of chromaffin cells. A number of recent studies carried out in different laboratories suggest that this classical concept of adrenal innervation and function needs to be modified. There is convincing new evidence that in addition to ACh, several peptides released in the adrenal medullary synapse are capable of affecting CA secretion, synthesis, and differentiation of chromaffin cells. This article briefly covers new evidence about peptidergic innervation of the rat adrenal medulla and the contribution of the peptides released from these neurons in the secretion and synthesis of CA.

A large number of experiments were carried out in isolated perfused adrenal gland of rat. A distinct advantage of this preparation was that secretion of CA could be evoked by excitation of splanchnic neurons as well as exogenous ACh (2). Perfusates of nonstimulated and stimulated adrenals were processed for peptide and CA estimations using radioimmunoassay and fluorometry or high-performance liquid chromatography coupled to electrochemical detection, respectively. Perfused adrenals were also used for *in situ* phosphorylation and activation of tyrosine hydroxylase (TH) experiments. Intact adrenal glands of rat were fixed for standard immunohistological detection of various peptides using specific antisera. In some experiments, rats were treated with capsaicin or monoclonal antibodies to acetylcholine esterase (AChE mab) prior to perfusion or fixation of adrenal glands.

Chromaffin cells of 3 to 4-week-old rats were cultured using standard methods and plated on glass coverslips for analysis of Ca^{2+} and exocytosis in a single cell. Cells loaded with Indo-1 dye were used for ratiometric estimation of intracellular Ca^{2+} by confocal microscopy. Carbon-fiber microelectrodes were used to record exocytosis from individual cells by amperometry. Details of all these techniques can be found in our previous publications.

Advances in Pharmacology, Volume 42

1054-3589/98 $25.00

I. Pharmacological and Physiological Evidence

If ACh was the only transmitter involved in the stimulation of CA secretion, then it was anticipated that antagonists of nicotinic and muscarinic receptors would block the secretion evoked by splanchnic nerve stimulation. Thus, 10 μM of mecamylamine plus 100 nM of atropine blocked CA secretion evoked by ACh. However, secretion evoked by splanchnic nerve stimulation was reduced but not blocked. For example, CA secretion evoked at high frequency (10 Hz) was reduced about 80%, but at low frequency (1Hz), the reduction was only 50% in the presence of antagonists. This substantial secretion in the face of effective cholinergic block was taken as a first indication that some noncholinergic transmitter released from neurons innervating the adrenal medulla was responsible for the secretion. The data also hinted that the contribution of noncholinergic substance was more prominent at lower than higher frequencies of stimulation (3).

Among a large list of potential candidates tested in the perfused adrenal gland, it was clearly found that only the members of secretin peptide family were effective in stimulating CA secretion. Thus, vasoactive intestinal polypeptide (VIP) and pituitary adenylate cyclase–activating polypeptide (PACAP) caused a marked increase in CA secretion. VIP was slightly less potent than ACh, but PACAP was almost 500 times more potent than ACh or VIP in stimulating the secretion.

II. Biochemical Evidence

Perfusates collected after various treatments were analyzed using radioimmunoassay for the peptides. Electrical stimulation of neurons produced an almost two- to threefold increase in VIP and PACAP contents of the perfusate over the nonstimulation period. When release of VIP was expressed as picomoles of peptide released per pulse, the frequency–release relationship showed more VIP release at low than high frequency. In chronically denervated adrenal glands, electrical stimulation of splanchnic nerves did not enhance the secretion of VIP. Further, stimulation of adrenal medulla by ACh did not secrete VIP. These data provided direct evidence for the secretion of VIP in the adrenal gland and that the origin of the secretory material was in the neuronal elements (4).

III. Immunohistochemical Evidence

Using specific antibodies for VIP and PACAP, it was possible to detect several immunoreactive nerve terminals that were positive for both peptides. Varicose and intense fluorescence were indicative of the presence of peptides in nerve terminals and not the chromaffin cells. Intravenous injection of AChE mab has been shown to cause destruction of peripheral autonomic cholinergic neurons. In such cholinergically denervated adrenals, VIP and PACAP immunoreactive nerve terminals could be still detected. These data indicated that ACh and peptidergic transmitters are most probably not located in the same neurons. Retrograde labeling with fluorogold combined with immunohistological studies showed that PACAP neurons originated in the dorsal root and nodose ganglia. This sensory origin of PACAP neurons was further supported using capsaicin to destroy sensory neurons. There was a marked reduction in PACAP immunoreactive fibers in adrenal glands of capsaicin-treated rats. These results showed for the first time that VIP- and PACAP-containing nerve terminals innervate chromaffin cells and that the cell bodies of these neurons are located in spinal cord as well as in sensory ganglia. Most important, ACh and these peptides are not colocalized in same splanchnic neurons (5).

IV. Physiological Actions of VIP and PACAP

Although VIP and PACAP stimulate CA secretion, the effect is not identical to that of ACh. ACh secretes epinephrine and norepinephrine in a ratio of about 4, but the peptides secrete mostly epinephrine. The ratio of epinephrine to norepinephrine is 10 for VIP and 7 for PACAP. More important, after electrical stimulation, the ratio was also around 6, indicating a significant contribution of peptides in secretion of epinephrine. These results taken together with earlier data suggest that peptides released during low neuronal activity (at rest) secrete epinephrine, whereas ACh secreted during high neural activity (stressful conditions) is responsible for norepinephrine secretion, which may be necessary to meet the demands of the stressful situation (6).

Biochemical studies aimed at examining phosphorylation and activation of TH, the rate-limiting enzyme of CA biosynthesis, showed that VIP enhanced phosphorylation of serine 41 residue of TH and also increased the catalytic activity of the enzyme. Although ACh caused phosphorylation of several serine moieties, there was not a concomitant increase in the activation of TH. Interestingly, electrical stimulation of splanchnic nerves resulted in phosphorylation of serine 41 and also activation of TH. PACAP has also been shown to behave similar to that of VIP in affecting the biochemical properties of TH. These studies offer new and clear evidence for a major role of adrenal neuropeptides in regulating the synthesis of CA (7).

V. Molecular Mechanism of CA Secretion by ACh and Peptide

Primary cultures of rat adrenal chromaffin cells were used to discover the mechanism of CA secretion by peptidergic and cholinergic agents. A brief application (500 ms) of nicotine immediately caused a rise in $[Ca^{2+}]$i and a burst of exocytotic events lasting for about 1 min. Muscarine application resulted in a delay in onset of both the rise in $[Ca^{2+}]$i and exocytosis. Muscarine-evoked responses lasted for several minutes. When Ca^{2+} was omitted from the ejection pipette, muscarine still increased $[Ca^{2+}]$i and exocytosis, but nicotine responses were completely blocked. Thse data provided direct support for the notion that muscarinic receptor stimulation mobilizes the internal pool of $Ca^{2+}$and nicotinic receptors increase the influx of external Ca^{2+} in chromaffin cells (8). After PACAP application, there was a consistent delay of about 7 sec in the onset of increase in [Ca^{2+}]i and exocytosis. A similar delay was observed for a maximum elevation of cyclic adenosine monophosphate (cAMP) in chromaffin cells treated with PACAP. The extent and duration of exocytosis was far greater than that of any other agent. PACAP utilized external Ca^{2+} for secretion. Ca^{2+} channel blocking agents, such as omega conotoxin, nifedipine, or cadmium, had very little effect in blocking PACAP responses, but a nicotine-mediated rise in [Ca^{2+}]i and exocytosis were fully blocked. On the other hand, zinc (100 uM) almost completely blocked exocytosis and [Ca^{2+}]i by PACAP but not by nicotine (9). These results suggest that PACAP stimulates secretion of CA by increasing the influx of extracellular Ca^{2+} involving the cAMP pathway. Although, the nature of Ca^{2+} channels involved in the influx of Ca^{2+} remains unknown at the present time, it is clear that ACh and PACAP utilize different mechanisms to evoke secretion of adrenal medullary hormones. It will be important to discover the alternate route used by the peptide to increase [Ca^{2+}]i and exocytosis.

References

1. Feldberg, W., Minz, B., and Tsudzimura, H. (1934). The mechanism of nervous discharge of adrenaline. *J. Physiol.* (London) **81**, 286–304.
2. Wakade, A. R. Studies on secretion of catecholamines evoked by acetylcholine or transmural stimulation of the rat adrenal gland. (1981). *J. Physiol.* **313**, 463–480.
3. Malhotra, R. K., and Wakade, A. R. Non-cholinergic component of rat splanchnic nerves predominates at low neuronal activity and is eliminated by naloxone. *J. Physiol.* (London) **383**, 639–652.
4. Wakade, T. D., Blank, M. A., Malhotra, R. K., Pourcho, R., and Wakade, A R. (1991). The peptide VIP is a neurotransmitter in rat adrenal medulla: physiological role in controlling catecholamine secretion. *J. Physiol.* (London) **444**, 349–362.
5. Dun, N. J., Tang, H., Dun, S. L., Huang, R., Dun, E. C., and Wakade, A. R. (1996). Pituatary adenylate cyclase activating polypeptide-immunoreactive sensory neurons innervate rat adrenal medulla. *Brain Res.* **716**, 11–21.
6. Xi, Guo, and Wakade, A. R. Differential secretion of catecholamines in response to peptidergic and cholinergic transmitters in rat adrenals. (1994). *J. Physiol.* (London) **475·3**, 539–545.
7. Haycock, J. W., and Wakade, A. R. Activation and multiple site phosphorylation of tyrosine hydroxylase in perfused rat adrenal gland. *J. Neurochem.* **58**, 57–64.
8. Xi, G., Przywara, D. A., Wakade, T. D., and Wakade, A. R. (1996). Exocytosis coupled to mobilization of intracellular calcium by muscarine and caffeine in rat chromaffin cells. *J. Neurochem.* **67**, 155–162.
9. Przywara, D. A., Xi, G., Angelilli, L., Wakade, T. D., and Wakade, A. R. (1996). A noncholinergic transmitter, pituatary adenylate cyclase activating polypeptide, utilizes a novel mechanism to evoke catecholamine secretion in rat adrenal chromaffin cells. *J. Biol. Chem.* **271**, 10545–10550.

Toshio Nishikimi
Division of Hypertension
National Cardiovascular Center
Osaka 565, Japan

Adrenomedullin in Cardiovascular Disease

The 52-residue vasoactive peptide adrenomedullin (AM), first isolated from human pheochromocytoma, has potent and long-lasting hypotensive effects (1). Immunoreactive AM has been detected in human plasma by a specific radioimmunoassay, and vascular smooth-muscle cells possess specific AM receptors that are functionally coupled to adenylate cyclase. AM mRNA is reportedly highly expressed in the adrenal gland, lung, kidney, and heart (1), suggesting that these organs actively produce and secrete AM in humans. However, whether these organs actually release AM into the circulation remains unknown. In study 1, to investigate the sites of production and degradation of AM in human subjects, we obtained blood samples from various sites during cardiac catheterization in 15 patients with ischemic heart disease and in five hypertensive patients who were suspected of having renovascular hypertension and measured immunoreactive AM concentrations. We also measured plasma AM levels in two patients with pheochromocytoma at rest and during hypertensive attacks. In study 2, to investigate the pathophysiological significance of AM in cardiovascular diseases, we measured AM concentrations in essential hypertension (n = 35), renal failure (n = 29), and heart failure (n = 66). Other humoral factors that were also measured simultaneously included norepinephrine (NE), atrial natriuretic peptide (ANP), and brain natriuretic peptide (BNP). In eight patients with severe heart failure, plasma AM levels were measured before and after the treatment. Study 3 was designed to determine the pathophysiological significance of AM in the pulmonary circulation in heart failure. To that end, we measured plasma concentrations of AM in the femoral veins, pulmonary artery, left atrium, and the aorta from 23 consecutive patients with rheumatic mitral stenosis who were undergoing percutaneous transvenous mitral commissurotomy (PTMC). We also simultaneously measured plasma concentrations of ANP and BNP and compared them with plasma concentrations of AM to characterize this peptide. In study 4, to investigate the pathophysiological significance of AM in acute myocardial infarction, we serially measured plasma AM levels in 31 patients with acute myocardial infarction over the time course of 4 wk.

In study 1, there were no significant differences in plasma AM concentrations in various sites of the right-sided circulation. There was no step-up of plasma AM levels in the coronary sinus. However, the plasma concentration of AM in aorta was slightly but significantly lower than in pulmonary artery (Fig. 1A). There was no difference of plasma AM levels between right and left renal veins and aorta.

Advances in Pharmacology, Volume 42

1054-3589/98 $25.00

The plasma concentration of AM in left adrenal vein tended to be higher than in aorta, but it was not significant (Fig. 1B). Plasma AM concentration did not increase at rest or even during a hypertensive attack in patients with pheochromocytoma, even though epinephrine and norepinephrine rose markedly.

In study 2, the plasma level of AM in normal subjects was 2.5 ± 0.3 pmol/liter (mean ± SD). Plasma AM levels were increased by 25% ($p < .05$) in essential hypertensive patients without organ damage and by 45% ($p < .01$)

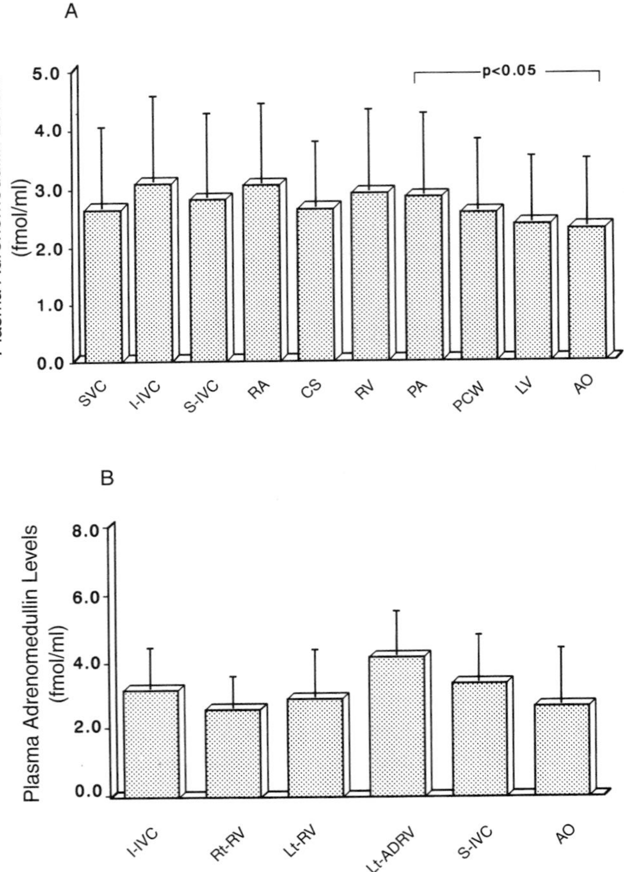

FIGURE 1 (*A*) Plasma AM levels in various sites in 15 patients with ischemic heart disease. Values are expressed as means ± SD. SVC, superior vena cava; I-IVC, infrarenal inferior vena cava; S-IVC, suprarenal inferior vena cava; RA, right atrium; CS, coronary sinus; RV, right ventricle; PA, pulmonary artery; PCW, pulmonary capillary wedge; LV, left ventricle; AO, aorta. (*B*) Plasma AM levels in the renal and adrenal veins in five patients with hypertension. Rt-RV, right renal vein; Lt-RV, left renal vein; Lt-ADRV, left adrenal vein. (Reprinted with permission from American Heart Association Scientific Publishing, Vol. 24, No. 5, 1994 by T. Nishikimi *et al.*)

in those with organ damage (Fig. 2). The increase of AM was more prominent in chronic renal failure than in hypertension. Renal failure patients with plasma creatine 1.5–.03, 3–6, and higher than 6 mg/dl had higher plasma AM levels than healthy subjects by 78% ($p < .05$), 131% ($p < .001$), and 214% ($p < .001$), respectively (see Fig. 2). Moreover, AM showed intimate correlations with norepinephrine ($r = .63, p < .001$), ANP ($r = .66, p < .001$), and BNP ($r = .46, p < .001$) in hypertension and renal failure. In heart failure, plasma levels of AM were unaffected in patients with heart failure in New York Heart Association (NYHA) class I (2.9 ± 0.6), but there was a significant and proportionate increase in the level in patients with heart failure more severe than NYHA class II (NYHA II: 3.5 ± 0.8; NYHA III: 4.8 ± 1.2; HYHA IV: 8.7 ± 3.4) (see Fig. 2). There were significant correlations between the plasma level of AM and those of norepinephrine, ANP, and BNP (norepinephrine: $r = .62, p < .001$; ANP: $r < .70, p < .001$; BNP: $r = .69, p < .001$). Plasma adrenomedullin levels significantly decreased after the treatment in severe heart failure patients (from 7.40 ± 3.40 to 3.98 ± 1.00, $p < .05$).

In study 3, patients with mitral stenosis had significantly higher AM concentrations than age-matched normal controls (3.9 ± 0.3 pmol/liter, $p < .001$). There was a significant reduction of plasma AM concentrations from the pulmonary artery to the left atrium (3.8 ± 0.2 vs 3.2 ± 0.4, pmol/liter, $p < .001$). These concentrations were significantly correlated with ANP ($r = .54, p < .01$), BNP ($r = .46, p < .05$), mean pulmonary artery pressure ($r = .65, p < .001$), total pulmonary vascular resistance ($r = .83, p < .0001$), and pulmonary vascular resistance ($r = .65, p < .001$). Plasma AM levels did not change 20 min after PTMC.

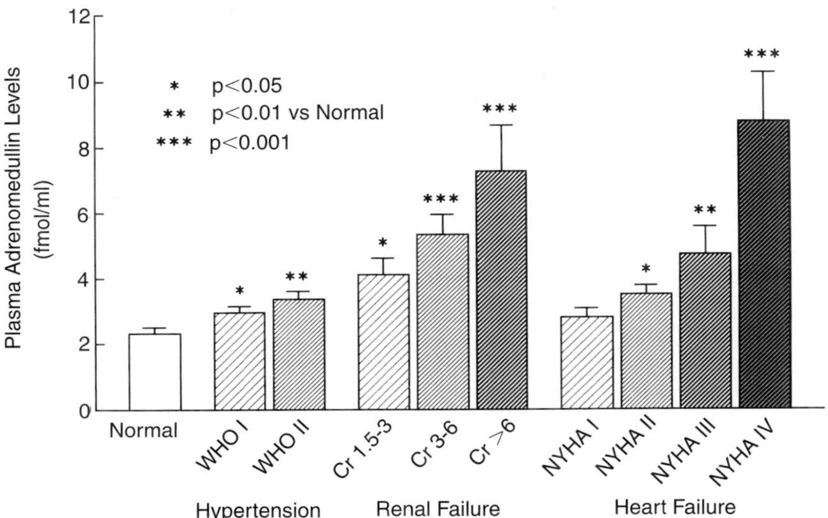

FIGURE 2 Plasma AM levels in normal subjects and patients with hypertension, renal failure, and heart failure.

In study 4, in patients with acute myocardial infarction, plasma AM levels reached their peak (7.0 ± 0.9) at 24 hr, remained increased within 1 wk, and gradually decreased thereafter. Plasma AM levels returned to normal levels at 4 wk after the onset of acute myocardial infarction. Plasma AM levels positively correlated with ANP ($r = .40$, $p < .001$) and BNP ($r = .53$, $p < .001$). The plasma AM levels in Killip's class II, III, and IV were higher than that of Killip's class I. Plasma AM levels on admission positively correlated peak creatinine kinase levels and left ventricular end-diastolic volume index and negatively correlated with left ventricular ejection fraction. Plasma AM levels from 12 to 48 hr negatively correlated with systemic vascular resistance index.

The results in study 1 suggest that the adrenal glands, lungs, kidneys, and heart may not be the main sources of circulating AM, despite the fact that these organs highly express AM mRNA. The pulmonary circulation, however, may be one of the sites of AM clearance (2). Adrenal glands do not greatly contribute to circulating AM levels in humans, despite AM having been discovered in pheochromocytoma and named accordingly (2). These findings were supported by the later report, which showed that vascular endothelium and smooth-muscle cell actively synthesize and secrete AM *in vitro,* suggesting that vascular wall is a main production site of AM (3, 4).

The results in study 2 suggest that plasma levels of AM are elevated in hypertension, renal failure, and heart failure in proportion to each condition's severity. Furthermore, plasma AM is supposed to increase in association with increased plasma volume and an activated sympathetic nervous system in these disorders. Considering that AM is synthesized and secreted by vascular endothelial cells and vascular smooth-muscle cell *in vitro,* and that it exerts potent cardiovascular effects, a possible role of this peptide is suggested in the pathophysiology of hypertension, renal failure, and heart failure (5, 6).

The results in study 3 suggest that plasma concentrations of AM were significantly elevated in patients with mitral stenosis compared with normal controls, and we observed significant relationships between plasma concentrations of AM and mean pulmonary artery pressure, pulmonary vascular resistance, and total pulmonary resistance. Findings suggest the involvement of AM in the pathophysiology of pulmonary hypertension in patients with mitral stenosis. Given that the vascular wall mainly produces AM, that AM-specific receptors are present on smooth-muscle cells, and that AM acts on the pulmonary vascular bed, increased plasma levels of AM in patients with mitral stenosis may help to prevent the increase in pulmonary arterial resistance in patients with secondary pulmonary hypertension.

The results in study 4 suggest that plasma AM levels are increased markedly in the early phase of acute myocardial infarction in proportion to the clinical severity. Increased plasma AM levels in patients with acute myocardial infarction may be involved in the defense mechanism against further elevation in peripheral vascular resistance.

In conclusion, plasma AM levels are increased in proportion to each condition's severity in many cardiovascular diseases, such as hypertension, renal failure, heart failure, and acute myocardial infarction. Although the precise mechanism by which AM is synthesized and released remains to be elucidated, increased AM may be involved in the defense mechanism against further elevation of blood pressure or fluid retention.

Acknowledgments

This work was supported in part by Special Coordination Funds for Promoting Science and Technology (Encouragement System of COE) from the Science and Technology Agency of Japan, the Ministry of Health and Welfare, and the Human Science Foundation of Japan.

References

1. Kitamura, K., Kangawa, K., Kawamoto, M., Ichiki, Y., Nakamura, S., Matsuo, H., and Eto, T. (1993). Adrenomedullin: A novel hypotensive peptide isolated from human pheochromocytoma. *Biochem. Biophys. Res. Commun.* **192,** 553–560.
2. Sugo, S., Minamino, N., Kangawa, K., Miyamoto, K., Kitamura, K., Sakata, J., Eto, T., and Matsuo, H. (1994). Endothelial cells actively synthesize and secrete adrenomedullin. *Biochem. Biophys. Res. Commun.* **201,** 1160–1166.
3. Sugo, S., Minamino, N., Shoji, H., Kangawa, K., Kitamura, K., Eto, T., and Matsuo, H. (1994). Production and secretion of adrenomedullin from vascular smooth cells: Augmented production by tumor necrosis factor α. *Biochem. Biophys. Res. Commun.* **203,** 719–726.
4. Nishikimi, T., Kitamura, K., Saito, Y., Shimada, K., Ishimitsu, T., Takamiya, M., Kangawa, K., Matsuo, H., Eto, T., Omae, T., and Matsuoka, H. (1994). Clinical studies for the sites of production and clearance of circulating adrenomedullin in human subjects. *Hypertension* **24,** 600–604.
5. Ishimitsu, T., Nishikimi, T., Saito, Y., Kitamura, K., Eto, T., Kangawa, K., Matsuo, H., Omae, T., and Matsuoka, H. (1994). Plasma levels of adrenomedullin, a hypotensive peptide, in patients with hypertension and renal failure. *J. Clin. Invest* **94,** 2158–2161.
6. Nishikimi, T., Saito, Y., Kitamura, K., Ishimitsu, T., Eto, T., Kangawa, K., Matsuo, H., Omae, T., and Matsuoka, H. (1995). Increased plasma levels of adrenomedullin in patients with heart failure. *J. Am. Coll. Cardiol.* **26,** 1424–1431.

Gal Yadid,* Yigal Fraenkel,* and Eliahu Golomb†

*Department of Life Science
Bar-Ilan University
Ramat-Gan, Israel

†Department of Pathology
Tel-Aviv University
Tel Aviv, Israel

Strychnine, Glycine, and Adrenomedullary Secretion

The convulsive alkaloid strychnine has been established as a specific antagonist of the postsynaptic receptor for the inhibitory neurotransmitter, glycine, in the central nervous system. A strychnine recognition site appears to be located on the 48-kDa α subunit of the glycine receptor (GlyR) (1), which is homologous to the nicotinic receptor family (2). The two tyrosine residues separated by four amino acids (tyr 197 and 202), which have been predicted to participate in strychnine binding to the GlyR (1), parallel the two tyrosines (tyr 198 and 203) included in the nicotinic acetylcholine receptor (AChR) of many species, suggesting that strychnine might also bind to this site (3). It has been shown in different natural as well as recombinant systems that strychnine depresses nicotinic–cholinergic responses through interaction with nicotinic receptors (4).

The chromaffin cell has been extensively studied as a cholinergic neurosecretory system, and activation of the nicotinic receptor in these cells leads to the Ca^{2+} dependent exocytotic release of catecholalmines, adenosine triphosphate (ATP), chromogranins, and enkephalins. These cells possess a high-affinity [^{3}H]strychnine binding site (5), and glycine can evoke catecholamine release from adrenomedullary tissue *in vitro* (8) and *in vivo,* as demonstrated by the use of microdialysis techniques (6). The α_3 subunit of glycine receptors (but not the α_1 or α_2) is expressed in rat adrenal and may mediate this effect (6).

Strychnine inhibits the nicotinic stimulation of catecholamine release from cultured bovine adrenal chromaffin cells in a concentration-dependent (1–100 μM) manner. At 10 μM of nicotine, the IC 50 value for strychnine is approximately 30 μM. Strychnine also inhibits the nicotine-induced membrane depolarization and increase in intracellular Ca^{2+} concentration. The inhibitory action of strychnine is reversible and is selective for nicotinic stimulation, with no effect observed on secretion elicited by a high external K^+ concentration, histamine, or angiotensin II (4).

Nuclear magnetic resonance (NMR) techniques, based on the averaging properties brought about by rapid chemical exchange, allow one to obtain quantitative information concerning receptor–agonist complexes. NMR selective relaxation times (T_{1s}) for monitoring ligand interactions are particulary sensitive to binding of small ligands to macromolecules (7). In a system in

Advances in Pharmacology, Volume 42

1054-3589/98 $25.00

which the total concentration of ligand is in a large excess relative to the receptor concentration, the proportion of ligand bound is very small. In such systems, one can express the longitudinal relaxation (T_1) of the ligand's protons by the following equation:

$$1/T_{1obs} - 1/T_{1free} = f/(T_{1bound} + \tau_{1bound}),$$

where T_{1obs} is the observed relaxation time, T_{1free} is the relaxation time of the free ligand (no binding), f is the fraction of bound ligand, and T_{1bound} and τ_{1bound} are the relaxation and lifetime of the bound state, respectively. This equation is correct for both nonselective (T_{1ns}) and selective (T_{1s}) longitudinal relaxation times. It is apparent that the expressions rates for the two distinct physical situations differ in the frequency-independent term τ_c, the tumbling rate, which is a determinant of the relaxation of T_{1s} (3, 7).

All T_{1s} measurements were done on a VARIAN XL series 300-Mhz NMR instrument using the inversion-recovery pulse sequence. The desired resonance was inverted with a 7-ms pulse from the decoupler. The accumulative values from 16 pulses were determined at intervals of 5 times the estimated T_{1s}. This procedure was repeated at eight different intervals between the inversion and reading pulses. From these data, T_{1s} was determined for the acetyl group of ACh in the absence of membrane, after addition of bovine adrenal medulla membrane and after the addition of the indicated ligand. ACh concentration was 4 mM, and the concentration of the displacing ligands varied in the different experiments: glycine in the milimolar range and strychnine and hexamethonium in the micromolar range. Each experiment was ended by adding a threefold excess of hexamethonium to determine nonspecific binding (T_{1free}).

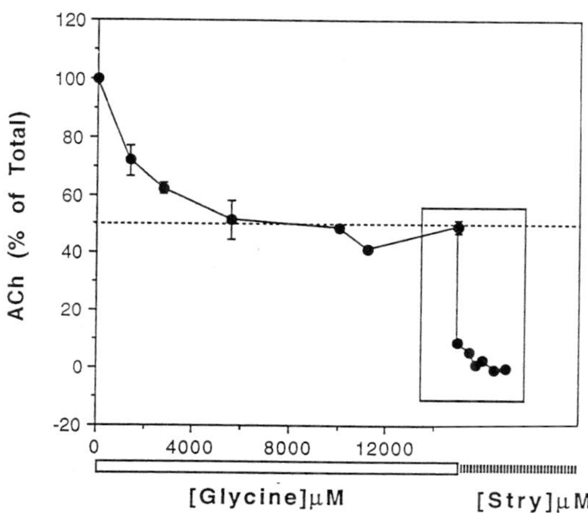

FIGURE 1 The effect of glycine and strychnine on the specific ACh binding to the adrenal medulla membranes by monitoring the T_{1s} of ACh. Glycine (up to a concentration of 15 mM) displaced 50% of the specifically bound ACh. Further addition of strychnine (20–80 μM) displaced ACh totally.

Strychnine and glycine specifically interact with ACh-binding site(s) in bovine adrenal medulla membranes. Strychnine completely displaced ACh from its binding sites with an apparent Km of 1.8 μM and a Hill coefficient (n_H) of 0.97, compared with hexamethonium, which showed two apparent Km values (1.72 and 5.6 μM) and an n_H of 2.1. Glycine was able to displace only 50% of the specific bound ACh, with a Km of 5.9 mM and an n_H of 0.65 (Fig. 1).

These data suggest that strychnine and glycine interact with the agonist-binding site of the nicotinic AChR in chromaffin cells, thus exerting a pharmacological effect that may have a modulatory role on the AChR.

References

1. Ruiz-Gomez, A., Morato, E., Garcia-Calvo, M., Valdivieso, F., and Mayor, F. Jr. (1990). Localization of strychnine binding site on the 48-kilodalton subunit of the glycine receptor. *Biochemistry* **29**, 7033–7040.
2. Grenningloh, G., Rienitz, A., Schmitt, B., Methfessel, C., Zensen, M., Beyreuther, K., Gundelfinger, E. D., and Betz, H. (1987). The strychnine-binding subunit of the glycine receptor shows homology with nicotinic acetylcholine receptor. *Nature* **328**, 215–220.
3. Fraenkel, Y., Navon, G., Aronheim, A., and Gershoni, J. M. (1990). Direct measurements of agonist binding to genetically engineered peptides of the acetylcholine receptor by selective T1 NMR relaxation. *Biochemistry* **29**, 2617–2622.
4. Kuijpers, G. A. J., Leoncio, A., Soledad, C., and Yadid, G. (1994). Strychnine inhibits acetylcholine receptor activation in bovine adrenal medullary chromaffin cells. *Br. J. Pharmacol.* **113**, 471–478.
5. Yadid, G., Youdim, M. B. H., and Zinder, O. (1989). High affinity strychnine binding to adrenal medulla chromaffin cell membranes. *Eur. J. Pharmacol.* **175**, 365–366.
6. Yadid, G., Goldstein, D. S., Pacak, K., Kopin, I. J., and Golomb, E. (1995). Functional α_3-glycine receptors in rat adrenal. *Eur. J. Pharmacol.* **288**, 399–401.
7. Valensin, G., Kushnir, T., and Navon, G. (1982). Selective and nonselective proton spin-lattice relaxation studies of enzyme-substrate interactions. *J. Magn. Reson.* **46**, 23–29.
8. Yadid, G., Zinder, O., and Youdim, M. B. H. (1992). Preferred release of epinephrine by glycine from adrenal chromaffin cells. *Eur. J. Pharamacol.* **221**, 389–391.

J. W. M. Lenders,* M. C. Jacobs,* Th. Thien,* and D. S. Goldstein†

*Department of Medicine
Division of General Internal Medicine
University Hospital Nijmegen
Nijmegen, The Netherlands 6525 GA

†Clinical Neuroscience Branch
National Institute of Neurological Disorders and Stroke
National Institutes of Health
Bethesda, Maryland 20892

Catecholamines and Neurocardiogenic Syncope

The common faint is the most frequent cause of neurocardiogenic or vasodepressor syncope. Examples of other types of vasodepressor syncope are the carotid sinus syndrome, cough syncope, and micturation syncope. The common faint mostly occurs in otherwise healthy people during exposure to emotional stress, heat, prolonged upright posture, or pain. The hemodynamic alterations at the time of fainting are characterized by systemic vasodilation, with or without vagally mediated bradycardia. Cardiac output does not increase but may even decrease slightly despite the systemic vasodilation. This vasodilation is caused by a sudden decrease in sympathetic nervous outflow, as has been shown by the microneurographic technique. Several studies have shown that this sympathoinhibition is reflected by a decrease in plasma norepinephrine levels or by an absence of an appropriate increase in plasma norepinephrine levels at the time of fainting. It has also been shown previously that this is caused by a decrease in the neuronal release of norepinephrine (1). In contrast, plasma epinephrine levels are frequently increased at the time of fainting.

Much less is known about what happens with the plasma or urinary catecholamine levels before the onset of fainting. This is probably due to the unexpectedness and unpredictability of vasodepressor reactions. Chosy and Graham (2) demonstrated higher epinephrine excretions in urine of blood donors that was collected before they had a faint at the time of blood donation. Vingerhoets (3) showed an increased plasma epinephrine and norepinephrine level before a syncope in one subject. In another case report, Goldstein *et al.* (4) could also demonstrate an increase in plasma epinephrine long before a faint. In a larger study, Sra *et al.* (5) reported on increased plasma epinephrine levels before the actual onset of the syncope during head-up tilt testing. At the time of syncope, the plasma epinephrine levels increased to an even larger extent. We have studied the plasma catecholamine responses and the plasma norepinephrine kinetics in healthy subjects who were exposed to low- and high-intensity lower-body negative pressure (LBNP) (6). Low-intensity LBNP at less

Advances in Pharmacology, Volume 42

1054-3589/98 $25.00

than −20 mm Hg reduces central venous pressure without altering arterial blood pressure or pulse pressure, indicating relative selective inhibition of cardiac baroreceptors, whereas high-intensity LBNP of more than −20 mm Hg decreases blood pressure and pulse pressure, thereby unloading both the cardiac and arterial baroreceptors. We compared the responses of plasma catecholamines and plasma norepinephrine kinetics to low-intensity LBNP in subjects who did versus those who did not develop a vasodepressor reaction during high-intensity LBNP, in order to identify possible neurochemical alterations before the onset of vasodepressor syncope. The radiotracer dilution technique with infusion of tritiated norepinephrine was used to calculate total body and forearm spillover and clearance of norepinephrine. High-performance liquid chromatography with fluorometric detection was used as assay for the endogenous and radiolabeled catecholamines

Of the 26 subjects who participated in the study, 15 subjects underwent 30 min of LBNP at −15 mmHg as well as at −40 mm Hg without any signs of a vasodepressor spell. The remaining 11 subjects developed a vasodepressor reaction during LBNP at −40 mm Hg but also showed no symptoms of vasodepression during LBNP at −15 mm Hg. Baseline levels of venous and arterial plasma norepinephrine were similar in the two groups, as were the baseline arterial plasma epinephrine levels. During LBNP at −15 mm Hg, venous plasma norepinephrine levels increased significantly in both groups, with a tendency for smaller venous plasma norepinephrine responses in the group with subsequent syncope. In both groups, arterial plasma epinephrine levels increased significantly during low-intensity LBNP at −15 mm Hg; however, in the group that did not faint at high-intensity LBNP at −40 mm Hg, arterial epinephrine levels plateaued at about 0.22 nmol · liter^{-1} after 10 min, whereas in the group that did faint, arterial epinephrine levels progressively rose to a peak value of 0.34 ± 0.05 nmol · liter^{-1} (Fig. 1). The absolute increments in arterial epinephrine levels were larger in the group with than in the group without subsequent

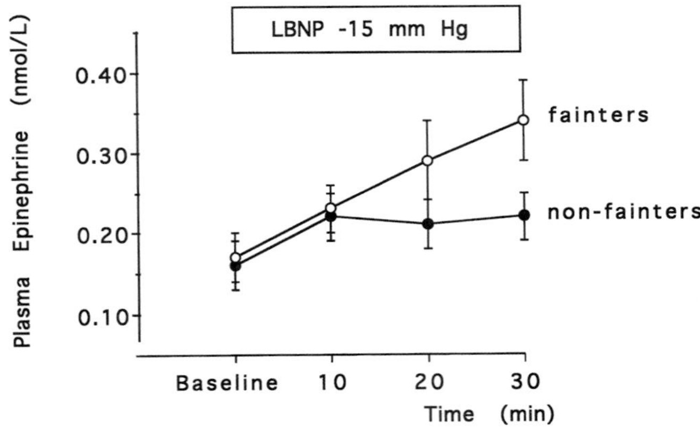

FIGURE 1 The plasma epinephrine concentrations (mean ± SEM) in the fainters and nonfainters at baseline and at 10, 20, and 30 min of low-intensity LBNP (LBNP at −15 mm Hg).

syncope. In the 16 out of 26 subjects who received a [^{3}H]norepinephrine infusion, the average response of forearm norepinephrine spillover across all time points during LBNP at −15 mm Hg was larger in the non-fainting subjects (0.31 ± 0.13 pmol · min^{-1} · 100 ml^{-1}) than in the fainting subjects (-0.06 ± 0.04 pmol · min^{-1} · 100 ml^{-1}, $p < .05$). There were no differences for total body norepinephrine spillover or clearance between both groups. During high-intensity LBNP at −40 mm Hg, venous and arterial plasma norepinephrine levels increased significantly in both groups, and the responses were not different between the groups. Arterial plasma epinephrine levels increased in the non-fainting subjects, from 0.21 ± 0.02 nmol · liter^{-1} to 0.39 ± 0.04 nmol · liter^{-1} ($p < .01$), and increased markedly in the subjects who fainted, from 0.30 ± 0.04 nmol · liter^{-1} to 1.24 ± 0.44 nmol · liter^{-1} ($p < .01$), after 10 minutes of LBNP. Total body and forearm norepinephrine spillovers increased significantly during LBNP at −40 mm Hg in the non-fainters, while these increments were not significant in the group that had a faint.

These findings indicate that the simultaneous occurrence of attenuated noradrenergic and exaggerated adrenomedullary activity to a mildly decreased cardiac filling (LBNP at −15 mm Hg) is characteristic for subjects who are prone to vasodepressor reactions during exposure to more severely decreased cardiac filling. It is unclear whether the increased arterial plasma epinephrine levels in the fainters, at a time when there is not yet any sign of syncope, elicits arterial vasodilation and contributes in that way to the later development of the syncope. The attained plasma epinephrine levels seem, however, not high enough to induce vasodilation, but it is possible that vasodilation by epinephrine becomes manifest when there is a simultaneous attenuation of the noradrenergic responsiveness or even sympathoinhibition. In addition, venous vasodilation in the splanchnic and skeletal muscle vascular bed would enhance the decrease in cardiac filling. An alternative way by which epinephrine could play a role in the pathogenesis of a syncope is by its inotropic effects on the ventricular wall, thus stimulating the inhibitory afferent C-fibers. It has been shown in animal experiments that stimulation of these fibers will lead to vasodilation, because of sympathoinhibition, and to bradycardia, both being the major hemodynamic alterations during vasodepressor syncope. On the other hand, infusion of a vasodilator has been reported to evoke a vasodepressor syncope in a heart transplant recipient who lacked cardiac innervation. Heart transplant recipients do not differ from healthy control subjects in their susceptibility to LBNP-induced vasodepression. This suggests that the occurrence of vasodepressor syncope in humans may not require neuronal afferent input from ventricular baroreceptors.

There are two other arguments that support a key role for epinephrine. First, it is well documented that the β-adrenergic agonist isoproterenol can enhance the diagnostic sensitivity of head-up tilt testing. This suggests that either the β_2-adrenoceptor–mediated vasodilation or the β_1-adrenoceptor–mediated inotropic effect of isoproterenol is a requirement for the development of syncope. Second, it has been shown the β-adrenoceptor antagonists are efficient in preventing syncopes.

An important question that arises is: what is the cause of the simultaneous occurrence of enhanced adrenomedullary and attenuated noradrenergic responses before there is any sign of syncope? One explanation is a central neural process, arising from sudden resetting of baroreceptors. Central processes involving vasopressin or endogenous opioids may also operate in some vasodepressor reactions. Vasovagal syncope frequently occurs in emotionally distressing circumstances. Individuals who developed a vasodepressor reaction had enhanced plasma epinephrine responses. A sudden central resetting of baroreflexes may also be responsible for the abrupt sympathoinhibition. One possibility is that hypotension fails to release sympathoneural outflows from baroreceptor restraint. Because endogenous opioids augment the extent of sympathoinhibition during a hypotensive stimulus, sudden release of endogenous opioids in the brain could contribute to baroreflex resetting and, thereby, produce sympathoinhibition despite hypotension. Vasopressin also augments baroreflex restraint of sympathoneural outflows, and extremely high circulating vasopressin levels often accompany vasodepressor syncope. Other mechanisms for the sudden sympathoinhibition include stimulation of ventricular baroreceptors, which increase activity of inhibitory afferent C-fibers, and collapse-firing of low pressure baroreceptors in the right atrium and great veins (7). The final feature of the central and peripheral hypothesized mechanisms is sympathoinhibition despite hypotension.

In summary, the combination of enhanced adrenomedullary and attenuated sympathoneural responses during slight reductions in cardiac filling antecede the syncope. The findings are consistent with the view that sudden, central resetting of baroreflex function can evoke a pattern of sympathoinhibition combined with adrenomedullary and vagal activation. The role of epinephrine in the development of vasodepressor reactions deserves further research.

References

1. Esler, M., Jennings, G., Lambert, G., Meredith, I., Horne, M., and Eisenhofer, G. (1990). Overflow of catecholamine neurotransmitters to the circulation: Source, fate, and functions. *Physiol. Rev.* **70**, 963–985.
2. Chosy, J. J., and Graham, D. T. (1965). Catecholamines in vasovagal fainting. *J. Psychosom. Res.* **9**, 189–194.
3. Vingerhoets, A. J. J. M. (1984). Biochemical changes in two subjects succumbing to syncope. *Psychosom. Med.* **46**, 95–102.
4. Goldstein, D. S., Spanarkel, M., Pitterman, A., Toltzis, R., Gratz, E., Epstein, S., and Keiser, H. R. (1982). Circulatory control mechanisms in vasodepressor syncope. *Am. Heart. J.* **104**, 1071–1075.
5. Sra, J. S., Murthy, V., Natale, A., Jazayeri, M. R., Dhala, A., Deshpande, S., Sheth, M., and Akhtar, M. (1994). Circulatory and catecholamine changes during head-up tilt testing in neurocardiogenic (vasovagal) syncope. *Am. J. Cardiol.* **73**, 33–37.
6. Jacobs, M. C., Goldstein, D. S., Willemsen, J. J., Smits, P., Thien, Th., Dionne, R. A., and Lenders, J. W. M. (1995). Neurohumoral antecedents of vasodepressor reactions. *Eur. J. Clin. Invest.* **25**, 754–761.
7. Dickinson, C. J. (1993). Fainting precipitated by collapse-firing of venous baroreceptors. *Lancet* **342**, 970–972.

Gloria Feuerstein and Robert R. Ruffolo, Jr.

Cardiovascular Pharmacology
Smith Kline Beecham Pharmaceuticals
King of Prussia, Pennsylvania 19406–0939

β-Blockers in Congestive Heart Failure: The Pharmacology of Carvedilol, a Vasodilating β-Blocker and Antioxidant, and Its Therapeutic Utility in Congestive Heart Failure

Congestive heart failure (CHF) is the most rapidly growing cardiovascular disorder in developed countries, and it is the leading cause of hospitalization in the elderly. In the United States, it is estimated that 3–4 million people have CHF, which is a major cause of cardiovascular morbidity and mortality. The prognosis for patients with CHF is poor, with an overall survival rate of 50% in 5 years, which decreases to less than 1 year in those patients with New York Heart Association class III and IV CHF. Current therapy for CHF involves the administration of several drugs, including diuretics, digitalis, and angiotensin–converting enzyme inhibitors, with only the latter producing a significant, but modest, reduction in mortality.

The pathological abnormalities that occur in CHF are multiple and complex. As the left ventricle fails and cardiac output decreases, a series of reflexes are activated that result in sodium and water retention, peripheral vasoconstriction, and activation of several neurohormonal systems, including, among others, the renin-angiotensin system and the sympathetic nervous system. The ability of the angiotensin-converting enzyme inhibitors, which block the renin-angiotensin system, to reduce morbidity and mortality, albeit to only a limited extent, has generated interest in inhibiting neurohormonal systems as a strategy to treat CHF. Of the neurohormonal systems that are activated in CHF, the sympathetic nervous system may have the broadest implications. This hypothesis is based on the direct toxicity that the catecholamines released by the sympathetic nervous system have on the myocardium, as well as the ability of norepinephrine to increase peripheral vascular resistance and to activate the renin-angiotensin system. The generalized activation of the sympathetic nervous system in CHF is inferred from the high levels of catecholamines that occur in the circulation of patients with CHF and that are highly correlated with disease severity and the probability

Advances in Pharmacology, Volume 42

1054-3589/98 $25.00

of mortality. In addition, the chronic activation of the sympathetic nervous system in CHF results in an increase in myocardial work and oxygen demand, which may trigger serious and life-threatening ventricular arrhythmias. Furthermore, the ischemia induced by activation of the sympathetic nervous system may also enhance myocardial apoptosis, or programmed cell death, which is thought to be involved in the progression of CHF.

These deleterious actions of the sympathetic nervous system in CHF provided the original rationale for clinical trials with β-blockers in CHF. The early clinical trials by Waagstein and coworkers (1) provided encouraging results with respect to the tolerability and potential clinical benefits of β-blockers in CHF, and led to subsequent evaluations of various β-blockers in CHF (2, 3). Most of these trials were small, short in duration, poorly controlled, and inconsistent with respect to the clinical endpoints that were evaluated. Most importantly, a majority of these trials were not designed to explore the effects of the β-blockers on mortality.

More recently, a large, multicenter, double-blind, placebo-controlled clinical trial of carvedilol, a vasodilating β-blocker with antioxidant activity, was conducted in patients with mild-to-severe CHF (4). In these studies, carvedilol was associated with a 65% reduction in mortality ($p < .0001$), and a significant reduction in hospitalization (4).

I. Pharmacology of Carvedilol

Carvedilol (Fig. 1) is a multiple-action drug that has the capacity to suppress a variety of neurohormonal systems. The pharmacological actions of carvedilol have been studied in detail and reviewed previously (5). The major pharmacological actions of carvedilol can be summarized as follows: (1) Carvedilol is a potent competitive antagonist of both β_1 and β_2 adrenoceptors, with no intrinsic

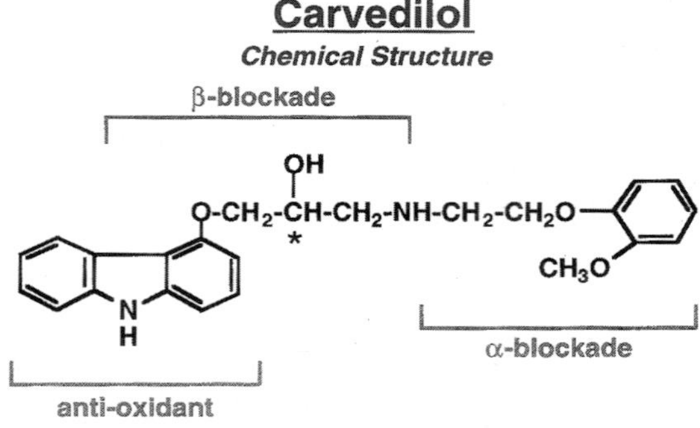

FIGURE 1 Chemical structure of carvedilol.

sympathomimetic activity; (2) carvedilol is a potent antagonist of α_1 adrenoceptors that is responsible for the vasodilating effects of the drug; and (3) carvedilol is an extremely potent antioxidant as a result of its carbazole moiety (see Fig. 1). The antioxidant activity of carvedilol may be of particular importance in CHF inasmuch as oxygen radical stress has been shown to occur in CHF and may be responsible for producing damage to the myocardium and inducing apoptosis.

The cardioprotective effects of carvedilol have been studied extensively in a variety of experimental animal models of myocardial ischemia. The results of these investigations can be summarized as follows: (1) Carvedilol consistently reduced myocardial infarction induced by permanent or transient cardiac ischemia, and this effect was noted when carvedilol was administered either before or after the ischemic insult; (2) carvedilol produced cardioprotection in no less than 12 separate studies involving five different species (i.e., rat, rabbit, cat, dog, pig); (3) in all studies in which a direct comparison was made between carvedilol and propranolol (a prototypic nonselective β-blocker comparable in potency to carvedilol), carvedilol consistently produced greater cardioprotection at comparable β-blocking doses (5); in fact, in some studies (e.g., cat, pig), carvedilol nearly completely (90–98%) prevented ischemia-induced cardiac necrosis and arrhythmias; and (4) the unusually high degree of cardioprotection produced by carvedilol strongly suggests that actions of the drug in addition to β blockade (i.e., antioxidant effect) may contribute to the cardioprotective effects of carvedilol (5).

The ability of carvedilol to produce cardioprotection to a degree that is greater than that achieved with β-blockers has been attributed to the antioxidant effects of the drug (5). The antioxidant properties of carvedilol have been characterized in detail and may be summarized as follows: (1) Carvedilol is a direct antioxidant that blocks free electron transfer in both lipid and aqueous media; (2) carvedilol prevents lipid peroxidation in biological membranes (e.g., cardiac membranes) subjected to oxidative stress induced by chemical, enzymatic, and cellular sources of oxygen radicals (3) carvedilol prevents the consumption of endogenous antioxidants, such as vitamin E and glutathione, in organs subjected to oxidative stress; (4) carvedilol is metabolized to several hydroxycarbazole derivatives that are 50 to 100-fold more potent than carvedilol as antioxidants and approximately 1000-fold more potent than vitamin E; (5) carvedilol produces antioxidant effects at concentrations that are consistent with the blood levels of the drug obtained clinically; and (6) carvedilol has been shown to be an antioxidant *in vivo* in both animals and humans. The antioxidant effects described for carvedilol are not observed with any other β-blocking agents.

In addition to the antioxidant effects of carvedilol that likely contribute to the cardioprotective effects of the drug, carvedilol has also been shown to produce a number of other actions that may be secondary to the antioxidant effects of the drug. Thus, carvedilol has the capacity to (1) suppress leukocyte infiltration into injured myocardium by blocking the gene expression of ICAM-1, (2) inhibit ischemia-induced apoptosis in the heart, (3) suppress life-threatening ventricular arrhythmias, (4) inhibit left ventricular remodeling, and (5) block vascular remodeling through the inhibition of vascular smooth-muscle cell migration and proliferation. These additional actions of carvedilol may all be relevant to the marked reduction in morbidity and mortality produced by the drug in patients with CHF.

II. Clinical Efficacy of Carvedilol in CHF

The clinical effectiveness of carvedilol when added to standard therapy for CHF has been demonstrated consistently in many single and multicenter clinical trials (6–9). Clinical and hemodynamic benefits produced by carvedilol have been demonstrated in patients with mild, moderate, and severe CHF and in patients with ischemic or nonischemic origins of heart failure. Moreover, carvedilol produced dose-dependent reductions in morbidity and mortality (4). Most notably, in the large U.S. multicenter clinical trial of carvedilol in CHF, the drug was associated with a reduction in all-cause mortality of 65%, as well as a significant reduction in hospitalization (4). These observations represent a breakthrough in the therapy of CHF and may open a new era in the pharmacological management of this progressive disease.

The positive outcome of the clinical trials of carvedilol in CHF cannot be dissociated from the relatively unique pharmacological profile of the drug. Thus, other β-blockers that lack the additional actions of carvedilol, such as metoprolol (2) and bisoprolol (3), have not, to date, been associated with the degree of efficacy in CHF that has been observed with carvedilol. The apparent ability of carvedilol to slow the progression of heart failure may reflect its unique actions in suppressing many of the pathological processes that contribute to the progression of CHF. In this respect, the ability of carvedilol to suppress oxygen radical stress in the ischemic myocardium, inhibit inflammatory cell infiltration and activation, block apoptosis in the myocardium, and suppress many of the neurohormonal systems and their mediators, which have been implicated in myocardial damage and cardiac remodeling, may contribute to the overall effects of carvedilol on the failing myocardium.

References

1. Waagstein, F., Hjalmarson, A., Varnauskas, E., and Wallentin, I. (1975). Effect of chronic β-adrenergic receptor blockade in congestive cardiomyopathy. *Br. Heart J.* **37,** 1022–1036.
2. Waagstein, F., Bristow, R. M., Swedeberg, K., Camerini, F., Fowler, M. B., Silver, M. A., Gilbert, E. M., Johnson, M. R., Goss, F. G., and Hjalmarson, A. (1993). Beneficial effects of metoprodol in idiopathic dilated cardiomyopathy. *Lancet* **342,** 1441–1446.
3. CIBIS. (1994). A randomized trial of β-blockade in heart failure. The Cardiac Insufficiency Bisopriol Study (CIBIS). *Circulation* **90,** 1765–1773.
4. Packer, M., Bristow, M. R., Cohn, J. N., Colucci, W. S., Fowler, M. B., Gilbert, E. M., and Shusterman, N. H. (1996). The effect of carvedilol on morbidity and mortality in patients with chronic heart failure. *N. Engl. J. Med.* **334,** 1349–1355.
5. Feuerstein, G. Z., Poste, G., and Ruffolo, R. R. Jr. (1995). Carvedilol Update III: Rationale for use in congestive heart failure. *Drugs Today* **31,** 307–326.

6. Olsen, S. L., Gilbert, E. M., Renlund, D. G., Taylor, D. O., Yanowitz, F. D., and Bristow, M. R. (1995). Carvedilol improves left ventricular function and symptoms in chronic heart failure: A double-blind randomized study. *J. Am. Coll. Cardiol.* **25**, 1225–1231.
7. Metra, M., Nardi, M., Giubbini, R., and Cas, L. D. (1994). Effects of short- and long-term carvedilol administration on rest and exercise hemodynamic variables, exercise capacity and clinical conditions in patients with idiopathic dilated cardiomyopathy. *J. Am. Coll. Cardiol.* **24**, 1678–1687.
8. Krum, H., Sackner-Bernstein, J. D., Goldsmith, R. L., Kukin, M. L., Schwartz, B., Penn, J., Medina, N., Yushak, M., Horn, E., Katz, S. D., Levin, H. R., Neuberg, G. W., DeLong, G., and Packer, M. (1995). Double-blind, placebo-controlled study of the long-term efficacy of carvedilol in patients with severe chronic heart failure. *Circulation* **92**, 1499–1506.
9. Australia-New Zealand Heart Failure Research Collaborative Group. (1995). Effects of carvedilol, a vasodilator–β-blocker, in patients with congestive heart failure due to ischemic heart disease. *Circulation* **92**, 212–218.

David S. Goldstein,* Courtney Holmes,*
Richard O. Cannon, III,† Graeme Eisenhofer,*
and Irwin J. Kopin*

*Clinical Neuroscience Branch
National Institute of Neurological Disorders and Stroke
†Cardiology Branch
National Heart, Lung, and Blood Institute
National Institutes of Health
Bethesda, Maryland 20892

Sympathetic Cardioneuropathy in Dysautonomias

Dysautonomias, derangements of sympathetic or parasympathetic nervous system function, occur fairly commonly in neurology and cardiology. Most attention has been paid to autonomic hypofunction or failure, causes of which include drugs and disease-associated polyneuropathy (e.g., diabetes, amyloid). Less commonly, autonomic failure occurs without an identifiable cause or in association with a disease for which the pathophysiological basis for dysautonomia remains obscure.

The classification of dysautonomias has been confusing and the pathophysiology obscure (1). We examined sympathetic innervation of the heart in patients with acquired, idiopathic dysautonomias, using thoracic 6-[^{18}F]fluorodopamine positron emission tomographic scanning (2) and assessments of the entry rate of norepinephrine, the sympathetic neurotransmitter, into the cardiac venous

Advances in Pharmacology, Volume 42

1054-3589/98 $25.00

drainage (cardiac norepinephrine spillover); and we related the laboratory findings to signs of sympathetic neurocirculatory failure (orthostatic hypotension and abnormal blood pressure responses associated with the Valsalva maneuver), central neural degeneration, and responsiveness to L-dopa/carbidopa (Sinemet).

One might expect absent myocardial 6-[^{18}F]fluorodopamine-derived radioactivity in dysautonomia patients with diffuse sympathetic denervation. Because blockade of ganglionic neurotransmission increases concentrations of myocardial 6-[^{18}F]fluorodopamine-derived radioactivity (2), one might expect increased myocardial 6-[^{18}F]fluorodopamine-derived radioactivity in dysautonomia patients with functionally intact sympathetic terminals but absent sympathetic nerve traffic.

Scans were obtained after intravenous 6-[^{18}F]fluorodopamine in 26 dysautonomia patients (3). Fourteen had sympathetic neurocirculatory failure: three with no signs of central neurodegeneration (pure autonomic failure), two with Sinemet-responsive parkinsonism, and nine with Sinemet-unresponsive central neurodegeneration (the Shy-Drager syndrome). Cardiac norepinephrine spillover rates were estimated based on intravenously infused [^{3}H]norepinephrine during right-heart catheterization.

Sympathetic neurocirculatory failure was diagnosed from persistent orthostatic hypotension and characteristic blood pressure abnormalities during and after performance of the Valsalva maneuver: a progressive fall in blood pressure during phase II (normally mean arterial pressure increases from its nadir by the end of phase II) and an absence of an overshoot in systolic pressure during phase IV. The Shy-Drager syndrome was diagnosed from sympathetic neurocirculatory failure and progressive central neural degeneration: Sinemet-resistant parkinsonism, progressive cerebellar ataxia, or supranuclear or bulbar palsy. Pure autonomic failure was diagnosed by sympathetic neurocirculatory failure without signs of central neural degeneration. Multiple-system atrophy with parasympathetic autonomic failure was diagnosed by central neural degeneration and persistent impotence, constipation, urinary incontinence, urinary retention, or decreased sweating, without evidence specifically of sympathetic neurocirculatory failure.

Myocardial perfusion was assessed by thoracic positron emission tomographic scanning for 20 min after a 1-min infusion of 5 mCi of [^{13}N]ammonia. After at least 1 hr, 6-[^{18}F]fluorodopamine (dose in most cases, 1.0 mCi) was infused intravenously at a constant rate for 3 min, with continuous thoracic positron emission tomographic scanning for up to 3 hr. Radioactivity concentrations in two circular regions of interest each in the left ventricular free wall and septum were averaged. Decay-corrected concentrations (in nCi/cc) were adjusted for the dose of radioactive drug per kilogram of body weight (in mCi/kg) and expressed in units of nCi-kg/cc-mCi. During right heart catheterization, a tracer amount of [^{3}H]norepinephrine was infused intravenously, with coronary sinus blood flow measured by thermodilution, and arterial and coronary sinus blood sampled after at least 20 min.

None of the three patients with pure autonomic failure had detectable myocardial 6-[^{18}F]fluorodopamine-derived radioactivity (Fig. 1), cardiac norepinephrine spillover, or cardiac arteriovenous increments in plasma levels of L-dopa, dihydroxyphenylglycol, or dihydroxyphenylacetic acid (Table I). All

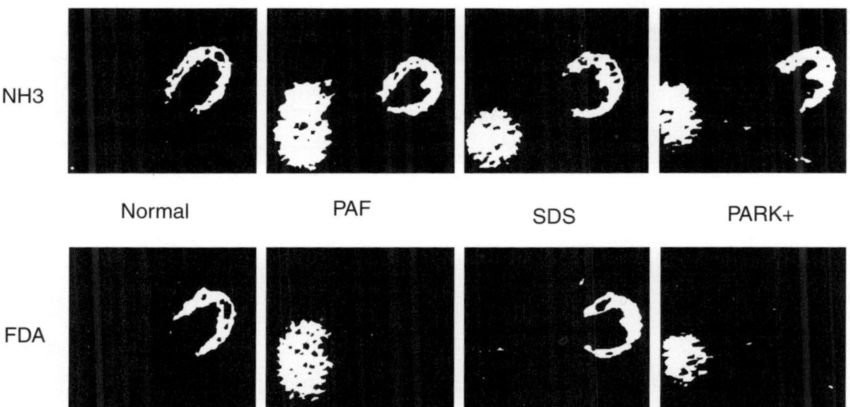

FIGURE 1 Thoracic positron emission tomographic scans after intravenous injection of 5 mCi of [^{13}N] ammonia (NH3) and 1 mCi of 6-[^{18}F]fluorodopamine (FDA) into a healthy volunteer (Normal), a patient with pure autonomic failure (PAF), a patient with the Shy-Drager syndrome (SDS), and a patient with parkinsonism and sympathetic neurocirculatory failure (Park+). The pictures represent time-averaged, non-gated. White is the maximum and black is the minimum radioactivity concentration in each picture. The right side of each picture corresponds to the left side of the subject.

nine patients with the Shy-Drager syndrome had clearly visible left ventricular myocardial 6-[^{18}F]fluorodopamine-derived radioactivity (see Fig. 1) and increased myocardial mean concentrations of 6-[^{18}F]fluorodopamine-derived radioactivity, compared with values in patients with multiple-system atrophy without sympathetic neurocirculatory failure or in healthy volunteers. Patients with multiple-system atrophy without sympathetic neurocirculatory failure had normal myocardial concentrations of 6-[^{18}F]fluorodopamine-derived radioactivity, normal cardiac norepinephrine spillover, and in most cases, normal cardiac arteriovenous increments in plasma levels of L-dopa, dihydroxyphenylglycol, and dihydroxyphenylacetic acid. Both patients with Sinemet-responsive parkinsonism and sympathetic neurocirculatory failure had undetectable myocardial 6-[^{18}F]fluorodopamine-derived radioactivity, cardiac norepinephrine spillover, or cardiac arteriovenous increments in plasma levels of L-dopa, dihydroxyphenylglycol, or dihydroxyphenylacetic acid.

The lack of myocardial 6-[^{18}F]fluorodopamine-derived radioactivity in the patients with pure autonomic failure probably reflected loss of postganglionic cardiac sympathetic terminals. Decreased activity of a sympathoneuronal membrane transporter can also produce this result; however, the same patients had strong neurochemical evidence of cardiac sympathetic denervation—virtually no cardiac spillover of norepinephrine and no cardiac production of L-dopa, dihydroxyphenylglycol, or dihydroxyphenylacetic acid.

In contrast, patients with the Shy-Drager syndrome (defined here as multiple-system atrophy with sympathetic neurocirculatory failure) had clearly depicted myocardial 6-[^{18}F]fluorodopamine-derived radioactivity, normal car-

TABLE I Mean Values (±SEM) for Coronary Sinus Blood Flow and Concentrations of Norepinephrine and Other Catechols in Patients with Dysautonomias.

	Diagnosis				
Variable	*PAF*	*SDS*	*Park+*	*Other*	*Normal*
Q_{CS} (cc/min)	142 ± 14 (3)	184 ± 18 (7)	205 ± 20 (2)	144 ± 27 (8)	184 ± 15 (32)
NE_A (pg/ml)	107 ± 44 (3)	235 ± 20 (9)	732 ± 483 (2)	331 ± 57 (13)	226 ± 15 (32)
E_{NE} (%)	6 ± 3 (3)	70 ± 4* (8)	12 ± 4 (2)	68 ± 5* (9)	81 ± 2* (32)
$NESO_{CS}$ (ng/min)	0 ± 0 (3)	21 ± 4* (7)	4 ± 1 (2)	27 ± 12* (8)	18 ± 2* (32)
$V\text{-}A_{DOPA}$ (pg/ml)	7 ± 39 (3)	120 ± 33* (8)	−52 ± 43 (2)	155 ± 24* (7)	236 ± 18* (32)
$V\text{-}A_{DHPG}$ (pg/ml)	−17 ± 11 (3)	472 ± 106* (8)	18 ± 2 (2)	633 ± 105* (9)	711 ± 35* (32)
$V\text{-}A_{DOPAC}$ (pg/ml)	91 ± 140 (3)	117 ± 26* (8)	−38 ± 32 (2)	152 ± 35* (7)	331 ± 50* (32)

* Significantly different from zero.
Numbers in parentheses indicate numbers of subjects.
To convert concentrations of NE from pg/ml to nmoles/liter, divide by the molecular weight of NE, 169; for DOPA, divide by 197; for DHPG, divide by 170; and for DOPAC, divide by 168.
PAF, pure autonomic failure; SDS, Shy-Drager syndrome; Park+, parkinsonism with sympathetic neurocirculatory failure; Q_{CS}, coronary sinus blood flow; NE_A, arterial plasma norepinephrine concentration; E_{NE}, cardiac extraction fraction of norepinephrine; $NESO_{CS}$, norepinephrine spillover into coronary sinus plasma; $V\text{-}A_{DOPA}$, arteriovenous increment in plasma L-dopa levels in patients not treated with L-dopa/carbidopa; $V\text{-}A_{DHPG}$, arteriovenous increment in plasma dihydroxyphenylglycol levels; $V\text{-}A_{DOPAC}$, arteriovenous increment in plasma dihydroxyphenylacetic acid levels.

diac spillovers of norepinephrine, and significant cardiac production of L-dopa, dihydroxyphenylglycol, and dihydroxyphenylacetic acid, confirming the presence of functionally intact cardiac sympathetic terminals. In fact, the Shy-Drager patients had increased myocardial concentrations of 6-[^{18}F]fluorodopamine-derived radioactivity, compared with values in healthy volunteers or in multiple-system atrophy patients without sympathetic neurocirculatory failure. Because the loss of 6-[^{18}F]fluorodopamine-derived radioactivity depends partly on ongoing sympathoneural traffic (2), decreased or absent sympathetic outflow to the heart can explain the increased myocardial 6-[^{18}F]fluorodopamine-derived radioactivity in Shy-Drager patients. The results suggest that the Shy-Drager syndrome differs pathophysiologically from multiple-system atrophy without sympathetic neurocirculatory failure, in that only the former is associated with decreased or absent sympathetic nerve traffic.

The present results highlight the apparent paradox of normal entry of norepinephrine into the bloodstream in the setting of apparently decreased or absent postganglionic sympathetic nerve traffic in Shy-Drager patients. "Constitutive neurosecretion"—spontaneous release of norepinephrine independent of sympathetic nerve traffic—may explain this phenomenon. Mechanisms of constitutive neurosecretion in Shy-Drager patients, if it occurs, are unknown.

The differential diagnosis between the Shy-Drager syndrome and parkinsonism with autonomic failure has proven particularly challenging, because both entities feature progressive central neural degeneration, neurogenic orthostatic hypotension, and a failure to increase plasma norepinephrine levels during orthostasis. In the present study, two patients with parkinsonism and sympathetic neurocirculatory failure had no myocardial 6-[^{18}F]fluorodopamine-derived radioactivity, in contrast with increased myocardial concentrations of 6-[^{18}F]fluorodopamine-derived radioactivity in Shy-Drager patients, providing a clear laboratory distinction between these entities and supporting the separate classification of Parkinson's disease with autonomic failure.

The results lead us to propose a pathophysiological classification of dysautonomias, in which sympathetic neurocirculatory failure results from peripheral sympathetic denervation or from decreased or absent sympathoneural traffic, with or without signs of central neural degeneration; and in which both parkinsonism with sympathetic neurocirculatory failure and multiple-system atrophy without sympathetic neurocirculatory failure differ from the Shy-Drager syndrome.

Refined clinical laboratory means to identify pathophysiological mechanisms of dysautonomias should enable future development of more effective means to diagnose the type of dysautonomia in individual patients, establish prognosis, and predict responses to therapy.

References

1. The Consensus Committee of the American Autonomic Society and the American Academy of Neurology. (1996). Consensus statement on the definition of orthostatic hypotension, pure autonomic failure, and multiple system atrophy. *Neurology* **46**, 1470.
2. Goldstein, D. S., Eisenhofer, G., Dunn, B. B., Armando, I., Lenders, J., Grossman, E., Holmes, C., Kirk, K. L., Bacharach, S., Adams, R., Herscovitch, P., and Kopin, I. J. (1993). Positron emission tomographic imaging of cardiac sympathetic innervation using 6-[^{18}F]fluorodopamine: Initial findings in humans. *J. Am. Coll. Cardiol.* **22**, 1961–1971.
3. Goldstein, D. S., Holmes, C., Stuhlmuller, J. E., Lenders, J. W. M., Kopin, I. J. (1997). 6-[^{18}F]Fluorodopamine positron emission tomographic (PET) scanning in the assessment of cardiac sympathoneural function—studies in normal humans. *Clin. Auton. Res.* **7**, 17–29.

Philip E. Cryer

Division of Endrocrinology, Diabetes, and Metabolism
Washington University School of Medicine
St. Louis, Missouri 63110

Hypoglycemia-Associated Autonomic Failure in Insulin-Dependent Diabetes Mellitus

Were it not for the devastating effects of hypoglycemia, particularly on the brain, diabetes mellitus would be easy to treat. One would simply use enough insulin to lower plasma glucose concentrations to or below the normal range. That would eliminate symptoms caused by hyperglycemia as well as ketoacidosis and the nonketotic hyperosmolar syndrome, prevent the specific long-term complications (retinopathy, nephropathy, and neuropathy), and likely reduce atherosclerotic risk to baseline. But, hypoglycemia is a reality, indeed, a fact of life for most people with insulin-dependent diabetes mellitus (IDDM) (1). Those attempting to achieve glycemic control suffer untold numbers of episodes of asymptomatic hypoglycemia and an average of two episodes of symptomatic hypoglycemia per week. At a minimum, 25% suffer an episode of severe, temporarily disabling hypoglycemia, often with seizure or coma, in a given year. Iatrogenic hypoglycemia causes recurrent physical and psychosocial morbidity and some mortality and precludes achievement of euglycemia in the vast majority of patients with IDDM. If we are to achieve euglycemia safely, we must learn to replace insulin in a much more physiological fashion, to prevent, correct, or compensate for compromised glucose counterregulation, or both (1).

In defense against falling plasma glucose concentrations, decrements in insulin, increments in glucagon, and, absent the latter, increments in epinephrine normally prevent or correct hypoglycemia (1). In addition to these physiological defenses, warning symptoms (largely autonomic in origin) normally prompt a behavioral defense (food ingestion). All of these defenses are compromised in established IDDM (1). As the patient becomes totally insulin-deficient over the first few years of clinical IDDM, insulin levels do not decrease as glucose levels fall. Because it is passively absorbed from subcutaneous injection sites, the appearance of insulin in the circulation is unregulated. Over the same time frame, the glucagon secretory response to falling glucose levels is lost. In this setting, the development of reduced epinephrine secretory responses to falling glucose levels is a critical pathophysiological event. Compared with those with absent glucagon but normal epinephrine responses, patients with combined deficiencies of their glucagon and epinephrine responses are at 25-fold or greater increased risk for severe iatrogenic hypoglycemia during intensive therapy of

Advances in Pharmacology, Volume 42

1054-3589/98 $25.00

their IDDM. They have the syndrome of *defective glucose counterregulation.* A related disorder, associated with about a sixfold increased risk for severe hypoglycemia, is the syndrome of *hypoglycemia unawareness* (loss of the warning symptoms of developing hypoglycemia). Awareness of hypoglycemia is normally the result of the perception of physiological changes caused by the autonomic (both sympathetic neural and adrenomedullary) discharge triggered by falling glucose levels. Hypoglycemia unawareness is best attributed to loss of these neurogenic (autonomic) symptoms. Thus, the first symptoms are neuroglycopenic, and it is often too late for the patient to abort the episode by eating.

Iatrogenic hypoglycemia in IDDM is the result of the interplay of relative or absolute insulin excess, which must occur from time to time because of the gross imperfections of all currently available insulin-replacement regimens, and compromised glucose counterregulation (1). Reduced autonomic, including epinephrine, and symptomatic responses to falling glucose levels play central roles in the pathogenesis of the syndromes of compromised glucose counterregulation, defective glucose counterregulation, and hypoglycemia unawareness. What is the mechanism (or mechanisms) of the reduced autonomic responses that occur commonly in IDDM?

Recent antecedent hypoglycemia causes reduced autonomic, including epinephrine, and symptomatic responses to subsequent hypoglycemia in nondiabetic individuals (2) and patients with IDDM (3). This finding led to the concept of *hypoglycemia-associated autonomic failure,* a functional disorder distinct from classical diabetic autonomic neuropathy (4). As illustrated in Figure 1, the hypoglycemia-associated autonomic failure concept posits that episodes of iatrogenic hypoglycemia lead to reduced autonomic responses to subsequent hypoglycemia and, thus, to both hypoglycemia unawareness and defective glucose counterregulation. Among the growing body of evidence in support of this concept, perhaps the most compelling is the finding, from three different laboratories,

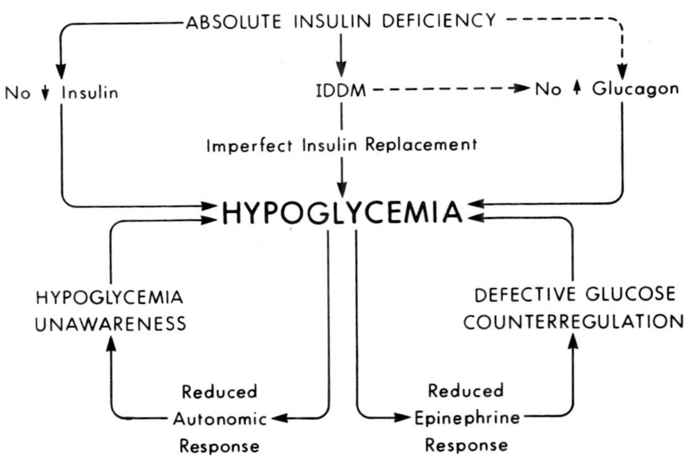

FIGURE 1 Hypoglycemia-associated autonomic failure concept. (Reprinted from *Hypoglycemia* Oxford University Press, 1997, with permission.)

that scrupulous avoidance of iatrogenic hypoglycemia completely reverses the syndrome of hypoglycemia unawareness and at least partially reverses the reduced epinephrine component of defective glucose counterregulation (5).

The specific mechanism(s) of the reduced autonomic responses induced by recent antecedent hypoglycemia remains to be established but it may involve increased blood-to-brain glucose transport. Regardless of the mechanism, hypoglycemia unawareness and, at least in part, the reduced epinephrine component of defective glucose counterregulation are reversible by short-term scrupulous avoidance of iatrogenic hypoglycemia in IDDM.

References

1. Cryer, P. E. (1994). Hypoglycemia: The limiting factor in the management of IDDM. *Diabetes* **43**, 1378–1389.
2. Heller, S. R., and Cryer, P. E. (1991). Reduced neuroendocrine and symptomatic responses to subsequent hypoglycemia after one episode of hypoglycemia in nondiabetic humans. *Diabetes* **40**, 223–226.
3. Dagogo-Jack, S. E., Craft, S., and Cryer, P. E. (1993). Hypoglycemia-associated autonomic failure in insulin-dependent diabetes mellitus. *J. Clin. Invest.* **91**, 819–828.
4. Cryer, P. E. (1992). Iatrogenic hypoglycemia as a cause of hypoglycemia-associated autonomic failure in IDDM: A vicious cycle. *Diabetes* **41**, 255–260.
5. Dagogo-Jack, S. E., Rattarasarn, C., and Cryer, P. E. Reversal of hypoglycemia unawareness, but not defective glucose counterregulation, in IDDM. *Diabetes* **43**, 1426–1434.

D. G. Maggs and R. S. Sherwin

Section of Endocrinology
Yale University School of Medicine
New Haven, Connecticut 06520

Mechanisms of the Sympathoadrenal Response to Hypoglycemia

Hypoglycemia is characterized by a neurohormonal response that facilitates glucose recovery, and a critical component of this response is an increase in sympathoadrenal activity. The neural origin of the sympathoadrenal response to hypoglycemia has been an area of some interest because it is undoubtedly the critical site where glucose sensing occurs. A role for both cerebral and extracerebral glucose sensors has been described, and although there is some

Advances in Pharmacology, Volume 42

1054-3589/98 $25.00

evidence that the liver may play some role in activating sympathoadrenal activity during hypoglycemia, most data point to the central nervous system as the key center for sensing hypoglycemia. This view is supported by studies showing that infusion of glucose into carotid and vertebral arteries (maintaining brain glucose levels during systemic hypoglycemia) abolishes the neurohormonal response to hypoglycemia. These studies, however, do not define the precise region in the brain involved in glucose sensing. Although data exist that suggest that important glucose sensing occurs in the hindbrain or spinal cord, recent studies employing lesioning and microdialysis techniques provide strong evidence that the ventromedial hypothalamus (VMH) is essential for glucose sensing and, therefore, the launching of the neurohormonal response to hypoglycemia. Hypoglycemic clamp studies were done in conscious rats with bilateral VMH lesions produced by local ibotenic acid injection [2 wk earlier], and rats with lesions at other neural sites served as controls. In the control rats, robust catecholamine responses were observed during hypoglycemia, however, catecholamine responses in the VMH-lesioned rats were markedly inhibited. Employing local microdialysis techniques, subsequent studies further determined the role of the VMH in triggering counterregulatory responses to hypoglycemia. In the VMH of awake rats, 2-deoxyglucose (2-DG) was perfused via bilateral microdialysis probes to produce localized cellular glucopenia. In control studies, probes were perfused with glucose instead of 2-DG under identical experimental conditions. Perfusion of 2-DG into the VMH caused a striking increase in plasma epinephrine (sixfold) and norepinephrine (fourfold) in association with a prompt elevation of circulating blood glucose, evidence that effective glucose counterregulation had occurred. As expected, in the control studies, there was no sympathoadrenal response when glucose was delivered into the VMH. To follow the findings from this particular study, further studies were done using a reversed experimental paradigm: inducing systemic hypoglycemia in awake rats but preventing local glucopenia in the VMH by perfusing glucose via bilateral microdialysis probes. Results indicate that the rats receiving glucose via the microdialysis probes, therefore protecting the VMH from the effects of systemic hypoglycemia, had markedly diminished sympathoadrenal responses, again suggesting that the VMH is the dominant glucose sensor site for activation of the sympathoadrenal response to hypoglycemia.

The sympathoadrenal response comprises two distinct components working in tandem: increased neural activity at sympathetic nerve endings and the humoral effects of the adrenomedullary response. Measurement of circulating levels of catecholamines during hypoglycemia would suggest that epinephrine is the predominant catecholamine, and, therefore, the humoral effects of the adrenomedullary response outweigh the neuronal effects of increased sympathetic neural activity. However, in a group of healthy human subjects, local tissue levels of catecholamines were measured by microdialysis in skeletal muscle and adipose tissue during controlled hypoglycemic (plasma glucose 50 mg/dl) clamp conditions. As expected, there was a much greater relative increase in plasma epinephrine than norepinephrine, such that epinephrine comprised approximately 85% of the total molar increment in circulating catecholamines, but a different picture emerged in muscle and adipose tissue extracellular fluid (ECF). Norepinephrine comprised approximately 35% of the total catecholamine increment in adipose tissue and approximately 50% of the total increment in muscle. Although this finding highlights the importance of local norepinephrine release, the actual increase in local sympathetic activity may have been greater still. The increment in tissue epinephrine is almost entirely due to the passage of epinephrine from the circulation into the ECF space, whereas, the rise in local ECF norepinephrine is due to overspill from the neuroeffector junction, where a significant fraction of locally released norepinephrine is cleared by an efficient re-uptake mechanism or broken down enzymatically. Therefore, the role of peripheral sympathetic activation in hypoglycemic counterregulation is underestimated when based on blood sampling alone. Furthermore, these data suggest that the importance of local hypoglycemia-induced sympathetic activation, as manifested by local norepinephrine release at the neuroeffector junction, may be much greater than previously thought.

The net systemic effects of the sympathoadrenal response are wide and varied, most important, having powerful effects on the cardiovascular system and fuel metabolism. It is through these effects that the sympathoadrenal response is critical in facilitating glucose recovery, and these far-reaching effects can best be described in three areas.

First, the systemic effects of increased sympathoadrenal activity result in important subjective symptoms that alert subjects that they are hypoglycemic, therefore facilitating the ingestion of rapidly absorbed carbohydrate. In light of the disconcerting phenomenon of "hypoglycemic unawareness" described in insulin-dependent diabetes—a phenomenon in which patients no longer experience the normal warning symptoms of hypoglycemia—the subjective symptoms of hypoglycemia have become an area of clinical interest. At present, it is surmised that the spectrum of symptoms that occur during hypoglycemia manifest either as a consequence of the neurohormonal or, in other words, autonomic response to hypoglycemia or as a consequence of neuroglycopenia. More importantly, autonomic symptoms seem to be especially critical because they tend to manifest at a higher glycemic level than those attributed to neuroglycopenia, therefore allowing the earliest warning of hypoglycemia. It is also important to note that although hypoglycemia causes a generalized autonomic activation, the predominant autonomic symptoms have sympathoadrenal origins (i.e., sweating, tremor, pounding heart). Therefore, the sympathoadrenal response is critical in allowing normal subjective awareness of hypoglycemia.

Second, the sympathoadrenal response has very powerful effects on fuel metabolism, essentially opposing the systemic effects of insulin: (a) inhibiting glucose consumption in extraneural tissues, thereby sparing glucose for brain utilization; (b) generating gluconeogenic precursors from peripheral energy stores (lactate, glycerol and alanine); and (c) generating alternate fuels (lactate and fatty acids) for consumption by extraneuronal tissues (skeletal muscle and myocardium). In normal glucose counterregulation to acute hypoglycemia, glucose recovery is dependent on the hormonal release of both glucagon and epinephrine, and, through the powerful effects of glucagon on hepatic glucose production, epinephrine action is of secondary importance. However, in insulin-dependent diabetes, in which the glucagon response to hypoglycemia is blunted or absent after approximately 2 years' diabetes duration, the epinephrine, and therefore sympathoadrenal, response to hypoglycemia and its action on fuel metabolism becomes of critical importance.

Third, and finally, the sympathoadrenal response causes a thermoregulatory disturbance, resulting in a fall in core temperature, which can be interpreted as an important defensive or protective mechanism. Since the early days of insulin use, it has long been reported that hypoglycemia is invariably accompanied by a fall in core temperature. Recent studies have shown that this occurs through responses initiated by increased sympathoadrenal activity: increased evaporative heat loss through sympathetic-mediated sweating and increased limb muscle blood flow presumably raising the temperature of the deeper limb tissues, thereby conducting heat to the skin surface and sustaining evaporative heat loss (1) (Fig. 1). The net effect is a fall in body temperature and, more importantly, brain temperature. The physiological importance of the hypothermic response during hypoglycemia is questionable, but recent studies in animals

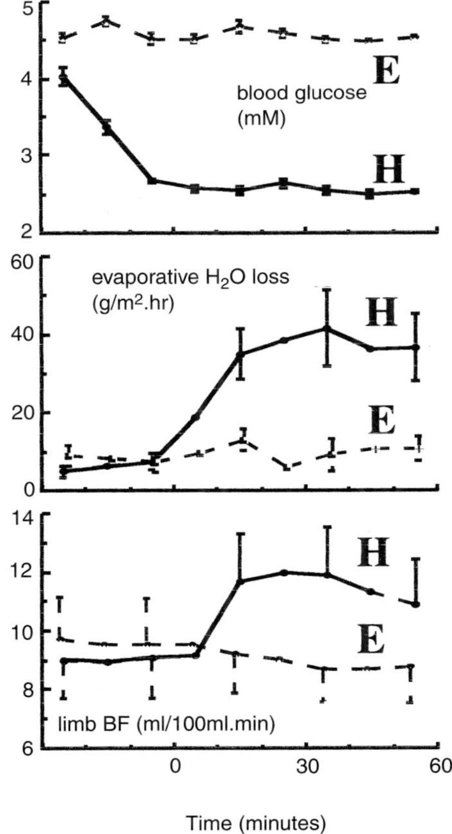

FIGURE 1 Blood glucose, evaporative water loss from skin surface (sweating), and limb blood flow during euglycemic (E) and hypoglycemic (H) insulin clamps in healthy humans.

demonstrate that brain damage or mortality rate during hypoglycemia is greatly increased if the natural fall in core temperature that occurs during hypoglycemia is artificially prevented. It is theorized that, by lowering core temperature, metabolic substrate requirements are reduced, and, of particular reference to the central nervous system, this protective mechanism seems to spare neuronal tissue from excitotoxic damage. However, it is difficult to equate this important observation in animals, where a prolonged and severe hypoglycemic insult was employed, to hypoglycemia occurring in humans. Studies have attempted to address this issue in human subjects by demonstrating that the magnitude of the hypothermic response is porportional to the severity of the hypoglycemic insult (glucose nadir and duration) and that the fall in core temperature during hypoglycemia is aided by a suppression of normal shivering. Relating these findings to the clinical setting, where diabetic patients experience severe hypoglycemia relatively infrequently, it is arguable whether less severe hypoglycemia causes potential neurologic harm and whether the accompanying fall in core temperature serves any protective benefit. Nevertheless, in the early days of insulin-shock therapy, it was observed that patients recovered quicker from hypoglycemia if their reaction was accompanied by a fall in core temperature, and, more recently, it has also been shown that insulin-treated diabetic patients with an impaired sympathoadrenal response to hypoglycemia are unable to dissipate heat effectively and are, therefore, likely to suffer euthermic hypoglycemia. It is under these conditions that the integrity of the hypothermic response may indeed become important in at-risk patients.

To summarize, the sympathoadrenal response to hypoglycemia plays a key role in glucose counterregulation. In insulin-dependent diabetes, its role becomes even more critical because the glucagon response to hypoglycemia is lost after a short duration of diabetes. Moreover, the sympathoadrenal responses may also become blunted in some patients, rendering them without appropriate subjective warning symptoms, without effective glucose counterregulation, and at potential risk of a greater insult to the central nervous system.

Reference

1. Maggs, D. G., Scott, A. R., and Macdonald, I. A. (1994). Thermoregulatory responses to hyperinsulinemic hypoglycemia and euglycemia in man. *Am. J. Physiol.* **267**, R1266–1272.

Geremia B. Bolli

Department of Internal Medicine and
Endocrinological and Metabolic Sciences
University of Perugia
06126 Perugia, Italy

Importance of Catecholamines in Defense against Insulin Hypoglycemia in Humans

It has been suggested that, in defense against hypoglycemia in humans, secretion of glucagon plays the dominant counterregulatory role (1). According to the view, secretion of adrenaline is usually not critical as long as the responses of glucagon are appropriate, and it becomes important only when secretion of glucagon is prevented (1) or impaired, as in type I diabetes mellitus (2).

These principles have recently been reexamined. A major criticism is that they have been derived from a model of acute hypoglycemia in which the i.v. insulin bolus produces a pharmacological increase in plasma insulin with a subsequently rapid decrease in plasma insulin over a few minutes' time (1). In common clinical situations, such as insulinoma, sulphonylurea-induced hypoglycemia, and insulin-dependent diabetes mellitus, hypoglycemia develops gradually, is less severe, and is reversed more slowly than that of the acute experimental models (1).

We have developed a model of prolonged hypoglycemia that mimics the clinical situation better than the acute model (1). In this model, insulin is infused to slowly increase the plasma insulin concentration to a plateau in a physiological range (two- to threefold above baseline), and the plasma glucose concentration gradually decreases into the hypoglycemic range (50–60 mg/dl). Using this model, it has already been shown (1) that mechanisms that were not found to be important for counterregulation in the acute model, such as suppression of endogenous insulin secretion and glucose utilization, do play important counterregulatory role; (2) that growth hormone and cortisol, which were not considered as counterregulatory hormones in the acute model, do indeed play an important role in defense against clinical hypoglycemia; and (3) that neither the failure of glucagon to respond to hypoglycemia nor that of growth hormone or cortisol is fully compensated by a larger secretion of epinephrine and norepinephrine.

The several differences in the counterregulatory mechanisms already observed between the acute (1) and the clinical model of hypoglycemia (3) suggest that, in the clinical model, the secretion of epinephrine also may play a much more important counterregulatory role as compared with that envisaged on the basis of the acute hypoglycemia model.

In this chapter, the contribution of the adrenergic mechanisms to counterregulation in an experimental model of prolonged hypoglycemia closely mimick-

Advances in Pharmacology, Volume 42

1054-3589/98 $25.00

ing the clinical situation will be reviewed. For this purpose, the pancreatic-adrenocortical-pituitary clamp, a technique that allows one to examine the role of a single counterregulatory hormone without confounding changes in the plasma concentrations of other hormones, was used.

In an initial series of studies to determine the overall counterregulatory contribution of adrenergic response to hypoglycemia, insulin (15 mU · m^{-2} · min^{-1} for 12 hr) was infused subcutaneously either alone or in combination with propranolol and phentolamine (3). Exogenous glucose was infused whenever needed in the α-, β-blockade experiments to match studies for plasma glucose concentrations. In both studies, plasma insulin slowly increased to a plateau of approximately 25 μU/ml, while plasma glucose decreased to a plateau of approximately 50 mg/dl. However, during catecholamine blockade, hepatic glucose production increased less both in an early phase and throughout. In addition, after 5 hr, glucose utilization was less suppressed than in control experiments. However, plasma C-peptide was more suppressed and plasma growth hormone less stimulated when catecholamine receptors were not blocked. These studies provided the first evidence of an important counterregulatory role of catecholamines in defense against hypoglycemia in humans. Interestingly, the effect was evident in both an early and a late phase and located both at the hepatic and muscular level. Finally, the role of catecholamines in defense against hypoglycemia was so important that other counterregulatory hormones, including glucagon, did not substitute for blockade of catecholamine effects.

However, these studies were not matched for plasma C-peptide (possible overestimation of the role of catecholamines because of greater portal plasma insulin concentrations during α,β blockade) nor for plasma growth hormone (which could have underestimated the role of catecholamines in a late phase). Thus, a second series of studies were designed to assess the extrapancreatic counterregulatory role of catecholamines without the confounding changes in plasma C-peptide and growth hormone. The pancreatic-adrenocortical-pituitary clamp was preformed either alone or in combination with i.v. propranolol and phentolamine to block the receptors of plasma and tissue catecholamines. In these experiments, studies were matched for portal plasma insulin concentrations and peripheral plasma counterregulatory concentrations, except for greater plasma adrenaline during α,β blockade. The results were similar to those observed during α,β blockade in the absence of pancreatic-adrenocortical-pituitary clamp; that is, blockade of catecholamines resulted in an early and sustained suppression of hepatic glucose production, and later in less suppression of peripheral glucose utilization.

In a final series of studies to quantitate the severity of hypoglycemia that would develop in the absence of the contribution of catecholamines to counterregulation, no exogenous glucose was infused in the experiments of adrenergic blockade. However, these experiments had to be interrupted early because plasma glucose decreased below 40 mg/dl, despite greater secretion of plasma glucagon, growth hormone, and cortisol.

One interesting question regarding the counterregulatory role of catecholamines is the contribution of *direct* versus *indirect* effects. For example, suppression of endogenous insulin secretion is a well-known indirect effect of catecholamines that contributes to counterregulation by limiting portal hyperinsulinemia.

However, there are, potentially, additional indirect mechanisms by which catecholamines may contribute to counterregulation. For example, in response to hypoglycemia, after initial suppression, lipolysis is activated. The following rebound increase in plasma free fatty acids (FFAs), glycerol, and ketone bodies, which are largely mediated by adrenergic activation, might both stimulate gluconeogenesis and suppress peripheral glucose utilization.

To test the hypothesis that catecholamine-mediated lipolysis plays a physiological role in counterregulation to prolonged insulin-induced hypoglycemia in humans, two series of experiments were performed (4, 5). In an initial series of experiments, hypoglycemia (plasma glucose ~55 mg/dl) was induced by low-dose, continuous i.v. insulin infusion (plasma insulin ~30 μU/ml) on one occasion (study 1). In a second study, α,β blockade was superimposed to i.v. insulin to block adrenergic-mediated lipolysis. In a third study, heparin and 10% Intralipid werer infused at rates to mimic the posthypoglycemic increase in plasma FFAs and glycerol of the control study. In the second and third studies, exogenous glucose was infused whenever needed to match the control study for plasma glucose concentration. During the adrenergic blockade of study 2, plasma FFAs and glycerol concentrations and rates of lipid oxidation were suppressed. This was associated with a decrease in hepatic glucose output and less suppression of peripheral glucose utilization. When plasma FFAs and glycerol were increased in study 3 to the values of the control study 1, lipid oxidation increased to the values of study 1, whereas hepatic glucose production increased by at least 50%, as compared with study 2, and peripheral glucose utilization was suppressed 85% more than in study 2. These studies (4) have been the first to demonstrate the critical role of lipolysis in glucose counterregulation. In particular, these studies (4) have shown that plasma FFAs contribute to 50% of stimulated hepatic glucose production and to at least 85% suppression of peripheral glucose utlization during hypoglycemia. Notably, this impressive contribution of the FFA substrate occurs during adrenergic blockade, a condition that, if anything, underestimates the role of FFAs in counterregulation.

To examine the contribution of lipolyis to counterregulation in the presence of catecholamine effects, in a second series of experiments, hypoglycemia was induced during selective pharmacological blockade of lipolysis with acipimox (5). As compared with a control study in which lipolysis was not blocked, blockade of lipolysis resulted in an approximate 40% decrease in overall hepatic glucose production, an approximate 70% decrease in gluconeogenesis (from alanine) as well as an approximate 15% increase in glucose utilization. These studies (5) have confirmed that the contribution of lipolysis to glucose counterregulation is important. Notably, failure of FFAs to increase during hypoglycemia results in severe impairment of hepatic and peripheral mechanisms of defense against hypoglycemia, despite physiological activation of the adrenergic system. Thus, a large part of the counterregulatory effects of catecholamines is indirect, because they are mediated by the substrate FFAs.

In conclusion, catecholamines contribute to human glucose counterregulation since an early phase of hypoglycemia throughout; they exert effects both at the liver and at the muscle level; their counterregulatory effects are not compensated by other hormones, in particular glucagon; thus, the failure of catecholamines to exert their effects in hypoglycemia, such as during α,β pharmacological blockade, results in severe hypoglycemia, despite increase in other counterregulatory hormones. Finally, catecholamines largely work during counterregulation via indirect mechanisms such as stimulation of lipolysis, which enhances gluconeogenesis and hepatic glucose production, and suppress oxidation and utilization of glucose.

The findings of these studies are that catecholamines are in the very first line of defense against hypoglycemia along with glucagon and that most of their effects are indirect (i.e., mediated by stimulation of lipolysis and mobilization of FFAs to the liver and muscle.

References

1. Rizza, R. A., Cryer, P. E., and Gerich, J. E. (1979). Role of glucagon, catecholamines, growth hormone in human glucose counterregulation. *J. Clin. Invest.* **64**, 64–71.
2. Gerich, J., Langlois, M., Noacco, C., Karam, J., and Forsham, P. (1973). Lack of glucagon response to hypoglycemia in diabetes: Evidence for an intrisinc pancreatic alpha-cell defect. *Science* **182**, 171–173.
3. De Feo, P., Perriello, G., Torlone, E., *et al.* (1991). The contribution of adrenergic mechanisms to glucose counterregulation in humans. *Am. J. Physiol.* **261**, E725–E736.
4. Fanelli, C., De Feo, P., Porcellati, F., *et al.* (1992). Adrenergic mechanisms contribute to the late phase of hypoglycemic glucose counterregulation in humans by stimulating lipolysis. *J. Clin. Invest.* **89**, 2005–2013.
5. Fanelli, C., Calderone, S., Epifano, L., *et al.* (1993). Demonstration of a critical role for free fatty acids in mediating counterregulatory stimulation of gluconeogenesis and suppression of glucose utilization in humans. *J. Clin. Invest.* **92**, 1617–1622.

M. Vaz, M. D. Esler, H. S. Cox, G. L. Jennings, D. M. Kaye, and A. G. Turner

Baker Medical Research Institute
Melbourne 3181, Australia

Sympathetic Nervous Activity and the Thermic Effect of Food in Humans

Studies have demonstrated that a reduced rate of energy expenditure is a risk factor for body weight gain and that the sympathetic nervous system (SNS) is a determinant of energy expenditure, although racial differences in

Advances in Pharmacology, Volume 42

1054-3589/98 $25.00

this relationship may be present. The SNS has indeed been implicated in all the major components of daily energy expenditure, including resting metabolic rate, energy expenditure related to physical activity, the thermic effect of food (TEF), cold-induced thermogenesis, and the thermogenic response to daily stimulants such as caffeine and nicotine. A reduction in sympathetic nervous activity is, therefore, a potential mechanism predisposing to weight gain.

TEF accounts for approximately 10% of daily energy expenditure. TEF is the increment in energy expenditure above resting values that occurs following food intake. TEF has been divided into an obligatory component that relates to the digestion, absorption, and processing of nutrients and a regulatory or facultative component that is due to energy expended in excess of the obligatory demands. Alterations in the facultative component could potentially result in weight gain over a period of time. The demonstration that β-adrenergically mediated SNS activity was responsible for almost the entire rise in energy expenditure in excess of the obligatory requirements during glucose-induced thermogenesis focused attention on the role of the SNS in the facultative component of TEF (1).

The most widely used method for assessing SNS activation following the consumption of a meal is the estimation of the primary neurotransmitter of the SNS in plasma, norepinephrine (NE). There are, however, a number of issues that need to be addressed in the interpretation of plasma NE concentrations as an index of SNS activity. First, several earlier studies employed forearm venous sampling, which reflects local SNS activation in the forearm vascular bed. Second, any major change in clearance of NE from plasma would affect the interpretation of plasma NE concentration as an index of SNS activation. There is an increase in cardiac output following the ingestion of food and a redistribution of blood flow into the splanchnic vascular bed. These changes could potentially alter the clearance of NE from plasma. Finally, the estimation of NE from plasma does not provide any information about the regional sites of SNS activation.

Studies carried out in our laboratory addressed these issues. We were able to demonstrate that there were indeed significant increases in the whole body clearance of NE from plasma following the ingestion of a 750-kcal mixed meal (2). This increase, however, was transient, confined to the first 30-min postprandial period, and coincided with the reported peaks in postprandial cardiac output and regional blood flow changes. The transient nature of plasma NE clearance changes postprandially is unlikely, by itself, to limit the interpretation of plasma NE concentrations in relation to TEF, which proceeds for many hours following the consumption of food. However, the role of the size of the meal and the macronutrient composition of the meal, both likely to influence plasma NE clearance, remain uninvestigated.

The limitations of venous sampling have been well established. This is particularly relevant when assessing SNS activity postprandially, because skeletal muscle is known (from microneurographic studies) to be a major site of postprandial SNS activity following the ingestion of glucose. Our studies corroborated this earlier evidence. There was a greater than 100% increase in forearm NE spillover following the ingestion a 750-kcal mixed meal (3). This large increase in forearm NE spillover is likely to obscure the contributions of increased SNS activity in regions distant from the venous sampling site.

There has been a growing impetus to determine the regional sites of thermogenesis. The contribution of skeletal muscle to TEF has been investigated by combining arteriovenous oxygen differences across a largely skeletal muscle vascular bed, such as the forearm or calf, with regional blood flow measurements. Earlier studies suggested that skeletal muscle was an important site of thermogenesis following the ingestion of glucose or largely carbohydrate meals. We were unable to demonstrate any increase in forearm oxygen consumption over a 2-hr postprandial period following a mixed meal, despite a significant elevation in forearm NE spillover (3). The application of central venous catheterization techniques to metabolic studies has allowed for the quantification of visceral contributions to whole body thermogenesis. Wahren *et al.* have elegantly demonstrated the importance of splanchnic thermogenesis following protein but not carbohydrate ingestion (4). Our own studies demonstrated a significant increase in splanchnic oxygen consumption following the ingestion of a mixed meal. In addition, there was also a significant increase in postprandial renal oxygen consumption, although this did not contribute substantially to the increase in whole body oxygen consumption. Cardiac oxygen consumption was unchanged from fasting levels (5).

The regional pattern of postprandial SNS activity was obtained using isotope dilution methodology coupled with regional venous sampling and plasma flows. Our studies indicate that the kidneys and skeletal muscle are the major sites of postprandial sympathetic nervous activity following the ingestion of a mixed meal, accounting for over 50% of the increment in whole body NE spillover postprandially (Fig. 1).

Although studies such as ours allow a clear delineation of the regional pattern of both oxygen consumption and SNS activity postprandially, they do not establish a causal relationship between the increment in SNS activity and the

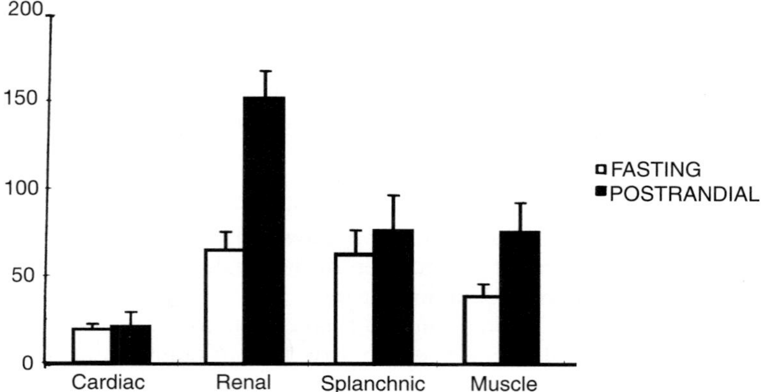

FIGURE 1 Regional postprandial NE spillover following the ingestion of a mixed meal. (Reprinted from *J. Aut. Ner. System,* 56, Vaz *et al.,* 1995, with kind permission of Elsevier Science, Sara Burgerhartstraat 25, 1055 KV, Amsterdam, The Netherlands.)

increment in oxygen consumption within a regional vascular bed. Sympathetic blockade studies have been used to better understand the role of SNS activity on TEF at a whole body level. These studies have shown that sympathetic nervous blockade significantly attenuates the TEF following glucose or carbohydrate ingestion but that it is without effect following protein ingestion. A similar approach to regional studies would provide valuable information.

In conclusion, we have attempted to establish the regional sites of both thermogensis and sympathetic nervous activation following the ingestion of food. We have demonstrated that while forearm NE spillover increases significantly following ingestion of a mixed meal, it is not associated with any local increase in oxygen consumption. Cardiac NE spillover and oxygen consumption were both unaltered postprandially. The high level of postprandial renal NE spillover may have significance in obesity-related hypertension. Further studies elucidating the pattern of regional SNS activity and oxygen consumption following the ingestion of different macronutrients and in the context of obesity need to be done.

References

1. Acheson, K. J., Ravussin, E., Wahren, J., and Jequier, E. (1984). Thermic effect of glucose in man. Obligatory and facultative thermogenesis. *J. Clin. Invest.* **74**, 1572–1580.
2. Vaz, M., Cox, H. S., Kaye, D. M., Turner, A. G., Jennings, G. L., and Esler, M. D. (1995). Fallibility of plasma noradrenaline measurements in studying postprandial sympathetic nervous responses. *J. Auton. Nerv. Syst.* **56**, 97–104.
3. Vaz, M., Turner, A., Kingwell, B., Chin, J., Koff, Cox, H., Jennings, G. and Esler, M. (1995). Postprandial sympatho-adrenal activity: Its relation to metabolic and cardiovascular events and to changes in meal frequency. *Clin. Sci.* **89**, 349–357.
4. Wahren, J. (1993). The biochemical basis of splanchnic thermogenesis. *Int. J. Obesity* **15**, S56–S59.
5. Cox, H. S., Kaye, D. M., Thompson, J. M., Turner, A. G., Jennings, G. L., Itsiopoulos, C., and Esler, M. D. (1995). Regional sympathetic nervous activation after a large meal in humans. *Clin. Sci.* **89**, 145–154.

Eva Hagström-Toft

Department of Medicine
Karolinska Institute
Huddinge Hospital
S-141 86 Huddinge, Sweden

Microdialysis for the Assessment of Catecholamine-Induced Lipolysis in Adipose and Skeletal Muscle Tissue

Microdialysis was developed more than 25 years ago for studies of neuropharmacology in experimental animals, but has, since the 1980s, also been available for human research. During this century, adipose tissue, which was the initial site of investigation, has emerged from being merely inactive "insulation" to a role as one of the more important regulators of energy homeostasis in the body. For many years, research on adipose tissue was restricted to *in vitro* studies or whole body measurements. However, during the last two decades, *in situ* techniques for studies of *regional* adipose tissue lipolysis have been developed, such as the arteriovenous-metabolite difference technique over a subcutaneous fat depot and the microdialysis technique (1). Microdialysis also has been introduced for studies of skeletal muscle tissue as well as several other tissues in the body.

I. The Microdialysis Method

Several types of microdialysis devices have been used in investigations of peripheral tissues, but the principle is similar for all. The microdialysis device used by our group consists of a double-lumen catheter with a dialysis membrane (10–30 mm, 20,000 D cut-off) at the tip. The perfusion fluid enters the catheter through the outer cannula and streams slowly (0.3–5.0 μl/min) along the membrane, over which substances equilibrate. The perfusion fluid then leaves the catheter through the inner tubing and is collected and analyzed. The major factors influencing the recovery of a substance (the ratio of substance concentration in the dialysate vs the respective concentration in the extracellular fluid) are the length of the dialysis membrane (direct correlation) and the perfusate velocity (inverse correlation). In addition, recovery is influenced by factors such as membrane pore size and chemical properties, the composition of the perfusate fluid, the tissue temperature, and speed of diffusion through the tissue. The *in vivo* recovery of a certain substance can be determined by various calibration

Advances in Pharmacology, Volume 42

1054-3589/98 $25.00

procedures: no-net flux, 0-flow, and isotope (retro-) recovery. The recovery factor is subsequently used for calculation of the true tissue level. Using a membrane length of 30 mm and a very low perfusate speed (0.3–0.5 μl/min) it has been possible to obtain full recovery (near 100%) of glucose, glycerol, and lactate.

II. Lipolysis and Microdialysis

The major regulatory hormones for human adipose tissue lipolysis are insulin and catecholamines, which act on hormone-sensitive lipase. Binding sites for all adrenoceptor subtypes have been found in fat cells. It has to be pointed out that, with microdialysis, only one of the end-products of lipolysis, glycerol, can be monitored. Free fatty acids (FFAs) are hydrophobic and not dialyzable with the presently used membranes. Because glycerol is not reutilized by adipose tissue, it is considered an index of lipolysis. Of great importance is the evaluation of blood flow variations in conjunction to *in situ* glycerol measurements. This is because the interstitial concentration of a metabolite will be determined both by its production and uptake in the tissue and by the transportation away or delivery to the tissue by the capillary bloodstream. The content of the dialysis sample reflects the net sum of these events. Tissue blood flow can be estimated either by Xenon-133 washout or by addition of a flow marker (e.g., ethanol) to the microdialysis perfusate solvent. The microdialysis technique can be used in various types of lipolysis studies, which will be examplified:

- Mechanistic *in situ* studies
- Systemic hormonal challenge
- Combinations of mechanistic *in situ* studies and systemic hormonal challenge
- Hormonal interactions
- Investigations of various tissues

A. Mechanistic *In Situ* Studies

One advantage of the microdialysis technique is the possibility of studying effects of metabolically active drugs and substances locally, by retroperfusion, with no risk of systemic side effects. This was done to investigate β-adrenoceptor subtype function in adipose tissue *in vivo* (2). Non-obese healthy subjects were investigated in the postabsorptive state with microdialysis of abdominal, subcutaneous adipose tissue. In different microdialysis catheters, the tissue was perfused with agonists of the three β-adrenoceptors: dobutamine, terbutaline, and CGP 12177. All three agonists caused a rapid increase of the dialysate glycerol concentration. Pretreatment of the tissue with propranolol ($\beta_{1,2}$-adrenoceptor antagonist), via the microdialysis catheter, blocked the dobutamine and terbutaline effect on glycerol, whereas the effect of the β_3-agonist was unchanged.

Variations in adipose tissue nutritive blood flow was estimated by the ethanol perfusion technique (3). Blood flow increased when terbutaline was perfused, as shown by a decreasing ethanol outflow versus inflow concentration ratio. Taking the blood flow change into account when interpreting the data, it appears that the β_2-adrenoceptor is the dominant β-adrenoceptor as regards lipid mobilization in subcutaneous adipose tissue.

The blood flow aspect further illustrated by a mechanistic study in which the tissue was perfused by the α_2-agonist clonidine (3). α_2-adrenoceptor stimulation inhibits lipolysis, but *in situ,* a paradoxical increase in glycerol was induced at low clonidine concentrations. The effect was more pronounced in gluteal than in abdominal tissue. Pretreatment of the tissue by vasodilators, hydralazine or nitroprusside, abolished the dialysate glycerol increase. Thus, it was concluded that the effect of clonidine at low concentrations was caused by retention of lipid products in the tissue due to vasoconstriction.

B. Systemic Hormonal Challenge

Another way to investigate catecholamine-induced lipolysis is to induce a systemic increase of the hormone levels by a functional challenge (e.g., physical exercise, mental stress, surgical trauma, or insulin-induced hypoglycemia). Insulin-induced hypoglycemia evokes strong adrenergic counterregulatory mechanisms, mainly a release of epinephrine from the adrenal medulla. It is well known that hypoglycemia promotes lipolysis, leading to stimulation of gluconeogenesis and decreased peripheral glucose utilization due to increased circulating FFAs. Hypoglycemia was induced in normal subjects and type 1 diabetic patients (4). The 60-fold increase in epinephrine was accompanied by both a local and a systemic rise in glycerol. The lipolytic increase was higher in the insulin-dependent diabetes mellitus patients, compared with the healthy subjects, despite a reduced increase in epinephrine.

C. Combinations of Mechanistic *in situ* Studies and Systemic Hormonal Challenge

The adrenergic receptors involved in the lipolytic response to hypoglycemia were investigated by combining the hypoglycemic procedure with *in situ* perfusion by α- and β-adrenoceptor blocking agents. It could be demonstrated that the lipolytic response is mediated by both α- and β-adrenoceptors. The addition of propranolol to the perfusate inhibited the lipolytic response, and the addition of phentolamine increased the glycerol levels. Hypoglycemia induced an increased ethanol washout, indicating an increase in tissue blood flow. This was mediated by β-adrenoceptors, because propranolol prevented this effect. Phentolamine (α-adrenoceptor blockade) had no effect on blood flow.

D. Hormonal Interactions

Interactions of the hormones acting on lipolysis can be investigated with microdialysis, and this is exemplified by a study combining systemic insulin infusion (euglycemic, hyperinsulinemic clamp) with *in situ* isoprenaline perfusion (nonselective β-adrenoceptor stimulator). In adipose tissue, the antilipolytic effect of insulin was effectively counteracted by the β-adrenergic stimulation (5).

E. Hormonal Interactions and Investigations of Various Tissues

Hormonal interactions and the effect of a hormonal challenge on lipolysis in two different tissues (adipose tissue and skeletal muscle) was studied by inducing a hyperinsulinemic, hypoglycemic clamp (unpublished data). It is well known that the FFAs needed in the myocyte during exercise in part originate from hydrolysis of intracellular triglycerides. However, it was shown by Maggs *et al.* (p. 622) with microdialysis of *M. Vastus lateralis,* glycerol seems to be mobilized *from* the tissue. In the basal state, glycerol levels were found that were twofold those in plasma, indicating production of glycerol in the tissue. During insulin-induced hypoglycemia (hypoglycemic clamp), there was an initial insulin-induced suppression of interstitial glycerol with a subsequent rapid increase induced by the adrenergic counterregulatory response. The time course of lipolysis activation was different in adipose and skeletal muscle tissue in that the glycerol rise appeared earlier in the muscle.

In conformity with the adipose tissue findings, *in situ* perfusion with propranolol resulted in attenuation of glycerol levels. Phentolamine, on the other hand, did not affect glycerol levels in skeletal muscle. A different regulation of regional blood flow, compared with fat tissue, may be involved.

III. Summary

The results from studies performed in our group on catecholamine-induced lipolysis using microdialysis is summarized in Table I.

TABLE I Catecholamine-induced Lipolysis; Adrenergic Hormones and Receptors Involved

	Hormone		*Receptor*	
	A	*NA*	α_2	$\beta_{(1),2,(3)}$
Adipose tissue				
Basal, postabsorptive			+	
Physical exercise	+	++		+
Mental stress	++	+		+
Hypoglycemia; normal subjects	+++	+	+	+
Hypoglycemia; IDDM	++	+	+	+
Abdominal surgery	+	++	+	+
Skeletal muscle				
Hypoglycemia	+++	+		+

A (adrenaline), NA (noradrenaline)

References

1. Lafontan, M., and Arner, P. (1996). Application of in situ microdialysis to measure metabolic and vascular responses in adipose tissue. *TIPS* **17**, 309–313.
2. Enocksson, S., Schimizu, M., Lönnqvist, F., Nordenström, J., and Arner, P. (1995). Demonstration of an in vivo functional beta$_3$-adrenoceptor in man. *J. Clin. Invest.* **95**, 2239–2245.
3. Galitzky, J., Lafontan, M., Nordenström, J., and Arner, P. (1993). Role of alpha$_2$-adrenoceptors in regulating lipid mobilization from human adipose tissue. *J. Clin. Invest.* **91**, 1997–2003.
4. Bolinder, J., Sjöberg, S., and Arner, P. (1996). Stimulation of adipose tissue lipolysis following insulin-induced hypoglycemia: Evidence of increased beta-adrenoceptor-mediated lipolytic response in IDDM. *Diabetologia* **39**, 845–853.
5. Hagström-Toft, E., Arner, P., Johansson, U., Eriksson, L. S., Ungerstedt, U., and Bolinder, J. (1992). Effects of insulin on human adipose tissue in situ. Interactions with beta-adrenoceptors. *Diabetologia* **35**, 664–670.

Patrice G. Guyenet, Yu Wen Li, Donghai Huangfu, and Ann M. Schreihofer

Department of Pharmacology
University of Virginia School of Medicine
Charlottesville, Virginia 22902

Bulbospinal C1-Adrenergic Neurons: Electrophysiological Properties in the Neonate Rat

The rostral ventrolateral medulla (RVLM) contains the C1 cell group, neurons that express all the enzymes necessary for the synthesis of adrenaline with the possible exception of dopamine-β-hydroxylase. C1 cells selectively innervate brain structures involved in autonomic regulation and are believed to play a key role in generating sympathetic tone, especially its vasomotor component (1). C1 cells are scattered at low density within the RVLM, which may explain the scarcity of studies devoted to their electrophysiological properties. This drawback has now been remedied to a considerable extent by two recently introduced technical developments. The first is the use of thin slices of neonatal brain in which individual cell bodies can be visualized prior to patching. The second is the retrograde labeling of bulbospinal cells with fluores-

Advances in Pharmacology, Volume 42

1054-3589/98 $25.00

cent tracers injected into the thoracic cord (2). Providing that the coronal slice selected for study is prepared from a narrow region located just caudal to the facial motor nucleus, the targeted population of RVLM fluorescent cells consists of about 65% C1 cells (for experimental details see refs. 2, 3). This chapter briefly summarizes data from whole cell recordings of more than 250 RVLM bulbospinal C1 neurons using the thin slice approach in neonate rats (3–12 days after birth). All recordings were done at room temperature. The overlapping work of Kangrga and Loewy (2) is also mentioned when appropriate.

The vast majority of C1 cells are slowly firing at rest (0.5–5.0 Hz; mean, 2.4 Hz) (2, 3). The pattern of firing of the most active cells (>3 Hz) is regular, with single spikes separated by ramplike interspike trajectories (2, 3). Slower cells discharge irregularly, and their action potentials are triggered from slow oscillations of the resting membrane potential (3). C1 cells discharge spontaneously, even when recordings are made in a cell-attached mode (no or partial intracellular access) with patch electrodes filled with normal extracellular medium (3). Therefore, their spontaneous activity is not an artifact due to whole-cell intracellular recording. The discharge rate of C1 cells is affected very little by the addition to the superfusate of a mixture of receptor antagonists, which eliminates most observable postsynaptic potentials (the nonselective glutamate-receptor antagonist kynurenic acid, the $GABA_A$-receptor antagonist bicuculline, and strychnine, a blocker of glycinergic receptors) (3), which suggests that $GABA_A$, glycinergic, and glutamatergic ionotropic receptors do not contribute significantly to the firing rate of these cells *in vitro*. Furthermore, the rate of discharge of C1 cells is not reduced when synaptic transmission is attenuated by incubating the slice in a low Ca^{2+}/high Mg^{2+} medium (3). In fact, this procedure produces an increase in the cells' discharge rate, associated with and probably caused by a reduction in the afterhyperpolarization (3). As a whole, the data suggest that the spontaneous discharges of C1 cells *in vitro* are due predominantly to intrinsic properties rather than to synaptic inputs.

The action potentials of C1 neurons are triggered primarily from slow 2- to 6-mV oscillations of the resting membrane potential in the slower firing cells (<2 Hz). The same oscillations can be uncovered in the most active cells during injection of small amounts of hyperpolarizing current (3). These oscillations disappear when the cells are hyperpolarized below −70 mV (3) or when external sodium is reduced to 27 mM (Huangfu and Guyenet, unpublished). The membrane oscillations are more resistant to the fast sodium channel blocker tetrodotoxin (TTX, 1 μM) than the action potential (3). These observations suggest that the membrane oscillations are intrinsic to the C1 cells and are due to a voltage-activated persistent sodium current that is less sensitive to TTX than the fast Na^+ current.

At present, only a limited number of voltage-dependent conductances have been characterized in C1 cells. These neurons have a hyperpolarization-activated current (I_h) (2, 3) with a threshold for activation of about −65 to 70 mV, which is at or slightly more negative than the membrane potential reached during the peak of the afterhyperpolarization. Therefore the contribution of I_h to the autoactivity of the cells in the neonate slice is doubtful. An A-like current, a delayed rectifier potassium current, and an afterhyperpolarization that appears to be at least partially Ca^{2+}-dependent have been described (2, 3).

C1 cells have high-threshold calcium currents and, in most cells, a low-threshold calcium current can be also detected (Li, Guyenet, and Bayliss, unpublished data). The low-threshold calcium current could contribute to spike initiation.

Bulbospinal C1 cells have somatodendritic α_2-adrenergic receptors, as indicated by the following evidence. Catecholamines cause hyperpolarization and a reduction in membrane resistance. Their effect persists in the presence of TTX, is mimicked by selective α_2-adrenergic agonists such as clonidine and UK 14,104, and is eliminated by the selective α_2-adrenergic antagonist methoxyidazoxan (3). In voltage clamp (~ -60 mV), catecholamines and the aforementioned α_2-adrenergic agonists produce an outward current that is blocked by 0.2 mM of barium and reverses polarity at the equilibrium potential for potassium in the presence of a range of extracellular K^+ concentrations (2.5–10.0 mM) (3). $GABA_B$-receptor stimulation produces effects identical to those of α_2-adrenergic agonists on K^+ currents (4). Noradrenaline also reduces the peak calcium current evoked by a depolarization from −100–0 mV by up to 35% on bulbospinal C1 cells (Li, Guyenet, and Bayliss, unpublished results). The attenuation of high-threshold calcium currents is also a hallmark of the stimulation of α_2-adrenergic receptors in many central nervous system (CNS) neurons. The downregulation of Ca^{2+} channels by α_2-adrenergic receptors in C1 cells could reduce dendritic spiking or the dendritic release of transmitters (if present) and may provide a mechanism for presynaptic control of transmitter release at the terminals. These α_2 receptors probably contribute to the inhibition of C1 cells *in vivo* by low doses of clonidine and related antihypertensive agents and secondarily to the sympatholytic and hypotensive action of these agents.

In the neonate, C1 cells have muscarinic and nicotinic receptors, both of which cause depolarization in current clamp and an inward current in voltage clamp (5). The muscarinic response may contribute to the hypertension caused by intoxication with acetylcholinesterase inhibitors.

Angiotensin II (AngII) depolarizes C1 cells and increases their firing rate *in vitro* (6). These effects are mediated via the activation of postsynaptic type-1 angiotensin receptors (AT-1 receptors) and are due, at least in part, to the closure of a resting potassium conductance (6). These neurophysiological data reinforce prior anatomical evidence that, in the RVLM, AngI receptors may be predominantly associated with the C1 cells. This also raises the interesting possibility that converting-enzyme inhibitors and AT-1-receptor antagonist (e.g., losartan) could produce some of their clinical antihypertensive effects by reducing the excitatory effect of AngII within the RVLM.

Bulbospinal C1 cells are excited by substance P *in vitro*. This effect involves postsynaptic receptors and the closure of a resting potassium conductance, as in the case of AngII (Li and Guyenet, unpublished results). However, contrary to those of AngII, the effects of substance P are not restricted to the adrenergic cells but are found in most other bulbospinal neurons of the RVLM (Li and Guyenet, unpublished results). The effects of substance P are mediated by neurokinin-1 (NK1) receptors (Li and Guyenet, unpublished results). These results are in agreement with immunohistochemical and other histological evidence suggesting the presence of a high density of NK1 receptors in the RVLM. They are also consistent with the fact that the microinjection of substance P in the RVLM produces sympathoexcitation and an increase in arterial pressure in rats. The most likely source of substance P input to the C1 cells is the medullary raphe.

In conclusion, our studies in the neonate rat indicate that C1 adrenergic neurons have two basic characteristics that are reminiscent of other CNS catecholaminergic neurons, namely (1) a slow nonbursting pattern of discharge that is due to intrinsic properties of the neurons rather than synaptic drives and (2) somatodendritic catecholaminergic inhibitory receptors (autoreceptors). The autoactivity of C1 cells *in vitro* appears due, in large part, to a persistent sodium current, which leads to slow, irregular fluctuations of the membrane potential from which action potentials are triggered. This characteristic also may be present in mature C1 cells, because comparable membrane oscillations have been observed *in vivo* in adults rats (7). The presence of inhibitory α_2-adrenergic receptors on C1 cells probably accounts for their sensitivity to low doses of clonidine *in vivo*. The somatodendritic receptors so far characterized on C1 cells are in good agreement with the type of synaptic inputs previously described by anatomical means (GABA, acetylcholine, substance P, angiotensin) and are reminiscent of those present on the locus ceruleus. However, the similarity between C1 and other monoaminergic neurons, including the locus ceruleus, should not be overemphasized, because other evidence suggests that C1 cells may not have a high-affinity amine transporter, may use glutamate as their principal transmitter, and may play a specialized role in controlling the vasomotor component of the sympathetic outflow (1).

Acknowledgments

This work was supported by grants from the National Institutes of Health (RO1 HL 28785) and the American Heart Association.

References

1. Guyenet, P. G., Koshiya, N., Huangfu, D., Baraban, S. C., Stornetta, R. L., and Li, Y. W. (1996). Role of medulla oblongata in generation of sympathetic and vagal outflows. *Prog. Brain Res.* **107**, 127–144.
2. Kangrga, I. M., and Loewy, A. D. (1995). Whole-cell recordings from visualized C1 adrenergic bulbospinal neurons: Ionic mechanisms underlying vasomotor tone. *Brain Res.* **670**, 215–232.
3. Li, Y. W., Bayliss, D. A., and Guyenet, P. G. (1995). C1 neurons of neonatal rats: Intrinsic beating properties and α_2-adrenergic receptors. *Am. J. Physiol.* **269**, R1356–R1369.
4. Li, Y. W., and Guyenet, P. G. (1996). Activation of $GABA_B$ receptors increases a potassium conductance in rat bulbospinal neurons of the C_1 area. *Am. J. Physiol.* **271**, R1304–R1310.
5. Huangfu, D., Schreihofer, A. M., and Guyenet, P. G. (1997). Effect of cholinergic agonists on bulbospinal C_1 neurons in rats. *Am. J. Physiol.* **272**, R249–R258.
6. Li, Y. W., and Guyenet, P. G. (1996). Angiotensin II decreases a resting K^+ conductance in rat bulbospinal neurons of the C1 area. *Circ. Res.* **78**, 274–282.
7. Lipski, J., Kanjhan, R., Kruszewska, B., and Rong, W. F. (1996). Properties of presympathetic neurones in the rostral ventrolateral medulla in the rat: An intracellular study '*in vivo*'. *J. Physiol. (Lond.)* **490**, 729–744.

Virginia M. Pickel,* Sue A. Aicher,* Chiye Aoki,†, Peter Y. Cheng,* and Melissa J. Nirenberg*

*Department of Neurology and Neuroscience
Division of Neurobiology
Cornell University Medical College
New York, New York 10021

†Center for Neural Sciences
New York University Medical Center
New York, New York 10003

Catecholamines, Opioids, and Vagal Afferents in the Nucleus of the Solitary Tract

Catecholamines and/or opioids in the nucleus of the solitary tract (NTS) can modulate both cardiorespiratory and nociceptive reflexes. (See ref. 1 for discussion and references.) In addition, both noradrenergic and adrenergic neurons in this region contain endogenous opioid peptides and receive direct synaptic input from vagal afferents. The functional sites for storage and release of catecholamines and for the physiological effects of catecholamines and/or opioids in the NTS, however, are largely unknown. To identify these sites, we used previously characterized antipeptide antisera to examine the immunocytochemical localization of (a) the vesicular monoamine transporter-2 (VMAT2), the transporter responsible for the reserpine-sensitive vesicular uptake of monoamines in neurons (2); (b) the β_2-adrenergic receptor (βAR), a receptor subtype implicated in the presynaptic regulation of noradrenaline release (3, 4); and (c) the μ-opioid receptor (μOR), the major opioid-receptor subtype involved in modulation of vagal reflexes (5).

Sprague-Dawley rats were anesthetized with pentobarbital (100 mg/kg i.p.), and the brains were fixed by vascular perfusion with 2% paraformaldehyde and 3.8% acrolein. Coronal sections through the NTS at the level of the area postrema were cut with a Vibratome. These sections were processed for immunogold silver and/or immunoperoxidase labeling of rabbit antipeptide antisera against VMAT2, βAR, and μOR in three sets of experiments. The antisera were provided by the laboratories of Drs. R. H. Edwards (Department of Neurology and Physiology, University of California), C. D. Strader (Shering-Plough Research Institute), and L. Y. Liu-Chen (Department of Pharmacology, Temple University School of Medicine), respectively. We also used a commercial mouse monoclonal antiserum (Incstar) directed against tyrosine hydroxylase (TH) to identify the catecholaminergic neurons. Further details about the antisera, labeling protocols, and in-depth studies have been published (1–5).

Advances in Pharmacology, Volume 42

1054-3589/98 $25.00

In the first set of experiments directed toward identification of monoamine storage sites in the NTS, we showed the cellular and subcellular localization of VMAT2 (2). By light microscopy, VMAT2 was localized to the perinuclear region of TH-labeled perikarya, as well as to varicose and nonvaricose processes. Electron microscopy showed that in perikarya, VMAT2 was localized to the trans-Golgi lamellae and associated large dense-core vesicles (LDCVs) and tubulovesicular organelles. In dendrites, there was also extensive labeling of tubulovesicles and infrequent labeling of LDCVs. In axon terminals, there was preferential localization of VMAT2 to membranes of LDCVs distant from synaptic vesicles, less frequent labeling of small synaptic vesicles (SSVs), and only rare labeling of tubulovesicular structures. The prominent localization of VMAT2 to LDCVs in presumptive noradrenergic, adrenergic, and serotonergic terminals in the NTS is in marked contrast to the localization of the transporter in dopaminergic terminals in the striatum, where it is detected almost exclusively in SSVs (see Nirenberg *et al.,* in this volume). These results show differences in the subcellular location of VMAT2, suggesting that there are compartment, region, and cell-specific differences in the sorting of this protein within monoaminergic neurons. They also suggest that catecholamines may be released by exocytosis mainly from LDCVs to reach receptors at a distance from synaptic junctions.

In the second series of experiments, we established the localization of βAR in both neurons and glia in the NTS (3, 4). We showed that in neuronal perikarya and large dendrites in this region, βAR labeling was mainly associated with cytoplasmic membranes, Golgi lamellae, smooth endoplasmic reticulum, and/or tubulovesicles. In smaller dendrites and in axon terminals, however, βAR immunoreactivity was more prominently associated with pre- and postsynaptic membrane specializations. Some of the βAR-labeled neurons also contained TH immunoreactivity, supporting physiological evidence that this receptor is involved in autoregulation of catecholaminergic neurons. With one of the two βAR antisera, we observed prominent labeling of astrocytic processes apposed to catecholaminergic neurons (3). The results suggest that occupancy of βAR in the NTS can modulate not only the presynaptic release and postsynaptic responses of catecholaminergic neurons, but also the activity of neighboring glia. The findings support earlier studies showing that radiolabeled βAR ligands bind to astrocytes. (See ref. 3 for references to relevant literature.) Furthermore, because activation of βAR *in vivo* has been shown to alter the uptake of both glutamate and γ-aminobutyric acid (GABA), two of the neurotransmitters present in afferents to the catecholaminergic neurons, our results suggest that adrenergic modulation of catecholaminergic neurons in the NTS may also be mediated through astrocytes. (See ref. 4 for further discussion and additional references.)

Because opioid peptides are also stored in LDCVs and may be released at sites distant from typical synaptic junctions, in the third series of experiments, we examined the localization of μOR to determine whether this opioid-receptor subtype was present at extrasynaptic sites on neurons and/or glia in the NTS (5). Light microscopy showed prominent μOR labeling in the subpostrema region and in the dorsomedial and medial NTS. This distribution closely paralleled that of the vagal afferents, which were identified by anterograde labeling in the NTS following injections of biotinylated dextran amine (BDA) into the nodose ganglion. Electron microscopy showed that immunogold labeling for the μOR was prominently localized to extrasynaptic plasma membranes of dendrites, many of which received asymmetric, excitatory-type synapses. Unmyelinated axons and axon terminals showed μOR labeling associated with both the plasma membrane and the membranes of SSVs. Many of these μOR-labeled terminals were large and formed asymmetric synapses, as is characteristic of vagal afferents. Furthermore, in sections processed for μOR-immunogold labeling and for peroxidase detection of anterogradely transported BDA, some of these axons were dually labeled. Our results suggest that μOR agonists can produce direct extrasynaptic modulation of the excitability of neurons in the NTS. In addition, the findings provide ultrastructural evidence for involvement of μOR in the presynaptic release of neurotransmitters from vagal afferents to this region.

In summary, the results of these related studies suggest that there is prominent nonsynaptic vesicular release of catecholamines and/or opioid peptides in the NTS, and that this occurs mainly from LDCVs. These findings, as well as the presence of βAR on glia and μOR at extrasynaptic sites on dendrites and axons, suggest that catecholamines and opioids are involved in parasynaptic transmission. Our demonstration that VMAT2 is present in SSVs and that βARs are also localized within synaptic junctions suggests that catecholamines are also involved in classical synaptic transmission. Furthermore, the presence of μOR on vagal afferents suggests a major involvement of opiates in the presynaptic release of glutamate, one of the primary neurotransmitters in visceral afferents to the NTS.

Acknowledgments

V.M.P. receives salary support from a grant provided by the National Institute of Mental Health (MH00078) and research support for studies in the NTS from NIMH grant (MH48776) and from National Institutes of Health grant (HL-18974). P.Y.C. was supported in part by an Aaron Diamond Postdoctoral Fellowship.

References

1. Pickel, V. M., Chan, J., and Milner, T. A. (1989). Ultrastructural basis for interactions between central opioids and catecholamines. II. Nuclei of the solitary tracts. *J. Neurosci.* **9,** 2519–2535.
2. Nirenberg, M. J., Liu, Y. J., Peter, D., Edwards, R. H., and Pickel, V. M. (1995). The vesicular monoamine transporter 2 is present in small synaptic vesicles and preferentially localizes to large dense core vesicles in rat solitary tract nuclei. *Proc. Natl. Acad. Sci. U.S.A.* **92,** 8773–8777.
3. Aoki, C., and Pickel, V. M. (1992). C-terminal tail of β-adrenergic receptors: Immunocytochemical localization within astrocytes and their relation to catecholaminergic neurons in N. tractus solitarii and area postrema. *Brain Res.* **571,** 35–49.

4. Aoki, C., Zemcik, B. A., Strader, C. D., and Pickel, V. M. (1989). Cytoplasmic loop of β-adrenergic receptors: Synaptic and intracellular localization and relation to catecholaminergic neurons in the nuclei of the solitary tracts. *Brain Res.* **493,** 331–347.
5. Cheng, P. Y., Liu-Chen, L. Y., Chen, C., and Pickel, V. M. (1996). Immunolabeling of μ-opioid receptors in the rat nucleus of the solitary tract: Extrasynaptic plasmalemmal localization and association with Leu^5-enkephalin. *J. Comp. Neurol.* **371,** 522–536.

D. J. Reis and S. Regunathan

Division of Neurology
Cornell University Medical College
New York, New York 10021

Agmatine: A Novel Neurotransmitter?

Agmatine is a cationic amine formed by the decarboxylation of arginine by arginine decarboxylase (ADC). It is a normal constituent of bacteria, plants, and many invertebrates, in which it is primarily metabolized by hydrolysis by agmatine ureohydrolase (agmatinase) to putrescine, the metabolic precursor for polyamines, including spermine and spermidine. This represents an accessory pathway for putrescine biosynthesis to the decarboxylation by ornithine decarboxylase (ODC), the only pathway expressed in mammals.

Until recently, neither agmatine, ADC, or agmatinase were considered to be expressed in mammals. However, in 1994 (1), while searching for endogenous ligands for imidazoline receptors (IRs), we isolated agmatine from bovine brain (1) and discovered that it is widely expressed in various organs and tissues (1, 2). The observations indicated, for the first time, that agmatine is also present in mammalian species.

The presence of the amine in brain and the fact that it binds to α_2-adrenergic receptors and IRs raised the question of whether agmatine might, like other bioamines, have biological actions in its own right, possibly as a neurotransmitter–neuromodulator. We have pursued this hypothesis and summarize evidence to support the contention.

Agmatine

Advances in Pharmacology, Volume 42

1054-3589/98 $25.00

I. Neurochemistry

A. Distribution and Storage

Agmatine is present in whole bovine and rat brain in concentrations (2.5–15.5 ng/gm) comparable to that of the classic monoamine transmitters dopamine, adrenaline, or serotonin (2, 3). It is also, like many other transmitters, found in chromaffin cells of adrenal medulla.

Immunocytochemically, agmatine has been localized in brain and adrenal using polyclonal antibodies (4). While in untreated brain agmatine-like immunoreactivity (-LI) is detected in only a few perikarya, after pretreatment with intracerebroventricular (i.c.v.) colchicine, it can be seen in the cytoplasm of many neurons throughout the brain. These are most heavily concentrated in hypothalamus, neocortex, and periaquaductal regions of midbrain. By electron microscopy agmatine-LI can be detected in dense-core vesicles in axon terminals in the nucleus tractus solatarii and in the hilus of the hippocampal formation. Hence, its storage site is that which would be expected of a neurotransmitter candidate.

B. Biosynthesis

Brain agmatine appears to be locally synthesized, because rat brain expresses ADC activity. Unlike bacterial ADC, which is cytosolic, mammalian ADC is membrane associated, synaptosomal, and enriched in the innner membranes of mitochondria. This is probably why previous investigators failed to detect the enzyme in soluble extracts of mammalian tissue. ADC is unevenly distributed in brain, with activity greatest in cerebral cortex and hippocampus, in agreement with the immunocytochemical distribution of agmatine. The mammalian enzyme has other properties that distinguish it from the bacterial enzyme: Rat hepatic ADC utilizes arginine *and* ornithine as substrates (Kms of 0.75 and 0.25 mM) and has pH and thermal optima of 8.23 and 25°C, respectively. ADC activity is facilitated by Mg^{2+} and inhibited by Ca^{2+}, while other ions are ineffective. In contrast to most decarboxylases, ADC does not require pyridoxal phosphate as a cofactor for maximal activity. Difluoromethylarginine (DFMA) or -ornithine (DFMO), selective inhibitors of bacterial ADC or ODC, or inhibitors of nitric oxide synthase (NMA, NMMA), an enzyme also utilizing arginine as substrate, are ineffective as inhibitors. Whether ADC is expressed in neurons is not known. Agmatine can be synthesized in glia and might be taken up and stored (but not synthesized) in neurons. A precedent for the possibility exists in blood vessels, where agmatine is stored in endothelium and smooth-muscle cells, but only the former expresses ADC. The absence of specific inhibitors of mammalian ADC and agmatinase has limited the investigations into the function of the amine in brain.

C. Release

Like other transmitters, agmatine is released by depolarization of rat brain slices, synaptosome, or bovine adrenal chromaffin cells. Incubation of these tissues with guanido [^{14}C] or [^{3}H]agmatine or [^{3}H]putrescine releases radioactivity into the medium, as measured for 10 min after exposure to 55 mM of KCI (brain) or nicotine (100 μM) (adrenal) with or without Ca^{2+}. Depolarization with 55 mM of KCI results in significant ($p < .05$; n = 6) release of labeled agmatine from synaptosome (9.9 ± 0.36 vs 12.6 ± 0.37, expressed as percent of total radioactivity released) or slices (9.1 ± 0.78 vs 15.3 ± 2.5), while removal of Ca^{2+} from the incubation medium significantly ($p < .05$) reduces it. Because under these conditions [^{3}H]putrescine is not released, the radioactivity can not be attributed to metabolic conversion of agmatine to putrescine. Depolarization with KCI (55 mM) or nicotine (10 μM) also significantly releases [^{3}H]agmatine from adrenal chromaffin cells. These results indicate that, like other transmitters, agmatine can be released from presynaptic neurons and chromaffin cells in a calcium-dependent manner by depolarizing stimuli.

D. Inactivation

1. Reuptake

Agmatine can be biologically inactivated in mammalian brain by two mechanisms: re-uptake and enzymatic degradation. Uptake has been studied by measuring accumulation of guanido [^{14}C]agmatine (4.2 μM) in the P_2 fractions of synaptosome prepared from rat brain. The uptake of agmatine is temperature-dependent and saturable only at high concentrations (Km of 18.8 ± 3.3 mM and Vmax of 4.78 ± 0.67 nmol/mg protein/min). Treatment with ouabain (a Na^{+}/K^{+} adenosine triphosphatase inhibitor) or replacement of extracellular Na^{+} did not attenuate the uptake. Agmatine uptake was not inhibited by various amino acids, polyamines, or monoamines. Of various ion-channel modulators, only Ca^{2+} channel blockers inhibited, whereas reduction of extracellular Ca^{2+} increased uptake. Some drugs acting at imidazoline receptors (e.g., idazoxan [Ki = 0.24 mM] and phentolamine [Ki = 0.22 mM]), were strong noncompetitive inhibitors of uptake. These observations indicate that synaptosome can take up and concentrate agmatine by a mechanism differing from common amino acid, polyamine, or monoamine transporters and might be mediated or regulated by calcium channel activity.

2. Metabolism

In bacteria, the major metabolic pathway for agmatine is hydrolysis by agmatinase to putrescine and urea. We have discovered that agmatinase activity can be detected in mammalian brain. Agmatinase was assayed by measuring hydrolysis of guanido [^{14}C]agmatine to [^{14}C]urea and putrescine and subsequent trapping of $^{14}CO_2$ released by urease from [^{14}C]urea. Incubation of guanido [^{14}C]agmatine with rat brain homogenates resulted in substantial hydrolysis of agmatine (7.6–11.8 nmol/hr/mg protein). Activity was reduced (up to 75%) by boiling, while approximately 25% of total activity remained in (−) homogenate controls due to non-enzymatic degradation of agmatine. With subcellular fractionation of rat brain agmatinase, activity was maximal (48.1 nmol/hr/mg protein) in the soluble fraction of the P_2 pellet (synaptosomal–mitochondrial). Further fractionation of the P_2 pellet resulted in enrichment of agmatinase in the mitochondrial (327.8 nmol/hr/mg protein) versus synaptosomal (31.2 nmol/hr/mg protein) fractions. Agmatinase activity in the P_2 pellets varied regionally in brain: hypothalamus (133 nmol/hr/mg protein) > hippocampus (88) > medulla (64.5) > cerebellum (47.5) > striatum (35.8) > cerebral cortex (30.2). These observations indicate that rat brain expresses agmatinase, which can convert agmatine to putrescine and urea, and that the enzyme is soluble and associated with mitochondria. The importance of the finding is that not only can agmatine be metabolically degraded, but also the presence of an agmatine–putrescine pathway suggests a novel metabolic pathway for polyamine biosynthesis in mammals.

II. Receptors Mechanisms

Agmatine binds with high affinity to α_2-adrenergic receptors of all subclasses and to the imidazoline receptors of the I_1 and I_2 subclasses (5). Agmatine is not promiscuous at adrenergic receptors, for it does not bind to α_1- or β-adrenergic receptors, suggesting it might be a ligand for only the α_2 subclass. It is not known whether agmatine is an agonist and/or antagonist at any receptor, although most endogenous ligands are agonists.

However, agmatine may have unique features in so far as, as an organic cation, it may enter the cell via cation channels, including voltage-gated and ligand-gated Ca^{2+} channels. The latter include cholinergic, nicotinic, and N-methy-D-aspartate receptors. Thus, agmatine may be able to enter cells without metabolic conversion and not only interact with the channels, but also reach I_2 receptors that are appended to mitochondria.

III. Functions

That agmatine is biologically active has been demonstrated in a number of systems. It dose-dependently releases catecholamines from adrenal chromaffin cells, insulin from pancreatic islet cells exposed to glucose, and luteinizing hormone–releasing hormone from the hypothalamus of ovariectomized rats, and enhances gastric secretion. It blocks the stimulated proliferation of vascular smooth–muscle cells of rat aorta by interaction with I_2 receptors and, when administered intrathecally in mice, enhances morphine-induced analgesia ninefold via actions on κ_1 receptors without affecting pain thresholds, and acutely prevents the morphine tolerance at the same receptor. The cellular mechanisms of these effects, however, are not fully understood for, as yet, it is not certain that they are mediated by α_2-adrenergic receptors or IRs. Conceivably, some of the actions may relate to modulation by agmatine of ionic conductance, including competition at voltage-gated or ligand-gated cation channels.

Although agmatine binds to α_2-adrenoreceptors, its actions at this site are unclear. To date, no one has been able to show actions of agmatine attributable to agonism or antagonism at the site *in vitro*. We have, however, recently observed that in the rat tail artery, agmatine can inhibit the contractions elicited by transmural nerve stimulation, an effect antagonized by α_2-adrenergic receptor antagonists.

An additional issue of interest is whether agmatine is a precursor of polyamines in mammals, as in bacteria. While it is generally assumed that in mammals polyamines are generated from putrescine formed from ornithine by the action of ODC, we have reported that rat brain expresses agmatinase, indicating an alternative pathway for polyamine biosynthesis in mammals.

IV. Summary

Current evidence is consistent with an hypothesis that agmatine meets many criteria for a neurotransmitter–neuromodulator. It is synthesized, stored, and released in brain; is contained in neurons and axon terminals with a heterogeneous distribution; interacts with cell-specific receptors; and elicits biological actions within the central nervous system. Its role in normal brain function, however, has not yet been established, in part because of the absence of agents that selectively affect its biosynthesis or degradation.

References

1. Li, G., Regunathan, S., Barrow, C. J., Eshraghi, J., Cooper, R., and Reis D. J. (1994). Agmatine: An endogenous clonidine-displacing substance in brain. *Science* **263,** 966–969.
2. Raasch, W., Regunathan, S., Li, G., and Reis, D. J. (1995). Agmatine, the bacterial amine, is widely distributed in mammalian tissues. *Life Sci.* **56,** 2319–2330.
3. Feng, Y., Halaris, A. E., and Piletz, J. E. (1997). Determination of agmatine in brain and plasma using high-performance liquid chromatography with fluorescence detection. *J. Chromatogr.* **691,** 277–286.
4. Wang, H., Regunathan, S., Youngson, C., and Reis, D. J. (1995). An antibody to agmatine localizes the amine in bovine adrenal chromaffin cells. *Neurosci. Lett.* **183,** 17–21.
5. Regunathan, S., and Reis, D. J. (1996). Imidazoline receptors and their endogenous ligands. *Annu. Rev. Pharmacol. Toxicol.* **36,** 511–544.

M. Esler, G. Lambert, G. Jennings, A. Turner, and D. Kaye

Baker Medical Research Institute
Prahran 3181
Melbourne, Australia

Central and Peripheral Norepinephrine Kinetics in Heart Failure, Coronary Artery Disease, and Hypertension

Traditionally, the assessment of sympathetic nervous system function in human cardiovascular disorders has involved the measurement of plasma norepinephrine concentrations in antecubital venous plasma. This methodology has several deficiencies. One is the dependence of the plasma concentration on norepinephrine clearance, which is sometimes altered, such as in cardiac failure. The second drawback is that sympathetic nervous system responses are typically differentiated (regionalized), so that activation of one sympathetic outflow might be accompanied by no change or a reduction in sympathetic in other sites. One example of this is provided by the sympathetic nervous response to dietary sodium restriction, in which sympathetic stimulation is evident in the renal sympathetic nerves, while cardiac sympathetic tone is unchanged (1).

I. Essential Hypertension

Measurement of regional sympathetic nervous system activity in patients with essential hypertension, using powerful electrophysiological methods (clinical microneurography, which measures skeletal muscle sympathetic nerve firing rates) and neurochemical techniques (radiotracer-derived measurement of norepinephrine spillover from individual organs), has indicated that activation of the sympathetic outflows to skeletal muscle, the heart, and the kidneys is commonly present, particularly in younger patients (2). No consistent increase in cardiac sympathetic activity is evident in hypertensive patients when heart rate spectral analysis methods are used, utilizing 0.1-Hz spectral power as a surrogate measure of cardiac sympathetic tone, but this is due to the well-demonstrated inability of heart rate spectral analysis to validly quantify efferent sympathetic activity in the heart.

This sympathetic nervous activation no doubt contributes to the blood pressure elevation through neural influences on cardiac performance, vascular

Advances in Pharmacology, Volume 42

1054-3589/98 $25.00

resistance, and renal function (renin secretion, sodium reabsorption). In addition, it appears to have adverse consequences in hypertensive patients beyond blood pressure elevation. There is evidence that neural vasoconstriction has metabolic effects, in skeletal muscle impairing glucose delivery to muscle, causing insulin resistance and hyperinsulinemia. A trophic effect of sympathetic activation on cardiovascular growth is also probable, contributing to the development of left ventricular hypertrophy (2). An arrhythmogenic effect of the cardiac sympathetic activation is also likely, possibly accentuated by vasodilator antihypertensive drugs, which stimulate the cardiac sympathetic outflow.

II. Coronary Artery Disease

Sympathetic nervous stimulation also contributes importantly to coronary artery disease syndromes. Experimental studies in nonhuman primates incriminate sympathetic nervous activation in atherogenesis, perhaps due to increased splanchnic sympathetic tone that reduces visceral blood flows and retards postprandial clearing of plasma lipids or through sympathetic nervous stimulation of the cardiovascular system that increases arterial shear stress and endothelial damage. In patients with unstable angina, the cardiac sympathetic outflow is activated at rest; this is thought to contribute both to thrombogenesis and to the development of ventricular tachyarrhythmias. In such patients, drugs that stimulate the cardiac sympathetic outflow appear to predispose to arrhythmia development.

In patients with stable coronary artery disease who unexpectedly develop life-threatening ventricular arrhythmias, caridac sympathetic tone after recovery has been demonstrated to be markedly increased (3). A behavioral trigger for such arrhythmias is often evident, as has been demonstrated frequently in studies of nontraumatic sudden death during an earthquake and in patients with panic disorder. Mental stress studied in the laboratory has been shown to cause dramatic and relatively selective stimulation of the sympathetic nerves of the heart.

III. Cardiac Failure

The demonstration that the level of sympathetic nervous drive to the failing heart in patients with severe heart failure is a major determinant of prognosis (4). Mortality in heart failure is reduced by β-adrenergic blockade with carvedilol, indicating the clinical relevance of cardiac neuroscience research. Important initial findings were observations that the plasma concentration of sympathetic transmitter, norepinephrine, is elevated in heart failure and that clinical outcome overall is related to the plasma norepinephrine concentration (although here heart failure severity may be a confounder). Sympathetic nerve recording and radiotracer methods measuring regional sympathetic activity in the heart have now largely supplanted antecubital venous norepinephrine measurements as research tools, with the newer methods providing information on regional sympathetic function that was previously lacking.

The cardiac sympathetic nerves are preferentially stimulated in severe heart failure, with norepinephrine release from the failing heart at rest being increased as much as 50-fold, similar to the level seen in healthy people during maximum exercise (1). There is lesser stimulation of the sympathetic outflows to the kidneys and skeletal muscle. In early, mild heart failure it is *only* the cardiac sympathetic nerves that are activated. This preferential activation of the cardiac sympathetic outflow contributes to arrhythmogenesis and also probably to progression of the heart failure and has been linked to mortality in both mild and severe cardiac failure.

An additional neurophysiological abnormality that seems to be present in heart failure patients is presynaptic augmentation of transmitter release at the existing high rates of sympathetic nerve firing. The mechanism is uncertain but may involve regional release of adrenaline from the failing heart (5) acting on presynaptic β-adrenergic receptors on sympathetic nerves. The increased cardiac norepinephrine spillover could also possibly arise in part from impaired neuronal re-uptake of the neurotransmitter.

IV. Central Nervous System Control of Human Sympathetic Nervous Outflow

Although sympathetic nervous activation is present in patients with hypertension and with heart failure, the central nervous system (CNS) mechanisms involved are not entirely clear. In heart failure patients, increased intracardiac diastolic pressure seems to one peripheral reflex stimulus, and increased forebrain norepinephrine turnover an important central mechanism (6). Experiments in laboratory animals have challenged the conventional view that the dominant effects of CNS noradrenergic neurons in cardiovascular control are sympathetic nervous inhibition and blood pressure reduction, describing instead sympathetic activation. We have tested whether such a stimulant effect on sympathetic outflow is also evident in human hypertension and heart failure. CNS norepinephrine turnover was estimated from the combined overflow of norepinephrine, methoxyhydroxyphenylglycol (MHPG), and dihroxyphenylglycol into the internal jugular veins. Cerebral blood flow scans allowed differentiation between cortical and subcortical jugular venous drainage (7).

In patients with pure autonomic failure, jugular overflow of norepinephrine and metabolites was not reduced, indicating brain neurons and not cerebrovascular sympathetics were the source. In healthy men, CNS norepinephrine turnover and muscle sympathetic nerve activity were directly related ($p < .02$). Administration of the ganglion blocker, trimethaphan, caused a compensatory fivefold increase in jugular overflow of MHPG. Conversely, intravenous clonidine reduced CNS norepinephrine turnover by approximately 50%, this possibly representing a mechanism of drug action. In cardiac failure patients, sympathetic nervous activation was associated with a trebling of CNS norepinephrine turnover ($p < .01$). In untreated patients with essential hypertension, the sympathetic activation present was associated with 250% higher CNS norepinephrine turnover ($p < .01$) but only in subcortical brain regions.

A close and direct relation exists between brain norepinephrine turnover and human sympathetic nervous activity. CNS release of norepinephrine, presumably in the forebrain where noradrenergic neurons are sympathoexcitatory and pressor, appears to mediate increased sympathetic nerve firing in patients with essential hypertension and cardiac failure. Elucidation of the abnormalities in central nervous control of sympathetic outflow in heart failure in particular has become of major clinical relevance. Following the demonstration of the benefical effects of β-adrenergic blockade in heart failure, imidazoline receptor binding agents, centrally acting sympathetic nervous system suppressants similar to clonidine, are also under evaluation. Such drugs inhibit both firing of brain noradrenergic neurons and sympathetic nerve firing and are potentially cardioprotective, although this is yet to be established.

References

1. Esler, M., Jennings, G., Lambert, G., Meredith, I., Horne, M., and Eisenhofer, G. (1990). Overflow of catecholamine neurotransmitter to the circulation: Source, fate and functions. *Physiol. Rev.* **70,** 963–985.
2. Esler, M., Lambert, G., and Jennings, G. (1990). Increased regional sympathetic nervous activity in human hypertension: Causes and consequences. *J. Hypertens.* **8,** (Suppl 7), S53–S57.
3. Meredith, I. T., Broughton, A., Jennings, G. L., and Esler, M. D. (1991). Evidence for a selective increase in resting cardiac sympathetic activity in some patients suffering sustained out of hospital ventricular arrhythmias. *N. Eng. J. Med.* **325,** 618–624.
4. Kaye, D. M., Lefkovits, J., Jennings, G. L., Bergin, P., Broughton, A., and Esler, M. D. (1995). Adverse consequences of high sympathetic nervous activity in the failing human heart. *J. Am. Coll. Cardiol.* **26,** 1257–1263.
5. Kaye, D. M., Cox, H., Lambert, G., Jennings, G. L., Turner, A., and Esler, M. D. (1995). Regional epinephrine kinetics in severe heart failure: Evidence for extra-adrenal, non-neural release. *Am. J. Physiol.* **269,** H182–H188.
6. Kaye, D. M., Lambert, G. W., Lefkovits, J., Morris, M., Jennings, G. L., and Esler, M. D. (1994). Neurochemical evidence of cardiac sympathetic activation and increased central nervous system norepinephrine turnover in severe congestive heart failure. *J. Am. Coll. Cardiol.* **23,** 570–578.
7. Ferrier, C., Jennings, G. L., Eisenhofer, G., Lambert, G., Cox, H. S., Kalff, V., Kelly, M., and Esler, M. D. (1993). Evidence for increased noradrenaline release from subcortical brain regions in essential hypertension. *J. Hyperten.* **11,** 1217–1227.

PART F

CATECHOLAMINES IN THE CENTRAL NERVOUS SYSTEM

A. A. Grace,* C. R. Gerfen,† and G. Aston-Jones‡

*Departments of Neuroscience and Psychiatry
University of Pittsburgh
Pittsburgh, Pennsylvania 15260

†Laboratory of Systems Neuroscience
National Institute of Mental Health
Bethesda, Maryland 20892

‡Department of Psychiatry
Allegheny University
Philadelphia, Pennsylvania 19102

Overview

Studies concentrating on catecholaminergic systems in the central nervous system (CNS) have initially lagged far behind the more aggressive approaches used to examine their function in the periphery. However, the preponderance of studies over the past several decades demonstrating a role for these neurotransmitter systems in neurological and psychiatric disorders has been a driving force for studies focusing on central aspects of catecholaminergic function. Indeed, it is becoming increasingly clear that dopamine, norepinephrine (NE), and epinephrine exhibit essential roles in the regulation and coordination of activity among widespread central systems. A multitude of methodological approaches was applied to study a comparatively diverse collection of topic materials; nonetheless, there was a remarkable degree of convergence among the

Advances in Pharmacology, Volume 42

1054-3589/98 $25.00

experimental findings and conceptual models advanced. The papers presented in this part generally could be divided along two foci: the anatomy, pharmacology, and molecular regulation of catecholamines in the normal brain; and how these relationships may be altered in disease states.

I. Molecular Regulation of Dopaminergic Responses

An effective approach for studying the functional role of dopamine in the basal ganglia has been to measure changes in the expression levels of various genes and their products in the basal ganglia following pharmacological manipulations. Gerfen and his colleagues (p. 670) describe studies of dopamine regulation of striatal output neurons, which follow from their demonstration of the segregation of D1 and D2 dopamine receptor subtypes, respectively, to the so-called direct–striatonigral and indirect–striatopallidal striatal projection neurons. One issue surrounding the segregation of these receptor subtypes stems from models that synergy between the subtypes requires colocalization. To address this issue, single-cell measurements were made of changes in the immediate early gene zif 268, which has a constitutive level of expression and thus allows both increases and decreases in expression levels to be examined in identified direct (D1) or indirect (D2) striatal projection neurons in animals treated with a D1-receptor agonist alone or in combination with D2 agonists. Results demonstrated that combined treatment with D1 and D2 agonists results in a selective increase in gene regulation in D1-expressing "direct" projection neurons and decreased gene regulation in D2-expressing "indirect" projection neurons. Thus, the synergistic interaction between these receptor subtypes appears to occur through intercellular interactions between different neurons, rather than through intracellular interactions between receptors expressed by the same neurons. While a variety of gene-regulation effects are seen by manipulation of dopaminergic neurotransmission, the functional significance of these effects remains for the most part obscure. Work by Steiner and Gerfen was described, which has shown that elevated expression of the opiate neuropeptide dynorphin following repeated overstimulation of D1 receptors (e.g., by cocaine) is an adaptive response of these neurons to blunt the effects of such overstimulation.

While there are many effects of dopamine receptor–mediated action within the striatum, other parts of the basal ganglia circuitry are necessarily involved in the behavioral consequences of dopaminergic system manipulation. Chesselet and her coworkers (p. 674) describe studies in which such extrastriatal basal ganglia effects are examined by measurement of changes in GAD 67 mRNA in basal ganglia nuclei downstream from the striatum. Although it might be anticipated that dopamine-mediated changes in the output of the striatum would be reflected in changes in the target nuclei of that output, the direction of such changes does not always appear to follow from the expected changes in activity levels of the neurons. For example, elevated markers in striatopallidal neurons resulting from striatal dopamine depletion or neuroleptic treatment resulted in increased GAD 67 mRNA levels in globus pallidus neurons, which may be considered surprising because increased striatopallidal output would be expected to decrease activity in such pallidal neurons. A number of possible mechanisms are examined that might account for these results. Rather than simply altering the level of activity in pallidal neurons, changes in striatal outputs might alter the pattern of activity, which might result from altered membrane properties. Along these lines, results are described in which altered dopaminergic neurotransmission results in altered expression of the Kv3.1 "shaw-like" potassium channel. Moreover, nuclei that are connected with, but not necessarily considered part of, the basal ganglia are shown to display alterations in gene expression and are examined in dopaminergic manipulation paradigms. One result of interest is that of a change in the level of GAD 67 in the thalamic reticular nucleus. Such results are discussed in terms of the circuitry and mechanisms through which dopamine exerts effects on behavior.

Marshall and his colleagues (p. 678) describe studies from their laboratory that have examined D1 and D2 dopamine receptor regulation of striatal output neurons in paradigms utilizing psychostimulant drugs such as amphetamine. They provide a number of examples of an obligatory dependence on D2 receptor mechanisms of D1–receptor–mediated induction of Fos in striatonigral neurons. Such studies point out the importance of D1- and D2-receptor interactions in the effect of dopamine in the striatum, particularly in the normal striatum in animals treated with psychostimulant drugs. An additional set of studies is described in which the effects of alteration of the striatopallidal pathways on the target neurons in the globus pallidus are examined. The most interesting result from these studies is that distinct populations of pallidal neurons, characterized on the basis of their expression of the calcium-binding protein parvalbumin and on their axonal projections, display different immediate early gene responses to dopamine agonists or antagonists.

Liu and Graybiel (p. 682) describe work from their laboratory that examined dopamine- and calcium-signaling regulation of cyclic response element-binding (CREB) phosphorylation in striatal neurons. A developmental model was described in which organotypic cultures of the striatum were used as a substrate for dopaminergic- and calcium-signaling pharmacological treatment paradigms and in which the measured effect is that of phosphorylation of CREB. One of the intriguing results described was the difference in the kinetics of phosphorylation of CREB (pCREB) in different striatal compartments. Whereas D1 dopamine receptor agonist treatment for a short period (7 min) resulted in striosomal–patch and matrix pCREB, sustained agonist incubation resulted in only striosomal–patch labeling. A different pattern of pCREB kinetics was observed when an L-type calcium channel activator (BAY K 8644) was used. Similar to the D1 agonist, short-term treatment resulted in pCREB in neurons of both compartments; however, sustained treatment resulted in labeling primarily in the matrix. These results are discussed in terms of the convergence of different signaling pathways through CREB for gene regulation in striatal neurons.

II. Physiology of Dopaminergic Neurons

Studies were presented of the neurophysiological activity of dopaminergic neurons of the midbrain spanning the environmental and behavioral contexts that are represented in the activity of these neurons and the afferent glutamatergic, cholinergic, and GABAergic receptor–mediated membrane mechanisms responsible for such activity. Midbrain dopaminergic neurons in the substantia nigra and ventral tegmental area are generally accepted as playing a role in the motivational control of voluntary behavior. Schultz describes his group's work, which has helped define that role. Using the powerful approach of single-unit recording in awake behaving primates, Schultz (p. 686) has examined the relationship between the reward value of specific behavior-evoking stimuli and dopamine neuron activity. While most dopamine neurons respond to primary appetitive stimuli, it is the relationship between such reward and stimulus parameters that provides the most interesting insight into the function of dopamine neuronal activity. Among the observations that Schultz provides is the relationship between the novelty of a stimulus and dopamine neuron activity. Once a reward becomes fully predictable, responses of dopamine neurons diminish. However, this aspect of the novelty, or unpredictability, in the relationship between a stimulus and associated reward has an important corollary function in that it allows for the representation of a reward to be transferred to secondary, intrinsically neutral conditional stimuli. Such determinants of dopamine neuron activity are suggested by Schultz to indicate that dopamine neuron activity codes errors in reward prediction as an appropriate signal for appetitive learning.

In a series of three chapters, Johnson (p. 691), Tepper *et al.* (p. 694), and Kitai (p. 700) describe membrane physiological properties underlying glutamatergic, GABAergic, and cholinergic afferent synaptic regulation of dopamine neuron activity. Johnson has used a slice preparation to study the effects of synaptic inputs, membrane potentials, or currents on dopamine neurons using intracellular recording from a slice preparation. Results demonstrate that in dopamine neurons, both N-methy-D-aspartate (NMDA) and non-NMDA glutamatergic inputs evoke fast excitatory synaptic currents, whereas metabotropic glutamatergic receptors mediate slow-onset long-duration responses. GABAergic synaptic inputs evoke inhibitory responses, producing fast and slow IPSPs through $GABA_A$ and $GABA_B$ receptors, respectively. Tepper and his colleagues expand on the differential actions of GABAergic synaptic input through the A- and B-receptor subtypes on dopamine neuron activity. Their studies demonstrate that GABAergic inputs are provided to dopamine neurons through multiple sources, from the striatum, from the globus pallidus, and from the GABAergic neurons in the substantia nigra pars reticulata. $GABA_A$- and $GABA_B$-receptor subtypes appear to be involved not only in synaptic input from each of these sources, but also in regulating release through presynaptic mechanisms. Importantly, the two receptor subtypes appear to have distinct roles in determining transitions between burst and pacemaker patterns of activity: $GABA_A$ antagonists increase burstiness, whereas $GABA_B$ antagonists decrease burstiness and increase pacemaker activity. Examining the circuits involved in regulating afferent GABAergic regulation of dopamine neuron activity, Tepper and his group determined that altered patterns of pallidal output affect, at least in part, dopamine neuron activity through multisynaptic mechanisms involving the GABAergic neurons of the substantia nigra pars reticulata, which themselves provide inhibition to the dopamine neurons. In addition to the GABAergic input are other afferents thought to be responsible for excitatory responses of midbrain dopamine neurons.

Kitai describes work in his laboratory that has examined two such inputs: glutamatergic inputs from the subthalamic nucleus and cholinergic input from the pedunculopontine nucleus. Kitai and his colleagues have pioneered work examining the importance of glutamatergic input to both GABAergic and dopaminergic neurons of the substantia nigra. Such input is thought to be responsible, in large measure, for the tonic activity of these neurons. In this chapter, Kitai focuses on the contribution of such glutamatergic input to the dopamine neurons and on studies that examine the interaction of cholinergic receptor–mediated mechanisms on such excitatory input. Conclusions drawn from these studies suggest that such glutamatergic–cholinergic interactions may be involved in the modification of dopamine neuron activity patterns from an intrinsic, regular, rhythmic pattern to an irregular, bursty pattern.

III. Catecholamines in the Prefrontal Cortex

One focus of recent research efforts delves into the anatomy and pharmacology of neuronal interactions within the prefrontal cortex (PFC). This interest has been driven by recent studies implicating the PFC as an important limbic component related to the pathophysiology of schizophrenia. Details of the anatomy of dopamine input to the frontal cortex are provided in the chapter by Lewis *et al.* (p. 703). This group showed that the distribution of dopamine in the frontal cortex is dependent on the region of the cortex examined, that dopamine is selectively distributed within different laminae of different frontal cortical regions, and moreover that this innervation exhibits distinct maturational modifications throughout development. Thus, dopamine innervation is densest in the dorsomedial prefrontal cortex, particularly within the superficial and deep layers; in contrast, in area 9, the innervation is trilaminar. Moreover, the type of elements that these axons contact differs depending on the layer of innervation, reinforcing the proposal that dopamine exerts potent direct and indirect actions over the pyramidal cell population. Indeed, Goldman-Rakic (p. 707) suggests that the distinct types of dopamine innervation may play unique roles in the modulation of working memory function within the frontal cortex. She suggests that the direct dopaminergic input onto pyramidal neuron spines and distal dendrites may underlie its ability to intensify and focus inputs related to memory fields, as shown by her work on D1 agonist modulation of this behavior. In contrast, the more pervasive nonsynaptic actions of dopamine on pyramidal neurons may also modulate working memory via a unique interaction with D1 receptors. Finally, the ability of dopamine to indirectly modulate pyramidal cells via a D4-mediated action on GABAergic interneurons may act in concert with pyramidal neurons to hold information in working memory. These relationships may have important functional consequences during development, because Lewis *et al.* showed that these patterns of innervation reveal unique alterations during development in a region-specific manner.

In addition to dopamine's modulatory action over intrinsic neuronal elements within PFC, Thierry *et al.* (p. 717) provide evidence that dopamine also provides a modulatory influence over long-loop afferents that control activity states within this region. Stimulation of the mediodorsal nucleus was shown to exert both orthodromic and antidromic excitation of PFC neurons. However, it appears that dopamine selectively modulates only the local axon collaterals activated by the antidromic stimulus and, therefore, may enable a focusing of activity within this region by selectively blocking local interactions. Dopamine was also found to exert direct inhibitory actions within the PFC via D1 and D2 receptor stimulation. However, it appears that the D2-mediated response is dependent on intact noradrenergic α_1 transmission. Indeed, this interdependence of the dopaminergic and noradrenergic systems within the PFC is supported by studies by Tassin (p. 712). In a complex series of studies, he extends this correlation to suggest that even subcortical dopaminergic actions are dependent on intact α-adrenergic activation within the prefrontal cortex. This may occur via a tonic influence of norepinephrine over dopaminergic transmission in the PFC. Furthermore, the presence of this noradrenergic innervation may be essential for the expression of dopamine denervation-induced supersensitivity to D1 stimulation. This noradrenergic–dopaminergic interdependence may provide a cellular basis for the proposal by Breier *et al.* that clozapine exerts its unique antipsychotic actions via an effect on noradrenergic systems.

IV. Interaction among Glutamatergic Afferents and Dopamine in the Striatum

In addition to the effects of dopamine within the frontal cortex, an increasing amount of evidence has implicated the PFC in the regulation of subcortical dopaminergic transmission. In the chapter by Thierry *et al.*, findings are presented that dopamine agonists injected into the PFC exert an opposite action to those injected systemically, and that this appears to be mediated by a D1 action. Furthermore, the effects of prefrontal glutamatergic input onto subcortical structures are also modulated by dopaminergic agonists. Within the nucleus accumbens, Grace *et al.* (p. 721) have shown that dopamine acting at very low concentrations exerts a tonic inhibitory influence on PFC afferents. In contrast, in the dorsal striatum, Levine and Cepeda have shown that dopamine exerts a pharmacologically selective modulation of glutamate via distinct postsynaptic mechanisms. Thus, dopamine acting on D1 receptors potentiates postsynaptic glutamate responses mediated via the NMDA receptor, whereas D2 agonists attenuate non-NMDA-mediated glutamatergic excitation of striatal neurons. This was found to be a directly mediated action, because it was also observed in whole-cell clamp studies in the presence of tetrodotoxin. Indeed, Thierry *et al.* provide evidence that suggests that dopamine can modulate glutamatergic responses within the PFC as well. Konradi (p. 729) provides a molecular mechanism through which such a glutamatergic–dopaminergic interaction may take place. In an elegant series of studies, she shows that c-fos expression mediated by dopamine-receptor stimulation is dependent on functional NMDA receptors and is inhibited by the glutamate antagonist MK 801. This interaction is proposed to occur via a D1-stimulated, cyclic adenosine monophosphate–dependent phosphorylation of CREB that is additive with glutamate-mediated activation of CREB.

The relevance of dopaminergic-glutamatergic interaction within subcortical structures is further underscored in two models that examine the functional impact of these systems. Arbuthott *et al.* attempt to bridge studies examining the electrophysiological actions of dopamine in a model system that also examines ultrastructure. As they point out, it is clear from many studies that dopamine does not exert a simple inhibitory or excitatory action in the striatum. Indeed, studies by Schultz have shown that dopamine acts in reward-related events as a way to increase the probability that an unexpected reward will be generated by increasing the type of behavior that is associated with the rewarding event. Arbuthnott *et al.* (p. 733) show physiological evidence for such a role, in that repetitive corticostriatal stimulation tends to induce an LTD-like inhibition of responses to subsequent stimuli; however, in the presence of dopamine, repetitive responses are potentiated. Examined at the ultrastructural level, data are presented to suggest that dopamine may be involved in the selective survival of potentiated synaptic connections within this highly integrative structure. Another aspect of dopamine function as it relates to motor systems is examined by Rebec (p. 737), who used the elegant technique of combined extracellular recordings from striatal cells in freely moving animals with microiontophoresis. This approach allows his laboratory to examine dopamine–glutamate interactions at a single-cell level and to correlate these results with behavior. In these studies, locally applied amphetamine was found to selectively potentiate the activation of motor-related striatal units; furthermore, it was discovered that this activation is dependent on corticostriatal inputs. This work has important parallels with that of Levine and Cepeda (p. 724), in that it demonstrates a behavioral function for the potentiation of glutamatergic responses by dopamine that occurs at the cellular level. Finally, as Svensson *et al.* point out, an important component of glutamatergic actions on dopamine systems also occurs with respect to afferents to the dopamine cell body region, particularly when the drugs are administered systemically.

Dopaminergic systems may also play a role in modulating the interaction of other types of inputs within the striatum. In particular, Grace *et al.* have shown a unique type of modulatory interaction among afferents to the accumbens that may be related to the regulation of information flow within these structures. Thus, accumbens neurons receive overlapping inputs from the PFC, hippocampus, and amygdala. Moreover, the hippocampus and amygdala appear to exert a facilitatory action on prefrontal inputs, which they propose underlies selective gating of information flow from the PFC through the accumbens to eventually influence thalamocortical activity. In particular, the hippocampal input is proposed to provide a context-dependent bias over selective prefrontal cortical inputs, whereas the amygdala would provide input selection based on the affective state of the organism. The ability of dopamine agonists and phenycyclidine (PCP) to interfere with this information-gating process may, therefore, provide a mechanism by which these systems contribute to at least a subset of the cognitive deficits observed in schizophrenia.

V. Systems Level Analyses of Brain Noradrenergic Systems

There was notable coherence in the systems-level analyses of NE in the CNS. Regulation of behavioral state, sensory processing, attention, and memory were particularly noteworthy themes. Interestingly, these themes are closely related, interdependent functions of the forebrain, and, therefore, it should perhaps be no surprise that different investigators would conclude such related functions for NE in the CNS. Nonetheless, this convergence of results is encouraging and suggests that we may be closing in on fundamental aspects of the functions of this enigmatic brain system.

A. Behavioral State

Two reports deal directly with the role of forebrain NE in behavioral state regulation. Rajkowski *et al.* (p. 740) recorded from locus ceruleus (LC) neurons in monkeys during spontaneously occurring changes in alertness. As previously reported in New World monkeys as well as other species, this study in Old World (*Cynomolgus*) monkeys found that LC neurons dramatically decreased tonic impulse activity during drowsiness and slow-wave sleep. This effect appears to be especially pronounced in Old World monkeys because these cells became virtually quiescent, even during relatively early stages of drowsiness. As also previously found in other species, LC neurons in monkeys increased activity abruptly either just preceding or coincident with the transition from drownsiness or slow-wave sleep to waking. Recently, this group has also found that LC neurons in Old World monkeys are completely silent during paradoxical sleep, the first results for LC neurons in monkeys during this important stage of sleep, again similar to properties of these cells found in other species. Thus, these findings confirm for the Old World monkey properties that were also prominent for LC/NE neurons in rats and cats. These results indicate that, across species, LC neurons may play a permissive role in sleep but an executive role in waking from slow-wave sleep. Additional studies in monkeys indicate that such state control extends to the attentional state within continuous waking, so that regulation of sleep and waking is only a portion of the state regulation produced by this system.

Berridge (p. 744) reports studies in rats that were very compatible with the view that increased arousal and waking behaviors are associated with increased tonic LC impulse activity. In lightly anesthetized rats, he found that microinfusion of the β-adrenoceptor agonist isoproterenol (Iso) into the area of the medial septum reliably activated cortical and hippocampal electroencephalographic (EEG) signs of arousal. In addition, microinfusions of the β-adrenoceptor antagonist timolol into this same region were effective in shifting these EEG measures toward activities associated with less alert states. Additional studies in unanesthetized rats confirmed that Iso injections into the medial septal area increased waking, indicating that the former results are not limited to the anesthetized preparation. These findings are entirely compatible with those from LC recordings in behaving animals (as the LC densely innervates the medial septal area) and indicate that the LC system may regulate alertness, at least in part via β-adrenoceptor mechanisms in the medial septal area.

B. Sensory Processing

Waterhouse and colleagues (p. 749) report a series of anatomical and physiological experiments aimed at further delineating the role of the LC/NE system in somatosensory processing. LC neurons are perhaps best known for their highly divergent axonal projections. Recent anatomical studies from this group call for a re-examination of the resulting widely held assumption that these cells are also nonspecific in their wiring. Instead, Devilbiss *et al.* report the intriguing result that individual LC neurons preferentially innervate functionally connected circuit elements at different levels of the neuraxis. Thus, for example, neurons that project to the somatosensory cortex are more likely to also project to the somatosensory thalamus (ventroposterior nucleus) than to, for example, the visual thalamus (lateral geniculate nucleus). Thus, although LC neurons may project widely, they exhibit much greater specificity in their projections than typically assumed. These investigators also found evidence for substantial specificity in the physiological actions of NE on somatosensory cortical neurons. In experiments using the *in vitro* cortical slice preparation, they found that low concentrations of NE or α_1-adrenoceptor agonists facilitate spike production in layer 5 neurons in response to near threshold synaptic inputs, while in other neurons, β-adrenoceptor activation augments GABAergic synaptic responses recorded in similar neurons. Together, these results indicate that the LC efferent system exhibits substantial anatomical and physiological specificity in its impact on target cortical circuits. Thus, instead of a simple "gain" control over a process such as arousal, the LC may specifically and coordinately regulate processing of sensory information by virtue of actions on select neurons via different receptors along circuits that mediate responses to certain sensory modalities of stimulation.

C. Attention

Selective processing of sensory information is a hallmark of attention, and, therefore, the aforementioned findings for modulation of senory circuits would be consistent with a role for the LC in attentional activity. In fact, results reported by Aston-Jones and colleagues (p. 755) from their recording of LC neurons in monkeys during performance of a visual discrimination-attention task strongly implicate the LC in attention. This task required animals to stably foveate a central fixation spot on a video monitor and selectively release a lever after presentation of a target cue (10–20% of trials) to receive a drop of juice. Responses to nontarget cues (80–90% of trials) resulted in a 3-sec time-out. LC neurons exhibited tonic and phasic activities that varied in relation to task performance. Tonically, LC activity varied among low, intermediate, and high levels. Low activity (near zero) was strongly associated with drowsiness (as noted in the report by Rajkowski *et al.*) and poor task performance characterized by frequently missed stimuli. High tonic LC activity (~3–5 spikes/sec) was also associated with poor task performance, in this case characterized by frequent false alarms (incorrect responses to nontarget stimuli), poor foveation of fix spots, and increased scanning eye movements. Optimal performance was associated with an intermediate level of tonic LC discharge (~1–2 spikes/sec). Phasic LC activation was evoked at short latencies (~100 ms) selectively by target stimuli but not by other task events. Moreover, the phasic responses to target stimuli occurred only during epochs of excellent task performance associated with intermediate tonic LC activity. Aston-Jones *et al.* conclude that the LC exhibits two modes of activity. The *aphasic* mode is characterized by moderate tonic activity and robust phasic responses to attended sensory stimuli; this mode corresponds to focused (selective) attention and good task performance. These cells also exhibit a *tonic* mode characterized by high tonic activity and no phasic responsiveness; this mode corresponds to scanning, labile attention, and poor performance on this task. Additional studies by this group, using connectionist modeling, indicated that these different modes of LC operation may play a causal role in determining the associated level of behavioral performance. These modeling studies also indicate that these two LC modes may result from different levels of electrotonic coupling among LC neurons, consistent with recent results from Williams and colleagues for such coupling in the adult LC.

D. Memory

A system involved in sensory processing and attention may also be expected to play a role in memory. Indeed, two separate reports indicate that actions of NE in the dorsolateral PFC (area 46) of behaving monkeys may play a significant role in memory. Sawaguchi and Kikuchi (p. 759) used iontophoretic methods to study the effects of NE agents on activity of area 46 neurons in monkeys during a delay task. Previous studies by Fuster and colleagues and by Goldman-Rakic's group revealed that neurons in this area exhibit prolonged activity during the delay period in a delayed-response task. Goldman-Rakic trained monkeys on an oculomotor delayed-response task, in which they were required to remember the spatial location of a visual stimulus during a delay period and to subsequently move their center of gaze to that location in the absence of the stimulus. Local iontophoretic application of the α_2-adrenoceptor antagonist yohimbine attenuated delay period activity of area 46 neurons, whereas iontophoresis of the β-adrenoceptor antagonist propranolol had no consistent effects on such activity. Further analysis revealed that this α_2-adrenoceptor antagonist attenuated the sharpness of tuning of the delay-period activity (relative selectivity of delay-period activity for the specific location in space to be recalled) more than the baseline firing of delay-period activity. These results indicate that activation of α_2 adrenoceptors plays a modulatory role in the coding of impulse activity thought to underlie working memory processes in the primate PFC.

This proposed role of α_2-adrenoceptor activity in the PFC is consistent with behavioral studies in monkey carried out by Arnsten and her colleagues. They summarize experiments in which α_2-adrenoceptor agonists given systemically improved performance in a delayed-response task in aged monkeys with naturally occurring NE depletion. The beneficial effects of α_2-adrenoceptor agonists was particularly evident in conditions of high interference (e.g., distracting stimuli), and the pharmacologic profile indicated that these effects primarily involved actions at the α_{2A}-receptor subtype. Additional studies indicate that the α_2-adrenoceptor agonist clonidine is more effective in monkeys that have suffered presynaptic NE depletion (aged animals, reserpine treatment, 6-hydroxydopamine [6-OHDA], or MPTP lesions) than in normal animals, implicating postsynaptic rather than presynaptic α_2-adrenoceptor sites in these behavioral effects. Local microinfusion experiments with α_2-adrenoceptor agonists and antagonists support this general view. Recent studies by this group indicate that, in contrast to results with α_2 adrenoceptors, α_1-adrenoceptor activation in monkey PFC impairs working memory performance. This leads Arnsten *et al.* (p. 764) to propose that a balance of α_2 and α_1-adrenoceptor activation in the PFC may be involved in regulating the functional status of this area and its ability to maintain normal working memory activity.

E. Functional Organization of Inputs to the Locus Ceruleus

Ennis and colleagues (p. 767) describe neuroanatomical tract-tracing studies in which the organization of afferents to the core and pericellular shell of the LC has been examined. A number of afferents specifically target dendrites of noradrenergic neurons that extend into the pericellular shell region surrounding the LC core region, in which the cell bodies are located. Among these afferents, two specific systems, those from the midbrain periaqueductal gray and from the medial preoptic area, have been studied with both neuroanatomical tracing methods and functional activation studies. Specific pharmacological activation of each of these sources of afferent input to the pericellular shell of the LC resulted in induction of the immediate early gene Fos in the noradrenergic neurons. These functional studies demonstrate that the afferents to the pericellular shell into which noradrenergic neurons extend their dendrites may functionally activate these neurons.

Nakamura and his colleagues (p. 772) describe their neurophysiological studies of the noradrenergic neurons of the LC. First, they describe the development of sensory-evoked responses. Specifically, they relate the patterns of sensory-evoked activity to the postnatal development of a negative-feedback mechanism, which replaces the more predominant positive-feedback system of early development. Second, they examine the role of GABAergic inhibition of LC neurons, using caloric vestibular stimulation as a paradigm for examining this phenomenon. Their results suggest that vestibular-mediated inhibition of the LC is mediated by the ventral lateral medullary afferents, which because they are primarily excitatory, presumably activate GABAergic afferents connected with the noradrenergic neurons.

F. Ultrastructural Localization of NE and NE Receptors in Primate Cortex

Finally, detailed electron microscopic studies by Aoki *et al.* (p. 777) have begun to shed light on the possible microcircuitry involved in such NE functions in the PFC. These investigators used immunohistochemistry to localize NE fibers and receptors at the ultrastructural level in tissue from monkey area 46. Laminar analysis indicated that NE fibers form synapses most frequently in layers 2 through 5, and that such synapses occur at about every micrometer along the length of an NE fiber in this area. This reveals a high connectivity from the LC/NE system to the PFC. Staining, using antibodies against the subtypes of α_2 adrenoceptors in the PFC, revealed that the α_{2A} subtype was most prevalent in perikarya and proximal dendrites of neurons throughout layers, and that α_{2B} and α_{2C}-receptor staining was more scarce in area 46. These findings are consistent with the results of behavioral studies with preferential α_2-adrenoceptor subtype agents by Arnsten *et al.* These studies also revealed that α_{2A}-adrenoceptors are more frequently localized to presynaptic terminals than to postsynaptic sites, but that receptors associated with identified synapses were more often postsynaptic than presynaptic. These results may indicate that synaptically released NE may operate preferentially at postsynaptic sites but that exogenous as well as endogenously released NE may also have substantial effects mediated via presynaptic α_2-adrenoceptors. Immunohistochemical localization of β adrenoceptors revealed a surprisingly similar pattern for β and α_{2A}-adrenoceptor activation, which have opposite effects on the adenylate cyclase cascade. However, dual labeling was much more frequently observed for β adrenoceptors and GABA than for α_{2A}-adrenoceptors in GABA+ profiles, indicating that the cortical NE system (which is derived exclusively from the LC) interacts with GABA interneurons primarily via β adrenoceptors.

Thus, converging evidence from several levels of analysis indicates roles for forebrain NE in related functional processes. First, state control is an important factor in the control of sensory processing and memory. Similarly, regulation of sensory processing is at the heart of attentional function. Finally, attention is an essential prerequisite for proper memory functioning. Indeed, late attentional stage processing can be readily and closely related to working memory. This group of papers, then, provides a glimpse at how functional attributes of the NE system at different levels of analysis may be involved in integrative behavioral attributes of this system. It is notable, for example, that the aforementioned results may provide an anatomically and physiologically specific mechanism underlying the well-known linkage between attention and memory functions from psychological work.

Such brain substrates for psychological phenomena may also hold important insights for clinical application. Indeed, there are several clinical implications of this work. Perhaps most obvious are the parallels between LC function and attentional dysfunctions such as attention deficit hyperactivity disorder (ADHD). The results of these studies indicate that ADHD may be associated with the tonic mode of LC function (see Aston-Jones *et al.*), which may produce overstimulation of α_1 adrenoceptors in cortical sites (see Arnsten *et al.*). These results also indicate that NE in the PFC may play an important role in memory processes, much as it may play a role in sensory processing when acting in more sensory-related cortical areas. Roles of the LC system in the PFC in attention and in memory may become disrupted in clinical conditions associated with decreased NE availability in cortical regions, such as that found in aging and Alzheimer's disease.

VI. Catecholamines: Involvement in Clinical Disorders

A. Stress

Because of the diverse nature of catecholaminergic projections, these systems are believed to play a role in general activational processes related to adaptation and survival of the organism. Such responses are particularly evident in the CNS activation produced by stress-related stimuli, often referred to as the fight or flight response. However adaptive such a response may be in dealing with immediately threatening environments, the continuous presence of stress can also exert pathological changes within the CNS. Valentino *et al.* (p. 781) point to evidence that the noradrenergic system is a principal component of the acute stress response, which is likely to be mediated via an excitatory amino acid transmitter input to the LC. In contrast, Valentino *et al.* describe some convincing experiments to suggest that corticotropin-releasing factor (CRF) may exert a major role in the long-term activation of the noradrenergic system in both normal and pathological states. Thus, CRF has been shown to mediate stress-induced activation of the LC, particularly when certain classes of stimuli serve as triggers. Moreover, there is clear evidence that CRF afferents contact NE neuron dendrites near the LC. Finally, CRF injection into the LC exerts functional actions. Therefore, CRF exerts a clear activation of LC noradrenergic cell firing in a manner that is sufficient to stimulate postsynaptic targets. In doing so, this system may be capable of maintaining arousal states in threating environments and may serve to link certain peripheral effects on the autonomic nervous system with CNS activity.

Another treatment that appears to elevate noradrenergic activity is administration of the atypical antipsychotic drug clozapine, as outlined by Breier *et al.* (p. 785). Indeed, it appears that clozapine is unique in activating LC neuron firing. The ability of clozapine to alter this system could provide a functional link to account for the known involvement of stressful stimuli in the exacerbation of schizophrenia, and it provides a unique perspective on the potentially important tonic influence of NE on dopaminergic systems, as outlined by Thierry *et al.* and Tassin. Of course, it is also evident from studies by Svensson *et al.* that noradrenergic input to the ventral tegmental dopamine system plays an important role in modulating the activity of these neurons, and it also must be taken into consideration when analyzing the impact of noradrenergic–dopaminergic interactions in mediating CNS responses to stress.

B. Parkinson's Disease

Zigmond and Hastings (p. 788) review studies of the changes in various measures of dopamine neurotransmission following 6-OHDA lesions of the nigrostriatal dopaminergic pathway. These studies have attempted to address two main issues related to the pathophysiology of Parkinson's disease, for which 6-OHDA lesions are a useful animal model. One issue involves why the behavioral effects of dopamine depletion develop over an extended period, rather than abruptly with the onset of the lesion. Studies are described that demonstrate increases in dopamine release from remaining terminals as an example of mechanisms that are temporarily able to compensate for the early phases of dopamine afferent destruction. A second issue addressed is why dopamine neurons are susceptible to toxicity. Studies are described that raise the possibility that dopamine itself, like the analogue 6-OHDA, may be toxic to dopaminergic neurons.

Although L-dopa therapy has provided an effective treatment for Parkinson's disease, the fact that this pharmacological treatment does not remain effective indefinitely and that, despite this treatment, the degeneration of the dopaminergic system progresses, make clear the necessity for an alternative treatment approach. Bankiewicz *et al.* (p. 801) describe work from their group that is exploring the use of gene therapy for the treatment of Parkinson's disease. Studies are described in which a cell-based delivery system is being tested. The strategy is to deliver the protein machinery necessary to synthesize dopamine into the deafferented striatum. Two alternative strategies are discussed. One is based on using immortalized cell lines that have been genetically engineered to produce dopamine. Another is to used adeno-associated virus as a means of chronic delivery of therapeutic agents. Preliminary data using these approaches suggest that gene therapy may one day provide an effective strategy for the treatment of Parkinson's disease.

C. Schizophrenia

One issue raised by Lipska and Weinberger (p. 806) points out a major difficulty in this line of research—developing an animal model that can accurately reflect at least some aspects of a disorder that is uniquely human in nature. This is particularly true for a disorder such as schizophrenia, in which the neuropathological underpinnings of the disease are not obvious at the ultrastructural or neurochemical level, and the major components of the disorder involve disturbances of higher cognitive processes. Experimental approaches can nonetheless be employed toward this end. For example, one can focus on examining the basic physiological properties and types of network interactions that occur among neuronal systems that are likely to play a major role in a specific disorder. Such an approach has been described in the chapters by Thierry *et al.*, Goldman-Rakic, Lewis *et al.*, and Grace *et al.*, in which a systems-interaction approach is utilized to derive models related to the pathophysiology of schizophrenia.

One important aspect of models as they relate to schizophrenia is the concept that this disorder is developmental in origin. Indeed, the developmentally related changes that take place within the dopaminergic system and its innervation of the PFC, as outlined by Lewis *et al.*, demonstrate the complex nature and potential for pathological insult that is present throughout the formation and elaboration of the frontal cortical dopaminergic system. As such, there is the potential for a number of developmental insults to impact on the functional development of this system. Such an approach was used by Lipska and Weinberger to investigate potential developmental pathologies related to this disorder. Given the evidence that insults to fetal development that occur within the second trimester of pregnancy tend to increase the incidence of schizophrenic offspring, they embarked on a series of studies to examine potential pathologies produced at an equivalent rat developmental stage that may impact on limbic system function. By lesioning the hippocampus, Lipska and Weinberger found that, although the prepubertal rat appeared normal along several dimensions, when tested as an adult, the animals displayed distinct alterations in behavioral testing and pharmacological responsiveness that are consistent with some of the observations made in schizophrenic humans. Indeed, such alterations impact on the same systems that Grace *et al.* have shown to be involved in limbic interactions within the nucleus accumbens. As such, the model by Lipska and Weinberger is the first developmentally relevant animal model of schizophrenia advanced to date, and it is likely to have an important impact on our ability to examine systems-based developmental interactions as they relate to human diseases.

Another method that is often employed to utilize animal models as approaches to understanding the pathophysiology of schizophrenia is the use of pharmacological tools. This approach can be divided into two categories: (1) examining the mode of action of agents that are known to mimic schizophrenia-like symptoms and (2) examining the actions of antipsychotic drugs as a method for analyzing what actions may correspond to the clinical response profile of these compounds. One particularly important pharmacological model of schizophrenia relates to PCP. As reviewed by Jentsch *et al.* (p. 810), unlike other pharmacological models, PCP is a psychotomimetic that can reproduce both the positive and negative symptoms of schizophrenia at doses related to its NMDA-antagonistic properties. In an elegant series of studies, Jentsch *et al.* have investigated the short-term and long-term consequences of PCP administration on nonhuman primates. Their work has shown that PCP given acutely can exert significant influences on both prefrontal cortical and subcortical metabolism of dopamine. Furthermore, they have shown that agents that block stress-induced increases in dopamine also block these actions of PCP, again showing the importance of stress, norepinephrine–dopamine interactions, and frontal cortical function. Perhaps more importantly, repeated administration of PCP appears to induce long-term changes in both behavioral tests and the PFC that may be related to schizophrenic pathophysiology.

Svensson *et al.* have utilized a different approach to the study of schizophrenia: one that focuses on the mode of action of antipsychotic drugs. As Svensson *et al.* point out, dopamine D2-receptor occupancy cannot account for the efficacy of antipsychotic drugs, particularly as it relates to the atypical neuroleptics. Their work focuses on the importance of ventral tegmental area dopaminergic neuron activity in assessing pharmacological responses to drugs. In particular, they show that, in addition to actions in forebrain structures, as demonstrated by Jentsch *et al.* and Grace *et al.*, indirect glutamatergic blockers also have important and selective actions on dopamine neuron activity states. Perhaps more importantly, the research by Svensson *et al.* (p. 814) suggests that atypical antipsychotic drugs may achieve their unique therapeutic profile due to their actions on serotonin-1A- and noradrenergic α_1-receptor systems, which in turn produce potent modulatory actions on dopaminergic neuron activity.

VII. Summary

A goal of this part is to examine the functional impact of catecholaminergic systems on CNS function as it relates to normal and pathological states. The participants achieve their objectives individually by relating their high-quality work on this expansive topic, while maintaining a focus on the functional implications of their findings. Nonetheless, despite the necessarily broad nature of the topics presented, there is a remarkable degree of convergence of information. Several subthemes have emerged as a consequence of considering this work in its entirety. First is the importance of examining neurotransmitter effects, not in isolation, but in terms of interactions with other neurotransmitter systems. History has shown that a limited focus often produces confusing or inconsistent results, which become increasingly clear on consideration of the state of the organism. Second is the importance of examining pharmacological and pathophysiological interactions in light of the anatomy of the system and how developmental influences can alter this relationship. Such very general considerations have been found to provide an essential ingredient in understanding the nature of catecholamine function within this complex system.

Charles R. Gerfen, Kristen A. Keefe, and Heinz Steiner

Laboratory of Systems Neuroscience
National Institute of Mental Health
Bethesda, Maryland 20892

Dopamine-Mediated Gene Regulation in the Striatum

Dopamine's critical role in the basal ganglia is evident in clinical disorders such as Parkinson's disease and the implication of striatal dopamine effects of drugs of abuse, including cocaine. The response of genes encoding transcription factors and neuropeptides to manipulation of dopamine receptors may be used to study the dynamic modulatory role that dopamine plays in affecting basal ganglia function. Basal ganglia output is antagonistically determined by two separate striatal output systems, the direct and indirect pathways, which are oppositely modulated by their respective expression of the D1 and D2 dopamine receptor subtypes (1). Two sets of studies are reviewed that demonstrate how dopamine receptor–mediated gene-regulation effects provide insight into the functional role of dopamine in the striatum.

I. D1–D2 Dopamine Receptor Segregation in Direct and Indirect Striatal Output Pathways

While there is now considerable support for the original observations of the segregation of D1 and D2 dopamine receptor subtypes to the neurons contributing to the "direct" and "indirect" striatal output pathways (2), some physiological studies suggest a functional colocalization of these receptors. For

Advances in Pharmacology, Volume 42

1054-3589/98 $25.00

example, coactivation of D1 and D2 receptors results in a potentiated response in striatal neurons that suggests mechanisms involving synergistic interactions between D1 and D2 receptors colocalized in striatal neurons. In a series of studies, we have addressed the potentiated response to combined D1- and D2-agonist treatment compared with D1-agonist treatment alone (3, 4).

To examine D1- and D2-receptor interactions, we used a D1–D2 synergy paradigm in which a relatively low dose of a D1 agonist, SKF38393 (0.5–1.0 mg/kg), results in a small induction of the immediate early genes c-fos and zif 268 in the dopamine-depleted striatum, and co-treatment with a D2-selective agonist, quinpirole (1.0 mg/kg), greatly potentiates the immediate early gene induction. We determined that coactivation with D1- and D2-selective receptor agonists results in opposite effects on D1- and D2-containing striatal neurons, while nonetheless potentiating the response in D1 neurons. The level of the immediate early gene zif 268 mRNA was measured in identified D2- or D1-containing neurons, which had been marked with a second marker. In this case, a probe for the peptide enkephalin (ENK) was used, which provides a nearly 1:1 marker for D2-bearing neurons. D1-agonist treatment alone (SKF38393, 1.0 mg/kg) results in the induction of zif 268 selectively in D1 neurons, while not affecting zif 268 mRNA levels in D2 neurons. Importantly, nearly the entire population of striatal D1 (ENK-negative) neurons shows elevated zif 268 mRNA levels. The combination of the D2-receptor agonist quinpirole (1.0 mg/kg) with D1-receptor agonist (SKF38393, 1.0 mg/kg) treatment causes a decrease in the zif 268 mRNA levels in D2 (ENK-positive) neurons and a potentiated zif 268 mRNA response in D1 (ENK-negative) neurons. Thus, there is a potentiated induction of immediate early genes in D1 neurons with combined D2- and D1-agonist treatment, compared with D1-agonist treatment alone, which is coupled with a decrease of immediate early gene levels in D2 neurons. These data demonstrate that D1- and D2-receptor agonists selectively affect separate striatal neuron populations when administered alone or together, as expected if the receptors are segregated on different neurons. This leads to the further suggestion that the potentiated response of D1-containing striatal neurons to combined D1- and D2-agonist treatment results from interactions between neurons, rather than by interaction of receptors expressed on the same neurons. Thus, these functional data support the view that D1 and D2 dopamine receptors are segregated, respectively, to direct and indirect striatal output pathways.

II. Striatal Adaptive Responses: Dopamine–Opiate Interactions

One of the compelling characteristics of striatal organization is regional and compartmental variation in the distribution of neuropeptides such as dynorphin, substance P, and ENK in striatal neurons (1). A number of studies have shown that the level of expression of these peptides is regulated by dopamine receptor–mediated processes (2) and suggests that regional differences reflect ongoing differences in the patterns of activity in these regions.

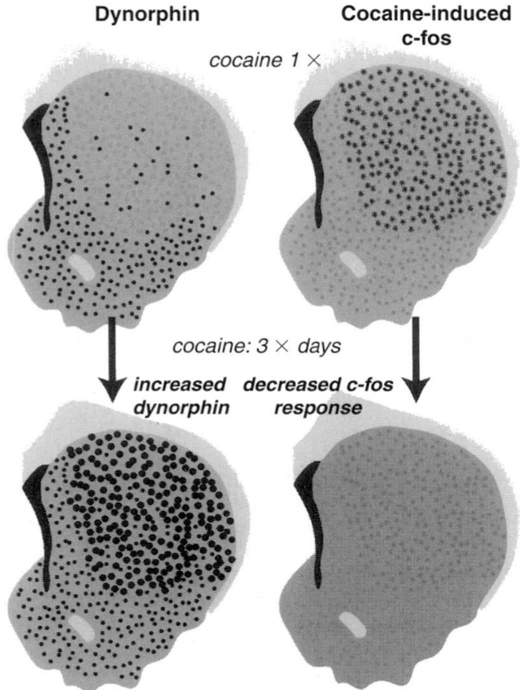

FIGURE 1 Diagram of the regional response within the striatum to the indirect dopamine agonist cocaine, which demonstrates the functional role of dynorphin in modulating this response. Basal levels of dynorphin are higher in ventral and medial striatal regions. A single injection of cocaine induces the immediate early gene c-fos, by a D1-mediated mechanism, in the dorsal striatum, complementary to areas with high dynorphin levels. Repeated cocaine treatment (single daily injections of 30 mg/kg for 3 days) results in increased dynorphin in the dorsal striatum, which has low basal expression, and a marked reduction of c-fos induction in this area, in which c-fos had previously been induced. These data suggest that dynorphin blunts the response of neurons of D1-receptor stimulation. Further studies have shown that this effect of dynorphin is mediated through kappa opiate receptors. From Steiner and Gerfen (5).

To study such regulatory processes as they are involved in the adaptive responses of striatal neurons to dopamine-receptor stimulation, the indirect dopamine agonist cocaine has been used (5, 6). In a first study, cocaine was administered at a dose of 20 mg/kg once a day for 4 days, and the levels of c-fos and peptide mRNAs were analyzed in the striatum on each of those days at a time point 45 min after drug administration (5). After the first injection of cocaine, there was a significant induction of c-fos mRNA in the dorsal striatum. A similar pattern and level of induction were also observed with a second-day treatment, but the level of induction decreased after the third- and fourth-day injections. We also observed that the regional pattern of cocaine-induced c-fos induction after the first-day injection was complementary to the pattern of dynorphin mRNA expression in the striatum. That is, dynorphin mRNA, which is expressed in direct projecting neurons throughout the striatum, has a higher basal level of expression on a per-cell basis in the ventral striatum, including the nucleus accumbens, than in the dorsal striatal region in which cocaine-induced c-fos mRNA levels were elevated. None of the other peptide mRNA levels measured, including ENK and substance P, displayed this complementary pattern. With successive cocaine injections, the levels of dynorphin mRNA were elevated in the dorsal striatal region, coincident with the decreasing response of c-fos mRNA to these injections. These results led us to propose that dynorphin functions to "blunt" the cocaine induction of c-fos and that repeated cocaine administration results in an adaptive response to increase dynorphin expression. As a test of this, spiradoline, an agonist of the kappa opiate receptor, which is the receptor dynorphin is thought to act on, was coadministered with cocaine systemically in drug-naive animals. In this case cocaine-induced c-fos in the dorsal striatum was markedly reduced. A similar result was obtained when spiradoline was infused directly into the striatum (6).

The results of these studies suggest a functional role for dynorphin in the striatum to modify the response of "direct" striatal projection neurons to excessive D1-receptor stimulation. Evidence in support of such a function comes first from the finding that the D1-mediated cocaine induction of immediate early genes in the striatum occurs in the dorsal striatal region in which dynorphin is normally at a relatively low level and does not occur in regions with constituatively high levels of dynorphin. Second, repeated cocaine treatment results in an adaptive response in the dorsal region that involves induction of dynorphin levels coincident with a decreased D1-mediated cocaine induction of immediate early genes. Third, pharmacological treatment with a dynorphin agonist is able to mimic the adaptive response of elevated dynorphin levels by blocking D1-mediated immediate early gene induction in the dorsal striatum.

References

1. Gerfen, C. R. (1992). Neostriatal mosaic: Multiple levels of compartmental organization. *TINS* **15**, 133–139.
2. Gerfen, C. R., Engber, T. M., Mahan, L. C., Susel, Z., Chase, T. N., Monsma, F. U., and Sibley, D. R. (1990). D1 and D2 dopamine receptor-regulated gene expression of striatonigral and striatopallidal neurons. *Science* **250**, 1429–1432.
3. Keefe, K. A., Gerfen, C. R. (1995). Synergistic response to combined D1- and D2-dopamine receptor stimulation in striatum: Immediate early gene response to intrastriatal drug administration. *Neuroscience* **66**, 903–913.
4. Gerfen, C. R., Keefe, K. A., and Gauda, E. B. (1996). D1 and D2 dopamine receptor function in the striatum: Co-activation of D1- and D2-dopamine receptors on separate populations of neurons results in potentiated immediate early gene response in D1-containing neurons. *J. Neurosci.* **15**, 8167–8176.
5. Steiner, H., and Gerfen, C. R. (1993). Cocaine-induced c-fos messenger RNA is inversely related to dynorphin expression in striatum. *J. Neurosci.* **13**, 5066–5081.
6. Steiner, H., and Gerfen, C. R. (1994). Kappa opioid receptor inhibition of D1 dopamine receptor mediated induction of immediate early genes in striatum. *J. Comp. Neurol.* **353**, 200–212.

M-F. Chesselet, J. M. Delfs, and L. Mackenzie

Department of Neurology
UCLA School of Medicine
Los Angeles, California 90095

Dopamine Control of Gene Expression in Basal Ganglia Nuclei: Striatal and Nonstriatal Mechanisms

Much attention has been devoted to the effects of dopamine on gene regulation in the striatum, the area of the basal ganglia that receives the densest dopaminergic innervation. In particular, numerous groups have shown that lesions of the nigrostriatal pathway or administration of blockers of dopamine D2 receptors induce an increase in expression of mRNA encoding enkephalin, a neuropeptide present in striatal efferent neurons projecting to the globus pallidus or external pallidum (1). Similarly, nigrostriatal lesions and D2 antagonists increase the expression of mRNA encoding glutamic acid decarboxylase (Mr 67,000: GAD 67), which is also present in striatal efferent neurons. Together with many data from electrophysiological and biochemical studies, these results suggest that the GABA/enkephalin-containing striatal output to the globus pallidus is activated after decreased dopaminergic transmission in striatum. It has been postulated that this results in inhibition of the globus pallidus and contributes significantly to the generation of the akinesia and catalepsy seen after decreased dopaminergic transmission in the striatum (2). However, when we examined the level of expression of either enkephalin or GAD 67 mRNA in the striatum in relation to behavior in rats with alterations of dopaminergic transmission, we noticed a striking absence of correlation between these effects and the motor abnormalities displayed by the animals. For example, sustained increase in enkephalin mRNA did not always occur in the striatum when rats were cataleptic, whereas increases in enkephalin and GAD mRNA were observed in the striatum even when catalepsy was not observed or was blocked (3). Furthermore, enkephalin and GAD mRNAs were still increased in the striatum of rats that displayed orofacial dyskinesia, rather than catalepsy, after long-term administration of the D2 antagonist haloperidol (4).

In an effort to examine changes in gene expression caused by alterations in dopaminergic transmission in regions of the basal ganglia that received striatal inputs, we examined the effects of nigrostriatal lesions and haloperidol administration on the expression of GAD 67 mRNA in the globus pallidus.

Advances in Pharmacology, Volume 42

1054-3589/98 $25.00

Measuring GAD mRNA offers a distinct opportunity to obtain information on the GABAergic efferent neurons of this region, because, in contrast to GAD or GABA, the mRNA is contained in their cell bodies, not in the terminals of striatal neurons projecting to the pallidus.

Despite converging evidence that the inhibitory, GABAergic input from the striatum to the globus pallidus is activated after decreased dopaminergic transmission, we observed an increased expression of GAD 67 mRNA in neurons of the globus pallidus, both after unilateral nigrostriatal lesions and after short-term (3–21 days) treatment with haloperidol (1 mg/kg s.c.) (3, 5). In contrast to changes in enkephalin and GAD 67 mRNA levels in the striatum, the increase in GAD 67 mRNA in the globus pallidus was correlated with the presence of catalepsy. In collaboration with Dr. Ian Creese (Rutgers University), we have found that this effect can be selectively reproduced by intracerebroventricular administration of antisense oligonucleotides against the dopamine D2 receptor, a treatment that also produces robust catalepsy. Furthermore, in rats with orofacial dyskinesia following haloperidol administration, GAD 67 mRNA levels were decreased rather than increased in the globus pallidus. These data suggest that despite the increased inhibitory input from the sriatum, GABA production may be increased in efferent neurons of the globus pallidus. We have hypothesized that this effect may be related to the increase in burst firing recorded in the globus pallidus after unilateral dopamine lesions in awake animals, and we tested the hypothesis that increased excitatory inputs from the subthalamic nucleus could mediate the effect. Indeed, a kainate-induced lesion of the subthalamic nucleus on the side of the nigrostriatal lesion abolishes the increase in GAD 67 mRNA seen in the globus pallidus after the nigrostriatal lesion alone (5).

In models of the basal ganglia (2), the globus pallidus is viewed mostly as a relay nucleus in the "indirect pathway," which conveys information from the striatum to the output regions of the basal ganglia, the internal pallidum, and the substantia nigra pars reticulata by way of the subthalamic nucleus. More recently, however, it has been recognized that the globus pallidus also influences these output regions through direct GABAergic projections. Furthermore, the globus pallidus also projects outside the basal ganglia, in particular to the reticular nucleus of the thalamus (6). Together with the changes in gene expression, these anatomical data suggest a much more central role for the globus pallidus in basal ganglia circuitry than was considered previously. Through its reciprocal connections with the globus pallidus, the subthalamic nucleus, which shows a marked increase in firing after dopamine depletion, may contribute to changes in neuronal activity in the globus pallidus as much as the striatal input. The long-term effects of the changes in activity in these two inputs to the globus pallidus are likely to depend on the time course of alterations in the neurotransmitter receptors and intracellular effectors, which mediate their effects, in response to prolonged changes in neurotransmission.

Although there is evidence that GABA and N-methyl-D-aspartate receptors decrease in the globus pallidus after prolonged alteration in dopamine transmission (7), less is known about other molecules that are likely to play a role in changes in neuronal activity. One intriguing possibility is that prolonged changes in neuronal firing are associated with changes in expression of ion channels, in particular, potassium channels, which are known to influence the pattern of neuronal firing. One advantage of addressing this question at the level of gene expression is that *in situ* hybridization histochemistry can provide information on relative changes in levels of mRNA expression at the single-cell level, providing a very high degree of anatomical resolution. Furthermore, specific probes can be designed to distinguish mRNAs encoding subtypes of channels for which specific antibodies may not yet be available. However, very little is known about the regulation of mRNAs encoding potassium channels in brain. In collaboration with Theresa Perney (Rutgers University), we have tested the hypothesis that alterations in dopamine transmission affect the expression of the mRNA encoding the shaw-like potassium channel Kv3.1, which is expressed at a high level in neurons of the globus pallidus. Both 6-hydroxydopamine lesions and treatment with haloperidol (2 mg/kg for 10 days) significantly decreased Kv3.1 mRNA levels in the globus pallidus. The effect of the nigrostriatal lesion was blocked by a subthalamic lesion, although the subthalamic lesion itself, unexpectedly, produced the same effect. In contrast to haloperidol (2 mg/kg), which affects both D1 and D2 dopamine receptors, treatment with 1 mg/kg, a dose that primarily blocks D2 receptors, for 7 days, did not affect Kv3.1 mRNA levels in the globus pallidus, suggesting that the effect may be mediated by D1 receptors. The data reveal a novel molecular consequence of alterations in dopamine transmision and suggest that further study of alterations in potassium channels may provide new clues for the understanding and perhaps the treatment of the long-term consequences of dopaminergic lesions or dopamine-receptor blockade in the basal ganglia.

Alterations of dopaminergic transmission in the basal ganglia result in changes in activity of neurons in the thalamus and cerebral cortex, which ultimately mediate the resulting abnormal motor behavior (2). Therefore, it is likely that in addition to the striatum and its output regions, including the globus pallidus, dopamine depletion will affect gene expression in other brain areas. Very little, however, is known about the molecular consequences of dopamine depletion in areas outside, but anatomically and functionally related to, the basal ganglia. We were intrigued by the possibility that nigrostriatal lesions may affect the thalamic reticular nucleus (RTN). This region consists of a band of GABAergic neurons that surrounds the thalamus, receives inputs from both the thalamocortical and corticothalamic neurons, and projects to the thalamus. In addition, the RTN receives direct inputs from the globus pallidus and the substantia nigra pars reticulata. After unilateral lesions of the nigrostriatal pathway in rats, the level of GAD 67 mRNA increased bilaterally in the ventrolateral RTN (6). This effect is very robust and was observed 2–3 wk after lesions made with a variety of experimental paradigms. It is intriguing that, in contrast to most reported effects on gene regulation within the basal ganglia, this effect was bilateral. Several anatomical circuits can be involved, including bilateral cortical projections or crossed inputs from the pedunculopontine nucleus (6). Interestingly, low doses of haloperidol (1 mg/kg), which have profound effects on gene expression within the basal ganglia, did not produce this effect. However, the effects of higher concentrations and of other drugs affecting D1 dopamine receptors remain to be examined.

In summary, examining the effects of dopamine depletion and dopamine-receptor blockade on gene expression, not only in the striatum, but also in other regions of the basal ganglia and the brain, revealed widespread effects of altered dopaminergic transmission at the molecular level. Futhermore, the data showed that dopamine depletion affects not only the expression of genes encoding transcription factors and neurotransmitters, but also other molecules, such as potassium channels, which may be involved in long-term alterations in neuronal function and provide new molecular targets for the treatment of movement disorders resulting from dopamine dysfunction.

References

1. Angulo, J. A., and McEwen, B. S. (1994). Molecular aspects of neuropeptide regulation and function in the corpus striatum and nucleus acumbens. *Brain Res. Rev.* **19**, 1–28.
2. Albin, R. L., Young, A. B., and Penney, J. B. (1990). The functional anatomy of basal ganglia disorders. *Trends Neurosci.* **12**, 366–375.
3. Delfs, J. M., Anegawa, N. J., and Chesselet, M-F. (1995). Glutamic acid decarboxylase mRNA in rat pallidum: Comparison of the effects of haloperidol, clozapine and combined haloperidol-scopolamine treatments. *Neuroscience* **66**, 67–80.
4. Delfs, J. D., Ellison, G. D. Mercugliano, M., and Chesselet, M-F. (1995). Expression of glutamic acid decarboxylase mRNA in striatum and pallidum in an animal model of tardive dyskinesia. *Exp. Neurol.* **133**, 175–188.
5. Delfs, J. D., Ciaramitaro, V. M., Parry, T. J., and Chesselet, M-F. (1995). Subthalamic nucleus lesions: Widespread effects on changes in gene expression induced by nigrostriatal dopamine depletion in rats. *J. Neurosci.* **15**, 6562–6575.

6 Delfs, J. D., Ciaramitaro, V. M., Soghomonian, J-J., and Chesselet, M-F. (1995). Unilateral nigrostriatal lesions induce a bilateral increase in glutamic acid decarboxylase mRNA in the reticular thalamic nucleus. *Neuroscience* **71**, 383–395.

John F. Marshall, David N. Ruskin, and Gerald J. LaHoste

Department of Psychobiology
University of California, Irvine
Irvine, California 92697

Dopaminergic Regulation of Immediate Early Gene Expression in the Basal Ganglia

Models of basal ganglia organization have emphasized that information flows from the neostriatum via two routes, the striatonigral and striatopallidal pathways (1). Correspondingly, the dopaminergic influences in neostriatum are thought to be mediated by different dopamine receptors having distinct locations on these two pathways. Yet several attributes of dopaminergic influences on basal ganglia function are not addressed by or seem contradictory to these models.

First, D1 (including D1 and D5 receptors) and D2 (including D2, D3, and D4 receptors) subfamilies of dopamine (DA) receptor synergize to influence the function of basal ganglia neurons in intact animals. D1–D2 synergy in the control of basal ganglia activity is evident using a variety of dependent measures, including motor behaviors, cerebral metabolism, cell firing, and gene expression (2). After extensive nigrostriatal injury, this property of D1–D2 synergy breaks down, such that independent D1- or D2-receptor agonism elicits many of the behavioral, physiological, and genomic influences observed following concurrent D1–D2 agonism in intact animals. The cellular bases for D1–D2 synergy in intact animals and D1–D2 independence after injury are presently unclear. Second, DA receptors located outside of the striatum contribute to the basal ganglia actions of exogenously applied DA agonists as well as to endogenous DA. At least one DA receptor (or mRNA) has been found in globus pallidus (GP), subthalamic nucleus (STN), and substantia nigra pars reticulata (SNr), where they may modulate the activity of excitatory or inhibitory basal ganglia circuits. Third, some recent findings concerning dopaminergic influences on basal ganglia cell function are opposite to predictions of the existing models. These considerations call for a reformulation of the underlying models.

The present studies use the expression of Fos, the protein product of the immediate early gene (IEG) *c-fos,* to characterize dopaminergic influences on cells of the direct (striato-SNr) and indirect (striato-GP-STN-SNr) pathways. IEGs were first characterized as a set of genes that were quickly transcribed following growth factor–induced re-entry of cultured cells into the proliferative cycle. Typically, IEGs are expressed at low levels basally, and their induced transcription is short-lived. By phosphorylation of a constitutively expressed transcription factor, cyclic response element binding (CREB), depolarization and adenylate cyclase activation can both induce IEG expression.

Advances in Pharmacology, Volume 42

1054-3589/98 $25.00

I. Direct Pathway

The ability of DA agonists to induce striatal c-*fos* has been a key observation in exploring the basal ganglia circuitry on which DA acts. The necessary role of D1 receptors in amphetamine (AMPH)- or cocaine (COC)- induced striatal c-*fos* has long been recognized, because the D1 antagonist SCH 23390 blocks the striatal IEG induction. The role of the D2 receptors, however, has been less clear. Varying degrees of inhibition of AMPH- and COC-induced IEG expression by D2 antagonists have been reported, and it has been difficult to draw conclusions because D2 antagonists alone cause considerable striatal IEG expression. It was reasoned that the problems of interpretation arising from the use of D2 antagonists might be minimized by analyzing the AMPH- or COC-induced striatal Fos in restricted cell populations. Because D2 antagonists induce Fos in striatopallidal neurons, we measured Fos in striatonigral cells, identified by labeling with the fluorescent retrograde tracer Fluorogold placed in the SNr.

Experiments used male Sprague-Dawley rats iontophoresed with Fluorogold in the SNr of each hemisphere. Five to 8 days later, the rats were given d-amphetamine (5 mg/kg) or COC (40 mg/kg) preceded by either the D2 antagonist eticlopride (ETIC; 0.5 mg/kg) or saline. Two hours later, the rats were perfused. Double fluorescence immunocytochemistry was used to visualize Fluorogold and CY-3-labeled Fos.

Numerous Fluorogold-positive striatal neurons were found. In rats given ETIC alone, there were almost no Fos immunofluorescent nuclei in identified striatonigral neurons of rats receiving either saline or ETIC alone. Substantial numbers of immunofluorescent nuclei were observed in striata of ETIC-treated animals, but these nuclei rarely coincided with the striatonigral population. In contrast, both AMPH and COC induced considerable Fos-like immunofluorescence in striatonigral neurons. Pretreatment with ETIC nearly abolished Fos-like immunofluorescence in striatonigral neurons, reducing it by 98% (for AMPH) and 94% (for COC).

While previous investigations have indicated the obligatory role of D1 receptors in the striatal c-*fos* induction due to AMPH or COC administration, the present findings indicate the necessity of concomitant D2-receptor stimulation for these effects. Thus, D1-receptor activation is *not sufficient* to mediate indirect DA agonist–induced striatal Fos, and coactivation of D1 and D2 classes of dopamine receptor are needed. These conclusions are in accordance with earlier results from this laboratory, in which direct DA agonists were used to stimulate striatal Fos expression (3).

II. Indirect Pathway

A. D1 Influences

The indirect pathway, involving striatopallidal projections and pallidal efferents to STN and SNr, contributes importantly to motor functions. While some evidence suggests that the striatopallidal enkephalinergic neurons contain

D2 but not D1 mRNA, electrophysiological studies have shown that D1 agonism can influence the activity of GP neurons. The current experiments were designed to characterize the roles of dopamine D1 and D2 receptors in GP and STN Fos expression, comparing intact and 6-hydroxydopamine (6-OHDA)-lesioned rats.

Rats were given unilateral 6-OHDA injections into the ventral tegmentum to damage extensively the ascending DA projections. Other rats were unlesioned. Five weeks later, rats were injected i.p. with saline or DA agonists and killed 2 hr later. After sectioning, tissue was processed for Fos immunocytochemistry using the ABC method.

In unlesioned rats, administration of SKF 38393 (20 mg/kg) plus quinpirole (0.5 mg/kg) produced substantial numbers of Fos-ir nuclei in GP and STN (126 ± 32 and 56 ± 13 per mm^2, respectively). Quinpirole alone had a much smaller effect in GP and no effect on STN Fos. Saline or SKF 38393 alone (20 mg/kg) failed to induce significant numbers of Fos-ir nuclei in either structure.

In the lesioned hemispheres of 6-OHDA-injected rats, either (SKF 38393 (4 or 20 mg/kg) or quinpirole alone elicited GP Fos. In STN, SKF 38393 at either dose, but not quinpirole, evoked significant numbers of Fos-ir nuclei. These effects of SKF 38393 were fully antagonized by pretreatment with SCH 23390 (0.1 mg/kg, i.p.), but not eticlopride (0.5 mg/kg i.p.), indicating that they were D1-mediated.

Current models of D1 versus D2 control of basal ganglia efferent pathways have difficulty explaining two aspects of these findings. First, it is not clear how D1 agonists synergize with D2 agonists to control GP Fos. Second, current models fail to predict that DA agonists can activate STN. Here, STN Fos expression was increased by combined D1- and D2-agonist treatment (in unlesioned rats) and to an even greater extent after D1-agonist treatment in 6-OHDA-injected rats. This novel finding is difficult to reconcile with the inhibitory nature of the GP–STN pathway. We hypothesize (1) that the D1 agonist may act directly on dopamine receptors in STN or (2) that the D1 agonist acts indirectly (e.g., via excitatory cortico-STN connections) to activate STN neurons in the 6-OHDA-injected hemisphere.

B. Controls of Distinct GP Cell Populations

GP Fos expression is increased both by DA agonists and by acute *reductions* in D2 receptor stimulation. A similar, apparently paradoxical Fos regulation is observed in striatum, where either DA agonists or D2 antagonists can induce IEG expression but do so in largely separate populations of neurons. A population-based distinction may also apply to GP, because GP neurons evidence both physiological and anatomical heterogeneities. Features of anatomical heterogeneity include GP neuron dendritic shapes and spininess, somatic sizes, destinations of axonal projections, and whether they express parvalbumin (PARV). The present studies were undertaken to determine which populations of GP neurons expressed Fos after treatment with DA agonists versus D2 antagonist. In the following experiments, GP cell populations were distinguished on the basis of (1) whether they were PARV+ or PARV− or (2) whether they were retrogradely labeled from one of four target sites.

Experiments used unlesioned male rats. Rats to be used for Fos–Fluorogold double labeling underwent surgery for intracerebral Fluorogold iontophoresis 4–6 days before sacrifice, during which tracer was applied in the SNr, entopeduncular nucleus (EP), STN, or caudate–putamen (CPu). Saline, DA agonists (SKF 38393 [20 mg/kg], quinpirole [0.5 mg/kg], or a combination of SKF 38393 plus quinpirole [20 + 0.5 mg/kg]), or D2 antagonist (eticlopride [1 mg/kg]) was administered, and the rats were killed 2 hr later. Sections through anterior GP were incubated for Fos immunoreactivity and visualized using vector SG for chromogen. PARV immunoreactivity was visualized using a monoclonal antibody (1 : 1000, Sigma) followed by incubation in CY-3-conjugated goat anti-mouse IgG. In separate groups of rats, Fos-positive nuclei were classified as being present in (1) PARV+ or PARV− cells or (2) Fluorogold+ or Fluorogold− cells.

Administration of either DA agonists or ETIC induced significant numbers of GP Fos-IR nuclei. Fos-IR nuclei were observed more frequently in PARV− than PARV+ GP cells. This difference may arise because PARV is a calcium-binding protein, which can lower intracellular calcium concentrations, thereby blunting calcium-dependent activation of CREB. Compared with saline injection, administration of ETIC, quinpirole, or SKF 38393 and quinpirole induced significant numbers of Fos-IR nuclei in the PARV− cells. By contrast, only combined SKF 38393 and quinpirole induced significant numbers of Fos-ir nuclei in the PARV+ cell population.

In rats having Fluorogold iontophoresis, the combination of SKF 38393 and quinpirole significantly increased Fos expression in cells projectig to cPu, SNr, EP, and STN, whereas ETIC increased Fos only in neurons projecting to CPu.

These findings suggest a parcellation of GP neurons based on their function. GP neurons activated by D2 antagonists represent the most distinct population, being only PARV− and projecting only to CPu. In contrast, DA agonist–activated neurons include both PARV+ and PARV− cells and project to all of the target nuclei examined in this study. However, it is not clear from the present data how many types of GP cells the agonist-activated GP neurons represent, because GP neurons can collateralize in their projections to CPu, SNr, EP, and STN. Assuming complete collateralization of the agonist-activated neurons, it appears that there is a minimum of two separate populations (collateralized PARV+ and collateralized PARV− cells). Revised models of basal ganglia organization will need to incorporate such heterogeneities in the GP neuron population.

References

1. Gerfen, C. R. (1992). The neostriatal mosaic: Multiple levels of compartmental organization in the basal ganglia. *Annu. Rev. Neurosci.* **15**, 285–320.
2. Marshall, J. F., Ruskin, D. N., and LaHoste, G. J. (1996). D1/D2 dopamine receptor interactions in basal ganglia functions. *In* The Dopamine Receptors. (K. A. Neve, and R. L. Neve, eds.) Human Press, Totowa, NJ, pp 193–219.
3. LaHoste, G. J., Yu, J., and Marshall, J. F. (1993). Striatal Fos expression is indicative of dopamine D1/D2 synergism and receptor supersensitivity. *Proc. Nat. Acad. Sci. U.S.A.* **90**, 7451–7455.

Fu-Chin Liu* and Ann M. Graybiel†

*Institute of Neuroscience and Department of Life Science
National Yang-Ming University
Taipei, Taiwan 11221

†Department of Brain and Cognitive Sciences
Massachusetts Institute of Technology
Cambridge, Massachusetts 02139

Dopamine and Calcium Signal Interactions in the Developing Striatum: Control by Kinetics of CREB Phosphorylation

Dopamine and Ca^{2+} are major signaling molecules in the striatum. Intracellular cascades transduced by dopamine and Ca^{2+} are known to be involved in neuronal plasticity and development in the striatum as well as several other regions of the brain. It is striking that striosomes and matrix both show compartmentalized developmental patterns of expression for their dopamine and Ca^{2+} systems (1). The concentration of dopamine- and Ca^{2+}-related signaling molecules in the striatum during development renders the developing striatum an attractive model system for studying the interactions between these two plasticity-related systems during development.

It is likely that gene regulation mediated by dopamine and Ca^{2+} signals is one of the molecular bases of neuronal plasticity and development. We have approached this issue by analyzing regulation of the cyclic adenosine monophosphate (cAMP) response element-binding protein (CREB). CREB is a transcription factor that is an important mediator in the plasticity underlying memory in both the invertebrate and vertebrate systems (2). Activation of CREB can occur through phosphorylation of CREB on its Ser^{133} residue by cAMP-dependent protein kinase or by Ca^{2+} influx through voltage-sensitive Ca^{2+} channels. Activated CREB can subsequently regulate cAMP-response element (CRE)-containing genes through the CRE. Accordingly, CREB has been proposed to be a convergence molecule for cAMP- and Ca^{2+}-mediated activation (3). As dopamine D1-class receptors are linked to the cAMP–adenylate cyclase pathway, dopamine and Ca^{2+} signals may share a common transduction pathway through CREB in regulating striatal gene expression.

Advances in Pharmacology, Volume 42

1054-3589/98 $25.00

We review here findings suggesting that the kinetics of CREB phosphorylation may be a critical factor in determining downstream gene activation by dopamine and Ca^{2+} in the developing striatum (4).

I. Strategy for Studying Phosphorylation of CREB in the Striatum

There were two key features to our study. The first was to use a phosphospecific CREB (PCREB) antiserum. We used the PCREB antiserum generated by Ginty *et al.*, which specifically recognizes phosphorylated CREB on Ser^{133} (5). We also used this group's CREB antiserum, which recognizes phosphorylated as well as nonphosphorylated CREB. With these antisera, we were able to detect CREB phosphorylation at the single-cell level of resolution. The second feature of our approach was to use organotypic slice cultures of the neonatal striatum, which allowed us to identify striatal compartment-specific cell types *in vitro*. We have found that striatal architecture is largely maintained in slice cultures and that many of the neurochemical distinctions between striosomes and matrix are maintained (4, 6). To prepare the slice cultures, the forebrains of newborn rat pups were cut into $^{300}\mu$m-thick slices that were then placed into Millipore culture plate inserts and cultured for 1–3 days (4, 6).

II. Induction of Sustained CREB Phosphorylation in Developing Striosomes by Activation of Dopamine D1-Class Receptors

In vehicle-treated slices, there were numerous CREB-positive nuclei distributed through the striatum. By contrast, PCREB immunoreactivity was nearly nil. These results indicated that CREB is constitutively expressed in the cultured striatum, but the basal level of CREB phosphorylation is very low under culture conditions. Incubation of the striatal cultures with SKF 81297 (100 nM), an agonist selective for dopamine D1-class receptors, induced a striking modular pattern of PCREB immunostaining in the striatum. PCREB-positive nuclei appeared in many patches scattered throughout the striatum. Double immunostaining for PCREB and DARPP-32 demonstrated that these PCREB patches corresponded to DARPP-32-positive striosomes. This induction of PCREB was blocked by pretreatment with the D1 antagonist SCH 23390 (1 μM), and no induction of PCREB occurred on treatment with the D2 agonist quinpirol (10 μM). These results suggested that the PCREB induction we found was selective for D1-class dopamine receptors.

III. Kinetics of CREB Phosphorylation: Transient versus Sustained Phosphorylation

A parameter that we found to be critical for the pattern of CREB phosphorylation was the length of time that the slices were incubated with the drugs. The striosomal pattern of PCREB induction occurred in slices incubated for 30 min with SKF 81297. However, when the incubation time was reduced to 7 min, SKF 81297 induced PCREB throughout the striatum, in both striosomes and matrix. Thus, stimulation of D1-class receptors for 7 min induced CREB phosphorylation in both striosomes and matrix, but by 30 min, the CREB phosphorylation was diminished in the matrix but was sustained in striosomes (Table I). These findings suggested that, in terms of the kinetics of CREB phosphorylation, there are two modes of CREB phosphorylation: transient and sustained.

IV. Induction of Sustained CREB Phosphorylation in DARPP-32-Negative Developing Matrix Neurons and in Scattered DARPP-32-Negative Cells in Striosomes by Activation of L-Type Voltage-Sensitive Ca^{2+} Channels

For the Ca^{2+} system, we studied the activation of L-type voltage-sensitive Ca^{2+} channels (L-type channels). We used BAY K 8644, an activator of L-type channels. Incubation of the slices with this L-type channel agonist for 30 min did not produce the striosome-selective pattern that we found with the D1/D5-receptor agonist. Instead, scattered PCREB-positive nuclei appeared throughout the striatum. Double immunostaining for PCREB and DARPP-32 showed that these PCREB-positive nuclei were mainly in the DARPP-32-negative matrix and some scattered DARPP-32-negative cells in striosomes (see Table I). Like SKF 81297 at 7-min incubation times, treatment with BAY K 8644 for 7 min induced PCREB in both striosomes and matrix. The specificity of BAY K 8644 was demonstrated by blockade induced by pretreatments with the L-type channel antagonists, nifedipine (10 μM) and nitrendipine (10 μM), and with the Ca^{2+} chelator, EGTA (10 mM).

TABLE I Substained CREB Phosphorylation, but Not Transient CREB Phosphorylation, Is Correlated with Subsequent Expression of Fos-like Protein in Organotypic Cultures of Neonatal Striatum

	Dopamine D1/D5 receptors (SKF 81297)		*L-type Ca^{2+} channels (BAY K 8644)*	
Compartmental phenotype	*DARPP-32-positive cells in striosomes*	*DARPP-32-negative cells in matrix*	*DARPP-32-positive cells in striosomes*	*DARPP-32-negative cells in matrix*
Transient PCREB	↑	↑	↑	↑
Sustained PCREB	↑	↓	↓	↑
Fos-like protein	↑	↓	↓	↑

Striatal slice cultures were treated with agonists for 7 min (transient PCREB), for 30 min (sustained PCREB), or for 2.5–3 hr (Fos-like protein).
↑, high levels of induction; ↓, low levels of induction.

V. Spatial Correlation between Induction of Sustained CREB Phosphorylation and Subsequent Fos Expression

These findings suggest that activation of dopamine D1/D5 receptors induces sustained PCREB in DARPP-32-positive cells in striosomes. By contrast, activation of L-type channels induces sustained PCREB in DARPP-32-negative cells in the matrix and in scattered DARPP-32-negative cells in striosomes. The immediate question of importance was whether these different kinetics of CREB phosphorylation would have different effects on downstream gene activation. To test this possibility, we studied the activation of the immediate-early gene, *c-fos,* which is known to be regulated by CREB phosphorylation (3). Our results showed that there was a covariation of PCREB and Fos expression induced by treatments with SKF 81297 and BAY K 8644. Sustained CREB phosphorylation and induction of Fos-like immunoreactivity were both pronounced in DARPP-32-positive striosomes with SKF 81297 treatment. In BAY K 8644–treated slices, the pattern was opposite, with sustained PCREB and Fos-like immunoreactivity being high in DARPP-32-negative neurons, concentrated in the matrix (see Table I). These findings indicate that sustained, but not transient, phosphorylation of CREB may be critical to activating downstream gene expression, at least as exemplified by the CRE-containing gene *c-fos.*

VI. Kinetics of CREB Phosphorylation: A Novel Mechanism in Control of Dopamine and Ca^{2+} Signal Interactions

Our findings suggest that although CREB has been proposed as a convergence molecule for integration of Ca^{2+} and cAMP signals, the convergence is only transient in the striatal dopamine and Ca^{2+} systems. With increasing of time, the transient convergent pattern evolves into the sustained divergent pattern. Thus, dopamine and Ca^{2+} signals may be processed convergently in striatal neurons for fast neurotransmission or short-term plasticity on the time scale of seconds and minutes. However, for signaling mechanisms that function at longer time frames, such as gene regulation, dopamine and Ca^{2+} may differentially regulate gene expression via CREB phosphorylation in different striatal compartments. Control of the kinetics of CREB phosphorylation may be a novel mechanism underlying the interaction between dopamine and Ca^{2+} systems in the developing striatum.

Acknowledgments

Supported by National Institutes of Health grant 1 R01 HD28341, and the Science Partnership Fund at the Massachusetts Institute of Technology.

References

1. Liu, F-C., and Graybiel, A. M. (1992). Heterogeneous development of calbindin-D_{28K} expression in the striatal matrix. *J. Comp. Neurol.* **320**, 304–322.
2. Frank, D. A., and Greenberg, M. E. (1994). CREB: A mediator of long-term memory from mollusks to mammals. *Cell* **79**, 5–8.
3. Sheng, M., and Greenberg, M. E., (1990). The regulation and function of c-fos and other immediate early genes in the nervous system. *Neuron* **4**, 477–485.
4. Liu, F-C., and Graybiel, A. M. (1996). Spatiotemporal dynamics of CREB phosphorylation: Transient vs. sustained phosphorylation in the developing striatum. *Neuron* **17**, 1133–1144.
5. Ginty, D. D., Kornhauser, J. M., Thompson, M. A., Bading, H., Mayo, K. E., Takahashi, J. S., and Greenberg, M. E. (1993). Regulation of CREB phosphorylation in the suprachiasmatic nucleus by light and a circadian clock. *Science* **260**, 238–241.
6. Liu, F-C., Takahashi, H., McKay, R. D. G., and Graybiel, A. M. (1995). Dopaminergic regulation of transcription factor expression in organotypic cultures of developing striatum. *J. Neurosci.* **15**, 2367–2384.

Wolfram Schultz

Institute of Physiology
University of Fribourg
1700 Fribourg, Switzerland

The Phasic Reward Signal of Primate Dopamine Neurons

Anatomical and psychopharmacological evidence indicates that mammalian dopamine systems play a major role in the motivational control of voluntary behavior. They appear to be particularly involved in the learning and execution of intentional, goal-directed behavior. However, very little is known about underlying cellular mechanisms and representations of behavioral events in these structures. To this end, we recorded the activity of single midbrain dopamine neurons in monkeys learning and performing controlled behavioral tasks. The results were evaluated according to concepts of current learning theories, which view primary liquid and food rewards as key determinants for controlling basic approach behavior. Neutral stimuli gain the capacity to elicit approach behavior after being paired with primary rewards. The rate of learning is directly related to the unpredictability of reward and reaches an asymptote when reward is entirely predicted by a conditioned stimulus.

Advances in Pharmacology, Volume 42

1054-3589/98 $25.00

I. Responses to Primary Rewards

A majority of dopamine neurons (75–85%) are activated by phasically occurring, primary appetitive stimuli, such as foods and liquids, whereas the remaining neurons are not influenced by any stimulus tested. Responses are very phasic, occurring with latencies of 50–110 ms and lasting less than 300 ms. Responses are observed when animals touch a small morsel of hidden food during exploratory movements in the absence of other phasic stimuli or when receiving a drop of liquid at the mouth outside of any behavioral task or while learning a task (1, 2). The responses do not discriminate between different rewards but distinguish rewards from nonreward objects. Neurons activated by primary rewards do not respond or are occasionally depressed in their activity when nonfood objects are touched, even when they are similarly shaped, or when a fluid valve is operated audibly without actually delivering liquid. Only a few dopamine neurons are activated by innocuous primary aversive stimuli, such as an air puff to the hand or a drop of hypertonic saline to the mouth. However, these responses are not strong enough to result in an average population response (2).

II. Responses to Conditioned, Reward-Predicting Stimuli

A majority of dopamine neurons (55–75%) are also activated by conditioned visual and auditory stimuli that have become valid reward predictors (1). Responses do not discriminate between visual and auditory stimuli, the same neurons responding to both modalities. Conditioned stimuli are generally slightly less effective than primary rewards, both in terms of response magnitude in individual neurons and in terms of fractions of neurons activated. In most situations, dopamine responses show an all-or-none discrimination between appetitive and neutral or aversive stimuli (1, 2). Only few neurons are also activated by conditioned aversive visual and auditory stimuli in innocuous air puff or saline avoidance tasks, without leading to an average population response (2). When neutral or aversive stimuli are presented in close temporal proximity and in random alternation with physically very similar appetitive stimuli, the discriminative capacity of dopamine neurons is expressed by quantitative preferences for appetitive stimuli, whereas the all-or-none discrimination is lost.

III. Responses to Novel Stimuli

Dopamine neurons are activated by novel stimuli as long as they elicit behavioral orienting reactions (e.g., ocular saccades). Neuronal responses subside together with orienting reactions after several tens of stimulus repetitions (1). Particularly salient stimuli are effective in over 1000 trials for both behavioral and neuronal responses.

IV. Coding an Error in the Prediction of Reward

The responses of dopamine neurons to primary rewards and reward-predicting stimuli do not occur unconditionally but depend entirely on the unpredictability of these events. Dopamine neurons respond to primary rewards only when the reward occurs unpredictably, either outside of a task or during learning (Fig. 1, top) (1, 2). By contrast, a fully predicted reward does not elicit a dopamine response, and the response is transferred to the earliest conditioned, reward-predicting stimulus in most neurons (Fig. 1, middle). During repeated learning of new contingencies within a given task structure, dopamine neurons are activated by primary reward during each learning phase but stop responding when the asymptote of each learning curve is reached and rewards are fully predicted (3). An inverse situation exists when a fully predicted reward fails to occur, because of an error of the animal or the deliberate withholding by the experimenter. Dopamine neurons are depressed in their activity exactly at the time at which the reward would have occurred (Fig. 1, bottom) (3). It, thus, appears that dopamine neurons code the deviation or "error" between the prediction and the actual occurrence of reward, according to

$$\text{Dopamine response} = \text{Reward occurrence} - \text{Reward prediction}$$

A comparable dependence on event unpredictability holds for conditioned, reward-predicting stimuli. Dopamine neurons respond only to the first of a series of consecutive conditioned stimuli, but not to the subsequent predicted stimuli nor the predicted primary reward (3). This suggests that stimulus unpredictability is a common requirement for all appetitive events activating dopamine neurons.

V. Using the Dopamine Reward Signal

Taken together, dopamine neurons respond to a limited range of stimuli. Most of them are specifically appetitive in nature, namely primary rewards and conditioned, rewarding-predicting stimuli. The remaining effective stimuli are potentially appetitive, namely novel or appetitive-resembling stimuli. Dopamine neurons, thus, report nearly exclusively the positive motivational value of environmental events.

The response transfer from primary rewards to conditioned, reward-predicting stimuli demonstrates the adaptive capacities of dopamine neurons that allow them to acquire responses to a wide variety of intrinsically neutral environmental stimuli, which become associated with rewards. The predictive reward signal provides advance information about appetitive events, which is useful for ameliorating approach behavior as part of behavioral learning.

The relationship to event unpredictability suggests that dopamine neurons report an error in the prediction of appetitive events. These characteristics follow basic assumptions of associative learning theories, in which learning directly depends on the unpredictability of the reinforcer (4). Thus, the dopamine response could serve as an appropriate teaching signal for appetitive learning. The message of dopamine neurons is broadcast as a global reinforcement signal to the large majority of neurons in the striatum (caudate nucleus

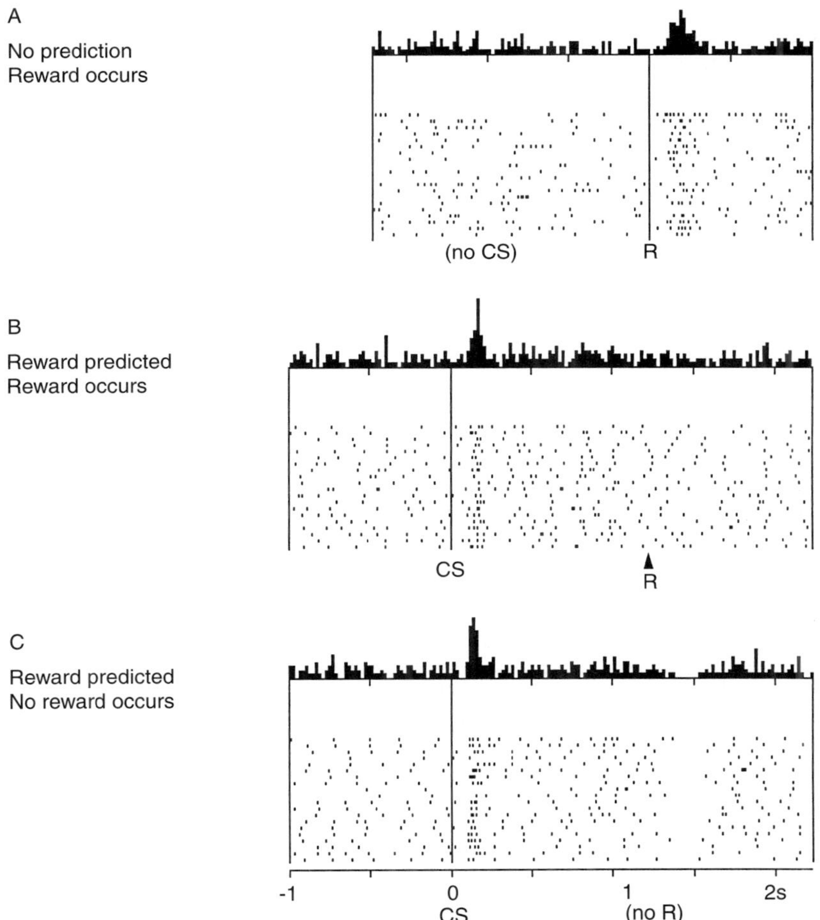

FIGURE 1 Dopamine neurons code an error in reward prediction. (A) A drop of liquid reward occurs in the absence of prediction, hence a deviation, or "error" in the prediction of reward. The dopamine neuron is activated by this unpredicted occurrence of liquid. (B) A conditioned stimulus (CS) predicts a reward (R), and the reward occurs according to the prediction, hence no error in the prediction of reward. The dopamine neuron is activated by the reward-predicting stimulus but fails to be activated by the predicted reward (*right*). (C) A conditioned stimulus predicts a reward, but the reward fails to occur because of lack of reaction by the animal. The activity of the dopamine neuron is depressed exactly at the time when the reward would have occurred.

and putamen) and to many neurons in the frontal cortex, where it is able to influence neuronal and synaptic activity occurring around the time of the dopamine signal.

The dopamine signal and the embedding basal ganglia architecture strongly resemble the Adaptive Critic module of Temporal Difference models of predictive reinforcement learning developed on the basis of behavioral theories (5). The Critic modifies the synaptic strength of a target Actor according to the predicted outcome of behavior after having learned the outcome prediction, very similar to the adaptive properties of dopamine neurons. These models have been used for implementing a variety of biologically plausible artificial networks that very efficiently learn even complex tasks. These comparisons suggest that the dopamine message could be effectively used as a global reinforcement signal for adapting the behavior according to the motivational value of environmental stimuli.

Acknowledgments

Recent financial support was obtained by the Swiss National Science Foundation, the Roche Research Foundation, the U.S. National Institute of Mental Health, the McDonnell Foundation and the British Council.

References

1. Ljungberg, T., Apicella, P., and Schultz, W. (1992). Responses of monkey dopamine neurons during learning of behavioral reactions. *J. Neurophysiol.* **67**, 145–163.
2. Mirenowicz, J., and Schultz, W. (1996). Preferential activation of midbrain dopamine neurons by appetitive rather than aversive stimuli. *Nature* **379**, 449–451.
3. Schultz, W., Apicella, P., and Ljungberg, T. (1993). Responses of monkey dopamine neurons to reward and conditioned stimuli during successive steps of learning a delayed response task. *J. Neuronsci.* **13**, 900–913.
4. Rescorla, R. A., and Wagner, A. R. (1972). A theory of Pavlovian conditioning: Variations in the effectiveness of reinforcement and nonreinforcement. *In* Classical Conditioning II: Current Research and Theory (A. H. Black and W. F. Prokasy, eds.), pp. 64–99. Appleton-Century-Crafts, New York.
5. Sutton, R. S., and Barto, A. G. (1981). Toward a modern theory of adaptive networks: Expectation and prediction. *Psychol. Rev.* **88**, 135–170.

Steven W. Johnson

Departments of Physiology and Pharmacology and Neurology
Oregon Health Sciences University
Portland, Oregon 97201

Afferent Control of Midbrain Dopamine Neurons: An Intracellular Perspective

Dopamine neurons in the substantia nigra and ventral tegmental area fire spontaneous action potential *in vitro* due to intrinsic membrane properties. However, afferent synaptic inputs regulate the firing frequency and pattern of these neurons *in vivo*. To study effects of synaptic inputs, membrane potentials or currents from single dopamine neurons were recorded with intracellular microelectrodes or patch pipettes in whole-cell configuration in the rat midbrain slice. The slices were cut in the horizontal plane and completely submerged and perfused with bicarbonate-buffered artificial cerebrospinal fluid saturated with oxygen. Synaptic potentials or currents were evoked with stimuli of constant voltage (0.1-ms duration) using bipolar tungsten electrodes placed in the slice within 500 μm of the recording electrode. Dopamine-containing neurons were identified by well-established electrophysiological criteria, which include the spontaneous firing of action potentials, a large postspike hyperpolarization, presence of *Ih* (hyperpolarization-activated current), and membrane hyperpolarization or outward current evoked by perfusing the slice with dopamine (30 μM).

A single electrical stimulus delivered to the slice evokes a biphasic potential composed of an early depolarization mediated by glutamate (fast excitatory postsynaptic potential [EPSP]) followed by a hyperpolarization mediated by synaptic release of GABA (fast inhibitory postsynaptic potential [IPSP]). Perfusing the slice with the $GABA_A$ antagonist bicuculline (30 μM) blocks the fast IPSP and permits the study of the fast EPSP in isolation. When recording in AP5 (50 μM), which blocks N-methyl-D-aspartate (NMDA) receptors, the EPSP peaks within 1–2 ms and decays within 50 ms. However, when recording in CNQX (10 μM) (6-cyano-7-nitroquinoxaline-2,3-dione), which blocks non-NMDA receptors, the EPSP peaks after 10–20 ms and has a duration greater than 100 ms. Thus, the fast EPSP is produced by currents activated by both NMDA and non-NMDA receptors (1).

Although these excitatory potentials can evoke action potentials, synaptic release of glutamate can also alter the firing pattern. When perfusing the slice with bicuculline (30 μM) and CNQX (10 μM), continuous electrical stimulation of the slice at 20 Hz converts single-spike firing to a burst-firing pattern (n = three of four neurons). The burst-firing pattern is caused by activation of NMDA receptors, because it is blocked by AP5 (50 μM), and perfusion with NMDA (10–30 μM) also evokes burst firing. Burst firing induced by NMDA

Advances in Pharmacology, Volume 42

1054-3589/98 $25.00

is dependent on extracellular Na^+ and K^+ but not extracellular or intracellular Ca^{2+}. It is also not blocked by the Ca^{2+}-activated K^+ channel blocker apamin (100 nM), which instead potentiates burst firing. However, burst firing was blocked by the Na^+/K^+–adenosine triphosphatase (ATPase) inhibitors ouabain or strophanthidin (2–3 μM) and was also blocked in perfusate containing no added Mg^{2+}. It was concluded that stimulation of NMDA receptors causes bursts of action potentials that are interrupted by membrane hyperpolarization caused by the electrogenic extrusion of Na^+ by Na^+/K^+–ATPase. Moreover, voltage-dependent block by Mg^{2+} facilitates the interburst membrane hyperpolarization by reducing conductance through NMDA receptor–channels (2).

When the fast EPSP is blocked by perfusing the slice with AP5 (50 μM) and CNQX (10 μM), the IPSP mediated by $GABA_A$ receptors can be studied in isolation. The IPSP has a latency-to-peak amplitude of 10 ms and lasts about 100 ms. At an initial holding potential of −60 mV, the IPSP is hyperpolarizing when potassium acetate (2 M)–filled electrodes are used, but the IPSP is depolarizing when recording with microelectrodes containing potassium chloride (2 M). Spontaneous tetrodotoxin-sensitive IPSPs can also be recorded in dopamine neurons when the extracellular K^+ concentration is raised from 2.5 to 6 or 8 mM. These $GABA_A$ IPSPs are blocked by [Met^5]enkephalin in a concentration-dependent manner (1–10 μM), yet μ-opioid agonists produce no direct change in postsynaptic membrane properties of dopamine neurons. In contrast, μ-opioid agonists hyperpolarize nondopamine secondary neurons in the substantia nigra compacta and ventral tegmental area. These neurons are presumed to contain GABA and make synaptic contact with dopamine neurons. Moreover, membrane hyperpolarization of secondary neurons occurs over the same concentration range as that which effectively reduces the frequency of spontaneous IPSPs recorded in dopamine neurons. Thus, by reducing GABA release from local interneurons, μ-opioid agonists may increase the firing rate of dopamine neurons (3).

When a brief train of stimuli is delivered to the slice, a long-duration "slow" IPSP follows the fast excitatory and inhibitory potentials. In the presence of AP5 (50 μM), CNQX (10 μM), and bicuculline (30 μM), the slow IPSP has a 250-ms latency-to-peak amplitude and a duration of about 1 sec. Its apparent reversal potential is −110 mV, which is consistent with current mediated by K^+. The slow IPSP is mediated by $GABA_B$ receptors because it is blocked by the $GABA_B$-receptor antagonist CGP 35348 (100 μM) (Fig. 1). 5HT reduces the $GABA_B$ IPSP amplitude by 40% at its IC 50 of 10 μM. This effect is mediated by $5HT_{1B}$ receptors, because the action of 5HT is blocked by cyanopindolol (100 nM) and mimicked by the $5HT_{1B}$ agonist TFMPP (300 nM). Furthermore, this action is presynaptic, because 5HT (30 μM) does not block the membrane hyperpolarization induced by stimulation of $GABA_B$ receptors by GABA (1 mM) recorded in bicuculline (30 μM). Despite a presynaptic site of action, 5HT (30 μM) does not reduce the $GABA_A$ IPSP. These data are most consistent with a segregation of GABA-containing inputs, such that those that release GABA onto $GABA_B$ receptors express $5HT_{1B}$ receptors on their nerve terminals, while those inputs that release GABA onto $GABA_A$ receptors do not (4). The striatonigral pathway is a likely candidate source for the pathway that selectively activates $GABA_B$ receptors, because the striatum but not the midbrain expresses mRNA for $5HT_{1B}$ receptors, and striatal lesions decrease the density of $5HT_{1B}$-binding sites in the midbrain.

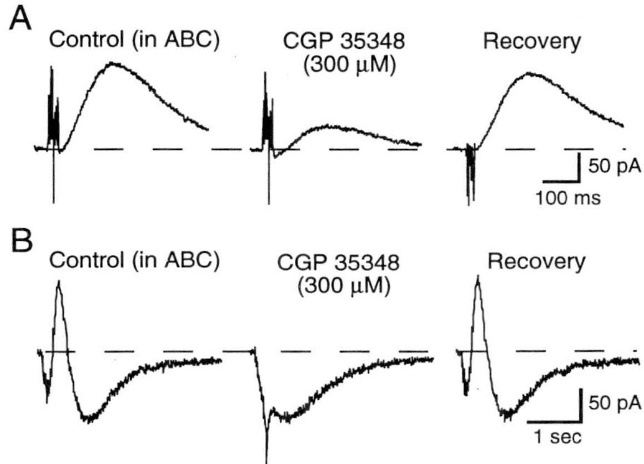

FIGURE 1 Focal electrical stimulation evokes a slow IPSC and slow EPSC recorded in dopamine neurons *in vitro*. (*A*) A brief train of stimuli evokes an outward current (slow IPSC), which is reduced by the $GABA_B$-receptor antagonist CGP 35348. (*B*) A train of stimuli also evokes a long-lasting, slowly developing inward current (slow EPSC), which is best seen when the slow IPSC is blocked by CGP 35348. Note the compressed time scale in (*B*) compared with (*A*). All recordings were made in perfusate containing AP5 (50 μM), CNQX (10 μM), and bicuculline (50 μM) to block ionotropic glutamate and $GABA_A$ receptors. Broken lines indicate zero current.

When stimulation electrodes are placed in the caudal region of the horizontal midbrain slice, a slow EPSC can be recorded in dopamine neurons in voltage clamp. The slow EPSC peaks 560 ms after onset and lasts several seconds, with a decay time constant of 630 ms (see Fig. 1). The slow EPSC is blocked when pipettes contain the G-protein inhibitor GDP-β-S (0.5 mM), and the slow EPSC amplitude is increased and its duration prolonged when pipettes contain the G-protein activator GTP-γ-S (0.5 mM). Metabotropic glutamate receptors mediate the slow EPSC because it is blocked by ($\pm$)-α-methyl-4-phosphonophenylglycine (MCPG) (300 μM) and is occluded by inward current produced by the metabotropic glutamate agonist *trans*-1-aminocyclo-pentane-1S,3R-dicarboxylic acid (ACPD) (300 μM). The finding that the slow EPSC is preferentially evoked by stimulation of caudal regions of the slice is consistent with the hypothesis that afferents arising from pontine nuclei selectively release glutamate onto metabotropic glutamate receptors on dopamine neurons (5).

Finally, an inhibitory synaptic potential mediated by dopamine can be evoked, but this synaptic potential is encountered rarely. The dopamine-mediated IPSP is similar in time course to the $GABA_B$ IPSP but is insensitive to CGP 35348. Its time course is prolonged by cocaine (3 μM) and blocked by the D2-receptor antagonist sulpiride (100 nM). Because the dopamine-mediated IPSP is recorded rarely, this suggests that stimulation of somatodendritic dopamine receptors by synaptic release of dopamine does not play an important role in regulating dopamine neuronal excitability under standard *in vitro* conditions (1).

These studies show that dopamine neurons receive inputs containing glutamate, which evokes fast excitatory synaptic currents mediated by NMDA and non-NMDA receptors, as well as a slow-onset long-duration synaptic current mediated by metabotropic glutamate receptors. Moreover, inputs containing GABA produce a fast IPSP mediated by $GABA_A$ receptors and a slow IPSP mediated by $GABA_B$ receptors. The use of ligands to activate or block presynaptic receptors on afferents containing glutamate and GABA may be an effective strategy for influencing dopamine neuronal activity and may ultimately be useful in the treatment of a variety of pathological human conditions in which dopamine release plays a role.

References

1. Johnson, S. W., and North, R. A. (1992). Two types of neurone in the rat ventral tegmental area and their synaptic inputs. *J. Physiol. (Lond)* **450**, 455–468.
2. Johnson, S. W., Seutin V., and North, R. A. (1992). Burst firing in dopamine neurons induced by N-methyl-D-aspartate: Role of electrogenic sodium pump. *Science* **258**, 665–667.
3. Johnson, S. W., and North, R. A. (1992). Opioids excite dopamine neurons by hyperpolarization of local interneurons. *J. Neurosci.* **12**, 483–488.
4. Johnson, S. W., Mercuri, N. B., and North, R. A. (1992). 5-Hydroxytryptamine$_{1B}$ receptors block the $GABA_B$ synaptic potential in rat dopamine neurons. *J. Neurosci.* **12**, 2000–2006.
5. Shen, K-Z., and Johnson, S. W. (1997). A slow excitatory postsynaptic current mediated by G-protein-coupled metabotropic glutamate receptors in rat ventral tegmental dopamine neurons. *Eur. J. Neurosci.* **9**, 48–54.

J. M. Tepper, C. A. Paladini, and P. Celada

Center for Molecular and Behavioral Neuroscience
Rutgers, The State University of New Jersey
Newark, New Jersey 07102

GABAergic Control of the Firing Pattern of Substantia Nigra Dopaminergic Neurons

In vivo, dopaminergic neurons fire spontaneously in three different patterns: a pacemaker-like regular firing pattern, a random mode, and a bursty mode. Whereas the random pattern is the most common firing pattern in anesthetized rats, dopaminergic neurons recorded *in vitro* fire spontaneously

Advances in Pharmacology, Volume 42

1054-3589/98 $25.00

almost exclusively in the pacemaker-like mode. This last fact suggests that afferent inputs play an important role in the control of the firing pattern of the nigrostriatal dopaminergic neurons. The origins of the different firing patterns are important, because increases in bursting have been correlated with increases in dopamine release.

Excitatory amino acid afferents originating in the frontal cortex, subthalamic nucleus, and pedunculopontine nucleus have been shown to play an important role in the modulation of dopaminergic neuron firing pattern (1), although these inputs may not be the only ones important for the control of the firing pattern *in vivo*. The predominant inputs to dopamine neurons in the substantia nigra are GABAergic, the best characterized of which originate from the neostriatum and globus pallidus. Data have shown that substantia nigra dopaminergic neurons also receive a significant input from the axon collaterals of substantia nigra pars reticulata projection neurons (2). Dopaminergic neurons have been shown to exhibit both $GABA_A$ and $GABA_B$ responses *in vitro* (3).

In vivo extracellular recordings from dopaminergic neurons in urethane-anesthetized rats following direct stimulation of striatum, globus pallidus, or antidromic stimulation of substantia nigra pars reticulata projection neurons revealed that all three afferents produce inhibition of dopaminergic neurons. Data from experiments using minute local pressure application of $GABA_A$ and $GABA_B$ antagonists via a picospritzer from antidromically identified nigrostriatal neurons show that the inhibition arising from each of the principal GABAergic inputs is completely and reversibly blocked by the $GABA_A$ antagonist, bicuculline (250 μM, 1 nl). This is illustrated for inhibition elicited from globus pallidus for one representative dopaminergic neuron (Fig. 1A, 2). The $GABA_B$ antagonist, saclofen (400 μM, 1 nl), not only failed to block these evoked inhibitions, but often facilitated them (Fig. 1A, 3). Thus, it appears that the monosynaptic inhibition arising from striatum, globus pallidus, and the axon collaterals of the pars reticulata GABAergic projection neuron is mediated by a $GABA_A$ receptor. Furthermore, these data also suggest that all three of these GABAergic afferents possess presynaptic inhibitory $GABA_B$ autoreceptors on their nerve terminals.

Local application of $GABA_A$ and $GABA_B$ antagonists differentially affect the firing pattern of dopaminergic neurons. The pattern of activity of substantia nigra dopaminergic neurons was quantified by computing autocorrelograms, coefficients of variation, and the percentage of spikes fired in bursts (2). Local nigral application of the $GABA_A$ antagonist, bicuculline, through the recording pipette markedly increases the proportion of neurons that fire in the bursty mode, or shifts single dopaminergic neurons firing in the pacemaker or random modes to bursty firing when applied by local pressure injection. Conversely, local application of the $GABA_B$ antagonist, saclofen, through the recording pipette (or excitotoxic lesion of globus pallidus) decreases bursty firing and produces a modest increase in the proportion of neurons that fire in a pacemaker-like mode, or shifts single dopaminergic neurons to pacemaker-like firing when applied by local pressure injection through a second pipette. This latter effect is apparently due to increased GABA release acting on postsynaptic $GABA_A$ receptors as a result of the blockade of inhibitory presynaptic $GABA_B$ autoreceptors. These data suggest that bursty firing in substantia nigra dopamin-

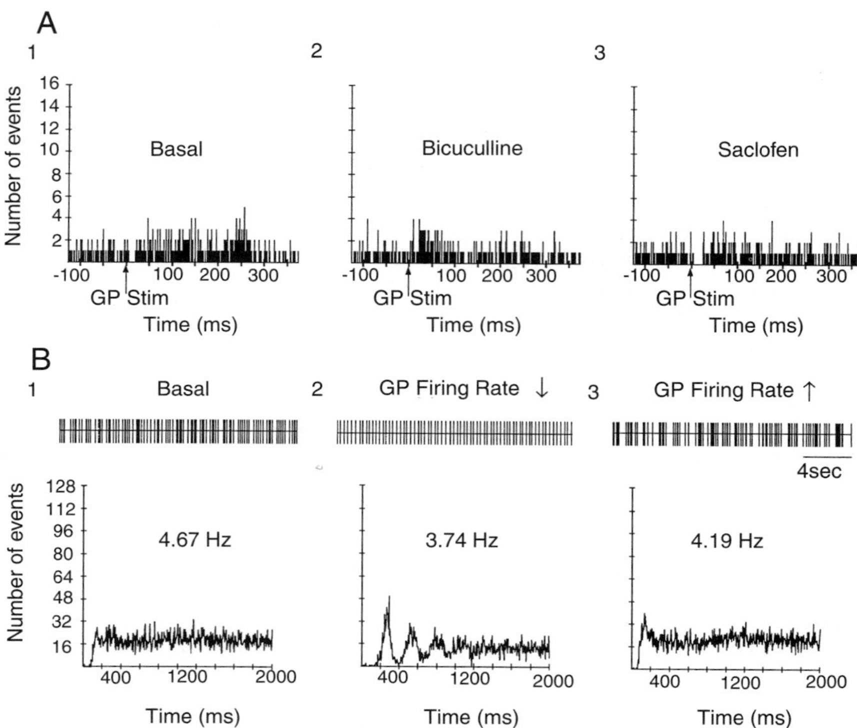

FIGURE 1 (*A*) Globus pallidus (GP) stimulation (1 mA, 0.5 ms, marked by arrow at time = 0) produces inhibition of an antidromically identified nigrostriatal dopaminergic neuron (*1*). Local application of bicuculline to the same neuron blocked the inhibition completely (*2*) while subsequent local application of saclofen (after recovery from bicuculline) increases the strength and duration of the inhibition (*3*). (*B*) Spike trains and autocorrelograms showing changes in the firing pattern of a single dopaminergic neuron as globus pallidus neurons are inhibited (*2*) and excited (*3*) by local infusion of muscimol and bicuculline, respectively, into globus pallidus. Inhibition of globus pallidus by muscimol leads to a decrease in spontaneous firing rate and a shift to pacemaker firing. Excitation of globus pallidus neurons by subsequent infusion of bicuculline leads to an increase in dopaminergic neuron firing rate and a shift to bursty firing.

C

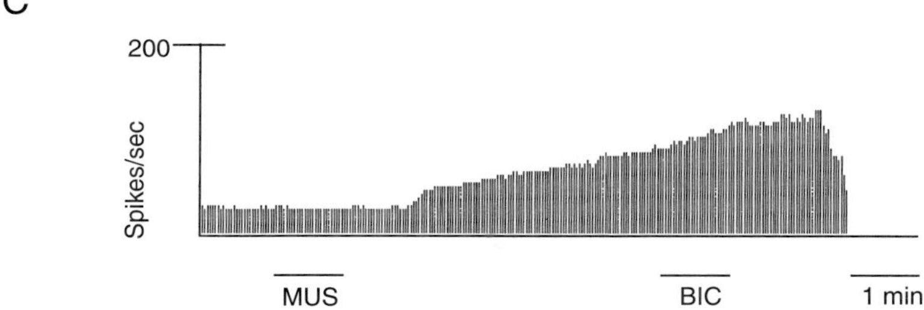

D

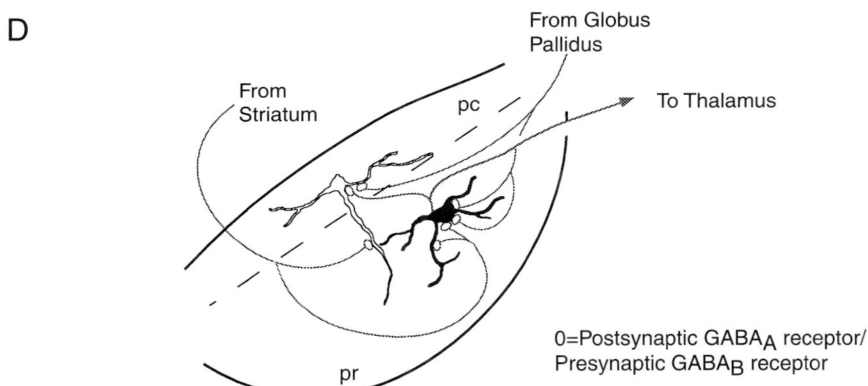

FIGURE 1 Continued. (*C*) Infusion of muscimol (MUS) into globus pallidus produces a dramatic increase in the spontaneous activity of an unidentified pars reticulata GABAergic neuron to over 100 spikes/sec. Subsequent infusion of bicuculline (BIC) reverses the excitation and produces complete cessation of firing. (*D*) A proposed model of interaction of the GABAergic inputs into the dopamine neurons in substantia nigra pars compacta.

ergic neurons is suppressed when $GABA_A$ inhibition is increased and is facilitated when $GABA_A$ input is decreased.

To determine the physiological relevance of these pharmacological phenomena with respect to the control of firing pattern of dopaminergic neurons *in vivo*, and to begin to identify the pathways involved, we examined the activity of substantia nigra dopaminergic neurons *in vivo* while manipulating globus pallidus neuron firing rates. Globus pallidus neuron firing rates were decreased by infusing muscimol (800 μM, 200 nl) or increased by infusing bicuculline (1000 μM 200 nl) into the globus pallidus through a 32-g cannula (4). Muscimol-induced inhibition of globus pallidus neurons, verified by extracellular recording, shifted dopaminergic neurons to the pacemaker firing pattern, accompanied by a slight but significant decrease in firing rate (86% of baseline), as shown for a typical dopaminergic neuron in Figure 1B, 2. Conversely, bicuculline-induced excitation of globus pallidus neurons shifted dopaminergic neurons to the bursty firing mode, with a concomitant increase in firing rate

(129% of baseline), as shown in Figure 1B, 3). Because the changes in firing rates of the dopaminergic neurons were opposite of what would be expected based on a monosynaptic GABAergic input from globus pallidus, the effects of manipulating the globus pallidus firing rate on the activity of nondopaminergic pars reticulata neurons were examined. These data revealed that nondopaminergic pars reticulata neurons were far more sensitive to the pharmacological manipulation of globus pallidus cells than were the dopaminergic neurons, and they responded in the opposite manner. That is, inhibition of globus pallidus activity produced a dramatic increase in the firing rate of the nondopaminergic pars reticulata neurons, while globus pallidus excitation decreased the nondopaminergic pars reticulata neuron firing rate (Fig. 1C). Thus, the changes in activity of substantia nigra dopaminergic neurons following manipulation of the globus pallidus neuron firing rate appear to be mediated, at least in part, indirectly through pars reticulata neurons. Although not conclusively identified, these nondopaminergic pars reticulata neurons exhibited spontaneous activity similar to that of identified GABAergic projection neurons (tonic firing at rates from 15 to 40/sec in a regular pattern), and some of these neurons were antidromically activated at short latency (0.8–3.0 ms) from ipsilateral thalamus.

These data suggest that GABAergic inputs can affect the activity of dopaminergic neurons *in vitro* in several ways. In addition to the well-known monosynaptic pathways originating in striatum and globus pallidus, there is a potent modulation of the firing pattern of dopaminergic neurons from globus pallidus that is mediated disynaptically through GABAergic neurons of the substantia nigra pars reticulata. The pars reticulata GABAergic projection neurons appear to exert a tonic inhibitory influence on burst firing. When they are inhibited by activation of globus pallidus neurons (or perhaps also is a result of increases in the activity of the direct striatonigral pathway), the dopaminergic neuron is released from tonic $GABA_A$ inhibition and burst firing ensues, perhaps mediated by a glutamatergic input from cortex, subthalamic nucleus, and/or pedunculopontine nucleus (1). Conversely, when the pars reticulata neurons are disinhibited, they act to depress bursty firing on dopaminergic neurons through activation $GABA_A$ receptors.

The physiological role of the postsynaptic $GABA_B$ receptor on dopaminergic neurons and the source of the $GABA_B$ input *in vivo* remain to be determined. These receptors do not appear to be stimulated *in vivo* by input from striatum, globus pallidus, or the axon collaterals of the pars reticulata projection neurons. However, these data do not rule out the possibility of an heretofore unidentified source of GABAergic input to dopaminergic neurons that may access the $GABA_B$ receptors. Spontaneous $GABA_B$-like synaptic responses have been observed in substantia nigra dopaminergic neurons *in vitro* (5), suggesting the possibility that there may be an intrinsic source of $GABA_B$ inputs to substantia nigra dopaminergic neurons.

Acknowledgments

Supported by N. I. H. grants NS30679 and MH45286. P. Celada was supported through Spanish and Catalan (CIRIT) Government Fellowships.

References

1. Johnson, S. W., Seutin, V., and North, R. A. (1992). Burst firing in dopamine neurons induced by N-methyl-D-aspartate: Role of electrogenic sodium pump. *Science* **258,** 665–667.
2. Tepper, J. M., Martin, L. P., and Anderson, D. R. (1995). $GABA_A$ receptor-mediated inhibition of rat substantia nigra dopaminergic neurons by pars reticulata projection neurons. *J. Neurosci.* **15,** 3092–3103.
3. Lacey, M. G., Mercuri, N. B., and North, R. A. (1988). On the potassium conductance increase activated by $GABA_B$ and dopamine D_2 receptors in rat substantia nigra neurones. *J. Physiol.* (*Lond.*) **401,** 437–453.
4. Celada, P., and Tepper, J. M. (1996). Role of globus pallidus in the regulation of the firing and pattern of nigral dopaminergic neurons. *Soc. Neurosci. Abstr.* **22,** 2026.
5. Koós, T., and Tepper, J. M., (1996). Spontaneous $GABA_B$ synaptic potentials in neonatal rat substantia nigra dopaminergic neurons. *Soc. Neurosci. Abstr.* **22,** 2026.

S. T. Kitai

Department of Anatomy and Neurobiology
University of Tennessee, College of Medicine
Memphis, Tennessee 38163

Afferent Control of Substantia Nigra Compacta Dopamine Neurons: Anatomical Perspective and Role of Glutamatergic and Cholinergic Inputs

The substantia nigra pars compacta (SNc) contains dopamine (DA) neurons that project mainly to the striatum, and the loss of these neurons gives rise to parkinsonian symptoms (1). In this report, we illustrate that the pedunculopontine nucleus (PPN) and the subthalamic nucleus (STN) could be a source of strong excitatory inputs that control firing patterns of SNc DA neurons.

Activities of SNc DA neurons of the rat were studied using a combination of intracellular electrophysiological recording *in vitro* and subsequent immunocytochemical double-labeling techniques. The neurons recorded in the SNc were identified as DA neurons if they were immunopositive for tyrosine hydroxylase (TH). Slices were prepared from male Sprague-Dawley rats (Harlan: 60–150 g) using previously described methods (2). Briefly, the rats were anesthetized with methoxyflurane and decapitated. The brain was removed and placed in ice-cold, artificial cerebrospinal fluid (aCSF), the composition of which was (mM) NaCl, 124; KCl, 2.5; $NaHCO_3$, 26; NaH_2PO_4, 1.24; $MgSO_4$, 1.3; $CaCl_2$, 2.4; and glucose, 10. Parasagittal sections (400–500 μm) were cut on a Vibratome and maintained in aCSF at room temperature for at least 1 hr prior to transferral to the recording chamber. Glass micropipettes filled with 0.5 M KCl in 0.05 Tris buffer and 1–3% biocytin (pH = 7.6, impedance 100–150 MΩ) were used for recording. A glass-coated, carbon-fiber electrode (tip diameter of 6 μm, impedance of 500 KΩ) was used to stimulate PPN and STN. Biocytin-filled SNc neurons were identified by Avidin-Texas Red histochemistry and were further processed for immunohistochemistry. DA and non-DA biocytin-filled neurons were stained with DAB-Ni using the ABC method for morphological analysis.

The DA neurons had longer-duration action potentials (mean, 1.49 ± 0.10 ms halfwidth), usually large monophasic spike afterhyperpolarization (AHP), a low-threshold calcium spike that could be activated from hyperpolarized potentials, and displayed anomalous inward and transient outward rectification.

Advances in Pharmacology, Volume 42

1054-3589/98 $25.00

PPN stimulation induced in DA neurons monosynaptic excitatory postsynaptic potentials (EPSPs) that consisted of an initial early transient followed by a late slow component. An application of antiglutamatergic agents (1 mM kynurenic acid and/or 30 μM CNQX) in the bathing media partially suppressed the EPSPs, indicating that PPN inputs to SNc DA neurons consist of glutamatergic and nonglutamatergic components. The early phase of both EPSPs was considered monosynaptic in nature because of stable latencies in spite of variation in stimulation intensities and the ability of EPSPs to follow high-frequency stimulation. The antiglutamatergic-resistant EPSPs were suppressed by applications of anticholinergic agents, such as atropine (50 μM), mecamylamine (10 μM), and pirenzepine (10 μM), indicating both the nicotinic and muscarinic nature of PPN inputs to SNc DA neurons. Bath superfusion of 10–100 μM of nicotine induced a depolarization and a decrease in input resistance of SNc DA neurons. The depolarization was usually accompanied by an increase in cell firing. Whether the nicotinic receptors in the SNc are pre- or postsynaptic was investigated by a combined immunocytochemical labeling and electron microscopy. The α_4 and β_2 nicotinic receptor subunits were immunohistochemically localized using mAB 299 and mAB 270 (gifts of Dr. Jon Lindstrom), respectively. In sections labeled for TH and the α_4 subunit, it was found that many TH-immunopositive neurons were labeled for the α_4 subunit. At the ultrastructural level, reaction product for the α_4 subunit was located in TH-immunopositive dendrites in the SNc. The monosynaptic inputs from the PPN to SNc neurons are substantiated by our intracellular labeling study of PPN neurons, which demonstrated that the collaterals of both choline acetyltransferase (CHAT)–positive and CHAT-negative neurons were traced into the substantia nigra. These results indicate a convergence of glutamatergic and cholinergic excitatory inputs from the PPN to SNc DA neurons. The monosynaptic latencies for cholinergic EPSPs (1.9–8.0 ms, n = 44) were quite similar to glutamatergic-induced EPSPs (1.6–8.0 ms, n = 33). These latencies correspond well with the latencies (0.5–12.0 ms) of antidromically activated action potentials recorded from PPN neurons (3).

Intracellular responses to STN stimulation consisted of initial fast monosynaptic EPSPs followed by slow, long-lasting EPSPs which were suppressed by application of kynurenic acid. Fast and slow EPSPs were blocked by CNQX and APV, respectively, indicating an involvement of both non-NMDA and NMDA receptors in this synaptic transmission.

The most characteristic property of a DA neuron is its regular pacemaker firing pattern. The sodium spikes are triggered from a peak of each rhythmically oscillating membrane wave, which is known to be calcium dependent. When a single spike is triggered from each summit of oscillatory wave, it promptly repolarizes, and each spike is followed by a strong spike-AHP. In our study of muscarine and carbachol effects on DA cell-firing patterns, the duration of the rising phase of each oscillatory wave was prolonged as a plateau-like depolarization, with a decrease in its amplitude by application of cholinergic agents (these effects were antagonized by atropine or pirenzepine). Under these conditions, an addition of strong excitatory inputs (e.g., glutamatergic inputs from the subthalamus and PPN) and/or disinhibition of GABAergic inputs (e.g., substantia nigra reticulata [4] and neostriatum [5]) could be able to increase the amplitude of a prolonged oscillatory wave sufficient enough to trigger a burst of action potential from the prolonged plateau of each wave. A functional significance of PPN and subthalamic cholinergic and glutamatergic inputs and nigral and striatal GABAergic inputs may, therefore, be to modify the firing pattern of DA neurons from an intrinsic regular rhythmic firing to an irregular burst firing.

Acknowledgments

The author is greatly indebted to the following individuals who contributed in making this presentation possible. These include Drs. T. Futami (PPN inputs), Y. Kang (DA neuron firing, cholinergic action, STN inputs), Y. Kuga (morphology and immunohistochemistry), C. Richards (DA neurons), T. Shiroyama (morphology, electron microscopy, and immunohistochemistry), E. Sorenson (nicotinic action), K. Takakusaki (PPN inputs and morphology), B. Teng (morphology and immunohistochemistry), and T. Yamamoto (morphology and immunohistochemistry). These studies were supported by U.S. Public Health Service grants NS20702 and NS23886 and the Human Frontier Science Program.

References

1. Gibb, W. R. and Lees, A. J. (1991). Anatomy, pigmentation, ventral and dorsal subpopulations of the substantia nigra, and differential cell death in Parkinson's disease. *J. Neurol. Neurosurg. Psychiatry* **54**,388–396.
2. Takakusaki, K., Shiroyama, T., Yamamoto, T., and Kitai, S. T. (1996). Cholinergic and noncholinergic tegmental pedunculopontine projection neurons in rats revealed by intracellular labeling. *J. Comp. Neurol.* **371**, 345–361.
3. Kang, Y., and Kitai, S. T. (1990). Electrophysiological properties of pedunculopontine neurons and their postsynaptic responses following stimulation of substantia nigra reticulata. *Brain Res.* **535**, 75–95.
4. Tepper, J. M., Martin, L. P., and Anderson, D. R. (1995). GABA(A) receptor-mediated inhibition of rat substantia nigra dopaminergic neurons by pars reticulata projection neurons. *J. Neurosci.* **15**, 3092–3103.
5. Preston, R. J., McCrea, R. A., Chang, H. T., and Kitai, S. T. (1981). Anatomy and physiology of substantia nigra and retrorubral neurons studied by extra and intracellular recording and by horseradish peroxidase labeling. *Neuroscience* **6**, 331–344.

David A. Lewis,*† Susan R. Sesack,*† Allan I. Levey,‡ and David R. Rosenberg*

Departments of *Psychiatry and †Neuroscience
University of Pittsburgh
Pittsburgh, Pennsylvania 15213

‡Department of Neurology
Emory University
Atlanta, Georgia 30322

Dopamine Axons in Primate Prefrontal Cortex: Specificity of Distribution, Synaptic Targets, and Development

Multiple lines of investigation have demonstrated that dopamine (DA) afferents from the mesencephalon play a critical role in the normal development of the prefrontal cortex (PFC), as well as in the regulation of neuronal activity in these regions of the adult brain. In addition, abnormalities in DA neurotransmission have been implicated in the pathophysiology of certain disorders, such as schizophrenia, that are associated with disturbances in the cognitive functions subserved by the PFC. Understanding how DA mediates these effects in the normal PFC and contributes to the dysfunction of the PFC in disease states requires knowledge of the organization of DA axons in the PFC and of the refinements that occur in this organization during development. Consequently, we have examined the regional and laminar innervation patterns of DA axons in monkey PFC, their synaptic targets, and their maturational changes during postnatal development. These studies have been conducted using light and electron microscopic immunocytochemical techniques with antibodies that recognize tyrosine hydroxylase (TH), the rate-limiting enzyme in DA biosynthesis, or the DA transporter (DAT), the uptake carrier that terminates the action of extracellular DA.

The specificity of the TH antibody and its tendency to selectively label DA axons in primate neocortex have been previously described (1). The specificity of the DAT antibody was demonstrated by immunoblot, immunoprecipitation, and preadsorption immunocytochemical experiments. In addition, antibody specificity was confirmed by the anatomical distribution of the labeled structures. For example, DA neurons in the ventral mesencephalon were intensively DAT-immunoreactive (IR), whereas the cell bodies of noradrenaline- or serotonin-containing neurons in the locus ceruleus and raphe nuclei, respectively, were unlabeled. These findings indicate that this antibody does in fact

Advances in Pharmacology, Volume 42

1054-3589/98 $25.00

recognize the DAT but does not cross-react with the norepinephrine or serotonin transporters. In addition, the differential labeling of monoamine-containing cell bodies did not appear to be a confound of limited antibody sensitivity, because DAT-labeled axon terminals were clearly evident within the locus ceruleus, consistent with the previously described DA innervation of this region.

Within the monkey cerebral cortex, the distribution of DAT-IR axons differed substantially across cytoarchitectonic regions. In general, the density of labeled axons was greatest in motor regions, intermediate in association regions, and lowest in primary sensory areas, although differences across regions were also present within each of these categories. In the frontal lobe, the density of DAT-IR axons decreased from the motor and premotor regions to the more rostrally located PFC. However, substantial differences in density were present across cytoarchitectonic regions of the PFC. For example, the density of labeled axons was greatest in the dorsomedial regions (areas 9 and 8), intermediate in the orbital and ventromedial regions (areas 11, 12, 14, and 32), and lowest in the dorsolateral (area 46) and frontal pole (area 10) regions.

In most areas, DAT-labeled axons in the PFC were distributed in a bilaminar fashion, with the density of labeled axons greatest in the superficial layers (I–IIIa), intermediate in the deep layers (Vb–VI), and very low in the middle layers (IIIb–Va). However, some regions had labeled axons distributed across all cortical layers. For example, area 46 had the typical bilaminar pattern, whereas the medially adjacent area 9 also had a dense band of labeled axons in the middle layers, giving rise to a trilaminar appearance.

These regional and laminar patterns were quite similar to those previously observed with an antibody against TH that appeared to selectively label DA axons in primate PFC (1). Indeed, dual-label studies revealed that over 95% of TH-IR axons were also labeled for DAT and that over 98% of DAT-IR axons were also labeled with TH. However, the relative intensity of TH versus DAT immunoreactivity differed across populations of DA axons, and some differences in the light microscopic morphological appearance of TH- and DAT-labeled axons were also observed. These findings may reflect differences in the subcellular localization of TH and DAT and the different role that these two proteins play in the regulation of DA neurotransmission.

The distinctive regional and laminar distribution patterns of DA afferents raise the question of whether these axons target specific subclasses of PFC neurons. In the monkey PFC, DA axon terminals are known to form symmetric contacts with the dendritic spines and shafts of pyramidal neurons, as well as with the dendrites of local circuit neurons that contain GABA (2). To determine whether DA synaptic inputs selectively target certain subpopulations of GABA neurons, we used an immunocytochemical electron microscopic approach that combined peroxidase staining for TH with a pre-embedding gold-silver marker for either parvalbumin (PARV) or calretinin (CR), calcium-binding proteins that are expressed in separate subpopulations of GABA-containing local circuit neurons in monkey PFC (3).

In the middle layers (IIIb–IV) of PFC area 9, TH-IR terminals were found in apposition to dendrites immunolabeled for PARV, and 40% of these contacts had the morphological features of symmetric synapses. In contrast, in the superficial layers (I–IIIa) of area 9, TH-positive terminals were much less frequently found in contact with PARV-IR dendrites, and no synaptic specializations were detected at these sites of apposition. This apparent specificity of the synaptic targets of DA axons was further demonstrated by the fact that TH-positive terminals did not form synapses with CR-containing dendrites in the superficial cortical layers of monkey PFC (4), although synaptic contacts between TH-IR terminals and GABA-labeled dendrites were clearly present in these layers (2).

Taken together, these findings reveal both the cell type and the laminar specificity of the synaptic targets of DA axon terminals in monkey PFC. That is, DA axons provide synaptic input to the dendrites of PARV- but not CR-containing local circuit neurons and to PARV-IR dendrites located in the middle but not the superficial cortical layers. The biochemical identity of the subclass(es) of GABA neurons in the superficial layers that do receive DA synaptic input remains to be determined, but this information may reveal additional specificity in the synaptic targets of DA axons. It should also be noted that although our findings demonstrate that the effects of DA on PARV-IR neurons are likely to be synaptically mediated, they do not exclude nonsynaptic effects of DA on PARV-containing or other subpopulations of GABA neurons.

The synaptic regulation of PARV-containing local circuit neurons by DA is particularly interesting, given the potent inhibitory control that these neurons may exert on pyramidal cells. PARV is expressed in both the wide arbor and the chandelier classes of GABA neurons, which provide inhibitory synaptic input to the soma and axon initial segments, respectively, of pyramidal neurons (3). Thus, the influence of DA on the activity of pyramidal neurons may be mediated both through direct synaptic inputs to these neurons, as well as through indirect, GABA-mediated pathways. Interestingly, both the density of dendritic spines on layer III pyramidal cells, the principal site of excitatory inputs to these neurons, and the expression of PARV in the chandelier class of local circuit neurons change substantially, and in a parallel fashion, during postnatal development. For example, the densities of both dendritic spines and PARV-positive chandelier neuron axon terminals in layer III decline markedly during adolescence. To determine whether theses changes were also associated with late developmental shifts in the DA innervation of the PFC, we quantified the densities of TH-IR varicosities in the PFC of a series of macaque monkeys ranging in age from birth to adulthood (5).

In layer III of PFC area 9, the number of TH-IR varicosities was relatively low and uniform during the first postnatal month but then increased threefold by the third postnatal month. The density of labeled varicosities continued to increase, reaching peak levels (sixfold greater than the newborn animals) in monkeys 2–3 years old, the typical age of onset of puberty in this species. The number of DA varicosities then declined to relatively stable adult levels by 5 years of age. Similar changes were also observed in PFC area 46, although the magnitude of these changes was substantially smaller than in area 9. Peak values in the density of DA varicosities appeared to be achieved prior to the decline in the densities of pyramidal neuron dendritic spines and PARV-IR chandelier neuron axon terminals and to persist until the adult levels of these markers of excitatory and inhibitory inputs to pyramidal neurons were achieved. These temporal patterns of change suggest that the neuromodulatory effects of DA may influence the adolescent refinement of excitatory and inhibitory inputs to layer III pyramidal neurons and that DA may have a particularly strong influence on cortical informational processing around the time of puberty.

The potential significance of these developmental changes is further illustrated by their laminar and regional specificities. In contrast to the marked changes in layer III, no significant developmental changes in the density of TH-IR axons or varicosities were observed in the superficial or deep layers of the PFC. Furthermore, within the entorhinal cortex of the same animals, developmental changes in the density of TH-IR varicosities were also restricted to the middle cortical layers. However, in contrast to the time course observed in the PFC, the density of TH-IR varicosities in the entorhinal cortex peaked at around 6 months of age before rapidly declining to levels that remained stable throughout adolescence and adulthood.

In summary, the substantial specificity in innervation patterns, synaptic targets, and postnatal development exhibited by DA axons may provide insight into the role of these afferents in the functional architecture of the PFC in both normal and disease states.

Acknowledgments

These studies were supported by National Institute of Mental Health Research Scientist Development Award MH00519 and grants MH43784, MH50314, and MH45156.

References

1. Lewis, D. A., Foote, S. L., Goldstein, M., and Morrison, J. H. (1988). The dopaminergic innervation of monkey prefrontal cortex: A tyrosine hydroxylase immunohistochemical study. *Brain Res.* **449**, 225–243.
2. Sesack, S. R., Snyder, C. L., and Lewis, D. A. (1995). Axon terminals immunolabeled for dopamine or tyrosine hydroxylase synapse on GABA-immunoreactive dendrites in rat and monkey cortex. *J. Comp. Neurol.* **363**, 264–280.
3. Condé, F., Lund, J. S., Jacobowitz, D. M., Baimbridge, K. G., and Lewis, D. A. (1994). Local circuit neurons immunoreactive for calretinin, calbindin D-28k, or parvalbumin in monkey prefrontal cortex: Distribution and morphology. *J. Comp. Neurol.* **341**, 95–116.
4. Sesack, S. R., Bressler, C. N., and Lewis, D. A. (1995). Ultrastructural associations between dopamine terminals and local circuit neurons in the monkey prefrontal cortex: A study of calretinin-immunoreactive cells. *Neurosci. Lett.* **200**, 9–12.
5. Rosenberg, D. R., and Lewis, D. A. (1995). Postnatal maturation of the dopaminergic innervation of monkey prefrontal and motor cortices: A tyrosine hydroxylase immunohistochemical analysis. *J. Comp. Neurol.* **358**, 383–400.

Patricia S. Goldman-Rakic

Section of Neurobiology
Yale University School of Medicine
New Haven, Connecticut 06510

The Cortical Dopamine System: Role in Memory and Cognition

Evidence from a variety of studies is accumulating to indicate that dopamine has a major role in regulating the excitability of the cortical neuron on which the working memory functions of the prefrontal cortex depend (1). Interactions between monoamines and a compromised cortical circuitry additionally may hold the key to the salience of frontal lobe symptoms in schizophrenia, in spite of widespread pathological changes (2). I will outline three distinct possible cellular mechanisms that have so far been identified for dopamine modulation of working memory function in the prefrontal cortex. These are (1) direct synaptic modulation of receptors on the distal dendrites and spines of pyramidal neurons, (2) direct nonsynaptic modulation of pyramidal neurons, and (3) indirect synaptic modulation of pyramidal neurons via feedforward inhibition from GABAergic interneurons (Fig. 1).

I. Direct Synaptic Modulation

One mode of action appears to be via direct synapses on cortical neurons in the prefrontal cortex (see Fig. 1A). In primate prefrontal cortex, synaptic complexes, termed *triads,* have been observed in which spines of pyramidal neurons are frequently postsynaptic to both a dopamine terminal and an excitatory terminal (3). Because pyramidal cells receive the major sensory inputs arriving at the cortex via spine synapses, this synaptic triad complex allows direct dopamine modulation of local spine responses to excitatory input, thereby regulating a pyramidal neuron's integration of its myriad inputs and ultimately affecting its output via axonal projections to various cortical and subcortical structures: the corticothalamic, corticostriatal, and corticocortical projections. Because the majority of the dopamine synapses appear to be formed on pyramidal neurons, dopamine axons are thus placed in *direct* contact with the major projection neurons of the prefrontal cortex. The dopamine receptors(s) responsible for these actions have not been definitively identified. Members of both the D1 and the D2 familes of dopamine receptors are prominent in the prefrontal cortex of primates (4), and D1, D3, D4, and D5 receptor proteins have been localized to the distal dendrites and spines of pyramidal cells that are postsynaptic to asymmetrical terminals, but none has yet been consistently found in apposition to a dopamine synaptic terminal (5–7). Studies in this laboratory

Advances in Pharmacology, Volume 42

1054-3589/98 $25.00

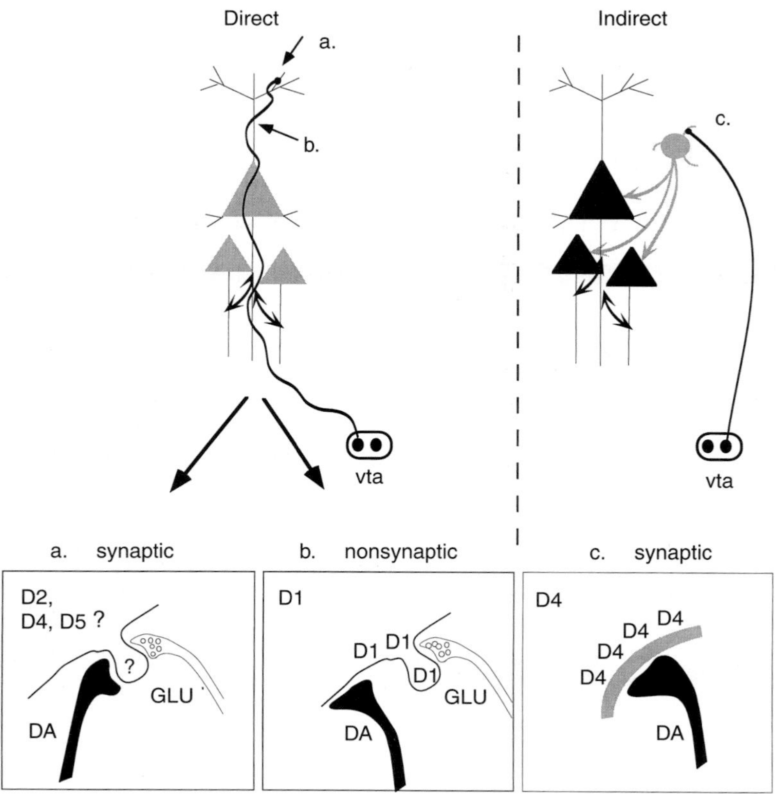

FIGURE 1 (*Upper panel*) Diagram of the anatomical arrangement of the simplest pyramidal and nonpyramidal cell interaction in the cortex and the localization of dopamine (DA) afferents on pyramidal neurons (Direct) and nonpyramidal cells (Indirect). vt a; visual target a. (*Lower panel*) (*A*) Synaptic triad with a dopamine (DA) and glutamate (Glu) terminal forming synapses on the same spine. The receptor subtype postsynaptic to the DA afferent has not been clearly established. (*B*) Nonsynaptic apposition of a DA terminal on a pyramidal cell. Both (*A*) and (*B*) present modes of regulating pyramidal cell firing directly. (*C*) DA terminal on a GABAergic interneuron, which can control pyramidal cell firing via feedforward inhibition (i.e., providing an indirect mode of regulation).

indicate that the D1 receptor may be importantly involved in regulating excitatory transmission in neurons with memory fields, defined as maximal firing of a neuron concomitant with the memory or recall of a visual target in one or a few locations of the visual field (1). D1 stimulation intensifies the neuron's memory field and reduces its reponse to memories of targets in the opponent direction (1). Different neurons encode different items of information, which, in the sphere of spatial processing, amounts to a memory map in the dorsolateral prefrontal cortex. Imaging studies of the human brain show that the homologous prefrontal area is activated in subjects engaged in similar tasks with similar spatial memory requirements.

II. Direct Nonsynaptic Modulation

A second mode of action in the prefrontal cortex is undoubtedly nonsynaptic (see Fig. 1B). Nonsynaptic neurotransmission may be the more pervasive means of altering pyramidal cell activity, because numerous dopamine varicosities are observed in nonsynaptic relationship to cortical elements (7, 8). Members of the D1 family of dopamine receptors have been found to be particularly prominent in the prefrontal cortex of primates (4), and both D1 and D5 receptor proteins have been localized to the distal dendrites and spines of pyramidal cells (5). It is notable that the D5-receptor protein is more prominent in the shafts of pyramidal neurons than in their spines, suggesting a high degree of specificity in their functions within the same neuron (5). Many receptors that have been localized to spines of pyramidal cells appear to lack a dopamine synaptic terminal apposition, although these receptors are often in close relationship to glutamatergic synapses on the same spine (7). Thus, actions at any and all of these receptors could be due to diffusion of transmitter from nearby varicosities or possibly at heteroreceptors. Whether the D1 influence on neurons that specifically subserve specific working memory functions occurs via synaptic or nonsynaptic transmission remains to be determined (1) (see following discussion).

III. Indirect Modulation

A third mechanism of dopamine action is an indirect one, appearing to involve feedforward inhibition on pyramidal neurons from nonpyramidal neurons (see Fig. 1C). The indirect action of dopamine on this circuit derives from the identification of dopamine synapses on nonpyramidal GABAergic neurons in the prefrontal cortex (8, 9) and the finding that the D4 member of the D2 family of dopamine receptors is localized postsynaptically on a subset of GABA interneurons (6). Evidence indicates that pyramidal–nonpyramidal interaction neurons interact physiologically, while monkeys hold a particular item of information in working memory (10). In particular, interneurons, like pyramidal neurons, express directional preferences; and the patterns of activity expressed by closely adjacent pyramidal and nonpyramidal neurons are often inverse, such that as a nonpyramidal neuron increases its rate of discharge, a nearby pyramidal neuron decreases its rate (10). These findings provide suggestive evidence that feedforward inhibition may play a role in the construction of a memory field in prefrontal neurons. Dopamine has been shown to both stimulate and inhibit interneurons in rat cortical neurons. The different effects of dopamine may depend on the subtype of interneuron engaged, and future studies may indicate the basis for the differential modulation of interneurons by dopamine.

IV. Serotonergic Influences on Cortical Circuits

Understanding the interactions between the major cellular constituents of cortical circuits—pyramidal and nonpyramidal cells—is considered a necessary step in unraveling the receptor mechanisms that could lead to an effective pharmacological treatment of negative and cognitive symptoms, as well as improved insight into the pathophysiological basis of schizophrenia and other disorders. It is highly relevant, therefore, that the soma and dendrites of nonpyramidal neurons in the cortex are a primary postsynaptic target of serotoninergic axons, although pyramidal neurons also receive some serotoninergic afferents (11, 12). Serotonin $5HT_2$ receptors have been reported in cortical interneurons by Morilak *et al.* (13). The presumed colocalization of D4 and serotonin receptors in nonpyramidal neurons could provide a basis for a synergistic action of these monoamines on cognitive function. Give that the atypical neuroleptic, clozapine, has a high affinity for the $5HT_2$ as well as the D4 receptor (14), the new D4 localization data focus attention on the nonpyramidal cell as a major target of pharmacological intervention and offer a possible neural explanation for the reported improvements in prefrontal functions (e.g., negative symptoms) by atypical neuroleptics. Because both dopamine and serotonin have complex effects—modulating pyramidal cell firing directly and indirectly through control of nonpyramidal cell firing—understanding the *relative* impact of direct and indirect actions of these neurotransmitters on pyramidal cell firing *in vivo* may hold the key to effective pharmacotherapy of all classes of symptoms in schizophrenia.

References

1. Williams, G. V., and Goldman-Rakic, P. S. (1995). Modulation of memory fields by dopamine D1 receptors in prefrontal cortex. *Nature* **376**, 572–575.
2. Selemon, L. D., Rajkowska, G., and Goldman-Rakic, P. S. (1995). Abnormally high neuronal density in two widespread areas of the schizophrenic cortex. A morphometric analysis of prefrontal area 9 and occipital area 17. *Arch. Gen. Psychiatry* **52**, 805–818.
3. Goldman-Rakic, P. S., Leranth, C., Williams, S. M., Mons, N., and Geffard, M. (1989). Dopamine synaptic complex with pyramidal neurons in primate cerebral cortex. *Proc. Natl. Acad. Sci. U.S.A.* **86**, 9015–9019.
4. Goldman-Rakic, P. S., Lidow, M. S., and Gallager, D. W. (1990). Overlap of dopaminergic, adrenergic, and serotonergic receptors and complementarity of their subtypes in primate prefrontal cortex. *J. Neurosci.* **10**, 2125–2138.
5. Bergson, C., Mrzljak, L., Smiley, J. F., Pappy, M., Levenson, R., and Goldman-Rakic, P. S. (1995). Regional, cellular, and subcellular variations in the distribution of D1 and D5 dopamine receptors in primate brain. *J. Neurosci.* **15**, 7821–7836.
6. Mrzljak, L., Bergson, C., Pappy, M., Levenson, R., Huff, R., and Goldman-Rakic, P. S. (1996). Localization of dopamine D4 receptors in GABAergic neurons of the primate brain. *Nature* **381**, 245–248.
7. Smiley, J. F., Levey, A. I., Ciliax, B. J., and Goldman-Rakic, P. S. (1994). D1 dopamine receptor immunoreactivity in human and monkey cerebral cortex: Predominant and extrasynaptic localization in dendritic spines. *Proc. Natl. Acad. Sci. U.S.A.* **91**, 5720–5724.

8. Smiley, J. F., and Goldman-Rakic, P. S. (1993). Heterogeneous targets of dopamine synapses in monkey prefrontal cortex demonstrated by serial section electron microscopy: A laminar analysis using the silver-enhanced diaminobenzidine sulfide (SEDS) immunolabeling technique. *Cereb. Cortex* **3**, 223–238.
9. Sesack, S. R., Snyder, C. L., and Lewis, D. A. (1995). Axon terminals immunolabeled for dopamine or tyrosine hydroxylase synapse on GABA-immunoreactive dendrites in rat and monkey cortex. *J. Comp. Neurol.* **363**, 264–280.
10. Wilson, F. A. W., O Scalaidhe, S. P., and Goldman-Rakic, P. S. (1994). Functional synergism between putative γ-aminobutyrate-containing neurons and pyramidal neurons in prefrontal cortex. *Proc. Natl. Acad U.S.A.* **91**, 4009–4013.
11. Smiley, J. F., and Goldman-Rakic, P. S. (1996). Serotonergic axons in monkey prefrontal cerebral cortex synapse predominantly on interneurons as demonstrated by serial section electron microscopy. *J. Comp. Neurol.* **367**, 431–443.
12. Jakab, R. L., and Goldman-Rakic, P. S. (1996). Serotonin innervation of neurochemically identified interneuron subtypes in the monkey prefrontal cortex. *Soc. Neurosci. Abstr.* **22**, 356.11.
13. Morilak, D. A., Garlow, S. K., and Ciaranello, R. D. (1993). Immunocytochemical localization and description of neurons expressing serotonin 2 receptors in the rat brain. *Neuroscience* **54**, 701–717.
14. Leysen, J. E., Schotte, A., Janssen, F., Gommeren, W., Van Gommeren, W., Van Gompel, P., Lesage, A. S., De Backer, M. D., Luytten, W. H. M. L., Amlaiky, N., and Megens, A. A. H. P. (1995). Interaction of new antipsychotics with neurotransmitter receptors *in vitro* and *in vivo:* Pharmacological to therapeutic significance. *In* Schizophrenia: An Integrated View (R. Fog, J. Gerlach, and R. Hemmingsen, eds.), pp. 344–356. The Alfred Benzon Foundation, Copenhagen.

Jean-Pol Tassin

INSERM U.114
Chaire de Neuropharmacologie
Collège de France
75231 Paris, France

Norepinephrine–Dopamine Interactions in the Prefrontal Cortex and the Ventral Tegmental Area: Relevance to Mental Diseases

Ascending noradrenergic (NE) neurons located in the locus ceruleus innervate most of the structures of the forebrain. Mesocortical dopaminergic (DA) neurons originate from the ventral tegmental area (VTA) and innervate different cortical structures, including the prefrontal cortex. The respective anatomical distributions of NE and mesocortical DA neurons suggest that interactions between these two systems may occur at the levels of either the DA nerve terminals (prefrontal cortex) or the DA cell bodies (VTA).

I. NE–DA Interactions in the VTA

Initially, a specific destruction of the ascending NE neurons innervating the VTA was performed by injection of 6-hydroxydopamine (6-OHDA) in the vicinity of the NE pathway connecting the locus coeruleus to the VTA (1). These lesions did not affect NE cortical innervation. Seven days later, a significant 38% decrease of the dihydroxyphenylacetic acid (DOPAC)/DA ratio was observed in the prefrontal cortex, with no change in the nucleus accumbens. These data suggested that NE neurons exert a specific tonic excitatory control on mesocortical DA neurons. This was further confirmed in microdialysis experiments performed on awake freely moving animals, where short-lasting, abrupt physiological increases of NE extracellular levels in the prefrontal cortex were correlated with those of cortical DA extracellular levels ($N = 21$, $r = .91$; $p < .0001$) (2).

II. NE–DA Interactions in the Prefrontal Cortex

The first indication of the presence of interactions between NE and DA neurons in the prefrontal cortex was obtained when it was found that the 6-

Advances in Pharmacology, Volume 42

1054-3589/98 $25.00

OHDA destruction of ascending NE neurons induced a collateral sprouting of mesocortical DA neurons (3). Later on, it was shown that the denervation supersensitivity of cortical postsynaptic D1 receptors occurred only if NE fibers were preserved by the lesion of mesocortical DA neurons. This permissive role of ascending NE fibers on the appearance of denervation supersensitivity of cortical D1 receptors was finally demonstrated by showing that rats with an electrolytic lesion of the VTA exhibited increases in cortical DA-sensitive adenylyl cyclase activity, whereas rats with both electrolytic VTA lesions of DA neurons and 6-OHDA lesions of ascending NE neurons did not (4, 5).

Two groups of experiments were then performed to investigate what type of adrenergic receptor was responsible for the heteroregulation of cortical D1 receptors by NE. First, EEDQ, an irreversible antagonist of monoaminergic receptors, was injected into rats at a dose that induced a three-fold higher mean decrease of cortical α_1-adrenergic receptors than of cortical D1 receptors. Four hours after EEDQ injection, we observed a significant increase (+25%) of DA-sensitive adenylyl cyclase activity in the prefrontal cortex. This increase was abolished when rats were preinjected with prazosin, an α_1-adrenergic antagonist (6). Second, the rate of resensitization of D1 receptors in embryonic cortical cell cultures previously incubated with 50 μM of DA was analyzed. We found that the rate of resensitization of the DA-sensitive adenylyl cyclase activity was almost doubled when cultured cells were incubated in the presence of methoxamine, an α_1-adrenergic agonist. This effect was abolished in the presence of 1 μM of prazosin (6).

Both types of experiments indicate that the stimulation of cortical α_1-adrenergic receptors inhibits the cortical DA transmission mediated by D1 receptors. Moreover, embryonic cell culture experiments suggest that α_1-adrenergic and D1 receptors are, at least partly, located on the same neurons in the rat cerebral cortex.

III. Behavioral Consequences of NE–DA Interactions in the Prefrontal Cortex

Bilateral electrolytic lesions of the rat VTA induce deficits such as locomotor hyperactivity and disappearance of spontaneous alternation (7). Correlation studies have indicated that the amplitude of locomotor hyperactivity is proportional to the extent of destruction of the DA fibers innervating the prefrontal cortex and to the development of a D1-receptor supersensitivity in this area (3, 4, 8).

Because destruction of ascending NE pathways downregulates cortical D1-receptor denervation supersensitivity induced by electrolytic lesion of the VTA, we have investigated whether the locomotor hyperactivity induced by this electrolytic lesion could be affected by chemical (6-OHDA) lesions of the NE innervation. Indeed, both deficits induced by the electrolytic lesion of the VTA—locomotor hyperactivity and disappearance of spontaneous alternation—were abolished by a superimposed 6-OHDA lesion of ascending NE neurons (9). This functional recovery provided new insights into the antagonistic properties of NE and DA neurons. A functional hierarchy may exist between these systems, because no significant modification of the locomotor activity or spontaneous alternation was observed in rats with NE lesions alone.

To confirm the hypothesis that the prefrontal cortex could be an important site of interaction between NE fibers and the target cells of the mesocortical DA neurons, two groups of experiments were performed with rats implanted with chronic bilateral cannulae in both the prefrontal cortex and the nucleus accumbens. First, we found that the locomotor hyperactivity induced by the infusion of amphetamine into the nucleus accumbens was inhibited when amphetamine was simultaneously injected into the prefrontal cortex. Complementary experiments have indicated that this cortical effect is mediated via D1 receptors, because the injection of SCH 23390, a D1 antagonist, into the prefrontal cortex, doubled the locomotor hyperactivity induced by the infusion of amphetamine into the nucleus accumbens. The injection of sulpiride, a D2 antagonist, was without effect (10). These results suggest that the inhibitory role of prefrontocortical DA innervation on locomotor behavior is mediated by D1 receptors when the DA transmission in the nucleus accumbens is activated. Second, we have shown that the injection of prazosin or WB 4101, two α_1-adrenergic antagonists, into the prefrontal cortex, completely reversed the locomotor hyperactivity induced by the injection of amphetamine into the nucleus accumbens. Both antagonists had no effect on locomotor activity when injected alone (11). Altogether, these results not only confirm that NE and DA neurons exert opposite functional roles in the prefrontal cortex, but also demonstrate that the stimulation by NE of cortical α_1-adrenergic receptors is necessary for a functional subcortical DA transmission.

Recent experiments, performed with minute-by-minute sampling microdialysis, have indicated that prazosin does not modify the increase of extracellular DA levels in the nucleus accumbens following a systemic injection of amphetamine. It seems rather that it changes the mode of firing of DA neurons following the amphetamine injection. Briefly, this was shown by perfusing by reverse microdialysis 3 μM of amphetamine into the nucleus accumbens and by observing that, in these conditions, the 60% increase in DA extracellular levels induced by a subsequent systemic injection of amphetamine was completely blocked by a previous systemic injection of prazosin (Darracq *et al.*, submitted). We propose, therefore, that stimulations of cortical α_1-adrenergic receptors by NE initiate, through an inhibition of cortical D1 transmission, the phasic release of DA in subcortical structures.

IV. Relevance to Mental Diseases of NE–DA Interactions

Up to now, the clinical efficacy of antidepressants has been correlated with their biochemical property to desensitize cortical β-adrenergic and/or serotonergic ($5HT_2$) receptors. This does not necessarily mean that these desensitizations are essential, but rather that both types of receptors are extremely sensitive to modifications of their respective NE and 5HT transmissions, and that the reactivation of these latter is critical for depression relief. NE and 5HT neurons are not, however, independent, and there are some indications of a synergy between both transmissions, the stimulation of $5HT_2$ receptors increasing, for example, the responses of NE neurons to sensory stimuli (12). On the other hand, among the different symptoms of depression, anhedonia and emotional blunting removal are probably two of the best indications of antidepressant efficacy, and these symptoms have been related to a reduction of DA transmission in subcortical structures.

Because the presence of a normal NE transmission is necessary to obtain a functional subcortical DA transmission, a physiological sequence (5HT/NE/subcortical DA) may link together monoaminergic cells activities and antidepressants would participate in the re-establishment of this sequence. Following the reactivation of 5HT neurons, antidepressants would, through NE cells, reinstate *in fine* a functional subcortical transmission. The therapeutic delay, generally observed with antidepressants, would then be due to the time necessary to obtain this readjustment and not only to the well-documented desensitization of autoreceptors. The final relief may, however, only occur following the recoupling of monoaminergic cells to neuronal networks responsible for sensory stimuli processing (13).

Patricia Goldman-Rakic and her colleagues have shown that the stimulation of the prefrontocortical D1 receptors by DA was necessary for the expression of a working memory (14). If we consider that the stimulation of cortical α_1-adrenergic receptors by NE exerts an antagonist effect on D1 cortical transmission, an activation of NE neurons should block the working memory. On the other hand, the activation by NE of mesocortical DA neurons in the VTA may initiate a working memory processing. We have proposed that two modes of brain functioning, analogical and cognitive, characterized by their rates of processing, exist in the awake state (15). At birth, during REM sleep, and following the ingestion of psychostimulants, the mode of brain functioning would be essentially rapid and analogical with a preponderance of subcortical DA function. In the awake state, oscillations between the two modes would be under the control of monoaminergic neurons, an increase in DA cortical release favoring a slow cognitive-processing mode, whereas intermittent activations of NE neurons would switch the brain into the analogical mode of processing. Schizophrenic patients with "positive" symptoms would suffer from an abnormal preponderance of the analogical mode while awake, whereas "negative" symptoms would be due to the excessive presence of cognitive mode caused by the inactivation of NE neurons blocked into an hyperdepolarization state. In normal subjects, the activation of mesocortical DA neurons would be due to the excitatory effect of NE neurons in the VTA following the emergence of an unexpected, new stimulus in the environment.

References

1. Hervé, D., Blanc, G., Glowinski, J., and Tassin J-P. (1982). Reduction of dopamine utilization in the prefrontal cortex but not in the nucleus accumbens after selective destruc-

tion of noradrenergic fibers innervating the ventral tegmental area in the rat. *Brain Res.* **237,** 510–516.

2. Gillibert, C. (1994). Libération des monoamines dans le cortex préfrontal du rat au cours du cycle veille-sommeil: analyse en microdialyse. DEA Thesis, University Paris VI
3. Tassin, J-P., Stinus, L., Simon, H., Blanc. G., Thierry, A-M., Le Moal, M., Cardo, B., and Glowinski, J. (1978). Relationship between the locomotor hyperactivity induced by A10 lesions and the destruction of the fronto-cortical dopaminergic innervation in the rat. *Brain Res.* **141,** 267–281.
4. Tassin, J-P., Simon, H., Glowinski, J., and Bockaërt, J. (1982). Modulations of the sensitivity of dopaminergic receptors in the prefrontal cortex and the nucleus accumbens. Relationship with locomotor activity. In *Brain Peptides and Hormones,* R. Collu *et al.,* eds. Raven Press, New York, pp. 17–30.
5. Tassin, J-P., Studler, J-M., Hervé, D., Blanc, G., and Glowinski, J. (1986). Contribution of noradrenergic neurons to the regulation of dopaminergic (D1) receptor denervation supersensitivity in rat prefrontal cortex. *J. Neurochem.* **46,** 243–248.
6. Trovero, F., Hervé, D., Blanc, G., Glowinski, J., and Tassin, J-P. (1992). *In vivo* partial inactivation of dopamine D1 receptors induces hypersensitivity of cortical dopamine-sensitive adenylate cyclase: permissive role of α1-adrenergic receptors. *J. Neurochem.* **59,** 311–337.
7. Le Moal, M., Stinus, L., Simon, H., Tassin, J-P., Thierry, A-M., Blanc, G., Glowinski, J., and Cardo, B. (1977). Behavioral effects of a lesion in the ventral mesencephalic tegmentum: Evidence for involvement of A10 dopaminergic neurons. *Adv. Biochem. Psychopharmacol.* **16,** 237–245.
8. Tassin, J. P., Trovero, F., Vezina, P., Blanc, G., Glowinski, J., and Hervé D. (1995). L'hétérorégulation des récepteurs ou la présence d'une relation fonctionnelle entre deux ensembles neuronaux. *Med. Sci.* **11,** 829–836.
9. Taghzouti, K., Simon, H., Hervé, D., Blanc, G., Studler, J-M., Glowinski, J., Le Moal, M., and Tassin, J-P. (1988). Behavioural deficits induced by an electrolytic lesion of the rat ventral mesencephalic tegmentum are corrected by a superimposed lesion of the dorsal noradrenergic system. *Brain Res.* **440,** 172–176.
10. Vézina, P., Blanc, G., Glowinski, J., and Tassin, J. P. (1992). Opposed behavioural outputs of increased dopamine transmission in prefrontocortical and subcortical areas: A role for the cortical D1 dopamine receptor. *Eur. J. Neurosci.* **3,** 1001–1007.
11. Blanc, G., Trovero, F., Vezina, P., Herve, D., Godeheu, A. M., Glowinski, J., and Tassin, J. P. (1994). Blockade of prefronto-cortical α1-adrenergic receptors prevents locomotor hyperactivity induced by subcortical D-amphetamine injection. *Eur. J. Neurosci.* **6,** 293–298.
12. Chiang, C., and Aston-Jones, G. (1993). A 5-hydroxytryptamine2 agonist augments γ-aminobutyric acid and excitatory amino acid inputs to noradrenergic locus coeruleus neurons., *Neuroscience* **54:**409–420.
13. Tassin, J. P. (1994). Interrelations between neuromediators implicated in depression and antidepressive drugs: *Encephale* **4,** 623–628.
14. Sawaguchi, T., and Goldman-Rakic, P. S. (1991). D1 receptors in prefrontal cortex: involvement in working memory. *Science* **251,** 947–950.
15. Tassin, J. P. (1992). NE/DA interactions in prefrontal cortex and their possible roles as neuromodulators in schizophrenia. *J. Neural Transm. Suppl.* **36,** 135–162.

A. M. Thierry, S. Pirot, Y. Gioanni, and J. Glowinski

INSERM U114
Chaire de Neuropharmacologie
Collège de France
75231 Paris, France

Dopamine Function in the Prefrontal Cortex

The prefrontal cortex (PFC), defined as the essential cortical projection area of the thalamic mediodorsal nucleus (MD), is implicated in the control of locomotor activity and plays a major role in cognitive processes as well as in affective and emotional behaviors. The dopaminergic (DA) innervation of the PFC, which originates from the ventral tegmental area (VTA), has been shown to be crucial for its functions. Locomotor hyperactivity induced by amphetamine is decreased by concomittant amphetamine injection into the PFC and, at the opposite, increased by intracortical injection of a D1 antagonist. The particularly high reactivity of the mesocortical DA system to stressful situations or anxiogenic drugs suggests that this system is involved in the control of emotional behavior. Finally, the DA neurotransmission in the PFC contributes to the regulation of cognitive functions. The depletion of DA or the local administration of DA antagonists into the PFC impaired tasks requiring delayed responses.

The PFC is likely a crucial target area for the action of antipsychotic drugs such as neuroleptics, which impair DA neurotransmission. In the rat, the mesocortical DA system exerts an inhibitory influence on the spontaneous activity of PFC cells. The characterization of the DA-receptor subtype involved and the contribution of GABA interneurons in this inhibitory effect will be discussed. We have shown that DA afferents exert an inhibitory control on excitatory inputs to the PFC that originate from the hippocampal formation (1). We will analyze here the influence of the mesocortical DA system on the excitatory responses induced in the PFC following electrical stimulation of the MD.

I. Pharmacological Characteristics of the Inhibitory Responses Induced by the Mesocortical DA System on PFC Neurons

Binding and *in situ* hybridization studies have revealed that the two main classes of DA receptors are present in the rat PFC, although the number of D1 receptors exceeds that of D2 receptors. D1 receptors have a more widespread laminar distribution (layers II–VI) than D2 receptors, the localization of which is restricted to deep layers, mainly layer V. The effects of selective D1 and D2 antagonists on the inhibitory responses evoked in PFC cells either by iontopho-

Advances in Pharmacology, Volume 42

1054-3589/98 $25.00

retic application of DA or by electrical stimulation of the VTA were examined in anesthetized rats, using extracellular single-unit recordings (2). Sulpiride, a D2 antagonist, contrarily to SCH 23390, a D1 antagonist, was effective in blocking the inhibitory effects evoked by iontophoretic application of DA or VTA stimulation. Results obtained with DA agonists further support the involvement of D2 receptors, because quinpirole, a D2 agonist, was more potent than SKF 38393, a D1 agonist, in inhibiting the spontaneous activity of PFC cells.

Surprisingly, haloperidol, a potent D2 antagonist, applied locally or systemically did not block the DA- or VTA-induced inhibitions in the PFC. When compared with sulpiride, the lack of effect of haloperidol could be due to the fact that haloperidol also has a high affinity for α_1-adrenergic receptors, whereas sulpiride is a specific D2 antagonist. A possible interaction between α_1-adrenergic and DA receptors is supported by recent data obtained in our laboratory by Y. Gioanni and J. P. Tassin (unpublished observations). It was observed that peripheral or local administration of prazosin, an α_1-adrenergic antagonist, blocked the VTA-induced inhibition of PFC cells and that this effect can be reversed by iontophoretic application of sulpiride. Thus, under the concomittant action of prazosin and sulpiride, the VTA-induced inhibition in PFC cells was observed, as well as after haloperidol administration. This observation suggests that blockade of α_1-adrenergic receptors results in a functional inactivation of D2-mediated DA transmission in the PFC. Because most of the antipsychotic drugs block both D2 and α_1-adrenergic receptors, the potential contribution of α_1-adrenergic receptors in their therapeutic effects has to be taken into account and may have relevance to the design of new antipsychotic drugs.

Differential modulatory actions of DA on responses evoked by N-methyl-D-aspartate (NMDA) and glutamate or quisqualate have been reported in a study preformed on neocortical slices. In recent *in vivo* experiments (S. Pirot, unpublished observations), we have investigated the effects of DA, quinpirole, and SKF 38393 on the excitatory responses evoked by iontophoretic application of AMPA and NMDA on PFC cells. Although, the primary action of DA as well as of quinpirole or SKF 38393 was to attenuate AMPA- and NMDA-evoked responses, AMPA responses were enhanced in a small contingent of PFC cells. Moreover, NMDA-evoked responses were enhanced by DA or quinpirole in a relatively large proportion of cells (34% and 27%, respectively). Whether these effects result from direct postsynaptic interactions between DA and excitatory amino acid receptors or involve local circuits remains to be established.

The DA- and VTA-induced inhibition of efferent PFC neurons may result from a direct action of DA on pyramidal cells and/or an indirect effect involving GABA interneurons. DA terminals establish symetrical synaptic contacts with dendritic shafts and spines of pyramidal neurons, suggesting that DA may exert a direct inhibitory effect on pyramidal cells. Several observations also support a possible involvement of GABA interneurons in the inhibitory effect of DA on pyramidal cells. DA varicosities have been observed in close nonsynaptic contiguity with GABA neurons, and DA application in PFC slices enhances the spontaneous release of GABA. Moreover, the inhibitory GABAergic postsynaptic potentials recorded in pyramidal cells are increased by DA application on

PFC slices. Our *in vivo* study also suggests that part of the inhibitory responses of PFC cells induced by the activation of the mesocortical DA system involves a local GABAergic component. Indeed, iontophoretic application of bicuculline blocked both DA- and VTA-induced inhibition in 57% and 51%, respectively, of the PFC neurons recorded. The contribution of GABA interneurons might explain that although the cortical DA innervation is relatively discrete, the mesocortical DA system influences the activity of a large population of PFC cells in deep cortical layers.

II. Influence of the Mesocortical DA System on the Excitatory Responses Evoked by MD Stimulation

The PFC and the MD are reciprocally connected and pyramidal cells projecting to the MD are located in layer VI and to a lesser extent in layer V, whereas nerve terminals originating from MD neurons are located in superficial layers, mainly in layer III. Electrical stimulation of the MD evokes excitatory responses in PFC neurons, which result not only from the activation of the MD–PFC pathways, but also from the activation of recurrent collaterals of the antidromically driven PFC fibers projecting to the MD. These two types of excitatory responses were discriminated mainly on the basis of the differential conduction times of these two pathways (3). Excitatory responses resulting from the activation of the MD–PFC pathway are characterized by their short latency (<4 ms) and are observed mainly in superficial layers. In contrast, excitatory responses resulting from the activation of recurrent collaterals of the PFC–MD pathway are characterized by their longer latency (>10 ms) and are recorded more frequently in deep than in superficial layers.

Local application of DA or VTA stimulation did not affect the short latency excitatory responses but blocked the long latency excitatory responses (Fig. 1). Thus, the mesocortical DA system does not influence markedly the excitatory responses evoked by the activation of the MD–PFC pathways, the main thalamic afferent input to the PFC. The lack of effect of DA may partly be related to the localization of MD terminals in superficial layers of the PFC, which receive a less dense DA innervation than deep layers. To the contrary, the mesocortical DA system exerts a potent inhibitory control on the excitatory responses induced by the activation of intracortical recurrent collaterals of the PFC pyramidal cells, which project to the MD. Because recurrent excitations in neocortical circuits may play an important role in amplifying excitatory inputs originating from subcortical structures, it can be proposed that, by suppressing the propagation of excitation through the recurrent collateral network, the mesocortical DA system allows a spatial focalization of afferent excitatory signals to a restricted cortical area or to a limited group of PFC cells.

The mesocortical DA system is particularly reactive to environmental conditions. Indeed, various stressful situations increase DA release, and, at the opposite, long-term isolation decreases DA metabolism in the rat PFC. As described in this brief review, the modulatory effect of DA on information processing in the PFC depends on multiple factors: the DA receptor subtype involved, the

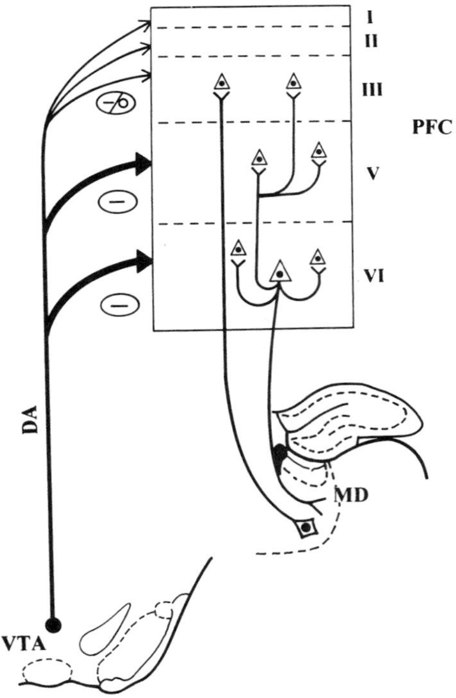

FIGURE 1 Diagram illustrating the influence of the mesocortical DA system on MD–PFC connections. The mesocortical DA system does not affect the excitatory responses evoked by the activation of the MD–PFC pathway, but exerts a marked inhibitory control on the excitatory responses resulting from the activation of the recurrent collaterals of the PFC neurons projecting to the MD. The inhibitory influence of the mesocortical DA system on the spontaneous activity of PFC cells is more potent in deep than in superficial layers.

afferent excitatory pathways and the excitatory amino acid receptor subtype implicated, and finally the activity of other afferent modulatory systems, such as the noradrenergic system. The complex but specific influence of DA on the transfer of information may explain the major role of the mesocortical DA system in the functions of the PFC.

References

1. Jay, T. M., Glowinski, J., and Thierry, A. M. (1995). Inhibition of hippocampo-prefrontal cortex excitatory responses by the mesocortical DA system. *Neuroreport* **6**, 1845–1848.
2. Godbout, R., Mantz, J., Pirot, S., Glowinski, J., and Thierry, A. M. (1991). Inhibitory influence of the mesocortical dopaminergic neurons on their target cells: Electrophysiological and pharmacological characterization. *J. Pharmacol. Exp. Ther.* **258**, 728–738.
3. Pirot, S., Jay, T., Glowinski, J., and Thierry, A. M. (1994). Anatomical and electrophysiological evidence for an excitatory amino acid pathway from the thalamic mediodorsal nucleus to the prefrontal cortex in the rat. *Eur. J. Neurosci.* **6**, 1225–1234.

Anthony A. Grace, Holly Moore, and Patricio O'Donnell

Departments of Neuroscience and Psychiatry
Center For Neuroscience
University of Pittsburgh
Pittsburgh, Pennsylvania 15260

The Modulation of Corticoaccumbens Transmission by Limbic Afferents and Dopamine: A Model for the Pathophysiology of Schizophrenia

Despite the substantial research effort directed toward ascertaining the etiology of schizophrenia, insight into the primary site of pathophysiology of this illness has remained elusive. Initial pharmacological studies focused on the dopamine (DA) system; however, there has been a remarkable lack of evidence to substantiate the existence of a primary deficit in this neurotransmitter system. In contrast, much recent work based on brain imaging studies has shifted the focus onto cortical regions, with a primary interest in the prefrontal cortex (PFC) and the hippocampus. More specifically, recent studies have concentrated on schizophrenia as an abnormality in cortical regulation of subcortical DA systems. A region in which the hippocampus, PFC, and the mesolimbic DA system overlap is the nucleus accumbens. We have used both *in vivo* and *in vitro* intracellular recording techniques in the rat to examine how a dysfunction in the interaction among these systems could lead to psychosis.

Our first study focused on the effects of DA agonists on accumbens neurons and on their activation by PFC afferents in an *in vitro* brain slice preparation. Slices containing the core region of the nucleus accumbens and the prelimbic PFC were cut in the sagittal plane (400 μm thick), and bipolar stimulating electrodes were placed on the subcortical white matter between the PFC and the rostral pole of the nucleus accumbens. The D1/D2 agonist apomorphine (50 μM), the D2-selective agonist quinpirole (1–5 μM) and antagonist sulpiride (1–10 μM), and the D1-selective agonist SKF 38393 (3–10 μM) and antagonist SCH 23390 (10 μM) were added to the superfusate. Apomorphine produced a depolarization of accumbens neurons combined with an increase in spike threshold; nonetheless, the net result was a decrease in excitability of the neuron. This decrease in excitability was reproduced by coadministration of both D1 and D2 agonists (1). In addition to these direct actions, DA agonists also affected excitatory input to accumbens neurons. Thus, stimulation of PFC afferents evoked monosynaptic excitatory postsynaptic potentials (EPSPs)

Advances in Pharmacology, Volume 42

1054-3589/98 $25.00

(8.8 ± 5.4 mV) in accumbens neurons. Administration of the D1 agonist did not significantly alter the amplitude of the evoked EPSP. However, administration of quinpirole reduced the EPSP amplitude in approximately half of the neurons tested; this was reversed by sulpiride. On the other hand, sulpiride administered alone caused an *increase* in the evoked EPSP in 80% of the cells tested. In contrast, sulpiride did not affect EPSP amplitude when the rats were first depleted of DA by prior administration of α-methy-*para*-tyrosine and reserpine; moreover, in these rats, quinpirole readily decreased EPSP amplitude in all cells tested (2). Therefore, DA exerts a tonic inhibition of PFC afferent excitation of accumbens neurons in addition to causing a direct decrease in cell excitability.

To examine the interaction among excitatory afferents in this system, *in vivo* intracellular recordings were made from accumbens neurons while stimulating afferents originating in the amygdala, the hippocampus, and the PFC. When recorded *in vivo* in chloral hydrate–anesthetized rats, nucleus accumbens neurons typically exhibited spontaneous spike discharge. Furthermore, most of the neurons exhibited a bistable membrane potential. This was characterized by an alternation between a hyperpolarized, inactive state and one of membrane depolarization and spontaneous spike discharge (3) (Fig. 1), with the hyperpolarizing phase exhibiting an average membrane potential of −77.3 ± 7.1 mV and the depolarized state occurring at −63.0 ± 7.4 mV.

Stimulation of accumbens afferents arising from the amygdala, the fornix, and the PFC consistently evoked short-latency EPSPs in most cells recorded. In fact, in 95% of the cases, the nucleus accumbens cells tested exhibited evoked responses to all three convergent inputs. However, the nature of the evoked response varied with the site of stimulation. Stimulation of the hippocampal afferents at the level of the fornix with single current pulses evoked short-latency EPSPs in accumbens cells, followed by depolarized plateaus similar to those occurring spontaneously. Furthermore, repetitive stimulation of the hippocampal afferents evoked a prolonged membrane depolarization in accumbens cells. Following transection of the fornix, none of the accumbens cells recorded exhibited this bistable state. Moreover, infusion of the local anesthetic lidocaine onto the fornix caused a reversible suppression of the depolarizing

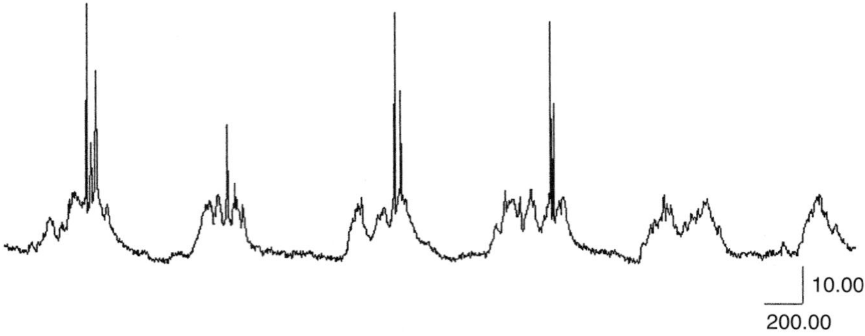

FIGURE 1 Bistable nucleus accumbens neuron recorded intracellularly *in vivo* (spikes truncated).

plateaus in single accumbens neurons. Therefore, the transition into the depolarized state was mediated by hippocampal afferents to the nucleus accumbens.

In contrast to the hippocampal afferents, stimulation of the PFC fibers typically evoked only an EPSP in the majority of accumbens neurons when they were in the hyperpolarized state. However, accumbens cells that were in the depolarized state responded to PFC stimulation with spike discharge; this was followed by a termination of the depolarization and a rapid transition of the membrane back to its hyperpolarized state. Moreover, although accumbens neurons in the hyperpolarized state typically exhibited only EPSPs upon PFC stimulation, during the sustained membrane depolarization produced by repetitive stimulation of the fornix, PFC stimulation readily evoked spike discharge. Therefore, a steady-state depolarization of the membrane of accumbens neurons by hippocampal afferents was necessary to enable PFC afferent input to evoke spike discharge. We interpret this as a hippocampal gating of PFC throughput via an interaction in the accumbens.

In summary, our data show that the PFC throughput in the accumbens is dependent on two systems: It is gated by depolarizations arising from the hippocampal input, and it is attenuated by DA acting presynaptically on PFC terminals and on accumbens neurons by decreasing excitability. Systemic administration of DA agonists also appears to interfere with PFC transmission in another manner—by affecting the transitions to the depolarized state. Thus, recent studies (4) show that systemic administration of D1 and D2 DA agonists causes a decrease in the frequency of transitions to the depolarized state in bistable neurons. Furthermore, systemic administration of the psychotomimetic phenycyclidine (PCP) also causes a potent suppression of the bistable state. Because the primary action of these drugs is a slowing of the frequency of the bistable state rather than a decrease in its amplitude, these drugs are most likely acting directly or indirectly on the hippocampus. Therefore, administration of DA agonists and PCP or damage to the hippocampus has a similar impact in the accumbens: an attenuation of the gating of PFC afferent inputs.

The ability of the hippocampus to gate the PFC has implications of relevance to schizophrenia. The hippocampus has been implicated in a diverse array of functions, among which is context-dependency. By depolarizing the membrane of a selective subset of accumbens neurons, the hippocampal input is positioned to selectively gate PFC throughput at the level of the accumbens for inputs that are congruent with a given context, while blocking transmission of input that is outside of this context. Several recent studies suggest that a deficit in hippocampal function may be involved in the etiology of schizophrenia. In our model, a dysfunction in the hippocampal afferent regulation of the accumbens would interfere with this proposed context-dependent gating function, allowing signals to pass through to the thalamocortical system unfiltered. This could provide a physiological correlate to the difficulty experienced by schizophrenics in the interpretation of stimuli based on the context in which they occur, and how psychotomimetic drugs may mimic these symptoms.

In addition, afferents from the amygdala to the nucleus accumbens also appear to exert a unique modulatory influence over the PFC inputs. During *in vivo* intracellular recording, stimulation of the amygdala evoked a long-latency (i.e., 17.1 $\pm$ 5.4 ms) depolarization of the membrane of accumbens neurons.

Using extracellular recording techniques, we found that brief trains of stimuli delivered to the amygdala facilitated PFC-evoked spiking as well; however, in this case, the facilitation occurred in a time-locked manner; that is, only when the amygdala stimulation preceded the PFC stimulus by 7–30 ms was the facilitatory effect noted (5). Therefore, unlike the long-duration gating of PFC input to the accumbens mediated by the hippocampus, the amygdala appears to exert a narrow, time-locked facilitation that may be selective for specific stimuli. These systems are, therefore, likely to exert both a context-dependent (hippocampal) and affect-selective (amygdalar) gating of PFC throughput in the accumbens.

References

1. O'Donnell, P., and Grace, A. A. (1996). Dopaminergic reduction of excitability in nucleus accumbens neurons recorded *in vitro*. *Neuropsychopharmacology* **15**, 87–97.
2. O'Donnell, P., and Grace, A. A. (1994). Tonic D_2-mediated attenuation of cortical excitation in nucleus accumbens neurons recorded *in vitro*. *Brain Res.* **634**, 105–112.
3. O'Donnell, P., and Grace, A. A. (1995). Synaptic interaction among excitatory afferents to nucleus accumbens neurons: Hippocampal gating of prefrontal cortical input. *J. Neurosci.* **15**, 3622–3639.
4. O'Donnell, P., and Grace, A. A. (1996). Hippocampal gating of cortical throughput in the nucleus accumbens: Modulation by dopamine. *Biol. Psychiatry* **39**, 632.
5. Moore, H., and Grace, A. A. (1996). Interactions between amygdala and prefrontal cortical afferents to the nucleus accumbens and their modulation by dopamine receptor activation. *Soc. Neurosci. Abstr.* **22**, 1088.

Michael S. Levine and Carlos Cepeda

Mental Retardation Research Center
University of California, Los Angeles
Los Angeles, California 90024

Dopamine Modulation of Responses Mediated by Excitatory Amino Acids in the Neostriatum

A growing body of evidence suggests that dopamine (DA) and excitatory amino acid (EAA) neurotransmitters are functionally coupled within the neostriatum. Corticostriatal and thalamostriatal excitatory glutamatergic afferents

Advances in Pharmacology, Volume 42

1054-3589/98 $25.00

project onto neostriatal medium-spiny neurons, while nigrostriatal DA modulates these EAA inputs and alters cellular responsiveness to glutamate (1–3). The present studies were designed to examine the electrophysiological interactions that occur when subtypes of DA and EAA receptors are activated simultaneously and to test the hypothesis that the direction of DA modulation is a function of both the subtype of EAA receptor and the subtype of DA receptor activated. Our findings demonstrate the direction of the modulatory effects of DA receptor agonists can be very predictable if the subtype of receptor activated is controlled. When N-methyl-D-aspartate (NMDA) receptors are activated, D1 DA receptor agonists potentiate excitatory responses. When non-NMDA receptors are activated, D2 DA receptor agonists attenuate excitatory responses.

Two types of experiments were performed using brain slices. In the first type, current-clamp techniques were used to examine DA-receptor modulation of responses induced by activation of NMDA or non-NMDA EAA receptors. The second type concentrated on DA-receptor modulation of NMDA-induced membrane currents using whole-cell voltage-clamp techniques in visually identified neostriatal cells. Sprague-Dawley rats (60–90 days old for current-clamp studies) or rat pups (12–18 days old for voltage-clamp studies) were used. For current-clamp recording studies, brains were extracted and cut, and slices were incubated in artificial cerebrospinal fluid (aCSF) and then transferred to an interface recording chamber for intracellular current-clamp recordings. We examined the interactions between two subtypes of EAA receptors, α-amino-3-hydroxy-5-methyl-4-isoxazole propionic acid (AMPA) (i.e., representing non-NMDA ionotropic receptors) and NMDA and the effects of D1- and D2-receptor agonists. Responses to EAA- and DA-receptor subtype agonists were examined either by iontophoretic application of EAA agonists or by synaptic activation using local stimulation and bath application of EAA antagonists to define the EAA-receptor subtype. DA-receptor agonists were applied either iontophoretically or in the bath. A five-barrelled pipette was positioned close to the recording electrode for iontophoresis. Pipettes contained AMPA, NMDA, and saline for current balancing and control. Each EAA was ejected iontophoretically to produce depolarizing responses. SKF 38393 (D1 agonist, 1–20 μM) was either applied in the bath or by iontophoresis. Quinpirole (D2 agonist, 5–10 μM) was always bath-applied. To test the effects of DA-receptor agonists, a single EAA ejection intensity was chosen. Responses induced by the EAA were characterized first, then reassessed in the presence of DA-receptor agonists and retested after application of the DA-receptor agonists.

Stimulating electrodes, placed close to the recording electrode, were used to evoke depolarizing synaptic responses (DPSPs). Stimuli were constant-current 100-μsec square wave pulses of varying amplitudes (100–1000 μA, 1 pulse/4–5 sec). Changes in response area determined the modulatory effects of DA agonists and were computed from averages of four to six responses. Cells were prebathed in Mg^{2+}-free aCSF (for at least 45 min) to unmask NMDA-receptor activation, then exposed to either 6-cyano-7-nitroquinoxaline-2,3-dione (CNQX, 5 μM, at least 30 min), a non-NMDA-receptor antagonist, or the selective NMDA-receptor antagonist, 2-amino-5-phosphonovalerate (AP5, 25 μM, at least 30 min) to induce DPSPs mediated by activation of either NMDA or non-NMDA receptors, respectively. In a number of experiments,

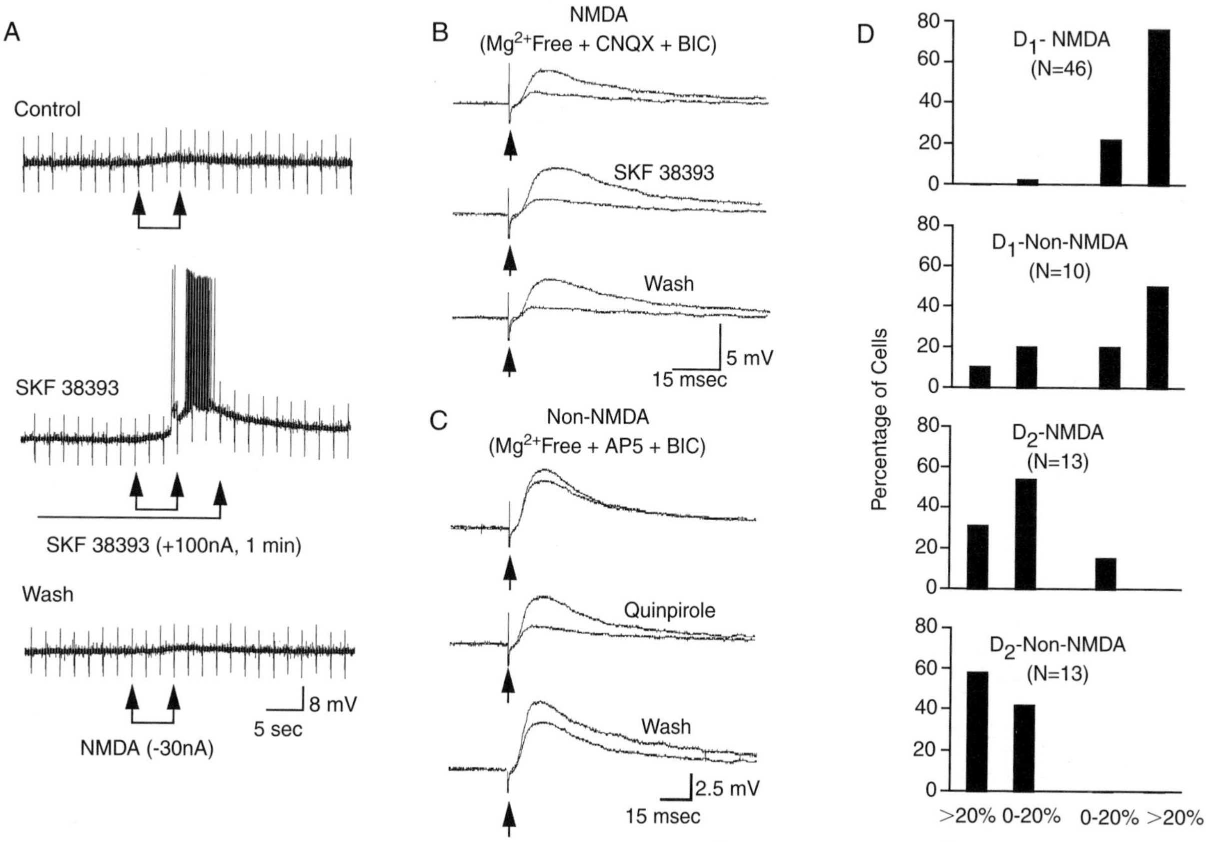

A
Control
SKF 38393
SKF 38393 (+100nA, 1 min)
Wash
NMDA (-30nA)
8 mV
5 sec
B
NMDA
(Mg2+Free + CNQX + BIC)
SKF 38393
Wash
5 mV
15 msec
C
Non-NMDA
(Mg2+Free + AP5 + BIC)
Quinpirole
Wash
2.5 mV
15 msec
D
D1- NMDA
(N=46)
D1-Non-NMDA
(N=10)
D2-NMDA
(N=13)
D2-Non-NMDA
(N=13)
Percentage of Cells
80
60
40
20
0
>20% 0-20% 0-20% >20%
Decrease
Increase

solutions also contained bicuculline (15 μM) to block activation of $GABA_A$ receptors.

SKF 38393 (applied iontophoretically or in the bath) produced a consistent enhancement of the iontophoretically induced NMDA response (Fig. 1A), increasing the amplitude of NMDA-evoked depolarizations and the firing frequency in 25 of 26 cells (one cell displayed a reduction). In these iontophoretic experiments, SKF 38393 effects were more variable on AMPA-induced responses; enhancement occurred in four of seven cells (three cells were decreased). When SKF 39393 was tested on DPSPs mediated by NMDA receptors (Mg^{2+}-free aCSF and CNQX), it produced a statistically significant increase in the average DPSP area (32.2 ± 6.1%; $p < .01$, t-test; data averaged over three concentrations of SKF 38393, 1, 5, and 10 μM) (see Fig. 1B). In contrast, SKF 38393 (10 μM) produced little net change in average DPSP area when responses were mediated by non-NMDA receptors (Mg^{2+}-free aCSF and AP5) (6.2 ± 5.4%, N = 5). Both increases (N = 3) and decreases (N = 2) in DPSP area occurred.

Quinpirole (5–10 μM) primarily decreased responses in iontophoretic experiments, reducing NMDA-evoked responses in nine of 10 cells (one cell increased) and diminishing AMPA-evoked responses in seven of seven cells. Quinpirole (10 μM) produced little net change in average response area (0.6 ± 8.3% increase) in three cells when DPSPs were mediated by NMDA receptors (Mg^{2+}-free aCSF and CNQX). When quinpirole was tested on DPSPs mediated by activation of non-NMDA receptors (Mg^{2+}-free aCSF and AP5), it produced a statistically significant decrease in area (22.6 ± 6.3%; $p < .01$ t-test) (see Fig. 1C). When DA agonists were applied alone, no consistent changes were produced in the membrane potential or the input resistance that could account for their effects on EAA-receptor–mediated responses.

To further examine the distribution of changes in direction of modulation of EAA responses by DA-receptor subtype agonists, responses were categorized according to the magnitude and direction of the change (see Fig. 1D). There were four categories: D1-NMDA, D1-non-NMDA, D2-NMDA, D2-non-NMDA. Data were collapsed across iontophoretically or synaptically evoked responses. In each category, responses were divided into two groups according to the magnitude of the change (0–20%, >20%). All but one NMDA-mediated

FIGURE 1 (*A*) Iontophoretically applied SKF 38393 enhances a response induced by iontophoretically applied NMDA. The top trace shows a control response to NMDA (−30 nA, applied between upward pointing arrows). The middle trace shows that SKF 38393 (applied iontophoretically, +100 nA for 1 min before NMDA) markedly enhanced the response. The bottom trace shows a wash 5 min after SKF 38393. (*B*) The top traces show NMDA-mediated DPSPs induced by two intensities of stimulation (200 and 450 μA). The slice was bathed in Mg^{2+}-free aCSF, bicuculline (BIC), and CNQX. The middle traces show that SKF 38393 (10 μM) increased response size. The bottom traces show a wash 30 min after SKF 38393. (*C*) The top traces show non-NMDA-mediated DPSPs induced by two intensities of stimulation (250 and 350 μA). The slice was bathed in Mg^{2+}-free aCSF, bicuculline, and AP5. The middle traces show that quinpirole (5 μM) decreased response size. The bottom traces show a wash 30 min after quinpirole. (*D*) Distributions of percentages of responses that displayed increases or decreases in DPSP area as a function of the subtypes of EAA and DA receptor examined.

response was enhanced by the D1 agonist; most increased by more than 20%. Although a majority of non-NMDA-mediated responses also were enhanced by SKF 38393, a number were decreased. Most of the NMDA-mediated responses were decreased by quinpirole, while all of the non-NMDA-mediated responses were decreased, most by more than 20%. Thus, the most predictable combinations of outcomes occurred when D1-receptor activation modulated NMDA-mediated responses and when D2-receptor activation modulated non-NMDA responses.

For voltage-clamp studies, brains were dissected and cut, and slices were incubated and then transferred to a perfusion chamber attached to the stage of a fixed-stage upright microscope. Slices were illuminated with near infrared light by placing an infrared filter in the light path. Cells were typically visualized from 30 to 100 μm below the surface of the slice. Patch electrodes (3–6 MΩ) were filled with standard internal solutions, depending on the purpose of the experiment. Tight seals (2–10 GΩ) from visualized medium-sized cells were obtained by applying negative pressure. The membrane was disrupted with additional suction to obtain the whole-cell configuration. Cells were held at −70 mV, which closely corresponded to the resting membrane potential of neostriatal neurons in the slice. Tetrodotoxin (1 μM) was added to the bath to block Na^+ currents after the whole-cell configuration was obtained. Drugs were applied in the bath or iontophoretically through a five-barrelled pipette placed approximately 30–50 μm from the recorded cell, as described previously. To test the effects of DA, D1, or D2 agonists, a single NMDA ejection intensity was chosen, and DA-, D1-, or D2-receptor agonists were applied and responses to NMDA reassessed. Three measures were recorded; maximum response amplitude, response duration at half maximum amplitude, and response area.

Inward currents were evoked by iontophoretic application of NMDA in all cells tested. These currents were voltage-dependent (displaying maximum amplitudes at holding potentials of −20 to −40 mV), enhanced in Mg^{2+}-free aCSF, and reduced by bath application of AP5. DA enhanced NMDA-induced currents. Application of DA (N = 15) produced statistically significant, reversible increases in the mean maximum amplitude (14 ± 6%), the mean half-amplitude duration (19 ± 7%), and the mean area (38 ± 14%) of NMDA-evoked current responses. Alone, DA produced no change in current. Application of D1-receptor agonists (SKF 38393 [N = 8] and A 77636 [N = 3]) also significantly and reversibly increased amplitude (26 ± 7%) and area (32 ± 9%) of the NMDA-evoked inward currents. Half-amplitude duration increased only slightly (5 ± 2%). Alone, these D1 agonists produced no change in current. The effects of SKF 38393 were blocked when cells were exposed first to SCH 23390 (N = 3), a D1-receptor antagonist. Application of the D2-receptor agonist, quinpirole (N = 6), produced inconsistent effects from cell to cell. Mean changes in amplitude (10 ± 11%), half-amplitude duration (−11 ± 5%), and area (−3 ± 10%) were small.

In conclusion, the present findings demonstrate that the complex modulatory actions of DA are dependent on combinations of coactivation of specific subtypes of EAA and DA receptors. Responses mediated by NMDA receptors are consistently potentiated by D1 DA agonists, while those mediated by non-NMDA receptors are consistently attenuated by D2 DA agonists. These findings are of clinical relevance since the actions of DA and EAAs have been implicated in neurological and affective disorders.

Acknowledgment

This work was supported by U.S. Public Health Service grant NS 33538.

References

1. Cepeda, C., Buchwald, N. A., and Levine, M. S. (1993). Neuromodulatory actions of dopamine in the neostriatum are dependent upon the excitatory amino acid receptor subtypes activated. *Proc. Natl. Acad. Sci. U.S.A.* **90**, 9576–9580.
2. Levine, M. S., Li, Z., Cepeda, C., Cromwell, H. C., and Altemus, K. L. (1996). Neuromodulatory actions of dopamine on synaptically-evoked neostriatal responses in slices. *Synapse* **24**, 65–78.
3. Smith, A. D., and Bolam, J. P. (1990). The neural network of the basal ganglia as revealed by the study of synaptic connections of identified neurones. *Trends Neurosci.* **13**, 259–265.

Christine Konradi

Laboratory of Molecular and Developmental Neuroscience
Massachusetts General Hospital
Charlestown, Massachusetts 02129

The Molecular Basis of Dopamine and Glutamate Interactions in the Striatum

One of the most interesting challenges in neuroscience research is the elucidation of neuronal plasticity. The striatum offers an excellent model to study plastic changes due to alterations in evironmental stimuli. Striatal neurons display a high degree of adaptability to changes in dopamine-receptor activation. For example, in Parkinson's disease (PD), striatal neurons are able to adjust to a decrease in dopaminergic input of 70–90%, before clinical symptoms become apparent. Moreover, supersensitivity to dopamine after unilateral removal of the dopaminergic input is observed in the 6-hydroxydopamine (6-OHDA) model of PD. Striatal neuroplasticity is also seen after stimulation of

Advances in Pharmacology, Volume 42

1054-3589/98 $25.00

dopamine receptors by psychostimulants: Drug dependence and withdrawal symptoms after cessation of chronic drug abuse are signs of adaptation to hyperstimulation of dopamine receptors.

Because increased RNA and protein synthesis may be one component of striatal neuroplasticity, it is crucial to study the effects of dopamine-receptor stimulation or inhibition on gene expression. We decided to investigate the cellular components linking dopamine-receptor stimulation with the expression of a particular target gene, *c-fos*. *C-fos* is an immediate early gene that is a perfect candidate for studies of neuroplasticity: It has low basal levels, which can be rapidly and highly induced when stimulated; the *c-fos* promoter is one of the best explored promoters; and the Fos protein is a DNA-binding transcription factor involved in the transactivation of AP1-responsive genes.

Psychostimulants like the indirect dopamine-receptor agonists cocaine or amphetamine stimulate *c-fos* synthesis (1, 2). In primary striatal cultures, direct activation of D1 receptors with dopamine or D1 agonists (SKF 38393; SKF 82958) also lead to an increase of *c-fos* levels (1, 3). *In vivo* as well as in culture, the induction of *c-fos* mRNA synthesis is prevented when D1 receptors are inhibited. Thus D1 receptors relay their state of activity to the cell nucleus to regulate *c-fos* mRNA synthesis. The presumed second-messenger pathway involves generation of cyclic adenosine monophosphate (cAMP), cAMP-mediated activation of protein kinase A (PKA), and phosphorylation of the transcription factor cAMP-responsive element-binding protein (CREB) at ^{133}Ser (Fig. 1A). CREB, which is constitutively bound to the *c-fos* calcium and cAMP response element (CaRE) site, binds to CREB-binding protein when phosphorylated on ^{133}Ser, thereby promoting recruitment of proteins of the transcription initiation complex. Indeed, in primary striatal culture, dopamine D1-receptor activation (1) increases cAMP levels, (2) leads to phosphorylation of CREB at ^{133}Ser, and (3) induces *c-fos* mRNA synthesis (see Fig. 1A) (1).

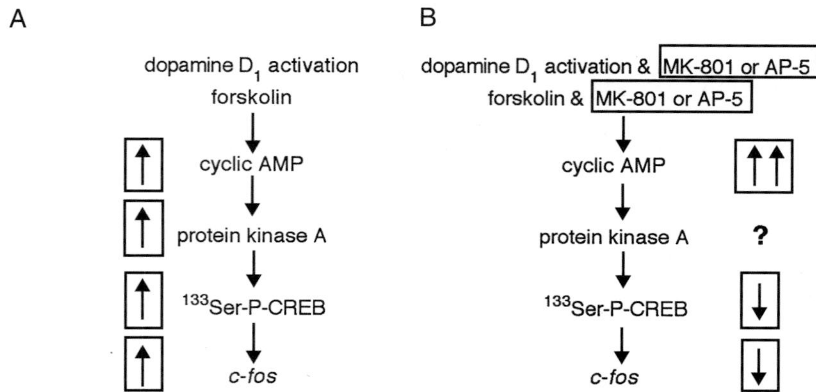

FIGURE 1 (*A*) Second-messenger pathway leading from D1-receptor stimulation or forskolin treatment to *c-fos* gene expression. (*B*) NMDA antagonists block D1 or forskolin–mediated *c-fos* induction while superstimulating cAMP levels.

Surprisingly, the NMDA antagonist MK 801 is able to inhibit amphetamine-stimulated gene expression *in vivo* (3). Thus, while the hypothetical D1-receptor stimulated second messenger pathway does not predict a role for glutamate–NMDA (see Fig. 1A), functional NMDA receptors are required for dopamine-mediated gene expression. *In vivo* studies cannot address the question of whether the interaction between the dopaminergic and glutamatergic–NMDA neurotransmitter systems are intraneuronal or transsynaptic. Because the striatum is part of a neuronal loop with glutamatergic components (e.g., cortex and thalamus), systemic administration of MK 801 may potentially be effective at several sites within this circuitry. For instance, release studies in striatal synaptosomal fractions show a stimulation of dopamine release by glutamate and NMDA (4), which may suggest a possible block of amphetamine-mediated dopamine release by NMDA antagonists.

Potential intraneuronal interactions of dopamine–cAMP and NMDA/Ca^{2+} pathways can be studied in primary striatal cultures, which lack presynaptic dopaminergic or glutamatergic inputs. In these cultures, like in the *in vivo* experiments, MK 801 and the competitive NMDA antagonist AP5 block dopamine D1-receptor–mediated *c-fos* expression (3). Thus, the D1-activated second-messenger pathway requires functional NMDA receptors within the same neuron to induce *c-fos* gene expression. D1-receptor–phosphorylation of ^{133}Ser CREB is also blocked in the presence of NMDA antagonists (3), while D1-stimulated cAMP levels are potentiated (see Fig. 1B). In effect, a dissociation of cAMP levels and CREB phosphorylation is observed. Replacing the cAMP stimulant forskolin for D1 agonists yields similar results at low forskolin concentrations (3), while higher concentrations of forskolin stimulate *c-fos* gene expression independently of the presence of NMDA-receptor antagonists.

Because one of the characteristics of NMDA receptors is their permeability for Ca^{2+} ions, it is imperative to examine the role of Ca^{2+} in D1-receptors–stimulated *c-fos* synthesis. Removal of Ca^{2+} ions from the medium or addition of EGTA blocks D1-mediated *c-fos* expression (3). Nevertheless, the Ca^{2+} dependence is NMDA-specific, because blockade of L-type Ca^{2+} channels does not affect D1-mediated *c-fos* expression and (2) KCl cannot overcome the NMDA-antagonist blockade of D1-mediated *c-fos* synthesis (3).

Addition of glutamate to dopamine agonists superstimulates *c-fos* expression. In the absence of D1 agonists or forskolin, glutamate does not significantly affect basal cAMP levels, but it potently induces ^{133}Ser CREB phosphorylation as well as *c-fos* gene expression. Thus, glutamate-induced ^{133}Ser CREB phosphorylation and *c-fos* gene expression is independent of cAMP increase.

Preliminary investigation into the regulation of a variety of cAMP-responsive striatal genes (e.g., prodynorphin, proenkephalin) yielded results similar to *c-fos* regulation. One possible model that explains our findings proposes a temporary modulation of NMDA receptors by a factor activated by the D1/cAMP pathway: PKA, activated by increased cAMP levels, phosphorylates NMDA receptors and removes the Mg^{2+} block (Fig. 2), an interaction already described for NMDA receptors with PKC (5). The ensuing Ca^{2+} influx activates protein kinases like Ca^{2+}–calmodulin kinases, which phosphorylate CREB at ^{133}Ser. However, we do not exclude an additional role for enhancer elements other than cAMP-responsive elements (e.g., the serum response element of the

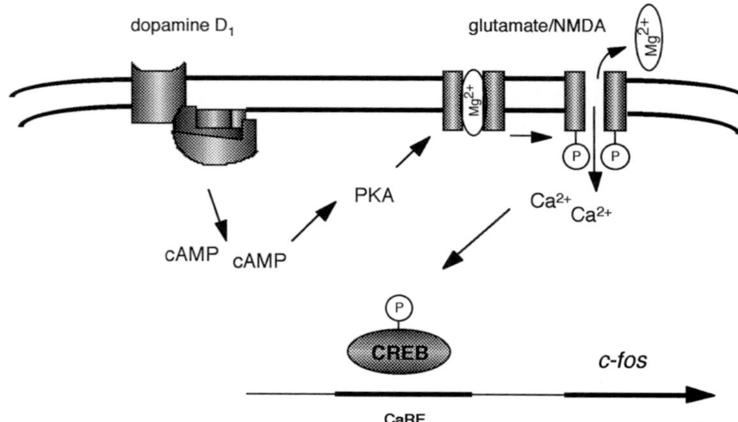

FIGURE 2 Possible mechanism of interaction between D1–cAMP second-messenger pathway and glutamate–NMDA receptors.

c-fos promoter). Neither do the data exclude additional interactions between dopamine and glutamate pathways, both intra- and extraneuronally.

In summary, D1 dopamine receptor stimulation leads to an increase in cAMP levels, phosphorylation of ^{133}Ser CREB, and induction of *c-fos* gene expression. Glutamate-receptor stimulation leads to phosphorylation of ^{133}Ser CREB and induces *c-fos* gene expression without increasing cAMP levels. The NMDA antagonist MK 801 prevents dopamine-mediated ^{133}Ser CREB phosphorylation and *c-fos* gene expression but does not inhibit cAMP induction. Dopamine-receptor–mediated *c-fos* expression is dependent on functional NMDA receptors and Ca^{2+} influx. The route of Ca^{2+} influx into the neuron may be crucial for gene expression, because neither L-type Ca^{2+} channel blockers nor KCl can reproduce the effects of NMDA antagonists or agonists. The inhibition of dopamine-mediated *c-fos* gene expression by MK 801 and the activation by glutamate reflect intrastriatal and intraneuronal rather than interneuronal events.

References

1. Konradi, C., Cole, R. L., Heckers, S., and Hyman, S. E. (1994). Amphetamine regulates gene expression in rat striatum via transcription factor CREB. *J. Neurosci.* **14**, 5623–5634.
2. Graybiel, A. M., Moratalla, R., and Robertson, H. A. (1990). Amphetamine and cocaine induce drug specific activation of the *c-fos* gene in striosome-matrix compartments and limbic subdivisions of the striatum. *Proc. Natl. Acad. Sci. U.S.A.* **87**, 6912–6916.
3. Konradi, C., Leveque, J. C., and Hyman, S. E. (1996). Amphetamine and dopamine-induced immediate early gene expression in striatal neurons depends upon postsynaptic NMDA receptors and calcium. *J. Neurosci.* **16**, 4231–4239.

4. Carrozza, D. P., Ferraro, T. N. Golden, G. T., Reyes, P. F., and Hare, T. A. (1992). In vivo modulation of excitatory amino acid receptors: Microdialysis studies on *N*-methyl-D-asparate-evoked striatal dopamine release and effects of antagonists. *Brain Res.* 574, 42–48.
5. Chen, L., and Huang, L. Y. M. (1992). Protein kinase C reduces Mg^{2+} block of NMDA receptor channels as a mechanism of modulation. *Nature* 356, 521–523.

G. W. Arbuthnott,* C. A. Ingham,* and J. R. Wickens†

*Department of Preclinical Veterinary Sciences
University of Edinburgh Centre for Neuroscience
Edinburgh EH9 1QH, Scotland

†Department of Anatomy and Structural Biology
University of Otago
Dunedin, New Zealand

Modulation by Dopamine of Rat Corticostriatal Input

I. A Continuing Conundrum

It is somewhat embarrassing to have to agree with a recent obituary for Paul Feltz that, in spite of the heroic efforts of Dr. Feltz and others like him, we still have no clear summary of how dopamine acts in the striatum. We will try to shed some light in this difficult place, but to do so, we have first to abandon some of the preconceptions that governed the field until recently. First, it seems that there is no evidence for a "simple" excitatory or inhibitory action of dopamine. Even the idea that the striatonigral cells are inhibited and the striatopallidal ones excited by dopamine cannot explain why the laboratories that report excitatory effects also find inhibitory ones in the same cells, nor does it help explain why so many laboratories have failed to see either effect. The results of attempts to identify the membrane current in striatal cells, which is modified by dopamine application, have also been contradictory. A recent paper has identified a Ca^{2+} current, which may be the target for dopamine action, but it is only likely to change the behavior of the cells during cortical activation. It has also been shown that dopamine changes the synaptic potentials on spiny neurones but in a way that depends on the glutamate-receptor subtype being activated. Similarly, although we could demonstrate an inhibition of the afterhyperpolarization following trains of action potentials, the effect was only visible in cells already depolarized from rest. So the effects of dopamine seem to depend on the recent history of the neuron's membrane potential.

Advances in Pharmacology, Volume 42

1054-3589/98 $25.00

II. A Different Kind of Electrophysiology

Our most recent physiological experiments were an attempt to understand the actions of dopamine at the single-cell level in the context of the suggested timing of its release. In the behaving monkey, dopamine cells fire bursts of action potentials as a consequence of stimuli that predict the likelihood of a rewarding event. Schultz and his colleagues (p. 686) suggest that dopamine is released in situations in which unexpected reward occurs. During training, this release coincides with the solution to a behavioral problem. The appropriate response is, "Do that again."

To make use of such an instruction, the animals need a record of what has just been done, and the arrival of the dopamine must make that sequence of events more likely to occur. If we suppose that the dopamine is acting on corticostriatal synapses, then the "trace" of what has just been done would correspond to a set of coincidentally activated cortical cells and their respective striatal partners. The arrival of dopamine must somehow make a memory trace as a record of this set of connections. Once such a set is "confirmed," then the actions that led to reward being obtained become more likely to recur in similar circumstances and continue for some time in the absence of further reward. Thus, the changes induced by dopamine must themselves be long term and outlast the actions of the dopamine release.

Long term potentiation (LTP) seemed likely to be a suitable model on which to build this conditional memory system. We studied the consequences of trains of stimuli to the corticostriatal system and found, like others, that the usual consequence of repetitive trains of stimulation to the cortex is the reduction of the size of the excitatory postsynaptic potential (EPSP) in the striatal cell (1). Usually cortical activation—if it is paired with postsynaptic depolarization in the striatum—leads to a reduction in the efficacy of the synaptic input to the striatal cell—a kind of long term depression (LTD). However, if dopamine is applied in a pulsatile manner intended to mimic the release of dopamine thought to follow a train of spikes in a dopamine cell, then the corticostriatal synapses can show LTP (Fig. 1.) (2). LTP is also a consequence of treating slices of the corticostriatal system in ways that enhance the N-methyl-D-aspartate component of transmission (e.g., with low Mg^{2+} solutions) and after blocking potassium channels with tetraethylammonium, which would also increase dopamine release in the slices. Dopamine, therefore, seems to have long-term consequences for the efficacy of corticostriatal synapses—not just their efficacy but perhaps even their survival.

III. Ultrastructual Consequences of Dopamine Damage

Ultrastructural studies have shown that dopamine synapses are often localized to the necks of dendritic spines on striatal neurons. We examined the rat neostriatum in the electron microscope to investigate whether the increase in enkephalin, which follows dopamine destruction, resulted from sprouting of enkephalin-containing terminal onto the spine neck sites vacated by dopamine. Enkephalin-immunoreactive synaptic boutons were 50% larger after dopamine

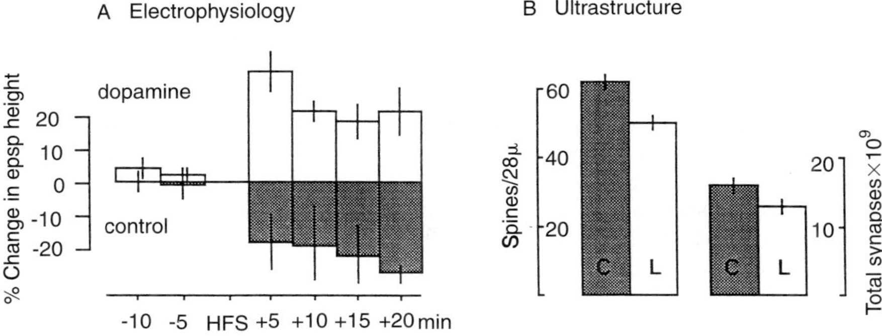

FIGURE 1 Both the electrophysiological results and the anatomical data are shown in summary form. (*A*) Each bar shows the mean peak height of the EPSP recorded in striatal cells after stimulation of the cortex. Heights were normalized to the size of the response immediately before the application of six conditioning trains of stimuli to the cortex coincident with a depolarizing pulse to the striatal cell (HFS). The shaded columns show the mean and SEM for seven cells recorded with no other treatment, while the clear bars show the results in seven cells to which dopamine was added during each train. Modified from Wickens *et al.* (2). (*B*) On the left is an analysis of the number of spines on Golgi-stained striatal cell dendrites, counted at the same distance from the beginning of spines on the dendrites. The filled column (C) represents cells on the control side; the clear one (L), those on the side of the brain with a 6-hydroxydopamine-induced lesion of the dopamine input. (Modified from Ingham *et al.* [3]). On the right is an estimate of the total number of asymmetric synapses in the striata of a similar set of animals. Again, the mean from the control side is shaded.

denervation, but they did not contact spines more frequently than in the control striatum. As a parallel study progressed, a possible explanation emerged.

Detailed spine counting revealed that the number of dendritic spines on the output cells of the striatum from which the dopamine has been removed by a 6-hydroxydopamine injection, is reduced (3). This loss of spines happens over the first 3 wk after the lesion, and the output cells are less spiny even after 1 year. Recently, we have used "unbiased" stereological methods to count the number of asymmetric synapses in the striatum on the side of a 6-hydroxydopamine lesion, compared with the control side. The total number of asymmetric synapses in the striatum on the lesioned side is reduced by about 18%, which is similar to the reduction in spine density at equivalent times after the lesion (4). Cortical terminals are one of the main sources of asymmetric synapses in the striatum, and they are particularly associated with dopamine input to the same spines (5). Therefore, it seems likely that the reduction in the number of synapses is the consequence of the loss of corticostriatal synapses. Could the lost spines be the ones that originally received dopamine input?

The percentage of loss is larger than the estimate of the number of spines receiving a dopamine synapse from a study of all the synapses in a small cube of striatal volume (<7%) but smaller than the estimate derived from the dendrite of an identified striatonigral cell (39%). The large range suggests that only some cells have a large percentage of spines innervated by dopamine, and our estimate is a global average of cells with dopamine synapses and those without.

Current experiments are designed to test the idea that the loss of synapses is specific to only one of the two sets of output cells of the striatum, but, of course, there may be a loss of spines that do not receive dopamine input.

There is evidence from other systems (e.g., retina) and recently from experiments on developing striatal cells in culture that dopamine might have a "morphogenic" effect. The synapses that disappear after dopamine depletion might be the most vulnerable of a system of corticostriatal synapses maintained somehow by dopamine. Indeed, one theory of the "purpose" of spines is to limit the toxic consequences of glutamate release to small parts of the cell, so that cells lose spines rather than die by excitotoxic action.

IV. A Synthesis

Of course, the preceding discussion is only a speculation, but it is highly plausible that a mechanism like LTP, which can change synaptic efficacy for weeks, should have ultrastructural consequences in the striatum, as it has been shown to have in hippocampus. Thus, taking the morphological and physiological experiments together suggests that LTP induced by dopamine has morphological consequences. It could be potentiated synapses survive and in the absence of dopamine, nonpotentiated synapses are eliminated. Testing these ideas will require more detail about how potentiation happens and some pharmacological tools with which to interfere with various stages of it, but as the molecular biology of LTP races ahead, it seems likely that suitable tools are not so very far away.

References

1. Calabresi, P., Maj, R., Mercuri, N. B., and Bernardi, G. (1992). Coactivation of D_1 and D_2 dopamine receptors is required for long-term synaptic depression in the striatum. *Neurosci. Lett.* **142**, 95–99.
2. Wickens, J. R., Begg, A. J., and Arbuthnott, G. W. (1996). Dopamine reverses the depression of rat cortico-striatal synapses which normally follows high frequency stimulation of cortex in vitro. *Neuroscience* **70**, 1–6.
3. Ingham, C. A., Hood, S. H., Weenink, A., van Maldegem, B., and Arbuthnott, G. W. (1993). Morphological changes in the rat neostriatum after unilateral 6-hydroxydopamine injections into the nigrostriatal pathway. *Exp. Brain Res.* **93**, 17–27.
4. Ingham, C. A., Hood, S. H., Taggart, P., and Arbuthnott, G. W. (1996). Synaptic plasticity in the rat neostriatum after unilateral 6-hydroxydopamine lesion of the nigrostriatal dopaminergic pathway. In: *The Basal Ganglia V,* edited by Ohye, C., Kimura, M., and McKenzie, J. S. Plenum Press: New York, p. 157–164.
5. Smith, Y., Bennett, B. D., Bolam, J. P., Parent, A., and Sadikot, A. F. (1994). Synaptic relationships beween dopaminergic afferents and cortical or thalamic input in the somatosensory territory of the striatum in monkey. *J. Comp. Neurol.* **344**, 1–19.

George V. Rebec

Program in Neural Science
Department of Psychology
Indiana University
Bloomington, Indiana 47405

Dopamine, Glutamate, and Behavioral Correlates of Striatal Neuronal Activity

As the main afferent structure of the basal ganglia, the striatum plays a key role in processing cerebrocortical information for motor output. Although both dopamine (DA) and glutamate (GLU) have been implicated in this role, it is unclear how these transmitters alter striatal function under naturally occurring behavioral conditions. To address this issue, a series of experiments recorded single-unit activity in the striatum of awake, unrestrained rats (1–5). DA transmission was altered either pharmacologically or via DA iontophoresis, and possible interactions with GLU were assessed. The results indicate that although the postsynaptic actions of DA are complex, they must be considered in conjunction with the level of GLU input from cerebrocortical afferents.

Male, Sprague-Dawley rats (300–400 g), approved for use by the Institutional Animal Care and Use Committee, were used for all recordings. The animals were anesthetized for stereotaxic surgery, and their skulls were prepared for freely moving electrophysiology. Some animals also received aspiration lesions of cereborcortex from 2.0 posterior to bregma to the frontal pole. Sham controls underwent the same surgical procedure, but the cerebrocortex remained intact. After an appropriate recovery period (4–8 days in most cases and 10–20 days for cortical and sham lesions), the animals were habituated to an open-field arena (1.2 m^2), and a custom-designed microdrive was secured to their skulls. For the first group of studies (1, 4, 5), an epoxy-insulated tungsten microelectrode (30–50 MΩ at 60 Hz) was lowered into the target site. Neuronal signals passed through a head-mounted preamplifier and then via shielded cable to an electronic swivel. Unit discharges were amplified, filtered (bandpass, 0.3–3.0 kHz), and displayed by conventional means. Spontaneously active units, isolated to a signal-to-noise ratio of at least 3 : 1, also were stored on an audio channel of a videocassette recorder, which was used in conjunction with a videotaping system to record behavioral activity. On completion of recording, rats were anesthetized, and the unit recording site was marked by passing current through the tungsten recording electrode. After a transcardial perfusion with formosaline, the brain was processed for histological analysis.

Unit activity was compared during periods of either quiet rest or active movement. A sample of more than 100 neurons revealed that approximately

Advances in Pharmacology, Volume 42

1054-3589/98 $25.00

80% were motor-related in that they increased spike activity in close temporal association with spontaneous movement. The remaining neurons responded with either no or inconsistent changes in rate. Systemic injection (s.c.) of 1.0 mg/kg of D-amphetamine, an indirect DA agonist, increased motor activity and activated motor-related neurons. Detailed analysis of videotape records of matched behavioral responses before and after the drug revealed an activating effect of amphetamine on unit activity well beyond that associated with movement alone (1). The results of this behavioral clamping analysis were confirmed when the drug was applied to striatal neurons via an adjacent infusion cannula (5). In this case, neurons previously characterized as motor-related increased activity in response to amphetamine several minutes before behavioral activation. It appears, therefore, that amphetamine acts directly in the striatum to accelerate motor-related neuronal activity, which may help shape subsequent behavioral changes. In addition, non-motor-related neurons were inhibited by either systemic or intrastrital amphetamine injections. Thus, in awake, behaving animals, amphetamine creates a divergence in firing rate between two behaviorally distinct neuronal populations.

That DA may mediate both types of neuronal responses is supported by evidence that they are reversed by a subsequent injection of a dopamine-receptor antagonist. Interestingly, however, the amphetamine-induced excitation of motor-related neurons is significantly attenuated in rats with cerebrocortical lesions, compared with sham controls (4). Such lesions disrupt the corticostriatal pathway, which provides GLU input to the striatum. Collectively, the results of these experiments suggest (1) both an excitatory and inhibitory role for DA in the neuronal effects of amphetamine on striatal neurons and; (2) that the role that DA assumes depends, in part, on neuronal responsiveness to movement; and (3) the amphetamine-induced striatal activation depends on GLU input from cerebrocortex.

These conclusions have been supported by experiments involving DA and GLU iontophoresis in freely moving rats (2, 3). In the first of these studies, Pierce and Rebec (3) classified striatal neurons as motor- or non-motor-related before iontophoresis to test for possible differential effects of DA. When the animals resumed a resting posture, DA or GLU was ejected for brief periods (15–30 sec), and dose-response effects (5–80 nA) were assessed. Consistent with the amphetamine data, iontophoretic DA typically excited motor-related but inhibited non-motor-related cells. Both types of responses were dose-dependent, with half-maximal effects typically observed at 20 nA. GLU had uniformly excitatory effects on both neuronal categories. The GLU response, which was evident at 10 nA, also was dose-dependent. Whenever possible, DA and GLU were coapplied at half-maximal doses to test for possible interactions. In 13 of 17 units, DA and GLU together increased firing rate by at least 50% more than the application of either substance alone, suggesting a strong supra-additive effect. The DA–GLU combination had additive effects on the remaining four neurons.

In a subsequent investigation, Kiyatkin and Rebec (2) tested separate applications of DA and GLU on a total of 88 spontaneously active neurons in anterior striatum and its ventral extent, the nucleus accumbens. GLU caused dose-dependent excitations (mean threshold dose of 22.2 nA) with relatively short onset and offset latencies (0.5–4.0 sec). DA, in contrast, had excitatory or inhibitory effects with onset and offset latencies that were highly variable (2–20 sec). Both types of responses were relatively weak. The DA-induced inhibition, moreover, disappeared during periods of spontaneous movement. Apart from a significantly higher level of basal firing rate in the nucleus accumbens during quiet rest, striatal and accumbal neurons responded similarly. In some cases, DA was applied at low currents for prolonged periods (1–2 min) to assess possible modulatory effects on brief applications of GLU. Under these conditions, DA had weakly inhibitory, excitatory, or no effects on basal levels of unit activity, but in 74% of the cases, the relative magnitude of the GLU response was enhanced. This enhancement occurred even in those cases in which the absolute magnitude of the GLU response, compared with pre-DA iontophoresis, was reduced. Thus, prolonged applications of DA increased the magnitude of the GLU signal relative to the level of background activity.

Taken together, the results of these experiments suggest that DA, either applied directly or released by amphetamine, excites or inhibits striatal neurons, depending on their sensitivity to overt movement. These effects, moreover, occur in conjunction with ongoing cortical input, which, in the case of amphetamine, is critical for the drug-induced excitation. It appears, therefore, that the synaptic action of DA depends, at least in part, on the level of cortical GLU input. As shown in Fig. 1, DA may exert excitatory or inhibitory effects on neurons receiving substantial or little activation respectively, from cerebrocortex. Another aspect of the striatal DA–GLU interaction is the ability of prolonged DA application at low ejection currents to enhance the magnitude of the GLU-induced excitation relative to the level of background activity. Thus, in a possible role as neuromodulator, DA may enhance the signal-to-noise ratio of the GLU signal. Further assessment of this role in ambulant animals is required to identify the mechanisms by which DA alters striatal function under naturally occurring behavioral conditions.

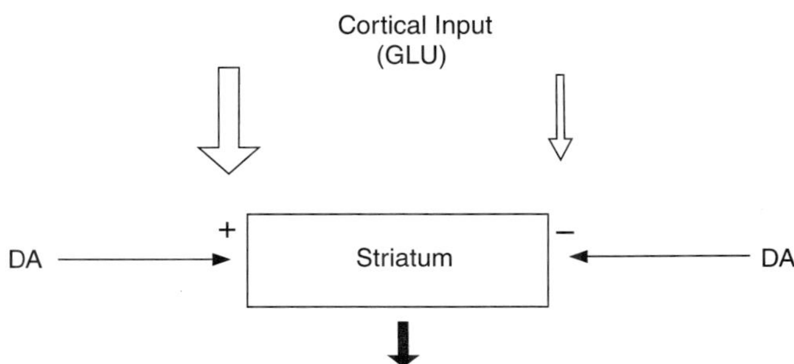

FIGURE 1 Model of DA-GLU interactions in the striatum of awake, unrestrained rats. DA appears to exert excitatory or inhibitory effects on striatal background activity, depending on the level of cortical input. In either case, moreover, the net result of DA is an increase in the relative magnitude of the GLU signal.

Acknowledgment

This research was supported by the National Institute on Drug Abuse (DA 02451). The technical and administrative assistance of P. Langley and F. Caylor, respectively, is gratefully acknowledged.

References

1. Haracz, J. L., Tschanz, J. T., Wang, Z., White, I. M., and Rebec, G. V. (1993). Striatal single-unit responses to amphetamine and neuroleptics in freely moving rats. *Neurosci. Biobehav. Rev.* **17**, 1–12.
2. Kiyatkin, E. A., and Rebec, G. V. (1996). Dopaminergic modulation of glutamate-induced excitations of neurons in the neostriatum and nucleus accumbens of awake unrestrained rats. *J. Neurophysiol.* **75**, 142–153.
3. Pierce, R. C., and Rebec, G. V. (1995). Iontophoresis in the neostriatum of awake unrestrained rats: Differential effects of dopamine, glutamate, and ascorbate on motor- and nonmotor-related neurons. *Neuroscience* **67**, 313–324.
4. Tschanz, J. T., Griffith, K. E., Haracz, J. L., and Rebec, G. V. (1994). Cortical lesions attenuate the opposing effects of amphetamine and haloperidol on neostriatal neurons in freely moving rats. *Eur. J. Pharmacol.* **257**, 161–167.
5. Wang, Z., and Rebec, G. V. (1993). Neuronal and behavioral correlates of intrastriatal infusions of amphetamine in freely moving rats. *Brain Res.* **627**, 79–88.

J. Rajkowski, P. Kubiak, S. Ivanova, and G. Aston-Jones

Department of Psychiatry
Allegheny University
Philadelphia, Pennsylvania 19102

State-Related Activity, Reactivity of Locus Ceruleus Neurons in Behaving Monkeys

Alertness varies dramatically between sleep and wakefulness, with variations in sustained attentiveness occurring during waking. Results presented here reveal that tonic activity of neurons of the brain noradrenergic nucleus locus ceruleus (LS) is highly correlated with changing degree of alertness, while at the same time, LC cells respond phasically to behaviorally significant stimuli. These results support the hypothesis that the LC–norepinephrine system regu-

Advances in Pharmacology, Volume 42

1054-3589/98 $25.00

lates states of alertness and significantly enhances the brain's ability to react adaptively to environmental stimuli (1).

LC neurons were recorded extracellularly from four cynomolgus monkeys performing a visual discrimination task (2). This task required the animal to press a level and foveate a central fix spot to initiate each trial, to release a lever within 650 ms of a target stimulus (10–20% of trials) to receive a juice reward, and to withhold responses to nontarget stimuli (80–90% of trials). Stimuli were horizontal or vertical bars presented in random order on a video screen. Eye position and pupil diameter were measured using a video-based eye-tracking system; in addition, an electroencephalogram (EEG), neuronal activity, and time marks for stimuli and behavioral responses were recorded. Monkeys were allowed to work uninterrupted for up to 3 hr. Fluctuations in task performance allowed us to identify states of alertness spanning the range between drowsiness and high arousal. Three general states of attentiveness were defined: Drowsiness (inattention), focused attention, and scanning (searching mode).

Drowsiness and early stages of sleep were observed in our experiments, as indicated by episodic bursts of slow-wave activity. During drowsiness, task performance deteriorated, as indicated by decreased lever releases for target stimuli (hit responses) and increased errors of omission. As previously reported

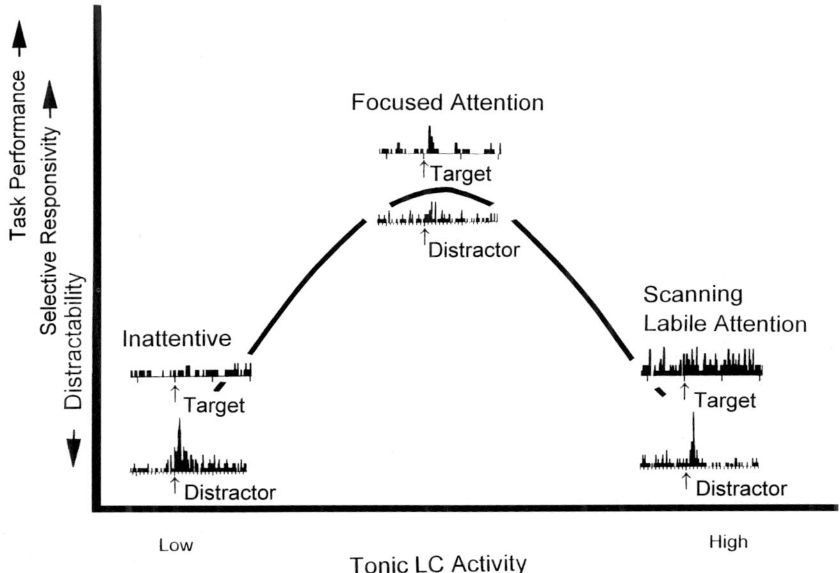

FIGURE 1 Performance on the visual discrimination–vigilance task. Performance on this task requires focused attention and is optimal with intermediate levels of tonic LC activity (*solid line*). This also corresponds to epochs of robust, selective LC responsivity (phasic responses to target stimuli). Conversely, phasic responses to unconditioned auditory stimuli (distractors) are weakest at this same time, corresponding to diminished distractability during focused attention.

(3), dramatic changes in LC activity were associated with transitions between sleep and waking. LC neurons became inactive during drowsiness and the initial phases of sleep. Usually, LC inactivity preceded eyelid closure or EEG synchronization, and long pauses (several seconds) in LC discharge were reliable predictors of impinging drowsiness. Conversely, LC neurons generated vigorous bursts prior to or at the onset of awakening, as indicated by EEG desynchronization and eye opening. These results reveal a role for the LC in sleep regulation such that LC inactivity may be a critically permissive condition for sleep to occur. In contrast, the LC may play a more executive role in awakening. Awakening-related LC activity may indicate a role for the LC in rapid restoration of vigilance after sleep.

LC neurons exhibit relatively little activity during low-vigilance behaviors such as grooming or food consumption, while LC tonic activity typically accelerates during exploratory activity. The task (visual discrimination–vigilance task) required that the animal foveate a central fix spot to initiate each trial. Such foveation is effortful and reflects attentiveness to the task. During task performance, the frequencies of both LC discharge and successful foveations fluctuated over short (10–30 sec) and long time intervals (10–30 min). These changes in LC discharge were consistently inversely correlated with attentiveness to the task, such that slightly elevated LC activity (by 0.5–1.0 spike/sec) was accompanied by decreased fixation frequency. Even short-term increases in LC tonic activity often corresponded to marked, short-lasting reductions in foveation frequency. Task performance was best during moderate LC neuronal activity, whereas high LC tonic activity was associated with behavioral hyperactivity and poor task performance (frequent false alarms). In epochs of moderate tonic LC discharge and good performance, animals exhibited high foveation frequency, presumably corresponding to focused attention, whereas during poor performance, in addition to frequent errors, the animals exhibited larger amplitude eye movements with shorter, less stable foveations (apparent scanning attention).

LC neurons were phasically activated at a short latency (about 100 ms) by target stimuli but not by other task events, such as bar releases or reward presentations signaled by target. Other sensory events (distractors, unconditioned stimuli) also phasically activated LC if they occurred unexpectedly or were of high intensity; they usually produced an orienting reaction.

The amplitude of LC responses to both targets and distractor stimuli showed remarkable, albeit converse, modulation with different alertness states. During drowsiness, the tonic discharge rate of LC neurons decreased and LC responses to target were significantly smaller. In contrast, LC responses to distractor stimuli (white noise bursts; 70 msec, 75 dB) were of relatively great magnitude during drowsiness, compared with epochs of good task performance. Similarly, LC responses decreased for target cues but increased (preliminary result) for distractor stimuli during epochs of poor performance associated with higher LC discharge rate and behavioral hyperresponsivity. These results show that a close relationship exists between attentiveness, LC tonic and phasic activities, and performance of a vigilance task. Phasic excitatory responsiveness of LC neurons may produce a transient, stimulus-oriented increase of attentiveness.

The rate of tonic LC discharge, changing inversely with focused attentiveness and quality of performance, correlated directly with tonic pupil diameter. Thus, during poor task performance, high tonic LC activity was associated with pupillary dilation. These results confirm for the monkey a close association between LC discharge and autonomic activity. It is noteworthy that stimuli that activate the LC also activate the sympathetic system. This coactivation may be the result of collateral projections from the nucleus paragigantocellularis to LC and spinal cord or hypothalamus (1). The present results show that while the sympathetic system prepares the animal physically for rapid responses to urgent stimuli, a parallel activation of the LC increases vigilance and attentiveness, preparing the animal cognitively for adaptive responsiveness to such stimuli.

The striking relationship between LC activity and reactions to stimuli with varying states of alertness and attentiveness may indicate a role for LC in controlling alertness. The role of LC in alertness is demonstrated by alterations in LC activity during sleep and waking. It may be argued that LC activity not only is unnecessary, but also may be disruptive for sleep. The strong burst of LC discharges that precedes or accompanies awakening may be interpreted as restorative to vigilance and would be behaviorally adaptive (e.g., when in danger of predation). Together, the LC and sympathetic system constitute a coherent system of adaptive responsiveness to urgent stimuli.

Additional data indicate that attentional regulation may not be the only role for the LC system. It is known that only through awakening from REM sleep, when most dreaming occurs, can the content of a dream be remembered. Our results indicate that such awakening typically is associated with a strong burst of LC discharge. In addition to the role proposed for restoring alertness, this discharge may be important for memorizing dream content (4). Indeed, a role for LC in memory and learning has been proposed before (see accompanying chapter by Kubiak *et al.* in Part H). Sleep-related silence of the LC (as well as in other structures of synergistic function, such as the raphe nucleus) may function in disallowing or disrupting memory formation during sleep. This explains why a dream can be remembered only when interrupted by awakening, when it can be properly recognized as a dream. Such a mechanism may help to prevent inclusions of unreferenced and unrealistic dream memories to real-life memories.

A role for LC in memory is also consistent with tonic and phasic LC activity observed within the alert state. The specific stimuli that activate LC neurons and increase alertness are also the events most likely to be remembered. This speculation links attention and memory via functions hypothesized for the LC. Interestingly, the same stimuli also have the capacity to activate emotional systems, acting as still another memory enhancer (4).

References

1. Aston-Jones, G., Chiang, C., and Alexinsky, T. (1991). Discharge of noradrenergic locus coeruleus neurons in behaving rats and monkeys suggests a role in vigilance. *Prog. Brain Res.* 88, 501–520.

2. Aston-Jones, G., Rajkowski, J., Kubiak, P., and Alexinsky, T. (1994). Locus coeruleus neurons in the monkey are selectively activated by attended stimuli in a vigilance task. *J. Neurosci.* **14**, 4467–4480.
3. Foote, S. L., Aston-Jones, G., and Bloom, F. E. (1980). Impulse activity of locus coeruleus neurons in awake rats and monkeys is a function of sensory stimulation and arousal. *Proc. Natl. Acad. Sci. U.S.A.* **77**, 3033–3037.
4. Aston-Jones, G., Rajkowski, J., Kubiak, P., Valentino, R., and Shipely, M. (1996). Role of the locus coeruleus in emotional activation. *Prog. Brain Res.* **107**, 379–402.

Craig W. Berridge

Psychology Department
University of Wisconsin
Madison, Wisconsin 53706

Modulation of Forebrain Electroencephalographic Activity and Behavioral State by the Locus Ceruleus–Noradrenergic System: Involvement of the Medial Septal Area

The locus coeruleus (LC)–noradrenergic system has long been postulated to be a critical component of an ascending activational system involved in the induction and maintenance of prolonged epochs of alert waking and/or in the modulation of state-dependent cognitive processes. To directly assess the causal relationship between LC neuronal discharge activity and forebrain EEG state, we previously conducted a series of studies in the halothane-anesthetized rat, that demonstrated substantial, tonic excitatory actions of the LC noradrenergic system on electroencephalographic activity (EEG). It was observed that unilateral enhancement of LC neuronal activity elicits bilateral activation of cortical (ECoG) and hippocampal (HEEG) EEG, whereas bilateral suppression of LC discharge activity increases EEG measures of sedation (1, 2). The EEG-activating effects of LC stimulation appear to involve noradrenergic β receptors (1). These results indicate that the LC–noradrenergic system is a potent modulator of forebrain EEG state via actions of β receptors in the halothane-anesthetized rat.

Advances in Pharmacology, Volume 42

1054-3589/98 $25.00

Possible candidate sites at which LC efferents might act to modulate forebrain EEG include portions of the basal forebrain containing the medial septum–diagonal band of Broca (collectively referred to as MS) and the substantia innominata–nucleus basalis (SI). MS and SI receive a dense noradrenergic innervation, the preponderance of which arises from LC. Further, both send efferent projections widely throughout the hippocampal formation and neocortex, respectively, and influence HEEG and ECoG. The present studies tested the hypothesis that noradrenergic β receptors located within MS and SI modulate ECoG and HEEG state in the halothane-anesthetized rat (3). The EEG effects of 150-nl infusions of a β agonist (isoproterenol [ISO]), a β antagonist (timolol), glutamate, and vehicle into sites within and adjacent to MS and SI were examined. In all experiments, male Sprague-Dawley rats were anesthetized with halothane and guide cannulae were stereotaxically implanted over MS, SI, or adjacent structures. Bipolar, EEG electrodes were implanted bilaterally into frontal cortex and hippocampus. EEG was recorded to polygraph and analogue recording tape. Power spectral analyses (PSAs) were conducted on selected pre-, post-, and recovery epochs of recorded EEG. Dye (2% Pontamine Sky Blue) was mixed in with infusate to facilitate histological determination of the infusion center.

Halothane was adjusted such that baseline (at least 30 min) ECoG was characterized by uninterrupted large-amplitude, slow-wave activity and a 2-sec tail pinch elicited robust ECoG desynchronization. Vehicle infusions into MS lacked obvious effects on either ECoG or HEEG. In contrast, 150-nl unilateral infusions of ISO (3.75 μg) into MS (n = 21) elicited robust ECoG desynchronization (low-amplitude, high-frequency) and nearly pure HEEG theta activity. EEG responses followed infusions by 3–8 min and were observed only when infusions were placed within a radius of approximately 500 μm of MS. PSA provided quantitative verification of these qualitative observations. ISO infusions into sites adjacent to MS did not alter either ECoG or HEEG.

ISO infusions into SI did not alter forebrain EEG. The ineffectiveness of SI ISO infusions might result from insufficient volume or properties unique to this experimental preparation that interfere with the normal EEG modulatory action of SI neurons. To assess this, glutamate infusions (150 nl) and larger ISO infusions (450 nl) were made into SI. Glutamate infusions elicited robust activation of ECoG and HEEG within 30–90 sec, whereas 450 nl infusions of ISO had no effect on either ECoG or HEEG.

To determine the degree to which endogenous norepinephrine (NE) contributes to the maintenance of forebrain EEG activation, the effects of bilateral MS infusions of the β antagonist, timolol, on ECoG and HEEG activity were examined (n = 8). Halothane was adjusted so that low-voltage, high-frequency activity in ECoG and theta activity in HEEG were consistently present. Under these conditions, bilateral infusion of timolol (3.75 μg/150 nl in each hemisphere) induced in a shift in ECoG from predominantly desynchronized to large-amplitude, slow-wave activity and a shift in HEEG from theta-dominated to mixed-frequency activity. The onset of these EEG responses occurred within 3–30 min (median, 10 min) of the infusions, with a duration from 60 min to greater than 120 min. Bilateral vehicle infusions (n = 7) lacked obvious effects on either ECoG or HEEG, as did bilateral timolol infusions placed adjacent to MS.

To summarize, these results indicate that, under these experimental conditions, noradrenergic efferents, presumably arising from LC, modulate forebrain EEG state via actions at MS β receptors. The anesthetized rat facilitates the study of LC modulation of EEG state. However, to obtain a clearer understanding of the behavioral functions of the LC–noradrenergic system, it is essential to determine the degree to which manipulations of this system, or selected LC terminal fields, modulate forebrain EEG in the unanesthetized animal and whether changes in EEG state are associated with concomitant changes in behavioral state. Therefore, additional studies were conducted to examine the bahavioral and EEG and electromyographic (EMG), effects of ISO infusions (3.75 μg/150 nl) into or immediately adjacent to MS in the undisturbed, resting rat (4).

In these studies, rats were implanted with a guide cannula aimed at MS or adjacent sites and, in a subset of animals, bipolar ECoG and EMG (implanted into neck muscle) electrodes. On the day of testing, a 33-gauge needle was connected to a liquid swivel and infusion pump contained outside the testing chamber and inserted into and secured to the cannula. This preparation permitted making infusions in the sleeping rat without disturbing the animal. Behavioral and EEG and EMG signals were continuously recorded on a polygraph and on videocassette recording tape. ECoG and EMG were scored for the following behavioral state categories: (1) slow-wave sleep, (2) REM sleep, (3) quiet waking, and (4) active waking. The time spent in each state across five 30-min epochs was measured. A variety of behaviors, determined from videotaped records, were scored for all animals: (1) asleep, (2) quiet awake, (3) body up, (4) grooming, (5) rears, (6) quadrant entries, (7) eating, (8) drinking, and (9) total time spent awake. Data collection began 60–120 min following needle insertion. All infusions were performed following the collection of at least 60 min of baseline (two 30-min epochs), and behavior and EEG and EMG were recorded for 90 min following infusions (three 30-min epochs).

ISO infusions into MS consistently resulted in a substantial increase in waking as compared with vehicle-infusion (Fig. 1). Similar to that observed in the anesthetized rat, the distribution of infusion sites that did or did not result in substantial time spent awake identifies a region of the basal forebrain, encompassing the medial septum, in which ISO infusions act to alter behavioral state. Quantitative analyses of the behavioral and EEG and EMG measures were conducted on those animals in which the ventral tip of the infusion needle was located within this region as determined from histological analyses.

ISO infusions (n = 26), but not vehicle (n = 12), significantly increased total time spent awake as well as time spent engaged in all of the behavioral categories listed here, except for drinking, with a response latency following infusions of 4–10 min. Total time spent awake was significantly increased for the three 30-min postinfusion epochs. A comparable activating effect on behavioral state was observed with EEG and EMG measures, with a significant decrease in slow-wave sleep and a significant increase in total time spent awake observed following MS ISO (n = 8) infusions in each postinfusion 30-min epoch (see Fig. 1). The EEG and EMG responses either coincided with behav-

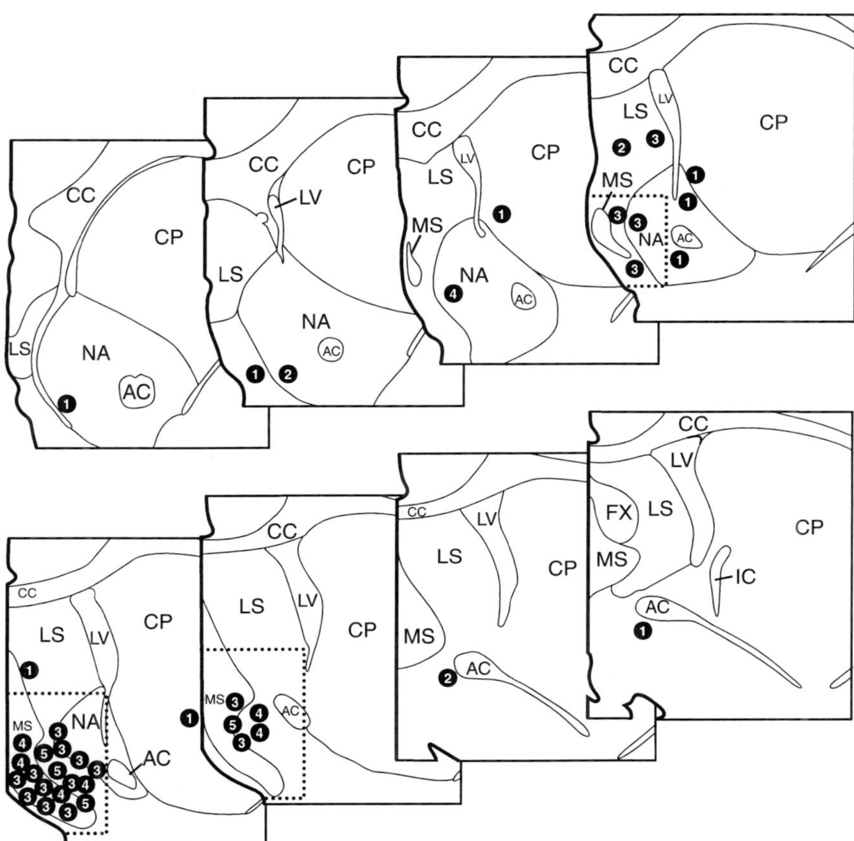

FIGURE 1 Schematic depiction of the location of each ISO infusion with a numeral indicating the effect of each infusion on total time spent awake. Numerals specify the appropriate time range (1 = 0–500 sec; 2 = 500–1000 sec, 3 = 1000–2000 sec; 4 = 2000–3000 sec; 5 = >3000 sec) for total time spent awake during the 60-min post-infusion interval for each animal. Vehicle-treated animals displayed a mean total time awake in this period of 302 ± 88 sec with a range of 125–1098 sec. These infusions identify a region within which ISO infusions increase waking, that encompasses the medial septum, the vertical limb of the diagonal band of Broca, the Island of Calleja, and posterior portions of the shell region of the nucleus accumbens. This region is identified by the dotted line border and is collectively referred to as MS. Infusions outside this region were substantially less effective at increasing waking beyond that observed in pre-infusion epochs or in vehicle-treated animals. AC = anterior commissure; CC = corpus callosum; CP = caudate-putamen; IC = internal capsule; LS = lateral septum; LV = lateral ventricle; MS = medial septum; NA = nucleus accumbens. Each level separated by 250 μm, with the most anterior section in the top left panel (adapted from Swanson, 1992).

ioral activation (n = 3) or preceded it by approximately 10–320 sec (n = 7). ISO infusions into MS resulted in a near-complete suppression of REM sleep. Vehicle infusions into MS (n = 5) lacked consistent, obvious effects on EEG, EMG, or behavioral indices of waking.

Based on these observations and those of others, it is posited that one function of LC is to facilitate the induction of a behavioral state appropriate for acquisition of sensory information (e.g., waking). This is proposed to be dependent on changes in the tonic activity of LC neurons associated with changes in behavioral state and, at least in part, involves actions of NE at β receptors located within the MS. Superimposed on these actions may be facilitatory actions on state-dependent cognitive processes, which may, in part, involve phasic fluctuations in LC neuronal activity and possibly actions of NE within cortical and thalamic terminal fields. Combined, these actions would serve to facilitate appropriate behavioral responding to sensory information.

References

1. Berridge, C. W., and Foote, S. L. (1991). Effects of locus coeruleus activation on electroencephalographic activity in neocortex and hippocampus. *J. Neurosci.* **11,** 3135–3145.
2. Berridge, C. W., Page, M., Valentino, R. J., and Foote, S. L. (1993). Effects of locus coeruleus inactivation on forebrain electroencephalographic activity. *Neuroscience* **55,** 381–393.
3. Berridge, C. W., Bolen, S. J., Manley, M. S. and Foote, S. L. (1996). Modulation of forebrain electroencephalographic (EEG) activity in the halothane-anesthetized rat via actions of noradrenergic β-receptors located within the medial septal region of the basal forebrain. *J. Neurosci.* **16,** 7010–7020.
4. Berridge, C. W., and Foote, S. L. (1996). Enhancement of behavioral electroencephalographic (EEG), and electromyographic (EMG) indices of waking following stimulation of noradrenergic β-receptors located within the medial septal region of the basal forebrain in the unanesthetized rat. *J. Neurosci.* **16,** 6999–7009.

Barry D. Waterhouse, David Devilbiss, Daniel Fleischer, Francis M. Sessler, and Kimberly L. Simpson

Department of Neurobiology and Anatomy
Allegheny University of the Health Sciences
Philadelphia, Pennsylvania 19129

New Perspectives on the Functional Organization and Postsynaptic Influences of the Locus Ceruleus Efferent Projection System

A variety of behavioral, electrophysiological, and neuroanatomical studies suggest that one role of the central noradrenergic system is to regulate the transfer of information through sensory circuits during periods of arousal and selective or sustained attention. Sensory signal–processing regions of the brain are densely innervated by locus ceruleus (LC) efferents, and investigations from our own laboratory (1) as well as others have shown that local administration of norepinephrine (NE) can increase the magnitude of individual sensory neuron responses to synaptic stimuli. However, many questions remain concerning the precise way in which NE influences the stimulus coding properties of individual cells and ultimately how output from the LC impacts on the operation of ensembles of neurons engaged in common sensory functions. In addition, there has been considerable recent interest in several peptides, such as galanin, that are colocalized with NE in subpopulations of LC neurons. Despite the potential for corelease of these neuroactive substances at noradrenergic synapses in sensory circuits, there have been few studies designed to characterize the organization of peptide-containing LC neurons with respect to modality-specific efferent targets of the LC nucleus. Finally, although many reductionist approaches have been used to advance our understanding of the receptor-linked second-messenger mechanisms underlying NE's actions at the single-cell and membrane level, it is well to remember that such *in vitro* studies are limited insofar as the cells and circuits under consideration are isolated from the neural networks in which they normally operate. As a result, the impact of NE or other putative LC transmitters on the coordinated activities of ensembles of neurons can only be indirectly inferred from studies of isolated cells. With these considerations in mind, we have focused our recent efforts on the design of anatomical and physiological experiments that are capable of bridging the gap between cellular-membrane studies of LC–NE attributes and more global hypotheses regarding

Advances in Pharmacology, Volume 42

1054-3589/98 $25.00

the impact of the LC projection system on sensory neural network function. The rat somatosensory system has been used throughout this work as a model sensory network for investigating these issues.

The specific goals have been to: (1) obtain a more sophisticated understanding of the anatomical relationship between the LC and the sensory system that relays tactile information from the mystacial vibrissae on the rat's snout to the contralateral somatosensory cortex, (2) determine the neurochemical composition of LC efferents that project to trigeminal somatosensory pathway in rat, and (3) further clarify the precise nature of NE's influences on membrane excitability and cellular-response properties in different morphologically and electrophysiologically identified classes of neurons within the rat cortical circuitry. We believe that by mapping the ceruleosomatosensory projection system and by determining the way in which specific cortical circuit elements are influenced by NE, it will be possible to begin predicting how sensory signal processing along afferent pathways and in cortical networks may be altered under conditions such as arousal and attention, where the output of the LC efferent system is increased.

I. Functional Organization of LC Efferents

Recent studies (2) using fluorescent retrograde tracers have shown that LC efferents to target structures along the trigeminal somatosensory pathway in rat exhibit an orderly projection with respect to the crossed trajectory of this ascending sensory system. Quantitative analysis of the distribution of LC cells labeled following unilateral injections of fluorogold, rhodamine-, or fluorescein-coated latex microspheres into the whisker-related region of the primary somatosensory cortex (i.e., barrel field cortex [BFC]), ventrobasal (VB) thalamus, and principal nucleus of V (PrV) revealed a bias such that output from one LC nucleus is heaviest to the contralateral PrV, ipsilateral VB thalamus, and ipsilateral BFC. Thus, based on the relative density of retrograde labeling, output from one LC nucleus appears capable of exerting its major influence on structures conveying tactile information from the contralateral side of the body.

A second finding of this study was that LC projection neurons to somatosensory structures are organized into overlapping subsets within the nucleus, thus raising the possibility that a significant proportion of these cells may project to multiple, common somatosensory targets via axon collaterals. Studies employing double fluorescent retrograde labeling strategies confirmed that single LC neurons preferentially collateralize to multiple sites that process the same sensory information (2B); that is, individual LC cells are more likely to coinnervate like modality structures (e.g., VB thalamus and barrel field cortex) than structures associated with different sensory modalities (e.g., dorsal lateral geniculate and BFC).

The results of this investigation contradict conventional wisdom and demonstrate that the LC is neither randomly organized nor without specificity in its projection of forebrain targets. Instead, the LC exhibits an internal organization and efferent projection pattern that is consistent with the idea that LC outputs can selectively and coordinately modulate the flow of somatosensory information through forebrain circuits.

II. Colocalization of Neuroactive Peptides and NE in LC Projection Neurons

To determine if specific subpopulations of LC neurons that project to anatomically discrete regions of the trigeminal somatosensory pathway differ with respect to the proportion of cells expressing the neuroactive peptide, galanin, brain stem tissue sections were processed for peptide immunoreactivity after retrograde transport of fluorescent tracers. Preliminary results indicate that galanin immunoreactivity was observed prominently within LC, with the heaviest concentration of galanin-positive cells in the dorsal aspect of the nucleus. Comparison of peptide expression in the LC with fluorogold labeling from the BFC revealed that many but not all ceruleosomatosensory cortical projection neurons contain galanin. These latter studies are important in that they begin to identify the neurochemical composition of an identified population of LC projection neurons.

III. Specificity of NE Actions on Sensory Cortical Neurons

In a recent *in vitro* extracellular recording study, cortical neuronal responses to uniform iontophoretic pulses (8-sec duration, every 30 sec) of glutamate were monitored before, during, and after incremental microiontophoretic doses of NE (1–50 nA). For each incremental dose of NE applied, perievent histograms of cell responses to glutamate were quantitatively analyzed to determine iontophoretic dose-response relationships for NE–glutamate interactions. In some cells, suppression of glutamate responses was observed at all NE doses tested. However, in other cells, we observed an "inverted U" dose-response relationship such that glutamate-evoked responses were progressively facilitated to an optimum as NE levels were increased and then progressively suppressed as NE levels were increased further (Fig. 1). Using a different experimental strategy, we noted that a fixed level of NE could facilitate cell responses to minimal ejection currents of glutamate more than responses to currents of glutamate that were initially near maximal for producing peak excitation. The demonstration that NE can either facilitate or suppress cortical neuronal responsiveness to glutamate emphasizes the intrinsic heterogeneity of cortical neurons with respect to noradrenergic modulatory actions. More importantly, these results suggest that under ideal conditions, modulatory interactions between NE and glutamate can be optimized. Furthermore, our finding of an inverted U dose-response relationship between NE and glutamate-evoked discharge is particularly exciting because it parallels a similar relationship between tonic rates of

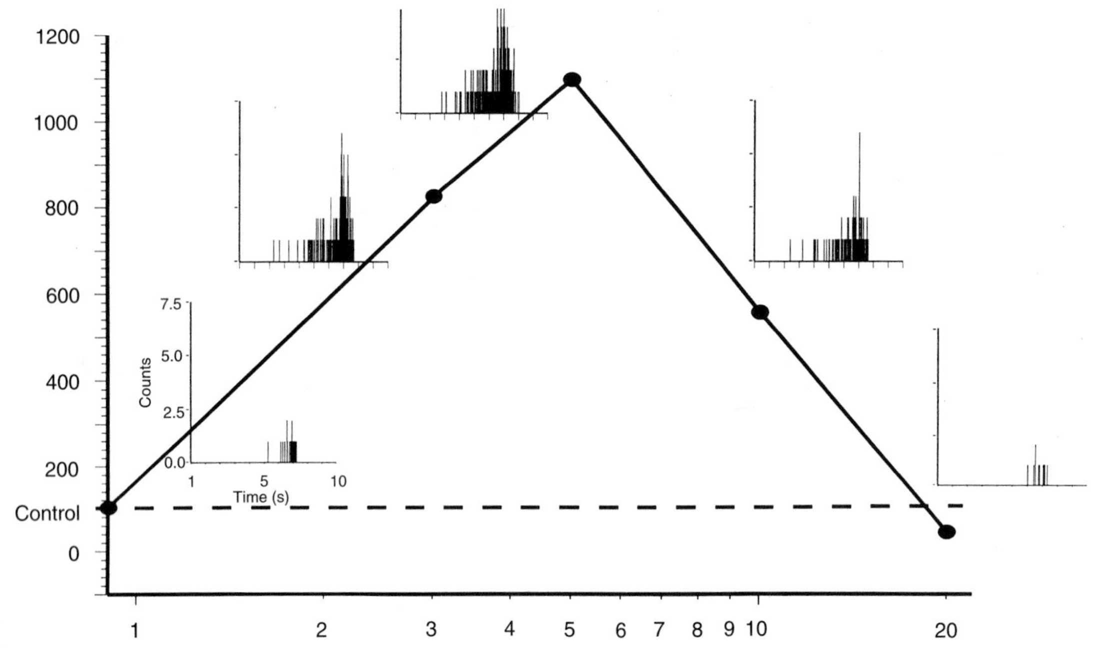

FIGURE 1 Biphasic dose-response relationship for NE modulation of glutamate-evoked responses in rat sensory cortical neurons. Each point on the graph and its associated perievent histogram represent the extracellularly recorded response of a single layer V somatosensory coritcal neuron (*in vitro* tissue slice preparation) to uniform iontophoretic pulses (20 nA, 8-sec duration, 40-sec cycle) of glutamate before (control, at left) and during administration of different tonic levels of iontophoretic NE (1–20 nA). Ech histogram sums unit activity during six consecutive glutamate applications.

LC neuron discharge and monkey performance on a sustained-attention task, as reported by Rajkowski and colleagues (3).

Intracellular studies in tissue slice preparations have shown that in the absence of a direct hyperpolarizing action and under current-clamp conditions, bath-applied NE (1–30 μM) can increase the probability of cortical neuronal spiking in response of *threshold, perithreshold, and subthreshold* activation of afferent synaptic inputs. This effect was reliably mimicked by α_1-receptor activation and was consistently observed in "regular" spiking but not "intrinsic burst"–type neurons of layer V somatosensory cortex. NE-induced increases in spike probability were observed both with and without concomitant increases in excitatory postsynaptic potential amplitude. In other cells that were morphologically identified as callosally projecting layer V neurons, NE (over the same dose range) consistently suppressed membrane responses to excitatory synaptic inputs. In regular spiking cells of layers II and III, we have observed both increases and decreases in rheobase and input resistance with NE administration.

Finally, we have shown that NE can enhance GABA-induced membrane conductance changes in most but not all layer V regular spiking neurons (4). Such GABA-potentiating actions were mimicked and blocked by β agonists and antagonists, respectively, and also mimicked by agents that elevate intracellular levels of cyclic adenosine monophosphate (cAMP) (e.g., forskolin and 8-bromo-cAMP). Thus, these findings indicate that NE can augment membrane responses to GABA via a β-receptor–linked/cAMP cascade of events. As such, these noradrenergic influences on putative transmitter-induced conductance changes are consistent with previously observed effects of iontophoretic NE on single-unit responses to GABA in anesthetized (5) and awake, behaving (6) animals.

Overall, the results obtained to date support a working hypothesis that NE differentially modulates the discharge properties and transmitter responsiveness of various well defined cell types within the sensory cortical circuitry.

IV. Implication of Recent Findings

The results reported here continue to support and refine the idea that the LC efferent system exhibits a high degree of anatomical and physiological specificity with respect to its potential impact on target neuronal circuits. Such findings are significant in that they run counter to the prevailing notion that the LC efferent system is a diffuse, nonspecific projection system that serves a generalized alerting function in the brain. For example, the demonstration of an ordered projection pattern and collateralization of LC efferents to the rat trigeminal somatosensory pathway suggests that under certain behavioral circumstances LC output may be capable of coordinately regulating the transfer of like-modality information through multiple levels of the neuraxis. Such a mode of operation may be advantageous in situations that require the organism to briefly focus its perceptual capabilities on a salient or imperative stimulus. By contrast, *en masse* activation of LC neurons and subsequent global release of NE throughout forebrain circuits may be more appropriate as a means of supporting generalized arousal or vigilance.

Identification of LC projection neurons that contain both NE and galanin raises several new questions. What would be the effect of simultaneously released NE and galanin on target neurons in sensory circuits? Under what conditions would such corelease be expected to occur? Likewise, the demonstration of cell-specific physiological actions of NE within the cortical circuitry implies that noradrenergic output from the LC can differentially regulate the operation of various functional components in neural networks. Moreover, depending on local tissue concentrations, NE appears capable of inducing a range of enhancement and suppression of responsiveness to excitatory amino acids in sensory cortical neurons. This finding further emphasizes the need to identify the physiological levels of NE that prevail in noradrenergic terminal fields as output from the LC fluctuates. Overall, these advances in our understanding of the organizational principles and physiological specificity of the LC-NE efferent system provide new insights into the role of this monoamine-containing brain–stem nucleus in behavior. While the LC has often been implicated in tasks that require orienting, alerting, or attending (7), the data emerging from these studies are beginning to identify neural substrates that can participate in achieving such complex perceptual and cognitive functions.

References

1. Waterhouse, B. D., Azizi, S. A., Burne, R. A., and Woodward, D. J. (1990). Modulation of area 17 simple and complex cell responses to moving visual stimuli during norepinephrine and serotonin iontophoresis. *Brain Res.* **514**, 276–292.
2. Simpson, K. L., Altman, D. W., Wang, L., Kirifides, M. L., Lin, R. C. S., and Waterhouse, B. D. (1997). Lateralization and functional organization of the locus coeruleus projection to the trigeminal somatosensory pathway in rat. *J. Comp. Neurol.* (in press)
3. Rajkowski, J., Kubiak, P., and Aston-Jones, G. (1992). Activity of locus coeruleus (LC) neurons in behaving monkeys varies with changes in focused attention. *Soc. Neurosci. Abstr.* **18**, 538.
4. Sessler, F., Liu, W., Kirifides, M. L., Mouradian, R., Lin, R. C. S., and Waterhouse, B. D. (1995). Noradrenergic enhancement of GABA-induced input resistance changes in layer V regular spiking pyramidal neurons of rat somatosensory cortex. *Brain Res.* **675**, 171–182.
5. Waterhouse, B. D., Moises, H. C., and Woodward, D. J. (1980). Noradrenergic modulation of somatosensory cortical neuronal responses to iontophoretically applied putative transmitters. *Exp. Neurol.* **69**, 30–49.
6. West, M. O., and Woodward, D. J. (1983). A technique for microiontophoresis in the freely moving rat. *Soc. Neurosci. Abstr.* **9**, 227.
7. Foote, S. L., Bloom, F. E., and Aston-Jones, G. (1983). Nucleus locus coeruleus: New evidence of anatomical and physiological specificity. *Physiol. Rev.* **63**, 844–914.

G. Aston-Jones,* J. Rajkowski,* S. Ivanova,* M. Usher,† and J. Cohen†

*Department of Psychiatry
Allegheny University
Philadelphia, Pennsylvania 19102

†Department of Psychology
Carnegie Mellon University
Pittsburgh, Pennsylvania 15260

Neuromodulation and Cognitive Performance: Recent Studies of Noradrenergic Locus Ceruleus Neurons in Behaving Monkeys

Locus ceruleus (LC) neurons were recorded in monkeys during performance of a visual discrimination task (1, 2). This task required animals to stably foveate a central fixation spot on a video monitor and selectively release a lever after presentation of a target cue (10–20% of trials) to receive a drop of juice. Responses to nontarget cues (80–90% of trials) resulted in a 3-sec time out. LC neurons exhibited phasic and tonic activities that varied in relation to task performance.

LC neurons were phasically activated at short latencies (~100 ms) selectively by target stimuli but not by other task events (Fig. 1) (1, 2). This response easily shifted to the new target (previous nontarget) cue after reversal training, confirming that these responses were driven by stimulus meaning (Fig. 2) (3). Tonically, LC activity varied among low, intermediate, and high levels. Low activity (near zero) was strongly associated with drowsiness and poor task performance with many misses. High tonic LC activity (~3–5 spikes/sec) was associated with poor task performance characterized by frequent false alarms (incorrect responses to nontarget stimuli), poor foveation of fix spots, and increased scanning eye movements (Fig. 3). Optimal performance (with few or no false alarms and misses, and with steady visual fixation of task events) was associated with an intermediate level of tonic LC discharge (~1–2 spikes/sec) (4). In addition, phasic responses to target stimuli (described previously) occurred only during epochs of excellent task performance associated with intermediate tonic LC activity. These data indicate that the LC in the alert monkey exhibits two modes of activity: (1) high phasic corresponding to focused attention and good task performance and (2) high tonic corresponding to scanning attention and poor performance on the task. Intra-LC injections of pilocarpine or clonidine to acutely activate or inactivate LC neurons, respectively, affected task performance as predicted by the aforementioned recordings.

Advances in Pharmacology, Volume 42

1054-3589/98 $25.00

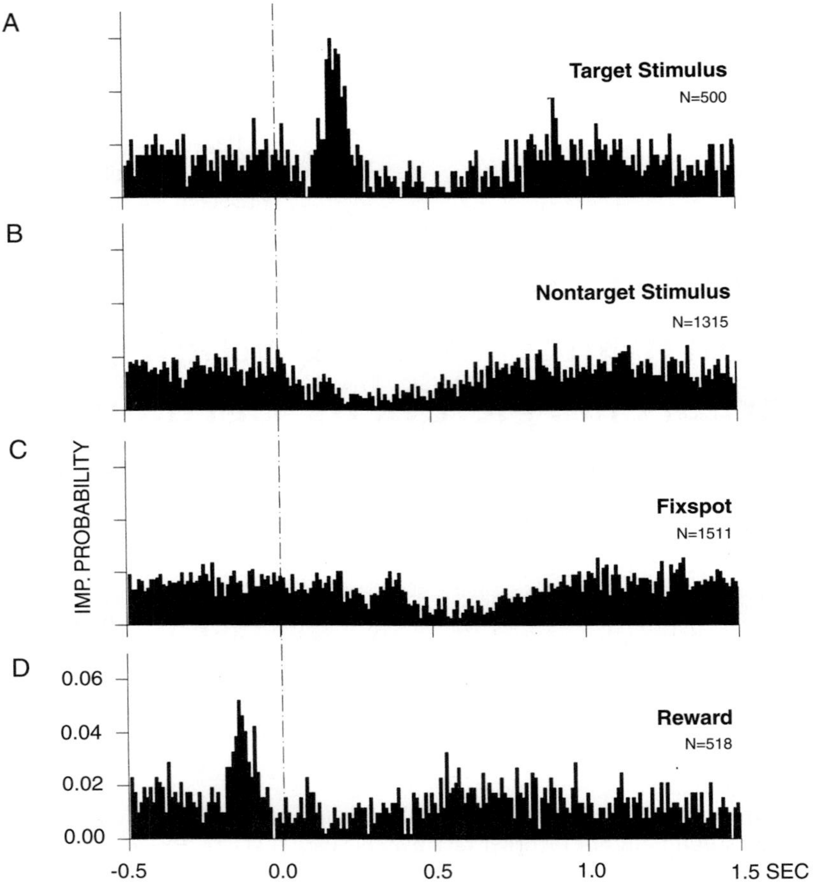

A

Target Stimulus
N=500

B

Nontarget Stimulus
N=1315

C

Fixspot
N=1511

D

Reward
N=518

IMP. PROBABILITY
0.06
0.04
0.02
0.00
-0.5 0.0 0.5 1.0 1.5 SEC

FIGURE 1 Peristimulus time histograms (PSTHs) for a typical individual LC neuron in response to various events during performance of the vigilance task. Normalized PSTHs are synchronized with (*A*) target stimuli, (*B*) nontarget stimuli, (*C*) fix spot presentation, and (*D*) juice solenoid activation. Note the selective activation by target stimuli (*A*). The activation seen before reward presentation (*D*) is due to activation by target cues. All LC neurons tested to date were activated by target stimuli; inhibition by nontargets (as here) occurs only in a subset of cells. Inhibition in the Fixspot PSTH is due to frequent nontargets, which follow fix spot presentation.

FIGURE 2 PSTHs for an LC multineuron recording (*A*) before and (*B*) after reversal of stimulus meaning. Note in the left panels (*A*) that neurons were activated by vertical stimuli when they were target cues and not by the nontarget horizontal stimuli. In contrast, after reversal, neurons were selectively activated by horizontal stimuli when they were target cues and not activated by the vertical stimuli when they were nontarget cues (*B* right panels). These response profiles were typical of LC neurons recorded during the respective stimulus contingencies. This indicates that LC neurons respond to the meaning of stimuli, not to their physical attributes.

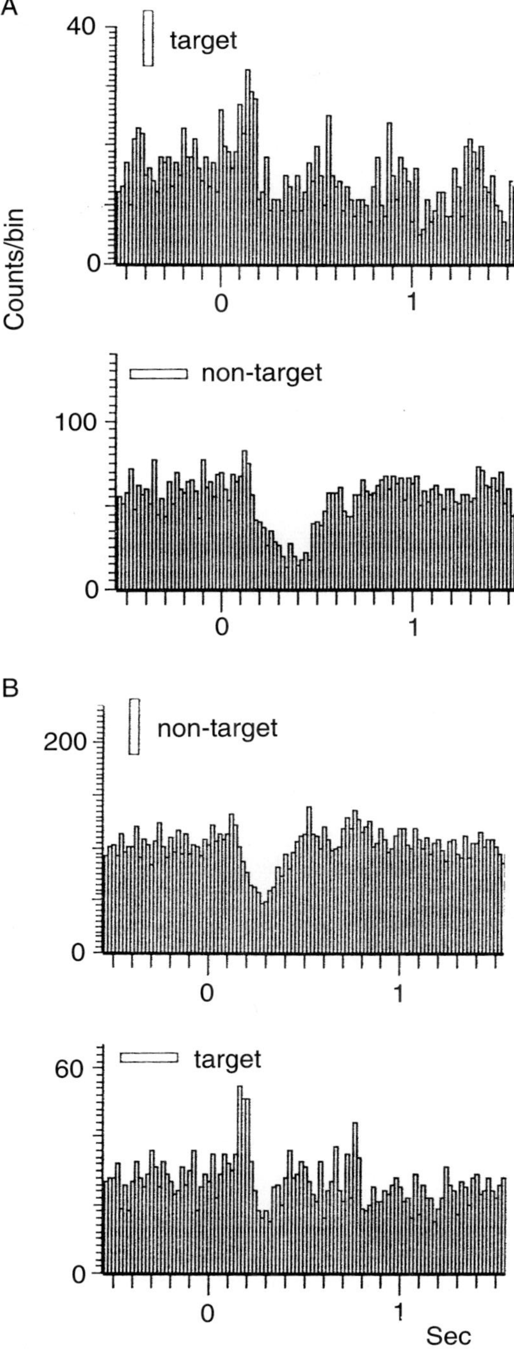
A
40
target
Counts/bin
0
0
1
non-target
100
0
0
1
B
non-target
200
0
0
1
target
60
0
0
1
Sec

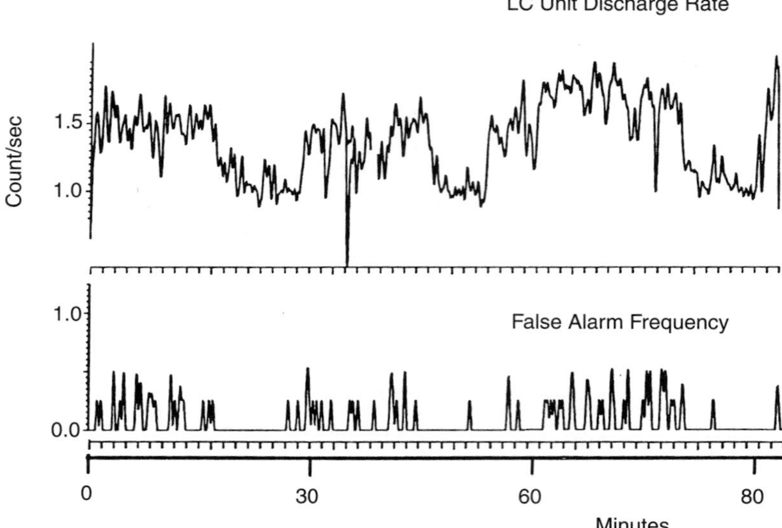

FIGURE 3 Plots of discharge frequency of an individual LC neuron (*upper*) and false alarm frequency (*lower*) from a typical recording session during the visual discrimination task. Note that activity of this neuron varies between two levels. The animal was alert throughout this recording. Thus the lower level of LC activity does not correspond to drowsiness (which would typically correspond to zero LC activity). Note the close correlation between the high tonic activity and more frequent false alarm errors (mistaken bar releases following nontarget stimuli). Optimal performance on this task, which requires focused attention, corresponds with the more moderate level of tonic LC discharge; this is also the state when LC neurons exhibit selective phasic responses to target stimuli (as shown in Figs. 1 and 2).

A computational model of the LC and of task performance was constructed that closely simulated LC activity and its relationship to task performance (5, 6). This model explains the relationship between patterns of LC firing and task performance by changes in electrotonic coupling among LC neurons. Specifically, it demonstrates that an increase in coupling results in greater phasic responses to target but not nontarget cues and also produces lower spontaneous (tonic) activity. This results in tighter coupling of cortical norepinephrine release to target cues, which in turn improves task performance (reduction in false alarms without an increase in response latency to targets). Thus, this model predicts that modulated electrotonic coupling among LC neurons may be an important factor that regulates LC activity and its role in behavioral performance. Overall, these results indicate a function for the LC system in regulation of the attentional state.

Acknowledgments

Supported by U.S. Public Health Service grant MH55309 and the Human Frontiers Science Program.

References

1. Aston-Jones, G., Rajkowski, J., Kubiak, P., and Alexinsky, T. (1994). Locus coeruleus neurons in the monkey are selectively activated by attended stimuli in a vigilance task. *J. Neurosci.* **14,** 4467–4480.
2. Rajkowski, J., Kubiak, P., and Aston-Jones, G. (1994). Locus coeruleus activity in monkey: Phasic and tonic changes are associated with altered vigilance. *Brain Res. Bull.* **35,** 607–616.
3. Kubiak, P., Rajkowski, J., and Aston-Jones, G. Responses of monkey locus coeruleus neurons anticipate acquisition of discriminative behavior in a vigilance task. *Neuroscience* (in press)
4. Aston-Jones, G., Rajkowski, J., Kubiak, P., Valentino, R., and Shipley, M. (1996). Role of the locus coeruleus in emotional activation. *Prog. Brain Res.* **197,** 379–402.
5. Usher, M., Cohen, J., Servan-Schreiber, D., Zemel, R., Kubiak, P., Rajkowski, J., and Aston-Jones, G. (1995). A computational model of locus coeruleus influence on performance in a visual discrimination task. *Soc. Neurosci. Abstr.* **21,** 2093.
6. Usher, M., Cohen, J. D., Rajkowski, J., Kubiak, P., and Aston-Jones, G. A computational model of locus coeruleus function and its influence on cognitive performance. (submitted)

T. Sawaguchi and Y. Kikuchi

Department of Psychology
Hokkaido University
N10 W7, Kita-ku
Sapporo 060, Japan

Noradrenergic Effects on Activity of Prefrontal Cortical Neurons in Behaving Monkeys

The dorsolateral prefrontal cortex of primates is involved in the working memory process for visuospatial information, and this area contains a subset of neurons that shows mnemonic coding for representing the visuospatial working memory process (1). During the 1980s, several lines of research suggested that noradrenaline might be involved in the working memory process of the prefrontal cortex. Both noradrenaline-containing fibers and noradrenaline receptors, including α_2 and β receptors, are prominent in this area (2, 3). Furthermore, a behavioral pharmacological study by Arnsten and Goldman-Rakic (4) demonstrated that intramuscular injections of the α_2-adrenergic receptor agonist clonidine improved performance on a delayed-response working memory task in aged monkeys. This improving effect was antagonized by the α_2-receptor antagonist yohimbine, and the effect of clonidine disappeared when the dorso-

Advances in Pharmacology, Volume 42

1054-3589/98 $25.00

lateral prefrontal cortex was surgically removed. These findings imply that α_2 receptors are involved in the working memory process mediated by the prefrontal cortex. However, it has remained unclear how α_2 receptors are involved in perfrontal neuronal activity for working memory process. To address this problem, we combined iontophoretic analysis of noradrenaline antagonists with single-neuron recording during an oculomotor delayed-response (ODR) task that requires visuospatial working memory. We report here that iontophoretic application of the α_2-receptor antagonist yohimbine to prefrontal neurons in the monkey attenuated delay-period activity, indicating that activation of α_2 receptors plays a critical role in maintaining the directional or mnemonic coding of prefrontal neurons.

Two rhesus monkeys (*Macaca mulatta*) were trained to perform the ODR task. The task was started by the monkey fixating on a central spot on a cathode ray tube. One second later, a visual cue was presented for 0.5 sec, followed by a delay period. The cue was presented randomly at one of four peripheral locations, which were separated by 90 degress with an eccentricity of 15 degrees. After a delay period of 4 sec, the fixation spot was then extinguished, which instructed the monkey to make a memory-guided saccade to the location that had been cued prior to the delay period. The correct response was rewarded by a drop of water 0.2 sec after the response. Trials were separated by an intertrial interval of 2 sec.

For the iontophoretic application of drugs and extracellular recording of neuronal activity, we employed exactly the same techniques as used for our previous studies (5). Briefly, multibarreled glass micropipettes were used for extracellular recording of neuronal activity and iontophoretic application of drugs. The central barrel of the micropipette, which contained a carbon-fiber filament (7 m in diameter), was filled with 0.9% saline and used to record neuronal activity. The surrounding barrels were filled with and used for iontophoretic application of solutions of the following drugs: yohimbine (0.01 M, pH = 5–6), propranolol (0.01 M, pH = 5–6), 0.9% saline (to balance the current), and other drugs, such as SCH 23390, for other studies. While the monkey performed the ODR task, the activity of a single neuron was recorded by the micropipette. The neuronal activity was converted from analogue to digital by a window discriminator for analyses with a personal computer. Raster displays and time histograms, aligned at onsets of task periods, were constructed

FIGURE 1 (*A*) The effect of iontophoretic application of SCH 23390 on a prefrontal neuron with the directional delay-period activity. Raster displays and averaged histograms, which are aligned by the events that comprise the task, for the activity before (control) and during the application of yohimbine with 50-nA current are shown. The activities associated with different target locations (down, left, upper, and right) are illustrated separately. F, fixation period; C, cue period; D, delay period; G, go period from the onset of the go signal to the onset of reward delivery. Vertical scale, 10 spikes/sec. (*B*) Tuning curves, fitted by a cosine function for the delay period activity shown in (*A*). The mean discharge rate during the delay period is plotted against the direction of the response. Note that the tuning curve with yohimbine was below the curve without the drug and that the tuning became less sharp during the application of yohimbine.

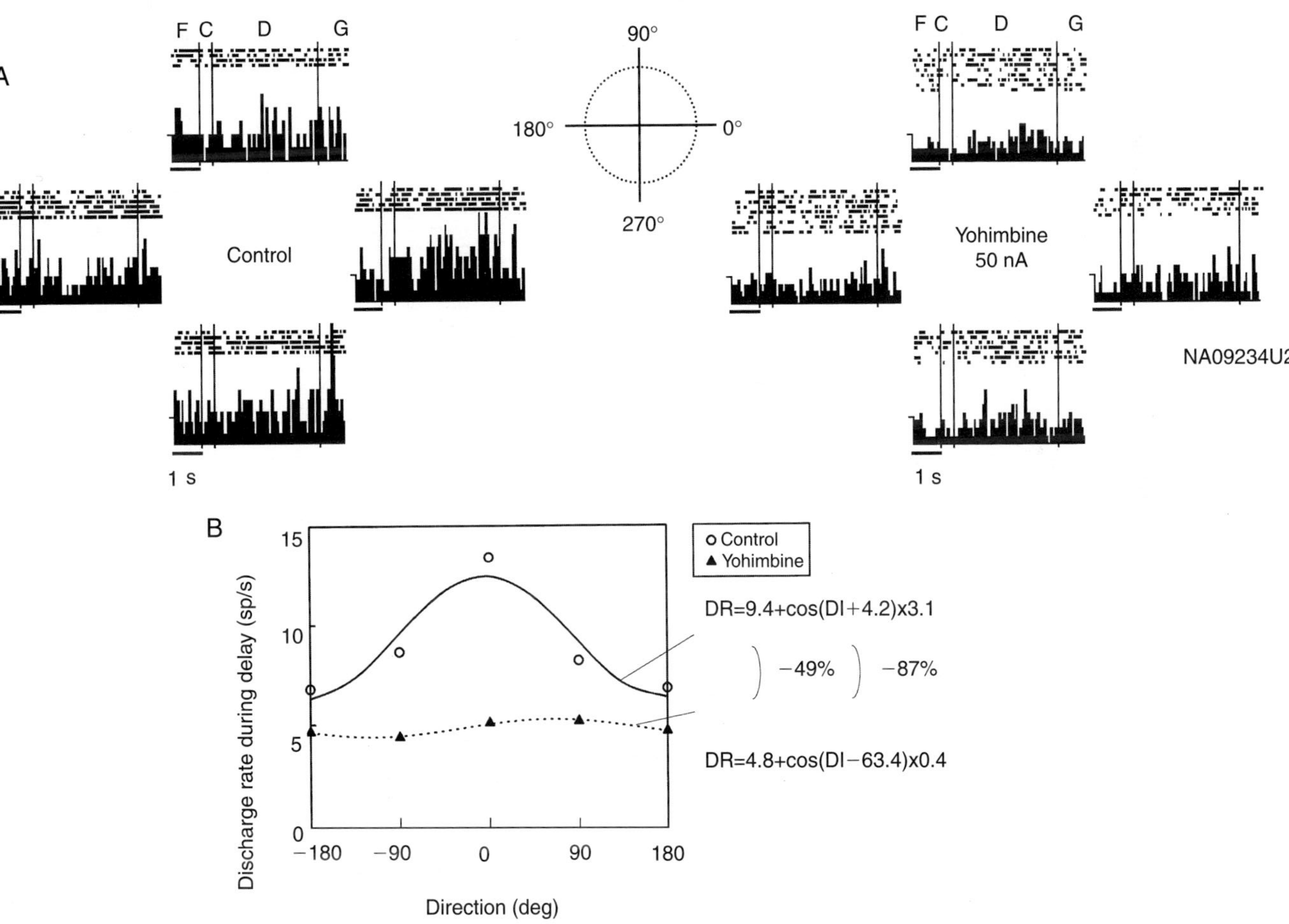
A
F C D G
Control
1 s
90°
180°
0°
270°
F C D G
Yohimbine
50 nA
1 s
NA09234U2S
B
Discharge rate during delay (sp/s)
15
10
5
0
−180
−90
0
90
180
Direction (deg)
Control
Yohimbine
DR=9.4+cos(DI+4.2)x3.1
−49%
−87%
DR=4.8+cos(DI−63.4)x0.4

by the computer (usually, with a sampling rate of 50 ms), and the discharge rate during each period of the task was compared with the background discharge rate during the 1-sec period prior to the cue (i.e., pre-cue period). When the discharge rate of a neuron during the delay period was significantly larger than that during the background activity (Mann-Whitney's U-test, $p < .05$), the neuron was judged to be activated during the delay period. When neuronal activity related to the task was encountered during the penetration of the micropipette, it was recorded for more than 20 successive trials. Each of the drugs was then applied with a current of 30–90 nA (usually 50 nA) for more than 20 successive trials to examine the influence of the drug on task-related neuronal activity. When the discharge rate of a neuron changed significantly during the application of a drug (t-test or Mann-Whitney's U-test, $p < .05$), the neuron was judged to be responsive to the drug. Throughout the experiment, the monkey was treated in accordance with the Guide for Care and Use of Laboratory Animals (National Institutes of Health, U.S.A.) and the Guide for Care and Use of Laboratory Primates (Primate Research Institute, Kyoto University, Japan).

Of 57 neurons recorded during ODR performance, we concentrate here on 31 neurons that showed delay-period activity (i.e., an increase in activity during the delay period), of which magnitude differed significantly with the direction of cue–response, because such directional delay-period activity has been suggested to play a central role in the working memory process for visuospatial information. These neurons were located in the principal sulcus or immediately adjacent cortex. Iontophoretic application of yohimbine with a 50-nA current attenuated the activities of most of these neurons (n = 24/31), and the activity during the delayed period appeared to become diffuse during the application. In contrast, iontophoretic application of propranolol did not have a clear effect on most of the neurons tested (n = 15/17). An example of the effect of yohimbine on such activity is shown in Figure 1A. This neuron showed sustained activation during the delay period, particularly for trials on the right side. The discharge rate during the delay period differed significantly with direction (ANOVA, $p < .001$). Iontophoretic application of yohimbine decreased this activity, and the directional preference of the delay period activity became insignificant ($p > .05$, NS).

To examine better the effect of yohimbine on directional tuning, we calculated a tuning curve by using a cosine function, as follows:

$$DR = b + a \times \cos (DI - c)$$

where DR is the averaged discharge rate during the delay period, DI is a direction, and c is the ideal preferred direction, b is an index of the baseline firing during the delay period, and a is an index of the sharpness of the tuning of the delay-period activity. Thus, this function represents the directional or mnemonic coding of delay-period activity. For the particular activity shown in Figure 1A, the cosine function showed a good fit to the actual data for both pre-drug control activity during the delay period (proportion, $r^2 = .92$) and the activity with yohimbine ($r^2 = .99$) (Fig. 1B). Both a and b decreased during the application of yohimbine, from 3.1 to 0.4 and from 9.4 to 4.8, respectively. It should be noted that the percentage decrease in a (−87%) was much larger than that in *b* (−49%). This indicates that the application of yohimbine attenuated the sharpness of the tuning more strongly than the baseline firing of delay-period activity. Exactly the same results were obtained for all of the 24 neurons that showed the directional delay-period activity.

Thus, iontophoretic application of the selective α_2-adrenergic receptor antagonist, yohimbine, attenuated the directional delay-period activity of prefrontal neurons during ODR performance. At the same time, yohimbine attenuated the sharpness of directional tuning, examined by fitting to a cosine function, of directional delay-period activity. These findings suggest that the activation of α_2 receptors plays a modulatory role in maintaining the directional tuning of delay-period activity of prefrontal cortical neurons during ODR performance. Furthermore, because the directional delay-period activity with tuning is considered to represent mnemonic coding for visuospatial working memory (1), we can conclude that the activation of α_2 receptor plays a modulatory role in maintaining mnemonic coding of neuronal activity for the working-memory process in primate prefrontal cortex.

References

1. Funahashi, S., Bruce, C., and Goldman-Rakic, P. S. (1989). Mnemonic coding of visual space in the monkey's dorsolateral prefrontal cortex. *J. Neurophysiol.* **61,** 331–349.
2. Goldman-Rakic, P. S., Lidow, M. S., and Gallager, D. W. (1990). Overlap of dopaminergic, adrenergic, and serotoninergic receptors and complementarity of their subtypes in primate prefrontal cortex. *J. Neurosci.* **10,** 2125–2138.
3. Levitt, P., Rakic, P., and Goldman-Rakic P. S. (19894). Region-specific distribution of catecholamine afferents in primate cerebral cortex: A fluorescence histochemical analysis. *J. Comp. Neurol.* **225,** 1–14.
4. Arnsten, A. F. T., and Goldman-Rakic, P. S. (1985). Alpha-2-adrenergic mechanisms in prefrontal cortex associated with cognitive decline in aged nonhuman primates. *Science* **230,** 1273–1276.
5. Sawaguchi, T., Matsumura, M., and Kubota, K. (1990). Catecholaminergic effects on neuronal activity related to a delayed response task in monkey prefrontal cortex. *J. Neurophysiol.* **63,** 1385–1400.

A. F. T. Arnsten,* J. C. Steere,* D. J. Jentsch,* and B. M. Li†

*Section on Neurobiology
Yale Medical School
New Haven, Connecticut 06510

†Shanghai Institute of Psychology
Shanghai, China 200031

Noradrenergic Influences on Prefrontal Cortical Cognitive Function: Opposing Actions at Postjunctional α_1 Versus α_2-Adrenergic Receptors

Accumulating evidence indicates that norepinephrine (NE) has an important influence on the spatial working-memory and attentional functions of the prefrontal cortex (PFC). Research in rodents and primates indicates that NE has beneficial effects on PFC function through actions at postsynaptic, α_2 receptors in the PFC but has detrimental actions through α_1-receptor mechanisms.

Several lines of research have shown that NE improves PFC function via postsynaptic, α_{2A}-receptor mechanisms (see ref. 1 for review). α_2-Adrenergic agonists such as clonidine, guanfacine, and UK14304 improve performance of cognitive tasks relying on the PFC (delayed response, delayed alternation, delayed match-to-sample with repeated stimuli, reversal of a visual discrimination problem) but produce little or no improvement performance of tasks that do not depend on the PFC (aquisition of a visual discrimination problem, delayed nonmatch-to-sample with trial unique stimuli, Morris water maze). The beneficial effects of guanfacine on delayed-response performance in aged monkeys with naturally occurring NE depletion can be seen in Figure 1B. Improvement with α_2 agonists is especially prominent under conditions of high interference (e.g., distracting stimuli, proactive interference) when the PFC is most needed. The beneficial effects of α_2 agonists on the delayed-response task have a pharmacological profile consistent with actions at the α_{2A}-receptor subtype. Low doses of agonists such as guanfacine and UK14304, with higher affinity for the α_{2A} subtype, can improve spatial working memory without sedative or hypotensive side effects, while compounds such as clonidine and BHT920, with equal or higher affinity for the α_{2B} or α_{2C} subtypes, only improve performance at higher doses, which also produce marked side effects. Furthermore, the beneficial effects of α_2 agonists on cognitive performance can be reversed by the $\alpha_{2A/B/C}$

Advances in Pharmacology, Volume 42

1054-3589/98 $25.00

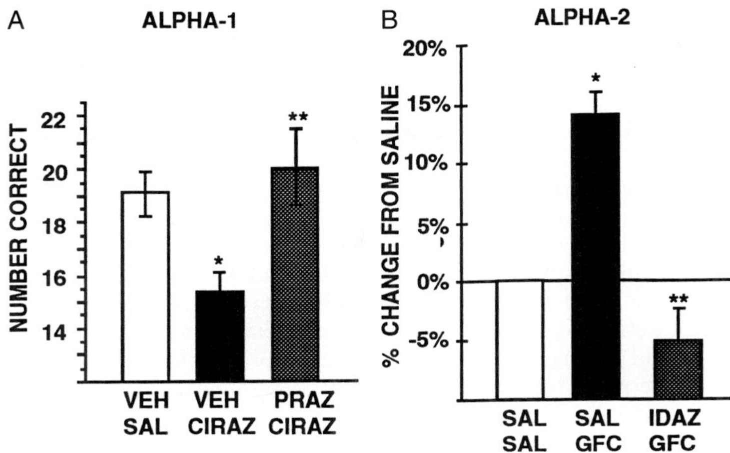

FIGURE 1 A comparison of the effects of α_1-versus α_2-adrenergic agonists on the delayed-response performance of aged monkeys. (A) The α_1-adrenergic agonist, cirazoline, significantly impairs delayed-response performance; the impairment is reversed by pretreatment with the α_1-receptor antagonist, prazosin. Results represent mean ± SEM number correct out of 30 trials; chance performance is 15. VEH, chocolate rice vehicle; SAL, saline; CIRAZ, cirazoline; PRAZ, prazosin; *, significantly different from vehicle + saline; **, significantly different from vehicle + cirazoline. Adapted from Arnsten and Jentsch, *Pharmacol. Biochem. Behav.* 1996 (in press). (B) The α_{2A}-adrenergic agonist, guanfacine, significantly improves delayed-response performance; the improvement is reversed by pretreatment with the α_2-receptor antagonist, idazoxan. Results represent mean ± SEM change from saline control performance. SAL, saline; GFC, guanfacine; IDAZ, idazoxan; *, significantly different from saline + saline; **, significantly different from saline + guanfacine. Adapted from Arnsten, A. F. T. *et al., J. Neurosci.* 1988; 8: 4287–4298.

antagonist, yohimbine, but not the α_1, $\alpha_{2B/C}$ antagonist, prazosin. Recently, the α_{2C}-receptor–preferring antagonist, MK912, has been shown *not* to reverse the beneficial effects of low doses of guanfacine (Arnsten, unpublished results), consistent with an α_{2A} mechanism. α_{2A} Receptors are localized both pre- and postsynaptically in the central nervous system, but evidence suggests that the beneficial effects of α_2 agonists on cognitive function result from actions at postsynaptic receptors. Thus, clonidine is *more* effective in monkeys with NE depletion produced by reserpine, 6-hydroxydopalmine, or MPTP, or in aged monkeys with naturally occurring NE loss. Further evidence indicates that α_2 agonists act at postsynaptic receptors in the PFC. Infusion of α_2 agonists into the PFC of aged monkeys or rats improves spatial working-memory performance. Conversely, infusion of the α_2 antagonist, yohimbine, into the PFC of young monkeys produces a delay-related impairment in working-memory performance. Taken together, these data indicate that NE improves PFC working-memory function at α_{2A} receptors localized postsynaptic to NE terminals in the PFC. α_{2A} Agonists are now being tested as potential cognitive enhancers in disorders with prominent PFC dysfunction, such as attention deficit hyperactivity disorder. Recent results from open trials suggest that α_{2A} agonists can improve PFC function in humans, as they do in animal models.

In contrast to the beneficial effects of α_2 agonists, recent evidence indicates that NE impairs spatial working memory function through α_1-receptor mechanisms. The effects of systemic administration of the α_1-adrenergic agonist, cirazoline, on spatial working-memory performance were examined in aged rhesus monkeys (2). Cirazoline has additional high affinity for imidazoline receptors but has good brain penetrance when administered systemically. As can be seen in Figure 1A, low doses of cirazoline (0.00001–0.001 mg/kg) significantly impaired delayed-response performance. This impairment did not appear to result from nonspecific changes in behavior, because cirazoline had no significant effect on performance of control trials where the delay was zero sec, and it had no significant effect on behavioral ratings. Impairment was reversed by pretreatment with the α_1-adrenergic antagonist, prazosin, consistent with drug actions at α_1-adrenergic receptors.

Further studies have examined the effects of infusing a more selective α_1-adrenergic receptor agonist, phenylephrine, into the PFC of rats or monkeys performing a spatial working-memory task (3). In monkeys, phenylephrine infusion (1–2 μg/1 μl $\times$ five infusions) into the principal sulcal PFC produced a delay-related impairment in delayed-response performance: performance was unaffected at short delays but impaired at higher delays. In a parallel study in rats, infusion of phenylephrine (0.1 μg/0.5 μl, bilateral) into the medial PFC significantly impaired delayed-alternation performance relative to vehicle infusion. The impairment induced by phenylephrine was blocked by coinfusion of the α_1-receptor antagonist, uripidil (0.1 μg phenylephrine + 0.01 μg uripidil/0.5 μl, bilateral), consistent with an α_1-receptor mechanism. Preliminary results indicate that the impairment induced by phenylephrine can also be reversed by pretreatment with lithium, using a dose regimen known to interfere with phosphoinositide (PI) turnover. These results suggests that PFC cognitive impairment induced by α_1-receptor stimulation results from excessive PI turnover, which can be countered by lithium pretreatment. These results may have clinical relevance to mania, a disorder commonly treated with lithium medication.

In summary, these data suggest that postsynaptic α_1 and α_2 receptors may have opposing roles in the PFC, as they do in the thalamus regulating arousal (4) and in the hypothalamus regulating ingestive behavior (5). Preliminary data comparing the binding of [^{3}H]NE to recombinant human α-adrenoceptors suggests that NE has higher affinity for the α_{2A} subtype than for either the α_{1A} or α_{1D} subtypes (J. P. Hieble, personal communication). Thus, α_2 mechanisms may predominate when basal NE release is moderate (e.g., normal, attentive waking) and PFC function is optimal, while α_1 mechanisms may predominate under conditons of higher levels of NE release (e.g., during stress), contributing to PFC cognitive impairment.

Acknowledgment

This work was supported by U.S. Public Health Service grant AG06036 to A. F. T. Arnsten.

References

1. Arnsten, A. F. T. Steere, J. C., and Hunt, R. D. (1996). The contribution of alpha-2 noradrenergic mechanisms to prefrontal cortical cognitive function: Potential significance to attention deficit hyperactivity disorder. *Arch. Gen. Psychiatry* 53, 448–455.
2. Arnsten, A. F. T. and Jentsch, J. D. (1997). The alpha-1 adrenergic agonist, cirazoline, impairs spatial working memory performance in aged monkeys. *Pharmacol. Biochem. Behav.* 57, 1–5.
3. Steere, J. C., Li, B-M., Jentsch, J. D., Mathew, R., Taylor, J. R. and Arnsten, A. F. T. (1996). Alpha-1 noradrenergic stimulation impairs, while alpha-2 stimulation improves, prefrontal cortex spatial working memory function. *Soc. Neurosci. Abstr.* 22, 1126.
4. Buzsaki, G., Kennedy, B., Solt, V. B., and Ziegler, M. (1991). Noradrenergic control of thalamic oscillation: The role of alpha-2 receptors. *Eur. J. Neurosci.* 3, 222–229.
5. Davies, B. T., and Wellman, P. J. (1992). Effects on ingestive behavior in rats of the alpha-1-adrenoceptor agonist cirazoline. *Eur. J. Pharmacol.* 210, 11–16.

M. Ennis,* M. T. Shipley,* G. Aston-Jones,† and J. T. Williams‡

*Department of Anatomy and Neurobiology
University of Maryland School of Medicine
Baltimore, Maryland 21201

†Department of Mental Health
Hahnemann University
Philadelphia, Pennsylvania 19102

‡Department of Biomedical Research
Oregon Health Sciences University
Vollum Institute
Portland, Oregan 97201

Afferent Control of Nucleus Locus Ceruleus: Differential Regulation by "Shell" and "Core" Inputs

Previous studies in our laboratories demonstrate that nucleus locus ceruleus (LC) receives restricted afferent input (1). The two major afferent inputs to LC arise from nucleus paragigantocellularis (PGi) and nucleus prepositus hypo-

Advances in Pharmacology, Volume 42

1054-3589/98 $25.00

glossi in the rostral medulla. A long-standing question about the afferent regulation of LC is the degree to which LC dendrites extend outside of nucleus proper, and whether such extranuclear dendrites are targeted by synaptic inputs different from those to LC proper. This is important because several brain areas project specifically to the pericerulear (pLC) region immediately adjacent to the LC nuclear core; such projections, therefore, are candidate afferent inputs to the extranuclear dendrites of LC neurons.

In the present studies, light and electron microscopic techniques were used to investigate the distribution and ultrastructure of LC processes immunocytochemically identified with antibodies for tyrosine hydroxylase or dopamine β-hydroxylase. Tract-tracing and electrophysiological methods were used to characterize projections from several brain areas to the pLC region.

I. Distribution of LC Extranuclear Processes

Light microscopic analysis (2) revealed that the vast majority of the extranuclear processes of LC neurons extended preferentially into two zones:

1. The region rostral and medial to LC proper, designated as the rostromedial pLC (pLCrm) region. Within this region, LC processes were focally aggregated immediately ventral to the ventrolateral margin of the fourth ventricle and dorsal to Barrington's nucleus. Some processes were present in Barrington's nucleus and the laterodorsal tegmental nucleus; however, the majority of labeled processes do not extend into the dorsal tegmental nucleus.
2. A narrow region adjacent to the fourth ventricle caudomedial to LC proper, designated as the caudal juxtaependymal pLC (pLCcje) region. Far fewer labeled processes extended into the lateral and ventral pericerulear regions.

II. Electron Microscopic Analysis of LC Extranuclear Processes

Electron microscopic analysis of the ultrastructure of labeled processes was conducted to determine the percentages of axons and dendrites in different pLC regions. The most striking findings were that all 188 labeled profiles in pLCrm were dendrites. In pLCcje, 84 of 89 (94%) of the labeled profiles examined were dendrites. By contrast, only 21% of those in the lateral parabrachial nucleus were dendrites. LC extranuclear dendrites were heavily targeted by synapses from unlabeled presynaptic axonal terminals. The majority of these were asymmetric synapses, although symmetric synapses also were observed.

It is conceivable that only neurons situated along the edge of LC proper extend dendrites into the pLC region. In this case, afferent inputs of pLC would influence only the activity of a subset of peripherally located LC neurons. To address this question, we reconstructed the dendritic arbors of individual LC neurons impaled and filled intracellularly with biocytin *in vitro*. This experiment demonstrated that the dendrites of individual neurons extended exclusively into the rostral, medial, and/or caudomedial pLC region. This pattern was exhibited by all LC neurons independent of their location within LC proper. Taken together with the aforementioned findings, these results suggest that extranuclear LC dendrites, particularly in the rostromedial and caudal juxtaependymal zones, comprise an extensive postsynaptic receptive surface of LC neurons.

III. Anatomical and Physiological Studies of Afferent Inputs to the PLC Region

Tract-tracing studies in our laboratories and others (1, 3, 4) demonstrated that the midbrain periaqueductal gray (PAG), medial preoptic area (MPO), medial prefrontal cortex, central nucleus of the amygdala (CNA), and nucleus tractus solitarius (NTS) send afferents to pLC but provide weak or no innervation of the LC nuclear core. In the experiments described next, we mapped the terminal distribution of projections from PAG and MPO to identified LC dendrites in pLC. Electrophysiological and Fos staining techniques were used to assess the functional influence of these inputs on LC neurons.

A. Periagueductal Gray

Injections of wheatgerm agglutinin-horseradish peroxidase encompassing pLCrm retrogradely labeled neurons in caudal PAG located lateral and ventrolateral to the cerebral aqueduct; injections restricted to LC proper did not consistently label PAG neurons. Deposits of the anterograde axonal tracer *Phaseolus vulgaris* leucoagglutinin into this same region of PAG-labeled axons indicate that they robustly innervated Barrington's nucleus and the pLCrm zone containing LC extranuclear dendrites (3). Only sparse fibers were observed in LC proper.

Stimulation of PAG produced weak activation of 69% of LC neurons. Typically, driven spikes from LC neurons were elicited on only 33% of the stimulus trials. Microinjection of small doses of glutamate (0.012–0.024 μmol in 60–240 nl) into the PAG produced modest increases in the spontaneous discharge of 12 LC neurons (147.4 $\pm$ 16.4% of control discharge rate).

B. Medial Preoptic Area

Retrograde labeling produced by tracer injection into pLCrm revealed that MPO neurons projecting to the dorsolateral pontine tegmentum are preferentially distributed in distinct subregions of MPO, including the sexually dimorphic medial preoptic nucleus (4). Anterograde tracing demonstrated considerable target specificity in projections from MPO to the dorsolateral pontine tegmentum. Barrington's nucleus, adjacent to pLCrm, receives a dense focal input along its entire rostrocaudal axis. In addition, pLCrm is heavily targeted by MPO inputs. Double-labeling studies demonstrated that there is a remarkable degree of spatial overlap between MPO terminals and LC dendrites in the pLCrm region; MPO terminals closely apposed labeled dendrites in pLCrm. By contrast, the laterodorsal tegmental nucleus and LC proper receive only sparse input from MPO.

Focal activation of MPO in anesthetized rats and Fos immunocytochemistry were used to map the locations of neurons activated by MPO inputs to pLCrm. Intermittent activation of MPO via electrical or chemical (microinjection of a 50-μM bicuculline methiodide, 10-mM d,1-homocysteate solution) stimulation of MPO induced moderate Fos expression in pLCrm and LC proper neurons. By contrast, Barrington's nucleus, which receives a dense input from MPO, was devoid of Fos-positive neurons. These results suggest that MPO projections exert an excitatory influence on local, unidentified neurons in pLCrm and on LC neurons; the influence of MPO on Barrington's neurons may be inhibitory or modulatory.

IV. Transmitter Inputs to Pericerulear LC Dendrites

Several inputs to the pLC region studied to date exert an excitatory influence on LC discharge, suggesting that these projections may be glutamatergic. Consistent with this hypothesis, focal application of glutamate onto LC extranuclear dendrites in pLCrm potently activates LC neurons *in vitro* (5); glutamate-evoked responses persisted in conditions that block synaptic transmission, indicating that these responses were not presynaptically mediated. Although glutamate has not been identified in axon terminals that synapse with LC extranuclear dendrites, N-methyl-D-aspartate receptors have been localized to identified LC dendrites in pLCrm. Other transmitter–receptors present at synapses onto identified LC extranuclear dendrites include corticotropin-releasing factor, met-enkephalin, and the μ-opiate receptor (6–9).

Taken together, these results demonstrate that the dendrites of LC neurons are polarized to extend preferentially into two pericerulear zones located rostromedial and caudomedial to LC proper. This dendritic shell pattern is characteristic of all or most LC neurons. LC extranuclear dendrites receive synaptic contacts and comprise a significant postsynaptic receptive surface of LC neurons. These shell dendrites are selectively targeted by a number of extrinsic afferent inputs that do not project appreciably into the LC core. By contrast, the nuclear core of LC receives two major afferent inputs; these inputs do not appreciably target the dendritic shell. Evidence to date suggests that several inputs to the shell region (i.e., PAG, MPO, medial prefrontal cortex) exert an excitatory influence of LC discharge. In conclusion, inputs to the core LC nucleus versus extranuclear dendrites may comprise functionally distinct channels for regulating LC neurons. How LC neurons integrate multiple afferent inputs impinging on their core and shell receptive surfaces is currently under investigation.

References

1. Aston-Jones, G., Shipley, M. T., and Grzanna, R. (1995). The locus coeruleus, A5 and A7 noradrenergic cell groups. *In* The Rat Nervous System, 2nd ed. (George Paxinos, ed.) pp. 183–213. Academic Press, San Diego.

2. Shipley, M. T., Fu L., Ennis, M.,. Liu, W-L., and Aston-Jones, G. (1996). Dendrites of locus coeruleus neurons extend preferentially into two pericoerulear zones, *J. Comp. Neurol.* **365**, 56–68.
3. Ennis, M., Behbehani, M. M., Van Bockstaele, E., Shipley, M. T., and Aston-Jones, G. (1991). Projections from the periaqueductal gray to the pericoerulear region and nucleus locus coeruleus. *J. Comp. Neurol.* **306**, 480–494.
4. Rizvi, T. A., Ennis, M., Aston-Jones, G., Jiang, M., Liu, W-L., Behbehani, M. M., and Shipley, M. T. (1994). Preoptic projections to Barrington's nucleus and the pericoerulear region: Architecture and terminal organization. *J. Comp. Neurol.* **347**, 1–24.
5. Aston-Jones, G., and Ivanov, A. Y. (1995). Extranuclear dendrites of locus coeruleus neurons: Activation by glutamate and modulation of activity by alpha receptors. *J. Neurophysiol.* **74**, 2427–2436.
6. Van Bockstaele, E. J., Branchereau, P., and Pickel, V. M. (1995). Morphologically heterogeneous met-enkephalin terminals form synapses with tyrosine-hydroxylase-containing dendrites in the rat locus coeruleus. *J. Comp. Neurol.* **363**, 423–438.
7. Van Bockstaele, E. J., and Colago, E. E. O. (1996). Selective distribution of the NMDA-R1 glutamate receptor in astrocytes and presynaptic axon terminals in the nucleus locus coeruleus of the rat brain: An immunoelectron microscopic study. *J. Comp. Neurol.* **369**, 483–496.
8. Van Bockstaele, E. J., Colago, E. E. O., Cheng, P., Moriwaki, A., Uhl, G. R., and Pickel, V. M. (1996). Ultrastructural evidence for prominent distribution of the m-opioid receptor at extrasynaptic sites on noradrenergic dendrites in the rat nucleus locus coeruleus. *J. Neurosci.* **16**, 5037–5048.
9. Van Bockstaele, E. J., Colago, E. E. O., and Valentino, R. J. (1996). Corticotropin-releasing factor-containing terminals synapse onto catecholamine dendrites and may presynaptically modulate other afferents in the rostral pole of the nucleus locus coeruleus of the rat brain. *J. Comp. Neurol.* **364**, 523–534.

S. Nakamura,* S. Nishiike,† Y. Fujii,* N. Takeda,† and T. Kubo†

* Department of Physiology
Yamaguchi University School of Medicine
Ube, Yamaguchi 755, Japan

† Department of Otolaryngology
Osaka University Medical School
Suita, Osaka 565, Japan

Sensory Response of the Locus Ceruleus: Neonatal and Adult Studies

The major afferent input to the locus ceruleus (LC) is known to come from peripheral sensory nerves, most of which exert an excitatory influence on LC neurons (1). In adult, anesthetized animals, LC neurons reveal a vigorous excitatory response to noxious sensory stimuli such as tail pinches, while non-noxious stimuli such as air puffs to the skin have little effect.

In this review, we describe two topics from our recent experiments concerning sensory response of LC neurons: 1) sensory response of LC neurons in fetal and neonatal rats (2) and 2) a long-lasting, GABAergic inhibition of LC neurons following caloric vestibular stimulation in adult rats (3, 4).

Male and female Sprague-Dawley rats, embryonic day (E) 18-E22 and postnatal day (PD)1-adults, were used. The animals were anesthetized with urethane (1.3 g/kg, i.p.), supplemented as necessary during the experiments. The adult rats were intubated with a tracheal cannula and then fixed to a conventional stereotaxic apparatus. The methods of fixing the head of fetal and neonatal rats were previously described (2). The body temperature was maintained at 37 ± 1°C with a heating pad. In the caloric-stimulation experiments, one polyethylene catheter was inserted into a femoral vein for drug injections, and another was inserted into a femoral artery for the continuous recording of mean blood pressure (MBP).

For the electrophysiological identification of the location of the LC, bipolar stimulating electrodes were implanted into the frontal cortex and dorsal noradrenergic bundle (DNB) ipsilateral to the LC recording, as described previously (2). The electrical activity of LC neurons was recorded extracellularly with glass micropipettes filled with 2 M of NaCl or 0.5 M of sodium acetate containing 2% pontamine sky blue. The location of the LC was determined by the appearance of a field response of the LC evoked by DNB stimulation. The single-unit activity of LC neurons was recorded superimposed on the field response when the recording electrode was adequately advanced. For microiontophoretic appli-

Advances in Pharmacology, Volume 42

1054-3589/98 $25.00

cation of drugs, multibarrel electrodes were used: one for the recording of LC units and the remaining electrodes for drug application.

For caloric vestibular stimulation, the bulla tympani of the left ear was opened by a retroauricular surgical approach. A polyethylene tube (outer diameter, 1 mm) was inserted into the middle ear cavity. Through this tube, the middle ear was irrigated with 5 ml of hot water (44°C) or cold water (30°C) at a rate of 0.1 ml/sec.

I. Sensory Response of the LC in Fetal and Neonatal Rats

The spontaneous activity of LC neurons was examined in fetal and neonatal rats (2). In fetal rats, most of the fetal LC neurons had no spontaneous activity. In neonatal rats (PD1–7), approximately 50% of LC neurons recorded had no spontaneous activity, while LC neurons showing spontaneous activity revealed firing rates of less than 1 spike/sec. As age increased, the number of silent cells decreased, and finally the firing rate of LC neurons reached the level of adults by PD 20.

Despite less spontaneous activity of fetal and neonatal LC neurons, sensory stimuli were very effective in producing the excitation of LC neurons in these immature animals (2). Even in fetal LC neurons, nonnoxious sensory stimuli such as air puffs to the skin caused a vigorous excitation. Although the sensory response of adult LC neurons in anesthetized animals is characterized by a transient excitation followed by an inhibition, most LC neurons in fetal and neonatal rats revealed a long-lasting (tonic) excitation in response to even one nonnoxious sensory stimulus. The tonic sensory-evoked excitation was more frequently observed in silent cells than in spontaneously active cells. Thus, the tonic sensory activation of LC neuronal firing at early developmental stages appears to be related to less spontaneous activity.

It has been shown that negative- and positive-feedback mechanisms operate to regulate the electrical activity of LC neurons (5). LC neurons are known to possess autoreceptors, α_1 and α_2 adrenoceptors, on the somatodendritic membrane of the neurons. These autoreceptors are activated by noradrenaline (NA) released from the terminals of recurrent axon collaterals and/or from presynaptic dendrites of LC neurons (6). Activation of α_1 adrenoceptors causes excitation of LC neurons (positive feedback), while that of α_2 adrenoceptors results in the suppression of the neurons (negative feedback). Of particular interest is the fact that the negative-feedback mechanism of LC neurons is not functional until about PD 9, while the positive-feedback mechanism is manifest predominantly in early developmental stages and becomes less apparent with age (2, 7). Furthermore, the occurrence of the negative- and positive-feedback mechanisms of LC neurons is dependent on the spontaneous activity of the neurons themselves or the amount of NA released from the terminals of axon collaterals of the neurons: When LC neurons are firing at a low rate, a small amount of NA is released from the terminal of axon collaterals. NA at a low concentration is only effective in activating α_1 adrenoceptors, resulting in an increase of firing (positive feedback). When LC neurons are firing at a relatively high rate, a large amount of NA is released from axon collaterals. NA at a high concentration produces the suppression of the neurons by activating α_2 adrenoceptors (negative-feedback mechanism) (2, 5).

Based on these findings, the tonic activation of fetal and neonatal LC neurons by sensory stimuli may be explained by the feedback mechanisms that are dependent on the spontaneous activity of the neurons. In early developmental stages, when LC neurons with less spontaneous activity are activated by a sensory stimulus, the positive mechanism operates to further increase the firing of the neurons, resulting in the occurrence of a tonic activation. As age increases, LC neurons become firing at a relatively high rate, and the negative-feedback mechanism becomes predominant. Therefore the sensory-evoked response of LC neurons in later developmental stages reveals the pattern of a transient excitation followed by a suppression.

II. Long-Lasting, GABAergic Inhibition of LC Neurons Following Caloric Vestibular Stimulation

The sensory-evoked responses of LC neurons are mainly excitatory, despite the difference in response pattern, tonic or transient (1). Recently, however, we have found an atypical sensory response of LC neurons in adult rats, LC neurons revealed a long-lasting inhibition of neuronal firing following caloric vestibular stimulation or electrical stimulation of the medial vestibular nucleus (3, 8).

First, responses of the neurons of the vestibular nuclear complex (VNC) to caloric vestibular stimulation were examined (8). About half of VNC neurons responding to caloric stimulation with both hot and ice (4°C) water (16/30) revealed an excitatory response to hot-water irrigation of the middle ear and an inhibitory response to cold-water irrigation. The opposite responses to both the caloric stimuli were observed in six of the 30 neurons. The remaining neurons showed either excitatory ($n = 7$) or inhibitory ($n = 1$) responses to both the caloric stimuli. The change in the firing of the VNC neurons occurred immediately after the start of caloric stimulation and returned to the prestimulus level immediately after the end of the stimulation.

Most LC neurons exhibited suppression of neuronal firing in response to both hot- and cold-water irrigation of the middle ear (3, 4). The LC inhibition occurred with a long delay (longer than 1 min) and lasted long (~3 min) after caloric stimulation. In many neurons, the suppression of LC neurons by caloric stimulation was preceded by a transient increase in firing. The transient excitation was not thought to be specific to vestibular afferent, because the transient excitation but not the long-lasting suppression could still be induced by caloric stimulation after the destruction of the middle ear. Furthermore, electrical stimulation of the medial vestibular nucleus produced a long-lasting suppression of LC neuronal firing similar to that caused by caloric stimulation. The onset-latency (mean, 24 sec) and the duration (mean, 68 sec) of the inhibition by electrical stimulation of the vestibular nucleus were shorter than those of the caloric stimulation–induced inhibition. It is noted that a transient excitation observed following caloric stimulation was not induced by electrical vestibular stimulation.

The long-lasting inhibition of LC neuronal firing following caloric vestibular stimulation was antagonized by the GABA antagonists picrotoxin (intravenous injection) and bicucullin (iontophoretic application) (3). The spontaneous firing of LC neurons was significantly increased by application of the GABA antagonists. Therefore, this finding indicates that the long-lasting suppression of LC neuronal firing following caloric vestibular stimulation is mediated by GABA.

The next question to be answered is what pathway is involved in the caloric stimulation–induced inhibition of the LC. There are two possible pathways responsible for the inhibition: (1) Vestibular inputs processed by some brain site(s) causes the LC neuronal inhibition and (2) changes of blood pressure by vestibular stimulation secondarily cause an inhibition of LC neuronal firing. Regarding the first possibility, one of the possible brain sites for the LC neuronal inhibition is the forebrain, which is reported to exert an inhibitory influence on LC neurons (9). In addition, Aston-Jones *et al.* have identified two major inputs to the LC: one from the nucleus prepositus hypoglossi (PrH) and the other from the nucleus paragigantocellularis in the rostral ventrolateral medulla (rostral VLM) (10). Thus these two nuclei are also the possible candidates for the LC inhibition. On the other hand, it has been reported that vestibular stimulation affects the sympathetic nervous system to increase blood pressure. Because it is known that an increase of blood pressure produces an inhibition of LC neuronal firing (11), it is likely that the increased blood pressure following caloric vestibular stimulation causes the LC neuronal inhibition.

We performed decerebration at the rostral pole of the superior colliculus for deafferentation of the forebrain and made electrical lesions of the PrH and PGi (4). Neither the deafferentation of the forebrain nor the PrH lesions altered the LC inhibition and the pressor effect following caloric stimulation. However, the lesions of the rostral VLM attenuated both the caloric stimulation–induced LC inhibition and the pressor effect. Chemical lesions of the VLM were made with microinjection of kainic acid into the VLM ipsilateral to caloric stimulation and LC recording. The unilateral lesion of the VLM made so could also attenuate the LC inhibition and pressor response to caloric stimulation. Neither the LC inhibition nor the pressor response was attenuated by deafferentation of baroreceptors (denervation of the vagal nerves, carotid sinus nerves, and aortic depressor nerves) and bilateral aspiration lesions of the nucleus tractus solitarius.

These results suggest that the caloric stimulation–induced LC inhibition is mediated by the VLM and may be independent of the VLM-mediated pressor effect (4). Because the VLM is the source of the major excitatory input to the LC (10), it is possible that VLM neurons receiving vestibular input exert excitatory influence on GABAergic neurons connecting to the LC.

References

1. Foote, S. L., Bloom, F. E., and Aston-Jones, G. (1983). Nucleus locus coeruleus: New evidence of anatomical and physiological specificity. *Physiol. Rev.* **63**, 844–914.

2. Nakamura, S., and Sakaguchi, T. (1990). Development and plasticity of the locus coeruleus: A review of recent physiological and pharmacological experimentation. *Prog. Neurobiol.* **34**, 505–526.
3. Nishiike, S., Nakamura, S., Arakawa, S., Takeda, N., and Kubo, T. (1996). GABAergic inhibitory response of locus coeruleus neurons to caloric vestibular stimulation in rats. *Brain Res.* **712**, 84–94.
4. Nishiike, S., Takeda, N., Kubo, T., and Nakamura, S. (1997). Neurons in rostral ventrolateral medulla mediate vestibular inhibition of locus coeruleus in rats. *Neuroscience* **77**, 219–232.
5. Nakamura, S., Sakaguchi, T., Kimura, F., and Aoki, F. (1988). The role of alpha adrenoceptor-mediated collateral excitation in the regulation of the electrical activity of locus coeruleus neurons. *Neuroscience* **27**, 921–929.
6. Aghajanian, G. K., Cedarbaum, J. M., and Wang, R. Y. (1977). Evidence for norepinephrine-mediated collateral inhibition of locus coeruleus neurons. *Brain Res.* **136**, 570–577.
7. Kimura, F., and Nakamura, S. (1987). Postnatal development of alpha-adrenoceptor-mediated autoinhibition in the locus coeruleus. *Dev. Brain Res.* **432**, 21–26.
8. Nishiike, S., Takeda, N., Nakamura, S., Arakawa, S., and Kubo, T. (1995). Responses of locus coeruleus neurons to caloric stimulation in rats. *Acta Otolaryngol.* (*Stockh.*) **520**, 105–109.
9. Sara, S. J., and Herve-Minvielle, A. (1995). Inhibitory influence of frontal cortex on locus coeruleus neurons. *Proc. Natl. Acad. Sci. U.S.A.* **92**, 6032–6036.
10. Aston-Jones, G., Shipley, M. T., Chouvet, G., Ennis, M., Van Bockstaele, E. Pieribone, V., Shiekahattar, R., Akaoka, H., Drolet, G., Astier, B., Charlety, P., Valentino, R. J., and Williams, J. T. (1991). Afferent regulation of locus coeruleus neurons: Anatomy, physiology and pharmacology. *Prog. Brain Res.* **88**, 47–75.
11. Svensson, T. H., and Thorén, P. (1979). Brain noradrenergic neurons in the locus coeruleus: Inhibition by blood volume load through vagal afferents. *Brain Res.* **172**, 174–178.

Chiye Aoki,* Charu Venkatesan,† and Hitoshi Kurose‡

*Center for Neural Science
New York University
New York, New York 10003

†Department of Neurobiology and Behavior
State University of New York
Stony Brook, New York 11794

‡University of Tokyo
‡Department of Toxicology and Pharmacology
Faculty of Pharmaceutical Science
Tokyo, 113, Japan

Noradrenergic Modulation of the Prefrontal Cortex as Revealed by Electron Microscopic Immunocytochemistry

The dorsolateral prefrontal cortex (PFC, area 46) is involved in the formation of spatial working memory. The locus ceruleus–noradrenergic system projects widely within brain, including the PFC, and has been shown to play an important role in modulating cognitive function requiring the PFC via activation of α_2-adrenergic receptors (α_2ARs) (see the review by Arnsten in this section). Other studies indicate that activation of β-adrenergic receptors (βARs) is necessary for certain forms of experience-dependent synaptic plasticity in the neonatal and the adult cerebral cortex (1). The goal of the studies presented here was to analyze the cellular and subcellular structures involved in noradrenergic modulation of PFC circuitry. To this end, electron microscopic immunocytochemical techniques were used to determine the laminar and cellular targets of noradrenergic fibers in the adult monkey PFC. In addition, antisera selective for the α_2- and βAR were used to determine the pharmacological profile of receptive sites.

Noradrenergic axons in area 46 of adult monkey PFC were identified immunocytochemically using a rabbit antiserum directed against dopamine β-hydroxylase (DBH) (Eugene Tech) and the avidin–biotin peroxidase complex (ABC) detection method. Ultrastructural analysis indicated that noradrenergic axons in this area form symmetric, asymmetric, and *en passant* synaptic junctions. However, only 9% of the axons encountered within single planes of section (34/374) form such morphologically identifiable synapses. Laminar analysis indicates that these morphologically identifiable synapases are nearly absent in layer 6 (0/46 terminals encountered), and of nearly equivalent

Advances in Pharmacology, Volume 42

1054-3589/98 $25.00

frequency (5–12%) in layers 1 through 5. It is often argued that, although axons may not appear to form synaptic junctions within the single planes of section sampled for ultrastructural analysis, synapses may occur in other planes of section along those axons. Indeed, we calculate that these synaptic junctions occur, on average, once every 1 μm along the length of single noradrenergic axons as they course within the PFC. This calculation is based on the assumption that the orientation of noradrenergic fibers is nearly anisotropic, so that sampling of a large area but within a single plane of section would be roughly equivalent to sampling a smaller area extensively through its z-axis (i.e., serially reconstructed neuropil). Thus, an encounter with morphologically identifiable synaptic junctions at a rate of 8% would be equivalent to an encounter with cross-sectional profiles of junctions once every 12 sections spanning a micron along the length of axons.

The segments of axons lacking morphologically identifiable synaptic junctions often are packed with small clear vesicles, suggesting that these portions also may participate in noradrenergic transmission. To investigate this possibility, we proceeded to analyze whether noradrenergic receptors might occur at apparently nonjunctional axonal and dendritic sites within the neuropil by using highly selective receptor antisera. Recent advances in molecular pharmacology have elucidated that the human genome expresses three subtypes of α_2AR, designated the A, the B, and the C subtypes. Antisera to each of these three subtypes of α_2AR were generated by Kurose *et al.* (2) and shown to exhibit minimal cross-reactivity across the subtypes.

Immunocytochemical results using the antisera directed against the three subtypes of α_2AR indicate that, within the monkey PFC, the A subtype is overwhelmingly more prevalent than the B or the C subtypes. Light microscopy shows that only the A subtype is present in perikarya and proximal dendrites of neurons. These α_2AR-immunoreactive neurons are distributed throughout the layers. Electron microscopy confirms that immunoreactivity for the B or C subtypes of α_2AR is scarce, although not entirely absent, from neuropil or perikarya of monkey PFC. In contrast, the A subtype occurs frequently along the plasma membrane of proximal dendrites, dendritic spines, within terminals, in astrocytic processes, and in the perikaryal cytoplasm associated with the endoplasmic reticulum (3). The antigen used to produce the antisera corresponds to the third intracellular loop portion of the receptor molecule. Accordingly, immunoreactivity is present along the intracellular surface of plasma membranes. Quantitative analysis of the 7985 immunoreactive profiles encountered from 82,700 μm^2 of two monkey PFCs indicates that, in all layers, immunoreactivity is more prevalent in axons than in dendrities. However, the proportion of immunoreactive axonal profiles forming morphologically identifiable synaptic junctions within the planes of sections is only 14%, as opposed to the corresponding value of 53.5% for the immunoreactive dendritic profiles. Thus, in all layers, the areal density of immunoreactivity *at morphologically identifiable synapses* is more prevalent postsynaptically than presynaptically (ca. 2.5-fold).

It is widely recognized that α_2ARs, operating as autoreceptors on noradrenergic axon terminals, cause reduction in noradrenaline release following activation. Thus, one might expect the laminar distribution of α_2AR-immunoreactive synaptic junctions to parallel that of the DBH-immunoreactive synaptic junctions. However, we noted that layers 1 and 6, which exhibit relatively low areal densities of DBH-immunoreactive synaptic junctions, are abundant in α_2AR-immunoreactive synapses. This difference suggests that α_2AR may also occur at non-noradrenergic synapses. Dual labeling of monkey PFC with α_2AR and GABA indicates that these receptors do, indeed, occur in GABAergic axon terminals, albeit infrequently (<10%).

The finding that the A subtype is more prevalent than the B or C subtypes within the PFC reflects a significant gain in our knowledge of the pharmacological profile of this cortical region. This is because previous studies using radioligand binding proved to be difficult, due to the unavailability of sufficiently selective radioligands that distinguish the binding to the three subtypes (4). Behavioral studies indicate that the rank order for improvement of PFC function corresponds to that for the drugs' relative selectivity to the A subtype over the B and C subtypes (5). Although this behavioral study involved systemic administration of α_2AR agonists, this finding, together with results from the present study, is consistent with the idea that α_2AR agonists act directly in the PFC to enhance cognitive functions. Moreover, prevalence of the A subtype of α_2AR-immunoreactivity within axons indicates that axons are major targets of noradrenaline released into the PFC neuropil. Finally, a measurable portion of immunoreactive sites consists of postsynaptic spines. While there is evidence that the firing rate of locus ceruleus neurons is strongly inhibited by activation of α_2AR, little has been reported about the physiological actions of α_2AR on cortical neurons. (However, see Sawaguchi and Kikuchi, 1997, in this volume.)

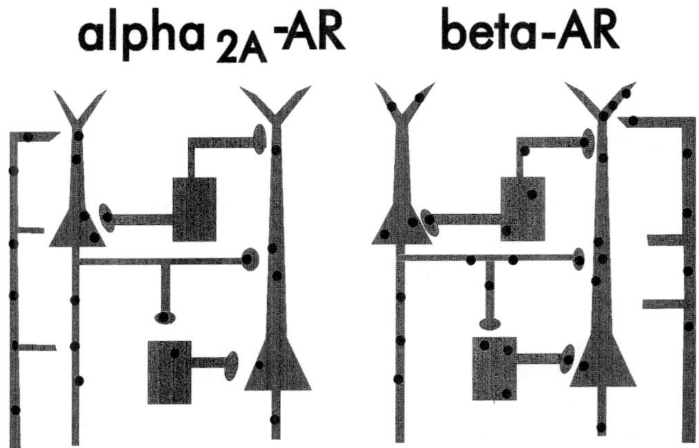

FIGURE 1 The subcellular distribution of α_2AR and βAR immunoreactivity (*dots*) within the microcircuitry of the PFC. The shown canonical microcircuitry is adpated from that for the sensory cortices. Triangles depict cell bodies of pyramidal neurons; rectangles depict cell bodies of GABAergic inhibitory interneurons; tapering thick lines depict apical dendritic processes; thick lines of constant width depict axonal profiles; and ovals depict terminal boutons. The multipronged elements to the extreme right and left depict afferents to the PFC.

One possibility is that α_2AR activates a biochemical cascade that has only subtle effects on the biophysical properties of neurons.

The distribution of βAR in the PFC has been analyzed using an antiserum generated against the β_2 subtype (hamster lung type). A quantitative ultrastructural analysis revealed a surprisingly similar pattern for β- and the A subtype of α_2AR. Thus, in all layers, immunoreactivity is most prevalent in axons, while *synaptic* labeling is more prevalent in dendrites than in axons. In particular, over 90% of all immunoreactive spines exhibit labeling directly over postsynaptic densities, while this value drops to 26% and 23% for axons and dendritic shafts, respectively. The prevalence of labeling in spines indicates that spiny neurons, such as the pyramidal neurons, are major targets of noradrenergic fibers. To determine whether the aspiny GABAergic interneurons also are targets for noradrenergic fibers, monkey PFC tissue was immunolabeled dually for GABA and βAR. Results indicate that, indeed, the βAR-immunoreactive axon terminals and dendritic shafts include those belonging to GABAergic interneurons, including the population that exhibits some spines.

The occurrence of neuronal profiles immunolabeled dually for βAR and GABA is much more frequent than that noted for α_2AR, thus suggesting that the noradrenergic system recruits the GABAergic interneurons into the PFC circuitary mainly via βAR (Fig. 1). βAR activation has an effect opposite of α_2AR activation, namely to stimulate adenylyl cyclase via its linkage to the G_s protein, which, in turn, promotes activation of the cyclic adenosine monophosphate–dependent protein kinase (PKA). The consequences of βAR activation can be numerous, due to the variety of proteins containing sites phosphorylatable by PKA. These include changes in the ion permeability of N-methyl-D-aspartate receptors and reduction of the afterhyperpolarization current. The latter effect, which can occur without the counteracting effect of α_2AR activation within GABAergic neurons, might lead to enhanced suppression of cortical circuitry. Another consequence of βAR is the depolymerization of microtubules following phosphorylation of MAP2, a microtubule-associated protein enriched in spines and dendrities. Phosphorylation of MAP2 can lead to destabilization of the cytoskeletal matrix, which may, in turn, allow for localized changes in the dendritic arbor and spine structure of activated neurons. It is intriguing to postulate that such a structural change may be linked to synaptic plasticity by altering the axospinous and axodendritic connectivity.

References

1. Kasamatsu, T. (1991). Adrenergic regulation of visuocortical plasticity: a role for the locus coerulens system. *Prog. Brain Res.* **88**, 599–616.
2. Kurose, H., Arriza, J. L., and Lefkowitz, R. J. (1993). Characterization of α_2-adrenergic receptor subtype-specific antibodies. *Mol. Pharmacol.* **43**, 444–450.
3. Aoki, C., Go, C.-G., Venkatesan, C., and Kurose, H. (1994). Perikaryal and synaptic localization of α_2-adrenergic receptor immunoreactivity in brain as revealed by light and electron microscopic immunocytochemistry. *Brain Res.* **650**, 181–204.
4. Jones, C. R., and Palacios, J. M. (1991). Autoradiography of adrenoceptors in rat and human brain: α-adrenoceptor and idazoxan binding sites. *Prog. Brain Res.* **88**, 271–292.
5. Arnsten, A. F. T., Steeve, J. C., and Hunt, R. D. (1996). The contribution of α2-noradrenergic mechanisms to prefrontal cortical cognitive function. *Arch. Gen. Psychiatry* **53**, 448–445.

Rita J. Valentino, Andre L. Curtis, Michelle E. Page, Luis A. Pavcovich, and Sandra M. Florin-Lechner

Department of Psychiatry
Allegheny University
Philadelphia, Pennsylvania 19102

Activation of the Locus Ceruleus Brain Noradrenergic System during Stress: Circuitry, Consequences, and Regulation

Neurochemical and neurophysiological evidence indicates that the locus ceruleus (LC)–noradrenergic (NE) system is activated by physiological challenges and stimuli that are considered stressors. One potential mediator of this activation is corticotropin-releasing factor (CRF), which was initially characterized as the primary neurohormone responsible for eliciting adrenocorticopin release from the anterior pituitary during stress (1). This chapter will: (1) review physiological evidence supporting the hypothesis that CRF serves as a neurotransmitter in the LC to mediate its activation by particular stimuli, (2) describe potential circuitry underlying CRF–LC interactions, (3) speculate on consequences of LC activation by CRF, and (4) discuss regulation of CRF–LC interactions.

Although it has long been accepted that the LC–NE system is activated by sensory stimuli, visceral stimuli, and diverse stressors, the neurotransmitters and circuitry underlying this activation remained unknown. Recent anatomical and physiological studies determined that LC activation by several stimuli, including opiate withdrawal, bladder distention, and sciatic nerve stimulation, is mediated by excitatory amino acid inputs to the LC (2). The identification of CRF-immunoreactive (CRF-IR) fibers within the LC and in pericerulear regions into which LC dendrites extend suggested that CRF might also act as a neurotransmitter in the LC to mediate its activation by certain stimuli, particularly stressors (2).

CRF increased LC discharge rates of both halothane-anesthetized and unanesthetized rats when administered intracerebroventricularly (i.c.v.). Similarly, local microinfusion of CRF (3–100 ng) into the LC increased discharge rate (28 ± 8% to 105 ± 26%), and the CRF dose-response curve generated by local microinfusion was shifted approximately 200-fold to the left of that generated by i.c.v. administration. Moreover, LC activation by maximally effective doses of i.c.v.-administered CRF were antagonized by local microinfusion of doses of CRF antagonists that do not alter LC activation by other stimuli. These findings provide strong evidence that exogenous CRF acts within the LC to activate these cells.

Advances in Pharmacology, Volume 42

1054-3589/98 $25.00

Pharmacological studies using i.c.v. or intracerulear administration of CRF antagonists into the LC have began to elucidate the stimuli that activate the LC via CRF release within the LC region (2). The stimuli that have been studied using this approach thus far include hypotensive challenge, bladder distention, colon distention, sciatic nerve stimulation, and opiate withdrawal. Of these, LC activation by hypotensive challenge and low magnitudes of colon distention appear to be mediated by CRF release within the LC, because intracerulear administration of CRF antagonists prevented or greatly attenuated the effect.

The hypotensive stress that has been examined is one that is sufficient to elicit CRF release into the hypophyseal portal system and activate the hypothalamic-pituitary-adrenal axis, and, therefore, may be classified as a stressor. CRF antagonists administered i.c.v. prevented LC activation by hypotensive stress in both anesthetized and unanesthetized rats, and antagonist IC 50s for antagonism of LC activation by hypotensive challenge were similar to their IC 50s for antagonism of i.c.v.-administered CRF, indicating that receptors involved in LC activation by hypotension and CRF are similar (2). The potential circuitry underlying LC activation by hypotension and possible consequences are discussed below.

LC activation by pelvic visceral stimuli has been previously described. Thus, both bladder and colon distention, in magnitudes that are thought to be nonnoxious, increase the LC discharge rates of anesthetized rats (3). Low magnitudes of distal colon distention produced an increase in LC discharge rate that was greatly attenuated by i.c.v. and intracerulear administration of CRF antagonists, but not by intracerulear administration of excitatory amino acid antagonists. The recent finding that Barrington's nucleus sends divergent projections to the sacral parasympathetic nucleus (which innervates the colon) and the LC, some of which are CRF-IR, implicate this nucleus as a source of CRF involved in LC activation by colon distention (discussion follows).

These physiological–pharmacological approaches have revealed, not surprisingly, that activation of the LC–NE system is mediated in a heterogenous manner, involving different neurotransmitters and presumably different afferents. Of the potential neuromediators, excitatory amino acids and CRF have been the most well studied thus far. However, with the elucidation of the neurochemical identity of LC afferents, it is likely that other neuromediators will be identified as integral to LC activation or inhibition by particular stimuli.

The anatomical substrates for endogenous CRF–LC interactions have begun to be investigated. Early studies by several groups identified CRF-IR varicose fibers within the LC, and autoradiographic studies revealed CRF-binding sites in this nucleus (2). Recent electron microscopy studies demonstrated synaptic contacts, the majority of which were asymmetric, between CRF-IR terminals, and tyrosine hydroxylase–immunoreactive (TH-IR) dendrites of LC neurons in the rostromedial dendritic zone (4). CRF-IR terminals also contacted unlabeled terminals that were apposed to TH-IR dendrites and unlabeled dendrites. These findings, providing morphological evidence for CRF–LC interactions, suggest a number of scenarios by which CRF in the LC region may impact on LC neurons, including direct interactions, effects on terminals of afferents to the LC, or effects on non-LC dendrites. The apparent lack of mRNA for CRF receptors in LC neurons favors an indirect action of CRF on these cells. However, this interpretation relies on the sensitivity of these approaches and on the assumption that the specific CRF receptor that is involved in LC activation by CRF is identical to that which is detected by the probes being used.

Anatomical studies combining retrograde tract-tracing from the nucleus LC and CRF immunohistochemistry implicate Barrington's nucleus, the nucleus paragigantocellularis (PGi), and the dorsal cap of the paraventricular nucleus of the hypothalamus as sources of CRF-IR fibers within the LC (2). However, because LC dendrites extend outside of the nucleus proper, these studies may underestimate the sources of CRF that can impact on the LC. Based on the different potential sources of CRF that can impact on LC discharge, it is likely that different stimuli that activate the LC via CRF release may do so via different pathways. For example, LC activation by hypotensive challenge, which activates neurons in the ventrolateral medulla, may be mediated by CRF inputs from the nucleus PGi. Because the PGi also projects to the intermediolateral column of the spinal cord, this nucleus could serve to coactivate the peripheral sympathetic and brain noradrenergic systems. On the other hand, LC activation by colon distention may be mediated by CRF–Barrington's neurons that also project to the sacral parasympathetic nucleus, thereby resulting in coactivation of the sacral parasympathetic system and brain noradrenergic system. Thus, CRF afferents to the LC may be organized in such a way as to integrate peripheral autonomic functions with forebrain activity. Future anatomical and physiological studies are necessary to begin to elucidate this circuitry and test this hypothesis.

The magnitude of LC activation produced by endogenous or exogenous CRF is relatively small compared with that produced by muscarinic agonists or excitatory amino acid agonists, (e.g., approximately a doubling of discharge rate). Nonetheless, neurochemical and physiological findings indicate that this is sufficient to impact on cortical targets. For example, microdialysis studies demonstrated that CRF microinfusion into the LC increases NE levels in prefrontal cortex and hippocampus. Moreover, LC activation elicited by exogenous CRF, or stimuli that activate the LC via endogenous CRF (e.g., hypotensive challenge, colon distention) is temporally correlated with forebrain electroencephalogram (EEG) activation (2). This effect has been investigated in detail using hypotensive challenge as the stimulus (2). Selective bilateral inactivation of the LC by intracerulear injections of clonidine prevented both LC activation and forebrain EEG activation produced by hypotensive challenge, suggesting that LC activation is necessary for forebrain activation. Importantly, bilateral intracerulear injections of CRF antagonists, in doses that prevented LC activation by hypotensive challenge, also prevented forebrain EEG activation by this stimulus. These results indicate that physiological CRF release in the LC region can impact on LC targets and suggest that one consequence of LC activation by endogenous CRF may be to increase or maintain arousal. Findings from other laboratories that have investigated the effects of local injection of CRF or CRF antagonists into the LC have suggested other consequences of CRF–LC interactions, including anxiogenic effects, increased colonic motility, and suppression of certain immune responses.

The neurohormone function of CRF is highly regulated by glucocorticoids, which affect CRF synthesis and release (5). This regulation is well demonstrated by the effects of adrenalectomy, which include enhanced synthesis of CRF in paraventricular hypothalamic neurons and release into the median eminence. Previous studies suggested that putative neurotransmitter effects of CRF are not regulated or are regulated differentially from neurohormone CRF. However, recent studies provide evidence for regulation of CRF effects within the LC by adrenalectomy. Thus, adrenalectomy was reliably associated with elevated mean basal LC discharge rates (2.3 $\pm$ 0.1 Hz vs 1.7 $\pm$ 0.1 Hz in adrenalectomized vs sham rats, respectively; $p < .001$) recorded 14 days after adrenalectomy, and intracerulear administration of the CRF antagonist D-PheCRF$_{12-24}$ (10 ng) decreased LC discharge rates of adrenalectomized rats to rates that were comparable to those recorded in sham-operated rats, while having no effect in sham-operated rats. These results suggested that CRF might be tonically released in the LC of adrenalectomized but not intact rats. Consistent with this, the CRF dose-effect curve in adrenalectomized rats was shifted in a manner that suggested that a proportion of CRF receptors were occupied prior to CRF administration and that low doses of CRF produced an additive effect. Together, these findings suggest that CRF–LC interactions and CRF neurohormone activity may be coregulated. Such parallel regulation could underlie the coexistence of neuroendocrine and behavioral dysfunction in certain psychiatric disorders (e.g., depression).

In summary, anatomical and physiological findings suggest that endogenous CRF, perhaps acting as a neurotransmitter within the LC, activates the LC–NE system under physiological conditions and that this activation is sufficient to impact on LC targets. CRF–LC interactions may be important components of the stress response and serve to link certain peripheral autonomic effects with forebrain activity. Finally, coregulation of CRF–LC interactions and neurohormone CRF may be important in symptoms of stress-related psychiatric disorders.

References

1. Vale, W., Spiess, J., Rivier, and C., Rivier, J. (1981). Characterization of a 41-residue ovine hypothalamic peptide that stimulates secretion of corticotropin and beta-endorphin. *Science* **213**, 1394–1397.
2. Valentino, R. J., Foote, S. L., and Page, M. E. (1993). The locus coeruleus as a site for integrating corticotropin-releasing factor and noradrenergic mediation of stress responses. *Ann. N. Y. Acad. Sci.* **697**, 173–188.
3. Elam, M., Thoren, T., and Svensson, T. H. (1986). Locus coeruleus neurons and sympathetic nerves: Activation by visceral afferents. *Brain Res.* **375**, 117–125.
4. Van Bockstaele, E. J., Colago, E. E. O., and Valentino, R. J. (1996). Corticotropin-releasing factor-containing axon terminals synapse onto catecholamine dendrites and may presynaptically modulate other afferents in the rostral pole of the nucleus locus coeruleus in the rat brain. *J. Comp. Neurol.* **364**, 523–534.
5. Dallman, M. F., Akana, S. F., Scriber, K. A., Bradbury, M. J., Walker, C. D., Strack, A. M., and Cascio, C. S. (1992). Stress, feedback and facillitation in the hypothalamo-pituitary-adrenal axis. *J. Neuroendocrinol.* **4**, 517–526.

Alan Breier,* Igor Elman,* and David S. Goldstein†

*Section on Clinical Studies
Experimental Therapeutics Branch
National Institute of Mental Health
Bethesda, Maryland 20892

†Clinical Neuroscience Branch
National Institute of Neurological Disorders and Stroke
Bethesda, Maryland 20892

Norepinephrine and Schizophrenia: A New Hypothesis for Antipsychotic Drug Action

Clozapine is an antipsychotic agent that was approved in the United States in 1989 for chronic schizophrenia. It marks a watershed in antipsychotic drug development because it was the first agent to demonstrate superior efficacy compared with traditional neuroleptic drugs and it is the only antipsychotic agent to date that is essentially void of extrapyramidal side effects. There is currently an intensive research focus attempting to uncover which of clozapine's many neurochemical effects is responsible for its superior clinical profile. This line of investigation is likely to lead to the development of even better agents for the treatment of schizophrenia.

Perhaps the most unique and consistently found neurochemical effect of clozapine is enhancement of noradrenergic activity. We found that clozapine produced fivefold increases in plasma norepinephrine (NE) levels, whereas traditional neuroleptics have essentially no effect on peripheral noradrenergic function (1) (Fig. 1). This finding has been replicated by several research groups. We also discovered that the clozapine-induced increases in NE levels were positively correlated to clinical improvement: Patients with the greatest increases had the best clinical responses (Fig. 2). These data suggested that clozapine's effects on noradrenergic function were related to its superior clinical efficacy. In an electrophysiological study of the locus ceruleus in rat brain, Ramirez and Wang (2) reported that chronic clozapine caused significant increases in the number of spontaneously active NE cells and in their average firing rates. In contrast, chronic haloperidol did not change the number of spontaneously firing NE cells and significantly *reduced* their average firing rates. In another study, clozapine, in comparison to other antipsychotic drugs, was the most potent agent to increase ^{3}H efflux from electrically stimulated cortical slices incubated with [^{3}H]NE.

The mechanism of clozapine-induced enhanced noradrenergic activity is currently not understood, although there are several possibilities. Clozapine antagonizes NE α_1 and α_2 receptors, which could account for enhanced NE outflow. In our earlier clinical study, in which we found fivefold increases in

1054-3589/98 $25.00

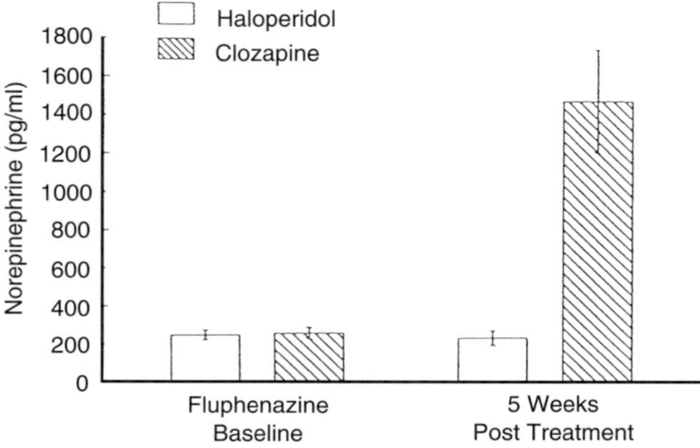

FIGURE 1 The effects of clozapine ($n = 11$) and haloperidol ($n = 15$) on plasma NE levels (mean ± SD).

plasma NE (1), we also discovered that there were no changes in plasma levels of 3,4-dihydroxyphenylglycol (DHPG). This was remarkable because DHPG is an intraneuronal metabolite of NE. When there is increased NE outflow, a compensatory increase in uptake occurs, which results in heightened intraneuronal NE metabolism and production of DHPG. One possible explanation of greatly increased NE release without concomitant increases in DHPG is NE uptake blockade. Another possible mechanism is that clozapine decreases NE clearance, which could account for elevated plasma NE levels.

To better understand the mechanism of clozapine's noradrenergic enhancement, we conducted a clinical study in schizophrenic patients to assess effects on NE uptake, clearance, and spillover. Nine clozapine-treated, six fluphenazine-treated, and five medication-free patients with chronic schizophrenia participated in the study. To assess NE uptake, we examined the conversion (i.e., ratio) of [^{3}H]NE to [^{3}H]DHPG. If clozapine blocked uptake, low [^{3}H]DHPG levels will be associated with [^{3}H]NE infusions. In addition, arterial levels of NE and DHPG were determined, and clearance and spillover were examined.

We found that patients with schizophrenia had significantly higher arterially derived plasma NE levels than fluphenazine-treated and medication-free patients (3). There were no significant differences in DHPG levels across the three groups. Moreover, [^{3}H]DHPG levels during a constant infusion of [^{3}H]NE were not different among the three groups, indicating clozapine was not blocking NE uptake (3). Lastly, NE clearance was not different across the three groups, while spillover paralleled the arterial NE levels, demonstrating significant increases in clozapine-treated patients (3).

These data replicate our earlier findings of robust increases in NE without changes in DHPG. We have not found support for clozapine-mediated blockade of the NE transporter and cannot explain elevated NE levels as secondary to changes in clearance. NE receptor antagonism is not an adequate explanation for these data in that highly potent α_1 and α_2 antagonists have not been shown

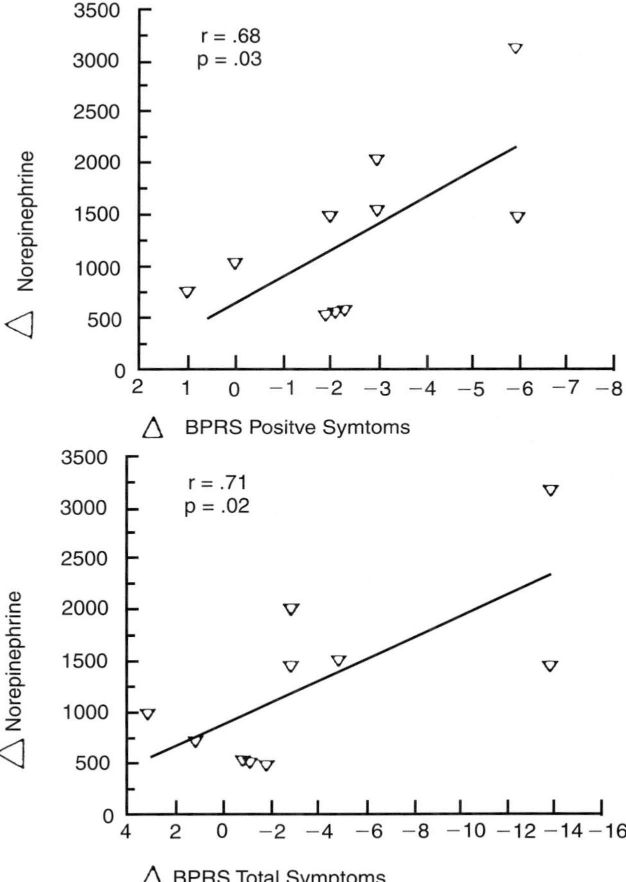

FIGURE 2 Relationship between clozapine-induced changes in plasma NE (pg/ml) and changes in positive and total BPRS symptoms. BPRS, Brief Psychiatric Rating Scale.

to produce the magnitude of NE effects or the NE/DHPG ratio observed in this study (4). We are currently examining the possibility that clozapine preferentially affects NE in deep vesicular storage pools, as opposed to vesicular pools near the synaptic cleft that are readily available for release, which might explain the findings reported here.

The noradrenergic-enhancing property of clozapine raises questions about the role of NE in the pathophysiology of schizophrenia. Numerous postmortem studies have presented evidence for cortical cytoarchitectural abnormalities in this illness. One functional implication of these data may be inefficient neurotransmission in affected synapses. NE, as demonstrated in long-term potentiation models and other models of neuronal plasticity, enhances synaptic strength and improves so-called signal-to-noise. We hypothesize that clozapine's superior clinical efficacy is related to its ability to enhance noradrenergic function, which in turn improves synaptic strength in cortical regions responsible for the symptoms of this illness.

References

1. Breier, A., Buchanan, R. W., Waltrip, R. W., Listwak, S., Holmes, C., Goldstein, D. S. (1994). The effect of clozapine on plasma norepinephrine: relationship to clincial efficacy. *Neuropsychopharmacology* **10**, 1–7.
2. Ramirez, O. A., Wang, R. Y. (1986). Locus coeruleus norepinephrine-containing neurons: effects produced by acute and subchronic treatment with antipsychotic drugs and amphetamine. *Brain Res* **362(1)**, 165–170.
3. Elman, I., Goldstein, D. S., Holmes, C., Pickar D., Folio, J., Adler C., Breier, A. (1996). Mechanism of plasma norepinephrine elevation in clozapine-treated patients. Proceedings the Eighth International Catecholamine Symposium, p. 34. Asilomar, CA
4. Goldstein, D. S. (1995). *In* stress catecholamines, and cardiovascular disease. pp. 234–286. Oxford University Press, New York, NY.

Michael J. Zigmond and Teresa G. Hastings

Departments of Neuroscience, Neurology, and Psychiatry
University of Pittsburgh
Pittsburgh, Pennsylvania 15260

Neurochemical Responses to Lesions of Dopaminergic Neurons: Implications for Compensation and Neuropathology

Parkinson's disease affects 1% of individuals above the age of 55. Its cause is unknown, and, although the symptoms can be treated, there is no cure. The disease is associated with a relatively selective loss of dopamine (DA) neurons of the nigrostriatal projection. A considerable amount of information regarding Parkinson's disease has come from animal studies using the selective neurotoxin 6-hydroxydopamine (6-OHDA). This compound is selectively accumulated in catecholamine neurons, where it autoxidizes to form such toxic species as

1054-3589/98 $25.00

superoxide anions, hydrogen peroxide, hydroxyl radicals, and 6-OHDA quinones. We generally have administered 6-OHDA via the lateral cerebroventricles. We have used this model to explore several issues regarding parkinsonism. In this review, we will briefly touch on two of them.

I. Why Is There an Extended Preclinical Phase in Parkinsonism?

6-OHDA–lesioned rats, like patients with Parkinson's disease, show a "preclinical phase" until the loss of striatal DA exceeds 80–90% (1). This absence of permanent gross behavioral deficits despite extensive lesions does not appear to be a result of regeneration, sprouting, or receptor supersensitivity. Moreover, because behavioral impairments can be elicited by drugs that interfere with dopaminergic transmission, the DA terminals that remain after 6-OHDA–induced lesions appear to play an important functional role (1).

The absence of gross neurological dysfunction despite extensive loss of DA neurons is associated with an increase in the capacity of the remaining DA neurons to deliver transmitter to denervated sites. Our group has examined this phenomenon in a variety of ways. Studies using neostriatal slices prepared from control and 6-OHDA–lesioned rats demonstrated that partial destruction of neostriatal DA terminals increased the fractional overflow of DA (i.e., the amount of electrically evoked DA efflux per apparent DA terminal remaining). The increased fractional overflow was related to lesion size, reaching sevenfold with losses of DA greater than 90%. Further studies have suggested that this apparent increase in DA efflux per terminal is due to a reduction in DA reuptake as well as an increase in DA release from residual terminals (1, 2). More recent observations by Denise Jackson and Li Ping Liang in our group indicate that the increases in DA synthesis and efflux develop gradually over at least a month and are accompanied by an increase in the availability of tyrosine hydroxylase within the remaining nerve terminals.

The impact of partial lesions on the concentration of DA in the extracellular fluid of neostriatum was examined using *in vivo* microdialysis, and, like others, we observed that the extracellular concentration of DA was not significantly different from normal in animals with less than an 80% loss of DA. Moreover, the capacity to maintain extracellular DA constant appears to serve a compensatory function. Our studies indicate that the ability of endogenous DA to exert an inhibitory influence on the overflow of GABA and acetylcholine from striatal neurons is not permanently disrupted until the loss of striatal DA exceeds 90% (1, 3).

II. What Is the Basis of the Selective Toxicity of DA Neurons?

The underlying mechanism for the loss of DA neurons in Parkinson's disease is unknown. However, it has long been suspected by many investigators that DA itself may be involved. The hypothesis that DA contributes to the etiology of Parkinson's disease was first suggested by Doyle Graham in 1978 on the basis of the observation that DA, like 6-OHDA, is a reactive molecule that will oxidize to form free radicals and quinones. The electrophilic quinones will react with nucleophilic sulfhydryl groups, a major source of which is provided by the cysteinyl residues of proteins. Thus, the binding of DA to cysteinyl residues may serve as an index of DA oxidation; it also may be a cytotoxic event.

To examine conditions under which such binding might occur, we incubated neostriatal slices with [^{3}H]DA under various buffer conditions and then determined the amount of radioactivity bound to the acid-precipitated protein. The amount of tritium bound to protein was greatly influenced by the concentration of reducing agent, ascorbate or reduced glutathione, that was present in the incubation buffer (4). Binding to protein could be completely inhibited by glutathione, suggesting that the thiol group on glutatione was competing with protein cysteinyl residues for the oxidized DA quinone.

When high concentrations of DA (0.1–1.0 μmol in 2 μl) were injected into neostriatum, evidence of *in vivo* toxicity was observed. First, 24 hr after the injection, protein-bound cysteinyl-DA and cysteinyl-dihydroxyphenylacetic acid were increased up to 100-fold in the tissue surrounding the injection. In addition, after 7 days, cellular damage and gliosis were observed at the injection site. This was surrounded by a region of specific loss of tyrosine hydroxylase immunoreactivity with no detectable loss of synaptophysin immunoreactivity or alterations in cellular architecture. Coadministration of DA with an equimolar concentration of ascorbate or glutathione greatly reduced both protein cysteinyl-catechol formation and the loss of tyrosine hydroxylase immunoreactivity (5). These findings suggest that DA can cause specific toxicity to dopaminergic neurons (also shown by Filloux and Townsend, 1993), that the binding of DA to protein is correlated with this toxicity, and that both result from the oxidation of DA.

III. Summary and Conclusions

As DA neurons degenerate, several neurological responses conspire to minimize the functional impairments that might otherwise occur (Fig. 1). The first of these is passive: a reduced capacity for the clearance of DA from extracellular fluid resulting from the loss of high-affinity DA transporters present on DA terminals. With larger lesions and over time, more active processes come into play, including an increase in the synthesis and release of DA from the remaining neurons. These events appear to be sufficient for a small number of DA terminals to maintain much of the control over striatal events that normally is exerted by the full complement of nigrostriatal neurons. On the other hand, like 6-OHDA, DA itself can oxidize and exert a selective, toxic effect on DA neurons. Thus, it is possible that the increase in the turnover of DA within remaining neurons contributes to the progressive neurodegenerative process while also reducing the symptoms of DA neuron loss.

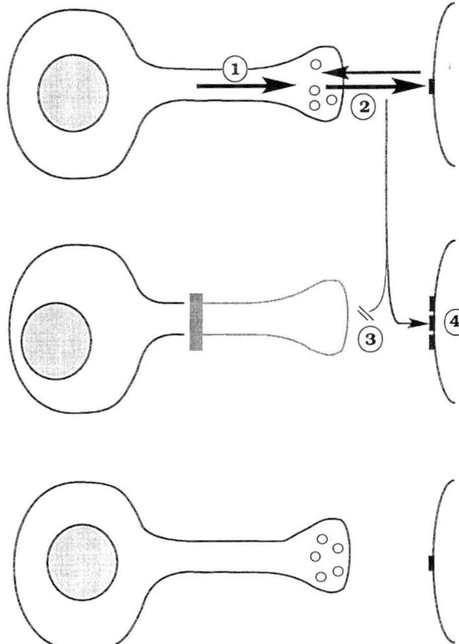

FIGURE 1 Model of compensatory changes after partial injury to the nigrostriatal dopamine (DA) projection. After injury (note middle neuron), residual neurons exhibit an increase in (1) DA synthesis and (2) DA release. We propose that this increased DA is able to diffuse to sites at which (3) denervation has occurred and thus can no longer take up DA and (4) DA receptor upregulation may have developed. We suggest that as long as the loss of DA terminals does not exceed approximately 80%, this combination of increased DA availability, decreased DA inactivation, and increased target cell sensitivity results in a "biochemical reinnervation" of the denervated site, thereby forestalling neurological deficits. We also suggest that the hyperactivity of DA neurons may produce reactive metabolites that contribute to the degenerative process.

Acknowledgments

Many colleagues have contributed to the research and concepts summarized in this review, including Elizabeth D. Abercrombie, Laszlo G. Harsing, Jr., David A. Lewis, Li Ping Liang, Ariel Rabinovic, Gretchen L. Snyder, Denise Jackson, Robert G. MacKenzie, and Michal K. Stachowiak. This work was supported in part by U.S. Public Health Service grants NS19608, MH43947, MH29670, MH45156, MH00058, and NS09076.

References

1. Zigmond, M. J., Abercrombie, E. D., Berger, T. W., Grace, A. A., and Stricker, E. M. (1993). Compensatory responses to partial loss of dopaminergic neurons: Studies with 6-

hydroxydopamine. *In* Current Concepts in Parkinson's Disease Research. (J. S. Schneider and M. Gupta, eds.), pp. 99–140. Hogrefe & Huber, Toronto.
2. Snyder, G. L., Keller, R. W., Jr., and Zigmond, M. J. (1990). Dopamine efflux from striatal slices after intracerebral 6-hydroxydopamine: Evidence for compensatory hyperactivity of residual terminals. *J. Pharmaol. Exp. Ther.* **253**, 867–876.
3. Harsing, L. G., and Zigmond, M. J. (1996). Dopaminergic inhibition of striatal GABA release after 6-hydroxydopamine. *Br. Res.* **738**, 142–145.
4. Hastings, T. G., and Zigmond, M. J. (1994). Identification of catechol-protein conjugates in neostriatal slices incubated with ^{3}H-dopamine: Impact of ascorbic acid and glutathione. *J. Neurochem.* **63**, 1126–1132.
5. Hastings, T. G., Lewis, D. A., and Zigmond, M. J. (1996). Role of oxidation in the neurotoxic effects of intrastriatal dopamine injections. *Proc. Natl. Acad. Sci. U.S.A.* **93**, 1956–1961.
6. Filloux, F. and Townsend, J. J. (1993). Pre- and post-synaptic neurotoxic effects of dopamine demonstrated by intrastriatal injection. *Exp. Neurol.* **119**, 79–88.

G. Andringa,*† R. J. Vermeulen,* B. Drukarch,*
J. C. Stoof,* and A. R. Cools†

*Department of Neurology
Research Institute for Neurosciences
Vrije Universiteit
Amsterdam, The Netherlands

†Department of Psychoneuropharmacology
Catholic University
Nijmegen, The Netherlands

Dopamine Receptor Subtypes as Targets for the Pharmacotherapy of Parkinson's Disease

The prevailing therapeutic strategy in Parkinson's disease (PD) has been confined to substitution for the dopaminergic deficit in the striatum by using L-dopa (precursor of dopamine [DA]) and/or D2-receptor agonists. Long-term treatment of PD patients with L-dopa and/or D2-receptor agonists often leads to wearing off, on-off effects, and/or dyskinesias. However, it is well known that apart from D2 receptors, D1 receptors also are targets for the action of DA in the striatum and are involved in the regulation of motor behavior. Thus, apart from D2 receptors, D1 receptors also might serve as pharmacotherapeutic targets in PD. We elaborated in this issue by testing two newly developed benzazepines in the pri-

Advances in Pharmacology, Volume 42

1054-3589/98 $25.00

mate MPTP model for PD. In both unilaterally and bilaterally MPTP-lesioned rhesus monkeys, we focused on the therapeutic and unwanted side effects of the two D1 agonists SKF 81297 and SKF 82958. In our behavioral analysis, we have tried to discriminate between real antiparkinsonian and "stimulating" effects of the compounds on various aspects of motor behavior.

I. Unilaterally Lesioned MPTP Monkey

By using SKF 81297 and the D2 agonist LY 171555, we and others have previously reported that these compounds stimulate several aspects of motor behavior, such as circling behavior and use of the invalidated forelimb (1). In the present analysis, in which we observed the monkeys in a cage of 1 × 1 m, we have tried to further refine the analysis of circling behavior by considering the following: *turning,* defined as rotating in circles with a diameter of more than 70 cm while walking on four limbs; *dragging,* defined as turning, with the exception that the dorsum of the outer hindlimb is dragging over the floor of the cage before it is lifted and correctly placed on the floor; *pivoting,* defined as rotating around the axis of the inner hindlimb in circles with a diameter of less than 40 cm—the inner hindlimb is carrying the weight, making small correcting steps during the rotation; and *shuffling,* defined as rotations around the axis of the trunk, while the monkey is sitting. Likewise, we have further refined the analysis of forelimb use by considering *goal-directed* movements, defined as normal hand movements directed either at the body itself or at objects such as bars, ball, and feeding-trough-, and *dyskinesias,* encompassing athetoid and choreiform movements. Finally, we defined *displacement activity* as any displacement (walking, climbing, etc.) of the animal over a minimum distance of 50 cm.

It should be realized that the monkeys that were used (n = 4) for this study had been pretreated with various dopaminergic compounds. On acute administration of SKF (0.3 mg/kg i.m.) or LY (0.01 mg/kg i.m.), the following behavioral effects could be observed. First, although LY 171555 induced slight contralateral turning behavior in the monkeys, the animals spent much more time in dragging, pivoting, and shuffling. Similarly, following SKF 81297 administration, the animals spent even less or virtually no time in turning behavior, but displayed predominantly the other mentioned categories of circling behavior. Second, saline-treated monkeys clearly displayed goal-directed forelimb movements with their unsevered (ipsilateral) arm, whereas drug-treated monkeys reduced the number of goal-directed movements of the ipsilateral arm and did not induce any goal-directed movements in the contralateral arm, but instead new and dyskinetic types of movements were observed in the contralateral arm. Third, whereas LY 171555 induced significant displacement activity in the cage, SKF 81297 did not. Finally, both compounds induced epileptoid attacks, of which the occurrence was confirmed by electroencephalographic analysis. With this epileptoid behavior, the animals actually went through a series of successively appearing phases.

- Phase 1: rapid opening and closing of the eyelids

- Phase 2: brief muscular contractions and relaxations of muscle groups in the ears, small sections of the face, and the mouth
- Phase 3: rapid flexions and extensions of the lower arm and upper arm, resulting in abrupt, staccato-like forelimb movements as well as abrupt, uncoordinated movements of the upper and lower torso
- Phase 4: contralateral turning of the eyeballs, followed by abrupt, contralateral rotations of the head, and then followed by the appearance of jerky, contralateral rotations with the whole body

We conclude on the basis of these findings that, although SKF 81297 apparently "stimulates" motor behavior, the effects are predominantly dyskinetic, abnormal, and clearly not goal directed and, thus, cannot be considered "real" antiparkinsonian. Because not only SKF81297, but also LY 171555 reduced the goal-directed movements in the intact side and induced epileptoid behavior, it is tempting to suggest that the unilaterally MPTP-lesioned monkey is not the best model for PD.

II. Bilaterally Lesioned MPTP Monkey

Based on our experiences with the unilaterally MPTP-lesioned model, we decided to investigate the effects of a D1 agonist in bilaterally MPTP-lesioned rhesus monkeys and especially the behavioral effects of chronic treatment. Because SKF 81297 induced epileptoid activity in the unilaterally lesioned monkey, we decided to use SKF 82958 instead and compared the behavioral effects of this compound with the effects of chronic L-dopa treatment. We investigated the following behavioral categories: goal-directed movements, dyskinesias, and displacement activity. Two groups of animals were used, four for the L-dopa treatment and four for the SKF treatment. All animals received (daily) either L-dopa in a dose of 10.0 mg/kg (i.m.) or SKF 82958 in a dose of 1.0 mg/kg (i.m.) in the early afternoon for a period of 25 days. Behavior was analyzed until 1 hr after the injection. L-dopa clearly stimulated, from the first day until the last day, the goal-directed forelimb movements. In contrast, SKF 82958 did not show any change in goal-directed forelimb movement on any particular day. The D1 agonist did induce, however, statistically significant displacement activity, from day 1 till day 25, comparable to the changes induced by L-dopa. Both compounds induced dyskinesias (see also Blanchet *et al.* [2]). On average, the dyskinesias started earlier in the SKF group than in the L-dopa group. Epileptoid activity could be observed in two out of four animals of the SKF group, whereas none of the animals in the L-dopa–treated group displayed this kind of activity.

On the basis of these findings, we conclude that

- The D1 agonists SKF 81297 and SKF 82958 "stimulate" to a certain extent motor behavior, but they lack a clear effect on goal-directed movements, which might be considered a real antiparkinsonian effect.
- Both agonists clearly induce dyskinetic effects.

- Both agonists induce epileptoid activity. Reduction of the seizure threshold by D1 agonists has been reported previously to be an inevitable consequence of D1-receptor stimulation (3).
- Summarizing, we have serious doubts concerning whether pursuing behavioral studies with these SKF analogues in MPTP-lesioned monkeys, aiming at the development of new antiparkinsonian drugs, is a fruitful endeavor.

III. *In Vitro* Experiments with SKF 83959

It has been reported that the effects of several benzazepines on motor behavior failed to correlate with their efficacy in stimulating adenylate cyclase activity (4), a paradigm generally considered the hallmark of D1-receptor efficacy. For instance, SKF 83959, which had been reported to display antiparkinsonian effects in MPTP-lesioned marmosets and 6-OH-DA rats upon acute administration, behaved *in vitro* not as an agonist on the D1 receptor. In rat and primate glial cells expressing D1 receptors, we found that the compound did not stimulate cyclic adenosine monophosphate (AMP) formation but instead blocked very potently the DA-induced AMP formation, with a similar potency as the D1-receptor antagonist SCH 23390. Because the compound does not display any significant D2-receptor affinity, the mechanism underlying its reported antiparkinsonian effect in the animal models is unclear. Therefore, we are further analyzing the pharmacological profile of this compound and will study the effects of chronic administration in our bilaterally MPTP-lesioned monkeys, especially to discover how the compound behaves with respect to goal-directed behavior, dyskinesias, and epileptoid activity.

Acknowledgment

The financial support of this work by the Prinses Beatrix Fonds (Den Haag, The Netherlands) is greatly acknowledged.

References

1. Vermeulen, R. J., Drukarch, B., Sahadat, M. C. R., Goosen, G., Wolters, E. G., and Stoof, J. C. (1993). The selective D1 receptor agonist SKF 81297 stimulates motor behavior of MPTP-lesioned monkeys. *Eur. J. Pharmacol.* **235**, 143.
2. Blanchet, P. J., Grondin, R., and Bedard, P. J. (1996). Dyskinesia and wearing-off following dopamine D1 agonist treatment in drug-naive 1-methyl-4-phenyl-1,2,3,6-tetrahydropyridine-lesioned primates. *Mov. Disord.* **11**, 91.
3. Starr, M. S., Starr, B. S. (1993). Seizure promotion by D1 agonists does not correlate with other dopaminergic properties. *J. Neural Transm. Park. Disease Dem. Sect.* **6**, 27.
4. Gnanalingham, K. K., Erol, D. D., Hunter, A. J., Smith, L. A., Jenner, P., and Marsden, C. D. (1995). Differential anti-parkinsonian effects of benzazepine D1 dopamine agonists with varying efficacies in the MPTP-treated common marmoset. *Psychopharmacology* **117**, 275.

Chuang C. Chiueh and Pekka Rauhala

Unit on Neurotoxicology and Neuroprotection
Laboratory of Clinical Science
National Institute of Mental Health
Bethesda, Maryland 20892

Free Radicals and MPTP-Induced Selective Destruction of Substantia Nigra Compacta Neurons

1-Methyl-4-phenyl-1,2,3,6-tetrahydropyridine (MPTP) is a synthetic neurotoxin that produces a parkinsonian syndrome in humans. In experimental animals, low doses of MPTP induce a selective destruction of the pigmented substantia nigra compacta neurons of primates, while large doses of MPTP are needed to induce a reversible dopamine depletion in rodents. MPTP-induced parkinsonian animal models are useful for evaluating new treatments, transplant, and pallidotomy procedures. Earlier findings that MPTP (<1.5 mg/kg i.v.) spares nonpigmented dopamine neurons, while inducing a selective destruction of pigmented dopamine neurons in rhesus monkeys, have stimulated current investigations of the possible role of free radicals in MPTP-induced selective dopaminergic neurotoxicity for supporting the oxidant stress hypothesis in Parkinson's disease (1).

I. Oxidative Stress and Formation of Dopamine Melanin

Dopamine neuromelanin pigments are generated by mixing together dopamine, iron, and oxygen; this free radical–mediated polymerization of neuromelanin is blocked by hydroxyl radical scavengers *in vitro* (2), leading to a speculation of a site-specific and age-dependent generation of cytotoxic hydroxyl radicals in these melanized dopamine neurons. Among brain dopamine neurons, iron-containing substantia nigra compacta neurons are subjected to a continuous oxidant stress. Hydroxyl radicals and semiquinone radicals are generated during iron-catalyzed dopamine auto-oxidation. We have developed an intracerebral microdialysis procedure for trapping generation of hydroxyl radicals *in vivo*. Furthermore, dopamine-derived oxidants propagate lipid peroxidation that leads to generation of lipid radicals and accumulation of malondialdehyde. Prolonged oxidative stress may cause neuronal degeneration when cellular repair mechanisms and antioxidant defense systems are weakened by aging, brain injury, and/or neurotoxic insult. Therefore, the accumulation of age pigments such as dopamine melanin in the midbrain substantia nigra compacta neurons may be a reliable biological marker for oxidative stress (3).

II. MPTP and Generation of Reactive Oxygen Species

A. Inhibition of Mitochondrial Complex I

Free radicals are unavoidable side products of redox cycling of oxygen. The inhibition of complex I may donate unpaired electron to oxygen, leading to the generation of reactive oxygen species. It has been shown that high levels of 1-methyl-4-phenylpyridinium ion (MPP^+, up to 100 mM) inhibit complex I activities in mitochondrial preparations. However, the highest brain levels of MPP^+ are located in the locus aeruleus noradrenergic neurons but not in the substantia nigra dopaminergic neurons. It is not known whether the substantia nigra compacta neurons can concentrate enough MPP^+ *in vivo* to inhibit complex I.

B. Generation of Hydroxyl Radicals by MPP^+

Earlier reports indicate that high levels of iron and dopamine in the basal ganglia may exaggerate iron-catalyzed dopamine oxidation and generation of cytotoxic free radicals, including superoxide anion, semiquinone, and hydroxyl radicals in nigrostriatal dopamine neurons. In cell cultures, dopamine causes oxidative injury and apoptosis in PC-12 cells. MPP^+-induced oxidative injury in the cultured midbrain dopamine neurons is suppressed by antioxidants, including lazaroid, trolox, selegiline, and melatonin. Additional *in vivo* studies further suggest that hydroxyl radicals and associated oxidative stress mediate MPTP-induced neurotoxicity (2).

1. Accumulation of Iron in Nigral Neurons

Iron is age-dependently accumulated in the basal ganglia, especially in the substantia nigra compacta dopamine neurons and globus pallidum. Most of the cellular iron is deposited as ferritin complexes. However, excess amounts of small molecular weight iron complexes (i.e., ferrous citrate) may generate reactive hydroxyl radical through redox cycling or iron–oxygen complexes. High levels of iron may contribute to progressive nigrostriatal degeneration during senescence and in Parkinson's disease. Furthermore, nigral uptake of iron is significantly augmented by MPTP in monkeys (4).

Iron is normally functioned as a catalyst for activating tyrosine hydroxylase. Because of a relative high concentration of transferrin transporter in the nigral neurons, they can take up more iron and, thus, synthesize greater levels of dopamine than other brain dopaminergic neurons. Intranigral infusion of ferrous citrate causes an acute increase in dopamine turnover and a chronic striatal dopamine depletion. This iron-induced dopamine depletion is mediated by lipid peroxidation caused by hydroxyl radicals generated by iron complexes (5). This site-specific iron-catalyzed dopamine auto-oxidation and oxidative stress are greatly augmented by MPTP that results in a selective injury to A9 dopaminergic neurons in the substantia nigra.

2. Role of Oxidative Stress in Site-Specific Dopamine Neurotoxicity Induced by MPTP

MPTP analogues (such as, 2′-methyl-MPTP, MPDP, and MPP^+) produce a sustained dopamine efflux from the nigrostriatal nerve terminals, lasting for hours. Most of released dopamine is oxidized either enzymatically or nonenzymatically, leading to generation of toxic species such as hydroxyl radicals, dopamine aldehyde, hydrogen peroxide, and semiquinone radicals inside and/or near the dopaminergic synapse. Moreover, intranigral infusion of MPP^+ causes not only the generation of cytotoxic hydroxyl radicals, but also an increase in brain lipid peroxidation—a hallmarker for free radical–induced chain reactions. Finally, it causes calcium overload and oxidative injury. Prevention of MPP^+ uptake into nigral neurons may suppress MPTP's neurotoxicity, whereas blockade of vesicular transporters by MPP^+ itself may augment dopaminergic toxicity because intracellular dopamine can no longer be protected and stored in synaptic vesicles. Therefore, MPTP's selective neurotoxicity may be mediated by dopamine-induced oxidative stress, especially in the substantia nigra compacta neurons associated with high iron levels (2).

3. Generation of Hydroxyl Radicals through Putative Peroxynitrite Pathway

Theoretically, peroxynitrite is generated when equal molars of nitric oxide and superoxide anion are mixed. It may decompose to form hydroxyl and nitrogen dioxide radicals, initiating lipid peroxidation. However, exogenously administered nitric oxide inhibits peroxynitrite-induced brain lipid peroxidation *in vitro* (P. Rauhala and C. C. Chiueh, unpublished observations). Surprisingly, nitric oxide inhibits oxidative stress caused by either peroxynitrite or iron. In addition, it is known that neuronal nitric oxide synthase is not located in the nigrostriatal dopaminergic neurons, and intranigral infusion of nitric oxide does not cause oxidative brain injury. Furthermore, the generation of nitrites and nitrates but not peroxynitrite has been demonstrated *in vivo*. Acidosis due to brain injury may convert peroxynitrite directly into nitrate. As compared with the pro-oxidant effects of iron, peroxynitrite is a relatively weak pro-oxidant for inducing brain lipid peroxidation both *in vitro* and *in vivo*. In neuronal nitric oxide synthase knockout mice, MPTP still causes significant nigral injury and dopamine depletion. MPTP produces more dopamine depletion in wild-type than in transgenic knockout animals. However, this interesting finding remains to be confirmed using transgenic mice overexpressing neuronal nitric oxide synthase.

III. Protection of Nigral Neurons from MPP^+ Neurotoxicity by Antioxidants

Selegiline's neuroprotective action is similar to that of U-78517, a potent antioxidant and inhibitor of brain lipid peroxidation, which protects nigral neurons from MPP^+-induced injury both *in vitro* and *in vivo*. Other atypical antioxidants (i.e., estrogen and melatonin) are being tested for their possible neuroprotective actions. Selegiline protects nigral neurons from dopaminergic toxicity caused by both MPTP and MPP^+. In addition to inhibiting monoamine oxidase B, selegiline also suppresses MPP^+-induced generation of hydroxyl radicals and lipid peroxidation (6). It protects brain neurons from oxidative injury caused by retrograde degeneration, ischemia–reperfusion injury, and MPP^+ neurotoxicity. In PC-12 cells, selegiline prevents p53-mediated apoptotic cell death. *In vivo* studies indicate that selegiline protects and/or rescues nigral neurons from mild-to-moderate but not severe oxidative injury induced by MPP^+ (6).

Based on the peroxynitrite hypothesis, attempts have been made to protect nigral neurons from injury mediated by toxic nitric oxide derivatives through inhibiting neuronal nitric oxide synthase. Neuronal nitric oxide synthase is blocked by both 7-nitroindazole and L-N^G-nitro-arginine methyl ester. However, only 7-nitroindazole protects nigral neurons from MPTP-induced dopamine depletion (7). In addition, our preliminary findings indicate that high doses of 7-nitroindazole may quench the accumulation of oxidized fluorescent products of 2′-methyl-MPTP or 2′-methyl-MPP^+ and lipid peroxidation (malondialdehyde) in rat brain homogenates (Table I). Therefore, similar to broad

TABLE I Effects of 7-Nitroindazole on the Bioactivation of 2′-methyl-MPTP (2′-methyl-MPP^+) and the Peroxidation of Brain Lipids (Malondialdehyde).

7-nitroindazole (μM)	*2′-methyl-MPP^+*	*Malondialdehyde*
	(relative fluorescent intensity)	
0	8.8 ± 0.22	1.472 ± 0.018
25	8.1 ± 0.37	
62.5		1.342 ± 0.041
100	6.3 ± 0.57	
125		1.049 ± 0.013
250		0.715 ± 0.015
400	5.27 ± 0.66	
500		0.331 ± 0.013
1000	2.87 ± 0.23	0.089 ± 0.007

The control sample (50 mg/ml rat brain homogenates) were treated with 1% DMSO, while other samples were treated with various concentrations of 7-nitroindazole in Ringer's solution containing 1% DMSO and incubated at 37°C for 2 hr. The effects of 7-nitroindazole on bioactivation of 2′-methyl-MPTP (100 μM) were assayed fluorometrically by measuring relative fluorescent intensity of 2′-methyl-MPP^+ (excitation/emission wavelength = 295/420 nm) in the aqueous phase (N = 3). Malondialdehyde fluorescent adducts (excitation/emission wavelength = 356/426 nm) were measured in chloroform and methanol extracts obtained from rat brain homogenates incubated with ferrous citrate (1 μM) for initiating lipid peroxidation (N = 4).

pharmacological actions of selegiline, 7-nitroindazole may have atypical antioxidant properties in addition to its known inhibition of the neruonal nitric oxide synthase and the MAO-B (8). In fact, recent reports indicate that selegiline and 7-nitroindazole protect brain neurons from oxidative injury caused by MPTP and ischemia–reperfusion procedure as well.

References

1. Fahn, S., and Cohen, G. (1992). The oxidant stress hypothesis in Parkinson's disease: Evidence supporting it. *Ann. Neurol.* **32,** 804–812.
2. Chiueh, C. C., Miyake, H., and Peng, M-T. (1993). Role of dopamine autoxidation, hydroxyl radical generation, and calcium overload in underlying mechanisms involved in MPTP-induced parkinsonism. *Adv. Neurol.* **60,** 251–258.
3. Chiueh, C. C., Murphy, D. L., Miyake, H., Lang, K., Tulsi, P. K., and Huang, S. J. (1993). Hydroxyl free radicals (·OH) formation reflected by salicylate hydroxylation and neuromelanin: In vivo marker for oxidant injury of nigral neurons. *Ann. N. Y. Acad. Sci.* **679,** 370–375.
4. Mochizuki, H., Imai, H., Endo, K., Yokomizo, K., Murata, Y., Hattori, N., and Mizuno, Y. (1994). Iron accumulation in the substantia nigra of 1-methyl-4-phenyl-1,2,3,6-tetrahydropyridine (MPTP)-induced hemiparkinsonian monkeys. *Neurosci. Lett.* **168,** 251–253.
5. Mohanakumar, K. P., de Bartolomeis, A., Wu, R-M., Yeh, J. J., Sternberger, L. M., Peng, S-Y., Murphy, D. L., and Chiueh, C. C. (1994). Ferrous-citrate complex and nigral injury: Evidence for free-radical formation and lipid peroxidation. *Ann. N. Y. Acad. Sci.* **738,** 392–399.
6. Wu, R. M., Murphy, D. L., and Chiueh, C. C. (1995). Neuronal protective and rescue effects of deprenyl against MPP^+ dopaminergic toxicity. *J. Neural Transm.* **100,** 53–61.
7. Hantraye, P., Brouillet, E., Ferrante, R., Palfi, S., Dolan, R., Matthews, R. T., and Beal, M. F. (1996). Inhibition of neuronal nitric oxide synthase prevents MPTP-induced parkinsonism in baboons. *Nat. Med.* **2,** 1017–1021.
8. Castagnoli, K., Palmer, S., Anderson, A., Bueters, T., and Castagnoli, N., Jr. (1997). The neuronal nitric oxide synthase inhibitor 7-nitroindazole also inhibits the monoamine oxidase-B-catalyzed oxidation of 1-methyl-4-phenyl-1,2,3,6-tetrahydropyridine. *Chem. Res. Toxicol.* **10,** 364–368.

K. S. Bankiewicz, J. R. Bringas, W. McLaughlin, P. Pivirotto, R. Hundal, B. Yang, M. E. Emborg, and D. Nagy

Somatix Therapy Corporation
Alameda, California 94501

Application of Gene Therapy for Parkinson's Disease: Nonhuman Primate Experience

Parkinson's disease (PD) is characterized by the progressive loss of dopaminergic cells and a deficiency of tyrosine hydroxylase (TH), which is necessary for the synthesis of dopamine (DA). As the degeneration of nigral dopamine neurons continues, intracellular biochemical responses occur in the presynaptic dopamine neurons and the postsynaptic striatal neurons that enable the nigrostriatal dopamine system to compensate for great depletions of DA prior to the onset of clinical symptoms. As the disease progresses, and up to 85% of nigral dopaminergic neurons have degenerated, the synthesis of TH and the release of DA from the presynaptic terminals is increased in a compensatory manner.

Early in the disease, this TH deficiency can be offset by oral administration of L-dopa. The exogenous L-dopa is taken into the cell via active amino acid transport, where it is acted on by L-AADC, thus circumventing the rate-limiting TH. Currently, it is thought that the exogenous L-dopa exerts its therapeutic effects by conversion to dopamine in the remaining striatonigral DA cells, which then release DA in a physiologically relevant manner.

Oral administration of L-dopa continues to be the mainstay of current therapy for PD but is limited by: (1) the inability to achieve site-specific delivery, which results in unwanted side effects and limits the amount of drug that can be given; (2) the inability to maintain a sustained drug level within the central nervous system (CNS), which is thought to contribute to unpredictable on-off effects in the treatment; and (3) the inability to prevent the progressive degeneration of dopamine-secreting nerve cells (1). Development of new therapeutic approaches to PD must address these issues.

I. Local Delivery of Therapeutic Agents into CNS

Delivery of macromolecular therapeutic agents into the CNS presents unique challenges. Systematically administered proteins do not enter the brain tissue due to exclusion by the brain capillary endothelial cells. Because the blood-

Advances in Pharmacology, Volume 42

1054-3589/98 $25.00

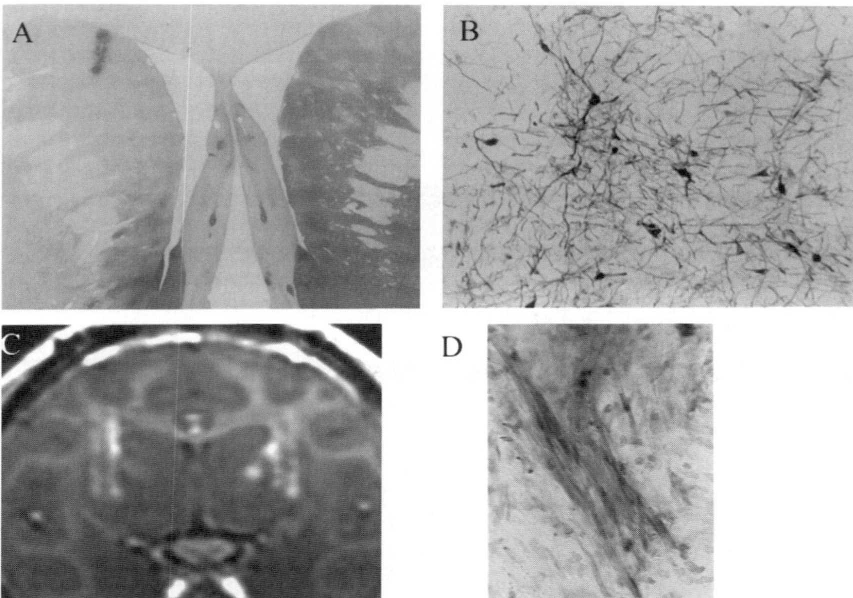

FIGURE 1 (*A*) Low-power photomicrograph of TH immunostain of MPTP-treated monkey striatum. The AAV-Th vector was infused into the caudate nucleus on the MPTP-treated side. (*B*) High-power view of TH-immunoreactive (TH-IR) cells infected by the AAV-TH vector at 21 days. (*C*) Magnetic resonance imaging of the monkey brain 5 days after implantation of fibroblasts transfected with the TH gene. (*D*) High-power view of TH-IR fibroblast graft in the monkey striatum at 4 mo.

brain barrier (BBB) is permeable to lipid-soluble compounds, enhancement of the lipid solubility of polypeptides has been attempted, but with limited success. The permeability through brain capillaries can also be increased with chemical modifications of proteins, such as cationization. Conjugation of peptides and proteins to antitransferrin-receptor antibodies also increases uptake into the brain, but this has resulted in systemic toxicity. Drug-impregnated polymer devices can provide sustained and local release into the CNS. However, such delivery is possible only for a limited period of time, and "empty" polymer capsules that have limited biodegradability remain in the CNS after therapy. Regardless of the method used for direct drug delivery, the ability of agents to distribute within the brain following introduction at a local site is a critical issue in determining drug efficacy and toxicity.

Direct administration of therapeutic agents into the CNS eliminates the need to bypass the BBB, thereby reducing systemic side effects. Delivery of proteins into the cerebral ventricles results in high protein concentration in the tissue adjacent to the ventricle, but there is limited penetration into surrounding regions, and clearance into the peripheral circulation is rapid. Direct injection into the brain parenchyma at the target site has been shown to be a more effective means of achieving high drug concentration at specific sites within the brain. Intraparenchymal cannulae and infusion pumps are required for chronic delivery, making this approach less desirable.

The short penetration distances of proteins pose an even greater problem in the human brain than in the relatively small rat brain. The treatment of many neurodegenerative disorders will require protocols that allow for delivery of sufficient levels of therapeutic agents over significant distances in brain parenchyma.

II. Cell-Based Delivery into CNS

Cellular implants can overcome the need for persistent and local physical delivery of therapeutic agents into the CNS. Limited but striking successes have been achieved through the implantation of DA-producing fetal cells in parkinsonian patients. The requirement of many embryos per patient is a major factor restricting the wide-scale adoption of this approach as a possible treatment for PD.

Several alternative approaches utilizing genetic engineering have been proposed by many investigators (2). Primary or immortalized cells have been engineered to produce a specific protein in culture, and the cells are then implanted into the host CNS either by direct cell transplantation or by encapsulating cells into semipermeable membranes. Other approaches consist of *in vivo* gene transfer based on direct introduction of genetic material into the CNS, using viral or synthetic vectors. *Ex vivo* methods using primary fibroblasts transduced with retroviral vectors and the *in vivo* method utilizing adeno-associated virus (AAV) are described here as possible means of chronic delivery of therapeutic agents into CNS. To develop a viable gene therapy for any neurological disorder, three requirements must be met: (1) demonstration of long-term expression, (2) lack of toxicity of the procedure, and (3) the ability to reverse the treatment if adverse events occur.

III. *Ex Vivo* Gene Therapy Using Autologous Cells

One method for delivery of L-dopa locally into the CNS is to transplant cells that synthesize and secrete L-dopa after *ex vivo* genetic modification. In this scenario, a PD patient's own cells would be genetically modified to express TH. Depending on the cell type being used as a vehicle, it may be necessary to express a second enzyme, guanosine triphosphate (GTP) cyclohydrolase I, to facilitate the production of the cofactor for TH activity, tetrahydrobiopterin (BH_4). These cells then would be grafted into appropriate sites in the striatum to provide a local supply of L-dopa at the sites in the brain that are normally innervated by dopaminergic neurons. The L-dopa secreted by these cells may then be taken up by remaining neuronal and nonneuronal cells, converted to DA, and released either in a normally regulated or in an unregulated fashion to ameliorate symptoms of the underlying parkinsonism in these patients.

IV. *In Vivo* Gene Transfer

Significant advances have recently been made regarding the introduction of genetic material into mammalian cells *in vivo*. Several vehicles have been used for *in vivo* transfer of cDNA, including herpes simplex viral vectors, adenoviral vectors, direct plasmid DNA transfer, and, most recently, retroviral vectors (lentivirus) (2). An additional promising vector system utilizes AAV, a nonpathogenic in human single-stranded DNA parvovirus (3). In addition to having no known pathology in humans, the wild-type virus is incapable of replication without helper functions provided by an adenovirus. Wild-type AAV has been shown to infect postmitotic mammalian cells and to insert its DNA into the host genome with some specificity for chromosome 19. It should be stressed, however, that at the present time, it is difficult to predict which of these *in vivo* delivery systems will prove to be the most optimal and clinically applicable.

V. Delivery of TH- AAV Vector into Monkey Striatum

In addition to the difficulty of introducing therapeutic agents into the CNS, another significant challenge is presented by PD. Because the disease affects mostly nigrostriatal and partially mesolimbic systems, therapeutic agents must be spread over a considerable area of the brain. This can be accomplished by utilizing multiple delivery sites. However, with the increasing number of penetration sites into the CNS, the risk of complication increases as well. At the present time, direct gene delivery into the CNS can be accomplished only by introduction of viral vectors into the brain parenchyma. Injection of the AAV vector results in a limited and highly concentrated infection of the brain located within close proximity to the injection cannulae. Such a limited area of coverage is not desirable when large regions of the human brain are to be targeted. In addition, due to the limited diffusion of therapeutic agents in the brain, infections are highly localized. An overproduction of a gene product in a limited area can result in local toxicity, which gives rise to unwanted side effects.

Attempts to spread the infection into larger areas of the brain, using a slow infusion technique and fused silica cannulae, resulted in a higher number of infected cells and a less concentrated area of infection. No signs of toxicity in the striatum related to the volumes or techniques used were observed (4).

VI. Gene Expression at Four Months

A. *In Vivo* Approach

Gene expression was detected by immunostaining in monkey striatum after 14–21 days and at 3 mo survival for both β-galactosidase (β-gal) and TH-AAV vectors. Gene expression after 3 mo appeared less robust than at the earlier time points; however, the level of gene expression over time is now under investigation. Based on double-labeling immunocytochemistry using antibodies against the transgene and neurofilament or glial fibrillary acidic protein (GFAP), it appears that the AAV vector used infects, almost exclusively, neurons in the monkey brain.

Some glialike cells were detected expressing β-gal. However, they constituted a minority of AAV-infected cells, because they were estimated to be less than 5% of all the cells detected to express β-gal. The number of β-gal-positive cells was greatest when the slow infusion methods were applied. In this case, it was estimated that there were between 14,000 and 31,000 cells per single delivery site (4).

B. *Ex Vivo* Approach

Expression of the TH gene was examined at 4 mo after the fibroblast implantation by *in situ* hybridization, immunostaining with a TH antibody, and Southern polymerase chain reaction. The application of the photographic emulsion combined with H&E counterstaining allows the cellular resolution of *in situ* hybridization. The TH mRNA–expressing cells were recognized by the accumulation of silver grains over the cell bodies. The grains were highly concentrated over the groups of elongated fibroblast inside the implants. Positive TH staining was observed in the fibroblasts implants as well.

So far, we have examined gene expression in genetically modified fibroblasts for up to 4 mo. We have demonstrated persistent expression in the grafted cells for this period. The addition of GTP-cyclohydrolase-I vectors, and, therefore, the production of BH_4, appears to facilitate TH immunostaining, perhaps by lengthening the half-life of the TH enzyme in coinfected fibroblasts. As has been shown in rat experiments, the combination of expressing TH and GTP-cyclohydrolase-I cDNAs in grafted cells also makes it possible to detect L-dopa by microdialysis.

The application of AAV-based vectors into the monkey striatum results in robust gene expression, which is limited mostly to neuronal populations of cells in the stratum. The mechanism for specific targeting of AAV vector expression to neurons is not clear at this time. However, we have observed this pattern, using both the cytomegalovirus immediate–early promoter and a Moloney Murine leukemia virus promoter driving expression in the AAV vector. A critical question still remains regarding the regulation of gene expression. The development of regulatable promoters might help to overcome this in the near future. Further technological advances are required to optimize gene delivery, regulation of gene expression, and testing in appropriate functional models before gene therapy can be used extensively.

References

1. Chase, T. N., Juncos, J., Serrati, C., Fabbrini, G., and Bruno, G. (1986). Fluctuations in response to chronic levodopa treatment of therapy: Pathogenic and therapeutic considera-

tions. *In* Advances in Neurology: Parkinson's Disease. (M. D. Yahr and K. J. Bergman, eds.), pp. 477–480. Raven Press, New York.
2. Freese, A., Stern, M., Kaplitt, M. G., O'Connor, W., Abbey, M., O'Connor, M. J., and During, M. J. (1996). Prospects for gene therapy in Parkinson's disease. *Mov. Disord.* 5, 469–488.
3. Muzyczka, N. (1992). Use of adeno-associated virus as general transduction vector for mammalian cells. *Curr. Top. Microbiol. Immunol.* 158, 97–129.
4. Bankiewicz, K. S., Snyder, R., Zhou, S. Z., Morton, M., Conway, J., and Nagy, D. (1996). Adeno-associated (AAV) viral vector-mediated gene delivery in non-human primates. *Soc. Neurosci. Abstr.* **22**, 768.

Barbara K. Lipska and Daniel R. Weinberger

Clinical Brain Disorders Branch
National Institute of Mental Health
Washington, D.C. 20032

Prefrontal Cortical and Hippocampal Modulation of Dopamine-Mediated Effects

Animal models of diseases are widely embraced as tools for studying pathophysiological mechanisms and developing effective treatments. However, a different standard tends to be applied to animal models of psychiatric diseases, which, in general, are thought to be of limited value. One reason for this is that mental diseases are uniquely human in the sense that they affect mood, thought, perception, and cognition, and, thus, cannot be reproduced in phylogenetically lower species. Another reason is that much of the data about mental diseases are inconclusive, and usually neither the basic inducing conditions nor the fundamental neurobiological mechanisms are known. This seems particularly true for schizophrenia, an illness characterized by dramatic abnormalities in perception and comportment, as well as in attention, memory, and executive functions of unknown origin and pathophysiology. The situation is changing, however, with the advent of new technologies and more sensitive tools to study brain function. For instance, there is rapidly growing evidence of structural abnormalities in brains of patients with schizophrenia. Both brain imaging and postmortem neuroanatomical studies show morphometric changes in the cerebral cortices of schizophrenics. In addition to structural abnormalities, functional imaging studies indicate cortical malfunction and suggest functional disorganization of prefrontal-temporal-limbic cortical connectivity, perhaps consistent with evidence of a developmental cortical defect. Other observations

Advances in Pharmacology, Volume 42

1054-3589/98 $25.00

also indicate that the disease process in patients with schizophrenia most likely occurs early in life, probably prenatally, and that the changes in brain morphology are nonprogressive and not related to the duration of the illness. Moreover, the beneficial effects of neuroleptic drugs that ameliorate some of the symptoms implicate the involvement of a dysfunctional dopaminergic system in schizophrenia. Overall, recent clinical and basic research data suggest that schizophrenia is related to subtle maldevelopment of prefrontal-temporal-limbic cortices and abnormal dopamine (DA) function.

In our experimental approach to model some aspects of schizophrenia in the animal, we have utilized these two seemingly divergent lines of evidence. In subsequent paragraphs, we will demonstrate that this approach led us to construct an animal model with a high degree of analogy to some aspects of schizophrenia, and indicated that a neurodevelopmental hypothesis of schizophrenia based on cortical maldevelopment and adult onset dopaminergic dysregulation is biologically plausible.

We have attempted to model some aspects of schizophrenia in the rat by reproducing what is presumably a primary defect in this illness—damage of a cortical region in a developing brain. We hypothesized that this primary defect would result in secondary effects analogous to those observed in schizophrenia, including dysregulation of the DA system and frontal cortical dysfunction. We chose the hippocampal formation as a target area to induce such a lesion. The hippocampus appears to be the region of the most reproducible anatomical pathology found in patients with this disorder. The hippocampal region constitutes, however, a large part of the rat brain and is anatomically and functionally heterogeneous. The results of anatomical studies led us to posit that the ventral hippocampus (VH) may make a predominant contribution to hippocampal regulation of the mesolimbic DA system. Abnormal function of this system is strongly implicated in schizophrenia. Moreover, the VH is the part of the rat hippocampal formation that, in terms of anatomical connections, corresponds to the human anterior hippocampus, the site of structural pathology in schizophrenic patients.

Although the psychotic symptoms of schizophrenia occur around the time of puberty, several lines of evidence suggest that the disorder arises from a neurodevelopmental event probably in utero around the second trimester of gestation. This developmental period is thought to be critical for migration and setting of young neurons into the cortical target areas. In contrast to primates, however, rats undergo considerable brain development postnatally. Accordingly, the first or second postnatal week in a rat's life roughly corresponds to the late second trimester of gestation in humans in terms of similar neurodevelopmental changes. Thus, we have chosen the seventh postnatal day (PD7) to induce an excitotoxic VH lesion in the newborn rat. It is widely recognized that the effects of early brain damage differ depending on age at which the lesion is induced. Based on other studies showing that the capacity for compensating for a loss of function diminishes with age, we hypothesized that VH damage produced early (PD3 and PD7) would lead to less severe abnormalities, perhaps with a different temporal profile of change than a similar lesion acquired later (PD14) or in adulthood.

Experimental damage to the VH of the neonatal rat was induced with ibotenic acid, an excitatory neurotoxin with presumably axon-sparing properties. The electron microscopic studies show, however, that ibotenic acid may cause more generalized damage, so it is likely that at least extrinsic myelinated axons were affected. Moreover, the cavitation commonly observed in the lesion site indicated as well that axons might have been damaged. Nevertheless, ibotenic acid is much less destructive than surgical or electrolytic techniques.

These neonatal excitotoxic lesions of the rat VH produced on PD7 resulted in a temporally specific pattern of abnormalities in a number of DA-related behavioral paradigms (1–3). When tested before puberty (PD35), rats with neonatal lesions of the VH did not demonstrate any apparent DA-linked behavioral abnormalities. For example, their locomotion after exposure to novelty, saline injection, swim stress, or amphetamine did not differ from that in control rats. Also, haloperidol-induced catalepsy, apomorphine-induced stereotypies, and prepulse inhibition of startle (PPI) tested in the lesioned rats at PD35 were comparable to controls. Thus, despite early induced damage, these rats seem to exhibit relative functional sparing. However, such sparing is transient because it fails around puberty. At a postpubertal age (PD56), the VH-lesioned animals displayed markedly increased locomotion compared with the sham-operated rats when exposed to novelty, swim stress, and after receiving saline or amphetamine (1.5 mg/kg, i.p.) injections. These behavioral changes are thought to be primarily linked to increased mesolimbic DA transmission. At PD56, these rats are also hyperresponsive to presumably postsynaptic DA manipulations within the nigrostriatal system. For instance, at PD56, reduced haloperidol-induced (1 mg/kg, i.p.) catalepsy and enhanced apomorphine-induced (0.75 mg/kg, s.c.) stereotypies are observed in these animals. Moreover, subchronic treatment with either haloperidol (0.1 mg/kg, i.p.) or clozapine (4 mg/kg, i.p.) suppresses novelty-induced hyperlocomotor activity in the lesioned animals, further supporting possible involvement of DA systems in the behavioral abnormalities. Neonatally lesioned rats also demonstrate reduced PPI when tested postpuberty (4). Because PPI has been shown to be reduced by D2 DA receptor activation and restored by antidopaminergic drugs, these data again suggest excessive dopaminergic activity associated with this lesion at PD56.

When tested postpubertally, neonatally lesioned rats also demonstrate impairments that may correspond to the negative symptoms of schizophrenia, such as working-memory deficits and deficits in social interactions. We have recently found that the neonatal VH lesion dramatically impairs performance of adult rats in the delayed-alternation test with a delay of 10 sec (Moghaddam *et al.*, unpublished observations). A majority of these neonatally lesioned rats were unable to learn the task, and those few animals that reached the criterion of 80% correct choices needed a significantly longer time to learn it as compared with the sham-operated controls. Interestingly, rats with analogous lesions produced in adulthood did not show imparments in this test, suggesting that the early but not the adult lesion affects the development of the brain region that is critically important for working memory, such as, for instance, the medial prefrontal cortex. We have also investigated social behaviors of neonatally lesioned animals and demonstrated that the lesion results in a significant reduction of social interactions both before and after puberty. Chronic treatment with clozapine did not improve these impairments (5). These results indicate that neonatal hippocampal lesions result in behavioral abnormalities that resemble negative symptoms of schizophrenia, such as cognitive impairments and social withdrawal, and that these deficits appear relatively early in development (i.e., before puberty).

In contrast to the delayed appearance of disturbances associated with the neonatal VH lesions, analogous excitotoxic lesions of the adult rat VH and lesions at PD14 produce marked behavioral changes that appear relatively shortly after the lesion (6). Rats receiving lesions of the VH as adults and at PD14 are hyperactive after amphetamine injections after 2 and 3 wk, respectively. Furthermore, their response to apomorphine is qualitatively different from the neonatally lesioned rats; that is, the adult- and PD14-lesioned animals show reduced rather than potentiated stereotypic behaviors following apomorphine administration. As mentioned before, in contrast to the neonatally lesioned rats, their performance in the working-memory test (delayed alternation) is not impaired. Adult-lesioned animals differ also from neonatally lesioned animals in their PPI (7). Their response is altered only after apomorphine, but not after saline injection as compared with sham-operated controls. Taken together, the results of these studies demonstrate that a similar lesion induced later in development (PD14) or in adulthood results not only in a different temporal profile of changes, but also in qualitative differences in behavioral abnormalities, as compared with the lesion induced early in life.

The results of this series of studies indicate that some behavioral abnormalities associated with the neonatal hippocampal lesion and corresponding to positive symptoms of schizophrenia remain relatively quiet until a certain time in development, while others (i.e., those that correspond to negative symptoms) appear soon after the lesion. The profile of changes depends on the age at which damage is induced.

References

1. Lipska, B. K., Jaskiw, G. E., and Weinberger, D. R. (1993). Postpubertal emergence of hyperresponsiveness to stress and to amphetamine after neonatal hippocampal damage: A potential animal model of schizophrenia. *Neuropsychopharmacology* **9**, 67–75.
2. Lipska, B. K., and Weinberger, D. R. (1993). Delayed effects of neonatal hippocampal damage on haloperidol-induced catalepsy and apomorphine-induced stereotypic behaviors in the rat. *Dev. Brain Res.* **75**, 213–222.
3. Lipska, B. K., and Weinberger, D. R. (1994). Subchronic treatment with haloperidol or clozapine in rats with neonatal excitotoxic hippocampal damage. *Neuropsychopharmacology* **10**, 199–205.
4. Lipska, B. K., Swerdlow, N. R., Geyer, M. A., Jaskiw, G. E., Braff, D. L., and Weinberger, D. R. (1995). Neonatal excitotoxic hippocampal damage in rats causes postpubertal changes in prepulse inhibition of startle and its disruption by apomorphine. *Psychopharmacology* **122**, 35–43.
5. Sams-Dodd, F., Lipska, B. K., Weinberger, D. R. Neonatal lesions of the net ventral hippocampus result in hyperlocomotion and deficits in social behavior in adulthood. *Psychopharmacology* (in press).
6. Lipska, B. K., Jaskiw, G. E., Braun, A. R., and Weinberger, D. R. (1995). Prefrontal cortical and hippocampal modulation of haloperidol-induced catalepsy and apomorphine-induced stereotypic behaviors in the rat. *Biol. Psychiatry* **38**, 255–262.
7. Swerdlow, N. R., Lipska, B. K., Weinberger, D. R., Braff, D. L., Jaskiw, G. E., and Geyer, M. A. (1995). Increased sensitivity to the sensorimotor gating-disruptive effects of apomorphine after lesions of frontal cortex or hippocampus in adult rats. *Psychopharmacology* **122**, 27–34.

J. David Jentsch, John D. Elsworth, Jane R. Taylor, D. Eugene Redmond, Jr., and Robert H. Roth

Departments of Neurobiology, Psychiatry, and Pharmacology
Yale University School of Medicine
New Haven, Connecticut 06510

Dysregulation of Mesoprefrontal Dopamine Neurons Induced by Acute and Repeated Phencyclidine Administration in the Nonhuman Primate: Implications for Schizophrenia

It has been more than 35 years since it was first noted that phencyclidine (PCP, or Sernyl) induces psychotomimetic reactions in normal humans (1). Since that time, increasing usage of PCP as a psychedelic drug by humans has provided more insights regarding PCP as a behavioral and neurological model of schizophrenia, but as of yet, the critical neurobiological effects of PCP underlying the induction of schizophrenic-like behavior remain unknown.

PCP induces several distinct behavioral effects in humans and animals that appear to represent schizophrenia-like symptomatology. In humans, acute PCP exposure can induce both positive and negative symptoms of schizophrenia (2). These acute effects, however, are usually transient. In contrast, repeated exposure to PCP in humans can lead to enduring presentation of both negative and positive symptoms of schizophrenia (2). Thus, repeated PCP administration may induce neurobiological defects that mimic those present in the brain of patients with schizophrenia.

Despite the vast amount of data suggesting that PCP models the primary behavioral symptoms of schizophrenia, it is still unclear what pathophysiology underlies the PCP-induced behavior. Several plausible possibilities exist. First, PCP's primary pharmacological action is noncompetitive antagonism of the N-methyl-D-aspartate (NMDA)–sensitive glutamate receptor; thus, a hypoglutamatergic mechanism has been posited to underlie these effects (3). A second possible mechanism responsible for PCP-induced psychosis involves the anatomically distinct neuropathological effects of this drug. Neurons within the corticolimbic axis appear to be sensitive to PCP, undergoing pathological changes or even necrosis following acute or repeated administration. This pattern of neural damage has been suggested to roughly parallel that observed in

Advances in Pharmacology, Volume 42

1054-3589/98 $25.00

schizophrenia (4). A final substrate by which PCP can induce psychotic-like symptoms in humans is dysregulation of mesotelencephalic dopamine (DA) systems. Acute PCP administration has been shown to activate the mesoprefrontal DA system in the rat (5).

In this chapter, we will discuss some of our recent studies on the effects of acute and repeated PCP administration on the rodent and nonhuman primate DA systems and how these neurobiological effects may underlie the psychotomimetic properties of PCP. Finally, we will evaluate the relevance of DAergic dysfunction for the hypoglutamatergic and neuropathological hypotheses of PCP's effects.

We have replicated the finding that PCP markedly increases prefrontal cortex (PFC) and nucleus accumbens DA metabolism in the rodent (Fig. 1) and have extended those findings by demonstrating that the isomers of 3-amino-1-hydroxypyrrolid-2-one (HA966) and the α_2-noradrenergic receptor agonist clonidine, which can block stress-induced increases in PFC DA metabolism, can also prevent the activation of mesoPFC DA neurons after acute PCP without altering DA turnover in that area on their own. These data suggest

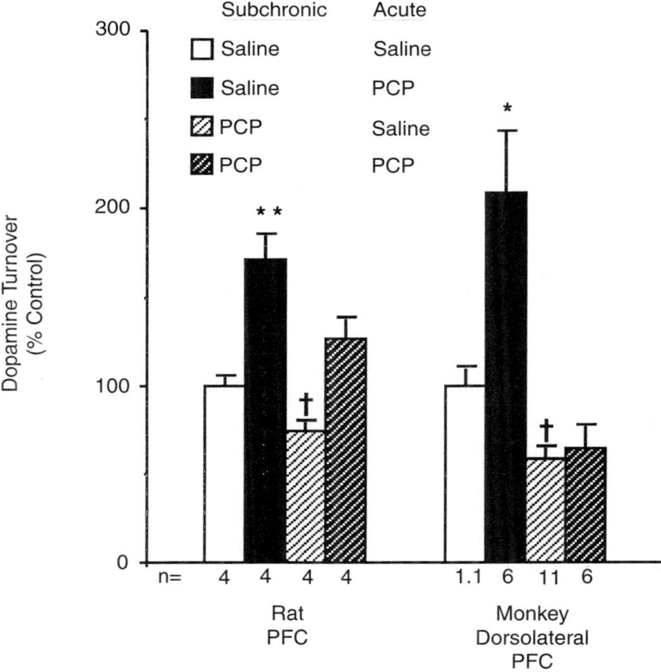

FIGURE 1 Acute PCP administration increases while subchronic PCP exposure decreases DA turnover in the rat PFC cortex (*left*) and monkey dorsolateral PFC (*right*). In addition, the effects of an acute PCP challenge dose on PFC DA turnover is reduced in the rat and eliminated in the monkey after subchronic PCP administration. Significantly increased versus saline/saline: $^{*}p < .05$; $^{**}p < .01$. †Significantly reduced versus saline/saline: $p < .05$.

that PCP-stimulated PFC DA metabolism is dependent on increased impulse flow in mesoPFC DA neurons, a conclusion supported by *in vivo* microdialysis studies demonstrating that PCP-induced PFC DA release is tetrodotoxin-sensitive (6), and that the enantiomers of HA966 and clonidine can selectively modulate stimulated states of these neurons.

In the primate, acute PCP administration causes a selective activation of DA metabolism in the frontal cortex (see Fig. 1), while sparing subcortical DA systems. In addition, as in the rodent, S-(-)HA966 blocks PCP-induced increases in DA turnover in the monkey frontal cortex (7). This finding demonstrates that the DAergic activation noted in rodent brain after PCP is conserved in the primate and that this effect can be pharmacologically modulated in a fashion similar to that observed in the rodent.

These neurochemical effects of acute PCP exposure do have relevance to observed behavioral effects of PCP in humans. Research from our laboratory has already demonstrated that increased DA turnover in rodent and monkey PFC impairs spatial working-memory function (8). Thus, it is likely that the impairments in cognitive functioning observed in normal human subjects after acute PCP exposure are due to a hyperdopaminergic state of the PFC and that compounds that attenuate this increased DA-receptor activation in the PFC will ameliorate the cognitive deficits.

The marked activation of mesoPFC DA systems after an acute PCP challenge is consistent with hypotheses involving hypoglutamatergic and neuropathological mechanisms in the PCP model of schizophrenia. First, there is developing evidence that NMDA receptors have a predominant role in regulating recurrent inhibition (RI) in the cortex (9) and that by blocking NMDA receptors, a preferential reduction in RI is lost, leading to increases in the firing of excitatory corticomesencephalic fibers. Thus, acute PCP may lead to a stimulation of DA neurons indirectly by activating cerebral cortical output circuits. This loss of RI in the cortex and stimulation of monoamine neuron firing may explain the almost paradoxical psychostimulant effects of this potent glutamatergic antagonist.

Increases in DA transmission by acute PCP challenge may also explain the neuropathological effects of this compound. Interestingly, the antipsychotic drugs haloperidol and clozapine, both of which are DA-receptor antagonists, and the α_2-noradrenergic agonist clonidine, which blocks the activation of PFC DA turnover after PCP exposure, all prevent the PCP-induced neural injury in rat brain (3). This suggests that activation of DA systems by PCP may be involved in the development of neuropathology after acute PCP administration.

Data from schizophrenics, however, support the idea that there is a hypoactivity, rather than hyperactivity, of the PFC DA innervation in schizophrenia. Interestingly, we have recently observed that subchronic PCP administration, a paradigm that induces enduring schizophreniform symptoms in humans, reduces basal DA turnover in the rat and monkey PFC (see Fig. 1). In addition, the response to an acute PCP challenge dose in the rat and monkey PFC is blunted after repeated PCP exposure (see Fig. 1). These data suggest that there is an inhibition of basal and stimulated DA metabolism in rat and monkey brain after repeated PCP administration (10).

To determine whether there are functional consequences to this DAergic depression, we tested monkeys on an object retrieval-detour task, the performance of which is sensitive to reductions in mesotelencephalic DA function. Successful performance of this task is dependent on the capacity for response inhibition and the absence of perseverative behavior. Normal, saline-treated monkeys perform almost perfectly on this task, while monkeys previously treated subchronically with PCP were significantly impaired, as measured by a significantly higher incidence of perseverative response than controls and increased context-specific inappropriate responses (10). Perseverative behavior and incapacity for response inhibition are behavioral deficits displayed by schizophrenics. Finally, we most recently observed that the cognitive deficits displayed by monkeys subchronically treated with PCP are partially ameliorated with the atypical antipsychotic drug clozapine (10), as are the negative symptoms of schizophrenia. Further, preliminary data suggest that a D4 mechanism is critically involved in the action of clozapine in this paradigm. These data suggest that subchronic PCP administration in the monkey induces a profound and chronic cognitive dysfunction similar to that manifested in schizophrenia and that the PCP-induced impairments respond to pharmacological treatment, as do the negative symptoms of schizophrenia.

Taken together, our data indicate that repeated PCP administration may model the cognitive dysfunction of schizophrenia and suggest that this dysfunction may be due to an inhibition of DAergic function in the PFC. Further, this paradigm may represent a step forward in biological psychiatric research by offering a tool for evaluating the pathophysiology of the deficits of schizophrenia and a mechanism for the evaluation of novel agents that may be able to ameliorate the typically treatment-refractory negative symptoms in this debilitating disorders.

References

1. Luby, E. D., Cohen, B. D., Rosenbaum, G., Gottleib, J. S., and Kelly, R. (1959). Study of a new schizophreniomimetic drug—Sernyl. *Arch. Neurol. Psychiatry* **81**, 363–369.
2. Pearlson, G. D. (1981). Psychiatric and medical syndromes associated with phencyclidine (PCP) abuse. *John Hopkins Med. J.* **148**, 25–33.
3. Olney, J. W., and Farber, N. B. (1995). Glutamate receptor dysfunction and schizophrenia. *Arch. Gen. Psychiatry* **52**, 998–1007.
4. Ellison, G. (1995). The N-methyl-D-aspartate antagonists phencyclidine, ketamine and dizocilpine as both behavioral and anatomical models of the dementias. *Brain Res. Rev.* **20**, 250–267.
5. Deutch, A. Y., Tam, S. Y., Freeman, A. S., Bowers, M. B., and Roth, R. H. (1987). Mesolimbic and mesocortical activation induced by phencyclidine: Contrasting pattern to striatal response. *Eur. J. Pharmacol.* **134**, 257–264.
6. Nishijima, K., Kashiwa, A., Hashimoto, A., Iwama, H., Umino, A., and Nishikawa T. (1996). Differential effects of phencyclidine and methamphetamine on dopamine metabolism in rat frontal cortex and striatum as revealed by *in vivo* dialysis. *Synapse* **22**, 304–312.
7. Jentsch, J. D., Elsworth, J. D., Redmond, D. E., and Roth, R. H. (1997). Phencyclidine increases monoamine turnover in rodent and monkey forebrain: Modulation by the isomers of HA966. *J. Neurosci.* **17**, 1769–1776.

8. Murphy, B. L., Arnsten, A. F. T., Goldman-Rakic, P. S., and Roth, R. H. (1996). Increased dopamine turnover in prefrontial cortex impairs spatial working memory in rats and monkeys. *Proc. Natl. Acad. Sci. U.S.A.* **93,** 1325–1329.
9. Grunze, H. C., Rainnie, D. G., Hasselmo, M. E., Barkai, E., Hearn, E. F., McCarley, R. W., and Greene, R. W. (1996). NMDA-dependent modulation of CA1 local circuit inhibition. *J. Neurosci.* **16,** 2034–2043.
10. Jentsch, J. D., Redmond, D. E., Jr., Elsworth, J. D., Taylor, J. R., Youngren, K. D., and Roth, R. H. (1997). Enduring cognitive dysfunction and cortical dopamine deficits in monkeys after long-term phencyclidine administration. *Science* (in press).

T. H. Svensson, J. M. Mathé, G. G. Nomikos, B. Schilström, M. Marcus, and M. Fagerquist

Department of Physiology and Pharmacology
Division of Pharmacology
Karolinska Institutet
S-171 77 Stockholm, Sweden

Interactions between Catecholamines and Serotonin: Relevance to the Pharmacology of Schizophrenia

The glutamate-dopamine (DA) hypothesis of schizophrenia is largely based on indirect, pharmacological evidence. This includes the schizophrenomimetic effects of various N-methyl-D-aspartate (NMDA)–receptor antagonists, such as phencyclidine (PCP), MK-801, or ketamine, which can appear even with acute drug administration to healthy volunteers, as well as the fact that all approved antipsychotic drugs have one effect in common—to impair brain DA neurotransmission, mostly via DA–D_2-receptor antagonism—and that paranoid schizophrenics tend to worsen on administration of indirectly or directly acting DA-receptor agonists, as well as on PCP-like agents. However, although most neuroleptics, including selective D_2 antagonists, in positron emission tomography studies display a D_2-receptor occupancy of about 75% in clinically effective doses, the atypical antipsychotic drug clozapine shows only about 50% D_2-receptor occupancy in the human brain, in spite of superior clinical efficacy. Therefore, it is of considerable interest to find out how other properties of clozapine and other potentially atypical neuroleptics, such as a potent antagonistic action at central 5-HT_{2A} and α_1-adrenoceptors, may contribute to the antipsy-

Advances in Pharmacology, Volume 42

1054-3589/98 $25.00

chotic efficacy. Here we report recent pharmacological results demonstrating beyond doubt the implication of the mesoaccumbens DA pathway in the behavioral effects of MK-801 in the rat, as well as the stabilizing effects of potent 5-HT_2 and α_1-adrenoceptor blocking agents on different dysfunctions in the mesocorticolimbic DA system produced by systemic administration of the NMDA-receptor antagonists, with associated behavioral correlates. In addition, a biochemical and behavioral analysis of the 5-HT_{2A}/DA-D_2-receptor antagonist combination will be provided. The methods used include single-cell recording techniques, *in vivo* voltammetry as well as microdialysis in freely moving animals together with high-performance liquid chromatography with electrochemical detection, and, finally, several behavioral techniques (1).

Systemic administration of low doses of MK-801 to rats, which induces profound locomotor stimulation, yet without significant ataxia (Fig. 1B), caused a moderate activation of the firing rate of ventral tegmental area (VTA) DA neurons. However, whereas cells in the parabrachial pigmented nucleus, a subdivision of the VTA that provides *inter alia* the prefrontal DA projection, displayed a reduced burst firing, cells in the paranigral nucleus, another subdivision of the VTA projecting subcortically (e.g. to the nucleus accumbens [NAC]), displayed a burst-like, high-frequency firing pattern relatively similar to that obtained by microiontophoretic application of kainate or quisqualate onto VTA cells. Previous results from our and other laboratories indicate that spontaneous burst firing in VTA DA cells is regulated by their excitatory amino acid (EAA) inputs. However, evidence also supports a differential regulation by glutamate of subdivisions of VTA cells, with the NMDA-receptor subtype preferentially controlling neurons projecting to the prefrontal cortex (PFC) and the kainate subtype largely modulating the mesoaccumbens neurons (2). We have now, accordingly, found that local administration of CNQX, which blocks AMPA/kainate receptors, into the VTA can even totally abolish the evoked DA release in the NAC caused by systemic MK-801, as well as the associated locomotor stimulation (Figs. 1A and B). In addition, our preliminary evidence indicates that systemic administration of MK-801 (0.1 mg/kg s.c.) causes an increased extracellular concentration of glutamate and asparate locally in the VTA. Thus, the locomotor stimulant action of low doses of MK-801 in the rat is, indeed, causally related to the increased neuronal activity of the mesoaccumbens DA neurons, which is generated within the VTA, probably as a consequence of enhanced release of EAAs acting at AMPA/kainate receptors. Notably, typical physiological bursts were rarely seen in the VTA cells following MK-801 administration. Previous results in primates indicate that VTA DA neurons are involved with phasic changes in activity in basic attentional and motivational processes underlying learning and cognitive behavior, and that DA in the PFC is important for behavioral performance (e.g., of delayed response tasks). Phasic DA responses appear to signal the salience of various stimuli. Moreover, our recent results demonstrate that, rather than the absolute mean discharge rate of midbrain DA cells, the temporal organization of the action potentials they generate conveys information to their target areas, as reflected in the expression of immediate-early genes in DA target areas (3). Thus, the distorted firing patterns of the VTA DA neurons caused by the psychotomimetic NMDA-

A

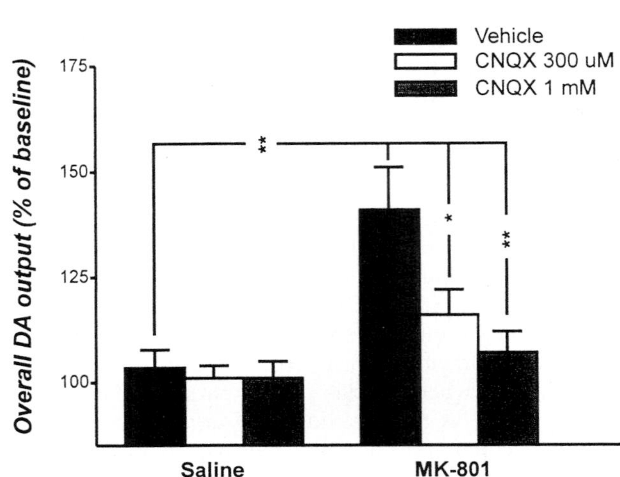

B

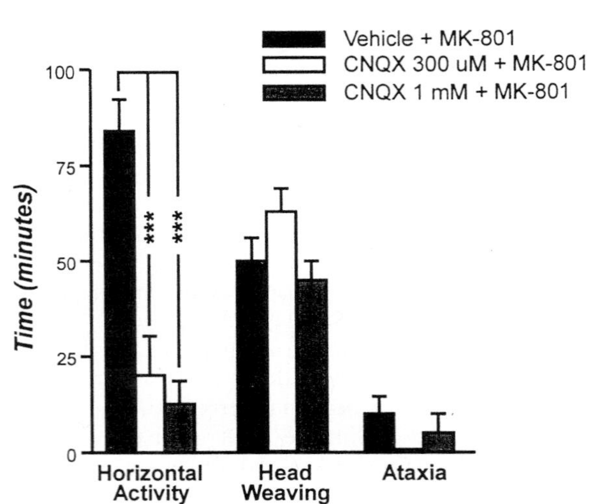

FIGURE 1 Effects of systemic administration of MK-801 on DA release in the NAC, as assessed by microdialysis, and behavior in rats during perfusion of the VTA with the AMPA/kainate-receptor antagonist CNQX or vehicle. CNQX, 300 μM or 1 mM, or vehicle was perfused for 240 min, starting 40 min before the subcutaneous injection of MK-801, 0.1 mg/kg. (*A*) Data represent means + SEM of the percent change of extracellular DA levels in the NAC during the 200-min observation period. (*B*) Data represent means + SEM of the total time during which the animals displayed the behavioral signs. * $p < .05$; ** $p < .01$; *** $p < .001$.

receptor antagonists should significantly impair attentional, motivational, and cognitive aspects of behavior.

Because schizophrenia frequently has been associated with relative prefrontal hypometabolism, so-called hypofrontality, we previously explored the effects of functional inactivation of the medial PFC (mPFC) of the rat, by means of local cooling or lidocaine application on the activity of the VTA DA neurons. The mPFC inactivation was also found to cause a marked distortion of the firing pattern (e.g., a selective reduction in burst firing), probably due to inhibition of an afferent EAA pathway from the mPFC, as well as regularization of firing (1). Notably, also PCP-intoxicated subjects may show prefrontal hypometabolism similar to that observed in schizophrenia, and the hypofrontality state has been claimed to correlate with reduced DA metabolism in brain as well with negative symptoms, such as avolition, attentional deficits, and emotional indifference.

In subsequent pharmacological experiments in rats, we observed that drugs with potent 5-HT_{2A}-receptor antagonistic action, which have been claimed to be particularly effective against negative symptoms in schizophrenia, effectively restored the impaired burst firing in DA cells caused by the psychotomimetic NMDA-receptor antagonists, as well as by the PFC inactivation. In other, biochemical studies utilizing microdialysis and *in vivo* voltammetry, we observed that administration of high doses of ritanserin or atypical neuroleptics with prominent 5-HT_2-receptor antagonistic action produced preferential activation of the PFC DA projection, whereas classical DA–D_2-antagonists, such as haloperidol, caused preferential activation of subcortical DA projections. In fact, adding ritanserin to raclopride markedly enhanced the raclopride-induced increase in DA release in the mPFC, probably mediated via augmentation of the raclopride-induced increase in burst firing of mesocortical DA neurons, but failed to affect the action of raclopride on striatal DA release. In other experiments, we found that drugs with potent 5-HT_2-receptor antagonistic action, such as amperozide, ritanserin, sertindole, and MDL 100, 907, as well as low doses of risperidone or clozapine, increased DA levels to a greater extent in the shell than in the core of the NAC, whereas drugs with a prominent D_2-receptor antagonistic action, such as haloperidol and raclopride, as well as high doses of risperidone or clozapine, elicited a larger DA increase in the core than in the shell of the NAC (4). Because the shell region is closely linked to the limbic system and the PFC, these results support our previous findings. Most recently, we have observed that adding ritanserin administration to raclopride in rats enhances the raclopride-induced suppression of conditioned avoidance response, without affecting the raclopride-induced catalepsy at either maximal or submaximal doses (5). Taken together, these data suggest that addition of 5-HT_2-receptor antagonism to a D_2-receptor antagonist can enhance the antipsychotic-like properties of the D_2-receptor antagonist, without increasing the propensity for extrapyramidal side effects.

In separate experiments in rats, the tentative, auxilliary antipsychotic activity of α_1-adrenoceptor antagonism was assessed within the framework of the PCP model of schizophrenia. Briefly, administration of prazosin antagonized the behavioral stimulation (i.e., locomotor hyperactivity) induced by systemic administration of MK-801. The α_1-adrenoceptor antagonist also selectively suppressed the evoked, burst-like firing of subcortically projecting DA cells in the paranigral subdivision of the VTA, as well as the evoked DA release in the NAC, associated with systemic MK-801 administration in low doses (6). This effect may, at least in the rat, be related to blockade of a previously demonstrated, noradrenergic excitatory input to the VTA DA neurons, mediated via α_1 adrenoceptors. Because prazosin suppressed only evoked but not basal DA release in the NAC, such an auxilliary, DA-inhibiting mechanism might be of particular clinical significance in association with stressful environmental challenge (i.e., conditions with enhanced central noradrenergic activity).

Taken together, this research clearly suggests that the potent antagonistic effect at central 5-HT_{2A} and α_1-adrenergic receptors of several atypical neuroleptics may well contribute significantly to their clinical profile. Specifically, 5-HT_{2A}-receptor antagonism can help to improve defective phasic DA signaling in the mesocortical DA projection, and may tentatively be of particular value against negative symptoms and cognitive dysfunction, whereas the α_1-adrenoceptor blockade, at a presynaptic level, might add to the postsynaptic D_2-receptor blockade in antagonizing a hyperactive or hyperresponsive mesolimbic DA projection, with a more efficient control of positive symptomatology. This hypothesis also fits another animal model of schizophrenia, created by ibotenic acid lesions of the ventral hippocampus, which likewise proposes an impaired DA function in the PFC concomitant with a hyperactive or hyperresponsive mesolimbic DA system (7). Thus, the potent 5-HT_{2A} and α_1-adrenergic receptor-blocking effects of a number of atypical neuroleptic drugs may, in different ways, act to stabilize several dysfunctions within the mesocorticolimbic DA system, which may contribute to the generation of psychotic symptomatology.

References

1. Svensson, T. H., Mathé, J. M., Andersson, J. L., Nomikos, G. G., Hildebrand, B. E., and Marcus, M. (1995). Mode of action of atypical neuroleptics in relation to the phencyclidine model of schizophrenia: Role of 5-HT_2- and α_1-receptor antagonism. *J. Clin. Psychopharmacol.* **15**, 11S–18S.
2. Kalivas, P. W., Duffy, P., and Barrow, J. (1989). Regulation of the mesocorticolimbic dopamine system by glutamic acid receptor subtypes. *J. Pharmacol. Exp. Ther.* **251**, 378–387.
3. Chergui, K., Nomikos, G. G., Mathé, J. M., Gonon, F., and Svensson, T. H. (1996). Burst stimulation of the medial forebrain bundle selectively increases Fos immunoreactivity in the limbic forebrain of the rat. *Neuroscience* **72**, 141–156.
4. Marcus, M. M., Nomikos, G. G., and Svensson, T. H. (1996). Differential actions of typical and atypical antipsychotic drugs on dopamine release in the core and shell of the nucleus accumbens. *Eur. Neuropsychopharmacology* **6**, 29–38.
5. Wadenberg, M-L., Salmi, P., Jimenez, P., Svensson, T. H., and Ahlenius, S. (1996). Enhancement of antipsychotic-like properties of the dopamine D_2 receptor blocking agent, ritanserin, in the rat. *Eur. Neuropsychopharmacology* **6**, 305–310.
6. Mathé, J. M., Nomikos, G. G., Hildebrand, B., Hertel, P., and Svensson, T. H. (1996). Prazosin inhibits MK-801 induced hyperlocomotion and dopamine release in the nucleus accumbens. *Eur. J. Pharmacol.* **309**, 1–11.
7. Lipska, B. K., Jaskiw, G. E., Chrapusta, S., Karoum, F., and Weinberger, D. R. (1992). Ibotenic acid lesion of the ventral hippocampus differentially affects dopamine and its metabolites in the nucleus accumbens and prefrontal cortex in the rat. *Brain Res.* **585**, 1–6.

PART G

NOVEL CATECHOLAMINERGIC SYSTEMS

David S. Goldstein

Clinical Neuroscience Branch
National Institute of Neurological Disorders and Stroke
National Institutes of Health
Bethesda, Maryland 20892-1424

Overview

Recent research about novel catecholaminergic systems has related to (1) catecholestrogens, (2) nonneuronal biosynthesis of catecholamines, (3) L-dopa as a neurotransmitter, and (4) dopamine (DA) as a renal autocrine–paracrine substance.

I. Catecholestrogens

Production of 2-hydroxylated catecholestrogens (2-OH-CEs) constitutes a major pathway for hepatic metabolism of estrone (E_1) and estradiol (E_2). Aromatic hydroxylation of E_2 by estradiol-2/4 hydroxylase yields the catecholestrogens 2-OH-E_2 and 4-OH-E_2. Because CEs undergo inactivation by catechol-O-methyltransferase (COMT), the high affinity of these compounds for the catechol-metabolizing enzyme led to the early suggestion that CEs could inhibit breakdown of catecholamines.

Little evidence has accrued for effects of CEs on catecholaminergic function; however, CEs have been implicated in carcinogenesis, by mediating actions of

Advances in Pharmacology, Volume 42

1054-3589/98 $25.00

primary estrogens. Liehr (p. 824) has put forth the hypothesis that 4-OH-E_2 (but not 2-OH-E_2) increases mutagenesis by generating free radicals consequent to peroxidase- or P450-induced formation of a quinone (CE-Q). COMT would, therefore, be protective here. Inactivation of the presumably toxic CE-Qs can also take place via quinone reductase, which regenerates CEs, and via S-transferase, which conjugates CE-Qs with glutathione. Weisz and coworkers (p. 828) have reported that in hamster kidney, a model for estrogen-induced cancer related to local generation of CEs, treatment with estrogen increases presumably protective COMT activity in both the cytoplasm and nucleus of proximal convoluted tubular cells. Stack and coworkers (p. 833) have studied reactions of the two forms of electrophilic *o*-quinones $E_1(E_2)$-2,3-Q and $E_1(E_2)$-3,4-Q, with DNA bases; the authors conclude that formation of 4-OH-E_1 and 4-OH-E_2 are carcinogenic, due to formation of a 4-OH-$E_1(E_2)$-1(α,β)-N7guanine depurinating adduct. Cavalieri and Rogan (p. 837) have also hypothesized that N7Gua adducts formed by reaction of estrogen-3,4-quinones with DNA initiate estrogen-associated cancers. In human breast cancer microsomes, 4-OH-E_2 constitutes a major catechol metabolite, consistent with a role of CEs in production of estrogen-dependent tumors.

CEs also appear to play an important role in blastocyst implantation. Paria and coworkers (p. 840) have studied a mouse model of "delayed implantation." Ovariectomy before preimplantation estrogen secretion on day 4 prevents implantation and produces blastocyst dormancy, and ovarian progesterone can maintain this condition. Estrogen injection in the progesterone-primed uterus initiates implantation, and Paria and coworkers propose that this occurs via rapid production of CEs by the blastocyst.

II. Nonneuronal Biosynthesis of Catecholamines

Based on Cannon's influential experiments and writings, the sympathoadrenal system was thought to function as a unit, preserving homeostasis during exposure to extreme environmental or physiological stimuli and during "fight or flight" responses. Until recently, investigators viewed DA as only an intermediary in the biosynthesis of norepinephrine (NE) in sympathetic nerves and of epinephrine (Epi) in adrenomedullary cells. Recent evidence has supported a physiological role for DA as an autocrine–paracrine hormone that influences Na^+ disposition in the periphery. At least some of this DA does not appear to derive from either the adrenal medulla or sympathetic nerves, but from catecholamine biosynthesis in nonneuronal cells.

Phenylethanolamine-*N*-methyltransferase (PNMT) catalyzes the conversion of NE to Epi. NE can undergo methylation also by the action of a less specific methyltransferase, NMT, which N-methylates many amines. Ziegler and coworkers (p. 843) have summarized evidence for the presence of PNMT in nonneuronal cells from several types of tissues. PNMT was distinguished from NMT by the ability of NMT to convert DA to epinine and by selective effects of PNMT blockers. Cardiac atrial PNMT increased in response to glucocorticoids in adrenal-demedullated rats and approximately doubled atrial Epi concentrations. Skeletal muscle also was found to possess PNMT, the activity of which increased with dexamethasone. Because sympathectomy did not alter

PNMT activity in masseter muscle of rats with superior cervical ganglionectomy, and 6-hydroxydopamine treatment increased atrial PNMT activity, PNMT can exist as a nonneuronal enzyme. Dexamethasone doubled rat pulmonary PNMT mRNA expression, whereas administration of the glucocorticoid antagonist RU 486 decreased pulmonary PNMT mRNA expression; and in cultured human bronchial epithelial cells, dexamethasone increased PNMT activity by about 20-fold.

Duodenal and gastric secretion of bicarbonate alkalinizes the mucus gel adherent to mucosal cells, protecting them from acidic injury. Flemstrom and coworkers (p. 846) have reported that DA- or D1-receptor stimulation increases duodenal bicarbonate secretion in rats and that in enterocytes, these agonists increase production of cyclic adenosine monophosphate (cAMP). Peripheral COMT inhibition by nitecapone also increases bicarbonate secretion. The results were consistent with participation by DA in a "brain-gut axis" that modulates bicarbonate secretion.

Nonneuronal cells may provide a source of DA in the gastrointestinal tract. In an oral presentation based on a submitted abstract, Mezey and coworkers have reported high concentrations of DA in gastric juice of intact or chemically sympathectomized rats. Gastric parietal cells and pancreatic exocrine cells had TH immunostaining and measurable TH activity and had detectable DA and positive immunostaining for the plasma membrane DA transporter and the vesicular monoamine transporters VMAT1 and VMAT2. Moreover, gastric mucosal cells took up [^{3}H]DA, and GBR 12909, a specific inhibitor of the DA transporter, blocked this uptake.

Local uptake of L-dopa and intracellular decarboxylation by nonneuronal cells constitutes another potential nonneuronal mechanism for DA production in the gastrointestinal tract. In an oral presentation based on a submitted abstract, Vieira-Coelho and Soares-da-Silva described apical and basolateral cell border outward-transfer of newly formed DA in CACO-2 cells, a cell line derived from human colon adenocarcinoma. The cells had saturable apical uptake of L-dopa but nonsaturable basolateral uptake; rapid, extensive DA production from applied L-dopa; and saturable apical extrusion of DA and nonsaturable basolateral extrusion of DA.

Given that at least some cell types produce DA from L-dopa after uptake of L-dopa from extracellular fluid, from whence does L-dopa in extracellular fluid arise? Eldrup and coworkers (p. 851) summarized studies about the origin and significance of plasma L-dopa levels. Because stimuli that produce large changes in plasma NE levels produce little or no changes in plasma L-dopa levels, plasma L-dopa does not reflect sympathetically mediated exocytosis. Administration of benserazide to inhibit L-aromatic-amino-acid decarboxylase (L-AAD) increased plasma L-dopa levels by almost eightfold, indicating substantial conversion of plasma L-dopa to DA in the body. After eating, plasma L-dopa levels decreased, implying that dietary intake does not maintain plasma L-dopa levels.

III. Is L-Dopa a Neurotransmitter?

L-Dopa undergoes conversion to DA in many types of cells, including neurons, and DA acts as a classical central neurotransmitter. Misu and coworkers (p. 855) have proposed that L-dopa itself may behave as a neurotransmitter. In the nucleus of the solitary tract (NTS), L-dopa may modulate the arterial baroreflex. Locally microinjected L-dopa dose-dependently decreased heart rate and blood pressure in rats, and DOPA ester antagonized the responses, whereas an inhibitor of L-AAD, NSD-1015, did not. The authors also provide evidence that gamma-aminobutyric acid (GABA), acting at $GABA_A$ receptors, inhibits presynaptic L-dopa release in the NTS and, thereby exerts an inhibitory modulatory effect on the baroreflex. Conversely, L-dopa administration increased local microdialysate GABA concentrations.

The presence of TH immunoreactivity does not necessarily imply the presence of catecholamine biosynthesis, because neurons that transiently have TH immunoreactivity may not have immunoreactivity for L-AAD. Moreover, TH absolutely requires the cofactor, tetrahydrobiopterin, for activity. Guanosine triphosphate cyclohydrolase I (GCH) catalyzes that first enzymatic step in the biosynthesis of tetrahydrobiopterin. Nagatsu and coworkers (p. 859) have reported findings using a polyclonal antibody against an oligopeptide of GCH. In postnatal mice, GCH immunoreactivity appeared in the intergeniculate leaflet and the internal and external lamellae of the ventral lateral geniculate nuclear region, but the cells did not have TH and did not produce DA or serotonin. The authors also report the presence of TH immunoreactivity with absent GCH immunoreactivity in the medial geniculate nuclear region, again in cells that did not belong to the catecholaminergic neuron system, because they did not produce DA. These results have led to the suggestion that TH and GCH have as yet unidentified additional functions besides mediating production of catecholamines.

L-3,4-dihydroxyphenylserine (L-*threo*-DOPS) undergoes enzymatic decarboxylation to form NE. As an amino acid, L-*threo*-DOPS can cross the blood-brain barrier, providing precursor for NE production. Since 2-fluoronorepinephrine (2-FNE) acts as a selective β-adrenoceptor agonist and 6-fluoronorepinephrine (6-FNE) as a selective α-adrenoceptor agonist, Chen and coworkers (p. 862) have studied methods to synthesize (2-FNE) and (6-FNE) from 2F- and 6F-*threo*-DOPS. Efforts to synthesize these compounds enzymatically using L-AAD failed, despite efficient decarboxylation of L-*threo*-DOPS, probably because (+)-*threo*-DOPS inhibited the enzyme. The authors therefore are pursuing enantioselective syntheses.

IV. Dopamine as a Renal Autocrine–Paracrine Substance

The renal DOPA–DA system constitutes the most well studied nonneuronal catecholaminergic system. L-dopa in the tubular filtrate undergoes Na^+-dependent uptake into proximal tubular cells. DA formed intracellularly can exit the cell to occupy basolateral membrane D1 and D2 receptors and apical membrane D1 receptors, the former inhibiting Na^+-K^+ adenosine triphosphatase (ATPase) and the latter inhibiting transmembrane Na^+-H^+ exchange. By both actions, renal DA acts as an autocrine–paracrine agent that augments natriuresis.

In isolated renal cortical tubules, Soares-da-Silva and Vieira-Coelho (p. 866) have found that L-dopa undergoes uptake by an active transport process inhibited by corticosterone and by organic cation transport inhibitors in a pH-dependent manner, the latter consistent with uptake by an organic cation-H^+ exchanger. Outward transport of the newly formed DA appears to depend on a Na^+-H^+ exchange, because amiloride reduced renal DA outflow considerably. Whereas DA could undergo basolateral re-uptake, the amine did not undergo apical re-uptake.

Greengard and coworkers (p. 870) have reviewed mechanisms by which renal DA inhibits Na^+-K^+ ATPase. The kidney expresses a cAMP-regulated phosphoprotein, termed DARPP32. Protein kinase A (PKA)–induced phosphorylation of DARPP32 inhibits a protein phosphatase, 1,PP1, which in turn dephosphorylates and activates Na^+-K^+ ATPase. α-Adrenoceptor agonists can dephosphorylate DARPP32 via calcineurin. Thus, neuronal NE and autocrine-paracine DA exert opposing effects on intracellular signaling mechanisms regulating Na^+-K^+ ATPase and sodium excretion. The authors suggest that DA-induced inhibition of Na^+-K^+ ATPase also can occur via a PKC pathway, where occupation of D2 receptors leads to activation of phopholipase A, releasing arachidonic acid and increasing formation of the arachidonic acid metabolite 20HETE, which inhibits Na^+-K^+ ATPase by a PKC-dependent mechanism.

The kidneys possess high concentrations of COMT, and as noted by Aperia (p. 870), administration of the COMT antagonist, nitecapone, exerts a pronounced natriuretic effect in anesthetized rats, consistent with increased intrarenal effects of locally generated DA.

Carey and coworkers (p. 873) have studied the renal DOPA–DA system *in vivo* using microdialysis in anesthetized rats. High salt intake increased urinary DA excretion and decreased microdialysate DA concentrations. Because renal denervation failed to affect either DA excretion or microdialysate DA, renal DA appears to derive mainly if not exculusively from nonneuronal sources. In response to glucodopa, renal DA excretion increased proportionately more than did interstitial DA, consistent with preferential release of renal DA into the tubular lumen.

In an oral presentation, Aherne and coworkers reported the cellular localization and regional distribution of DA-receptor subtypes. The kidney was found to possess D1A, D1B, and D3 receptors.

Classical endocrine systems may contribute to regulation of the renal DOPA–DA natriuretic system. In an oral presentation, Armando and coworkers noted effects of manipulations of the adrenocortical and thyroid axes. Glucocorticoid administration in general produced parallel changes in excretion rates of sodium, DOPA, and DA. Hypothyroid rats had decreased sodium and DOPA excretion, with normal DA excretion. The results supported the view that the renal DOPA–DA system mediates some hormonal influences on renal sodium handling.

V. Conclusions and Future Directions

Whether polymorphisms for COMT or treatments with COMT inhibitors relate to the incidence of estrogen-dependent tumors or to fertility merits further study. Establishment of L-dopa as a neurotransmitter will depend on more information about mechanisms regulating L-dopa release and determining L-dopa effects.

Three types of endogenous catecholaminergic systems operate outside the central nervous system: The sympathetic nervous system, the adrenomedullary hormonal system, and the DOPA–DA system exemplify the major known means by which chemical messengers act on cells. In particular, DA, the natriuretic catecholamine, is synthesized in, released in, and acts in the kidneys, gut, and probably elsewhere as an autocrine–paracrine substance. The functions of this "third catecholamine system" require further study. The relative roles of local TH activity in nonneuronal cells and of local uptake and decarboxylation of L-dopa in synthesis of nonneuronal DA should be compared. The sources and significance of plasma L-dopa and DA sulfate require fuller understanding.

Joachim G. Liehr

Department of Pharmacology and Toxicology
University of Texas Medical Branch
Galveston, Texas 77555

Catecholestrogens in the Induction of Tumors in the Kidney of the Syrian Hamster

Estrogens induce tumors in several rodent species. For instance, the administration of estradiol or estrone implants to male Syrian hamsters induces kidney tumors with 80–100% incidence within 6–8 mo (reviewed by Liehr and Sirbasku [1]). The estrogen-induced kidney tumor in hamsters has been investigated extensively as a mechanistic model for estrogen-associated cancers in humans, such as breast or uterine cancers (2, 3). In the hamster model, several synthetic steroidal estrogens have been identified, which were only weakly carcinogenic despite their potent hormonal activities. For example, 17α-ethinylestradiol, 2-fluoroestradiol, and 11β-methylestradiol are powerful estrogens but induced kidney tumors with only 0–10% incidence. It was concluded from these studies that the hormonal activity of estrogens is necessary but not sufficient for tumor development to occur. We, thus, developed a hypothesis of a dual role of estrogens as both tumor initiators inducing DNA damage and genetic alterations by metabolic activation and as hormones inducing cell transformation and growth by hormone receptor–mediated pathways. Specifi-

Advances in Pharmacology, Volume 42

1054-3589/98 $25.00

cally, the metabolic conversion of estrogens to catecholestrogens was thought to provide the substrates for metabolic redox cycling, which generates potentially mutagenic free radicals. In mechanistic studies of metabolic activation of catecholestrogens and induction of DNA damage, the increased efficiency of the 4- compared with the 2-hydroxylated catechols pointed to the former metabolites as possible endogenous procarcinogens.

A mechanism of metabolic activation of 4-hydroxyestrogens includes redox cycling between the catechol (hydroquinone) and the quinone-reactive intermediate, as shown in Figure 1. The oxidation to the quinone is catalyzed by organic hydroperoxide-dependent cytochrome P450 1A enzymes or other peroxidases, wheras the reduction of the quinone is catalyzed by NADPH-dependent cytochrome P450 reductase or other reductases. The quinone intermediates may bind covalently to DNA and, thus, induce mutations (see accompanying chapters by Stack *et al.* and Cavalieri and Rogan). In addition, the semiquinone intermediates in the metabolic redox cycle may react with molecular oxygen and generate superoxide radicals. The superoxide may be reduced to hydrogen peroxide and, in the presence of metal ions, further to hydroxy radicals. Thus, metabolic redox cycling of catecholestrogens is a mechanism of potentiation of DNA damage by potentially mutagenic free radicals generated by possibly small amounts of catecholestrogen substrate undergoing continuous oxidations and reductions (4).

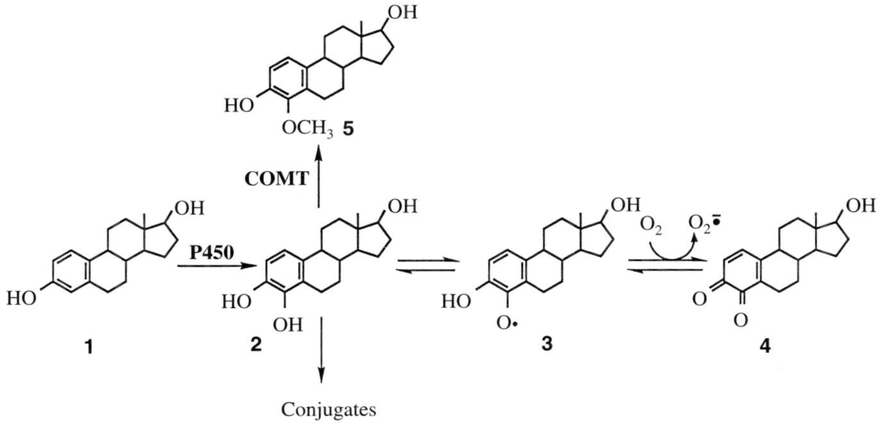

FIGURE 1 Estradiol (*1*) is converted by cytochrome P450 enzymes such as P450 1B1 to 4-hydroxyestradiol (*2*). Metabolic redox cycling between 4-hydroxyestradiol (*2*) and its corresponding quinone estradiol-3,4-quinone (*4*) may be catalyzed by organic hydroperoxide-dependent cytochrome P450 1A1 for the oxidation step and cytochrome P450 reductase for the reduction. The semiquinone intermediate in this cycle (*3*) is a free radical and may react with molecular oxygen to form superoxide radical and quinone. The conversions of 4-hydroxyestradiol to 4-methoxyestradiol (*5*) by catechol-O-methyltansferase (COMT) or by other phase II enzymes to other conjugates represent detoxification pathways. Although 2-hydroxyestradiol is also capable of metabolic redox cycling, its concentrations may be too low for a substantial contribution to free radical generation because of its more efficient conversion to phase II metabolites than has been observed with 4-hydroxyestrogens.

The following evidence supports the postulated dual role of 4-hydroxyestradiol in tumor initiation and growth in rodent models of estrogen-induced carcinogenesis:

4-Hydroxylation of estradiol is the predominant form of catechol metabolite formation by cytochrome P450 enzymes of hamster kidney, mouse uterus, or rat pituitary, the three major organ systems where estrogens induce tumors. In contrast, 2-hydroxylation predominates in rodent livers or other organs, where tumors are not induced by this treatment (see accompanying chapter by Weisz *et al.*).

4-Hydroxyestradiol is as carcinogenic as estradiol in the hamster kidney tumor model, whereas 2-hydroxyestradiol does not induce tumors when chronically administered by implant. Similarly, 4-hydroxyestrone is also carcinogenic, albeit at a lower incidence rate, while 2-hydroxyestrone does not induce any tumors.

The induction of DNA damage by 4-hydroxyestradiol is in agreement with its capability to induce tumors. For instance, in Syrian hamsters *in vivo,* 4-hydroxyestradiol induces DNA single-strand breaks and 8-hydroxylation of guanine bases of DNA. Both processes are initiated by hydroxy radicals. In incubations with microsomes *in vitro,* guanine bases of DNA are hydroxylated at C-8 only by 4-hydroxyestrogen metabolites or their quinone intermediates, but not by the parent hormones estrone or estradiol nor by their 2-hydroxylated metabolites. In addition, 4-hydroxyestrogen quinones form DNA adducts of a different type than do 2-hydroxyestrogens. This differential DNA adduct formation and its biological consequences are discussed by Stack *et al.* and Cavalieri and Rogan (see accompanying chapters).

The differential reactivity of 4- versus 2-hydroxyestrogens may be based on the more efficient oxidation of the former catechols to reactive quinone metabolites by microsomal enzymes and, thus, on more efficient metabolic redox cycling. *In vivo,* it may also be based on the preferential formation of 4-hydroxyestrogens in target organs, which results in their higher tissue concentrations, as discussed previously and also in the accompanying text of Weisz *et al.* In addition, elevated tissue concentrations of these metabolites may also be due to a preferential inhibition of the conversion of 4-hydroxyestrogens to phase II metabolites compared with 2-hydroxyestrogens. For instance, the catechol-*O*-methyltransferase–mediated methylation of 4-hydroxyestrogens is inhibited by 2-hydroxyestrogens. Thus, in organs in which both catechol metabolites are formed, such as in the hamster kidney, concentrations of 4-hydroxyestrogens are higher than would be expected based on their rates of formation, because 2-hydroxyestradiol inhibits the inactivation of 4-hydroxyestradiol by catechol-*O*-methyltransferase–mediated methylation. Other endogenous catechols, such as catecholamines, also inhibit more efficiently the methylation of 4- than of 2-hydroxyestrogens. In this context, it is noteworthy that concentrations of norepinephrine in hamster kidney or mouse uterus are higher than in livers of these rodents or in kidney or uterus of other rodent species, where estrogens do not induce tumors. Thus, it is possible that particularly high catecholamine concentrations in any organ of any species facilitate induction of tumors by estrogens by way of inhibiting the phase II catabolism of the procarcinogenic 4-hydroxyestrogen metabolites.

Tumor modulation by enzyme inhibition also supports the concept of estrogen-induced tumorigenesis mediated by reactive 4-hydroxyestrogen metabolites. The flavonoid quercetin enhances estrogen-induced tumorigenesis in hamsters and enhances 4-hydroxyestrogen metabolite concentrations in the kidney, the target of estrogen-induced carcinogenesis. The increase in metabolite concentrations likely is based on the known inhibition of catechol-O-methyltransferase by quercetin administration.

In contrast, α-naphthoflavone, an inhibitor of cytochrome P450 1A and 1B enzymes, completely suppresses estradiol-induced hamster kidney tumorigenesis. The basis of tumor inhibition likely is the inhibition of catechol metabolite formation and their metabolic redox cycling, because the oxidation step of the redox cycle (i.e., the conversion of the catechol to the quinone), is catalyzed mainly by cytochrome P450 1A1.

The inhibition of estrogen-induced carcinogenesis by free-radical scavengers such as ascorbic acid (vitamin C) or t-butyl hydroxyanisol (BHA) also supports a mechanism of tumor initiation by estrogen-mediated free-radical damage to DNA. This inhibition is not consistent with a mechanism of tumorigenesis postulated by others, which is based solely on estrogen receptor-mediated activation of genes or cell proliferation. Vitamin C or t-butyl hydroxyanisol are not known to have any estrogen-receptor agonist or antagonist activity and, thus, could not have modulated any processes mediated by hormone-receptor binding.

4-Hydroxyestradiol is known to be a long-acting estrogen because of its slower dissociation from the estrogen receptor compared with estradiol. The prolonged hormonal activity achieved by 4-hydroxylation of estradiol may also be necessary for tumor induction and may possibly complete the development of estrogen-induced tumors by receptor-mediated processes in cells damaged by events described previously.

Initial evidence indicates that human breast carcinogenesis may also be characterized by features of estrogen activation and DNA damage, as described previously. For instance, 4-Hydroxyestradiol formation is the predominant catechol metabolism in human breast cancer microsomes (5). The elevated formation of this metabolite corresponds to its predominant concentration in a breast cancer extract analyzed for estrogen metabolites. The elevated hydroxy radical damage to DNA of mammary tissue of breast cancer patients compared with controls is an indicator of metabolic redox cycling and free-radical generation by this estrogen metabolite, as described previously. Additional research is required to substantiate this postulated concept of estrogen-induced tumorigenesis in hamsters and to explore it applicability to human breast cancer development.

Acknowledgment

This study was supported by the National Institutes of Health, NCI (CA 59497).

References

1. Liehr, J. G., and Sirbasku, D. A. (1985). Estrogen-dependent kidney tumors. *In* Tissue Culture of Epithelial Cells. (M. Taub, ed.), pp. 205–234. Plenum Press, New York.
2. Yager, J. D., and Liehr, J. G. (1996). Molecular mechanisms of estrogen carcinogenesis. *Annu. Rev. Pharmacol. Toxicol.* **36**, 203–232.
3. Liehr, J. G. (1997). Hormole-associated cancer: Mechanistic similarities between human breast cancer and estrogen-induced kidney carcinogenesis in hamsters. *Environ. Health Perspect.* **105**, 565–569.
4. Liehr, J. G., and Roy, D. (1990). Free radical generation by redox cycling of estrogens. *Free Radic. Biol. Med.* **8**, 415–423.
5. Liehr, J. G., and Ricci, M. J. (1996). 4-Hydroxylation of estrogens as marker of human mammary tumors. *Proc. Natl. Acad. Sci. U.S.A.* **93**, 3294–3296.

J. Weisz,* G. A. Clawson,† and C. R. Creveling‡

Departments of *Obstetrics and Gynecology and †Pathology
The M. S. Hershey Medical Center
Pennsylvania State University
Hershey, Pennsylvania 17033

‡National Institute of Diabetes and Digestive and Kidney Diseases
National Institutes of Health
Bethesda, Maryland 20892

Biogenesis and Inactivation of Catecholestrogens

That catecholestrogens (CEs) may provide a functional link between estrogens (Es) and catecholamines was first suggested over three decades ago. It was based on the discovery that 2-hydroxylated CEs (2-OH-CEs), products of a major pathway of metabolic *inactivation* of estrone and estradiol in liver, were not only inactivated by catechol-O-methyltransferase (COMT), but also, by virtue of their high affinity for COMT, could inhibit inactivation of catecholamines by this enzyme (1). Interest in 2-hydroxylation of estrogens (Es) as a potential pathway of metabolic *activation* of Es was rekindled a decade later following reports that 2-OH-CEs may be generated in the central nervous system (CNS) (1). This finding led to the suggestion that CEs may provide a direct biochemical link between Es and catecholaminergic function restricted to sites in the CNS where they are produced. The demonstration that 2-OH-CEs can inhibit not only COMT, but also tyrosine hydroxylase lent further support to this hypothesis (1, 2). However, the initial enthusiasm for this

Advances in Pharmacology, Volume 42

1054-3589/98 $25.00

postulate waned as difficulties were encountered in obtaining conclusive, reproducible evidence for any essential role for CEs in the E-dependent neuronal functions selected for study (1). In retrospect, the frustrations caused by these early studies are not surprising considering how little was known at that time about the biogenesis and metabolism of CEs, the mechanism by which they may act, and the lack of specificity and precision of the indirect assays used in many studies. It is now clear that CEs can be generated by several discrete enzymatic pathways, that these can result in the formation of both 2- and 4-hydroxylated CEs (4-OH-CEs), and that these two classes of CEs differ in their properties, both as catechols and as Es (2). The value of these insights, availability of specific and sensitive assays for measuring CE synthase activities, and molecular biological tools now available for probing the role of CEs as local mediators of the *physiological* actions of Es are highlighted in this symposium by S. K. Day. A new dimension to the potential significance of CE formation within E-target organs was added by evidence, reviewed in this symposium by Liehr and Cavalieri, implicating CEs in the *pathopysiology* of E-action, notably carcinogenesis and neuronal degeneration.

I. Biogenesis of CEs

Hydroxylation of Es at C-2 and C-4 can be catalyzed by two distinct biochemical mechanisms (2). The two differ in their cofactor requirement and in the profile of CEs they generate. One utilizes NADPH as the cofactor, molecular oxygen as the oxygen source, and a cytochrome P450 as the catalyst. It generates predominantly either 2- *or* 4-OH-CEs, depending on the form of P450 involved. The second utilizes an organic hydroperoxide (OHP) as cofactor and oxygen source, can be catalyzed by some hemoprotein peroxidases as well as by some as yet unidentified cytochrome P450, and generates 2- and 4-OH-CEs in similar amounts. For reasons presented in this chapter, these differences in cofactor requirement and profile of CEs generated have functional implications for both the physiology and the pathophysiology of E action.

A. NADPH-Dependent 2-Hydroxylation of Es

Studies of NADPH-dependent aromatic hydroxylation of Es have until recently focused almost exclusively on 2-hydroxylation. Several forms of P450 have been identified that can catalyze 2-hydroxylation of Es. Specifically, these are two members of the P4501A subfamily (P4501A1 and -1A2), members of the P4503A subfamily, and, in rats, also the male-specific P4502C11. These P450s are expressed in many tissues besides the liver and, as shown in cytochemical studies, their expression in different organs is restricted to anatomically defined cell populations. Because P4501A1, P4501A2, and P4503A1 are also responsible for the phase I metabolism (metabolic activation–inactivation) of many prevalent lipophilic zenobiotics and drugs, and are inducible by them, these P450s provide an interface between environmental chemicals and Es and, by implication, also catecholaminergic function.

NADPH-dependent 4-hydroxylation of Es was first identified in rat anterior pituitary and then in hamster kidney (2). However, only recently has a form of P450 been identified that can catalyze selective 4-hydroxylation Es (3). It was discovered by investigators interested in mechanisms responsible for the apparent antiestrogenicity of TCDD, and in determining if altered metabolism of Es could contribute to the phenomenon. On treating MCF-7 breast cancer cells with TCDD, they noted a marked increase in both 2- and 4-hydroxylation of Es. The increase in 2-hydroxylation could be accounted for by induction of members of the P4501A subfamily. The catalyst for the 4-hydroxylation proved to be a previously unidentified member of the P450 superfamily. Based on its sequence homology with P4501A1 and -1A2, it was designated P4501B1. Because 4-OH-CEs are potent long-acting Es, the discovery of P4501B1 did not contribute to understanding TCDD's apparent antiestrogenic effects. It did, however, direct attention to 4-OH-CEs as potential local mediators of E-action. Although much remains to be learned about the properties of 4-OH-CEs, it is already evident that these are likely to differ in many important respects from those of 2-OH-CEs, both as Es and as catechols, differences with implications for both physiology and pathophysiology (discussion follows). P4501B1 appears to be expressed constitutively in many extrahepatic tissues, and its expression, like that of P450s mediating 2-hydroxylation of Es, can be shown to be restricted to defined anatomical loci and cell populations, a finding consistent with the concept that CEs act as autocrine–paracrine mediators of the action of primary Es (4).

B. Peroxidative CE Formation

Aromatic hydroxylation of ES can be catalyzed in the presence of an OHP by certain hemoprotein peroxidases as well as microsomes (2, 4). The mechanism involved is analogous to the OHP-dependent metabolism of polycyclic aromatic hydrocarbon. Cytochrome P450-dependence of microsomal OHP-dependent CE formation has been demonstrated using CO and pharmacological inhibitors of P450. However, the form(s) of P450 mediating this pathway CE formation is not known. None of the major, inducible hepatic forms of P450 appear to be responsible, because neither inducers of P450 administered *in vivo* nor inhibitory antibodies *in vitro* affected peroxidative activity of hepatic microsomes. Whether peroxidative activity is expressed under normal physiological conditions, and what the nature or source of OHPs might be under such conditions are also not known. However, OHPs can be generated in virtually unlimited amounts under conditions of oxidative stress. Under these conditions, peroxidative CE formation could be triggered and sustained unchecked and contribute to geno- and cytotoxicity by mechanisms described by others in this symposium. In particular, semiquinone and quinone E metabolites of CEs are highly reactive and can form DNA adducts. CEs can also become a source of reactive oxygen species (ROS) by entering into redox cycling (i.e., repeated cycles of oxidation and reduction between catechol and quinone Es). Because peroxidative activity generates 2- and 4-OH-CEs in similar amounts, and 2-OH-CEs can inhibit the inactivation of 4-OH-CEs by COMT, the peroxidative pathway of CE formation could be especially relevant to pathophysiology.

Enzymatic demethylation of O-methylated CEs by specific isoforms of P450 could provide an additional mechanism by which CEs could be generated at specific sites. This postulate is supported by evidence for demethylation of O-methylated CEs administered *in vivo* and reports of remarkably high potency of O-methylated CEs in some *in vitro* bioassays (4). This potential mechanism for targeting of CEs to specific cell populations has remained unexplored.

II. Differences in Properties of 2- and 4-OH-CEs; Estrogenic Potency and Inactivation

There is good evidence that 4-OH-CEs are potent long-acting Es (2, 4). In contrast, 2-OH-CEs have been generally considered to be weak Es based on the marked decrease in affinity for the E receptor caused by 2-hydroxylation and the weak estrogenicity of 2-OH-Es in *in vivo* bioassays. This conclusion may, however, have to be revised in light of the unexpectedly high potency of 2-OH-CEs in *in vitro* bioassays, in which target cells are exposed directly to the steroids. These differences between *in vivo* and *in vitro* potency could be due to differences in pharmacokinetics of the two classes of CEs and/or their further metabolism.

The 2- and 4-OH-CEs differ also in their properties as catechols. Thus, although 2- and 4-OH-CEs have similar apparent affinity for COMT and similar maximal velocities of O-methylation, 2-OH-CEs can inhibit the O-methylation of 4-OH-CEs (2, 4). Whether 2- and 4-OH-CEs differ also in their ability to inhibit tyrosine hydroxylase remains to be determined. Systematic comparisons of 2- and 4-OH-CEs as catechols are needed in light of evidence that 4-OH-CEs are not simply minor by-products of 2-hydroxylation, but that they may be generated in significant amounts in many extrahepatic tissues, either by some forms of P450 (e.g., P4501B1) or by a peroxidative mechanism.

III. Metabolism of CEs: The COMT Paradigm

Because CEs are highly reactive and are potential sources of excess ROS, there is an obvious need for their prompt inactivation. CEs' postulated role as local mediators of E-action also implies a need for inactivating enzymes close to sites of formation of CEs in order to restrict their actions spatially and temporally. Understanding of how CEs are inactivated, however, has lagged behind that of their biogenesis. Only COMT has been investigated to any extent within this context. COMT in liver and red blood cells plays a major role in bulk inactivation of catechols, including CEs generated in liver (2, 4). In *extrahepatic* organs, such as brain, uterus, kidney, and mammary gland, COMT has been shown to be expressed only in a few anatomically defined loci and cell populations (5). Such selective expression is consistent with the notion that COMT participates in regulation of local concentrations of catechols, including

CEs. Studies of expression of COMT in uterus in relationship to implantation also support this notion (6). Implantation is a critical physiological function in which locally generated CEs have been postulated to play an essential role. In rat uterus, COMT has neen shown to be induced at the time of implantation, specifically on the antimesometrial side where implantation is known to occur, and its induction has been shown to depend on progesterone, a key regulator and coordinator of implantation. We have examined the effect of Es on the expression of COMT in hamster kidney, a model for E-induced cancer in which CEs generated *in situ* are postulated to play an essential, initiating role. Using immunocytochemistry, we have shown COMT to be localized in control animals in epithelial cells of proximal convoluted tubules (PCTs), mainly in the deep juxtamedullary cortex. E treatment not only induced COMT in PCTs throughout the cortex, but also caused the appearance of nuclear COMT. By immunoblotting, the nuclear COMT was identified to be the smaller soluble form of COMT. The physiological role of COMT in PCTs is presumably to regulate levels of locally generated dopamine and, thereby, the rate of natruresis (7). The induction of COMT caused by E treatment, in particular its transport into the nucleus, implies a threat to the genome by increased catechol load, resulting, presumably, from CEs generated *in situ*. Interestingly, we also observed nuclear localization of COMT in human breast cancer cells, whereas in normal mammary ductal epithelial cells COMT is localized exclusively in cytoplasm. Clearly, such correlations, although suggestive, do not *prove* participation of CEs in either implantation or carcinogenesis. However, they underscore the need for a systematic examination of enzymes that can inactivate CEs, and they provide model systems for such studies. Ultimately, defining the role of CEs will require observing, and modulating their generation, using specific pharmacological inhibitors and molecular biological strategies.

Acknowledgment

Supported by National Institutes of Health grants R55 CA58739-01 and CA21141-17.

References

1. Catchol Estrogens. Proceedings of Fogarthy International Conference, Bethesda, MD. (G. R. Meriam and M. B. Lipsett, eds.). Raven Press, New York, 1983.
2. Weisz, J. (1994). Biogenesis of catecholestrogens: Metabolic activation of estrogens by phase I enzymes. Polycyclic Arom. Compd. 6, 241–251.
3. Hayes, C. L., Spink, D. C., Spink, B., Cao, J. Q., Walker, N. J., and Suter, T. R. (1996). 17β-Estradiol hydroxylation by human cytochrome P450 1B1. *Proc. Natl. Acad. Sci. U.S.A.* **93**, 9776–9781.
4. Weisz, J. (1991). Metabolism of estrogens in target cells. Diversification and amplification of hormone action and the catecholestrogen hypothesis. *In* New Biology of Steroid Hormones. (R. Hochberg and F. Naftolin, eds.), pp. 201–212. Raven Press, New York.
5. Creveling, C., and Hartman, B. K. (1982). Relationship between the cellular localization and physiological function catechol-O-methyltransferase. *In* Biochemistry of D-Adenosyl

L-Methionine and Related Compounds. (E. Usdin, R. T. Burchard, and C. R. Creveling, eds.). pp. 479–486. Macmillan Press, London.
6. Creveling, C. R., and Inoue K. (1995). Induction of catechol-O-methyltransferase in the luminal epithelium of rat uterus by progesterone. *Drug Metab. Dispos.* **23**, 430–432.
7. Meister, B., Bean, A. J., and Aperia, A. (1993). Catechol-O-methyltransferase mRNA in kidney and its appearance during ontogeny. *Kidney Int.* **44**, 726–733.

D. E. Stack, E. L. Cavalieri, and E. G. Rogan
Eppley Cancer Research Institute
University of Nebraska Medical Center
Omaha, Nebraska 68198

Catecholestrogens as Procarcinogens: Depurinating Adducts and Tumor Initiation

Excessive exposure to endogenous estrogens has been linked to cancer in both rodent models and human studies. While many studies have been aimed at understanding the relationship between estrogen exposure and cancer, the mechanism of estrogen-induced cancer is still poorly understood. The role of estrogens in receptor-mediated processes leading to the stimulation of cell proliferation has been widely studied to better understand the etiology of estrogen-induced cancers. More recently, the genotoxic properties of electrophilic estrogen metabolites have been investigated to help shed light on the mechanism of hormonal carcinogenesis (1).

Estradiol (E_2) and estrone (E_1), when hydroxylated at the 2- or 4-position, form catecholestrogens (CEs), which are among the major metabolites of E_1 and E_2. CEs are typically conjugated by catechol-*O*-methyltransferases (COMTs) to afford *O*-methylated CEs. Nonmethylated CEs can be oxidized to the reactive electrophilic *o*-quinones (CE-Qs) by peroxidases and cytochrome P450. Therefore, COMTs act as protective enzymes. With elevated rates of CE formation and/or deficient COMT activity, increased levels of CEs can occur, leading to the formation of CE-Qs. Once formed, CE-Qs can act as Michael acceptors, binding to cellular macromolecules, including DNA.

Malignant renal tumors are induced in Syrian golden hamsters by treatment with 4-OH-E_1 or 4-OH-E_2, whereas the corresponding 2-OH isomers are inactive (2, 3). Furthermore, elevated levels of the 4-OH isomer (relative to the 2-OH isomer) have been detected in tissues prone to estrogen-induced cancers, such as rat pituitary, mouse uterus, human MCF-7 breast cancer cells, human

Advances in Pharmacology, Volume 42

1054-3589/98 $25.00

uterine myometrial tumors, and breast cancer tissues (1). Because we hypothesize that CE-Qs are the ultimate carcinogenic forms of estrogens, we have investigated the chemical differences in the two CE-Qs generated from the isomeric CEs, estrogen-2,3-quinones [$E_1(E_2)$-2,3-Q] and estrogen-3,4-quinones [$E_1(E_2)$-3,4-Q]. The chemical properties of the two CE-Qs in relation to nucleosides and DNA are the subject of this presentation.

CE-Qs can be generated chemically by MnO_2 oxidation of the corresponding CEs in acetonitrile, chloroform, methylene chloride, or acetone. The stability of the two *o*-quinones, E_1-2,3-Q and E_1-3,4-Q, was measured in aqueous solutions at various pH. In neutral conditions (pH 6.7, no buffer), the two quinones have similar half-lives of 110 min. However, the E_1-3,4-Q showed an increased stability at lower pH (T½ of 190 and 390 min at pH 5 and 3, respectively), whereas the E_1-2,3-Q showed no increase in stability at lower pH (T½ of 110 and 70 min at pH 5 and 3, respectively). It has been reported that the 2,3-quinones of estrogens are less stable and hence more reactive than the 3,4-quinones; however, we have found that in competitive studies with glutathione (GSH) at various pH, the 3,4-quinones generate approximately 10 times more of the CE-Q-GSH adduct than the 2,3-quinones (the reaction between CE-Q and GSH occurs rapidly, with complete conversion of CE-Q-GSH adducts). Thus, because the 2,3-quinone and the 3,4-quinone have similar stabilities under physiological conditions, the 3,4-quinones possess more electrophilic character.

The reaction of CE-Qs with the nucleosides deoxyguanosine (dG) and deoxyadenosine (dA) was conducted to gain insight into the behavior of these electrophiles with DNA bases. In addition, the synthetic adducts would serve as standard compounds for *in vitro* and *in vivo* studies aimed at understanding the genotoxic effects of CE-Qs. When an acetonitrile solution of E_1-3,4-Q or E_2-3,4-Q was mixed with dG (dissolved in acetic acid-water, 50 : 50), one adduct, 4-OH-$E_1(E_2)$-1(α,β)-N7Gua, was formed. The adduct is a mixture of two conformational isomers resulting from the restricted rotation of the guanine moiety about the N7(Gua)–C1(estrogen) bond. Attack of the CE-Qs by dG at the N7-position results in loss of the deoxyribose moiety; hence, the 4-OH-$E_1(E_2)$-1(α,β)-N7Gua adducts are referred to as *depurinating* adducts. Reaction of $E_1(E_2)$-3,4-Q with dA afforded no adducts.

Reaction of E_1-2,3-Q with dG or dA generated a different adduct profile from that produced by E_1-3,4-Q. Reaction with dG resulted in the formation of 2-OH-E_1-6-N^2dG, whereas reaction with dA yielded 2-OH-E_1-6-N^6dA. In this case, the CE-Qs did not react as an *o*-quinone; instead, tautomerization to a CE-Q methide (CE-QM) occurred, followed by attack of the exocyclic amino group of the nucleoside at the 6-position of the CE-QM. This yields CE adducts with the deoxyribose moiety intact; hence, 2-OH-E_1-6-N^2dG and 2-OH-E_1-6-N^6dA are referred to as *stable* adducts that would remain in DNA, unless repaired.

The reaction of CE-Qs with DNA was undertaken to determine the type of DNA damage that would occur when these electrophiles bind to DNA. The formation of stable adducts was determined by the ^{32}P-postlabeling technique, whereas formation of depurinating adducts was measured by analysis of the reaction medium via high-performance liquid chromatography interfaced to an electrochemical detector. Reaction of E_1-2,3-Q and E_2-2,3-Q with calf thymus DNA produced 8.59 and 1.14 μmol stable adduct/mol DNA-P, respectively (Table I). Reaction of E_1-3,4-Q and E_2-3,4-Q generated 0.11 and 0.07 μmol stable adduct/mol DNA-P, respectively (4). Thus, as in the chemical reactions, the 2,3-quinones form stable adducts, whereas the 3,4-quinones have a low propensity for stable adduct formation. Reaction of E_1-3,4-Q nd E_2-3,4-Q also produced the depurinating adducts 4-OH-E_1(E_2)-1(α,β)-N7Gua at 59 and 213 μmol adduct/mol DNA-P, respectively. Thus, when both stable and depurinating adducts are considered, the 3,4-quinones produce a much higher level of binding with DNA.

Binding of CE-Qs via activation of the CE precursors is also examined (see Table I). When 4-OH-E_1 was activated with horseradish peroxidase (HRP) in the presence of DNA, 4-OH-E_1-1(α,β)-N7Gua was found at 50 μmol adduct/mol DNA-P. Likewise, HRP activation of 4-OH-E_2 produced 194 μmol adduct/mol DNA-P of the 4-OH-E_2-1(α,β)-N7Gua adduct. This level of binding was similar to the direct reaction of the corresponding CE-Qs. However, lactoperoxidase (LP) activation of 4-OH-E_1 generated approximately 9 times the amount of depurinating adduct compared with CE-Q or HRP activation of CE (see Table I). Phenobarbital-induced rat liver microsomes with cumene hydroperoxide as cofactor also activated 4-OH-E_1 to CE-Qs, resulting in 87 μmol adduct/mol DNA-P of the 4-OH-E_1-1(α,β)-N7Gua adduct.

Formation of the 4-OH-E_1-1(α,β)-N7Gua depurinating adduct was examined *in vivo* by treatment of the mammary glands of female Sprague-Dawley rats with 200 nmol/gland of 4-OH-E_1. After 24 hr, the rats were killed and the second and third mammary glands on both the right and the left sides were excised. The mammary glands were minced, ground in liquid nitrogen, pooled, and divided into two aliquots, one for analysis of depurinating DNA adducts and one for analysis of the stable DNA adducts. The 4-OH-E_1-1(α,β)-N7Gua depurinating adduct was detected at a level of 30 μmol adduct/mol DNA-P, whereas the level of stable adducts was less than 0.05 μmol adduct/mol DNA-P, the limit of detection. These results are in agreement with the properties of E_1-3,4-Q in both chemical and biochemical experiments, as reported previously.

TABLE I Reaction of CE-Qs and HRP-, LP-, or P-450-Activated CEs with DNA

Compound	*Depurinating adducts (μmol/mol DNA-P)*	*Stable adducts (μmol/mol DNA-P)*
E_1-3,4-Q	59	0.11
HRP-activated 4-OH-E_1	50	0.07
LP-activated 4-OH-E_1	440	0.06
PB-microsome/CuOOH-activated 4-OH-E_1	87	0.01
E_2-3,4-Q	213	0.07
HRP-activated 4-OH-E_2	194	0.10
E_1-2,3-Q		8.59[a]
HRP-activated 2-OH-E_1		3.00[a]
E_2-2,3-Q		1.14[a]
HRP-activated 2-OH-E_2		9.26[a]

[a] Data from Dwivedy *et al.* (4).

The CE-Qs derived from 4-OH-E_1(E_2) and 2-OH-E_1(E_2) display different chemical properties in their reaction with nucleosides and DNA. Notably, 2,3-quinones bind at the exocyclic amino group of dG and dA, allowing the deoxyribose moiety to remain attached to the adduct, whereas the 3,4-quinones form N7Gua adducts in which the glycosidic bond of dG is destabilized and cleaved. The implications for DNA damage are that 2,3-quinones will bind to form stable adducts, whereas the 3,4-quinones, via generation of the 4-OH-E_1(E_2)-1(α,β)-N7Gua adduct, will form predominantly depurinating adducts. Recent findings have established a strong correlation between the formation of depurinating adducts with polyaromatic hydrocarbons and induction of oncogenic mutations in mouse skin papillomas induced by these compounds (see chapter by Cavalieri and Rogan, this section). Because the 4-hydroxy isomers of E_1 and E_2 are carcinogenic, whereas the 2-hydroxy isomers are not, we hypothesize that the formation of the 4-OH-E_1(E_2)-1(α,β)-N7Gua depurinating adduct via the E_1(E_2)-3,4-Q is the critical step in the initiation of cancer by estrogens.

References

1. Stack, D. E., Byun, J., Gross, M. L., Rogan, E. G., and Cavalieri, E. L. (1996). Molecular characteristics of catechol estrogen quinones in reaction with deoxyribonucleosides. *Chem. Res. Toxicol.* **9**, 851–859.
2. Liehr, J. G., Fang, W. F., Sirbasku, D. A., and Ari-Ulubelen, A. (1986). Carcinogenicity of catechol estrogens in Syrian hamsters. *J. Steroid Biochem.* **24**, 353–356.
3. Li, J. J., and Li, S. A. (1987). Estrogen carcinogenesis in Syrian hamster tissues: Role of metabolism. *Fed. Proc.* **46**, 1858–1863.
4. Dwivedy, I., Devanesan, P., Cremonesi, P., Rogan, E., and Cavalieri, E. (1992). Synthesis and characterization of estrogen 2,3-quinones and 3,4-quinones. Comparison of DNA adducts formed by the quinones versus horseradish peroxidase-activated catechol estrogens. *Chem. Res. Toxicol.* **5**, 828–833.

E. L. Cavalieri and E. G. Rogan

Eppley Cancer Research Institute
University of Nebraska Medical Center
Omaha, Nebraska 68198

Role of Aromatic Hydrocarbons in Disclosing How Catecholestrogens Initiate Cancer

The estrogens 17β-estradiol (E_2) and estrone (E_1) are metabolized via two major pathways: 16α-hydroxylation (not shown) and formation of catechol estrogens (CEs) (Fig. 1). The catechols formed are the 2-hydroxy and 4-hydroxy derivatives. Generally, these two catechols are inactivated by O-methylation catalyzed by catechol-O-methyltransferases (COMT). If this conjugation is insufficient, there is the possibility that CEs are oxidized to semiquinones (not shown) and quinones (CE-Qs) (see Fig. 1).

The CE-Qs can conjugate with glutathione, catalyzed by S-transferases. Another inactivating process occurs by reduction of the CE-Qs to CEs, catalyzed by quinone reductase. If these two inactivating processes are insufficient, estrogen-2,3-quinones can react with DNA to form stable adducts that remain in DNA unless repaired. Estrogen-3,4-quinones, if not inactivated, react with DNA to form depurinating adducts, more specifically, two rotational conformers bound to the N7 of guanine (Gua); these adducts are lost from DNA by cleavage of the glycosidic bond, leaving apurinic sites in the DNA.

We hypothesize that formation of N7Gua adducts by reaction of estrogen-3,4-quinones with DNA is the tumor-initiating event in several human cancers, which include breast, prostate, ovarian, and, possibly, brain tumors. In fact, when tested in Syrian golden hamsters, 4-CEs are carcinogenic in the kidney, whereas 2-CEs are not (1, 2).

Comprehensive studies of carcinogenic polycyclic aromatic hydrocarbons (PAHs) have led to understanding of their mechanism of tumor initiation (3). This basic knowledge has been instrumental in delineating the DNA adducts formed by CE-Qs (depurinating adducts) that in our hypothesis would represent the events leading to tumor initiation.

PAHs are activated by two main pathways: one-electron oxidation to produce reactive intermediate radical cations and monooxygenation to yield bay-region diol epoxides (3). The reactive intermediates formed by the two mechanisms, radical cations and diol epoxides, can bind to DNA to form adducts that initiate the tumor process. PAH-DNA adducts are obtained by reaction of the metabolically activated PAHs with the nucleophilic groups of the two purine bases, adenine (Ade) and Gua. These adducts can be either stable adducts or depurinating adducts. The depurinating adducts are normally

Advances in Pharmacology, Volume 42

1054-3589/98 $25.00

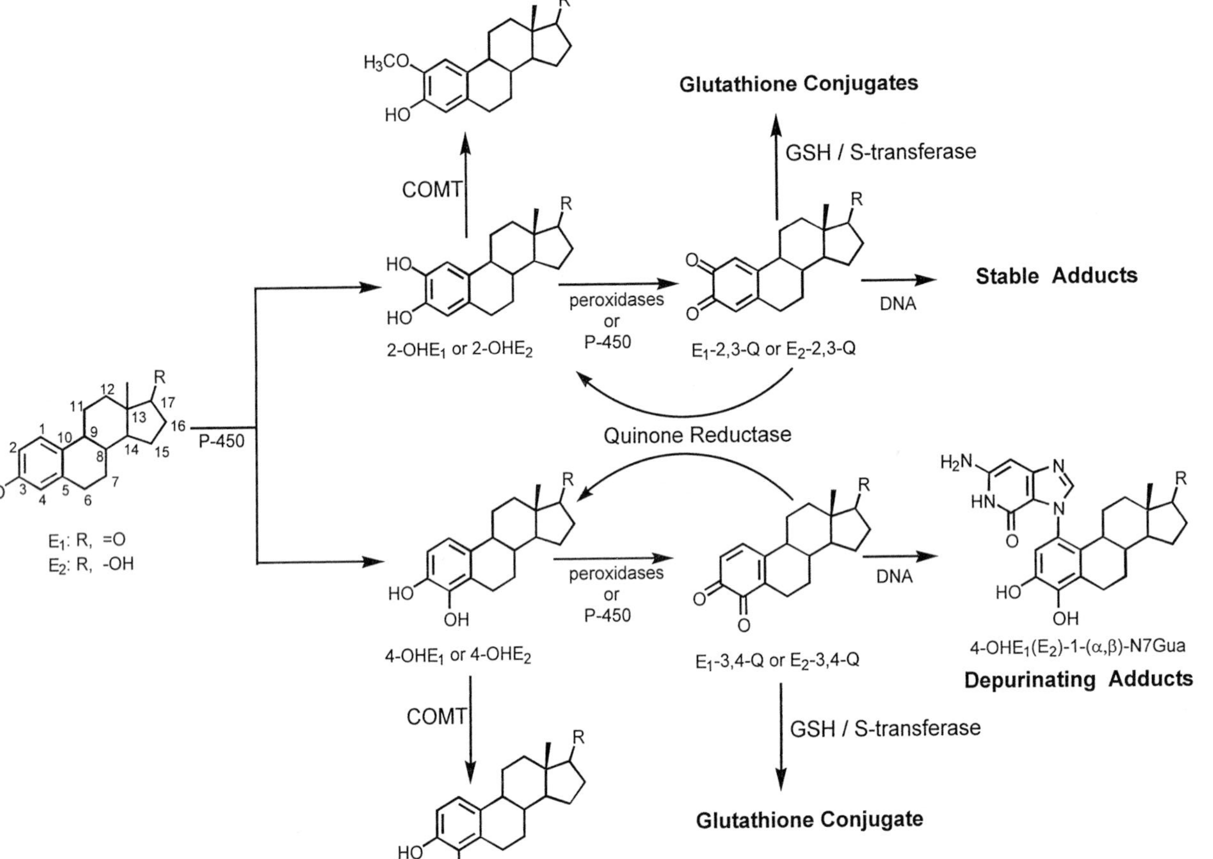

FIGURE 1 Metabolism and DNA adducts of estrogens.

formed by reaction of activated PAHs with the N3 and N7 of Ade, the N7 of Gua, and, sometimes, the C8 of Gua.

The pattern of adducts formed in the target-tissue mouse skin treated with the carcinogenic PAH benzo[*a*]pyrene, 7,12-dimethylbenz[*a*]anthracene (DMBA), or dibenzo[*a,l*]pyrene (DB[*a,l*]P) correlates with mutations found in the Harvey-*ras* oncogene in mouse skin papillomas induced by the PAH (4). When mouse skin is treated with DMBA, 79% of the adducts are depurinating Ade adducts, and 20% are depurinating Gua adducts (3). For DB[*a,l*]P, 81% are depurinating Ade adducts, and 18% are depurinating Gua adducts (3). In contrast, mouse skin treated with benzo[*a*]pyrene produced 46% depurinating Gua adducts and 25% depurinating Ade adducts (3). Examination of the Harvey-*ras* mutations in mouse skin papillomas induced with DMBA or DB[*a,l*]P demonstrates that in both cases an A→T transversion at codon 61 (CAA→CTA) consistently occurs (4). These mutations correlate with the predominant formation of depurinating Ade adducts in the mouse skin by these carcinogenic PAHs. About twice as many of the papillomas induced with benzo[*a*]pyrene exhibit a G→T transversion at codon 13 of the Harvey-*ras* oncogene (GGC→GTC), compared with the number of tumors having the codon 61 CAA→CTA mutation (4). This ratio of mutations is consistent with the profile of depurinating Gua and Ade adducts formed by benzo[*a*]pyrene in this target tissue.

This pattern of *ras* mutations suggests that the oncogenic mutations found in mouse skin papillomas induced by these PAHs are generated by misreplication of unrepaired apurinic sites derived from the loss of depurinating adducts (4). For example, an A→T transversion can be attributed to loss of a depurinating Ade adduct and generation of an apurinic site. If the apurinic site is not repaired, in the next round of DNA replication the most likely base to be inserted opposite the apurinic site is Ade. When the coding strand of DNA is then replicated, a thymine is inserted opposite the new Ade. This results in the A→T transversion observed in codon 61 of the Harvey-*ras* oncogene in tumors initiated by PAHs forming depurinating Ade adducts. When a Gua adduct is lost by depurination, leaving an apurinic site in the DNA, the preferential insertion of Ade in the opposite DNA strand leads to a G→T transversion at the site of the adduct.

Because thousands of apurinic sites spontaneously form and are repaired per cell per day, repair of the apurinic sites induced by depurinating adducts could be expected. The level of apurinic sites arising from treatment with PAHs is, however, 15 to 120 times higher than those formed spontaneously, suggesting that this large increase in apurinic sites overwhelms the capacity of the cell to repair them before DNA replication occurs. In addition, the apparent nonrepair of apurinic sites induced by treatment with PAHs may also be due to the presence of stable adducts that interfere with error-free repair of apurinic sites.

Because estrogen-3,4-quinones specifically form the depurinating N7Gua adducts and their precursor CEs are carcinogenic in male Syrian golden hamsters (1, 2), formation of these adducts is thought to be the initiating event that can lead to tumor development in several human cancers. Evidence collected thus far on the formation of estrogen-N7Gua adducts *in vitro* and *in vivo* (see chapter by Stack *et al.* in this section) supports the possible role of estrogens as tumor initiators.

References

1. Li, J. J., and Li, S. A. (1987). Estrogen carcinogenesis in Syrian hamster tissues: Role of metabolism. *Fed. Proc.* **46,** 1858–1863.
2. Liehr, J. G., Fang, W. F., Sirbasku, D. A., and Ari-Ulubelen, A. (1986). Carcinogenicity of catechol estrogens in Syrian hamsters. *J. Steroid Biochem.* **24,** 353–356.
3. Cavalieri, E., and Rogan, E. (1997). Mechanisms of tumor initiation by polycyclic aromatic hydrocarbons in mammals. *In* The Handbook of Environmental Chemistry, vol. 3: PAHs and Related Compounds. (A. H. Neilson, ed.), Springer-Verlag, Heidelberg. (in press)
4. Chakravarti, D., Pelling, J. C., Cavalieri, E. L., and Rogan, E. G. (1995). Relating aromatic hydrocarbon-induced DNA adducts and c-Harvey-*ras* mutations in mouse skin papillomas: The role of apurinic sites. *Proc. Natl. Acad. Sci. U.S.A.* **92,** 10422–10426.

B. C. Paria, S. K. Das, and S. K. Dey

Department of Molecular and Integrative Physiology
Ralph L. Smith Research Center
University of Kansas Medical Center
Kansas City, Kansas 66160

Embryo Implantation Requires Estrogen-Directed Uterine Preparation and Catecholestrogen-Mediated Embyronic Activation

In the mouse, both ovarian progesterone (P_4) and estrogen are essential for the preparation of the uterus for blastocyst implantation. Ovariectomy before the preimplantation estrogen secretion on day 4 results in implantation failure and blastocyst dormancy. This condition is termed *delayed implantation* and can be maintained with continued P_4 treatment. However, an injection of estrogen in the P_4-primed uterus initiates the implantation process (1). The mechanisms by which estrogen initiates the process of implantation are not known.

Using the delayed implantation model, blastocyst transfer experiments have established that the blastocyst's state of activity determines the "window" of implantation in the receptive uterus (1). A key finding of this investigation was that a factor(s) is produced very rapidly, but transiently (1 hr), in the P_4-primed

1054-3589/98 $25.00

uterus after an estradiol-17β (E_2) injection, which activates dormant blastocysts for implantation in the receptive uterus. The identity of such a factor(s) remains unknown. Furthermore, the blastocyst's state of activity is associated with its epidermal growth factor receptor (EGF-R) status: downregulation of the EGF-R gene is correlated with blastocyst dormancy, while its upregulation is associated with blastocyst activation *in utero,* but not *in vitro,* by E_2 (2). These observations suggest that uterine preparation and blastocyst activation by E_2 is differentially regulated in the P_4-primed uterus. The rapid and transient nature of the event also indicates that the synthesis of a new protein(s) via gene transcription may not be involved. However, involvement of a protein(s) translated from an existing mRNA(s) or posttranslational modification of an existing protein(s) cannot be ruled out. The present investigation provides evidence that while E_2 prepares the uterus to the receptive state, its metabolites, catecholestrogens (CEs), are involved in blastocyst activation. CEs (2-OH-E_2 and 4-OH-E_2) are formed as a result of aromatic hydroxylation of phenolic E_2 by estradiol-2/4-hydroxylase (E-2/4-H). Although CEs can function via the classical estrogen receptor, membrane binding sites for CEs have also been demonstrated. Because of their extremely rapid clearance rate, CEs are not likely to function as circulating hormones, but rather as local mediators of estrogen action.

Adult female CD-1 mice were mated with fertile or vasectomized males of the same strain to induce pregnancy or pseudopregnancy, respectively. The presence of vaginal plugs the next morning denoted day 1 of pregnancy or pseudopregnancy. For experiments requiring the depletion of endogenous ovarian steroids, females were ovariectomzied without regard to their stage of estrous cycle and rested for 15 days before receiving any treatment. Pregnant or pseudopregnant mice were ovariectomized on day 4 morning to induce the conditions of delayed implantation. These conditions were maintained with daily P_4 injections (2 mg/mouse) from days 5–7 (2). Pregnant and pseudopregnant mice served as blastocyst donors and recipients, respectively.

To determine the effects of various agents on EGF-R induction in dormant blastocysts, day 7 dormant blastocysts were cultured for 24 hr in Whitten's medium in the presence or absence of various factors. After termination of cultures, blastocysts were washed several times and used for either [^{125}I]EGF-binding studies or transfered to the P_4-treated, delayed pseudopregnant recipients after an injection of 2-Fl-E_2. This fluoroestradiol is a potent estrogen but is a poor substrate and an inhibitor of CE formation (3).

To confirm that 2-Fl-E_2 is estrogenic with respect to uterine growth and gene expression, nuclear [^{3}H]thymidine incorporation (DNA synthesis) and expression of lactoferrin, an estrogen-responsive gene, were monitored in ovariectomized mice after injections of various steroids (4, 5).

Our prediction was that the uterus hdyroxylates E_2 to CEs, which then activate the dormant blastocysts. Product isolation assay using high-performance liquid chromatography showed that luminal epithelial microsomes from P_4-primed delayed implanting uterus on day 7 can form CEs, mostly 4-OH-E_2. 2-Fl-E_2 inhibited this activity by threefold. [^{125}I]EGF-binding sites were induced in dormant blastocysts when cultured *in vitro* for 24 hr with CEs; E_2 was ineffective in this response. Furthermore, ICI-182780 (a pure antiestrogen) did not block EGF binding to dormant blastocysts exposed to 4-OH-E_2, suggesting the existence of an alternate mechanism for CE-mediated effects on blastocysts. Induction of EGF binding in dormant blastocysts in the presence of prostaglandins (PGs) *in vitro* suggests that CEs may act by generating PGs. Inhibition of EGF binding by indomethacin, an inhibitor of PG synthesis, in the presence of 4-OH-E_2 is consistent with the aforementioned results.

2-Fl-E_2 is a poor substrate but a potent inhibitor of E-2/4-H. However, it is a potent estrogen with respect to uterotrophic and gene expression. Thus, 2-Fl-E_2, like E_2, in the adult ovariectomized mice stimulated nuclear [^{3}H]thymidine uptake in the uterine epithelial cells, and in stromal cells when combined with P_4 (4). Further, E_2 or 2-Fl-E_2 upregulated the expression of lactoferrin, an estrogen-responsive gene in the mouse uterus. In spite of the estrogenic effects, 2-Fl-E_2 was not effective in inducing implantation in P_4-primed, delayed, implanting mice. It is possible that 2-Fl-E_2 prepared the uterus but failed to activate dormant blastocysts for implantation. This was tested by blastocyst transfer between delayed pregnant donors and pseudopregnant recipients. Normal day 4 blastocysts transferred into P_4-treated, delayed, pseudopregnant recipients implanted after an injection of 2-Fl-E_2. In contrast, dormant blastocysts transferred into similarly treated recipients failed to implant. Moreover, day 7 dormant blastocysts cultured for 24 hr for metabolic "activation" in the presence or absence of E_2 failed to implant when transferred to P_4-primed, delayed, pseudopregnant recipients 15 min after an injection of 2-Fl-E_2, while dormant blastocysts cultured with 4-OH-E_2 or PGE_2 for 24 hr implanted in similarly treated recipients.

These results suggest that estrogen differentially regulates uterine preparation and blastocyst activation for implantation. While primary estrogen is important for the preparation of the uterus, its catechol metabolites are important for blastocyst activation for implantation. Failure of either component results in implantation failure. Further, blastocyst's state of activity is associated with its EGF-R status: Dormancy is associated with EGF-R downregulation, whereas activation is associated with its upregulation. CEs appear to act on blastocyst functions via PGs, and CE-mediated effects do not operate through the nuclear estrogen receptor.

Acknowledgments

This research was supported in part by grants from the National Institutes of Health (HD 12304) and the National Cooperative Program on Markers of Uterine Receptivity for Nonhuman Blastocyst Implantation (HD 29968).

References

1. Paria, B. C., Huet-Hudson, Y. M., and Dey, S. K. (1993). Blastocyst's state of activity determines the "window" of implantation in the receptive mouse uterus. *Proc. Natl. Acad. Sci. U.S.A.* 90, 10159–10162.

2. Paria, B. C., Das, S. K., Andrews, G. K., and Dey, S. K. (1993). Expression of the epidermal growth factor receptor gene is regulated in mouse blastocysts during delayed implantation. *Proc. Natl. Acad. Sci. U.S.A.* **90,** 55–59.
3. Liehr, J. (1983). 2-fluoroestradiol: Separation of estrogenicity and carcinogenicity. *Mol. Pharmacol.* **23,** 278–281.
4. Huet, Y. M., Andrews, G. K., and Dey, S. K. (1989). Cell type-specific localization of c-myc protein in the mouse uterus: Modulation by steroid hormones and analysis of the periimplantation period. *Endocrinology* **125,** 1683–1690.
5. Wang, X-N., Das, S. K., Damm, D., Klagsbrun, M., Abraham, J. A., and Dey, S. K. (1994). Differential regulation of heparin-binding EGF-like growth factor in the adult ovariectomized mouse uterus by progesterone and estrogen. *Endocrinology* **135,** 1264–1271.

Michael G. Ziegler, Brian P. Kennedy, and Frederick W. Houts

Department of Medicine
Division of Nephrology/Hypertension
USCD School of Medicine
San Diego, California 92103-8341

Extra-Adrenal Nonneuronal Epinephrine and Phenylethanolamine-*N*-Methyltransferase

Epinephrine (Epi) is synthesized in the adrenal from norepinephrine (NE) by the enzyme phenylethanolamine-N-methyltransferase (PNMT), an enzyme that specifically methylates β-hydroxylated phenylethanolamines. NE can also be N-methylated by a less specific *N*-methyltransferse (NMT) which N-methylates many amines. Several groups have observed that both rats and people maintain nearly normal basal levels of blood and urinary Epi following adrenalectomy. We developed a sensitive assay to measure PNMT and NMT using NE as a substrate with [^{3}H]S-adenosylmethionine as the methyl donor. This assay technique has the advantage of sensitivity and of measuring actual N-methylation of NE to Epi. NMT is distinguished from PNMT by the ability of NMT to convert dopamine into [^{3}H]epinine and by the selective effect of PNMT inhibitors to block N-methylation by PNMT but not by NMT.

PNMT and NMT activity was measured in atrial and ventricular homogenates from the hearts of 12 male Sprague-Dawley rats. Cardiac atria contained PNMT with high affinity for NE, substrate specificity for NE over dopamine, and

Advances in Pharmacology, Volume 42

1054-3589/98 $25.00

inhibition by the PNMT inhibitor SKF 29661. Adrenal medullary PNMT activity is increased by glucocorticoid hormones. We surgically removed the adrenal medullae of rats and treated them with the glucocorticoid dexamethasone. PNMT levels in the cardiac atria increased 230% by 1 day and fivefold after 12 days of glucocorticoid treatment. In these animals, 12 days of glucocorticoid treatment nearly doubled levels of atrial Epi. Chemical sympathectomy with 6-hydroxydopamine actually increased PNMT levels. After dexamethasone treatment, greater volumes of anti-PNMT antiserum were needed to decrease PNMT enzymatic activity, indicating that dexamethasone treatment resulted in greater amounts of PNMT and did not simply activate existing PNMT molecules. PNMT activity was also measurable in skeletal muscle, and dexamethasone induced PNMT activity there as well (1). Sympathectomy of the masseter muscle of rats by unilateral superior cervical ganglionectomy had no effect on PNMT activity and depleted tissue Epi levels from 50% to 90%, less than the greater than 95% depletion of tissue NE levels. These findings suggest that PNMT is an extraneuronal enzyme in both cardiac and skeletal muscles (2).

Epi has a profound effect on pulmonary function. It stimulates α and β receptors in the lung and has a particularly high affinity for β_2 receptors, which comprise up to 70% of the β receptors in the lung. Stimulation of pulmonary adrenergic receptors by Epi causes bronchodilation. When we removed the adrenal medullae from rats, Epi levels in lung remained at 30% of the level of sham-operated rats. Treatment of these adrenal demedullated rats with 6-hydroxydopamine plus reserpine did not further reduce lung Epi. The Epi-forming capacity of lung homogenates was inhibited by the PNMT inhibitor SKF 29661 about half as well as in adrenal but better than in cardiac ventricle. It appeared that 50% of lung Epi-forming activity was due to PNMT and 50% from NMT. The NMT activity was unaffected by glucocorticoid, whereas PNMT activity was stimulated by dexamethasone. Administration of dexamethasone doubled levels of mRNA encoding for PNMT in rat lung. Administration of the glucocorticoid antagonist RU-486 after 7 days of dexamethasone treatment reduced PNMT mRNA levels by two-thirds within 24-hr, while Epi levels correlated with lung PNMT activity in both dexamethasone-treated and control rats. Administration of the PNMT inhibitor SKF 64139 (15 mg/kg t.i.d.) reduced *in vitro* lung PNMT activity in chronically dexamethasone-treated adrenalectomized rats by 96% and reduced lung Epi levels by 75%. The data suggest that a substantial fraction of lung Epi is locally synthesized. Dexamethasone was capable of tripling lung PNMT mRNA levels in 6 hr. The half-life of lung PNMT mRNA appears to be between 10 and 24 hr based on the rate of its disappearance after administration of the glucocorticoid receptor blocker RU-486.

Human bronchial epithelial cells (16 HBE14 0−) were grown to confluence in minimal essential culture medium with 10% fetal calf serum. Some of the cell cultures were then incubated for 7 days in 1 μmol of dexamethasone. PNMT activity in these cells was increased approximately 20-fold by treatment with dexamethasone. This PNMT activity was inhibited by anti-PNMT antibody.

We carried out a group of experiments to determine if rat and human kidneys can synthesize Epi. Rats that underwent adrenal demedullation decreased plasma Epi levels but did not decrease their renal Epi. Rat renal tissue contained PNMT and higher levels of the nonspecific methylating enzyme NMT. When an adrenalectomized rat received intravenous [^{3}H]methionine, its urine contained radioactivity that appeared to be [^{3}H]Epi with small amounts of the N-methylated metabolite of dopamine, [^{3}H]epinine. After [^{3}H]methionine was infused in a renal artery, the major urinary product appeared to be [^{3}H]epinine with a small amount of [^{3}H]Epi. We administered tracer amounts of intravenous [^{3}H]Epi to eight control human subjects and five older, heavier hypertensives for 3 hr to attain plateau levels of [^{3}H]Epi. Both groups had a clearance of H-Epi from arterial blood into urine of about 170 ml/min. We measured plasma Epi levels and calculated a predicted rate of clearance of Epi from arterial blood into urine. Among the control subjects, 43% ± 9% of urinary Epi appeared to be derived from the kidney and not from plasma. Among the hypertensives, 85% ± 4% of urinary Epi appeared to be derived from the kidneys ($p < .01$ when compared with control subjects) (3). We conclude that some urinary Epi is made by the kidney in both rats and humans. The renal conversion of NE to Epi by the enzymes PNMT and NMT seems a likely source of some urinary Epi. Rat kidney can convert NE to urinary Epi using two different N-methylating enzymes, and these enzymes are present in human kidney in even higher levels.

Epi can raise both blood pressure and blood glucose. Glucocorticoids induce PNMT in many tissues and can also induce hypertension and diabetes. We induced hypertension in adrenalectomized rats with the glucocorticoid dexamethasone and then administered the PNMT inhibitors SKF 29661 or SKF 64139. Both PNMT-inhibiting drugs lowered systolic blood pressure as measured by tail cuff in rats (2). We induced insulin resistance in rats by treatment with dexamethasone for 12 days. Dexamethasone rapidly increased tissue PNMT level and also increased muscle Epi levels. Dexamethasone-treated rats had elevated insulin levels after a glucose load, and the chronic administration of the PNMT inhibitor SKF 64139 returned insulin levels to normal. Chronic treatment with a PNMT inhibitor improved glucose tolerance in normal rats (1).

We studied Epi-forming activity in human tissues (4). PNMT and NMT activity was detectable in most tissues. PNMT activity predominated in heart and lung, whereas NMT was more prevalent in kidney, liver, bronchus, and trachea. Overall, tissue Epi levels correlated with PNMT activity ($r = .34$; $p = .025$) but not with NMT ($r = 11$). Because circulating Epi can be taken up by sympathetic nerves, it was of interest to inspect Epi levels in tissue with relatively poor sympathetic innveration, such as lung. In bronchus, the correlation coefficient between Epi and PNMT activity was .968 ($p = .007$), and between Epi and NMT it was .636. In trachea, the correlation between Epi and PNMT was .825 and between Epi and NMT, .864. The correlation between Epi and lung NMT was also highly significant ($r = .976$; $p < .001$).

We studied Epi-forming activity in six cell lines derived from human tissues. NMT activity was detectable in all of the cell lines; PNMT activity was present in three of the cell lines. PNMT was visualized in sections of human kidney, lung, and pancreas, using immunoperoxidase staining with an anti-PNMT antibody. In the lung, PNMT staining was most prominent in pulmonary alveolar cells. Staining of both parenchymal cells and islet cells was observed in the pancreas. There appeared to be inhibitors of PNMT activity in human tissue, because lung PNMT activity more than tripled after dialysis of lung homogenates. NMT activity was not changed by dialysis. PNMT was present in human red blood cells (RBCs). The RBC–PNMT activity in a population of healthy women was normally distributed and was about 7% higher than in men ($p = 0.14$ by Wilcoxin Rank Sum test) (4). PNMT activity in men did not appear to be normally distributed.

Human PNMT appears to be responsive to thyroid hormones. RBC–PNMT activity was below normal levels in hypothyroid patients and returned to normal levels when the patients became euthyroid after chronic thyroxin treatment. On the other hand, hyperthyroid subjects had elevated RBC–PNMT activity. NMT activity was not altered by either hypothyroidism or hyperthyroidism.

We conclude that PNMT is present and capable of synthesizing Epi in several extra-adrenal tissues. PNMT levels are increased by glucocorticoid treatment (5), and PNMT appears to partially mediate glucocorticoid hypertension and insulin resistance. Among both normal subjects and unmedicated hypertensives, the kidney appears to synthesize a significant fraction of urinary Epi.

References

1. Kennedy, B., Elayan, H., and Ziegler, M. G. (1993). Glucocorticoid induction of epinephrine synthesizing enzyme in rat skeletal muscle and insulin resistance. *J. Clin. Invest.* **92,** 303–307.
2. Kennedy, B., Elayan, H., and Ziegler, M. G. (1993). Glucocorticoid hypertension and nonadrenal phenylethanolamine N-methyltransferase. *Hypertension* **21,** 415–419.
3. Ziegler, M. G., Aung, M., and Kennedy, B. (1990). Sources of human urinary epinephrine. *J. Hypertens.* **8,** 927–931.
4. Kennedy, B., Bigby, T. D., and Ziegler, M. G. (1995). Nonadrenal epinephrine-forming enzymes in humans. Characteristics, distribution, regulation and relationship to epinephrine levels. *J. Clin. Invest.* **95,** 2896–2902.
5. Kennedy, B., Elayan, H., and Ziegler, M. G. (1993). Glucocorticoid elevation of mRNA encoding epinephrine-forming enzyme in lung. *Am. J. Physiol.* **265,** I117–I120.

Gunnar Flemström, Bengt Säfsten, and Lars Knutson

Department of Physiology and Medical Biophysics
Uppsala University
SE-751 23 Uppsala, Sweden

Dopamine and the Brain–Gut Axis

The bicarbonate secretion by the duodenal (and gastric) mucosa alkalinizes the viscoelastic mucus gel adherent to the mucosal surface and is one major mechanism in the protection of these epithelia against luminal acid. The charac-

Advances in Pharmacology, Volume 42

1054-3589/98 $25.00

teristics of the secretion have been reviewed (1). Duodenal secretion is deficient in patients with chronic and acute duodenal ulcer disease, and studies in animals have demonstrated that some stimuli of the duodenal mucosal bicarbonate secretion protect this epithelium against acid-induced injury. Stimulation of the secretion by vasoactive intestinal polypeptide (VIP), at doses not affecting duodenal mucosa blood flow, protects the duodenal mucosa in the rat and pig from acid-induced morphological changes. VIP protected against damage induced by 10 mM of hydrochloric acid in the rat and against that induced by 30 mM of hydrochloric acid in the pig. Some σ-receptor ligands, which stimulate duodenal mucosal bicarbonate secretion but do not inhibit gastric acid secretion, protect against duodenal but not gastric ulceration in animals.

Duodenal (and gastric) bicarbonate secretion is under neurohumoral control and also influenced by local mucosal production of prostaglandins. Peripheral sympathetic effects have been investigated by administration of adrenergic agonists and antagonists, by studying secretion in splanchnicotomized and/or adrenal-ligated animals, and by elicitation of sympathetic intestinointestinal and brain stem reflexes. The peripheral sympathetic influence is generally inhibitory and mediated by peripherally located α_2 adrenoceptors. Drugs acting at β adrenoceptors have been found without effect on duodenal secretion in the rat and on gastric secretions in the cat.

The α_2-adrenoceptor–mediated inhibition of the secretion contrasts to the stimulation mediated by the vagal nerves. The latter is transmitted peptidergically and by muscarinic M_3 receptors. Sham feeding, for example, stimulates the alkaline secretion by the duodenal as well as gastric mucosa in humans and conscious dogs. Electrical stimulation (in the peripheral direction) of the vagus nerves in cats and rats increases duodenal secretion, an effect abolished by hexamethonium but only partially blocked by atropine.

Central nervous administration of some agents have been used to further elucidate the central nervous influence on duodenal (and gastric) mucosal bicarbonate secretion. Intracerebroventricular infusion of corticotropin-releasing hormone in the rat increases secretion by the duodenal mucosa, and release of β-endorphin from the pituitary appeared to be part of the mediation of the response. Intracerebroventricular but not intravenous administration of thyrotropin-releasing hormone (TRH) increases duodenal secretion in anesthetized and conscious rats, and the transmission of this vagally mediated response probably involves VIP. In addition, TRH seems to cause a minor, adrenergically transmitted inhibition of the secretion. It should be noted that also some benzodiazepines influence the central nervous control of duodenal secretion. Thus, centrally elicited stimulation mediated by the vagal nerves was demonstrated with the benzodiazepines Ro 15-1788 and diazepam. The tricyclic antidepressant trimipramine was also tested but was without effect in this report.

Very potent stimulation (up fourfold increase) of duodenal mucosal bicarbonate secretion occurs on intracerebroventricular infusion of the α_1-adrenoceptor agonist phenylephrine in the rat (2). This stimulation is inhibited by central nervous (but not by intravenous) administration of the antagonist prazosin. Hexamethonium abolished stimulation, whereas cervical vagotomy, epidural blockade, and naloxone were without effect. The response was, thus, not mediated by the vagal nerves or the spinal cord, nor was it due to release of β-

endorphin. This may suggest that central nervous α_1-adrenoceptor stimulation induces the release of potent humoral stimuli of the alkaline secretion. The response involves nicotinic, possibly enteric nervous transmission. In contrast, intracerebroventricular infusion of the α_2-adrenoceptor agonist clonidine inhibits the secretion.

Some dopaminergic compounds are reported to be efficient in the treatment of duodenal ulcer disease in humans and prevent gastroduodenal mucosal damage in animal models of ulcer disease (1). These findings in humans and animals made it of interest to investigate the effects of dopamine and dopaminergic compounds on duodenal mucosal bicarbonate secretion. The secretion has been studied in anesthetized animals (3) and in human volunteers (4). Effects on duodenal mucosal production of cyclic adenosine monophosphate (cAMP) has been measured by use of isolated duodenal villus and crypt enterocytes (5). The results of these studies are summarized and some recent findings are reported here.

I. Studies in Animals

Studies of effects of dopamine and dopaminergic compounds have to consider that well-documented sites of action are the central nervous system as well as some peripheral tissues, including the intestinal tract and the kidney. It should also be kept in mind that dopamine may exert effects not only on dopamine receptors, but also on adrenoceptors.

Duodenum in the thiobarbiturate-anesthetized rat spontaneously secretes bicarbonate at a steady basal rate. Continuous intravenous infusion of dopamine dose-dependently increases the mucosal bicarbonate secretion, and the D_1 agonist SKF 38393 causes an increase in secretion of similar magnitude (Fig. 1). Dopamine D_1-receptor stimulation by bromocriptine, in contrast, caused a decrease in secretion. Catecholamine O-methyl transferase (COMT) inhibitors decrease tissue degradation of catecholamines, including dopamine, and effects of the peripherally acting inhibitor 3-[(3,4-dihydroxy-5-nitrophenyl) methylene]-2,4-pentanedione (nitecapone) have been studied. Nitecapone caused a dose-dependent increase in duodenal mucosal bicarbonate secretion of a magnitude similar to that observed with dopamine (see Fig. 1). Domperidone, a peripherally acting dopamine antagonist, inhibits the increase in secretion in response to the agonist SKF 38393 or nitecapone. Pretreatment with the adrenoceptor antagonists prazosin or phentolamine or the nicotinic antagonist hexamethonium, in contrast, has no such inhibitory effect. The combined findings strongly suggest that dopaminergic compounds stimulate duodenal mucosal bicarbonate secretion via an action on dopamine D_1 receptors located on the duodenal enterocytes.

Interestingly, the (centrally acting) D_1/D_2 antagonist haloperidol does not decrease but stimulates the duodenal secretion (see Fig. 1). This may suggest the presence of central nervous as well as peripheral dopaminergic modulation of the duodenal secretion.

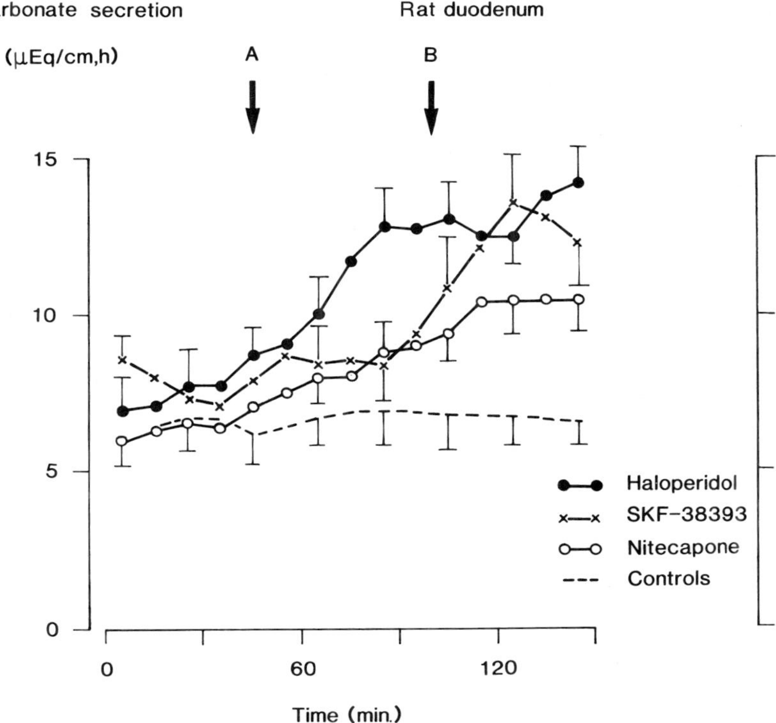

FIGURE 1 Intravenous bolus injection of haloperidol (5 and 25 μg/kg, n = 7), SKF-38393 (10 and 50 μg/kg, n = 6), or nitecapone (50 and 500 μg/kg, n = 12) stimulates duodenal mucosal bicarbonate secretion in the thiobarbiturate-anesthetized rat. The lower dose of each compound was injected at arrow *A*, and the higher, at arrow *B*. Means ± SEM are presented.

II. Studies of Isolated Duodenal Enterocytes

To confirm an action of dopamine on enterocyte receptors, cells were collected from the duodenum in rats by a combination of enzyme treatment and calcium chelation (5). Two major fractions, one mainly of villus and the other mainly of crypt origin, have been studied. Dopamine and the D_1 agonist SKF 39393 increased the accumulation of cAMP with a maximal response 5–15 min after start of incubation. In line with the effects on the bicarbonate secretion, dopamine D_2-receptor stimulation by quinpirole causes a slight decrease in the production of cAMP. There were no significant differences between villus and crypt cell fractions in respect to the effects of dopamine, SKF 38393, or quinpirole. These studies lend strong support to the suggestion that dopamine stimulates duodenal mucosal bicarbonate secretion by an action on the duodenal enterocytes and confirm that stimulation is D_1-receptor–mediated.

III. Studies in Human Volunteers

The distribution of dopamine in peripheral tissues varies considerably between species. The effects of the COMT inhibitor nitecapone on duodenal mucosal bicarbonate secretion in healthy volunteers was, thus, studied (4). The first 3 cm of the bulb was isolated using a pear-shaped balloon in the distal part of the stomach and two button-shaped balloons in the duodenum. Isotonic NaCl solution was infused into the duodenal test segment, and the effluent was collected by gravity drainage. Nitecapone added to the luminal perfusate (30–150 mg) caused a significant and dose-dependent increase in the bicarbonate secretion. The rise in secretion was similar in magnitude to that obtained with the prostaglandin E_1 analogue misoprostol. Peripheral COMT inhibition, thus, increases duodenal mucosal bicarbonate secretion in human volunteers as well as in animals.

IV. Conclusions

Duodenal mucosal secretion of bicarbonate is under peripheral as well as central nervous neurohumoral control, and strong evidence has been presented that stimuli of bicarbonate secretion protect the mucosa against acid-induced injury. Some dopaminergic compounds ameliorate mucosal damage in animal models of ulcer disease, and depletion of mucosal dopamine has been related to the appearance of duodenal ulcer disease in humans. Dopamine D_1-receptor agonists and a peripherally acting COMT inhibitor stimulate duodenal mucosal bicarbonate secretion in the rat, and similar stimulation by the COMT inhibitor was observed in human volunteers. These results, indicating that duodenal secretion is stimulated via peripheral dopamine D_1 receptors, are supported by the finding that D_1 (but not D_2) receptor agonists increase the production of cAMP in duodenal crypt and villus enterocytes. The stimulation of the secretion by haloperidol may suggest the presence of central nervous as well as peripheral dopaminergic modulation of duodenal bicarbonate secretion and mucosal protection.

Acknowledgment

This work was supported by the Swedish Medical Research Council (grant 04X-3515).

References

1. Flemström, G. (1994). Gastric and duodenal mucosal bicarbonate secretion. *In* Physiology of the Gastrointestinal Tract, 3rd ed. (L. R. Johnson, E. D. Jacobson, J. Christensen, D. Alpers, and J. H. Walsh, eds.), pp. 1285–1309. Raven Press, New York.
2. Larson, G. M., Jedstedt, G., Nylander, O., and Flemström, G. (1996). Intracerebral adrenoreceptor agonists influence rat duodenal mucosal bicarbonate secretion. *Am. J. Physiol.* **271**, G831–G840.

3. Flemström, G., Säfsten, B., and Jedstedt, G. (1993). Stimulation of mucosal alkaline secretion in rat duodenum by dopamine and dopaminergic compounds. *Gastroenterology* **104**, 825–833.
4. Knutson, L., Knutson, T. W., and Flemström, G. (1993). Endogenous dopamine and duodenal bicarbonate secretion in humans. *Gastroenterology* **104**, 1409–1413.
5. Säfsten, B., and Flemström, G. (1993). Dopamine and vasoactive intestinal peptide stimulate cyclic adenosine-3′,5′-monophosphate formation in isolated rat villus and crypt duodenocytes. *Acta Physiol. Scand.* **149**, 67–75.

Ebbe Eldrup,* Erik A. Richter,† Merete L. Hetland,* Jan Andreasen,* Jannik Hilsted,‡ Svend E. Møller,§ and Niels Juel Christensen*

* Department of Internal Medicine and Endocrinology
Herlev University Hospital
2730 Herlev, Denmark

† August Krogh Institute
University of Copenhagen
2100 Dopenhagen, Denmark

‡ Hvidore Hospital
2930 Klampenborg, Denmark

§ Clinical Research Laboratory
Sct. Hans Hospital
4000 Roskilde, Denmark

Origin and Significance of Plasma Dihydroxyphenylalanine

3,4-Dihydroxyphenylalanine (DOPA) is both a catecholamine and an aromatic amino acid. DOPA occupies a central position in the function of the sympathetic nervous system. DOPA is the immediate product of the rate-limiting step in the biosynthesis of catecholamines: dopamine, norepinehrine (NE), and epinephrine. Also, DOPA concentrations are relatively high, 1–3 ng/ml, in venous plasma. Regional release of DOPA into the bloodstream has been indicated by arteriovenous increments of plasma DOPA concentrations in the arm, leg, head, and heart.

There are some key questions regarding plasma DOPA that we hope this brief review will answer: Is plasma DOPA of neuronal origin? (1) Is plasma DOPA an index of sympathetic activity? (2) And finally, is it significant that DOPA is an amino acid?

Advances in Pharmacology, Volume 42

1054-3589/98 $25.00

In 1987, it was suggested by Goldstein and coworkers that plasma DOPA is of neuronal origin (3). This hypothesis was based on observations in animals and humans. An arteriovenous increase in plasma DOPA is normally observed across the forearm. In three of the four patients who had undergone a unilateral regional sympathectomy, an arteriovenous decrease in plasma DOPA was observed. Human plasma DOPA concentrations decreased after administration of clonidine. Plasma DOPA concentration did not change, however, in response to a tilt in humans or after adrenalectomy in monkeys. In dogs treated with an inhibitor of tyrosine hydroxylase, alpha-metyl-para-tyrosine, the concentrations of NE and DOPA decreased, but not at all in a parallel manner. In another study by the same group in healthy controls, in patients with multiple systemic atrophy and in patients with pure autonomic failure, plasma DOPA concentrations were lower in patients with autonomic failure, but no changes were seen in patients with multiple systemic atrophy. No convincing relationship, however, was found between plasma DOPA and plasma NE. The correlation coefficient was .38 when two patients with dopamine β-hydroxylase deficiency were excluded (4).

In a study of isometric exercise, mental stress, coffee drinking, and administration of the neuronal re-uptake inhibitor, desipramine, very small changes in forearm venous plasma DOPA concentrations were found (5). When these changes were related to NE spillover, a small but significant relationship was observed.

Devalon and coworkers (6) found a small increase in plasma DOPA after exercise. In another exercise study, Morita and coworkers found a very small increase in plasma DOPA (7). Statistical analyses revealed that time (i.e., also during exercise) only explained 5% of the variation in plasma DOPA. Again, no relationship between plasma NE and plasma DOPA was found. To summarize the results of studies in animals and humans from the groups of Kopin, Goldstein, Eisenhofer, Tyce, and Morita: Small changes in plasma DOPA have been observed despite tremendous changes in sympathetic activity, measured as plasma NE concentration or plasma NE spillover.

When we examined healthy individuals, patients with diabetic neuropathy and diabetics without neuropathy no changes in plasma DOPA were found, in contrast to the changes in plasma NE measured. In a clonidine experiment in humans, a decrease in plasma DOPA was found, but this was also observed in the control experiment, meaning that no changes in plasma DOPA were induced by clonidine, in contrast to the significant decrease in plasma NE. Thus, we have not observed changes in plasma DOPA in conditions with acutely or chronically changed sympathetic activity in humans (8).

Why these discrepancies? Can plasma DOPA be of nonneuronal origin? Banwart and coworkers observed that DOPA does not seem to originate from the food because no change in plasma DOPA was observed after feeding in dogs (9). DOPA can be synthesized by tyrosinase in the melanin synthesis, but in a study by Garty and coworkers in albinos, blacks, and caucasians, similar plasma DOPA levels were found, indicating that melanocytes are not the source of plasma DOPA (10). No one has found any apparent contribution to plasma DOPA from the central nervous system. In conclusion, no extra-neuronal source of plasma DOPA has been identified.

Where is DOPA located? In a study in rats, which we unilaterally sympathectomized in the lower limb, muscle content of DOPA did not change, whereas NE and dopamine content decreased as expected (11). The DOPA content in different tissues was measured. The gut, kidney, and muscles had a high DOPA content. From this study we concluded that DOPA is not accumulated in sympathetic neurons. Also, muscle cells seem to be an important DOPA reservoir.

We performed a kinetic study of plasma DOPA in humans (12). Plasma DOPA appearance rate was 1.1 μg/min. The extremities accounted for about 20%. DOPA is also extracted from plasma. In the lower arm, the extraction fraction is 24%. Across the heart, Goldstein and coworkers have measured that the extraction fraction is close to zero (13).

When a small dose of a decarboxylase inhibitor, benserazide, was administered, plasma DOPA concentration increased significantly (12). Plasma DOPA appearance rate increased almost eightfold. Plasma DOPA clearance, however, only decreased insignificantly. The normal DOPA clearance was 1 liter/min, of which 20% seem to be cleared in the kidneys.

The results from the kinetic study are compatible with the concept that substantial amounts of DOPA are normally decarboxylated in sympathetic nerves. Furthermore, DOPA spillover to plasma provided it is not decarboxylated to dopamine.

Recently, we have examined the influence of meals in a 25-hr study measuring DOPA, as well as the dopamine metabolites, dopaminesulphate and dihydroxyphenylacetic acid (14). Seven healthy young men were investigated on two different days, one with food intake and one without. When comparing the difference between the mean plasma DOPA values 3 hr after and 1 hr before each meal with the same difference before and after the time of the meals on the fasting day, a significant decrease was found after the 12 A.M. meal and after the 6 P.M. meal. No relationship, however, was found between plasma NE concentrations and plasma DOPA concentrations.

Plasma total dopamine, which in humans is dopaminesulphate, increased after lunch and after the evening meal. The increments in total dopamine concentrations were closely correlated to the food content of DOPA plus dopamine plus dopaminesulphate, because we observed no increase after breakfast, a small increase after the hot noon meal, and a huge increase after the evening open-sandwiches meal.

In conclusion, this study indicates that DOPA in ordinary meals is most likely metabolized in the gastrointestinal tract, and DOPA in plasma is probably taken up by muscles in response to the rise in plasma insulin after a meal. We also conclude that substantial amounts of plasma dopaminesulphate are derived from ordinary meals.

In a recent study of cardiac sympathetic nerve function, Eisenhofer and coworkers demonstrated that cardiac DOPA spillover is increased when cardiac sympathetic activity is increased (15). In the heart, the increase in DOPA spillover seem to be correlated to the increase in NE turnover. This close relationship between plasma DOPA spillover and sympathetic activity may be explained by the fact that in the heart, we have the more or less unique situation that DOPA is extracted only to a very small degree.

Our major conclusions are: (1) plasma DOPA seems to be of neuronal origin; (2) plasma DOPA is not an index of sympathetic activity; (3) plasma DOPA may be an index of NE turnover in the heart; and (4) DOPA in plasma behaves as an amino acid after a meal.

References

1. Kopin I. J. Origins and significance of dopa and catecholamine metabolites in body fluids (1992). *Pharmacopsychiat.* **25,** 33–36.
2. Anton, A. H. Is plasma Dopa a valid indicator of sympathetic activity? (1991). *J. Lab. Clin. Med.* **117,** 263–264.
3. Goldstein, D. S., Udelsman, R., Eisenhofer, G., Stull, R., Keiser, H. R., Kopin, I. J. Neuronal source of plasma dihydroxyphenylalanine (1987). *J. Clin. Endocinol. Metab.* **64,** 856–861.
4. Goldstein, D. S., Polinsky, R., Garty, M., Robertson, D., Brown, R., Biaggioni, I. (1987). *et al.* Patterns of plasma levels of catachols in neurogenic orthostatic hypotension. (1989). *Ann. Neurol.* **26,** 558–563.
5. Eisenhofer, G., Meredith, I. T., Ferrier, C., Cox, H. S., Lambert, G., Jennings, G. L., and Esler, M. D. (1991). Increased plasma dihydroxyphenylalanine during sympathetic activation in humans is related to increased norepinephrine turnover. *J. Lab. Clin. Med.* **117,** 266–273.
6. Devalon, M. L., Miller, T. D., Squires, R. W., Rogers, P. J., Bove, A. A., Tyce, G. M. Dopa in plasma increases during acute exercise and after exercise training. (1989). *J. Lab. Clin. Med.* **114,** 321–327.
7. Morita, K., Minami, M., Inagaki, H. (1992). Kinetics of plasma 3,4-dihydroxyphenylalanine (DOPA) during physical exercise: a comparative study with that of plasma noradrenaline. *Biogenic Amines* **8,** 431–441.
8. Eldrup, E., Christensen, N. J., Andreasen, J., and Hilsted, J. Plasma dihydroxyphenylalanine (DOPA) is independent of sympathetic activity in humans. (1989). *Eur. J. Clin. Invest.* **19,** 514–517.
9. Banwart, B., Miller, T. D., Jones, J. D., and Tyce, G. M. (1989). Plasma dopa and feeding. *Proc. Soc. Exp. Biol. Med.* **191,** 357–361.
10. Garty, M., Stull, R., Kopin, I. J., Goldstein, D. S. (1989). Skin color, aging, and plasma L-dopa levels. *J. Auton. Nerv. Syst.* **26,** 261–263.
11. Eldrup, E., E. A., Richter, and N. J., Christensen. (1989). Dopa, norepinephrine, and dopamine in rat tissues. no effect of sympathectomy on muscle DOPA. *Am. J. Physiol.* **256,** E284–E287.
12. Eldrup, E. M. L., Hetland, and N. J., Christensen. Increase in plasma 3,4-dihydroxyphenylalanine (DOPA) appearance rate after inhibition of DOPA decarboxylase in humans. (1994). *Eur. J. Clin. Invest.* **24,** 205–211.
13. Goldstein, D. S., Cannon, III R. O., Quyyumi, A., Chang, P., Duncan, M., Brush, Jr., J. E., and Eisenhofer, G. (1991). Regional extraction of circulating norepinephrine, DOPA, and dihydroxyphenylglycol in humans. *J. Auton. Nerv. Syst.* **34,** 17–36.
14. Eldrup, E., Møller, S. E., Andreasen, J., Christensen, N. J. (1997). Effects of ordinary meals on plasma concentrations of 3,4-dihydroxyphenylalanine, dopamine sulphate and 3,4-dihydroxyphenylacetic acid. *Clin. Sci.* **92,** 423–430.
15. Eisenhofer, G., Friberg, P., Rundqvist, B., Quyyumi, A. A., Lambert, G., Kaye, D. M., Kopin, I. J., Goldstein, D. S., and Esler, M. D. (1996). Cardiac sympathetic nerve function in congestive heart failure. *Circulation* **93,** 1667–1676.

Y. Misu, Y. Goshima, J-L. Yue, and T. Miyamae
Department of Pharmacology
Yokohama City University School of Medicine
Yokohama 236, Japan

Is L-Dopa a Neurotransmitter of the Primary Baroreceptor Afferents Terminating in the Nucleus Tractus Solitarii of Rats?

Dihydroxyphenylalanine (DOPA) is believed to be a precursor for dopamine (DA) converted by aromatic L-amino acid decarboxylase (AADC). Since 1986, we have proposed that DOPA is a neurotransmitter in the central nervous system (1–5). Transmitter-like DOPA is released from rat striata and blood pressure regulation centers in the lower brain stem. Exogenous DOPA itself elicits *in vitro* presynaptic and *in vivo* postsynaptic responses. All are stereoselective, and most are antagonized by DOPA methyl ester (DOPA ester), a competitive DOPA antagonist. DOPA did not displace selective binding of α_2, β, D_1, and D_2 ligands in brain membrane preparations. A recognition site for DOPA may exist. In striata, DOPA is a precursor for DA, an endogenous potentiator for activities of postsynaptic D_2 receptors and of presynaptic β adrenoceptors to facilitate evoked DA release, and an inhibitor of acetylcholine release. All may cooperate for effectiveness in Parkinson's disease. Meanwhile, DOPA induces DOPA ester–sensitive neuronal glutamate release from slices, even under inhibition of AADC. This may relate to neuroexcitatory side effects. DOPA further elicits postsynaptic decreases in blood pressure and heart rate (HR) in the nucleus tractus solitarii (NTS) and caudal ventrolateral medulla and increases in those in the rostral ventrolateral medulla. Immunocytochemically, tyrosine hydroxylase (TH) (+)-, AADC (−)-, DOPA (+)-, and DA (−)-neurons that may contain DOPA as an endproduct exist in various regions, including NTS (2).

The first topic is a DOPAergic system in NTS. NTS has the first synapase of the baroreceptor reflex arc. Neurotransmitters of the primary baroreceptor afferents have not been identified, although glutamate is one of candidates. DOPA is proposed to be a neurotransmitter (1–5). TH and DOPA, but not DA and DA-β-hydroxylase immunoreactivities decreased in the unilateral NTS/dorsal vagal nucleus area 7 days after ipsilateral aortic nerve denervation peripheral to the ganglion nodosum, suggesting a monosynaptic DOPAergic relay from baroreceptor to NTS area. Using high-performance liquid chromatography (HPLC) with an electrochemical detector, basal DOPA release during microdialysis of NTS in anesthetized rats was partially tetrodotoxin (TTX)-

Advances in Pharmacology, Volume 42

1054-3589/98 $25.00

sensitive, Ca^{2+}-dependent, and reduced by TH inhibitor (α-methyl-p-tyrosine [α-MPT] i.p.). The release by intermittent aortic nerve stimulation (100 Hz) was TTX-sensitive and that by high K^+ was Ca^{2+}-dependent (4). Ca^{2+} may be primarily involved in DOPA release process, because high K^+ elicited the release prior to enhancement of TH activity, $[^3H]H_2O$ formation, in $[^3H]$tyrosine-superfused brain slices. The activity was enhanced after the release ended (2). The release by baroreceptor stimulation with phenylephrine was abolished by bilateral sinoaortic denervation without modification of hypertension (Fig. 1A). DOPA (10–100 ng) microinjected into depressor sites of the medial NTS identified by prior glutamate led to dose-dependent decreases in blood pressure and HR in untreated rats and in those treated with central AADC inhibitor (NSD-1015 100 mg/kg i.p.) or with i.v.t. 6-hydroxydopamine, without the effect of D-DOPA, DA, or noradrenaline (100 ng) (2, 3). The decreases seem to be postsynaptic in nature, not due to conversion to DA but to activation of the DOPA recognition site, because the decreases were antagonized by DOPA ester (1 μg) ipsilaterally microinjected without modification of those by glutamate (30 ng). Left aortic nerve stimulation (20 Hz, 10 sec) elicited depressor and bradycardic responses similar to DOPA, which were antagonized by ipsilateral DOPA ester (4). DOPA evoked functions tonically to activate depressor sites, because DOPA ester bilaterally microinjected elicited hypertension and tachycardia, being abolished by α-MPT (Fig. 1B) with a decrease in basal DOPA release. Phenylephrine-induced reflex bradycardia was abolished by sinoaortic denervation and reduced by bilateral DOPA ester.

The second topic is interactions between DOPAergic and GABAergic systems. GABA is an inhibitory neuromodulator for baroreflex at the level of NTS. GABA functions tonically via $GABA_A$ receptors to elicit increases in blood pressure and HR and to inhibit decreases in those by electrical aortic nerve stimulation (2). Thus, if our DOPA neurotransmitter hypothesis is truly the case, GABA has to inhibit presynaptic and/or postsynaptic components of DOPA in NTS. GABA (3–300 ng) microinjected dose-dependently and nipecotic acid, a GABA uptake inhibitor (100 ng), elicited increases in blood pressure and HR and inhibited decreases in those by DOPA (30 ng) when the respective pressor responses returned to a base, while bicuculline, a $GABA_A$ antagonist (10 ng), elicited decreases in blood pressure and HR and potentiated those by DOPA when the depressor response returned to the base. DOPA ester (1 μg) unilaterally microinjected reduced by a half increases in blood pressure and HR by GABA (300 ng). DOPA ester up to 1 mM did not displace $[^3H]$GABA binding in brain membrane preparations. Muscimol, a $GABA_A$ agonist (10–30 μM), perfused via probes decreased basal DOPA release by 50–60%, with a peak 2 hr after perfusion, and the decrease was antagonized by bicuculline (10 μM), while bicuculline (30 μM) alone increased it by 40%, with a peak similar to that of muscimol. Using microdialysis and HPLC with fluorescence spectrophotometry, GABA was constantly detectable in dialysates. Under intact AADC, DOPA (10 nM) perfused for 20 min increased GABA release by 40% at the first 20 min. Under inhibited AADC (NSD-1015 200 μM), DOPA (1–100 nM) concentration-dependently increased GABA release by 35–60%, a DOPA (10 nM)-induced increase was Ca^{2+}-dependent and TTX-sensitive, and DA or D-DOPA elicited no release. GABA seems to act tonically on $GABA_A$

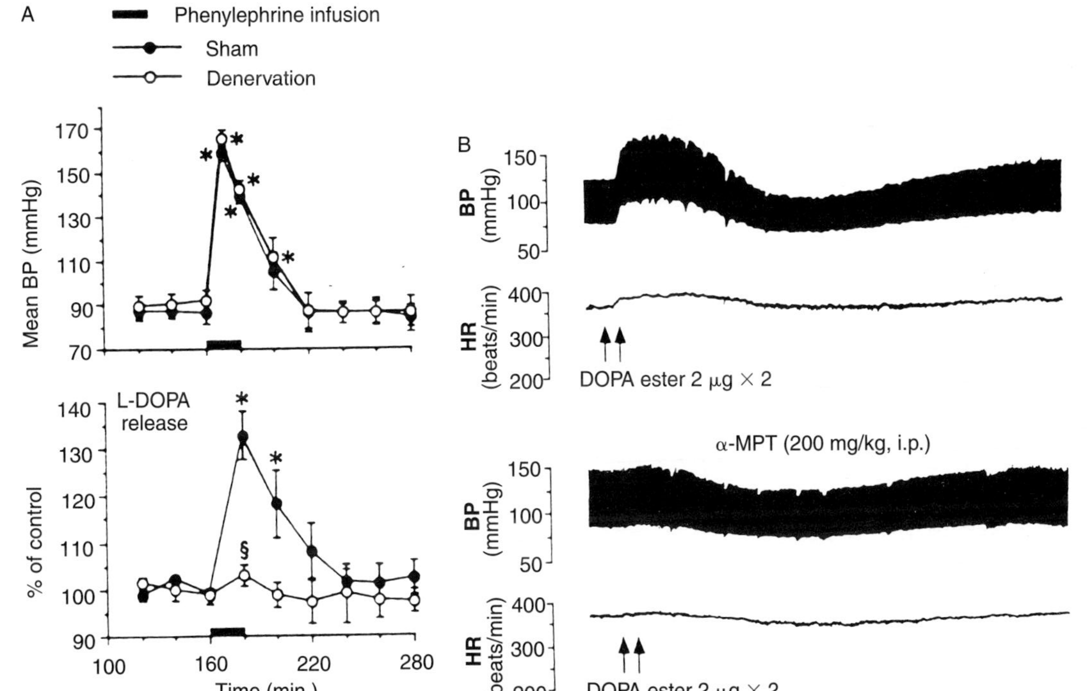

FIGURE 1 (*A*) Phenylephrine-induced hypertension and DOPA release during microdialysis of NTS and abolition of the release by bilateral sinoaortic denervation. (*B*) Initial hypertension and tachycardia by DOPA ester bilaterally microinjected in NTS and the abolition by α-MPT in anesthetized rats. (Modified from Yue *et al.* (4) with permission of International Brain Research Organization and *Neuroscience* **62**, 145–161, 1994.)

receptors to inhibit presynaptic DOPA release and postsynaptic depressor responses to DOPA released in NTS. This inhibition seems to be in part involved in GABA-induced hypertension in NTS. DOPA itself releases neuronal GABA, being a reflection of a mutual compensatory mechanism between both systems within the NTS of the baroreflex. These findings further support our DOPA neurotransmitter hypothesis.

The third topic is altered functions of the DOPA system in the NTS of adult spontaneously hypertensive rats (SHR) (5). Basal DOPA release was lower in SHR than age-matched Wistar-Kyoto (WKY) rats. This release was reduced by TTX to the same absolute levels in the two strains. Tonic neuronal DOPA release was impaired in SHR. This impairment is secondarily neither due to a decrease in TH activity nor due to an increase in AADC activity, because the former was increased in SHR compared with WKY, while no difference of the latter was seen in the NTS area punched out, and may be involved in maintenance of hypertension in SHR.

References

1. Misu, Y., Ueda, H., and Goshima, Y. (1995). Transmitter-like actions of L-DOPA. *Adv. Pharmacol.* **32,** 427–459.
2. Misu, Y., Goshima, Y., Ueda, H., and Okamura, H. (1996). Neurobiology of L-DOPAergic systems. *Prog. Neurobiol.* **49,** 415–454.
3. Kubo, T., Yue, J.-L., Goshima, Y., Nakamura, S., and Misu, Y. (1992). Evidence of L-DOPA systems responsible for cardiovascular control in the nucleus tractus solitarii of the rat. *Neurosci. Lett.* **140,** 153–156.
4. Yue, J.-L., Okamura, H., Goshima, Y., Nakamura, S., Geffard, M., and Misu, Y. (1994). Baroreceptor-aortic nerve-mediated release of endogenous L-3,4-dihydroxyphenylalanine and its tonic depressor function in the nucleus tractus solitarii of rats. *Neuroscience* **62,** 145–161.
5. Yue, J.-L., Okumura, Y., Miyamae, T., Ueda, H., and Misu Y. (1995). Altered tonic L-3,4-dihydroxyphenylalanine systems in the nucleus tractus solitarii and the rostral ventrolateral medulla of spontaneously hypertensive rats. *Neuroscience* **67,** 95–106.

I. Nagatsu,* M. Sakai,* N. Karasawa,* T. Takeuchi,* R. Arai,* K. Yamada,* and T. Nagatsu†

*Department of Anatomy
†Institute for Comprehensive Medical Science
School of Medicine
Fujita Health University
Toyoake, Aichi 470-11, Japan

Immunocytochemical Evidence of Novel Catecholamine- or Biopterin-Related Neurons of Mammalian Brain

Since the first discovery of tyrosine hydroxylase (TH) by Nagatsu *et al.* in 1964, TH has been used as a marker enzyme of catecholamine (CA) biosynthesis, such as dopamine (DA), noradrenaline, and adrenaline. Our extensive immunocytochemical studies previously revealed that transiently TH-immunoreactive (IR) neurons were not stained, using antiserum against aromatic L-amino acid decarboxylase (AAD or DDC), dopamine β-hydroxylase, phenylethanolamine-N-methyl transferase, DA, or serotonin (5HT) in some noncatecholaminergic (non-CAnergic) neurons, during pre- and postnatal development, in the anterior olfactory nucleus, striatum, cerebellum, and spinal trigeminal nucleus.

We produced for the first time a polyclonal antibody with highly sensitive immunoreactivity against an oligopeptide of rat GTP cyclohydrolase I (GCH): GFPERELPRPGA. GCH is the first enzyme for biosynthesis of tetrahydrobiopterin (BH4), the cofactor of phenylalanine hydroxylase, TH, tryptophan hydroxylase, and nitric oxide synthase. GCH is selectively stained in the monoaminergic neurons in the mouse brain as well as in the liver cells, adrenal medullary cells (1), and in eccrine sweat gland cells (Sakai, preliminary results).

TH and GCH, AAD, or 5HT were detected on the serial sections of the mouse brain by use of an avidin–biotin peroxidase complex (ABC) method. Our rabbit polyclonal antibody directed against GCH-glutaraldehyde–hemocianin conjugate (1), or antibody against bovine adrenal TH (2), L-dopa (2), AAD (2), DA (2), or 5HT (2) were used as primary antibodies.

The transient appearance of GCH-IR cells in the ventral lateral geniculate (VLG) nuclear region of mice was detected by immunocytochemistry. We found for the first time novel GCH-IR cells already at postnatal day 1 (P1). The numbers reached maximum at P7 to P14 and decreased until P29, and then mostly disappeared at P56. These cells were distributed in the intergeniculate leaflet and internal and external lamellae of the VLG. These GCH-IR cells were TH-negative, AAD-negative, 5HT-negative, and did not belong to the monoaminergic neuron system, because they lacked dopamine or serotonin production (3).

Advances in Pharmacology, Volume 42

1054-3589/98 $25.00

TABLE I Analysis of the Transgene Expression in the Brain of hTH Tg Mice in Comparison with nTg Mice

Region	*Tg*	*nTg*					
Introduced gene	*hTh*						
protein detected	*hTH*	*GCH*	*mTH*	*DOPA*	*DDC*	*DA*	*5HT*
Non-CAnergic Region							
Ventral lateral geniculate nucleus	+	+/−	−		−	−	−
Medial geniculate nucleus	+	−	+/−		−	−	−
Spinal trigeminal nucleus	+	−	+/−	−	−	−	−
Nucleus parabigeminalis	++	−	−		−	−	−
Accessory olfactory bulb (gr)	−		−	−	−	−	−
Septum	+	−	−	−	−	−	−
Nucleus accumbens	+	−	−	−	+	−	−
Caudoputamen	+	−	−	−	−	−	−
Stria terminalis (D14)	+	−	−	−	+	−	−
Amygdala (medial nucleus)	++	−	−	−	+	−	−
Nucleus Supraopticus (human)	+	−	+/− (+)	−	−	−	−
Main olfactory bulb (except A16)	+	−	−	−	−	−	−
Nucleus olfactorius anterior	++	−	+/−	−	−	−	−
Piriform cortex	+++	−	−	−	−	−	−
Hippocampus	++	−	−	−	−	−	−
Parasubiculum	+	−	−	−	−	−	−
Entorhinal cortex	+++	−	−	−	−	−	−
Pineal body	+	+			+	−	+
Nucleus suprachiasmaticus (D13)	+	+/−	−	−	+	+/−	−
Zona incerta (D10)	+	−	−	−	+	−	−
Paraventricular nucleus (human)	+	−	+/− (+)	−	−	−	−
Lateral habenula (D6)	+	−	−	−	+	−	−
Habenula (suncus)		−	−	−(+)	−	−	−
Mammillary body	++	−	−	−	+	−	+
Nucleus raphe dorsalis (B7)	++	++	+	−	++	−	+
Inferior colliculus (D4)	++	−	+/−	−	+	−	−
Nucleus parabrachialis (D3)	+	−	−	−	+	−	−
Nucleus tractus solitarius (D2) (A2m)(suncus)	++	−	−	− (+)	+	−	−
Substantia gelatinosus of nucleus tractus solitarius (human)	++		+				(DBH−)
Area postrema	++	+	+	−	+	+/−	−
CAnergic Region							
A16	++	+	++	−	+	+	−
A15 ~A14	++	+	++	−	++	+	−
A13	++	+	++	+/−	++	+	

(continues)

TABLE I *Continued*

Region		*Tg*	*nTg*					
	Introduced gene	*hTh*						
	protein detected	*hTH*	*GCH*	*mTH*	*DOPA*	*DDC*	*DA*	*5HT*
A12 (suncus)		++	+	++	−(+)	++	+	
A11		++	+	++	+/−	++	+	
A10		++++	+	+++	+/−	+++	++	−
A9 ~A8		++++	+	+++	+/−	+++	++	−
A7 ~A6		+++	+	++	+	++	+	−
A5 ~A4		+	+	+	+/−	+	+	−
A2/C2		++	+	+	+/−	+	+	−
A1/C1		+	+	+	+/−	+	+	−

The signs represent the abundance of GCH-, TH-, DOPA-, DDC-, DA-, and 5HT-immunoreactive cells in the different brain regions: ++++ ~ +++, areas where most neurons show strong reactivity; ++, some cells show strong reactivity; +, a few cells show strong reactivity; +/−, transiently appeared rare cell reactivity of any type; −, no reactivity in any cells. CAnergic regions are classified as described by Hökfelt *et al.* (1984) and DDC-containing neurons by Nagatsu *et al.* (2) in mice. The hTH-1 mice were generated and analyzed in previous papers (2). Tg, transgenic; nTg, non-Tg; hTH, human TH, mTH, mouse TH.

The transient appearance of TH-IR cells in the medial geniculate (MG) nuclear region of mice was investigated at P7. The numbers of TH-IR cells reached maximum at P14 to P21 and decreased at P29. These TH-IR cells were GCH-negative, AAD-negative, DA-negative, and did not belong to the CAnergic neuron system, because they did not produce DA. The novel TH-IR/GCH-negative cells were distributed from MG to nucleus brachium inferior colliculus (4).

These transiently TH-only-IR neural regions of the postnatal mouse brain showed continuous TH immunoreactivity in the adult brain of human TH transgenic (hTH Tg) mice (5), such as anterior olfactory nucleus, cerebral cortex, hippocampus, striatum, MG nucleus, nucleus parabigeminalis, spinal trigeminal nucleus, and cerebellum, which contain CAnergic terminals but not cell bodies in the adult brain of wild-type nontransgenic (nTg) mice (Table I).

Novel GCH-IR cells in the VLG were TH-negative, and novel TH-IR cells in the MG were GCH-negative, suggesting that both GCH and TH have new independent functions in the GCH-IR or TH-IR cells.

References

1. Nagatsu, I., Ichinose, H., Sakai, M., Titani, K., Suzuki, M., and Nagatsu, T. (1995). Immunocytochemical localization of GTP cyclohydrolase I in the brain, adrenal gland, and liver of mice. *J. Neural Transm.* **102**, 175–188.
2. Nagatsu, I., Yamada, K., Karasawa, N., Kaneda, N., Sasaoka, T., Kobayashi, K., Fujita, K., and Nagatsu, T. (1993). Non-catecholaminergic neuronal expression of human tyrosine hydroxylase in the brain of transgenic mice with special reference to aromatic L-amino acid

decarboxylase. *In* M. Naoi and S. H. Palvez (eds.) Tyrosine Hydroxylase pp. 37–57. VSP Science Press, Zeist, The Netherlands.
3. Nagatsu, I., Takeuchi, T., Sakai, M., Arai, R., Karasawa, N., and Nagatsu, T. (1996). Transient appearance of GTP cyclohydrolase I-positive non-monoaminergic neurons in the ventral lateral geniculate nucleus of postnatal mice. *Neurosci. Lett.* **215**, 79–82.
4. Nagatsu, I., Takeuchi, T., Sakai, M., Karasawa, N., Yamawaki, Y., Arai, R., and Nagatsu, T. (1996). Transient appearance of tyrosine hydroxylase–immunoreactive non-catecholaminergic neurons in the medial geniculate nucleus of postnatal mice. *Neurosci. Lett.* **211**, 183–186.
5. Nagatsu, I., Yamada, K., Karasawa, N., Sakai, M., Takeuchi, T., Kaneda, N., Sasaoka, T., Kobayashi, K., Yokoyama, M., Nomura, T., Katsuki, M., Fujita, K., and Nagatsu, T. (1991). Expression in brain sensory neurons of the transgene in transgenic mice carrying human tyrosine hydroxylase gene. *Neurosci. Lett.* **127**, 91–95.
6. Hökfelt, T., Martensson, R., Björklund, A., Kleinau, S. and Goldstein, M. (1984). Distributional maps of tyrosine-hydroxylase-immunoreactive neurons in the rat brain. In A. Björklund and T. Hökfelt (Eds.), *Classical Transmitters in the CNS. Part I. Handbook of Chemical Neuroanatomy, Vol. 2.* pp. 277–379. Elsevier, Amsterdam.

B-H. Chen,* J-Y. Nie,* M. Singh,* R. Davenport, V. W. Pike,† and K. L. Kirk*

* Laboratory of Bioorganic Chemistry
NIDDK
National Institutes of Health
Bethesda, Maryland 20892

† Cyclotron Unit
MRC CSC
Hammersmith Hospital
London W12 ONN, England

Fluorinated Dihydroxyphenylserines as Potential Biological Precursors of Fluorinated Norepinephrines

Recent attention has been given to the pharmacological properties and therapeutic potential of L-*threo*-(3,4-dihydroxyphenyl)serine (L-*threo*-DOPS) (1a). Of special interest is evidence that administered L-*threo*-DOPS crosses the blood-brain barrier and is subsequently decarboxylated to produce norepinephrine (NE) in the central nervous system, particularly in situations wherein

Advances in Pharmacology, Volume 42

1054-3589/98 $25.00

catecholamine deficiencies are indicated. Indeed, several clinical trials suggest that L-*threo*-DOPS may be beneficial in treating disorders of both the central and sympathetic nervous systems that are characterized by NE deficiencies. For example, Tohgi and coworkers found a dose-dependent increase in cerebrospinal fluid NE concentrations in six advanced parkinsonian patients, and in three of six patients the "freezing phenomenon" in gait and speech improved. Because such symptoms, in advanced cases, become unresponsive to L-dopa treatment, the hypothesis has been made that these symptoms may be caused by NE deficiency (1). Therapeutic potential also was shown by *threo*-DOPS–mediated improvement of certain memory functions in patients with Korsakoff's disease (amnesia induced by chronic alcoholism) and by its effectiveness in treatment of orthostatic hypotension associated with Shy-Drager syndrome and in familial amyloid neuropathy. There is also evidence that L-*threo*-DOPS functions directly, as the amino acid, in certain systems (2).

The biological interest and therapeutic potential of L-*threo*-DOPS suggested that ring fluorinated analogues of this amino acid, as precursors of the corresponding fluorinated norepinephrine (FNE), would be useful tools to study mechanistic aspects of these actions. We have shown previously that 2-fluoronorepinephrine (2-FNE) is a selective β-adrenergic agonist and 6-fluoronorepinephrine (6-FNE) is a selective α-adrenergic agonist (3). If 2- and 6-F-*threo*-DOPS serve as precursors for the biosynthesis of the corresponding FNE, this would provide a pro-drug strategy to deliver the selective α- and β-adrenergic agonists, 2-FNE and 6-FNE, to the central nervous system. *In vivo* studies with these analogues of DOPS could provide insights into the pharmacological effects of activating only α-adrenergic pathways or β-adrenergic pathways, systems that DOPS-derived NE would activate concomitantly. It is possible that 2- or 6-F-*threo*-DOPS could be more potent and/or more selective than *threo*-DOPS in clinical applications.

The synthesis of these analogues appeared at the outset to be straightforward. Aldol condensation of a glycine equivalent with the appropriately substituted benzaldehyde was an obvious approach, because several diastereoselective and enantioselective procedures had been published for the preparation of simpler α-amino-β-hydroxy acids, including phenylserines. However, several attempts using base-catalyzed aldol condensations that were well precedented proved to be fruitless. Although we have no explanation for this recalcitrant behavior, in previous synthetic projects we had experienced difficulties in effecting base-catalyzed condensations with fluorine-substituted protected procatechualdehydes. In contrast, Lewis acid-catalyzed condensations have been facile. Consistent with this, condensation of 6-fluoro-3,4-dibenzyloxybenzaldehyde (2c) with the trimethylsilyl ketene acetal 3 derived from the benzophenone imine of glycine ethyl ester in the presence of 5 mol% of $ZnCl_2$ afforded a mixture of *threo*- and *erythro*-condensation products, with the *threo*-diastereomer 4c present as the major product (6 : 1) (Equation 1). A similar result with predominant formation of the *threo*-diastereomer 4b (7 : 1) was obtained using 3,4-dibenzyloxy-2-fluorobenzaldehyde (2b). After separation of the major diastereomer in each series, the imine and silyl ether functionalities were removed with dilute acid to give the esters (5b,c). Saponification of the ester and hydrogenolysis produced racemic 2- and 6-F-*threo*-DOPS (1b,c). As

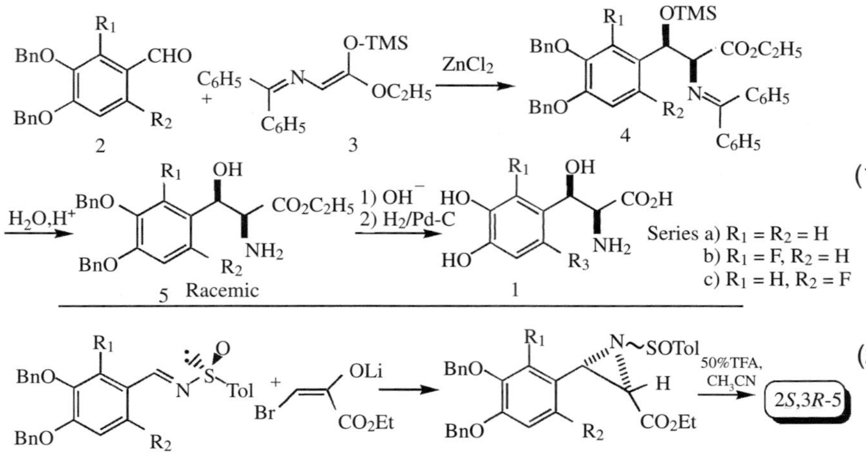

(1)

(2)

verification of the final stereochemical assignments, a similar sequence starting with 3,4-dibenzyloxybenzaldehyde (2a) produced *threo*-DOPS (1a), the NMR spectrum of which was identical in all respects to that of authentic *threo*-DOPS (4).

Transport across the blood-brain barrier and enzymatic decarboxylation are required for elaboration of 2- and 6-FNE from 2- and 6-F-*threo*-DOPS in the CNS. Unfortunately, conditions under which L-*threo*-DOPS was efficiently decarboxylated, using dopa decarboxylase or tyrosine decarboxylase, gave no conversion of 2- and 6-F-*threo*-DOPS to the corresponding FNEs. In this regard, it is important to note that (+)-*threo*-DOPS has been shown by others to be an inhibitor of rat kidney aromatic amino acid decarboxylase–catalyzed decarboxylation of (−)-*threo*-DOPS. Thus, to get a valid assessment of the biological potential of fluoro-*threo*-DOPS, it becomes essential to have the pure enantiomers in hand. Progress toward that end is summarized next.

Extensive research in organic synthesis in recent years has increased enormously the available strategies for enantioselective syntheses. Included in this work are elegant new routes to serine derivatives. For example, *threo* or *erythro* α-amino-β-hydroxy acids, including phenylserines, are produced by base-induced condensations of aldehydes with a glycine equivalent comprised of a Ni(II) complex of a chiral proline-derived Schiff base. The major diastereomer formed can be reversed by changing reaction conditions. Using *threo*-selective conditions, condensation of the Ni(II) complex with fluoroveratraldehydes was facile, but removal of the O-methyl groups from the catechol ring proved problematic due to the strenuous conditions required (BBr_3 or HBr). On the other hand, benzyl-protected fluorinated procatechulaldehydes were unreactive in the condensation step. A dicarboethoxymethylenedioxy (DCEM)-protecting group was found to facilitate condensation and is readily removed under mild acid conditions. However, the acid conditions required to decompose the Ni(II) complex resulted in epimerization of the product to the *erythro*-isomer, apparently due to the acid lability of the benzylic OH group in the intermediate possessing the free catecholic hydroxyl groups (4). A similar acid-catalyzed epimerization occurred with intermediates formed in the $ZnCl_2$-catalyzed condensations of DCEM-protected aldehydes, indicating that in any synthetic sequence, neutral or basic removal of the catechol-protecting group will be, of necessity, the final step. Benzyl-protected intermediates have obvious advantages, because the benzyl group can removed in a final step by hydrogenolysis.

We have explored several other new approaches with this constraint in mind. For example, condensation of aldehydes with α-isothiocyanatoacetyl derivatives of chiral oxazolidones occurs with good diastereoselectivity, and the resulting thiazolidones formed *in situ* can be cleaved under conditions that produce 2*R*,3*S*-α-amino-β-hydroxy acids in excellent enantiomeric excess. However, fluorinated dibenzyloxybenzaldehydes gave disappointing yields in the initial condensation. Because there are several subsequent steps required to give the final product, and because the "conservation of atoms" in the overall sequence is quite low, we have abandoned this approach for now.

Aziridines are versatile intermediates that can be used for the preparation of amines through reductive ring opening and of hydroxy amines by hydrolytic ring opening. Recent attention has been given to the preparation of homochiral aziridines as intermediates for the synthesis of chiral amines and amino alcohols. In a recent elegant advance, Davis and coworkers (5) have shown that chiral sulfinimines (chirality resident on sulfur) derived from aldehydes condense with the lithium enolate of methyl bromoacetate to give, following *in situ* ring closure, chiral *Z*-2-substituted-1-carbomethoxy sulfinylaziridines. The reaction occurs with excellent stereoselectivity. Further, hydrolytic ring opening and removal of the sulfinate group occur with only minimal loss of stereochemical integrity to give hydroxyamino acid esters (including the methyl ester of 2*S*,3*R*-phenylserine) (5). We have now carried out this sequence using sulfoximines derived from 3,4-dibenzyloxy-6-fluorobenzaldehyde. Condensation with the lithium salt of ethyl bromo acetate produces the sulfinyl aziridine. Hydrolytic ring opening (water, trifluoroacetic acid, acetonitrile) produces a dibenzyl-protected 6-fluoro-*threo*-DOPS ethyl ester that is identical, except for optical rotation, with that prepared as described previously using the $ZnCl_2$-catalyzed reaction (Equation 2).

References

1. Kondo, T. (1993). L-threo-DOPS in advanced parkinsonism. *Adv. Neurol.* **60,** 660–665.
2. Yue, J-L., Goshima, Y., Nakamura, S., and Misu, Y. (1992). L-Dopa-like regulatory actions of L-*threo*-3,4-dihydroxyphenylserine on the release of endogenous noradrenaline via presynaptic receptors in rat hypothalamic slices. *J. Pharm. Pharmacol.* **44,** 990–995.
3. Kirk, K. L. (1995). Chemistry and pharmacology of ring-fluorinated catecholamines. *J. Fluorine Chem.* **72,** 26–266.
4. Chen, B-H., Nie, J-Y., Singh, M., Pike, V. W., and Kirk, K. L. (1995). Synthesis of 2- and 6-fluoro analogues of *threo*-3-(3,4-dihydroxyphenyl)serine (2- and 6-fluoro-*threo*-DOPS). *J. Fluorine Chem.* **75,** 93–101.
5. Davis, F. A., Zhou, P., and Reddy, G. V. (1994). Asymmetric synthesis and reactions of cis-N-(p-toluene-sulfinyl)aziridine-2-carboxylic acids. *J. Organ. Chem.* **59,** 3243–3245.

P. Soares-da-Silva and M. A. Vieira-Coelho

Institute of Pharmacology and Therapeutics
4200 Porto, Portugal

Nonneuronal Dopamine

The best example of nonneuronal dopamine (DA) is that acting in the kidney. This nonneuronal DA has a role in the renal handling of Na^+, and a close relationship has been demonstrated to occur between urinary DA and Na^+. The renal delivery of Na^+ also has been demonstrated to positively influence the activity of renal aromatic L-amino acid decarboxylase (AAD) and to increase the renal delivery of L-dopa. Although DA has been shown in the kidney to produce renal vasodilatation, the effects of renal endogenous DA appear to occur mainly at the tubular level. Renal DA is also an important issue because a deficiency in its production has been suggested to occur in salt-sensitive hypertensives. Figure 1 is a schematic representation of a renal proximal tubule epithelial cell taking up L-dopa and converting it up to DA. The major source of L-dopa appears to be that in the tubular filtrate, and filtered L-dopa enters the cells from the apical side of the cell. Newly formed DA has the ability to leave the cell and activate DA receptors in the external surface of the membrane. Simultaneous activation of D_1 and D_2 receptors located in the basolateral side of the cell results in Na^+-K^+ adenosine triphosphatase (ATPase) inhibition, whereas activation of D_1 receptors located at the apical side of the cell inhibits Na^+/H^+ exchange. Natriuretic effects of DA appear to result from both inhibition of Na^+-K^+ ATPase and Na^+/H^+ exchange (1).

This work is aimed to review some basic aspects related to the physiology of these nonneuronal dopaminergic systems, namely the mechanisms involved in DA formation (uptake of L-dopa and its decarboxylation), the metabolic degradation of the amine, and the mechanisms regulating the amine outflow.

When we started looking at the mechanisms regulating DA formation in the kidney, we first considered previous information that high Na^+ diet resulted in an increase in the urinary excretion of DA. One of the first things we studied was the influence of extracellular Na^+ on the uptake and decarboxylation of L-dopa in rat kidney slices. The data obtained under *in vitro* conditions, as have been described under *in vivo* experimental conditions, suggested that the uptake of L-dopa and formation of DA is a Na^+-dependent process requiring the integrity of mechanisms governing transtubular transfer of Na^+ and Na^+-K^+ ATPase (4). This fitted well the evidence of a close relationship between Na^+ and DA; DA produces natriuresis, and, on the other hand, a high renal delivery of Na^+ increases DA formation as a protective mechanism in order to avoid excess Na^+ reabsorption.

The uptake of L-dopa in isolated renal cortical tubules, incubated in the presence of benserazide (50-μM) in order to inhibit AAD, was found to be a saturable process, with a Km of about 100 μM. Using this procedure, we decide to evaluate in more detail the pharmacological characteristics of L-dopa uptake.

Advances in Pharmacology, Volume 42

1054-3589/98 $25.00

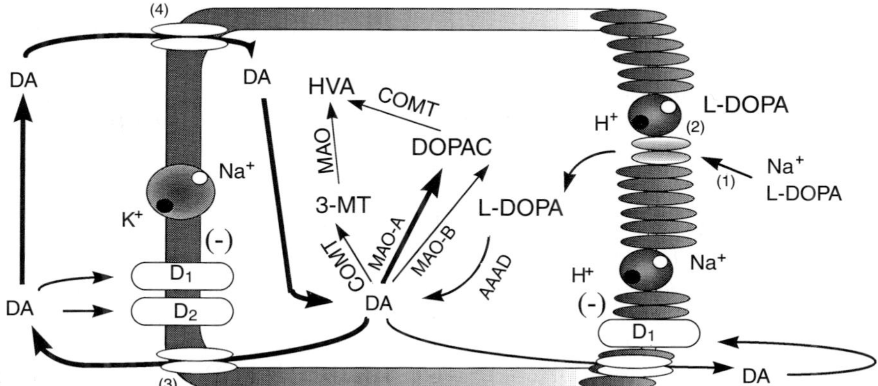

FIGURE 1 A renal proximal tubular epithelial cell taking up L-dopa through the apical cell border and converting it to DA. The uptake of L-dopa may be promoted by an Na^+-dependent transporter (1) or by the organic cation/H^+ exchanger (2). Newly formed DA is rapidly metabolized to inactive metabolites (DOPAC, 3MT, and HVA) by types A and B monoamine oxidase (MAO-A and MAO-B) and catechol-O-methyltransferase (COMT). Intracellular DA can leave the cell through the apical or the basolateral cell border by saturable mechanisms (3); this is under the influence of the Na^+/H^+ exchanger. The apical surface is impermeable to DA, whereas at the basolateral side, extracellular DA can be taken by a saturable process (4).

The uncoupling agent, 2,4-dinitrophenol, produced a concentration-dependent inhibition of L-dopa uptake, with an $IC/_{50}$ of 12 μM, indicating that we were dealing with an active transporter. The neuronal uptake blocker, cocaine, produced a nonsignificant decrease in L-dopa uptake, whereas the extraneuronal uptake blocker, corticosterone, produced a pronounced inhibitory effect on L-dopa uptake with an $IC/_{50}$ of 11 μM. The organic anion transport inhibitors probenecid and 4,4′-diisothiocyanotostilbene-2,2′-disulphonic acid (DIDS) produced no effect on L-dopa uptake, whereas the organic cation transport inhibitor cyanine 863 was found to potently inhibit L-dopa uptake with an $IC/_{50}$ of 2 μM. Compound decynium 24 (1,1′-diethyl-2,4′-cyanine) is a chemically related cyanine derivative that has also been found to inhibit the organic cation transporter; both decynium 24 and decynium 22 (1,1′-diethyl-2,2′-cyanine) were found even more potent than cyanine 863 in inhibiting L-dopa uptake. The findings that all three cyanine derivatives and corticosterone produce a noncompetitive type of inhibition, and because the magnitude of these inhibitory effect is pH dependent, led us to suggest that L-dopa is transported by the organic cation/H^+ exchanger (2). Because these experiments were performed in isolated renal tubules, we cannot be sure where this transporter is located. But the classical view about the localization of the organic cation/H^+ exchanger favors its position at the apical side of the cell. How is this compatible with the results showing that L-dopa is Na^+-dependent? It is quite possible that changes in transepithelial Na^+ fluxes, as induced during incubation in the presence of different extracellular concentrations of Na^+ and in the presence of ouabain, would produce changes in membrane potential, and this obviously may alter the uptake of L-dopa.

Once inside the cell, L-dopa is rapidly decarboxylated to DA, and this represent the second major event in the process of DA formation. Decarboxylation of taken-up L-dopa is usually a rapid and efficient process, most probably because renal proximal tubular epithelial cells are endowed with one of the highest activities in the body.

These epithelial cells are also endowed with high monoamine oxidase (MAO) and catechol-O-methyltransferase activities, the main enzymes involved in the metabolic degradation of DA. Figure 1 emphasizes that metabolic degradation of newly formed DA is a matter of considerable importance, being DA rapidly converted to inactive metabolites. DA can be methylated to 3-MT or deaminated to DOPAC; 3-MT and DOPAC can be converted to HVA by deamination and methylation, respectively. Although HVA is classically assumed the final product of DA metabolism, deamination of renal DA is probably the most important mechanism of amine inactivation. Renal tissues are endowed with two types of MAO, MAO-A and MAO-B. Experiments performed in the rat kidney suggested that DA is predominantly deaminated by MAO-A (the most abundant and easily accessible enzyme) and to a minor extent by MAO-B (the less abundant and not so easily accessible enzyme).

Newly formed DA can leave the cellular compartment where synthesis takes place. This is of particular importance for the actions of the amine, because DA receptors are located in the external surface of the cells. A considerable amount of renal DA must come out of the cell by crossing the apical cell border, because it appears in the urine, but there is evidence suggesting that DA can also leave the cell by crossing the basolateral side. Previous work in the rat and canine kidney has shown a saturable DA outward transporter with a Km of 340 $\pm$ 41 and 396 $\pm$ 45 μM, respectively. This DA outward transporter is age-dependent and can be modulated pharmacologically. Inhibition of the Na^+/H^+ exchanger by amiloride and ethylisopropylamiloride (EIPA) results in considerable reduction in the outflow of renal DA, whereas activation of this mechanism by both phenylephrine and UK 14,304 results in facilitation of the outflow of DA (3). Altogether, the data available suggest that the outflow of DA from these epithelial cells is under the control of the Na^+/H^+ exchanger. The precise mechanisms are not yet fully clarified, and it is not known whether this outward DA transfer system operates at both sides of the cell or is just located at the apical border. Released DA, on the other hand, can enter the cell through the basolateral border, but the apical cell border is impermeable to the amine. This suggests that there is no return for that DA leaving the cell through the apical cell border, whereas the intracellular DA crossing the cell membrane at the basolateral side can be taken up.

Any of these processes involved in the formation and fate of nonneuronal DA may rate-limit the availability of renal DA. Several studies, namely those by the Goldstein group (5), have established that nonneuronal renal DA is originated from circulating or filtered L-dopa. Because L-dopa plasma levels are fairly stable and in the pmol/ml range, it is possible that the renal delivery of L-dopa may not rate-limit the formation of DA. In studies performed in OK and LLC-PK_1 cells incubated in the absence of benserazide, it was observed that the formation of DA was a rapidly saturable process, and a substantial amount of taken-up L-dopa was not converted to DA. This clearly suggested that the uptake process is not the rate-limiting step in the formation of DA. However, this does not occur in isolated renal proximal tubules and in intestinal Caco-2 cells; the intracellular concentrations of DA are greater than those of L-dopa; that is, most of taken-up L-dopa is rapidly converted to DA, and only residual amounts of L-dopa escape decarboxylation. One of the reasons we think that in some epithelial cells L-dopa uptake rate-limits the formation of DA has to do with the amount of AAD present. Our hypothesis is that if cells have enough AAD, then the rate-limiting step is L-dopa uptake.

Acknowledgment

This work was supported by grant SAU 20/95.

References

1. Lee, M. R. (1993). Dopamine and the kidney: Ten years on *Clin. Sci.* **84,** 357–375.
2. Pinto-do-O, P. C., and Soares-da-Silva, P. (1996). Studies on the pharmacology of the inward transport of L-DOPA in rat renal tubules. *Br. J. Pharmacol.* **118,** 741–747.
3. Soares-da-Silva, P. (1993). Renal tubular dopamine outward transfer during Na^+-H^+ exchange activation by α_1- and α_2-adrenoceptor agonists. *Br. J. Pharmacol.* **109,** 569–576.
4. Soares-da-Silva, P., Fernandes, M. H., and Pestana, M. (1993). Studies on the role of sodium on the synthesis of dopamine in the rat kidney. *J. Pharmacol. Exp. Ther.* **264,** 406–414.
5. Wolfovitz, E., Grossman, E., Folio, C. J., Keiser, H. R., Kopin, I. J., and Goldstein, D. S. (1993). Derivation of urinary dopamine from plasma dihydroxyphenylalanine in humans. *Clin. Sci.* **84,** 549–557.

Anita Aperia,* Ann-Christine Eklöf,* Ulla Holtbäck,*
Susana Nowicki,* Martin Sundelöf,* and Paul Greengard†

*Department of Woman and Child Health
Pediatric Unit
St. Göran's Children's Hospital
Karolinska Institute
11281 Stockholm, Sweden

†Department of Molecular and Cellular Neuroscience
Rockefeller University
New York, New York 10021

The Renal Dopamine System

It is well documented that intrarenally formed dopamine plays a central role in the regulation of sodium metabolism (1). Thus, the natriuresis induced by a high salt diet is attenuated by dopamine antagonists as well as by dopamine depletion. Locally formed dopamine inhibits the activity of Na/KATPase, the enzyme responsible for active Na^+ reabsorption from the renal tubular lumen. Several clinical observations as well as studies on animals with various forms of genetic hypertension suggest that defects in the renal dopamine system may, by causing salt retention, contribute to the development of hypertension. A better understanding of the regulatory steps in the renal dopamine system may, therefore, lead to improved early diagnosis and more directed treatment of various forms of salt-sensitive hypertension. This review will deal both with the regulation of renal Na/KATPase by dopamine and with the regulation of dopamine availability in the renal tubular cells.

I. Dopamine Regulation of Renal Na/KATPase Occurs via a Complex Signal Transduction Network

Dopamine inhibits proximal tubular Na/KATPase in a dose-dependent manner. Incubation of proximal tubular segments with the dopamine precursor L-dopa will also lead to inhibition of Na/KATPase, but this inhibition is abolished if the tubules have been incubated with an inhibitor of aromatic acid decarboxylase, the enzyme that converts L-dopa to dopamine. Furthermore, the Na/KATPase activity in proximal tubules from rats that have received gludopa, a kidney-specific dopamine prodrug, is significantly lower than in tubules from control rats. This inhibitory effect is abolished if the rats have also been treated with a dopamine antagonist.

In the proximal tubule, Na/KATPase activity is synergistically inhibited by D1 and D2 agonists. Dopamine inhibition of tubular Na/KATPase appears to require both activation of adenylyl cyclase, as well as arachidonic acid release with subsequent formation of 20 HETE and protein kinase C activation.

Advances in Pharmacology, Volume 42

1054-3589/98 $25.00

A. Role of the Adenylyl Cyclase–PKA Pathway

The kidney expresses a dopamine and cyclic adenosine monophosphate (cAMP)-regulated phosphoprotein DARPP32. This phosphoprotein, which was originally discovered in cells containing dopamine 1 receptors in the striatum, will, following protein kinase A (PKA)-induced phosphorylation, be a potent inhibitor of protein phosphatase 1 (PP1). PP1 can dephosphorylate and activate Na/KATPase. By incubating transiently permeabilized cells with a peptide corresponding to the active domain of phospho-DARPP32, it was demonstrated that phospho-DARPP32, but not dephospho-DARPP32, dose-dependently inhibited the activity of proximal tubular Na/KATPase (2). Interestingly, DARPP32 is dephosphorylated and inactivated by calcineurin, a calcium-dependent protein phosphatase.

α-Adrenergic receptors stimulate renal Na/KATPase and have a salt-retaining effect. The α-adrenergic stimulation of Na/KATPase is mediated by calcineurin. This has led us to propose that in the kidney, dopamine and α-adrenergic receptors bidirectionally regulate renal tubular Na/KATPase activity by exerting opposing forces on a common intracellular signaling system (3).

B. Role of the Phospholipase A2–Arachidonic Acid–20 HETE–PKC Pathway

Renal Na/KATPase is dose-dependently inhibited by the arachidonic acid metabolite 20HETE. Indirect lines of evidence have suggested that dopamine inhibition of tubular Na/KATPase may also involve activation of phospholipase A, release of arachidonic acid, and formation of 20HETE. There is also evidence that dopamine inhibition of renal tubular Na/KATPase most likely also involves PKC. PKC phosphorylates and inactivates rat Na/KATPase (4). Taken together, these observations prompted us to examine whether 20HETE may mediate its effects via PKC activation and subsequent phosphorylation and inactivation of Na/KATPase. This turned out to be the case. 20HETE inhibition of rat renal Na/KATPase is abolished in the presence of specific PKC inhibitors. 20HETE inhibition of Na/KATPase is also abolished if the PKC regulatory site has been deleted by site-directed mutagenesis (5). Studies in cell lines have suggested that the D2 receptors can be coupled to PLA_2. It is not unlikely that they also do so in the kidney. In fact, we have observed that 20HETE and D1, as well as D2 and D1, synergistically inhibit proximal tubular Na/KATPase.

II. The Effects of the Renal Dopamine System May Be Influenced by the Availability and Metabolism of Dopamine: Role of Catechol-O-Methyltransferase

Dopamine is formed in proximal tubular cells. The renal source of dopamine is the proximal tubular cell. The expression of aromatic acid decarboxylase is high in the proximal convoluted tubule, considerably lower in the proximal straight tubule, and not detectable in other renal cells. Tyrosine hydroxylase is not present in proximal tubular cells. L-dopa enters proximal tubular cells in an Na^+-coupled transport and is rapidly converted to dopamine. The concentration of dopamine in renal cortical tissue, which mainly consists of proximal tubular cells, is higher than in blood. This may suggest that dopamine is stored in these cells in such a way that it is protected from being metabolized.

Catechol-O-methyltransferase (COMT) may play an important role in modulating the effects of intrarenal dopamine. The kidney has a very high COMT activity. *In situ* hybridization studies have shown that COMT mRNA is present in almost all renal tubular cells, but the signal is particularly strong in proximal straight tubular cells (6). To examine the physiological role of COMT, we have used a specific inhibitor of the activity of peripheral COMT, nitecapone. Nitecapone was administered orally by gavage to anesthetized rats. This resulted in an almost complete inhibition of renal COMT activity and a significant increase of urinary HVA excretion. In some of the protocols, the rats were also given a D1 antagonist. Nitecapone caused a dramatic increase in urinary salt excretion, which was abolished when a D1 antagonist was also administered. Nitecapone caused a marked inhibition of Na/KATPase activity, both in proximal convoluted and in straight tubules. Simultaneous treatment with a D1 antagonist abolished this inhibitory effect. The effect of nitecapone was somewhat enhanced in rats that had received a high-salt diet. The natriuretic effect of nitecapone was more pronounced than the natriuretic effect of the dopamine precursor, glu-dopa. Nitecapone and gludopa had an additative effect on urinary sodium excretion.

III. Summary

Intrarenally formed dopamine induces natriuresis by inhibiting the activity of renal tubular Na/KATPase. This effect is mediated via a complex signal network, which includes inhibition of PP1 via the adenylyl cyclase–PKA-DARPP32 pathway and activation of PKC via the PLA2–arachidonic acid–20HETE pathway. The renal dopamine availability is a major determinant of the natriuretic effect of dopamine and is to a large extent modulated by the activity of COMT. The possibility that regulation of dopamine storage and release influences renal dopamine effects should be considered.

References

1. Aperia, A. (1994). Dopamine action and metabolism in the kidney. *Curr. Opin. Nephrol. Hypertens.* **3**, 39–45.
2. Meister, B., Fryckstedt, J., Schalling, M., Cortés, R., Hökfelt, T., Aperia, A., Hemmings, H. C., Jr., Nairn, A. C., Ehrlich, M., and Greengard, P. (1989). Dopamine- and cAMP-regulated phosphoprotein (DARPP-32) and dopamine DA1 agonist-sensitive Na^+, K^+-ATPase in renal tubule cells. *Proc. Natl. Acad. Sci. U.S.A.* **86**, 8068–8072.
3. Aperia, A., Ibarra, F., Svensson, L.-B., Klee, C., and Greengard P. (1992). Calcineurin mediates α-adrenergic stimulation of Na^+, K^+-ATPase activity in renal tubule cells. *Proc. Natl. Acad. Sci. U.S.A.* **89**, 7394–7397.

4. Logvinenko, N. S., Dulubova, I., Fedosova, N., Larsson, S. H., Nairn, A. C., Esmann, M., Greengard, P., Aperia, A. (1996). Phosphorylation by protein kinase C of serine-23 of the α-1 subunit of rat Na^+, K^+-ATPase affects its conformational equilibrium. *Proc. Natl. Acad. Sci. U.S.A.* **93**, 9132–9137.
5. Nowicki, S., Chen, S., Aizman, O., Cheng, X., Li, D., Nowicki, C., Nairn, A., Greengard, P., and Aperia, A. 20 HETE activates protein kinase C. Role in regulation of rat renal Na^+, K^+ ATPase *J. Clin. Invest.* **99**, 1224–1230.
6. Meister, R., Bean, A. J., and Aperia, A. (1993). Catechol-O-methyltransferase mRNA in the kidney and its appearance during ontogeny. *Kidney Int.* **11**, 726–733.

Robert M. Carey,* Zhi-Qin Wang,* Helmy M. Siragy,* and Robin A. Felder†

Departments of Medicine* and †Pathology
University of Virginia Health Sciences Center
Charlottesville, Virginia 22908

Renal Dopamine Production and Release in the Rat: A Microdialysis Study

The renal dopaminergic system plays an important role in the regulation of blood pressure, sodium homeostasis and kidney function. Proximal tubule cells, rich in aromatic amino acid decarboxylase, can take up circulating and/or filtered L-dopa for decarboxylation to dopamine (DA). Most of the DA appearing in urine is of proximal tubule origin. However, the extent to which the urinary excretion of DA reflects the amount of the amine that has been synthesized in tubule cells, and the fate and outflow of newly formed DA from the proximal tubules remains unknown. D1-like receptors are localized in renal blood vessels, the juxtaglomerular apparatus, the proximal tubule (both apical and basolateral), cortical collecting duct, and medullary thick ascending limb. D2-like receptors are described in the endothelial and adventitial layers of renal vasculature and the glomerulus. DA generated intrarenally has a well-documented natriuretic effect, and infusion of DA antagonists decreases sodium excretion independently of hemodynamic changes. Because DA is synthesized, stored, and released within the kidney in close proximity to its target site(s), it is believed that DA produced by the renal proximal tubule serves as an intrarenal paracrine hormone mediating diuresis and natriuresis. The precise mechanism by which DA exerts its cell-to-cell action is not fully understood. In the present study, renal interstitial (RIF) DA (by *in vivo* microdialysis tech-

Advances in Pharmacology, Volume 42

1054-3589/98 $25.00

nique) and urinary DA excretion ($U_{DA}V$) were investigated in anesthetized rats on either normal (0.28% NaCl, NS) or high (4.0% NaCl, HS) sodium balance and in response to acute administration of a DA prodrug, γ-L-glutamyl-L-dopa (gludopa). Responses of renal sodium excretion and intrarenal blood flow distribution to gludopa also were examined.

Microdialysis probes were constructed as previously described (1, 2). Studies of *in vitro* recovery of DA showed that the best relative recovery was observed with a perfusion rate of 1 μl/min and was 92.7 ± 2.0%, compared with 73.9 ± 4.9% and 44.3 ± 3.3% at 3 and 5 μl/min, respectively. *In vivo* renal microdialysis was performed in anesthetized Sprague-Dawley rats (1, 2). *In vivo* recovery of i.v. infused [^{3}H]inulin has demonstrated that [^{3}H]inulin appearing in the RIF sample is 2% of urinary [^{3}H]inulin (2), indicating that the dialysate is not significantly contaminated by renal tubule fluid. Using a gradient equilibrium dialysis (2), the theoretically derived point (4.1 pg/μl) at which there was no net transmembrane flux was similar to the interstitial DA concentration (2.9 pg/μl) measured directly during normal salt intake.

Urine flow rate (UV) and sodium excretion ($U_{Na}V$) in HS (n = 9) were greater than in NS (n = 9) rats (UV 7.2 ± 0.6 vs 3.8 ± 0.3 μl/min, $p < .01$; $U_{Na}V$ 497 ± 66 vs 265 ± 27 nmol/min, $p < .01$). $U_{DA}V$ increased in HS compared with NS rats (601 ± 68 vs 420 ± 37) pg/min, $p < .05$). In contrast, RIF DA was significantly lower in HS than NS rats (1.3 ± 0.4 vs 3.7 ± 0.5 pg/min, $p < .01$). Basal $U_{DA}V$ and RIF DA from rats with prior bilateral renal denervation (n = 10) were similar to rats with intact renal innervation (n = 14) ($U_{DA}V$ 476 ± 30 vs 394 ± 28 pg/min, $p > 0.5$; RIF DA 2.9 ± 0.2 vs 3.4 ± 0.3 pg/min, $p > .05$). UV tended to be higher in rats with prior bilateral renal denervation (4.1 ± 0.2 vs 3.4 ± 0.2 μl/min, $p = .05$) while U_{Na} was not significantly different between the two groups (286 ± 13 vs 265 ± 16 nmol/min, $p > .05$).

In rats with intact renal innervation, i.v. injection of gludopa at 3, 5, and 7.5 nmol/kg (n = 7 in each group) produced a larger increase in $U_{DA}V$ than RIF DA. Only the highest dose of gludopa (7.5 nmol/kg), which resulted in a 7.3-fold increase in $U_{DA}V$ and 1.7-fold increase in RIF DA, was associated with significant diuresis and natriuresis (UV 3.4 ± 0.4 vs 7.4 ± 0.5 μl/min, $p < .01$; $U_{Na}V$ 265 ± 27 vs 711 ± 120 nmol/min, $p < .01$). Renal blood flow of the cortex and medulla recorded by a laser-Doppler flowmeter did not change in rats receiving gludopa at 7.5 nmol/kg (n = 6), while angiotensin II (100 ng/kg/min) induced significant reduction in cortical (42.6%) and medullary (28.1%) blood flow, which gradually returned toward preangiotensin levels when the angiotensin infusion was stopped (n = 6). In rats with prior bilateral renal denervation (n = 5 in each group), gludopa at 7.5 nmol/kg produced a significant increase in $U_{DA}V$ (8.3-fold) and RIF DA (1.8-fold), accompanied by significant diuresis and natriuresis (UV 3.9 ± 0.2 vs 8.2 ± 0.5 μl/min, $p < .01$; $U_{Na}V$ 289 ± vs 778 ± 56 nmol/min, $p < .01$).

In rats on a normal salt diet in the present study, renal interstitial fluid DA levels were much lower than $U_{DA}V$, suggesting that intrarenally produced DA is released preferentially into the tubule lumen rather than in the peritubular space. Cellular mechanisms of renal DA secretion are not understood. DA immunoreactive granules and L-dopa–induced fluorescence are mainly concentrated in the region of the apical membrane of proximal tubule cells. There may be lumenal organic cation transporters with high affinity for DA in porcine proximal tubule (LLC-PK_1) cells.

Our study confirmed that $U_{DA}V$ increases in response to chronic salt loading. The mechanism underlying the observed reduction in RIF DA with chronic salt loading is not apparent. RIF DA may derive from tubular outward transport of DA through the basolateral membrane. Histofluorescent and neurochemical findings also suggest the presence of dopaminergic neurons in the kidney, and adrenergic nerves may become dopaminergic under certain circumstances. Vagal afferents have been shown to stimulate renal release of DA and produce a neurogenically mediated natriuresis. Our data confirm that the main source of DA in the urine is nonneuronal, and further demonstrate that renal nerve activity does not contribute significantly to either urinary or interstitial fluid DA. During chronic salt loading, intrarenally produced DA is released preferentially into the luminal space, where it may act on apical DA receptors as an autocoid or paracrine factor.

Significant natriuresis and renal vasodilation occurs both in the whole animal and humans and in the isolated perfused rat kidney following pharmacological increase of renal DA production by gludopa. In the present study, much smaller quantities of gludopa (7.5 nmol/kg) resulted in a physiological increase in $U_{DA}V$ (7.3-fold), accompanied by a slight increase in renal interstitial fluid DA (1.7-fold), and produced significant diuresis and natriuresis without detectable changes in intrarenal blood flow. These results also support preferential secretion of DA generated by proximal tubule cells into the tubule lumen. Our previous study in the uninephrectomized conscious dog demonstrated that blockade of the renal DA-1 receptor with intrarenal infusion of SCH 23390 produced a 50% decrease in urine flow rate and sodium excretion, but there were no renal hemodynamic changes accompanying the antinatriuresis (3). A 5.2-fold increase in kidney DA content with gludopa administration was associated with decreased renal cortical brush border Na^+/H^+ antiporter activity and significant natriuresis (4.9-fold) and diuresis (2.6-fold) without any change in glomerular filtration rate in anesthetized rats (4). Low-dose L-dopa infusion, which resulted in a five- to eightfold increase in urine DA excretion, had a natriuretic effect without any change in blood pressure or glomerular filtration rate in healthy subjects of a low-sodium diet (5). These results strongly support that intrarenal DA plays a role in the control of natriuresis through a tubule mechanism.

The renal sympathetic nerve endings are closely related to the juxtaglomerular apparatus and proximal tubule cells. Whether the sympathetic nervous system influences intrarenal DA generation and regulates renal DA-mediated natriuresis and diuresis is unknown. Basal DA production (as reflected by $U_{DA}V$ and renal interstitial fluid DA) and $U_{Na}V$ and their responses to gludopa were similar between rats with and without prior renal denervation. These results demonstrate that intrarenal production of DA and its renal effects are not significantly influenced by renal sympathetic nerve activity.

Our data demonstrate that renal interstitial microdialysis can be used to monitor DA levels in the renal interstitial fluid in anesthetized rats. DA produced in the kidneys is released preferentially into the tubule lumen and exerts a direct tubule effect in the control of sodium excretion. Renal DA production and its renal effects were not significantly influenced by renal sympathetic nerve activity.

Acknowledgments

The authors wish to thank Dawn M. Fultz and Nancy L. Howell for their skillful technical assistance. This work was supported in part by National Institutes of Health grant RO1-HL-49575 to Dr. R. M. Carey.

References

1. Siragy, H. M., and Carey, R. M. (1996). The subtype-2 *AT_2) angiotensin receptor regulates renal cyclic guanosine 3′, 5′-monophosphate and AT_1 receptor-mediated prostaglandin E_2 production in conscious rats. *J. Clin. Invest.* **97**, 1978–1982.
2. Siragy, H. M., and Linden, J. (1996). Sodium intake markedly alters renal interstitial fluid adenosine. *Hypertension* **27**, 404–407.
3. Siragy, H. M., Felder, R. A., Howell, N. L., Chevalier, R. L., Peach, M. J., and Carey, R. M. (1989). Evidence that intrarenal dopamine acts as a paracrine substance at the renal tubule. *Am. J. Physiol.* **257**, F469–F477.
4. Jose, P. A., Eisner, G. M., Drago, J., Carey, R. M., Felder, R. M. (1996). Dopamine receptor signaling defects in spontaneous hypertension. *Am. J. Hypertens.* **9**, 400–405.
5. Barendregt, J. N. M., Muizert, Y., van Nispen tot Pannerden, L. L. A. M., and Chang, P. C. (1995). Intrarenal production of dopamine and natriuresis following dopa and saline infusion in healthy human volunteers. *J. Human. Hypert.* **9**, 187–194.

PART H

DEVELOPMENT AND PLASTICITY

Joan P. Schwartz

Molecular Genetics Section
Clinical Neuroscience Branch
National Institute of Neurological Disorders and Stroke
National Institutes of Health
Bethesda, Maryland 20892

Overview

Much of what is presented in the chapters comprising this section that is new and exciting with respect to development and plasticity of catecholaminergic systems can be summed up by the term *neurotrophic factors*. There were reports of novel proteins, of new members, and even new gene families of factors, active during development in both the central and peripheral nervous systems but also capable of restoring function to injured neurons. Orphan receptors have been identified as mediating actions of some of these factors. Treatment modalities, from genetic therapy, to cell transplants, to infusion of neurotrophic factors, are looking more promising than ever. The role of catecholamines in learning and memory promises further insights into their possible function in plasticity responses of the central nervous system. Finally, the contributions of invertebrate neurobiology to the field are leading us to further understanding about the diverse functions of catecholamines, including actions as trophic factors.

I. Development and Plasticity of Peripheral Catecholaminergic Systems

This section runs the gamut from factors that determine the fate of neural crest cells to factors that determine which of those cells become catecholaminergic. Bronner-Fraser (p. 883) discusses two classes of factors that affect the origins of neural crest cells from neural tube, induction factors, and dorsalizing signals. The neural crest arises as a result of inductive interactions between neural and nonneural ectoderm; among the factors involved are sonic hedgehog (SHH), BMP-4 and BMP-7 (bone morphogenetic proteins), and a novel gene cloned in her laboratory, named NP-1. The crest cells are still multipotent, but dorsalizing signals, genes expressed only in the dorsal tube, act to determine which neural crest cells populate the roof plate or become commissural or other association neurons. Among these dorsalizing signals is dorsalin-1, a member of the transforming growth factor-β (TGFβ) family.

Weston (p. 887) then describes factors involved in determining the fate of trunk neural crest cells, which can become either neurons and glia or melanocytes. He discusses a hypothesis that expression of specific receptor tyrosine kinases determines which pathway a given cell will follow, while specific neurotrophic factors ultimately affect survival and differentiation once the cell has reached its final destination. Thus, steel factor and its receptor c-kit appear to be required transiently for melanocyte migration and differentiation, while the neurotrophin receptor trk C apparently determines migration of neuroglial precursors and its ligand, neurotrophin-3 (NT-3), their differentiation. In addition, he throws in the concept that a nonfunctional competition for a given factor between functional and nonfunctional receptors might influence the pathway chosen by a specific population of cells.

The BMPs, also members of the TGFβ family, not only affect neural crest cell induction, but also are involved in induction of tyrosine hydroxylase (TH) in the neural crest–derived sympathetic neurons. Rohrer and Ernsberger (p. 891) present evidence that choline acetyltransferase and TH are both detectable within some crest cells well before innervation. The factors responsible are thought to derive from surrounding tissues; the best candidates for induction of the noradrenergic phenotype are BMP-2, -4, and -7. Rohrer and Ernsberger suggest that an as yet unidentified factor, perhaps from the CNTF/LIF family, causes further maturation of the cholinergic cells to express vasoactive intestinal peptide (VIP) also. Guidry and Landis (p. 895) present evidence that the cytokine LIF (leukemia inhibitory factor) can mediate the switch from catecholaminergic to cholinergic phenotype of the salivary gland innervation, as can other members of this neuropoietic cytokine family, such as ciliary neurotrophic factor (CNTF). However, data from the LIF knockout mouse rule out LIF and suggest that a novel, not yet identified member of the family is responsible for this specific action. Zigmond *et al.* (p. 899) present data that indirectly support a comparable role for LIF, with the same downregulation effect on TH, in sympathetic ganglia. Axotomy of the superior cervical ganglion (SCG) results in loss of TH and neuropeptide Y (NPY) with concomitant increased expression of VIP, substance P, and galanin, as well as increased levels of LIF. The effects of LIF are the exact opposite of those for NGF on SCG and suggest that LIF and other members of the CNTF family may play a significant role in plasticity responses of the SCG neurons to injury, whereas NGF and the other neurotrophins are presumably required for restoration of their normal function. Further evidence that the cholinergic phenotype may be coexpressed before innervation and that the switch from noradrenergic to cholinergic phenotype may actually be a switching off of the NE system came from studies on the ontogeny of expression of the vesicular monoamine transporter (VMAT) and vesicular acetylcholine transporter (VAChT) genes reported by Schütz *et al.* (p. 903).

II. Development and Plasticity of Central Catecholaminergic Systems

Rosenthal (p. 908) reports that SHH, acting via its recently discovered receptor, *patched,* is released by ventral floorplate cells and induces the dopaminergic neurons of the ventral mesencaphalon. Thus, SHH can act first in the generation of neuronal progenitor cells and then in the differentiation of a population of those cells to become dopaminergic. In culture, those neurons become dependent on a member of the TGFβ family, possibly GDNF (glial-derived neurotrophic factor), neurturin (a newly cloned member of the family), or TGFβ2 or -β3. All of these related proteins can support survival of dopaminergic neurons in culture, but whether any of them play a comparable role *in vivo* remains to be determined. The ability of these factors to substitute for one another may explain results from knockout animals, because dopaminergic neurons survive in GDNF-knockout mice. Alternatively, the results from knockout mice may simply indicate that these neurotrophic factors are not required for either generation or phenotypic differentiation of DA neurons and, therefore, are not active during development. Results presented by Hoffer (Gash *et al.*) (p. 911) complement those of Rosenthal with respect to GDNF-knockout animals but go further in terms of the effects of GDNF in animal models of Parkinson's disease. In rats lesioned unilaterally with 6-OHDA as well as mice and rhesus monkeys lesioned with MPTP, intracerebroventricular injections of GDNF increased numbers of TH-positive neurons in the nigra as well as enhancing recovery of motor activity. Most remarkably, a single injection of GDNF, even when given 2 to 3 wk after the lesion, was effective for at least 9 wk post-lesion. GDNF injection was also effective in aged rats (24 mo old) in enhancing release and uptake of DA and in restoring motor activity. Because GDNF increased DA content and turnover rate in normal rats and monkeys, Hoffer (Gash *et al.*) consider GDNF to be not only neurotrophic but also neurorestorative, a useful term for many of the effects being studied.

In contrast to these very long-lived effects of GDNF on plasticity of the dopaminergic system, the other two chapters in this topic area emphasize short-term neuromodulatory effects of neurotrophins. Altar *et al.* (p. 915) present two sets of data as support for the idea that brain-derived neurotrophic factor (BDNF) could have acute effects. BDNF infusions into the nigra affected DA content and turnover in striatum, as determined by microdialysis; however, because the measurements were carried out 10–14 days after the start of the BDNF infusion, it is difficult to argue that these necessarily represent acute effects of BDNF. However, based on results that show that there is no BDNF mRNA in the striatum but high levels of BDNF, which are depleted by decortication, they suggest the possibility that BDNF is coreleased from cortical glutamatergic synapses within the striatum and may function as a cotransmitter. Levine *et al.* (p. 921) provide more direct support for the idea that BDNF could affect both the presynaptic cell's firing rate as well as the amplitude of the postsynaptic cell's response. The studies, carried out in hippocampal cultures, demonstrate BDNF enhancement of glutamate, but not of acetylcholine, within 3–5 min of application of BDNF. If, as Altar *et al.* argue, BDNF or other neurotrophic factors can be coreleased with a neurotransmitter, such acute effects of the neurotrophins could have significant effects on the strength and efficacy of synaptic transmission.

III. Genetic Diseases Involving Catecholamine Systems and the Potentials of Gene Therapy

The apparent stability of and/or the long-lasting response to a neurotrophic factor such as GDNF, as well as its ability to affect an injured neuron 2–3 wk following injury, have profound implications for the potential of gene therapy. However, the realization that no therapy can be devised until the primary neural deficit is understood is emphasized by Axelrod *et al.* (p. 925) They discuss the genotype and phenotype of familial dysautonomia in terms of new findings that suggest that it can now be distinguished from other forms of neurogenic orthostatic hypotension on the basis of the pattern of plasma catechols. The gene for this autosomal-recessive disorder has been localized to chromosome 9q31, but only its identification will allow the possibility of gene therapy. Despite indications that the defect might involve a neurotrophic factor, to date, all the known factors and their receptors have been ruled out.

Zurn *et al.* (p. 929) discuss the type of gene therapy, using neurotrophic factors, that is currently being tested for treatment of both amyotrophic lateral sclerosis (ALS) and Parkinson's disease. The approach involves the encapsulation of a cell line, baby hamster kidney (BHK) cells engineered to express either CNTF or GDNF, in a polymer that allows ready release of the factor without permitting immunological recognition of the BHK cells as foreign. Both CNTF and GDNF have been shown in animal models to support survival of motor neurons and are being tested on ALS patients by implantation of the polymer intrathecally. Whereas in an earlier clinical trial, systemic injection of CNTF that achieved plasma levels on the order of 3 ng/ml were not tolerated by patients, the polymers are generating cerebrospinal fluid levels of 300–6000 pg/ml, which are well tolerated but to date have had no effect on disease progression. The BHK cells continue to proliferate and produce CNTF for at least 3 mo. The same GDNF-producing BHK-containing polymer has been tested in rats with a medial forebrain bundle lesion; the animals showed decreased turning behavior, increased numbers of TH-positive neurons, but no changes in DA content. These GDNF-producing polymers are presumably on their way to a clinical trial for Parkinson's disease.

Unsicker (Deimling *et al.*, p. 932) discuss adrenal medullary chromaffin cell function in terms of the requirement for glucocorticoids, using as a model a glucocorticoid receptor knockout mouse line. The experiments confirm that glucocorticoids are not required for survival of chromaffin cells, but rather that they are needed for induction of the phenotypic marker, phenylethanolamine-N-methyltransferase, responsible for synthesis of epinephrine. In addition, glucocorticoids are not responsible for determination of the epinephrine versus norepinephrine sublineages of chromaffin cells. These observations may change the thinking about treatment of patients with adrenal medulla dysfunctions.

IV. Evolution and Phylogeny

Cardinaud *et al.* (p. 936) discuss the evolution of an ancestral dopamine D1-receptor gene as it diverged in vertebrates from agnathans up to primates. The goal of the analysis was to identify events that led to the appearance of new D1-receptor genes during evolution. They conclude that there had been two steps of diversification, the first occurring at the level of chordates, which generated the D1A subtype, while further diversification between the agnathans and fish led to the D1B and D1C subtypes. Each subtype shares functional properties among species, which allows one to define functional homologies.

Whereas Cardinaud *et al.* dealt with what homologies at the level of the gene can tell us about function, the following three chapters address what the shared properties and functions of a given gene between invertebrate and vertebrate species might say. Rand *et al.* (p. 940) discuss recent data on VMAT and VAChT gene expression in *C. elegans*. The VAChT gene was first cloned from *C. elegans*, and that gene was used to fish out the mammalian gene, while the VMAT gene was first cloned from mammalian tissue and the *C. elegans* gene recognized as the homologue in that genome. This is a powerful example of the application of knowledge obtained from invertebrate species to mammalian neurobiological problems and vice versa. Mutants in both the VAChT and VMAT genes exist in *C. elegans*, and analysis of their behavioral phenotypes can be used to determine the contribution of these proteins to behavior, thus providing clues for vertebrate biologists to consider in their correspondingly more complex organisms.

Hirsh (p. 945) demonstrates how one can use behavioral analyses of *Drosophila* to dissect the roles of the various aminergic receptors present. As in vertebrates, DA receptors are involved in control of locomotion but in addition play an indirect role in courtship behavior, as indicated by pharmacological studies. He presents a novel model in which the biogenic amines are applied directly to decapitated flies and behavior–activity analyses carried out. The power of *Drosophila* for genetic approaches is contrasted with that of the grasshopper, which is significantly larger than the fly, making cellular studies easier to carry out. Condron and Zinn (p. 949) discuss the use of the grasshopper central nervous system to carry out an analysis of the role of DA in its development. One can perturb biochemical pathways in individual cells through the use of antibodies or antisense oligonucleotides, for example. With such techniques, Condron and Zinn demonstrate that DA has two effects on the development of a serotonergic interneuron, increasing serotonin synthesis while inhibiting neurite extension. Such results are not readily obtained in vertebrate systems because of the complexity of the nervous system, but they offer important clues for vertebrate neurobiologists to pursue. In addition, they underscore the ever-broadening definition of neurotrophic factors, which is now recognized to include neurotransmitters and neuropeptides; thus, DA functions not only as a neurotransmitter but can have neurotrophic effects as well.

V. Catecholamines and Learning

In the final section, four chapters address specifically the mechanisms by which NE affects attention, learning, and memory. Four different experimental paradigms were utilized. Harly (p. 952) presents evidence that NE projections from the locus ceruleus (LC) induce a form of long-term potentiation in the dentate gyrus of the hippocampus that is independent of that induced by glutamate via NMDA receptors and requires high-frequency bursting of the LC neurons to generate the relatively high concentrations of NE necessary for the effect. In the other three chapters, NE is shown to be required for specific forms of learning in the "whole animal," specifically for acquisition of a new task contingency by Cynomolgus monkeys (Kubiak *et al.*) (p. 956), in the mediation by tactile stimuli that is required in the development of the olfactory attraction used by infants to recognize their mothers (Leon) (p. 961), and for the storage of memory in the amygdala related to highly emotional experiences (Cahill) (p. 964). Thus, catecholamines play a fundamental role in a variety of forms of learning and memory, which, of course, are the ultimate examples of plasticity responses in the nervous system.

Marianne Bronner-Fraser

Division of Biology 139-74
California Institute of Technology
Pasedena, California 91125

Inductive Interactions Underlie Neural Crest Formation

I. The Neural Crest

Towards the end of gastrulation in vertebrates, an embryo consists of three "germ" layers: the ectoderm, mesoderm, and endoderm. Of these, the ectoderm gives rise to the entire nervous system, in addition to the epidermis, which will eventually cover the whole surface of the embryo and contribute to the skin. At some point prior to or during neurulation, a divergence has been thought to occur between the central and peripheral nervous systems. During neurulation, a region of central ectoderm becomes thickened, and this neural plate folds upon itself to generate the neural tube, from which all neurons and glial cells of the central nervous system (CNS) arise. In contrast, the peripheral nervous system (PNS) is derived largely from the neural crest, with the remaining contribution coming from the sensory epidermal placodes.

The neural crest is a transient population of cells defined in terms of their ectodermal origin, their migratory behavior, and their derivatives. Neural crest cells classically have been thought to be a segregated population within the neural plate, bordered laterally by presumptive epidermis and medially by prospective CNS. However, our recent cell lineage analyses have indicated a common origin within the neural folds for epidermal, neural crest, and neural tube derivatives (1). After they emigrate from the neural tube, neural crest cells give rise to multiple derivatives. Cell lineage analyses *in vitro* (2, 3) and *in vivo* (4, 5) have demonstrated that individual neural crest cells are multipotent. In fact, they have the properties of stem cells with at least a limited ability to self-renew (3). Interestingly, this stem cell population appears to remain within the dorsal neural tube for some time, because older neural tubes are also able to produce "neural crest" cells after the predominant period of cell migration (6).

II. Generation of the Neural Crest

It has often been assumed that the neural crest is a population segregated between the presumptive epidermis and neural plate. However, our cell lineage analyses of the neural folds show that the progeny of some single cells contribute to ectodermal, neural crest, and neural tube derivatives when injections are performed prior to neural tube closure. Only after neural tube closure does the

Advances in Pharmacology, Volume 42

1054-3589/98 $25.00

epidermal lineage segregate from the neural lineage (1). This suggests that the epidermal lineage only segregates from the neural tube lineage after the neural folds have approximated and fused. Thus, individual ectodermal cells are multipotent, such that a precursor within the neural folds has the ability to form epidermis, neural tube (CNS), and neural crest (PNS and other) derivatives. We are currently exploring the function of a novel gene, *NP-1,* in neural crest formation, because its distribution correlates with the potential to form neural crest cells. In our preliminary "gain-of-function experiments," transfection of embryonic cells with *NP-1* leads to ectopic neural crest formation. On the other hand, loss of *NP-1* function by antisense oligonucleotides leads to abnormalities in the neural tube and neural crest.

Our recent experiments demonstrate that interactions between the presumptive neural plate and the nonneural ectoderm lead to induction of the avian neural crest at the interface between these tissues. This supports the idea that an inductive signal travels through the epidermis to generate neural crest cells in the gastrulating embryo. At slightly later stages, inductive interactions between the presumptive epidermis and neural plate result in "dorsalization" of the neural plate, as assayed by the expression of the *Wnt-1* gene (7). These experiments suggest that inductive signals result in neural crest formation prior to formation of other dorsal neural tube derivatives. On the other hand, the competence of the neural plate to form different types of neural crest derivatives appears to change with time. The finding that avian neural crest cells form by inductive interactions of ectodermal tissues is in agreement with experiments performed in amphibian embryos.

Recently, BMP4 and BMP7 (bone morphogenetic proteins) have been shown to be sufficient to substitute for the nonneural ectoderm in inducing neural crest cells (8). Although it remains to be demonstrated whether BMPs are necessary for neural crest induction and if their effects are primary, this class of molecules represents a good candidate for a neural crest inducer.

III. Rostrocaudal and Dorsoventral Patterning of the Neuraxis

The neural tube has a characteristic polarity along the rostrocaudal and the dorsoventral axes. Rostrocaudal regionalization is manifested by the formation of subdivisions in the neural tube, such as the forebrain, midbrain, hindbrain, and spinal cord. Neural crest cells arising from different axial levels contribute to derivatives characteristic of the axial level of origin. Based on the results of quail/chick chimeric grafting experiments (9), it appears that the neural crest is regionalized such that cells derived from different axial levels follow distinct migratory pathways and give rise to a stereotyped set of derivatives. From rostral to caudal along the neuraxis, the levels have been designated as cranial, vagal, trunk, and lumbosacral. Cranial neural crest cells contribute to the connective tissue and periocular skeleton of the face, Schwann cells, and ciliary and cranial sensory ganglia (9–11). Vagal neural crest cells populate the enteric nervous system, first in the rostral and then in progressively more caudal regions of the gut. In the trunk region, neural crest cells form skin melanocytes, sensory and sympathetic ganglia, Schwann cells, and adrenomedullary cells. Sacral neural crest cells, like vagal neural crest cells, contribute to the enteric nervous system. Thus, neural crest cells from different axial levels contribute to some distinct and some overlapping derivatives.

Along the dorsoventral axis, different cell types arise from different portions of the neural tube. Dorsal structures include the roof plate, commissural neurons, other association neurons and neural crest cells, which subsequently emigrate to form the PNS and other derivatives (9). Ventral structures include the floor plate, which acts as a chemoattractant for commissural axons via production of netrins (12–14), and the motor neuron columns that form lateral to the floor plate and project their processes to targets in the periphery. The notochord is thought to play an important role in establishing the dorsoventral polarity of the neural tube. A notochord grafted lateral to the neural tube induces an extra floor plate and motor neuron pools (15–17) via a sonic hedgehog (SHH)-mediated signal (18, 19). SHH protein is expressed in the floor plate and notochord (20), and application of either SHH-expressing cells or purified peptides to intermediate neural plate explants leads to induction of floor plate and/or motor neuron differentiation in a dose-dependent fashion (21).

Grafting a notochord lateral to the closing neural folds juxtaposes dorsal and ventral cells types by inducing floor plate cells and motor neurons dorsally. By performing single-cell lineage analysis using the vital dye, lysinated rhodamine dextran, we have found that both "dorsal" and "ventral" neural tube derivatives can arise from a single precursor in the dorsal neural tube adjacent to the grafted notochord. Cells as diverse as sensory neurons, presumptive pigment cells, roof plate cells, motor neurons, and floor plate cells were observed in the same clone. The presence of such diversity within single clones indicates that the responses to dorsal and ventral signals are not mutually exclusive; in the closing neural tube, neuroepithelial cells are multipotent rather than restricted to form only dorsal or ventral neural tube derivatives (22).

There is increasing evidence that dorsalizing signals influence development of not only the neural crest, but also the dorsal neural tube (23). Various genes are selectively expressed in the dorsal, but not in the ventral, portion of the neural tube. These include: *Wnt-1, Wnt-3a, Pax-3, Pax-7, dorsalin-1,* and *Slug.* Grafting a notochord is known to suppress the dorsal expression of some of these genes (e.g., *Pax-3*) (24) and *dorsalin-1* (23). However, a notochord grafted into the closing neural tube does not suppress the formation of neural crest cells or commissural neurons (25). This suggests that some dorsal properties may be independent of ventral signal by the time of neural tube closure.

IV. Summary

The data summarized here indicate that (1) the neural crest can arise by means of inductive interaction between the neural and nonneural ectoderm; (2) initially, progenitor cells are multipotent, having the potential to form multiple ectodermal derivatives (epidermis, neural crest, and neural tube derivatives); and (3) with time, the precursors become progressively restricted to form neural crest derivatives and eventually to individual phenotypes.

References

1. Selleck, M. A. J., and Bronner-Fraser, M. (1995). Origins of the avian neural crest: The role of neural plate/epidermal interactions. *Development* **121,** 525–538.
2. Baroffio, A., Dupin, E., and Le Douarin, N. M. (1991). Common precursors for neural and mesectodermal derivatives in the cephalic neural crest. *Development* **112,** 301–305.
3. Stemple, D. L., and Anderson, D. J. (1992). Isolation of a stem cell for neurons and glia from the mammalian neural crest. *Cell* **71,** 973–985.
4. Bronner-Fraser, M., and Fraser, S. (1988). Cell lineage analysis shows multipotentiality of some avian neural crest cells. *Nature* **355,** 161–164.
5. Bronner-Fraser, M., and Fraser, S. (1989). Developmental potential of avian trunk neural crest cells in situ. *Neuron* **3,** 755–766.
6. Sharma, K., Korade, Z., and Frank, E. (1995). Late-migrating neuroepithelial cells from the spinal cord differentiate into sensory ganglion cells and melanocytes. *Neuron* **14,** 143–152.
7. Dickinson, M., Selleck, M., McMahon, A., and Bronner-Fraser, M. (1995). Dorsalization of the neural tube by the non-neural ectoderm. *Development* **121,** 2099–2106.
8. Liem, K. F. Jr., Tremml, G., Roelink, H., and Jessell, T. M. (1995). Dorsal differentiation of neural plate cells induced by BMP-mediated signals from epidermal ectoderm. *Cell* **82,** 969–979.
9. LeDouarin, N. M. (1982). The Neural Crest. Cambridge University Press, London.
10. Noden, D. M. (1975). An analysis of the migratory behavior of avian cephalic neural crest cells. *Dev. Biol.* **42,** 106–130.
11. D'Amico-Martel, A., and Noden, D. (1983). Contributions of placodal and neural crest cells to avian cranial peripheral ganglia. *Am. J. Anat.* **166,** 445–468.
12. Tessier-Lavigne, M., Placzek, M., Lumsden, A., Dodd, J., Jessell, T. M. (1988). Chemotropic guidance of developing axons in the mammalian central nervous system. *Nature* **366,** 775–778.
13. Placzek, M., Tessier-Levigne, M., Jessell, T., and Dodd, J. (1990). Orientation of commissural axons in vitro in response to a floor plate derived chemoattractant. *Dev. Biol.* **110,** 19–30.
14. Kennedy, T. E., Serafini, T., de la Torre, J. R., and Tessier-Lavigne, M. (1994). Netrins are diffusible chemotropic factors for commissural axons in the embryonic spinal cord. *Cell* **78,** 425–435.
15. van Straaten, H. W. M., Hekking, E. J. L. M., Wiertz-Hoessels, F. T., and Drukker, J. (1988). Effects of the notochord on the differentiation of a floor plate area in the neural tube of the chick embryo. *Anat. Embryol.* **177,** 317–324.
16. Smith, J., and Schoenwolf, G. C. (1989). Notochordal induction of cell wedging in the chick neural plate and its role in neural tube formation. *J. Exp. Zool.* **25,** 49–62.
17. Yamada, T., Placzek, M., Tanaka, H., Dodd, J., Jessell, T. M. (1991). Control of cell pattern in the developing nervous system: Polarizing activity of the floor plate and notochord. *Cell* **64,** 635–647.
18. Echelard, Y., Epstein, D. J., St-Jacques, B., Shen, L., Mohler, J., McMahon, J. A., and McMahon, A. P. (1993). Sonic hedgehog, a member of a family of putative signaling molecules, is implicated in the regulation of CNS polarity. *Cell* **75,** 1417–1430.
19. Roelink, H., Porter, J. A., Chiang, C., Tanabe, Y., Chang, D. T., Beachy, P. A., and Jessell, T. M. (1995). Floor plate and motor neuron induction by different concentrations of the amino-terminal cleavage product of sonic hedgehog autoproteolysis. *Cell* **81,** 445–455.

20. Marti, E., Takade, R., Bumcrot, D. A., Sasaki, H., and McMahon, A. P. (1995). Distribution of sonic hedgehog peptides in the developing chick and mouse embryo. *Development* **121**, 2537–2547.
21. Marti, E., Bumcrot, D. A., Takada, R., and McMahon, A. P. (1995). Requirement of 19K sonic hedgehog for induction of distinct ventral cell types in CNS explants. *Nature* **375**, 322–325.
22. Artinger, K. B., and Bronner-Fraser, M. (1992). Partial restriction in the developmental potential of late emigrating avian neural crest cells. *Dev. Biol.* **149**, 149–157.
23. Basler, K., Edlund, T., Jessell, T. M., and Yamada, T. (1993). Control of cell pattern in the neural tube: Regulation of cell differentiation by dorsalin-1, a novel TGF beta family member. *Cell* **73**, 687–702.
24. Goulding, M. D., Lumsden, A., and Gruss, P. (1993). Signals from the notochord and floor plate regulate the region-specific expression of two Pax genes in the developing spinal cord. *Development* **117**, 1001–1016.
25. Artinger, K., Fraser, S., and Bronner-Fraser, M. (1995). Dorsal and ventral cell types can arise from common neural tube progenitors. *Dev. Biol.* **172**, 591–601.

James A. Weston
Institute of Neuroscience
University of Oregon
Eugene, Oregon 97403

Lineage Commitment and Fate of Neural Crest–Derived Neurogenic Cells

Trunk neural crest cells of vertebrate embryos give rise to neuronal and nonneuronal derivatives in precise embryonic locations. Crest cells in the dorsal part of the embryonic neural tube undergo an epithelial–mesenchyme transition and segregate from the neuroepithelium. When they do so, crest cells enter an interstitial matrix-filled "migration staging area" (MSA) bounded by the neural tube, the epithelial somite, and the overlying embryonic epithelium (1). They eventually leave the MSA on two spatially and temporally distinct migration pathways. The first crest cells to leave the MSA enter a ventromedial ("medial") pathway into the sclerotome or between adjacent somites. Cells on this pathway give rise to both neuronal and nonneuronal (glial and Schwann sheath cell) derivatives. After the medial migration has been underway for about a day,

Advances in Pharmacology, Volume 42

1054-3589/98 $25.00

some of the crest cells that remain in the MSA begin to disperse on a ventrolateral ("lateral") pathway between the dermatome and the overlying ectodermal epithelium. The cells that undergo this delayed dispersal never give rise to neuronal derivatives but ultimately enter the skin and its derivatives and produce melanocytes. Thus, in general, sensory and autonomic ganglia are formed in the embryonic trunk from crest cells that disperse early in the medial pathway, whereas melanocytes arise primarily from cells that disperse later on the lateral pathway.

I. Regulation of Crest Cell Migration and Survival Patterns in the Embryonic Trunk

We have undertaken to understand the mechanism(s) by which the various crest derivatives differentiate in precise embyronic locations. Several alternatives can be imagined: First, developmental cues might be differentially localized on the two pathways, so that developmentally labile crest cells in the staging area disperse as the pathways become available to them, and differentiate in response to the distinct environmental cues that they encounter in the different embyronic locations. Alternatively, the cues required for the development of specific crest derivatives might be more generally distributed in the embryo, and developmentally distinct crest-derived subpopulations differentially exploit the spatially and temporally distinct pathways. Finally, some combination of the two mechanisms might operate, so that developmentally distinct crest-derived subpopulations migrate selectively on the two pathways in response to localized environmental cues. Recent progress in analyzing the complexion of the crest cell population in the MSA during the early stages of crest cell dispersal, and in characterizing the distribution of specific growth factor activities known to be required for survival and differentiation of specific crest derivatives, have allowed us to elucidate these alternatives.

II. Developmentally Distinct Subpopulations of Crest-Derived Cells Arise in the MSA and Respond to Growth Factor Activities Differentially Localized on the Crest Migration Pathways

We and others have shown that developmentally distinct subpopulations are present in the MSA prior to the onset of crest cell dispersal. These include both neurogenic and nonneurogenic cells that express characteristic receptor tyrosine kinases (RTKs) and other early cell type–specific markers. Our analysis of melanocyte precursor dispersal and survival on the lateral pathway of the mouse embyro will be discussed first, because it suggested a novel mechanism for regulating these processes. Then, results will be presented that suggest that at least some crest-derived neurogenic precursors use similar mechanisms to regulate their dispersal and ultimate survival on the medial migration pathway.

Melanocyte precursors express the Steel factor (SlF) receptor, c-*kit,* and depend on localized SlF activity for dispersal and survival on the lateral migration pathway. Murine crest–derived melanocyte precursors transiently require SlF, the ligand for the RTK, c-*kit,* for differentiation and survival *in vivo* and *in vitro* (see ref. 2 and references therein). *In vivo,* after most of the crest cells have departed on the medial migration pathway, some of the cells remaining in the MSA begin to express mRNAs encoding c-*kit* and tyrosinase-related protein-2 (TRP-2) characteristic of melanocyte precursors. At about the same time, mRNA for the c-*kit* ligand, SlF, begins to be expressed by cells of the epithelial dermatome that border the lateral migration pathway, and shortly thereafter, the melanocyte precursors leave the MSA on this pathway. In mouse embyros homozygous for the *Steel* mutation, which lack SlF, melanocyte precursors (c-kit^{+}/TRP-2^{+} cells) appear in the MSA but fail to disperse on the lateral pathway. We have concluded, therefore, that an SlF signal on the lateral pathway is required for crest cells to disperse there (3).

Steel factor is normally produced as two variants of a transmembrane molecule. One variant contains an extracellular protease-susceptible site that is normally cleaved to release a diffusible ("secreted") form of SlF. The other variant lacks the protease cleavage site and remains predominantly at the cell surface. Additional analysis of melanocyte precursor behavior in mouse embryros homozygous for a *Steel* allele (Sl^{d}) that lacks cell-surface–associated SlF, but produces soluble SlF, has revealed that the soluble and cell-bound forms of SlF have different functions. Thus, in these mutant homozygotes, melanocyte precursors leave the MSA on schedule and disperse normally on the lateral pathway. This suggests that the soluble form of SlF is sufficient to promote or permit dispersal on the lateral pathway. However, after initial dispersal, melanocyte precursors on the lateral pathway disappear, suggesting that the cell-bound form of SlF is necessary for ultimate differentiation and survival of these precursors (3) These inferences are supported by observations in yet another mouse mutant, *Patch,* a recessive lethal, coat color pattern mutant. Melanocyte precursors from embyros carrying the *Patch* mutation express c-kit and undergo melanogenesis in the presence of SlF *in vitro. In vivo,* however, these cells leave the MSA but do not persist in the dermis. In these mutants, although SlF is present on the lateral migration pathway, c-kit is ectopically expressed in the lateral somitic structures. This results in a clear reduction in SlF immunoreactivity in the dermal mesenchyme. The resulting pigment phenotype in these mutants seems to be the result of the reduction of localized SlF activity, which in turn, results in failure of melanocytes to thrive (see ref. 4 and references therein).

Some crest-derived neurogenic precursors express the neurotrophin receptor, *trkC,* and depend on its ligand NT-3 for differentiation and survival on the medial migration pathway. The control of pathway selection, migration, and survival of melanocyte precursors by c-kit/SCF signaling represents a clear example of RTK involvement in the regulation of neural crest cell behavior. We suggest that the differential expression of RTKs in neural crest subpopulations, combined with localized growth factor activity, may also provide a mechanism for the specific migration and localization of other neural crest–derived subpopulations. To determine if crest-derived *neuronal* cells differentially expressed RTKs and responded to localized neurotrophin activity, we examined whether neurotrophin-3 (NT-3), a survival factor for subsets of peripheral neurons, was involved in the regulation of neurogenesis by neural crest cells (5). First, we found that cells in the neural tube and some premigratory and migrating avian neural crest cells on the medial migration pathway express mRNAs encoding multiple isoforms of the NT-3 receptor, *trk*C. Later in development, a subpopulation of neurons in nascent sensory ganglia express *trk*C message. Second, we learned that *trk*C mRNA is only expressed in neural crest cell populations that possess neurogenic potential, and that the presence of NT-3 is required during the initial development of cultured neural crest cells for neuronal development by such cells. These results suggest that, *in vivo,* at least a subpopulation of neurogenic neural crest–derived cells expresses functional *trk*C receptors and requires the timely availability of NT-3 for normal gangliogenesis.

NT-3 is produced by the neural tube and possibly by other embryonic tissues and is presumed to be present in somitic tissues (6) and, therefore, available on both the medial and lateral crest migration pathways. Because *trk*C-containing crest-derived cells are found primarily, if not exclusively, on the medial pathway, it seems unlikely that differential localization or limited availability of NT-3, similar to what has been described for Steel factor, could explain the distribution of neurogenic crest-derived subpopulations solely on the medial migration pathway. It is noteworthy, however, that during crest cell dispersal, the only cells outside of the neural tube, other than crest-derived cells, that express *trk*C mRNA are found in the epithelial dermamyotome (5, 6). It is tempting to suggest, therefore, that this "paradoxical" expression of *trk*C receptors by somite cells bordering the lateral crest migration pathway plays a role in the exclusive development of crest-derived neuronal derivatives on the medial pathway. Regardless of its function in the dermatome, *trk*C would compete with dispersing crest cells for available ligand on the lateral pathway, thereby assuring that responsive neurogenic cells would only disperse, survive, and undergo gangliogenesis on the medial pathway.

In summary, the differentiation of neural crest derivatives in distinct embryonic locations in the vertebrate embryo has stimulated many attempts to understand the underlying mechanism of specific pathway choices made by migrating neural crest cells. We now know that developmentally distinct subpopulations of crest-derived cells arise from crest cell precursors prior to their dispersal from an MSA. The MSA would, therefore, be the embyronic location where neural crest cells make their first specific pathway choices. We suggest that the expression of specific receptors (RTKs) in developmentally distinct subpopulations allows them to respond to differentially localized ligands in the embyro and that the responses to specific growth-factor activities mediate the onset of dispersal and subsequent survival of crest-derived subpopulations in specific embryonic locations (7).

Acknowledgment

Supported by U.S. Public Health Service grants DE04316 and NS29438.

References

1. Weston, J. A. (1991). Sequential segregation and fate of developmentally restricted intermediate cell populations in the neural crest lineage. *Curr. Top. Dev. Biol.* **25,** 133–153.
2. Morrison-Graham, K., and Weston, J. A. (1993). Transient Steel factor dependence by neural crest-derived melanocyte precursors. *Dev. Biol.* **159,** 346–352.
3. Wehrle-Haller, B., and Weston, J. A. (1995). Soluble and cell-bound forms of steel factor activity play distinct roles in melanocyte precursor dispersal and survival on the lateral neural crest migration pathway. *Development* **121,** 731–742.
4. Wehrle-Haller, B., Morrison-Graham, K., and Weston, J. A. (1996). Ectopic c-kit expression affects the fate of melanocyte precursors in *Patch* mutant embryos. *Dev. Biol.* **177,** 463–474.
5. Henion, P. D., Garner, A. S., Large, T. H., and Weston, J. A. (1995). *trk*C-mediated NT-3 signalling is required for the early development of a subpopulation of neurogenic neural crest cells. *Dev. Biol.* **172,** 602–612.
6. Brill, G., Kahane, N., Carmeli, C., vonSchack, D., Barde, Y-A., and Kalcheim, C. (1995). Epithelial-mesenchymal conversion of dermatome progenitors requires neural tube-derived signals: Characterization of the role of Neurotrophin-3. *Development* **121,** 2583–2594.
7. Wehrle-Haller, and Weston, J. A. (1997). Receptor tyrosine kinase-dependent neural crest migration in response to differentially localized growth factors. *BioEssays* **19,** 337–345.

Hermann Rohrer and Uwe Ernsberger

Department of Neurochemistry
Max-Planck-Institute for Brain Research
D-60528 Frankfurt/M., Germany

The Differentiation of the Neurotransmitter Phenotypes in Chick Sympathetic Neurons

The postganglionic neurons in the sympathetic nervous system use either noradrenaline or acetylcholine as neurotransmitter (1). Because sympathetic ganglia are composed of only two major transmitter phenotypes, noradrenergic and cholinergic neurons, the sympathetic nervous system serves as a model for the analysis of neurotransmitter phenotype development. Important issues in this analysis are the identification of cell interactions involved in the induction of a specific neurotransmitter phenotype and characterization of the growth factors that are mediating these inductive interactions.

Advances in Pharmacology, Volume 42

1054-3589/98 $25.00

I. The Noradrenergic Differentiation

The analysis of the noradrenergic differentiation of postganglionic sympathetic neurons has been accelerated by the availability of markers indicating catecholamine biosynthesis and storage in embryonic tissue (2). In the chick embryo, tyrosine hydroxylase (TH), the rate-limiting enzyme in catecholamine biosynthesis, is expressed after the neural crest–derived precursors of the sympathetic neurons have gathered at the dorsal aorta to form the primary sympathetic ganglia (3). The start of TH expression at the end of the third day of development (stage 18) coincides with the appearance of catecholamine histofluorescence in the primary sympathetic ganglia, indicating that the machineries for catecholamine synthesis and storage are expressed coordinately in the developing sympathetic neurons. In addition, catecholamine uptake is detectable at the same time in these cells. Interestingly, not only characters of a noradrenergic neurotransmitter phenotype are expressed at that stage. The expression of a set of early neuronal markers and of the transcription factor Phox2 starts at the same time. These findings indicate that an important inductive process leads to the expression of transcription factors and proteins characteristic of the noradrenergic neurotransmitter phenotype (3). There is at present no evidence, however, to suggest that these transcription factors, such as Phox2, are involved in the regulated expression of the noradrenergic characteristics.

The appearance of many characteristic properties of the noradrenergic phenotype shortly after neural crest-derived sympathetic precursors have reached the dorsal aorta indicates that the dorsal aorta may play an important role in this process. This notion is strongly corroborated by the finding that dorsal aorta explanted from the chick embryo dramatically increases the number of TH expressing cells in cultures of neural crest (4). In addition to the dorsal aorta, several embryonic tissues have been demonstrated to promote catecholaminergic differentiation in developing sympathetic neurons or their precursors (2). In particular, ventral neural tube and notochord are necessary for catecholaminergic differentiation of sympathetic neurons *in vivo*. It is currently unclear, whether aorta, notochord, and neural tube exert their effects on catecholaminergic differentiation in a cooperative manner or independently. Because neural crest cells first migrate past ventral neural tube and notochord before they reach the dorsal aorta, it is conceivable that signals from the axial structures have a conditioning effect and induce competence to respond to signals from the dorsal aorta. Another possibility is that the axial structures notochord and neural tube exert their effects indirectly, for example, by inducing the aorta to express growth factors, which in turn induce catecholaminergic differentiation in the developing sympathetic neurons.

The search of identified growth factors that mediate the induction of noradrenergic differentiation in sympathetic neurons revealed that bone morphogenetic proteins (BMPs) are effective in inducing TH expression *in vitro* and *in vivo* (4). BMP-2, -4, and -7 lead to TH expression in cultured neural crest cells, with BMP-2 and -4 acting at lower concentration than BMP-7. Interestingly, the effect of BMP-2 and -4 vanishes at higher concentrations of the recombinant growth factors in the culture media. In addition to its effect in cell culture,

BMP-4 has been demonstrated to promote catecholaminergic differentiation *in vivo*. Inducing ectopic expression of BMP-4 via infection of chick embryos with a RCAS virus containing a mBMP-4 insert leads to an increase in the number of TH-positive cells in the area of the sympathetic ganglia as well as to the expression of TH in ectopically located cells. This observation indicates that BMPs are able to promote catecholaminergic differentiation in neural crest–derived sympathetic precursors *in vitro* as well as *in vivo*. Significantly, BMP-4 and -7 are expressed in the dorsal aorta before and during noradrenergic differentiation of sympathetic neurons (4). Thus, it is highly likely that BMPs are the growth factors responsible for the effect of dorsal aorta on noradrenergic differentiation in sympathetic neurons and play a critical role in sympathetic neuron development.

II. The Cholinergic Differentiation

The analysis of the developing sympathetic innervation of the rat sweat gland has led to the conclusion that the cholinergic neurotransmitter phenotype of sympathetic neurons is induced by interactions with the target tissues, such as the sweat gland (5). The neurons innervating the sweat glands initially display a noradrenergic neurotransmitter phenotype. Markers of the cholinergic neurotransmitter phenotype, such as the enzymes choline acetyltransferase (CHAT), acetylcholine esterase (AChE), and the neuropeptide vasoactive intestinal peptide (VIP), were detectable only after sweat gland innervation. In parallel to the induction of cholinergic characteristics, the expression of noradrenergic characters was reduced after target encounter. Transplantation of the target tissues clearly demonstrated a causal role of the target in the change of transmitter phenotype *in vivo*. In addition, extracts containing soluble proteins from developing and adult target tissues induced cultured noradrenergic sympathetic neurons to exhibit a similar increase in the expression of cholinergic characters, as can be observed *in vivo*. Although the results indicate that the development of the cholinergic neurotransmitter phenotype occurred under the influence of certain sympathetic target tissues, it remained unclear which signals led to the initial expression of the genes characteristic for the cholinergic phenotype.

In the 18-day-old chick embryo, at a time when sympathetic neurons have already established connections with their target tissues, noradrenergic and cholinergic subpopulations of neurons are detectable (U. Ernsberger, H. Patzke, and H. Rohrer, in preparation). The cholinergic neurons express CHAT and VIP but are negative for TH, which is expressed in the noradrenergic cell population. The expression of VIP mRNA is detectable from embryonic day 10 onward. The peripheral targets of these VIP-expressing neurons are arteries in skin and muscle (Zechbauer and Rohrer, unpublished). Sympathetic innervation of these blood vessels, as revealed by catecholamine fluorescence, was observed during development several days before the appearance of VIP-positive fibers. These findings are in agreement with the notion that in chick sympathetic neurons, VIP expression is induced by target interaction. Surprisingly, CHAT mRNA was detectable already at embryonic day 6 at a time when sympathetic fibers are only starting to grow out from the just-formed secondary ganglia. A few CHAT-expressing cells were even detected in the primary sympathetic ganglia. The data strongly indicate that the initial expression of CHAT occurs independent of target contact. A second, later maturation of the cholinergic neurons to express both CHAT and VIP occurs during a time when the neurites have reached their targets.

Both the target-derived differentiation factors responsible for the late differentiation step and the factor(s) leading to the initial expression of CHAT remain to be indentified. *In vitro,* the change in transmitter properties observed *in vivo* can be elicited by members of a family of related differentiation factors, leukemia inhibitory factor (LIF), ciliary neurotrophic factor, (CNTF), oncostatin M, cardiotrophin-1, and growth-promoting activity (GPA). Transgenic mice deficient in LIF or CNTF displayed normal development of cholinergic sympathetic differentiation. Thus, it remains to be analyzed whether the factors inducing cholinergic differentiation *in vivo* are members of this family.

III. Conclusions

The analysis of the expression of enzymes involved in the biosynthesis of noradrenaline and acetylcholine in sympathetic neurons of the chick embryo revealed an early induction of noradrenergic as well as cholinergic properties. TH and CHAT are first expressed when the precursor cells are still located in the primary sympathetic ganglia before finishing the migration to their final sites in secondary sympathetic ganglia. Several days later, when neurites have grown from the secondary ganglia to the sympathetic target tissues, cholinergic neurons mature to express VIP in addition to CHAT. Whereas BMPs have been shown to be prime candidates for the growth factors mediating the induction of noradrenergic differentiation of sympathetic neurons in the primary sympathetic ganglia, the factors responsible for the induction of CHAT expression are unknown. Several candidate growth factors involved in the later target-dependent maturation of the cholinergic properties have been described *in vitro,* but the factor responsible *in vivo* is still to be defined. The results favor a model of neurotransmitter phenotype differentiation whereby the initial expression of the key properties is under the control of adjacent embryonic tissues, such as aorta, notochord, and neural tube. In a later maturation step, target tissues of the neurons induce further properties of the neurotransmitter phenotype or may even lead to a switch from a noradrenergic to a cholinergic phenotype.

Acknowledgments

Work from our group referred to in this publication has been supported by grants from the Deutsche Forschungsgemeinschaft (SFB 269) and the Fonds der chemischen Industrie.

References

1. Elfvin, L-G., Lindh, B., and Hökfelt, T. (1993). The chemical neuroanatomy of sympathetic ganglia. *Annu. Rev. Neurosci.* **16,** 471–507.

2. Ernsberger, U., and Rohrer, H. (1996). The development of the noradrenergic transmitter phenotype in postganglionic sympathetic neurons. *Neurochem. Res.* **21**, 829–835.
3. Ernsberger, U., Patzke, H., Tissier-Seta, J. P., Reh, T., Goridis, C., and Rohrer, H. (1995). The expression of tyrosine hydroxylase and the transcription factors cPhox-2 and Cash-1: Evidence for distinct inductive steps in the differentiation of chick sympathetic precursor cells. *Mech. Dev.* **52**, 125–136.
4. Reissmann, E., Ernsberger, U., Francis-West, P. H., Rueger, D., Brickell, P. M., and Rohrer, H. (1996). Involvement of bone morphogenetic proteins-4 and -7 in the specification of the adrenergic phenotype in developing sympathetic neurons. *Development* **122**, 2079–2088.
5. Rao, M. S., and Landis, S. C. (1993). Cell interactions that determine sympathetic neuron transmitter phenotype and the neurokines that mediate them. *J. Neurobiol.* **24**, 215–232.

Guy L. Guidry and Story C. Landis

Neural Development Section
National Institute of Neurological Disorders and Stroke
National Institutes of Health
Bethesda, Maryland 20892

Developmental Regulation of Neurotransmitters in Sympathetic Neurons

The developmental mechanisms that regulate the expression of neurotransmitters and neuropeptides in autonomic neurons are incompletely understood. Previous studies of developing neural crest cells *in vivo* and *in vitro* have provided evidence that during migration or shortly after arrival at the presumptive ganglia, neural crest cells receive an instructive signal that induces noradrenergic properties (1). The notochord and/or the ventral neural tube have been suggested as a source of the noradrenergic signal, and two candidate molecules, sonic hedgehog and bone morphogenetic proteins, have been put forward. Examination of cervical and thoracic ganglia from embryonic rats between embryonic days 14 and 16 reveal that all neuronal cell bodies contain immunoreactivity for tyrosine hydroxylase, the rate-limiting enzyme for catecholamine synthesis.

Although all cervical and thoracic sympathetic ganglia are noradrenergic during embryonic development, in adult rats a minority population of mature sympathetic neurons are cholinergic. We have focused our attention on the development of this subpopulation (2). Sweat glands are innervated exclusively by cholinergic sympathetic neurons, and, therefore, we have analyzed the devel-

Advances in Pharmacology, Volume 42

1054-3589/98 $25.00

oping innervation in this target tissue. In adult rats, the sweat gland innervation contains choline acetyltransferase activity (ChAT), acetylcholinesterase activity (AChE), and immunoreactivity for vasoactive intestinal peptide (VIP). Previous studies of the developing sweat gland innervation of rat hind footpads demonstrated that the axons that initially associate with sweat gland primordia at postnatal day 4 contain intense catecholamine histofluorescence and lack AChE, VIP, and detectable ChAT activity. These results provide evidence that the neurons that innervate eccrine sweat glands undergo a switch from noradrenergic to cholinergic and VIP-containing. The recent development of antisera that recognize the vesicular acetylcholine transporter (VAChT) offers the possibility of a more sensitive assay for the expression of cholinergic markers in the early sweat gland innervation. Because the coding sequence for VAChT is contained within the first intron of the gene encoding ChAT and expression of the two genes is coordinately regulated, VAChT is an excellent marker for cholinergic function. Our preliminary data indicate that while there are many tyrosine hydroxylase immunoreactive axons associated with the developing sweat glands from postnatal days 2 through 14 in hind footpads, significant VAChT immunoreactivity is not evident until postnatal day 6. VAChT immunoreactivity is present 2 days earlier in the innervation of front footpads, presumably due to the earlier arrival of sympathetic axons in this more anterior target. Consistent with the postulated switch in transmitter phenotype from noradrenergic to cholinergic, treatment of neonatal rat pups with the adrenergic neurotoxin, 6-hydroxydopamine, resulted in the absence of cholinergic sympathetic innervation of sweat glands.

The fact that only specific targets are innervated by cholinergic sympathetic neurons and the temporal correlation between arrival in the target tissue and the acquisition of cholinergic and peptidergic properties suggested that sweat glands provide a retrograde instructive signal that induces cholinergic differentiation. To test this possibility, sweat gland primorida were replaced with parotid gland, a noradrenergic target. We found that the presumptive sweat gland innervation remained catecholaminergic and did not acquire ChAT activity and VIP immunoreactivity. Evidence for an inductive role of sweat glands was obtained when gland primordia transplanted to hairy skin caused noradrenergic sympathetic innervation to lose catecholamines and become cholinergic and peptidergic. In these experiments, we could not, however, exclude the possibility that the parotid gland inhibited an intrinsically regulated transmitter switch.

We have reexamined the role of the target sweat glands in altering the transmitter properties of sympathetic neurons. First, we took advantage of the fact that sweat glands fail to form in *Tabby* mutant mice (3). *Tabby* is an X-linked mutation affecting morphogenesis of epidermal derivatives, including sweat glands. The mutation delays eye opening and tooth eruption and completely prevents sweat gland development, but it does not directly affect sympathetic neuron development. The *Tabby* sequence is unknown, but the gene is believed to play an important role in dermal induction of epidermal derivatives. Despite the absence of sweat glands, we find that noradrenergic sympathetic axons pathfind correctly to the footpads and remain in the presumptive target area for approximately 10 days. Because sympathetic axon pathfinding was not dependent on the target tissue, we compared the development of transmitter properties by sweat gland axons in *Tabby* footpads with that in control footpads. We found that sympathetic axons containing catecholamines and lacking VIP were still present in *Tabby* footpads between P10 and P14, when catecholamines have declined and VIP has appeared in wild-type footpads.

To determine whether cholinergic and peptidergic properties would appear in the presumptive sweat gland innervation in *Tabby* mice with additional development, we crossed *Tabby* mutant mice with a transgenic line in which nerve growth factor (NGF) expression is driven by the promoter of the keratin 14 gene. As a result, NGF is expressed earlier than normal in the basal cells of keratinized epithelia and at much higher levels than normal. In K14-NGF transgenic mice, sympathetic and sensory neurons are significantly increased in number, as is the density of sensory and sympathetic innervation of hairy and footpad skin. It seemed likely to us that the excess NGF produced in the skin of K14-NGF transgenic mice would promote the survival of the presumptive sweat gland innervation in *Tabby* mice. When we analyzed the footpads of adult double-mutant mice, we found brightly fluorescent tyrosine hydroxylase immunoreactive fascicles that extended through the glandless pad core and formed a dense plexus in the dermis and at the dermal–epidermal junction. In contrast to the robust tyrosine hydroxylase immunoreactivity, neither VAChT or VIP immunoreactivities were detected in sympathetic axons. Despite the presence of the dense sympathetic plexus in the footpads of the *Tabby*/K14-NGF double mutants, ChAT activity was not greater than that present in the footpads of *Tabby*.

Our studies of the *Tabby* and *Tabby*/K14-NGF mice indicate that in the absence of sweat glands, the gland-targeted innervation does not alter its neurotransmitter properties and confirm the interpretation of our previous transplantation studies. Because the footpad tissues remaining in the single- and double-mutant mice appear normal, these findings also identify sweat glands as the source for the differentiation factor in footpads. This identification is consistent with the reduction of cholinergic-inducing activity in *tabby* footpads and the production of cholinergic differentiation activity by sweat gland cells when they are cocultured with sympathetic neurons.

Several lines of evidence indicate that the cholinergic differentiation factor produced by sweat glands is a novel member of the neuropoietic cytokine family. The changes in neurotransmitter properties that occur during normal development of the sweat gland innervation can be induced in cultured sympathetic neurons by members of the neuropoietic cytokine family, including leukemia inhibitory factor (LIF), ciliary neurotrophic factor (CNTF), and cardiotrophin-1 (CT-1) but not by other cytokines and growth factors. Further, monoclonal antibodies that block activation of LIFRβ, one of the receptor subunits utilized by members of the neuropoietic cytokine family, block the induction of cholinergic function in sweat gland/sympathetic neuron cocultures. Although footpads contain low levels of mRNA encoding each of these factors, examination of adult mice deficient in LIF, CNTF, or LIF and CNTF revealed that the sweat gland innervation possessed cholinergic and peptidergic properties (4). Similarly, the induction of cholinergic and peptidergic properties in sympathetic neurons cocultured with sweat gland cells was not prevented by the presence of antibodies that block CT-1 function. Thus, the known members of the neuropoietic cytokine family do not appear to account for the cholinergic induction by sweat glands. While the sweat gland cholinergic differentiation activity remains to be identified, we do know that its production is regulated by gland innervation both *in vivo* and *in vitro* (5). In addition, culture studies suggest that the catecholamines produced by the early innervation are important in its regulation.

We have recently identified a new cholinergic sympathetic target, the periosteum or connective tissue covering of the bone, and have examined the development of the innervation of the periosteum of the sternum. Like the developing sweat gland innervation, periosteal innervation initially expresses noradrenergic properties, which are suppressed as cholinergic and peptidergic ones appear, and this change is regulated by interactions with the target. In contrast to the developing sweat gland innervation, however, a significant proportion of the sympathetic axons that reach the sternum already expresses immunoreactivity for the VAChT and VIP in addition to catecholaminergic properties. This difference in the time at which cholinergic and peptidergic properties are expressed could reflect the fact that some sympathetic neurons are exposed to cholinergic signals in the ganglion before they extend axons or as they grow to the sternum through the rib periosteum. Thus, while target interactions are required for the induction of cholinergic properties in neurons that innervate sweat glands, target signals may not be necessary for induction of cholinergic properties in sympathetic neurons that innervate other cholinergic targets. It is possible, however, that target-derived factors play a role in maintaining cholinergic function in such neurons.

References

1. Ernsberger, U., and Rohrer, H. (1996). The development of the noradrenergic transmitter phenotype in postganglionic sympathetic neurons. *Neurochem. Res.* **21,** 823–829.
2. Landis, S. C. (1990). Target regulation of transmitter phenotype. *Trends Neurosci.* **12,** 344–350.
3. Guidry, G., and Landis, S. (1995). Sympathetic axons pathfind successfully in the absence of target. *J. Neurosci.* **16,** 2179–2190.
4. Francis, N. J., Asmus, S. E., and Landis, S. C. (1997). CNTF and LIF are not required for the target-directed acquisition of cholinergic and peptidergic properties by sympathetic neurons in vivo. *Dev. Biol.* **182,** 76–87.
5. Habecker, B., Tresser, S., Rao, M., and Landis, S. C. (1995). Production of sweat gland differentiation factor depends on innervation. *Dev. Biol.* **167,** 307–316.

R. Zigmond, R. Mohney, R. Schreiber, A. Shadiack, Y. Sun, Y. S. Vaccariello, and Y. Zhou

Department of Neurosciences
Case Western Reserve University
Cleveland, Ohio 44106

Changes in Gene Expression in Adult Sympathetic Neurons after Axonal Injury

Cheah and Geffen reported in 1973 that tyrosine hydroxylase (TH) activity decreases in neurons in lumbar sympathetic ganglia 3 days after transection of their axons in the sciatic nerve. The authors proposed that axotomy may "reorder the priorities of protein synthesis in the neurons to favor the production of proteins necessary for regenerative rather than transmitter functions" (1). This hypothesis has withstood over 20 years of investigations in three different types of peripheral neurons: sympathetic, sensory, and motor (3, 5); however, until recently little has been known about the mechanisms underlying such phenotypic changes. This brief chapter will summarize advances in our knowledge of the phenotype of axotomized peripheral neurons since 1973, concentrating on the phenotypic changes that occur in axotomized sympathetic neurons and the signals that trigger these changes.

The decrease in TH activity in axotomized sympathetic neurons is accompanied by a decrease in TH mRNA (2) (Fig. 1). Changes can also be found in the mRNA levels of other proteins involved in synaptic transmission. For example, in addition to synthesizing catecholamines, about 60% of the noradrenergic postganglionic neurons in the rat superior cervical ganglion (SCG) express neuropeptide Y (NPY)-like immunoreactivity (IR), and this peptide is thought to play an important role in sympathetic neural regulation of the vascular system. Interestingly, following axotomy, NPY mRNA decreases with a time course similar to that for the decrease in TH mRNA (2) (Fig. 1). Decreases in transmitter synthesis also occur in axotomized sensory and motor neurons. For example, substance P expression decreases in small-diameter sensory neurons that normally use the peptide as a neurotransmitter, and the enzyme choline acetyltransferase decreases in motor neurons (3).

In addition to changes in neurotransmitter–neuromodulator levels in peripheral neurons after axotomy, changes have been found in their complement of receptors for these signaling molecules. For example, a decrease in muscarinic receptors was reported by Rotter *et al.* in 1977 in facial motoneurons after transection of the facial nerve (3). Recently, it has been shown that axotomy produces a dramatic decrease in the mRNA for several subunits of the nicotinic receptors in the SCG (4).

Advances in Pharmacology, Volume 42

1054-3589/98 $25.00

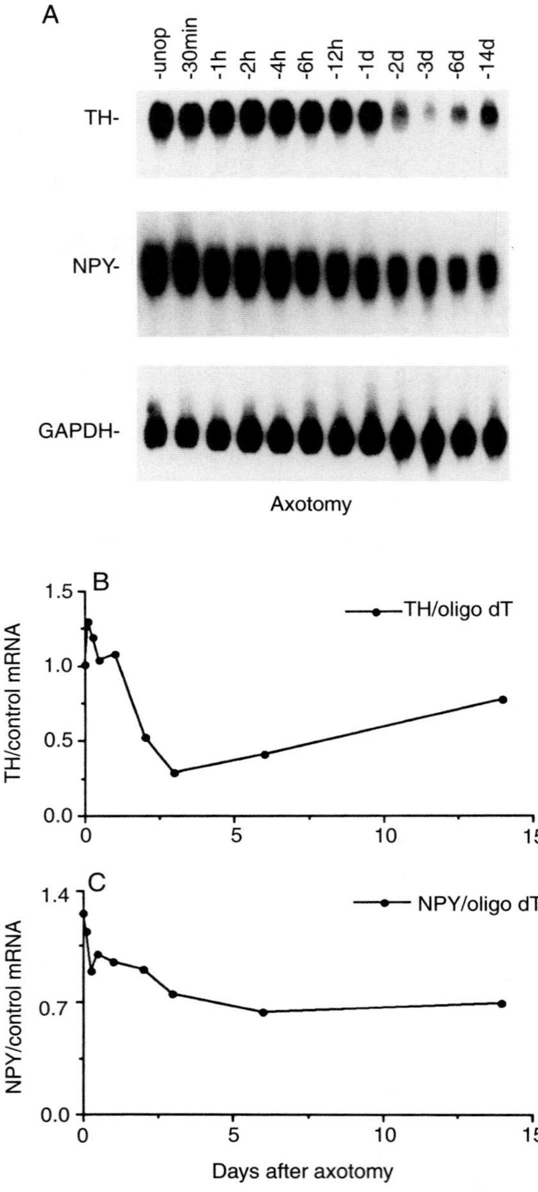

FIGURE 1 (*A*) Northern blot of SCG taken from unoperated animals (unop) and from animals at various times (30 min to 14 days) after transection of the postganglionic internal and external carotid nerves. Decreases were seen in the levels of TH and NPY mRNA but not in glyceraldehyde-3-phosphate dehydrogenase (GAPDH) mRNA. The data plotted in (*B*) and (*C*) are normalized by the total mRNA hybridizing to an oligo dT probe for each sample.

As predicted by Cheah and Geffen, axotomy of peripheral neurons also produces increases in expression of certain proteins known to be involved in regeneration, such as tubulin and growth-associated protein (GAP)-43 (3). More recent evidence demonstrates that, in addition, several neuropeptides, not normally expressed by these neurons, are induced in response to axotomy. Thus, very few, if any, sympathetic neurons in the SCG normally express substance P, vasoactive intestinal peptide (VIP), or galanin; however, within 48 hr after transection of the two major postganglionic trunks of the SCG, significant numbers of neurons express these peptides. Increases in the levels of the peptides are accompanied by increases in their corresponding mRNAs. Interestingly, galanin and VIP, which are not normally expressed by dorsal root ganglion neurons or motoneurons, are expressed after these neurons are axotomized. (3). The functional significance of these changes in neuropeptide expression is not yet known. It has been speculated, however, that at least some of these peptides might promote neuronal survival and/or fiber outgrowth under conditions in which the neurons are deprived of their target-derived trophic factors (3, 5).

We have identified two events involved in triggering these phenotypic changes following axotomy: (1) the induction and release of leukemia inhibitory factor (LIF) in nonneuronal cells within the SCG and/or at the site of nerve transection and (2) the reduction in levels of nerve growth factor (NGF) in the SCG as a consequence of the disconnection of the neurons from their target tissues (3, 5). Part of the evidence for a role for LIF comes from experiments in which the phenotypic changes after axotomy, at both the mRNA and peptide levels, were compared in wild-type mice and in transgenic LIF-minus mice. The magnitude of the changes in peptide expression are significantly reduced in the transgenic animals. The influence of LIF on these mRNA levels is particularly large with regard to the increases in galanin and VIP, although the decreases in TH and NPY are also significantly reduced in the transgenic mice (2). While LIF mRNA is not detectable in the SCG under normal conditions, it is seen within 1–2 hr after axotomy and reaches peak levels at 6 hr (6). Interestingly, in dissociated cell cultures, levels of VIP increase substantially more in mixed cultures of neurons and nonneuronal cells than in cultures in which the nonneuronal cells are largely removed by preplating (5). Medium conditioned for 48 hr by nonneuronal cells from the SCG increases levels of VIP in neuron-enriched cultures. The influence of this conditioned medium can be removed if the medium is treated first with an antiserum of LIF that immunoprecipitates LIF (5).

Other evidence confirms that dramatic changes occur in the nonneuronal population of the SCG after axotomy (5). Changes in the satellite–Schwann cell population can be followed by increases in glial fibrillary acidic protein expression, and changes in the macrophage population by increases in immunostaining for ED-1, a marker for infiltrating macrophages. The influx of macrophages is of interest because it has been suggested that interleukin 1β (IL-1β) released by macrophages might be the trigger for the induction of LIF mRNA after axotomy. However, time-course studies indicate that the peak in LIF mRNN occurs many hours before the infiltrating macrophages are seen in the SCG (6). Studies with a second macrophage marker, ED-2, demonstrate that the SCG, like other tissues, has a population of resident macrophages, which could serve as a source for LIF-inducing factors.

Evidence for an endogenous LIF-inducing factor has recently been reported (6). This factor is released if either excised ganglia or segments of sympathetic nerve trunks are incubated *in vitro* for an hour. The endogenous factor is heat- and protease-sensitive. The activity was not affected by a function-blocking antiserum to IL-1β, and its presence is not affected by including cycloheximide in the medium. We hypothesize that the LIF-inducing factor is a protein other than IL-1β, that it preexists in neural tissue under normal conditions, and that it is activated and/or released following axotomy.

Changes in galanin, VIP, and NPY still occurred in the LIF-minus transgenic mice, although the changes were significantly smaller than those seen in wild-type mice (2). The data suggest that the changes in neuropeptide expression have both a LIF-dependent and a LIF-independent component. The possibility that, in addition to LIF induction, a reduction in NGF levels in the SCG also plays an important role in triggering these changes comes from recent experiments from our laboratory in which normal (i.e., nonlesioned) rats received daily systemic injections of an antiserum raised against NGF. The antiserum caused decreases in the levels of NPY and its mRNA and increases in the levels of galanin and VIP and their mRNAs, though it produced no change in substance P expression. Considering the relative magnitudes of the effects of the NGF antiserum, it appears that removal of NGF plays a larger role in the decrease of NPY expression than it does in the increases in expression of galanin and VIP.

Acknowledgment

The research from our laboratory described in this paper was supported by the U.S. Public Health Service (NS 17512 and NS 12651).

References

1. Cheah, T. B., and Geffen, L. B. (1973). Effects of axonal injury on norepinephrine, tyrosine hydroxylase and monoamine oxidase levels in sympathetic ganglia. *J. Neurobiol.* **4,** 443–452.
2. Sun, Y., and Zigmond, R. E. (1996). Involvement of leukemia inhibitory factor in the increases in galanin and vasoactive intestinal peptide mRNA and the decreases in neuropeptide Y and tyrosine hydroxylase mRNA after axotomy of sympathetic neurons. *J. Neurochem.* **67,** 1751–1760.
3. Zigmond, R. E., Hyatt-Sachs, H., Mohney, R. P., Schreiber, R. C., Shadiack, A. M., Sun, Y., and Vaccariello, S. A. (1996). Changes in neuropeptide phenotype after axotomy of adult peripheral neurons and the role of leukemia inhibitory factor. *Perspect. Dev. Neurobiol.* **4,** 75–90.
4. Zhou, Y., Deneris, E., and Zigmond, R. E. (1996). Regulation of nicotinic receptor subunit transcripts in the superior cervical ganglion (SCG) after axotomy. *Soc. Neurosci. Abstr.* **22,** 1961.

5. Zigmond, R. E. (1996). Retrograde and paracrine influences on neuropeptide expression in sympathetic neurons after axonal injury. *In* Cytokines and the CNS: Development, Defenses and Disease. (R. M. Ransohoff and E. N. Benveniste, eds.), pp. 169–186. CRC Press, Boca Raton.
6. Sun, Y., Landis, S. C., and Zigmond, R. E. (1996). Signals triggering the induction of leukemia inhibitory factor in sympathetic superior cervical ganglia and their nerve trunks after axonal injury. *Mol. Cell. Neurosci.* 7, 152–163.

B. Schütz,* M. K-H. Schäfer,* L. E. Eiden,† and E. Weihe*

* Department of Anatomy and Cell Biology
Philipps University
Marburg, Germany

† Section on Molecular Neuroscience
Laboratory of Cellular and Molecular Regulation
National Institute of Mental Health
National Institutes of Health
Bethesda, Maryland

Ontogeny of Vesicular Amine Transporter Expression in the Rat: New Perspectives on Aminergic Neuronal and Neuroendocrine Differentiation

The vesicular monoamine transporter VMAT1 is restricted to neuroendocrine cells, and only the VMAT2 isoform is expressed in neurons, in adult rats (1, 2). The vesicular acetylcholine transporter (VAChT) is expressed only in cholinergic neurons throughout the adult rat central and peripheral nervous systems, including neurons of the sympathetic nervous system that during early development coexpress noradrenergic traits (3–7). The patterns of expression of VMAT1, VMAT2, and VAChT during ontogenesis of monoaminergic and cholinergic neurons and endocrine cells, therefore, should provide insight into pathways of lineage commitment leading to the final chemical coding of aminergic neurotransmission in the adult rat.

Advances in Pharmacology, Volume 42

1054-3589/98 $25.00

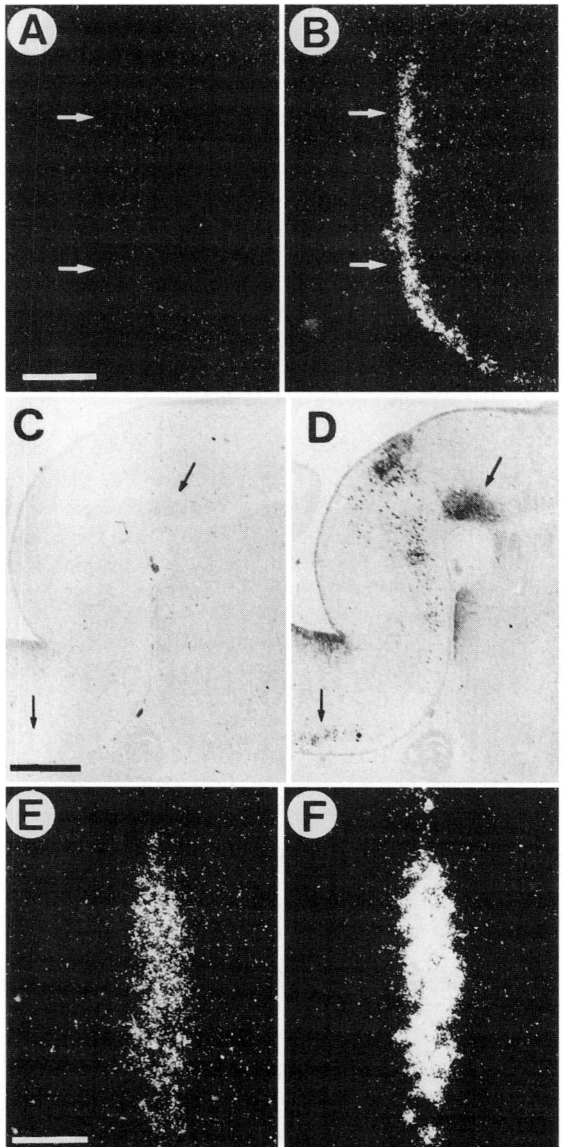

FIGURE 1 Expression of VMAT1, VMAT2 and VAChT in the developing central and peripheral nervous systems. (*A, B*) Adjacent sections of embryonic day 12 (E12) neuroepithelium processed for (*A*) VMAT1 and (*B*) VMAT2 *in situ* hybridization histochemistry (ISHH). Scale bar in *A* (also for *B* and *I*) = 200 μm. Arrows mark identical locations within the two sections. (*C, D*) Adjacent sections of brainstem at E16 processed for (*C*) VMAT1 and (*D*) VMAT2 immunohistochemistry (IHC). Scale bar in *C* (also for *D*) = 500 μm. Arrows mark identical locations within the two sections. (*E, F*) Adjacent sections of E12 sympathetic chain, processed

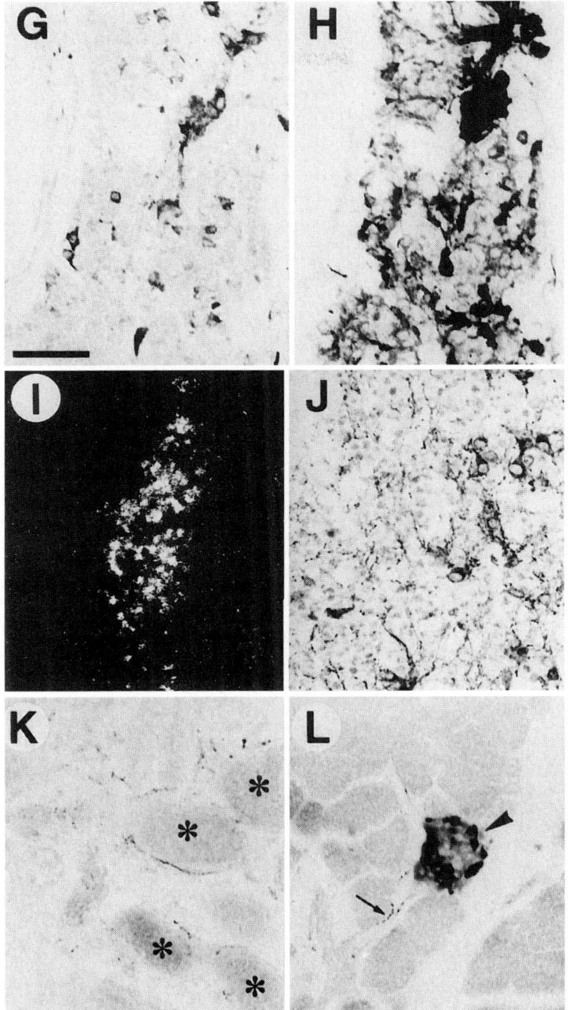

for (*E*) VMAT1 and (*F*) VMAT2 ISHH. Scale bar in *E* (also for *F*) = 100 μm. (*G, H*) Adjacent sections of E16 stellate ganglion processed for (*G*) VMAT1 and (*H*) VMAT2 IHC. Scale bar in *G* (also for *H, J, K,* and *L*) = 50 μm. (*I*) VAChT ISHH in E19 stellate ganglion. (*J*) VAChT IHC in E19 stellate ganglion. (*K*) VAChT IHC in postnatal day 4 (P4) sweat glands. Asterisks mark sweat gland coils. (*L*) Adult pancreas, VMAT2 IHC. Arrowhead marks endocrine pancreatic islet; arrow marks blood vessel innervated with VMAT2-positive fibers. Methods: *In situ* hybridization and immunohistochemistry was carried out in embryonic and neonatal rat tissues using riboprobes for rat VMAT1, VMAT2, and VAChT, and rabbit antisera raised against C-terminal peptides of the corresponding rat proteins, as previously described. (Parts *J* and *K* reproduced from Ref. 4, Copyright 1996 National Academy of Sciences, U.S.A.)

Neuroepithelial cells expressing exclusively VMAT2 appear around embryonic day 12 (E12), and are the presumptive neuroblasts giving rise to the monoaminergic cell groups of the central nervous system (CNS) (Figs. 1A, B). These cell groups, including serotonergic neurons of the raphe nucleus, dopaminergic neurons of the substantia nigra, noradrenergic neurons of the locus ceruleus, and histaminergic neurons of the hypothalamus, all express VMAT2 in cell bodies and fibers from around day E14 (illustrated for brainstem monoaminergic cell groups in Figs. 1C, D). VMAT1 expression at both protein and mRNA levels is absent from CNS neurons at all stages of development examined (e.g., Figs. 1A, C). Like the monoaminergic cell groups of the CNS, a population of intrinsic neurons of the gut and histaminocytes (ECL cells) of the stomach that express VMAT2 in the adult rat appear to express only VMAT2 during development, beginning early postnatally (not shown). Pancreatic islet cells also express VMAT2 in the adult rat (Fig. 1L) and during development (not shown). Whether the endocrine cells of the pancreas are functionally monoaminergic, either during development or in the adult, is not known, although transient expression of tyrosine hydroxylase in α, β, and δ cells of the endocrine pancreas has been reported (8). Neither VMAT1 nor VMAT2 mRNA or protein could be detected in exocrine pancreas in contrast to a recent report of VMAT1 and VMAT2 mRNA expression in exocrine pancreas in the adult rat (9). Enterochromaffin (EC) cells, which express only VMAT1 in the adult rat, also express only VMAT1 during development. Thus, monoaminergic neuronal and endocrine cells of the central nervous and gastrointestinopancreatic systems appear to develop via an initial commitment to exclusive expression of one or the other VMAT isoform.

In contrast, VMAT1 and VMAT2 are coexpressed during development in adrenomedullary and carotid body neuroendocrine cells and in principal ganglion cells of the sympathetic chain (Figs. 1E–H). In these sympathoadrenal cell groups, VMAT1 expression becomes progressively higher and VMAT2 expression lower in neuroendocrine cells postnatally, while VMAT2 expression remains high, with VMAT1 expression progressively diminishing in developing neurons. Thus, sympathoadrenal monoaminergic cells that ultimately differentiate into VMAT1-expressing neuroendocrine and VMAT2-expressing neuronal cells proceed through a developmental phase that includes expression of both transporters.

A subpopulation of postganglionic neurons of the paravertebral sympathetic chain, responsible for innervation of skeletal muscle vasculature and sweat glands, is cholinergic. It has been postulated, both from experimental observations *in vivo* and from using superior cervical ganglion cells in culture as a model system, that these sympathetic cholinergic neurons become so postnatally by switching from a noradrenergic to a cholinergic phenotype after establishing synapses on the sweat gland epithelia (10, 11). However, the stellate ganglion, which supplies cholinergic innervation to sweat glands and other peripheral targets, contains a population of VAChT-positive neurons by embryonic day 19 (Figs. 1I, J), and sweat gland innervation is already VAChT-positive as early as postnatal day 4 when sweat gland coils are first forming in the footpads of the forepaw (Fig. 1K). These VAChT-positive terminals are also VMAT2- and VMAT1-negative and, thus, unlikely to be functionally noradrenergic during early development of the sweat gland. Cholinergic sympathetic differentiation within the sympathetic chain could conceivably occur at the same time as sympathetic noradrenergic differentiation, as reported in other vertebrate species (12), although this clearly remains to be determined. Based on these observations, factors controlling cholinergic sympathetic differentiation might productively be sought in the vicinity of the developing sympathetic chain or the migrating neural crest, rather than in the developing sweat gland. Additional factors may be present in and secreted from the sweat gland that mediate attenuation of noradrenergic traits, including tyrosine hydroxylase expression, in postganglionic sympathetic cholinergic neurons.

In summary, progressive commitment of VMAT1/VMAT2 coexpressing cells to predominantly VMAT1 (endocrine) or VMAT2 (neuronal) expression is a feature of sympathoadrenal differentiation. Other monoaminergic cells commit directly to a VMAT1 (EC cells) or VMAT2 (ECL cells, CNS neurons) phenotype without an intermediate developmental stage of VMAT1/VMAT2 coexpression. Postganglionic cholinergic neurons of the sympathetic nervous system are functionally cholinergic (i.e., express both VAChT and choline acetyltransferase) early in sympathoadrenal development. It remains to be determined if the cholinergic phenotype is established concommitantly with noradrenergic traits whose expression is later attenuated in these neurons.

References

1. Weihe, E., Schäfer, M. K-H., Erickson, J. D., and Eiden, L. E. (1994). Localization of vesicular monoamine transporter isoforms (VMAT1 and VMAT2) to endocrine cells and neurons in rat. *J. Mol. Neurosci.* **5**, 149–164.
2. Peter, D., Liu, Y., Sternini, C., de Giorgio, R., Brecha, N., and Edwards, R. H. (1995). Differential expression of two vesicular monoamine transporters. *J. Neurosci.* **15**, 6179–6188.
3. Schäfer, M. K-H., Weihe, E., Varoqui, H., Eiden, L. E., and Erickson, J. D. (1994). Distribution of the vesicular acetylcholine transporter (VAChT) in the central and peripheral nervous systems of the rat. *J. Mol. Neurosci.* **5**, 1–18.
4. Weihe, E., Tao-Cheng, J-H., Schäfer, M. K-H., Erickson, J. D., and Eiden, L. E. (1996). Visualization of the vesicular acetylcholine transporter in cholinergic nerve terminals and its targeting to a specific population of small synaptic vesicles. *Proc. Natl. Acad. Sci. U.S.A.* **93**, 3547–3552.
5. Gilmor, M. L., Nash, N. R., Roghani, A., Edwards, R. H., Yi, H., Hersch, S. M., and Levey, A. I. (1996). Expression of the putative vesicular acetylcholine transporter in rat brain and localization in cholinergic synaptic vesicles. *J. Neurosci.* **16**, 2179–2190.
6. Schäfer, M. K-H., Schütz, B., Erickson, J. D., Eiden, L. E., and Weihe, E. (1996). Visualization of vesicular monoamine and acetylcholine transporters in developing neurons and neuroendocrine cells. *Soc. Neurosci. Abstr.* **22**, 29.
7. Landis, S. C. (1988). Neurotransmitter plasticity in sympathetic neurons and its regulation by environmental factors in vitro and in vivo. *In* The Peripheral Nervous System. (A. Björklund, T., Hökfelt, and C. Owman, eds.), pp. 65–115. Elsevier, Amsterdam.
8. Cochard, P., Goldstein, M., and Black, I. B. (1978). Ontogenetic appearance and disappearance of tyrosine hydroxylase and catecholamines in the rat embryo. *Proc. Natl. Acad. Sci. U.S.A.* **75**, 2986–2990.

9. Mezey, E., Eisenhofer, G., Harta, G., Hansson, S., Gould, L., Hunyady, B., and Hoffman, B. J. (1996). A novel nonneuronal catecholaminergic system: Exocrine pancreas synthesizes and releases dopamine. *Proc. Natl. Acad. Sci. U.S.A.* **93**, 10377–10382.
10. Landis, S. C. (1990). Target regulation of neurotransmitter phenotype. *TINS* **13**, 344–350.
11. Habecker, B. A., and Landis, S. C. (1994). Noradrenergic regulation of cholinergic differentiation. *Science* **264**, 1602–1604.
12. Reissmann, E., Ernsberger, U., Francis-West, P. H., Rueger, D., Brickell, P. M., and Rohrer, H. (1996). Involvement of bone morphogenetic protein-4 and bone morphogenetic protein-7 in the differentiation of the adrenergic phenotype in developing sympathetic neurons. *Development* **122**, 2079–2088.

Arnon Rosenthal

Department of Neuroscience
Genentech, Inc.
South San Francisco, California 94080

Specification and Survival of Dopaminergic Neurons in the Mammalian Midbrain

Midbrian dopaminergic (DA) neurons, constitute a small group of cells that innervate the striatum, limbic system, and neocortex. These cells play a seminal role in postural reflex, reward-associated behavior, and learning, and their loss or abnormal function has been linked to Parkinson's disease, schizophrenia, and drug addiction. Despite the physiological and clinical importance of DA neurons, the mechanisms that control their survival and genesis are not well understood. We have shown that several members of the transforming growth factors (TGF) TGFβ protein family including, in addition to glial-cell line–derived neurotrophic factor (GDNF), TGFβ2 and -β3 can prevent the death of cultured rat embryonic midbrain DA neurons at picomolar concentrations. Furthermore, we found that TGFβ2 and -β3 and GDNF are expressed sequentially as local and target-derived trophic factors and that subpopulations of DA neurons projecting to the striatum would have access to GDNF whereas DA neurons projecting to the cortical and limbic system will have access mainly to the TGFs. These findings lead us to propose that multiple members of the TGF protein family may serve as physiological survival factors for distinct subpopulations of DA neurons (1). Surprisingly, given the potent survival activity of TGFβ2 and -β3 on DA neurons *in vitro,* we found that the only member of this protein family that appears to be efficacious in preventing the death of DA neurons in lesioned animal models *in vivo* was GDNF (2). We, therefore,

Advances in Pharmacology, Volume 42

1054-3589/98 $25.00

undertook to explore the mechanism of action of this protein. We found that physiological responses to GDNF require the presence of a novel glycosylphosphatidyl inositol–linked protein (designated GDNFR) that is expressed on GDNF-responsive cells and binds GDNF with a high affinity. We further demonstrated that GDNF promotes the formation of a physical complex between GDNFR and the orphan tyrosine kinase receptor Ret, thereby inducing its tyrosine phosphorylation. These findings support the hypothesis that GDNF utilizes a multisubunit receptor system in which GDNFR and Ret function as the ligand-binding and signaling components, respectively (3).

Because GDNF appears to be the most potent and efficacious survival factor for DA neurons known so far, we have explored whether it is indeed the long sought after physiological survival factor for DA neurons during normal development. For this aim, we have generated mice that are deficient in GDNF. Surprisingly, we found that at postnatal day 1, no deficiency in the number of DA neurons in the midbrain or in the density of DA nerve fibers was detected in the GDNF-deficient mice. Furthermore, we have previously demonstrated that GDNF is 75-fold more potent than any other neurotrophic factor in supporting the survival of purified embyronic rat motoneurons in culture and that GDNF mRNA was found in the immediate vicinity of motoneurons during the period of cell death in development. *In vivo,* GDNF was found to rescue and prevent the atrophy of facial motoneurons that have been deprived of target-derived survival factors by axotomy (4). Despite this, no significant deficit in the number or morphology of motoneurons was found in the GDNF-deficient mice. In addition, we have found that GDNF is a potent survival factor for distinct populations of sensory and sutonomic nervous system neurons (5). Consistent with this finding, GDNF-deficient mice displayed partial deficits in superior cervical sympathetic neurons as well as in subpopulations of dorsal root and nodose sensory ganglia neurons (6).

Most surprisingly, we find that GDNF-deficient mice completely lack ureters and kidneys as early as embyronic day 10.5. The metanephric kidney develops by reciprocal inductive interactions between the ureteric bud, an evagination of the mesonephric-wolffian duct, which gives rise to the collecting ducts–ureter and the metanephric blastema, a caudal intermediate mesodermal precursor of the renal parenchyma, which gives rise to the renal tubules of the metanephric (definitive) kidney. The pattern of expression of GDNF in the metanephric blastema, combined with the absence of a morphologically defined ureteric bud, raises the possibility that GDNF may participate in the induction of the ureteric bud by the mesonephric duct. Alternatively, GDNF may control the proliferation or survival of ureteric bud cells after they have been specified. In the absence of a ureteric bud, the mesenchyme will not develop into renal tubules, and no kidney will be formed. In addition, these mice fail to develop an enteric nervous system, suggesting that GDNF may be essential for the commitment, migration, or differentiation of ENS neuron precursors (6).

The discrepancy between the efficacy of GDNF as a survival factor for multiple neuronal populations in culture and the mild neuronal deficient observed in GDNF-deficient mice suggested that *in vivo* GDNF-like proteins may compensate for its absence. Consistent with this possibility is the recent isolation by the groups of Millbrant and Johnson of a novel protein designated Neurturin, which is structurally related to GDNF. Using survival assays in culture, we have demonstrated that Neurturin-like GDNF is a potent survival factor for DA neurons. Moreover, we have identified a multi-component receptor complex for Neurturin composed of a novel, GPI-linked protein designated Neurturin Rα and the tyrosine kinase receptor Ret.

In addition to the studies on survival factors for DA neurons, we have been exploring the mechanisms by which these cells acquire their particular cell fate during normal development. We show that DA neurons develop in the midbrain in vicinity to the floor plate, a specialized group of cells that transiently occupy the ventral nerual tube through most of its extant. We further demonstrated, using explant culture *in vitro,* that the floor plate–produced signals induce the formation of DA neurons in ectopic locations in midbrain explants. In addition, we showed that transgenic mice that harbor a supernumerary floor plate in the dorsal midbrain region develop a duplicated cluster of DA adjacent to the ectopic floor plate (7). Subsequent to these studies, we were able to identify the molecular nature of the floor plate–derived inducer and provided evidence that it is sonic hedgehog (SHH), a secreted protein produced by the floor plate and notochord. Addition of recombinant SHH to midbrain explants led to the formation of DA neurons as well as to the differentiation of floor plate cells, motoneurons, and HNF3β-positive cells. The types of cell that developed in these midbrain explant appeared to be dependent on the concentration of SHH. The concentration of SHH required to induce these different cell types is lowest for motoneurons, about fivefold higher for HNF3β, and higher yet for DA neurons and for FP3/4+ cells. These findings, combined with discoveries from the laboratories of Drs. Jessell and MacMahon, suggest that SHH may function as a morphogene to specify distinct cell types along the dorsoventral axis of the neural tube in a concentration-dependent manner (7).

Because SHH appeared to be an important inducer of DA neurons, we undertook the elucidation of its mechanism of action. We provided evidence that SHH mediates its function through a multi-component receptor system composed of 12 transmembrane ligand binding protein patched and a 7 transmembrane signalling component designated smoothened. In sum, the genesis and survival of DA neurons are complex processes regulated by mutliple growth factors that utilize diverse signaling systems.

References

1. Poulsen, K. T., Armanini, M. P., Klein, R. D., Hynes, M. A., Phillips, H. S., and Rosenthal, A. (1994). TGFβ2 and TGFβ3 are potent survival factors for midbrain dopaminergic neurons. *Neuron* **13,** 1245–1252.
2. Beck, K. D., Valverde, J., Alexi, T., Poulsen, K., Moffat, B., Vandlen, R. A., Rosenthal, A., and Hefti, F. (1995). Mesencephalic dopaminergic neurons protected by GDNF from axotomy-induced degeneration in the adult brain. *Nature* **373,** 339–341.
3. Treanor, J., Goodman, L., de Sauvage, F., Stone, D., Poulsen, K., Beck, K., Gray, C., Armanini, M., Pollock, R. A., Hefti, F., Phillips, H., Goddard, A., Davies, A., Asai, N., Takahashi, M., Vandlen, R., Henderson, C., and Rosenthal, A. (1996). Characterization of a multicomponent receptor for GDNF *Nature* **382,** 80–83.
4. Henderson, C. E., Phillips, H. S., Pollock, R. A., Davies, A. M., Lemeulle, C., Armanini, M., Simmons, L., Moffet, B., Vandlen, R. A., Koliatsos, V. E., and Rosenthal, A. (1994).

GDNF: A potent survival factor for motoneurons present in peripheral nerve and muscle. *Science* **266**, 1062–1064.
5. Buj-Bello, A., Horton, A., Rosenthal, A., and Davies, A. M. (1995). GDNF is an age-specific survival factor for sensory and autonomic neurons. *Neuron* **15**, 821–828.
6. Moore, M. W., Klein, R. D., Fariñas, I., Sauer, H., Armanini, M., Phillips, H., Reichardt, L. F., Ryan, A. M., Carver-Moore, K., and Rosenthal, A. (1996). Renal and neuronal abnormalities in mice lacking GDNF. *Nature* **382**, 76–79.
7. Hynes, M., Porter, J. A., Chiang, C., Chang, D., Tessier-Lavigne, M., Beachy, P. A., Rosenthal, A. (1995). Induction of midbrain dopaminergic neurons by sonic hedgehog. *Neuron* **15**, 35–44.

Don M. Gash,* Greg A. Gerhardt,† and Barry J. Hoffer†

*Department of Anatomy and Neurobiology and Research Magnetic Resonance Imaging and Spectroscopy Center
University of Kentucky College of Medicine
Lexington, Kentucky 40536

†Departments of Pharmacology and Psychiatry
Neuroscience Training Program and Rocky Mountain Center for Sensor Technology
University of Colorado Health Sciences Center
Denver, Colorado 80262

Effects of Glial Cell Line-Derived Neurotrophic Factor on the Nigrostriatal Dopamine System in Rodents and Nonhuman Primates

Glial cell line–derived neurotrophic factor (GDNF) is a glycosylated, disulfide-bonded homodimer distantly related to the transforming growth factor-β superfamily, which might have a therapeutic potential for treating Parkinson's disease (1). While GDNF has been found to play an important role in the survival and development of a variety of cell types throughout the body, our studies have focused on analyzing its effects on the nigrostriatal dopaminergic system in rodents and nonhuman primates.

The *in vivo* effects of GDNF on midbrain dopamine (DA) neurons in normal adult rats and rhesus monkeys have been shown to be dramatic. In normal Fischer 344 rats, a single intranigral injection of 10 μg of GDNF (2), increased nigral DA levels over threefold by 3 wk postinjection (Table I). Striatal DA levels on the injected side were either unchanged (1 wk posttreatment) or significantly lower (3 wk posttreatment), while striatal DA turnover was

Advances in Pharmacology, Volume 42

1054-3589/98 $25.00

TABLE I Effects of GDNF in Normal Animals

Animal	*Dose*	*Striatum*	*Substantia nigra*
Rat	vehicle	102%	84%
	10μg GDNF	74%	367%
Rhesus monkey	vehicle	86%	97%
	150μg GDNF	89%	240%

Intranigral injections of GDNF consistently increased nigral levels of DA in the substantia nigra in the normal animals tested (2, 3). In contrast, striatal DA levels were either unaffected or significantly lower.

significantly elevated. Both spontaneous and amphetamine-induced activity levels were significantly higher in the first week following GDNF treatment but returned to control levels after that. The other behavioral effect observed was a reduction in food consumption for the first 7 days following intranigral GDNF administration.

Using the data from the normal rat study to scale up for the larger nonhuman primate brain, the response to a single intranigral injection of 150 μg of GDNF was evaluated in normal adult female rhesus monkeys whose substantia nigra was 12–15 times larger than that of the rat (3). Over the 3-wk period following GDNF administration, five of the six treated animals experienced some weight loss, and four monkeys displayed a small increase in daytime activity. No other behavioral effects were observed. At 3 wk postadministration, nigral DA levels on the injected side of GDNF recipients were nearly two and one-half times higher than the vehicle recipients (see Table I). Striatal DA levels were not elevated in the trophic factor recipients, but electrochemical measurements revealed a significant increase in DA release in both the caudate nucleus and putamen in response to potassium stimulation. Thus, the response to intranigral GDNF injections was similar in both Fischer 344 rats and rhesus monkeys. Both species showed some weight reductions and increased activity levels. Nigral DA levels were elevated, but striatal DA levels tended to be lower.

GDNF also displays neuroprotective properties when given prior to the administration of dopaminergic toxins. In mice given intrastriatal GDNF 24 hr prior to systemic 1-methyl-4-phenyl-1,2,3,6-tetrahydropyridine (MPTP), striatal and nigral DA levels were partially protected from depletion (4). Intranigral GDNF administration prevented nigral DA depletion, but striatal DA levels were not significantly different than those of animals receiving vehicle intranigral injections followed by a systemic MPTP challenge. When Fischer 344 rats were administered 10 μg of GDNF intranigrally 24 hr prior to either an intranigral or an intrastriatal 6-hydroxydopamine (6-OHDA) lesion, nigral DA neuronal loss was significantly attenuated (5). Adjacent dopamine neurons in the ventral tegmental area were also protected from the neurotoxin.

In addition to pronounced neuroprotective effects, GDNF also exhibits neurorestorative properties. By a week following a single intranigral injection of 100 μg of GDNF, rats with a 95% or higher reduction of tyrosine hydroxylase (TH)–positive nigral DA neurons from a unilateral 6-OHDA lesion demonstrated an 85% or greater decrease in apomorphine-induced rotation behavior (6, 7). Five weeks after the intranigral injections, nigral DA levels in GDNF recipients were threefold higher compared with cytochrome-C treated controls (Table II). TH-positive fiber density was significantly increased in the region of the injection site, suggesting that GDNF can induce neurite growth in adult DA neurons or that TH levels were upregulated in processes that previously contained undetectable levels (7).

Because of its antiparkinsonian properties in rodent model systems, we have proceeded to analyze the behavioral, anatomical, and neurochemical effects of GDNF in rhesus monkeys with MPTP-induced parkinsonian features (8). A unilateral infusion of MPTP through the right carotid artery was used to produce a severe loss of DA neurons and their processes on the right side of the brain. The left side of the brain was less severely affected. GDNF administration was made at least 3 mo post-MPTP treatment at a time when the animals displayed stable, moderate hemiparkinsonian signs. In contrast to the modest behavioral effects seen in normal monkeys, the behavioral effects of GDNF in the parkinsonian monkeys were pronounced. Two weeks after a single injection of GDNF, either into the substantia nigra (150 μg), caudate nucleus (450 μg), or lateral ventricle (100 μg or 450 μg) on the lesioned side, there was significant improvement in parkinsonian features, which lasted for several more weeks (8). Four of the five cardinal features of parkinsonism were improved: bradykinesia, rigidity, posture, and balance. The action tremor exhibited by the parkinsonian animals did not appear to be diminished by trophic factor treatment. In animals given lateral ventricular injections of GDNF, behavioral improvements were maintained by repeated monthly administrations of GDNF. The magnitude of the improvement in parkinsonian features was similar to that seen after treatment with levodopa. However, the effects of a single treatment with GDNF lasted for up to 4 wk (the longest time point examined), while improvement following a single levodopa treatment lasted for only a few hours.

TABLE II Dopamine Levels in Parkinsonian Animals following GDNF Treatment

Animal (model)	*Treatment (site)*	*Striatum (%)*	*Substantia nigra (%)*
Mouse (MPTP)	Vehicle (SN)	26	97
	10 μg GDNF (SN)	23	158
Mouse (MPTP)	Vehicle (STR)	21	92
	10 μg GDNF (STR)	36	178
Rat (6-OHDA)	Vehicle (SN)	0.2	21
	100 μg GDNF (SN)	0.3	70
Rhesus monkey (MPTP)	Vehicle (LV)	0.5	7
	100–450 μg GDNF (LV)	0.4	14

The levels of DA in the lesioned striatum and substantia nigra in comparison to normal controls are shown for three animal models of Parkinson's disease in which GDNF treatment has been evaluated. In these animal models, GDNF has consistently increased nigral levels of DA regardless of the intracerebral site of injection (4, 6, 8). In contrast, striatal DA levels either were unaffected or, in STR injections in the mouse, showed a relatively small increase.
SN, substantia nigra; STR, striatum; LV, lateral ventricle.

Similar to the normal animals, intracerebral administration of GDNF to MPTP-treated monkeys led to significant increases in midbrain DA neuron soma size and in TH-positive fiber area. MPTP-treated monkeys that received a single injection of GDNF or vehicle into the caudate nucleus or substantia nigra were killed 4 wk after trophic factor administration and their midbrains processed for quantitative image analysis using unbiased stereological procedures (8). On the less severely lesioned left side of the brain, there were significant increases in nigral cell size (18%), nigral cell number (50%), and TH-positive fiber area (40%). On the more severely lesioned right side, there were significant increases in nigral cell size (20%) and TH-positive fiber area (189%). Nigral cell number on the severely lesioned side tended to increase by an average of 76%; however, this increase was not statistically significant due to the variance of the data for this parameter. As with the normal monkeys, the bilateral effects observed following unilateral administration of trophic factor suggest that GDNF may be able to diffuse effectively throughout the brain.

The MPTP-treated monkeys that received three lateral ventricular injections of GDNF or vehicle at monthly intervals were killed 3 wk after the last dose and multiple tissue punches were taken. In the GDNF-treated animals (100 or 450 μg per dose, n = 3 in each group), DA levels were increased by approximately 200% on the lesioned side in the substantia nigra, ventral tegmental area, and globus pallidus compared with the lesioned side of three vehicle-treated animals (see Table II). However, while DA levels in these three regions were increased, they were still significantly lower than DA levels in normal animals. In addition, DA levels were not increased in the striatum. In fact, this was the pattern observed in all three animal models evaluated: The effects of GDNF were most pronounced on midbrain DA neurons, while the DA levels in the striatum either were less affected or showed no significant change.

In conclusion, these studies have demonstrated that GDNF is a potent dopaminergic trophic factor for midbrain DA neurons in both normal and parkinsonian animals. GDNF's long-term effects on the nigrostriatal dopaminergic system suggest that it may prove useful in therapeutic strategies for the treatment of Parkinson's disease.

Acknowledgments

This work was supported in part by grants from the U.S. Public Health Service and by contracts from Synergen, Inc. (Boulder, CO) and AMGEN, Inc. (Thousand Oaks, CA).

References

1. Lin, L-F. H., Doherty, D. H., Lile, J. D., Bektesh, S., and Collins, F. (1993). GDNF: A glial cell line-derived neurotrophic factor for midbrain dopaminergic neurons. *Science* **260,** 1130–1132.
2. Hudson, J., Granholm, A-C., Gerhardt, G., Henry, M. A., Hoffman, A., Biddle, P., Leela, N. S., Mackerlova, L., Lile, J. D., Collins, F., and Hoffer, B. J. (1995). Glial cell line-derived

neurotrophic factor augments midbrain dopaminergic circuits in vivo. *Brain Res. Bull.* **36,** 425–432.
3. Gash, D. M., Zhang, Z., Cass, W. A., Ovadia, A., Simmerman, L., Martin, D., Russell, D., Collins, F., Hoffer, B. J., and Gerhardt, G. A. (1995). Morphological and functional effects of intranigrally administered GDNF in normal rhesus monkeys. *J. Comp. Neurol.* **363,** 345–358.
4. Tomac, A., Lindqvist, E., Lin, L-F. H., Ogren, S. O., Young, D., Hoffer, B. J., and Olson, L. (1995). Protection and repair of the nigrostriatal dopaminergic system by GDNF in vivo. *Nature* **373,** 335–339.
5. Kearns, C. M., and Gash, D. M. (1995). GDNF protects nigral dopamine neurons against 6-hydroxydopamine in vivo. *Brain Res.* **672,** 104–111.
6. Hoffer, B. J., Hoffman, A., Bowenkamp, K., Huettl, P., Hudson, J., Martin, D., Lin, L-F. H., and Gerhardt, G. A. (1994). Glial cell line-derived neurotrophic factor reverses toxin-induced injury to midbrain dopaminergic neurons in vivo. *Neurosci. Lett.* **182,** 107–111.
7. Bowenkamp, K. E., Hoffman, A. F., Gerhardt, G. A., Henry, M. A., Biddle, P., Hoffer, B. J., and Granholm, A-C. E. (1995). Glial cell line-derived neurotrophic factor supports survival of injured midbrain dopaminergic neurons. *J. Comp. Neurol.* **355,** 479–489.
8. Gash, D. M., Zhang, Z., Ovadia, A., Cass, W. A., Yi, A., Simmerman, L., Russell, D., Martin, D., Lapchak, P. A., Collins, F., Hoffer. B. J., and Gerhardt, G. A. (1996). Functional recovery in GDNF-treated parkinsonian monkeys. *Nature* **380,** 252–255.

C. Anthony Altar, Michelle Fritsche, and Ronald M. Lindsay

Regeneron Pharmaceuticals, Inc.
Tarrytown, New York 10591

Cell Body Infusions of Brain-Derived Neurotrophic Factor Increase Forebrain Dopamine Release and Serotonin Metabolism Determined with *in Vivo* Microdialysis

Brain-derived neurotrophic factor (BDNF) and neurotrophin-3 (NT-3) each promote the survival and phenotypic expression of a variety of cultured central nervous system neurons, including dopaminergic and serotonergic neu-

Advances in Pharmacology, Volume 42

1054-3589/98 $25.00

rons (1). Chronic infusions of BDNF or NT-3 above the rat pars compacta elevate spontaneous or amphetamine-stimulated locomotor behaviors and increase neostriatal dopamine and serotonin metabolism (2–5). BDNF or NT-3 infusions to the rat dorsal raphe nucleus elevate nociceptive thresholds to heat and increase serotonin turnover in the nucleus accumbens, caudate-putamen, and other forebrain regions (6, 7).

Only indirect support exists for a monoamine release–promoting action of BDNF or NT-3. *In vivo* microdialysis was used here to determine whether chronic nigral infusions of either factor can elevate the release or metabolism of dopamine in the caudate–putamen, where infusions of BDNF produces large and consistent increases in dopamine metabolites, including the dopamine release–specific metabolite, 3-methoxytyramine (3-MT) (2, 3). Infusions of BDNF into the dorsal raphe were evaluated for effects on serotonin release and metabolism in the nucleus accumbens–ventral striatal area, which shows large elevations of 5-HIAA levels and serotonin turnover during raphe infusions of BDNF (7).

Alzet 2002 minipumps were filled with the sterile phosphate-buffered saline (PBS) vehicle (12 μl/day) or with BDNF or NT-3 (AMGEN-Regeneron Partnership) at concentrations of 0.90 to 1.0 mg/ml to deliver about 12 μg/day. Male Sprague-Dawley rats (200–240 g) were housed and treated in compliance with guidelines set forth in the Public Health Service (PHS) manual, *Guide for the Care and Use of Laboratory Animals.* During general anesthesia, a cannula attached to the osmotic pump was mounted on the skull to terminate above the right substantia nigra or right raphe nucleus–midbrain tegmentum. A guide cannula for a 3.0-mm microdialysis probe (BAS/CMA/12; Bioanalytical Systems, Layfayette, IN) was implanted at the same time over the right neostriatum. When inserted, the tip of the dialysis probe terminated in the center of the caudate–putamen in nigra-infused animals or in the nucleus accumbens of dorsal raphe–infused animals.

Dialysis sampling in the striatum (8) was conducted 10–14 days after the start of neurotrophin infusions. Samples were collected every 30 min for 80–120 min before and for 150 min after i.p. injections of 3.3 mg/kg of the dopamine-releasing drug, *d*-amphetamine (nigral infused animals) or 2.5 mg/kg of the serotonin-releasing drug, p-chloramphetamine (PCA; raphe-infused animals). Dialysis probe recoveries of 10–30% (dopamine) and 23–51% (serotonin) were used to adjust extracellular monoamine and amino acid levels determined by high-performance liquid chromatography (9). Statistical analyses were performed using JMP version 3.15 (SAS Institute, Cary, NC). The statistical significance of changes in neurochemical measurements was assessed with a two-way analysis of variance followed by a paired *t*-test or with contrast tests.

I. Dopamine Dialysis in the Caudate–Putamen

Nigral infusions of BDNF produced a 42% increase in the spontaneous release of dopamine above that of PBS-infused animals and smaller (16–22%), nonsignificant elevations of dihydroxyphenylacetic acid (DOPAC) and homovanillic acid (HVA) (Table I). Nigral NT-3 infusions produced a 22%, nonsignificant elevation in dopamine release and 17–18%, nonsignificant increases in DOPAC and HVA compared with animals infused with PBS (see Table I).

TABLE I Baseline Dopamine, Serotonin, and Metabolite Concentrations in Striatal Dialysates of Rats with Midbrain Infusions of Vehicle, BDNF, or NT-3

Supranigral Infusion		
	Vehicle (n = 27)	BDNF (n = 25)
Dopamine (pg/20 μl)	12.1 ± 0.87	17.2 ± 2.0* (42%)
DOPAC (ng/20 μl)	8.2 ± 0.78	9.5 ± 1.1 (16%)
HVA (ng/20 μl)	4.1 ± 0.38	5.0 ± 0.48 (22%)
	Vehicle (n = 24)	NT-3 (n = 27)
Dopamine (pg/20 μl)	9.4 ± 0.80	11.4 ± 1.3 (22%)
DOPAC (ng/20 μl)	9.6 ± 0.57	11.3 ± 0.9 (17%)
HVA (ng/20 μl)	6.5 ± 0.40	7.7 ± 0.58 (18%)
Dorsal Raphe Infusion		
	Vehicle (n = 13)	BDNF (n = 13)
Serotonin (pg/20 μl)	3.9 ± 0.62	4.4 ± 0.92 (13%)
5-HIAA (ng/20 μl)	2.2 ± 0.19	4.3 ± 0.57* (91%)

Male rats received supranigral or dorsal raphe infusions of the PBS vehicle, BDNF, or NT-3 (12 μg/day) for 10–14 days. Sequential dialysis samples from cannulae that terminated in the striatum (supranigral groups) or nucleus accumbens and ventral striatum (dorsal raphe infusion) were analysed for monoamine contents, and these values were averaged for each animal. Group means ± SEM were analyzed by Dunnet's *t*-test; $^*p < .01$ versus vehicle-infused animals. Percentages in parentheses are the elevations above vehicle group values, regardless of statistical significance.

As expected, *d*-amphetamine treatment elevated dopamine release by 20-fold over time (F {4, 16} = 13, $p < .001$). Neither these increases nor the suppressions of DOPAC (F {4, 16} = 14, $p < .001$) or HVA (F {4, 16} = 20, $p < .001$) produced by *d*-amphetamine were altered by BDNF. NT-3 also did not alter the dopamine response {F [1, 27] = 0.94, NS}. During the second hour after amphetamine, DOPAC was 23–45% greater {F [1, 30] = 4.7, $p < .04$} and HVA was 30–42% greater in NT-3–treated animals compared with PBS-treated animals.

BDNF infusions doubled basal tyrosine dialysis levels from 5.9 ± 0.85 ng/20 μl to 11.4 ± ng/20 μl (n = 11–14/group) and did not change the basal levels of alanine, glutamate, glutamine, glycine, histidine, serine, taurine, or threonine. Amphetamine injections decreased by at least half dialysis levels of tyrosine, glutamate, serine, taurine, and threonine and did so to the same extent when vehicle or BDNF was infused into the nigra. Tyrosine was the one exception to this finding, as the suppression in tyrosine produced by *d*-amphetamine was completely prevented by BDNF.

II. Serotonin and 5-HIAA Dialysis in the Nucleus Accumbens

Raphe infusions of BDNF produced a 13% nonsignificant increase in nucleus accumbens serotonin dialysis and a doubling of 5-HIAA (see Table I).

Treatment of the PBS-infused animals with PCA elevated serotonin release by fivefold and decreased 5-HIAA by 50%, compared with baseline levels (Fig. 1). BDNF infusions produced a nonsignificant 30% elevation in the release of serotonin and significantly attenuated the suppression of 5-HIAA, such that steady-state 5-HIAA levels were 80% greater than those of PBS-infused animals (see Fig. 1).

The 42% increase in the basal release of dopamine from nigrostriatal nerve terminals in BDNF-infused animals is consistent with the ability of nigral infusions of BDNF to elevate by 38% the accumulation of striatal 3-MT after monoamine oxidase-B inhibition with pargyline (3), to increase by 32% the spontaneous firing rate of BDNF-treated pars compacta dopamine neurons, and to double the number of spontaneously active pars compacta neurons (10). Thus, nigral infusions of BDNF elevate striatal dopamine release in intact animals. The smaller elevations in striatal dopamine metabolism (3) and lesser induction of rotational behavior after amphetamine in both intact (2) and lesioned animals (11) treated with intranigral NT-3 are consistent with the smaller effects of NT-3 on striatal dopamine release and extracellular metabolites found here with dialysis.

Nigral infusions of BDNF or NT-3 showed only small, nonsignificant increases in striatal dialysis levels of DOPAC and HVA. The greater (eg., 30–50%) and significant increases in striatal *homogenate* levels of these metabolites in BDNF and NT-3 treated rats (2) suggest that much of these increases are due to an intracellular formation and accumulation of metabolites, which would not be detected by *in vivo* dialysis. Because nigral infusions of BDNF but not NT-3 elevate dopamine release, supranigral infusions of BDNF probably increase intracellular dopamine metabolite levels at least in part in response to elevated dopamine release. In the case of NT-3, metabolite elevations appear to result exclusively from increased intracellular dopamine catabolism of nonreleased transmitter.

Because nigral infusions of NT-3 do not augment striatal dopamine release, and because BDNF augments basal but not amphetamine-stimulated dopamine release, the effects of these nigral infusions on spontaneous and amphetamine-stimulated behaviors (2–5, 11) cannot be explained solely by actions on nigrostriatal dopamine neurons. A recruitment of peptidergic, GABAergic (12), or serotonergic neurotransmitters (4, 6, 7) may explain the ability of BDNF and NT-3 to amplify amphetamine's behavioral effects.

The doubling of striatal tyrosine during nigral exposure to BDNF may contribute to the elevated release and turnover of dopamine by pars compacta neurons that have been firing at an accelerated rate (10). Treatment of BDNF-infused animals with tyrosine might reveal whether dopamine synthesis can be augmented to further elevate dopamine content or produce greater increases in dopamine turnover or release. Such a dependence on exogenous tyrosine may be even more dramatic for the hyperfunctioning pars compacta neurons of BDNF-treated animals with partial dopamine lesions.

Midbrain infusions of BDNF had little if any effect on serotonin release, even though they doubled serotonin metabolism before and after amphetamine challenge. These changes are consistent with the failure of midbrain infusions of BDNF to alter most of the measured parameters of dorsal raphe neuron cell

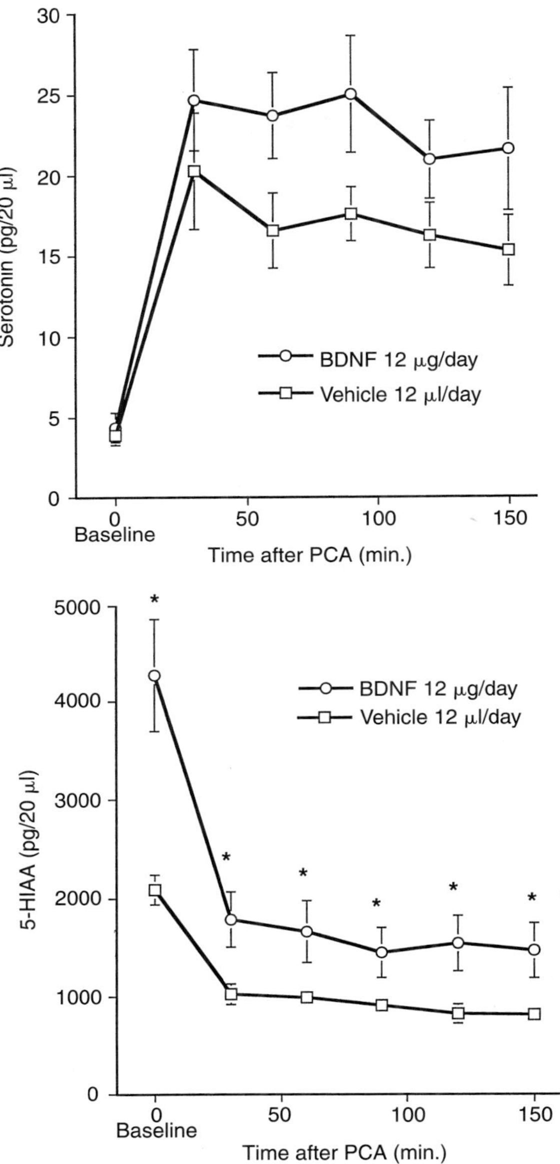

FIGURE 1 Appearance of serotonin and 5-HIAA in the dialysate of rats infused for 10–14 days with the PBS vehicle (12 μl/day; n = 13) or with BDNF (12 μg/day; n = 13) before (basline) or after a 2.5 mg/kg injection of PCA. ANOVA for BDNF versus vehicle treatment interactions: serotonin, F [1, 13] = 3.2, ns; 5-HIAA, F [1, 16] = 5.4, $p < .03$. $^*p < .05$ versus vehicle, paired t-test.

body electrical activity, except to make firing patterns more random (13). BDNF infusions produced a somewhat greater elevation in serotonin release during PCA. In the basal state, these increases may be masked by efficient serotonin reuptake and intracellular degradative mechanisms, which would explain the large increase in 5-HIAA produced by BDNF. The PCA block of serotonin reuptake and heightened serotonin release could, thus, explain the trend toward greater serotonin release associated with BDNF in PCA-treated animals.

Acknowledgments

The expert assistance with the statistical analysis of these data by Nancy Stambler is greatly appreciated, as were the helpful suggestions of Dr. Richard Kellar concerning the serotonin dialysis method.

References

1. Lindsay, R. M., Weigand, S. J., Altar, C. A., and DiStefano, P. S. (1994). Neurotrophic factors: From molecule to man. *Trends Neurosci.* **17,** 182–190.
2. Altar, C. A., Boylan, C. B., Jackson, C., Hershenson, S., Miller, J., Wiegand, S. J., Lindsay, R. M., and Hyman, C. (1992). Brain-derived neurotrophic factor augments rotational behavior and nigrostriatal dopamine turnover in vivo. *Proc. Natl. Acad. Sci. U.S.A.* **89,** 11347–11351.
3. Altar, C. A., Boylan, C. B., Fritsche, M., Jackson, C., Hyman, C., and Lindsay, R. M. (1994). The neurotrophins NT-4/5 and BDNF activate serotonin, dopamine, and GABA neurons following behaviorally effective infusions to the substantia nigra. *Exp. Neurol.* **130,** 31–40.
4. Martin-Iverson, M. T., Todd, K. G., and Altar, C. A. (1994). Brain-derived neurotrophic factor and neurotrophin-3 activate striatal dopamine and serotonin metabolism and related behaviors: Interaction with amphetamine. *J. Neurosci.* **14,** 1262–1270.
5. Martin-Iverson, M. T., and Atlar, C. A. (1996). Spontaneous behaviors are differently affected by substantia nigra infusions of brain-derived neurotrophic factor and neurotrophin-3. *Eur. J. Neurosci.* **8,** 1696–1706.
6. Siuciak, J. A., Altar, C. A., Wiegand, S. M., and Lindsay, R. M. (1994). Antinociceptive effects of brain-derived neurotrophic factor and neurotrophin-3 *Brain Res.* **633,** 326–330.
7. Siuciak, J. S., Boylan, C., Fritsche, M., Altar, C. A., and Lindsay, R. M. (1996). BDNF increases monoaminergic activity in rat brain following intraventricular or intraparenchymal administration. *Brain Res.* **710,** 11–20.
8. Hernandez, L., Lee, F., and Hoebel, G. B. (1987). Simultaneous microdialysis and amphetamine infusion in the nucleus accumbens and striatum of freely moving rats: Increase in extracellular dopamine and serotonin. *Brain Res. Bull.* **19,** 623–628.
9. Gamache, P. H., Ryan, E., and Svendsen, C. N. (1993). Simultaneous measurement of the monoamines, metabolites, and amino acids in brain tissue and microdialysis perfusates. *J. Chromatography* **614,** 221–220.
10. Shen, R-Y., Altar, C. A., and Chiodo, L. A. (1994). Brain-derived neurotrophic factor increases the electrical activity of pars compacta dopamine neurons in vivo. *Proc. Natl. Acad. Sci. U.S.A.* **91,** 8920–8924.
11. Altar, C. A., Boylan, C. B., Fritsche, M. A., Jackson, C., Jones, B. E., Weigand, S. W., Lindsay, R. M., and Hyman, C. (1994). Efficacy of BDNF and NT-3 on neurochemical and

behavioral deficits associated with partial nigrostriatal dopamine lesions. *J. Neurochem.* 63, 1021–1032.
12. Arenas, E., Äkerud, P., Wong, V., Boylan, C. B., Persson, H., Lindsay, R. M., and Altar, C. A. (1996). Effects of BDNF and NT-4/5 infusions in striatonigral neuropeptides and nigral GABA neurons in rat brain. *Eur. J. Neurosci.*
13. Celada, P., Siuciak, J. A., Tran, T. M., Altar, C. A., and Tepper, J. M. (1996). Local infusion of BDNF modifies the firing pattern of dorsal raphe serotonergic neurons. *Brain Res.* 712, 293–298.

Eric S. Levine,* Ira B. Black,* and Mark R. Plummer†

* Department of Neuroscience and Cell Biology
Robert Wood Johnson Medical School
University of Medicine and Dentistry of New Jersey
Piscataway, New Jersey 08854

† Department of Biological Sciences
Rutgers University
Piscataway, New Jersey 08855

Neurotrophin Modulation of Hippocampal Synaptic Transmission

A central question in neuroscience concerns the cellular basis of learning and memory. At a mechanistic level, this is a question of how brain activity itself modifies the communicative connections among neurons, the synapses. Activity-dependent modulation of synaptic transmission is thought to play an important role in these processes. The synaptic apparatus both subserves rapid interneuronal communication and stores long-term information, thereby participating in the mnemonic process. Although there are a large number of transmitters, peptides, and hormones that function as synaptic messengers, little is known about the signals that regulate the strength and efficacy of synaptic transmission.

One approach is to focus on the signals that regulate nervous system development when maximal synaptic and neural plasticity occurs. During development, the functional wiring of the brain is established through processes of axonal and dendritic outgrowth, synapse formation and stabilization, and the loss of neurons that fail to make appropriate contacts. Neuronal growth and trophic factors are important classes of signals involved in these events. In particular, nerve growth factor (NGF) and related members of the neurotrophin

Advances in Pharmacology, Volume 42

1054-3589/98 $25.00

gene family are essential for the selective survival and differentiation of specific neurons within the peripheral and central nervous systems (1). Our recent studies suggest a novel role for these trophic factors in the dynamic modulation of synaptic efficacy, both during development and in maturity. Thus, signaling mechanisms that participate in the formation of neuronal circuits during development appear to operate throughout life to modulate interneuronal communication.

The neurotrophin gene family includes NGF, brain-derived neurotrophic factor (BDNF), neurotrophin-3 (NT-3), and neurotrophin-4/5 (NT-4/5). Three *trk* family members, *trkA, trkB,* and *trkC,* encode essential components of receptors for these neurotrophins (2). NGF acts on trkA, BDNF and NT-4/5 on trkB, and NT-3 primarily on trkC (but also interacts with trkA and trkB). All neurotrophins also bind to p75, a widely distributed receptor of uncertain function. Neurotrophin receptors exist in several isoforms. The full-length, active forms of the trk family are necessary for the normal development and maintenance of specific neuronal populations. It remains to be determined whether the truncated trkB and trkC receptors, which lack tyrosine kinase catalytic domains, play roles in neurotrophin-mediated function.

We examined trophic modulation of synaptic transmission in the hippocampus. Several forms of activity-dependent synaptic plasticity have been demonstrated in this brain structure, including long-term potentiation, considered to be a cellular model of memory mechanisms. The hippocampus is also the site of highest expression in the brain for neurotrophins and their receptors during development and maturity. Thus, trophin modulation of synaptic transmission in the hippocampus may have consequences for plasticity underlying mnemonic processes.

Patch-clamp recordings were made from neuronal cultures obtained from embryonic rat hippocampus grown in serum-free, fully-defined medium. Under these conditions, neurons form extensive connections, display spontaneous synaptic activity, and fire action potentials in response to synaptic inputs. In the first set of studies, impulse activity was recorded using the whole-cell current-clamp configuration. Application of BDNF caused an immediate twofold increase in the firing rate of hippocampal neurons (3). This increase was sustained in the presence of BDNF, and firing rate returned to baseline levels 10–15 min following removal of the factor.

To investigate underlying synaptic mechanisms, the BDNF effect was characterized in whole-cell voltage-clamp experiments. Recordings of excitatory postsynaptic currents (EPSCs) revealed a marked enhancement within 2–3 min of BDNF exposure. These synaptic currents were not spontaneous miniature potentials and were action potential-dependent, because their occurrence was eliminated by the sodium channel blocker tetrodotoxin. Currents were quantified by measuring the integrated current, or synaptic charge. BDNF application elicited a twofold increase in synaptic charge with a time course similar to the increase in action potential activity. In contrast, treatment with heat-inactivated BDNF had no effect on synaptic charge. Additionally, bath application of the specific *trk* receptor tyrosine kinase inhibitor K-252a, which prevents biological responses to neurotrophins, completely blocked the effect of BDNF on synaptic charge (3).

Related members of the neurotrophin gene family often have overlapping effects on responsive populations. To examine the specificity of the effect on synaptic currents, we examined responses to other trophic factors. TrkB-receptor activation appears critical for the synaptic enhancement, because NT-4, another trkB ligand, elicits effects similar to those of BDNF. In contrast, the related neurotrophins, NGF and NT-3, which primarily interact with other trk receptors, did not share this effect. Additionally, we found that the unrelated growth factors, epidermal growth factor and basic fibroblast growth factor, were also without effect (4).

We investigated the role of postsynaptic changes in trophin-induced synaptic plasticity. These experiments took advantage of the ability to inject drugs selectivity into the postsynaptic cell by inclusion of the drug in the internal solution of the patch pipette. Intracellular injection of K-252a significantly decreased the magnitude of the effect compared with BDNF alone. Conversely, intracellular injection of the phosphatase inhibitor, okadaic acid, which prolongs phosphorylation events, enhanced the magnitude of the effect (3). Baseline synaptic charge was not affected by inclusion of either K-252a or okadaic acid in the internal solution. These results indicate that phosphorylation-dependent changes in postsynaptic responsiveness are required for the trophin effect on synaptic currents.

Exposure to BDNF, in fact, produced increases in both the frequency and amplitude of synaptic currents. Because the EPSCs were action potential–dependent, the frequency of synaptic currents was increased as a direct consequence of increased impulse activity of the presynaptic inputs to the recorded cell. Intracellular injection of K-252a or okadaic acid directly into the postsynaptic cell, therefore, had no effect on BDNF-induced increase in frequency. Conversely, the increase in synaptic current amplitude was enhanced by intracellular injection of the phosphatase inhibitor okadaic acid and completely abolished by injection of the tyrosine kinase inhibitor K252a. These results indicate that the effect of BDNF on EPSC amplitude, in particular, is a direct result of postsynaptic modulation.

In an additional set of studies, Western blot analyses revealed that full-length, functional trkB receptors are an intrinsic component of the postsynaptic density (PSD), a specialization of the postsynaptic membrane, in both the developing and adult hippocampus (5). The truncated, nonsignaling form of the trkB receptor, although present in the synaptic membrane fraction, was undetectable in the PSD. Several characteristics suggest that the PSD plays a pivotal role in synaptic function and synaptic plasticity. Neurotransmitter receptors, protein kinases, and ion channel proteins are anchored to the PSD, suggesting that the PSD participates in signal transduction and, potentially, receptor regulation. In fact, recent studies revealed that transmitter receptors, including the NMDA and AMPA glutamate receptor subtypes, can be phosphorylated by intrinsic PSD kinases, resulting in enhanced synaptic transmission. Many of the same protein kinases are known to be involved in trkB signal transduction pathways. Thus, neurotrophins may modulate synaptic transmission via phosphorylation of these and other receptors in the PSD.

In summary, the neurotrophins, BDNF and NT-4, ligands at the trkB receptor, rapidly increase both the frequency and amplitude of excitatory synaptic currents in hippocampal neurons. The potentiation of synaptic current amplitude, in particular, results from a phosphorylation-dependent change in postsynaptic responsiveness. Neurotrophins have also been shown to enhance synaptic transmission by increasing presynaptic transmitter release. These results suggest that trophic factors may play a central role in the dynamic modulation of synaptic efficacy. Acute modulation by neurotrophins may constitute an intermediary stage, preceding more stable changes (e.g., in synaptic morphology or ion channel expression) that depend on alterations in gene expression and/or protein synthesis. Our findings represent a first step in understanding the role of neurotrophin signaling in brain synaptic communication.

References

1. Thoenen, H., Bandtlow, C., and Heumann, R. (1987). The physiological function of nerve growth factor in the central nervous system: Comparison with the periphery. *Rev. Physiol. Biochem. Pharmacol.* **109,** 145–178.
2. Meakin, S. O., and Shooter, E. M. (1992). The nerve growth factor family of receptors. *Trends Neurosci.* **15,** 323–331.
3. Levine, E. S., Dreyfus, C. F., Black, I. B., and Plummer, M. R. (1995). Brain-derived neurotrophic factor rapidly enhances synaptic transmission in hippocampal neurons via postsynaptic tyrosine kinase receptors. *Proc. Natl. Acad. Sci. U.S.A.* **92,** 8074–8077.
4. Levine, E. S., Dreyfus, C. F., Black, I. B., and Plummer, M. R. (1996). Selective role for trkB neurotrophin receptors in rapid modulation of hippocampal synaptic transmission. *Mol. Brain. Res.* **38,** 300–303.
5. Wu, K., Xu, J. L., Suen, P. C., Levine, E. S., Huang, Y. Y., Mount, H. T. J., Lin, S. Y., and Black, I. B. (1996). Functional trkB neurotrophin receptors are intrinsic components of the adult brain postsynaptic density. *Mol. Brain Res.* **43,** 286–290.

F. B. Axelrod,* D. S. Goldstein,† C. Holmes,† and I. J. Kopin†

* Dysautonomia Treatment and Evaluation Center
New York University Medical Center
New York, New York 10016

† Clinical Neuroscience Branch
National Institute of Neurological Disorders and Stroke/NIH
Bethesda, Maryland 20892

Genotype and Phenotype in Familial Dysautonomia

Familial dysautonomia (FD) is a genetic disorder that is classified within the broader category of hereditary sensory and autonomic neuropathy (HSAN) (1). Each of the HSANs is phenotypically distinct based on clinical assessments and neuropathologic findings. It is now appreciated that they are also genetically distinct (Table I). The gene for FD, an autosomal recessive disorder affecting Ashkenazi Jewish individuals, is located at 9q31 (2). However, an understanding of the genotype and the expressed phenotype in this rare disorder still eludes us.

Clinical features of FD encompass sensory and autonomic disturbances. Consistent peripheral neuropathologic findings suggest arrested development of the unmyelinated neuronal population, as well as progressive neurological deterioration (1). Disturbances in levels of catecholamine metabolites were thought secondary to decreased neuronal number. Neurochemical findings suggested a problem in norepinephrine (NE) synthesis, release, or degradation. They included abnormal ratio of urinary catecholamine metabolites (twice normal levels of homovanillic acid [HVA] and normal to low levels of vanillylmandelic acid [VMA], resulting in elevated HVA/VMA ratios) (3), exaggerated response to both NE and mecholyl infusions, and lack of an appropriate rise in NE and dopamine β-hydroxylase (DBH) when the FD patient went from the supine to the erect position (4).

We investigated plasma levels of catechols, as well as their metabolites, to define better the neurochemical phenotype and elucidate possible pathophysiological mechanisms. Ten FD patients and eight control subjects were studied. Venous blood samples were obtained with the subjects supine and after 5 min upright. Plasma levels of catechols—dihydroxyphenylalanine (DOPA), NE, epinephrine (Epi), dopamine (DA), dihydroxyphenylglycol (DHPG), and dihydroxyphenylacetic acid (DOPAC)—were assayed. Catechol concentrations were assayed using liquid chromatography with electrochemical detection after batch alumina extraction, as previously described. (5)

Consistent with previous studies (4), FD subjects had significantly higher mean plasma levels of NE, mean blood pressures, and higher supine heart rates than control subjects. Individual supine mean blood pressure in FD subjects correlated positively with plasma NE levels (r^2 = .494, $p < .0005$). Individual

Advances in Pharmacology, Volume 42

1054-3589/98 $25.00

TABLE I Hereditary Sensory and Autonomic Neuropathies (HSANs)

Common name	*HSAN type*	*Transmission*	*Chromosomal location/gene*
Hereditary sensory radicular neuropathy	I	Autosomal dominant	9q22.1–22.3; gene?
Congenital insensitivity to pain	II	Autosomal recessive?	?
Familial dysautonomia	III	Autosomal recessive	9q31; gene?
Congenital insensitivity to pain with anhydrosis	IV	Autosomal recessive	1q21–22; TRKA/NGF
Congential insensitivity to pain with partial anhydrosis	V	Autosomal recessive?	?
Congenital autonomic dysfunction with universal pain loss		Autosomal recessive?	?
Progressive pandysautonomia		Autosomal recessive?	?

supine heart rates did not correlate with supine plasma catechol levels in FD or control subjects.

In addition, we noted that supine FD subjects had significantly higher mean plasma levels of NE and DOPA and lower mean levels of DHPG. Low DHPG levels and elevated DOPA levels in FD subjects resulted in markedly elevated DOPA/DHPG ratios (Fig. 1). All FD DOPA/DHPG ratios were above the

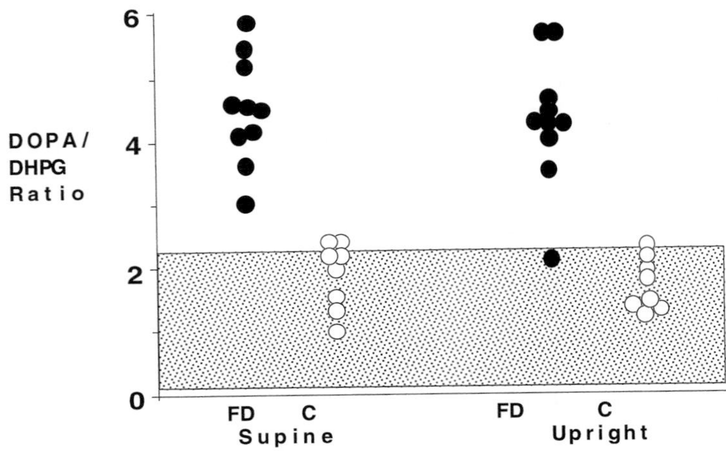

FIGURE 1 DOPA/DHPG ratios for nine FD and eight control subjects. One FD subject was not included because DHPG values were not available. FD values (*filled circle*) are averages from two to three testing sessions. Control values (*open circle*) are absolute values. Grey area indicates reported normal range of this ratio in plasma (0.13–2.28) (Reprinted with permission from *Clinical Autonomic Research,* Vol. 6, pp. 205–209, 1996.)

previously published normal range of 0.13–2.28 (6). The groups did not differ significantly in supine plasma levels of Epi, DA, or DOPAC. The range of values for DOPAC was very broad, with one FD subject having extraordinarily high values on all testing sessions. This was the oldest FD subject and the only one who was not independent in his ambulation. In general, there were parallel decreases in DOPAC and DHPG ratios in FD patients, and DOPAC/DHPG ratios remained normal.

With tilt, FD subjects consistently demonstrated orthostatic hypotension with a range of mean blood pressure decrease from 24 to 109 mm Hg within 5 min. In contrast, control subjects either maintained or increased their mean blood pressures. At 5 min upright, the FD group continued to have significantly higher plasma levels of DOPA and lower levels of DHPG than normal control subjects, as well as consistently elevated DOPA/DHPG ratios (see Fig. 1). FD and control groups did not differ significantly in plasma levels of NE, Epi, DA, or DOPAC. When erect, individual FD mean blood pressure no longer correlated with plasma NE levels, but highly significant correlations were noted between mean blood pressure and plasma DA levels ($r^2 = .63$, $p = .0001$) and between mean blood pressure and plasma DOPAC levels ($r^2 = .43$, $p < .001$).

Our data demonstrate that FD patients have a distinctive pattern of plasma levels of catechols. Regardless of posture, FD patients have disproportionately high plasma levels of DOPA and low plasma levels of DHPG, with strikingly elevated plasma DOPA/DHPG ratios. This pattern of plasma catechols is distinctly different than that found in other disorders associated with neurogenic orthostatic hypotension (7). Patients with acquired dysautonomia due to multiple system atrophy have normal supine NE and normal DOPA/DHPG ratios. Patients with DBH deficiency had barely detectable levels of NE when supine, no detectable DHPG, markedly elevated DOPA and DOPAC levels, and extremely high DOPA/DHPG and DOPAC/DHPG ratios.

The normal supine NE levels but low DHPG levels in FD are intriguing. Because plasma DHPG levels reflect oxidative deamination of NE in the axoplasm of sympathetic nerve terminals, low levels of DHPG could be a consequence of either decreased availability of axoplasmic NE or decreased sequential activity of monoamine oxidase (MAO) and aldehyde reductase on NE. Decreased MAO activity should also result in decreased formation of DOPAC, the MAO metabolite of DA. Plasma levels of DOPAC were not consistently decreased in FD patients; about half of FD patients had decreased DOPAC levels. If the singular FD patient with an extremely high DOPAC level is omitted, then the mean plasma levels of DOPAC for the FD patient group are significantly lower than normal.

The normal supine plasma NE levels in FD patients are also consistent with decreased neuronal re-uptake of released NE due to sparse sympathetic innervation and perhaps ongoing constitutive neurosecretion. Moreover, if sympathoneural traffic in the surviving neurons were already functioning at maximum levels in the supine position, then plasma NE levels would fail to increase during standing.

A decrease in available NE could induce a compensatory increase in tyrosine hydroxylation and enhanced DOPA production. This could explain the high plasma DOPA levels in patients with FD and would be consistent with Pearson's description of large amounts of tyrosine hydroxylase in the superior cervical ganglia by monoclonal antibody stains (8).

In summary, FD patients have a distinctive pattern of plasma catechols, which could distinguish FD from other forms of neurogenic orthostatic hypotension and may guide genetic investigators toward factors that would influence neurodevelopment, as well as differentiation of catecholaminergic systems.

References

1. Axelrod, F. B. (1996). Autonomic and sensory disorders. *In* Principles and Practice of Medical Genetics, 3rd ed. (A. E. H. Emory and D. L. Rimoin, eds.), pp. 397–411. Churchill Livingstone, Edinburgh.
2. Blumenfeld, A., Slaugenhaupt, S. A., Axelrod, F. B., Lucente, D. E., Maayan, Ch., Lieberg, C. B., Ozelius, L. J., Trofatter, J. A., Haines, J. L., Breakefield, X. O., and Gusella, J. F. (1993). Localization of the gene for familial dysautonomia on chromosome 9 and definition of DNA markers for genetic diagnosis. *Nat. Genet.* **4,** 160–164.
3. Smith, A. A., Taylor, T., and Wortis, S. B. (1963). Abnormal catecholamine metabolism in familial dysautonomia. *N. Engl. J. Med.* **268,** 705–707.
4. Ziegler, M. G., Lake, R., and Kopin, I. J. (1976). Deficient sympathetic nervous response in familial dysautonomia. *N. Engl. J. Med.* **294,** 630–633.
5. Eisenhofer, G., Goldstein, D. S., Stull, R., Keiser, H. R., Sunderland, T., Murphy, D. L., and Kopin, I. J. (1986). Simultaneous liquid chromatographic determination of 3,4-dihydroxyphenylglycol, catecholamines, and 3,4-dihydroxyphenylalanine in plsama and their responses to inhibition of monoamine oxidase. *Clin. Chem.* **32,** 2030–2033.
6. Kaler, S. G., Goldstein, D. S., Holmes, C., Salerno, J. A., and Gahl, W. A. (1993). Plasma and cerebrospinal fluid neurochemical pattern in Menkes disease. *Ann. Neurol.* **33,** 171–175.
7. Goldstein, D. S., Polinksy, R. J., Garty, M., Robertson, D., Brown, R. T., Biaggioni, I., Stull, R., and Kopin, I. J. (1989). Patterns of plasma levels of catecholamines in neurogenic orthostatic hypotension. *Ann. Neurol.* **26,** 558–563.
8. Pearson, J., Goldstein, M., and Brandeis, L. (1979). Tyrosine hydroxylase immunohistochemistry in human brain. *Brain Res.* **165,** 333–337.

A. D. Zurn, J. L. Tseng, N. Déglon, J. M. Joseph, and P. Aebischer

Gene Therapy Center and Surgical Research Division
Lausanne University Medical School
CH-1011 Lausanne, Switzerland

A Gene Therapy Approach for the Treatment of Amyotrophic Lateral Sclerosis and Parkinson's Disease

Many human neurodegenerative diseases are the result of dysfunction and death of specific populations of neurons in the central and peripheral nervous systems. In amyotrophic lateral sclerosis (ALS), subsets of neurons degenerate in the spinal cord, brain stem, and motor cortex. This results in progressive muscle weakness, paralysis, and death within 3–5 years. Most of the ALS cases are sporadic and have an unknown etiology. Parkinson's disease (PD) is characterized by major clinical disturbances such as resting tremor, bradykinesia, and rigidity. These clinical symptoms are caused by dopamine depletion in the striatum resulting from neuronal loss in the substantia nigra. Both genetic factors and environmental toxins have been implicated in the etiology of PD, but no cause for the selective degeneration of dopaminergic neurons in the substantia nigra has yet been identified. Although symptomatic treatment of PD with L-dopa to substitute for the missing neurotransmitter dopamine efficiently restores neurological function early in the disease, the efficacy of the drug decreases after 5–6 years and side effects appear. The goal would, therefore, be to find a neuroprotective therapy for PD (i.e., find factors that protect the dopaminergic neurons from the degenerative process).

Neurotrophic factors are naturally occurring proteins that are essential for neuronal survival, neurite outgrowth, and neurotransmitter expression during development, as well as for the maintenance of normal neuronal function in the adult. These factors, however, can also prevent neuronal degeneration after injury-induced cell death. For instance, ciliary neurotrophic factor (CNTF) and glial cell line-derived neurotrophic factor (GDNF) have been shown to rescue motoneurons from axotomy-induced cell death and to slow down the degeneration of motoneurons in a mouse mutant with progressive motor neuronopathy (animal models of ALS) (1–4). In addition, GDNF has been demonstrated to protect dopaminergic neurons from cell death in several animal models of PD (5–7). Neurotrophic factors may, therefore, be of therapeutic importance for the treatment of neurodegenerative diseases such as ALS and PD.

Two of the major difficulties encountered in the use of neurotrophic factors for the treatment of neurodegenerative diseases are (1) the large amounts needed

Advances in Pharmacology, Volume 42

1054-3589/98 $25.00

to have a biological effect in animal models and (2) the instability of some of these factors. Therefore, the transplantation of polymer-encapsulated cells genetically engineered to continuously release neurotrophic factors at a relatively low dose is an extremely useful tool. Polymer capsules consist of a permselective membrane that sequesters the transplanted cells but allows diffusion of small molecules, such as nutrients and trophic factors, into and out of the polymer envelope (8). Furthermore, this surrounding membrane prevents rejection of the transplanted cells by the host immune system and therefore permits the use of cell lines from a xenogeneic origin. This encapsulation technology has been used to deliver neurotrophic factors in neonatal rats following facial nerve axotomy and in mutant mice with progressive motor neuronopathy (4).

Cell lines genetically engineered to release the neurotrophic factors CNTF or GDNF have been constructed. Baby hamster kidney (BHK) cells were transfected with the mutant dihydrofolate reductase–based expression vector containing the CNTF gene or the GDNF cDNA using standard calcium phosphate transfection procedures, and were selected for CNTF and GDNF expression using increasing concentrations of methotrexate (2). BHK-CNTF and BHK-GDNF cells release 5 μg and 20 ng of neurotrophic factor/10^6 cells/day, respectively, as measured by ELISA assay. Devices consisting of hollow-fiber membranes with an inner diameter of 500 μm have been sterilized and aseptically filled with genetically engineered cells.

The capacity of GDNF to slow down the degeneration of dopaminergic neurons in a rat model of PD (medial forebrain bundle axotomy) was investigated (9). Encapsulated BHK cells genetically engineered to release GDNF were implanted unilaterally close to the substantia nigra. One week later, the medial forebrain bundle was axotomized ipsilateral to the capsule. Seven days later, the animals were tested for amphetamine-induced rotation and killed GDNF reduced the number of turns from 6.2 $\pm$ 2.3 turns/min to 1.2 $\pm$ 1.0 turns/min. In addition, BHK-GDNF–implanted animals had 64.9 $\pm$ 7.2 tyrosine hydroxylase–positive cells in the substantia nigra on the lesioned compared with the nonlesioned side, while control animals had 27.2 $\pm$ 3.5% tyrosine-positive cells on the lesioned side (9). However, GDNF did not prevent the loss of dopamine in the striatum as analyzed by high-performance liquid chromatography. These results indicate that the continuous release of low levels of GDNF close to the substantia nigra is capable of protecting nigral dopaminergic neurons from axotomy-induced cell death and improving amphetamine-induced rotation by a mechanism other than dopaminergic striatal reinnervation.

A phase I clinical trial using intrathecal delivery of human CNTF has been performed in six patients suffering from ALS (10). This trial was designed as an open-label safety study. BHK cells transfected with the CNTF gene were loaded in 5-cm-long polyethersulfone fibers and surgically placed within the lumbar intrathecal space. The patients were first implanted with a device releasing 0.5–1.0 μg of CNTF per day. According to the protocol, the implant was retrieved after 3–4 mo. Nanogram quantities of CNTF were detected in the cerebrospinal fluid (CSF) of all implanted patients: 170–810 pg CNTF/ml CSF, 12 wk after implantation (patients 1–3); 666–2165 pg/ml, 10 wk after implantation (patients 4–6). No CNTF was detected in the CSF of patients prior to capsule implantation. Viable BHK cells were observed in all retrieved implants, which released from 0.125–0.833 μg of CNTF per day. Patients were evaluated on a monthly basis using the following tests: pulmonary capacity (forced vital capacity), muscle strength (Tuft's Quantitative Neuromuscular Examination), and neurological status (Norris score). None of these tests showed any improvement after capsule implantation. However, because of the small number of patients evaluated, it is too early to assess a potential slowing of the degenerative process. The present study indicates that significant doses of hCNTF can be delivered directly into the central nervous system, and that, in contrast to the side effects associated with systemic delivery of CNTF in ALS patients, no limiting adverse effects are observed in any of the patients exposed to low intrathecal doses of CNTF.

References

1. Sendtner, M., Kreutzberg, G. W., and Thoenen, H. (1990). Ciliary neurotrophic factor prevents the degeneration of motoneurons after axotomy. *Nature* **345,** 440–441.
2. Zurn, A. D., Baetge, E. E., Hammang, J. P., Tan, S. A., and Aebischer, P. (1994). Glial cell line-derived neurotrophic factor (GDNF), a new neurotrophic factor for motoneurons. *Neuroreport* **6,** 113–118.
3. Yan, Q., Matheson, C., and Lopez, O. T. (1995). In vivo neurotrophic effects of GDNF on neonatal and adult facial motor neurons. *Nature* **373,** 341–344.
4. Sagot, Y., Tan, S. A., Baetge, E. E., Schmalbruch, H., Kato, A. C., and Aebischer, P. (1995). Polymer encapsulated cell lines genetically engineered to release CNTF can slow down progressive motor neuronopathy in the mouse. *Eur. J. Neurosci.* **7,** 1313–1322.
5. Tomac, A., Lindquist, E., Lin, L. F., Ögren, S. O., Young, D., Hoffer, B. J., and Olson, L. (1995). Protection and repair of the nigrostriatal dopaminergic system by GDNF in vivo. *Nature* **373,** 335–339.
6. Sauer, H., Rosenblad, C., and Björklund, A. (1995). GDNF but not TGF-β3 prevents delayed degeneration of nigral dopaminergic neurons following striatal 6-hydroxydopamine-lesion. *Proc. Natl. Acad. Sci. U.S.A.* **92,** 8935–8939.
7. Gash, D. M., Zhang, Z., Ovadia, A., Cass, W. A., Yi, A., Simmerman, L., Russel, D., Martin, D., Lapchak, P. A. Collins, F., Hoffer, B. J., and Gerhardt, G. A. (1996). Functional recovery in parkinsonian monkeys treated with GDNF. *Nature* **380,** 252–255.
8. Aebischer, P., Winn, S. R., Presco, P. A., Greene, L. A., and Jaeger, C. B. (1991). Long-term cross-species brain transplantation of a polymer encapsulated dopamine-secreting cell line. *Exp. Neurol.* **103,** 269–275.
9. Tseng, J. L., Baetge, E. E., Zurn, A. D., and Aebischer, P. (1997). GDNF reduces drug-induced rotational behavior following medial forebrain bundle transection by a mechanism not involving striatal dopamine. *J. Neurosci.* **17,** 325–333.
10. Aebischer, P., Schluep, M., Déglon, N., Joseph, J. M., Hirt, L., Heyd, B., Goddard, M., Hammang, J. P., Zurn, A. D., Kato, A. C., Regli, F., and Baetge, E. E. (1996). Intrathecal delivery of CNTF using encapsulated genetically modified xenogeneic cells in amyotrophic lateral sclerosis patients. *Nat. Med.* **2,** 696–698.

Frauke Deimling,* Susetta Finotto,* Karin Lindner,* Barbara Brühl,* Jose L. Roig-Lopez,† Jose E. Garcia-Arraras,† Christo Goridis,‡ Kerstin Krieglstein,* and Klaus Unsicker*

* Departments of Anatomy and Cell Biology
University of Heidelberg
D-69120 Heidelberg, Germany

† Department of Biology
University of Puerto Rico
San Juan, Puerto Rico 00931

‡Developmental Biology Institute of Marseilles
F-13288 Marseille, France

Characterization of Adrenal Chromaffin Progenitor Cells in Mice

Molecular cues underlying the diversification of neural gene expression are a focus of current research in developmental neurobiology. The neural crest and its derivatives have occupied and continue to play a paradigmatic role in the eludication of mechanisms that determine cell fate in the nervous system. Current hypotheses and text books maintain that chromaffin cells, the neuroendocrine cells of the adrenal medulla and paraganglia, and sympathetic neurons share a common precursor, the sympathoadrenal (SA) progenitor cell. An essential role in triggering chromaffin as opposed to sympathetic neuronal fates has been attributed to glucocorticoid hormones (1, 2). Evidence based on *in vitro* studies with SA and chromaffin progenitor cells has suggested that glucocorticoids are necessary to, first, suppress neuronal markers in SA progenitors, channeling them toward a chromaffin cell phenotype, and, subsequently, to induce the adrenaline synthesizing enzyme, phenylethanolamine-N-methyltransferase (PNMT), in a majority of chromaffin cells. Analysis of mice lacking the glucocorticoid receptor (3) has generated the unexpected result that chromaffin cells expressing tyrosine hydroxylase (TH) and showing normal ultrastructural features typical of chromaffin cells are present inside the adrenal gland. However, they do not form a solid adrenal medulla, lack PNMT and adrenaline, and seem to be reduced in their number. These data suggested that glucocorticoid signaling is essential for the induction of PNMT but not for triggering other features typical for the chromaffin cell phenotype. Beyond catecholamine synthesis and their ultrastructural appearance, chromaffin cells are characterized by numerous other markers. Such markers include a variety of neuropeptides, granins, monoamine transporters, and several other mole-

Advances in Pharmacology, Volume 42

1054-3589/98 $25.00

cules, such as L1, GAP43, and acetylcholinesterase. Some of these markers, such as various neuropeptides, secretogranin II, chromogranin B, L1, and GAP43 have been described to specifically recognize subpopulations of chromaffin cells, at least in rat. This makes them potentially attractive for addressing the problem of whether particular subpopulations of chromaffin cells might be deleted in glucocorticoid receptor-deficient mice. However, most of these markers have not been applied to mouse chromaffin tissue. The present study was, therefore, designed to reveal the pattern of expression of several candidate markers in the embryonic adrenal gland of wild-type mice.

NMR-I 32 mice were used throughout this study. Mice were mated overnight and designated day 0.5 of pregnancy (E0.5) following plug discovery. Pregnant mice were killed at E12.5, E13.5, E15.5, and E18.5 by cervical dislocation. Embryos were rapidly removed and fixed by immersion in ice-cold-phosphate-buffered paraformaldehyde (4%) for various periods of time, depending on antibodies used. Adrenal anlagen or glands, respectively, were cryostate-sectioned, and sections were processed for immunocytochemistry. Localization of primary antibodies was visualized by indirect immunofluorescence using one of the following components: Cy3, DTAF, or streptavidin-Cy3/DTAF.

I. TH and the Transcription Factor Phox 2

Adrenal anlagen containing immigrating cells that were positive for TH and the transcription factor Phox 2 were first recognized at E13.5. Colabeling for TH and Phox 2 revealed some cells at E13.5 that were positive for Phox 2 (a general marker for catecholaminergic cells), without staining for TH. At E15.5 and E18.5, TH and Phox 2 immunoreactivities were strictly colocalized. Taking these data together supports the notion that Phox 2 is a marker for both differentiated and presumptive catecholaminergic adrenal chromaffin cells prior to reaching detectable levels of TH immunoreactivity.

II. Numeric Development, Cell Proliferation, and Death

Proliferation, survival, and death of adrenal chromaffin progenitor cells have not been studied *in vivo* as yet. We found that numbers of Phox 2–positive adrenal cells increase approximately threefold between E13.5 and E18.5, suggesting a substantial proliferative activity during this period. In accordance with these data, double-labeling adrenal cells at E15.5 for TH and proliferation cell nuclear antigen revealed that about 40% of the TH-positive cells also display the nuclear marker. Using the *in situ* apoptosis technique (TUNEL) in conjunction with TH immunostaining at E15.5 showed that about 1% of adrenal TH-positive cells undergo apoptosis at this age.

III. Adrenergic Subpopulation of Adrenal Chromaffin Cells

PNMT immunoreactivity first appeared at E15.5 and was found in approximately two-thirds of TH-positive cells at E18.5. Secretogranin II, a member of the granin family, also appeared at E15.5 and colocalized with PNMT. Chromogranin B is localized predominantly in adrenergic chromaffin cells of the early postnatal mouse adrenal gland. In contrast to PNMT and secretogranin II, chromogranin B was detectable in a few adrenal cells as early as E13.5, becoming more prominent and localized in more cells with embryonic age.

IV. Noradrenergic Subpopulation of Adrenal Chromaffin Cells

Noradrenergic chromaffin cells possess the full repertoire of chromaffin cell functions except the capacity to synthesize adrenaline. Adult noradrenergic chromaffin cells exhibit at least three markers that they do not share with adrenergic chromaffin cells, GAP43, the adhesion molecule L1, and vesicular monoamine transporter-1 (VMAT2). GAP43 is detectable in all adrenal TH-positive cells at E13.5. Numbers of GAP43-immunoreactive cells decrease with embryonic age. With the appearance of PNMT, some cells transiently express both markers, PNMT and GAP43, before most PNMT-positive cells become GAP43-negative at E18.5. Cells immunoreactive for the VMAT2 (postnatally expressed in noradrenergic chromaffin cells) are detectable as early as E13.5, subsequently increase in number, but remain a minor subpopulation (approximately 20%) of TH-positive adrenal cells throughout embryonic development.

V. Pan-chromaffin Markers and Neuropeptides

In addition to TH, Phox 2, peanut agglutinin (PNA) and neuropeptide Y (NPY) are detectable in all postnatal chromaffin cells. Whereas antibodies to NPY recognize all TH-positive cells from E13.5 onward, PNA staining does not become detectable until E15.5. VMAT, are detectable in a marker found in both noradrenergic and adrenergic chromaffin cells, appears at E18.5. Cells that are positive for metenkephalin (Met-Enk) appear at E15.5 but constitute a very small group of adrenal cells throughout embryonic development. Galanin immunoreactivity is undetectable in the embryonic mouse adrenal gland; very few cells with a small intensely fluorescent (SIF) cell-like morphology are labeled at postnatal day 6.

VI. Conclusions

The present data (summarized in Table I) provide a useful basis for further analyzing the phenotypes of adrenal chromaffin cells in glucocorticoid-receptor knockout mice. Furthermore, our results suggest that one marker found predominantly or exclusively in mature adrenergic chromaffin cells in mice, chro-

TABLE I Markers for Mouse Chromaffin Progenitor Cells

Marker	*E13.5*	*E15.5*	*E18.5*	*Source*	*Dilution*
TH	+	+	+	Boehringer Mannheim	1:2000
Phox-2	+	+	+	Goridis	1:1000
NPY	+	+	+	Garcia-Arraras	1:500
Met-Enk	—	+	+	Garcia-Arraras	1:500
Galanin	ND	ND	—	Garcia-Arraras	1:1000
Chromogranin B	+	+	+	Rosa *et al.*	1:600
PNMT	—	+	+	Incstar	1:1000
Secretogranin II	—	+	+	Rosa *et al.*	1:600
VMAT1	–	–	+	M. Hannah	1:2000
VMAT2	+	+	+	M. Hannah	1:100
GAP43	+	+	+	Sigma	1:100
PNA	—	+	+	Vector	1:100

ND, not determined.

mogranin B, can be detected in chromaffin progenitor cells prior to the expression of other key markers for the adrenergic phenotype, PNMT, and secretogranin II. This suggests that differentiative properties of adrenergic chromaffin cells can be regulated sequentially and are probably, in part, independent from glucocorticoid signaling. Several markers of the mature noradrenergic chromaffin cell, including GAP43 and VMAT2, are found at very early stages of adrenal chromaffin cell development. While GAP43 is initially found in a majority of chromaffin progenitor cells and becomes restricted to PNMT-negative cells late in development, VMAT2 is always restricted to a small subpopulation of chromaffin cells. Thus, regulation of both adrenergic and noradrenergic phenotypic features of adrenal chromaffin cells is not fully synchronized, suggesting that several distinct mechanisms rather than glucocorticoid signaling alone must be involved in the generation of subpopulations of chromaffin cells.

References

1. Unsicker, K., Krisch, B., Otten, U., and Thoenen, H. (1978). Nerve growth factor-induced fiber outgrowth from isolated rat adrenal chromaffin cells: Impairment by glucocorticoids. *Proc. Natl. Acad. Sci. U.S.A.* 75, 3498–3502.
2. Anderson, D. J., and Axel, R. (1986), A bipotential neuroendocrine precursor whose choice of cell fate is determined by NGF and glucocorticoids. *Cell* 47, 1079–1090.
3. Cole, T. J., Blendy, J. A., Monaghan, A. P., Krieglstein, K., Schmid, W., Aguzzi, A., Fantuzzi, G., Hummler, E., Uniscker, K., and Schütz, G. (1995). Targeted disruption of the glucocorticoid receptor gene blocks adrenergic chromaffin cell development and severely retards lung development. *Genes Dev.* 9, 1608–1621.

B. Cardinaud,* J-M. Gilbert,* F. Liu,† K. S. Sugamori,†
J-D. Vincent,* H. B. Niznik,† and P. Vernier*

*Institute Alfred Fessard, CNRS
F91198 Gif-sur-Yvette, France

†Clarke Institute of Psychiatry
Toronto, Ontario, M5T 1T8 Canada

Evolution and Origin of the Diversity of Dopamine Receptors in Vertebrates

Dopamine, like the other bioamines, elicits very diverse effects in the nervous system of vertebrates. This physiological diversity relies both on a widespread anatomical distribution and on a multiplicity of membrane receptors (1). A major challenge for modern molecular pharmacology and neurobiology is to understand the role of each of the receptor subtypes in the multiple actions of dopamine neurotransmitters. In this respect, the notions of the "gene family" and "homology" are particularly useful. Indeed, the bioamine receptors belong to the very broad G-protein–coupled receptor superfamily. This "family concept" implicates that all the members of the defined family originate from a common ancestor gene. Then, the multiplicity of receptors was generated by gene duplications followed by mutations and conservation of important functional peculiarities (physiological constraints) in the species of the lineage where the duplications occurred. During evolution, these events of duplication and conservation of specific genes are infrequent and generally associated with identifiable changes in the physiology of the corresponding species. To understand the physiological consequences of such changes, it is necessary to define whether they correspond to characters conserved in most of the species sharing the same ancestor or if they are only derived features specific to a given species. In other words, one needs to depict the nature of homologies shared by the different bioamine receptor subtypes. Homology defines the characteristics inherited from common ancestry, and it remains the most useful concept to account for the mix of conservation and changes that characterizes evolution-dependent processes such as gene diversification. In the case of bioamine receptors, phylogenetic analysis points to the key events that could have justified the conservation of such a high number of receptors for just one endogenous ligand in most of the vertebrate species

As a first step in the understanding of the diversification of bioamine receptors in vertebrates, we used the methods of molecular phylogeny to obtain a "natural" classification of receptors. These methods describe as accurately as possible the structural, and therefore genetic, relationships of the receptor

Advances in Pharmacology, Volume 42

1054-3589/98 $25.00

sequences. They calculate and display in a dichotomic fashion the sequence similarities of the studied receptors, and the topology of the branching should correspond to the phylogeny of the gene family. In particular, this analysis allows the precise depiction of the paralogous or orthologous nature of the receptors. Pharmacological differences between receptors reflect in part this evolutionary diversification, but the pharmacological analysis itself cannot unravel the nature of meaningful receptor homologies, both between vertebrates and invertebrates and even among vertebrates (2).

Important results of this phylogenetic analysis are the following: (1) The phylogenetical classification essentially fits with the pharmacological classification but allows it to assign still unclassified receptor sequences to well-defined classes or subtypes of bioamine receptors. (2) The rate of receptor sequence evolution is highly irregular in this protein family (no molecular clock), indicating that the functional constraints exerted on the receptor sequences have varied during the course of evolution of this membrane receptor family. (3) Saturation in the sequence similarities, which marks the strong structural requirements of bioamine receptors, precludes any robust prediction of the ancient relationship of bioamine receptors. Nevertheless, tree topologies suggest that the main classes of bioamine receptors (D1, D2, α_1, α_2, $5HT_{1A}$, $5HT_2$, etc.) were generated before or at least close to the divergence of cephalochordates from vertebrates. These methodological limitations rendered it essential to look for bioamine receptor diversity directly in the species whose lineages diverged at a different period of time during vertebrate evolution, in order to gain robust information on the events leading to receptor diversity.

A comparative approach of the dopamine D1-receptor class was undertaken to analyze in detail the dopamine-receptor multiplicity of D1-like receptors in vertebrates by cloning the corresponding genes in most of the main groups of vertebrates, with the aim to identify the events that led to the emergence of a "new" D1-receptor gene in the evolution of vertebrates. The choice of the D1-receptor family was justified by the fact that the corresponding genes have no introns in the coding regions, facilitating the isolation of all the members of the family from genomic DNA in a given species.

In brief, we cloned members of the D1-receptor family (in a manner as exhaustive as possible) from agnathans (jawless fish: hagfish and lampreys [*Petromyzon* and *Lampetra*]), whose ancestors diverged close to the "emergence" of vertebrates; a chondrychtian (cartilaginous fish: the electric ray *Torpedo marmorata*); an actinopterygian (bony fish: the teleost European eel); an amphibian (*Xenopus laevis*); and one representative of birds (chicken, *Gallus gallus*). Mammalian sequences from rodent (mouse and rat), marsupials (opposum), and primates (humans and apes) but also from the fish Fugu and Tilapia were already available. The description and interpretation of the receptor diversification in vertebrates required comparison with a nonvertebrate species to be used as a satisfactory outgroup, and we cloned a D1-like sequence from amphioxus, a cephalocordate proposed to be the closest sister group of vertebrates.

This phylogenetical approach provided conclusive evidence for the existence of two other subtypes of D1 receptors, one named D1C, found in all the jawed vertebrates except mammals (3, 4), and one termed D1D, which as of

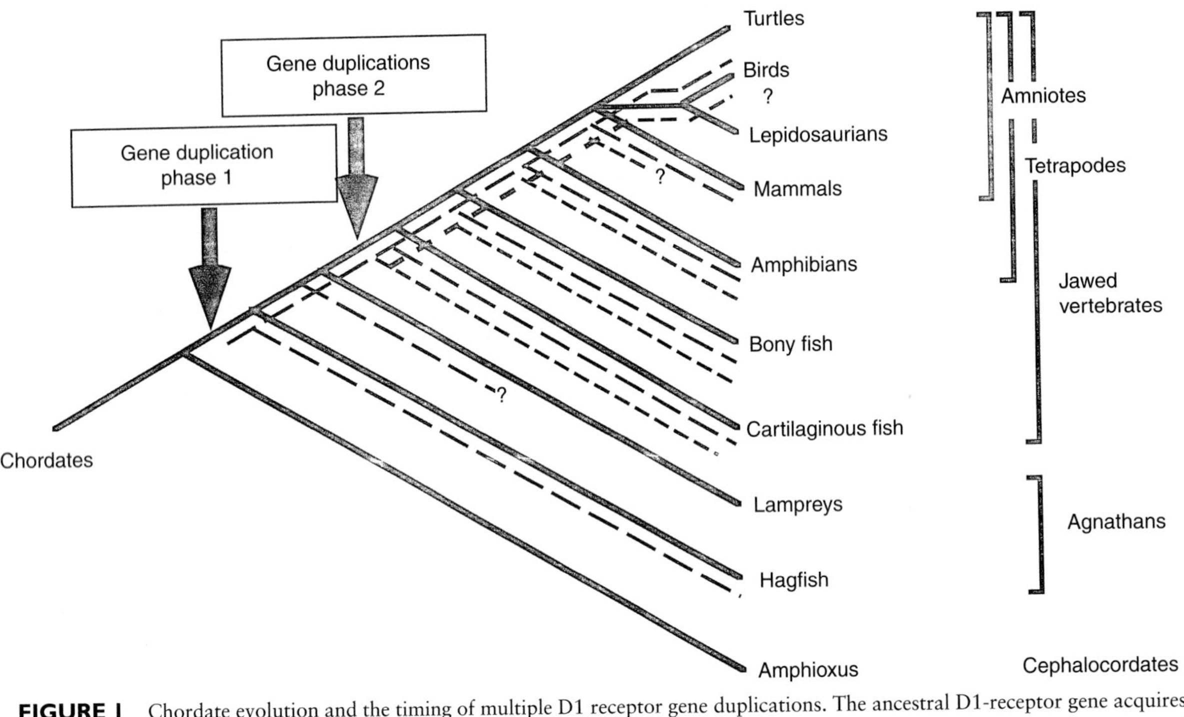

FIGURE 1 Chordate evolution and the timing of multiple D1 receptor gene duplications. The ancestral D1-receptor gene acquires a D1A character (*solid lines*) with the emergence of craniates. It duplicated a first time before hagfish appeared (*long interrupted lines*) and then a second time before fish emergence (*short interrupted lines*). (Modified and reprinted with permission from Holland and Garcia-Fernàndez, *Developmental Biology* 173, pp. 382–395, 1996.)

now appears only to be found in birds (5). Therefore, the D1-receptor class comprises three to four subtypes in "higher" vertebrates, virtually indentical to other catecholamine receptor families, indicating that the gene duplications at the origin of this receptor multiplicity occurred before or concomitantly with the appearance of Chondrychtians (cartilaginous fish). In contrast, hagfish express two D1-like receptors, one being orthologous to the vertebrate D1A receptor and the other exhibiting a sequence related to D1B- and D1C-receptor subtypes. Two lamprey species have surprisingly only one D1 receptor, which is clearly of the D1A subtype, but lampreys are known to have lost genes as compared with hagfish. Finally, amphioxus appears to contain only one D1-like receptor that cannot be assigned to any of the known D1-receptor subtypes in vertebrates, as is also the case for *Drosophila* and other invertebrate D1-like receptors.

Our current view of the evolution of the D1-receptor class in vertebrates is that two main steps of diversification surely occurred during evolution (probably through genome tetraploidization) to generate the present state of receptor diversity. The first of these events occurred before the emergence of vertebrates from a chordate ancestor and allows the fixation of the D1A-receptor subtype. The second led to the appearance of the D1B and D1C subtypes and occurred between the separation of fish ancestors from the agnathans group. This picture strongly resembles the one proposed for the emergence of vertebrates and more recently for the evolution of families of developmental genes (Fig. 1) (6). In addition, whereas three or four different D1 receptors belonging to the D1A, D1B, D1C, and D1D subtypes are found in most vertebrate groups, eutherian mammals appear, surprisingly, to have only two of them (D1A and D1B). Finally, clearly shared functional peculiarities are observed for most of the dopamine-receptor subtypes, leading to a definition of significant functional homologies for each receptor subtype (see chapter by Niznik *et al.* in Part D).

Because all of the known bioamine receptor subtypes are expressed in the central nervous system—and in general with a discrete distribution—it is tempting to propose that the bioamine and dopamine D1-receptor diversification accompanies in some respect the genetic mechanisms leading to the encephalization of the vertebrate nervous system, which could possibly support novel functions. Indeed, once a duplicated receptor becomes expressed as a functional protein, changes in function may be favored, provided that a new expression territory is found, mediated in part by means of tissue-specific regulatory trans-acting factors. Acquisition and changes in the expression pattern of bioamine receptors is likely to have been the major factor of duplicated gene conservation in vertebrates.

References

1. Smeets, W. J. A. J., and Reiner, A. (eds.). (1994). Phylogeny and Development of Catecholamine Systems in the CNS of Vertebrates. Cambridge University Press.
2. Vernier, P., Cardinaud, B., Valdenaire, O., Philippe, H., and Vincent, J-D. (1995). An evolutionary view of drug-receptor interactions: the example of the bioamine receptor family. *Trends Pharmacol. Sci.* **16**, 375–381.

3. Sugamori, K. S., Demchyshyn, L. L., Chung, M., and Niznik, H. B. (1994). D_{1A} D_{1B} and D_{1C} dopamine receptors from Xenopus laevis. *Proc. Natl. Acad. Sci.* U.S.A. **91**, 10536–10540.
4. Cardinaud, B., Sugamori, K. S., Coudouel, S., Vincent, J-D., Niznik, H. B., and Vernier, P. (1997). Early emergence of three dopamine D1 receptor subtypes in vertebrates: Molecular phylogenetic, pharmacological and functional criteria defining D1A, D1B and D1C receptors in european eel Anguilla anguilla. *J. Biol. Chem.* **272**, 2778–2787.
5. Demchyshyn, L. L., Sugamori, K. S., Lee, F. J., Hamadanizadeh, S. A., and Niznik, H. B. (1995). The dopamine D_{1D} receptor: cloning and characterization of three pharmacologically distinct D_1-like receptors from Gallus domesticus, *J. Biol. Chem.* **270**, 4005–4012.
6. Holland, P. W. H., and Garcia-Fernandez, J. (1996). Hox genes and chordate evolution. *Dev. Biol.* **173**, 382–395.

James B. Rand, Janet S. Duerr, and Dennis L. Frisby

Program in Molecular and Cell Biology
Oklahoma Medical Research Foundation
Oklahoma City, Oklahoma 73104

Neurogenetics of Synaptic Transmission in *Caenorhabditis elegans*

The nematode *Caenorhabditis elegans* has a number of advantages for the analysis of synaptic molecules. These include ease of manipulation, a simple nervous system, and powerful tools for the analysis of mutants and genes.

Genetically, *C. elegans* is advantageous because of its short generation time (3 days), its prolific progeny yield (280 per parent), its small size (1.5 mm long), and its ease of laboratory culture (on *E. coli* lawns on agar Petri dishes). There are now thousands of mutant strains of *C. elegans,* exhibiting a wide variety of behavioral, morphological, and developmental phenotypes, and hundreds of genes have now been mapped on the animals' six linkage groups.

Cellularly, *C. elegans* is remarkably simple. At hatching, there are 550 somatic cells (1), and over the course of the next 48 hr, this number increases to produce an adult total of 959 somatic cells (2). The adult contains 302 neurons, and reproducibility of neuron structure and connectivity has been demonstrated by serial section electron microscopy for most portions of the nervous system (3).

Molecular biology using this organism is simplified by its relatively small genome size of 10^8 bp, the availability of a transposon for gene cloning (4), and the near completion of the physical map (5) and sequence of the entire genome (6).

Advances in Pharmacology, Volume 42

1054-3589/98 $25.00

I. Mammalian Biology and Nematode Biology

From the perspective of a mammalian neurobiologist, there are two basic strategies that may profitably be employed using *C. elegans*. In one case, basic information derived from *C. elegans* biology and molecular genetics is applied to mammalian systems; that is, genes first characterized in *C. elegans* are used to identify mammalian homologues. The other powerful use of *C. elegans* is to start with known mammalian biology and genes and then identify the *C. elegans* homologues of these genes. This permits *C. elegans* cell biology and mutant analysis to be applied. We will present one example of each strategy. Both examples involve synaptic vesicle neurotransmitter transporters: The vesicular acetylcholine transporter (VAChT) was first identified and characterized in *C. elegans* and then used to identify the mammalian gene and protein; the vesicular monoamine transporters (VMATs) were first identified in mammals, and we have recently started to analyze a *C. elegans* homologue.

II. Cholinergic Function in *C. elegans*

Acetylcholine is the major excitatory neurotransmitter at nematode neuromuscular junctions. The transmitter is synthesized by choline acetyltransferase (ChAT). We identified ChAT-deficient mutants and demonstrated that they defined the ChAT structural gene, which we named *cha-1* (7). During our genetic analysis of *cha-1* mutants, we found that the *cha-1* gene lay very close to the *unc-17* gene, which was noteworthy because mutations in the two genes led to very similar phenotypes (discussion follows). After *cha-1* was cloned, it became clear that *unc-17* and *cha-1* are both parts of a complex transcription unit (8, 9). The two genes share a common 5′-exon, with the remainder of the *unc-17* gene nested within the long first intron of *cha-1*. The two transcripts apparently arise by alternative splicing of a common precursor.

III. *Unc-17* Encodes VAChT

When it was cloned, the *unc-17* gene product (UNC-17) had no homologues in the databases. In an effort to understand more about this transcription unit and its proteins, we raised antibodies to *C. elegans* ChAT and UNC-17. Immunocytochemical studies demonstrated that the two proteins were colocalized both at the cellular and the subcellular levels. Immunoreactivity to both proteins is present in synaptic regions of approximately 100 neurons, and additional experiments showed that UNC-17 was associated with synaptic vesicles. Almost all of the ChAT- and UNC-17–positive cells are (apparently excitatory) motor neurons. When the cloning of the rat VMATs was published (10, 11), it became clear that UNC-17 was related to the VMATs, and it was, therefore, likely to be the *C. elegans* VAChT (12). The *C. elegans unc-17* gene was then used to isolate vertebrate homologues, and the identity of the corresponding gene products as VAChTs was confirmed by binding and uptake studies (13, 14). An unexpected result was that the rat and human VAChT genes had a gene structure similar to *C. elegans,* with the VAChT gene nested within the first intron of the ChAT gene (14, 15). Thus, in both mammals and nematodes, the synthesis and the vesicular transport of acetylcholine are coupled at the genomic level. This suggests that the organization of this "cholinergic locus" is somehow important for its function.

The availability of a large collection of *C. elegans* mutants has permitted careful *in vivo* analysis of cholinergic regulation and the behavioral consequences of cholinergic hypofunction. Animals completely lacking the VAChT protein (null mutants) cannot grow or survive after hatching. Animals with milder mutations in the *unc-17* gene (i.e., partially VAChT-deficient animals) are small, slow-growing, and display a number of neuromuscular deficits (jerky, uncoordinated locomotion; reduced pharyngeal pumping, and reduced defecation). They are also quite strongly resistant to inhibitors of acetylcholinesterase. All of these phenotypes (including the lethality of null mutations) are shared by *cha-1* (ChAT-deficient) mutants, which argues that vesicular transport is necessary for cholinergic function.

IV. Catecholamines in *C. elegans*

Three amine neurotransmitters have been identified thus far in *C. elegans.* Dopamine was originally identified in eight sensory neurons using the technique of formaldehyde-induced fluorescence (FIF) (16). Exogenous dopamine inhibits locomotion and egg laying. Serotonin (5HT) has been identified in *C. elegans* neurons by FIF (17) and by anti-5HT immunostaining (18, 19). In *C. elegans,* exogenous 5HT stimulates egg laying and pharyngeal pumping and inhibits locomotion and defecation (17, 20). 5HT is also required for male mating behavior (21). There are approximately 11 neurons in *C. elegans* with anti-5HT immunoreactivity (18, 21), but little is known about the role of 5HT in these cells. Horvitz *et al.* (17) detected neither epinephrine nor norepinephrine in *C. elegans;* however, octopamine (*p*-hydroxyphenylethanolamine) is present in *C. elegans* homogenates, and exogenous octopamine inhibits egg laying and stimulates locomotion.

V. *C. elegans* VMAT

The *C. elegans* Genome Sequencing Project identified a genomic sequence that appeared to encode part of a VMAT-like protein. Using standard methods, we have now cloned and sequenced the complete cDNA corresponding to this gene; the predicted *C. elegans* protein is 47% identical to rat VMAT1 and 49% identical to rat VMAT2. Preliminary experiments (in collaboration with Jeffrey Erickson and Lee Eiden) suggest that the *C. elegans* protein is capable of mediating time-dependent, reserpine-sensitive, and tetrabenazine-sensitive transport of serotonin.

We have obtained antibodies specific for this protein, and immunolocalization studies indicate that the VMAT protein is primarily localized to synaptic regions of a subset of neurons. There are approximately 25 VMAT-positive neurons; these include the eight putative dopaminergic neurons and the 11 putative serotonergic neurons, as well as a few, still unidentified cells. These neurons are (with at least two clear exceptions) distinct from those that are immunopositive for VAChT. All eight of the dopaminergic cells are sensory neurons, but the serotonergic neurons include sensory cells, interneurons, motor neurons, and secretory cells.

The *C. elegans* VMAT homologue appears to be encoded by the previously identified *cat-1* gene. In *cat-1* mutants, dopamine (visualized by FIF) (16) and 5-HT (visualized by immunocytochemistry) (21) are no longer localized to neuronal processes but are found only in cell bodies. Furthermore, the apparent abundance of dopamine is decreased by approximately 60% (16). It is noteworthy that this is the same phenotype obtained by treating wild-type animals with reserpine. The *cat-1* mutants are also deficient in dopamine- and 5HT-mediated behaviors, such as slowing their rate of locomotion in the presence of food. In addition, *cat-1* mutants are completely deficient for VMAT immunoreactivity.

Using these transporter genes and mutants, together with standard *C. elegans* tools, it will be now possible to analyze the differentiation of particular neuronal cell types and also the contribution of particular protein domains to cellular and overall behavioral function.

Acknowledgments

The early molecular analysis and sequencing of the *cha-1–unc-17* cholinergic locus was performed by Aixa Alfonso, Kiely Grundahl, and John McManus. The kinetic analysis of the *C. elegans* VMAT and VAChT transport activities were performed in collaboration with Jeff Erickson and Lee Eiden, at the National Institute of Mental Health. We are also grateful to Angie Duke and Jennifer Gaskin for technical assistance with the sequencing of VMAT cDNAs and the VMAT immunohistochemical analysis, respectively. These studies were supported by research grants from the National Institutes of Health and the Oklahoma Center for the Advancement of Science and Technology.

References

1. Sulston, J. E., Schierenberg, E., White, J. G., and Thomson, J. N. (1983). The embryonic cell lineage of the nematode *Caenorhabditis elegans. Dev. Biol.* **100,** 64–119.
2. Sulston, J. E., and Horvitz, H. R. (1977). Post-embryonic cell lineages of the nematode *Caenorhabditis elegans. Dev. Biol.* **56,** 110–156.
3. White, J. G., Southgate, E., Thomson, J. N., and Brenner, S. (1986). The structure of the nervous system of the nematode *Caenorhabditis elegans. Phil. Trans. R. Soc. Lond. B* **314,** 1–340.
4. Herman, R. K., and Shaw, J. E. (1987). The transposable genetic element Tc1 in the nematode *C. elegans. Trends Genet.* **3,** 222–225.

5. Coulson, A., Sulston, J., Brenner, S., and Karn, J. (1986). Toward a physical map of the genome of the nematode *Caenorhabditis elegans. Proc. Natl. Acad. Sci. USA* **83,** 7821–7825.
6. Waterston, R., and Sulston, J. (1995). The genome of *Caenorhabditis elegans. Proc. Natl. Acad. Sci. USA* **92,** 10836–10840.
7. Rand, J. B., and Russell, R. L. (1984). Choline acetyltransferase-deficient mutants of the nematode *Caenorhabditis elegans. Genetics* **106,** 227–248.
8. Alfonso, A., Grundahl, K., McManus, J. R., and Rand, J. B. (1994a). Cloning and characterization of the choline acetyltransferase structural gene (*cha-1*) from *C. elegans. J. Neurosci.* **14,** 2290–2300.
9. Alfonso, A., Grundahl, K., McManus, J. R., Asbury, J. M., and Rand, J. B. (1994b). Alternative splicing leads to two cholinergic proteins in *Caenorhabditis elegans. J. Mol. Biol.* **241,** 627–630.
10. Erickson, J. D., Eiden, L. E., and Hoffman, B. J. (1992). Expression cloning of a reserpine-sensitive vesicular monoamine transporter. *Proc. Natl. Acad. Sci. USA* **89,** 10993–10997.
11. Liu, Y., Peter, D., Roghani, A., Schuldiner, S., Privé, G. G., Eisenberg, D., Brecha, N., and Edwards, R. H. (1992). A cDNA that suppresses MPP^+ toxicity encodes a vesicular amine transporter. *Cell* **70,** 539–551.
12. Alfonso, A., Grundahl, K., Duerr, J. S., Han, H-P., and Rand, J. B. (1993). The *Caenorhabditis elegans unc-17* gene: A putative vesicular acetylcholine transporter. *Science* **261,** 617–619.
13. Varoqui, H., Diebler, M.-F., Meunier, F.-M., Rand, J. B., Usdin, T.B., Bonner, T. I., Eiden, L. E., and Erickson, J. D. (1994). Cloning and expression of the vesamicol binding protein from the marine ray *Torpedo:* Homology with the putative vesicular acetylcholine transporter UNC-17 from *Caenorhabditis elegans. FEBS Lett.* **342,** 97–102.
14. Erickson, J. D., Varoqui, H., Schäfer, M. K.-H., Modi, W., Diebler, M., Weihe, E., Rand, J., Eiden, L., Bonner, T., and Usdin, T. (1994). Functional identification of a vesicular acetylcholine transporter and its expression from a "cholinergic" gene locus. *J. Biol. Chem.* **269,** 21929–21932.
15. Bejanin, S., Cervini, R., Mallet, J., and Berrard, S. (1994). A unique gene organization for two cholinergic markers, choline acetyltransferase and a putative vesicular transporter of acetylcholine. *J. Biol. Chem.* **269,** 21944–21947.
16. Sulston, J., Dew, M., and Brenner, S. (1975). Dopaminergic neurons in the nematode *Caenorhabditis elegans. J. Comp. Neurol.* **163,** 215–226.
17. Horvitz, H. R., Chalfie, M., Trent, C., Sulston, J. E., and Evans, P. D. (1982). Serotonin and octopamine in the nematode *C. elegans. Science* **216,** 1012–1014.
18. Desai, C., Garriga, G., McIntire, S. L., and Horvitz, H. R. (1988). A genetic pathway for the development of the *Caenorhabditis elegans HSN motor neurons. Nature* **336,** 638–646.
19. McIntire, S. L., Garriga, G., White, J., Jacobson, D., and Horvitz, H. R. (1992). Genes necessary for directed axonal elongation of fasciculation in *C. elegans. Neuron* **8,** 307–322.
20. Ségalat, L., Elkes, D. A., and Kaplan, J. M. (1995). Modulation of serotonin-controlled behaviors by G. in *Caenorhabditis elegans. Science* **267,** 1648–1651.
21. Loer, C. M., and Kenyon, C. J. (1993). Serotonin-deficient mutants and male mating behavior in the nematode *Caenorhabditis elegans. J. Neurosci.* **13,** 5407–5417.

Jay Hirsh

Department of Biology
University of Virginia
Charlottesville, Virginia 22903

Decapitated *Drosophila*: A Novel System for the Study of Biogenic Amines

It is now clear that the basic principles and genes governing neural development and functioning were in place before the evolutionary divergence of vertebrates and invertebrates. These conserved aspects include the basic mechanisms governing anterior–posterior differentiation of the nervous system, synaptic release mechanisms, and neurotransmitters and receptors, and the fundamental mechanisms governing learning and memory. The molecular and genetic approaches possible in *Drosophila* have contributed much to this understanding (1), and the ability to assess complex behaviors, including learning and memory, in this animal make it a favorable system for future studies. The genes and enzymes of biogenic amine metabolism are also well conserved between *Drosophila* and vertebrates, as are a number of amine receptors that have been isolated from *Drosophila* (2–4). The amines dopamine, serotonin, and octopamine are found in the *Drosophila* central nervous system (CNS). The behavioral and/or developmental roles for these compounds in the insect CNS are not well defined, but the recent finding of a dopamine receptor localized to the mushroom bodies, a brain structure involved in olfactory learning and memory, suggests roles for dopamine in these processes (4).

In this chapter, we review recent findings from my laboratory that imply that the linkage to dopamine receptors to specific types of behaviors is also conserved between higher vertebrates and insects. We have obtained these findings utilizing decapitated adult fruit flies, *Drosophila melanogaster*. This preparation can survive for days if kept moist and shows a strong righting response and both a basal level of grooming as well as a provoked grooming response following stimulation of a sensory bristle. With this preparation, we add to the end of the cut nerve cord biogenic amines and drugs that interact with vertebrate amine receptors and observe the behavioral consequences. Data supporting the findings summarized herein are contained in Yellman *et al.* (5).

Serotonin, dopamine, and octopamine show distinguishable effects when added to the decapitated preparations at millimolar concentrations. All three amines stimulate locomotion and hindleg grooming. Whereas the decapitated flies show a basal level of grooming before addition of amines, locomotion is never observed. The relative level of locomotor activity stimulated by these amines varies greatly, with serotonin the least effective and octopamine the most

Advances in Pharmacology, Volume 42

1054-3589/98 $25.00

effective. Exposure to octopamine stimulates flies to locomote approximately 2.5 cm in a 2-min observation period. Additional behaviors stimulated selectively by dopamine versus octopamine allow one to further distinguish these responses. These responses are further potentiated by concurrent addition of hydrazaline, an inhibitor of vertebrate monoamine oxidases. Hydrazaline is presumably acting similarly in this system to increase the effective concentration of both endogenously and exogenously added amines. Many of the hydrazaline-induced effects are sexually dimorphic, with males showing greater responses than females. These results are striking in that there is only one report that detected low levels of monoamine oxidase (MAO) activity in *Drosophila* (6), and MAO activity is undetectable in many other insects.

Behaviors similar to those induced by dopamine can be induced by application of the vertebrate dopamine D2-like receptor agonist quinpirole. A decapitated fly locomoting following addition of quinpirole is shown in Figure 1. The effects of quinpirole on locomotion are also sexually dimorphic, with males preferentially stimulated at low concentrations of quinpirole, whereas both sexes are stimulated similarly at high concentrations. Other vertebrate dopamine agonists and antagonists show effects distinguishable from quinpirole: vertebrate D2-like and D1-like dopamine antagonists result in akinesia, and D1-like agonists result in a selective stimulation of hindleg grooming without locomotion. The akinesias induced by the antagonists are also distinguishable as a function of vertebrate receptor-class specificity: The D1-like antagonists SCH 23390 and SKF 85366 lead to akinesia with associated 10- to 15-Hz tremor and an extended body posture, whereas the D2-like antagonists eticlopride and raclopride lead to akinesia with no tremor and a rather contracted posture.

Identification of the relevant receptors will be crucial to a more thorough understanding of the behaviors induced in this system. We suspect that none of the heretofore identified amine receptors in *Drosophila* is involved in the nerve cord responses assayed in the decapitated preparations. The two D1-like dopamine receptors found to date in *Drosophila* are both localized primarily if not exclusively to the brain, thus excluding them from consideration. An octopamine–tyramine receptor has been isolated from *Drosophila,* but it responds more strongly to tyramine than octopamine for adenylyl cyclase inhibitory responses and equally to both compounds when cytosolic Ca^{2+} release is assayed, results not consistent with the weak effects of tyramine relative to octopamine in the decapitated flies.

Are the vertebrate dopamine-receptor agonists and antagonists interacting with *Drosophila* dopamine receptors or with other types of amine receptors? Our best evidence on this issue comes from comparing behaviors stimulated by the dopamine-receptor agonists with the behaviors stimulated by the amines. The similarities between the dopamine agonist–induced behaviors and those induced by dopamine argue in favor of specific interactions with dopamine receptors. However, until the receptor(s) mediating these responses are identified, we cannot eliminate the possibility that multiple amines could be interacting with a single novel receptor subtype and activating it in different manners.

The responses of decapitated *Drosophila* show many resemblances to the effects of vertebrate dopamine-receptor agonists and antagonists after injection into rodents: both D1-like and D2-like receptor agonists stimulate locomotion

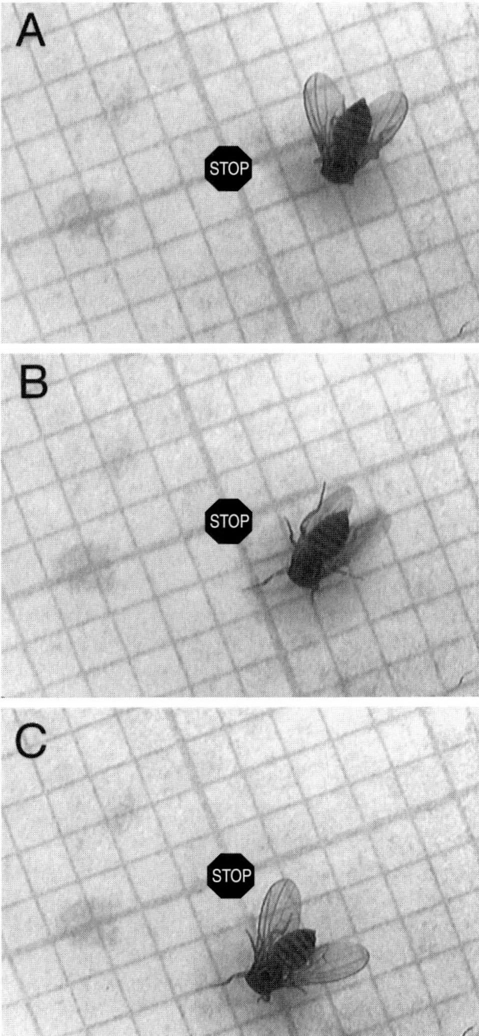

FIGURE 1 Locomotion of a decapitated female fly following exposure to 5 mM of quinpirole. (*A–C*) Images captured at 2-sec intervals from a fly locomoting on 0.05-inch gridded graph paper. The STOP Sign marks a fixed reference point on the field. The wings in (*B*) are blurred due to wing activity stimulated by the quinpirole.

and sterotype behaviors, with differences in the types of stereotypies induced, and both D1-like and D2-like antagonists lead to an akinesic state (7). Stereotyped behaviors include grooming, sniffing, chewing, and posturing, behaviors that are controlled by spinal cord neural oscillators. These phenotypic similarities are striking, indicating that if these compounds are interacting with receptors analogous to the vertebrate dopamine receptors, the link to locomotor and stereotyped grooming behaviors occurred very early in evolution, before the split between *Drosophila* and vertebrates. One important and obvious distinction between the results obtained in vertebrates and in the decapitated *Drosophila* is that the brain has been removed from these preparations. The vertebrate behavioral responses to dopamine depend on striatal dopamine receptors. It, thus, seems possible that the insect nerve cord contains functions that have been taken over by the forebrain in higher vertebrates. Further resolution of these issues will await identification and detailed study of the *Drosophila* receptors involved in these responses.

Modulation of seven-transmembrane–helix receptor sensitivity by phosphorylation is a hallmark of this family of receptors, and this modulation can change affinities for agonist. Changes in amine receptor sensitivity or in types of receptors have not been previously linked to sexual dimorphisms. Our results lead to the hypothesis that differences in amine receptor sensitivity or in types of receptors expressed will be associated with sexually dimorphic behaviors in *Drosophila*. This is a readily testable hypothesis, because the genes involved in *Drosophila* sexual differentiation have been well characterized and can be used to separate the behavioral from the phenotypic aspects of sex.

Acknowledgment

This work was supported by National Institutes of Health grant R01 GM 27318.

References

1. Thor, S. (1995). The genetics of brain development: conserved programs in flies and mice. *Neuron* **15,** 975–977.
2. Jackson, F. R. (1990). Prokaryotic and eukaryotic pyridoxal-dependent decarboxylases are homologous. *J. Mol. Evol.* **31,** 325–329.
3. Nagatsu, T. (1991). Genes for human catecholamine-synthesizing enzymes. *Neurosci. Res.* **12,** 315–341.
4. Han, K.-A., Millar, N. S., Grotewiel, M. S., and Davis, R. L. (1996). DAMB, A novel dopamine receptor expressed specifically in *Drosophila* mushroom bodies. *Neuron* **16,** 1127–1135.
5. Yellman, C., Tao, H., He, B., and Hirsh, J. (1997). Conserved and sexually dimorphic behavioral responses to biogenic amines in decapitated *Drosophila.* Proc Nat Acad Sci. USA **94,** 4131–4136.
6. Dewhurst, S. A., Croker, S. G., Ikeda, K., and McCaman, R. E. (1972). Metabolism of biogenic amines in *Drosophila* nervous tissue. *Comp. Biochem. Physiol.* **43B,** 975–981.
7. Waddington, J. L., and Daly, S. A. (1993). Regulation of unconditioned motor behaviour by D1:D2 interactions. *In* D1:D2 Dopamine Receptor Interactions. pp. 51–78. Academic Press, London England.

Barry Condron*† and Kai Zinn*

*Division of Biology
California Institute of Technology
Pasadena, California 91125

†Department of Neuroscience
School of Medicine
University of Virginia
Charlottesville, Virginia 22908

Dopaminergic Control of Serotonergic Neuron Development in the Grasshopper Central Nervous System

The embryonic central nervous system (CNS) of the grasshopper (*Schistocerca americana*) provides a model system in which to study the basic mechanisms of neural development. The cells are large and are relatively easy to manipulate. In addition, the grasshopper CNS is remarkably similar to that of the fruit fly, both anatomically and in gene expression patterns. This has allowed many neuron-specific markers to be identified in the grasshopper. While the fruit fly is a powerful organism in which to conduct genetic studies of neural development, the grasshopper is very useful for cellular studies.

The CNS of the grasshopper consists of a brain and a set of segmental ganglia, which together make up the ventral nerve cord. Each ventral nerve cord ganglion develops very similarly during early embryogenesis. Neural development in the grasshopper begins with the selection of the ventral ectoderm to become neural ectoderm (1). Through a process mediated by direct cell–cell interactions, certain members of the neural ectoderm are chosen to become neuroblasts. A set of 30 neuroblasts develop in each hemiganglion—each with a distinct fate—which is thought to be determined by their position in the two-dimensional array. Each of these divide to produce stereotypical families of neurons and/or glia. Each precursor cell in these neuroblast families is thought to acquire identity by virtue of its birth order and the identity of the neuroblast that gave rise to it. Neuronal members of these lineages send axons along very stereotypical pathways to reach their distinct targets. The availability of cells that make stereotypical decisions in the developing CNS and that can be identified in the living embryo makes the grasshopper an attractive system for researching specific questions about neural development.

We have developed a number of techniques to perturb cells in the developing grasshopper CNS in very specific ways. Embryos can be partially

Advances in Pharmacology, Volume 42

1054-3589/98 $25.00

dissected, specific cells microinjected, and the whole embryo cultured for a number of days, such that neural development will occur normally. This has allowed lineage analysis of neuroblasts (2). In addition, we have inhibited the expression of a specific gene by microinjection of antisense oligonucleotides into cells (3). We can also manipulate signal transduction pathways in specific cells (4). We are now using these techniques to study the development of the one local serotonergic interneuron that innervates each ventral nerve cord ganglion.

Serotonin is a very important neuromodulator in invertebrates. Application of this compound to small neuronal circuits or whole nervous systems profoundly affects their physiological behavior. Invertebrates have small numbers of dopaminergic and serotonergic neurons that send extensive processes that innervate most of the central neuropil (5). Like raphe neurons in vertebrates, the invertebrate serotonergic neurons examined have an autonomous oscillatory type of electrical activity that is automodulated by serotonin (6).

In insects, each CNS hemiganglion has two serotonergic neurons (5). Only one of these innervates the neuropil extensively, while the other provides connections between segments, possibly to synchronize serotonergic modulation throughout the CNS. This CNS neuropilar innervation cannot be well visualized in the *Drosophila* due to its small size but can be seen in some peripheral extensions. Neurite outgrowth in these peripheral extensions is significantly amplified in flies that cannot synthesize dopamine (7). Thus, dopamine appears to inhibit neurite outgrowth of this insect serotonergic neurite, as in mollusks (8). This recapitulates the effects seen with vertebrate heterotypic outgrowth after dopaminergic neuron ablation (9).

Dopamine and serotonin receptors have been characterized in insects and mollusks, both pharmacologically and molecularly (10). These are remarkably conserved between vertebrates and invertebrates. It is likely that the serotonergic and dopaminergic neuronal systems were formed before the divergence of vertebrates and invertebrates and have been, possibly due to their central roles, conserved during evolution. Thus, experimental observations of these cells made in insects may also be applicable to vertebrates. The experimental access afforded by insects makes such studies an attractive approach to understanding the much more complex vertebrate system.

Serotonergic innvervation of the neuropil of each ventral nerve cord hemiganglion derives from one local serotonergic interneuron, s1. This neuron extends an axon early in development. It expresses serotonin uptake activity before making serotonin. Serotonin synthesis begins when the early branches reach the target neuropil. Once in the neuropil, s1 neurites form a set of stereotypical secondary branches. All of the neurite extension and neurotransmitter synthesis can occur in the correct manner in cultured embryos. We have conducted a screen for reagents that can control both neurotransmitter synthesis and secondary branch formation when applied to the incubation medium. Dopamine seems to be required both for the serotonin synthesis and for controlling the rate of neurite extension. The receptors controlling these processes have been pharmacologically characterized.

The stereotypy and manipulability of the grasshopper serotonergic neurons means that it should now be possible to study in great detail the mechanisms by which dopamine controls both neurotransmitter synthesis and neurite extension.

References

1. Doe, C. Q., Kuwada, J. Y., and Goodman, C. S. (1985). From epithelium to neuroblasts to neurons: The role of cell interactions and cell lineage during insect neurogenesis. *Philos. Trans. R. Soc. Lond. B* **312,** 67–81.
2. Condron, B. G., and Zinn, K. (1994). The grasshopper median neuroblast is a pluripotential stem cell that generates both neurons and midline glia. *J. Neurosci.* **14,** 5766–5777.
3. Condron, B. G., Patel, N. H., and Zinn, K. (1994). Engrailed controls glial/neuronal cell fate decisions at the midline of the central nervous system. *Neuron* **13,** 541–554.
4. Condron, B. G., and Zinn, K. (1995). Activation of cAMP-dependent protein kinase triggers a glial-to-neuronal cell-fate switch in an insect neuroblast lineage. *Curr. Biol.* **5,** 51–61.
5. Nassel, D. R. (1988). Serotonin and serotonin-immunoreactive neurons in the nervous system of insects. *Prog. Neurobiol.* **30,** 1–87.
6. Ma, P. M., and Weiger, W. A. (1993). Serotonin-containing neurons in lobsters: The actions of g-Aminobutyric acid, octopamine, serotonin and proctolin on activity of a pair of identified neurons in the first abdominal ganglion. *J. Neurophysiol.* **69,** 2015–2029.
7. Budnik, V., Wu, C-F., and White, K. (1989). Altered branching of serotonin-containing neurons in Drosophila mutants unable to synthesize serotonin and dopamine. *J. Neurosci.* **9,** 2866–2877.
8. Haydon, P. G., McCobb, D. P., and Kater, S. B. (1986). The regulation of neurite outgrowth, growth cone motility, and electrical synaptogenesis by serotonin. *J. Neurobiol.* **18,** 197–215.
9. Jacobs, B. L., and Azmitia, E. C. (1992). Structure and function of the brain serotonin system. *Physiol. Rev.* **72,** 165–229.
10. Roeder, T. (1994). Biogenic amines and their receptors in insects. *Comp. Biochem. Physiol.* **107C,** 1–12.

Carolyn W. Harley

Departments of Psychology and Basic Sciences
Memorial University of Newfoundland
St. John's, Newfoundland, Canada A1B 3X9

Noradrenergic Long-Term Potentiation in the Dentate Gyrus

I. Historical Perspective

We first described noradrenergic (NE)-induced long-lasting (or long-term) potentiation (LTP) in the dentate gyrus in 1983, a decade after the initial report of high-frequency–induced LTP in the same structure. While noradrenergic-induced LTP has not received the same attention as high-frequency–induced LTP, we suggest it is of similar relevance for learning and memory.

In parallel with the early studies of high-frequency–induced LTP, NE-LTP was first observed *in vivo* in anesthetized animals using 0.1-Hz stimulation of the perforant path, the major input to the dentate gyrus of the hippocampus. The population spike of the perforant path evoked potential, an index of granule cell firing to the perforant path glutamatergic input, was consistently potentiated by NE iontophoresis in the granule cell layer. Both short-term and long-term (>30 min) potentiations were observed. In one experiment NE-LTP was followed for 11 hr after a brief iontophoretic application.

While NE potentiation of glutamatergic inputs had previously been described in the sensory cortices and cerebellum, potentiation at those sites was transient and mediated by activation of α_1 receptors. *In vitro* pharmacological studies of NE-LTP in dentate gyrus revealed its dependence on β_1-receptor activation, implicating cyclic adenosine monophosphate (cAMP) as the second messenger. *In vitro* work also suggested that NE-LTP was related to dose and duration of NE superfusion.

Although it was clear that exogenous NE could induce long-term changes in dentate gyrus circuitry, it remained to be proven that synaptic release of NE would be effective in inducing NE-LTP. We had observed potentiation of perforant path evoked potentials by electrical stimulation in the locus ceruleus (LC), the primary source of dentate gyrus NE, which became long-term with repeated pairings, but we were unable to block such potentiation with a β antagonist. Because electrical stimulation is likely to activate many systems in the LC region, we adopted micropressure ejection of glutamate locally in the LC as an alternative strategy for LC activation. Glutamatergic activation of LC induced both short- and long-term potentiation of the perforant path population spike, and these effects were attenuated by systemic administration of a β antagonist or by local ejection of a β antagonist in the dentate gyrus.

Advances in Pharmacology, Volume 42

1054-3589/98 $25.00

This earlier work was reviewed in Harley (1). More recently, we have examined the relationship of NE-dependent potentiation to N-methy-D-aspartate (NMDA)–dependent potentiation, the control of NE-dependent potentiation by NE levels *in vivo,* the influence of LC-NE on interneurons in the dentate gyrus circuit, and the occurrence of NE-dependent potentiation in awake animals.

II. Relationship between NMDA-Dependent LTP and NE-LTP

In vivo it appears that NMDA blockade does not affect LC-NE–induced potentiation. Using two neighboring micropipettes with 40-μm tips in the dentate gyrus to permit the diffusion of either saline or the NMDA blocker, ketamine, Frizzell and Harley (2) showed that ketamine effectively blocked perforant path high-frequency–induced LTP but did not attenuate LC-NE–induced potentiation, either short- or long-term. More recently, we have found that the β-blocker timolol in one pipette slows but does not prevent the development of, maximal high-frequency–induced LTP. Potentiation produced by stimulating the major glutamatergic input to LC from the paragigantocellularis nucleus (PGi) is blocked, as expected, on the intradentate timolol pipette. More importantly, when high-frequency LTP was saturated and could no longer be increased, stimulation of the PGi produced additional potentiation on the saline, but not the timolol, micropipette. These results suggest that NE-mediated potentiation is distinct from that initiated by NMDA receptor activation *in vivo.*

III. Estimating the Synaptic Concentration of NE That Induces Potentiation *in Vivo*

In both anesthetized and awake animals, we have demonstrated that NE administered intracerebroventricularly (i.c.v.) can produce NE potentiation. The duration of the potentiation is related to the dose of NE, although there is inherent variability in the levels of NE that reach the recording site in the dentate gyrus. This inherent variability of the i.c.v. administration route provided us with an opportunity to correlate extracellular NE concentrations at the dentate gyrus recording site, measured by microdialysis, with the occurrence of potentiation. These experiments were done in collaboration with Dr. Nutt's group at the University of Bristol. Although only 5 μg of NE were administered i.c.v., a dose we expected to give short-term effects, long-term changes were seen in the majority of experiments. The lowest increase in NE associated with LTP at the recording site was 30 times the basal NE levels. Levels of 6 times or less did not alter the evoked potential. In one experiment, 8 times the basal level was associated with a weak transient potentiation. Using these data and other data from Dr. Nutt's laboratory measuring the extracellular NE synaptic spillover associated with NE potentiation induced by the α_2 antagonist idazoxan, we estimated that synaptic concentrations of 0.4 μM or higher would be required to engage LTP mechanisms (Fig. 1). This is considerably higher than levels estimated for effective α_2-receptor activation and suggests that NE-LTP is likely to depend on higher frequency bursts of NE activity at targeted sites (3).

IV. LC-NE Modulation of the Dentate Gyrus Circuitry

Our most recent investigations have been directed at identifying the cellular changes produced by NE in dentate gyrus and relating those changes to the potentiation effects we first observed more than a decade ago. In these studies, we use glutamate ejections in LC to activate the LC and simultaneously monitor evoked potentials, electroencephalogram, and spontaneous unit activity in the dentate gyrus. We had earlier demonstrated, in collaboration with Dr. Susan Sara, that glutamate ejection in the LC produces a brief (< 0.5 sec) but dramatic burst of LC firing, followed by minutes of silence and a subsequent return to baseline rates. Curiously, the potentiation observed in the dentate gyrus lags the LC burst by 10–20 sec. Our preliminary unit data may provide an explanation of this delay.

We have found three distinct populations of nonbursting cells that are likely to be GABAergic neurons in the dentate hilus (the site of densest LC-NE innervation). One group, found in the immediate subgranular zone, was immediately, and briefly (10–20 sec), excited by LC activation. A second group, farther from the granular layer, was immediately inhibited by LC activation for a somewhat longer period (80–90 sec), and a third subgranular population was unchanged by LC activation. In all experiments, potentiation of the population spike was observed, and in eight of 14 experiments, NE-LTP occurred. In relating cell behavior to the evoked potential changes, it seemed evident that

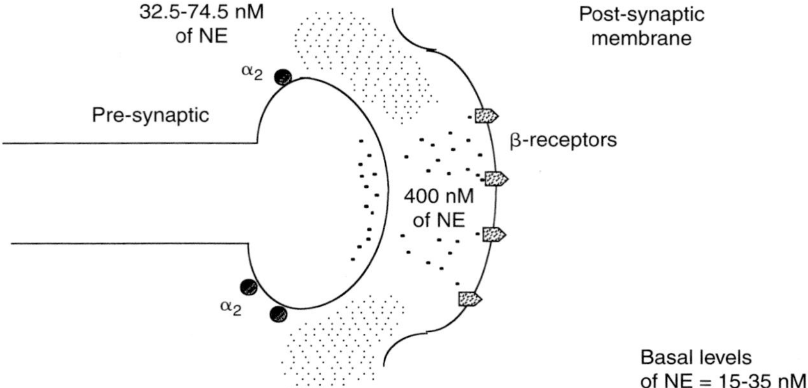

FIGURE 1 Estimates of the NE synaptic concentration required to activate β-mediated LTP suggest that higher NE levels (400 nM) must be reached to recruit NE-LTP than to engage α_2 autoreceptors or heteroreceptors, which appear responsive at normal extracellular levels.

the initial 10- to 20-sec period of unchanged population spike responses reflects mutually antagonistic influences of both increased and decreased GABA tone from different populations. Temporally, the initial phasic increase in the population spike after the brief null period relates well to the period of inhibition of the GABAergic neurons, which outlasts the excitatory burst. This disinhibition might be expected to promote NMDA-mediated plasticity change or calcium entry in the granule cells. The tonic potentiation that then follows, however, occurs after the firing rates of both GABAergic groups have returned to normal. Theta activity also appears to be briefly promoted by glutamate-induced LC-NE activation, while the longer period of LC silence (5–6 min) seemed to be accompanied by theta decreases.

Taken together, there are multiple ways in which NE may promote long-term functional change in the dentate gyrus: (1) by a direct cAMP-mediated effect on granule cells or other cells or inputs of the dentate gyrus, (2) by producing a transient but marked disinhibition of granule cells, and (3) by enhancing theta rhythmicity. These last effects would increase the likelihood of NMDA-mediated plasticity or possibly enhance voltage-mediated calcium entry.

V. Occurrence of NE Potentiation in the Awake Animal

In collaboration with Dr. Susan Sara's laboratory at the University of Paris VI, we examined the dentate gyrus perforant path evoked potential in awake rats exploring novel objects in a hole board apparatus. Dr. Sara had previously demonstrated that LC cells burst transiently during the first inspection of a hole or novel object. As predicted, we found potentiation of the population spike amplitude after the rats inspected novel holes or objects but not familiar holes or objects. An LTP was seen when animals were first placed in the hole board environment. All novelty-induced potentiation effects were blocked by prior administration of propranolol (4).

VI. LC-NE Potentiation: A Continuum from Attention through Memory

While it is likely that even transient LC-NE activation will invariably be accompanied by potentiation effects in the awake animal, as we have shown during novel object inspection, it is predicted that more pronounced or repeated LC activation will produce NE-LTP, as appeared to happen with the first exposure to a novel environment. Memory and learning mediated by LC-NE would, thus, be at one end of a continuum of attentional effects promoted by the LC-NE system, with either level or duration or repetition of NE exposure as the controlling variables.

References

1. Harley, C. W. (1991). Noradrenergic and locus coeruleus modulation of the perforant path-evoked potential in rat dentate gyrus supports a role for the locus coeruleus in attentional and memorial processes. *Prog. Brain Res.* **88**, 307–321.
2. Frizzell, L. M., and Harley, C. W. (1994). The N-methyl-D-aspartate channel blocker ketamine does not attenuate, but enhances, locus coeruleus-induced potentiation in rat dentate gyrus. *Brain Res.* **663**, 173–178.
3. Harley, C. W., Lalies, M. D., and Nutt, D. J. (1996). Estimating the synaptic concentration of norepinephrine in dentate gyrus which produces β-receptor mediated long-lasting potentiation in vivo using microdialysis and intracerebroventricular norepinephrine. *Brain Res.* **710**, 293–298.
4. Kichigina, V., Vankov, A., Harley, C. W., and Sara, S. J. (1997). Novelty-elicited, noradrenaline-dependent enhancement of excitability in the dentate gyrus. *Eur. J. Neurosci.* **9**, 41–47.

P. Kubiak, J. Rajkowski, S. Ivanova, and G. Aston-Jones

Department of Psychiatry
Allegheny University
Philadelphia, Pennsylvania 19102

Rapid Acquisition of Discriminative Responding in Monkey Locus Coeruleus Neurons

In 1970 Kety (1) proposed that activity of the central norepinephrine (NE) system enhances attention and memory for affectively important events. Indeed, in accordance with this hypothesis, improved learning and memory has been observed after stimulation of the locus coeruleus (LC)–NE system (2). In addition, several studies have shown that NE, or increased LC activity, can enhance long-term potentiation, indicating a possible cellular substrate for LC-NE involvement in learning (3). Moreover, depletion of NE decreased long-term potentiation in the dentate gyrus (3). Lesions of LC neurons have produced less consistent results. Learning deficits were found in some studies after lesions of the LC nucleus or efferent LC axons, but other investigations reported no effect of LC lesions on learning of olfactory or visual discrimination (4). The reason for these discrepancies is unknown, but the plasticity and compensation that occur following damage to the LC-NE system may confound lesion results.

Thus, although results with lesions have been variable, previous evidence consistently indicates that increased LC-NE activity may facilitate learning. It

Advances in Pharmacology, Volume 42

1054-3589/98 $25.00

follows that central NE may contribute most to learning about behaviors or stimuli that are associated with activation of LC-NE neurons. Recordings in behaving animals revealed that increased LC activity occurs during exploration or behavioral orientation (5). In addition, LC neurons are phasically activated by many sensory stimuli in unanesthetized, freely behaving animals (5); such responses are largest in magnitude for salient stimuli that cause behavioral orienting or waking responses. We reported that LC neurons in behaving monkeys are selectively activated by CS+ stimuli in a visual discrimination–vigilance task (6). The use of CS+ and CS− stimuli enabled us to reverse the task contingency and examine LC activity during the acquisition of new stimulus meaning. This approach offered an opportunity to reinvestigate the role of the LC in attention and learning. Specifically, this study investigated the hypothesis that LC activity can influence the acquisition of a new task contingency. An important advantage of our approach was that, after some experience, reversal of the task contingency can be accomplished during a single experimental session. This allowed study of LC activity in response to a change in stimulus meaning and task contingency, and permitted comparison of the response of LC neurons to stimuli in the original contingency with that for the same cells at various times after reversal.

Three adult Cynomolgus monkeys (*Macaca fasicularis*) were trained to perform a visual discrimination task. In this task, after the monkey depressed a nearby lever, a small rectangular spot (fixation spot) appeared in the center of the video monitor. The animal was required to visually fixate this spot for at least 230 ms, after which it was replaced by either a vertical or horizontal line segment (7×2.5 mm, 1.7-degree visual angle; white on a dark background). One orientation was initially used as the target (CS+) stimulus and the other as the nontarget (CS−) stimulus. The animal had to release the lever within 665 ms of the onset of the target stimulus to receive a juice reward. Incorrect releases within 665 ms of nontarget stimuli resulted in 3-sec time out. Target stimuli occurred on 20% of trials and were randomly dispersed among nontarget stimuli. Performance was measured in terms of hits (releases to target stimuli), misses (nonreleases to target stimuli), rejections (nonreleases to nontarget stimuli), and false alarms (releases to nontarget stimuli). Monkeys were overtrained for several months on one stimulus contingency in the task. Subsequently, reversal was instituted unexpectedly by abruptly reversing the meaning of the target and nontarget stimuli (i.e., the previous target stimulus became nontarget, and the previous nontarget became target). The overtrained contingency (original task) was employed for at least three experimental sessions between sessions that contained reversal to the alternative contingency (reversed task).

Extracellular recordings from individual or multiple LC neurons were obtained from microwire electrodes with beveled tips (25-μ-diameter stainless steel, Teflon insulated). The distinctive electrophysiological properties of LC neurons (6) were used to help identify LC neurons during recording sessions. Electroencephalographic signals were recorded differentially between two scull screws.

In this task, LC neurons were selectively and phasically activated by CS+ stimuli, as previously reported (6) (onset latency $\sim$ 140 ms in the overtrained contingency). Other task events (nontarget stimuli, fix spot stimuli, lever release) elicited no consistent responses from these neurons. This selective activation by target stimuli was true for whichever orientation was chosen as target. These findings indicate that phasic responses to CS+ stimuli are specifically related to their meaning and not driven by their sensory characteristics.

After reversal of the stimulus–response contingency, all LC neurons tested reversed their stimulus–response profile and became selectively responsive to the new target (old nontarget) stimuli and were unresponsive to the old target stimuli (new nontarget). In all 11 LC recordings examined in this manner, neural excitation was detected within the first 100 new target stimuli, while PSTHs revealed that each of these neurons had extinguished responses to nontarget stimuli within the first 100 nontarget stimuli of the new contingency. The magnitudes of phasic responses (Rmags) for all LC neurons for the first 100 target stimuli of the new contingency were slightly smaller but were not significantly different from the Rmags for 100 stimuli of the original contingency just preceding reversal.

After each reversal of the task contingency, monkeys initially performed the task with numerous errors. Within each session, an improvement in accuracy of behavioral discrimination with continued performance of the reversed task was observed. However, behavioral acquisition of the new contingency to a criterion of fewer than 5% errors was not reached for over 1000 trials with the reversed task (n = 11 recordings).

To analyze the temporal relationship between LC and behavioral responses during acquisition of a new stimulus contingency, the rate of reversal of LC phasic responses was compared with the rate of behavioral reversal for the same task epochs. As shown in Table I, early after reversal (within the first 500 trials of the new contingency), LC neurons exhibited very good discrimination, and Rmags for LC responses to the new target and nontarget stimuli did not differ significantly from values for corresponding stimuli of the original contingency. In particular, by this point after reversal, LC neurons exhibited very little response to old target (new nontarget) stimuli. In contrast, animals did not behaviorally discriminate well at this time and exhibited significantly more errors (especially false alarms for old target stimuli) than before reversal (see Table I). Introduction of the new stimulus–response contingency also rapidly and transiently increased the tonic activity of LC neurons (~25% increase for at least 10 min; $p < .05$).

These data indicate that conditioned responses of LC neurons reflect stimulus meaning and cognitive processing and are not driven by physical sensory attributes. Especially noteworthy is the finding that following task reversal, the LC acquires the new response profile before accurate behavioral performance on the new task is established. The early changes that occur in LC tonic and stimulus-evoked activity after task reversal indicate that the LC may play an active role in acquisition of stimulus significance.

The widespread efferent anatomy of the NE-LC system indicates that it may be involved in a very general aspect of learning rather than for learning one type of task only. The LC has been associated with several functions, perhaps most strongly with attentional processes. This view does not conflict with a role of this system in learning. Indeed, attentiveness is an important

TABLE I Reversal of LC Responses to Task Stimuli Anticipates Reversal of Lever Release Responses

	LC RMag (vertical stimuli)	*LC RMag (horizontal stimuli)*	No. *lever releases (vertical stimuli)*	No. *lever releases (horizontal stimuli)*
Before reversal (target = vertical)	56.8 ± 17.1 (targets)	0.3 ± 3.9 (nontargets)	99.3 ± 0.4* (100 targets)	3.4 ± 1.5** (400 nontargets)
After reversal (target = horizontal)	11.5 ± 4.9 (nontargets)	52.1 ± 18.0 (targets)	115.3 ± 27.0** (400 nontargets)	88.5 ± 3.9* (100 targets)

Data are expressed as average ±SEM response magnitudes (Rmags) for LC single neurons in response to vertical or horizontal stimuli, and as the average ±SEM number of lever releases for these same stimuli. Data are for 500 consecutive trials (100 target and 400 nontarget stimuli), both before and after reversal, as indicated. LC Rmags for new targets and nontargets at this point after reversal did not differ statistically from the corresponding values for old targets and nontargets before reversal (n = 11 recordings, paired *t*-tests). However, false alarm behavioral responses were significantly greater ($^{**}p < .002$) and hits were significantly fewer ($^{*}p = .02$) than before reversal. Thus, by this point, there is no LC response to new nontargets, while behavioral responses to these same stimuli (false alarms) are still frequent, indicating that LC responses extinguished before behavioral responses. The latter required many more than the 500 trials analyzed here to extinguish (see text). Rmags were calculated for 100 target and nontarget stimuli (including only every fourth nontarget stimulus).

prerequisite for learning. The brain substrate for the connection between attentiveness and learning has not been identified, but these results indicate that it may involve LC. Specifically, LC could play a role in acquisition of a new contingency by altering the relative salience of target and nontarget stimuli. Increased salience for a new target stimulus (caused in part by association with phasic LC responses) may alter the response probabilities for task stimuli (e.g., modulate attention) and facilitate processes involved in learning the new reward contingency. Alternatively, the LC could directly enhance the conditioned relationship among circuit elements that represent the new task contingency. In this view, diminished LC responses to the previous target stimuli may withdraw support from the circuit connections that represented the prior contingency, while the onset of LC responses to the new target stimulus may reinforce and strengthen over trials the relationship among circuit elements that represent the new contingency.

References

1. Kety, S. S. (1970). The biogenic amines in the central nervous system: Their possible roles in the arousal, emotion and learning. *In* The Neurosciences: Second Study Program, F. O. Schmitt, ed.), pp. 324–336. University Press, New York.
2. Velley, L., and Cardo, B. (1982). Facilitation of acquisition and extinction of an operant task four weeks after stimulation of brainstem aminergic nuclei of the rat brain. *Behav. Neural Biol.* **35**, 395–407.
3. Harley, C. (1991). Noradrenergic and locus coeruleus modulation of the perforate path-evoked potential in rat dentate gyrus supports a role for locus coeruleus in attentional and memorial processes. *Prog. Brain Res.* **88**, 307–321.
4. Mason, S. T. (1984). Catecholamines and Behavior. Cambridge University Press, London.
5. Aston-Jones, G., Chiang, C., and Alexinsky, T. (1991). Discharge of noradrenergic locus coeruleus neurons in behaving rats and monkeys suggest a role in vigilance. *Prog. Brain Res.* **88**, 501–520.
6. Aston-Jones, G., Rajkowski, J., Kubiak, P., and Alexinsky, T. (1994). Locus coeruleus neurons in monkey are selectively activated by attended stimuli in a vigilance task. *J. Neurosci.* **14**, 4467–4480.

Michael Leon

Department of Psychobiology
University of California
Irvine, California 92697

Catecholaminergic Contributions to Early Learning

The developing brain has evolved unique capacities for dealing with the specific demands of the perinatal world. These capacities typically are transitory but confer a critical ability for the developing organism to survive as it and its world rapidly change. One capacity that is critical for perinatal mammals is the identification of the mother, because she holds the vital elements for the continued survival of the infants. In the period immediately following birth, there is a need to identify and approach the mother, and olfaction seems to play a key role in mediating this attraction.

Human infants use an early capacity to learn to develop an early attraction to such odors (1). On their first day of life, infants were taken from their mothers and exposed to a neutral odor, either in the presence of concurrent tactile stimulation that mimicked maternal care, odor alone, tactile stimulation alone, or odor followed by tactile stimulation. The next day, the infants were brought again to the same clean environment, where they were allowed to demonstrate a preference for the trained odor by turning their head toward it. Only those infants that had received concurrent odor and tactile stimulation had developed a preference for that odor. A second, neutral odor did not evoke the preferential response evoked by the trained odor.

As it happens, laboratory rats also have infants that become attracted to the odor of their mothers (2). Moreover, infant rats will acquire a preference for a neutral odor when it is paired either with tactile stimulation or maternal care. Similarly, infant rats will not be attracted to a second odor after training with a first. Unlike human infants, however, we have been able to identify clear changes in the brains of rat infants that are required for the olfactory preference acquisition.

It seemed likely that the tactile stimulation that was critical for the development of the olfactory attraction was mediated by a centrifugal system that interacted with the olfactory system. The major centrifugal system projecting to the olfactory bulb is the noradrenergic system, and we considered its role in the development of a learned attraction. Indeed, monitoring the extracellular space in the olfactory bulb revealed that there is a sharp increase in norepinephrine in response to early olfactory preference training (3).

To test the possible functional significance of this phenomenon, we blocked the binding of β-noradrenergic receptors during training and found that it prevented the induction of acquired olfactory preferences (4). The drugs that

Advances in Pharmacology, Volume 42

1054-3589/98 $25.00

successfully blocked the formation of the learned preference did not interfere with the expression of the acquired response, indicating that the drugs did not block the ability to sense an odor (5). In addition, destruction of the locus ceruleus prevented the formation of early olfactory preferences (6).

It was possible, of course, that the administration of β-adrenergic antagonists, or the destruction of the noradrenergic fibers interfered with the acquisition of the preference by a mechanism that was not associated with learning. To test that possibility, as well as the hypothesis that the effects of the tactile stimulation were mediated by noradrenergic action, we replaced the tactile stimulation with a noradrenergic agonist and paired it with an olfactory cue (1). Such a pairing successfully induced learning in the young rats. These data indicate that norepinephrine may well be involved in the primary mechanism of preference acquisition. Moreover, the effects of tactile stimulation appear to be mediated by the action of norepinephrine.

It was also the case that combining odor, tactile stimulation, and a noradrenergic agonist blocked the acquisition of an olfactory preference (1). These findings suggested the possibility that the combination of exogenous and endogenous noradrenergic stimulation effectively blocked the early preference acquisition. If that were the case, then a dose–response study with odor and a noradrenergic agonist might reveal an inverted U-shared function. Indeed, it did, and this type of relationship between noradrenergic stimulation and learned behavior has been reported repeatedly in other situations involving memory.

In an initial effort to determine the mechanism or mechanisms underlying the action of olfactory bulb norepinephrine in supporting the acquisition of early olfactory preferences, we determined the spatial localization of β-noradrenergic receptors in the bulb (7). These receptors localize in both the granule cell layer and the glomerular layer of the bulb. While the spatial distribution of these receptors was fairly uniform throughout the granule cell layer, there was a remarkable distribution of the receptors in the glomerular layer. Specifically, there was clear localization of the receptors to relatively few glomeruli, and these glomeruli were localized principally on the lateral aspect of the lamina. These data are particularly interesting, because the learned odor induces an enhanced response (as assessed by 2-deoxyglucose uptake or Fos immunostaining) (9, 10) in glomeruli only within the lateral aspect of the glomerular layer. It seems possible that the presence of the β-noradrenergic receptors may confer a spatial specificity to the neural plasticity within the olfactory bulb.

If the β-noradrenergic receptors are exposed to repeated high levels of norepinephrine during daily preference training, perhaps there would be a downregulation of these receptors in the bulb. In fact, the density of the β-noradrenergic receptors decreases with early olfactory preference training (8). Conversely, the loss of norepinephrine might be expected to upregulate the expression of these receptors. Destruction of the noradrenergic fibers that project to the bulb increases the binding of these receptors in that structure (11).

What might be the mechanism of noradrenergic action? Given the presence of the β receptors in two lamina, there may be two modes by which norepinephrine affects the formation of early olfactory preferences. The granule cells serve to inhibit the firing of the mitral cells, which are the dominant projection neurons of the bulb, and norepinephrine blocks this inhibition. If one records from a population of mitral cells during a period of noradrenergic stimulation, these cells continue to fire in response to an odor, while the mitral cells of unstimulated pups, or of pups in which the action of norepinephrine has been blocked, decrease their firing during successive odor presentations (12). These data suggest that the tactile stimulation may allow a unique, continuing response by mitral cells in response to trained odors by increasing bulb norepinephrine. Such a response may be critical in establishing the odor as one that subsequently will be preferred.

In both the glomerular and granule cell layers, we considered the possibility that β-noradrenergic receptors might affect early learning by affecting the resident astrocytes. A distinguishing characteristic of these cells in the brain is that they contain glycogen. Both laminae have very high levels of glycogen phosphorylase, the rate-limiting enzyme in the mobilization of glycogen (13). Moreover, the bulb has extraordinarily high levels of glycogen, particularly during the first week of life, when the pups are particularly sensitive to acquiring learned olfactory preferences (14, 15). During the first week of life, olfactory bulb slices respond to β- but not α-noradrenergic agonists by mobilizing glycogen, even at very low doses (14). The mobilization of glycogen may support a critical function of glia, such as potassium modulation, or it may supply local neurons with a continued energy supply during a period of high activity.

Acknowledgment

This work was supported by grant PO1 24326 from the National Institute of Child Health and Human Development.

References

1. Sullivan, R. M., McGaugh, J., and Leon, M. (1991). Norepinephrine-induced plasticity and one-trial olfactory learning in neonatal rats. *Dev. Brain Res.* **60,** 219–228.
2. Leon, M. (1992). Neuroethology of olfactory preference development. *J. Neurobiol.* **23,** 1557–1573.
3. Rangel, S., and Leon, M. (1995). Early olfactory preference training increases olfactory bulb norepinephrine. *Dev. Brain Res.* **85,** 187–191.
4. Sullivan, R. M., Wilson, D. A., and Leon, M. (1989). Norepinephrine and learning-induced plasticity in infant rat olfactory system. *J. Neurosci.* **9,** 3998–4006.
5. Sullivan, R. M., and Wilson, D. A. (1991). The role of norepinephrine in the expression of learned olfactory neurobehavioral responses in infant rats. *Psychobiology* **19,** 308–312.
6. Sullivan, R. M., Wilson, D. A., Lemon, C., Pham, C., and Gerhardt, G. A. (1994). Bilateral 6-OHDA lesions of the locus coeruleus impair associative olfactory learning in newborn rats. *Brain Res.* **643,** 306–309.
7. Woo, C. C., and Leon, M. (1995). Distribution and development of beta-adrenergic receptors in the rat olfactory bulb. *J. Comp. Neurol.* **352,** 1–10.
8. Woo, C. C., and Leon, M. (1995). Early olfactory enrichment and deprivation both decrease beta adrenergic receptor density in the main olfactory bulb of the rat. *J. Comp. Neurol.* **360,** 634–642.
9. Johnson, B. A., Woo, C. C., Duong, H-C., Nguyen, V., and Leon, M. (1995). A learned odor evokes an enhanced Fos-like glomerular response in the olfactory bulb of the young rats. *Brain Res.* **699,** 192–200.

10. Johnson, B., and Leon, M. Spatial distribution of ^{14}C 2-deoxyglucose uptake in the glomerular layer of the rat olfactory bulb following early odor preference learning. (1996). *J. Comp. Neurol.* **376,** 557–566.
11. Woo, C. C., Wilson, D. A., Sullivan, R. M., and Leon, M. Early locus coeruleus lesions increase the density of beta adrenergic receptors in the main olfactory bulb of rats. *Int. J. Dev. Neurosci.* **14,** 913–919.
12. Wilson, D. A., and Sullivan, R. M. (1991). Olfactory associative conditioning in infant rats with brain stimulation as reward II; norepinephrine mediates a specific component of the bulb response to reward. *Behav. Neurosci.* **105,** 843–849.
13. Coopersmith, R., and Leon, M. (1987). Glycogen phosphorylase activity in the olfactory bulb of the young rat. *J. Comp. Neurol.* **261,** 148–154.
14. Coopersmith, R., and Leon, M. (1995). Olfactory bulb glycogen metabolism: Noradrenergic modulation in the young rat. *Brain Res.* **674,** 230–237.
15. Woo, C. C., and Leon, M. (1987). Sensitive period for neural and behavioral response development to learned odors. *Dev. Brain Res.* **36,** 309–313.

Larry Cahill

Center for the Neurobiology of Learning and Memory
University of California
Irvine, California 92697

Interactions between Catecholamines and the Amygdala in Emotional Memory: Subclinical and Clinical Evidence

A large and diverse body of evidence now suggests that memory storage processes can be modulated by endogenous stress hormones released during emotionally arousing events (1, 2). Additional evidence suggests that stress hormones influence memory via the amygdaloid complex (AC), the brain region most consistently implicated in emotional memory processes (2). The evidence indicates that the AC has a particular role in brain memory mechanisms: It interacts with stress hormones released during and after emotional events to modulate memory for those events (2).

The effect of endogenous stress hormones (in particular the adrenergic hormones epinephrine and norepinephrine) on memory follow a typical pattern (1, 2). In general, stimulation at relatively low doses enhances and at higher doses impairs memory (the classic inverted-U dose–response function). Also, the effects are time-dependent (i.e., most effective when given soon after train-

Advances in Pharmacology, Volume 42

1054-3589/98 $25.00

ing). The time-dependent nature of the effects is generally considered evidence that the hormones act on memory consolidation processes.

The AC is the brain region most consistently implicated in memory enhancement by endogenous stress hormones. Many studies have shown that both memory enhancement and memory impairment resulting from peripheral injection of drugs and hormones are blocked by lesions of the AC and its related circuitry (2). Extensive evidence also indicates that treatments influencing noradrenergic functioning within the AC affect memory formation. For example, intra-AC infusions of the β-adrenergic antagonist propranolol impair retention. Retention is enhanced by infusions of low doses of norepinephrine and impaired by high doses (1). Also, AC stimulation with the indirect catecholaminergic agonist amphetamine affects memory storage in several water maze tasks (2). Finally, intra-AC injection of various drugs modulates the effects of systemically administered drugs and hormones on memory (1, 2).

Storage of Emotional Memory in Humans: Catecholamines and the Amygdala

Recent studies involving human subjects confirm the view derived from animal subject studies that memory for emotionally arousing events requires activation of β-adrenergic receptors and the AC. We examined the effect of β-adrenergic impairment (using the β-adrenergic blocker propranolol) on enhanced memory associated with emotional arousal in healthy humans (3). Subjects received either a placebo or 40 mg of propranolol 1 hr before viewing either a relatively neutral or a more emotionally arousing short story. Memory for the story was assessed 1 wk later. Propranolol significantly impaired memory only in subjects who viewed the emotional story, primarily for the story phase in which the emotional story elements were introduced (Fig. 1A, phase 2). Because propranolol selectively affected only the emotional story, its impairing effect on memory cannot be attributed to any general effects of the drug on attention or level of sedation. Also, because the propranolol subjects did not differ from placebo subjects in their self-assessed emotional reaction to either story, the memory impairment cannot easily be attributed to reduced emotional responsiveness of the subjects to the emotional story.

Cahill *et al.* (4) examined memory of a rare patient (B.P.) with bilateral, selective damage to the AC using the same emotional memory paradigm already shown to be sensitive to β-adrenergic blockade (3). They found that B.P. exhibited a memory deficit similar to that of healthy subjects under β-adrenergic blockade—a selective impairment of enhanced memory normally associated with emotional arousal (Fig. 1B). This finding indicates that the AC is involved in long-term, emotionally influenced, declarative memory in humans. The fact that both the β-adrenergic system (3) and the AC (4) are implicated in memory within a single experimental paradigm further argues for an interaction between the two in storing memory for emotional events.

We have also demonstrated selective involvement of the amygdala in enhanced long-term, emotionally influenced, declarative memory in healthy hu-

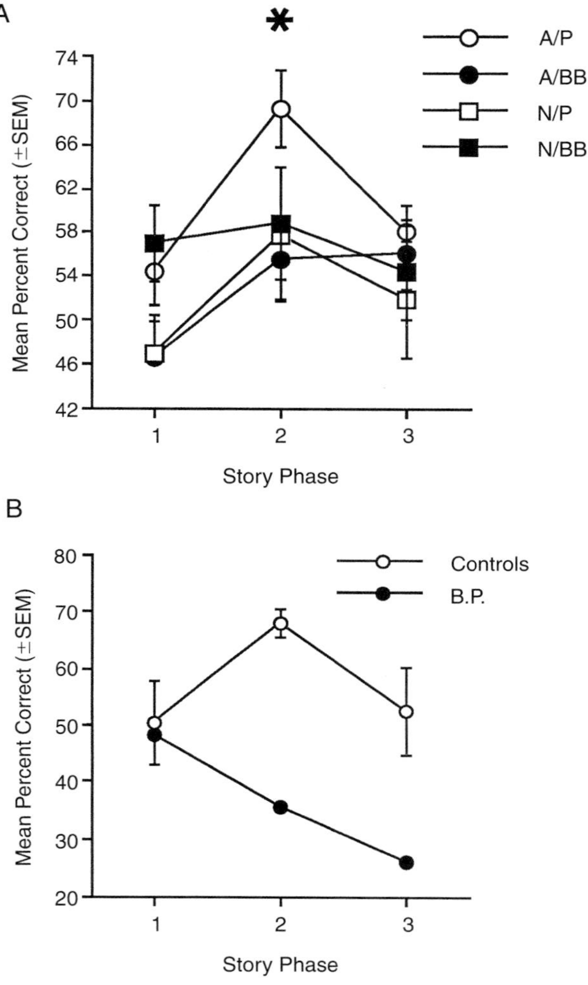

FIGURE 1 (*A*) Performance on a multiple-choice memory test for subjects taking either a placebo or propranolol 1 hr before viewing either an emotionally arousing or neutral story. The arousing story elements were introduced into the middle phase of the arousing story (phase 2). Enhanced memory for these elements was seen only in the placebo group. $^{*}p < .02$ compared with the arousal/β-blocker (A/BB) group for story phase 2. Propranolol had no effect on memory for the neutral story. A/P, arousal story/placebo; A/BB, arousal story/β-blocker; N/P, neutral story/placebo; N/BB, neutral story/β-blocker. (Reprinted with permission from *Nature*, Vol. 371, 1994.) (*B*) Performance of a patient (B.P.) with bilateral amygdala damage on the emotionally arousing story. Unlike healthy controls, B.P. failed to show enhanced memory for the material associated with emotional arousal (story phase 2). In this respect, his performance was similar to that of healthy controls subjects given propranolol. (Reprinted with permission from *Nature*, Vol. 377, 1995, Macmillan Magazines, Ltd.)

mans using positron emission tomography (PET) (5). Subjects in this study received two PET scans: one while viewing a series of relatively emotionally arousing films, and a second while viewing a series of relatively emotionally neutral films. The results showed that activity of the right amygdala while viewing the films correlated highly ($r = +.93$) with recall of the emotional but not neutral films. Thus, as found in the study of patient B.P. (4), these PET findings indicate that the amygdala is not required for declarative memory in all circumstances; rather, it is selectively important for the formation of enhanced memory normally associated with emotional arousal.

To summarize, converging evidence from animal and human subject studies suggests that memory can be modulated by endogenous hormones—in particular, the catecholamines—known to be released during and after stressful events. These hormones are thought to affect memory via the AC. Collectively, these experiments strongly support the view that memory for emotionally arousing events is modulated by endogenous, hormonally based systems (1, 2).

Acknowledgment

The author wishes to thank Dr. James L. McGaugh for his support during these studies.

References

1. McGaugh, J. L. (1989). Involvement of hormonal and neuromodulatory systems in the regulation of memory storage. *Annu. Rev. Neurosci.* **12**, 255–287.
2. Cahill, L., and McGaugh, J. L. (1996). Modulation of memory storage. *Curr. Opin. Neurobiol.* **6**, 237–242.
3. Cahill, L., Prins, B., Weber, M., and McGaugh, J. L. (1994). β-adrenergic activation and memory for emotional events. *Nature* **371**, 702–704.
4. Cahill, L., Babinsky, R., Markowitsch, H., and McGaugh, J. L. (1995). The amygdala and emotional memory. *Nature* **377**, 295–296.
5. Cahill, L., Haier, R., Fallon, J., Alkire, M., Tang, C., Keator, D., Wu, J., and McGaugh, J. L. (1996). Amygdala activity at encoding correlated with long-term, free recall of emotional information. *Proc. Natl. Acad. Sci. U.S.A.* **93**, 8016–8021.

PART I

DRUG ABUSE AND ALCOHOLISM

G. F. Koob

Department of Neuropharmacology
The Scripps Research Institute
La Jolla, California 92037

Overview

Significant advances have been made in the past few years regarding the mechanism of action of drugs of abuse that influence catecholamines at three levels of analysis: system, cellular, and molecular. The focus at the VII International Catecholamine Symposium was on psychostimulant drugs such as cocaine and amphetamine that act on the mesolimbic dopamine system and that act through a molecular transduction system that includes certain dopamine receptor subtypes and immediate early genes. This overview provides an integration of the chapters of this theme by level of analysis: system, cellular, and molecular. First, there is an overall synthesis, then there are brief synopses of each chapter, and finally the hypothesized implications of this body of work.

At the system level, the focus on the mesolimbic dopamine system has led to the identification of neural circuits involved in drug reward and to a new concept of the extended amygdala (Koob, p. 978). Dopamine elements of this extended amygdala, notably the shell or medial aspect of the nucleus accumbens, appear to be particularly sensitive to drug and natural rewards (Di Chiara *et al.*, p. 983). Chronic stimulant administration leads to a progressive increase in stimulant-induced activity, and this sensitization is environment-dependent (Robinson and Badiani, p. 987). Repeated administration of stimulant drugs also results in the development of a conditioned locomotor response in the absence of drug, and this conditioned effect appears to depend on mesolimbic

Advances in Pharmacology, Volume 42

1054-3589/98 $25.00

dopamine activation in the nucleus accumbens (Pert, p. 991). Substitution of a slow-onset dopamine agonist is hypothesized to be one potential approach for drug treatment for cocaine dependence based on the hypothesis whereby the abuse liability of a substance is directly related to the rapidity with which it reaches psychotropic elements in the brain (Gorelick, p. 995).

At the cellular level, mesolimbic system dopamine neurons respond to chronic cocaine administration by increasing their activity while their activity decreases during withdrawal (Diana, p. 998). At the level of the ventral tegmental area dopamine cell bodies, repeated administration of psychostimulants results in an inhibition of GABA release that contributes to the activation of mesolimbic neurons caused by increases in adenosine tone (Fiorillo *et al.*, p. 1002). Nucleus accumbens neurons show enhanced responsiveness to the inhibitory actions of D1-receptor stimulation with repeated administration, and during withdrawal show a decreased responsiveness (White *et al.*, p. 1006). *In vivo* microdialysis has shown increased extracellular levels of dopamine in the nucleus accumbens during cocaine self-administration, and more recently increased extracellular DA levels in the amygdala (Hurd *et al.*, p. 1010). Neuropeptides such as neurotensin are released by low doses of methamphetamine (Hanson *et al.*, p. 1014). Cocaine and amphetamine activate preprodynorphin and substance P in the striatum as well as glutamate and acetylcholine (McGinty and Wang, p. 1017). Using a highly sensitive polymerase chain reaction (PCR) amplification, dopamine-receptor subtypes have been shown to have specific distributions within the striatal medium spiny neurons with high levels of enkephalin associated with high levels of D2 receptors, with high levels of substance P containing neurons associated with high levels of D1A receptors, and the other dopamine receptor subtypes (D3, D4, and D1B) being widely distributed but at very low levels (Surmeier *et al.*, p. 1020).

At the molecular level, manipulations of specific neuropharmacological elements of dopaminergic neurotransmission, such as knockout mice, are hypothesized to provide insights into the genetic vulnerability to substance dependence (Uhl *et al.*, p. 1024). The genetic basis for traits linked to substance dependence that are affected by multiple genes can be localized on the genome using quantitative trait loci analysis (Crabbe *et al.*, p. 1033). The transcription of receptor activation within the dopaminergic system appears to lead to activation of specific immediate early genes (C-Fos) that in turn activate proteins (Fos-related proteins) that may play a role in neuroadaptation to repeated drug administration that is associated with sensitization (Hiroi and Nestler, p. 1037). Finally, dopamine transporters appear to have a mechanism by which they can be phosphorylated and dephosphorylated, providing a potentially new substrate for regulation of dopaminergic synaptic transmission (Vaughan *et al.*, 1042).

I. System Level of Analysis

The circuits implicated in the psychostimulant and reinforcing actions of drugs of abuse have long been associated with the mesolimbic dopamine system and its connections, as discussed by Koob. While the dopamine system is critical for the actions of indirect sympathomimetics, such as amphetamine and cocaine, the stimulant and reinforcing actions of opiates and sedative–hypnotics involve not only activation of the mesolimbic dopamine system, but also activation of opioid peptide receptors, opioid peptides, or GABAergic receptors.

An extension of the mesolimbic dopamine system in drug reinforcement has been the recruitment of the extended amygdala. Characterized as a continuum of similar morphology, immunohistochemistry, and connectivity, the extended amygdala is composed of the shell (medial part) of the nucleus accumbens, the central nucleus of the amygdala, and the bed nucleus of the stria terminalis. Recent studies have suggested that elements of the extended amygdala not only have a role in the acute positive reinforcement associated with drugs of abuse, but also in the motivational aspects of drug withdrawal, such as activation of the stress hormone corticotropin-releasing factor.

Di Chiara and coworkers discuss motivational substrates within the extended amygdala as they relate to drugs of abuse. In rodents, drugs of abuse share the ability to stimulate dopamine transmission in the shell (medial aspect) of the nucleus accumbens, compared with the core (lateral aspect), as observed using *in vivo* microdialysis. A nondrug reinforcer (a highly palatable cheese-based snack called Fonzies) produced similar effects in the shell. These data and the anatomical differences in the projections of the shell versus the core provide evidence for the hypothesis that the shell may be more involved in the central integration of emotions, and the core may be more involved in the extrapyramidal control of motor functions. Interestingly, dopamine release in the shell of the nucleus accumbens produced by nonpsychostimulant drugs of abuse appears to be tonically under control of opioid peptides, because pretreatment with the selective μ-opioid receptor antagonist, naloxonazine, blocks this dopamine release.

The similarity between conventional rewards and drugs of abuse separate at the level of dopamine release in the prefrontal cortex. Fonzies stimulate dopamine release in the prefrontal cortex, and repeated eating of Fonzies causes a rapid habituation, which can be reversed by exposure to stimuli that have been conditioned to Fonzie feeding. The nucleus accumbens responds to these stimuli only under the condition of deprivation, suggesting that nucleus accumbens dopamine is activated by vital, survival-related stimuli under deprivation conditions, while prefrontal cortex dopamine is activated by motivationally relevant stimuli in general. Under nondeprivation conditions, the nucleus accumbens may be related to the acquisition rather than the expression of motivated behavior important for incentive learning.

One form of drug-induced adaptation that involves catecholamines is sensitization, as presented by Robinson and Badiani. Here, intermittent administration of amphetamine or cocaine results in a progressive increase in the ability of these drugs to produce locomotor hyperactivity, stereotyped behavior, or rotational behavior. The neural basis of sensitization appears to include, among other changes, a persistent increase in the ability of amphetamine or cocaine to release dopamine and/or a decrease in responsiveness of dopamine D1 receptors in the nucleus accumbens.

Sensitization may have a role in addiction and, thus, it is important to determine the conditions that lead to sensitization and promote its expression. The expression of sensitization can be eliminated if the animal is tested in an environment other than the treatment environment, suggesting that sensitization has taken place but is not expressed. In contrast, drug treatment in a novel environment produced greater sensitization. This suggests that the circumstances surrounding drug administration may critically determine whether a given dose of a drug is capable of inducing sensitization and may have relevance for individual variations in drug responses, including addiction, in the human population.

Pert discussed another neuroadaptive change associated with repeated drug administration: conditioning. Repeated pairing of cocaine with a particular environment produces a robust conditioned locomotor response in the absence of drugs. The development of this conditioned locomotor response is blocked by lesions of the mesolimbic dopamine system or pretreatment with dopamine antagonists prior to cocaine administration. Nucleus accumbens dopamine-depleting lesions are more effective than dopamine-depleting lesions in the amygdala, but both block conditioning.

The expression of conditioned locomotor activity also appears to depend on an intact mesolimbic dopamine system but is more resistant to blockade. The expression of conditioned locomotion is accompanied by increased extracellular levels of dopamine in the nucleus accumbens and is blocked by dopamine-depleting nucleus accumbens lesions made after conditioning is established.

A role of limbic afferents to the nucleus accumbens in drug-conditioned locomotion has been hypothesized based on data showing that excitatory amino acid antagonists can block the acquisition of conditioned locomotor activity to cocaine. The expression of conditioned locomotion to cocaine may also depend on intact AMPA excitatory amino acid transmission in the ventral tegmental area.

Gorelick presented findings relating to the success of agonist substitution in the treatment of drug addiction, notably with oral methadone for opiate addiction and nicotine slow release for tobacco addiction. One hypothesis to explain the success of these substitution approaches is the so-called rate hypothesis. Here, the strength of the positive psychological effects (and by extrapolation, the abuse liability) is proportional to the rate of the drug binding to its site of action. Studies of experimental cocaine administration in humans are consistent with the rate hypothesis in that the degree of subjective liking parallels the latency in achieving peak drug effects, with intravenous and smoked cocaine having a very fast latency and oral cocaine a slow latency to achieve peak blood concentrations. Also consistent is the human data showing that cocaine addicts in treatment reported a faster progression to dependence and treatment-seeking when they began to use cocaine by smoking rather than by insufflation. Recent experimental studies show that chronic treatment with oral cocaine significantly reduced the acute psychological effects of cocaine taken intravenously. These findings are consistent with the concept that substitution with a slow-onset agonist drug may be an effective treatment for cocaine addiction.

II. Cellular Level of Analysis

Working at the cellular level, Diana has shown that psychostimulant drugs of abuse decrease dopaminergic neuronal activity, while nonstimulant drugs increase dopamine neuron firing rates and patterns. The psychostimulant effects probably reflect a neuronal attempt to counteract the powerful dopamine-releasing effects of these drugs, while the activation by other drugs probably is the neuronal basis of the dopamine transmission–potentiating effects of nonstimulant drugs.

Repeated administration of drugs of abuse increases the basal activity of dopaminergic neurons, perhaps due to the subsensitivity of somatic autoreceptors. In contrast, withdrawal from two major drugs of abuse, morphine and ethanol, is associated with a marked reduction in the spontaneous activity of antidromically identified mesolimbic dopaminergic neurons. In unanesthetized animals, this was not due to a depolarization block but rather a membrane hyperpolization. This reduction in dopamine activity has recently been shown to outlast the somatic withdrawal produced by these substances and may reflect a dopamine component of protracted abstinence. Repeated psychostimulant administration involves an adenosine-mediated change in dopamine-GABA neurotransmission in the ventral tegmental area.

Sensitization to psychomotor stimulant drugs is associated with an enhanced release of dopamine in the nucleus accumbens in response to the psychostimulant drug and a hypersensitive mesolimbic dopamine system. There are multiple mechanisms that may be involved, but one may be a change in the responsiveness of the dopamine cell bodies of origin in the ventral tegmental area (White, *et al.*). D1 agonists appear to act presynaptically to enhance the amplitude of GABA-B inhibitory postsynaptic potentials in the ventral tegmental area dopamine cells. However, in animals treated chronically with cocaine or morphine, D1 agonists inhibited the amplitude of inhibitory postsynaptic potentials. This effect appears to be mimicked by activation of adenylyl cyclase and compounds that increase the formation of adenosine. Thus, in drug-naive animals, D1 dopamine receptors facilitate GABA release in the ventral tegmental area, but repeated drug administration shifts the balance toward an inhibition of GABA release by increases in adenosine tone.

White and coworkers also revealed that neuroadaptions within the nucleus accumbens itself appear to be responsible for the expression of many of the effects associated with cocaine withdrawal, including behavioral sensitization. Nucleus accumbens neurons show enhanced responsiveness to the inhibitory actions of D1-receptor stimulation that follows the time course of behavioral sensitization over days and weeks.

However, acute withdrawal is also accompanied by significant membrane alterations, including a hyperpolarization of the resting membrane potential, and as a result, a decrease in the amplitude of current-induced action potentials. This decreased excitability of nucleus accumbens neurons appears to be mediated by an increased basal state of phosphorylation, resulting in a decrease in function of voltage-sensitive sodium channels. This marked reduction of voltage-sensitive sodium channels produces a profound decrease in responsiveness of the nucleus accumbens to excitatory commands and, thus, may be directly related to other effects of acute cocaine withdrawal, such as anergia, anhedonia, and depression.

Evidence supporting a critical role for dopamine in the acute actions of stimulant drugs has been obtained by Hurd and colleagues using *in vivo* microdialysis. The neuropharmacological mechanism for these effects appears to be, for cocaine, a marked potentiation of extracellular levels of dopamine originating from exocytotic release because of the absence of calcium-blocked cocaine-induced overflow. For amphetamine, the absence of calcium had no effect on amphetamine-induced overflow, and the enhanced dopamine overflow appears via a reversal of the dopamine transport carrier similar to the release of dopamine produced by low extracellular sodium concentrations.

During intravenous cocaine self-administration, extracellular levels of dopamine were increased not only in the nucleus accumbens, but also in the amygdala. Administration of a dopamine D1-receptor antagonist into the amygdala doubled the amount of cocaine self-administered and consequently doubled the amount of dopamine overflow in the nucleus accumbens, suggesting an important contribution of dopamine transmission in the amygdala to cocaine reinforcement.

The neuropeptide neurotensin has been reciprocally linked by Hanson *et al.* to dopaminergic activity in the nucleus accumbens and caudate nucleus. Depending on the dose, methamphetamine has been shown to alter the functional activity of neurotensin in the nucleus accumbens and caudate nucleus. Low doses of methamphetamine produce enhanced neurotensin release, and this appears to be mediated predominantly by dopamine D2 receptors. In contrast, high doses of methamphetamine produce increased tissue levels of neurotensin and increased mRNA formation, but no change in release of neurotensin. This effect is mediated by dopamine D1 receptors. These alterations in neurotensin function with high doses of methamphetamine may contribute to the significant motor and emotional consequences of high-dose methamphetamine use.

McGinty and Wang demonstrate that drugs such as cocaine induce the expression of neuropeptides such as preprodynorphin and substance P (SP) as well as immediate early genes such as c-fos in the medium spiny neurons of the striatum. These changes can be attenuated by N-methyl-D-aspartate antagonists. Acetylcholine also appears to modulate striatal gene expression but opposite that of stimulation of dopamine D1 receptors in the striatum.

Based on these data, a sequence of events has been hypothesized to occur within the striatum. Stimulation of D1 dopamine receptors on the soma and dendrites of striatonigral neurons results in multiple cellular responses, including release of acetylcholine, SP, GABA, and dynorphin, and a compensatory increase in preprodynorphin and SP expression in the striatum. Acetylcholine would bind to muscarinic receptors and inhibit preprodynorphin–SP expression in the striatum but would increase expression of preproenkephalin in striatopallidal neurons. At the level of the substantia nigra, D1-receptor stimulation may result in disinhibition of glutamate release in the thalamostriatal and corticostriatal pathways and lead to further stimulation of striatal acetylcholine release. According to this schema, the cholinergic neuron could act as a feedforward inhibitor of psychostimulant-induced striatonigral activity and a facilitator of striatopallidal activity. The ability of glutamate and acetylcholine to modulate the responses of medium spiny neurons in the striatum to dopamine provides a potentially useful neuropharmacological site for the modulation of actions of drugs of abuse.

The distribution of dopamine-receptor subtypes among the principal neurons of the neostriatum is important information necessary for understanding how drugs like cocaine and amphetamine influence basal ganglia function (Surmeier *et al.*). Medium spiny neurons having detectable levels of enkephalin, but not substance P mRNA, and express high levels of D2-receptor mRNA, as determined using a single-cell reverse transcription–PCR technique. Medium spiny neurons have detectable levels of SP mRNA, but not enkephalin mRNA, and expressed high levels of D1A-receptor mRNA. D3 mRNA was also seen in this population of cells. Neurons coexpressing detectable levels of SP and enkephalin mRNA (a group that composed about 20% of the total medium spiny neurons) consistently coexpressed D1A and D2 mRNAs. D3, D4, and D1B receptors are also expressed in many of these cell types, using a two-stage multiplex PCR amplification but in much lower abundance. The single-cell PCR technique reveals results similar to those observed for *in situ* hybridization for the most abundant receptor mRNAs (D1A and D2) but shows that lower-abundance dopamine receptors (D3, D2, and D1B) are also present.

III. Molecular Level of Analysis

Genetic contributions to drug abuse remain largely to be elucidated, but population comorbidity, twin studies, adoption, and family studies suggest important genetic and environmental interactions. Molecular genetic studies of variants in genes important for dopaminergic neurotransmission provide a rich substrate for individual differences in drug abuse vulnerabilities.

Uhl and coworkers discuss their findings of transgenic mice overexpressing a dopamine transporter gene variant that manifests increased psychostimulant and conditioned place preferences to cocaine. In contrast, D1-receptor and dopamine-transporter knockout mice show substantial blockade of the psychostimulant actions of cocaine. Interestingly, heterozygote brain vesicular monoamine transporter (VMAT2) knockout mice show a decreased conditioned place preference to amphetamine but a normal conditioned place preference to cocaine.

In human studies, some differences in the catechol *O*-methyltransferase gene have been linked to drug use, and recent studies have identified an association between pleasant emotional states and dopamine D4 polymorphisms. These studies in animals and humans point to a potentially prominent role for dopaminergic gene variants in contributing to the individual differences in vulnerability to drug use.

Crabbe and colleagues point out that the genetic contributions to complex traits such as those associated with drug abuse and dependence have been difficult to analyze using classical genetic methods because most responses to drugs are multigenic and polygenic. A recent development called quantitative trait locus (QTL) mapping is being successfully employed to identify genetic markers near genes that influence drug sensitivity. By studying specific animal populations, genetically segregating for a number of genes and markers that are polymorphic (where multiple allelic forms of the gene are maintained in the population), an association between drug effects and a particular allele can be established.

A number of interesting associations have been found using recombinant inbred strains. The sensitivity to the climbing response to a high dose of methamphetamine in mice has been linked to markers on several specific chromosomes. Ethanol withdrawal, as measured by handling induced convulsions, has also been linked to several mouse chromosomal regions. Ethanol preference has been linked to markers near the serotonin 5HT-1B receptor subtype, leading to the study of 5HT-1B knockouts. These animals drank twice as much ethanol as their congenic controls. QTL mapping provides a promising method for identifying genetic influences on complex pharmacobehavioral responses, and because of the high homology between the mouse and human genome (approximately 85%), such identified genes can be roughly mapped to humans.

Intermediate early genes, such as C-Fos, can be activated by a variety of drug treatments, and when activated, these genes transcribe their protein products (Hiroi and Nestler). This gene activation has been used to map neuronal activity; however, which cell systems are activated can be idiosyncratic. Perhaps more importantly, these genes also may play a crucial role in regulating particular cell functions. C-Fos proteins bind to a DNA sequence by forming heterodimers with protein products of another immediate early gene family, Jun. These Fos-Jun dimers bind to AP1 (activator protein-1) and as a result can activate or repress the transcription of other genes.

For example, cocaine acutely induces C-Fos but when administered chronically, loses its ability to induce C-Fos. However, the elevated AP1 binding persists with chronic cocaine treatment. Associated with this elevated AP1 binding was the presence of unique fos-related antigens (FRAs) Termed *chronic FRAs* because they have been induced in several brain regions following chronic treatment with several drugs, these proteins have been hypothesized to play critical roles in neuronal and behavioral plasticity. Further study revealed that all the FRAs induced are delta FosB variants. Mutant mice lacking FosB showed greater acute responses to cocaine but less subsequent sensitization, suggesting the hypothesis that the normal development of cocaine sensitization depends on transcription regulation by delta FosB variants and perhaps FosB.

Dopamine transporters are integral membrane neuronal proteins, which function to terminate dopaminergic transmission by rapid re-uptake of synaptic dopamine into presynaptic terminals. The dopamine transporter is the primary site of action of cocaine and amphetamine and the means by which neurotoxins such as 6-hydroxydopamine enter the presynaptic terminal. Consensus phosphorylation sites for numerous protein kinases have been identified on the dopamine transporter, and evidence provided by Vaughn and colleagues suggests that the dopamine transporter can be regulated by phosphorylation.

Synaptosomes treated with a protein kinase C activator increased phosphorylation of the dopamine transporter, and synaptosomes treated with a phosphatase inhibitor produced robust dephosphorylation. It is currently unknown what the relationship of phosphorylation of the dopamine transporter is to transporter function, but one hypothesis is that neurons utilize dopamine transporter phosphorylation as a mechanism for fine temporal and spatial regulation of synaptic dopamine levels.

IV. Implications

Significant progress has been made at every level of neuroscience research in understanding the mechanisms of action of drugs of abuse following both acute and chronic administration. The focus has been on psychostimulant drugs and the mesolimbic dopamine system, but the information gained within this context will be valuable for other drugs and insults to brain catecholamine systems. Notable new findings are the identification of the subregions of the projections of the mesolimbic dopamine systems as critical parts of drug reward circuitry, including the shell of the nucleus accumbens and the central nucleus of the amygdala. The other major new direction is that sophisticated behavioral, electrophysiological, neurochemical, and molecular techniques are being applied to determine the nature of the neuroadaptive processes that lead to changes in sensitivity to the dopamine system and its connection with repeated drug use. Changes in mesolimibic dopamine neuronal firing, the response of dopamine receptive elements in the terminal areas of the dopamine system, and molecular changes in elements of dopaminergic transmission, such as the dopamine transporter or in transcription of dopamine receptor activation (e.g., C-Fos), have been implicated in this plasticity. New genetic analyses and approaches, such as molecular genome manipulations and quantitative trait loci, provide hope for identifying elements that may confer vulnerability to substance abuse and dependence. The past several years have produced a rich harvest of advances in neuroscience research directed at understanding the role of brain catecholamines in drug abuse and alcoholism.

George F. Koob

Department of Neuropharmacology
The Scripps Research Institute
La Jolla, California 92037

Circuits, Drugs, and Drug Addiction

I. Neurobiological Substrates for Drug Reinforcement–Psychomotor Stimulants, Opiates, and Sedative–Hypnotics

Drug addiction or substance dependence is defined by the loss of control over drug intake and a compulsion to take the drug, as well as a characteristic withdrawal syndrome that results in physical and motivational signs of discomfort when the drug is removed. The concepts of reinforcement, reward, or motivation are crucial parts of both of these definitions, and the neural circuits involved in the acute reinforcing effects of drugs have begun to be elucidated. Early work focused on the role of the medial forebrain bundle in reward, and studies of drug reward subsequently implicated the mesolimbic dopamine system. More recent studies have begun to link subregions of the terminal projections of the mesolimbic dopamine system with the classical reward system of the medial forebrain bundle.

The neurobiological substrates for the stimulant and reinforcing properties of psychomotor stimulants that have an indirect sympathomimetic action (e.g., drugs such as cocaine and amphetamine which indirectly release catecholamines) have long been associated with a particular part of the forebrain dopamine projections, the mesolimbic dopamine system. A critical role for mesolimbic dopamine in the psychomotor stimulant effects of amphetamine and cocaine was provided by the observation that microinjections of the dopamine antagonist haloperidol into the nucleus accumbens blocked amphetamine-induced locomotor activity (1). Locomotor activation produced by amphetamine and cocaine was also blocked by 6-hydroxydopamine (6-OHDA) lesions of the nucleus accumbens (2). Similar 6-OHDA lesions of the nucleus accumbens blocked the locomotor stimulant effects of *d*-amphetamine but not of caffeine or scopolamine (3), heroin, or corticotropin-releasing factor (2).

Studies of intravenous self-administration of drugs have strongly implicated dopamine in the reinforcing effects of cocaine and amphetamine. Rats implanted with intravenous catheters and trained to self-administer cocaine with limited access (3 hr/day) will show a stable and regular drug intake over each daily session. The rats appear to regulate the amount of drug self-administered. Lowering the dose from the training level of 0.75 mg/kg per injection increases the number of self-administered infusions and vice versa.

Low doses of dopamine-receptor antagonists, when injected systemically, reliably increase cocaine and amphetamine self-administration in rats (4). The

Advances in Pharmacology, Volume 42

1054-3589/98 $25.00

animals appear to compensate for decreases in the magnitude of reinforcement with an increase in cocaine self-administration (or a decrease in the interinjection interval), a response similar to lowering the dose of cocaine. This suggests that a partial blockade of dopamine receptors produced a partial blockade of the reinforcing actions of cocaine.

In general, experiments investigating the effects of antagonists selective for D1, D2, and D3 receptors (4) on cocaine self-administration suggest that all three can decrease the reinforcing properites of cocaine. More recent data show that microinjections of the selective D1 antagonist SCH 23390 is particularly effective in blocking cocaine self-administration after injection into the shell of the nucleus accumbens and the central nucleus of the amygdala (5). These results suggest that D1 receptors in a subregion of the nucleus accumbens, and perhaps the amygdala, may be particularly important for the reinforcing properties of cocaine.

Opiate drugs such as heroin produce a different pattern of behavior, activation, and neuropharmacological effects from that observed with psychomotor stimulants (6). There is some initial suppression of activity followed by a sustained hyperlocomotion characterized by bursts of activity as opposed to continuous sniffing, rearing, and locomotion, as observed with stimulants. This opiate-induced locomotor activity is not blocked by mesolimbic dopamine 6-OHDA lesions but is blocked by microinjections of the opiate antagonist methylnaloxonium into the nucleus accumbens at doses significantly lower than those required by intracerebroventricular administration. An identical pattern of results was observed in rats self-administering heroin where nucleus accumbens 6-OHDA lesions failed to alter heroin self-administration, and methylnaloxonium microinjections into the nucleus accumbens were particularly effective in antagonizing heroin self-administration (6).

Ethanol, barbiturates, and benzodiazepines all have classic sedative hypnotic actions, as measured in pharmacological studies, including euphoria, disinhibition, anxiety reduction, sedation, and hypnosis. All of these drugs produce a release of punished responding in conflict situations, which correlates well with their ability to act as anxiolytics in the clinic (7). This anxiolytic or tension-reducing property of sedative hypnotics may be a major component of the reinforcing actions of these drugs, and the neurobiological basis of their anxiolytic properties may provide clues to their reinforcing properties and their abuse potential. GABAergic mechanisms have been hypothesized to be involved in the anxiolytic actions of sedative hypnotics as a result of studies showing that GABA antagonists block their anti-conflict effects and GABA agonists potentiate their effects (7). The brain sites for these anti-conflict actions may involve limbic regions such as the amygdala, because microinjection of the neurosteroid allopregnanolone, a compound that has ethanol-like actions on the GABA-benzodiazepineionophore complex, into the amygdala produces an anti-conflict action (8).

Support for a role of brain GABA in ethanol reinforcement is provided by the observation that the partial inverse benzodiazepine agonist R0 15-4513, which has been shown to reverse some of the behavioral effects of ethanol, produces a dose-dependent reduction of oral ethanol (10%) self-administration in nondeprived rats in an operant free-choice situation (9). Consistent with a role for the amygdala in the reinforcing, and by extrapolation intoxicating, actions of ethanol is the observation that very low doses of a selective GABA antagonist, SR 95531, potently decreases ethanol self-administration when it is microinjected into the central nucleus of the amygdala (10).

Opioid peptide antagonists such as naloxone have long been known to antagonize ethanol self-administration in the rat, suggesting a role for opioid peptides in the reinforcing actions of ethanol (9). Recent data with microinjections of the opioid antagonist methylnaloxonium suggest that the site of action may also be the central nucleus of the amygdala (11).

Dopamine-receptor antagonists also have been shown to reduce lever pressing for ethanol in nondeprived rats, and dopamine antagonists injected into the nucleus accumbens decrease oral ethanol self-administration in non-water-deprived rats trained in a two-lever, free-choice, self-administration task (9). Thus, the reinforcing actions of ethanol may involve multiple neurotransmitter systems that intersect with the brain reward circuitry at multiple sites, including the ventral tegmental area, nucleus accumbens, and amygdala (9).

II. Functional Circuitry Common to Drug Reinforcement: The Extended Amygdala

The neural circuits involved in drug reinforcement have used the mesolimbic dopamine system as a focal point; however, its role in reinforcement cannot be conceptualized to occur in isolation, even if one subscribes to the hypothesis that dopamine is the critical substrate for all reward. Clearly, the processing of the reinforcing stimulus requires connections with the terminal regions of the mesolimbic dopamine system, such as the nucleus accumbens, including both afferents and efferents to this region (6). This connectivity provides multiple sites for access of drugs and other reinforcers that can act independently of the mesolimbic dopamine system, thus providing a neural substrate for dopamine-independent reward (6).

Neuroanatomical data and new functional observations suggest the intriguing hypothesis that the neuroanatomical substrates for the reinforcing actions of drugs involve a neural circuitry that forms a separate entity within the basal forebrain, termed the *extended amygdala* (12). This large forebrain continuum is hypothesized largely on the basis of similarities between these structures in morphology, immunohistochemistry, and connectivity (12) and is composed of several basal forebrain structures: the bed nucleus of the stria terminalis, the centromedial amygdala and medial part of the nucleus accumbens (e.g., shell), and continuous cell columns in the sublenticular substantia innominata. Afferent connections to this complex include mainly limbic regions, while efferent connections from this complex include medial aspects of the ventral pallidum and substantia innominata, as well as a considerable projection to the lateral hypothalamus (12).

Functional studies support the hypothesis that this medial part of the ventral forebrain, described as the extended amygdala, may have a role in drug reinforcement. Microinjections of a D1 dopamine antagonist in animals trained to self-administer cocaine show that two particularly effective sites for blocking cocaine self-administration appear to be the shell of the nucleus accumbens and the central nucleus of the amygdala (5). In addition, lesions of the posterior medial ventral pallidum are particularly effective in blocking the motivation to work for intravenous cocaine (13). Further, data with *in vivo* microdialysis show that acute administration of drugs of abuse preferentially enhance dopamine release in the shell of the nucleus accumbens (14). Perhaps more intriguing is the observation that increased release of corticotropin-releasing factor, a neuropeptide implicated in behavioral response to stress, may be increased in the central nucleus of the amygdala during drug withdrawal (15).

These results suggest that certain neurochemical elements of the projection of extended amygdala may be important for the reinforcing effects of drugs. Clearly, the mesolimbic dopamine system is critical for psychomotor stimulant activation and psychomotor stimulant reinforcement, and plays a role in the reinforcing actions of other drugs. However, as has been discussed by others (1), the functions of the mesolimbic dopamine system may be determined largely by its specific innervations and not by any intrinsic functional attributes. The heterogeneity of connectivity within subregions of the nucleus accumbens suggests that the medial nucleus accumbens (shell), together with the rest of the extended amygdala, may provide a critical link between the terminals of the mesolimbic dopamine system and other forebrain circuitry involved in drug reinforcement.

References

1. Le Moal, M., and Simon, H. (1991). Mesocorticolimbic dopaminergic network: Functional and regulatory roles. *Physiol. Rev.* **71**, 155–234.
2. Amalric, M., and Koob, G. F. (1993). Functionally selective neurochemical afferents and efferents of the mesocorticolimbic and nigrostriatal dopamine system. *In* Chemical Signalling in the Basal Ganglia. (G. W. Arbuthnott and P. C. Emson eds.), pp. 209–226. Elsevier, Amsterdam.
3. Joyce, E. M., and Koob, G. F. (1981). Amphetamine-, scopolamine-, and caffeine-induced locomotor activity following 6-hydroxydopamine lesions of the mesolimbic dopamine system. *Psychopharmacology* **73**, 311–313.
4. Koob, G. F., Parsons, L. H., Caine, S. B., Weiss, F., Sokoloff, P., and Schwartz, J-C. (1996). Dopamine receptor subtype profiles in cocaine reward. *In* Dopamine Disease States. (R. J. Beninger, T. Palomo, and T. Archer, eds.), pp. 433–445. Editorial CYM, Madrid.
5. Caine, S. B., Heinrichs, S. C., Coffin, V. L., and Koob, G. F. (1995). Effects of the dopamine D-1 antagonist SCH 23390 microinjected into the accumbens, amygdala or striatum on cocaine self-administration in the rat. *Brain Res.* **692**, 47–56.
6. Koob, G. F. (1992). Drugs of abuse: Anatomy, pharmacology, and function of reward pathways. *Trends Pharmacol. Sci.* **13**, 177–184.
7. Koob, G. F., and Britton, K. T. (1996). Neurobiological substrates for the anti-anxiety effects of ethanol. *In* The Pharamcology of Alcohol and Alcohol Dependence (H. Begleiter and B. Kissin, eds.), pp. 477–506. Oxford University Press, New York.
8. Britton, K. T., Akwa, Y., Purdy, R. H., and Koob, G. F. (1996). The amygdala mediates the anxiolytic-like activity of the neurosteroid allopregnanolone. *Soc. Neurosci. Abstr.* **22**, 1292.

9. Koob, G. F., Rassnick, S., Heinrichs, S., and Weiss, F. (1994). Alcohol: The reward system and dependence. *In* Toward a Molecular Basis of Alcohol Use and Abuse. (B. Jansson, H. Jornvall, U. Rydberg, L. Terenius, B. L. Vallee, ed.). pp. 103–114. Birkhauser Verlag, Boston.
10. Hyytia, P., and Koob, G. F. (1995). $GABA_A$ receptor antagonism in the extended amygdala decreases ethanol self-administration in rats. *Eur. J. Pharmacol.* **283**, 151–159.
11. Heyser, C. J., Roberts, A. J., Schulteis, G., Hyytia, P., and Koob, G. F. (1995). Central administration of an opiate antagonist decreases oral ethanol self-administration in rats. *Soc. Neurosci. Abstr.* **21**, 1698.
12. Alheid, G. F., and Heimer, L. (1988). New perspectives in basal forebrain organization of special relevance for neuropsychiatric disorders: The striatopallidal, amygdaloid, and corticopetal components of substantia innominata. *Neuroscience* **27**, 1–39.
13. Robledo, P., and Koob, G. F. (1993). Two discrete nucleus accumbens projection areas differentially mediate cocaine self-administration in the rat. *Behav. Brain Res.* **55**, 159–166.
14. Pontieri, F. E., Tanda, G., and Di Chiara, G. (1995). Intravenous cocaine, morphine, and amphetamine preferentially increase extracellular dopamine in the "shell" as compared with the "core" of the rat nucleus accumbens. *Proc. Natl. Acad. Sci. U.S.A.* **92**, 12304–12308.
15. Merlo-Pich, E., Lorang, M., Yaganeh, M., Rodriguez de Fonseca, F., Raber, J., Koob, G. F., and Weiss, F. (1995). Increase of extracellular corticotropin-releasing factor-like immunoreactivity levels in the amygadala of awake rats during restraint stress and ethanol withdrawal as measured by microdialysis. *J. Neurosci.* **15**, 5439–5447.

Gaetano Di Chiara, Gianluigi Tanda, Cristina Cadoni, Elio Acquas, Valentina Bassareo, and Ezio Carboni

Department of Toxicology and CNR Centre for Neuropharmacology
University of Cagliari
09100 Cagliari, Italy

Homologies and Differences in the Action of Drugs of Abuse and a Conventional Reinforcer (Food) on Dopamine Transmission: An Interpretative Framework of the Mechanism of Drug Dependence

Microdialysis studies have shown that drugs and substances of abuse increase the extracellular concentration of dopamine (DA) in the nucleus accumbens, a terminal area of the mesolimbic dopamine system (1). This effect, although resulting from different mechanisms, depending on the pharmacological class of the drug, shows some homologies with a conventional reinforcer, such as palatable food (Fonzies). Thus, drugs of abuse and Fonzies share the properties of stimulating DA transmission within the nucleus accumbens "shell," as compared with the "core" (2, 3). These two compartments show distinct differences with regard to their input–output relationships and anatomical–functional significance. While the nucleus accumbens shell projects to limbic areas as the "extended amygdala" (central amygdala, bed nucleus of stria terminalis, sublenticular substantia innominata), lateral hypothalamus, and periacqueductal grey, areas involved in visceral, hormonal, and autonomic functions, the core projects to the ventral pallidum. Therefore, while the shell might be involved in the central integration of emotions, the core might be involved in the extrapyramidal control of motor functions. In the rat, drugs of abuse share the ability to stimulate DA transmission in the shell of the nucleus accumbens and to activate energy metabolism selectively in this area (Fig. 1) (2, 3). These observations, together with the aforementioned anatomical findings, are consistent with the idea that stimulation of DA transmission in the shell of the nucleus accumbens plays a role in the mechanism of drug abuse and dependence.

Nonpsychostimulant drugs of abuse and food depend for their property of stimulating DA release in the shell of the nucleus accumbens on an activation of the μ_1 subtype of opioid receptors. Thus, naloxonazine, a slowly reversible

Advances in Pharmacology, Volume 42

1054-3589/98 $25.00

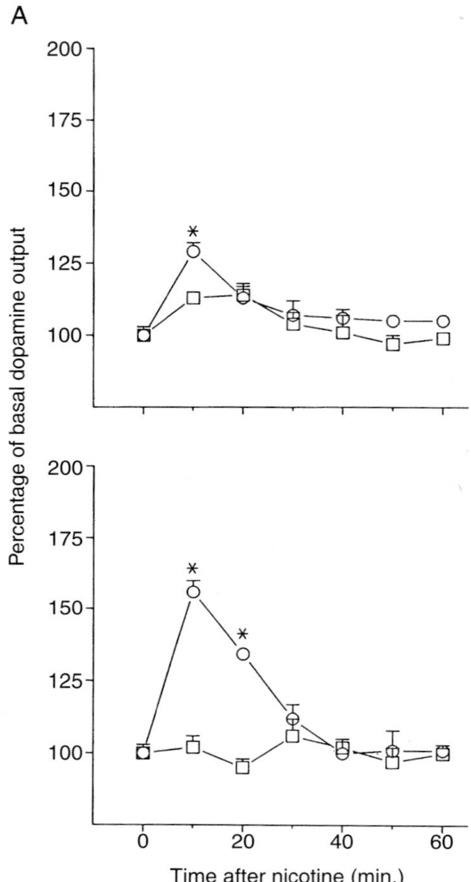

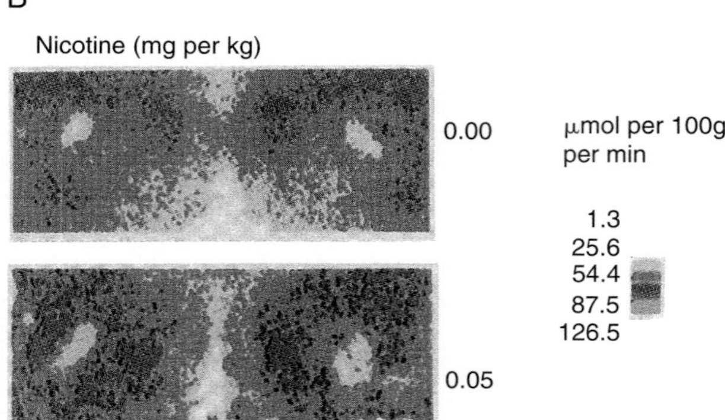

FIGURE 1 Effects of intravenous administration of nicotine on cerebral glucose utilization in the rat nucleus accumbens. Gray-coded transformations of autoradiographs of coronal brain sections at the level of the nucleus accumbens are from a control, saline-treated rat (A) and a rat treated with 0.05 mg/kg nicotine, free base (B), in which gray tones represent rates

μ_1 antagonist, blocks the stimulation of DA release in the nucleus accumbens shell by morphine, nicotine, and ethanol, as well as by highly palatable food (Fonzies) (Tanda and Di Chiara, in preparation).

In contrast with these homologies in regard to their effects on DA transmission in the nucleus accumbens, substances of abuse and Fonzies show differences in their ability to influence DA transmission in the prefrontal cortex. Thus, Fonzies activate DA release in the prefrontal cortex, but this effect, in contrast to that in the nucleus accumbens, is not blocked by naloxonazine. Drugs of abuse, as a whole, do not activate DA transmission in the prefrontal cortex; thus, morphine and ethanol fail to do so over a large range of doses; nicotine does it eventually at high doses but not consistently (4). Amphetamine and cocaine do increase extracellular DA in the prefrontal cortex. However, this effect is likely to be the result of a blockade of the noradrenaline carrier rather than of the dopamine carrier and is unrelated to their reinforcing properties. These results indicate that DA in the prefrontal cortex does not play an important role in the reinforcing properties of drugs of abuse.

Studies on Fonzies show that DA plays a different role in the prefrontal cortex and in the nucleus accumbens. In rats fed ad libitum with standard food, feeding of a novel palatable food like Fonzies phasically stimulates DA transmission in the nucleus accumbens and in the prefrontal cortex; on repeated Fonzies feeding, this effect undergoes habituation in the nucleus accumbens but not in the prefrontal cortex; moreover, presentation of incentive stimuli conditioned to Fonzies feeding phasically stimulates DA release in the prefrontal cortex but not in the nucleus accumbens; finally, exposure to an incentive stimulus conditioned to Fonzies feeding prevents the activation of DA release in the nucleus accumbens but not in the prefrontal cortex.

Under food deprivation, adaptive modulation of phasic activation of DA release in the nucleus accumbens by food is modified; most notably, conditional stimuli will eventually become effective in activating DA transmission also in the nucleus accumbens. It appears, therefore, that while prefrontal DA transmission reacts to generically salient stimuli independently of motivational state, nucleus accumbens DA reacts to these stimuli only under a deprivation state; under a nondeprivation state, nucleus accumbens DA reacts only to strong, unconditioned, novel, and unpredicted stimuli. We conclude that while prefrontal DA codes for motivationally relevant stimuli in general, nucleus accumbens DA codes for vital, survival-related stimuli.

---

of glucose utilization. Intravenous administration of nicotine increases energy metabolism in the shell of the nucleus accumbens without affecting values in the core (*Top right*). Effects of intravenous administration of nicotine on dopamine output in dialysates from the shell (*circles*) and core (*squares*) subdivisions of the nucleus accumbens in the rat. Percentage variations in dopamine output were produced by nicotine (free base) at 0.025 mg/kg (*a*) and 0.050 mg/kg (*b*). Post hoc analysis indicated a significant increase over basal values in the shell at 10 and 20 min after administration of 0.05 mg/kg nicotine and at 10 min after administration of 0.025 mg/kg nicotine (asterisk indicates $p < .05$ in the shell compared with basal and to the corresponding values in the core) (*Left*). Reproduced from Pontieri, Tanda, Orzi and Di Chiara, Nature *382,* 255–257, 1996, with permission.

A potentially critical difference between drugs of abuse and food is that while stimulation of dopamine release in the shell of the nucleus accumbens is prevented by previous exposure to a significant food stimulus, no such inhibition is observed in the case of drugs of abuse. In fact, repeated exposure to drugs of abuse does not result in a process of habituation of their ability to increase extracellular DA in the nucleus accumbens. The significance of this difference between drugs of abuse and conventional reinforcers such as food can be understood if one considers the possible role of the stimulation of DA release in the shell in motivation. Phasic stimulation of dopamine release in the nucleus accumbens can be clearly dissociated from incentive reactions to a secondary (conditioned) stimulus and from consummatory behavior itself (i.e., feeding); thus, under nondeprivation conditions, stimulation of DA release in the shell takes place only in response to novel or relatively novel primary food stimuli, habituates rapidly, and recovers slowly. These observations suggest that phasic stimulation of DA release in the shell of the nucleus accumbens by a conventional reinforcer such as food is not essential for the expression of incentive behavior and for the hedonic sensations arising from consumption of the reward. This, however, does not exclude that DA phasically released in the nucleus accumbens under the influence of survival-related stimuli can facilitate the expression of incentive responding.

We speculate that phasic release of DA in the nucleus accumbens, while not essential for incentive responding in general, might become important for incentive responding and survival in deprivation conditions. Therefore, under nondeprivation conditions, stimulation of dopamine release in the shell of the nucleus accumbens might be related to the acquisition rather than the expression of motivated behavior, being instrumental for incentive learning; by this process, otherwise neutral stimuli that predict the occurrence of positive (rewards) or negative (aversive) primary motivational stimuli, themselves, become powerful (secondary) motivational stimuli. Incentive learning might be critically dependent on the action of dopamine on D1 receptors. Consistent with this view is the observation that the D1 antagonist SCH 23390 and its analogues block not only acquisition of conditional place-preference and place-aversion induced by various drugs, but also the acquisition of conditioned taste aversion induced by lithium. These effects are stereospecific and take place at very low doses of the D1 antagonists (5).

Substances of abuse, in contrast to food, activate the release of DA in the shell of the nucleus accumbens independently of past experience of the same or related stimuli; the reiterated activation of D1 receptors in this area would result in a pathological strengthening of incentive learning, which is expressed in the compulsive reactivity of drug users toward secondary stimuli that predict the availability of the drug. Within this framework, craving for drugs and substances of abuse is hypothesized to be the result of a pathological facilitation of incentive learning resulting from the nonadaptive property of drugs to stimulate DA transmission in the nucleus accumbens shell.

References

1. Di Chiara, G. (1995). The role of dopamine in drug abuse viewed from the perspective of its role in motivation. *Drug Alcohol Depend.* **38**, 95–137.

2. Pontieri, F. E., Tanda, G., and Di Chiara, G. (1995). Intravenous cocaine, morphine, and amphetamine preferentially increase extracellular dopamine in the "shell" as compared with the "core" of the rat nucleus accumbens. *Proc. Natl. Acad. Sci. U.S.A.* **92,** 12304–12308.
3. Pontieri, F. E., Tanda, G., Orzi, F., and Di Chiara, G. (1996). Effects of nicotine on the nucleus accumbens and similarity to those of addictive drugs. *Nature* **382,** 255–257.
4. Bassareo, V., Tanda, G., Petromilli, P., Giua, C., and Di Chiara, G. (1996). Non-psychostimulant drugs of abuse and anxiogenic drugs activate with differential selectivity dopamine transmission in the nucleus accumbens and in the medial prefrontal cortex of the rat. *Psychopharamcology* **124,** 293–299.
5. Acquas, E., and Di Chiara, G. (1994). D_1 receptor blockade stereospecifically impairs the acquisition of drug-conditioned place-preference and place-aversion. *Behav. Pharmacol.* 5, 555–569.

Terry E. Robinson and Aldo Badiani

Department of Psychology (Biopsychology Program)
The University of Michigan
Ann Arbor Michigan 48109

Drug-Induced Adaptations in Catecholamine Systems: On the Inevitability of Sensitization

It is well established that psychostimulant drugs produce many of their effects via actions on brain catecholamine (CA) systems. When these drugs are administered repeatedly, their effects often change, and these changes are thought to be due in part to drug-induced adaptations in CA systems. One form of drug-induced adaptation is manifest behaviorally as the phenomenon of sensitization (1). For example, the repeated intermittent administration of amphetamine (AMPH) or cocaine (COC) results in a progressive increase in the ability of these drugs to produce locomotor hyperactivity, stereotyped behaviors, or rotational behavior. Once induced, sensitization may persist for very long periods, perhaps even for years. The neural basis of behavioral sensitization is not known, but both presynaptic and postsynaptic adaptations have been described in CA systems (1, 2). Presynaptic adaptations include (1) a persistent increase in the ability of AMPH or COC to enhance the overflow of dopamine (DA) in the caudate and nucleus accumbens and (2) a persistent increase in the overflow of norepinephrine (NE) in the hippocampus of rats

Advances in Pharmacology, Volume 42

1054-3589/98 $25.00

sensitized with AMPH, both under resting conditions and after an AMPH challenge (3). An example of a postsynaptic adaptation is a sensitization-related increase in the responsiveness of DA D1 receptors in the nucleus accumbens (2).

Sensitization of DA systems has attracted much of the attention in this field, perhaps because of the well-known role of these neurons in mediating the rewarding effects of drugs of abuse and of natural rewards. The fact that drugs of abuse can produce long-lasting alterations in a neural system involved in reward suggests that "neural sensitization" may play a role in the process of addiction. Indeed, it is difficult to imagine that if repeated exposure to drugs renders DA systems persistently hypersensitive to activating stimuli, including drugs themselves, that such neuroadaptations would not alter the process of reward and, thus, in some way contribute to addiction. Consistent with this idea, there is experimental evidence that repeated exposure to psychostimulant drugs results not only in sensitization to their psychomotor activating effects, but also to their rewarding effects (1). Exactly how sensitization of CA systems might contribute to addiction has been the subject of much speculation and discussion, and Kent Berridge and I (T. E. R.) have suggested (1) that neural sensitization may lead to pathological "wanting" (because of excessive incentive salience attribution), dissociated from the ability of drugs to produce subjective pleasurable effects ("liking").

Whatever the mechanism(s) by which psychostimulants produce adaptations in CA systems and alter the behavioral effects of drugs, it is important to determine the conditions that lead to neural sensitization and promote its expression. Sensitization is often thought of as an inevitable consequence of the pharmacological actions of a drug on a sensitive neural substrate; that is, when a sensitive neural substrate is repeatedly exposed to a drug (or other ligand), the substrate adapts. The behavioral effects of drugs are, however, powerfully modulated by the circumstances surrounding drug administration. In the remainder of this essay, we will consider two examples of how the circumstances surrounding drug administration can alter the induction and expression of sensitization. These examples illustrate that sensitization is *not* an inevitable consequence of exposure to psychostimulant drugs. In the first example, drug administration is thought to induce neural sensitization, but the circumstances surrounding drug administration and readministration determine whether neural sensitization is *expressed* in behavior. In the second example, the circumstances surrounding drug administration may determine whether neural sensitization is induced at all.

I. Modulation of Expression

There are many reports that under some circumstances the expression of behavioral sensitization is context-specific (1). If animals are given repeated treatments with AMPH or COO in one environment and then later a challenge injection of the drug in a different (test) environment ("unpaired" animals), behavioral sensitization may not be expressed in the test environment (i. e., it is specific to the treatment context). Despite this powerful conditioned stimulus control over the expression of behavioral sensitization, there are two reasons to believe that even unpaired animals undergo neural sensitization (4). One, unpaired animals *develop* behavioral sensitization in the drug treatment environment. This suggests that the neuroadaptations responsible for behavioral sensitization are induced in these animals, even though they will not be expressed when later the animals are given a drug challenge in a unique test environment. Two, examples of neural sensitization have been described under conditions that preclude the influence of contextual stimuli on the expression of the drug response, for example, in striatal tissue slices *in vitro* and in anesthetized animals (1, 2).

These studies indicate that behavioral sensitization does not simply represent unconditional neurobiological adaptations produced by the actions of drugs on a neural substrate, because the major determinant of whether sensitization is expressed at any particular place or time is the context in which the drug is readministered. Exactly how contextual stimuli and other circumstances surrounding drug administration and readministration modulate the *expression* of sensitization is not known, although Anagnostaras and Robinson (4) discuss the pros and cons of three different models (an excitatory conditioning model, an inhibitory conditioning model, and an occasion-setting model).

II. Modulation of Induction

Not only might the circumstances surrounding drug administration determine whether neural sensitization is expressed in behavior, but also whether neural sensitization is induced in the first place, or at least, the rate and extent of sensitization induced by a particular dose of a drug. Previous studies on context-specific sensitization do not address this issue because in none of these studies was the development of sensitization assessed, just the extent to which stimuli previously associated with drug administration influenced the subsequent expression of sensitization on a test day (5).

The first study to report that an environmental manipulation alters the *induction* of sensitization was by Badiani *et al.* (5). In these experiments, rats received repeated i.p. injections of AMPH under one of two conditions. One group received drug treatments in the cage in which they lived (at "home"). Another group was transported from distinct cages in which they lived to test cages that were physically identical to the other groups home cages, where they then received the drug. For the latter group, therefore, drug treatments were given in a "novel" context. In these experiments, the rate and extent of psychomotor sensitization was greater when AMPH was given in the novel situation, relative to the home situation. More recently, we reported that the unsignaled intravenous infusion of 1.0 mg/kg of AMPH at home (using a remotely activated syringe pump) failed to induce behavioral sensitization (6). When the same treatment was signaled by placement in a physically identical but novel environment, animals did develop behavioral sensitization. These studies suggest, therefore, that the circumstances surrounding drug administration may determine whether a given dose of a drug is capable of inducing neural sensitization.

As in the case of contextual control over the expression of neural sensitization, it is not known how environmental and psychological factors gain access to the neural substrate that is sensitized by psychostimulant drugs. Indeed, it is not even known what the critical variables are. We have speculated that exposure to a relatively novel environment may promote sensitization either because of some action as a stressor or because it facilitates associative learning, but these hypotheses remain to be tested (5). Whatever the mechanism, to fully understand the phenomenon of sensitization (and other drug-induced neuroadaptations), it will be important to determine how seemingly simple manipulations of the circumstances surrounding drug administration can have such profound effects on the ability of drugs to change the nervous system.

In humans, the experience associated with the administration of drugs of abuse varies enormously. This variation exists because a given drug effect is not a simple consequence of the pharmacological actions of a drug, but is due to complex *interactions* among pharmacological, environmental, and psychological factors, for example, whether drugs are taken in a drug-associated environment, what expectations the person brings to the situation, and so on. It is suggested here that similar complex interactions also determine the probability that psychostimulant drugs will produce adaptations in neural systems involved in psychomotor activation and reward, and in this way may contribute to susceptibility to sensitization and addiction. The two examples discussed, of how the circumstances surrounding drug administration can modulate the expression or induction of psychostimulant sensitization, may provide, therefore, a powerful animal model to study the nature of the environmental and psychological factors that influence the ability of drugs to change the nervous system, as well as a way to begin to explore the neurobiological mechanisms by which they act.

References

1. Robinson, T. E., and Berridge, K. C. (1993). The neural basis of drug craving: An incentive-sensitization theory of addiction. *Brain Res. Rev.* **18**, 247–291.
2. White, F. J., and Wolf, M. E. (1991). Psychomotor stimulants. *In* The Biological Bases of Drug Tolerance and Dependence. (J. Pratt, ed.), pp. 153–197. Academic Press, New York.
3. Camp, D. M., DeJonghe, D. K., and Robinson, T. E. (1997). Time-dependent effects of repeated amphetamine treatment on norepinephrine in the hypothalamus and hippocampus assessed with *in vivo* microdialysis. *Neuropsychopharmacol.* (in press).
4. Anagnostaras, S. G., and Robinson, T. E. (1996). Sensitization to the psychomotor stimulant effects of amphetamine: Modulation by associative learning. *Behav. Neurosci.* **110**, 1397–1414.
5. Badiani, A., Anagnostaras, S. G., and Robinson, T. E. (1995). The development of sensitization to the psychomotor stimulant effects of amphetamine is enhanced in a novel environment. *Psychopharmacology* **117**, 443–452.
6. Crombag, H. S., Badiani, A., and Robinson, T. E. (1996). Signalled versus unsignalled intravenous amphetamine: Large differences in the acute psychomotor response and sensitization. *Brain Res.* **722**, 227–231.

Agu Pert

Biological Psychiatry Branch
National Institute of Mental Health
National Institutes of Health
Bethesda, Maryland 20892

Neurobiological Substrates Underlying Conditioned Effects of Cocaine

A variety of neuroadaptive changes occur in the central nervous system following repeated exposure to addictive drugs such as cocaine and amphetamine (1–3). Such adaptations in cellular function have been postulated to underlie some aspects of the addictive process (3). Another form of adaptation that accompanies repeated drug exposure is associative in nature; that is, it is established through learning processes. Pavlov (4) was probably the first to recognize, for example, that drugs could act as unconditioned stimuli. In these early studies, it was found that discrete visual and auditory stimuli associated contiguously with injections of morphine and apomorphine developed over time the ability to elicit emesis. Thus, such classical conditioning can confer to neutral stimuli the ability to elicit certain pharmacological actions of a drug. We have argued previously (5) that the classical conditioning of drug effects to environmental cues underlies the development of incentive motivation, which is probably the neurobehavioral substrate responsible for craving. Understanding the circuitry that mediates conditioned drug effects may aid in the development of appropriate pharmacotherapeutic adjuncts for the treatment of drug addiction.

We have employed a relatively simple design in some of our studies to evaluate the behavioral and neurobiological variables regulating the acquisition and expression of cocaine-induced conditioned effects. Basically, three groups of rats are employed in this paradigm. In its simplest form, on day 1 the first group of rats (PAIRED) is injected with cocaine (30–40 ng/kg, i.p.) prior to placement in locomotor activity chambers for 30 min. One hour following return to their home cages, these rats are injected with saline. The second group (UNPAIRED) is treated in a similar fashion but receives saline prior to placement in the locomotor activity chambers and cocaine in the home cage. The third group (CONTROL) receives saline in both environments. On day 2, all rats are challenged with either saline or 10 ng/kg of cocaine prior to placement in the locomotor activity chamber. We have shown significant conditioned effects of cocaine using this design, which is reflected by dramatic increases in locomotor output in the PAIRED group on the test day relative to the other two groups. In some of our studies, the training sessions were increased to 7 days when it was necessary to increase the persistence and strength of conditioning.

Advances in Pharmacology, Volume 42

1054-3589/98 $25.00

One of our interests has been to define the neurobiological substrates underlying the conditioned effects of cocaine (6–8). We have found, for example, that dopamine (DA)-depleting lesions of the nucleus accumbens and amygdala prevented cocaine-induced conditioning after one training session. The amygdala lesions, however, were not effective in preventing conditioning when more extensive training was employed. The dopaminergic components of the amygdala appear to play a more subtle role in the formation of cocaine-conditioned behaviors than those in the nucleus accumbens. DA-depleting lesions of the frontal cortex and striatum were not effective in preventing the conditioned locomotor effects of cocaine. Neurotoxin-induced lesions of the raphe and locus ceruleus were equally ineffective. Radiofrequency lesions of the dorsal and ventral hippocampus as well as the cerebellum also had little effect on the establishment of cocaine-induced conditioning after 1 day of training. Such findings strongly suggest that intact DA function in the nucleus accumbens and, to a lesser degree, in the amygdala, is necessary for the formation of cocaine-conditioned behaviors.

It has been suggested that different neurobiological processes are involved in the acquisition and expression of psychomotor stimulant-induced conditioned increases in motor behavior. With lesions made prior to training, it is not possible to determine whether the deficit seen is related to disruptions to the acquisition process or to the expression of the behavior. Because DA is involved in the stimulatory and appetitive properties of psychomotor stimulants, it is not surprising that blockade of DA function would lead to decreases in the acquisition of conditioning. For example, neuroleptics coadministered with either amphetamine (9) or cocaine (10, 11) have been found to prevent the development of conditioned locomotor behaviors. More recently, we have found that D1 and D2 DA receptor antagonists are equally effective in preventing the formation of cocaine-induced conditioning (8). Likewise, conditioning in the 1-day design was found only following administration of a combination of D1 and D2 agonists during training, and not when either was administered separately. This would suggest that concurrent D1 and D2 DA receptor occupation is necessary for conditioning to occur.

There are a variety of mechanisms by which DA antagonists could disrupt the acquisition of cocaine conditioning (7). It is most likely, however, that the ability of these drugs to decrease or prevent conditioning to psychomotor stimulants is related to their ability to attenuate the unconditioned effects of the drugs, which are critical in forming the conditioned association. We have suggested that the ability of DA blockers to prevent conditioning is related to their ability to decrease the motivational significance of the unconditioned stimulus (e.g., cocaine). It is well established that the strength of conditioning is directly related to the intensity of the unconditioned stimulus in other conditioning paradigms (5).

Although mixed D1–D2 and selective D1 and D2 antagonists have been found to prevent the establishment of conditioning to cues associated with cocaine, they have been reported to be relatively ineffective in preventing expression once established. Early studies by Beninger and Hahn (9) and Beninger and Herz (10) found that pimozide did not eliminate the behavioral differential between cocaine-conditioned animals and their controls. We have recently extended these findings by demonstrating that neither D1 nor D2 antagonists are effective in altering the differential in performance between the conditioned and unconditioned rats during the test phase.

On the surface, these findings, together with the ability of DA antagonists to block the acquisition of conditioned behaviors, appear to suggest that while intact DA function is critical for the development of conditioning to cocaine-associated cues, it is not necessary for the expression of the conditioned response once established. It is possible that nondopaminergic pathways acquire the ability to elicit such conditioned reactions. The second alternative is that DA is involved in the expression of the conditioned behavior, and that the differential in activity seen between the conditioned and unconditioned groups is determined and maintained by increased activity of mesolimbic DA pathways in the former group, despite similar partial blockade of DA receptors in all experimental groups.

Using the 1-day conditioning paradigm described previously, we have evaluated the ability of stimuli associated with cocaine to increase mesolimbic DA function (6). Microdialysis procedures revealed significant increases in extracellular DA in the nucleus accumbens in the PAIRED group relative to the control groups during the test day. Kalivas and Duffy (12) also have reported increases in mesolimbic DA elicited by stimuli associated with cocaine. More recent studies in our laboratory have not found such conditioned increases in DA overflow in either the amygdala or striatum, suggesting some regional specificity in the effects of conditioned stimuli on DA function. Lesion studies also appear to support these findings. DA-depleting lesions of the nucleus accumbens made immediately after 7 days of conditioning were able to prevent the expression of the conditioned response when rats were tested 7 days postoperatively.

If mesolimbic DA is involved in mediating the expression of cocaine-induced conditioning, it is not working in isolation. Sensory information from conditioned stimuli, for example, needs to gain access to mesoaccumbens DA neurons. Likewise, these neurons need to ultimately activate motor pathways either directly or indirectly. It should be possible to disrupt the expression of cocaine-induced conditioned increases in locomotor behaviors by altering functional activity at any number of relays or integrative centers of this circuit.

Although neither the nucleus accumbens (DA terminals) nor the ventral tegmental area (VTA) DA perikarya) receives input from primary sensory cortical regions, DA activity in this system could be influenced indirectly through other structures, such as the amygdala or frontal cortex. The amygdala, for example, receives polysensory information from cortical sensory association areas and projects in turn to the nucleus accumbens and VTA (13). Because corticofugal neurons are predominantly glutamatergic in nature, it should be possible to disrupt both the acquisition and expression of cocaine-conditioned behaviors with excitatory amino acid antagonists. We, as well as others (14, 15), have found that MK-801 administered concurrently with psychomotor stimulants during training was effective in preventing the development of conditioning. In subsequent studies, we have found that injections of MK-801 into the VTA, but not the amygdala or nucleus accumbens, are also more effective in disrupting the acquisition of cocaine-induced conditioning. Thus, it appears that intact glutamate function in the VTA is necessary for the formation of cocaine-conditioned behaviors.

Despite the ability of MK-801 to disrupt the acquisition of cocaine-conditioned behavior, we have failed to find effects of systemically administered MK-801 and CPP (a competitive N-methyl-D-aspartate [NMDA] antagonist) on expression once established. Injections of MK-801 into the VTA were also ineffective in disrupting the expression of cocaine-conditioned increases in locomotor behavior. Intracerebral injections of DNQX, on the other hand, completely prevented the expression of conditioned behavior, indicating that NMDA receptor function in the VTA is necessary for the development of conditioning, while AMPA receptors apparently mediate the expression once established.

Considerable effort has been devoted to defining the neuropharmacological and neurobiological mechanisms underlying the acquisition of psychomotor stimulant-induced conditioning. Nothing is known, however, regarding the processes involved in the extinction of such behavior once established. An understanding of the processes involved in extinction may be useful for developing strategies to eliminate drug seeking or other aberrant behaviors based on conditioning. One possibility is the development of drugs that may facilitate the extinction process. We have recently evaluated the effects of dopaminergic drugs on the extinction of cocaine-induced conditioning established with five training sessions. The most significant finding to emerge from the series of studies is related to the actions of eticlopride. Although the D2 antagonist was able to completely eliminate the conditioned response during the extinction session, it had virtually no effect on the extinction process. It seems that for extinction to proceed, D2-receptor activation is necessary during unreinforced presentation of previously significant stimuli. If an extension is made to other behaviors, this would suggest that pharmacological elimination of physiological or behavioral symptoms established initially through conditioning may not necessarily facilitate the extinction of such symptoms during the course of treatment with behavioral therapeutic approaches.

References

1. Kalivas, P. W., and Stewart, J. (1991). Dopamine transmission in the initiation and expression of drug- and stress-induced sensitization of motor activity. *Brain Res. Rev.* **16**, 223–244.
2. Robinson, T. E., and Becker, J. B. (1986). Enduring changes in brain and behavior produced by chronic amphetamine administration: A review and evaluation of animal models of amphetamine psychosis. *Brain Res. Rev.* **11**, 157–198.
3. Nestler, E. J. (1995). Molecular basis of addictive states. *Neuroscientist,* 212–220.
4. Pavlov, I. P. (1927). Conditioned Reflexes: An Investigation of the Physiological Activity of the Cerebral Cortex. Dover, New York.
5. Pert, A. (1994). Neurobiological mechanisms underlying the acquisition and expression of incentive motivation by cocaine-associated stimuli: Relationship to craving. *In* Neurobiological Models of Drug Addiction. (L. Erinoff, ed.), pp. 163–190. NIDA Res. Monographs. U.S. Government Printing Office, Washington, D.C.
6. Fontana, D. J., Post, R. M., and Pert, A. (1993). Conditioned increases in mesolimbic dopamine overflow by stimuli associated with cocaine. *Brain Res.* **629**, 31–39.

7. Fontana, D. J., Post, R. M., Weiss, S. R. B., and Pert, A. (1993). The role of D1 and D2 dopamine receptors in the acquisition and expression of cocaine induced conditioned increases in locomotor behavior. *Behav. Pharmacol.* **4**, 375–387.
8. Pert, A., Post, R. M., and Weiss, S. R. B. (1990). Conditioning as a critical determinant of sensitization induced by psychomotor stimulants. *In* Neurobiology of Drug Abuse: Learning and Memory. (L. Erinoff, ed.), pp. 208–241. NIDA Res. Monographs. U.S. Government Printing Office, Washington, D.C.
9. Beninger, R. J., and Hahn, B. C. (1983). Pimozide blocks establishment but not expression of amphetamine-produced environment-specific conditioning. *Science* **220**, 1304–1306.
10. Beninger, R. J., and Herz, R. S. (1986). Pimozide blocks establishment but not expression of cocaine-produced environmental specific conditioning. *Life Sci.* **38**, 1424–1431.
11. Weiss, S. R. B., Post, R. M., Pert, A., Woodward, R. and Murman, D. (1989). Context-dependent cocaine sensitization: differential effects of haloperidol on development versus expression. *Pharmacol. Biochem. Behav.* **34**, 655–661.
12. Kalivas, P. W., and Duffy, P. (1990). Effects of acute and daily cocaine treatment on extracellular dopamine in the nucleus accumbens. *Synapse* **5**, 48–58.
13. Ono, T., Nishiyo, N., and Umano, T. (1995). Amygdala role in conditioned associative learning. *Prog. Neurobiol.* **46**, 401–422.
14. Druhan, J. D., Jakob, A., and Stewart, S. (1993). The development of behavioral sensitization to apomorphine is blocked by MK-801. *Eur. J. Pharmacol.* **243**, 73–77.
15. Wolf, W. E., and Khanra, M. R. (1991). Repeated administration of MK-801 produces sensitization to its own locomotor stimulant effects but blocks sensitization to amphetamine. *Brain Res.* **562**, 164–168.

David A. Gorelick

Treatment Branch
Division of Intramural Research
National Institute on Drug Abuse
National Institutes of Health
Baltimore, Maryland 21224

The Rate Hypothesis and Agonist Substitution Approaches to Cocaine Abuse Treatment

Successful agonist substitution approaches exist for treatment of several drug addictions (e.g., oral methadone for heroin, transdermal or transbuccal nicotine for cigarettes). Two of their shared favorable characteristics are moderate agonist action, which provides some positive psychological effects and improves compliance (as compared with antagonist approaches, e.g., naltrexone for heroin) and slow onset of action. The latter characteristic takes advantage of the so-called rate hypothesis of psychoactive drug effect, which holds that the strength of positive psychological effects (and, therefore, abuse liability) is proportional to the rate of drug binding to its site of action (or to onset of drug effect, a surrogate variable used in clinical research). Thus, a drug with slower onset of action would have less abuse liability. In principle, these two favorable characteristics could be achieved either pharmacokinetically, such as by giving the drug via a slowly absorbed route of administration (e.g., transdermal nicotine), or pharmacodynamically, such as by giving an analogue with slow, long-lasting binding to the site of action (e.g., oral methadone).

Evidence in humans for the rate hypothesis of cocaine effects is largely circumstantial, based on pharmacokinetic differences among routes of administration. Retrospective self-reports by experienced cocaine users and studies of experimental cocaine administration are consistent with the rate hypothesis, in that the degree of subjective liking (and positive psychological effects) associated with cocaine taken by various routes of administration parallels the latency to achieving peak blood concentration (presumably reflecting different rates of cocaine reaching the brain): smoked and i.v. faster than intranasal, which is faster than oral. Abuse liability also parallels route of administration, with oral cocaine associated with little abuse (e.g., coca leaf chewing or coca tea drinking in South American countries, where it is legal), while smoked and i.v. cocaine are associated with severe abuse. Cocaine addicts in treatment report a faster progression to dependence and treatment seeking when they began cocaine use by smoking than do those who began cocaine use by insufflation (intranasal) (1). These findings, while circumstantial, are consistent, with a slow-onset formulation of cocaine (e.g., oral or transdermal) meeting some of the feasibility criteria as an agonist-substitution approach.

Direct experimental evidence of the rate hypothesis for cocaine is limited to a study in rhesus monkeys that found increased cocaine self-administration produced by increased i.v. infusion rate across two different cocaine doses (2). Two current human studies are also evaluating the rate hypothesis. Gorelick and colleagues at the National Institute on Drug Abuse (NIDA) are administering i.v. cocaine (10, 25, and 50 mg) at three infusion durations (10, 30, and 60 sec) to experienced cocaine users, using a double-blind, crossover design in which each subject receives every dose–duration combination plus a saline placebo session. Preliminary results from four subjects show expected dose-dependent increases in positive psychological (e.g., craving for cocaine, "rush," measured by visual analogue scales) and cardiovascular (heart rate, blood pressure) effects, with indication of duration (i.e., rate)-dependent effects on some measures, consistent with the rate hypothesis. Bigelow and colleagues at Johns Hopkins University have administered i.v. cocaine (30 mg/70 kg) at three infusion durations (2, 15, and 60 sec) to nine experienced cocaine users, also using a double-blind, crossover design. Preliminary results show significant duration-dependent effects on psychological measures (e.g., rush) but less influence on cardiovascular measures.

Recent experimental human studies with oral cocaine (salt form in capsules) are also consistent with the rate hypothesis. Preston and colleagues at NIDA found that 25–50 mg (roughly comparable to estimate daily doses absorbed by coca leaf chewers in South America) was psychoactive in experienced cocaine users but did not cause a substantial positive psychological effect (rush or "high"). Subjects could learn to accurately discriminate cocaine from placebo and had increased response rates on a computerized fixed-ratio performance task, without significant increases in heart rate. Bigelow and colleagues at Johns Hopkins University recently found that chronic (2 wk) treatment with oral cocaine (up to 100 mg four times daily) significantly reduced the acute psychological effects of an i.v. cocaine (25 or 50 mg) challenge, with little change in cardiovascular effects (3). These findings are consistent with the concept of substitution with a slow-onset agonist drug as treatment for cocaine addiction (and with anecdotal case reports from South America that oral cocaine, in the form of coca tea, may reduce craving for and use of smoked cocaine).

Several medications that bind to the presumed major site of action for cocaine's psychoactive effects (the presynaptic dopamine transporters [DAT]) are being studied as possible agonist substitution agents. These agents have some mild stimulant-like agonist properties, with little apparent human abuse potential when taken in slow-onset form (orally). Both mazindol, an appetite suppressant, and bupropion, an antidepressant, have been ineffective in double-blind clinical trials. This may be related to dose-limiting side effects, which have kept doses in the range that probably occupies less than half of brain presynaptic DAT sites (4). The investigational compound GBR-12909, developed as an antidepressant, had sedative (rather than stimulant) psychological effects in the one published human study, suggesting low abuse potential. In animal studies, it has reduced the increased brain extracellular dopamine concentration produced by cocaine administration and reduced cocaine self-administration at doses that did not influence food intake (5). No clinical trials with GBR-12909 have been done.

In summary, agents with slow onset of effect, including possibly the abused drug itself by another route of administration, may be useful in implementing the agonist-substitution approach to treatment of cocaine abuse, consistent with the rate hypothesis of psychoactive drug effect.

References

1. Gorelick, D. A. (1992). Progression of dependence in male cocaine addicts. *Am. J. Drug Alcohol Abuse* **18,** 13–19.
2. Balster, R., and Schuster, C. R. (1973). Fixed-interval schedule of cocaine reinforcement: Effect of dose and infusion duration. *J. Exp. Anal. Behav.* **20,** 119–129.
3. Haberny, K., Walsh, S. L., Ginn, D. H., Liebson, I. A., and Bigelow, G. E. (1996). Modulation of intravenous cocaine effects by chronic oral cocaine in human drug abusers. *NIDA Res. Monogr.* **162,** 273.
4. Wong, D. F., Babington, Y., Dannals, R. F., Shaya, E. K., Ravert, H. T., Chen, C. A., Chan, B., Folio, T., Scheffel, U., Ricaurte, G. A., Neumeyer, J. L., Wagner, Jr., H. N., and Kuhar, M. J. (1995). In vivo imaging of baboon and human dopamine transporters by position emission tomography using [^{11}C]WIN 35, 428. Synapse **15,** 130–142.
5. Glowa, J. R., Wojnicki, F. H. E., Matecka, D., Rice, K. C., and Rothman, R. B. (1995). Effects of dopamine reuptake inhibitors on food- and cocaine-maintained responding: II. Comparisons with other drugs and repeated administration. *Exp. Clin. Psychopharmacol.* **3,** 232–239.

Marco Diana

Department of Drug Sciences
University of Sassari
07100 Sassari, Italy

Drugs of Abuse and Dopamine Cell Activity

Electrophysiological investigations provide a powerful method for assessing neuronal effects of addicting substances in the central nervous system. Perhaps more importantly, electrophysiology stands above other means commonly employed, for it allows the experimenter to monitor real neuronal activity, although under limited circumstances (anesthesia and/or stressful situations such as the use of neuromuscular blocking agents). Unfortunately, routine use of these methods in freely moving, undrugged animals, which will be desirable to study interactions of chemicals with dopaminergic systems, is still hampered by many technical difficulties, and studies have appeared only scattered through the years in the scientific literature (1, 2). Understanding of the physiological properties and pharmacological responsiveness of dopaminergic systems in the brain, however, has gained great benefit from the use of extracellular and intracellular recording techniques.

Specifically, in the field of drug addiction, amphetamine's electrophysiological effects on dopamine neurons were described in the early 1970s by the original studies of Bunney *et al.* (3). Although the interpretation of the inhibitory effect of intravenous amphetamine on dopaminergic neurons was not shared by Groves and colleagues (4), the first stone was posed in the comprehension of mechanisms related to the interaction of drugs of abuse with dopaminergic cells. Acute effects of major addicting drugs such as cocaine, nicotine, ethanol, and morphine were shortly thereafter described. The overall picture, as it appears today, is that with the exception of psychostimulants (i.e., amphetamines and cocaine), which produce a decrease in dopaminergic activity due to an increase in dopamine outflow (both terminal and dendritic), other substances potentiate central dopaminergic transmission as measured by electrophysiological recordings. Drugs of abuse can, thus, be divided, on the basis of their electrophysiological effects, into two classes: (1) psychostimulants, which decrease dopaminergic activity, and (2) others, the acute administration of which increases dopamine firing rate and pattern. While the reduction in firing rate that follows amphetamine and cocaine administration can be viewed as a neuronal attempt to counteract the powerful dopamine-releasing properties of these drugs (neuron's life-saving device), the increment in electrical activity observed after administration of other drugs may form the neuronal basis of the dopamine transmission-potentiating effect of drugs of abuse, widely recognized as the leading candidate in the search for the brain system that

Advances in Pharmacology, Volume 42

1054-3589/98 $25.00

mediates the reinforcing properties of drugs, which in turn is a key factor in the phenomenon of drug addiction and dependence.

Drug addiction, however, is a chronic phenomenon, a pathology that requires a fair amount of time (and drug administration) to develop and to produce behavioral alterations that are likely the result of a neuronal effect. Consequently, acute studies, although extremely informative on the one hand, are not very helpful in the context of drug dependence, because they do not meet one of the essential criteria in the definition of drug addiction. Chronic administration of drugs may thus mimic better neuronal alterations relevant for understanding mechanisms of drug dependence.

Not surprisingly, the mesolimbic dopamine system responds similarly to a chronic regimen of various addicting drugs. Repeated amphetamine, cocaine, and morphine all seem to increase the basal activity of dopaminergic neurons (5). In the case of psychostimulants, an additional factor has been described: subsensitivity of somatic autoreceptors, which may be the result of the increased dendritic dopamine release, which in turn is the cause of firing reduction after acute administration (5). In the case of ethanol, no increase in basal activity was reported after chronic administration, but a lack of tolerance to acute ethanol's administration has been described (6). Collectively, these results point to an increased dopaminergic tone after chronic administration of an addicting substance. This fact indicates once again that the mesolimbic dopamine system is a major target of drugs of abuse, not only after acute administration, but also after a chronic challenge. While the chronic intake of drugs of abuse represents an essential step in the development of drug addiction, the withdrawal syndrome is often the painful end. It is a phase in which a somatic suffering typical for each abused drug is associated with a common psychological discomfort, often termed *dysphoria*, that frequently triggers subsequent drug-seeking and taking behavior, with final relapse into drug dependence. The withdrawal syndrome is, thus, an extremely important phase in the process of drug dependence because it combines the positive and negative reinforcing properties of drugs of abuse (7).

We began a series of experiments aimed at investigating the physiological status of mesolimbic dopaminergic neurons during and after withdrawal from two of the most abused drugs: ethanol and morphine. In the first experiment (8), we found that rats made dependent on ethanol showed, on suspension of treatment (12 hr), a marked reduction in the spontaneous activity of antidromically identified mesolimbic dopaminergic neurons. Subsequently (9), the reduction in spontaneous activity was found to correlate with a decrease in dopamine outflow as measured by the microdialysis technique in the nucleus accumbens. Careful analysis of the electrophysiological parameters revealed a decrement in spontaneous firing rate, which was accompanied by an apparently normal (within control values) percentage of burst firing. Because the percentage of burst firing is related to basal neuronal activity (firing rate), we introduced (9) a different measure of burst firing: burst rate, which averages the burst firing per unit of time. In contrast to the apparently normal burst firing, burst rate disclosed another factor: a reduction in the number of spikes delivered in burst, which may be viewed as the main factor in the reduction of dopamine outflow. The number of action potentials within bursts was similarly reduced (9). Furthermore, evaluation of the number of spontaneously active dopaminergic neurons, through the cells/track method (8, 9), did not show variations between ethanol-dependent and controls (saline-treated rats), respectively. In spite of this finding, a subsequent report (10) described a reduction of the cells/track index in ethanol-dependent (chloral hydrate-anesthetized) rats at 24 hr after last ethanol administration, thus favoring the possibility of an ethanol withdrawal–induced depolarization blockade of dopamine neurons. This prompted us to directly compare the effects of chronic ethanol regimen, and subsequent withdrawal, in chloral hydrate–anesthetized and unanesthetized rats (11). The reduction in the cell–track ratio was found restricted to the group of rats anesthetized and withdrawn from ethanol but not in their unanesthetized counterpart. This fact, together with other ancillary findings (similar reduction in firing rate, pattern, and apomorphine response) (11) led us to the conclusion that the apparent depolarization block requires an essential ingredient, such as the use of chloral hydrate, and, thus, is extremely unlikely to be the effect of ethanol withdrawal. The presence, instead, of a higher number of silent and/or slowly firing dopamine neurons (9) was confirmed (11) in ethanol-withdrawn rats (both anesthetized and unanestethized). This result, once again, does not support the existence of a depolarization inactivation, but rather it suggests an opposite condition (hyperpolarization of the membrane). Although this possibility needs verification through intracellular investigations, it seems to be the most realistic possibility to explain the reduction in spontaneous activity affecting mesolimbic dopamine neurons after ethanol withdrawal.

Similarly, dopamine neurons recorded from rats made dependent on morphine while experiencing a measurable withdrawal syndrome (12) showed electrophysiological features reminiscent of those observed in ethanol-withdrawn rats. Firing rate and burst rate were found reduced, together with the number of spikes/burst. In sharp contrast, the number of spontaneously active neurons, as measured by the cells–track ratio, was found unaltered in morphine-withdrawn rats as compared with relative controls (12). The striking similarity of results obtained with two different chemicals, such as ethanol and morphine, reinforces the view that the mesolimbic dopamine system is profoundly affected by chronic administration and withdrawal of addicting drugs and consequently may be viewed as the target for potential new therapies aimed at ameliorating the psychological discomfort produced by drug withdrawal.

However, these experiments (9, 11, 12) were performed early in the withdrawal phase, a time window in which somatic manifestations of abstinence are still evident and measurable. No firm conclusion can, thus, be drawn from these results about possible relevance of dopamine in the slowly developing and long-lasting psychological aspects (dysphoria) of drug withdrawal. Therefore, we decided to study dopaminergic neuronal activity in the late phases (beyond resolution of somatic signs of withdrawal) of ethanol and morphine withdrawal to verify the possibility of a long-lasting hypodopaminergism. A similar result would allow: (1) exclusion of dopamine involvement in the somatic aspects of withdrawal and (2) pursuance of the idea of a reduced dopamine activity as an important factor in the long-lasting effects of drug dependence and design of new experiments aimed at further evaluating this hypothesis. For the time being, the hypothesis has been confirmed by the finding that dopamine neuronal activity reduction outlasts both ethanol (13) and morphine (14) somatic withdrawal.

Acknowledgments

The author is grateful to A. L. Muntoni, M. Pistis, M., Melis, and G. L. Gessa for continuous and informative discussions on the material discussed in this paper. The author also wishes to extend his thanks to Stefano Aramo for his expert technical assistance.

References

1. Freeman, A. S., Meltzer, L. T., and Bunney, B. S. (1985). Firing properties of substantia nigra dopaminergic neurons in freely moving rats. *Life Sci.* **36,** 1983–1994.
2. Diana, M., Garcia-Munoz, M., Richards, J. B., and Freed, C. R. (1989). Electrophysiological analysis of dopamine cells from the substantia nigra pars compacta of circulating rats. *Exp. Brain Res.* **74,** 625–630.
3. Bunney, B. S., Aghajanian, G. K., and Roth, R. H. (1973). Comparison of effects of L-DOPA, amphetamine and apomorphine on the firing rate of rat dopaminergic neurons. *Nature* **245,** 123–125.
4. Groves, P. M., Wilson, C. J., Young, S. J., and Rebec, G. V. (1975). Self-inhibition by dopamine neurons. *Science* **190,** 522–529.
5. White, F. J. (1996). Synaptic regulation of mesocorticolimbic dopamine neurons. *Annu. Rev. Neurosci.* **19,** 405–436.
6. Diana, M., Rossetti, Z. L., and Gessa, G. L. (1992). Lack of tolerance to ethanol-induced stimulation of mesolimbic dopamine system. *Alcohol Alcohol.* **27,** 329–333.
7. Wise, R. A. (1988). The neurobiology of craving: Implications for the understanding and treatment of addiction. *J. Abnorm. Psychol.* **97,** 118–132.
8. Diana, M., Pistis, M., Muntoni, A. L., Rossetti, Z. L., and Gessa, G. L. (1992). Marked decrease of A10 dopamine neuronal firing during ethanol withdrawal syndrome in rats. *Eur. J. Pharmacol.* **221,** 403–404.
9. Diana, M., Pistis, M., Carboni, S., Gessa, G. L., and Rossetti, Z. L. (1993). Profound decrement of mesolimbic dopaminergic neuronal activity during ethanol withdrawal syndrome in rats: Electrophysiological and biochemical evidence. *Proc. Natl. Acad. Sci. U.S.A.* **90,** 7966–7969.
10. Shen, R. Y., and Chiodo, L. A. (1993). Acute withdrawal after repeated ethanol treatment reduces the number of spontaneously active dopaminergic neurons in the ventral tegmental area. *Brain Res.* **622,** 289–293.
11. Diana, M., Pistis, M., Muntoni, A. L., and Gessa, G. L. (1995). Ethanol withdrawal does not induce a reduction in the number of spontaneously active dopaminergic neurons in the mesolimbic system. *Brain Res.* **682,** 29–34.
12. Diana, M., Pistis, M., Muntoni, A. L., and Gessa, G. L. (1995). Profound decrease of mesolimbic dopaminergic neuronal activity in morphine withdrawn rats. *J. Pharmacol. Exp. Ther.* **272,** 281–285.
13. Diana, M., Pistis, M., Muntoni, A. L., and Gessa, G. L. (1996). Mesolimbic dopaminergic reduction outlasts ethanol withdrawal syndrome: Evidence of protracted abstinence. *Neuroscience* **71,** 411–415.
14. Diana, M., Muntoni, A. L., Pistis, M., Palomba, F., and Gessa, G. L. (1995). Long-lasting changes in mesolimbic dopaminergic functioning after drug withdrawal. *Soc. Neurosci. Abstr.* **21,** 731.

C. D. Fiorillo, J. T. Williams, and A. Bonci

Vollum Institute
Oregon Health Sciences University
Portland, Oregon 97201

D1-Receptor Regulation of Synaptic Potentials in the Ventral Tegmental Area after Chronic Drug Treatment

A common feature of addictive drugs is their ability to enhance dopaminergic transmission. Activation of dopamine receptors appears to be both necessary and sufficient for the reinforcing properties of stimulants such as cocaine and amphetamine, and it is thought to be vital to the reinforcing effects of morphine, ethanol, and nicotine as well. The nucleus accumbens (NAc), which receives a dense dopaminergic innervation from the ventral tegmental area (VTA), is the primary locus of dopamine's reinforcing and locomotor activating effects. Repeated administration of amphetamine, cocaine, or morphine has been reported to result in sensitization to both reinforcing and locomotor stimulant effects on further exposure to any of the three drugs. Sensitization lasts long after withdrawal and serves as a model of the craving that persists in addicts and promotes relapse even after prolonged periods of abstinence. The phenomenon of cross-sensitization between psychostimulants and opiates suggests that these drugs produce common neural adaptations, despite having different pharmacological targets.

Behavioral sensitization has been found to correspond to a hypersensitive mesolimbic dopamine system (1). Two general mechanisms of sensitization have been identified using *in vivo* techniques: an enhanced release of dopamine in the NAc in response to drug and a sensitized response of NAc neurons to dopamine agonists. A particularly relevant adaptation at the cellular level is the enhanced inhibition of NAc neurons by D1-receptor activation in brain slices or *in vivo* after repeated administration of psychostimulants, though little is known about the mechanisms involved.

The VTA contains the somata of mesolimbic and mesocortical dopamine neurons. D1 receptors in the VTA are localized to GABAergic synaptic terminals originating in the NAc and ventral pallidum. D1 agonists act presynaptically to enhance the amplitude of $GABA_B$ (but not $GABA_A$) inhibitory postsynaptic potentials (IPSPs) in dopamine cells of guinea pig VTA (2). Forskolin, which activates adenylyl cyclase, mimics the effect of D1 agonists, while the selective D1 antagonist SCH 23390 by itself reduces the amplitude of the IPSP. This suggests that dopamine acts tonically at D1 receptors in the VTA to facilitate release of GABA.

Advances in Pharmacology, Volume 42

1054-3589/98 $25.00

We have investigated D1-receptor modulation of GABA transmission in slices of the VTA following a treatment regimen designed to elicit behavioral sensitization (3). Guinea pigs were given daily injections (i.p.) of cocaine (10 mg/kg) or saline for 14 days or morphine (10 mg/kg) for 7 days, and brain slices (300 μm thick) of VTA were prepared 7–10 days after the last injection. Intracellular recordings were made in dopamine cells identified by their distinct membrane properties. Pharmacologically isolated $GABA_B$ IPSPs were evoked once per minute by stimulating (10 stimuli of 500 μs at 70 Hz) with bipolar tungsten electrodes.

In slices from saline-treated animals, D1-receptor activation augmented the IPSP (~25%), as previously reported, while the D1 antagonist SCH 23390 reduced the amplitude of the IPSP (~20%). However, in slices from drug-treated animals, the effect of D1 activation was reversed (Fig. 1). D1 agonists (dopamine, SKF 38393, SKF 82958) depressed the IPSP (~20%), while D1 antagonists (SCH 23390, cis-flupenthixol) enhanced it (~25%). In all cases, activation of adenylyl cyclase by forskolin mimicked the effect of D1 agonists.

It is known that cyclic adenosine monophosphate (cAMP) can be metabolized to adenosine, which acts at A1 receptors to inhibit transmission in many areas of the central nervous system, including the $GABA_B$ IPSP in dopamine cells of the VTA. D1 activation could, therefore, produce an inhibition of transmitter release mediated by adenosine. To test this hypothesis, slices were treated with the A1 antagonists 8-CPT or DPCPX. After blockade of A1 receptors, activation of adenylyl cyclase by D1 agonists or forskolin in slices from drug-treated animals now produced an augmentation of the IPSP identical to

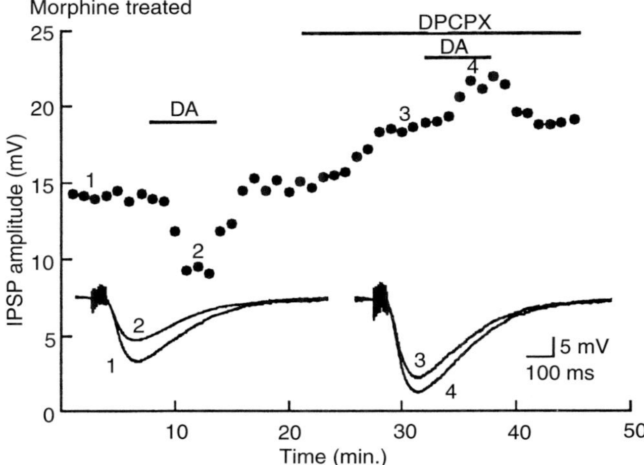

FIGURE 1 Dopamine depressed the amplitude of the $GABA_B$ IPSP in a slice taken from a morphine-treated animal. Application of the adenosine antagonist, DPCPX (1 μM), increased the amplitude of the IPSP, and subsequent application of dopamine caused a further increase in the IPSP. The amplitude of the IPSP is plotted as a function of time (1 IPSP/min). Selected IPSPs are illustrated in lower portion of graph.

that observed in slices from saline-treated animals (see Fig. 1), while SCH 23390 caused an identical depression of the IPSP. To further investigate the involvement of adenosine in the altered D1 response, several other agents were used: Ro 201724, an inhibitor of cAMP-dependent phosphodiesterase; probenicid, an inhibitor of cAMP transport; and adenosine deaminase, an enzyme that converts adenosine to inosine. In the presence of any of these agents, D1-receptor modulation of the IPSP in slices from drug-treated animals was the same as that seen in control slices. In addition, when applied by themselves, all of these agents as well as A1 antagonists (see Fig. 1) produced a significantly greater enhancement of the IPSP (30–40%) in slices from drug-treated animals than those from saline-treated controls (10–15%). A1 antagonists were without effect in the presence of the D1 antagonist SCH 23390. This suggests that adenosine tone is dependent on D1-receptor activation, is elevated in drug-treated animals, and is responsible for the reversal of dopamine's effect.

The sensitivity of D1- and A1-mediated responses was investigated in the presence of Ro 201724 to reduce endogenous adenosine levels. Concentration-response curves revealed that neither the sensitivity to nor the maximum effect of (±)SKF 38393 or the selective A1 agonist N6-CPA was altered in drug-treated animals, indicating that the drug-induced increase in adenosine tone results from an enhanced transport or metabolism of cAMP to adenosine rather than a change in the potency or efficacy of D1 or A1 agonists.

Because the actions of addictive drugs have been studied very little in guinea pigs, we have examined D1 and A1 receptors in rat VTA 10–20 days following a 2-wk period of injections. The A1 antagonist DPCPX (200 nM) enhanced the $GABA_B$ IPSP by 15 ± 7% (n = 8) in slices from saline-treated animals. After repeated injections of cocaine (14 × 20 mg/kg), amphetamine (5 × 2 mg/kg), or morphine (7 × 10 mg/kg), DPCPX enhanced the IPSP by 69 ± 23% (n = 8), 78 ± 24% (n = 7), and 55 ± 1% (n = 3), respectively, significantly more than in saline-treated animals (unpaired t-tests, $p < .05$). The D1 agonist SKF 82958 (1 μM) caused a 101 ± 34% (n = 4) increase in the amplitude of the IPSP in slices from saline-treated rats, but only a 32 ± 7% (n = 8; unpaired t-test, $p < .01$) increase in cocaine-treated rats. It can, therefore, be concluded that D1 receptors also facilitate GABA release in rat VTA, and that repeated drug administration, through a long-lasting increase in adenosine tone, shifts the balance toward inhibition of GABA release.

Studies utilizing repeated, moderate doses of psychostimulants have found no change in D1-receptor binding, while measurements of adenylyl cyclase activity have yielded mixed results. The present results indicate a lack of any long-lasting functional changes in D1 receptors or adenylyl cyclase following repeated injections of cocaine or morphine. Rather, the observed change occurred "downstream" of adenylyl cyclase and might involve transport of cAMP or adenosine or the metabolism of cAMP by enzymes such as phosphodiesterase or ecto 5′ nucleotidase. The long-lasting nature of this synaptic change suggests a persistent alteration in gene regulation. Psychostimulant-mediated induction of immediate early genes (IEGs) is observed in a large number of neurons of the NAc projecting to the midbrain. Chronic cocaine treatment induces IEGs in the NAc that are unique and long-lasting relative to the IEGs induced by acute treatment, and may underlie persistent changes in gene expression (4).

Increased adenosine inhibition of GABA transmission in the VTA would presumably result in disinhibition of dopamine neurons and enhanced dopamine levels in projection areas such as the NAc. One or more weeks after repeated treatment with opiates or psychostimulants, basal levels of dopamine in the NAc are unchanged, while drug-induced increases in dopamine are significantly enhanced. This might be explained by the present results, assuming that GABAergic afferents to the VTA are relatively silent under control conditions but are stimulated by drug administration. There is indeed evidence that activation of $GABA_B$ receptors on dopamine cells is not tonic (5), but it is unknown under what conditions these afferents are active.

The D1-receptor–dependent increase in adenosine tone reported here is likely to contribute to the persistent sensitization of the mesolimbic dopamine system observed following repeated administration of psychostimulants or opiates. It also demonstrates that cocaine and morphine, despite very different pharmacological targets, produce a common synaptic change that implicates the mesolimbic dopamine system in their chronic as well as their acute effects.

References

1. Kalivas, P. W., and Stewart, J. (1991). Dopamine transmission in the initiation and expression of drug- and stress-induced sensitization of motor activity. *Brain Res. Rev.* **16,** 223–244.
2. Cameron, D. L., and Williams, J. T. (1993). Dopamine D1 receptors facilitate transmitter release. *Nature* **366,** 344–347.
3. Bonci, A., and Williams, J. T. (1996). A common mechanism mediates long-term changes in synaptic transmission after chronic cocaine and morphine. *Neuron* **16,** 631–639.
4. Self, D. W., and Nestler, E. J. (1995). Molecular mechanisms of drug reinforcement and addiction. *Annu. Rev. Neurosci.* **18,** 463–495.
5. Westerink, B. H. C., Kwint, H. F., and de Vries, J. B. (1996). The pharmacology of mesolimbic dopamine neurons: A dual probe microdialysis study in the ventral tegmental area and nucleus accumbens of the rat brain. *J. Neurosci.* **16,** 2605–2611.

Francis J. White, Xiu-Ti Hu, and Xu-Feng Zhang

Neuropsychopharmacology Laboratory
Department of Neuroscience
Finch University of Health Sciences/Chicago Medical School
North Chicago, Illinois 60064

Neuroadaptations in Nucleus Accumbens Neurons Resulting from Repeated Cocaine Administration

During withdrawal, cocaine addicts experience a number of characteristic symptoms, including anergia, anxiety, anhedonia, cocaine craving, and sensitization to environmental cues associated with cocaine use. Animal studies of such behavioral alterations have identified a number of underlying neuronal correlates within the mesocorticolimbic dopamine (DA) system, which appears to be primarily responsible for the positive reinforcing (rewarding) efficacy of cocaine and other drugs of abuse. Using extracellular single-cell recording techniques, we have identified time-dependent alterations in the activity of neurons within several mesocorticolimbic structures (1). For example, increased basal activity of ventral tegmental area DA neurons appears to be necessary for the induction of behavioral sensitization, whereas decreases in the activity of this neuronal population may be associated with cocaine withdrawal.

Neuroadaptations within the nucleus accumbens (NAc) appear to be most responsible for the expression of many indices of cocaine withdrawal, including behavioral sensitization. We have demonstrated a close temporal relationship between the persistence of behavioral sensitization and enhanced responsiveness of NAc neurons to DA D1-receptor stimulation. Thus, both sensitized locomotor responses to cocaine and enhanced inhibitory responses of NAc neurons to DA D1-receptor stimulation are evident at withdrawal times of 1 day, 1 wk, and 1 mo, but not 2 mo (1). However, additional NAc neuronal alterations are also evident at early withdrawal times, including enhanced inhibitory effects of GABA and serotonin and markedly attenuated excitatory responses to glutamate (1). Because there is no evidence to suggest that repeated cocaine administration alters receptors for all of these various neurotransmitters within the NAc, such observations have led us to consider the possibility that repeated cocaine administration may alter the membrane properties that control the excitability of NAc neurons.

To examine the membrane properties of NAc neurons in cocaine-pretreated rats, we made intracellular recordings from rat brain slices *in vitro*. Male Sprague-Dawley rats (150–175 g) received once-daily injections of either (−)co-

Advances in Pharmacology, Volume 42

1054-3589/98 $25.00

caine HCl (15.0 mg/kg) or saline for 5 days. After a 3-day withdrawal period, the rats were decapitated (under halothane anesthesia) and 400-μM slices containing the NAc were cut in the coronal plane. Standard *in vitro* recordings were made using current-clamp techniques. Neurons within the core region of the NAc were studied beginning 2 hr after preparation of the brain slices.

Cocaine-pretreated NAc neurons exhibited a number of significant alterations in passive and active membrane properties, as compared with saline-pretreated controls. The resting membrane potential was significantly more hyperpolarized (-83.6 ± 0.7 vs -78.7 ± 0.8 mV), but there was no significant difference in input resistance.

Depolarization of NAc neurons by intracellular current injection revealed additional significant differences between cocaine-pretreated and control neurons. The current required to generate an action potential was significantly increased in the cocaine-pretreated neurons (0.98 ± 0.06 vs 0.65 ± 0.06 nA), as was the threshold for action potential generation (-40.4 ± 1.8 vs -46.5 ± 1.7 mV), whereas the action potential amplitude was significantly reduced (51.9 ± 1.8 vs 60.4 ± 1.1 mV). Repetitive firing during depolarization was also significantly reduced in the cocaine-pretreated neurons.

Taken together, these findings indicate that during the early period of withdrawal from repeated cocaine administration, NAc neurons are significantly less excitable than normal. Because we have demonstrated that D1-receptor–mediated responses are significantly enhanced by this cocaine-pretreatment regimen, we next determined whether endogenous DA within the slice was rendering the neurons less excitable by stimulating supersensitive D1 receptors. Bath perfusion with the selective DA D1-receptor antagonist SCH 23390 did not alter the passive or active membrane properties of NAc neurons, indicating that there was no basal activation of D1 receptors in our slices. The alterations could not be attributed to the local anesthetic properties of cocaine because it is unlikely that cocaine or its metabolites were present in the slice following a 3-day withdrawal period. Moreover, repeated administration of the local anesthetic lidocaine (50 mg/kg) failed to mimic the effects of repeated cocaine.

Several of the effects produced by repeated cocaine administration, in particular the increased threshold, decreased action potential amplitude, and reduction in repetitive firing, might be indicative of alterations in voltage-sensitive sodium channels (VSSCs). In view of previous research indicating that Na^+ currents in striatal neurons are modulated by DA receptors (2), we hypothesized that if similar modulation occurred with the NAc, then the repeated increase in DA-receptor occupation resulting from cocaine administration might have altered VSSCs in NAc neurons. To test this possibility, we made whole-cell recordings of Na^+ currents from freshly dissociated NAc neurons obtained from cocaine-pretreated or saline-pretreated control rats (as previously discussed).

As in the dorsal striatum, medium spiny neurons (capacitance <10 pF) within the NAc exhibit DA-receptor modulation of VSSCs. In particular, D1 receptor stimulation with SKF 38393 (100 nM–1 μM) suppressed whole-cell, tetrodotoxin-sensitive Na^+ current in approximately 80% of the neurons we tested. This effect was completely prevented by SCH 23390 (1 μM). The effect was produced through activation of the cyclic adenosine monophosphate (cAMP) signal transduction cascade, because it was mimicked by either bath application of the membrane permeable cAMP analogue 8-Br-cAMP or intracellular dialysis (via the patch-clamp electrode) with the catalytic subunit of cAMP-dependent protein kinase (PKA), and it was prevented by intracellular dialysis with an inhibitor of PKA.

In rats that had been pretreated with cocaine and withdrawn for 3 days, whole-cell Na^+ current, evoked by stepping the membrane voltage from a holding potential of −70 mV to a test potential of −20 mV, was significantly reduced; peak current (nA/pF) was reduced by approximately 33%. This effect was accompanied by a depolarizing shift in the voltage dependence of activation (Fig. 1). There was no significant alteration in the voltage dependence of inactivation. As in our current-clamp recordings from NAc slices, pretreatment with the sodium channel blocker lidocaine failed to reproduce the alterations in whole-cell Na^+ currents observed after repeated cocaine. In addition, recordings from acutely dissociated cortical neurons failed to reveal changes in VSSCs, such as those seen in the NAc. Thus, the effect is region-specific and cannot be attributed to the local anesthetic properties of cocaine.

The reduction in peak Na^+ current produced by repeated cocaine treatment is consistent with enhanced phosphorylation of VSSCs (3). However, debate exists regarding phosphorylation-induced alterations in voltage-dependent kinetics (3). The depolarizing shift in the voltage dependence of activation that we observed has been reported by some investigators to result from phosphorylation by protein kinase C (4), but others report reductions in peak current accompanied by a slowing of inactivation (3). Therefore, the mechanism underlying the change in voltage dependence of activation produced by repeated cocaine administration is presently unclear. What does appear to be clear is

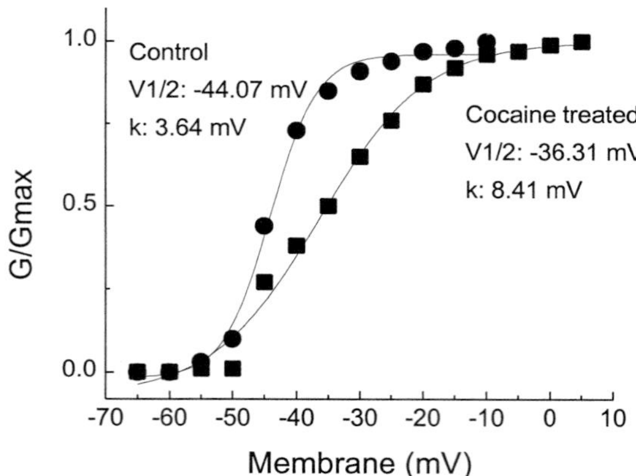

FIGURE 1 Examples of voltage-dependent activation of whole-cell Na^+ current obtained from representative saline-pretreated and cocaine-pretreated NAc neurons. Current-voltage (I-V) plots were generated by stepping the membrane voltage from the holding potential of −70 mV to various test potentials. Steady-state activation curves were constructed from I-V curves and fitted with a Boltzmann equation.

that repeated cocaine decreases the excitability of NAc neurons by increasing the basal state of phosphorylation of VSSCs.

The level of VSSC activity is subject to tonic modulation by the levels of PKA, an enzyme that is known to be increased within the NAc by repeated cocaine administration (5). Phosphorylation by PKA is also regulated by prior phosphorylation by PKC at serine 1506 within the intracellular loop connecting homologous domains III and IV of the α-subunit of brain type IIA Na^+ channels (3). Accordingly, neuromodulation of PKC activity might also be altered by repeated cocaine administration in such a way as to enhance basal states of phosphorylation. Given that there exists both a substantial range over which VSSCs can be modulated by phosphorylation and a broad array of potential neurotransmitters that can influence levels and activity of both PKA and PKC, any number of mechanisms could be responsible for the down-modulation of VSSCs in cocaine-pretreated NAc neurons. Whatever the mechanism underlying the effects of repeated cocaine administration, our findings indicate a novel form of plasticity. Although alterations in transmission at specific synapses (synaptic plasticity) are certain to modify the responsiveness of NAc neurons to selected inputs, the marked reduction of VSSC function (nonsynaptic or whole-cell plasticity) will produce a more profound and indiscriminate decrease in the responsiveness of the NAc to exitatory commands. Given that the major excitatory inputs into the NAc originate within prefrontal cortex, hippocampus, and amygdala, structures known to be involved in motivated behavior and the learned associations between such behavior and specific environmental setting and emotional states, the reduced excitability of NAc neurons may be directly related to cocaine withdrawal effects such as anergia, anhedonia, anxiety, and depression.

References

1. White, F. J., Hu, X.-T., Henry, D. J., and Zhang, X.-F. (1995). Neurophysiological alterations in the mesocorticolimbic dopamine system during repeated cocaine administration. *In* The Neurobiology of Cocaine: Cellular and Molecular Mechanisms. (R. P. Hammer, Jr., ed.), pp. 95–115. CRC Press, Boca Raton.
2. Surmeier, D. J. Eberwine, J., Wilson, C. J., Cao, Y., Stefani, A., and Kitai, S. T. (1992). Dopamine receptor subtypes colocalize in rat striatonigral neurons. *Proc. Natl. Acad. Sci. U.S.A.* **89,** 10178–10182.
3. Catterall, W. A. (1996). Molecular properties of sodium and calcium channels. *J. Bioenerg. Biomembr.* **28,** 219–230.
4. Dascal, N., Lotan, I. (1991). Activation of protein kinase C alters voltage dependence of a Na^+ channel. *Neuron* **6,** 165–175.
5. Nestler, E. J. (1994). Molecular neurobiology of drug addiction. *Neuropsychopharmacology* **11,** 77–87.

Y. L. Hurd,* M. Pontén,*† A. McGregor,† T. Guix,† and U. Ungerstedt†

Karolinska Institute
*Department of Clinical Neuroscience
Psychiatry Section
†Department of Physiology and Pharmacology
Karolinska Hospital S-17176
S-17177 Stockholm, Sweden

Dopamine Efflux Studies into *in Vivo* Actions of Psychostimulant Drugs

To date, no specific treatment exists for psychostimulant drug abuse. Effective therapeutic interventions may be obtained by understanding not only the long-term neurobiological effects of psychostimulant drugs, but also their acute pharmacological properties, because acute neural actions of addictive drugs initiate the chain of events that leads ultimately to neuroadaptive states driving drug abuse behaviors. It is has been very well documented that dopamine (DA) is a critical neurochemical substrate for the acute actions of psychostimulant agents such as cocaine and amphetamine (1). A large body of the *in vivo* evidence supporting the critical role of DA systems in drug abuse has been obtained within recent years using the microdialysis technique. By monitoring DA overflow (into the extracellular space) in response to drugs that are frequently abused, it has been feasible to (1) determine the distinct *in vivo* origin of DA efflux induced by different drugs, (2) follow the time course of drug action on DA efflux, (3) correlate *in vivo* DA levels with the presence of the drug in brain, and (4) correlate drug-induced behaviors with *in vivo* fluctuations of DA levels. All of these parameters have been helpful in providing clues about the chain of neural events involved in different stages of psychostimulant abuse.

This chapter outlines some of our *in vivo* microdialysis DA studies relating to the acute pharmacological actions of psychomotor stimulant drugs, and to adaptations in DA overflow following their repeated use. Dialysate samples were obtained from microdialysis probes inserted into either the caudate–putamen, nucleus accumbens, or amygdala. *In vivo* concentrations of DA were determined by high-performance liquid chromatography (HPLC) combined with electrochemical detection, and the levels of cocaine were determined by HPLC coupled to ultraviolet detection. In an attempt to examine the mechanism by which amphetamine and cocaine potentiate DA overflow, the microenvironment around the dialysis probe was modified by altering the concentration of various ions (mainly calcium and sodium) in the artificial cerebrospinal fluid perfusion media. Calcium is critical for exocytotic release of DA from nerve terminals,

Advances in Pharmacology, Volume 42

1054-3589/98 $25.00

while sodium concentrations influence the re-uptake of the DA back into the presynaptic nerve terminals via the DA transport carrier. *In vivo* microdialysis was also carried out in behavioral studies in which rats were trained to self-administer intravenous (i.v.) cocaine under a fixed ratio schedule. In this model, animals freely regulate their cocaine-intake pattern and, thus, provide information relevant to the human drug abuse condition.

Similar to earlier findings *in vitro,* cocaine and amphetamine had distinct *in vivo* pharmacological actions on striatal DA efflux. Cocaine potentiated extracellular levels of DA originating from exocytotic release, as evidenced by a marked attenuation of cocaine-induced DA overflow in the absence of calcium, which had no effect on DA levels potentiated by amphetamine (2). In contrast, amphetamine administration increased DA overflow via reversal of the DA transport carrier similar to the release of DA induced by low extracellular sodium concentrations (3). Transporter-mediated release was also evidenced by the fact that potent long-acting DA transport inhibitors, such as nomifensine and Lu 19005, were able to dramatically block the substantial release of DA caused by amphetamine (3), but these agents were not as effective at blocking the rapid rise and fall of DA overflow potentiated by i.v. cocaine. Although the DA transport carrier is undoubtedly an important target for treatment interventions of psychostimulant abuse, an important consideration is that blockade of the DA transport carrier will have different effectiveness for the different classes of stimulant drug abuse. Another consideration is that pharmacological blockade of the DA transport carrier will still lead to an increase of DA overflow (which might be "rewarding"), though with a much lower magnitude than that induced by cocaine or amphetamine. However, such a treatment might serve as a "substitution" therapy (e.g., methadone management for heroin abuse).

Although *in vivo* DAergic response to the acute pharmacological actions of cocaine are short-acting (30 min i.v. to 2 hr s.c. or i.p. elevation of DA levels, depending on the route of administration), the long-term consequences of this DA potentiation are quite marked. Just a single injection of cocaine (20 mg/kg s.c.) caused different effects on nucleus accumbens DA efflux, depending on the duration of time that had elapsed until the second drug injection. For example, sensitization (increased responsivity to the same dose of the drug) of DA overflow occurred during the first 2 days to a challenge injection of cocaine. However, 10 days after the first injection, an attenuation of the DA response to the same dose of cocaine was observed following re-exposure to the stimulant. Twenty days after the single cocaine injection, the DA response had renormalized, with an increase of DA similar to that observed during the initial drug administration. Thus, depending on the time period that has elapsed after just one injection of cocaine and the second drug exposure, sensitization, attenuation, or no change of the DAergic response can be observed.

As a psychomotor stimulant drug, cocaine produces marked effects on motor behavior. Chronic use in humans can cause stereotyped repetitive behavior, tics, and uncontrollable tremor. The sensitization of locomotor and stereotyped behaviors, which are also observed in animals, has been used as a model reflecting neuroadaptations in DA systems induced by repeated cocaine use. Sensitization of nucleus accumbens DA overflow was observed after 10 days of repeated cocaine (20 mg/kg s.c.) administration, as compared with effects induced on day 1 of treatment. The increased DA overflow was accompanied with a concomitant increase of locomotor and stereotyped behavior, confirming previous results concerning such sensitization (4). It has been speculated that increased brain concentrations of cocaine might underlie the increasing behavioral and DAergic effects observed following repeated cocaine administration since brain cocaine levels are linearly related to brain DA levels after acute injection of the drug. However, while we found that the DA and behavioral responsivity varied between 1 and 10 days of repeated cocaine injections, brain cocnentrations of the drug remained virtually the same under these conditions. Thus, although brain DA levels fluctuate with repeated use of the drug, the absolute amount of brain cocaine does not appear to be directly correlated with the neuroadaptations observed on DA efflux and behavior.

Although sensitization of motor behavior and drug-induced psychosis is one end of the spectrum of effects caused by repeated cocaine use in humans, this spectrum also encompasses depression and anhedonia. It was observed that if the 10-day administration of repeated cocaine is followed by a 20-day drug-free period, then the DA response to a challenge injection of cocaine is reduced. The attenuation of the DA overflow during this cocaine-free period might relate to the state of depressed mood noted in human cocaine users during abstinence. Sensitization and attenuation of the dopamine response observed at different phases of the drug abuse cycle might reflect the wide range of emotions and behaviors that cocaine induces in humans, from an initial high to severe anxiety and depression.

In an attempt to mimic more closely the human cocaine abuse behavior, DA levels were studied in the drug self-administration model. While our initial studies showed attenuation of nucleus accumbens DA response during cocaine self-administration compared with initial exposure to the drug (5), under different methodological conditions, we have recently observed potentiated DA overflow 400–500% of baseline during cocaine self-administration following 5–10 days of 3-hr limited access to cocaine (6). We have also found that during the self-administration of cocaine (1.5 mg/kg/injection), extracellular levels of DA are not only elevated in the nucleus accumbens, but also in the amygdala, though with a comparatively lower magnitude and a slower latency in response to cocaine (6). Despite these lower DA concentrations, amygdala DA transmission, mediated via D1 DA receptors, contributed significantly to the cocaine self-administration behavior. This was concluded because, consistent with previous behavioral findings, an intra-amygdala infusion of the D1 antagonist, SCH 23390, was found to cause an immediate concentration-dependent increase in the cocaine intake behavior. Moreover, at the highest concentration tested, 1.5 μg of SCH 23390, animals had to double their rate of cocaine intake, which doubled the amount of DA overflow in the nucleus accumbens, to compensate for the amygdala D1 blockade. While nucleus accumbens DA has been shown to be critical for cocaine reinforcement, it is apparent that amygdala DA transmission also provides an important contribution to cocaine reinforcement. It can be speculated that the subjective and motivational aspects associated with cocaine self-administration may be mediated by interactive DAergic alterations between the amygdala and nucleus accumbens or that drug-seeking and drug-taking behaviors, which are distinct components of drug self-administration, might be differentially regulated by these mesolimbic regions.

It is apparent that dose and route of administration, duration (acute, jshort-, and long-term) and pattern (passive, binge, self-regulated) of drug administration, withdrawal time period between last drug administration and testing, and subregion of specific brain regions in which testing is carried out can all influence conclusions drawn about the neurochemical–neuroadaptive state of the brain induced by psychostimulant drugs. These variables underlie the diversity of results noted in the literature regarding the state of *in vivo* DA transmission induced by psychostimulant use. Nevertheless, all of these studies emphasize that neuroadaptive responses to stimulants occur quickly and are long lasting. It is also evident that any therapeutic design should consider varying the pharmacological intervention for each specific stage of the drug abuse cycle.

Acknowledgment

This work was supported by grants from the Swedish Medical Research Council.

References

1. Koob, G. F. (1992). Dopamine, addiction and reward. *Semin. Neurosci.* **4,** 139–148.
2. Hurd, Y. L., and Ungerstedt, U. (1989). Cocaine: An in vivo microdialysis evaluation of its acute action on the dopamine transmission in rat striatum. *Synapse* **3,** 48–54.
3. Hurd, Y. L., and Ungerstedt, U. (1989). Influence of a carrier transport process on in vivo release and metabolism of dopamine: Dependence on extracellular Na^+. *Life Sci.* **45,** 283–293.
4. Robinson, T. E., Jurson, P. A., Bennett, J. A., and Bentgen, K. M. (1988). Persistent sensitization of dopamine neurotransmission in ventral striatum (nucleus accumbens) produced by prior experience with (+)-amphetamine: A microdialysis study in freely moving rats. *Brain Res.* **462,** 211–222.
5. Hurd, Y. L., Weiss, F., Andén, N. E., Koob, G. F., and Ungerstedt, U. (1989). Cocaine reinforcement and extracellular dopamine overflow: An in vivo microdialysis study. *Brain Res.* **498,** 199–203.
6. Hurd, Y. L., McGregor, A., and Pontén, M. *In vivo* amygdala dopamine levels modulate cocaine self-administration behavior in the rat: Dopamine D1 receptor involvement. *Eur. J. Neuroscience.* (in press).

Glen R. Hanson,* John D. Wagstaff,* Kalpana Merchant,† and James W. Gibb*

*Department of Pharmacology and Toxicology
University of Utah
Salt Lake City, Utah 84112

†Upjohn Laboratories/Central Nervous System Disease Research
Kalamazoo, Michigan 49001

Psychostimulants and Neuropeptide Response

The neuropeptides, neurotensin (NT), dynorphin, and substance P, appear to be reciprocally linked to dopaminergic activity related to emotional and motor functions. Because of these interactions, we tested the possibility that potent central nervous system stimulants, such as methamphetamine (METH) and cocaine, influence neuropeptide systems associated with extrapyramidal and limbic structures. We observed that stimulants of abuse dramatically increase the tissue content of these neuropeptides in the caudate nucleus, nucleus accumbens, and substantia nigra (1, 2). To understand the significance of the neuropeptide changes caused by these drugs, we studied in detail the response of NT to METH treatment. NT was selected for characterization because intracerebroventricular (i.c.v.) administration of this peptide has neuroleptic properties and antagonizes the behavioral effects of METH. In addition, NT receptors are predominantly located on caudate and accumbens dopamine terminals, suggesting they play a significant role in the regulation of the associated nigral–striatal and mesolimbic dopaminergic pathways (3). Identification of the mechanisms whereby METH treatment influences NT systems should elucidate the role of this and other peptides in mediating the motor and emotional effects of such stimulants, as well as provide insight as to the nature of interactions between the peptide and dopamine pathways in extrapyramidal and limbic functions.

The responses of NT systems to METH in caudate nucleus and nucleus accumbens of male Sprague-Dawley rats were assessed by measuring: (1) tissue NT content with radioimmunoassays employing a highly selective antiserum; (2) NT synthesis by determining the levels of mRNA for its precursor, NT/neuromedin N with *in situ* hybridization; and (3) NT release from midcaudate and nucleus accumbens employing microdialysis coupled with a solid-phase, disequilibrium radioimmunoassay. The role of NT systems in METH-mediated responses was determined by blocking extracellular NT activity with either NT-directed antibodies or a selective NT antagonist and measuring the effect on METH-induced behavior.

We observed interesting dose-dependent responses suggesting distinctive high- and low-dose effects by METH on NT systems. High doses of METH

Advances in Pharmacology, Volume 42

1054-3589/98 $25.00

(15 mg/kg) temporarily increased NT tissue content predominantly in the medial caudate and nucleus accumbens. The elevation in NT levels was selectively blocked by dopaminergic D1-, but not D2-, receptor antagonists. *In situ* analysis revealed a regionally specific increase in mRNA for the NT precursor also in the medial caudate and nucleus accumbens, which approximately correlated with the increased NT content. Microdialysis assessment of extracellular NT levels revealed that this high dose of METH did not significantly alter release of this peptide for up to 3 hr (Fig. 1).

In contrast, a low dose of METH (0.5 mg/kg) decreased NT tissue content in lateral caudate and in the nucleus accumbens. These changes are consistent with a D2-receptor mechanism, because the selective D2 agonist, quinpirole, similarly reduces NT tissue content. To date, there are no published studies concerning the effects of a low dose of METH on the mRNA for the NT precursor. As with the high-METH doses, we employed microdialysis to assess the effects of low-dose METH on extracellular NT levels. In contrast to a high dose (15 mg/kg) of METH, 0.5 mg/kg of this drug increased NT release approximately 200% in both the caudate and nucleus accumbens (see Fig. 1): The low-dose METH effect was completely blocked by a D2 antagonist (4).

Because the low dose of METH increased NT release, we examined the role of extracellular NT in behavioral effects caused by 0.5 mg/kg of this stimulant. Both an i.c.v. infusion of a selective NT antibody and an accumbens infusion of the NT antagonist, SR 48692 dramatically increased METH-induced motor behavior and dopamine release (5).

These data suggest that extrapyramidal and limbic NT systems are differentially influenced by low and high doses of METH. Low doses of METH increase NT release in both the caudate and nucleus accumbens, which reciprocally antagonizes METH-induced dopamine release and related behavior. The increased release of NT likely leads to a decrease in NT tissue content in lateral caudate and nucleus accumbens due to the increased neuropeptide turnover.

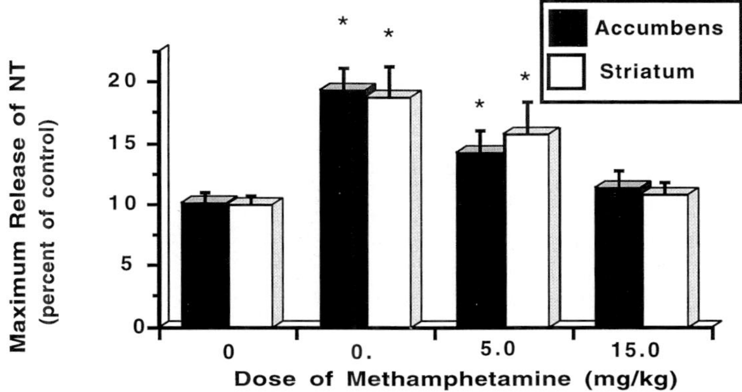

FIGURE 1 Effects of various doses of METH on the maximal release of neurotensin as measured by microdialysis. The values represent a 25-min fraction collected between 25 and 75 min after drug administration. $^{*}p < 0.05$ versus corresponding control values.

The low-dose METH effect appears mediated predominantly by D2 receptors, which likely interact with NT systems located in the lateral caudate. These data suggest that endogenous NT systems prevent excessive extrapyramidal and limbic DA responses during low-to-moderate stimulation of DA pathways.

In contrast, high-dose METH dramatically increases NT tissue content in medial caudate and nucleus accumbens by activating dopamine D1 receptors. The elevation in tissue levels likely is mediated by increased synthesis, because this METH treatment also increases the mRNA for the precursor in medial caudate and accumbens regions. Interestingly, release of NT is not altered in these structures after a high dose of METH and appears not to exert the regulatory impact observed with low doses of METH. This lack of NT release may lead to an exaggerated reaction of dopaminergic pathways to METH treatment and contribute to the significant motor and emotional consequences of high doses of this potent stimulant of abuse.

Acknowledgment

This research was supported by National Institutes of Health grants DA 00869 and DA 09407.

References

1. Hanson, G. R., Singh, N., Merchant, K., Johnson, M., and Gibb, J. W. (1995). The role of NMDA receptor systems in neuropeptide responses to stimlulants of abuse. *Drug Alcohol Depend.* **37**, 107–110.
2. Ritter, J. K., Schmidt, C. J., Gibb, J. W. and Hanson, G. R. (1984). Increases in substance P-like immunoreactivity within striatal-nigral structures following subacute methamphetamine treatment. *J. Pharmacol. Exp. Ther.* **229**, 487–492.
3. Bissett, G., and Nemeroff, C. (1995). The neurobiology of neurotensin. *In* Psychopharmacology: The Fourth Generation. (F. Bloom and D. Kupefer, eds.), pp. 573–583. Raven Press, New York.
4. Wagstaff, J. D., Gibb, J. W., and Hanson, G. R. (1996). Microdialysis assessment of methamphetamine-induced changes in extracellular content in the striatum and nucleus accumbens. *J. Pharmacol. Exp. Ther.* **278**, 547–554.
5. Wagstaff, J. D., Bush, L., Gibb, J. W., and Hanson, G. R. (1994). Endogenous neurotensin antagonizes methamphetamine-enhanced dopaminergic activity. *Brain Res.* **665**, 237–244.

Jacqueline F. McGinty and John Q. Wang

Department of Anatomy and Cell Biology
East Carolina University School of Medicine
Greenville, North Carolina 27858

Drugs of Abuse and Striatal Gene Expression

Mixed D1/D2-receptor agonists and indirect dopamine agonists, such as cocaine and amphetamine, but not the partial D1 agonist, SKF 38393, induce the expression of neuropeptides, such as preprodynorphin (PPD) and substance P (SP), and immediate early genes (IEGs), such as c-*fos* and *zif/268,* in medium spiny neurons of the normosensitive striatum. Induction of a constellation of such genes is thought to contribute to the molecular plasticity that underlies long-term adaptations in cellular physiology and behavior elicited by drugs of abuse. D1-receptor antagonists block dopamine agonist–induced gene expression in striatonigral neurons. However, D2 antagonists also significantly attenuate these effects, contributing to the idea that both D1 and D2 receptors must be stimulated to evoke gene expression in striatonigral neurons. How these D1/D2 interactions are accomplished is unclear in light of the controversy over the proposed segregation of D1 and D2 receptors on striatonigral and striatopallidal neurons, respectively. Furthermore, psychostimulants have been reported to increase preproenkephalin (PPE) mRNA levels in striatopallidal neurons, an effect that is mediated by D1, but not D2, receptors. Why D1-receptor stimulation should induce PPE mRNA in striatopallidal neurons if they do not express D1 receptors is also a mystery.

In addition to the strong regulation of gene expression in medium spiny neurons by dopamine receptors, recent studies indicate that glutamatergic and cholinergic neurotransmission strongly modulates the effects of dopamine on striatal gene expression. Because excitatory amino acid receptor antagonists and muscarinic receptor agonists are effective in blocking psychostimulant-induced behaviors and striatonigral gene expression, these two systems must participate cooperatively with dopamine to mediate the actions of these drugs. The final outcome of drug stimulation, then, is determined by interactions among these systems in the striatum as well as in related extrastriatal sites.

Glutamatergic regulation of the alterations in expression of IEGs and neuropeptide mRNAs and their protein levels in the striatum may be essential to the neuronal adaptation that underlies delayed and long-term effects of psychostimulants. Stimulant-induced IEG and neuropeptide expression in striatal neurons has been linked both to N-methyl-D-aspartate (NMDA) and non-NMDA receptor activation. A series of studies in our laboratory (1) has demonstrated that a single injection of amphetamine or methamphetamine is capable of stimulating mRNA expression of the IEGs, c-*fos,* and *zif/268* and prolonged induction of

Advances in Pharmacology, Volume 42

1054-3589/98 $25.00

opioid peptide mRNA in the striatum. The NMDA antagonists, MK 801, CPP, and the kainate/AMPA antagonist, DNQX, attenuated the increases in striatal *zif/268*, PPD, and PPE expression induced by acute amphetamine. Similar results were observed subsequently with acute methamphetamine, except that strong induction of *zif/268* was still present after pretreatment with DNQX. Because excitatory amino acid antagonists do not block the acute effects of amphetamines in rats but do block the behavioral effects of repeated amphetamine administration, these data indicate that NMDA and/or kainate/AMPA receptors may preferentially modulate those cellular responses that mediate the long-term effects of dopamine-receptor stimulation indicative of drug sensitization or addiction.

Acetylcholine also regulates striatal gene expression. Using *in situ* hybridization, we recently found that systemic or intrastriatal administration of the nonselective muscarinic receptor antagonist, scopolamine, upregulated PPD and SP mRNA expression in the dorsal and ventral striatum in a dose-dependent fashion (2). However, the muscarinic receptor agonist, oxotremorine, did not alter constitutive expression of PPD and SP mRNAs. In addition, blockade of muscarinic receptors substantially potentiated the ability of amphetamine or the full D1-receptor agonist, SKF 82958 to stimulate IEG (c-*fos* and *zif/268*) and peptide gene (PPD and SP) expression in intact rats. Conversely, oxotremorine attenuated amphetamine-stimulated PPD and SP gene expression. Based on these and other data, we and others (3) have postulated that endogenous inhibitory cholinergic tone protects striatonigral neurons from overexcitation by D1 receptors and potentially normalizes gene expression after D1 stimulation.

Acetylcholine and dopamine also exert opposite effects on striatopallidal gene expression. Although scopolamine does not alter the basal level of striatal PPE mRNA expression, systemic or intrastriatal scopolamine blocks the increase in PPE mRNA induced by either amphetamine or the full D1 agonist, SKF 82958 (2, 4). In contrast, oxotremorine increases basal PPE expression in the dorsal striatum but has no significant effect on PPE expression when coadministered with amphetamine (2, 4). Thus, stimulation of muscarinic neurotransmission in the dorsal striatum tonically facilitates PPE gene expression.

Based on this information, a detailed sequence of events between striatal medium spiny and cholinergic neurons is hypothesized to occur in response to dopamine stimulation (1, 3). This cell-to-cell sequence starts with stimulation of D1 receptors located on the soma and dendrites of striatonigral neurons as well as on their terminals in the substantia nigra. Activation of somatodendritic D1 receptors results in multiple cellular responses that include release of acetylcholine, SP, GABA, and dynorphin and a compensatory increase in PPD–SP expression. Acetylcholine released via this route would bind to muscarinic, probably M4, receptors and inhibit PPD–SP gene expression in striatonigral neurons by decreasing adenylate cyclase–dependent transcription. In contrast, acetylcholine would enhance the expression of PPE in striatopallidal neurons, possibly via M1 receptors coupled to phosphoinositide hydrolysis. At the level of the substantia nigra, D1-receptor stimulation, which facilitates local GABA release, may result in disinhibition of glutamate release in the thalamostriatal and corticostriatal pathways and further stimulation of striatal acetylcholine release (5). In this manner, the cholinergic neuron would be in a strategic position to serve as a feedforward inhibitor of psychostimulant-induced striatonigral activity and a facilitator of striatopallidal activity, as exemplified by scopolamine's ability to enhance D1-stimulated PPD and SP mRNA expression and attenuate PPE mRNA expression.

The inhibitory influence of cholinergic interneurons on striatonigral neurons also provides an intrastriatal mechanism for some D1/D2 interactions. Although D1 receptors are thought to control striatonigral gene expression, several studies have demonstrated that D2-receptor antagonists block stimulant-induced dynorphin immunoreactivity and PPD gene expression (1). Regardless of the controversy over D1/D2-receptor segregation on striatonigral and striatopallidal neurons, the D2 contribution to the full expression of D1-mediated function *in vivo* is likely to include an indirect, transynaptic mechanism. Because D2 tone is an important inhibitory force on acetylcholine release, it is possible that concomitant stimulation of D2 receptors, by minimizing acetylcholine release, decreases cholinergic neurotransmission, and, thus, synergistically enhances D1 stimulation of striatonigral neurons. In contrast, D2-receptor blockade, by increasing acetylcholine release, would attenuate D1-stimulated gene expression in these neurons. To support this putative acetylcholine-dependent pathway, we recently found that the D2-selective antagonist, eticlopride, blocked D1 agonist–stimulated PPD and SP mRNA expression in the rat striatum and that the effect of eticlopride was prevented by scopolamine (6).

In conclusion, glutamate and acetylcholine regulate the responses of medium spiny neurons to dopamine. Thus, both muscarinic receptor agonists and excitatory amino acid receptor antagonists are potentially useful tools in the search for novel therapeutic approaches to drug abuse.

Acknowledgments

The studies of the authors were supported by DA03982.

References

1. Wang, J. Q., and McGinty, J. F. (1996). Glutamatergic and cholinergic regulation of immediate early gene and neuropeptide gene expression in the striatum. *In* Pharmacological Regulation of Gene Expression in the CNS. (K. Merchant, ed.), pp. 81–113. CRC Press, Boca Raton, FL.
2. Wang, J. Q., and McGinty, J. F. (1996). Muscarinic receptors regulate striatal neuropeptide gene expression in normal and amphetamine-treated rats. *Neuroscience* **75**, 43–56.
3. Di Chiara, G., Morelli, M., and Consolo, S. (1994). Modulatory functions of neurotransmitters in the striatum: ACH/dopamine/NMDA interactions. *TINS* **17**, 228–233.
4. Wang, J. Q., and McGinty, J. F. (1997). Intrastriatal injection of a muscarinic receptor agonist and antagonist regulates striatal neuropeptide mRNA expression in normal and amphetamine-treated rats. *Brain Res.* **748**, 62–70.
5. DeBoer, P., and Abercrombie, E. D. (1994). Further characterization of the role of substantia nigra in the modulation of striatal acetylcholine in awake rats. *Soc. Neurosci. Abstr.* **20**, 285.
6. Wang, J. Q., and McGinty, J. F. (1997). The full D1 dopamine receptor agonist SKF-82958 induces neuropeptide mRNA in the normosensitive striatum of rats: Regulation of D1/D2 interactions by muscarinic receptors. *J. Pharmacol. Exp. Ther.* **281**, 972–982.

D. James Surmeier, Zhen Yan, and Wen-Jie-Song

Department of Anatomy and Neurobiology
University of Tennessee
Memphis, Tennessee 38163

Coordinated Expression of Dopamine Receptors in Neostriatal Medium Spiny Neurons

A clear picture of how dopamine (DA) receptor subtypes are distributed among the principal neurons of the neostriatum is of obvious importance to understanding how drugs like cocaine and amphetamine influence basal ganglia function. In recent years, this distribution has been the subject of debate (1, 2). For example, conventional anatomical and physiological approaches have yielded starkly different estimates of the extent to which D1- and D2-class DA receptors are colocalized. One plausible explanation for the discrepancy is that some DA receptors are present in physiologically significant numbers, but the mRNA for these receptors is not detectable with conventional techniques. To test this hypothesis, the expression of DA receptors in individual neostriatal neurons was examined using patch-clamp and single-cell reverse transcription–polymerase chain reaction (RT-PCR) techniques (3).

As a first step in the study of cellular localization of DA receptor mRNAs, tissue expression was examined. Conventional RT-PCR analysis of dorsal neostriatal mRNA isolated from coronal slices revealed that all five receptor mRNAs were present, and the expression of D3, D4, and D1b mRNAs was not a consequence of *de novo* transcription.

To determine the molecular identity of the receptors mediating the responses to the dopaminergic ligands in individual neurons, single-cell RT-PCR techniques were used. Neurons were initially divided into three groups on the basis of enkephalin (ENK) and substance P (SP) mRNA expression. The expression of these releasable peptides is strongly correlated with the axonal projection pattern of medium spiny neurons (1).

An obvious problem in attempting to determine how the expression of several mRNAs is coordinated in the same cell is that low-abundance mRNAs may be inadvertently missed when the products from a single cell are broken up into different reaction vessels. To address this issue, PCR analysis of DA-receptor mRNAs used either a small fraction of the total cellular cDNA (1/10), an intermediate level (one-fourth), or the product of a preamplification stage using multiplex amplification of all five DA-receptor mRNAs (3).

Medium spiny neurons having detectable levels of ENK but not SP mRNA (ENK+/SP−) expressed high levels of D2-receptor mRNA (long splice variant).

Advances in Pharmacology, Volume 42

1054-3589/98 $25.00

With a 1/10 aliquot, D2 cDNA was detected in nearly 90% of this group. The other DA-receptor mRNAs were rarely detected with this small aliquot. As the fraction of the total cellular cDNA used in the PCR reactions increased, other DA-receptor mRNAs emerged. Doubling the template concentration for D1a, D1b, D3, and D4 reactions led to a modest increase in the detection frequency of D1b mRNA (to 20% of the sample). The short isoform of the D2 mRNA was consistently seen in this condition. Using a two-stage multiplex procedure to maximize the detectability of low-abundance mRNAs increased the detection rate of D3 and D4 mRNAs primarily. An example of an ENK+/SP− cell in which D2, D3, and D4 mRNAs were seen is shown in Figure 1C. The data for ENK+/SP− neurons is summarized in Figure 1D. In this panel (and in subsequent summary diagrams), the extent to which particular mRNAs were found together is coded by the extent to which their lanes are shaded at similar points along the abscissa. The failure to detect D1b mRNA in as large a subject of neurons after multiplex amplification as after conventional PCR (0% vs 20%) probably reflects sampling variation.

Medium spiny neurons having detectable levels of SP but not ENK mRNA (ENK−/SP+) expressed high levels of D1a-receptor mRNA. With 1/10 aliquots, D1a mRNA was detected in 80% of the sample. D3 mRNA was detected in half of this sample, but the other receptor mRNAs were either not seen or seen rarely. Doubling the aliquot size increased the D1a detection rate to 100% and did not change the detection rate for the D3 mRNA. Using the multiplex protocol, the detection frequencies for D1a and D3 mRNAs were similar to those seen with smaller aliquots. However, the detection frequency of D4 mRNA increased to 25% (as seen with one-fifth aliquots) (Figs. 1A, B). In addition, D2 mRNA, which had not been seen with 1/10 or one-fifth aliquots, was seen in about 20% of the sample, suggesting that D2 mRNAs were of relatively low abundance in this group.

Neurons coexpressing detectable levels of SP and ENK (ENK+/SP+) consistently coexpressed D1a and D2 mRNAs. This group comprised about 20% of our total population of neurons. With 1/10 aliquots, D2 mRNA was the most commonly detected in this sample (78%). D1a-receptor amplicons were seen less frequently, and both amplicons were detected in about one-fifth of the sample. The other receptor mRNAs were either not detected or seen rarely with the small aliquot. Doubling the aliquot, did not change the D1a and D2 detection frequencies significantly. However, using the multiplex protocol, the detection rate for D1a mRNA increased to 90% (Fig. 1E). As a consequence, D1a and D2 mRNAs were detected in 70% of this group (Figs. 1E, F). The detection frequency of D3, D4, and D1b mRNAs also increased with the size of the aliquot used, suggesting that they were expressed at relatively low levels in 20–30% of ENK+/SP+ neurons.

In an attempt to determine whether these transcripts could give rise to functional protein, whole-cell recording was combined with conventional single-cell RT-PCR. Several neurons were found to express detectable levels of D2 and D1b mRNA, but not D1a mRNA; these cells consistently had robust responses to D1-class agonists. As shown, in other cells, D3 or D4 mRNA was detected but not D2 mRNA; these cells consistently had responses to D2-class agonists. While these results do not prove that D3-, D4-, and D1b-receptor

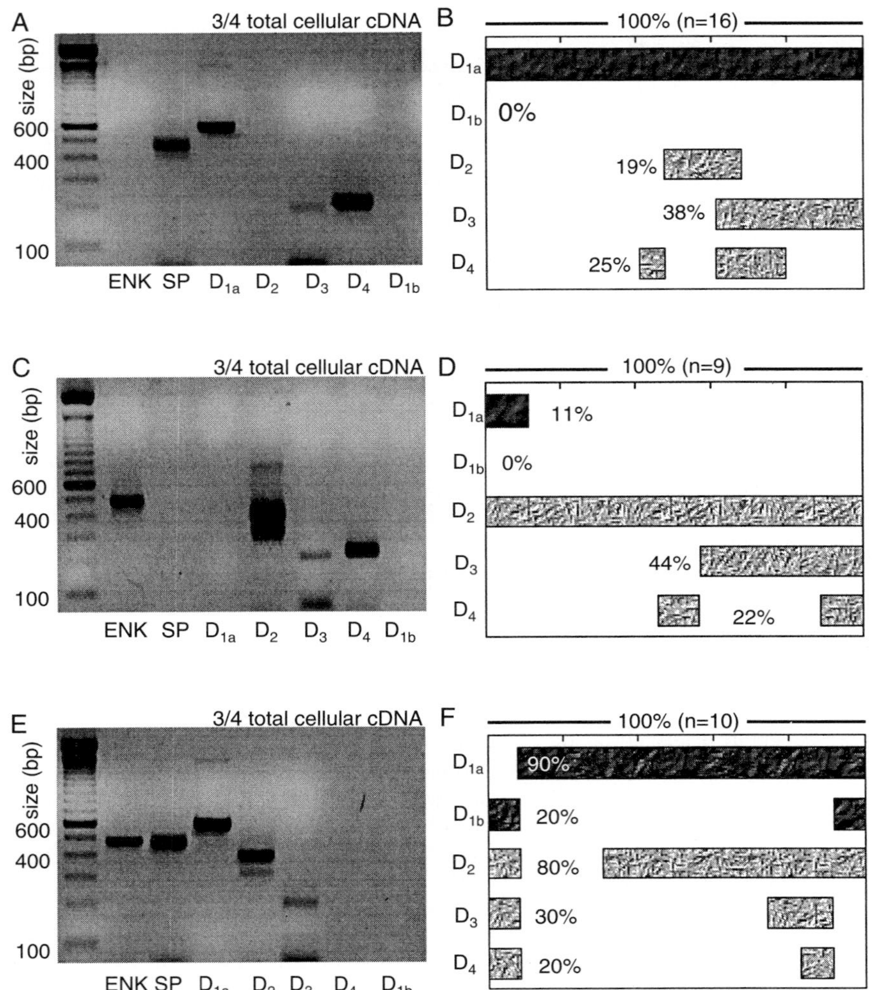

FIGURE 1 Single-cell RT-PCR revealed that neostriatal neurons coexpress mRNAs for several dopamine receptors. (*A*) Photograph of a gel containing amplicons from an SP+/ENK- medium spiny neuron in which a multiplex procedure employing three-fourths of the total cellular cDNA was used for detection of DA-receptor mRNAs. Note that D1a and D3 and D4 amplicons were detected. (*B*) Summary of coexpression detected with the multiplex procedure in 16 ENK-/SP+ neurons (*C*) Photograph of a gel containing amplicons from an SP-/ENK+ medium spiny neuron in which a multiplex procedure was employed as before. (*D*) Summary of coexpression detected with the multiplex procedure in nine ENK+/SP- neurons. (*E*) Photograph of a gel containing amplicons from an SP+/ENK+ medium spiny neuron in which the multiplex procedure was employed. (*F*) Summary of coexpression detected with the multiplex procedure in 10 ENK+/SP+ neurons. (Reprinted with permission from *The Journal of Neuroscience* **16**, 6579–6591, 1996.)

mRNAs invariably give rise to functional protein, they do suggest that these mRNAs are capable of being translated and properly processed.

The principal goal of this study was to understand why anatomical and functional studies of DA-receptor localization in the neostriatum have been so discordant (1, 2). It was our working hypothesis that at the heart of the dispute was the assumption that *only* D1a and D2 receptors were functionally significant in the workings of the dorsal neostriatum. This inference allowed the equation of responsiveness to D1-class agonists with the presence of D1a receptors and responsiveness to D2-class agonists with the presence of D2 receptors. Our results argue that, in fact, this assumption is incorrect. Significant levels of D3-, D4-, and D1b-receptor mRNAs are present in medium spiny neurons.

The demonstration that D1- and D2-class receptors are coexpressed by a significant percentage of medium spiny neurons, particularly those within the striatonigral pathway, establishes a cellular and molecular foundation for the well-known physiological and biochemical responses of these neurons to D1- and D2-class agonists. Our results also show that single-cell RT-PCR techniques yield a picture compatible with that derived from *in situ* hybridization if attention is restricted to only the most abundant receptor mRNAs. Because of their ability to evoke functionally significant physiological effects, the impact of the lower abundance DA receptors (D3, D4, D1b) must be factored into models of drug action within the neostriatum.

References

1. Gerfen, C. R. (1992). The neostriatal mosaic: Multiple levels of compartmental organization. *TINS* **15**, 133–139.
2. Surmeier, D. J., Reiner, A., Levine, M. S., and Ariano, M. A. (1993). Are neostriatal dopamine receptors co-localized? *TINS* **16**, 299–305.
3. Sibley, D. R. (1995). Molecular biology of dopamine receptors. *In* Cellular and Molecular Mechanisms of the Neostriatum. (D. J. Surmeier, and M. A. Ariano, eds.), R. G. Landes, Austin, TX.
4. Surmeier, D. J., Song, W-J., and Yan, Z. (1997). Coordinated expression of dopamine receptors in neostriatal medium spiny neurons. *J. Neurosci.* **16**, 6579–6591.

George R. Uhl,*† David J. Vandenbergh,*
Lawrence A. Rodriguez,* Lucinda Miner,*
and Nobuyuki Takahashi*

*Molecular Neurobiology Branch
Intramural Research Program
National Institute on Drug Abuse
National Institutes of Health
Baltimore Maryland 21224

†Departments of Neurology and Neuroscience
The John Hopkins University School of Medicine
Baltimore, Maryland 21205

Dopaminergic Genes and Substance Abuse

The use of addictive substances is a complex behavioral disorder likely to represent interactions between genetic and environmental factors (1). Evidence from twin, adoption, and family studies indicates that drug abuse phenotypes including quantity–frequency of use and features responsible for *Diagnostic and Statistical Manual* (DSM) diagnoses of substance abuse–dependence are each likely to reflect significant genetic contributions (1–3). Population comorbidity studies indicate that relative risk of a drug abuse disorder is enhanced substantially in individuals with comorbid antisocial personality disorder, alcoholism, depression, or the adult residua of attention deficit–hyperactivity disorder (1). Other personality traits have also been postulated to predispose to vulnerability to specific aspects of drug abuse, ranging from initiation of use to resistance to quitting to development of the drug's control over behavior that represents a hallmark of dependence.

Dopaminergic brain systems play prominent roles in drug reward (4), focusing attention on genes expressed in these circuits as candidates to contribute to substance abuse vulnerability. Psychostimulants engage these circuits with especial power, as evinced by a number of lines of evidence, including the magnitude of drug-induced dopamine spillover from nucleus accumbens dopaminergic synapses following psychostimulant treatments (5) and the results of lesion studies, in which dopaminergic lesions dampen brain reward systems and locomotor functions in both experimental animals and humans (6–8). A number of studies in experimental animals now suggest that manipulations of the levels of expression of several of the genes that encode proteins important for dopaminergic neurotransmission, or of genes regulated by psychostimulants, can have powerful influences on animal models of drug reward. Studies of the functional allelic variants that have been identified in several human dopaminergic genes, or of polymorphic markers that could tag heretofore underscribed

Advances in Pharmacology, Volume 42

1054-3589/98 $25.00

functional variation at other dopaminergic gene loci, have each indicated that expression of human genes with different levels of activity might contribute to differential responses to abused substances, different drug-abuse comorbidities, or different personality features of possible importance for substance abuse. In this chapter, we review several recent cases in which transgenic mouse and human data provide increasingly compelling evidence that variants in genes important for dopaminergic neurotransmission are strong candidates to contribute to interindividual differences in drug abuse vulnerabilities.

I. Transgenic Mice with Dopamine Transporter Over and Underexpressed

Studies of cocaine's primary site for reward and reinforcement in the brain have focused on the dopamine transporter (DAT) (9). DAT is a member of the 12-transmembrane domain sodium and chloride-dependent neurotransmitter transporter family, the primary structure of which has been elucidated by cDNA cloning (10). In studying this transporter, the effect of several mutations on transporter function have been characterized. Replacement of putative transmembrane 7 serines 350 and 353 with alanines, for example, results in a transporter variant that exhibits 30% increases in dopamine transport and 30% reductions in affinity for the cocaine analog 2β-carbomethoxy-3β-(t-fluorophenyl) tropane (CFT) (11). To model the effects that a DAT gene variant might have on drug abuse phenotypes, we constructed transgenic animals that express this DAT variant in catecholaminergic neurons (12). We utilized sequences from the 5′ flanking promoter region of the rat tyrosine hydroxylase (TH) gene because of their well-characterized ability to specifically mediate gene expression in catecholaminergic neurons (13, 14). Transgenic strains THDAT2 and THDAT4 each stably pass more than 10 copies of an 8.2-kb trangene construct made up of 4.8 kb of rat TH 5′ flanking promoter sequence linked to the serine 350 and 353 alanine substituted rat DAT variant, and express the transgene mRNA and encoded protein in a region-specific manner with high levels of expression in ventral midbrain, adrenal gland, and olfactory bulb, consistent with patterns of expression previously described as mediated by the TH promoter.

Heterozygous transgenic animals from each of the two lines expressing this construction display behavioral differences from wild-type littermate controls. During repeated exposure to novel testing apparatus during testing sessions spaced at 3 to 4-day intervals, both THDAT lines habituate to the testing environment more rapidly. THDAT mice are 150–230% as active following 30 mg/kg of cocaine i.p. More striking effects were observed when the reinforcing and rewarding properties of cocaine were tested using conditioned place preference to assess drug reward (15). Both THDAT2 and THDAT4 animals displayed significantly greater cocaine-induced conditioned place preferences than littermate control mice. Transgenic mice with region-specific overexpression of a DAT gene variant can manifest subtle but interesting differences in specific features related to higher order control of locomotor systems and major models of drug responsiveness.

The results on cocaine locomotion obtained in these mice complement those obtained by Giros *et al.* in DAT knockout mice (10). In these animals, although baseline locomotor activities are altered, cocaine induces virtually no excess locomotion. However, data from studies of cocaine reward in DAT knockout mice constructed independently in our laboratory suggest that cocaine-induced conditioned placed preference might also be less altered in these mice (Sora *et al.*, in preparation). These data are also in accord with studies of the locomotor responses in dopamine D1 receptor and DAT knockout mice, in which cocaine-stimulated locomotor functions were virtually eliminated (17, 18).

II. Transgenic Mice with Vesicular Monoamine Transporter Underexpressed

The brain vesicular monoamine transporter (VMAT2) pumps monoamine neurotransmitters from neuronal cytoplasm into synaptic vesicles. Vesicular monoamine stores accumulated by normal VMAT2 function may play significant roles in the locomotor stimulation and/or the behavioral reward produced by amphetamines, drugs increasingly abused in certain areas of the United States. Amphetamines dissipate proton gradients across the membranes of synaptic vesicles, disrupt VMAT2 function, enhance cytoplasmic monoamine concentrations, and cause calcium-independent, nonvesicular monoamine release into synapses (19). They also act like cocaine in blocking plasma membrane neurotransmitter transporters that use transmembrane sodium and chloride gradients to pump dopamine, norepinephrine, and serotonin from extracellular spaces into presynaptic neurons. Because little preexisting evidence documented which amphetamine action provided which contribution to amphetamine-induced locomotion or behavioral reward, we constructed transgenic VMAT2 knockout mice using a 12-kb targeting factor homologous to sequences flanking the first three exons of the murine 129 strain VMAT2 gene (20).

Heterozygotes VMAT 2 knockout mice are viable into adult life and display VMAT2 levels half of wild-type values. They also display alterations in several monoaminergic markers, heart rate, and blood pressure. However, their weight gain, fertility, habituation, passive avoidance, and locomotor activities are similar to wild-type littermates (20).

In these heterozygote VMAT2 knockout mice, 1- and 3-mg/kg intraperitoneal amphetamine doses enhanced locomotor activity, with increases at 1 mg/kg 146% of those in wild-type mice. However, amphetamine-conditioned place preferences in the heterozygotes were less than those in wild-type mice. Conversely, cocaine (5 mg/kg), a blocker of the plasma membrane DAT devoid of significant activity on vesicular monoamine stores, induced conditioned place preferences in heterozygotes that were indistinguishable from those in wildtype mice (20). These data support the idea that plasma membrane transporter blockade by amphetamine contributes primarily to its locomotor stimulation, but that normal function of vesicular monoamine stores is necessary for full amphetamine-induced behavioral reward. They are consistent with the results from studies that document differences between the effects of dopamine receptor knockouts on psychostimulant-induced locomotion and reward (18) and with the differences in dose-response relationships for amphetamine-induced locomotion and reward (19).

III. Human Studies of Catechol-O-Methyltransferase Functional Alleles in Polysubstance Abusers and Nonusers

Catechol-O-Methyltransferase (COMT) is expressed in dopaminergic brain regions, where its activity provides a major pathway by which extraneuronally released dopamine is inactivated (21). Three- to fourfold differences in human COMT activities are attributed to codon 158 polymorphisms that encode a valine, producing a high enzyme activity, or a methionine, yielding low activity (22–24, 38, 41). To seek influences of this COMT allelic variation on substance abuse vulnerability, we compared COMT genotypes in groups of unrelated polysubstance abuser research volunteers defined on the bases of: (1) quantity–frequency of drug use or (2) *Diagnostic and Statistical Manual of Mental Disorders III-Revised* (DSMIII-R) criteria for substance abuse/dependence to those of control research volunteers free of significant use of addictive substances (25, 26).

Significant differences in the distributions of both COMT genotypes and allele frequencies between controls and substance abusers, defined on the basis of self-reported peak lifetime use, and between these controls and volunteers who had also been administered the Diagnostic Interview Schedule (DIS) for DSMIII-R diagnoses of substance abuse or dependence were identified (25, 26).

Detection of single gene effects in complex disorders can be facilitated by examining phenotypes associated with the genotypes encoding extreme values; vulnerability could be enhanced in individuals with the highest or the lowest COMT activities, for example. The proportions of high activity G/G homozygotes and low activity A/A homozygotes were also examined: G/G homozygotes were nearly twice as frequent in volunteers who report high quantity–frequency drug use than in controls free of such use (25, 26). The sample defined on the basis of DSM diagnoses also shows a similar difference from control values. These data, thus, fit best with the hypothesis that high COMT activities might contribute to the genetic underpinnings of drug abuse vulnerability.

IV. Human Studies of DRD4 Functional Alleles in "Novelty Seeking" and Related Personality Substance Abuse Predispositions

Two recent studies have suggested an association between D4 dopamine receptor polymorphisms and measures designed to assess novelty-seeking behaviors, characterized by "exhilaration or excitement in response to novel stimuli" (27, 28). Such an association was plausible for several reasons. This personality dimension has been postulated to be substantially influenced by dopaminergic neurotransmission in one major formulation of personality typology and neurochemical correlates. The dopaminergic D4 receptor is expressed in human brain regions implicated in reward and mood. This receptor displays functional allelic variants that alter its efficiency in G-protein coupling and, thus, change features of dopaminergic neurotransmission in these circuits. However, the data from Benjamin *et al.,* suggest that only some individual item scores on the NEO-PI-R correlate with D4 genotype. These results were at least as consistent with the idea that happy, positive feelings or a ratio between positive and conscientious feelings could represent the underlying DRD4-associated phenotype.

We approached this question by assessing DRD4 genotypes in a sample taken from the Baltimore Longitudinal Study of Aging (29), a well-characterized population of research volunteers whose characterized personality types allowed us to select the individuals with the highest and with the lowest values on novelty-seeking scales and measures of "happiness," defined as ratios between measurements of excitement seeking to neuroticism (E/N).

We found only limited generalization of DRD4 genotype associations with specific personality features contributing to measures of novelty seeking; only the E/N ratio correlated with genotype (25). While our failure to identify the association to the original novelty-seeking item scores in this population strongly suggest that the findings of Benjamin *et al.* and Ebstein *et al.* cannot be generalized to all populations, the items positively correlated with DRD4 genotype (warmth, excitement seeking, and positive emotions) and the item that provided a negative correlate, deliberation, provide a ratio whose value did correlate weakly with DRD4 genotype.

These results support the idea that a more modest association with an assessment of a ratio of positive and negative feelings might represent a result of differential expression of D4 receptor genotypes. Indeed, given the relatively low levels of expression of the dopamine DRD4 gene, its limited brain distribution, and the substantially greater number of dopamine D2 receptors, even in areas of which the D4 receptors expressed, biological fit with this modest receptor expression and a more subtle personality influence on personality make some sense.

V. Other Dopamine Receptors

The status of dopamine D2-receptor polymorphisms in drug abuse has been reviewed (1, 42). More recent studies of a polymorphic marker at the dopamine D3 receptor have shown no association with substance abuse vulnerabilities (Rodriguez *et al.,* in preparation).

VI. Dopamine Transporter

Examination of association between variable number tandem repeat (VNTR) markers at the human DAT locus revealed no assocation (30). However, more recent observations may indicate association between polymorphic markers at this locus and two drug-associated conditions. Individuals who report paranoid experiences during cocaine administration display the DAT allele marked by a nine-copy VNTR marker more frequently than control subjects (31). Attention deficit–hyperactivity disorder (ADHD) risk is also increased in individuals with the DAT allele marked by the 10-copy VNTR (32). Because individuals with ADHD can display psychostimulant responses that differ from those of normal individuals, this evidence also plausibly links DAT allelic variants with alterations in human drug responses.

VII. Future Directions

This current review supports potentially prominent roles for dopaminergic gene variants in contributing to individual differences in vulnerability to drug abuse. However, other genes are quite likely also to be involved.

Seeking drug-regulated genes provides one avenue to find other classes of drug-abuse vulnerability genes in humans. New approaches, such as subtracted differential display, will allow us to identify hundreds of genes regulated by drug administration in experimental animals (33). Polymorphic markers at these gene loci can provide new tools to identify possible gene substrates for human individual-to-individual differences in drug abuse vulnerability.

Positional cloning of drug abuse vulnerability genes may also now become possible. Recent initial attempts to model this process (34) have been based on genetic heritability h^2 estimates of 0.45 and assumptions that environmental influences on drug abuse vulnerability are largely nonshared, both based on twin study data (1). An initial estimate of the relative risk for developing a substantial drug abuse disorder in the sibling of an abuser (λs) is three to fourfold, based on a group of family studies (1). The two to fourfold risk in excess of that in the general population can then be modelled as due to both some oligogenic and some polygenic influences. In quantitative trait locus studies in mice, the genetics of drug responsiveness can often be modelled as based on the detectable contributions of a limited number of genes, with a residual genetic variance assumed to be polygenic, due to effects of genes whose influences may be too small to determine individually (35–37). If these indications from mouse genetics are applicable to humans, a case in which five loci equally contribute to a 75% oligogenic influence could represent a plausible starting point for beginning to model the human condition. This would provide a locus-specific λs value of about 1.4 for the human condition.

Power calculations based on the work of Reisch (1990) and others indicate that evaluating 300 polymorphic markers spaced at 10 centimorgan intervals across the human genome in 400 nuclear affected sib pair pedigrees would provide an 80% opportunity to detect a locus of such an effect size (34). Although many of the assumptions used to make these power calculations require better and better estimates over time, these values also suggest that an affected sib pair approach to identification of drug abuse vulnerability genes is possible, with multicenter collaboration and powerful genotyping assistance.

As genome scaning–positional cloning approaches to detection of genes contributing to human drug abuse vulnerability become available, however, the study of candidate loci such as those expressed in the dopaminergic systems important for reward processes appears to provide a valuable initial route to identification of genes that contribute in both animal models and humans to interindividual differences in drug abuse vulnerability.

Acknowledgment

This work was largely supported by the Intraumural Research Program, National Institute on Drug Abuse, National Institutes of Health.

References

1. Uhl, G. R., Elmer, G. I., Labuda, M. C., Pickens, R. W. (1995). Genetic influences in drug abuse. Psychopharmacology: The Fourth Generation of Progress (F. E. Bloom and D. J. Kupfer, eds.). Raven Press, New York.
2. Goldberg, J., Lyons, M. J., Eisen, S. A., True, W. R., and Tsuang, M. (1993). Genetic influence on drug use: A preliminary analysis of 2674 Vietnam era veteran twins. *Behav. Genet.* **23,** 552.
3. Johnson, E. O., van den Bree, M. B. M., Uhl, G. R., and Pickens, R. W. (1996). Indicators of genetic and environmental influences in drug abusing individuals. *Drug Alcohol Depend.* **41,** 17–23.
4. Kuhar, M. J., Ritz, M. C., and Boja, J. W. (1991). The dopamine hypothesis of the reinforcing properties of cocaine. *Trends Neurosci.* **14,** 299–302.
5. Di Chiara, G., and Imperato, A. (1988). Drugs abused by humans preferentially increase synaptic dopamine concentrations in the mesolimbic system of freely moving rats. *Proc. Natl. Acad. Sci. U.S.A.* **85,** 5274–5278.
6. Roberts, D. C. S., and Koob, G. F. (1982). Description of cocaine self-administration following 6-hydroxydopamine lesions of the ventral tegmental area in rats. *Pharmacol. Biochem. Behav.* **17,** 901–904.
7. Pettit, H. O., Ettenberg, A., and Bloom, F. E. (1984). Destruction of dopamine in the nucleus accumbens selectively attenuates cocaine but not heroin self-administration in rats. *Psychopharmacology* **84,** 167–173.
8. Roberts, D. C. S., Corcoran, M. E., and Fibiger, H. C. (1977). On the role of ascending catecholamine systems in intravenous self-administration of cocaine. *Pharmacol. Biochem. Behav.* **6,** 615–620.
9. Ritz, M. C., Lamb, R. J., Goldberg, S. R., and Kuhar, M. J. (1987). Cocaine receptors on dopamine transporters are related to self-administration of cocaine. *Science* **237,** 1219–1223.
10. Shimada, S., Kitayama, S., Lin, C.-L., Patel, A., Nanthakumar, E., Gregor, P., and Uhl, G. R. (1991). Cloning and expression of a cocaine sensitive dopamine transporter cDNA. *Science* **254,** 576–578.
11. Kitayama, S., Wang, J.-B., and Uhl, G. R. (1993). Dopamine transporter mutants selectively enhance MPP^+ transport. *Synapase* **15,** 58–62.
12. Miner, L. L., Donovan, D. M., Perry, M., Revay, R., Sharpe, L. G., Przedborski, S., Isenwasser, S., Rothman, R. R., Schindler, C., and Uhl, G. R. Cocaine reward: Alteration by regional variant dopamine transporter overexpression. *J. Neurosci.* (submitted).

13. Harrington, C. A., Lewis, E. J., Krzemien, D., and Chikaraishi, D. M. (1987). Identification and cell type specifically of the tyrosine hydroxylase gene promoter. *Nucleic Acids Res.* **15,** 2363–2384.
14. Banerjee, S. A., Hoppe, P., Brilliant, M., and Chikaraishi, D. M. (1992). 5′ Flanking sequences of the rat tyrosine hydroxylase gene target accurate tissue-specific, developmental, and transsynaptic expression in transgenic mice. *J. Neurosci.* **12,** 4460–4467.
15. Carr, G. D., Fibiger, H. C., and Phillips, A. G. (1984). Conditioned place preference as a measure of drug reward. *In* Neuropharmacological Basis of Reward, pp. 264–319. Oxford New York.
16. Giros, B., Jaber, M., Jones, S. R., Wightman, R. M., and Caron, M. G. (1996). Hyperlocomotion and indifference to cocaine and amphetamine in mice lacking the dopamine transporter. *Nature* **379,** 606–612.
17. Drago, J., Gerfen, C. R., Lachowicz, J. E., Steiner, H., and Hollon, T. R. (1994). Altered striatal function in a mutant mouse lacking D1A dopamine receptors. *Proc. Natl. Acad. Sci.* **91,** 12564–12568.
18. Miner, L. L., Drago, J., Chamberlain, P. M., Donovan, D., and Uhl, G. R. (1995). Retained cocaine conditioned place preference in D1 receptor deficient mice. *Neuroreport* **6,** 2314–2316.
19. Cho, A. K., and Segal, D. S., eds. (1994). Amphetamine and its Analogs. Academic Press, San Diego.
20. Takahashi, N., Miner, L., Sora, I., Ujike, H., Revay, R., Kostic, V., Przedborski, S., and Uhl, G. R. VMAT2 knockout mice: Heterozygotes display reduced amphetamine-conditioned reward, enhanced amphetamine locomotion and enhanced MPTP toxicity. *Proc. Nat. Acad. Sci. USA* (in press).
21. Cooper, J. R., Bloom, F. E., and Roth, R. H. (1991). The Biochemical Basis of Neuropharmacology. Oxford Press, New York.
22. Aksoy, S., Klener, J., and Weinshilboum, R. M. (1993). Catechol-O-methyltransferase pharmacogenetics: Photoaffinity labelling and Western blot analysis of human liver samples. *Pharmacogenetics* **3,** 116–122.
23. Bertocci, B., Miggiano, V., Da Prada, M., Dembric, Z., Lahm, H.-W., and Malherbe, P. (1991). Human catechol-O-methyltransferase: Cloning and expression of the membrane associated form. *Proc. Natl. Acad. Sci.* **88,** 1416–1420.
24. Tenhunen, J., Salminen, M., Lundstrom, K., Kiviluoto, T., Savolainen, R., and Ulmanen, I. (1994). Genomic organization of the human catechol-O-methyltransferase gene and its expression from two distinct promoters. *Eur. J. Biochem.* **223,** 1049–1054.
25. Vandenbergh, D. A., Zonderman, A., Wang, J., Uhl, G. R., and Costa, P. A. Failure to replicate the association between long-repeat alleles of the novelty seeking and dopamine D4 receptor polymorphisms: Evidence for limited generalization. *Mol. Psychia.* (in press).
26. Vandenbergh, D. J., Rodriguez, L. A., Miller, I., Uhl, G. R., and Lachman, H. M. A High-activity catechol-O-methyltransferase allele is more prevalent in polysubstance abusers. *Psych. Genetics* (in press).
27. Benjamin, J., Li, L., Patterson, C., Greenberg, B. D., Murphy, D. L., and Hamer, D. H. (1996). Population and familial association between the D4 dopamine receptor gene and measures of novelty seeking. *Nat. Genet.* **12,** 81–84.
28. Ebstein, E. P., Novick, O., Umansky, R., Priel, B., Osher, Y., Blaine, D., Bennett, E. R., Nemanov, L., Katz, M., Belmaker, R. H. (1996). Dopamine D4 receptor (D4DR) exon III polymorphism associated with the human personality trait of novelty seeking. *Nat. Genet.* **12,** 78–80.
29. Shock, N. W., *et al.* Normal Human Aging: The Baltimore Longitudinal Study of Aging (1984). NIH Publication No. 84-2450, U.S. Government Printing Office, Washington, DC.
30. Persico, A. M., Vandenbergh, D. J., Smith, S. S., and Uhl, G. R. (1993). Dopamine transporter gene polymorphisms are not associated with polysubstance abuse. *Biol. Psychiatry* **34,** 265–267.

31. Gelernter, J., Kranzler, H. R., Satel, S. L., and Rao, P. A. (1994). Genetic association between dopamine transporter protein alleles and cocaine-induced paranoia. *Neuropsychopharmogy* **11,** 195–200.
32. Cook, E. H., Stein, M. A., Krasowski, M. D., Cox, N. J., Olken, D. M., Kieffer, J. E., and Leventhal, B. L. (1995). Association of attention deficit disorder and the dopamine transporter gene. *Am. J. Hum. Genet.* **56,** 993–998.
33. Wang, X.-B., Funada, M., Imai, Y., Revay, R., Ujike, H., and Uhl, G. R. rGβ1: A psychostimulant-regulated gene essential for establishing cocaine sensitization *J. Neurosci.* (submitted)
34. Uhl, G. R., Gold, L. H., Risch, N., and Wilson, A. Analyses of complex behavioral disorders. *Proc. Natl. Acad. Sci. USA,* **94,** (in press).
35. Crabbe, J. C., Belknap, J. K., and Buck, K. J. (1994). Genetic animal models of alcohol and drug abuse. *Science* **264,** 1715–1723.
36. Miner, L. L., and Marley, R. J. (1995). Chromosomal mapping of loci influencing sensitivity to cocaine-induced seizures in BXD recombinant inbred strains of mice. *Psychopharmacology* **117,** 62–66.
37. Miner, L. L., and Marley, R. J. (1995). Chromosomal mapping of the psychomotor stimulant effects of cocaine in BXD recombinant mice. *Psychopharmacology* **122,** 209–214.
38. Lachman, H., Papolos, D. F., Saito, T., Yu, Y. M., Szumlanski, C. L., and Weinshilboum, R. M. (1996). Human catechol-O-methyltransferase pharmacogenetics: Description of a functional polymorphism and its potential application to neuropsychiatric disorders. *Pharmacogenetics* **6,** 243–250.
39. Spielman, R. S., and Weinshilboum, R. M. (1981). Genetics of red cell COMT activity: Analysis of thermal stability and family data. *Am. J. Med. Genet.* **10,** 279–290.
40. Weinshilboum, R. M., and Raymond, F. A. (1977). Inheritance of low erythrocyte catechol-O-methyl transferase activity in man. *Am. J. Hum. Genet* **29,** 125–135.
41. Uhl, G., Blum, K., Noble, E., and Smith, S. (1994). Substance abuse vulnerability and D2 receptor gene. *Trends Neurosci.* **16,** 83–88.

John C. Crabbe, John K. Belknap, Pamela Metten, Judith E. Grisel, and Kari J. Buck

Portland Alcohol Research Center and Department of Behavioral Neuroscience
Oregon Health Sciences University and Department of Veterans Affairs Medical Center
Portland, Oregon 97201

Quantitative Trait Loci: Mapping Drug and Alcohol-Related Genes

Individual differences in most behavioral and pharmacological responses to drugs of abuse are dependent on both genetic and environmental factors; however, the genetic contributions to complex traits have been difficult to analyze using classical genetic methods. Most responses to drugs are multigenic (i.e., affected by multiple genes), are probably polygenic (i.e., each gene confers a relatively small influence on the trait), and are quantitative traits (i.e., they show a continuous distribution of individual differences). Recent developments in molecular biology and molecular genetics are beginning to allow the identification of specific genes that influence complex traits such as drug responses.

Quantitative trait locus (QTL) mapping is a method that has been successfully employed to identify genetic markers near genes that influence drug sensitivity. By studying specific animal populations, genetically segregating for a number of genes and markers that are polymorphic (i.e., where multiple allelic forms of the gene are maintained in the population), an association between drug sensitivity (or resistance) and possession of a particular allele is established. Because of the relatively precise localization in the mouse genome of such markers, such an association demonstrates linkage between the marker and a QTL; each such QTL represents a rough genomic map location for a gene (or genes) of influence. One method for accomplishing QTL mapping will be described, and its use to map genes influencing drug response will be shown in selected examples.

Our group (among others) has employed a multistage method for QTL mapping. We use a set of recombinant inbred (RI) strains to nominate QTL locations and follow this first-stage mapping effort with a verification in segregating populations, such as an F2 cross (1, 2). Within any inbred strain, all animals possess two copies of the same allele (i.e., are homozygous) for each gene; therefore, with the exception of sex chromosomal genes, they are monozygotic twins. Twenty-six recombinant inbred strains (BXD RI strains) have been derived from the cross of C57BL/6J (B6) and DBA/2J (D2) progenitor inbred strains by re-inbreeding from the genetically segregating F2 generation. At each gene and marker, each such BXD RI strain possesses a pair of identical alleles derived from either the B6 or D2 progenitor strain. The advantages of the BXD RI strain panel are that the progenitor B6 and D2 strains are genetically diverse by pedigree, therefore differing in allelic state for many genes, and that the

Advances in Pharmacology, Volume 42

1054-3589/98 $25.00

genomic locations of more than 1600 polymorphic markers have previously been mapped in each of the BXD RI strains. To map QTLs for a drug response trait, it is, therefore, only necessary to determine the phenotype (i.e., average magnitude of drug response) for each RI strain. By comparing the RI strain means with the allelic status of each strain for each previously mapped genetic marker, associations between particular alleles and drug sensitivity can be directly ascertained.

For example, we tested the B6 and D2 progenitors and the 26 BXD RI strains for their sensitivity to 16 mg/kg of methamphetamine by measuring the stereotyped climbing response (3). The RI strains showed a spectrum of climbing responses as compared with saline-treated control groups. These results indicated that there was significant genetic influence on the trait. QTL mapping analysis showed the presence of $p < .01$ genetic associations with markers on chromosomes 1, 2, 4, 5, 8, 12, 14, and 17. The associations on chromosomes 1 and 14 reached $p < .001$. Each such association suggests that there is a gene (or genes) nearby that affects methamphetamine sensitivity.

One powerful feature of QTL mapping is that the indication of a gene's location may suggest a candidate gene that is responsible for the drug sensitivity. For example, some of the markers on chromosome 1 associated with methamphetamine-induced climbing map to a position very close to that of the genes encoding the delta and gamma subunits of the nicotinic cholinergic receptor. Both biochemical and histochemical evidence suggests that the B6 and D2 strains differ in cholinergic activity (4), and striatal dopamine and acetylcholine systems are highly interactive (5). Together, this pattern of results is consistent with the hypothesis that the nicotinic receptor delta or gamma subunit gene may actually be the gene responsible for the QTL association with sensitivity to this form of methamphetamine stereotypy. However, much additional work would be necessary to accept this hypothesis, for provisional QTL mapping at the level of statistical stringency discussed here is relatively crude, and there may be 100 or more genes that fall within the confidence limits for this map location. Such a candidate gene hypothesis does, however, allow direct tests of the role of the nicotinic receptor in mediating methamphetamine responses in these strains.

An additional feature of large-scale QTL mapping projects using the BXD RI strains is that they are cumulative. Because the initial screen involves testing the same set of RI strains, data from one study can readily be compared with those from another. Our group has also examined several other responses to methamphetamine (e.g., stimulated homecage activity, alterations in body temperature, stereotyped chewing, and stimulated locomotor activity and its sensitization with repeated injections). BXD RI strains differ in their sensitivity to methamphetamine in all of these responses. Figure 1 shows that the body temperature, chewing, and home cage activity variables respond to methamphetamine in dose-dependent fashion. Of these, all but chewing also were significantly correlated with markers in the same region of chromosome 1 containing the nicotinic receptor subunit genes (3). Should these provisional associations be verified with further testing, this could offer clear evidence of genetic pleiotropism (i.e., the effects of a single gene on multiple traits). Such

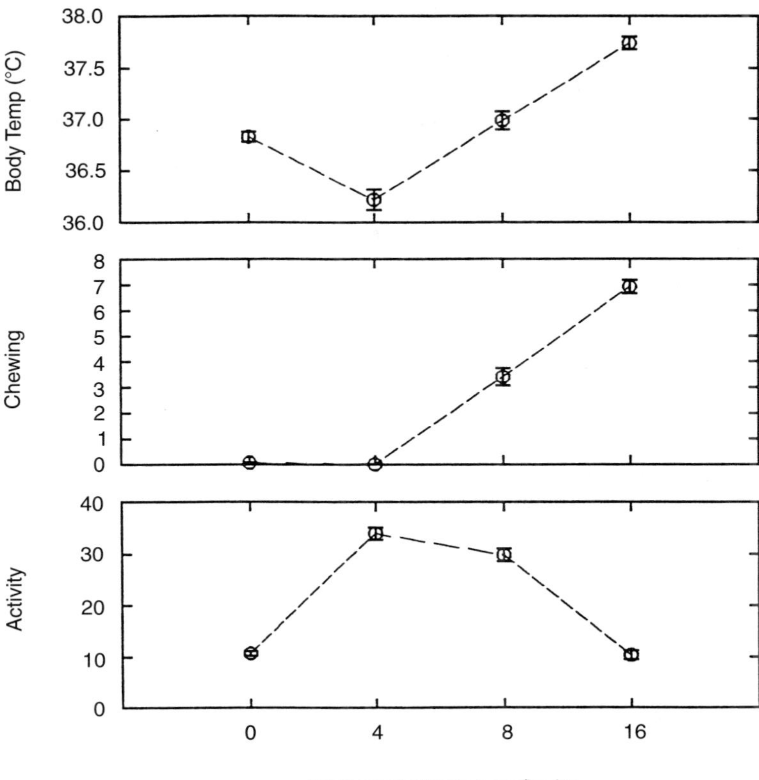

FIGURE 1 Dose-response curves for methamphetamine sensitivity measures across all mice tested (N = 213–217 per data point, N = 8 mice/strain/dose in most cases). Each point shows the mean ± SE for each dose. Body temperature (change from preinjection temperature 48 min after injection); stereotyped chewing (scored as number of incidences during a 1-min observation); and home cage activity (quadrant crossings during a 1-min observation). Data from Grisel *et al.* (3).

pleiotropic genes should be good candidates for more intensive investigation because of their broad spectrum of influence.

In an early review of provisional drug-related QTLs nominated by mapping studies ongoing in several laboratories (6), several drug response traits were associated with markers near the serotonin $5HT_{1B}$ receptor subtype. Among these were alcohol preference drinking and sensitivity and tolerance to alcohol's ataxic effects. The availability of a $5HT_{1B}$ knockout mouse developed by homologous recombination led us to study alcohol preference drinking and ataxia in the knockouts. These null mutants drank twice as much ethanol as their congenic controls (but equivalent amounts of saccharin, sucrose, and quinine) and had reduced sensitivity and tolerance to ethanol-induced ataxia (7). The increasing availability of transgenic preparations may make possible, in some cases, direct tests of candidate genes suggested by provisional QTL mapping efforts.

Provisional genetic map locations, such as those detected by a screen of BXD RI strains, require verification by additional genetic methods, such as analysis of segregating (e.g., F2) populations. For the methamphetamine traits described here, we have not yet progressed to this stage. However, another trait we have been pursuing offers an example of the subsequent stages of QTL verification. Acute injection of 4 g/kg of ethanol is followed by significant withdrawal, as assessed by the handling-induced convulsion (HIC), in some (e.g., D2) but not all (e.g., B6) strains of mice. An examination of the RI strains revealed significant variability among strains, and these results were used to nominate several mouse chromosome regions as provisional QTLs for alcohol withdrawal severity (1).

The great limitation of the BXD RI strains is that there are only 26 data points (genotypes) available for mapping. When these strain means are compared with the 1600+ available markers, these multiple tests of association result in an increased rate of false-positives (type I error). While this can be addressed by increasing markerwise alpha level, the power of the RI battery to detect true associations with QTLs is low, and the difficulty then becomes failure to detect true QTLs at the inflated alpha rates (type II error). We, therefore, advocate replication by turning to additional populations to increase the power to detect true associations while maintaining statistical rigor.

Of the seven ethanol withdrawal QTLs provisionally identified in the RI strain panel alone, subsequent tests with genetically segregating F2 mice showed that three were also significantly asssociated with allelic state at nearby markers. As a final test, we selectively bred animals for high or low withdrawal HICs, starting from the F2 population, and found that genotype (marker allele frequency) increased in parallel with HIC score over two generations of selective breeding. The combined estimates of statistical significance for these three loci were all $p \leq .00001$ (log of the odds favoring association, or LOD scores of 4.1–5.6) (8). For these rigorously verified QTLs, we are confident that a nearby gene affects alcohol withdrawal. The QTL on chromosome 11 is closely linked to a cluster of genes that encode the α_1, α_6 and γ_2 subunits of the $GABA_A$ receptor. Because GABA is the principal inhibitory neurotransmitter, its receptor is a highly plausible candidate gene for the chromosome 11 QTL. This region of mouse chromosome 11 maps to human chromosome 5 in the region 5q34–q35, suggests that the homologous human gene would be found in this region.

In conclusion, QTL mapping is a promising method for identifying genetic influences on complex pharmacobehavioral responses. The high degree of homology between mouse and human genome, currently thought to be in the vicinity of 85%, suggests that genes identified in mice can be immediately roughly mapped in humans.

Acknowledgments

These studies were supported by the Department of Veterans Affairs and by U.S. Public Health Service grants AA10760 and DA05228

References

1. Belknap, J. K., Metten, P., Helms, M. L., O'Toole, L. A., Angeli-Gade, S., Crabbe, J. C., and Phillips, T. J. (1993). QTL applications to substances of abuse: Physical dependence studies with nitrous oxide and ethanol. *Behav. Genet.* **23**, 213–222.
2. Johnson, T. E., DeFries, J. C., and Markel, P. D. (1992). Mapping quantitative trait loci for behavioral traits in the mouse. *Behav. Genet.* **22**, 635–653.
3. Grisel, J. E., Belknap, J. K., O'Toole, L. A., Helms, M. L., Wenger, C. D., and Crabbe, J. C. (1997). Quantitative trait loci affecting methamphetamine responses in BXD recombinant inbred mice. *J. Neurosci.* **17**, 745–754.
4. Durkin, T., Ayad, G., Ebel, A., and Mandel, P. (1977). Comparative study of acetylcholinesterase and choline acetyltransferase enzyme activity in brains of DBA and C57 mice. *Nat. New Biol.* **242**, 56–58.
5. Groves, P. M., Garcia-Munoz, M., Linder, J. C., Manley, M. S., Martone, M. M., and Young, S. J. (1995). Elements of the intrinsic organization and information processing in the neostriatum. *In* Models of Information Processing in the Basal Ganglia. J. C. Houk, J. L. Davis, and D. G. Beiser, (eds.), pp. 51–96. MIT Press, Cambridge, MA.
6. Crabbe, J. C., Belknap, J. K., and Buck, K. J. (1994). Genetic animal models of alcohol and drug abuse. *Science* **264**, 1715–1723.
7. Crabbe, J. C., Phillips, T. J., Feller, D. J., Wenger, C. D., Lessov, C. N., and Schafer, G. L. (1996). Elevated alcohol consumption in null mutant mice lacking 5-HT_{1B} receptors. *Nat. Genet.* **14**, 98–101.
8. Buck, K. J., Metten, P., Belknap, J. K., and Crabbe, J. C. Quantitative trait loci involved in genetic predisposition to acute alcohol withdrawal in mice. *J. Neurosci.* **17**, 3946–3955.

N. Hiroi and E. J. Nestler

Division of Molecular Psychiatry
Departments of Psychiatry and Pharmacology
Yale University School of Medicine
New Haven, Connecticut 06508

Nuclear Memory: Gene Transcription and Behavior

There is an ever-increasing list of stimuli and treatments that induce a particular set of genes, called immediate early genes. Perphaps the most widely studied immediate early genes belong to the Fos family (1). This family includes c-*fos*, *fos*B, *fra*1, and *fra*2. When activated, these genes transcribe their protein products, designated c-Fos, FosB, FRA1, and FRA2, respectively.

The initial impetus to examine the induction of these genes and proteins was the assumption that they could be used to "map" neurons and neuronal

Advances in Pharmacology, Volume 42

1054-3589/98 $25.00

networks activated by biologically significant stimuli. In some cases, this approach proved to be a useful one from a heuristic point of view (1). For example, noxious stimuli induce c-*fos* (mRNA) and c-Fos (protein) and upregulate prodynorphin mRNA in the spinal cord (2). The upregulation of prodynorphin mRNA is blocked by infusion of c-fos antisense oligonucleotides (3). However, c-*fos*/c-Fos turned out to be an erroneous marker for activated neurons in other cases. Acute treatment with the dopaminergic agonist apomorphine induces stereotyped behaviors in rodents, and this behavior has been shown to be dependent on activation of striatal neurons (4). Yet, the very same treatment does not induce c-Fos in normal striatum (5). Acute treatment with a psychomotor stimulant (e.g., cocaine and amphetamine) inhibits physiological activity of neurons in the nucleus accumbens (6) but potently induces c-Fos and other Fos family proteins in these cells (7). Conversely, chronic treatment with a psychomotor stimulant can result in potentiated physiological responses by these cells (6), while c-*fos* induction diminishes after repeated treatment (8). These are some examples of the numerous false-positive and false-negative cases for a relationship between neuronal activation and induction of Fos family member proteins. As such, these genes and their protein products cannot be used as reliable markers of activated neurons and networks.

This does not diminish the physiological importance of Fos family member proteins, however. These proteins play crucial roles in regulating particular cell functions. These proteins bind to a DNA sequence by forming heterodimers with protein products of another immediate early gene family, Jun. The Fos-Jun dimers bind to the consensus sequence TGACTCA, called activator protein-1 (AP1) sequence. When they bind to this DNA sequence, they activate or repress the transcription of other genes. Through this mechanism, numerous external and internal stimuli exert potent effects on the properties of specific target neurons.

We previously reported, as mentioned previously, that cocaine loses it ability to induce c-fos when repeatedly given to animals. However, we found that elevated AP1 binding activity persisted with chronic treatment when c-fos/c-Fos was not present (8). We later showed that unique Fos-related antigens (FRAs) are associated with this long-lasting AP1 binding activity (9). These FRAs were recognized by an "anti-FRA" antiserum raised against a highly conserved amino acid sequence of all Fos family member proteins and appeared to have the molecular weights of 35–37 kDa. These FRAs have been shown to be induced in distinct brain regions by a number of chronic treatments, including cocaine, morphine, electroconvulsive seizure, apomorphine, and various lesions (9–11), and, for this reason, they were termed *chronic FRAs* (9). Because these treatments induce dramatic behavioral changes ranging from sensitization–withdrawal to seizure, it has been hypothesized that chronic FRAs could play critical roles in neuronal and behavioral plasticity (9–11).

However, the identity of chronic FRAs remained unknown. Because chronic FRAs were recognized by an anti-FRA antiserum, raised against a highly conserved amino acid sequence of all Fos family members, they could conceivably be any Fos member protein or its variants. One Fos family member protein that has a predicted molecular weight around 35 kDa is an alternatively spliced form of FosB, termed ΔFosB, which lacks a portion of exon 4. Despite its truncated nature, however, ΔFosB seems to have functional activity as a transcription factor (12).

To further study the nature of the chronic FRAs, we used an antiserum raised against an N-terminus portion of FosB, which is present in full-length FosB and ΔFosB, and an antiserum raised against a C-terminus portion of full-length FosB, which is absent in ΔFosB. We demonstrated that the N-terminus, but not the C-terminus, anti-FosB antiserum recognized several bands around 35 kDa in the striatum of animals treated repeatedly with cocaine (9). This finding provided evidence that at least some of the chronic FRAs are ΔFosB, but the pattern of immunoreactive bands was not identical to that of chronic FRAs. This could be due to the fact that even if raised against the same protein, different antibodies do not produce identical staining patterns. Alternatively, this could indicate that not all chronic FRAs are ΔFosB. This latter possibility was further supported by the finding that the anti-FRA antiserum does not recognize FRAs at exactly the same positions when extracts of cells transfected with the Δ*fos*B gene and extracts of striatum of animals treated repeatedly with cocaine were compared (13). This finding suggested two possibilities. One is that some but not all of the chronic FRAs are ΔFosB. Alternatively, the induction of chronic FRAs, if they were all ΔFosB, could require cell surface stimulation that would lead to posttranslational modification of ΔFosB, which is absent in transfected cell lines that overexpress ΔFosB.

We have now obtained definitive evidence that all of the chronic FRAs are indeed ΔFosB variants: All of the chronic FRAs were absent in mice with targeted disruption of the *fos*B gene (14). Two-dimensional gel electrophoresis and FRA Western blotting were used for a more complete analysis. Chronic cocaine treatment was found to induce five distinct FRAs in the 35- to 37-kDa range. Most of these proteins were present under basal conditions, and their induction was enhanced by repeated cocaine treatment. In addition, one protein became detectable uniquely after repeated cocaine treatment. None of these cocaine-regulated FRAs was present in the striatum of *fos*B mutant mice (14). The absence of the 35- to 37-kDa FRAs in *fos*B mutant mice was accompanied by complete loss of the long-lasting increase in AP1-binding activity. These findings demonstrated that ΔFosB variants constitute the long-lasting AP1 activity induced in the striatum after repeated cocaine treatment. Related work indicated that the other constituent of this AP1 activity is JunD (13).

A remaining question regarding the chronic FRAs concerned their physiological roles. Repeated cocaine treatment induces a host of behavioral changes. One such behavioral change is locomotor sensitization. When animals are treated repeatedly with cocaine or amphetamine, their locomotor activation gradually increases. Behavioral studies have shown that locomotor sensitization to cocaine is mediated by the mesolimbic dopamine pathway projecting to the nucleus accumbens, a ventral portion of the striatum (15).

Three correlations prompted us to examine the physiological role of ΔFosB variants in the sensitization paradigm. First, ΔFosB variants are induced after repeated cocaine treatment in parallel to the development of behavioral sensitization. Second, ΔFosB variants are induced in the nucleus accumbens, a brain region implicated in cocaine sensitization. Third, the induction of ΔFosB variants and the development of cocaine sensitization are both blocked by a D1 dopamine-receptor antagonist, SCH 23390 (16, 17).

We used *fos*B mutant mice to further study their relationship (14). Cocaine acutely induced more robust behavioral activation in *fos*B mutant mice than in wild-type littermates. However, wild-type littermates exhibited greater locomotor sensitization: they increased their locomotor activity by 300% during 6 days of testing, while *fos*B mutant mice increased their locomotor activity by 150%. This was not due to focused stereotypy, which would behaviorally compete with and replace hyperactivity. Despite the fact that *fos*B mutant mice showed attenuated cocaine sensitization, they exhibited normal conditioned locomotor activity when placed in the same test boxes with saline injections.

These findings suggest several intriguing possibilities. First, the normal development of cocaine sensitization depends on transcriptional regulation by ΔFosB variants and perhaps FosB. Second, a certain element of sensitization, expressed as a low level of sensitization of *fos*B mutant mice, does not seem to require *fos*B gene products. Third, the development and expression of conditioned locomotor activity do not require *fos*B gene products, consistent with the view that sensitization and conditioned locomotor activity occur via distinct mechanisms.

The present set of findings shows that the long-lasting AP1 complex contains ΔFosB variants and that *fos*B gene products could play a crucial role in cocaine sensitization. In more general terms, this set of findings suggests that transcriptional regulation in the cell nucleus could play a critical role in the formation of certain types of memory.

References

1. Hughes, P., and Dragunow, M. (1995). Induction of immediate-early genes and the control of neurotransmitter-related gene expression within the nervous system. *Pharmacol. Rev.* **45**, 133–178.
2. Fitzgerald, M. (1990). c-Fos and the changing face of pain. *Trends Neurosci.* **13**, 439–440.
3. Lucas, J. J., Mellstrom, B., Colado, M. I., and Naranjo, J. R. (1993). Molecular mechanisms of pain: Serotonin$_{1A}$ receptor agonists trigger transactivation by c-*fos* of the prodynorphin gene in spinal cord neurons. *Neuron* **10**, 599–611.
4. Kelly, P. H., Seviour, P. W., and Iversen, S. D. (1975). Amphetamine and apomorphine responses in the rat following 6-OHDA lesions of the nucleus accumbens septi and corpus striatum. *Brain Res.* **94**, 507–522.
5. Pennypacker, K. R., Zhang, W. Q., Ye, H., and Hong, J. S. (1992). Apomorphine induction of AP-1 DNA binding in the rat striatum after dopamine depletion. *Mol. Brain Res.* **15**, 151–155.
6. Henry, D. J., and White, F. J. (1995). The persistence of behavioral sensitization to cocaine parallels enhanced inhibition of nucleus accumbens neurons. *J. Neurosci.* **15**, 6287–6299.
7. Graybiel, A. M., Moratalla, R., and Robertson, H. A. (1990). Amphetamine and cocaine induce drug-specific activation of the c-*fos* gene in striosome-matrix and limbic subdivisions of the striatum. *Proc. Natl. Acad. Sci. U.S.A.* **87**, 6912–6916.
8. Hope, B. T., Kosofsky, B., Hyman, S. E., and Nestler, E. J. (1992). Regulation of immediate-early gene expression and AP-1binding in the rat nucleus accumbens by chronic cocaine. *Proc. Natl. Acad. Sci. U.S.A.* **89**, 5764–5768.

9. Hope, B. T., Nye, H. E., Kelz, M. B., Self, D., Iadarola, M. J., Nakabeppu, Y., Duman, R. S., and Nestler, E. J. (1994). Induction of a long-lasting AP-1 complex composed of altered Fos-like proteins in brain by chronic cocaine and other chronic treatments. *Neuron* **13,** 1235–1244.
10. Pennypacker, K. R., Hong, J. S., and McMillian, M. K. (1995). Implications of prolonged expression of Fos-related antigens. *Trends Pharmacol. Sci.* **16,** 317–321.
11. Hiroi, N., Chen, J. S., Nye, H. E., and Nestler, E. J. (1996). Chronic FRAs: Novel transcription factors regulated in the basal ganglia by chronic neuronal perturbations. *In* The Basal Ganglia. V. (C. Ohye, M. Kimura, and J. S. McKenzie, eds.). Plenum Press, New York.
12. Mumberg, D., Lucibello, F. C., Schuermann, M., and Muller, R. (1991). Alternative splicing of fosB transcripts results in differentially expressed mRNAs encoding functionally antagonistic proteins. *Genes Dev.* **5,** 1212–1223.
13. Chen, J. S., Nye, H. E., Kelz, M. B., Hiroi, N., Nakabeppu, Y., Hope, B. T., and Nestler, E. J. (1995). Regulation of ΔFosB and FosB-like proteins by electroconvulsive seizure and cocaine treatments. *Mol. Pharmacol.* **48,** 880–889.
14. Hiroi, N., Brown, J. R., Haile, C., Greenberg, M. E., and Nestler, E. J. (1996). FosB mutant mice: Lack of chronic FRAs and abnormalities in cocaine-regulated behaviors. *Soc. Neurosci. Abstr.* **22,** 386.
15. Hemby, S. E., Jones, G. H., Justice, J. B. Jr., and Neill, D. B. (1992). Conditioned locomotor activity but not conditioned place preference following intra-accumbens infusions of cocaine. *Psychopharmacology* **106,** 330–336.
16. McCreary, A. C., and Marsden, C. A. (1993). Cocaine-induced behavior: Dopamine D1 receptor antagonism by SCH23390 prevents expression of conditioned sensitization following repeated administration of cocaine. *Neuropharmacology* **32,** 387–391.
17. Nye, H. E., Hope, B. T., Kelz, M., Iadarola, M., and Nestler, E. J. (1995). Pharmacological studies of the regulation of chronic Fos-related antigen induction by cocaine in the striatum and nucleus accumbens. *J. Pharmacol. Exp. Ther.* **275,** 1671–1680.

Roxanne A. Vaughan, Robin A. Huff, George R. Uhl, and Michael J. Kuhar

National Institute on Drug Abuse Intramural Research Program
Baltimore, Maryland 21224

Phosphorylation of Dopamine Transporters and Rapid Adaptation to Cocaine

Dopamine transporters (DATs) are integral membrane neuronal proteins that function to terminate dopaminergic neurotransmission by the rapid re-uptake of synaptic dopamine into presynaptic neurons. As the primary mechanism for the clearance of synaptic dopamine, DAT regulates the intensity and duration of dopaminergic neurotransmission. DAT is the primary site of action of cocaine and amphetamine and a mode of entry for the neurotoxins 6-hydroxydopamine and MPP^+, implicating it in mechanisms of drug abuse and neurodegeneration. DAT is a member of a class of neurotransmitter and amino acid transporters that drives re-uptake of transmitter by cotransport of Na^+ and Cl^- down electrochemical gradients. Molecular cloning of DAT reveals consensus phosphorylation sites for numerous protein kinases (PKs) (1). The identification of these potential phosphorylation sites on DAT and the other transporters of this class has increased the level of interest in the possibility that these proteins undergo functional regulation by phosphorylation, and evidence from a variety of studies indicates that this may be the case.

Treatment of striatal synaptosomes or DAT heterologous expression systems with phorbol esters results in reduced levels of dopamine transport (2, 3), and transporters for several other neurotransmitters including carriers for norepinephrine, serotonin, GABA, glycine, and glutamate also undergo functional regulation in response to treatment with PK activators. These studies have produced a variety of results, including instances of increased and decreased transport, and time courses of effects ranging from minutes to days. Mechanisms invoked to explain these results include transcriptional regulation of transporter expression, membrane trafficking, and direct effects of phosphorylation, although the precise mechanisms of action of these effects remain unknown. The potential involvement of transporter phosphorylation in mediating these effects is currently supported experimentally only for dopamine and glutamate transporters expressed in cultured cells (3, 4). Thus, while several studies have demonstrated functional effects induced by PK activators, only a small number of studies utilizing heterologous expression systems have demonstrated direct transporter phosphorylation, and little work has been done examining these parameters in the brain.

To examine the possibility that DATs undergo phosphorylation in the brain, rat striatal synaptosomes were incubated in Krebs-bicarbonate buffer

(25 mM Na_2HCO_3, 124 mM NaCl, 5 mM KCl, 1.5 mM $CaCl_2$, 5 mM $MgSO_4$, 10 mM glucose, pH 7.3) saturated with 95% O_2/5% CO_2, and containing 1.5 mCi/ml [^{32}P]orthophosphate. Test compounds were added, and samples were incubated at 30°C for 45 min. Incubation buffer was removed and tissue was disrupted and solubilized with sodium dodecyl sulfate (SDS). The solubilized samples were immunoprecipitated with antibody 16, directed against amino acids 42–59 of the deduced DAT primary sequence (5), and analyzed by SDS polyacrylamide gel electrophoresis and autoradiography.

The DAT was demonstrated to be a phosphoprotein based on the immunoprecipitation of a ^{32}P-labeled 80-kDa protein from rat striatal synaptosomes using a serum specific for DAT. Preimmune serum did not recognize the protein, and the ^{32}P-labeled protein exactly comigrated on gels with authentic [^{125}I]DEEP photoaffinity–labeled DAT (Fig. 1). The phosphorylation level of the protein was increased about threefold when the synaptosomes were treated with the PKC activator, phorbol 12-myristate, 13-acetate (PMA). The effect was dose-dependent between 0.1 and 1.0 μM, and the time course of PMA stimulation showed observable increases in phosphorylation by 5–10 min, with maximum levels reached by 15–20 min. Treatment of synaptosomes with the inactive phorbol ester 4α-phorbol 12,13, didecanoate (4αPDD) at 10 μM had no effect on DAT phosphorylation, while two other PKC activators, (−)indolactam V and OAG (1-oleoyl-2-acetyl-*sn*-gycerol), increased DAT phosphorylation. PMA-stimulated DAT phosphorylation was blocked by the PKC inhibitors staurosporine and bisindoylmaleimide. These results demonstrate that DAT undergoes phosphorylation in synaptosomes and indicate that activation of PKC increases the phosphorylation level of DAT.

Treatment of synaptosomes with the phosphatase inhibitor okadaic acid (OA) during ^{32}P labeling also resulted in a substantial increase in the level of DAT phosphorylation, indicating that DAT undergoes robust dephosphorylation *in vivo*. Time course studies showed increased levels of phosphorylated DAT by 5 min of treatment, with a maximum reached by 15–20 min. This time course

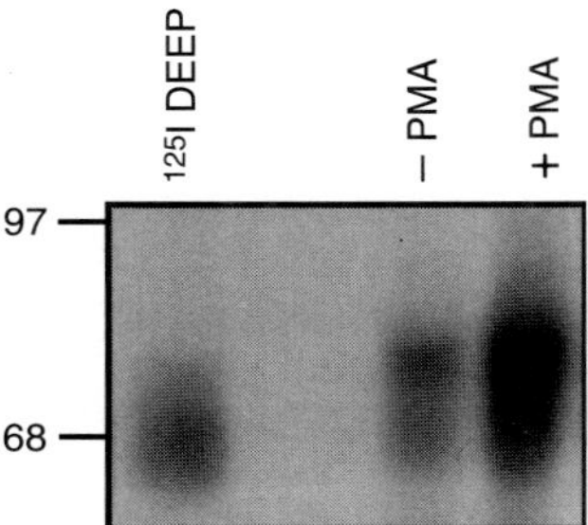

FIGURE 1 Phorbol ester–dependent phosphorylation of DATs. Rat striatal synaptosomes were incubated for 45 min with [^{32}P]orthophosphate and 10 μM of okadaic acid, in the presence or absence of 10 μM of PMA, as indicated. Tissue was solubilized and immunoprecipited with serum 16. [^{125}I]DEEP photoaffinity-labeled DATs were immunoprecipitated in parallel. Molecular mass markers are shown in kilodaltons.

is a reflection of the endogenous rate of DAT dephosphorylation and indicates that DATs undergo rapid and constitutive dephosphorylation.

This is the first demonstration that DAT or any of the Na^+-Cl^-–coupled neurotransmitter transporters undergoes phosphorylation in brain tissue. These results also demonstrate that activation of PKC promotes DAT phosphorylation, although it is not known if DAT is directly phosphorylated by PKC or by a downstream phosphorylation event.

Characterization of the DAT as a phosphoprotein has now been accomplished for neuronal DATs and for DATs expressed heterologously. In transfected rDAT-LLC-PK_1 cells, dopamine uptake is reduced 30–40% by phorbol esters via a reduction in DAT Vmax, and DATs undergo PMA-regulated *in vivo* phosphorylation with activator, inhibitor, and kinetic characteristic that parallel characteristics of the PMA-induced reduction of transport (3). Furthermore, the results found for both phosphorylation and uptake in cells are similar to the results found for the synaptosomal DATs. This indicates that heterologous expression systems provide valid models for examining DAT phosphorylation, and that results from studies such as site-directed or deletion mutagenesis using transfected DATs will relevant to the neuronal form of the protein.

The relationship of phosphorylation to transporter function currently remains obscure. While reduced dopamine uptake correlates exactly with conditions that increase DAT phosphorylation, the mechanisms underlying this effect remain to be elucidated. Reduction in dopamine transport could be produced by a direct effect on transporter activity or indirectly through changes in surface transporter number. Other aspects of DAT function, such as ion flux or binding of uptake blockers such as cocaine, may also be affected by phosphorylation.

The endogenous pathways involved in DAT phosphorylation also remain to be elucidated, although the kinetics of phosphorylation and dephosphorylation are compatible with receptor-mediated mechanisms acting though PKC. There is some evidence that dopamine receptors are coupled to breakdown of phosphatidylinositolbisphosphate and/or regulation of PKC action. This could provide a positive- or negative-feedback mechanism related to DAT phosphorylation, depending on whether the receptors are positively or negatively coupled to PKC. Other neurotransmitter receptors coupled to phospholipase C are also candidates for involvement in this phenomenon.

The demonstration that DATs undergo endogenous phosphorylation in synaptosomes under conditions that alter dopamine transport provides direct evidence in support of the possibility that neurons may utilize DAT phosphorylation as a mechanism for fine temporal and spatial regulation of synaptic dopamine levels. Such functional regulation could have profound effects on the intensity and duration of dopaminergic synaptic transmission, actions of psychostimulant drugs, and mechanisms of neurotoxicity and neurodegeneration.

References

1. Amara, S., and Kuhar, M. (1993). Neurotransmitter transporters: Recent progress. *Annu. Rev. Neurosci.* **16**, 73–79.

2. Copeland, B. J., Neff, N. H., and Hadjiconstantinou, M. (1995). Protein kinase C activators decrease dopamine uptake into striatal synaptosomes. *Soc. Neurosci. Abstr.* **21,** 1381.
3. Huff, R. A., Vaughan, R. A., Kuhar, M. J., and Uhl, G. R. (1997). Dopamine transporter: Phorbol esters enhance phosphorylation and decrease Vmax. *J. Neurochem.* **68,** 225–232.
4. Casado, M., Bendahan, A., Zafra, F., Danbolt, N. C., Aragon, C., Gimenez, C., and Kanner, B. I. (1993). Phosphorylation and modulation of brain glutamate transporters by protein kinase C. *J. Biol. Chem.* **268,** 27313–27317.
5. Vaughan, R. A. (1995). Photoaffinity-labeled ligand binding domains on dopamine transporters identified by peptide mapping. *Mol. Pharmacol.* **47,** 956–964.

Index

Contents of Previous Volumes

Volume 32

Volume 33

Volume 34

Volume 35

Volume 36

Volume 37

Volume 38

Volume 39

Volume 40

Volume 41

Biology of Cell Death

Apoptosis under Physiologic Conditions

Apoptosis in Pathologic States